POWERP

POWERPLANT TECHNOLOGY

M. M. El-Wakil

Professor of Mechanical and Nuclear Engineering
University of Wisconsin

McGraw-Hill Book Company

New York St. Louis San Francisco Auckland Bogotá Hamburg
London Madrid Mexico Montreal New Delhi
Panama Paris São Paulo Singapore Sydney Tokyo Toronto

POWERPLANT TECHNOLOGY
INTERNATIONAL EDITION 1984

30 29 28 27 26
20 09 08 07 06
BJE

This book was set in Times Roman by J. M. Post Graphics, Corp.
The editors were Anne Murphy and J. W. Maisel.

Library of Congress Cataloging in Publication Data

El-Wakil, M. M. (Mohamed Mohamed), date
Powerplant technology.

Bibliography: p.
Includes index.
1. Electric power-plants. I. Title.
TK1001.E39 1984 621.31'21 83-14944
ISBN 0-07-019288-X

When ordering this title use ISBN 0-07-066274-6

Printed in Singapore

He, who through vast immensity can pierce,
See worlds on worlds compose one universe,
Observe how system into system runs,
What other planets circle other suns,
What varied Being peoples every star,
May tell why Heaven has made us as we are.

. . . .

Who sees with equal eye, as God of all,
A hero perish, or a sparrow fall,
Atoms or systems into ruin hurled,
And now a bubble burst, and now a world.

Alexander Pope
(1688-1744)
An Essay on Man

CONTENTS

PREFACE

This book is the outgrowth of lecture notes used in a course on powerplant technology and engineering and given to a class of senior graduate students over several years. The author is indebted to many colleagues in the academic and industrial community who have made many helpful suggestions and supplied much needed material; and to his own students, whose give and take has helped in the evolution of the lecture notes into the present manuscript.

A course on powerplants was a popular and integral part of engineering curriculums, especially mechanical engineering, in the 1940s and 1950s. The material covered then was largely nonanalytical, preferring to concentrate on system and component description and operation. The course fell out of favor with the trend toward more fundamentalism in engineering education that started in the 1960s, and because the supply of cheap fossil fuels seemed so assurred that development in the field slowed down considerably.

With the turnaround in the energy picture in the 1970s, and the need for new solutions to efficient power-generation problems, the course experienced a remarkable revival with a strong student and graduate engineer demand. That demand, which is expected to continue for the indefinite future, relied on available books, most of which lack some of the modern aspects of power generation, concentrate on problems of energy supply, demand, conservation and economics; or address a wide variety of topics that are not all relevant to the subject matter.

It has become apparent that a modern up-to-date text is sorely needed. This book was written with this goal in mind. It attempts to focus on the treatment and synthesis of electric-generating powerplant technology and engineering, with a balance between the analytical and technological aspects of powerplant design, systems, and effects. The old powerplant systems, which were almost exclusively fossil-fired, have given way to a wide spectrum of systems including improved fossil and nuclear plants and plants that rely on renewable energy sources for their input. While the author has certain opinions on the viability of some of these as serious contributors to the large demands for electric energy in the near and long term, he has attempted to give most of them more or less equal emphasis, at least for the sake of completeness. The subject of energy, in general, generates much discussion, some of it emotional. The author has attempted to keep editorializing under control, however; but in a few instances, his opinions surfaced. Some readers may or may not agree with them.

The book is designed mainly for the use of mechanical, nuclear, and electrical engineers, but also for other engineering, energy, and applied-science majors. To accommodate a variety of backgrounds of readers, the book begins with introductory material that is necessary for the understanding and analysis of most powerplant systems. It then proceeds to cover fossil-fired powerplants of the Rankine, Brayton, and combined types. This is followed by introductory nuclear material and the current and most-promising fission-type nuclear powerplants. Fusion powerplants are not covered except in an introductory manner, since an electric-generating fusion powerplant is not now, and not expected to be, a reality in a well-defined form for decades to come. The book then proceeds to cover powerplants that rely upon renewable energy sources, such as geothermal, solar, wind, ocean temperature difference, and tide and wave energy. It terminates with presentations of various energy storage systems, most of which are still under development; and environmental aspects of electric-power generation, both fossil and nuclear.

Now a word about what this book does not cover. Probably the most important is hydroelectric energy, a major topic in its own right that is taught in separate courses and is adequately covered in many textbooks. Focusing on electric-power generation, the book does not cover areas of domestic and municipal heating and cooling, which are also covered elsewhere. It also does not cover the small but well-established electric-generating systems by internal-combustion engines which are also covered elsewhere. Power-generation systems from such sources as biomass and solid wastes, while environmentally helpful, are not expected to ever become major contributors to the electric-power-supply picture, and are covered with less emphasis than are other types.

Because of the expected variety in readers' backgrounds, a problem encountered in the author's own classes, some of the material in the book may not be necessary for some readers and may be bypassed by them, or used as a refresher. Also, because of the wide variety of topics covered, the book has been written with flexibility in mind, so that readers may change some of the order of presentation, or omit some of the material to suit their interests and purposes without loss of continuity. The book can be covered in three to five semester hours depending upon background and extent of coverage. While it is designed primarily as a college textbook with solved examples within and problems at the end of chapters, a conscious effort has been made to make it usable by graduate engineers in industry.

An engineering book that is published these days must face the dilemma of what system of units to use. Industry, particularly in the United States, still relies largely on the U. S. Customary System. In addition, many of the established heat-transfer, fluid-flow, and design correlations for powerplant equipment are based on it. On the other hand, the trend in the scientific and world communities is toward use of the S. I. (Le Système International d'Unités), adopted in 1960. This book uses both systems, often together, and trusts that the reader can, or can learn to, convert easily from one to the other. It is important to keep in mind that mass is mass, force is force, and weight is a force. The dimensional conversion factor g_c used whenever mass and force appear in the same equation is used throughout. It does not vanish in the S. I.

system but simply assumes a numerical value of unity. The Appendix contains an extensive number of conversion tables, the last column of each of which is reserved for the S. I. system.

The author is indebted to colleagues, too numerous to mention, who made many suggestions and gave much encouragement, and to his students, who by their probing questions have helped put proper emphasis where it was most needed. He is also indebted to many who have contributed to the preparation and typing of the manuscript, in particular, Mary Jo Biechler, who competently and cheerfully had to beat numerous deadlines on such a difficult and long manuscript.

The author is also grateful to the College of Engineering, University of Wisconsin, for the creative and challenging atmosphere in which he has been fortunate to work, and to his family and friends in both Madison and Alexandria for their love, support, and patience. It was on the shores of the latter that much of the inspiration for this work has come.

M. M. El-Wakil

CHAPTER

ONE

A THERMODYNAMICS REVIEW

1-1 INTRODUCTION

The design, operation, and performance of electric-generating powerplants are largely dependent upon the science of thermodynamics. It is assumed that the reader is familiar with the basics of that science. The material in this chapter, then, will not be a comprehensive treatment of thermodynamics but will instead review some fundamentals and present some aspects that are of particular relevance to power-generation systems.

The material in this chapter will include parts of both basic and applied thermodynamics. The first and second laws will be reviewed as well as processes, property relations, and cycles. The concepts of reversibility and entropy and the roles they play in the performance of power-generating cycles will be discussed.

Chapter 2 will be devoted to the analysis of the Rankine cycle, which plays the major role in power systems today. Other cycles and aspects, such as the Brayton cycle, heat transfer, and fluid flow in powerplant components, will be presented where appropriate and applicable throughout the book.

1-2 THE FIRST LAW AND THE OPEN SYSTEM

The first law of thermodynamics is the law of *conservation of energy,* which states that energy can neither be created nor destroyed. The energy of a system undergoing change (process) can be increased or decreased by exchange with the surroundings and converted from one form to another within that system. This is, therefore, simply a "bookkeeping" law that declares the exchange and convertibility of energy and sees to it that all energy is accounted for when a change occurs. The first law does *not* indicate whether conversions of energy from one form to another are or are not performed perfectly or whether some forms may be completely converted to others. Such limitations are left to the second law.

The most general mathematical expression of the first law is that for an open system undergoing change in a transient state. A *system* is a specified region, not necessarily of constant volume or fixed boundaries, where transfers and conversions of energy and mass are to be studied. An *open system* is one where energy and mass cross the boundaries of that system. An *open system in the transient state* is one in which the mass inflow and outflow are not equal or vary with time and in which the mass within the system changes with time. The transient system will not be covered here, although an example (a pressurizer in a nuclear pressurized-water reactor) will be presented in Chap. 10.

We will now discuss the *steady-state open system,* also called the *steady-state steady-flow (SSSF) system*. This is one in which mass and energy flows across the boundaries do *not* vary with time and in which the mass within the system remains constant. The SSSF system, applicable to mechanical energy (i.e., ignoring electrical, magnetic, chemical, and other effects) is shown schematically in Fig. 1-1. The first-law equation for that system is

$$\mathrm{PE}_1 + \mathrm{KE}_1 + \mathrm{IE}_1 + \mathrm{FE}_1 + \Delta Q = \mathrm{PE}_2 + \mathrm{KE}_2 + \mathrm{IE}_2 + \mathrm{FE}_2 + \Delta W_{\mathrm{sf}} \qquad (1\text{-}1a)$$

where the subscripts 1 and 2 indicate the inlet and exit stations of the open system. Equation (1-1*a*) assumes for simplicity that only one inlet and one exit exist, although the SSSF equation can be written easily enough for multiple inlets, exits, or both. In Eq. (1-1*a*)

$$\mathrm{PE} = \text{potential energy} = mz\frac{g}{g_c} \qquad (1\text{-}2)$$

where m is the mass of a quantity of matter or fluid entering and leaving the system (both equal in SSSF), z is the elevation of stations 1 or 2 above a common datum, g is the gravitational acceleration, and g_c is a conversion factor numerically equal to 32.2 $\mathrm{lb}_m \cdot \mathrm{ft}/(\mathrm{lb}_f \cdot \mathrm{s}^2)$ or 1.0 $\mathrm{kg} \cdot \mathrm{m}/(\mathrm{N} \cdot \mathrm{s}^2)$

$$\mathrm{KE} = \text{kinetic energy} = m\,\frac{V_s^2}{2g_c} \qquad (1\text{-}3)$$

where V_s is the velocity of the mass at 1 or 2.

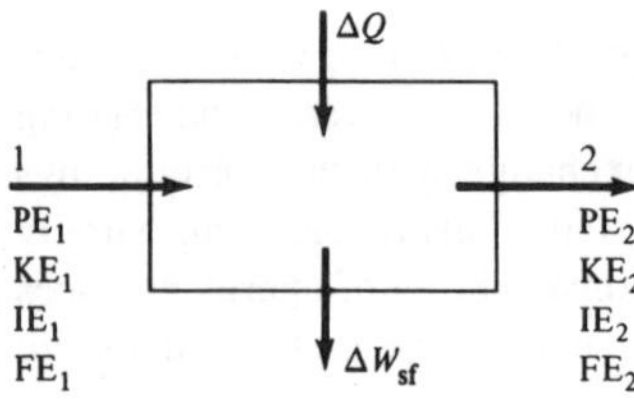

Figure 1-1 Schematic of a steady-state steady-flow (SSSF) system with one inlet and one outlet.

$$\text{IE} = \text{internal energy} = U \tag{1-4}$$

Internal energy is a sole function of temperature for perfect gases and a strong function of temperature and weak function of pressure for nonperfect gases, vapors, and liquids. It is a measure of the internal (molecular) activity and interaction of the fluid.

$$\text{FE} = \text{flow energy} = PV = Pmv \tag{1-5}$$

Flow energy, or flow work, is the work done by the flowing fluid to push the quantity represented by mass m into, and out of, the system. Mathematically it is equal to the product of pressure P and volume V.

$$\Delta Q = \text{net heat added} = Q_A - |Q_R| \tag{1-6}$$

where Q_A is heat *added* to (entering) and Q_R is the heat *rejected* by (leaving) the system across its boundaries. It is convenient to consider that heat added is positive and heat rejected is negative. Mathematically

$$\Delta Q = mc_n(T_2 - T_1) \tag{1-7}$$

where c_n is a specific heat that depends upon the process taking place between 1 and 2. Values for c_n for various processes are given in Table 1-1.

$$\begin{aligned} \Delta W_{sf} &= \text{net steady-flow mechanical work done by the system} \\ &= W_{by} - |W_{on}| \end{aligned} \tag{1-8}$$

where W_{by} is the work done by the system and W_{on} is the work done on the system. The convention is that the work done by the system is positive and the work done on the system is negative. Mathematically the steady-flow work is given by

$$\Delta W_{sf} = -\int_1^2 V\,dP \tag{1-9}$$

Equation (1-9) requires a relationship between pressure P and volume V for evaluation. The most general relationship is given by

$$PV^n = \text{constant} \tag{1-10}$$

where n is called the *polytropic exponent* and varies from zero to infinity. Its value for certain processes is given in Table 1-1.

Table 1-1 Values of c_n and n for various processes

Process	c_n	n
Constant pressure	c_p	0
Constant temperatures	∞	1
Adiabatic reversible	0	$k = \dfrac{c_p}{c_v}$
Constant volume	c_v	∞
Polytropic	$c_v \dfrac{k - n}{1 - n}$	$0 - \infty$

Equation (1-1*a*) may now be written for mass *m* entering and leaving the system as

$$mz_1\frac{g}{g_c} + m\frac{V_{s1}^2}{2g_c} + U_1 + P_1V_1 + \Delta Q$$

$$= mz_2\frac{g}{g_c} + m\frac{V_{s2}^2}{2g_c} + U_2 + P_2V_2 + \Delta W_{sf} \qquad (1\text{-}1b)$$

and for a unit mass

$$z_1\frac{g}{g_c} + \frac{V_{s1}^2}{2g_c} + u_1 + P_1v_1 + \Delta q$$

$$= z_2\frac{g}{g_c} + \frac{V_{s2}^2}{2g_c} + u_2 + P_2v_2 + \Delta w_{sf} \qquad (1\text{-}1c)$$

where the lowercase symbols represent specific values of the uppercase ones, i.e., per unit mass. Thus u = specific internal energy = U/m, v = specific volume = V/m, etc.

Table 1-2 Some common thermodynamic symbols

c_p	= specific heat at constant pressure, Btu/(lb$_m$ · °F) or J(kg · K)
c_v	= specific heat at constant volume, Btu/(lb$_m$ · °F) or J(kg · K)
h	= specific enthalpy, Btu/lb$_m$ or J/kg
H	= total enthalpy, Btu or J
J	= energy conversion factor = 778.16 ft · lb$_f$/Btu or 1.0 N m/J
M	= molecular mass, lb$_m$/lb·mol or kg/kg·mol
n	= polytropic exponent, dimensionless
P	= absolute pressure (gauge pressure + barometric pressure), lb$_f$/ft^2; unit may be lb$_f$/in^2, commonly written psia; or Pa
Q	= heat transferred to or from system, Btu or J, or Btu/cycle or J/cycle
R	= gas constant, lb$_f$ · ft/(lb$_m$ · °R) or J(kg · K) = $\bar{R}/M$
$\bar{R}$	= universal gas constant = 1.545.33, lb$_f$ · ft/(lb·mol · °R) or 8.31434 × 10^3 J/(kg·mol · K)
s	= specific entropy, Btu/(lb$_m$ · °R) or J/(kg · K)
S	= total entropy, Btu/°R or J/kg
t	= temperature, °F or °C
T	= temperature on absolute scale, °R or K
u	= specific internal energy, Btu/lb$_m$ or J/kg
U	= total internal energy, Btu or J
v	= specific volume, ft^3/lb$_m$ or m^3/kg
V	= total volume, ft^3 or m^3
W	= work done by or on system, lb$_f$ · ft or J, or Btu/cycle or J/cycle
x	= quality of a two-phase mixture = mass of vapor divided by total mass, dimensionless
k	= ratio of specific heats, c_p/c_v, dimensionless
η	= efficiency, as dimensionless fraction or percent
	Subscripts used in vapor tables
f	refers to saturated liquid
g	refers to saturated vapor
fg	refers to change in property because of change from saturated liquid to saturated vapor

The units of all the terms in Eqs. (1-1*a*) and (1-1*b*) are those of energy, such as Btu, ft · lb_f, or Joule (J). The units in Eq. (1-1*c*) are Btu/lb_m, ft · lb_f/lb_m or J/kg. A list of these and other common thermodynamic symbols is given in Table 1-2.

The Enthalpy

The sums $U + PV$ and $u + Pv$ appear together very frequently in thermodynamics. The combination, therefore, has been given the name *enthalpy* (with the stress on the middle syllable) and the symbols H and h, where h = specific enthalpy = H/m. Thus

$$H = U + PV \tag{1-11a}$$

and

$$h = u + Pv \tag{1-11b}$$

Equation (1-1*b*) can now be written, using H, as

$$mz_1\frac{g}{g_c} + m\frac{V_{s1}^2}{2g_c} + H_1 + \Delta Q$$

$$= mz_2\frac{g}{g_c} + m\frac{V_{s2}^2}{2g_c} + H_2 + \Delta W_{sf} \tag{1-1d}$$

and Eq. (1-1*c*) can be similarly written using h.

Enthalpies and internal energy are *properties* of the fluid, which means that each would have a single value at any given state of the fluid. They are defined as

$$c_v \equiv \left(\frac{\partial u}{\partial T}\right)_v \tag{1-12}$$

and

$$c_p \equiv \left(\frac{\partial h}{\partial T}\right)_P \tag{1-13}$$

where c_p and c_v are the specific heats at constant pressure and constant volume, respectively. They have units of Btu/(lb_m · °R) or J/(kg · K). They are related by

$$c_p - c_v = R \tag{1-14}$$

where R is the gas constant. For ideal gases

$$du = c_v\, dT \tag{1-15a}$$

and

$$dh = c_p\, dT \tag{1-16a}$$

where c_v and c_p are constant and independent of temperature for monatomic gases such as helium but increase with temperature for diatomic gases such as air and more so for triatomic gases such as CO_2, etc. For helium c_v = 0.753 Btu/(lb_m · °R) = 3.153 kJ/(kg · K), c_p = 1.250 Btu/(lb_m · °R) = 5.234 kJ/(kg · K), both independent of temperature. For air at low temperatures, c_v = 0.171 Btu/(lb_m · °R) = 0.716 kJ/(kg · K), c_p = 0.240 Btu/(lb_m · °R) = 1.005 kJ/(kg · K). For constant specific heats, or for small changes in temperature, Eqs. (1-15*a*) and (1-16*a*) may be written as

$$\Delta u = c_v\, \Delta T \tag{1-15b}$$

and $$\Delta h = c_p \, \Delta T \tag{1-16b}$$

Where there is a large increase in temperature

$$\Delta u = \int_1^2 c_v(T)\, dT \tag{1-17}$$

and $$\Delta h = \int_1^2 c_p(T)\, dT \tag{1-18}$$

Expressions for $c_p(T)$ in terms of temperatures for various gases (other than monatomic) may be found in the literature [4, 5]. $c_v(T)$ is found from $c_p(T)$ by subtracting the gas constant R for the particular gas, Eq. (1-14). For pure air

$$c_p(T) = 0.219 + 3.42 \times 10^{-5}T - 2.93 \times 10^{-9}T^2 \tag{1-19}$$

where $c_p(T)$ is in Btu/(lb$_m$ · °R) and T in °R. $c_v(T)$ for pure air is obtained by subtracting R for air or 53.34/778.16 = 0.0685 Btu/(lb$_m$ · °R) from Eq. (1-19).

It should be added here that the combination $U + PV$ is enthalpy H whether the system is open or closed (below). In the open system PV is the flow energy. In the closed system PV is simply the product of pressure times volume.

Equations (1-1*a*) through (1-1*d*) are used to solve problems of open systems. In considering a particular problem, it is often found that some terms drop out, some are unchanged between stations 1 and 2, or that the change is negligibly small compared with those of other terms. Some examples of open systems are

1. A steam generator

$$\Delta W_{sf} = 0$$

$$PE_2 - PE_1 = \text{negligible}$$

$$KE_2 - KE_1 = \text{negligible}$$

Thus $$\Delta Q = H_2 - H_1 \tag{1-20a}$$

and $$\Delta q = h_2 - h_1 \tag{1-20b}$$

2. A gas or steam turbine

$$\Delta Q = \text{negligible}$$

$$PE_2 - PE_1 = \text{negligible}$$

$$KE_2 - KE_1 = \text{negligible}$$

thus $$\Delta W_{sf} = H_1 - H_2 \tag{1-21a}$$

and $$\Delta w_{sf} = h_1 - h_2 \tag{1-21b}$$

3. A water (or incompressible fluid) pump

$$\Delta Q = \text{negligible}$$

$$PE_2 - PE_1 = 0 \qquad \text{(considering immediate inlet and exit)}$$

$$KE_2 - KE_1 = \text{negligible}$$

$$U_2 = U_1$$

And, because water is essentially incompressible and little or no change in temperature or volume takes place,

$$V_2 \approx V_1 = V$$

Thus $$\Delta W_{sf} = FE_1 - FE_2 = V(P_1 - P_2) \tag{1-22a}$$

and $$\Delta w_{sf} = v(P_1 - P_2) \tag{1-22b}$$

Both should be negative.

4. A nozzle

$$\Delta Q = 0$$

$$\Delta W_{sf} = 0$$

$$PE_2 - PE_1 = 0 \qquad \text{(considering immediate inlet and exit)}$$

$$KE_1 = \text{usually negligible compared with } KE_2$$

Thus $$V_{s2} = \sqrt{2g_cJ(h_1 - h_2)} \tag{1-23a}$$

$$= \sqrt{2g_cJc_p(T_1 - T_2)} \qquad \text{ideal gas} \tag{1-23b}$$

$$= \sqrt{2g_cv(P_1 - P_2)} \qquad \text{incompressible fluid} \tag{1-23c}$$

5. Throttling

$$\Delta Q = 0$$

$$\Delta W_{sf} = 0$$

$$PE_2 - PE_1 = 0 \qquad \text{or negligible}$$

$$KE_2 - KE_1 = \text{negligible}$$

Thus $$H_1 = H_2 \tag{1-24a}$$

and $$h_1 = h_2 \tag{1-24b}$$

1-3 THE FIRST LAW AND THE CLOSED SYSTEM

The open system, discussed above, is one in which mass crosses the boundaries. A *closed system,* by contrast, is one in which only energy and not mass may cross the boundaries. A third system of some interest is the *isolated system,* a special instance of the closed system. It is one in which neither mass or energy cross the boundaries but in which energy transformations may take place within the boundaries.

Because mass does not cross the boundaries in a closed system, the potential, kinetic, and flow energy terms of Eqs. (1-1*a*) through (1-1*d*) drop out and the first-law equation for the closed system is simplified to

$$U_1 + \Delta Q = U_2 + \Delta W_{nf} \tag{1-25a}$$

or

$$\Delta Q = \Delta U + \Delta W_{nf} \tag{1-25b}$$

and

$$\Delta q = \Delta u + \Delta w_{nf} \tag{1-25c}$$

In the case of the open system, Eqs. (1-1*a*) through (1-1*d*), 1 and 2 refer to positions in space with respect to the system. In the case of the closed system they instead refer to differences in the time domain, i.e., before and after the process in question has taken place.

ΔW_{nf} is called the *nonflow work*. Mathematically it is given by

$$\Delta W_{nf} = \int_1^2 P\,dV \tag{1-26}$$

which, as does Eq. (1-9), requires a relationship between P and V for evaluation, such as that given by Eq. (1-10).

Some examples of the closed system are:

1. A closed rigid tank

$$\Delta W_{nf} = 0$$

Thus

$$\Delta Q = U_2 - U_1 = \Delta U \tag{1-27}$$

2. An insulated cylinder in which a fluid expands behind a piston (or is compressed by it)

$$\Delta Q = 0$$

Thus

$$\Delta W_{nf} = U_1 - U_2 = -\Delta U \tag{1-28}$$

1-4 THE CYCLE

In order to convert forms of energy, particularly heat, to work on an extended or continuous basis (our main objective), one needs to operate on a cycle. A *process* begins at one state of the working fluid and ends at another, and that is that. A *cycle,* on the other hand, is a series of processes that begins and ends at the same state and thus can repeat indefinitely, or as long as needed. An example is the ideal diesel cycle,* shown on the $P - V$ and $T - s$ diagrams in Fig. 1-2. It is composed of an

* Rudolf Diesel (1858–1913) was born in Paris of German parents and moved to London in 1870 because of the Franco-German War. He was educated in Germany, where in 1893 he obtained a patent on the engine that bears his name. Originally his idea was to inject coal dust instead of liquid fuel into compressed air at high enough temperatures for the dust to ignite. He narrowly escaped death when his first attempt resulted in the engine's blowing up at the first injection of fuel. A successful engine was produced after some 4 years of tedious and costly work. Diesel disappeared and presumably drowned while crossing the English Channel during a storm.

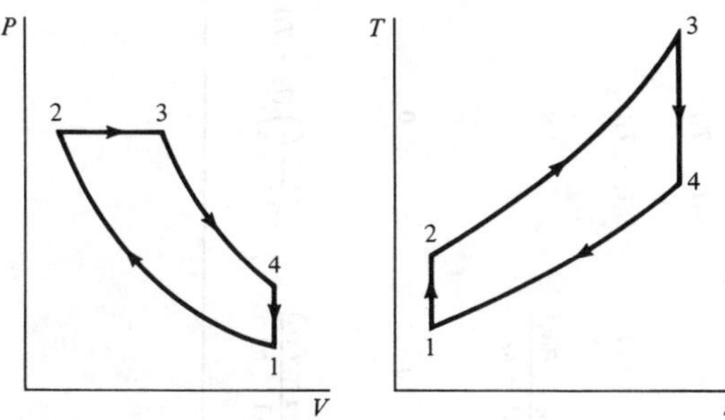

Figure 1-2 Pressure-volume and temperature-entropy diagrams of an ideal diesel cycle.

ideal and adiabatic (no heat exchanged) compression process 1-2, a constant-pressure heat addition process 2-3, an ideal and adiabatic expansion process 3-4, and a constant-volume heat rejection process 4-1, which returns the cycle back to 1. Because the beginning and the end of the cycle is 1 (or any other point), a thermodynamic cycle is a closed system where

$$\Delta U = U_1 - U_1 = 0$$

and the first law for this, and all other cycles, becomes

$$\Delta Q_{\text{net}} = Q_A - |Q_R| = \Delta W_{\text{net}} \qquad \text{(for a cycle)} \tag{1-29}$$

1-5 PROPERTY RELATIONSHIPS

Perfect Gases

Property relationships for perfect gases for different processes are given in Table 1-3. A *perfect (or ideal) gas* is one which, at any state, obeys the *equation of state for perfect gases*

$$PV = mRT \tag{1-30a}$$

$$Pv = RT \tag{1-30b}$$

$$pV = nR_oT \tag{1-30c}$$

where R = specific gas constant. Different gases have different values of R; for air $R = 53.3$ ft · lb$_f$/(lb$_m$ · °R), 286.8 J/(kg · K)

n = number of moles = m/M, where M is the molecular mass of the gas = 28.97 for air

R_o = universal gas constant = RM, the same for all perfect gases = 1545.33 ft · lb$_f$/(lb · mol · °R) = 8314.34 J/(kg · mol · K)

T = absolute temperature in degrees Rankine or Kelvin

Table 1-3 Perfect-gas relationships (constant specific heats)

Process	P, v, T relationships	$u_2 - u_1$	$h_2 - h_1$	$s_2 - s_1$	w (nonflow)	w (flow)	Q
Isothermal	T = constant $P_1/P_2 = v_2/v_1$	0	0	$(R/J)\ln(v_2/v_1)$	$(P_1v_1/J)\ln(v_2/v_1)$	$(P_1v_1/J)\ln(v_2/v_1)$	$(P_1v_1/J)\ln(v_2/v_1)$
Constant pressure	P = constant $T_2/T_1 = v_2/v_1$	$c_v(T_2 - T_1)$	$c_P(T_2 - T_1)$	$c_P\ln(T_2/T_1)$	$P(v_2 - v_1)/J$	0	$c_P(T_2 - T_1)$
Constant volume	v = constant $T_2/T_1 = P_2/P_1$	$c_v(T_2 - T_1)$	$c_P(T_2 - T_1)$	$c_v\ln(T_2/T_1)$	0	$v(P_1 - P_2)/J$	$c_v(T_2 - T_1)$
Isentropic (Adiabatic reversible)	s = constant $P_1v_1^k = P_2v_2^k$ $T_2/T_1 = (v_1/v_2)^{k-1}$ $T_2/T_1 = (P_2/P_1)^{(k-1)/k}$	$c_v(T_2 - T_1)$	$c_p(T_2 - T_1)$	0	$\dfrac{(P_2v_2 - P_1v_1)}{J(1 - k)}$	$\dfrac{k(P_2v_2 - P_1v_1)}{J(1 - k)}$	0
Throttling	h = constant T = constant $P_1/P_2 = v_2/v_1$	0	0	$(R/J)\ln(v_2/v_1)$	0	0	0
Polytropic	$P_1v_1^n = P_2v_2^n$ $T_2/T_1 = (v_1/v_2)^{n-1}$ $T_2/T_1 = (P_2/P_1)^{(n-1)/n}$	$c_v(T_2 - T_1)$	$c_P(T_2 - T_1)$	$c_v\ln(P_2/P_1) + c_P\ln(v_2/v_1)$	$\dfrac{(P_2v_2 - P_1v_1)}{J(1 - n)}$	$\dfrac{n(P_2v_2 - P_1v_1)}{J(1 - n)}$	$c_v\left(\dfrac{k - n}{1 - n}\right)(T_2 - T_1)$

Imperfect Gases

Property relationships for nonperfect gases and vapors are more complex than those for perfect gases. Property values for these fluids are facilitated by the use of charts and tables [4, 5, 6].

A *nonperfect gas* is one in which the molecules are close enough to exert forces on each other, as when a perfect gas is highly compressed and/or highly cooled with respect to its critical conditions. It is often described by modifying Eqs. (1-30*a*, *b*, and *c*) to the form

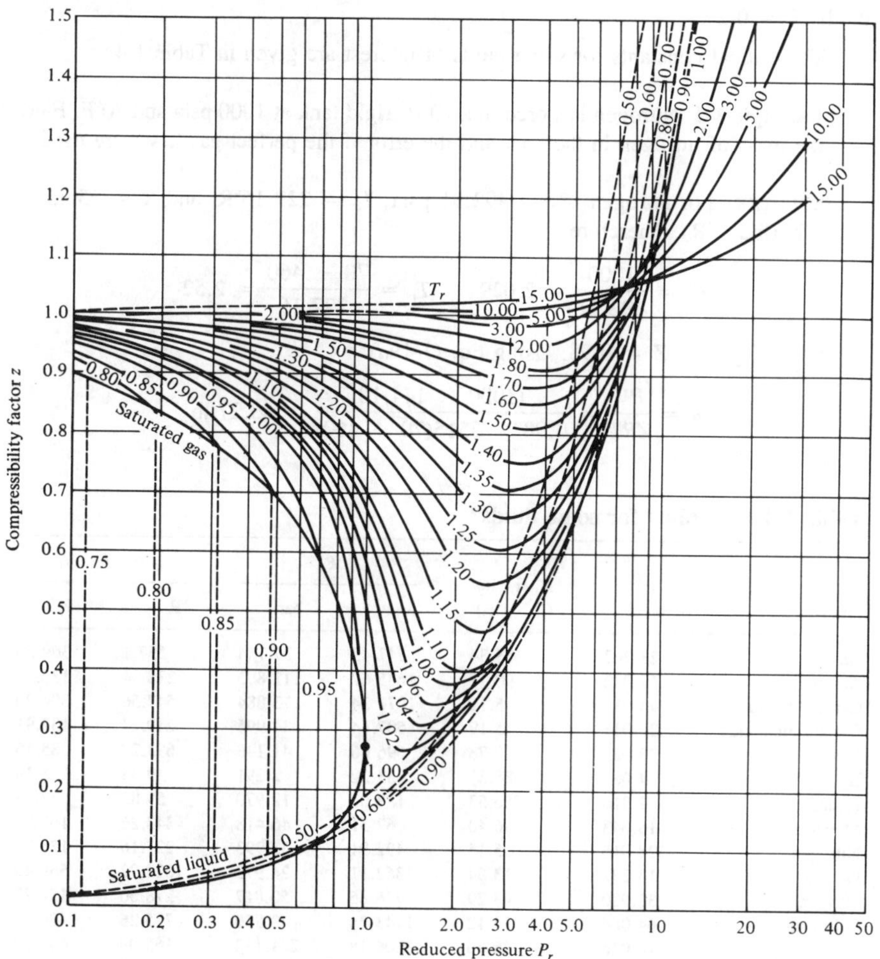

Figure 1-3 Generalized compressibility factor chart.

$$PV = mZRT \tag{1-31}$$

where Z is a compressibility factor that depends upon P, T, and the gas itself. Z is given in various compressibility charts. A generalized compressibility chart (see Fig. 1-3) gives Z for all gases as a function of the *reduced pressure* P_r and *reduced temperature* T_r, where

$$P_r = \frac{P}{P_c} \quad \text{and} \quad T_r = \frac{T}{T_c} \tag{1-32}$$

where P_c and T_c are the pressure and temperature, respectively, at the thermodynamic critical point for each gas. Note that at $P_c = 1$ and $T_c = 1$, the critical point for all fluids, $Z = 0.27$.

The critical constants for some fluids of interest are given in Table 1-4.

Example 1-1 Nitrogen is stored in a 10-ft^3 rigid tank at 1000 psia and 70°F. Find the mass of nitrogen in the tank and the error if the perfect gas law were used.

SOLUTION For nitrogen, $P_c = 492.91$ psia, $T_c = 227.16$°R, and $R = 55.15$ ft · lb$_f$/(lb$_m$ · °R). Therefore

$$P_r = \frac{1000}{492.91} = 2.029 \qquad T_r = \frac{70 + 460}{227.16} = 2.52$$

From Fig. 1-3: $Z = 0.98$. Using Eq. (1-19*d*)

$$m = \frac{PV}{ZRT} = \frac{(1000 \times 144)10}{0.98 \times 5515(70 + 460)} = 50.27 \text{ lb}_m$$

Table 1-4 Constants for some fluids*

Fluid	M	R, ft · lb$_f$/(lb$_m$ · °R)	P_c		T_c	
			psia	bar	°R	K
Air	28.967	53.34	547.43	37.744	557.1	309.50
Ammonia	17.032	90.77	1635.67	112.803	238.34	132.41
Carbon dioxide	44.011	35.12	1071.34	73.884	547.56	304.20
Carbon monoxide	28.011	55.19	507.44	34.995	239.24	132.91
Freon-12	120.925	12.78	596.66	41.148	693.29	385.16
Helium	4.003	386.33	33.22	2.291	9.34	5.19
Hydrogen	2.016	766.53	188.07	12.970	59.83	33.24
Methane	16.043	96.40	67.31	46.418	343.26	190.70
Nitrogen	28.016	55.15	492.91	33.993	227.16	126.20
Octane	114.232	13.54	362.11	24.973	1024.92	569.40
Oxygen	32.000	48.29	736.86	50.817	278.60	154.78
Sulfur dioxide	64.066	24.12	1143.34	78.850	775.26	430.70
Water	18.016	85.80	3206.18	221.112	1165.09	647.27

*Data from Ref. 4.

Using the perfect gas law, $m = PV/RT = 49.265 \text{ lb}_m$ with an error of about −2 percent.

Vapors

Property values for vapors, such as steam, are more complex than those for nonperfect gases and are given in the tables in App. A through F and in charts such as the *T-s*, *P-v*, and *h-s* (Mollier) diagrams (*s* is entropy, Sec. 1-8). They include data for the saturated liquid, the saturated vapor, the difference between these two, and for the superheated vapor. By the use of the familiar parameter *quality x,* which is the ratio of mass of vapor to mass of vapor and liquid in a two-phase mixture, one is able to obtain the properties of such a mixture. Recall that a two-phase mixture in equilibrium includes saturated liquid and saturated vapor and that neither can be subcooled or superheated. Thus the specific enthalpy of a two-phase mixture is given by

$$h = h_f + xh_{fg} \tag{1-33}$$

where h_f is the enthalpy of the saturated liquid and h_{fg} is the difference between the enthalpy of the saturated vapor, h_g, and h_f; that is, $h_{fg} = h_g - h_f$, all obtained at the pressure of the system. Similar equations are written for specific volume and entropy as

$$v = v_f + xv_{fg} \tag{1-34}$$

and

$$s = s_f + xs_{fg} \tag{1-35}$$

Subcooled Liquids

The subcooled liquid is the least understood. A *subcooled liquid* is one at a temperature below the saturation temperture at the given pressure. Examples are water at 70°F and 14.696 psia where the saturation temperature is 212°F, or water at 640°F and 2500 psia where the saturation temperature is 668.11°F. The term *compressed liquid* is synonymous with subcooled liquid. If we change our point of view from pressure to temperature and use the same examples above, we find that water at 70°F, if saturated, would have a pressure of 0.36292 psia. Because it is at 14.696 psia, it is said to be compressed. Likewise, water at 600°F has a saturation pressure of 2059.9 psia. Being at 2500 psia, it is said to be compressed. Hence, subcooled and compressed liquid have the same meaning.

Property data for the subcooled liquid may be found in tables [6], charts (Fig. 1-4), or by approximation from the saturated temperature steam tables. This works as follows. Data for a subcooled (or compressed) liquid are obtained as if the liquid were saturated at its given temperature, and the pressure is ignored. This approximation is reasonable, provided the actual and saturation pressures are not too far apart, within a few hundred psia for water, for example.

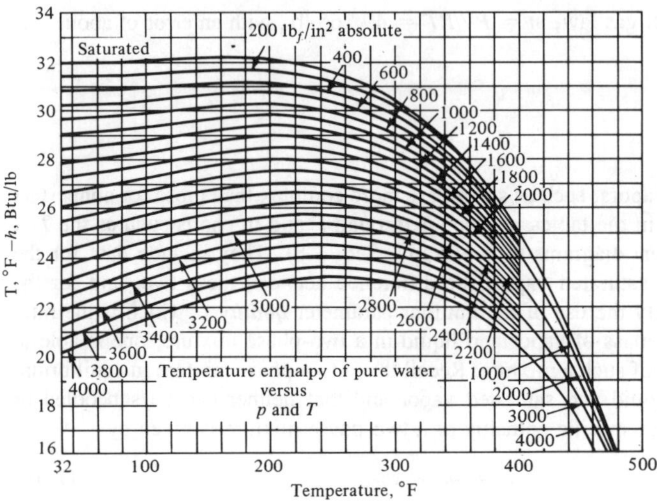

Figure 1-4 Enthalpy of subcooled water.

Example 1-2 Find the enthalpy of water at (1) 70°F and 1 atm and at (2) 640°F and 2500 psia.

SOLUTION

1. $h = h_f$ at 70°F $= 38.052$ Btu/lb$_m$
2. $h = h_f$ at 640°F $= 679.1$ Btu/lb$_m$

The American Society of Mechanical Engineers (ASME) tables for the compressed liquid give enthalpies of 38.09 and 673.47 Btu/lb$_m$, respectively. The approximations have resulted in errors of less than -0.1 percent in condition 1 and less than $+1$ percent in condition 2. Some subcooled water data is included in the superheat steam table, A-3, Appendix.

1-6 THE SECOND LAW OF THERMODYNAMICS

Whereas the first law of thermodynamics was one of conservation of energy, declaring that all forms of energy are convertible to one another, the second law puts a limitation on the conversion of some forms of energy to others. We are most concerned with two forms, heat and work. The second law does not negate the equivalence of conversion of these two, only the extent. Work is the more valuable commodity. It can

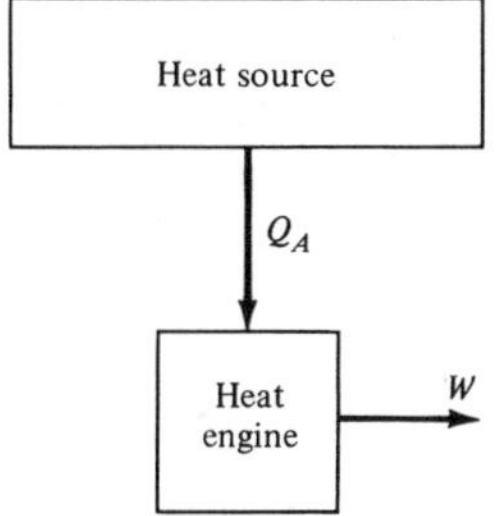

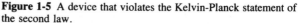

Figure 1-5 A device that violates the Kelvin-Planck statement of the second law.

be completely and continuously converted to heat. The opposite is not true. Heat cannot be completely and continuously converted to work. In other words, heat is not entirely *available* to do work on a continuous basis, i.e., in a cycle (though it may be in a process).

The portion of heat that cannot thus be converted to work, called *unavailable energy,* has to be rejected as low-grade heat after the work has been done. Thus, while energy is conserved, availability is not. The availability of a system always decreases. Another way of phrasing the second law is that the thermal efficiency of continuously converting heat to work, in a heat engine, must be less than 100 percent. The Carnot cycle (Sec. 1-9) represents an ideal heat engine that gives us an upper value of that efficiency between any two temperature limits.

There are a few historical statements of the second law that convey the above thoughts. Two are:

1. The Kelvin-Planck statement.* *It is impossible to construct a device which will operate in a cycle and produce no effect other than the raising of a weight and the exchange of heat with a single reservoir.* In this statement "operate in a cycle" means operate continuously, "raising of a weight" means doing work, and "exchange of heat with a single reservoir" means heat is only added, not rejected, and that there is a thermal efficiency of 100 percent. Figure 1-5 shows a device that violates the Kelvin-Planck statement.

* This statement is credited to both William Thompson, later Lord Kelvin (1824–1907), and Max Planck (1858–1947), though each stated it in a somewhat different way. Kelvin, who was knighted for helping lay the first transatlantic cable, was a professor of physics at Glasgow University, an excellent mathematician, an inventor, and a designer and was interested in athletics, the arts, and music. He contributed most to thermodynamics, establishing the thermodynamic temperature scale which is independent of the properties of matter. He also helped establish the first law on a firm foundation and contributed to the statement of the second law. Max Karl Ernst Ludwig Planck was a German professor who studied in Munich and taught in the universities of Munich, Kiel, and Berlin. One of his great contributions was in wavelength radiation and the definition of the "black body." He postulated the quantum theory and Planck's constant h, which has continued to influence physics and related sciences to a degree well beyond its original intent. Planck had wide interests in fields other than physics, including philosophy, religion, and social and political matters.

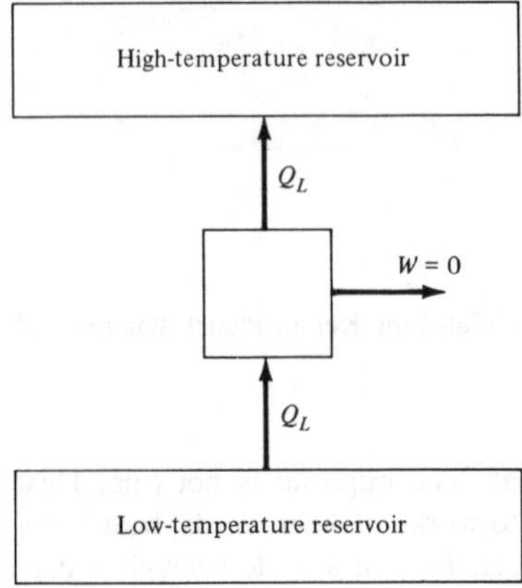

Figure 1-6 A device that violates the Clausius statement of the second law.

2. The Clausius statement.* *It is impossible to construct a device which operates in a cycle and produces no effect other than the transfer of heat from a cooler to a hotter body.*

These two statements, though they say different things, can be shown to be equivalent. Figure 1-6 shows a device that operates between a high-temperature reservoir and a low-temperature reservoir. A *reservoir* is a source of heat or a heat sink large enough that it does not undergo a change in temperature when heat is added or subtracted from it. The device in Fig. 1-6 produces no effect other than the transfer of heat from the low-temperature reservoir to the high-temperature reservoir; hence, it produces (or absorbs) zero work. By the first law, with this device as a system, the heat Q_L received from the low-temperature reservoir is equal to that delivered to the high-temperature reservoir.

Let us now add to the above a device that does not violate the Kelvin-Planck statement (Fig. 1-7*a*). Let us also choose that second device so that it rejects to the low-temperature reservoir the same Q_L as the first device. When we combine both devices (Fig. 1-7*b*), the result will be a device that receives $Q_H - Q_L$ from the single reservoir, the high-temperature one, and produces work W. This violates the Kelvin-Planck statement.

* Rudolf Julius Emmanuel Clausius (1822–1888) was a German professor of physics and a mathematical genius who worked in optics, electricity, and electrolysis and is credited with founding the kinetic theory of gases. Clausius elaborated and restated the work of Carnot (Sec. 1-9) and thus deduced his famous principle of the second law. He wrote an exhaustive treatise on the steam engine in which he emphasized the then new concept of entropy.

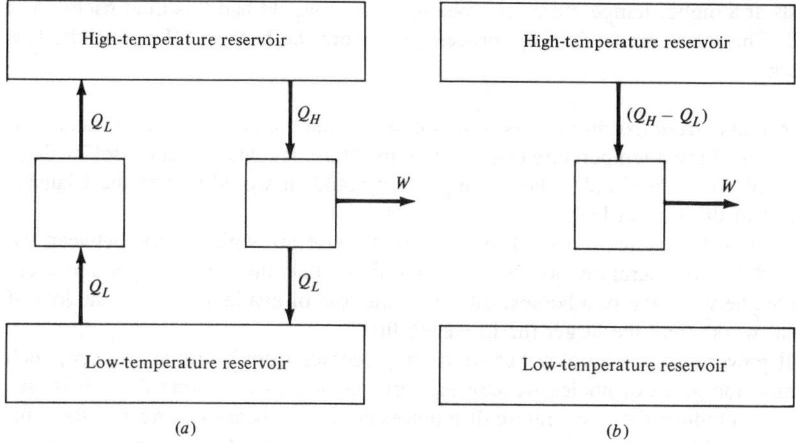

Figure 1-7 Conversion of device in Fig. 1-6 to device in Fig. 1-5.

1-7 THE CONCEPT OF REVERSIBILITY

The concept of reversibility was first introduced in 1824 by Sadi Carnot, who laid the foundations of the second law and introduced the concept of a cycle. Reversibility applies to processes. A cycle can be reversible, but only if *all* its processes are reversible. One irreversible process in a cycle renders the whole cycle irreversible.

A *reversible process,* also called an *ideal process,* is one which can reverse itself exactly by following the same path it undertook in the first place and thus restore to the system or the surroundings the same heat and work previously exchanged. For example, process 1-2 in Fig. 1-2, which was called ideal and is now called reversible, is so only when it can trace its exact path back from 2 to 1 and thus give back the exact work of compression done upon it from 1 to 2.

Needless to say there are no ideal, and hence no reversible, processes in the real world. Real processes are *irreversible,* although the degree of irreversibility varies between processes. There are many sources of irreversibility in nature. For our purposes, four are most important. These are *friction, heat transfer, throttling,* and *mixing*.

Friction *Mechanical friction* is one in which mechanical work is dissipated into a heating effect, such as in the case of a shaft rotating in a bearing. It is not possible to add the same heat to the bearing and expect the shaft to rotate. Another example is the rubbing of one's hands. Thus friction is the dissipation of energy that otherwise could have been transformed to useful work.

Fluid friction is similar to mechanical. A fluid expanding behind a piston or through a turbine undergoes internal friction, resulting in the dissipation of part of its energy into heating itself at the expense of useful work. The fluid then does less work and exhausts at a higher temperature (or enthalpy) than it would had this fluid friction not existed. The more irreversible the process, the more the heating effect and the less the work.

Heat transfer *Heat transfer* in any of its forms, conduction, convection, or radiation, occurs from a higher temperature to a lower temperature. It cannot reverse itself without the help of an external aid (a heat pump). If it could, it would violate the Clausius statement of the second law.

Heat transfer causes a loss of availability because no work is done between the high- and low-temperature bodies. It also follows that the larger the temperature difference between the two bodies, the larger the loss of availability (i.e., the loss of potential work), and the larger the irreversibility.

All power systems employ heat-transfer processes from a primary source, such as combustion gases or nuclear-reactor primary coolant, to a working fluid. It is thus important to reduce the temperature differences across the heat exchanger to the minimum practicable to reduce this source of irreversibility, which is a primary cause of reduction in efficiency of real systems.

Throttling *Throttling* is an uncontrolled expansion process of a fluid from a high-pressure (and occasionally high-temperature) region to a low-pressure region. Examples are leakage from a steam pipe or the flow through a valve. No work is done and heat transfer across the narrow opening is negligible. The resulting high kinetic energy is dissipated in fluid friction to restore the enthalpy of the fluid to its original value. In Eq. (1-1*d*) all terms drop out except enthalpy, resulting in Eq. (1-24*a*), which is repeated here.

$$H_1 = H_2 \tag{1-24a}$$

A throttling process, therefore, is a form of fluid friction, one in which the enthalpy is constant. Such a process is also called a *Joule-Thompson expansion*. Throttling is irreversible because flow cannot be reversed from the low-pressure region to the high-pressure region. It results in the loss of availability, i.e., the loss of work that could have been otherwise obtained if expansion between the two regions occurred ideally behind a piston or through a turbine.

It follows that the greater the pressure difference in throttling, the greater the irreversibility. Throttling should be avoided or minimized in power systems, though it is necessary in some applications.

Mixing When two or more separate fluids or gases are made to mix or diffuse into each other, they cannot unmix without external aid. Hence *mixing* is an irreversible process that results in the loss of availability of the constituent fluids. Mixing is unavoidable in many cases, such as when fuels and air are prepared for combustion or when steam and colder water mix in certain devices, such as open feedwater heaters. It is not, however, as major a concern in power systems as fluid friction or heat transfer.

External and Internal Irreversibilities In power systems, irreversibilities are sometimes classified as external and internal.

External irreversibilities are those that occur across the boundaries of the system. The primary source of external irreversibility in power systems is heat transfer, both at the high-temperature end, the heat source, and the low-temperature end, the heat sink. Another source of external irreversibility in power systems is mechanical friction in bearings of rotary machines such as turbines, compressors, pumps, and generators and electrical losses in the latter and in distribution systems.

Internal irreversibilities are those that occur within the boundaries of the system. The primary source of internal irreversibility in power systems is fluid friction in rotary machines such as turbines, compressors, and pumps and in pipes and valves. Other sources are throttling and mixing.

1-8 THE CONCEPT OF ENTROPY

Entropy, first introduced by Clausius in 1865, is a property, as are pressure, temperature, internal energy, and enthalpy. It is given the symbol S (and sometimes Φ) and has the units Btu per degree Rankine (Btu/°R) or the units joule per kelvin (J/K). Specific entropy s has the units Btu/(lb$_m$ · °R) or J/(kg · K). The physical meaning of entropy will be apparent later. It is convenient to introduce it first as a mathematical convenience.

An analog is made with nonflow work

$$\Delta W_{\text{nf}} = \int_1^2 P\,dV \tag{1-26}$$

This expression strictly gives the nonflow work for a reversible or ideal process because it is the only case where there is pressure equilibrium, i.e., when the pressure at the face of the piston is the same as in the bulk of the fluid, a situation that does not happen in nature, especially for high-speed machines. Its use, however, is extended to real processes because the pressure differential is small.

Equation (1-26) shows that the nonflow work is, graphically, the area under the process when plotted on a P-V diagram (Fig. 1-8a). Because heat, like work, is an

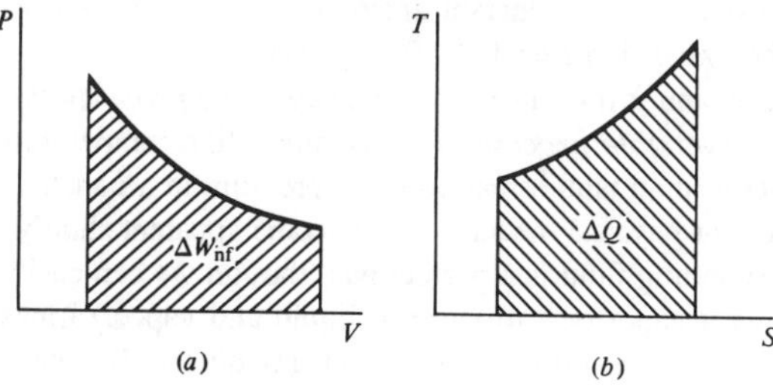

Figure 1-8 Areas under process 1-2 on P-V and T-S diagrams.

important energy term in energy systems, we ask ourselves: Wouldn't it be convenient to have a similar graph where areas represent heat? It is natural to have temperature as one of the ordinates. We invent the second and call it *entropy*. As $\Delta W_{nf} = \int_1^2 P\, dV$ strictly for a reversible process, then we can write

$$\Delta Q = \int_1^2 T\, dS \qquad \text{(reversible process only)} \tag{1-36}$$

and the area under process 1-2 on the *T-S* diagram equals ΔQ (Fig. 1-8*b*). As we shall see, although we can waive the reversibility requirement for Eq. (1-26) with little error, we cannot do the same for Eq. (1-36), which is true strictly for reversible processes.

In both Eqs. (1-26) and (1-36), if the integral is positive, i.e., we move from left to right, work is done by and heat is added to the process. In cycles, the net work and net heat of the cycles are represented by the enclosed areas on the *P-V* and *T-S* diagrams, provided the cycle is reversible (i.e., all its processes are reversible) in the case of the *T-S* diagram.

It also follows that the net work of a cycle, such as the diesel cycle (Fig. 1-2), is equal to the enclosed area of the cycle on the *P-V* diagram. A cycle is a power cycle, i.e., the net work is positive, if the cycle is clockwise. It is a reversed cycle or a heat pump if the net work is negative and the cycle is counterclockwise on the *P-V* diagram. It also follows from Eq. (1-29) that the enclosed area on the *T-S* diagram is also equal to the net work if the cycle is reversible. The enclosed areas on the *P-V* and *T-S* diagrams for the same reversible cycle are therefore equal in sign and magnitude, taking scales and conversion factors into account, of course.

We now examine (1-36) further. A *reversible adiabatic process* (already mentioned in the case of the ideal diesel cycle), while not existing in nature, is a most important process in cycle analysis. For such a process, being adiabatic, $\Delta Q = 0$, and being reversible, it obeys Eq. (1-36). Thus, for an adiabatic reversible process,

$$\int T\, dS = 0 \tag{1-37a}$$

and because T cannot equal zero, $dS = 0$

$$\begin{aligned} \text{or} \quad & S = \text{constant} \\ \text{and} \quad & s = \text{constant} \qquad \text{(adiabatic reversible process)} \end{aligned} \tag{1-37b}$$

Thus an adiabatic reversible process is one of constant entropy, i.e., vertical on the *T-S* diagram, as represented by the expansion line 1-2_s (Fig. 1-9*a*).

All this leads to one physical meaning for entropy: *Entropy is the property that remains constant in an adiabatic reversible process*. This is much like temperature being the property that stays constant in an isothermal process; pressure in a constant-pressure process, etc. A more important physical meaning will become apparent shortly.

Let us now assume that the expanding fluid is a perfect gas (though we can easily come to the same conclusions using a vapor or a mixture of liquid and vapor.) Lines P_1 and P_2 (Fig. 1-9*a*) are constant-pressure lines for a perfect gas on the *T-s* plane where $P_1 > P_2$. Their shape can be ascertained from the perfect-gas relationship

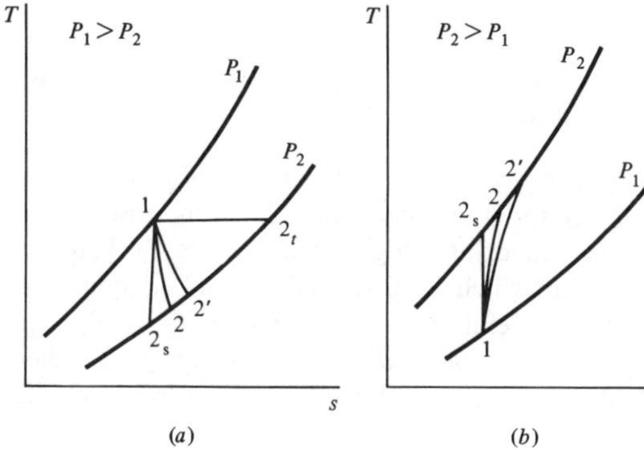

Figure 1-9 Expansion (*a*) and compression (*b*) of a gas from P_1 to P_2 on the *T-S* diagram. 1-2_s adiabatic reversible, 1-2 adiabatic irreversible, 1-2_t throttling.

$s_2 - s_1 = c_p \ln (T_2/T_1)$ (Table 1-3). Let us further assume that the gas starts at 1 with T_1, s_1 (and corresponding P_1 and v_1) and expands *adiabatically* to pressure P_2. We have already seen that if the process is reversible as well as adiabatic, entropy remains constant and the process is represented by 1-2_s.

The question is: What does the process look like if it were adiabatic but irreversible? We have already noted that irreversibility would manifest itself in an increase in temperature of the gas leaving at P_2, point 2, beyond that for the adiabatic reversible process. Thus $T_2 > T_{2s}$. On the P_2 line, which curves upwards, this can only occur if the entropy at 2 is greater than at 2_s. A more irreversible expansion results in greater self-heating of the gas and the process would be 1-2', etc. In other words: *The greater the irreversibility, the greater the increase in entropy in an adiabatic process.*

If the expansion were to occur in a turbine, the work is obtained from the first law (which applies equally to reversible and irreversible processes) for the three processes as $H_1 - H_{2s}$, $H_1 - H_2$, and $H_1 - H_{2'}$, respectively. Because the $T_{2'} > T_2 > T_{2s}$ and $dh = c_p\, dT$ for gases, then $H_{2'} > H_2 > H_{2s}$. Thus the work produced by the turbine W_T for three cases is given by

$$H_1 - H_{2s} > H_1 - H_2 > H_1 - H_{2'}$$

The adiabatic reversible turbine produces the most work. The greater the irreversibility, the less the work.

The process 1-2_t is one of constant temperature and hence, for a gas, constant enthalpy. This is a throttling process [Eq. (1-11)], where the work ΔH is zero and all energy is dissipated in fluid friction. This is the most irreversible process and the one with the most increase in entropy (Fig. 1-9*a*).

The degree of irreversibility is given by an expansion or turbine efficiency called the *polytropic turbine efficiency* η_T (sometimes called the isentropic or adiabatic turbine efficiency), which is equal to the ratio of actual work to ideal work and is given by

$$\eta_T = \frac{H_1 - H_2}{H_1 - H_{2s}} = \frac{h_1 - h_2}{h_1 - h_{2s}} \tag{1-38}$$

and for constant specific heats

$$\eta_T = \frac{T_1 - T_2}{T_1 - T_{2s}} \tag{1-39}$$

In the case of compression from P_1 to P_2 where $P_2 > P_1$ (Fig. 1-9*b*), an adiabatic reversible compression follows the constant entropy path 1-2_s. In the case of adiabatic irreversible compression the gas leaves at a higher temperature T_2, here because the fluid absorbs some work input which is dissipated in fluid friction. The greater the irreversibility, the greater the exit temperature $T_{2'} > T_2 > T_{2s}$ and the greater the increase in entropy. Again, since $dh = c_p\, dT$ for gases, then $H_{2'} > H_2 > H_{2s}$ and the work absorbed in compression $|W_c|$ increases with irreversibility or

$$H_{2'} - H_1 > H_2 - H_1 > H_{2s} - H_1$$

The degree of irreversibility here is given by a compressor efficiency, called the *polytropic compressor efficiency* η_c (and sometimes the isentropic or adiabatic compressor efficiency) equal to the ratio of ideal work to actual work (the reverse of that for expansion) and given by

$$\eta_c = \frac{H_{2s} - H_1}{H_2 - H_1} = \frac{h_{2s} - h_1}{h_2 - h_1} \tag{1-40}$$

and for constant specific heats

$$\eta_c = \frac{T_{2s} - T_1}{T_2 - T_1} \tag{1-41}$$

We can now state that *the change of entropy is a measure of the unavailable energy*, which leads us to an important physical meaning of entropy, namely that entropy is a measure of irreversibility, or more generally: *Entropy is a measure of disorder*. This is a concept that is used in sciences other than thermodynamics, such as social sciences that deal with society disorders.

Because the universe is an isolated system it does not exchange energy across its boundaries; i.e., it is adiabatic. Because it is full of irreversible processes, it follows that: *The entropy of the universe is continually on the increase* and the end of the universe would occur when entropy is at maximum, i.e., when all energy has been dissipated to a bottom state, when all availability is lost, when all matter is at the same temperature and no life, as we know it, is possible.

In the case of vapors, when expansion between P_1 and P_2 ends in the two-phase region, shown in Fig. 1-10 on both *T-s* and *h-s* (Mollier) charts, the same observation as for the gases applies except that the exit temperature is the same for the adiabatic reversible and adiabatic irreversible processes because they are both in the two-phase region. The exit enthalpy, however, is greater in the case of the irreversible process $h_2 > h_{2s}$ and the work is less: $h_1 - h_2 < h_1 - h_{2s}$. The degree of irreversibility here is given by a turbine efficiency, the same as in Eq. (1-38).

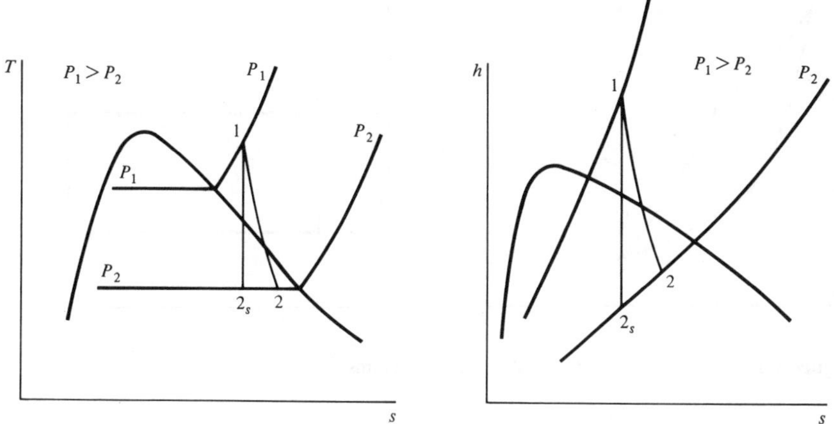

Figure 1-10 Expansion of a vapor from P_1 to P_2 on the T-s and Mollier (h-s) diagrams. 1-2_s adiabatic reversible, 1-2 adiabatic irreversible.

In the case of pumping liquids, a pump efficiency η_p, given by the ratio of ideal work to actual work, is used to obtain the actual work. Using Eq. (1-22a), actual work

$$|W_p| = \frac{H_{2s} - H_1}{\eta_p} \cong \frac{V(P_1 - P_2)}{\eta_p} \tag{1-42}$$

1-9 THE CARNOT CYCLE

Sadi Carnot* laid the foundations of the second law of thermodynamics, introduced the concepts of reversibility and cycles, and introduced the principle that the temperatures of the heat source and heat sink determined the thermal efficiency of a reversible cycle. He also postulated that because all such cycles must reject heat to the heat sink, efficiency is never 100 percent [7]. To show this, and to show that the effect of the working fluid on the thermal efficiency of a reversible cycle is nonexistent, Carnot invented his famous, though hypothetical (one cannot build a reversible engine), *Carnot cycle*.

The Carnot cycle, shown in Fig. 1-11 on the P-V and T-S diagrams, is composed of four processes:

* Nicolas Leonard Sadi Carnot (1796–1832), despite his profound and lasting effect on the science of thermodynamics, was a quiet, unassuming Frenchman who lived during the turbulent Napoleonic years and had an unspectacular life. One of his mottoes, "Speak little of what you know and not at all of what you do not know," reflects something of his demeanor.

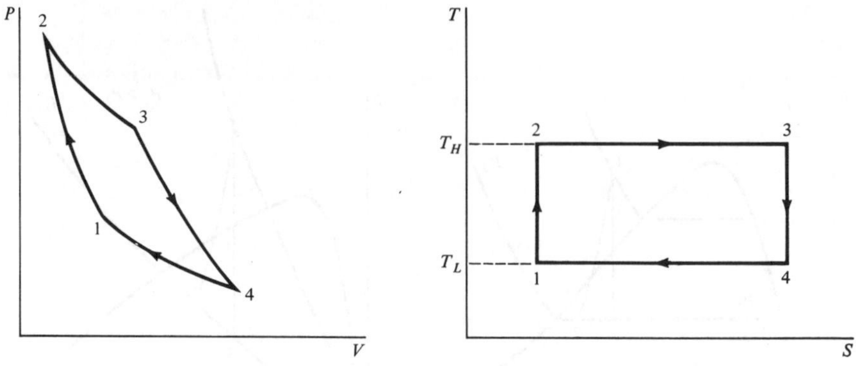

Figure 1-11 Carnot cycle on the *P-V* and *T-S* diagrams.

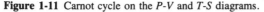

1. *1-2:* reversible adiabatic compression
2. *2-3:* reversible constant-temperature heat addition
3. *3-4:* reversible adiabatic expansion
4. *4-1:* reversible adiabatic heat rejection

The thermal efficiency of the Carnot cycle η_C can now be easily obtained, noting that the change in entropy during heat addition and rejection are equal in magnitude. Thus

$$Q_A = T_H (S_3 - S_2)$$

$$Q_R = T_L (S_1 - S_4)$$

or

$$|Q_R| = T_L (S_4 - S_1) = T_L (S_3 - S_2)$$

where T_H and T_L are the heat source and heat sink absolute temperatures, respectively. Note that the Carnot cycle is both externally and internally reversible. Thus heat transfer between the heat source and the working fluid occurs, hypothetically, across a zero temperature difference. Hence $T_H = T_2$. Similarly $T_L = T_1$.

For *all* power cycles the net work and the thermal efficiency are defined by

$$\Delta W_{\text{net}} = Q_A - |Q_R| \tag{1-43}$$

and

$$\eta_{\text{th}} = \frac{\Delta W_{\text{net}}}{Q_A} \tag{1-44}$$

Thus, the thermal efficiency of the Carnot cycle η_C is given by

$$\eta_C = \frac{T_H - T_L}{T_L} \tag{1-45}$$

This relationship can also be easily obtained from the net work of the different processes using the appropriate relationships in Table 1-1 (Prob. 1-11).

Equation (1-45) shows that the thermal efficiency of the Carnot cycle is only a function of the heat source and heat sink temperatures and is independent of the working fluid. Indeed this conclusion is true of other reversible cycles (Stirling, Ericsson) that receive and reject heat at constant temperature.

Because the Carnot cycle is reversible, it produces the maximum work possible between the given temperature limits T_H and T_L and, hence, a reversible cycle operating between given temperature limits has the highest possible thermal efficiency of all cycles operating between these same temperature limits.

The Carnot efficiency, therefore, is to be considered an upper limit, an ideal, a goal to strive toward when building real cycles, but one that is impossible to exceed or equal.

Another observation of importance is that the trends predicted by the Carnot cycle apply to real cycles. Thus real cycles will have higher efficiencies, the higher the temperature at which they receive heat and the lower the temperature at which they reject heat.

Example 1-3 Air expands in a gas turbine from 10 atm and 2000°F to 1 atm. The exhaust temperature is 1050°F. Assuming a constant specific heat of 0.240 Btu/(lb$_m$ · °F), find the turbine work in Btu/lb$_m$ and the turbine adiabatic efficiency.

SOLUTION $W_T = h_1 - h_2 = c_p(T_1 - T_2) = 0.24(2000 - 1050)$

$= 228$ Btu/lb$_m$

Referring to Table 1-3 and Fig. 1-9*a*

$$\frac{T_{2s}}{T_1} = \left(\frac{P_2}{P_1}\right)^{(k-1)/k} \quad \text{or} \quad \frac{T_{2s}}{(2000 + 460)} = \left(\frac{1}{10}\right)^{(1.4-1)/0.4}$$

Therefore $T_{2s} = 1274.2°\text{R} = 814.2°\text{F}$

$$\eta_T = \frac{h_1 - h_{2s}}{h_1 - h_2} = \frac{T_1 - T_{2s}}{T_1 - T_2} \quad \text{(for constant } c_p\text{)}$$

$$= \frac{2000 - 1050}{2000 - 814.2} = 0.801 = 80.1\%$$

Example 1-4 Find the work of the turbine of Example 1-1 but use variable specific heats.

SOLUTION Using Eq. (1-19)

$$W_T = \int_1^2 c_p\, dT = \int_{1510}^{2460} (0.219 + 3.42 \times 10^{-5} T - 2.93 \times 10^{-9} T^2)\, dT$$

$$= \left[0.219T + \frac{3.42 \times 10^{-5}}{2} T^2 - \frac{2.93 \times 10^{-9}}{3} T^3\right]_{1510}^{2460}$$

$$= 0.219(2460 - 1510) + \frac{3.42 \times 10^{-5}}{2}(2460^2 - 1510^2) - \frac{2.93 \times 10^{-9}}{3}(2460^3 - 1510^3)$$

$$= 208.05 + 64.49 - 14.54$$

$$= 258 \text{ Btu/lb}_m$$

as compared with $W_T = 228$ Btu/lb$_m$ when a constant specific heat at low temperature [0.24 Btu/(lb$_m \cdot$ °R)] was used.

To get the adiabatic efficiency, it is necessary to calculate T_{2s} and h_{2s} using average values of c_p and k between T_1 and the unknown T_{2s}. The solution may be obtained by trial and error, keeping in mind that because both c_p and c_v increase with temperature, and the difference between them is a constant, $k = c_p/c_v$ decreases with temperature. Thus one expects that T_{2s} would be greater than in Example 1-1. Assuming it to be 950°F, find an average $\overline{c_p}$ between 2000 and 950°F using Eq. (1-19)

$$\overline{c_p} = \frac{\int_{1410}^{2460} c_p \, dT}{2000 - 950} = \frac{287.64}{1050} = 0.2739 \text{ Btu/(lb}_m \cdot \text{°R)}$$

$$\overline{c_v} = \overline{c_p} - R = 0.2739 - \frac{53.34}{778.16} = 0.2739 - 0.0685 = 0.2054 \text{ Btu/(lb}_m \cdot \text{°R)}$$

Therefore $\bar{k} = \dfrac{0.2739}{0.2054} = 1.3335$

and $$T_{2s} = T_1 \left(\frac{1}{10}\right)^{\frac{0.3335}{1.3335}} = 2460 \times 0.5622 = 1383\text{°R} = 923\text{°F}$$

A second trial and error shows that $\overline{c_p} = 0.2736$ and $T_{2s} = 1382$°R $= 922$°F compared with 814.2°F for the adiabatic reversible work.

The ideal work is

$$h_1 - h_{2s} = \int_{1382}^{2460} c_p \, dT = \overline{c_p}(T_1 - T_{2s}) = 295.4 \text{ Btu/lb}_m$$

The turbine adiabatic efficiency is

$$\frac{258.0}{295.4} = 0.873 = 87.3\%$$

Compared with 80.1 percent when calculated using $c_p = 0.24$ Btu/(lb$_m \cdot$ °R). In the case of air with products of combustion, as in the case with gas turbines and other devices, there are available property tables [8] that take into account the temperature, the fuel-air ratios, and the effects of chemical equilibrium.

Example 1-5 A water pump compresses saturated water at 1 to 2500 psia. The pump efficiency is 0.70. Find the ideal and actual pump and the water exit temperature.

SOLUTION We will use two methods.

1. Assume water to be incompressible so that v = constant = v_1 at 1 psia = 0.016136 ft^3/lb$_m$, and no temperature rise across the pump,

$$\text{Ideal pump work } |W_{p,s}| = v_1(P_2 - P_1) = \frac{0.016136(2500 - 1)144}{778.16}$$

$$= 7.46 \text{ Btu/lb}_m$$

$$\text{Actual pump work } |W_p| = \frac{\Delta W_{p,s}}{\eta_p} = \frac{7.46}{0.7} = 10.66 \text{ Btu/lb}_m$$

2. Use the ASME steam tables [6]

$$T_1 = 101.74\text{°F} \qquad s_1 = 0.1326 \text{ Btu/lb}_m\text{°R} \qquad h_1 = 69.73 \text{ Btu/lb}_m$$

Using the compressed-liquid tables (a portion of which is reproduced in Table 1-5), for the ideal case $s_{2s} = s_1 = 0.1326$. By interpolation T_{2s} = 102.63°F, v_{2s} = 0.01602 ft^3/lb$_m$, h_{2s} = 77.17 Btu/lb$_m$. Thus, the temperature rises by 0.89°F and, in the ideal case, the specific volume decreases by 0.72 percent.

$$\text{Ideal work } |W_{p,s}| = 77.17 - 69.73 = 7.44 \text{ Btu/lb}_m$$

$$\text{Actual work } |W_p| = \frac{7.44}{0.7} = 10.63 \text{ Btu/lb}_m$$

The actual temperature and specific volume changes may be obtained by interpolation at h_2 = 69.73 + 10.63 = 80.36 Btu/lb$_m$, as 105.86°F and 0.01603 ft^3/lb$_m$.

It can be seen that the two methods are fairly comparable even for the large pressure difference chosen for this example. Thus method 1 is sufficiently accurate for most calculations, with the possible exception of exit water temperature.

Table 1-5 Liquid data used in calculation of example 1-5*

Pressure, psia	Temperature, °F	Specific volume, ft^3/lb$_m$	Enthalpy, Btu/lb$_m$	Entropy, Btu/(lb$_m$ · °R)
1.0	101.74 (saturated)	0.016136	69.73	0.1326
2500	100	0.01601	77.57	0.1280
	110	0.01605	84.45	0.1455

*Data from Ref. 6.

PROBLEMS

1-1 A perfect gas that has a constant specific heat at constant pressure c_p = 0.26 Btu/lb$_m$ undergoes an expansion process in a steady-flow machine with a mass flow rate of 100 lb$_m$/h. The machine is water-cooled. The water mass flow rate is 10 lb$_m$/h. During the process the gas temperature drops from 200 to 100°F and the water temperature rises from 70 to 100°F. Ignoring changes in potential and kinetic energies for the gas, calculate the work of the gas in Btus per hour and watts. [c_p for water = 1.0 Btu/(lb$_m$·°R).]

1-2 Air at 140°F and 100 psia is confined in an uninsulated 10-ft^3 vessel. A propeller is driven inside the vessel by a 50-W electric motor. After a period of 1 h the air temperature dropped to 100°F. Find the heat transfer in Btu per hour.

1-3 A gas has a molecular mass of 30 and a specific heat at constant pressure of 0.25 Btu/(lb$_m$·°R). It undergoes a nonflow polytropic compression during which its temperature increases from 100 to 200°F. The polytropic exponent is 1.3. Calculate the work done and the heat transfer in Btus per pound mass.

1-4 1000 lb$_m$/h of pure air enters a gas turbine at 2000°F and leaves at 1000°F. Find the work in kilowatts using variable specific heat for air, and the error if a constant specific heat equal to that at low temperatures is used.

1-5 Calculate the average specific heats for air in Btus per pound mass per degree Rankine between 1000 and 2000°F.

1-6 Steam is confined in a 100-ft^3 rigid vessel at 5000 psia and 1000°F. Find its mass in pounds mass using (*a*) data from the steam tables and (*b*) the compressibility chart.

1-7 A 10-m^3 rigid tank contains steam at 30 bar and 400°C. It is left to cool down until its pressure drops to 5 bar. Find (*a*) the final condition of the steam and (*b*) the heat transfer, in kilojoules.

1-8 A rigid 10-ft^3 vessel contains air at 15 psia and 1000°F. Heat is added until the air temperature reaches 2000°F. Assuming variable specific heats, calculate (a) the heat added, in Btus, and (*b*) the final pressure, pounds force per square inch absolute.

1-9 Steam at 30 bar and 400°C expands behind a piston in an insulated cylinder to 5 bar and 20 ft^3. Find the work done in kilojoules.

1-10 10 lb$_m$/h of liquid ammonia at 87°F and 250 psia is throttled into a flash tank to 100 psia. Ammonia vapor is drawn out at the top of the tank while liquid is discharged at the bottom. What are the temperature in degrees Fahrenheit, mass and volume flow rates in pounds mass per hour and cubic feet per hour of the two streams?

1-11 Derive the expression for the Carnot cycle efficiency [Eq. (1-45)] using the appropriate work relation for gases from Table 1-3.

1-12 An inventor claims to have built an engine that operates on a cycle, receives 1000 kJ at 500°C, produces work, and rejects 350 kJ at 50°C. Is this claim valid? Why?

1-13 The Carnot cycle is rectangular on the *T-s* diagram. Consider another cycle that is rectangular, but on the *P-v* diagram. Draw that cycle on both the *T-s* and *P-v* diagrams, labeling corners correspondingly, and name all its processes.

1-14 Derive expressions for the efficiency of the cycle in Prob. 1-13 in terms of its high and low temperatures T_H and T_L, and constant specific heats c_p and c_v of a gas working fluid for the cases of equal (*a*) temperature rises and (*b*) heats added at constant volume and pressure. Is this a good cycle? Why?

1-15 Using the expression for the change in entropy of gases for a polytropic process in Table 1-3, derive similar expressions in terms of changes in (*a*) pressure and temperature and (*b*) volume and temperature.

1-16 A reversible cycle consists of an isentropic compression from an initial temperature T_1 to 1000°R, a constant-volume process from 1000 to 1500°R a reversible adiabatic expansion to 1000°R a constant-pressure expansion from 1000 to 1500°R, and a constant-volume process to the initial temperature. Draw the cycle on the *P-v* and *T-s* diagrams, and calculate the initial temperature if the working fluid is a gas with k = 1.40.

1-17 Air expands from 10 bar and 1000°C to 1 bar and 500°C in an insulated turbine. Calculate (*a*) the turbine polytropic efficiency, (*b*) the change in entropy, in kilojoules per kilogram per Kelvin, and (*c*) the work, kilojoules per kilogram, and (*d*) the polytropic exponent n. Assume a constant specific of 1.005 kJ/kg·K.

1-18 Helium is compressed from 15 psia and 40°F to 60 psia. The compressor adiabatic efficiency is 0.70. Find (*a*) the helium exit temperature, in degrees Fahrenheit, (b) the work done in Btu per pound mass, and (c) the change in entropy, Btu/lb_m·°R.

1-19 1000 kg/h of water at 60°C and 1 bar are pumped to 100 bar. The pump efficiency is 0.65. Find the work in kilowatts.

1-20 Saturated Freon-12 vapor at 215 psia expands in a nozzle to 72.433 psia. The nozzle has an efficiency of 0.95 and an exit area of 10 in^2. Find the mass flow rate in pounds mass per hour.

1-21 10^6 lb_m/h of steam at 2500 psia and 1000°F expand in a turbine to 1 psia. The turbine has adiabatic and mechanical efficiencies of 0.90 and 0.95, respectively. It drives an electric generator that has an efficiency of 0.96. Calculate the output power of the generator in megawatts.

CHAPTER

TWO

THE RANKINE CYCLE

2-1 INTRODUCTION

When the Rankine* cycle was devised, it was readily accepted as the standard for steam powerplants and remains so today. Whereas the ideal diesel cycle (Fig. 1-2) is a gas cycle and the Carnot cycle (Fig. 1-11) is a cycle for all fluids, the Rankine cycle is a vapor-and-liquid cycle.

The real Rankine cycle used in powerplants is much more complex than the original, simple ideal Rankine cycle. It is by far the most widely used cycle for electric-power generation today and will most certainly continue to be so in the future. It is the backbone of much of the work presented in this book.

This chapter is devoted exclusively to the Rankine cycle, from its simplest ideal form to its more complex nonideal form with modifications and additions that render it one of the most efficient means of generating electricity today.

2-2 THE IDEAL RANKINE CYCLE

Because Rankine is a vapor-liquid cycle, it is most convenient to draw it on both the *P-V* and *T-S* diagrams with respect to the saturated-liquid and vapor lines of the working fluid, which usually, but not always, is H_2O. Figure 2-1 shows a simplified flow

* William John M. Rankine (1820–1872) was a professor of civil engineering at Glasgow University. He was an engineer and scientist of many talents which, besides civil engineering, included shipbuilding, waterworks, singing, and music composition. He was one of the giants of thermodynamics and the first to write formally on the subject.

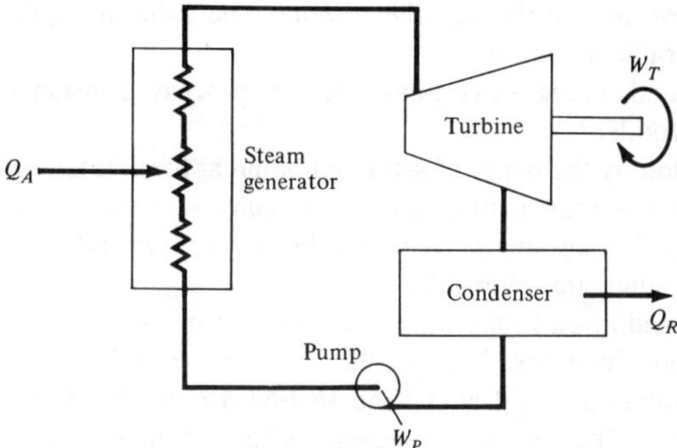

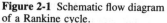

Figure 2-1 Schematic flow diagram of a Rankine cycle.

diagram of a Rankine cycle. Figure 2-2*a* and *b* shows ideal Rankine cycles on the (*a*) *P-v* and (*b*) *T-s* diagrams. The curved lines to the left of the *critical point* (CP) on both diagrams are the loci of all saturated-liquid points and are the *saturated-liquid* lines. The regions to the left of these are the *subcooled-liquid regions*. The curved lines to the right of CP are the loci of all saturated-vapor points and are the *saturated-vapor* lines. The regions to the right of these lines are the *superheat regions*. The regions under the domes represent the *two-phase* (liquid and vapor) *mixture region,* sometimes called the *wet region*.

Cycle 1-2-3-4-*B*-1 is a saturated Rankine cycle, meaning that saturated vapor enters the turbine. 1′-2′-3-4-*B*-1′ is a superheat Rankine cycle, meaning that superheated vapor enters the turbine. The cycles, being reversible, have the following processes.

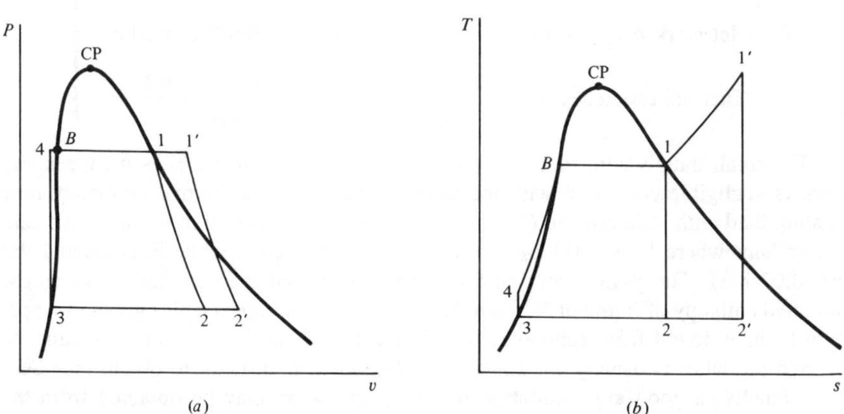

Figure 2-2 Ideal Rankine cycles of the (*a*) *P-v* and (*b*) *T-s* diagrams. 1-2-3-4-*B*-1 = saturated cycle. 1^1-2^1-3-4-*B*-1^1 = superheated cycle. CP = critical point.

1-2 or 1′-2′: adiabatic reversible expansion through the *turbine*. The exhaust vapor at 2 or 2′ is usually in the two-phase region.

2-3 or 2′-3: constant temperature and, being a two-phase mixture process, constant-pressure heat rejection in the condenser.

3-4: adiabatic reversible compression by the *pump* of saturated liquid at the condenser pressure, 3, to subcooled liquid at the steam-generator pressure, 4. Line 3-4 is vertical on both the *P-V* and *T-S* diagrams because the liquid is essentially incompressible and the pump is adiabatic reversible.

4-1 or 4-1′: constant-pressure heat addition in the *steam generator*. Line 4-*B*-1-1′ is a constant-pressure line on both diagrams. The portion 4-*B* represents bringing the subcooled liquid, 4, to saturated liquid at *B*. The section 4-*B* in the steam generator is called an *economizer*. The portion *B*-1 represents heating the saturated liquid to saturated vapor at constant pressure and temperature (being a two-phase mixture), and section *B*-1 in the steam generator is called the *boiler* or *evaporator*. Portion 1-1′, in the superheat cycle, represents heating the saturated vapor at 1 to 1′. Section 1-1′ in the steam generator is called a *superheater*.

The cycles as shown are internally reversible so that the turbine and pump are adiabatic reversible and hence vertical on the *T-S* diagram; no pressure losses occur in the piping so that line 4-*B*-1-1′ is a constant-pressure line.

The analysis of either cycle is straightforward. Based on a unit mass of vapor in the saturated cycle

$$\left.\begin{aligned}
&\text{Heat added } q_A = h_1 - h_4 \qquad \text{Btu/lb}_m \text{ or J/kg}\\
&\text{Turbine work } w_T = h_1 - h_2 \qquad \text{Btu/lb}_m \text{ or J/kg}\\
&\text{Heat rejected } |q_R| = h_2 - h_3 \qquad \text{Btu/lb}_m \text{ or J/kg}\\
&\text{Pump work } |w_p| = h_4 - h_3\\
&\text{Net work } \Delta w_{\text{net}} = (h_1 - h_2) - (h_4 - h_3) \qquad \text{Btu/lb}_m \text{ or J/kg}\\
&\text{Thermal efficiency } \eta_{\text{th}} = \frac{\Delta w_{\text{net}}}{q_A} = \frac{(h_1 - h_2) - (h_4 - h_3)}{(h_1 - h_4)}
\end{aligned}\right\} \tag{2-1}$$

For small units where P_4 is not too large compared with P_3, $h_4 \approx h_3$, the pump work is negligible compared with the turbine work, and the thermal efficiency may be simplified with little error to $(h_1 - h_2)/(h_1 - h_3)$. This is not true for modern steam powerplants where P_4 is 1000 lb_f/in^2 (about 70 bar) or higher, while P_3 is about 1 lb_f/in^2 (0.07 bar). The pump work in this case may be obtained by finding h_3 as the saturated enthalpy of liquid at P_3 from the steam (or other vapor) tables given in Apps. A to F. h_4 is found from subcooled liquid tables at T_4 and P_4. T_4 is nearly equal to T_3, and the latter is usually used in lieu of T_4, which is difficult to obtain (see Sec. 1-5). Finally, a good approximation for the pump work may be obtained from the change in flow work (Example 3, Sec. 1-2). Thus

$$|w_p| = v_3(P_4 - P_3) \tag{2-2}$$

which should be converted to the same units as in Eq. (2-1) by the use of proper conversion factors, such as multiply by 144 to convert psia (pounds force per square inch absolute) to pounds force per square foot absolute and divide by 778.16 to convert foot pounds force to Btu.

Another parameter of interest in cycle analysis is the *work ratio* WR, which is defined as the ratio of net work to gross work. For the simple Rankine cycle the work ratio is simply $\Delta w_{net}/w_T$.

The superheat cycle 1′-2′-3-4-*B*-1′ is analyzed by use of Eqs. (2-1) and (2-2), except 1′ is to be substituted for 1.

Because of the information it readily gives regarding the turbine and pump processes, the *T-S* diagram is more useful than the *P-V* diagram and is usually preferred when only one is used. The Mollier, or enthalpy-entropy, diagram is another useful diagram. Its utility, however, is restricted to processes involving the turbine because it gives little or no information of the liquid region.

2-3 THE EXTERNALLY IRREVERSIBLE RANKINE CYCLE

External irreversibility, we are reminded, is primarily the result of the temperature differences between the primary heat source, such as the combustion gases from the steam generator furnace or the primary coolant from a nuclear reactor, and the working fluid; and the temperature differences between condensing working fluid and the heat sink fluid, usually the condenser cooling water.

In Fig. 2-3, line *ab* represents the primary coolant in a counterflow heat exchanger with the working fluid 4-*B*-1 in a saturated Rankine cycle. Line *cd* represents the heat sink fluid (condenser cooling water) in a counterflow or parallel-flow heat exchanger with the condensing working fluid 2-3; both types are the same because the latter is at constant temperature.

As can be seen, the temperature differences between line *ab* and 4-*B*-1-1′ and

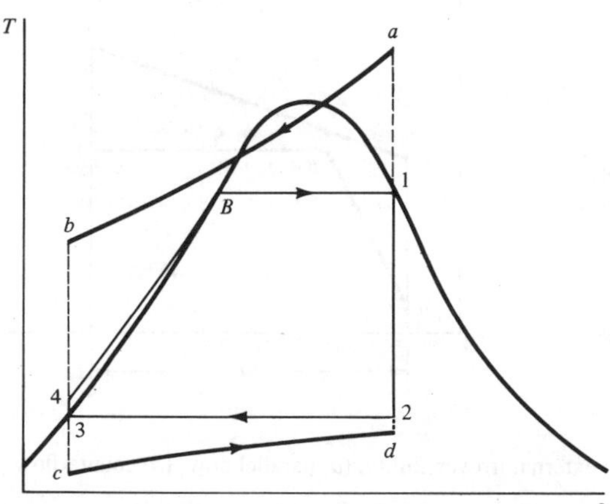

Figure 2-3 External irreversibility with Rankine cycle.

between 2-3 and line *cd* are not constant. We shall evaluate the effects of these differences beginning with the upper end. Figure 2-4 shows temperature-heat exchanger path length diagrams for (*a*) parallel-flow and (*b*) counterflow heat exchangers (steam generators) and the effect of flow directions in the heat exchanger. The minimum approach point between the two lines, called the *pinch point,* represented by *b*-1 and *e*-*B*, must be finite. Too small a pinch-point temperature difference results in low overall temperature differences and, hence, lower irreversibilities, but in a large and costly steam generator; too large a pinch-point temperature difference results in a small, inexpensive steam generator but large overall temperature differences and irreversibilities and, hence, reduction in plant efficiency. The most economical pinch-point temperature difference is obtained by optimization that takes into account both fixed charges (based on capital costs) and operating costs (based on efficiency and, hence, fuel costs).

Figure 2-4, in addition, clearly shows that the overall temperature differences between the heat source and the working fluids are greater in the case of the parallel-flow than counterflow heat exchangers; the result is a less efficient plant if parallel flow is used. Heat-transfer considerations also favor counterflow, resulting in higher overall heat-transfer coefficients and hence small heat exchanger. Thus counterflow is favored over parallel flow from both thermodynamic and heat-transfer considerations.

We will now examine the effect of the type of heat source fluid. Such a fluid may be a gas, such as the combustion gases in a fossil-fueled powerplant, the primary coolant in a gas-cooled reactor, such as CO_2 or He (Sec. 10-11), the water from a pressurized-water reactor (Sec. 10-2), or the molten sodium from a liquid-metal fast-breeder reactor (Chap. 11). This variety of fluids has different specific heats and mass-flow rates. Water from a pressurized-water reactor has a higher specific heat c_p than gases but also a higher mass-flow rate $\dot{m}$ because an effort is made to limit the temperature rise of water through the reactor to maintain nearly even moderation of

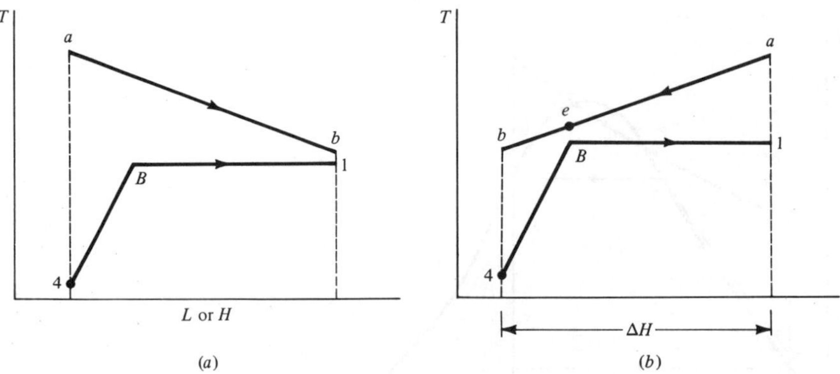

Figure 2-4 Effect of flow direction on external irreversibility: (*a*) parallel flow, (*b*) counterflow.

the neutrons (Sec. 9-8). Thus the product $\dot{m}c_p$ is greater in the case of water than in the case of gases.

Assuming that a differential amount of heat dQ exchanged between the two fluids is proportional to a path length dL and that $dQ = \dot{m}c_p\, dT$, where dT is the change in primary-fluid temperature in dL, the slope of line *ab* is then proportional to the reciprocal of $\dot{m}c_p$ or

$$\frac{dT}{dL} \propto \frac{1}{\dot{m}c_p} \qquad \text{(primary fluid)} \tag{2-3}$$

Hence the slope of line *ab* for water is much less than that for gases. Liquid sodium falls in between, though closer to gases than to water. This state of affairs is shown in Fig. 2-5 for a counterflow heat exchanger. It can be seen that for a given pinch-point temperature difference, the overall temperature differences between the primary and working fluids are greater in the case of gases than water, in particular in the boiler section, between *ae* and *B*-1.

This brings us to an important deduction, namely the determination of whether or not superheat (and reheat) is advantageous. We note that there are two distinct regions where the external irreversibility exists at the higher-temperature end of the cycle. These are: (1) between the primary fluid and the working fluid in the boiler section, i.e., between *ae* and *B*-1, and (2) between the primary fluid and the working fluid in the economizer section, i.e., between *be* and 4-*B*. We shall deal with these in turn in the next two sections.

There is little that can be done to improve things in the low-temperature end of the cycle, i.e., between 2-3 and *cd* in the condenser (Fig. 2-3), short of optimizing the condenser to obtain the lowest temperature differences between the two lines. Remember, however, that the lower the temperature of the cooling water at *c*, the lower the condenser steam temperature and the higher the cycle efficiency.

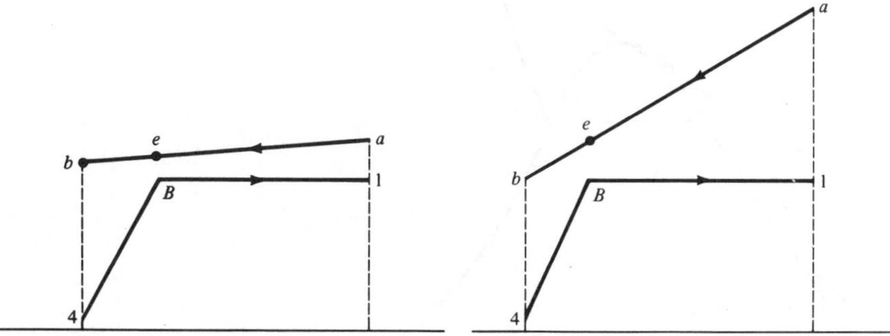

Figure 2-5 Effect of primary fluid type on external irreversibility: (*a*) water, (*b*) gases or liquid metal.

2-4 SUPERHEAT

In this section we will deal with the temperature differences between *ae* and *B*-1 (Fig. 2-5). It can be seen that these for a given pinch-point temperature difference ΔT_{e-B}, gases (and liquid metals) exhibit larger and increasing temperature differences as the working fluid boils from *B*-1 than is the case of water where the slope of line *ae* is much lower.

Although the temperature levels are not the same in the two cases, the gases are

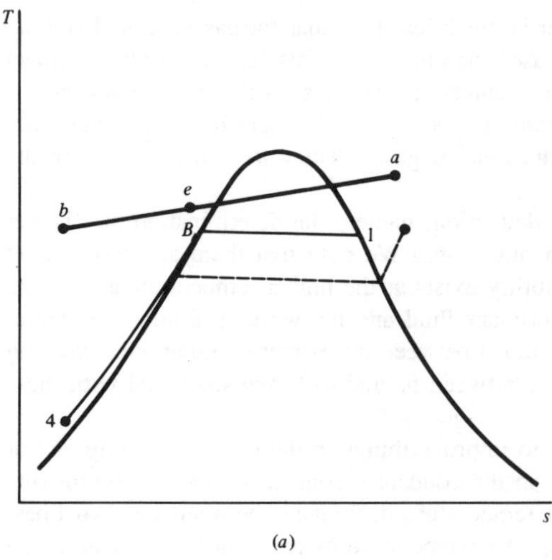

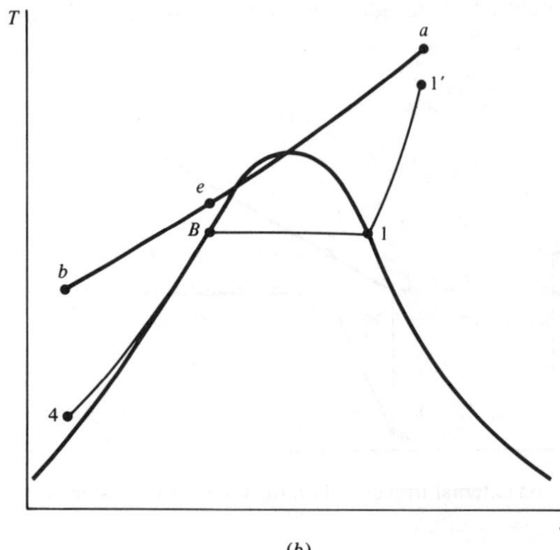

Figure 2-6 Superheat with (*a*) water as primary fluid, (*b*) gases or liquid metal as primary fluid.

usually at higher temperatures, the irreversibility in the case of gases can be reduced by the use of superheat (Fig. 2-6) by bringing the two lines back together again at *a* and 1′ and thus reducing the overall temperature differences between *ae* and *B*-1-1′ (line 4-*B*-1-1′ is a constant-pressure line). Thus superheat would improve the cycle thermal efficiency. Looking at it another way, superheat allows heat addition at an average temperature higher than using saturated steam only. From the Carnot analogy, this should result in higher cycle efficiency.

In the case of water, superheat is not practical because the differences between *ae* and *B*-1 vary little. Actually, if we were to fix the temperature at 1 and use superheat, we would need to lower the boiling temperature (and hence pressure) in *B*-1, as seen by the dashed line in Fig. 2-6*a*. This increases rather than decreases the overall temperature differences and results in reducing rather than increasing cycle efficiency. This is the reason why fossil-fuel and gas-cooled and liquid-metal-cooled nuclear powerplants employ superheat, while pressurized-water-cooled reactors do not. (A boiling-water reactor, Sec. 10-7, produces only saturated steam within the reactor vessel.)

Superheat has an additional beneficial effect. It results in drier steam at turbine exhaust 2′ as compared with 2 for saturated steam (Fig. 2-2 and Example 2-1). A turbine operating with less moisture is more efficient and less prone to blade damage.

Example 2-1 Consider three Rankine steam cycles, all exhausting to 1 psia. Cycle *A* operates at 2500 psia and 1000°F; cycle *B* operates with 2500 psia saturated steam; and cycle *C* operates with superheated steam at a temperature equal to that of cycle *B* but with a pressure of 1000 psia. Calculate the efficiencies and exhaust steam qualities of the three cycles.

SOLUTION Using Eqs. (2-1) and (2-2), and the steam tables, and referring to Fig. 2-2, calculations for cycle *A* are

$$h_{1'} = 1457.5 \text{ Btu/lb}_m \qquad s_{1'} = 1.5269 \text{ Btu/(lb}_m \cdot {}^\circ\text{R)}$$

Because the turbine is reversible adiabatic, its expansion line is isentropic, or $s_{2'} = s_{1'}$. Thus

$$s_{2'} = (s_f + x_2 s_{fg})_{1 \text{ psia}}$$

$$1.5269 = 0.1326 + x_2(1.8455)$$

From which quality of turbine exhaust $x_2 = 0.7555$

$$h_{2'} = (h_f + x_2 h_{fg})_{1 \text{ psia}} = 69.73 + 0.7555 \times 1036.1$$

$$= 852.5 \text{ Btu/lb}_m$$

$$h_3 = 69.73 \text{ Btu/lb}_m$$

$$|w_p| = h_4 - h_3 = v_3(P_4 - P_3) = \frac{0.016136(2500 - 1) \times 144}{778.16}$$

$$= 7.46 \text{ Btu/lb}_m$$

$$h_4 = 69.73 + 7.46 = 77.19 \text{ Btu/lb}_m$$

$$w_T = h_{1'} - h_{2'} = 1457.5 - 852.5 = 604.98 \text{ Btu/lb}_m$$

$$\Delta w_{\text{net}} = w_T - |w_p| = 604.98 - 7.46 = 597.52 \text{ Btu/lb}_m$$

$$q_A = h_{1'} - h_4 = 1457.5 - 77.19 = 1380.31 \text{ Btu/lb}_m$$

$$|q_R| = h_{2'} - h_3 = 852.5 - 69.73 - 782.77 \text{ Btu/lb}_m$$

$$\eta_{\text{th}} = \frac{\Delta w_{\text{net}}}{q_A} = \frac{597.52}{1380.31} = 0.4329 = 43.29\%$$

$$\text{WR} = \frac{\Delta w_{\text{net}}}{w_T} = \frac{597.52}{604.98} = 0.9877$$

Table 2-1 lists the results for cycle *A* and, using a similar procedure, for cycles *B* and *C*. Cycle *D* is a superheat-reheat cycle that will be discussed in Sec. 2-5. Cycle *E* is a nonideal cycle that will be discussed in Sec. 2-7.

Note that cycle *C* is actually less efficient than cycle *B*, which proves that superheat is not beneficial if the upper temperature is limited.

2-5 REHEAT

An additional improvement in cycle efficiency with gaseous primary fluids as in fossil-fueled and gas-cooled powerplants is achieved by the use of *reheat*.

Figures 2-7 and 2-8 show simplified flow and T-*s* diagrams of an internally reversible Rankine cycle (i.e., one with adiabatic reversible turbine and pump and no pressure drops) that superheats and reheats the vapor.

Table 2-1 Solutions for Examples 2-1, 2-2, and 2-3

	Cycle				
Data	*A* Superheat 2500/1000	*B* 2500 Saturated	*C* Superheat 1000/668.11	*D* 2500/ 1000/1000	*E* 2500/1000 Nonideal
Turbine inlet pressure, psia	2500	2500	1000	2500	2500
Turbine inlet temperature, °F	1000	668.11	668.11	1000	1000
Condenser pressure, psia	1	1	1	1	1
Inlet steam enthalpy, Btu/lb$_m$	1457.5	1093.3	1303.1	1457.5	1457.5
Exhaust steam enthalpy, Btu/lb$_m$	852.52	688.36	834.44	970.5	913.02
Turbine work, Btu/lb$_m$	604.98	404.94	468.66	741.8	544.48
Pump work, Btu/lb$_m$	7.46	7.46	2.98	7.46	11.52
Net work, Btu/lb$_m$	597.52	397.48	465.68	734.34	532.96
Heat added, Btu/lb$_m$	1380.31	1061.11	1230.39	1635.10	1376.25
Exhaust steam quality	0.7555	0.5971	0.7381	0.8694	0.8139
Cycle efficiency, %	43.29	39.12	37.85	44.91	38.73

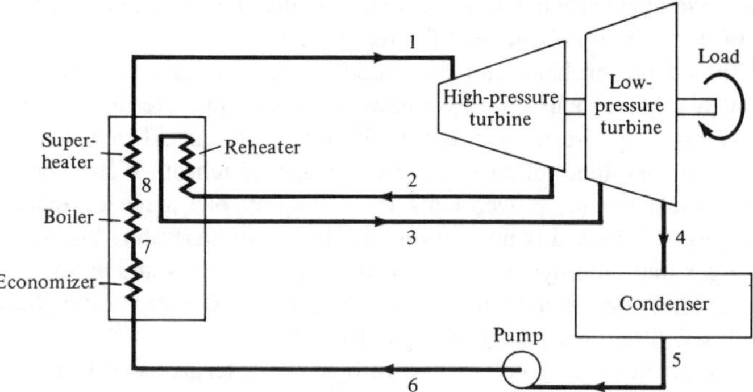

Figure 2-7 Schematic of a Rankine cycle with superheat and reheat.

In the reheat cycle, the vapor at 1 is expanded part of the way in a high-pressure section of the turbine to 2, after which it is returned back to the steam generator, where it is reheated at constant pressure (ideally) to a temperature near that at 1. The reheated steam now expands in the low-pressure section of the turbine to the condenser pressure.

As can be seen reheat allows heat addition twice: from 6 to 1 and from 2 to 3. It results in increasing the average temperature at which heat is added and keeps the boiler-superheat-reheat portion from 7 to 3 close to the primary fluid line *ae*, which

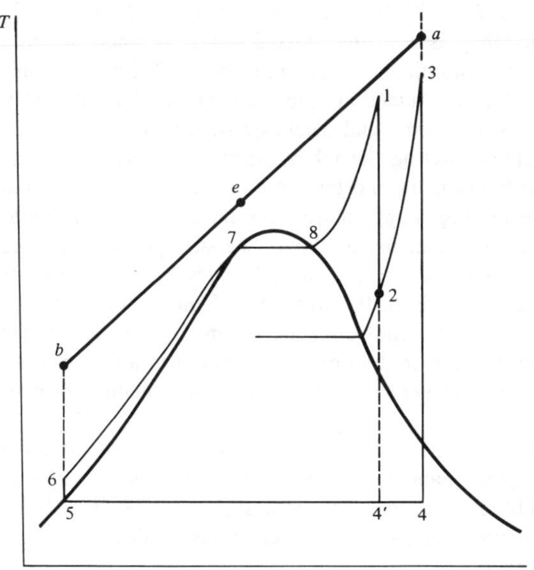

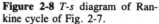

Figure 2-8 *T-s* diagram of Rankine cycle of Fig. 2-7.

results in improvement in cycle efficiency. Reheat also results in drier steam at turbine exhaust (4 instead of 4′), which is beneficial for real cycles.

Modern fossil-fueled powerplants employ superheat and at least one stage of reheat. Some employ two. More than two stages, however, results in cycle complication and increased capital costs that are not justified by improvements in efficiency. Gas-cooled nuclear-reactor powerplants often employ one stage of reheat. Water-cooled and sodium-cooled nuclear-reactor powerplants often employ one stage of reheat, except that the steam to be reheated is not returned to the steam generator. Instead, a separate heat exchanger that employs a portion of the original steam at 1 is used to reheat the steam at 2. That portion condenses and is sent to a feed water heater (Sec. 2-6). Examples of this will be presented in Chaps. 10 and 11.

The analysis of a reheat cycle involves two turbine work terms as well as two heat addition terms. Referring to Fig. 2-8

$$\left.\begin{aligned} w_T &= (h_1 - h_2) + (h_3 - h_4) \\ |w_p| &= h_6 - h_5 \\ \Delta w_{\text{net}} &= (h_1 - h_2) + (h_3 - h_4) - (h_6 - h_5) \\ q_A &= (h_1 - h_6) + (h_3 - h_2) \\ \eta_{\text{th}} &= \frac{\Delta w_{\text{net}}}{q_A} \end{aligned}\right\} \quad (2\text{-}4)$$

The pressure P_2 at which the steam is reheated affects the cycle efficiency. Figure 2-9 shows the change in cycle efficiency $\Delta\eta$ percent as a function of the ratio of reheat pressure to initial pressure P_2/P_1, for $P_1 = 2500$ psia, $T_1 = 1000$°F, and $T_3 = 1000$°F. $P_2/P_1 = 1.0$ is the case where no reheat is used and hence $\Delta\eta = 0$. A reheat pressure too close to the initial pressure results in little improvement in cycle efficiency because only a small portion of additional heat is added at high temperature. The efficiency improves as the reheat pressure P_2 is lowered and reaches a peak at a pressure ratio P_2/P_1 between 20 and 25 percent. Lowering the reheat pressure further causes the temperature differences between the primary and the working fluids to increase and begin to offset the addition of heat at high temperature, thus causing the efficiency to decrease again. Too low a reheat pressure, in the above case at a pressure ratio of about 0.025, actually results in a negative $\Delta\eta$, i.e., an efficiency below the case of no reheat. The optimum at a pressure ratio of 0.2 to 0.25, calculated for the above conditions, actually holds for most modern powerplants. Figure 2-9 also shows the value of T_2 and x_4. Note that reheat results in drier exhaust steam. Too low a pressure ratio may even result in superheated exhaust steam, an unfavorable situation for condenser operation.

A superheat-reheat powerplant is often designated by $P_1/T_1/T_3$ in pounds force per square inch absolute and degrees Fahrenheit. The above case, for example, is 2500/1000/1000, whereas a double-reheat plant may be designated 2400/1000/1025/1050. The following example shows a sample of the calculations conducted for Fig. 2-8, near the optimum pressure ratio.

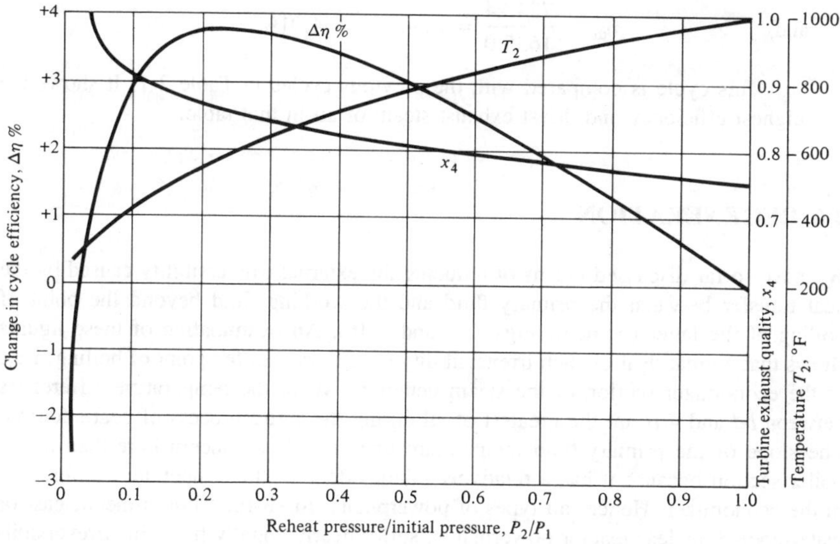

Figure 2-9 Effect of reheat-to-initial pressure ratio on efficiency, high-pressure turbine exit temperature, and low-pressure turbine exit quality. Data for cycle of Fig. 2-7 with initial steam at 2500 psia and 1000°F, and steam reheat to 1000°F (2500/1000/1000).

Example 2-2 Calculate the efficiency and exhaust steam quality of a 2500 psia/1000°F/1000°F internally reversible steam Rankine cycle (cycle D, Table 2-1). The reheat pressure is 500 psia. The condenser pressure is 1 psia.

SOLUTION Referring to Fig. 2-8

$$h_1 = 1457.5 \text{ Btu/lb}_m \qquad s_1 = 1.5269 = s_2 > s_g \text{ at 500 psia}$$

Therefore point 2 is in the superheat region. By interpolation

$$T_2 = 547.8\text{°F} \qquad h_2 = 1265.6 \text{ Btu/lb}_m$$

At 500 psia and 1000°F

$$h_3 = 1520.3 \text{ Btu/lb}_m \qquad s_3 = 1.7371 = s_4$$

Therefore $\qquad x_4 = 0.8694 \qquad h_4 = 970.5 \text{ Btu/lb}_m$

As in Example 2-1, $|w_p| = 7.46$ Btu/lb$_m$ and $h_6 = 77.19$ Btu/lb$_m$. Using Eqs. (2-4) gives

$$w_T = 191.5 + 549.8 = 741.7 \text{ Btu/lb}_m$$

$$\Delta w_{\text{net}} = 741.7 - 7.46 = 734.24 \text{ Btu/lb}_m$$

$$q_A = 1380.3 + 254.7 = 1635.0 \text{ Btu/lb}_m$$

and $$\eta_{th} = \frac{734.24}{1635.0} = 0.4491 = 44.91\%$$

This cycle is compared with the previous cycles in Table 2-1. It shows the highest efficiency and driest exhaust steam of all in that table.

2-6 REGENERATION

We have so far discussed means of reducing the external irreversibility caused by the heat transfer between the primary fluid and the working fluid beyond the point of boiling of the latter (point *B*, Figs. 2-3 and 2-4*b*). An examination of these figures shows that a great deal of such irreversibility occurs prior to the point of boiling, i.e., in the economizer section of the steam generator where the temperature differences between *bd* and 4-*B* are the greatest of all during the entire process of heat addition. The slope of the primary-fluid temperature line is of less concern here than in the boiler section because it has a relatively minor effect on the temperature differences in the economizer. Hence, all types of powerplants, fossil-fuel, liquid-metal, gas- or water-cooled nuclear-reactor powerplants, suffer nearly equally from this irreversibility.

This irreversibility can be eliminated if the liquid is added to the steam generator at *B* rather than at 4. This can be done by the process of *regeneration,* in which internal heat is exchanged between the expanding fluid in the turbine and the compressed fluid before heat addition. A well-known gas cycle that uses regeneration is the Stirling cycle, shown on the T-*s* diagram of Fig. 2-10. The ideal Stirling cycle is composed of heat addition at constant temperature 2-3 and heat rejection at constant temperature 4-1. Regeneration or heat exchange occurs reversibly between the constant volume processes 3-4 and 1-2, i.e., between portions of each curve that are at the same temperature. This heat exchange does not figure in the cycle efficiency because it is not obtained from an external source. The areas under 3-4 and 1-2 denoting heat lost by the expanded fluid and gained by the compressed fluid are equal in magnitude, though not in sign. The ideal Stirling cycle has the same efficiency as the Carnot cycle operating between the same temperature limits. This would not have been the case had heat been added from an external source during 1-2 and 2-3 and rejected to an external sink between 3-4 and 4-1.

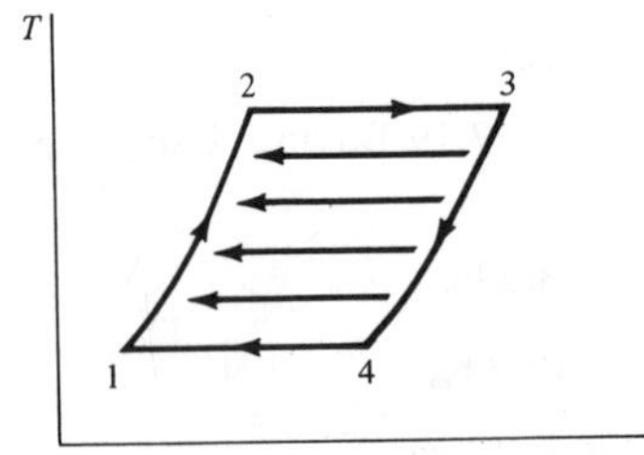

Figure 2-10 *T-s* diagram of Stirling cycle. Regeneration occurs between 3-4 and 1-2. Arrows indicate heat exchange.

Adopting the same procedure to a Rankine cycle, i.e., internal and reversible heat exchange from the expanding working fluid in the turbine and the fluid in the economizer section, would necessitate flow and T-s diagrams as shown in Fig. 2-11 for a saturated Rankine cycle. The compressed liquid at 4 would have to be carefully passed around the turbine to receive heat from the expanding vapor in the turbine reversibly at all times (i.e., with zero temperature difference) until it enters the steam generator at B. The steam generator would have no economizer and the irreversibility during heat addition to the economizer would be eliminated. The resulting Rankine cycle would receive and reject heat at constant temperature and, in the absence of other external irreversibilities, would also have the same efficiency as the Carnot cycle operating between the same temperature limits. Hence the great need for eliminating or minimizing the economizer irreversibility.

The ideal procedure of Fig. 2-11 is not practically possible. The vapor making its way through blade passages cannot be made to have adequate heat-transfer surface between it and the compressed liquid, which by necessity would have to be wrapped around the external turbine casing. Even if an adequate surface were possible, the mass-flow rates are so large that the effectiveness of such a heat exchanger would be low. Further, the vapor leaving the turbine would have an unacceptably high moisture content (low quality) for proper turbine operation and efficiency.

Feedwater Heating

A compromise that would reduce rather than eliminate the economizer irreversibility is accomplished by the use of *feedwater heating* (the more general term feed liquid heating that would apply to fluids other than H_2O is seldom used). Feedwater heating involves normal adiabatic (and ideally also reversible) expansion in the turbine. The compressed liquid at 4 is heated in a number of finite steps, rather than continuously, by vapor bled from the turbine at selected stages. Heating of the liquid takes place in heat exchangers called *feedwater heaters*. Feedwater heating dates back to the early 1920s, around the same time that steam temperatures reached about 725°F. Modern

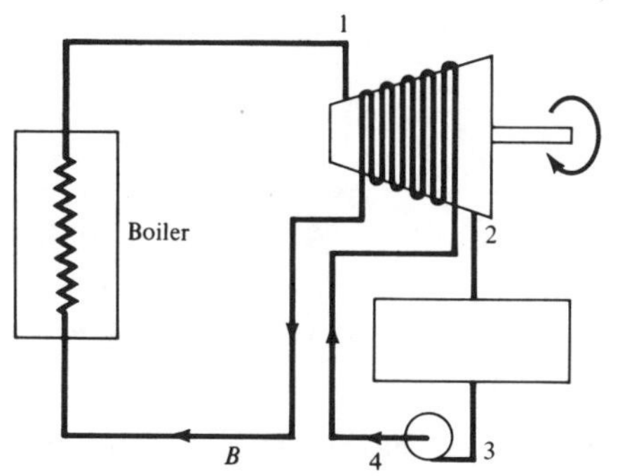

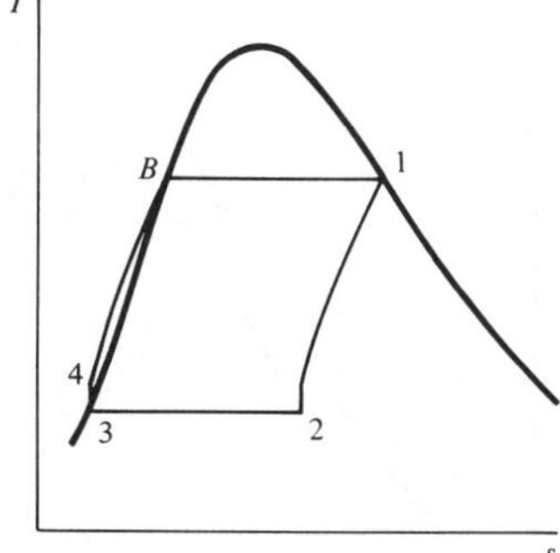

Figure 2-11 Ideal regeneration of a Rankine cycle.

large steam powerplants use between five and eight feedwater heating stages. None is built without feedwater heating.

Because of the finite number of feedwater heating stages, the liquid enters the steam generator at a point below *B*, necessitating an economizer section, though one that is much smaller than if no feedwater heating were used. Because of this, and because the feedwater heaters have irreversibilities of their own, the ideal situation of Fig. 2-11 is not attained and the Rankine cycle cannot attain a Carnot efficiency. A well-designed Rankine cycle, however, is the closest practical cycle to Carnot, and hence its wide acceptance for most powerplants.

There are three types of feedwater heaters in use. These are:

1. Open or direct-contact type
2. Closed type with drains cascaded backward
3. Closed type with drains pumped forward

These types will be discussed and analyzed in detail in this chapter beginning with Sec. 2-8. Their physical design will be described in Chap. 6.

2-7 THE INTERNALLY IRREVERSIBLE RANKINE CYCLE

Internal irreversibility is primarily the result of fluid friction, throttling, and mixing. The most important of these are the irreversibilities in turbines and pumps and pressure losses in heat exchangers, pipes, bends, valves, etc.

In the turbine and pumps, the assumption of adiabatic flow is still valid because the flow rates are so large that the heat losses per unit mass is negligible. However, they are no longer adiabatic reversible, and the entropy, in both, increases. This is shown in Fig. 2-12.

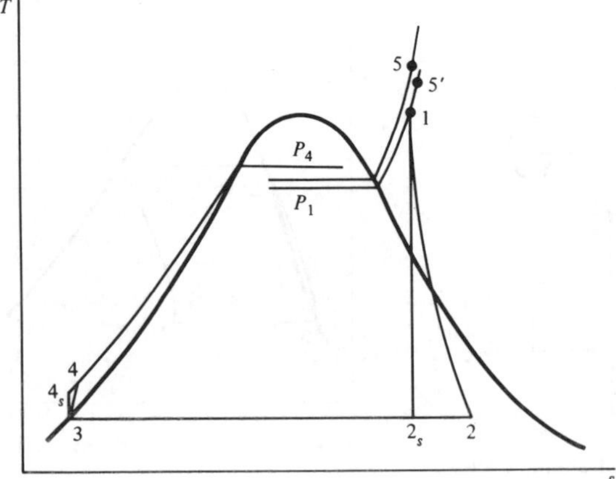

Figure 2-12 A *T-s* diagram of an internally irreversible superheat Rankine cycle.

The entropy increase in the turbine, unlike that in a gas turbine (Fig. 1-8), does not result in a temperature increase if exhaust is to the two-phase region, the usual case. Instead it results in an increase in enthalpy. Thus the ideal expansion, if the turbine were adiabatic reversible, is $1\text{-}2_s$, but the actual expansion is 1-2. The irreversible losses in the turbine are represented by a turbine efficiency η_T, called the *turbine polytropic efficiency* (and sometimes the adiabatic or isentropic efficiency). This is not to be confused with the cycle thermal efficiency. η_T is given by the ratio of the turbine actual work to the ideal, adiabatic reversible work. Hence

$$\eta_T = \frac{h_1 - h_2}{h_1 - h_{2s}} \tag{2-5}$$

Well-designed turbines have high polytropic efficiencies, around 90 percent. η_T usually increases with turbine size and suffers from moisture in the steam. η_T as given above is an overall polytropic efficiency. However, individual turbine stages have different efficiencies, being higher for early stages where the steam is drier. There will be more on turbines in Chap. 5.

No pressure losses are encountered in the condenser process 2-3 (Fig. 2-12) because it is a two-phase condensation process.

The pump process, being adiabatic and irreversible, also results in an increase in entropy. A single-phase (liquid) process, it results in an increase in temperature and enthalpy. Thus the actual work $h_4 - h_3$ is greater than the adiabatic reversible work $h_{4_s} - h_3$. In other words, one pays a penalty for irreversibility: the turbine produces less work, the pump absorbs more work. The pump irreversibility is also represented by a pump efficiency η_p, also called a *pump polytropic efficiency* (and sometimes adiabatic or isentropic efficiency). η_p is given by the ratio of the ideal work to the actual work, the reverse of that for the turbine. Thus

$$\eta_p = \frac{h_{4s} - h_3}{h_4 - h_s} \tag{2-6}$$

In both Eqs. (2-5) and (2-6), the smaller quantity is in the numerator. The actual pump work may now be obtained by modifying Eq. (2-6) to

$$|w_p| = \frac{h_{4s} - h_3}{\eta_p} \approx \frac{v_3(P_4 - P_3)}{\eta_p} \tag{2-7}$$

The liquid leaving the pump must be at a higher pressure than at the turbine inlet because of the friction drops in heat exchangers, feedwater heaters, pipes, bends, valves, etc. Thus P_4 represents the exit pump pressure, P_1 represents the turbine inlet pressure, and P_5 represents the steam-generator exit pressure. The steam leaves the generator at 5 and enters the turbine at 1. The path 5-1 is the result of the combined effects of friction and heat losses. Point 5′ at pressure P_1 represents frictional effects in the pipe connecting steam generator and turbine, including turbine throttle valve, if any. Heat losses from that pipe cause a decrease in entropy to 1. Pressure losses between 4 and 1 could be of the order of a few hundred pounds force per square inch.

Example 2-3 A superheat steam Rankine cycle has turbine inlet conditions of 2500 psia and 1000°F. The turbine and pump polytropic efficiencies are 0.9 and 0.7, respectively. Pressure losses between pump and turbine inlet are 200 psi. Calculate the turbine exhaust steam quality and cycle efficiency.

SOLUTION Referring to Fig. 2-12

$$h_1 = 1457.5 \qquad h_{2s} = 852.52 \text{ Btu/lb}_m \text{ (as in Example 2-1)}$$

$$w_T = \eta_T(h_1 - h_{2s}) = 0.9 \times 604.98 = 544.48 \text{ Btu/lb}_m$$

Therefore $$h_2 = h_1 - w_T = 913.02 \text{ Btu/lb}_m$$

At 1 psia $$913.02 = 69.73 + x_2(1036.1)$$

$$\therefore x_2 = 0.8139$$

$$P_4 = P_1 + 200 = 2700 \text{ psia}$$

Thus $$|w_p| \approx \frac{v_3(P_4 - P_3)}{\eta_p} = \frac{0.016136(2700 - 1) \times 144}{778 \times 0.7}$$

$$= 11.52 \text{ Btu/lb}_m$$

$$h_4 = h_3 + |w_p| = 69.73 + 11.52 = 81.25 \text{ Btu/lb}_m$$

$$\Delta w_{\text{net}} = w_T - |w_p| = 532.96 \text{ Btu/lb}_m$$

$$q_A = h_1 - h_4 = 1376.25$$

Therefore $$\eta_{\text{th}} = \frac{\Delta w_{\text{net}}}{q_A} = 0.3873 = 38.73\%$$

Thus the internal irreversibilities have resulted in reducing the cycle efficiency from 43.29 percent (Example 2-1) to 38.73 percent, but in an increase in exhaust steam quality from 0.7555 to 0.8139, one beneficial effect of an imperfect turbine. This example is listed as cycle *E* in Table 2-1.

2-8 OPEN OR DIRECT-CONTACT FEEDWATER HEATERS

In the open- or direct-contact-type of feedwater heater the extraction steam is mixed directly with the incoming subcooled feedwater to produce saturated water at the extraction steam pressure. Figure 2-13*a* and *b* shows a schematic flow diagram, and the corresponding T-*s* diagram for a Rankine cycle using, for simplicity of illustration, two such feedwater heaters, one low-pressure and one high-pressure (normally one open-type feedwater heater and between four and seven other heaters are used in modern large powerplants). The physical construction of such a feedwater heater is covered in Chap. 7. A typical open-type feedwater heater is shown in Fig. 6-15.

The condensate water leaves the condenser saturated at 5 and is pumped to 6 to

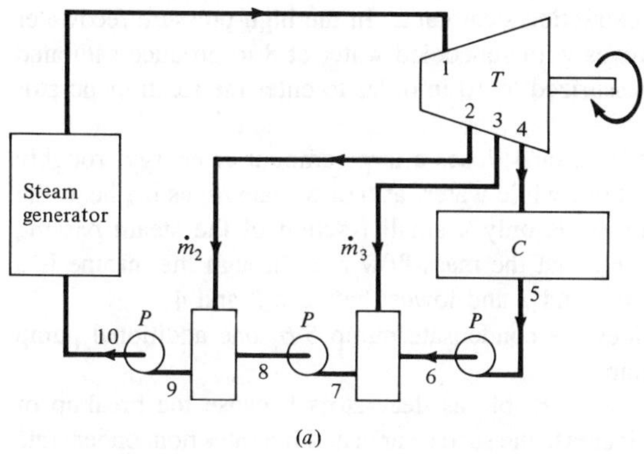

(a)

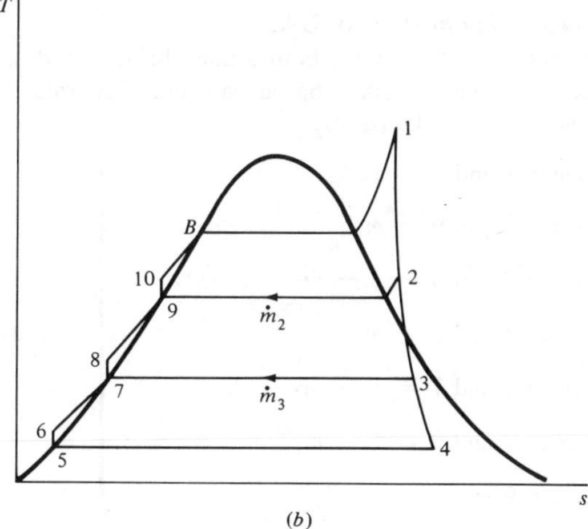

(b)

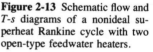

Figure 2-13 Schematic flow and T-s diagrams of a nonideal superheat Rankine cycle with two open-type feedwater heaters.

a pressure equal to that of the extraction steam at 3. The now-subcooled water at 6 and wet steam at 3 *mix* in the low-pressure feedwater heater to produce saturated water at 7. Thus the amount of bled steam $\dot{m}_3$ is essentially equal to that that would saturate the subcooled water at 6. If it were much less, it will result in a much lower temperature than that corresponding to 6, which would partially negate the advantages of feedwater heating. If it were more, it would result in unnecessary loss of turbine work and in a two-phase mixture that would be difficult to pump.

Line 6-7 in Fig. 2-13*b* is a constant-pressure line. (In practice some pressure drop is encountered.) The difference between it and the saturated liquid line 5-*B* is exaggerated for illustration purposes.

The pressure at 6-7 can be no higher than the extraction steam pressure at 3 (or else reverse flow of condensate water would enter the turbine at 3). A second pump must therefore be used to pressurize the saturated water from 7 to a subcooled condition at 8, which is at the pressure of extraction steam at 2. In the high-pressure feedwater heater, superheated steam at 2 mixes with subcooled water at 8 to produce saturated water at 9. This now must be pressurized to 10 in order to enter the steam generator at its pressure.

Because the extracted steam, at 2 or 3, loses a large amount of energy, roughly equal to its latent heat of vaporization, while water, at 6 or 8, gains sensible heat, the amount of extracted steam $\dot{m}_2$ or $\dot{m}_3$ is only a small fraction of the steam passing through the turbine. Note, however, that the mass-flow rate through the turbine is a variable quantity, highest between 1 and 2 and lowest between 3 and 4.

It can also be seen that besides the condensate pump 5-6, one additional pump per open feedwater heater is required.

Open-type feedwater heaters also double as deaerators because the breakup of water in the mixing process helps increase the surface area and liberates noncondensible gases (such as air, O_2, H_2, CO_2) that can be vented to the atmosphere (Sec. 6-7). Hence they are sometimes called *deaerating heaters,* or DA.

In order to analyze the system shown in Fig. 2-13, both a mass balance and an energy balance must be considered. The mass balance, based on a unit-flow rate (1 lb_m/h or kg/s) at throttle (point 1) is given, clockwise, by

$$\left.\begin{aligned}
\text{Mass flow between 1 and 2} &= 1 \\
\text{Mass flow between 2 and 9} &= \dot{m}_2 \\
\text{Mass flow between 2 and 3} &= 1 - \dot{m}_2 \\
\text{Mass flow between 3 and 7} &= \dot{m}_3 \\
\text{Mass flow between 4 and 7} &= 1 - \dot{m}_2 - \dot{m}_3 \\
\text{Mass flow between 7 and 9} &= 1 - \dot{m}_2 \\
\text{Mass flow between 9 and 1} &= 1
\end{aligned}\right\} \tag{2-8}$$

where $\dot{m}_2$ and $\dot{m}_3$ are small fractions of 1. Energy balances are now done on the high- and low-pressure feedwater heaters, respectively

$$\dot{m}_2(h_2 - h_9) = (1 - \dot{m}_2)(h_9 - h_8) \tag{2-9}$$

and

$$\dot{m}_3(h_3 - h_7) = (1 - \dot{m}_2 - \dot{m}_3)(h_3 - h_7) \tag{2-10}$$

where h is the enthalpy per unit mass at the point of interest. Equations (2-9) and (2-10) show that there are two equations and only two unknowns, $\dot{m}_2$ and $\dot{m}_3$, if the pressures at which steam is bled from the turbine (Sec. 2-13), and therefore the enthalpies, are all known. For any number of feedwater heaters there will be as many equations as there are unknowns, so solutions are always possible. A large number of feedwater heaters would, of course, require the solution of an equal number of simultaneous linear algebraic equations on a digital computer. The pertinent cycle parameters are now obtained, as energy per unit mass-flow rate at turbine inlet (point 1)

$$\left.\begin{aligned}
&\text{Heat added } q_A = (h_1 - h_{10}) \\
&\text{Turbine work } w_T = (h_1 - h_2) + (1 - \dot{m}_2)(\mathrm{h}_2 - \mathrm{h}_3) \\
&\qquad + (1 - \dot{m}_2 - \dot{m}_3)(h_3 - h_4) \\
&\text{Pump work } |\Sigma w_p| = (1 - \dot{m}_2 - \dot{m}_3)(h_6 - h_5) + (1 - \dot{m}_2)(h_8 - h_7) \\
&\qquad + (h_{10} - h_9) \approx (1 - \dot{m}_2 - \dot{m}_3)\frac{v_5(P_6 - P_5)}{\eta_p J} \\
&\qquad + (1 - \dot{m}_2)\frac{v_7(P_8 - P_7)}{\eta_p J} + \frac{v_9(P_{10} - P_9)}{\eta_p J} \\
&\text{Heat rejected } |q_R| = (1 - \dot{m}_2 - \dot{m}_3)(h_4 - h_5) \\
&\text{Net cycle work } \Delta w_{\text{net}} = w_T - |w_p| \\
&\text{Cycle thermal efficiency } \eta_{\text{th}} = \frac{\Delta w_{\text{net}}}{q_A} \\
&\text{Work ratio WR} = \frac{w_{\text{net}}}{w_T}
\end{aligned}\right\} \qquad (2\text{-}11)$$

where η_p is the pump efficiency and $J = 778.16$ ft·lb$_f$/Btu.

Example 2-4 An ideal Rankine cycle operates between 2500 psia and 1000°F at throttle and 1 psia in the condenser. One open-type feedwater heater is placed at 200 psia. Assuming 1 lb$_m$/h flow at turbine throttle and no flow pressure drops, calculate the mass-flow rate in the heater and the pertinent parameters for the cycle and compare them with those of the cycle in Example 2-1, which has the same conditions except that no feedwater heater was used.

SOLUTION Referring to Fig. 2-14 and the steam tables

$$h_1 = 1457.5 \text{ Btu/lb}_m \qquad s_1 = 1.5269 \text{ Btu/(lb}_m \cdot {}^\circ\text{F)}$$

At 200 psia

$$s_2 = s_1 = 1.5269 = 0.5438 + x_2(1.0016)$$

Therefore

$$x_2 = 0.9815 \qquad h_2 = 355.5 + 0.9815\,(842.8) = 1182.7 \text{ Btu/lb}_m$$

At 1 psia

$$s_3 = s_1 = 1.5269 = 0.1326 + x_3(1.8455)$$

Thus

$$x_3 = 0.7555 \qquad h_3 = 69.73 + 0.7555(1036.1) = 852.2 \text{ Btu/lb}_m$$

$$h_4 = 69.73 \text{ Btu/lb}_m \qquad v_4 = 0.016136 \text{ ft}^3/\text{lb}_m$$

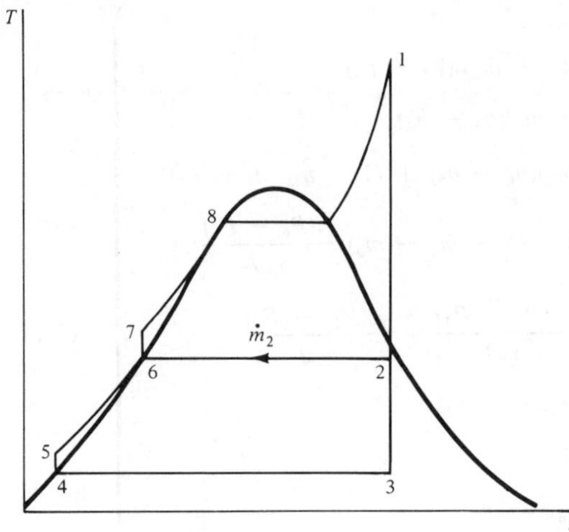

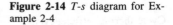

Figure 2-14 *T-s* diagram for Example 2-4

$$h_5 = 69.73 + \frac{0.016136 \times (200 - 1) \times 144}{778.16} = 69.73 + 0.59$$

$$= 70.32 \text{ Btu/lb}_m$$

$$h_6 = 355.5 \text{ Btu/lb}_m \qquad v_6 = 0.01839 \text{ ft}^3/\text{lb}_m$$

$$h_7 = 355.5 + 0.01839 + \frac{(2500 - 200) \times 144}{778.16} = 355.5 + 7.83$$

$$= 363.3 \text{ Btu/lb}_m$$

$$\dot{m}_2(h_2 - h_6) = (1 - \dot{m}_2)(h_6 - h_5)$$

$$\dot{m}_2(1182.7 - 355.5) = (1 - \dot{m}_2)(355.5 - 70.32)$$

$$\therefore \dot{m}_2 = 0.2564$$

$$w_T = (h_1 - h_2) + (1 - \dot{m}_2)(h_2 - h_3)$$

$$= (1457.5 - 1182.7) + (1 - 0.2564)(1182.7 - 852.5)$$

$$= 274.77 + 245.57 = 520.34 \text{ Btu/lb}_m$$

$$|\Sigma w_p| = (1 - \dot{m}_2)(h_5 - h_4) + (h_7 - h_6)$$

$$= (1 - 0.2564)(0.59 + 7.83) = 8.27 \text{ Btu/lb}_m$$

$$\Delta w_{\text{net}} = w_T - |\Sigma w_p| = 520.34 - 8.27 = 512.07 \text{ Btu/lb}_m$$

$$q_A = h_1 - h_7 = 1457.5 - 363.3 = 1094.2 \text{ Btu/lb}_m$$

$$|q_R| = (1 - \dot{m}_2)(h_3 - h_4) = (1 - 0.2564)(852.5 - 69.73)$$

$$= 582.1 \text{ Btu/lb}_m$$

$$\eta_{th} = \frac{\Delta w_{net}}{q_A} = \frac{512.07}{1094.2} = 0.468 = 46.8\%$$

$$WR = \frac{\Delta w_{net}}{w_T} = \frac{512.07}{520.34} = 0.984$$

Compare this with cycle *A* (Table 2-1), which had no feedwater heater. Note that the turbine work is decreased for the same mass-flow rate at throttle because of reduced turbine mass-flow rate after bleeding and that the pump work is increased. Note also the greater decrease in heat added, which more than makes up for the loss of net work, resulting in a marked improvement in cycle efficiency. This improvement increases as the number of feedwater heaters is increased. The number of feedwater heaters can be as high as seven or eight. An increase beyond that causes little increase in efficiency but adds complications and increased capital costs and thus diminishes returns.

As seen above, open feedwater heaters require, in addition to the condensate pump, as many additional pumps as there are feedwater heaters. Each of these pumps carries nearly full flow, or more accurately full flow minus the bled steam following it. For example, pump 7-8 (Fig. 2-13) carries $(1 - \dot{m}_2)$ lb_m/lb_m at throttle. In powerplants such large flow pumps are the source of operational, service, and noise problems and increase plant complexity and cost. In general only one open-type feedwater heater is used, which doubles up as a deaerating heater, followed by a pump called the *boiler feed pump*. (In some nuclear powerplants no open feedwater heaters are used and degassing is done elsewhere.) Other feedwater heaters in the system are therefore of the closed type.

2-9 CLOSED-TYPE FEEDWATER HEATERS WITH DRAINS CASCADED BACKWARD

This type of feedwater heater, though it results in a greater loss of availability than the open type, is the simplest and most commonly used type in powerplants. As in the case of the closed-type feedwater heater with drains pumped forward (Sec. 2-10), it too is a shell-and-tube heat exchanger but differs because of the lack of any moving equipment.

In a closed-type feedwater heater (of either type), feedwater passes through the tubes, and the bled steam, on the shell side, transfers its energy to it and condenses. Thus they are, in essence, small condensers that operate at pressures more elevated than those of the main plant condenser. Because the feedwater goes through the tubes in successive closed feedwater heaters, it does not mix with bled steam and therefore can be pressurized only once by the first condensate pump, which then doubles as a boiler feed pump, though often there is one condensate pump and a boiler feed pump placed downstream to reduce the pressure rise in each pump. A boiler feed pump is automatically required and placed after the deaerating heater if one is used in the plant.

Figure 2-15 shows a simplified flow diagram and corresponding T-s diagram of a nonideal superheat Rankine cycle showing, for simplicity, two feedwater heaters of this type. One pump, 5-6, pressurizes the condensate to a pressure sufficient to pass through the two feedwater heaters and enter the steam generator at 8. Again the difference between the high-pressure line 6-B and the saturated-liquid line 5-B is exaggerated for illustration purposes.

As the bled steam condenses in each feedwater heater, it cannot, of course,

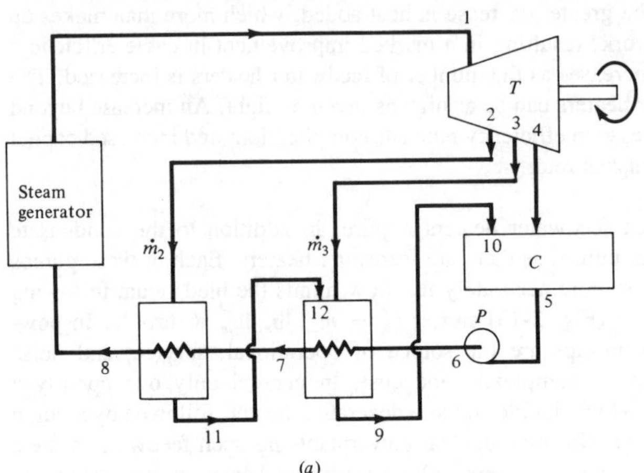

(a)

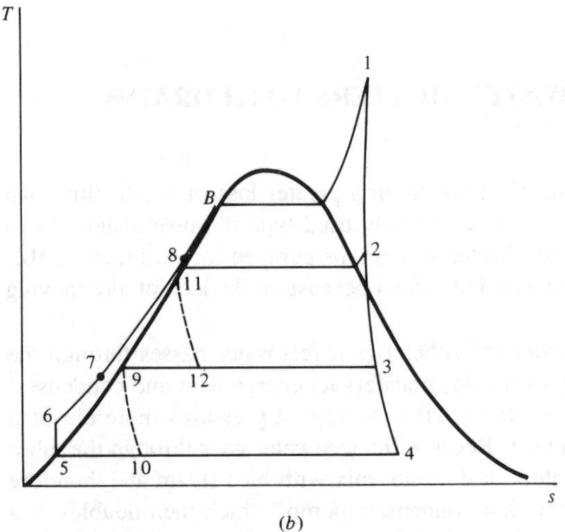

(b)

Figure 2-15 Schematic flow and T-s diagrams of a nonideal superheat Rankine cycle with two closed-type feedwater heaters with drains cascaded backward.

accumulate there and must be removed and fed back to the system. In this type of feedwater heater, the condensate is fed back to the next lower-pressure feedwater heater. The condensate of the lowest-pressure feedwater heater is (though not always) led back to the main condenser. One can imagine, then, a cascade from higher-pressure to lower-pressure heaters; hence, the name of this type of feedwater heater.

Again starting with the low-pressure feedwater heater, wet steam at 3 is admitted and transfers its energy to high-pressure subcooled water at 6. The events in that heater can be represented by the temperature-length diagram shown in Fig. 2-16*a*. The water exit temperature at 7 cannot reach the inlet bled steam temperature at 3. A difference called the *terminal temperature difference* (TTD, sometimes simply TD) is defined for all closed feedwater heaters as

$$\text{TTD} = \textit{saturation}\text{ temperature of bled steam} - \text{exit water temperature} \tag{2-12}$$

The value of TTD varies with heater pressure. In the case of low-pressure heaters, which receive wet or at most saturated bled steam, the TTD is positive and often of the order of 5°F. This difference is obtained by proper heat-transfer design of the heater. Too small a value, although good for plant efficiency, would require a larger heater than can be justified economically. Too large a value would hurt cycle efficiency. In some heaters, the drain at 9 is slightly subcooled. This will be shown later.

The drain from the low-pressure heater is now led to the condenser and enters it as a two-phase mixture at 10. This is a throttling process from the pressure corresponding to 9 to that of the main condenser, and hence there is loss of some availability, as alluded to earlier. There is also some loss of availability as a result of heat transfer. Process 9-10 is a throttling process and hence is a constant enthalpy one.

A closed feedwater heater that receives saturated or wet steam can have a drain cooler and thus be physically composed of a condensing section and a drain cooler section (Fig. 2-16*b*).

Returning to the system of Fig. 2-15, the high-pressure feedwater heater receives superheated steam bled from the turbine at 2 that flows on the shell side at the rate $\dot{m}_2$ and transfers its energy to subcooled liquid entering the tubes at 7. The events

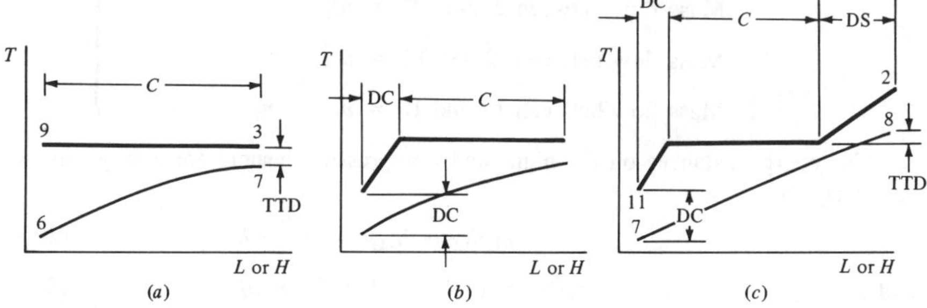

Figure 2-16 Temperature-enthalpy diagrams of (*a*) and (*b*) low-pressure and (*c*) high-pressure feedwater heaters of Fig. 2-15. TTD = terminal temperature difference, DS = desuperheater, C = condenser, DC = drain cooler.

there are shown by the temperature-path length diagram in Fig. 2-16*c*. Note here that because the inlet steam is superheated at 2, the exit water temperature at 8 can be higher than the saturation temperature of that steam and the TTD, defined by Eq. (2-12), can be negative. The TTD values for high-pressure heaters, therefore, range between 0 and -5°F, being more negative the higher the pressure, and hence the greater the degree of superheat of the entering steam.

Note also that the drain in this heater is slightly subcooled and hence imparts more energy to the water and thus reduces the loss of availability due to its throttling to the low-pressure heater. The heater is physically composed of a desuperheating section, a condensing section, and a drain cooler section (Fig. 2-16*c*).

Thus there are four physical possibilities of closed feedwater heaters composed of the following sections or zones (Sec. 6-5):

1. Condenser
2. Condenser, drain cooler
3. Desuperheater, condenser, drain cooler
4. Desuperheater, condenser

The drain at 11 is now throttled to the low-pressure heater entering it at 12 as a two-phase mixture where it joins with the steam bled at 3 and thus aids in the heating of the water in the low-pressure heater. The combined $\dot{m}_2 + \dot{m}_3$ constitutes the low-pressure heater drain, which is throttled to the main condenser at 10. The high-pressure heater exit water at 8 is led into the steam generator. Again, to analyze the system, both a mass and an energy balance are required. A mass balance, also based on a unit-flow rate at turbine inlet, point 1, is given, clockwise, by

$$\left.\begin{aligned}
&\text{Mass flow between 1 and 2} &&= 1\\
&\text{Mass flow between 2 and 3} &&= 1 - \dot{m}_2\\
&\text{Mass flow between 3 and 10} &&= 1 - \dot{m}_2 - \dot{m}_3\\
&\text{Mass flow between 10 and 1} &&= 1\\
&\text{Mass flow between 2 and 12} &&= \dot{m}_2\\
&\text{Mass flow between 3 and 12} &&= \dot{m}_3\\
&\text{Mass flow between 12 and 10} &&= \dot{m}_2 + \dot{m}_3
\end{aligned}\right\} \tag{2-13}$$

The energy balances on the high- and low-pressure heaters are now given, respectively, by

$$\dot{m}_2(h_2 - h_{11}) = h_8 - h_7 \tag{2-14}$$

and

$$\dot{m}_3(h_3 - h_9) + \dot{m}_2(h_{12} - h_9) = h_7 - h_6 \tag{2-15}$$

Recalling that a throttling process is a constant enthalpy process so that

$$h_{12} = h_{11} \qquad \text{and} \qquad h_{10} = h_9$$

and knowing the pressures at which steam is bled from the turbine (Sec. 2-13) so that the enthalpies in Eqs. (2-14) and (2-15) are all known, we again have two equations and two unknowns, $\dot{m}_2$ and $\dot{m}_3$. Or, in general, we will have as many equations as there are unknowns making a solution possible. The pertinent cycle parameters are now obtained, again as energy per unit mass flow rate at turbine inlet (point 1)

$$\left.\begin{aligned}
&\text{Heat added } q_A = h_1 - h_8 \\
&\text{Turbine work } w_T = (h_1 - h_2) + (1 - \dot{m}_2)(h_2 - h_3) \\
&\qquad\qquad + (1 - \dot{m}_2 - \dot{m}_3)(h_3 - h_4) \\
&\text{Pump work } |w_p| = h_6 - h_5 \approx \frac{v_5(P_6 - P_5)}{\eta_p J} \\
&\text{Heat rejected } |q_R| = (1 - \dot{m}_2 - \dot{m}_3)(h_4 - h_5) + (\dot{m}_2 + \dot{m}_3)(h_{10} - h_5) \\
&\text{Net cycle work } \Delta w_{\text{net}} = w_T - |w_p| \\
&\text{Cycle thermal efficiency } \eta_{\text{th}} = \frac{\Delta w_{\text{net}}}{q_A} \\
&\text{Work ratio WR} = \frac{\Delta w_{\text{net}}}{w_T}
\end{aligned}\right\} \quad (2\text{-}16)$$

Example 2-5 An ideal Rankine cycle operates with 1000 psia, 1000°F steam. It has one closed feedwater heater with drain cascaded backward placed at 100 psia. The condenser pressure is 1 psia. Use TTD = 5°F. The heater has a drain cooler resulting in DC (drain cooler temperature difference) = 10°F.

Solution Referring to Fig. 2-17, the enthalpies, all in Btu/lb$_m$, found by the usual procedure are

$$h_1 = 1505.4 \qquad h_2 = 1228.6 \qquad h_3 = 923.31 \qquad h_4 = 69.73 \qquad h_7 = 298.5$$

$$h_5 = h_4 + v_4(P_5 - P_4) = 69.73 + 2.98 = 72.71 \qquad \text{corresponding to } 104.72°\text{F}$$

For TTD = 5°F

$$t_6 = t_7 - 5 = 327.82 - 5 = 322.82°\text{F}$$

Therefore

$$h_6 = 293.36 \qquad \text{(by interpolation)}$$

For DC = 10°F

$$t_8 = t_5 + 10 = 104.72 + 10 = 114.72°\text{F}$$

Thus $h_8 = 82.69$ (by interpolation)

$$\dot{m}_2(h_2 - h_8) = h_6 - h_5$$

$$\dot{m}_2 = \frac{393.36 - 72.71}{1228.6 - 82.69} = 0.1926$$

$$
\begin{aligned}
w_T &= (h_1 - h_2) + (1 - \dot{m}_2)(h_2 - h_3) \\
&= (1505.4 - 1228.6) + (1 - 0.1926)(1228.6 - 923.31) \\
&= 276.8 + 246.49 = 523.29 \\
|w_p| &= (h_5 - h_4) = 2.98 \\
\Delta w_{\text{net}} &= 520.31 \\
q_A &= h_1 - h_6 = 1505.4 - 293.36 = 1212.04 \\
|q_R| &= (1 - \dot{m}_2)(h_3 - h_4) + \dot{m}_2(h_9 - h_4) \\
&= 689.18 + 2.50 = 691.68 \\
\eta_{\text{cycle}} &= \frac{520.31}{1212.04} = 0.4293 = 42.93\% \\
\text{WR} &= \frac{\Delta w_{\text{net}}}{w_T} = \frac{520.31}{523.29} = 0.9943
\end{aligned}
$$

Table 2-2 contains other solutions for ideal Rankine cycles with 1000 psia steam. The cycle in Example 2-5 is cycle *D* in that table. Again note the reduction in work but the improvement in η_{th} over the cycle with no feedwater heating. As stated for the open feedwater heaters, this improvement increases with the number of feedwater heaters until increases in complexity and capital cost make the addition of further heaters, beyond about seven or eight, unprofitable.

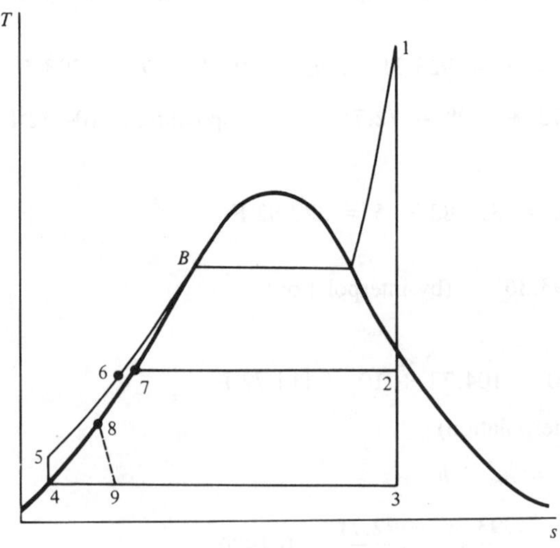

Figure 2-17 *T-s* diagram of Example 2-5.

Table 2-2 Results of example calculations for ideal Rankine cycles*

Cycle	Particulars	Δw_{net}	q_A	η%	$\lvert q_R \rvert$	WR
A	No superheat; no fwh†	413.72	1120.19	36.93	706.49	0.9928
B	Superheat; no fwh	579.11	1432.69	40.42	853.58	0.9949
C	Superheat; one open fwh	519.3	1203.95	43.13	685.25	0.9939
D	Superheat; one closed fwh; drains cascaded; DC	520.31	1212.04	42.93	691.68	0.9943
E	Superheat; one closed fwh; drains pumped; DC	529.85	1245.63	42.54	715.73	0.9945
F	Superheat; one closed fwh; drains pumped; no DC	520.59	1210.48	43.01	689.95	0.9943
G	Superheat; reheat; one open fwh	641.59	1447.44	44.33	805.83	0.9951
H	Superheat; reheat; two closed fwh; drains cascaded	609.83	1351.0	45.14	727.62	0.9952
I	Supercritical; double reheat; no fwh; 3500/1000/1025/1050	861.95	1831.92	47.05	969.97	0.9880

* All values in Btu/lb_m; all examples, except for cycle *A* which is saturated, and cycle I, at 1000 psia/1000°F. All at 1 psia condenser pressure.

† fwh = feedwater heater.

Although this type of feedwater heater is the most common, it causes some loss of availability because of throttling and, to a lesser extent, heat transfer.

2-10 CLOSED-TYPE FEEDWATER HEATERS WITH DRAINS PUMPED FORWARD

This second closed-type feedwater heater avoids throttling but at the expense of some added complexity because of the inclusion of a small pump. It also allows some flexibility to the plant cycle designer who prefers a mix of feedwater heater types that would be deemed most suitable.

As with the previous closed-type feedwater heater, it is a shell-and-tube heat exchanger in which the feedwater passes through the tubes and the bled steam, on the shell side, transfers its energy to it and condenses. They do not mix and the feedwater may be pressurized only once, although a deaerating heater followed by boiler feed pump are usually inserted into the system.

The drain from this type of heater, instead of being cascaded backward, is pumped forward into the main feedwater line. Figure 2-18 shows a simplified flow diagram and corresponding T-*s* diagram for a nonideal superheat Rankine cycle showing, for simplicity, two heaters of this type. Although this system requires one additional pump per heater, it differs from the system using open-type feedwater heaters in that the pumps this time are small and, rather than nearly full feedwater flow, carry only fractional flows corresponding to the bled steam $\dot{m}_2$ and $\dot{m}_3$.

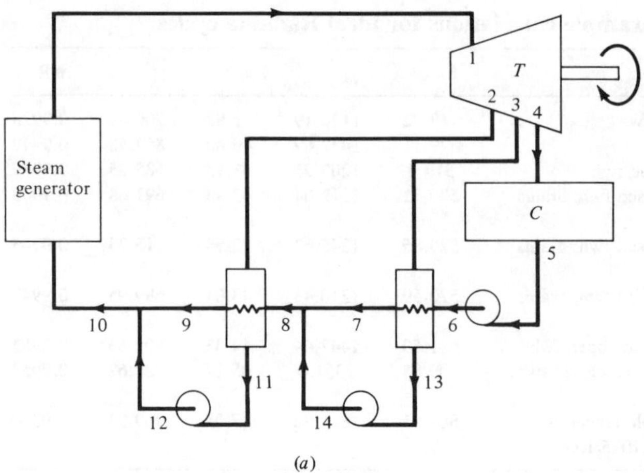

(a)

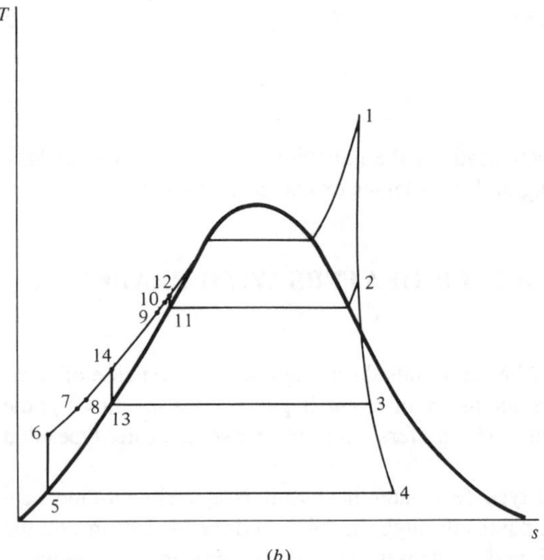

(b)

Figure 2-18 Schematic flow and *T-s* diagrams of nonideal superheat Rankine cycle with two closed-type feedwater heaters with drains pumped forward.

Starting with the low-pressure heater, the drain at 13 is pumped forward to the main feedwater line, enters it at 14, and mixes with the exit water from that heater at 7, resulting in a mixture at 8. Point 8 is closer to 7 than 14 on the *T-s* diagram because the main feedwater flow at 7 is greater than the drain flow $\dot{m}_3$.

The water at 8 enters the high-pressure heater and is heated to 9. The drain leaves the heater at 11, is pumped to 12, and mixes with the feedwater at 9, resulting in full feedwater flow at 10 which now goes to the steam generator.

A mass balance, based on a unit mass-flow rate at turbine inlet, point 1, is given, clockwise, on the *T-s* diagram by

$$\left.\begin{array}{ll} \text{Mass flow between 1 and 2} & = 1 \\ \text{Mass flow between 2 and 12} & = \dot{m}_2 \\ \text{Mass flow between 2 and 3} & = 1 - \dot{m}_2 \\ \text{Mass flow between 3 and 14} & = \dot{m}_3 \\ \text{Mass flow between 3 and 7} & = 1 - \dot{m}_2 - \dot{m}_3 \\ \text{Mass flow at 14} & = \dot{m}_3 \\ \text{Mass flow between 8 and 9} & = 1 - \dot{m}_2 \\ \text{Mass flow at 12} & = \dot{m}_2 \\ \text{Mass flow between 10 and 1} & = 1 \end{array}\right\} \tag{2-17}$$

The energy balances on the high- and low-pressure heaters are given, respectively, by

$$\dot{m}_2(h_2 - h_{11}) = (1 - \dot{m}_2)(h_9 - h_8) \tag{2-18}$$

and

$$\dot{m}_3(h_3 - h_{13}) = (1 - \dot{m}_2 - \dot{m}_3)(h_7 - h_6) \tag{2-19}$$

The values of h_9 and h_7 are obtained from the temperatures t_9 and t_7, which are equal to the saturation temperature of the steam in each heater minus its terminal temperature difference or

$$t_9 = t_{11} - \text{TTD} \qquad \text{hp heater} \tag{2-20a}$$

$$\text{and } t_7 = t_{13} - \text{TTD} \qquad \text{lp heater} \tag{2-20b}$$

h_{10}, needed for q_A, and h_8, to be used in Eq. (2-18), are obtained from h_{12} and h_{14}, respectively. The latter are given by

$$h_{12} = h_{11} + v_{11}\frac{P_{12} - P_{11}}{\eta_p} \tag{2-21a}$$

and

$$h_{14} = h_{13} + v_{13}\frac{P_{14} - P_{13}}{\eta_p} \tag{2-21b}$$

Thus

$$h_{10} = \dot{m}_2 h_{12} + (1 - \dot{m}_2)h_9 \tag{2-22a}$$

and

$$(1 - \dot{m}_2)h_8 = \dot{m}_3 h_{14} + (1 - \dot{m}_2 - \dot{m}_3)h_7 \tag{2-22b}$$

The turbine work

$$w_T = (h_1 - h_2) + (1 - \dot{m}_2)(h_2 - h_3) + (1 - \dot{m}_2 - \dot{m}_3)(h_3 - h_4) \tag{2-23}$$

$$\text{Pump work } |\Sigma w_p| = (1 - \dot{m}_2 - \dot{m}_3)(h_6 - h_5) + \dot{m}_3(h_{14} - h_{13}) + \dot{m}_2(h_{12} - h_{11}) \tag{2-24}$$

$$\text{Heat added } q_A = h_1 - h_{10} \tag{2-25}$$

$$\text{Thermal efficiency } \eta_{\text{th}} = \frac{w_T - |\Sigma w_p|}{q_A} \tag{2-26}$$

Example 2-6 Repeat Example 2-5 but for one closed-type feedwater heater with drain pumped forward. TTD = 5°F.

SOLUTION Refer to Fig. 2-19. h_1, h_2, h_3, h_4, h_5, h_6, h_7 are all the same as in Example 2-5

$$h_8 = h_7 + v_7 \frac{(P_8 - P_7) \times 144}{778.13} = 298.5 + 0.017740 \frac{(1000 - 100) \times 144}{778.17}$$

$$= 298.5 + 2.95 = 301.45 \text{ Btu/lb}_m$$

$$h_6 \text{ (as before)} = 293.36 \text{ Btu/lb}_m$$

$$\dot{m}_2(h_2 - h_7) = (1 - \dot{m}_2)(h_6 - h_5)$$

$$\dot{m}_2(1228.6 - 298.5) = (1 - \dot{m}_2)(293.36 - 72.71)$$

$$\therefore \dot{m}_2 = 0.1917$$

$$h_9 = \dot{m}_2 h_8 + (1 - \dot{m}_2)h_6 = 57.79 + 237.12 = 294.91 \text{ Btu/lb}_m$$

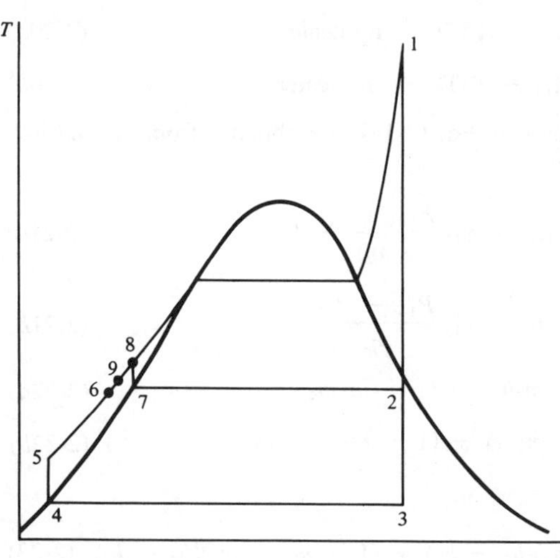

Figure 2-19 *T-s* diagram of Example 2-6.

$$w_T = (h_1 - h_2) + (1 - \dot{m}_2)(h_2 - h_3) = 276.8 + 246.77$$
$$= 523.57 \text{ Btu/lb}_m$$
$$\Sigma w_p = (1 - \dot{m}_2)(h_5 - h_4) + \dot{m}_2(h_8 - h_7) = 2.41 + 0.57$$
$$= 2.98 \text{ Btu/lb}_m$$
$$\Delta w_{\text{net}} = 520.59 \text{ Btu/lb}_m$$
$$q_A = h_1 - h_{10} = 1505.4 - 294.92 = 1210.48 \text{ Btu/lb}_m$$
$$|q_R| = (1 - \dot{m}_2)(h_3 - h_4) = 689.95 \text{ Btu/lb}_m$$
$$\eta_{\text{cycle}} = \frac{520.59}{1210.48} = 0.4301 = 43.01\%$$
$$\text{WR} = \frac{520.59}{523.57} = 0.9943$$

This example is listed as cycle F in Table 2-2.

As indicated earlier, the type of closed feedwater heater that has drains pumped forward avoids the loss of availability due to throttling inherent in the previous closed feedwater heater with drains cascaded backward. This, however, is done at the expense of the complexity of adding a drain pump following each heater. Note, however, that unlike the open feedwater heater the drain pump is a low-capacity one because its flow is only that of the bled steam being condensed in the heater. It must however pressurize that condensate to the full feedwater line pressure.

This type of feedwater heater results in a slightly better cycle efficiency if used without a drain cooler because energy transferred from the heater drain in the drain cooler lowers the point in the feedwater line at which energy is to be added from the primary heat source or from a higher pressure feedwater heater. Compare cycle F in Table 2-2 with cycle E, which is identical except that there is a drain cooler with DC = 10°F.

One other advantage of pumped drains is that, when used as the lowest-pressure feedwater heater in an otherwise all-cascaded system, or with all-cascaded feedwater heaters between it and an open feedwater heater, it prevents the throttling of the combined cascaded flows to the condenser pressure where the energy left in that combined flow is lost to the environment.

2-11 THE CHOICE OF FEEDWATER HEATERS

In general the choice of feedwater heater type depends upon many factors, including designer optimization and preference, practical considerations, cost, and so on, and one sees a variety of cycle designs. There are, however, features that are rather common.

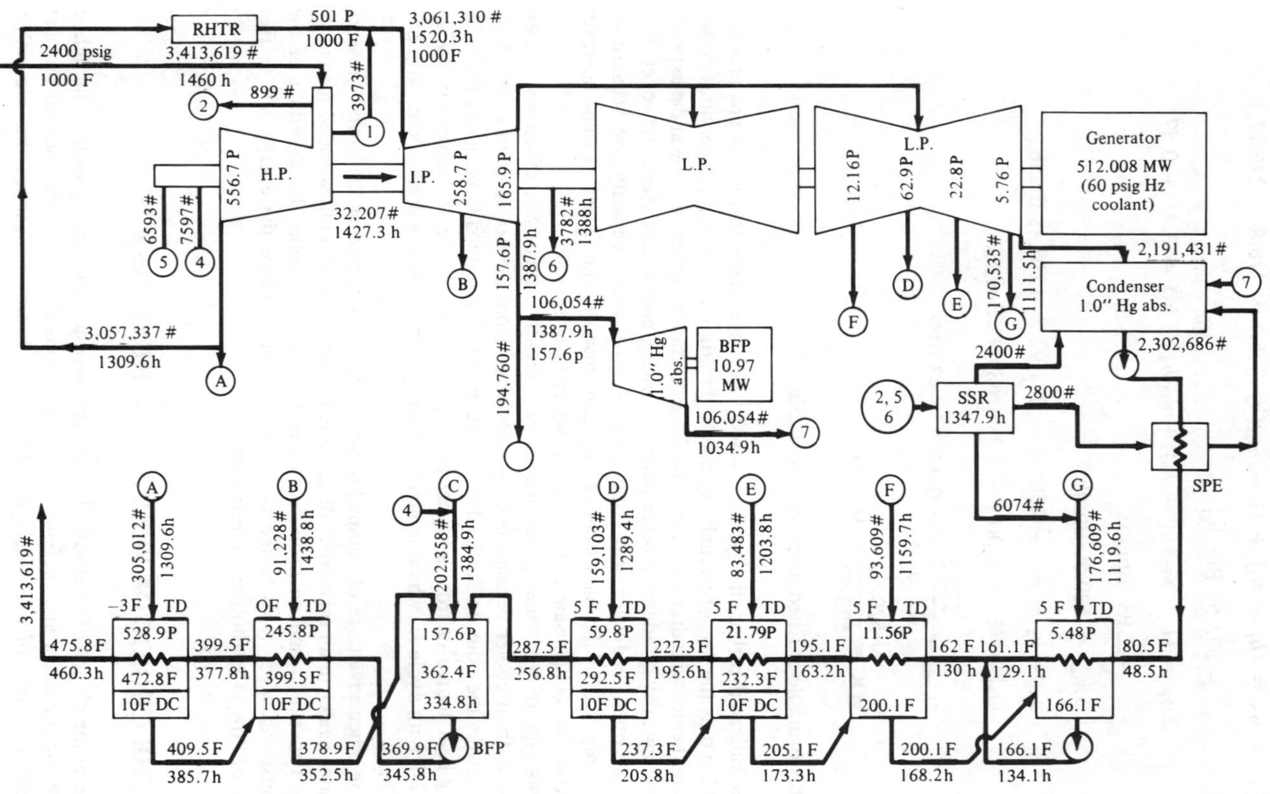

Figure 2-20 Flow diagram of an actual 512-MW 2400 psig/1000°F/1000°F reheat powerplant with seven feedwater heaters. *(Courtesy Wisconsin Power & Light Co.)*

1. One open-type feedwater heater, which doubles as a deaerator and is thus called the DA (deaerating) heater, is used in fossil-fueled powerplants. It is not yet the practice to use it in water-cooled-and-moderated nuclear powerplants because of the concern regarding radioactivity release with deaeration. This type of heater is usually placed near the middle of the feedwater system, where the temperature is most conducive to the release of noncondensables.
2. The closed-type feedwater heater with drains cascaded backward is the most common type, used both before and after the DA heater. It usually has integral desuperheating and drain cooler sections in the high-pressure stages but no superheating section in the very low-pressure stages because the bled steam is saturated or wet. A separate drain cooler is sometimes used for the lowest-pressure heater.
3. One closed feedwater heater with drains pumped forward is often used as the lowest-pressure feedwater heater to pump all accumulating drains back into the feedwater line, as indicated above. Occasionally one encounters one more feedwater of this type at a higher-pressure stage.

Table 2-2 is a compilation of the results of calculations similar to and including those in the previous examples. They all have 1000 psia, 1000°F steam at turbine inlet, except for cycle *A*, which is saturated. Cycles *G* and *H* have reheat to 1000°F. Cycles *A*, *B*, and *I* have no feedwater heaters. The rest have one feedwater of various types except for cycle *H*, which has two. All cycles are ideal, meaning that they are internally reversible with adiabatic reversible turbines and pumps.

Comparison shows large efficiency increases as a result of superheat, reheat, and the use of even one feedwater heater. The differences between different types of feedwater heaters are small. It is to be noted, however, that even a fraction of a percent difference in efficiency can mean a very large difference in annual fuel costs, especially in a fossil powerplant, where the fuel cycle costs are a large portion of the total cost of electricity. (Other costs are the fixed charges on the capital cost and the operation and maintenance cost, O & M.) Differences in efficiency also mean differences in plant size (heat exchangers, etc.) for a given plant output and hence differences in capital cost. Although the cycles summarized in Table 2-2 are ideal, the trends they exhibit are applicable to nonideal cycles, so one should expect the same relative standings in both cases.

Figure 2-20 shows a flow diagram of an actual 512-MW powerplant with superheat, reheat, and seven feedwaters: one DA, five closed with drains cascaded backward, and one, the lowest pressure, closed with drains pumped forward. In such diagrams, there are standard notations (not all to be found in Fig. 2-20), such as

AE	Available energy or isentropic enthalpy difference, Btu/lb_m
BFP	Boiler feed pump
DC	Drain cooler terminal temperature difference (Fig. 2-16*b* and *c*), °F
EL	Exhaust loss, Btu/lb_m
ELEP	Expansion line end-point enthalpy, Btu/lb_m
h	Enthalpy, Btu/lb_m

P	Pressure, psia
RHTR	Reheater
SGFP	Steam generator feed pump
SJAE	Steam-jet air ejector condenser
SPE	Steam packing exhaust condenser
SSR	Steam seal regulator
TD or TTD	Terminal temperature difference (Fig. 2-16), °F
UEEP	Used energy end point, Btu/lb$_m$
#	Mass-flow rate, lb$_m$/h

2-12 EFFICIENCY AND HEAT RATE

In the thermodynamic analysis of cycles and powerplants, the thermal efficiency and the power output are of prime importance. The *thermal efficiency* is the ratio of the net work to the heat added to the cycle or powerplant. The thermal efficiencies of powerplants are less than those computed for cycles as above because the analyses above failed to take into account the various auxiliaries used in a powerplant and the various irreversibilities associated with them. A complete analysis of a powerplant must take into account all these auxiliaries, the nonidealities in turbines, pumps, friction, heat transfer, throttling, etc., as well as the differences between full-load and partial-load operation. Such analyses are quite complex and require the use of high-capacity computers.

The *gross efficiency* is the one calculated based on the gross work or power of the turbine-generator. This is the work or power, MW gross, produced before power is tapped for the internal functioning of the powerplant, such as that needed to operate pumps, compressors, fuel-handling equipment, and other auxiliaries, labs, computers, heating systems, lighting, etc. (Fig. 2-21). The *net efficiency* is calculated based on

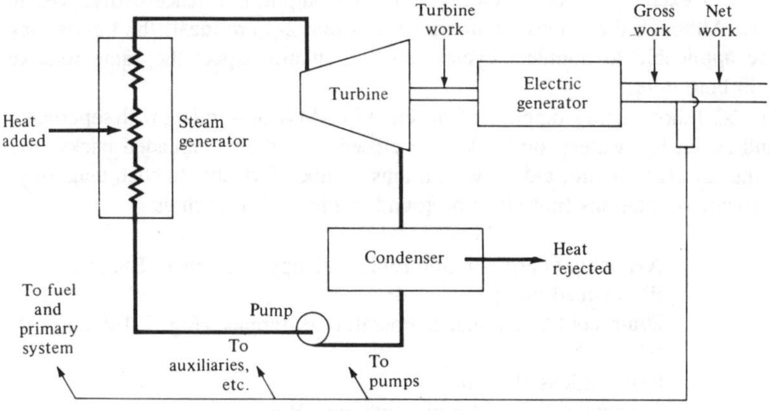

Figure 2-21 Schematic of a powerplant showing turbine, gross and net work.

the net work or power of the plant, i.e., the gross power minus the tapped power, above, or the power leaving at the station bus bars.

Powerplant designers and operators are interested in efficiency as a measure of the economy of the powerplant because it affects capital, fuel, and operating costs. They use in addition another parameter that more readily reflects the fuel economies. That parameter is called a *heat rate* (HR). It is the amount of heat added, usually in Btu, to produce a unit amount of work, usually in kilowatt hours (kWh). Heat rate thus has the units Btu/kWh. The HR is inversely proportional to the efficiency, and hence the lower its value, the better. There are various heat rates corresponding to the work used in the denominator. For example

$$\text{Net cycle HR} = \frac{\text{heat added to cycle, Btu}}{\text{net cycle work kWh}}$$

$$= \frac{\text{rate of heat added to cycle, Btu/h}}{\text{net cycle power, kW}}$$

$$\text{Gross cycle HR} = \frac{\text{rate of heat added to cycle, Btu/h}}{\text{turbine power output, kW}}$$

$$\text{Net station HR} = \frac{\text{rate of heat added to steam generator, Btu/h}}{\text{net station power, kW}}$$

$$\text{Gross station HR} = \frac{\text{rate of heat added to steam generator, Btu/h}}{\text{gross turbine-generator power, kW}}$$

and there are as many thermal efficiencies as there are heat rates. Because 1 kWh = 3412 Btu, the heat rate of any kind is related to the corresponding thermal efficiency by

$$\text{HR} = \frac{3412}{\eta_{\text{th}}} \tag{2-27}$$

Example 2-7 A coal-fired powerplant has a turbine-generator rated at 1000 MW gross. The plant requires about 9 percent of this power for its internal operations. It uses 9800 tons of coal per day. This coal has a heating value of 11,500 Btu/lb_m, and the steam generator efficiency is 86 percent. Calculate the gross station, net station, and the net steam cycle heat rates.

SOLUTION

$$\text{Rate of coal burned} = 9800 \times \frac{2000}{24} = 816{,}667 \ \ \text{lb}_m/\text{h}$$

$$\text{Gross station HR} = \frac{816{,}667 \times 11{,}500}{1000 \times 1000} = 9391.67 \text{ Btu/kWh}$$

$$\text{Station net power output} = (1 - 0.9) \times 1000 = 910 \text{ MW}$$

$$\text{Net station HR} = \frac{816{,}667 \times 11{,}500}{910 \times 1000} = 10{,}320.5 \text{ Btu/kWh}$$

$$\begin{aligned}\text{Heat added to steam generator} &= 816{,}667 \times 11{,}500 \times 0.86\\ &= 8.07683 \times 10^9 \text{ Btu/h}\end{aligned}$$

$$\text{Net steam cycle HR} = \frac{8.07683 \times 10^9}{0.91 \times 10^6} = 8875.64 \text{ Btu/kWh}$$

The corresponding thermal efficiencies are

$$\text{Gross station efficiency} = \frac{3412}{9391.67} = 36.33\%$$

$$\text{Net station efficiency} = \frac{3412}{10{,}320.5} = 33.06\%$$

$$\text{Net cycle efficiency} = \frac{3412}{8875.64} = 38.44\%$$

When the efficiency and heat rate of a powerplant are quoted without specification, it is usually the net station efficiency and heat rate that are meant. A convenient numerical value to remember for heat rate is 10,000 Btu/kWh. Usually large modern and efficient powerplants have values less than 10,000, while older plants, gas-turbine plants, and alternative power systems such as solar, geothermal, and others, exceed this value.

Figure 2-22, originally published in 1954 [9], contains a history of steam cycles since 1915 and an interesting prediction of things to come, up to 1980. It gives the average overall (net) HR range or band as a function of steam conditions, shown above the band. The heat rates are in turn dependent upon metallurgical constraints and development. The available materials are shown below the band. A landmark station was the 325-MW Eddystone unit I of the Philadelphia Electric Company, a double-reheat plant designed for operation with supercritical steam (Sec. 2-14) at 5000 psig/1200°F/1050°F/1050°F (about 345 bar, 650°C/565°C/565°C). Its actual operation was at 4700 psig and 1130°F turbine inlet (325 bar, 610°C). Built in 1959, it had the highest steam conditions and lowest HR of any plant in the world, and its power output was equal to the largest commercially available plant at the time.

Figure 2-22 is shown to predict conditions far beyond what has been achieved to date. The material *X* needed to raise the pressures and temperatures to the 7500 psig and 1400°F level, for example, remains to be developed. The most common steam conditions remain at 2400 to 3500 psia (165 to 240 bar) and 1000 to 1050°F (540 to 565°C). The 1960s and 1970s saw little improvements because there was no motivation to lower heat rates with the then-cheap fossil fuels and the advent of nuclear power. In fact, recent years have seen a rise in heat rates as a result of environmental restrictions on cooling and the increased use of devices to reduce the environmental impact of power generation (cooling towers, electrostatic precipitators, desulfurization, etc.).

Figure 2-22, however, correctly predicts advancements such as single and double

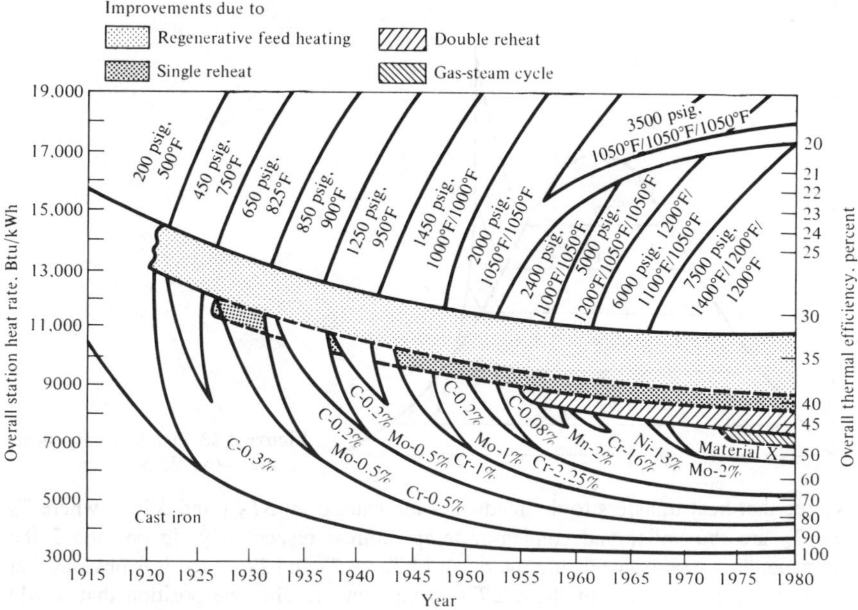

Figure 2-22 The evolution of the steam cycle as predicted in 1954 [9].

reheat and combined gas-turbine-steam-turbine cycles (Sec. 8-10). An advanced 773-MW plant design utilizing double reheat, supercritical steam at 4500 psig/1100/1050/1050°F (310 bar, 593/565/565°C), 10 feedwater heaters, and other novel features, and yielding a heat rate of 8335 Btu/kWh, has recently been proposed [10].

2-13 THE PLACEMENT OF FEEDWATER HEATERS

A natural question arises as to where to place the feedwater heaters (of any kind) in the cycle. In other words: What are the pressures at which steam is to be bled from the turbine that will result in the maximum increase in efficiency (or maximum reduction in heat rate)? It is expected that the answer to this question can be obtained most accurately by a complete optimization of the cycle, a job that entails large, complex, and usually not readily available computer programs.

There is, however, a simple answer based on physical reasoning. As indicated previously, the role of feedwater heaters is to bring the temperature of the feedwater as close as possible to that of the steam generator before the feedwater enters that steam generator. If we were to assume first for simplicity that only one feedwater heater (the type is not important for this discussion) is to be used, we may consider placing it in positions 1, 2, or 3 with respect to the cycle (Fig. 2-23). In position 1

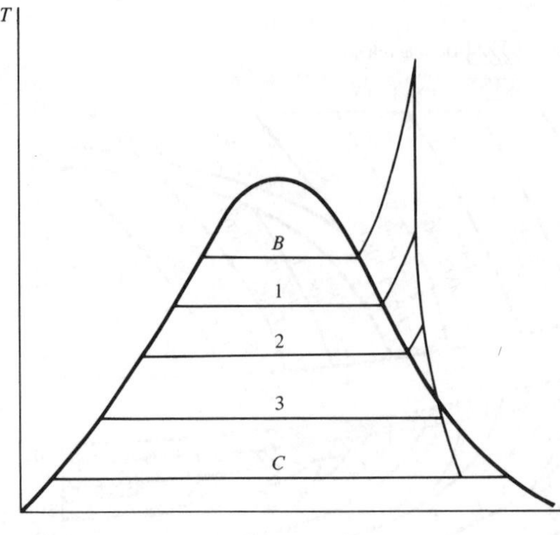

Figure 2-23 One feedwater heater in three possible positions.

we see that heat transfers to the feedwater are caused by ΔT_{B-1} and ΔT_{1-C}, where T_B and T_C are the boiler and condenser temperatures, respectively. In position 3 the corresponding heat transfers are the result of $T_B - T_3$ and $T_3 - T_C$. It is obvious that in both these cases one of these ΔT's is very large. The one position that would minimize both temperature differences is in the middle, position 2, where $T_B - T_2 = T_2 - T_C$. Thus the optimum, from an efficiency point of view, of the *pressure* at which the one feedwater heater is to be placed is obtained by finding the *temperature* that is half way between T_B and T_C and then obtaining the *saturation pressure* corresponding to that temperature. Note that the temperature at which steam is actually bled from the turbine may be in the superheat region at that pressure and thus higher than T_2.

If two feedwater heaters are to be used, the optimum placement is at temperatures that would divide $T_B - T_C$ into three equal parts. In general, then, for n feedwater heaters (Fig. 2-24), the optimum temperature rise per heater would be given by

$$\Delta T_{\text{opt}} = \frac{T_B - T_C}{n + 1} \tag{2-28}$$

Example 2-8 The Rankine cycle shown in Fig. 2-24 has an ideal turbine that operates between 1000 psia and 1000°F, and 1 psia. It has seven feedwater heaters. Find the optimum pressure and inlet temperature for the high- and the low-pressure feedwater heaters.

SOLUTION Referring to Fig. 2-24 and the steam tables

$$T_B = 544.58°\text{F} \qquad T_C = 101.74°\text{F} \qquad s_1 = 1.6530$$

$$\Delta T_{\text{opt}} = \frac{544.58 - 101.74}{7 + 1} = 55.36°\text{F}$$

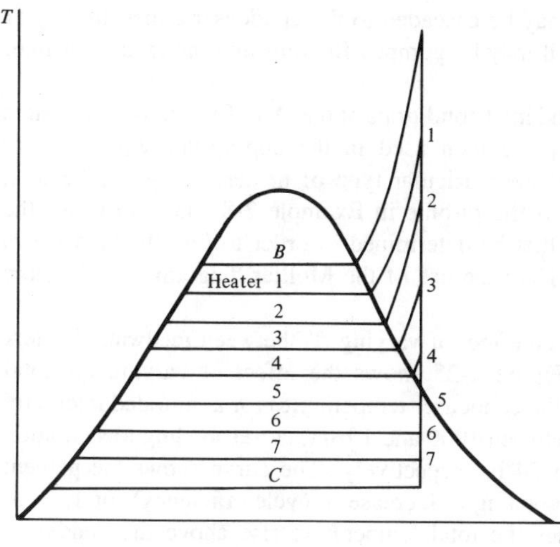

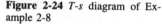

Figure 2-24 *T-s* diagram of Example 2-8

The low-pressure heater

$$T_7 = T_C + \Delta T_{\text{opt}} = 101.74 + 55.36$$
$$= 157.10°\text{F, corresponding to } P_7 = 4.422 \text{ psia}$$

Because s_g at $P_7 \approx 1.806 > s_1$, the bled steam to heater 7 is, as expected, in the two-phase region, for which

$$s_7 = s_1 = 1.6530 = (s_f + x_7 s_{fg})_{4.422 \text{ psia}}$$
$$= 0.2266 + x_7(1.6277)$$

Thus $\quad x_7 = 0.876$

and $\quad h_7 = 125.05 + 0.876 \times 1003.9 = 1004.5 \text{ Btu/lb}_m$

The high-pressure heater

$$T_{\text{sat},1} = T_B - \Delta T_{\text{opt}} = 544.58 - 55.36$$
$$= 489.22°\text{F, corresponding to } P_1 = 617.04 \text{ psia}$$

Because at P_1 $s_g = 1.4433 < s_1$, the bled steam to heater 1 is superheated. The inlet temperature, found by interpolation from the steam tables, is 850.0°F with a degree of superheat of 360.8°F, corresponding to an enthalpy of 1435.05 Btu/lb$_m$.

Heater 1, the high-pressure heater, receives highly superheated steam and thus would be constructed with a desuperheater zone, a condensing zone, and most likely, a drain cooler. Its TTD is most likely negative. Heater 7, the low-pressure heater, on the other hand, receives wet steam and will have no desuperheating zone. It will have a condensing section and may not have an integral drain cooler. If not, its drain may be cascaded to the condenser either directly or via a separate drain cooler, or it may be pumped forward into the feedwater line.

The temperatures, pressures, and inlet conditions of the other five feedwater heaters are found in a like manner. They are then used in the appropriate equations for determining the mass-flow rates in the particular type of heater, or mix of heaters, and the various cycle parameters. If the turbine in Example 2-8 were not ideal, the exact turbine expansion line must first be determined in order to find the bled steam inlet temperatures and enthalpies. Here the use of the Mollier diagram may be more useful than the *T-s* diagram.

It is now instructive to show the effect of varying ΔT between feedwater heaters from ΔT_{opt} on cycle efficiency. Figure 2-25 shows the effect of varying the total feedwater temperature rise (above the condenser temperature) for a saturated internally reversible steam cycle operating between 1000 and 1 psia, corresponding to saturation temperatures of 544.58°F and 101.74°F, respectively. The curve shows the percent decrease in cycle heat rate (corresponding to increase in cycle efficiency) for 1, 2, 3, 4, and 10 feedwater heaters versus the total temperature rise above the condenser temperature.

It can be seen, as expected, that the curve for a single feedwater heater peaks at a temperature rise halfway between the above saturation temperatures; i.e., it peaks at ΔT of 0.5(544.58 − 101.74), or about 222°F. For two feedwater heaters, the peak occurs at $\frac{2}{3}$(544.58 − 101.740), or about 295°F. It can also be seen that the curves are relatively flat about the optimum values, which indicates that small departures from these optimum values have no serious effect on heat rate. In actual powerplants, the feedwater heaters are not positioned necessarily at their optimum positions. Other considerations may dictate the exact positions. These considerations include the placement of the deaerating heater for best deaeration and the relative positions of the closed heaters before and after it, the existence of a convenient point at which steam is bled such as the crossover between turbine sections or at the steam outlet to the reheater, the design of the turbine casings, and others.

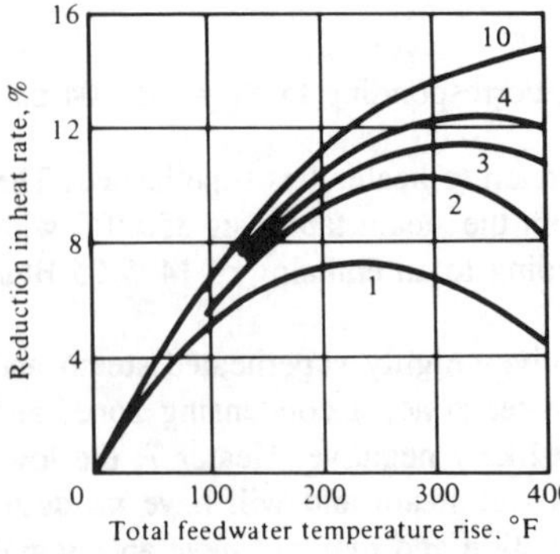

Figure 2-25 Effect of ΔT between feedwater heaters on cycle heat rate.

2-14 THE SUPERCRITICAL-PRESSURE CYCLE

In Fig. 2-26 the feedwater is pressurized at 8 to a pressure beyond the critical pressure of the vapor (3208 psia for steam). The feedwater heating curve shows a gradual change in temperature and density but not in phase to the steam temperature at 1. Such heating can be made to be closer to the heat source temperature than a subcritical cycle with the same steam temperature that shows an abrupt change in temperature within the two-phase region. Looking at it another way, the supercritical-pressure cycle receives more of its heat at higher temperatures than a subcritical cycle with the same turbine inlet steam temperature.

Because of the gradual change in density, supercritical-pressure cycles use once-through steam generators instead of the more common drum-type steam generators (Chap. 3).

A disadvantage of the supercritical-pressure cycle, however, is that expansion from point 1 to the condenser pressure would result in very wet vapor in the latter stages of the turbine. Hence, supercritical-pressure cycles invariably use reheat and often double reheat. A popular base design for a supercritical powerplant used 3500 psia and initial 1000°F steam with reheats to 1025°F and 1050°F (3500/1000/1025/1050). The higher temperatures after reheat were tolerated by the reheater tubes because of the much lower pressures in them.

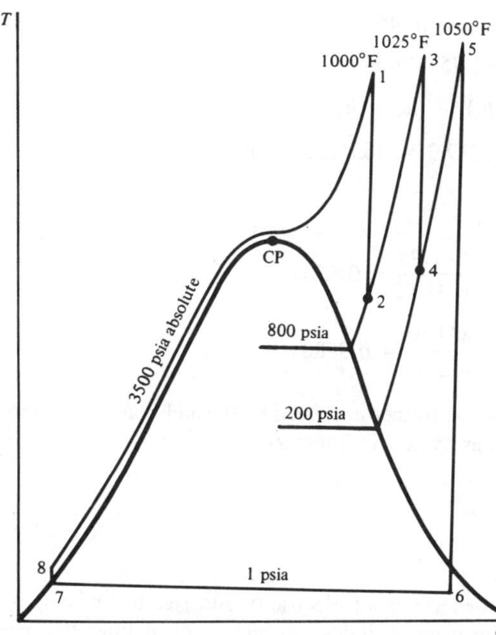

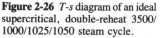

Figure 2-26 *T-s* diagram of an ideal supercritical, double-reheat 3500/1000/1025/1050 steam cycle.

Example 2-9 Calculate the net work, heat added, efficiency, and work ratio of an internally reversible supercritical double-reheat 3500/1000/1025/1050 cycle. Reheats occur at 800 and 200 psia. Condensing is at 1 psia.

SOLUTION Referring to Fig. 2-26 and the steam tables with h values in Btu/lb$_m$ and s values in Btu/(lb$_m$ · °R)

$$h_1 = 1422.2 \qquad s_1 = 1.4709$$

$$s_2 = 1.4709 \qquad h_2 = 1254.5$$

$$h_3 = 1525.3 \qquad s_3 = 1.69015$$

$$s_4 = 1.69015 \qquad h_4 = 16336.3$$

$$h_5 = 1555.4 \qquad s_5 = 1.8603$$

$$s_6 = 1.8603 \qquad x_6 = 0.936 \qquad h_6 = 1039.7$$

$$h_7 = 69.73$$

$$h_8 = 69.73 + \frac{0.016136(3500 - 1)144}{778.16} = 69.73 + 10.45 = 80.18$$

$$\begin{aligned}\Delta w_{\text{net}} &= (1422.20 - 254.5) + (1525.3 - 1336.3) \\ &\quad + (1555.4 - 1039.7) - 10.45 \\ &= 167.7 + 189 + 515.7 - 10.45 \\ &= 872.4 - 10.45 = 861.95 \text{ Btu/lb}_m\end{aligned}$$

$$\begin{aligned}q_A &= (h_1 - h_8) + (h_3 - h_2) + (h_5 - h_4) \\ &= 1342.02 + 270.8 + 219.3 = 1831.92 \text{ Btu/lb}_m\end{aligned}$$

Therefore

$$\eta_{\text{th}} = \frac{861.95}{1831.92} = 0.4705$$

and

$$\text{WR} = \frac{861.95}{872.41} = 0.9880$$

The efficiency, of course, would be further improved by the addition of feedwater heaters. This example is listed as cycle *I* in Table 2-2.

2-15 Cogeneration

Cogeneration is the simultaneous generation of electricity and steam (or heat) in a single powerplant. It has long been used by industries and municipalities that need process steam (or heat) as well as electricity. Examples are chemical industries, paper mills, and places that use district heating. Cogeneration is not usually used by large utilities which tend to produce electricity only. Cogeneration is advisable for industries and municipalities if they can produce electricity cheaper, or more conveniently, than that brought from a utility.

From an energy resource point of view, cogeneration is beneficial only if it saves primary energy when compared with separate generation of electricity and steam (or heat). The *cogeneration plant efficiency* η_{co} is given by

$$\eta_{co} = \frac{E + \Delta H_s}{Q_A} \tag{2-29}$$

where E = electric energy generated

ΔH_s = heat energy, or heat energy in process steam

= (enthalpy of steam entering the process) − (enthalpy of process condensate returning to plant)

Q_A = heat added to plant (in coal, nuclear fuel, etc.)

For separate generation of electricity and steam, the heat added per unit *total* energy output is

$$\frac{e}{\eta_e} + \frac{(1-e)}{\eta_h}$$

where e = electrical fraction of total energy output $= \dfrac{E}{(E+\Delta H_s)}$

η_e = electric plant efficiency

η_h = steam (or heat) generator efficiency

The *combined efficiency* η_c *for separate generation* is therefore given by

$$\eta_c = \frac{1}{(e/\eta_e) + [(1-e)/\eta_h]} \tag{2-30}$$

and cogeneration is beneficial if the efficiency of the cogeneration plant Eq. (2-29) exceeds that of separate generation, Eq. (2-30).

Types of Cogeneration

There are two broad categories of cogeneration:

1. *The topping cycle,* in which primary heat at the higher temperature end of the Rankine cycle is used to generate high-pressure and -temperature steam and electricity in the usual manner. Dependng on process requirements, process steam at low-pressure and temperature is either (*a*) extracted from the turbine at an intermediate stage, much as for feedwater heating, or (*b*) taken at the turbine exhaust,

in which case it is called a *back pressure* turbine. Process steam pressure requirements vary widely, between 0.5 and 40 bar.

2. *The bottoming cycle,* in which primary heat is used at high temperature directly for process requirements. An example is the high-temperature cement kiln. The process low-grade (low temperature and availability) waste heat is then used to generate electricity, obviously at low efficiency. The bottoming cycle thus has a combined efficiency that most certainly lies below that given by Eq. (2-30), and therefore is of little thermodynamic or economic interest.

Only the topping cycle, therefore, can provide true savings in primary energy. In addition, most process applications require low grade (temperature, availability) steam. Such steam is conveniently produced in a topping cycle. There are several arrangements for cogeneration in a topping cycle. Some are:

(*a*) Steam-electric powerplant with a back-pressure turbine.
(*b*) Steam-electric powerplant with steam extraction from a condensing turbine (Fig. 2-27).
(*c*) Gas-turbine powerplant with a heat-recovery boiler (using the gas turbine exhaust to generate steam).
(*d*) Combined steam-gas-turbine cycle powerplant (Secs. 8-8 and 8-9). The steam turbine is either of the back-pressure type (*a*) or of the extraction-condensing type (*b*), above.

The most suitable electric-to-heat generation ratios vary from type to type. The back-pressure steam turbine plant (*a*) is most suitable only when the electric demand is low compared with the heat demand. The combined-cycle plant (*d*) is most suitable only when the electric demand is high, about comparable to the heat demand or higher, though its range is wider with an extraction-condensing steam turbine than with a back-pressure turbine. The gas-turbine cycle (*c*) lies in between. Only the extraction-condensing plant (*b*) is suitable over a wide range of ratios.

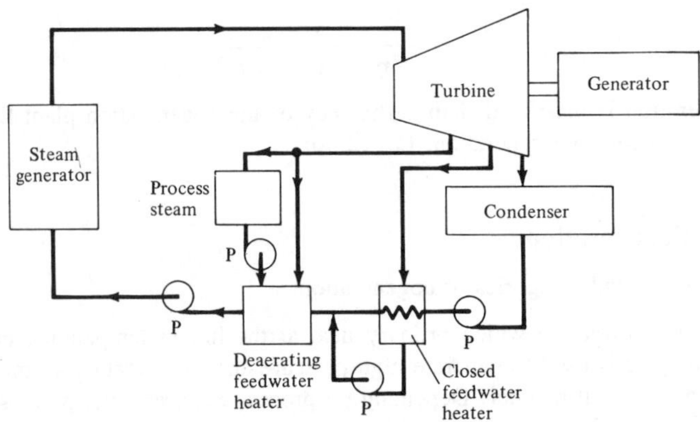

Figure 2-27 Schematic of basic cogeneration plant with extraction-condensing turbine.

Economics of Cogeneration

A privately or municipally owned cogeneration plant is advisable from an economic point of view if the cost of electricity generated by it is less than if purchased from a utility. (If a utility is not available, cogeneration becomes necessary, irrespective of economics.) In general, very low fractions of electric to total energy are not considered economical for cogeneration.

Since the main incentive of cogeneration is process steam (or heat), the economics of cogeneration are sharply influenced by the additional cost of generating electricity.

Powerplant costs are of two kinds: capital costs and production costs. *Capital costs* are given in total dollars or as *unit capital costs* in dollars per kilowatt net. Production costs are calculated annually, or more frequently if desired, and given in mills per kilowatt hour. A mill is one one-thousandth of a United States dollar. Capital costs determine whether a given utility or industry is sound enough to obtain financing and thus able to pay the fixed charges against these costs. *Production costs* are the true measure of the cost of power generated. They are composed of:

a. The fixed charges against the capital costs
b. The fuel costs
c. Operation and maintenance costs

all in mills per kilowatt hour. They are therefore given by:

$$\text{Production costs} = \frac{\text{total } (a+b+c) \text{ \$ spent per period} \times 10^3}{\text{KWh (net) generated during same period}} \tag{2-31}$$

where the period is usually taken as one year.

For a congeneration plant, it is important to calculate the production costs of electricity as an excess over the generating cost of steam alone, and to compare it with the cost of electricity when purchased from a utility. It is now necessary to introduce the *plant operating factor* POF, defined for all plants as

$$\text{POF} = \frac{\text{total net energy generated by plant during a period of time}}{\text{rated net energy capacity of plant during same period}} \tag{2-32}$$

where the period is again usually taken as one year. For estimation purposes, it is common to take POF = 0.80. A plant operating with POF = 0.8 is the same as if it operated only at rated capacity for 80 percent of the time or for 0.8(365 × 24) = 7008 h/yr, which is usually rounded out to 7000 h/yr.

The excess cost of electricity for a cogeneration plant may now be obtained from

$$\text{Electric cost} = [(C_{co}-C_h)r + (\text{OM}_{co}-\text{OM}_h) + (F_{co}-F_h)] \frac{10^3}{7000\,P} \text{ mills/kWh} \tag{2-33}$$

where C = capital costs, \$

r = annual fixed charges against the capital cost, fraction of C

OM = annual operation and maintenance costs, \$/yr

F = annual fuel costs, \$/yr

P = electric plant net power rating, kW

and the subscripts *co* and *h* indicate cogeneration and process heat plants, respectively.

Cogeneration plants, built mostly by industries or municipalities, are smaller than utility electric-generating plants and therefore tend to have higher unit capital and operating costs. They have not usually been considered for operation with coal or nuclear energy as a primary heat source, though this picture is slowly changing.

PROBLEMS

2-1 A simple ideal saturated Rankine cycle turbine receives 125 kg/s of steam at 300°C and condenses at 40°C. Calculate (*a*) the net cycle power, in megawatts, and (*b*) the cycle efficiency.

2-2 A simple nonideal saturated Rankine cycle turbine receives 125 kg/s of steam at 300°C and condenses at 40°C (same conditions as Prob. 2-1). This cycle has turbine and pump polytropic efficiencies of 0.88 and 0.75, respectively, and a total pressure drop in the feedwater line and steam generator of 10 bar. Calculate (*a*) the net cycle power, in megawatts, and (*b*) the cycle efficiency.

2-3 Analyze the ideal Rankine cycle *C* in Table 2-2 if the feedwater heater is placed at 100 psia.

2-4 Compare the inlet steam mass and volume flow rates in pound mass per second and cubic feet per second of (*a*) a fossil-fuel powerplant turbine having a polytropic efficiency of 0.90 and receiving steam at 2400 psia and 1000°F and (*b*) a nuclear powerplant turbine having a polytropic efficiency of 0.88 and receiving saturated steam at 1000 psia. Each turbine produces 1000 megawatts, and exhausts to 1 psia.

2-5 To reduce the volume flow rate and hence turbine physical size, powerplants that operate with low initial temperature water as a heat source, such as some types of geothermal (Chap. 12) and ocean temperature energy conversion, OTEC (Chap. 15), powerplants, use working fluids other than steam, such as Freon-12, ammonia, and propane. Compare the mass flow rates, pound mass per hour, volume flow rates, cubic feet per second, and boiler and condenser pressures of (*a*) Freon-12, (*b*) propane, and (*c*) steam, if all cycles operate with adiabatic reversible turbines that receive saturated vapor at 200°F and condense at 70°F, and each produces 100 kW.

2-6 In Prob. 2-5, why do the cycles operate with saturated vapor?

2-7 Consider three nonideal saturated Rankine cycles operating between 200 and 70°F using Freon-12, propane, and steam as working fluids. Each has turbine and pump polytropic efficiencies of 85 and 65 percent, respectively, and produces net work of 100 kW. Calculate (*a*) the mass flow rate in pound mass per hour, (*b*) the volume flow rate in cubic feet per second, (*c*) the heat added, in Btus per hour, and (*d*) the cycle efficiency.

2-8 Consider an ideal saturated steam Rankine cycle with perfect regeneration (Fig. 2-11) operating between 1000 and 1.0 psia. Neglecting pump work, calculate (*a*) the quality of the turbine exhaust steam, (*b*) the turbine work in Btus per pound mass, (*c*) the heat added in Btus per pound mass, and (*d*) the cycle efficiency. Compare that efficiency to that of a similar cycle but without regeneration, and a Carnot cycle, all operating between the same temperature limits.

2-9 Compare the net works, in Btus per pound mass, and efficiencies of two ideal saturated Rankine cycles using Freon-12 as a working fluid and operating between 200 and 72°F. One cycle has no feed heaters and the other has one open-type feed heater placed optimally. Why is feed heating not usually resorted to in such cycles?

2-10 A Rankine cycle with inlet steam at 90 bar and 500°C and condensation at 40°C produces 500 MW. It has one stage of reheat, optimally placed, back to 500°C. One feedwater of the closed type with drains cascaded back to the condenser receives bled steam at the reheat pressure. The high- and low-pressure turbine sections have polytropic efficiencies of 92 and 90 percent, respectively. The pump has a polytropic efficiency of 0.75. Calculate (*a*) the mass flow rate of steam at turbine inlet in kilograms per second, (*b*) the cycle efficiency, and (*c*) the cycle work ratio. Use TDD = − 1.6°C.

2-11 An ideal Rankine cycle operates with turbine inlet steam at 90 bar and 500°C, and a condenser temperature of 40°C. Calculate the efficiency and work ratio of this cycle for the following cases: (*a*) no feedwater heating, (*b*) one open-type feedwater heater, (*c*) one closed-type feedwater heater with drains cascaded back to the condenser, and (*d*) one closed feedwater heater with drains pumped forward. In each case the feedwater heater is optimally placed. Use TDD = 2.5°C.

2-12 A superheated nonideal steam cycle operates with inlet steam at 2400 psia and 1000°F and condenses at 1 psia. It has five feedwater heaters, all optimally placed. Assume the polytropic efficiencies of the turbine sections before, between, and after the bleed points to be all the same and equal to 0.90. Calculate (*a*) the specific enthalpies of the extraction steam to each feedwater heater, in Btus per pound mass and (*b*) the turbine overall polytropic efficiency; and (*c*) estimate the terminal temperature difference for each feedwater heater.

2-13 An 850-MW Rankine cycle operates with turbine inlet steam at 1200 psia and 1000°F and condenser pressure at 1 psia. There are three feedwater heaters placed optimally as follows: (*a*) the high-pressure heater is of the closed type with drains cascaded backward; (*b*) the intermediate-pressure heater is of the open type; (*c*) the low-pressure heater is of the closed type with drains pumped forward. Each of the turbine sections have the same polytropic efficiency of 90 percent. The pumps have polytropic efficiencies of 80 percent. Calculate (*a*) the mass flow rate at the turbine inlet in pound mass per hour, (*b*) the mass flow rate to the condenser, (*c*) the mass flow rate of the condenser cooling water, in pound mass per hour, if it undergoes a 25°F temperature rise, (*d*) the cycle efficiency, and (*e*) the cycle heat rate, in Btus per kilowatt hour.

2-14 If the Rankine cycle is to be used in outer space, heat rejection can be done only by thermal radiation to space which has an effective temperature of 0 absolute. To reduce the size and mass and hence lifting weight of the condenser, condensation has to be at temperatures much higher than those used in land-based Rankine cycles. Condensing temperatures of 1000 to 1500°F are considered. These are higher than the critical temperature of water. This also means a much higher turbine inlet temperature. Thus a liquid metal such as sodium must be used as the working fluid. Consider a 100-kW (thermal) Rankine cycle using sodium, operating with 24.692 psia and 2400°R sodium vapor at turbine inlet and condensing at 1500°R. The turbine and pump polytropic efficiencies are 0.85 and 0.65, respectively. For no feed heaters and ignoring pressure drops, calculate (*a*) the cycle efficiency and (*b*) the heat transfer area of the condenser-radiator if it has an overall heat transfer coefficient of 5 Btu/ft^2 · h · °F.

2-15 Calculate the gross heat rate, in Btus per kilowatt hour, and the gross efficiency of the powerplant shown in Fig. 2-20.

2-16 A 100-MW (thermal) binary-vapor cycle uses saturated mercury vapor at 1600°R at the top turbine inlet. The mercury condenses at 1000°R in a mercury-condenser–steam-boiler in which saturated steam is generated at 400 psia. It is further superheated to 1160°R in the mercury-boiler–steam superheater. The steam condenses at 1 psia. Assume both mercury and steam cycles to be ideal, and ignoring the pump work (*a*) draw flow and T-s diagrams of the binary cycle numbering points correspondingly, (*b*) calculate the mass flow rates of mercury and steam, and (*c*) calculate the heat added and heat rejected, in Btus per hour, and the cycle efficiency.

2-17 An advanced-type supercritical powerplant has turbine inlet steam at 7000 psia and 1400°F, double reheat at 1600 psia and 400 psia, both to 1200°F, and condenser at 1 psia. The three turbine sections have polytropic efficiencies of 0.93, 0.91, and 0.89 in order of descending pressures. The pump has a polytropic efficiency of 0.75. The plant receives one unit train of coal daily, which is composed of 100 cars carrying 110 short tons each. The coal has a heating value of 11,000 Btus/lb$_m$. The turbine-generator combined mechanical and electrical efficiency is 0.90. The steam-generator efficiency is 0.87. 8 percent of the gross output is used to run plant auxiliaries. Ignoring, for simplicity, all steam-line pressure drops and all feedwater heaters, calculate (*a*) the plant gross and net outputs, in megawatts, (*b*) the plant cycle, gross and net efficiencies, and (*c*) the cycle, and station gross and net heat rates, in Btus per kilowatt hour.

2-18 Draw flow diagrams of cogeneration plants of (*a*) the topping back-pressure steam turbine type and (*b*) the bottoming-condensing steam turbine type.

2-19 A cogeneration steam plant of the extraction-condensing steam turbine type has turbine inlet flow of 12.5 kg/s at 50 bar and 400°C, extraction for process steam at 220°C and condensation for both turbine and process steam at 40°C. Assuming ideal turbine and pump, no feedwater heating and 50 percent fraction of electricity to total energy output, calculate (*a*) the cogeneration plant efficiency, (*b*) the combined efficiency if separate electric and steam generation plants producing the same outputs as above are used. For all cases take the steam generator efficiency as 85 percent.

CHAPTER

THREE

FOSSIL-FUEL STEAM GENERATORS

3-1 INTRODUCTION

Steam generators are used in both fossil- and nuclear-fuel electric-generating powerplants. The most modern steam generators produce high-pressure (2400 to 3500 psia, 165 to 240 bar) superheated steam, the exception being pressurized-water reactor steam generators, which produce lower-pressure (1000 psia, 70 bar) saturated steam. The steam is invariably used in a Rankine cycle. Steam generators represent by far the greatest energy source for powerplants in the world today.

This chapter will cover fossil-fuel steam generators. Nuclear-fuel steam generators are of radically different design and will be covered in examples in Chaps. 10 and 11.

A *steam generator* is a complex combination of economizer, boiler, superheater, reheater, and air preheater. In addition, it has various auxiliaries, such as stokers, pulverizers, burners, fans, emission control equipment, stack, and ash-handling equipment. A *boiler* is that portion of the steam generator where saturated liquid is converted to saturated steam, although it may be difficult to separate it, physically, from the economizer. The term "boiler" is often used to mean the whole steam generator in the literature, however. Steam generators are classified in different ways. They may, for example, be classified as either (1) utility or (2) industrial steam generators.

Utility steam generators are those used by utilities for electric-power generating plants and are our main concern in this book. Modern utility steam generators are essentially of two basic kinds: (1) the subcritical water-tube drum type and (2) the supercritical once-through type. The supercritical units usually operate at about 3500 psia (240 bar) and higher, above the steam critical pressure of 3208.2 psia. The subcritical drum group usually operate at either 1900 psig (about 130 bar) or 2600 psig (180 bar). The majority of utility steam generators purchased in the 1970s and 1980s are of the 2600 psig water-tube drum variety, which produce superheated steam at about 1000°F (540°C) with one or two stages of reheat. They have the ability to burn coal in pulverized form or oil, although oil is being gradually retired as a fuel because of rising costs and supply problems. Natural gas, although still used in certain parts of the world, is also costly and is now being conserved for domestic uses in the United States. Gas, however, is a clean burning, relatively pollution-free fuel. The steam capacities of modern utility steam generators are high, ranging from 1 to 10 million lb_m/h (125 to 1250 kg/s). They power electric powerplants ranging in output from 125 to 1300 megawatts (MW).

Industrial steam generators, on the other hand, are those used by industrial and institutional concerns and are of many types. These include water-tube pulverized-coal units similar to those used by utilities, but they also may burn stoker (lump) coal, oil, or natural gas, often in combination, as well as municipal refuse and process wastes or by-products. Some even use electric heating. Some are heat-recovery types that use waste heat from industrial processes. They may also be of the fire-tube variety. Industrial steam generators usually do not produce superheated steam. Rather, they usually produce saturated steam, or even only hot water (in which case they should not be called steam generators). They operate at pressures ranging from a few psig to as much as 1500 psig (105 bar) and steam (or hot water) capacities ranging from a few thousand to 1 million lb_m/h (125 kg/s).

Fossil-fueled steam generators are more broadly classified [12, 13] as those having the following components or characteristics.

1. Fire-tube boilers
2. Water-tube boilers
3. Natural-circulation boilers
4. Controlled-circulation boilers
5. Once-through flow
6. Subcritical pressure
7. Supercritical pressure

3-2 THE FIRE-TUBE BOILER

Fire-tube boilers have been used in various early forms to produce steam for industrial purposes since the late eighteenth century. They are no longer used in large utility powerplants. They are covered here, however, for historical reasons and, by contrast, to emphasize the modern water-tube variety. Fire-tube boilers are still used in industrial plants to produce saturated steam at the upper limits of 250 psig (about 18 bar) pressure and 50,000 lb_m/h (6.3 kg/s) capacity. Although their size has increased, their general design has not changed appreciably in the past 25 years.

The fire-tube boiler is a special form of the shell-type boiler. A *shell-type boiler* is a closed, usually cylindrical, vessel or shell that contains water. A portion of the

Figure 3-1 The Stead fire-tube steam generator [11].

shell, such as its underside, is simply exposed to heat, such as gases from an externally fired flame. The shell boiler evolved into more modern forms such as the *electric boiler,* in which heat is supplied by electrodes embedded in the water, or the *accumulator,* in which heat is supplied by steam from an outside source passing through tubes within the shell. In both cases the shell itself is no longer exposed to heat.

The shell boiler evolved into the *fire-tube boiler*. Hot gases, instead of steam, were now made to pass through the tubes. Because of improved heat transfer the fire-tube boiler is much more efficient than the original shell boiler and can reach efficiencies of about 70 percent.

The fire tubes were placed in horizontal, vertical, or inclined positions. The most common was the horizontal-tube boiler. Figure 3-1 shows such an early fire-tube boiler. Figure 3-2 shows a simplified sketch of such a boiler. The furnace and grates are located underneath the front end of the shell. The gases pass horizontally along its underside to the rear, reverse direction, and pass through the horizontal tubes to the stack at the front.

There are two types of fire-tube boilers: (1) the fire-box and (2) the scotch marine. In the *fire box boiler,* the furnace, or fire box, is located within the shell, together with the fire tubes. In the *scotch marine boiler* (Fig. 3-3), combustion takes place within one or more cylindrical chambers that are usually situated inside and near the bottom of the main shell. The gases leave these chambers at the rear, reverse direction, and return through the fire tubes to the front and out through the stack. Scotch marine boilers are usually specified with liquid or gas fuels.

Because boiling occurs in the same compartment where water is, fire-tube boilers are limited to saturated-steam production. They are presently confined to relatively small capacities and low steam pressures, such as supplying steam for space heating

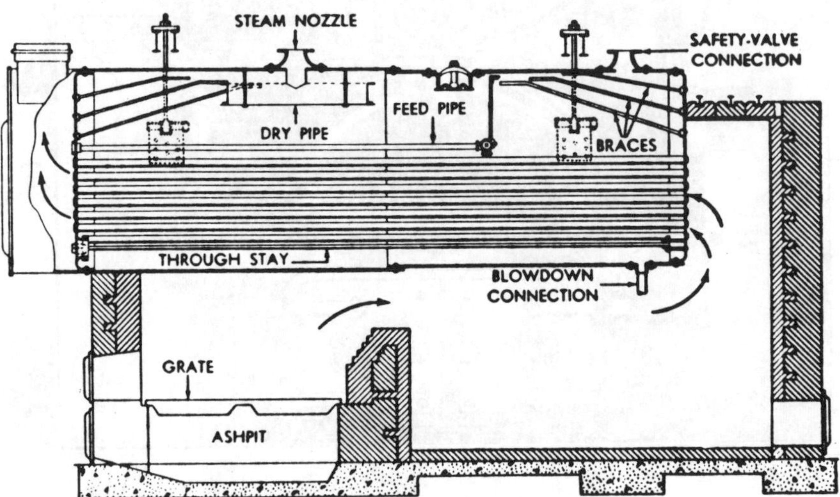

Figure 3-2 Schematic of an early horizontal-tube fire-tube boiler [12].

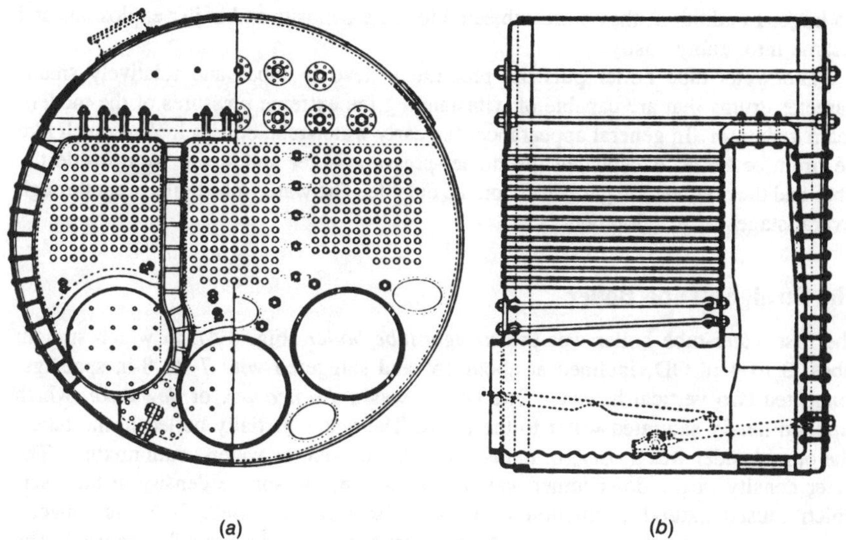

Figure 3-3 Schematic of an early scotch marine boiler.

and, in decreasing numbers, for railroad locomotive service. The largest scotch marine boiler offered in the United States today is rated at 2000 boiler horsepower (blhp),* contains two combustion chambers within a 13-ft diameter, 30-ft-long shell.

3-3 THE WATER-TUBE BOILER: EARLY DEVELOPMENTS

The forerunner of the modern steam generator was a water-tube boiler developed by George Babcock and Stephen Wilcox in 1867. Babcock and Wilcox called it the "nonexplosive" water-tube boiler, an allusion to disastrous boiler explosions that were frequent at the time. It was not, however, until early in the twentieth century, with the advent of the steam turbine and its requirement for large steam pressures and flows, that commercial development of the water-tube boiler became a reality.

With higher steam pressures and capacities, fire-tube boilers would need large-diameter shells. With such large diameters, the shells would have to operate under such extreme pressure and temperature stresses that their thicknesses would have been

* Boiler horsepower was originally used to indicate the size of the boiler. One blhp was defined as 10 ft^2 of boiler heating surface. This was later changed to the amount of heat required to evaporate 30 lb_m/h of water at 100°F to saturated steam at 70 psia. Thus, using enthalpies from the steam tables, this was equal to $30(1180.6 - 68) = 33{,}378$ Btu/h. Later on this was changed to the amount of heat necessary to evaporate 34.5 lb_m/h of saturated water to saturated steam at 1 atm or $34.5 \times 970.3 = 33475.35$ Btu/h. This is now rounded so that the modern value is 1 blhp = 33480 Btu/h.

too large. In addition, they were subjected to scale deposits and boiler explosions and became intolerably costly.

The water-tube boiler puts the pressure instead in tubes and relatively small-diameter drums that are capable of withstanding the extreme pressures of the modern steam generator. In general appearance, the early water-tube boilers looked much like the fire-tube boiler except that the higher-pressure water and steam were inside the tubes and the combustion gases were on the outside. The water-tube boiler went through several stages of development.

The Straight-Tube Boiler

The first water-tube boiler was the *straight-tube boiler* (Fig. 3-4), in which straight tubes, 3 to 4 in OD, inclined at about 15° and staggered with 7- to 8-in spacings, connected two vertical headers. One header was a *downcomer,* or *downtake,* which supplied nearly saturated water to the tubes. The water partially boiled in the tubes. The other header was a *riser,* or *uptake,* which received the water-steam mixture. The water density in the downcomer was larger than the two-phase density in the riser, which caused natural circulation in a clockwise direction (Sec. 3-5). As capacity increased, more than one header each and more than one tube "deck" were used. The two-phase mixture went into an upper drum that was arranged either parallel to the tubes (the *longitudinal drum,* Fig. 3-4*a*) or perpendicular to them (the *cross drum,* Fig. 3-4*b*). These drums received the feedwater from the last feedwater heater and supplied saturated steam to the superheater through a steam separator within the drum which separated steam from the bubbling water. The lower end of the downcomer was connected to a *mud drum,* which collected sediments from the circulating water.

A single longitudinal drum, usually 4 ft in diameter, can allow only a limited number of tubes and hence a limited heating surface. Longitudinal-drum boilers were built with one or more than one parallel drums, depending upon capacity. They were built with heating surfaces of 1000 to 10,000 ft^2 (93 to 930 m^2) and were limited to low pressures of 175 to 340 psia (12 to 23 bar) and steam capacities from 5000 to 80,000 lb_m/h (0.63 to 10 kg/s).

Cross-drum boilers, because of geometry, could accommodate many more tubes than longitudinal-drum boilers and were built with heating surfaces of 1000 to 25,000 ft^2 (93 to 2300 m^2), pressures of 175 to 1465 psia (12 to 100 bar), and steam capacities of 5000 to 500,000 lb_m/h (0.63 to 63 kg/s).

Baffles were installed across the tubes in both kinds to allow for up to three gas passes to ensure maximum exposure of the tubes to the hot combustion gases and minimal gas dead spots.

The Bent-Tube Boiler

There were many versions of the bent-tube boiler. In general, a *bent-tube boiler* used bent, rather than straight, tubes between several drums or drum and headers. The tubes were bent so that they entered and left the drums radially. The number of drums usually

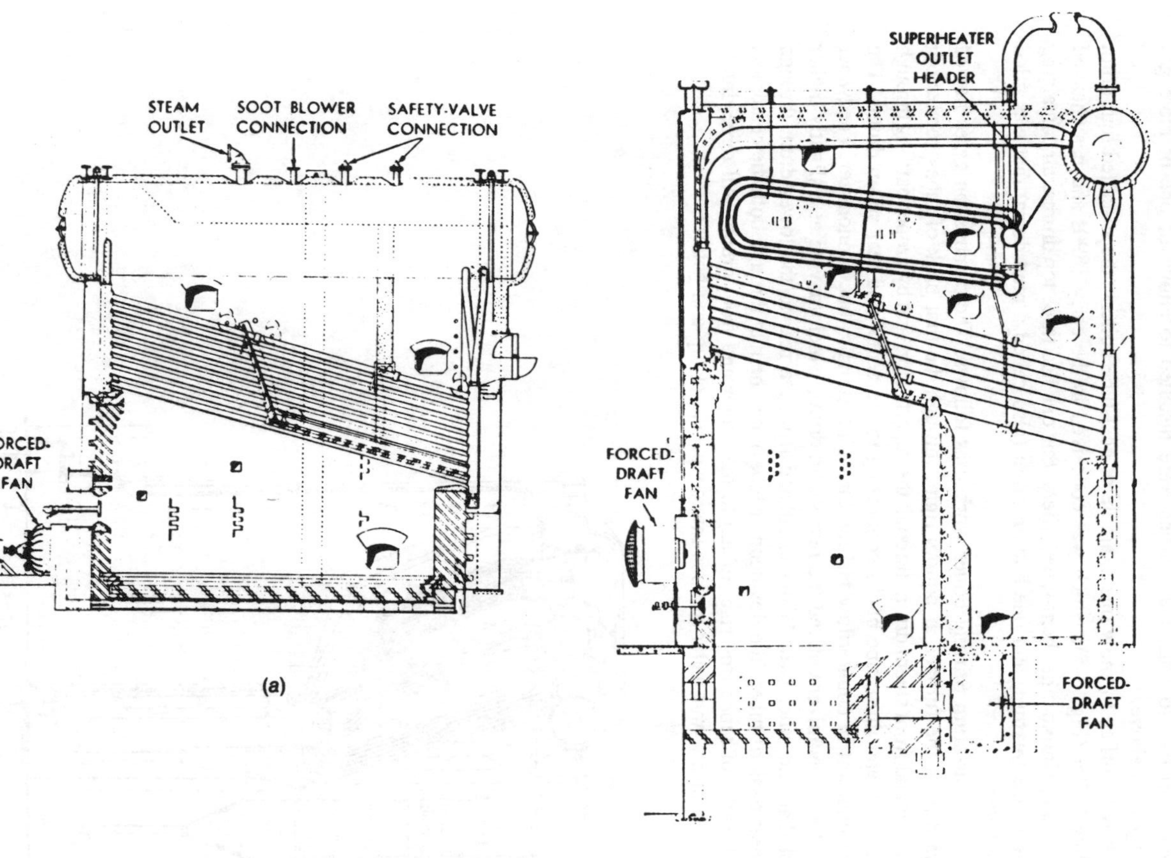

Figure 3-4 Early water-tube boilers: (*a*) longitudinal and (*b*) cross-drum [12].

varied from two to four. Gas baffles were installed to allow for one or more gas passages, as above.

It will suffice to show one example of bent-tube boilers, the so-called four-drum Stirling boiler (Fig. 3-5), which was conceived in the early 1890s and has changed little since. Unlike other bent-tube boilers, this one had three top drums, all containing a two-phase mixture, and one bottom drum (also called a mud drum) that was filled with water.

The four-drum Stirling boiler worked as follows. The combustion gases flowed upward from the furnace at bottom right, through the first bank of tubes connecting the water and front steam drum, through the superheater, and by proper baffling through the second and third tube banks connecting to the center and rear steam drums. The gases then left in counterflow fashion through a straight-tube economizer. Feedwater from the economizer entered the rear steam drum, which may be at a slightly higher level than the other two. Water circulated from the rear to the lower drum through the rear bank of tubes (the downcomer tubes) and then up through both the center and front banks of tubes (the riser tubes) to the center and front drums. All three drums had their steam and water regions interconnected at top and bottom.

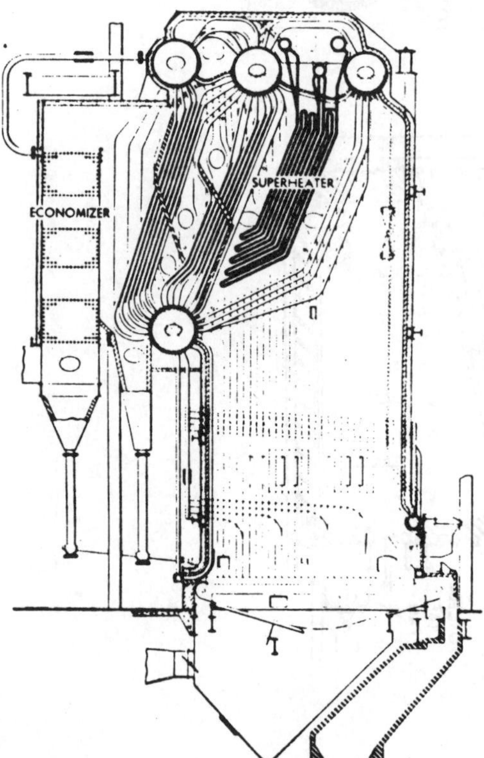

Figure 3-5 A four-drum early Stirling boiler [12].

The tubes were typically 3 to $3\frac{1}{2}$ in OD and spaced 5 and 7 in on centers, with back spacing decreased to maintain gas velocity of the cooler and denser gases. The spacings allowed for replacement of defective tubes without removing neighboring tubes.

The four-drum Stirling boiler was superseded by a simpler two-drum design in which the steam drum was directly above the water drum with one bank of bent tubes to the front, i.e., on the side of the incoming gas and another to the rear. Later designs of the two-drum Stirling boiler used a single gas path. More recent designs of Stirling boilers used cooled furnace walls by lining the interior of the walls with tubes carrying the same boiler water from the plant as seen in Fig. 3-5. These added to the heat-absorbing surface and protected the refractory lining of the walls from excessive temperature. The result was higher rates of combustion and higher steam-flow rates.

The Stirling boiler was generally capable of meeting conditions of rapidly varying loads, was useful where high water quality was difficult to maintain, and was adaptable to various fuels. It found use in both stationary and marine applications.

3-4 THE WATER-TUBE BOILER: RECENT DEVELOPMENTS

The advent of the water-cooled furnace walls, called *water walls,* eventually led to the integration of furnace, economizer, boiler, superheater, reheater, and air preheater into the modern steam generator. Water cooling is also used for superheater and economizer compartment walls and various other components, such as screens, dividing walls, etc. The use of a large number of feedwater heaters (up to seven or eight) means a smaller economizer, and the high pressure means a smaller boiler surface because the latent heat of vaporization decreases rapidly with pressure. Thus a modern high-pressure steam generator requires more superheating and reheating surface and less boiler surface than older units. Beyond about 1500 psia, the water tubes represent the entire boiler surface and no other tubes, such as those seen in the earlier designs of the previous two sections, are required.

Figure 3-6 shows a schematic flow diagram of a common steam-generator system. Water at 450 to 500°F from the plant high-pressure feedwater heater enters the economizer and leaves saturated or as a two-phase mixture of low quality. It then enters the steam drum at midpoint. Water from the steam drum flows through insulated downcomers, which are situated outside the furnace, to a header. The header connects to the water tubes that line the furnace walls and act as risers. The water in the tubes receives heat from the combustion gases and boils further. The density differential between the water in the downcomer and that in the water tubes helps circulation. Steam is separated from the bubbling water in the drum and goes to the superheater and the high-pressure section of the turbine. The exhaust from that turbine returns to the reheater, after which it goes to the low-pressure section of the turbine.

Atmospheric air from a *forced-draft (FD) fan* is preheated by the flue gases just before they are exhausted to the atmosphere. From there it flows into the furnace, where it mixes with the fuel and burns to some 3000°F. The combustion gases impart portions of some of their energy to the water tubes and then the superheater, reheater,

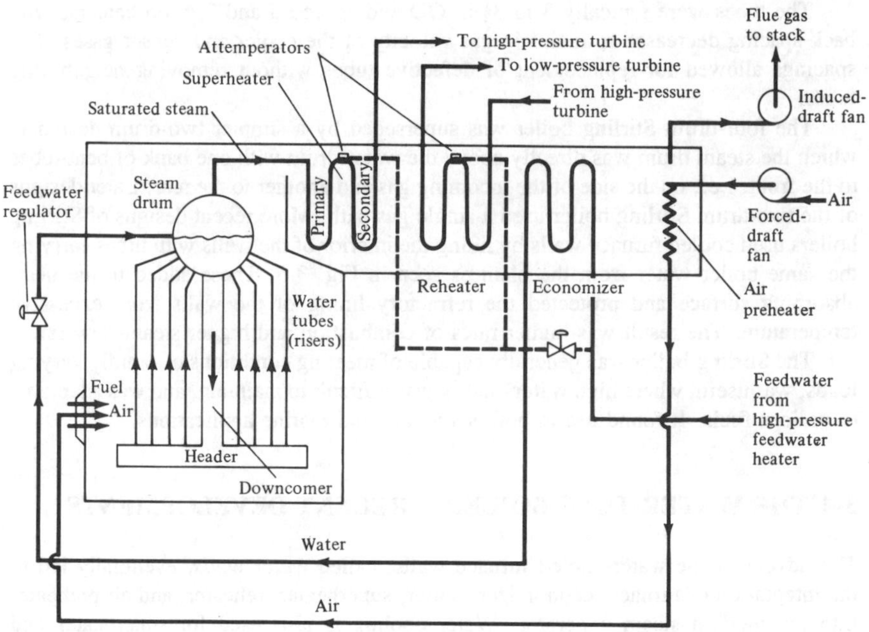

Figure 3-6 Schematic flow diagram of a modern steam generator.

and economizer, and leave the latter at about 600°F. From there they reheat the incoming atmospheric air in the air preheater, leaving it at about 300°F. An *induced-draft (ID) fan* draws the flue gases from the system and sends them up the stack. The temperature of about 300°F of the exiting flue gas represents an availability loss to the plant. This, however, is deemed acceptable because (1) the gas temperature should be kept well above the dew point of the water vapor in the gases (equal to the saturation temperature of water at the partial pressure of the water vapor) to prevent condensation which would form acids that would corrode metal components in its path, and (2) the flue gases must have enough buoyancy to rise in a high plume above the stack for proper atmospheric dispersion.

Example 3-1 A steam generator burns fuel oil with 20 percent excess air. The fuel oil may be represented by $C_{12}H_{26}$. Determine the minimum stack temperature needed to avoid condensation. Assume the flue gas pressure is leaving the air preheater at 45 psia.

SOLUTION Recalling that there are 3.76 mol N_2/mol O_2 in atmospheric air, the combustion equation for stoichiometric (chemically correct mixture) is

$$C_{12}H_{26} + 18.5O_2 + 69.56N_2 \rightarrow 12CO_2 + 13H_2O + 69.56N_2$$

With 20 percent excess air

$$C_{12}H_{24} + 22.2N_2 + 83.473N_2 \rightarrow 12CO_2 + 13H_2O + 3.7O_2 + 83.472N_2$$

The partial pressure of any component in a gas mixture is equal to the total pressure times the mole fraction of that component. Thus partial pressure of H_2O in products is

$$45 \times \frac{13}{12 + 13 + 3.7 + 83.472} = 5.215 \text{ psia}$$

This corresponds to a saturation temperature of 164°F. The average temperature of the gases is kept much higher to avoid local cool spots that might cause condensation.

The Boiler Walls

The water tubes that cool the water walls are closely spaced for maximum heat absorption. Tube construction has varied over the years (Fig. 3-7) from *bare* tubes (*a*) tangent to or (*b*) embedded in the refractory, to (*c*) *studded* tubes, to the now-common *membrane* design (*d*). The membrane design consists of tubes spaced on centers slightly wider than their diameter, connected by bars or membranes welded to the tubes at their centerlines. The membranes act as fins to increase the heat transfer as well as to afford a continuous rigid and pressure-tight construction for the furnace. No additional inner casing is required to contain the combustion gases. Insulation and metal lagging to protect it are provided on the outer side of the wall. One manufacturer has standardized its design on 3-in-diameter tubes on 3.75 in between centers, another on 3 in on 4 in, and yet a third on 2.75 in on 3.75 in.

The Radiant Boiler

Heat is transferred from the combustion gas to the water walls by both radiation and convection. A *radiant boiler,* as the name implies, receives most of its heat by radiation.

The combustion gases have characteristics that depend upon the fuel used, the combustion process, and the air-fuel ratio. They may be *luminous,* i.e., emit all wavelengths and hence strong visible radiation if there are particulates such as soot particles during the combustion process. This is the case with coal and oil. They may be *nonluminous,* in which case they burn cleanly without particulates, as is the case of gaseous fuels. No combustion gases are truly nonluminous because the heavier gases in the combustion products, in particular the triatomic CO_2 and H_2O, but also SO_2, ammonia, and sulfur dioxide, are selective radiators that emit (and absorb) radiation in certain wavelengths, mostly outside the visible range. The portion of radiation within the visible range is small but gives the combustion gases a green-blue appearance. Lighter gases such as the monatomic and diatomic gases are poor radiators.

The radiant energy emitted by the combustion gases depends upon the gas temperature (to the fourth power), the partial pressures of the individual constituent ra-

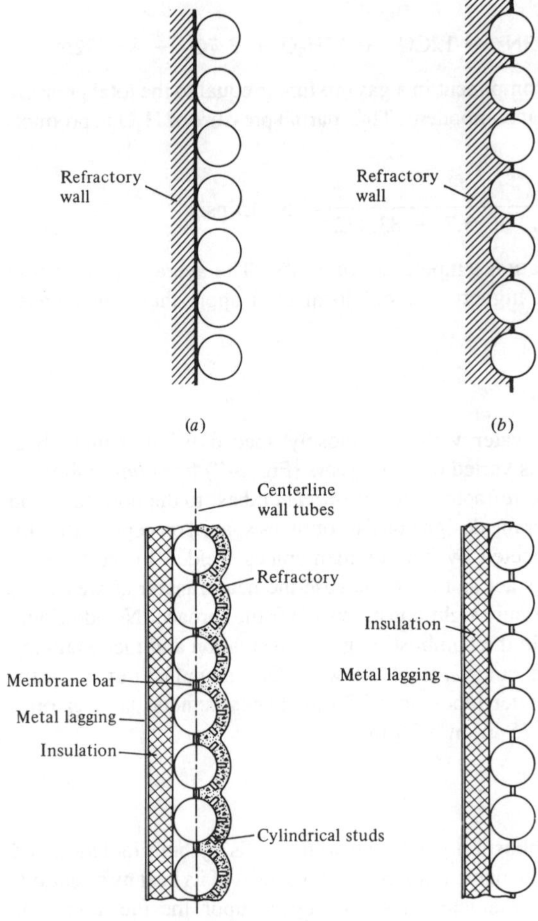

Figure 3-7 Plan views of different types of water tubes.

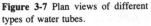

diating gases, the shape and size of the gases, their proximity to the absorbing body, and the temperature of that body (to the fourth power) [14].

The convective portion of the heat transfer follows the usual Nusselt-Reynolds turbulent forced-convection relationship. It is smaller than the radiant portion because radiation is caused by a thick body of gas, whereas convection is localized near the tube surface.

Heat received by the water walls is conducted through the membranes and tube walls and is then convected to the two-phase mixture inside the tubes by forced-convection nucleate-boiling heat transfer. The heat-transfer resistance of the latter is much smaller than the others so that it may be neglected in design calculations with little error.

Radiant boilers are designed for electric-generating stations to use coal or lignite for pulverized or cyclone furnace applications, oil, or natural gas. They are built to supply a wide range of steam pressures and temperatures, but usually around 1000°F (about 540°C) and steam capacities up to 10×10^6 lb_m/h (1260 kg/s). They are limited to subcritical pressures, usually 1800 to 2500 psig (about 125 to 170 bar).

3-5 WATER CIRCULATION

Water circulates from the steam drum via downcomer pipes to a bottom header, up the water tubes (which act as risers), where it partially boils, and back to the steam drum. Full boiling to 100 percent quality in the tubes is avoided because it would lead to tube burnout or failure as a result of *departure from nucleate boiling* (DNB). The density of the saturated water in the downcomers is greater than the average density of the two-phase mixture in the risers.

Natural circulation is dependent upon the difference between these two densities and the height of the drum above the bottom headers. Most large steam-generator boilers have sufficient natural-circulation driving force and are called *natural-circulation boilers*. Some require additional help by pumping the single-phase flow and are called *controlled-* or *forced-circulation boilers*. For a number of years, forced circulation was favored over natural circulation for relatively high subcritical pressures of about 2300 psig (160 bar) and higher. At such pressures the difference between water and steam densities rapidly decreases (becoming zero at the critical point, 3208 psia, 221 bar). This reduced natural circulation and required a margin of safety against tube failure and burnout due to DNB, i.e., reaching the critical heat flux (also a major concern in water-cooled nuclear-reactor design), if a particular tube receives reduced flow. A pump assist supplied this margin of safety.

More recently the water tubes in high-heat-absorbing areas of the furnace were provided with such devices as internal twisters and springs that would break the vapor film and thus inhibit or retard the onset of DNB. Another approach was to use tubes that were grooved, corrugated, or ribbed on their inside surface. The most recent and satisfactory answer has been tubes that were ribbed (rifled) helically on their inside surface. This ribbing creates a centrifugal action that directs water droplets to the vapor film clinging to the surface and provides a much greater margin of safety against DNB problems than do smooth tubes. Modern large steam-generator boilers can now be designed with natural circulation.

Natural-circulation driving forces are explained with the simplified flow diagram of Fig. 3-8. The *driving pressure* caused by natural circulation, Δp_d, is given by

$$\Delta p_d = (\rho_{dc} - \bar{\rho}_r)H\frac{g}{g_c} \tag{3-1}$$

where Δp_d = driving pressure, lb_f/ft^2 or N/m^2 (Pa)

ρ_{dc} = density of water in the downcomer, nearly saturated at the system pressure, lb_m/ft^3 or kg/m^3

$\bar{\rho}_r$ = average density of steam-water mixture in the riser, lb_m/ft^3 or kg/m^3

H = height of drum-water level above bottom header, ft or m

g = gravitational acceleration ft/s^2 or m/s^2

g_c = conversion factor 32.2 $lb_m \cdot ft/(lb_f \cdot s^2)$ or 1 $kg/(N \cdot s^2)$

The most difficult of the parameters in Eq. (3-1) to obtain is $\bar{\rho}_r$, the average density in the riser. It is a function of the *void fraction* distribution along the riser height. The void fraction α of a two-phase mixture is a volumetric quality defined as

$$\alpha = \frac{\text{volume of vapor}}{\text{volume of vapor + liquid}} \tag{3-2}$$

as opposed to the quality x, which is a mass quality, α and x are related by [2]

$$\alpha = \frac{1}{1 + [(1 - x)/x]\psi} \tag{3-3a}$$

and

$$x = \frac{1}{1 + [(1 - \alpha)/\alpha]\frac{1}{\psi}} \tag{3-3b}$$

where

$$\psi = \frac{v_f}{v_g}S \tag{3-3c}$$

where v_f and v_g are the specific volumes of the saturated liquid and vapor, respectively, at the system pressure, and S is the *slip ratio* of the two-phase mixture. The two phases of that mixture do not travel at the same speed. Instead there is a slip between them, which causes the vapor to move faster than the liquid. S is a dimensionless number, greater than 1, defined as

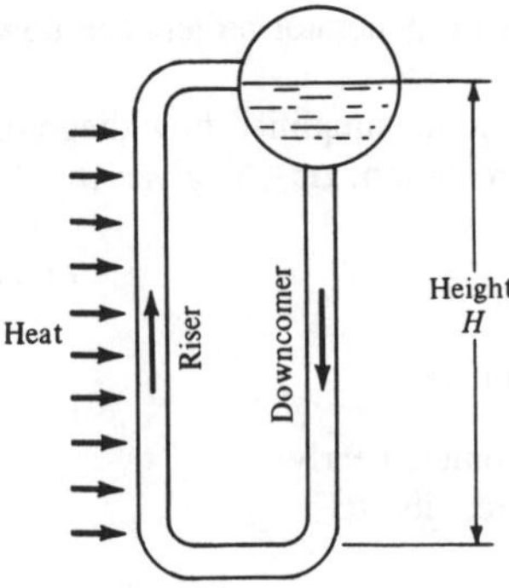

Figure 3-8 A natural circulation loop.

$$S = \frac{\overline{V}_{s,g}}{\overline{V}_{s,f}} \tag{3-4}$$

where $\overline{V}_{s,g}$ and $\overline{V}_{s,f}$ are the average vapor and average liquid velocities at any one cross section of the riser. S has been measured experimentally and found to vary between 1 to less than 10 in most systems, approaching 1 at high pressures (where the liquid and vapor densities approach each other). It is, in general, fairly constant along the path length.

The axial heat flux distribution to the riser determines the quality distribution. In turn, using a reasonable value for S (between 1 and 2), a void fraction distribution is obtained. A mixture density distribution ρ_m is now found from

$$\rho_m = (1 - \alpha)\rho_f + \alpha\rho_g \tag{3-5}$$

where ρ_f and ρ_g are densities (reciprocals of the specific volumes) of the saturated liquid and vapor, respectively. The average mixture distribution in the riser $\overline{\rho}_r$ is now obtained from

$$\overline{\rho}_r = \frac{\int_o^H \rho_m(z)\, dz}{H} \tag{3-6}$$

where z is the axial distance from the bottom of the riser. In the case of uniform axial heating, the solution of the above integral is [2]

$$\overline{\rho}_r = \rho_f - \frac{\rho_f - \rho_g}{1 - \psi}\left\{1 - \left[\frac{1}{\alpha_e(1 - \psi)} - 1\right] \ln\frac{1}{1 - \alpha_e(1 - \psi)}\right\} \tag{3-7}$$

where α_e is the riser exit void fraction.

The driving pressure given by Eq. (3-1) should balance the pressure losses of the single- and two-phase fluids in the loop, which are proportional to their mass-flow rate to the second power. The system seeks its own equilibrium; that is, if the driving pressure is greater than the losses, more fluid is pushed through, causing more losses but also causing x and α to become lower (for the same heat flux) and $\rho_{dc} - \overline{\rho}_r$ to be reduced until equilibrium is reached. The reverse is also true. If the driving pressure is too low for the desired flow rate, a pump is added to assist in circulation.

Example 3-2 A 40-ft-high downcomer-riser system operates at 2500 psia. The riser receives uniform heat flux and saturated water. The exit quality is 50 percent. Calculate the driving pressure. Take $S = 1.2$.

SOLUTION Using Eq. (3-3*c*)

$$\psi = \frac{v_f}{v_g}S = \frac{0.02859}{0.13068} \times 1.2 = 0.2625$$

Using Eq. (3-3*a*)

$$\alpha_e = \frac{1}{1 + [(1 - x_e)/x_e]\psi} = \frac{1}{1 + (0.5/0.5) \times 0.2625} = 0.8879$$

Using Eq. (3-7)

$$\bar{\rho}_r = 34.977 - \frac{34.977 - 7.652}{1 - 0.2625}\left\{1 - \left[\frac{1}{0.8879(1 - 0.2625)} - 1\right]\right.$$

$$\left.\ln \frac{1}{1 - 0.8879(1 - 0.2625)}\right\} = 18.701 \text{ lb}_m/\text{ft}^3$$

$$\rho_{dc} = \rho_f = 34.977 \text{ lb}_m/\text{ft}^3$$

Use Eq. (3-1) to find the driving pressure

$$\Delta p_d = (34.977 - 18.701)(40)\left(\frac{32.2}{32.2}\right) = 651.06 \text{ lb}_f/\text{ft}^2$$

$$= 4.521 \text{ psi}$$

3-6 THE STEAM DRUM

The steam drum provided in all modern steam generators except once-through types (Sec. 3-8) is where feedwater from the economizer is fed, saturated steam is separated from the boiling water, and the remaining water is recirculated as discussed above. The drum may also be used for chemical water treatment and blowdown to reduce solids in the water. The drum also must be of sufficient volume to accommodate water level changes caused by load changes (assisted by proper steam generator controls) and to prevent a dangerously low level or the "carryover" of water toward the superheater. This would cause deposits of entrained solids in the superheater tubes and thus materially increase their temperature, which would lead to their distortion or burnout. Carryover of solids with the steam has also been known to have far-reaching effects, such as deposit problems on turbine blades (the most troublesome being silica deposits, which are not easily removed by water washing).

The most important steam-drum function is separating the steam from the boiling water. The simplest method is *gravity separation* (Fig. 3-9*a*). If the steam velocity

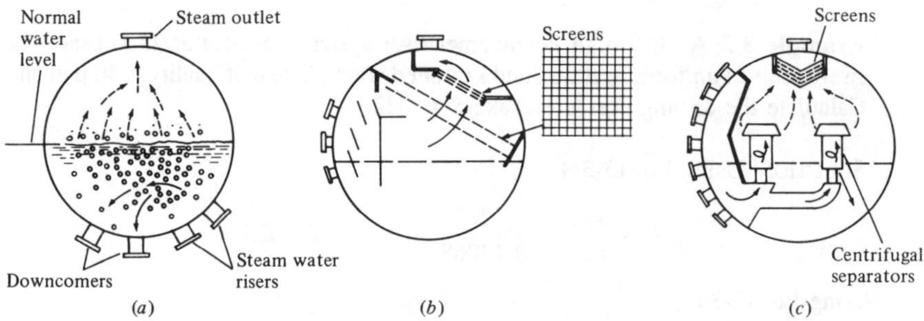

Figure 3-9 Steam drum separation: (*a*) gravity, (*b*) mechanical primary (baffler) and secondary (screen), (*c*) centrifugal.

leaving the water surface is low enough (below about 3 ft/s), the steam bubbles separate naturally without entraining water droplets and the solids they carry (carryover), and without being drawn with the recirculating water into the downcomers (carryunder). Factors other than velocity that affect this process are the positions of the downcomer and riser nozzles with respect to the steam outlet, usually at the top of the drum, and the operating pressure. Gravity separation is strongly affected by the difference in steam and water densities. The higher the pressure, the less the difference between these densities and the less effective the separation. Thus gravity separation, while requiring a simple drum, is economical only for low-steam-capacity, low-pressure service.

In modern high-capacity, high-pressure boilers, *mechanical separation* that assists or supplements gravity separation takes place in two steps: primary and secondary. *Primary separation* removes most of the water from the steam and prevents the carryunder of steam with the recirculating water to downcomers and risers. *Secondary separation,* also called *steam scrubbing* or *drying,* removes remaining mist or fine droplets and the solids they carry from the steam, which results in pure, or "dry and saturated," steam going to the superheater. Mechanical separation is accomplished by fitting the drums with baffles, screens, bent or corrugated plates, and centrifugal separators.

Baffle plates act as primary separators. They change or reverse the steam-flow direction (Fig. 3-9*b*), thus assisting gravity separation, and act as impact plates that cause the water to drain off. Screens made of wire mesh act as secondary separators where the individual wires attract and intercept the fine droplets, much like fabric filters attract dust from gases (Chap. 17). The accumulating drops then fall by gravity back to the main body of water. Bent or corrugated plates are used for both primary and secondary separation. Their effectiveness derives from their large ratio of surface to projected areas. The above plates and screens are used in many other configurations to maximize gravity separation [12, 13].

At high pressures, where the density differential between water and steam diminishes, centrifugal forces, much greater than the gravity forces, are used. *Centrifugal separation* devices are also called *cyclone* or *turbo separators*. They provide separation at pressures nearing critical. In a typical centrifugal separator (Fig. 3-9*c*), the mixture coming in from the risers is deflected tangentially downward into the main body of water. It then enters the separators, which are arranged along the length of the main steam drum. Guide vanes within the separators impart a spinning motion to the mixture, which causes the heavier water droplets to move radially through the lighter steam, to impinge on the separator wall, and to discharge downward below the water surface through an outer concentric cylinder. The separator may be equipped with a corrugated plate at its exit to provide further separation. Finally, screens, just under the drum exit, provide the final drying action.

Typical utility steam drums range in length to more than 100 ft, in diameter to more than 15 ft, and in mass to a few hundred tons. They contain as many as 30 outlet nozzles and many more riser and downcomer nozzles. The larger drums are usually constructed in cylindrical sections, called *courses,* which are welded together, and two hemispherical heads which are welded to the ends.

3-7 SUPERHEATERS AND REHEATERS

Superheaters and reheaters in utility steam generators are made of tubes of 2 to 3 in OD (smaller sizes, about half these dimensions, are used in marine service). The smaller diameters have lower pressure stresses and withstand them better. The larger diameters have lower steam-flow pressure drops and are easier to align. Finning on the outside surface of the tubes is avoided because it increases thermal stresses and makes cleaning difficult. Internal ribbing, like that used in boiler water tubes, is unnecessary because no DNB problems arise here. Adequate heat-transfer design is based on gas flow inside the tubes, which has a much lower conductance than nucleate boiling in the boiler tubes. Because the tubes are subjected to high temperatures, pressures, and thermal stresses, their materials of construction must be carefully selected. Below 850°F carbon steel is adequate. Modern superheaters and reheaters operating at about 1000°F, however, are usually made of special high-strength alloy steels chosen for both strength and corrosion resistance [15]. The exact alloy depends upon steam conditions and the types of fuel, especially if it contains undesirable impurities. The allowable stresses for materials drop drastically as the temperature increases.*

Convection Superheater

Early superheater designs placed them above or behind banks of water tubes to protect them from combustion flames and high temperatures. The main mode of heat transfer between the combustion gases and the superheater tubes, therefore, was convection, and that type of superheater became known as the *convection superheater*. The main distinguishing characteristic is its response to load changes. As demand for steam increases, fuel- and airflow, and hence combustion-gas flow, are increased. The convective heat-transfer coefficients increase both inside and outside the tubes, increasing the overall heat-transfer coefficient between gas and steam faster than the increase in mass-flow rate of the steam alone. (The combustion temperatures do not materially change with load.) Thus the steam receives greater heat transfer per unit mass-flow rate and its temperature increases with load (Fig. 3-10).

Radiant Superheater

Because of the need for greater heat absorption, superheaters were eventually placed nearer higher-temperature, in view of the combustion flames. Steam-flow velocities were increased to increase the overall heat-transfer coefficients, and overall superheater designs were improved to overcome expected higher metal temperatures.

This placement of superheater results in the main heat transfer between the hot gases and flame, and the tube outer walls, to be accomplished by radiation. This design

* For example, a material called Croloy 2¼, specification number SA213, Grade T22, has maximum allowable design stresses of 13,100, 11,000 and 7,800 psi at 900, 950, and 1000°F, respectively. Carbon steel SA210-A1 has allowable stresses of only 5000 and 3000 psi at 900 and 950°F, respectively.

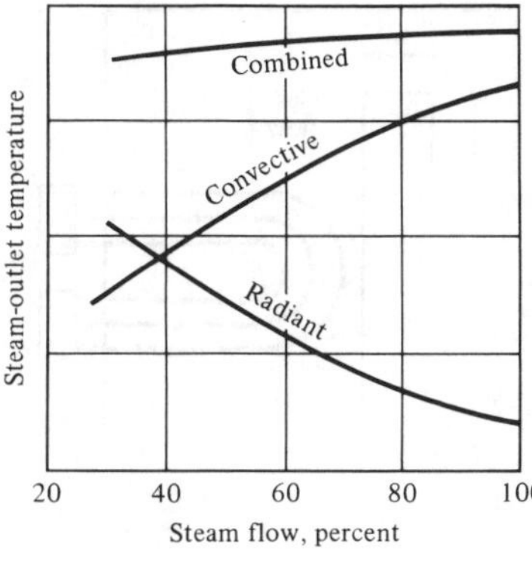

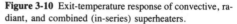

Figure 3-10 Exit-temperature response of convective, radiant, and combined (in-series) superheaters.

has come to be known as a *radiant superheater*. Radiation heat transfer is proportional to $T_f^4 - T_w^4$, where T_f and T_w are the flame and tube wall absolute temperatures, respectively. Because T_f is much greater than T_w, the heat transfer is essentially proportional to T_f^4. Because T_f is not strongly dependent on load, the heat transfer per unit mass flow of steam decreases as the steam flow increases. Thus an increase in steam flow due to an increased load demand would result in a reduction in exit steam temperature, the opposite effect of a convection superheater (Fig. 3-10).

Design considerations for reheaters are similar to those for superheaters except that, although the steam outlet temperatures are about the same, the overall temperatures are lower and the steam pressures are about 20 to 25 percent of those in the superheaters. The pressure stresses are therefore lower and a lower grade steel alloy is tolerated. In addition, larger tubing with higher stresses may be used, which has the additional beneficial effect of reducing the pressure losses in the reheater.

Convection superheaters alone are used with low-temperature steam generators. Radiant and convection superheaters and reheaters are used for high-temperature service. The radiant units are arranged in flat panels or platen sections with wide spacings of several feet to permit radiation through. These are usually followed downstream by sections on a narrower spacing that permit both radiation and convection. Mechanical construction of the sections are of three kinds: *pendant, inverted,* and *horizontal*.

Pendant-type superheaters and reheaters are those that are hung from above (Fig. 3-11*a*). They have the advantage of firm structural support but the disadvantage of flow blockage by condensed steam after a cold shutdown, which necessitates slow restart to purge the water that accumulates in the bottom. *Inverted-type* units, on the other hand, are supported from below (Fig. 3-11*b*). They have proper drainage of the condensed steam but lack the structural rigidity of the pendant type, especially in high-speed gas flow. The inverted type is not commonly used. *Horizontal-type* units (Fig. 3-11*c*) are usually supported horizontally in the vertical gas ducts parallel to the main furnace and receive the hot gases after a U-turn at the top. They do not view the flame directly and hence are mainly of the convection type. They have both proper drainage and good structural rigidity. Figure 3-12 shows a typical arrangement of superheaters

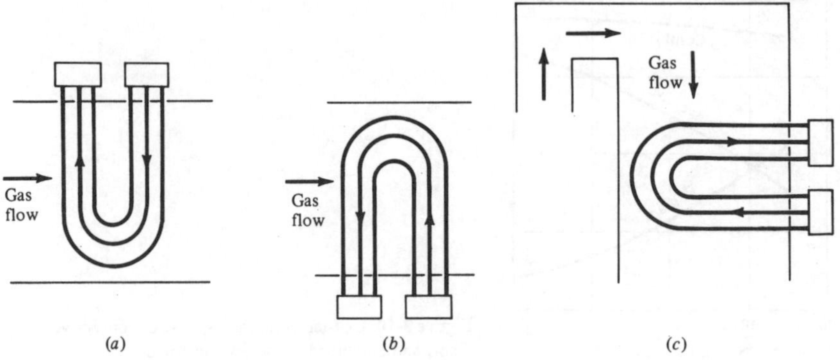

Figure 3-11 Schematic diagram showing (*a*) pendant, (*b*) inverted, and (*c*) horizontal superheaters and reheaters.

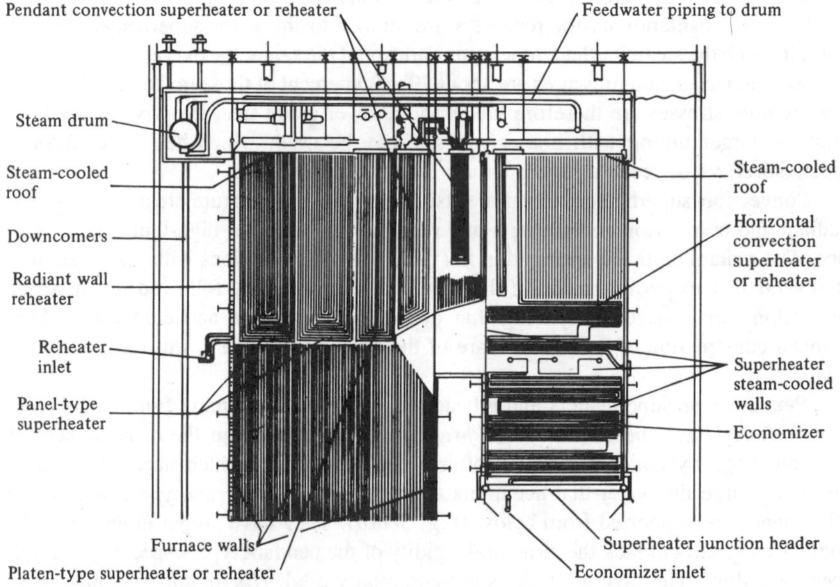

Figure 3-12 Superheater (SH), reheater (RH), economizer, and air preheater arrangements in a drum-type steam generator with cyclone furnace. (*Courtesy Babcock and Wilcox.*)

and reheaters. Superheaters and reheaters are often split into primary and secondary units for control purposes (Sec. 3-13).

3-8 ONCE-THROUGH BOILERS

The *once-through* boiler, or steam generator, is also called the *forced-circulation, Benson,* or *universal-pressure* boiler. The latter name is because it is applicable to all temperatures and pressures, although economically it is suited to large sizes and pressures in the high subcritical and supercritical range. In contrast to the drum type (Fig. 3-13*a*), the feedwater goes through the economizer, furnace walls, and superheater sections, changing sequentially to saturated water, saturated steam, and superheated steam in one continuous pass (Fig. 3-13*b*). No steam drum is required to separate saturated steam from boiling water and no water recirculation takes place. Reheat of steam after it is expanded in the high-pressure turbine is accomplished by a reheater in the usual manner. Because of the once-through mode of operation, very high purity feedwater is a requirement. Figure 3-14 shows a typical once-through steam generator.

The once-through boiler is the only type suited to supercritical-pressure operation (above 3208 psia, 221 bar, for steam) because the latent heat of vaporization at and beyond the critical pressure is zero and liquid and vapor are one and the same, so no separation in a drum is possible or necessary. While particularly applicable to supercritical pressures, once-through steam generators are used economically for high-

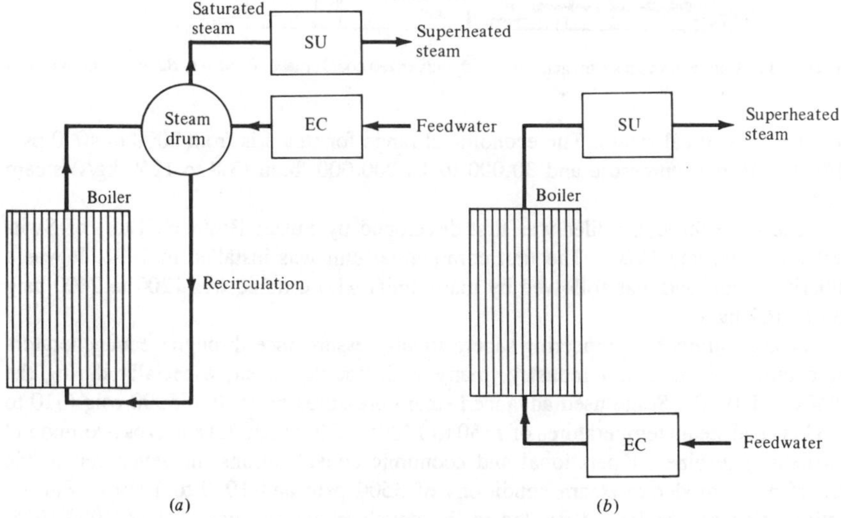

Figure 3-13 Schematic flow diagrams of (*a*) drum type and (*b*) once-through steam generators. SU = superheater, EC = economizer.

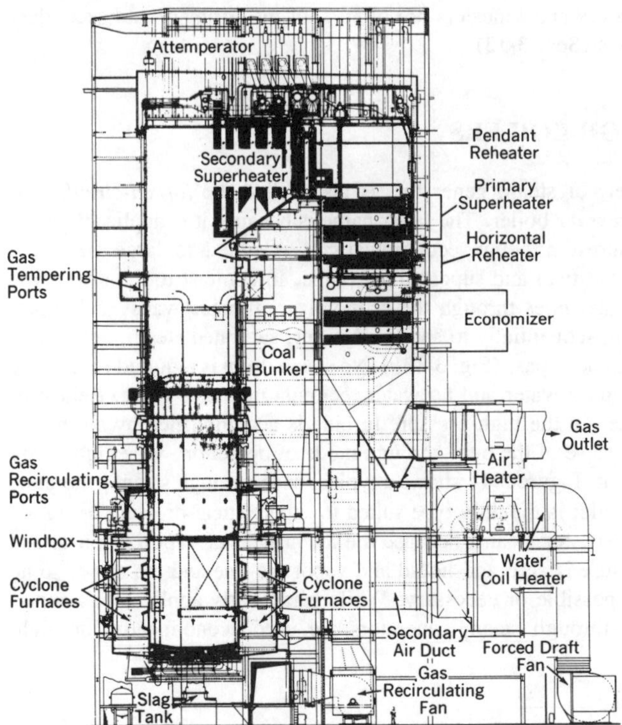

Figure 3-14 A once-through steam generator with pulverized coal furnace. (*Courtesy Babcock and Wilcox.*)

pressure subcritical steam. The economical range for steam is from 2000 to 4000 psia (138 to 276 bar) pressure and 30,000 to 10,000,000 lb_m/h (3.8 to 1260 kg/s) steam output.

The once-through boiler was first developed by Sulzer Brothers, Ltd., of Switzerland in the late 1920s. The first commercial unit was installed in 1932. It was a subcritical unit and was followed by many units whose range was 1200 to 2400 psig (84 to 167 bar).

A large number of pioneering supercritical-pressure once-through steam generators were built for the utility industry, many with double reheat, especially during the 1940s and 1950s. Some used advanced steam pressures of 4500 to 5000 psig (310 to 345 bar) and steam temperatures of 1150 to 1200°F (620 to 650°C) and cross-compound (two-shaft) turbines. Operational and economic considerations, however, led to the use of more moderate steam conditions of 3500 psig and 1000 to 1050°F. Further optimization by the late 1960s led to the development and use of 3500/1000/1025/1050 plants with single-shaft four-casing steam turbines. During the 1970s however, such units fell out of favor because of equipment unreliability, significantly mostly in the low-pressure components of the plants, as well as operational complexity. It is

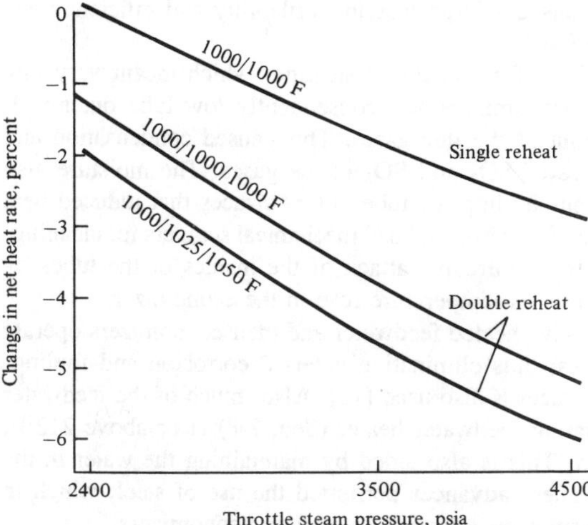

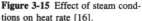

Figure 3-15 Effect of steam conditions on heat rate [16].

now believed that newly developed sophisticated schemes for boiler-turbine operation will largely overcome these difficulties [16] and that the large improvements in efficiency and reduction in heat rate possible with supercritical-double-reheat plants (Fig. 3-15) may reverse the downward trend of the 1970s. Under development, for example, is a 900-MW plant with steam conditions of 4500/1000/1025/1050 that will have a 6 percent improvement in heat rate over the currently popular 2400/1000/1000 plant.

Capital costs for a supercritical steam generator are a few percent higher than those of a drum-type subcritical one of the same capacity, but because of the increased efficiency of the powerplant the capital costs of the turbogenerator as well as the balance of plant (condenser, feedwater heaters, cooling towers, etc.) are lower. The net effect is that the total production costs of electricity in mills per kilowatt hour are lower for the supercritical cycle, particularly one with double reheat.

A *combined-circulation* boiler, developed by Combustion Engineering, Inc., combines once-through flow with circulation for use in supercritical-pressure operation [13]. Recirculation is used only during start-up and low loads in a spherical mixing vessel to protect the furnace walls. A stop-check valve assures once-through flow alone at high loads. A typical unit is 805 MW single reheat 3590/1005/1005.

3-9 ECONOMIZERS

The *economizer* (EC) is the heat exchanger that raises the temperature of the water leaving the highest-pressure feedwater heater to the saturation temperature corresponding to the boiler pressure. This is done by gases leaving the last superheater or reheater. These gases, at high enough temperatures to transfer heat to the superheater-reheaters, enter the economizer at 700 to 1000°F. Part of their energy is used to heat the feedwater. The term "economizer" historically was used because the discharge of such high-temperature gases would have caused a large loss in availability and efficiency and hence loss in economy of operation.

Economizers were introduced before feedwater heating, which meant very low inlet water temperatures to the economizers and consequently low tube outer wall temperatures, below the dew point of the flue gases. This caused condensation and corrosion because of the presence of SO_2 and SO_3 in the gases. The moisture also aided in the collection of ash, thus fouling the tube outer surfaces that reduced heat transfer. Economizers were made of cast iron and had mechanical scrapers for cleaning. Early in its use, steel suffered from corrosive attack of the insides of the tubes by oxygen freed from the feedwater as its temperature rose in the economizer.

Modern steam generators receive heated feedwater and their economizers operate above the dew point of the gases, thus eliminating external corrosion and fouling. Chemical cleaning of internal surfaces is also used [12]. Also, much of the feedwater oxygen is removed in the deaerating feedwater heater (Sec. 2-8) at or above 212°F, which reduces internal corrosion. This is also aided by maintaining the water in the economizer at a pH of 8 to 9. These advances permitted the use of steel, which in turn is suitable for the high pressures encountered in modern economizers.

Modern economizers are designed to allow some boiling of the feedwater in the outlet sections, up to 20 percent quality at full power, less at part loads.

Economizer tubes are commonly 1.75 to 2.75 in OD and are made in vertical sections of continuous tubes, between inlet to outlet headers, with each section formed into several horizontal paths connected by 180° vertical bends for proper draining. Sections are placed side by side on 1.75- to 2-in minimum spacings (edge to edge). The exact spacing depends upon the type of fuel and ash characteristics, which are smaller the cleaner the fuel, such as natural gas. When high-ash fuels are used, the water-soluble ashes accumulating on the economizer are dissolved and washed off during plant shutdown. In that case the economizer is usually located above a hopper, which receives the dissolved deposits. Steam or air-jet cleaning is also used in addition to washing.

Economizers have been built with plain or extended surface tubes. Extended surface tubes with fins or studs on their outer surface have higher heat-transfer characteristics and thus require smaller space. On balance, they have lower capital cost. They are, however, more suited to clean-burning gaseous fuels and situations in which no air preheaters are used, such as in combined steam-gas-turbine cycles (Chap. 8).

Economizers are generally placed between the last superheater-reheater and the air preheater. In some cases, a low-temperature economizer is placed after the air preheater. Such an economizer is called a *stack cooler* and acts as a low-pressure feedwater heater except that the heating medium is the flue gas instead of steam bled from the turbine.

3-10 AIR PREHEATERS

Like economizers, *air preheaters* (or simply air heaters) utilize some of the energy left in the flue gases before exhausting them to the atmosphere. They receive 600 to

800°F gases. As indicated previously, these gases are cooled to only 275 to 350°F to avoid gas condensation and corrosion problems and to allow for proper dispersion in the atmosphere. The air is heated from forced-draft outlet temperatures, not far from atmospheric to 500 to 650°F and sometimes higher. Preheating the air saves fuel that would otherwise be used for that heating. The fuel savings (and hence increase in plant efficiency) are nearly directly proportional to the air temperature rise in the preheater. Typical fuel savings are 4 percent for a 200°F air temperature rise and about 11 percent for a 500°F air temperature rise in the preheater.

In addition to fuel savings, preheated air is a requirement for the operation of pulverized-coal furnaces. Air in the 300 to 600°F range is needed for drying that fuel. Air is also used for transporting it to the furnace and then burning it there. Small stoker-fired units do not require preheated air. However, large stoker-fired bituminous-coal steam generators benefit from preheated air, but only up to about 350°F, to prevent damage to the stoker moving parts.

Air preheaters, like economizers, were first developed in Europe. The first unit commercially built in the United States was a flat-plate heat exchanger in which adjacent steel plates formed alternate air and gas passages. Present day higher air and gas pressures, however, use tubular or regenerative designs that better withstand these pressures. There are two general types of air preheaters: *recuperative* and *regenerative*.

Recuperative air preheaters have heat transferred directly from the hot gases to the air across the heat-exchange surface. They are commonly tubular, although some plate types are still used. Tubular units are essentially counterflow shell-and-tube heat exchangers in which the hot gases flow inside vertical or horizontal straight tubes and the air flows outside. Figure 3-16 shows a vertical preheater. Baffles are provided to maximize air contact with the hot tubes. The tubes are mechanically expanded into top and bottom tube sheets. Thermal expansion is provided by a bellows-type expansion unit shown at the bottom sheet. A hopper is provided below the tubes to collect soot and dust that deposit on the inside tube surfaces. Care is taken to avoid leakage between the air and the hot gases. Leakage would occur from the air that is at a higher pressure to the gas. A leak would result in short-circuiting of combustion air, which would increase both forced- and induced-draft-fan power consumption in direct proportion to the leakage. It also reduces the heater effectiveness.

Tubular preheaters are built in a variety of designs to suit particular steam generators spaces and duct layout. They may provide for one or more passes for both air and gas in counter- or crossflow, in vertical or horizontal arrangements. Tube sizes vary from 1.5 to 4 in OD. The smaller the diameter, the larger the number of tubes and the greater the surface area for a given overall size. Thus smaller diameters result in more compact heaters of a given heat load. Cost, cleaning requirements, and fuel type determine the diameter to be used in a given situation. Modern large steam generators use 2 to 2.5 in OD heater tubes. Diameters of 1.5 to 2 in are used in marine service where space and weight limitations are important.

Regenerative air preheaters are those in which heat is transferred from the hot flue gases, first to an intermediate heat-storage medium, then to the air. The most common is the rotary air preheater, known as the *Ljungstrom* preheater, first developed in Europe in 1920 and first installed in the United States in 1923. It is composed of

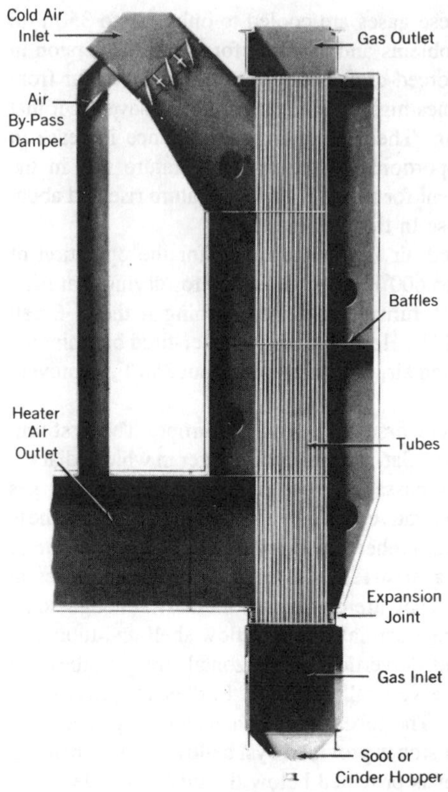

Figure 3-16 a tubular counterflow air preheater. Air bypass is used to control metal temperatures at air inlet end. (*Courtesy Babcock and Wilcox.*)

a rotor driven by an electric motor through reduction gearing so that it rotates slowly and continuously within a housing at 1 to 3 r/min, depending on diameter. The rotor has between 12 and 24 radial members that form sectors (Fig. 3-17). The sectors are filled with a heating surface composed of steel sheets that are flat or form-pressed with corrugated, notched, or undulated ribbing and formed into baskets. They constitute the heat-storage medium of the preheater. A stationary seal covers the equivalent of two opposite sectors. Half the remaining sectors are exposed at any one instant to the hot gases, which are moving in one direction, the other half are exposed to the air, which is moving in the opposite direction.

As the rotating sectors enter the hot-gas zone, they are progressively heated by the gas. They store that heat as sensible heat. When they enter the air zone, they progressively give up this heat to the air. The seal system reduces leakage.

Rotary air preheaters are designed with a vertical or horizontal shaft, depending upon layout and ducting. They are also designed as *laminar-* or *turbulent-surface* types. In the laminar type, the heat-storage elements are compactly spaced so that flow through them is laminar. They are used with clean-burning gaseous fuels. The turbulent

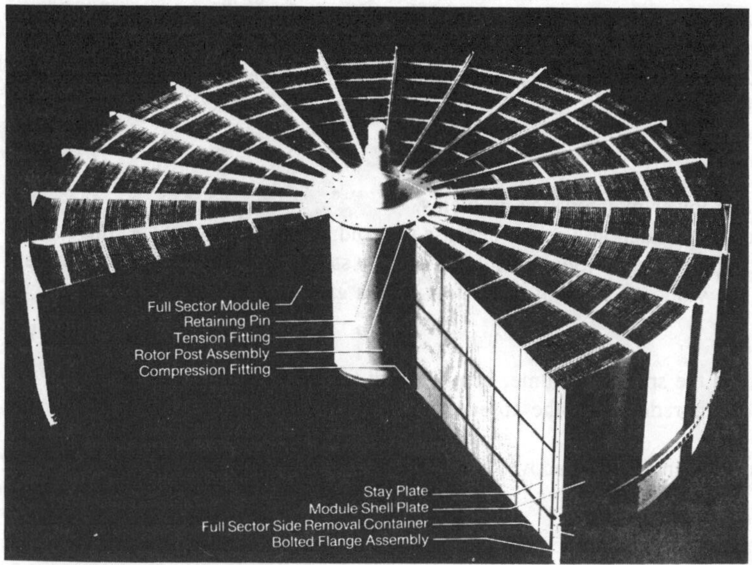

Figure 3-17 A vertical shaft Ljungstrom air preheater (*Courtesy Combustion Engineering, Inc.*)

type has elements with wider spacings (to facilitate cleaning of ash deposits) and turbulent flow in between. It is more suited to coal- and oil-fired systems. To get the same mass of heat-storage elements, the turbulent-type rotor is several times longer than the laminar-type and is, in general, vertically mounted, whereas the laminar type is usually horizontally mounted. Cleaning is accomplished by air or steam jets.

Another type of rotary air preheater, recently developed in Europe, is one in which the storage elements are stationary and the gas and air ducts are connected to two rotating segments each.

3-11 FANS

Early and small-capacity steam generators relied upon natural draft for the combustion gases. Proper design of gas passages must be provided to have the driving pressure between the atmosphere and the gases within the generator, caused by the density difference between atmospheric air and the average gas density, overcome the various pressure losses within the generator and supply the required air for combustion. Analogy here may be made with natural circulation in the boiler (Sec. 3-5).

Large steam generators require an assist to push the air in, pull the gas out, or both. For this, they use large fans [17, 18]. There are two types of fans in use today: *forced-draft (FD)* and *induced-draft (ID)* fans. When either one is used alone, it should overcome the total air and gas pressure losses within the generator.

Forced-draft fans, used alone as in many large steam generators and practically all marine applications, are placed at the air entrance to the air preheater and put the entire system up to the stack entrance under positive gage pressure. Because they handle only cold air, they have several advantages over induced-draft fans. Some advantages:

1. They have lower maintenance problems.
2. They consume much less power because the cold air has the lowest specific volume in air-gas path. Recall (Sec. 1-2) that a fan is a steady-flow thermodynamic system so that the work w_{sf} per unit mass-flow rate is given by

$$w_{sf} = \int v \, dP \tag{1-9}$$

   where v is the specific volume, being lowest for the cold air entering.
3. Their load is reduced by the absence of the additional gas equivalent of the fuel added.
4. As a consequence of the above, their capital and operating costs are lower.

Disadvantages result from the fact that they put the furnace under pressure, in which case it is called a *pressure furnace:*

1. Leakage of noxious gases from the furnace walls would be to the outside, necessitating a gas-tight furnace construction.
2. Special attention must be given to the design of inspection doors, soot blower boxes, and fuel-igniter openings.

For good reliability two forced-draft fans, operating in parallel are usually used, each one capable of at least 60 percent of full load flow when the other is out of service.

Induced-draft fans are located in the gas stream between the air preheater and the stack, either before or after the dust collector (Chap 18). They discharge essentially at atmospheric pressure and place the entire system under negative gage pressure. They must handle hot gases, including the original air, the gas equivalent of the fuel added, and leakages into the system. Their power requirements are therefore greater than forced-draft fans. In addition they must cope with corrosive combustion products and ash. Induced-draft fans are seldom used alone.

When both forced- and induced-draft fans are used in a steam generator, the FD fans push atmospheric air through the air preheater, dampers, various air ducts, and burners into the furnace. The ID fans pull the combustion gases from the furnace, through the heat-transfer surfaces in the superheaters, reheaters, economizer, and gas side air preheater and into the stack. (Sometimes they are built into the stack base.) The stack, because of its height, adds a natural driving pressure of its own (Sec. 3-12). The furnace in this case is said to operate with *balanced draft,* meaning that the pressure in it is approximately atmospheric. Actually, it is kept at a slightly negative gage pressure to ensure that any leakage would be inward.

Steam generators that use low-ash fuels such as gas or oil are usually designed with pressurized firing. Coal-fired generators may be designed with either pressurized- or balanced-draft firing. Modern systems seem to favor the latter.

Fans used in electric-generating plants are among the largest made: capacities of 1.5 million ft^3/min (700 m^3/s) and 60-in water static pressures (about 2.2 psi, 0.15 bar) are common. Because they operate continuously for long periods (up to 1 or $1\frac{1}{2}$ years), the fans must be well designed, ruggedly constructed, well balanced, and highly efficient over a wide range of outputs. There are two types of fans in common use: *centrifugal* and *axial*. In the centrifugal fan, the gases are accelerated radially through curved or flat impeller blades from rotor to a spiral or volute housing. In the axial fan, gases are accelerated parallel to the rotor axis. This is similar to a desk fan, but here the fan is housed in a casing to develop static pressure. Axial fans (with variable-pitch moving blades) maintain high efficiencies over a wider range of loads than constant-speed centrifugal fans but have higher capital costs.

In general, centrifugal fans with backward-curved blading are used for FD fans and with flat or forward-curved blading (Fig. 3-18) for ID fans. (Occasionally backward-curved blading but with curvature less than that for FD fans is used.) The lesser curvature results in lower tip speed and allows less dirt to cling to the backside of the blades, thus minimizing the erosive effect of ash. Low-speed fans with flat blades are used with particularly dirty or corrosive gases.

Because the pressure differential ΔP across a fan is usually small, the airflow or gas flow may be considered incompressible, that is v = constant, and Eq. (1-9) may be modified to

$$|w_{\text{sf}}| = \frac{v\,\Delta P}{\eta_f} \qquad \text{ft} \cdot \text{lb}_f/\text{lb}_m \text{ or J/kg} \tag{3-8}$$

The power would be given by

$$\dot{W}_{\text{sf}} = \frac{\dot{m}v\,\Delta P}{\eta_f} \qquad \text{ft} \cdot \text{lb}_f/\text{s or W} \tag{3-9}$$

where v = specific volume of air or gas, obtained from the perfect gas equations (1-30), ft^3/lb_m or m^3/kg

ΔP = pressure rise across fan, lb_f/ft^2 or N/m^2 (Pa)

η_f = fan efficiency, dimensionless

$\dot{m}$ = mass-flow rate of air or gas, lb_m/s or kg/s

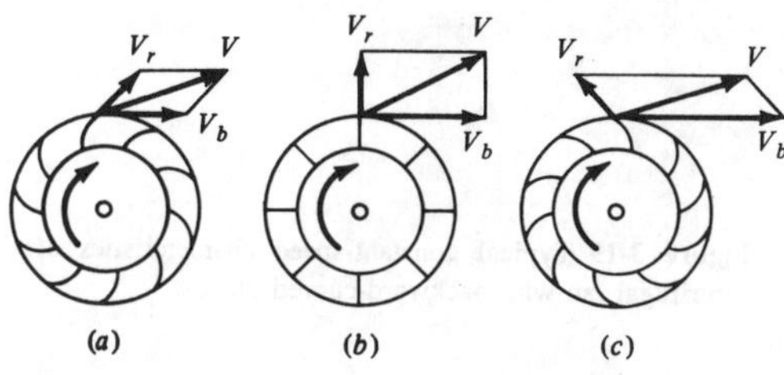

Figure 3-18 Centrifugal blading: (*a*) forward (*b*) flat, and (*c*) backward-curved. Vector diagrams show blade tip velocity V_b, air velocity relative to blade, and V absolute velocity of air leaving blade. V is the same in all cases.

The horsepower may be obtained by dividing $\dot{W}_{sf}$ in ft · lb$_f$/s by 550 or in W by 745.7. The assumption of incompressibility results in errors less than 3 percent for pressure ratios across the fan up to 1.10. Higher pressure ratios would require the use of a P-V relationship of the type $PV^n = C$ (Table 1-3) to evaluate the exact work and power.

Figure 3-19 shows typical characteristics of a centrifugal fan with backward-curved blades. The characteristics usually show pressure and other parameters based on static pressure. *Static pressure* is the force per unit area exerted on a wall by an adjacent fluid that is at rest with respect to the wall (the velocity at the wall in a boundary layer is zero). A fluid in motion, in addition, has a *velocity pressure*. The sum of the static and velocity pressures is called the *total pressure*. Static and velocity pressures are obtained correspondingly from the kinetic and flow energy terms in the general energy equation rewritten for no heat, work, or friction and for incompressible flow (the *Bernoulli equation*). Ignoring the potential energy term

$$Pv + \frac{V_s^2}{2g_c} = \text{constant} \tag{3-10}$$

where each term has units of energy. Multiplying by the density $\rho = 1/v$ gives

$$P + \rho\frac{V_s^2}{2g_c} = \text{constant} \tag{3-11}$$

where each term has units of pressure, P is the static pressure, and $\rho V_s^2/2g_c$ is the velocity (or kinetic) pressure. Dividing Eq. (3-11) by the weight density $\rho g/g_c$ gives

$$h + \frac{V_s^2}{2g} = \text{constant} \tag{3-12}$$

where the terms have the units of length, h here is called the *static head*, $V_s^2/2g$ is called the *velocity head*. The term head, in feet (ft) or meters (m) is often used in fluid technology. Static horsepower is that calculated from Eq. (3-9), where ΔP is the difference between static pressures.

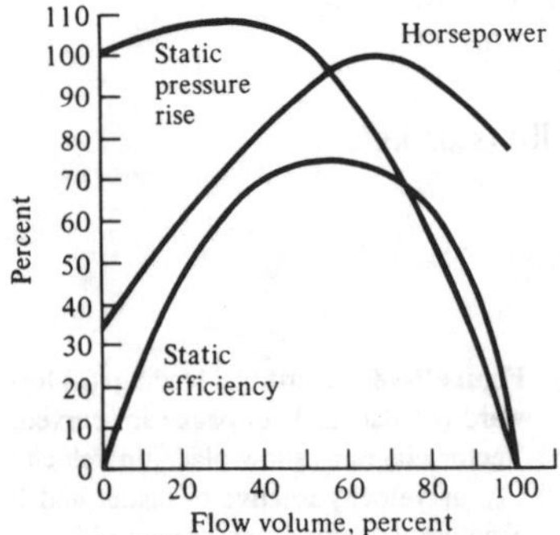

Figure 3-19 Typical constant-speed characteristics of a centrifugal fan with backward-curved blades.

Fan control There are two common methods of controlling the output of fans: *damper control* and *variable-speed* control.

Damper control has the advantage of low capital costs for the damper mechanism itself and for the fan drive motor, which would be a simple constant-speed induction ac motor. It is easily adaptable to automation and allows continuous rather than stepwise control. It suffers, however, from the disadvantage that it adds additional flow resistance that the fan must overcome by increasing its power input. Dampers are usually put on the outlet side of the fan, although inlet dampers, called *inlet vanes,* are sometimes used. They consume less power than outlet dampers but are only effective for moderate load changes near full load. When used with FD fans, they are usually used in combination with outlet dampers. Another combination that results in power savings is the use of a two-speed ac drive motor in conjunction with damper control. Two-speed motors are less expensive than variable-speed ac drives.

Variable-speed control has the advantage of reduced power consumption and is the most efficient method of fan control. The effect of speed on fan performance is that flow, pressure, and power input are directly proportional to N, N^2, and N^3, respectively, where N is the speed of the fan in revolutions per minute (r/min). Thus, reducing speed by say 70 percent reduces the capacity to 70 percent, the pressure to about 50 percent, and the power input to about 35 percent. The actual relationships are dependent upon changes in the effectiveness of the variable-speed drive used. Figure 3-20 shows the effect of speed on pressure and power for a centrifugal fan. The types of drives are (1) variable-speed steam turbine, (2) hydraulic coupling, (3) magnetic coupling, (4) variable-speed dc motor, (5) multiple-speed ac motor, and (6) electronically adjustable motor drive. The main disadvantage of variable-speed control is that all these methods involve higher capital costs than damper control, though the latter is less efficient.

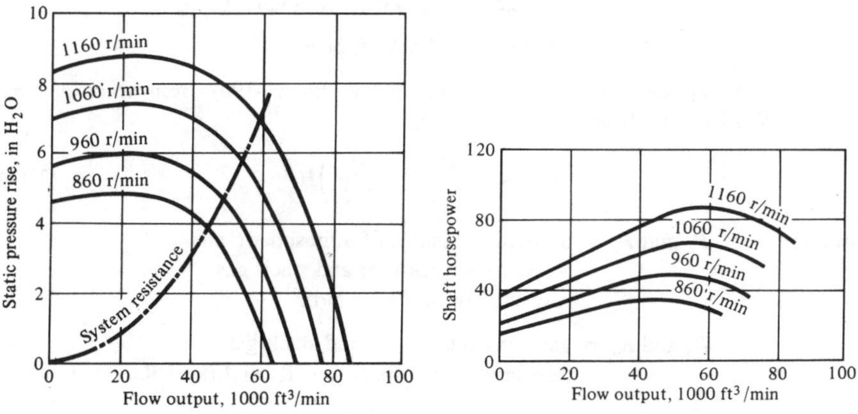

Figure 3-20 Typical centrifugal fan performance curves showing effect of varying speed, r/min [12].

Two other types of fans are used in powerplants: primary-air fans and gas-recirculation fans. *Primary-air fans* supply air to dry and transport pulverized coal to the furnace or to a storage bunker. *Gas-recirculation fans* recirculate gas from a point between the economizer and air preheater back to the bottom of the furnace as part of a steam-temperature control system.

Fans are a major source of noise in powerplants. To reduce that noise, they are often housed in thick masonry acoustical enclosures or equipped with inlet silencers (FD fans), or both [19].

3-12 THE STACK

Tall and conspicuous from a distance, stacks are used in nearly all powerplants. Early steam generators relied solely on stacks to meet the total pressure losses (draft) at the required gas flows. Modern ones have high gas-flow requirements and, because of the various heat exchangers (superheaters, reheaters, economizers, air preheaters), have such high pressure losses that stacks alone are insufficient and fans are added. Stacks have two functions: (1) to assist the fans in overcoming the pressure losses and (2) to help disperse the gas effluent into the atmosphere.

Driving pressure The *driving pressure* Δp_d, in lb_f/ft^2 or N/m^2, supplied by a stack is given by an equation similar to Eq. (3-1)

$$\Delta p_d = (\rho_a - \bar{\rho}_s)H\frac{g}{g_c} \tag{3-13}$$

where ρ_a = atmospheric air density, lb_m/ft^3 or kg/m^3

$\bar{\rho}_s$ = average stack gas density, lb_m/ft^3 or kg/m^3

H = height of the stack, ft or m

Because both air and gas obey the perfect-gas law, Eq. (1-19*a*), from which $\rho = m/V = P/RT$, Eq. (3-13) becomes

$$\Delta p_d = \left(\frac{P_a}{R_aT_a} - \frac{P_s}{R_s\bar{T}_s}\right)H\frac{g}{g_c} \tag{3-14}$$

where P_a and P_s = absolute pressures of atmospheric air (barometric pressure) and stack gas, respectively, lb_f/ft^2 or N/m^2

R_a and R_s = gas constants for air and stack gas, respectively, $ft \cdot lb_f/(lb_m \cdot °R)$ or $J/(kg \cdot K)$

T_a and $\bar{T}_s$ = air and average stack gas temperatures respectively, °R or K

Table 3-1 The variation of barometric pressure with altitude

	Altitude, ft									
P_a:	0	1000	2000	3000	4000	5000	6000	7000	8000	10,000
inHg	29.92	28.86	27.82	26.82	25.84	24.90	23.98	23.09	22.22	20.58
psia	14.70	14.17	13.66	13.17	12.69	12.23	11.78	11.34	10.91	10.11
bar	1.013	0.977	0.942	0.908	0.875	0.843	0.812	0.782	0.752	0.697

P_a and P_s differ only slightly. R_a = 53.34 ft · lb$_f$/(lb$_m$ · °R). R_s depends upon the gas composition* and hence upon the fuel used and is about 52.2 for bituminous coal, 53.2 for fuel oil, and 55.6 for natural gas. Equation (3-14) can then be rewritten with good accuracy into the form

$$\Delta P_d = \frac{P_a}{R_a}\left(\frac{1}{T_a} - \frac{1}{\overline{T}_s}\right)H\frac{g}{g_c} \tag{3-15}$$

In any one location P_a varies daily according to weather conditions. It is also a function of the altitude of the location. Standard values at different altitudes are given in Table 3-1. T_a, of course, varies seasonally, daily, hourly, etc. $\overline{T}_s$ is dependent upon the temperature variation of the gas along the stack, which is dependent upon the heat losses through the stack walls and any infiltration of cold outside air. The exact value of $\overline{T}_s$ is obtained by integrating the stack local temperature as a function of height and dividing by H, in the same manner as finding the average density in a two-phase channel [Eq. (3-6)]. This function is difficult to evaluate exactly, and arithmetic averaging, implying nearly constant heat loss along the stack, is accepted. Thus

$$\overline{T}_s = \frac{T_o + T_H}{2} \tag{3-16}$$

where T_o and T_H are the stack inlet and exit temperatures, respectively, in °R or K. Values of T_H depend upon T_o, the stack height H and internal diameter D, and the outside atmospheric conditions such as temperature and wind velocity. As expected, T_H increases as T_o and D increase and decreases as H increases.

Stacks, of course, introduce pressure drops of their own. These are caused by wall friction and the pressure equivalent to the kinetic energy of the gases leaving the stack. The latter is usually a few times greater than the former. The total is only a few percent of the driving pressure.

Dispersion The second important function of a stack, *dispersion* of the flue gases into the atmosphere, is defined as the movement of the flue gases horizontally and vertically

* R for a mixture is equal to the sum of the products of the mass fraction in the mixture times the gas constant for each individual constituent.

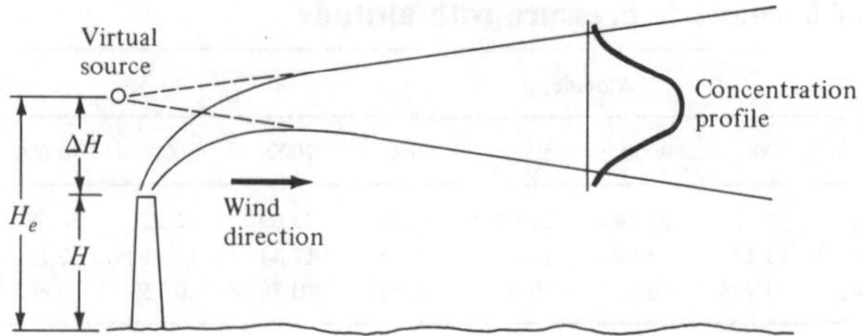

Figure 3-21 Dispersion model forms a stack of height H and plume ΔH.

and their dilution by the atmosphere. The horizontal motion is the result of existing wind. The vertical motion results from the upward motion of high-velocity warm stack-exit gases to much higher elevations. The stack-exit velocity results in a *plume* rise ΔH above the actual stack (Fig. 3-21). The gases bend in the direction of wind flow. According to most models, the *plume height* ΔH is the height of a virtual point source above the stack, obtained by extending the lines of dispersion backward. ΔH is therefore obtained some distance downwind from the stack, where the plume has reached its maximum height. There, thus, is an *effective stack height* H_e given by

$$H_e = H + \Delta H \tag{3-17}$$

In specific situations the proper design of a stack is dependent upon the local topography and airflow patterns. Valleys, for example, can concentrate emissions to unacceptable levels. Model studies are often necessary.

One of the most severe hindrances to dispersion is atmospheric *inversion,* which occurs when the temperature of the atmosphere increases with elevation. It is a condition of little wind and strong stability that results in the reduction of vertical dispersion and therefore the trapping and concentration of local emissions.*

There are several analytical methods of calculating ΔH [20]. Most of them utilize a momentum term that accounts for the vertical momentum of the gas caused by the stack exit velocity and a buoyancy term that accounts for the difference between stack gas and atmospheric densities. They yield different correlations and widely differing values for ΔH. Some of these correlations, giving ΔH in meters, are:

1. Carson and Moses [21]

$$\Delta H = -0.029 \frac{V_s D}{V_w} + 2.62 \frac{(Q_e)^{0.5}}{V_w} \tag{3-18}$$

where

V_s = stack-gas exit velocity, m/s

D = stack diameter, m

* There are two kinds of inversion. The first, called ***subsidence inversion,*** is due to the descent of a layer of air within a high-pressure air mass. The second, called ***radiation inversion,*** is due to thermal radiation from the earth's surface to the atmosphere above at night.

V_w = wind velocity at stack exit, m/s

Q_e = heat emission, J/s, given by

$$Q_e = \dot{m}c_p(T_s - T_a) \qquad (3\text{-}19)$$

where $\dot{m}$ = gas mass-flow rate, kg/s

c_p = specific heat of gas = 1005 J/(kg · K) for dry air at low temperature)

T_s = gas temperature at stack exit, K

T_a = air temperature at stack exit, K.

2. Briggs [22]

$$\Delta H = \frac{114CF^{1/3}}{V_w} \qquad (3\text{-}20)$$

where

C = dimensionless temperature gradient parameter = $1.58 - 41.4(\Delta\theta/\Delta z)$

$\Delta\theta/\Delta z$ = air potential temperature gradient, K/m = 0 for neutral atmospheric stability conditions*

F = buoyancy flux = $gV_sD^2(T_s - T_a)/4T_a$, m^4/s^3

g = gravitational acceleration = 9.8 m/s^2

3. TVA model [24]

$$\Delta H = \frac{173F^{1/3}}{V_w} e^{-(64\ \Delta\theta/\Delta z)} \qquad (3\text{-}21)$$

Equation (3-18) is based on many observations and is applicable to all atmospheric stability conditions. Equation (3-20) takes account of the various stability conditions via C and was based on values of $\Delta\theta/\Delta z$ between -0.001 to $+0.013$ K/m. Equation (3-21) was proposed to yield a value of ΔH some 2 km away from the stack where

* *Stability* is defined as the tendency to resist vertical motion or to suppress turbulence. When the atmosphere is considered to be an adiabatic system, free of frictional or inertial effects, it can be shown [23] that $dT_a/dz \approx -0.01$ K/m for dry air. The negative of this temperature gradient is called the *adiabatic lapse rate* $\Gamma = 0.01$ K/m. Because the standard atmosphere does not meet the above restrictions, a value, based on meterological data, of $dT_a/dz = -0.0066$ K/m has been adopted as an international standard. The air average temperature in the earth's middle latitudes decreases linearly with elevation in the troposphere from sea level to nearly 10,760 m (35,300 ft), above which it remains constant at 216.7 K (−67°F) in the stratosphere. A *stable atmosphere* is one that does not exhibit much vertical motion or mixing, and hot gases emitted near the earth's surface tend to remain there. A *strongly stable atmosphere* is one where dT_a/dz is positive. This is the case of *inversion* referred to earlier. A *weakly stable atmosphere* is one where dT_a/dz is negative but greater than the environmental standard. A *neutral atmosphere* is one where dT_a/dz is negative and equal to the environmental standard. In neutral conditions, air that is carried rapidly upward

the maximum plume rise was believed to have taken place. It takes into account the stability conditions via $\Delta\theta/\Delta z$, directly built into the equation. There are numerous other correlations for ΔH, some including the horizontal distance from the stack as a parameter. Finally, there are theories based on the assumption that the concentration of dispersed material in the wake is a three-dimensional gaussian distribution around the axis of the wake (Fig. 3-21). A formulation by Cramer [25] is presently accepted for predictions of particulate distributions of the atmosphere.

The nature and environmental effects of pollutants from powerplants are covered in Chap. 17.

3-13 STEAM-GENERATOR CONTROL

Steam-generator, and indeed total, powerplant control is a rather broad subject that includes instrumentation, data processing and controls for combustion, steam flow, temperature and pressure, drum level, burner sequencing, desulfurization precipitators, ash handling, system integration, start-up and shutdown, and automation. It obviously cannot be covered fully in this textbook. In this section, therefore, we will cover, in a simplified manner, only a few basic control systems that apply to the steam generator. These are: feedwater and drum-level control, steam-pressure control, and steam-temperature control.

Feedwater and Drum-Level Control

Feedwater (and therefore steam) flow is controlled to meet load demand by the turbine and at the same time maintain the level of water in the steam drum within relatively narrow limits. It is common to maintain the normal drum level at "half-a-glass,"

---

will have the same temperature as the environment and there is no further movement. An *unstable atmosphere* is one where dT_a/dz is negative and less than the environmental standard. It is the case where motion in the vertical direction is enhanced.

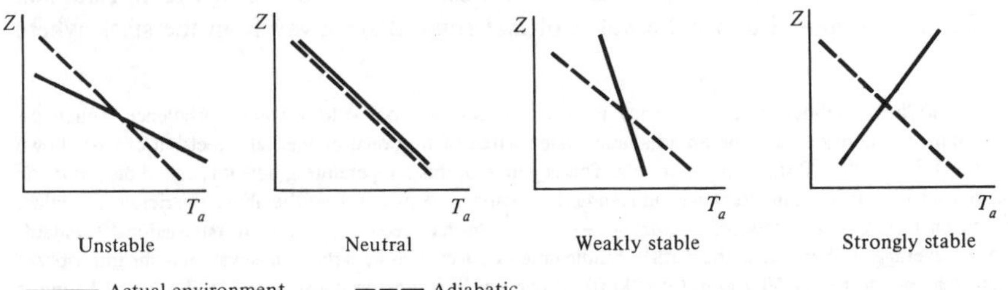

The parameter $\Delta\theta/\Delta z$ used in Eqs. (3-20) and (3-21) is called the potential temperature gradient and is equal to the difference between the environmental and adiabatic temperature gradients, $\Delta\theta/\Delta z = (\Delta T/\Delta z)_{env} - (\Delta T/\Delta z)_{adiab}$. Thus it is zero for a neutral atmosphere, positive for a strongly stable atmosphere, and negative for an unstable atmosphere.

meaning half full, in reference to sight glass tubes used outside the drum. A high steam consumption by the turbine, combined with low feedwater supply, for example, would lower the water level in the drum. Figure 3-22 shows a three-element automatic control system, of which drum level is one element.

The drum-level sensor responds to the error between actual drum and its set point, such as in the case of high steam consumption and low feedwater supply, and acts on the controller to increase the feedwater valve opening to meet the steam-flow demand. This action may be too slow and is supplemented by sensors for feedwater and steam flow. The difference between the signals from these two sensors anticipates changes in drum level and sends a signal to the controller to actuate the valve in the proper direction.

Steam-Pressure Control

The steam-pressure control system (Fig. 3-23), sometimes called the "boiler master," maintains steam pressure by adjusting fuel and combustion airflows to meet the desired pressure. When pressure drops, the flows are increased. A steam-pressure sensor acts directly on fuel- and airflow controls, such as the pulverized-coal power drives and the forced-draft fan, to affect the desired changes. A trimming signal from fuel- and airflow sensors maintains the proper fuel-air ratio. Because it is often difficult to obtain accurate fuel flows, a steam-flow sensor is sometimes substituted for the fuel-flow sensor.

Usually about a 5-s delay is allowed when changing coal flow and airflow to ensure the prevention of a momentary rich mixture (high fuel-air ratio) and thus assure smoke-free combustion.

Steam-Temperature Control

Steam-generator outlet temperature control within narrow limits is important to powerplant operation. Steam temperatures may fluctuate because of the buildup of slag or

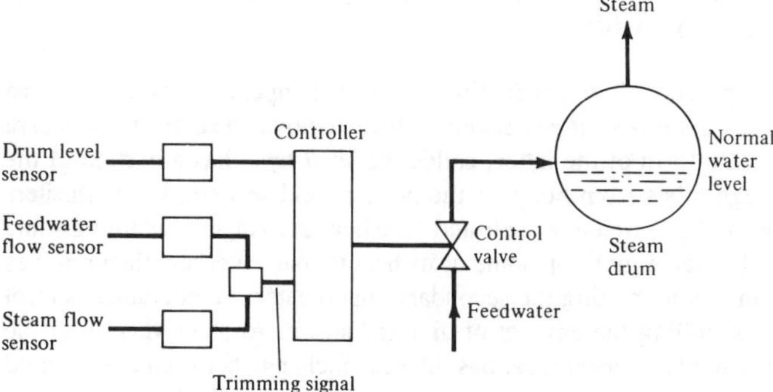

Figure 3-22 Schematic of a three-element feedwater control system.

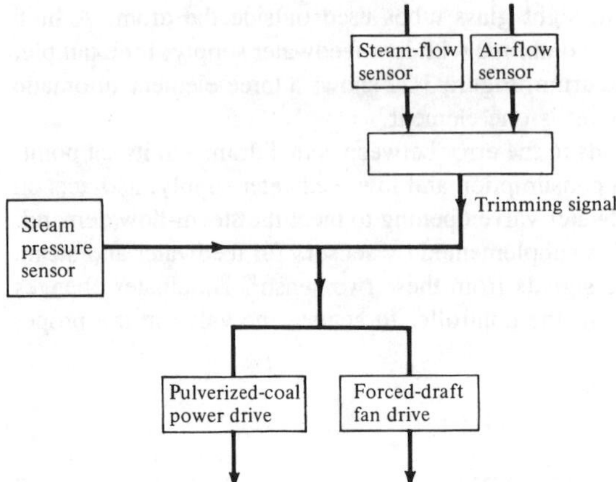

Figure 3-23 Schematic of a steam-pressure control system.

ash on heat-transfer surfaces. The main fluctuations, however, occur because of changes in load. Recall that radiant-type superheaters and reheaters have a drooping temperature-load characteristic, whereas convection-type units have a rising temperature-load characteristic (Fig. 3-10).

A reduction in steam temperature results in loss in plant efficiency. For example, a drop of 35 to 40°F results in about 1 percent increase in heat rate. On the other hand, an increase in steam temperature above design may result in overheating and failure of superheater and reheater tubes and turbine blades.

The temperature of the saturated steam leaving the drum corresponds to the system pressure and thus remains constant as long as the steam-pressure controls are in working order. It is the superheater-reheater responses to load changes that have to be corrected. There are several ways of adjusting the temperature, some of which are discussed below.

Combined radiant-convective superheaters In certain cases radiant and convective superheaters are arranged in series to yield a relatively flat final steam temperature over a wide load range (Fig. 3-10).

Attemperation Attemperation is the reduction of steam temperature by one of two methods. The first uses a *surface attemperator,* which removes heat from the steam in a heat exchanger. One form of the latter, called the *shell type,* has a portion of the steam taken out through tubes from between the primary and secondary superheaters by an automatic valve and diverted to a shell-and-tube heat exchanger containing some of the boiled water. The steam gives up some of its heat to that water and then remixes with the primary steam upon entering the secondary superheater. Temperature control is accomplished by controlling the amount of diverted steam. Another version of the surface attemperator, called the *drum type,* has the heat exchange between the diverted steam and the boiler water occur within the main steam drum, which must now be made larger to accommodate the attemperator tubes.

The second method of attemperation uses a device called the *spray*, or *direct contact, attemperator* (Fig. 3-24). It reduces the steam temperature by spraying low-temperature water from the boiler or economizer exit into the line between the primary and secondary superheaters, Fig. 3-6. The spray nozzle injects water into the throat of a mixing venturi, where the water mixes with high-velocity steam in the throat, vaporizes, and cools the steam. The venturi and a thermal sleeve also protect the main steam pipe from thermal shock caused by any unvaporized water droplets that might otherwise impact on the pipe. The water used must be of high purity to avoid adding deposits on the superheater tubes, pipes, and turbine blades. The spray attemperator has been satisfactory in service. It provides a rapid and sensitive means of temperature control. Steam temperature is controlled by regulating the amount of spray water to produce a flat temperature curve beyond point *a* (Fig. 3-25). A simple energy balance illustrates the principle. For no work, heat, or changes in potential and kinetic energies, the enthalpy entering the system equals that leaving it. Thus

$$\dot{m}_s h_{s1} + \dot{m}_w h_w = (\dot{m}_s + \dot{m}_w) h_{s2} \tag{3-22}$$

where $\dot{m}_s$ and $\dot{m}_w$ = mass-flow rates of steam and water, respectively, lb_m/h or kg/s

h_s and h_w = specific enthalpies of steam and water, respectively, Btu/lb_m or J/kg

subscripts 1 and 2 refer to steam inlet and exit, respectively

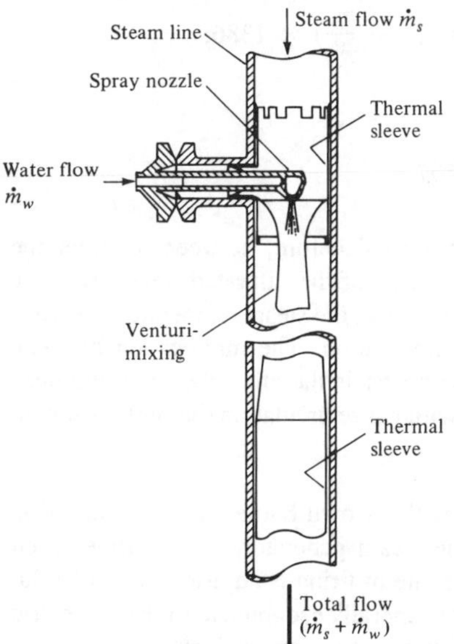

Figure 3-24 A spray attemperator for steam temperature control.

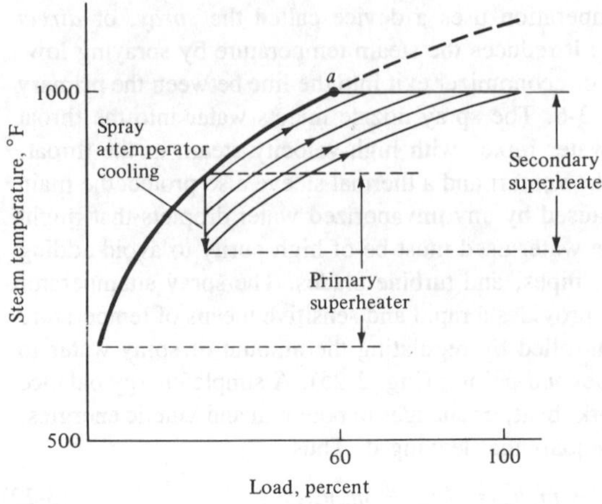

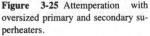

Figure 3-25 Attemperation with oversized primary and secondary superheaters.

Example 3-2 Steam enters a spray attemperator at 2500 psia and 950°F. The spray water comes from the boiler drum, which operates at 2600 psia. Calculate the mass of spray water that must be added per unit mass of steam to reduce its temperature to 900°F.

SOLUTION From the steam tables: $h_{s1} = 1423.1$, $h_{s2} = 1386.7$, $h_w = h_f$ at 2600 psia $= 744.47$, all in Btu/lb$_m$. Using Eq. (3-22)

$$1423.1 + \frac{\dot{m}_w}{\dot{m}_s} \times 744.47 = \left(1 + \frac{\dot{m}_w}{\dot{m}_s}\right) \times 1386.7$$

Therefore

$$\frac{\dot{m}_w}{\dot{m}_s} = 0.0567$$

The attemperator, of either type, may be located before, between, or after the superheater. The first choice will cause condensation of the saturated steam from the boiler before it enters the superheater. In the last choice, the steam temperature exceeds the final desired steam temperature before attemperation. The midlocation between the primary and secondary sections of the superheater is therefore the preferred one.

Attemperation is sometimes used in series with gas recirculation (below), as shown in Fig. 3-26.

Separately fired superheater A superheater with its own burner, fans, combustion chamber, controls, etc., all independent of the steam generator, is sometimes used and may serve more than one steam drum. The rate of firing is adjusted to yield a flat steam temperature-load curve. This system is not generally economical for large electric generating systems and is usually used in the chemical process industry.

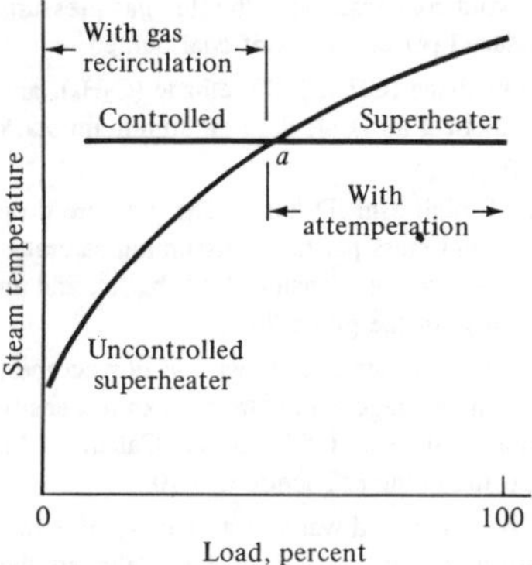

Figure 3-26 Gas recirculation and attemperation in series.

Gas recirculation In this system, gas from some point downstream of the superheater-reheaters, mostly from the economizer outlet but sometimes from the air preheater outlet, is recirculated back to the furnace by means of a gas-recirculation fan (mentioned at the end of Sec. 3-10). The term gas recirculation is restricted to the case where the gas is introduced back to the burning zone, such as before the burners, at the bottom hopper, etc. Gas recirculation to a point downstream of the burning zone is called *gas tempering*. Varying the percentage of recirculated gas alters the heat-absorbing characteristics of the various heat-absorbing surfaces in the steam generators to yield the desired effect and is taken into account in the initial design of the system.

Other types of steam-temperature control are *selecting burners* that give the desired gas temperature, using *tiltable burners* to shift the flame zone in the furnace, *bypassing* a portion of the hot gas around the superheater by dampers, and others.

Regulating reheat outlet steam temperature is necessary for the same reasons as those for regulating superheater outlet temperatures, and the methods used are generally the same.

PROBLEMS

3-1 Calculate the percent energy distribution in a steam generator that goes into steam, losses to flue gases, refuse and heat losses. The steam generator receives 480 short tons of coal per day. The heating value of coal is 13,000 Btu/lb_m. 375,000 lb_m/h of water enters the steam generator at 2500 psia and 450°F and leaves as steam at 2400 psia and 1000°F. Combustion air enters at 80°F and the flue gases leave at 350°F. The refuse, generated at the rate of 45 short tons per day, has an internal energy of 800 Btu/lb_m. The air/fuel ratio by mass is 20:1. Assume that the flue gases have the same variable specific heat as air.

3-2 A steam generator burns a fuel oil that has the following chemical analysis by mass: Carbon 85.3%, hydrogen 14.1%, sulfur 0.5%, and nitrogen 0.1%. Combustion takes place in 125% theoretical air. The flue gases leave the air preheater at 25 psia. Find (*a*) the minimum stack temperature to avoid condensation and (*b*) the maximum mass of sulfurous acid (H_2SO_3) that may be deposited per unit mass of fuel oil burned.

3-3 A West Virginia coal has the following mass analysis: Carbon 75.0%, oxygen 6.7%, hydrogen 5.0%, water 2.5%, sulfur 2.3%, nitrogen 1.5%, and ash 7.0%. It burns in a steam generator with 125 percent theoretical air. Calculate (*a*) the minimum stack temperature to avoid condensation if the flue gas pressure leaving the air preheater is 12 psia and (*b*) the mass of SO_2 produced per unit mass of coal burned.

3-4 A gaseous fuel has the following volumetric analysis: 90% methane (CH_4), 5.0% ethane (C_2H_6), and 5% nitrogen. It burns in a steam generator with 115 percent theoretical air. Calculate the minimum stack temperature if the flue gases leaving the air preheater are at 15 psia.

3-5 A water-tube steam generator operating at 2400 psia has 50-ft-high 3-in ID tubes. The pressure drop is given by $3.25 \times 10^{-6}\ \dot{m}^2$, where $\dot{m}$ is the mass flow rate, in pound mass per hour. Assuming saturated water in the downcomer, a water-steam mixture in the tubes with an average density of 18 lb_m/ft^3, and an inlet tube velocity of 3.0 ft/s, calculate the pump pressure necessary for the given flow.

3-6 A drum-type steam generator operates at 160 bars. 1260 kg/s of water goes down the downcomer, 7.32°C subcooled. The downcomer and risers are all 12 m high. The average water-steam mixture density in the risers is 350 kg/m^3. The pressure losses in the downcomer and risers total 0.5 bar. Calculate the power, in kilowatts, needed to drive a forced circulation pump if the pump efficiency is 0.70.

3-7 A 40-ft-long 3-in-diameter steam generator water tube receives saturated water at a velocity of 2 feet per second and a pressure of 2400 psia. Heat is added to it uniformly. The slip ratio is 1.8. Calculate the maximum heat added to the tube in Btus per foot if the exit void fraction is not to exceed 0.80.

3-8 Steam enters a convective-type superheater saturated at 160 bars. At a given load it leaves at 480°C. The convective heat transfer coefficients inside h_i and outside h_o of the tubes are proportional to $\dot{m}_s^{0.8}$ and $\dot{m}_g^{0.6}$, respectively, where $\dot{m}_s$ and $\dot{m}_g$ are the mass flow rates of steam and gas. Find the exist gas temperature in degrees centigrade if both steam and gas flow are doubled. The gas temperature T_g remains constant at 2000°C. Assume for simplicity that heat transferred between gas and steam is proportional to the product of an overall heat transfer coefficient U and $(T_g - \overline{T}_s)$, where $\overline{T}_s$ is the average steam temperature, and also proportional to $(T_{s2} - T_{s1})$, where T_{s2} and T_{s1} are the exit and inlet steam temperatures. Ignore the conductive resistance in the tube walls and recall from heat transfer that U is inversely proportional to $(1/h_o + 1/h_i)$.

3-9 Steam enters a radiant-type superheater in a dry and saturated condition at 2400 psia. The convective heat-transfer coefficient inside the tubes is proportional to $\dot{m}^{0.8}$, where $\dot{m}$ is the steam mass flow rate. The radiative heat transfer between the combustion gases and the tubes outer surface is proportional to $(T_f^4 - T_w^4)$, where T_f and T_w are the flame and tube wall temperatures, respectively. At a given load, the superheater exit steam temperature is 1000°F, and the external heat-transfer coefficient is five times the internal heat-transfer coefficient. Ignoring the conductive resistance in the tube walls and the tube thickness, estimate the temperature of the steam leaving the superheater if the steam mass flow rate is doubled.

3-10 A 24-sector Ljungstrom air preheater contains 1/16-in-thick corrugated steel sheets and rotates at 2 r/min. The hot gases enter at 600°F and leave at 350°F. The air enters at 70°F. For simplicity assume that the gases and air have equal mass flow rates of 100,000 lb_m/h each, and the same specific heat of 0.243 Btu/lb_m°R. The preheater has an effectiveness of 0.80 (ratio of actual air temperature rise to theoretical maximum temperature rise). Ignore short-circuiting in the preheater and assume perfect heat absorption and release by the steel sheets. Steel has a density of 490 $lb_m/ft^3$and a specific heat of 0.1 Btu/lb_m°R. Calculate (*a*) the air exit temperature and (*b*) the total preheater steel sheet area, in square feet.

3-11 A powerplant receives 92,000 lb_m/h of coal which burns in 125 percent theoretical air. The coal is assumed, for simplicity, to be composed of 100 percent carbon. The air-flue gas pressure losses in the steam generator total 60 in water. Find the power, in megawatts and horsepower, necessary to drive (*a*) a forced-draft fan with air at 60°F and (*b*) an induced-draft fan with flue gases at 300°F if either is used alone. The fans have efficiencies of 0.7. Use c_p for cold air and flue gases of 0.24 and 0.243 Btu/lb_m°R, respectively. Assume a 2 percent inleakage in the case of the induced-draft fan, and negligible change in specific volume across the fans.

3-12 Find the total power consumed, in megawatts and horsepower, by forced- and induced-draft fans if both are used on a steam generator operating on a balanced draft. Assume that the pressure losses before and after the furnace are 20 and 40 in water, respectively. The steam generator uses 800,000 lb_m/h of fuel oil with air-fuel mass ratio of 18.75. Take leakage into the steam generator at 1 percent of total flow. The outside air is at 40°F and the flue gases are at 325°F. The FD and ID fan efficiencies are 0.8 and 0.7, respectively.

3-13 A centrifugal fan equipped with damper control has the damper wide open at full flow. The damper is partly closed to reduce flow to 80 percent of full load. Estimate the changes, in percent, in static pressure, and the horsepower requirement.

3-14 A centrifugal fan equipped with variable speed control produces a flow of 20,000 ft^3/min at 860 r/min. Its speed is increased at 1160 r/min. Estimate the flow in cubic feet per minute, shaft horsepower and static pressure rise, in H_2O, if its speed is increased to 1160 r/min.

3-15 Combustion gases at 800°F enter a cross-flow-type economizer of a steam generator and leave at 600°F. The steam pressure is 1200 psia. 10^6 lb_m/h of feedwater leave the high pressure feedwater heater at 400°F and leave the economizer 7.2°F subcooled. The overall heat transfer coefficient U (which takes into account convection of both gases and water and radiation between the gases and the tubes) is 13.5 Btu/h·ft^2·°F. Calculate the total tube length required if the tubes are 2 in OD.

3-16 A powerplant is situated at an altitude of 1000 ft above sea level. Flue gases enter the stack at 285°F and leave at 230°F. (*a*) Find the height of stack, in feet, necessary for a driving pressure of 0.1 psia if the outside air is at 50°F and (*b*) the change in flue gas flow if the outside air temperature changes to 60°F.

3-17 A powerplant is situated at sea level. 1650 kg/s of flue gases enter a 5-m-diameter stack at 140°C and leave at 110°C. The outside air temperature is at 10°C. The stack is designed for a driving pressure of 0.007 bar. Using the Carson and Moses correlation, calculate the flue gas plume height if the prevailing winds are at 80 km/h.

3-18 A 200-m-high 4-m-diameter stack emits 1000 kg/s of 100°C gases into 5°C air. The prevailing wind velocity is 50 km/h. The atmosphere is in a condition of neutral stability. Calculate the height of the gas plume by two methods.

3-19 A 200-m-high 4-m-diameter stack emits 1000 kg/s of 100°C gases into 5°C air. The prevailing wind velocity is 50 km/h. Using the TVA model, calculate the height of the gas plumes if the atmosphere is (*a*) unstable where $dT/dz = -0.010$ K/m, (*b*) weakly stable where $dT/dz = 0.0050$ K/m, and (*c*) strongly stable where $dT/dz = +0.010$ K/m.

3-20 The steam outlet temperature from a superheater may be given by $t(°F) = 850 + 25\,L^{1/2}$, where L = load, percent. That temperature should not be allowed to exceed 1000°F. A spray-type attemperator is used to regulate that temperature, using feedwater at 400°F. Calculate (*a*) the percent load at which the attemperator is activated, and (*b*) the pound mass of water injected per pound mass of steam entering the attemperator at half load and at full load. Steam pressure is assumed constant at 1000 psia.

3-21 A spray attemperator receives 125 kg/s of steam at 80 bar and 540°C. It uses spray water from the feedwater line at 200°C. The attemperator exit is at 78 bar and 480°C. How much feedwater is used in kilograms per second?

3-22 A superheat-reheat Rankine cycle has attemperators on both the superheater and reheater which are of the convective type. Flows in the superheater and reheater are 1.1×10^6 and 1.0×10^6 lb_m/h, respectively. The maximum pressure is 1000 psia and the reheater is placed optimally in the cycle. At 25 percent load the exit steam from both is 1000°F. As load is increased to 50 percent, these temperatures (without attemperators) would become 1200 and 1150°F. How much water must be added in pounds mass per hour in the attemperators to maintain 1000°F exit temperatures in both if water is available at 400°F?

CHAPTER
FOUR
FUELS AND COMBUSTION

4-1 INTRODUCTION

In Chap. 3 we covered those components of fossil-fueled steam generators that dealt with the working fluid (water and steam) and with the air and flue gases. We deferred discussing the fuel aspects because they require independent treatment. There is a rather wide variety of fuels. Their preparation and feeding, often outside the steam generators, and their methods of firing deserve special attention.

The increasing worldwide demand for energy has focused attention on fuels, their availability (Table 4-1), and environmental effects. The fuels available to utility industry are largely nuclear and fossil, both essentially nonrenewable. Nuclear fuels originated with the universe, and it takes nature millions of years to manufacture fossil fuels.

Fossil fuels originate from the earth as a result of the slow decomposition and chemical conversion of organic material. They come in three basic forms: *solid* (coal), *liquid* (oil), and *natural gas*. Coal represents the largest fossil-fuel energy resource in the world. In the United States today (1983), it is responsible for about 50 percent of electric-power generation. Oil and natural gas are responsible for another 30 percent. The remaining percentage is mostly due to nuclear and hydraulic generation. Natural gas, however, is being phased out of the picture in the United States because it must be conserved for essential industrial and domestic uses.

New combustible-fuel options include the so-called synthetic fuels, or synfuels, which are liquids and gases derived largely from coal, oil shale, and tar sands. A tiny fraction of fuels used today are industrial by-products, industrial and domestic wastes, and biomass.

Table 4-1 World energy reserves

Fuel	Type	Energy, Q*
Fossil	Coal	32
	Oil and gas	6
Fissile	Uranium and thorium	600
Fusile	Deuterium	10^{10}

* 1 Q = 10^{18} Btu or roughly 10^{21} J, not to be confused with a quad q = 10^{15} Btu.

This chapter will cover the combustible fuels available to the utility industry, both natural (fossil) and synthetic, and their preparation and firing systems. Nuclear fuels and renewable energy sources, and the environmental aspects of power generation in general, will be covered later in this text.

4-2 COAL

Coal is a general term that encompasses a large number of solid organic minerals with widely differing compositions and properties, although all are essentially rich in amorphous (without regular structure) elemental carbon. It is found in stratified deposits at different and often great depths, although sometimes near the surface. It is estimated that in the United States there are 270,000 million tons of recoverable reserves (those that can be mined economically within the foreseeable future) in 36 of the 50 states. This accounts for about 30 percent of the world's total.

There are many ways of classifying coal according to its chemical and physical properties. The most accepted system is the one used by the American Society for Testing and Materials (ASTM), which classifies coals by grade or rank according to the degree of metamorphism (change in form and structure under the influences of heat, pressure, and water), ranging from the lowest state, lignite, to the highest, anthracite (ASTM D 388). These classifications are briefly described below in a descending order.

Anthracite This is the highest grade of coal. It cotains a high content, 86 to 98 mass percent, of fixed carbon (the carbon content in the elemental state) on a dry, mineral-matter-free basis and a low content of volatile matter, less than 2 to 14 mass percent (chiefly methane, CH_4). Anthracite is a shiny black, dense, hard, brittle coal that borders on graphite* at the upper end of fixed carbon. It is slow-burning and has a heating value just below that of the highest for bituminous coal (see below). Its use

* Graphite is a moderately soft allotropic form of carbon. Carbon crystallizes perfectly into diamond, imperfectly into graphite, and is amorphous (having no regular structure, noncrystalline) in anthracite and charcoal.

in steam generators is largely confined to burning on stokers, and rarely in pulverized form. In the United States it is mostly found in Pennsylvania.

The anthracite rank of coal is subdivided into three groups. In descending order of fixed-carbon percent, they are *meta-anthracite,* greater than 98 percent; *anthracite,* 92 to 98 percent; and *semianthracite,* 86 to 92 percent.

Bituminous coal The largest group, bituminous coal is a broad class of coals containing 46 to 86 mass percent of fixed carbon and 20 to 40 percent of volatile matter of more complex content than that found in anthracite. It derives its name from *bitumen,* an asphaltic residue obtained in the distillation of some fuels. Bituminous coals range in heating value from 11,000 to more than 14,000 Btu/lb$_m$ (about 25,600 to 32,600 kJ/kg). Bituminous coals usually burn easily, especially in pulverized form.

The bituminous rank is subdivided into five groups: *low-volatile, medium-volatile,* and *high-volatile A, B,* and *C*. The lower the volatility, the higher the heating value. The low-volatility group is grayish black and granular in structure, while the high-volatility groups are homogeneous or laminar.

Subbituminous coal This is a class of coal with generally lower heating values than bituminous coal, between 8300 to 11,500 Btu/lb$_m$ (about 19,300 to 26,750 kJ/kg). It is relatively high in inherent moisture content, as much as 15 to 30 percent, but often low in sulfur content. It is brownish black or black and mostly homogeneous in structure. Subbituminous coals are usually burned in pulverized form. The subbituminous rank is divided into three groups: *A, B,* and *C*.

Lignite The lowest grade of coal, lignite derives its name from the Latin *lignum,* which means "wood." It is brown and laminar in structure, and remnants of wood fiber are often visible in it. It originates mostly from resin-rich plants and is therefore high in both inherent moisture, as high as 30 percent, and volatile matter. Its heating value ranges between less than 6300 to 8300 Btu/lb$_m$ (about 14,650 to 19,300 kJ/kg). Because of the high moisture content and low heating value, lignite it is not economical to transport over long distances and it is usually burned by utilities at the mine site. The lignite rank is subdivided into two groups: *A* and *B*.

Peat Peat is not an ASTM rank of coal. It is, however, considered the first geological step in coal's formation. Peat is a heterogeneous material consisting of decomposed plant matter and inorganic minerals. It contains up to 90 percent moisture. Although not attractive as a utility fuel, it is abundant in many parts of the world. Several states in the United States have large deposits. Because of its abundance, it is used in a few countries (Ireland, Finland, the USSR) in some electric generating plants and in district heating.

4-3 COAL ANALYSIS

There are two types of coal analysis: *proximate* and *ultimate,* both done on a mass-percent basis. Both of these methods may be based on: an *as-received* basis, useful for combustion calculations; a *moisture-free* basis, which avoids variations of the moisture content even in the same shipment and certainly in the different stages of pulverization; and a *dry mineral–matter-free* basis, which circumvents the problem of the ash content's not being the same as the mineral matter in the coal.

Proximate Analysis

This is the easier of two types of coal analysis and the one which supplies readily meaningful information for coal's use in steam generators. The basic method for proximate analysis is given by ANSI/ASTM* Standards D 3172. It determines the mass percentages of fixed carbon, volatile matter, moisture, and ash. Sulfur is obtained in a separate determination.

Fixed carbon is the elemental carbon that exists in coal. In proximate analysis, its determination is approximated by assuming it to be the difference between the original sample and the sum of volatile matter, moisture, and ash.

The *volatile matter* is that portion of coal, other than water vapor, which is driven off when the sample is heated in the absence of oxygen in a standard test (up to 1750°F for 7 min). It consists of hydrocarbon and other gases that result from distillation and decomposition.

Moisture is determined by a standard procedure of drying in an oven. This does not account for all the water present, which includes combined water and water of hydration. There are several other terms for moisture in coal. One, *inherent moisture,* is that existing in the natural state of coal and considered to be part of the deposit, excluding surface water.

Ash is the inorganic salts contained in coal. It is determined in practice as the noncombustible residue after the combustion of dried coal in a standard test (at 1380°F).

Sulfur is determined separately in a standard test, given by ANSI/ASTM Standards D 2492. Being combustible, it contributes to the heating value of the coal. It forms oxides which combine with water to form acids. These cause corrosion problems in the back end of steam generators if the gases are cooled below the dew point, as well as environmental problems (Chap. 17).

Ultimate Analysis

A more scientific test than proximate analysis, ultimate analysis gives the mass percentages of the chemical elements that constitute the coal. These include carbon, hydrogen, nitrogen, oxygen, and sulfur. Ash is determined as a whole, sometimes in a separate analysis. Ultimate analysis is given by ASTM Standards D 3176.

Heating Value

The heating value, Btu/lb_m or J/kg of fuel, may be determined on as-received, dry, or dry-and-ash-free basis. It is the heat transferred when the products of complete

* American National Standards Institute/American Society for Testing and Materials.

combustion of a sample of coal or other fuel are cooled to the initial temperature of air and fuel. It is determined in a standard test in a *bomb calorimeter* given by ASTM Standards D 2015. There are two determinations: the *higher* (or *gross*) *heating value* (HHV) assumes that the water vapor in the products condenses and thus includes the latent heat of vaporization of the water vapor in the products; the *lower heating value* (LHV) does not. The difference between the two is given by

$$\text{LHV} = \text{HHV} - m_w h_{fg} \tag{4-1a}$$

or

$$\text{LHV} = \text{HHV} - 9m_{H_2} h_{fg} \tag{4-1b}$$

where m_w = mass of water vapor in products of combustion per unit mass of fuel (due to the combustion of H_2 in the fuel, i.e., not including initial H_2O in fuel)

m_{H_2} = mass of original hydrogen per unit mass of fuel, known from ultimate analysis

h_{fg} = latent heat of vaporization of water vapor at its partial pressure in the combustion products, Btu/lb$_m$ H_2O or J/kg H_2O

The partial pressure of water vapor in the products of combustion is obtained by multiplying the mole faction of H_2O in the products, which is obtained from the combustion equation in the usual manner, by the total pressure of the products. The 9 in Eq. (4-1*b*) is the ratio of the molecular masses of H_2O and H_2 and represents the mass of H_2O vapor obtained from a unit mass of H_2.

Because gases are not usually cooled down below the dew point in steam generators (or engines), it does not seem fair to charge them with the higher heating value in calculating energy balances and efficiencies of cycles or engines. Some, however, argue that they should be charged with the total energy content of the fuel. A uniform standard had to be agreed upon, whereupon everybody uses the HHV in energy balances and efficiency calculations. (The LHV is the standard used in European practice, however.)

As indicated above, heating values are obtained by testing. However, a formula of the *Dulong* type (which does not include the effects of dissociation) is used to give approximate higher heating values of anthracite and bituminous coals in Btu/lb$_m$.

$$\text{HHV} = 14{,}600\text{C} + 62{,}000\left(\text{H} - \frac{\text{O}}{8}\right) + 4050\text{S} \tag{4-2}$$

where C, H, O, and S are the mass fractions of carbon, hydrogen, oxygen, and sulfur, respectively, in the coal. For lower-rank fuels, the above formula usually underestimates the HHV.

Table 4-2 gives the proximate and ultimate analyses of some typical U.S. coals.

Calculations of heating value for fuels of known composition will be covered in Sec. 4-14.

Table 4-2 Proximate and ultimate analyses of some U.S. coals

Analysis, mass percent	Anthracite	Bituminous, medium volatility	Subbituminous	Lignite
		Proximate		
Fixed carbon	83.8	70.0	45.9	30.8
Volatile matter	5.7	20.5	30.5	28.2
Moisture	2.5	3.3	19.6	34.8
Ash	8.0	6.2	4.0	6.2
		Ultimate		
C	83.9	80.7	58.8	42.4
H_2	2.9	4.5	3.8	2.8
S	0.7	1.8	0.3	0.7
O_2	0.7	2.4	12.2	12.4
N_2	1.3	1.1	1.3	0.7
H_2O	2.5	3.3	19.6	34.8
		HHV		
Btu/lb_m	13,720	14,310	10,130	7,210

Example 4-1 Write the complete combustion equation, calculate the HHV and LHV, Btu/lb_m, of the medium volatility subbituminous coal in Table 4-2 by using the Dulong-type formula and find the dew point, °F. Assume stoichiometric, i.e. chemically correct, combustion air. Total pressure = 1 atm.

SOLUTION

$$\text{HHV} = 14{,}600 \times 0.807 + 62{,}000\left(0.045 - \frac{0.024}{8}\right) + 4050 \times 0.018$$

$$= 11{,}782 + 2976 + 73 = 14{,}831 \text{ Btu/lb}_m$$

Note that the actual value, 14,310, is slightly lower by about 3.5 percent because dissociation and other effects are not taken into account by the Dulong-type formula.

The relative mole fractions of the fuel constituents from the ultimate analysis are obtained by dividing the mass fractions by their molecular masses. The equivalent molecule of the coal, therefore, is

$$\frac{80.7}{12}\text{C} + \frac{4.5}{2}\text{H}_2 + \frac{1.8}{32}\text{S} + \frac{2.4}{32}\text{O}_2 + \frac{1.1}{28}\text{N}_2 + \frac{3.3}{18}\text{H}_2\text{O}$$

or $$0.725\text{C} + 2.250\text{H}_2 + 0.05625\text{S} + 0.075\text{O}_2 + 0.03929\text{N}_2 + 0.1833\text{H}_2\text{O}$$

A hydrogen balance gives $2.250H_2O$ from the H_2 in the fuel + $0.1833H_2O$ originally in the fuel = $2.4333H_2O$ in the products.

A sulfur balance gives $0.05625SO_2$ in the products.

An oxygen balance thus requires $0.725 + (2.250/2) + 0.05625 = 1.90625O_2$ for combustion, but only $1.90625 - 0.075$ originally in the fuel $= 1.83125O_2$ from the atmosphere.

A nitrogen balance (the atmosphere contains 3.76 mol N_2 per mol O_2) gives $3.76 \times 1.83125 = 6.8855N_2$ from the atmosphere + 0.03929 originally in the fuel $= 6.92479N_2$ in the products.

The complete combustion equation, therefore, is

$$(0.725C + 2.250H_2 + 0.05625S + 0.075O_2 + 0.03929N_2 + 0.1833H_2O) + 1.83125O_2 + 6.8855N_2 \rightarrow 0.725CO_2 + 2.4333H_2O + 0.05625SO_2 + 6.92479N_2$$

The latent heat of vaporization, necessary for the calculation of the LHV, is due to the partial pressure of the H_2O that was formed in the combustion process only, because the moisture originally in the fuel receives and gives off the same heat upon vaporization and condensation. The partial pressure of the combustion H_2O is equal to its mole fraction in the products

$$\frac{2.250}{0.725 + 2.4333 + 0.05625 + 6.92479} \times 14.696 = 3.705 \text{ psia}$$

From the steam tables, h_{fg} at 3.705 psia (by interpolation) is 1008.6 Btu/lb$_m$. Using Eq. (4-1*b*)

$$\text{LHV} = 14{,}831 - 9 \times 0.045 \times 1008.6 = 14{,}422.5 \text{ Btu/lb}_m$$

Note that the differences between HHV and LHV for coals are small (less than 3 percent in the above example) because of their low H_2 content. This is not the case with liquid or gaseous hydrocarbons, which contain a large portion of hydrogen in the C-H molecules.

The dew point of the products is the saturation temperature corresponding to the partial pressure of the total H_2O in the products, or

$$\frac{2.4333}{0.725 + 2.4333 + 0.05625 + 6.92479} \times 14.696 = 3.5268 \text{ psia}$$

This corresponds to a saturation temperature of about 148°F, which is the dew point.

There will be more discussion of heating value in Sec. 4-14.

4-4 COAL FIRING

Since the old days of feeding coal into a furnace by hand, several major advances have been made that permit increasingly higher rates of combustion.

The earliest in the history of steam boilers were *mechanical stokers,* and several types are still being used for small- and medium-sized boilers. All such stokers are designed to continuously feed coal into the furnace by moving it on a grate within the furnace and also to remove ash from the furnace.

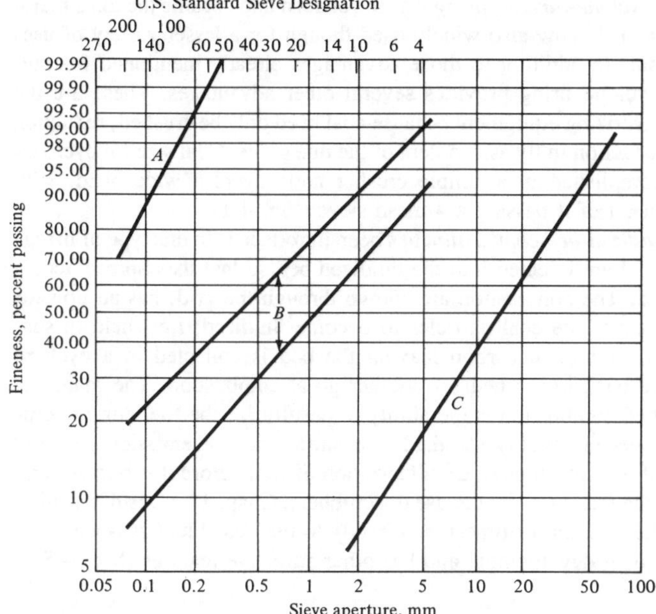

Figure 4-1 Coal sieve analysis. (*A*) pulverized-coal sample; (*B*) coal range for cyclone firing; (*C*) coal as fired. Plot based on graphical system proposed in Ref. 26.

Pulverized-coal firing was introduced in the 1920s and represented a major increase in combustion rates over mechanical stokers. It is widely used today. To prepare the coal for use in pulverized firing, it is crushed and then ground to such a fine powder that approximately 70 percent of it will pass a 200-mesh sieve* (Fig. 4-1). It is suitable for a wide variety of coal, particularly the higher-grade ones. Advantages of pulverized-coal firing are the ability to use any size coal; good variable-load response; a lower requirement for excess air for combustion, resulting in lower fan power consumption; lower carbon loss; higher combustion temperatures and improved thermal efficiency; lower operation and maintenance costs; and the possibility of design for multiple-fuel combustion (oil, gas, and coal).

* There are some seven screen, or sieve, standards in the United States and Europe. The one used here is the U.S. Standard Sieve, in which the number of openings per linear inch designates the mesh. A 100-mesh screen has 100 openings to the inch, or 10,000 openings per square inch. The higher the mesh, the finer the screen. The diameter of the wire determines the opening size. The U.S. Standard Sieve mesh and opening size in inches and millimeters are given below.

Mesh	20	30	40	50	60	100	140	200	325	400
in	1.0331	1.0234	0.0165	0.0117	0.0098	0.0059	0.0041	0.0029	0.0017	0.0015
mm	0.840	0.595	0.420	0.297	0.250	0.149	0.105	0.074	0.044	0.037

In the late 1930s *cyclone-furnace firing* was introduced and became the third major advance in coal firing. It is now also widely used though for a lesser variety of uses than is pulverized coal. In addition to those advantages already mentioned for pulverized-coal firing, cyclone firing provides several other advantages. These are the obvious savings in pulverizing equipment because coal need only be crushed, reduction in furnace size, and reduction in fly ash content of the flue gases. Coal size for cyclone-furnace firing is accomplished in a simple crusher and covers a wide band, with approximately 95 percent of it passing a 4-mesh sieve (Fig. 4-1).

Most recently, *fluidized-bed combustion* has been introduced. In this type of firing, crushed particles of coal are injected into the fluidized bed so that they spread across an air distribution grid. The combustion air, blown through the grid, has an upward velocity sufficient to cause the coal particles to become fluidized, i.e., held in suspension as they burn. Unburned carbon leaving the bed is collected in a cyclone separator and returned back to the bed for another go at combustion. The main advantage of fluidized-bed combustion is the ability to desulfurize the fuel during combustion in order to meet air quality standards for sulfur dioxide emissions. (Other methods are the use of low-sulfur coal, desulfurization of coal before it is burned, and removal of SO_2 from the flue gases by the use of scrubbers, Chap. 17.) Desulfurization is accomplished by the addition of limestone directly to the bed. Fluidized-bed combustion is still undergoing development and has other attractive features (Sec. 4-8).

4-5 MECHANICAL STOKERS

Almost all kinds of coal can be fired on stokers. Stoker firing, however, is the least efficient of all types of firing except hand firing. Partly because of the low efficiency, stoker firing is limited to relatively low capacities, usually for boilers producing less than 400,000 lb_m/h (50 kg/s) of steam, though designers are limiting stoker use to around 100,000 lb_m/h (12.6 kg/s). These capacities are the result of the practical limitations of stoker physical sizes and relatively low burning rates which require a large furnace width for a given steam output. Pulverized and cyclone firing, on the other hand, have higher burning rates and are flexible enough in design to meet the millions of pounds per hour of steam requirements of modern steam generators with narrower and higher furnaces. Stokers, however, remain an important part of steam-generator systems in their size range.

Mechanical stokers are usually classified into four major groups, depending upon the method of introducing the coal into the furnace. These are *spreader stokers, underfed stokers, vibrating-grate stokers,* and *traveling-grate stokers*.

The *spreader stoker* is the most widely used for steam capacities of 75,000 to 400,000 lb_m/h (9.5 to 50 kg/s). It can burn a wide variety of coals from high-rank bituminous to lignite and even some by-product waste fuels such as wood wastes, pulpwood, bark, and others and is responsive to rapid load changes. In the spreader stoker coal is fed from a hopper to a number of feeder-distributor units, each of which has a reciprocating feed plate that transports the coal from the hopper over an adjustable spill plate to an overthrow rotor equipped with curved blades. There are a number of such feeder-distributor mechanisms that inject the coal into the furnace in a wide uniform projectile over the stoker grate (Fig. 4-2). Air is primarily fed upward through the grate from an air plenum below it. This is called *undergrate air*. The finer coal particles, between 25 and 50 percent of the injected coal, are supported by the upward airflow and are burned while in suspension. The larger ones fall to the grate and burn in a relatively thin layer. Some air, called *overfire air*, is blown into the furnace just above the coal projectile. Forced-draft fans are used for both undergrate and overfire air. The unit has equipment for collecting and reinjecting dust and controls for coal flow and airflow to suit load demand on the steam generator.

The problem with stationary spreader stokers was the removal of ash, which was first done manually and then by shutting off individual sections of grates and their air supply for ash removal without affecting other sections of the stoker. The spreader stoker became widely accepted only after the introduction of the *continuous-ash-discharge traveling-grate* stoker in the late 1930s. Traveling-grate stokers, as a class, also include the so-called *chain-grate* stoker. They have grates, links, or keys joined in an endless belt that is driven by a motorized sprocket drive at one end and over an idle shaft sprocket mechanism at the other. Coal may be injected in the above manner or fed directly from a hopper onto the moving grate through an adjustable gate that regulates the thickness of the coal layer. Ash is discharged into an ash pit at either end, depending upon the direction of motion of the traveling grate.

Continuous-cleaning grates that use reciprocating or vibrating designs have also been developed, as have underfeed stokers that are suitable for burning special types

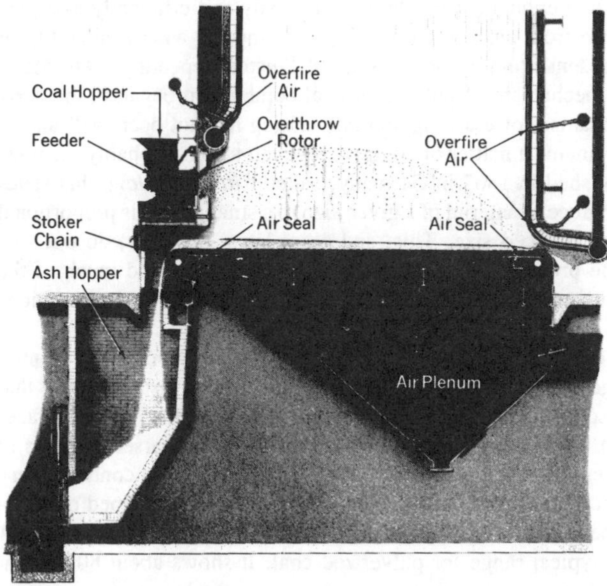

Figure 4-2 A traveling-grate spreader stoker with front ash discharge. *(Courtesy Babcock and Wilcox.)*

of coals. The continuous-ash-removal traveling-grate stoker, however, has high burning rates and remains the preferred type of stoker.

Ignition of the fresh coal in stokers, as well as its combustible volatile matter, driven off by distillation, is started by radiation heat transfer from the burning gases above. The fuel bed continues to burn and grows thinner as the stoker travels to the far end over the bend, where ash is discharged to the ash pit. Arches are sometimes built into the furnace to improve combustion by reflecting heat onto the coal bed.

4-6 PULVERIZED-COAL FIRING

The commercial development of methods for firing coal in pulverized form is a landmark in the history of steam generation. It made possible the construction of large, efficient, and reliable steam generators and powerplants. The concept of firing "powdered" coal, as it was called in earlier times, dates back to Carnot [7], whose idea envisaged its use for the Carnot cycle; to Diesel, who used it in his first experiments on the engine that now bears his name [27]; to Thomas Edison, who improved its firing in cement kilns, thus improving their efficiency and production; and to many others. It was not, however, until the pioneering efforts of John Anderson and his associates and the forerunner of the present Wisconsin Electric Power Company that pulverized coal was used successfully in electric generating powerplants at their Oneida Street and Lakeside Stations, Milwaukee, Wisconsin [28].

The impetus for the early work on coal pulverization stemmed from the belief that, if coal were made fine enough, it would burn as easily and efficiently as a gas. Further inducements came from an increase in oil prices and the wide availability of coal, which makes the present situation sound rather like history repeating itself. Much theoretical work on the mechanism of pulverized-coal combustion began in the early 1920s [29,30]. The mechanism of crushing and pulverizing has not been well understood theoretically and remains a matter of controversy even today. Probably the most accepted law is one published in 1867 in Germany, called *Rittinger's law,* that states that the work needed to reduce a material of a given size to a smaller size is proportional to the surface area of the reduced size. This, and other laws, however, do not take into account many of the processes involved in coal pulverization, and much of the progress in developing pulverized-coal furnaces relies heavily on empirical correlations and designs.

To burn pulverized coal successfully in a furnace, two requirements must be met: (1) the existence of large quantities of very fine particles of coal, usually those that would pass a 200-mesh screen, to ensure ready ignition because of their large surface-to-volume ratios and (2) the existence of a minimum quantity of coarser particles to ensure high combustion efficiency. These larger coarse particles should contain a very small amount larger than a given size, usually that which would be retained on a 50-mesh screen, because they cause slagging and loss of combustion efficiency. Line *A* in Fig. 4-1 represents a typical range for pulverized coal. It shows about 80 percent of the coal passing a 200-mesh screen that corresponds to a 0.074-mm opening and about 99.99 percent passing a 50-mesh screen that corresponds to a 0.297-mm opening, i.e., only 0.1 percent larger than 0.297 mm.

The size of bituminous coal that is shipped as it comes from the mine, called *run-of-mine coal,* is about 8 in. Oversized lumps are broken up but the coal is not screened. Other sizes are given names like *lump* (5 in), which is used in hand firing and domestic applications, *egg* (5 × 2 in), *nut* (2 × $1\frac{1}{4}$ in), *stoker* ($1\frac{1}{4}$ × $\frac{3}{4}$ in), and *slack* ($\frac{3}{4}$ × 0 in, meaning $\frac{3}{4}$ in or less). [Anthracite coal has similar designations, ranging from *broken* ($4\frac{3}{8}$ × 3 in) to *buckwheat* ($\frac{4}{10}$ × $\frac{5}{16}$ in) and *rice* ($\frac{5}{16}$ × $\frac{3}{16}$ in), ASTM D 310].

Coal is usually delivered to a plant site already sized to meet the feed size required by the pulverizing mill or the cyclone furnace (Sec. 4-7). If the coal is too large, however, it must go through *crushers,* which are part of the plant coal-handling system and are usually located in a crusher house at a convenient transfer point in the coal-conveyor system. The feed size required in pulverizing mills is designated at $1\frac{1}{4}$ × 0 in; that required for cyclone furnaces is $\frac{1}{4}$ × 0 in.

Crushers

Although there are several types of commercially available coal crushers, a few stand out for particular uses. To prepare coal for pulverization, the *ring crusher,* or *granulator,* (Fig. 4-3) and the *hammermill* (Fig. 4-4) are preferred. The coal is fed at the top and is crushed by the action of rings that pivot off center on a rotor or by swinging hammers attached to it. Adjustable screen bars determine the maximum size of the discharged coal. Wood and other foreign material is also crushed, but a trap is usually provided to collect tramp iron (metal and other hard-to-crush matter.) Ring crushers and hammermills are used off or on plant site. They reduce run-of-mine coals down to sizes such as $\frac{3}{4}$ × 0 in. Thus they discharge a large amount of fines suitable for

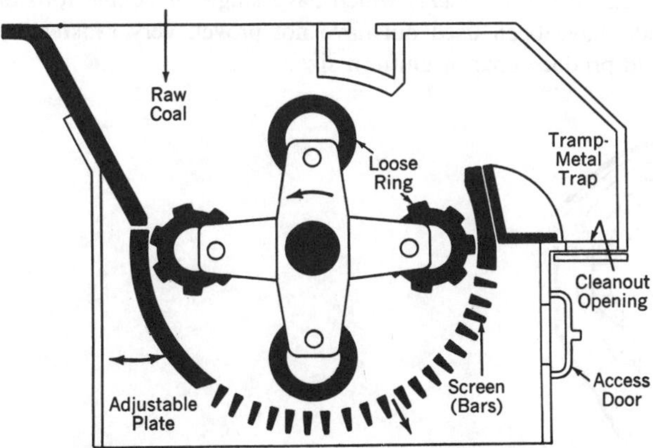

Figure 4-3 A ring-type coal crusher. *(Courtesy Babcock and Wilcox.)*

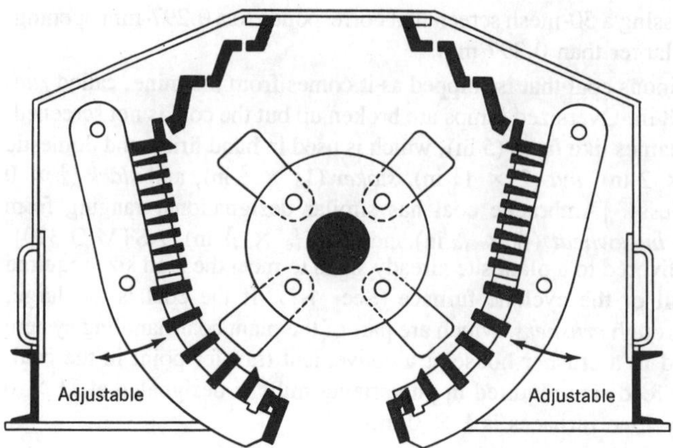

Figure 4-4 A hammer-mill coal crusher. *(Courtesy Babcock and Wilcox.)*

further pulverization, but not for cyclone-furnace firing. For the latter, a crusher type called the *reversible hammermill* is preferred.

A third type, the *Bradford breaker* (Fig. 4-5), is used for large-capacity work. It is composed of a large cylinder consisting of perforated steel or screen plates to which lifting shelves are attached on the inside. The cylinder rotating slowly at about 20 r/min receives feed at one end. The shelves lift the coal, and the breaking action is accomplished by the repeated dropping of the coal until its size permits it to be discharged through the perforations, whose size determines the size of the discharged coal. The quantity of fines is limited because the crushing force, due to gravity, is not large. Bradford breakers easily reject foreign matter and produce relatively uniform size coals. They are usually used at the mine but may also be used at the plant.

Other simple devices called *roll crushers,* which have single or double rolls or rotors equipped with teeth, have been used but have not proven very satisfactory because of their inability to produce coal of uniform size.

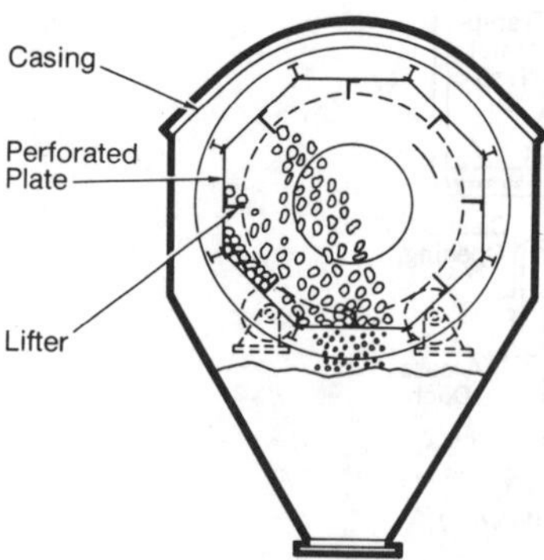

Figure 4-5 A Bradford breaker. *(Courtesy Combustion Engineering, Inc.)*

Pulverizers

The pulverizing process is composed of several stages. The first is the *feeding* system, which must automatically control the fuel-feed rate according to the boiler demand and the air rates required for drying (below) and transporting pulverized fuel to the burner (primary air). The next stage is *drying*. One important property of coal being prepared for pulverization is that it be dry and dusty. Because coals have varying quantities of moisture and in order that lower-rank coals can be used, *dryers* are an integral part of pulverizing equipment. Part of the air from the steam-generator air preheater, the primary air, is forced into the pulverizer at 650°F or more by the primary-air fan. There it is mixed with the coal as it is being circulated and ground.

The heart of the equipment is the *pulverizer,* also called grinding mill. Grinding is accomplished by impact, attrition, crushing, or combinations of these. There are several commonly used pulverizers, classified by speed: (1) low-speed (below 75 r/min): the ball-tube mill; (2) medium-speed (75 to 225 r/min): the ball-and-race and roll-and-race mill; and (3) high-speed (above 225 r/min): the impact or hammermill, and the attrition mill.

The low-speed *ball-tube mill,* one of the oldest on the market, is basically a hollow cylinder with conical ends and heavy-cast wear-resistant liners, less than half-filled with forged steel balls of mixed size. Pulverization is accomplished by attrition and impact as the balls and coal ascend and fall with cylinder rotation. Primary air is circulated over the charge to carry the pulverized coal to classifiers (below). The balltube mill is dependable and requires low maintenance, but it is larger and heavier in construction, consumes more power than others, and because of poor air circulation, works less efficiently with wet coals. It has now been replaced with more efficient types.

The medium-speed *ball-and-race* and *roll-and race* pulverizers are the type in most use nowadays. They operate on the principles of crushing and attrition. Pulverization takes place between two surfaces, one rolling on top of the other. The rolling elements may be balls or ring-shaped rolls that roll between two races, in the manner of a ball bearing. Figure 4-6 shows an example of the former. The balls are between a top stationary race or ring and a rotating bottom ring, which is driven by the vertical shaft of the pulverizer. Primary air causes coal feed to circulate between the grinding elements, and when it becomes fine enough, it becomes suspended in the air and is carried to the classifier. Grinding pressure is varied for the most efficient grinding of various coals by externally adjustable springs on top of the stationary ring. The ball-and-race pulverizer has ball circle diameters varying between 17 and 76 in and capacities between $1\frac{1}{2}$ to 20 tons/h. The roll-and-race pulverizer is operated at lower speeds and larger sizes. A typical one has an 89-in ball circle diameter, a 12-ft diameter, and a 22.5-ft height overall, weighs 150 tons, and is driven by a 700-hp motor [12]. Both types are suitable for direct-firing systems (see below).

High-speed pulverizers use *hammer beaters* that revolve in a chamber equipped with high-wear-resistant liners. They are mostly used with low-rank coals with high-moisture content and use flue gas for drying. They are not widely used for pulverized coal systems.

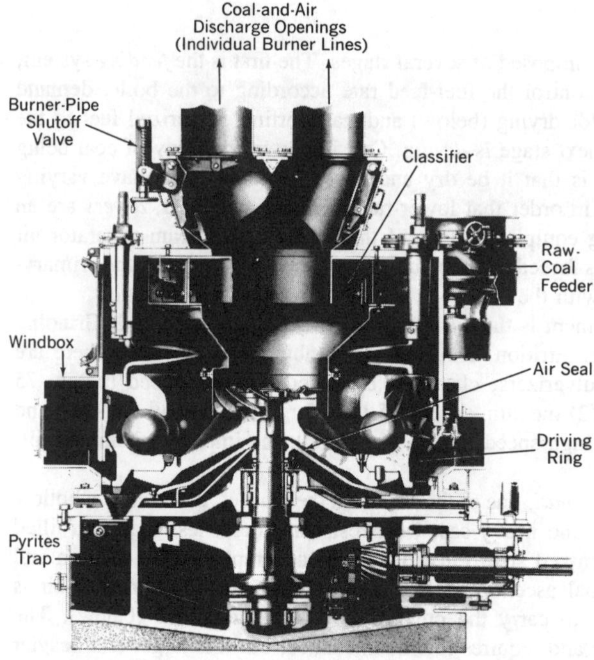

Figure 4-6 A single-row ball-and-race coal pulverizer. *(Courtesy Babcock and Wilcox.)*

The *classifier* referred to above is located at the pulverizer exit. It is usually a cyclone with adjustable inlet vanes. The classifier separates oversized coal and returns it to the grinders to maintain the proper fineness for the particular application and coal used. Adjustment is obtained by varying the gas-suspension velocity in the classifier by adjusting the inlet vanes.

The Pulverized-Coal System

A total pulverized-coal system comprises pulverizing, delivery, and burning equipment. It must be capable of both continuous operation and rapid change as required by load demands. There are two main systems: the bin or storage system and the direct-firing system.

The *bin system* is essentially a batch system by which the pulverized coal is prepared away from the furnace and the resulting pulverized-coal–primary-air mixture goes to a cyclone separator and fabric bag filter that separate and exhaust the moisture-laden air to the atmosphere and discharge the pulverized coal to storage bins (Fig. 4-7). From there, the coal is pneumatically conveyed through pipelines to utilization bins near the furnace for use as required. The bin system was widely used before pulverizing equipment became reliable enough for continuous steady operation. Be-

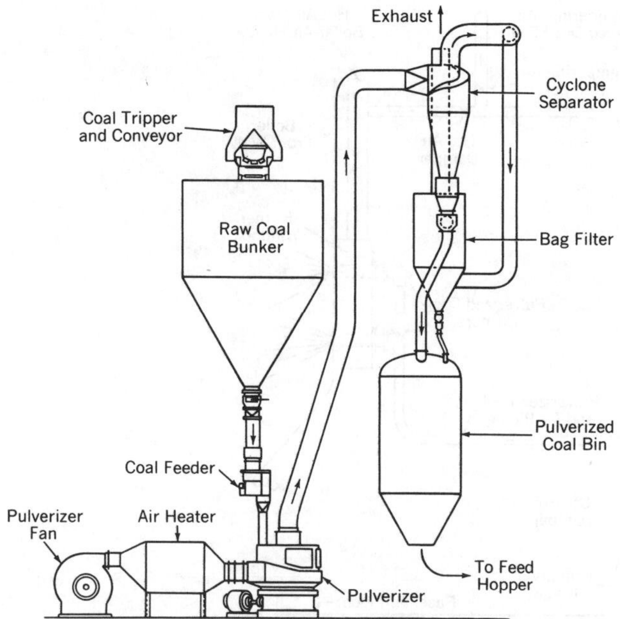

Figure 4-7 Pulverized-coal bin system. *(Courtesy Babcock and Wilcox.)*

cause of the many stages of drying, storing, transporting, etc., the bin system is subject to fire hazards. Nevertheless, it is still in use in many older plants. It has, however, given way to the direct-firing system, which is used exclusively in modern plants.

Compared with the bin system the *direct-firing system* has greater simplicity and hence greater safety, lower space requirements, lower capital and operating costs, and greater plant cleanliness. As its name implies, it continuously processes the coal from the storage receiving bunker through a feeder, pulverizer, and primary-air fan, to the furnace burners (Fig. 4-8). (Another version of this system, less used, places the fan on the outlet side of the pulverizer.) Fuel flow is suited to load demand by a combination of controls on the feeder and on the primary-air fan in order to give air-fuel ratios suitable for the various steam-generator loads. The control operating range on any one direct-firing pulverizer system is only about 3 to 1. Large steam generators are provided with more than one pulverizer system, each feeding a number of burners, so that a wide control range is possible by varying the number of pulverizers and the load on each.

Burners A pulverized-coal *burner* is not too dissimilar to an oil burner. The latter must atomize the liquid fuel to give a large surface-to-volume ratio of fuel for proper interaction with the combustion air. A pulverized-coal burner already receives dried pulverized coal in suspension in the primary air and mixes it with the main combustion

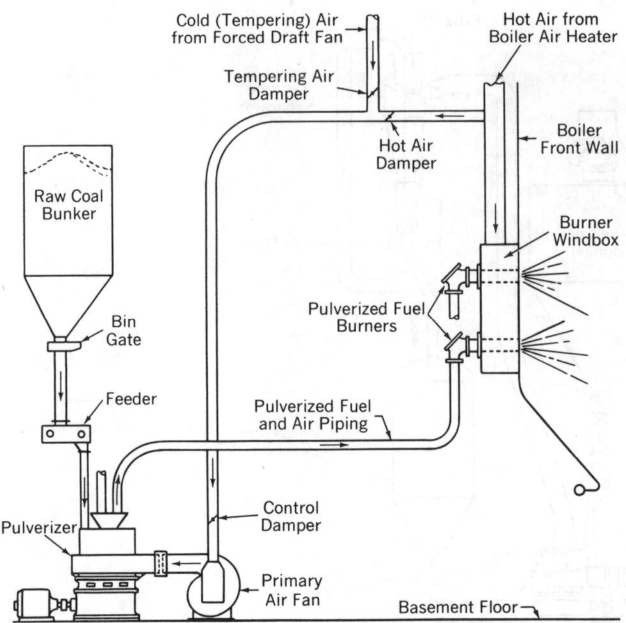

Figure 4-8 Pulverized-coal direct-firing system. *(Courtesy Babcock and Wilcox.)*

air from the steam-generator air preheater. The surface-to-volume ratio of pulverized coal or fineness requirements vary, though not too greatly, from coal to coal (the higher the fixed carbon, the finer the coal). For example, pulverized coal with 80 percent passing a 200-mesh screen and 99.5 percent passing a 50-mesh screen possesses a surface area of approximately 1500 cm^2/g with more than 97 percent of that surface area passing the 200-mesh screen.

The fuel burners may be arranged in one of two configurations. In the first, individual burners, usually arranged horizontally from one or opposite walls, are independent of each other and provide separate flame envelopes. In the second, the burners are arranged so that the fuel and air injected by them interact and produce a single flame envelope. In this configuration the burners are such that fuel and air are injected from the four corners of the furnace along lines that are tangents to an imaginary horizontal circle within the furnace, thus causing a rotative motion and intensive mixing and a flame envelope that fills the furnace area. Vertical firing is also used but is more complex and used only for hard-to-ignite fuels.

The burners themselves can be used to burn pulverized coal only (Fig. 4-9) or all three primary fuels, i.e., pulverized coal, oil, or gas (Fig. 4-10). In Fig. 4-9, the coal impeller promotes the mixing of fuel with the primary air and the tangential doors built into the windbox provide turbulence of the main combustion, or secondary, air to help mix it with the fuel–primary-air mixture leaving the impeller. The total air-

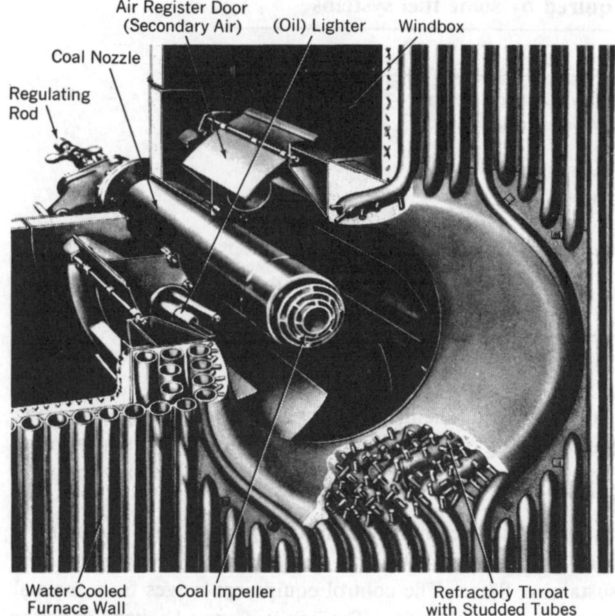

Figure 4-9 A pulverized-coal burner. *(Courtesy Babcock and Wilcox.)*

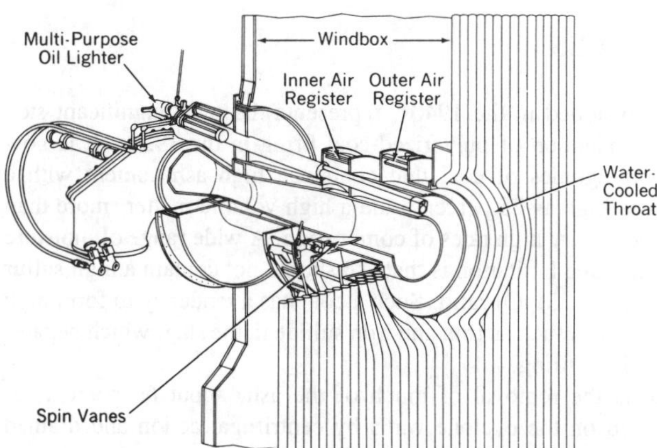

Figure 4-10 Multifuel burners for pulverized coal, oil, and gas. *(Courtesy Babcock and Wilcox.)*

Table 4-3 Excess air required by some fuel systems

Fuel	System	Excess air, %
Coal:	Pulverized, completely water-cooled furnace	15–20
	Pulverized, partially water-cooled furnace	15–40
	Spreader stoker	30–60
	Chain grate and traveling stoker	15–50
	Crushed, cyclone furnace	10–15
Fuel oil:	Oil burners	5–10
	Multifuel burners	10–20
Gas:	Gas burners	5–10
	Multifuel burners	7–12

fuel ratio is greater than stoichrometric (chemically correct) but just enough to ensure complete combustion without wasting energy by adding too much sensible heat to the air. Table 4-3 gives the range of excess air, percent of theoretical, necessary for good combustion of some fuels.

Initial ignition of the burners is accomplished in a variety of ways including a light-fuel oil jet, itself spark-ignited (Fig. 4-9). This igniter is usually energized long enough to ensure a self-sustaining flame. The control equipment ranges from manual to a remotely operated programmed sequence. The igniters may be kept only for seconds in the case of fuel oil or gas. In the case of pulverized coal, however, they are usually kept much longer, sometimes for hours, until the combustion-zone temperature is high enough to ensure a self-sustaining flame. It may also be necessary to activate the igniter at very light loads, especially for coals of low volatility. The impeller is the part of the burner that is subject to severe maintenance problems and is usually replaced once a year or so.

4-7 CYCLONE FURNACES

Cyclone-furnace firing, developed in the 1940s, represents the most significant step in coal firing since the introduction of pulverized-coal firing in the 1920s. It is now widely used to burn poorer grades of coal that contain a high ash content with a minimum of 6 percent to as high as 25 percent, and a high volatile matter, more than 15 percent, to obtain the necessary high rates of combustion. A wide range of moisture is allowable with predrying. One limitation is that ash should not contain a high sulfur content or a high $Fe_2O_3/(CaO + MgO)$ ratio. Such a coal has a tendency to form high ash-fusion temperature materials such as iron and iron sulfide in the slag, which negates the main advantage of cyclone firing.

The main advantage is the removal of much of the ash, about 60 percent, as molten slag that is collected on the cyclone walls by centrifugal action and drained off the bottom to a slag-disintegrating tank below. Thus only 40 percent ash leaves with the flue gases, compared with about 80 percent for pulverized-coal firing. This materially reduces erosion and fouling of steam-generator surfaces as well as the size of dust-removal precipitators or bag houses (Chap. 17) at steam-generator exit. Other advantages are that only crushed coal is used and no pulverization equipment is needed and that the boiler size is reduced. Cyclone-furnace firing uses a range of coal sizes averaging 95 percent passing a 4-mesh screen (region B, Fig. 4-1).

The disadvantages are higher forced-draft fan pressures and therefore higher power requirements, the inability to use the coals mentioned above, and the formation of relatively more oxides of nitrogen, NO_x, which are air pollutants, in the combustion process.

The cyclone is essentially a water-cooled horizontal cylinder (Fig. 4-11) located outside the main boiler furnace, in which the crushed coal is fed and fired with very high rates of heat release. Combustion of the coal is completed before the resulting hot gases enter the boiler furnace. The crushed coal is fed into the cyclone burner at left along with primary air, which is about 20 percent of combustion or secondary air. The primary air enters the burner tangentially, thus imparting a centrifugal motion to the coal. The secondary air is also admitted tangentially at the top of the cyclone at high speed, imparting further centrifugal motion. A small quantity of air, called *tertiary air,* is admitted at the center.

The whirling motion of air and coal results in large heat-release-rate volumetric densities, between 450,000 and 800,000 Btu/(h · ft^3) (about 4700 to 8300 kW/m^3), and high combustion temperatures, more than 3000°F (1650°C). These high temperatures melt the ash into a liquid slag that covers the surface of the cyclone and eventually drains through the slag-tap opening to a slag tank at the bottom of the boiler furnace,

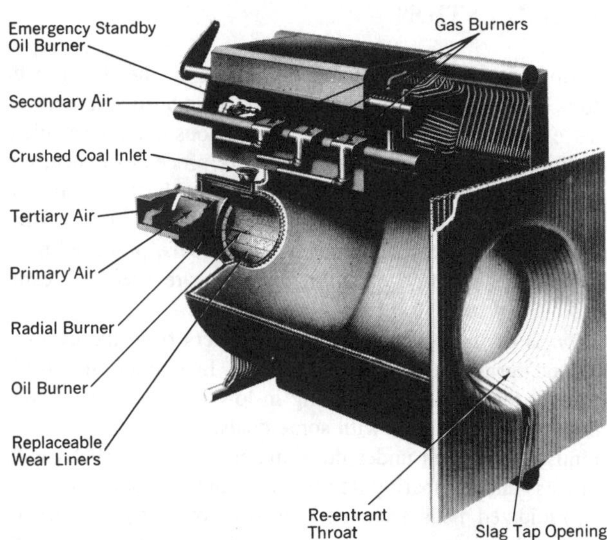

Figure 4-11 A cyclone furnace. *(Courtesy Babcock and Wilcox.)*

where it is solidified and broken for removal. The slag layer that forms on the walls of the cyclone provides insulation against too much heat loss through the walls and contributes to the efficiency of cyclone firing. The high temperatures also explain the large production of NO_x in the gaseous combustion products. These gases leave the cyclone through the throat at right and enter the main boiler furnace. Thus combustion takes place in the relatively small cyclone, and the main boiler furnace has the sole function of heat transfer from the gases to the water-tube walls. Cyclone furnaces are also suitable for fuel-oil and gaseous-fuel firing.

Initial ignition is done by small retractable oil or gas burners in the secondary air ports.

Like pulverized-coal systems, cyclone firing systems can be of the bin, or storage, or direct-firing types, though the bin type is more widely used, especially for most bituminous coals, than in the case of pulverized coal. The cyclone system uses either one-wall, or opposed-wall, firing, the latter being preferred for large steam generators. The size and number of cyclones per boiler depend upon the boiler size and the desired load response because the usual load range for good performance of any one cyclone is from 50 to 100 percent of its rated capacity. Cyclones vary in size from 6 to 10 ft in diameter with heat inputs between 160 to 425 million Btu/h (about 47,000 to 125,000 kW), respectively [12].

The cyclone component requiring the most maintenance is the burner, which is subjected to erosion by the high velocity of the coal. Erosion is minimized by the use of tungsten carbide and other erosion-resistant materials for the burner liners, which are usually replaced once a year or so.

4-8 FLUIDIZED-BED COMBUSTION

We have already noted various attempts at reducing pollutants in the flue gas of powerplant steam generators. The most common are the postcombustion processes that utilize such devices as electrostatic precipitators and baghouses for particulate matter and scrubbers for sulfur dioxide (Chap. 17). Others are processes concurrent with combustion such as cyclone-furnace combustion, in which much of the ash, and hence the particulate matter, is removed during the combustion process. Others, still, are precombustion processes in which "clean" gaseous or liquid fuels, produced from coal by gasification or liquefaction and free of sulfur and ash, are used as steam-generator fuels (Sec. 4-11).

Fluidized-bed combustion is of the concurrent type. It differs from the cyclone furnace in that sulfur is removed during the combustion process. In addition, fluidized-bed combustion occurs at lower temperatures, resulting in lower production of NO_x as well as the avoidance of slagging problems with some coals.

The fluidized-bed combustor has been under development since the 1950s with the aim of perfecting the process and comparing its reliability and economy with SO_2 postcombustion scrubbing. Fluidized beds have been in use for many decades in chemical-industry applications where intimate mixing and contact between reactants are desired. Such contact in a fluidized turbulent state increases heat and mass transfer and reduces time of reaction, plant size, and power requirement. Fluidized beds have even been proposed for use in nuclear-power reactors [1,3]. When used for coal, fluidized-bed combustion results in high combustion efficiency and low combustion temperatures.

Fluidization

A *fluidized bed* is one that contains solid particles which are in intimate contact with a fluid passing through at a velocity sufficiently high to cause the particles to separate and become freely supported by the fluid (Fig. 4-12). A fixed bed, on the other hand, is one in which the velocity of the fluid is too slow to cause fluidization. The minimum fluid velocity necessary for fluidization may be calculated by equating the drag force on a particle due to the motion of the fluid to the weight of the particle. Thus

$$C_D A_c \rho_f \frac{V_s^2}{2g_c} = V\rho_s \frac{g}{g_c} \tag{4-3}$$

where C_D = drag coefficient, a function of the particle shape and the Reynolds number, dimensionless

A_c = cross-sectional area of the particle = πr^2 for a spherical particle, ft² or m²

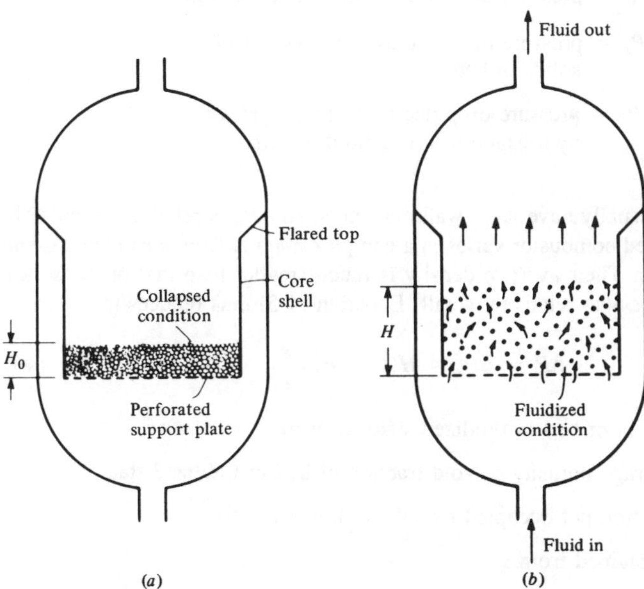

Figure 4-12 Schematic of a fluidized bed: (*a*) particles in collapsed state, (*b*) particles in fluidized state.

r = radius of the particle, if considered spherical, ft or m

ρ_f = density of the fluid, lb_m/ft^3 or kg/m^3

V_s = velocity of the fluid, ft/h or m/s

V = volume of the particle $= \frac{4}{3}\pi r^3$ if spherical, ft^3 or m^3

ρ_s = density of the solid particle, lb_m/ft^3 or kg/m^3

g = gravitational acceleration, ft/h^2 or m/s^2

g_c = conversion factor, 4.17×10^8 $lb_m ft/(lb_f \cdot h^2)$ or $1\ kg/(N \cdot s^2)$

For a spherical particle, Eq. (4-3) yields

$$V_s = \sqrt{\frac{8}{3C_D}\frac{\rho_s}{\rho_f} r g} \tag{4-4}$$

The total pressure drop ΔP, lb_f/ft^2 or Pa, in a fluidized bed is composed of three components

$$\Delta P = \Delta P_w + \Delta P_s + \Delta P_f \tag{4-5}$$

where ΔP_w = pressure drop due to friction at the wall

ΔP_s = pressure drop due to static weight of solids in bed

ΔP_f = pressure drop due to static weight (or hydrostatic head) of fluid in bed

Fluidized beds usually have large wall diameters, so ΔP_w is relatively small. The fluid in a fluidized-bed combustor varies in a complex manner from air to hot gaseous combustion products. Their average density is much smaller than that of the solids, and ΔP_f, therefore, is also relatively small. Equation (4-5) thus reduces to

$$\Delta P = \Delta P_s = H(1 - \alpha)\rho_s \frac{g}{g_c} \tag{4-6}$$

where H = height of bed in fluidized state, ft or m

α = average porosity or void fraction of bed in fluidized state

= fraction not occupied by solids, dimensionless

The value of α is obtained from

$$\frac{1 - \alpha}{1 - \alpha_o} = \frac{H_o}{H} \tag{4-7}$$

where H_o = height of the bed in the collapsed state, i.e., when all solids are randomly packed at bottom

and α_o = porosity in the collapsed state; the value of α_o of a bed of spheres of equal diameter, randomly packed is very nearly equal to 0.40 [31] but lower for spheres of different diameters

In fluidized-bed combustion, crushed coal, between 1/4 and 3/4 in maximum (6 to 20 mm), is injected into the bed just above an air-distribution grid in the bottom of the bed (Fig. 4-13). The air flows upwards through the grid from the air plenum into the bed, which now becomes the furnace where combustion of the swirling mixture takes place. The products of combustion leaving the bed contain a large proportion of unburned carbon particles that are collected in a cyclone separator, which separates these particles from the gas by imparting a centrifugal acceleration on the mixture. They are then returned back to the fluidized bed to complete their combustion. The boiler water tubes are located in the furnace.

As indicated previously, the most important advantage of fluidized-bed combustion is the concurrent removal of the sulfur dioxide that results normally from the combustion of the sulfur content of the coal. Desulfurization is accomplished by the addition of limestone directly to the bed together with the crushed coal. Limestone is a sedimentary rock composed mostly of calcium carbonate ($CaCO_3$) and sometimes some magnesium carbonate ($MgCO_3$). Limestone absorbs the sulfur dioxide with the help of some oxygen from the excess air, according to

$$CaCO_3 + SO_2 + \frac{1}{2}O_2 \rightarrow CaSO_4 + CO_2 \qquad (4\text{-}8)$$

The rate of this reaction is maximum at bed temperatures between 1500 and 1600°F (815 to 870°C), though a practical range of operation of fluidized beds of 1380 to 1740°F (750 to 950°C) is common.

The calcium sulfate $CaSO_4$ produced in this process is a dry waste product that is either regenerated or disposed of. Technical problems for handling this calcium sulfate are under study. Reductions in sulfur dioxide emissions of up to 90 percent have been achieved in fluidized-bed pilot plants.

There are other advantages as a result of the low combustion temperatures of fluidized-bed combustors. They allow inferior grades of coal to be used without slagging problems; the carbon and ash carryover in the flue gases does not reach temperatures at which they become soft and foul heat transfer surfaces; the low temperature combustion results in substantial reductions in the emission of oxides of nitrogen NO_x, a serious air pollutant; and cheaper alloy materials are possible, resulting in economy of construction.

Another advantage is the absence of pulverization equipment, resulting in further economies. Also, a fluidized-bed combustor can be designed to incorporate the boiler within the bed, resulting in volumetric heat-transfer rates that are 10 to 15 times higher and surface heat-transfer rates that are 2 to 3 times higher than a conventional boiler.

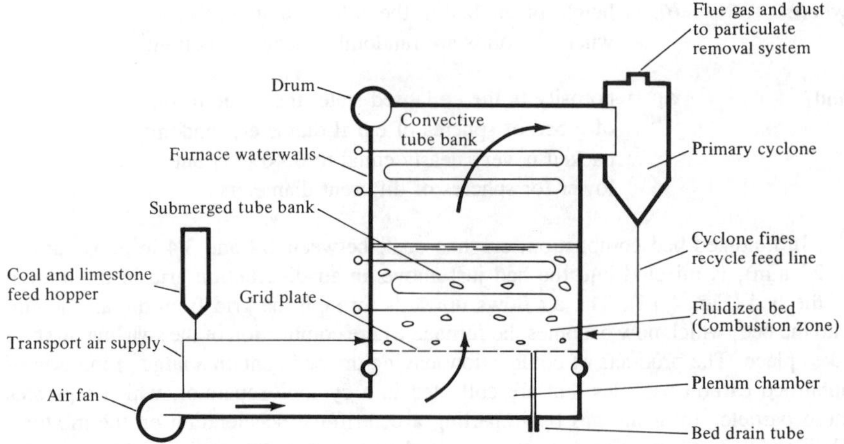

Figure 4-13 Schematic of a fluidized-bed-combustion steam generator. *(Courtesy Combustion Engineering, Inc.)*

A fluidized-bed steam generator is therefore much more compact than a conventional one of the same capacity.

In addition, reductions in sulfur dioxide (and trioxide) in the flue gas means that lower stack-gas temperatures can be tolerated because less acid is formed as a result of the condensation of water vapor. Lower stack-gas temperatures result in an increase in overall plant efficiency.

The problems facing the development of fluidized-bed combustors include those that are associated with feeding the coal and limestone into the bed, the control of carbon carryover with the flue gases, the regeneration or disposal of calcium sulfate, the quenching of combustion by the cooler water tubes within the bed, and variable-load operation.

Fluidized-bed combustors may be used with either a conventional steam powerplant (Rankine cycle) or a combined gas-steam powerplant (Brayton-Rankine) cycle (Sec. 8-6). In the conventional powerplant application, the bed can be of the *atmospheric*, or *pressurized*, type. Atmospheric beds utilize both forced- and induced-draft fans. Pressurized beds use compressors to supply combustion air at pressures of up to 10 bar, which results in slightly higher heat-transfer rates. Electrostatic precipitators, fly-ash removal, air preheaters, and other auxiliaries as are found in conventional steam generators are used. In the combined gas-steam cycle pressurized beds are used. They receive compressed air from the gas-turbine cycle compressor.

By the end of 1981 more than 20 fluidized-bed-combustion water-tube steam generators were contracted for worldwide. Their ratings extended up to 330,000 lb_m/h (about 42 kg/s) capacity, 2450 psig (about 170 bar), and 1000°F (about 540°C) steam conditions. They now are considered by some as competitive with conventional steam generators with gas scrubbing when high-sulfur coals or waste fuels are burned.

4-9 LIQUID FUELS

Technically, liquid fuels are an excellent energy source. They are easy to handle, store, and burn and have nearly constant heating values. They are usually a mixture of hydrocarbons that may be represented by the molecule C_nH_m, where m is a function of n that depends upon the "family" of the hydrocarbon. Table 4-4 gives the families of hydrocarbons found in crude and refined oils.

The number of carbon atoms in a hydrocarbon molecule is identified by

1 meth	6 hex	11 undec	16 hexadec	21 heneicos
2 eth	7 hept	12 dodec	17 heptadec	22 docos
3 prop	8 oct	13 tridec	18 octadec	23 tricos
4 but	9 non	14 tetradec	19 nonadec	30 triacont
5 pent	10 dec	15 pentadec	20 eicos	100 hect

and so on. Straight chain molecules are often called *normal* and are preceded by the letter *n*. Some chain molecules, called *isomers* of the original molecule, are branched and use the prefix *iso*. Some examples are:

1. Methane, CH_4:

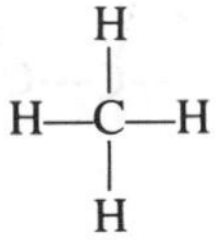

```
     H
     |
  H—C—H
     |
     H
```

2. n-Octane, C_8H_{18}:

```
 |   |   |   |   |   |   |   |
—C—C—C—C—C—C—C—C—
 |   |   |   |   |   |   |   |
```

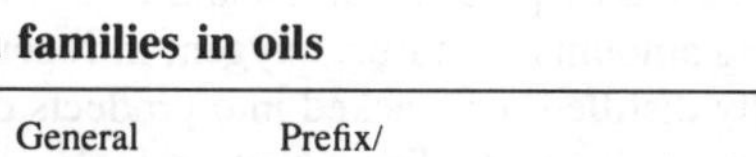

where the H symbol has been dropped for simplicity.

Table 4-4 Hydrocarbon families in oils

Family	General formula	Prefix/ suffix	Structure
Paraffin* (alkanes)	C_nH_{2n+2}	-ane	Chain, saturated
Olefin	C_nH_{2n}	-ene, -ylene	Chain, unsaturated, one double bond
Diolefin	C_nH_{2n-2}	-diene	Chain, unsaturated, two double bonds
Naphthene* (cycloparaffin)	C_nH_{2n}	Cyclo–ane	Ring, saturated
Aromatic*			
Benzene	C_nH_{2n-6}		Ring, unsaturated
Naphthalene	C_nH_{2n-12}		Double ring, unsaturated

* The primary hydrocarbons found in crude oil

3. Isooctane, C_8H_{18}:

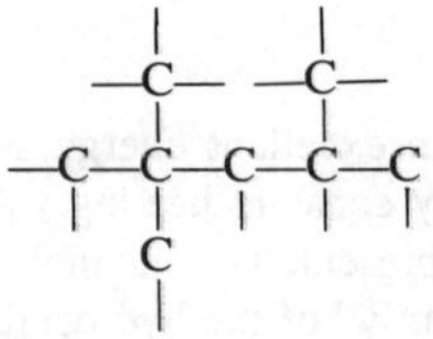

```
      |     |
    —C—   —C—
      |     |
 —C—C—C—C—C
  |  |  |  |  |
     C
     |
```

The latter molecule is also called 2-2-4-trimethyl pentane because it has the alkyl (C_nH_{2n+1}) methyl radical (CH_3) attached to the numbers 2, 2, and 4 carbon atoms on a pentane base. It still has the formula C_8H_{18}. This hydrocarbon, because of its structure, is hard to "break" and resists detonation in a gasoline engine. It is this isooctane, rather than normal octane, that is used as a standard with a scale of 100 for detonation-resistant gasoline fuels. Other examples are:

4. Butadiene, C_4H_6:

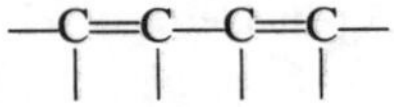

```
—C=C—C=C—
 |  |  |  |
```

where the double bonds indicate unsaturation, meaning only one hydrogen atom is attached to each of the adjacent carbon atoms.

5. Cyclopentane, C_5H_{10}:

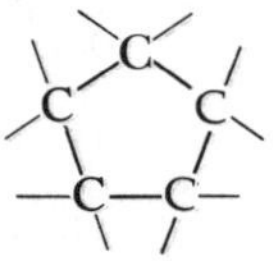

6. Benzene, C_6H_6:

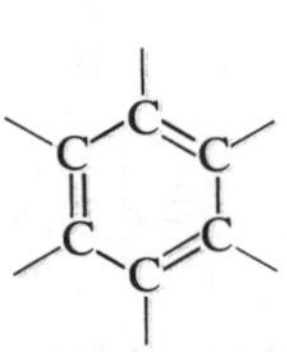

Crude oil is a mixture of an almost infinite number of hydrocarbons, ranging from light gaseous (low n) to heavy (high n) tarlike liquids and waxes of complex molecular structure. They average 83 to 87 percent carbon and 11 to 14 percent hydrogen. Crude oil also contains varying amounts of sulfur, oxygen, nitrogen, particulates, and water. It is refined, fractionally distilled, or cracked into products of narrower range suitable for various applications, such as gasoline, aviation fuels, diesel fuel, fuel oil, lubricating oil, etc.

The fuel most suitable for utility powerplants and industrial uses is fuel oil. It comes in various grades, from light to heavy. They are:

Distillates: No. 1, No. 2, and No. 4
Residual oils: No. 5 (light), No. 5 (heavy), and No. 6

No. 4 could be either a distillate or a mixture of refinery products. The latter grades require preheating for burning and handling. Like coals, oils are classified according

Table 4-5 Average characteristics of fuel oils

Grade	1	2	4	5	6
Analysis, mass %					
C	86.50	86.40	86.10	85.55	85.70
H_2	13.20	12.70	11.90	11.70	10.50
O_2 and N_2	0.20	0.20	0.48	0.70	0.92
S	0.10	0.4–0.7	0.4–1.5	2.0 maximum	2.8 maximum
Sediment and water	Trace	Trace	0.5 maximum	1.0 maximum	2.0 maximum
Ash	Trace	Trace	0.02	0.05	0.08
Density (60°F)					
lb_m/ft^3	51.46	53.98	57.87	59.43	61.50
lb_m/U.S. gal	6.870	7.206	7.727	7.935	8.212
Viscosity					
cSt (100°F)	1.60	2.68	15.0	50.0	360
Pour point, °F	< 0	<0	10	30	65
Atomizing temperature, °F	Atm	Atm	25 minimum	130	200
Higher heating value					
Btu/lb_m	19,940	19,570	18,900	18,650	18,260
Btu/U.S. gallon	137,000	141,000	146,000	148,000	150,000

to physical characteristics by an ASTM specification (ASTM Standards D 396), see for example Table 4-5.

Crude oil has been burned in some boilers and gas turbines. However, because it contains light fractions, its flash point* is low and presents a fire hazard and thus requires special handling procedures. The crude most suitable for direct burning is the so-called *sweet* or low-sulfur crude because it is a low air pollutant. However, it is usually reserved for use as refinery feedstock rather than as a fuel.

In many countries, including the United States, oil is becoming a scarce or valuable commodity and often results in extremely high plant fuel costs. In such countries, new powerplants are often nuclear or coal-fired. For existing oil-fired plants or where no large coal reserves are available, alternatives to straight oil burning are being sought. Some of these are described below.

Emulsion firing An *emulsion* is a suspension of a finely divided fluid in another, in this case water emulsified in heavy oil. Research has shown that, when atomized, the drops of such an emulsion undergo microexplosions of the entrained water as they enter the hot combustion chamber. This causes additional atomization and increased fuel surface-to-volume ratio. In turn, this reduces carbon loss to the combustion process, thus reducing soot and preventing deposit buildup on heat-transfer surfaces. It also reduces excess air requirements and improves combustion efficiency, thus reducing

* The *flash point* is the lowest temperature that allows inflammable vapors to be formed. It is found by heating the fuel slowly, sweeping a flame across the liquid surface, and noting the temperature at which a distinct flash occurs.

oil requirements. This reduction, although technically meaningful, is, however, much less than needed for oil conservation on a national scale.

Coal-oil and coal-water mixtures (COM and CWM) The main advantage of COMs and CWMs is their ability to reduce or replace oil as fuel in oil-fired utility steam generators. They are currently receiving increasing attention as the cost differential between coal and oil increases.

COMs usually contain about 50 percent coal on a heating-value basis. Their main attraction is that the technology for producing and burning them exists (the first tests on COMs were run as far back as 1880), and several fuel suppliers are offering them commercially, though in limited quantities.

CWMs, preferred by some over COMs because of their greater ability to replace oil, are 70 to 80 percent coal, the rest water, plus a fraction of a percent of a stabilizer. CWMs are used in the form of a slurry, that is, a relatively fine coal suspended in the water. They can thus be handled, stored, transported, and fired like oil. CWMs are expected to result in derating of the steam generators. Limited testing of CWMs as fuel is taking place in the time being.

4-10 LIQUID, GAS, AND SOLID BY-PRODUCTS

Combustible industrial wastes have received increased attention as steam-generator fuels, beginning in the decade of the 1970s. So used, they serve the double purpose of disposing of them and reducing the use of oil. The liquids include solvents, waste oil and oil sludges, oil-water emulsions, polymers, resins, chlorinated hydrocarbons, phenols, tars, combustible chemicals, greases, and fats. The obvious main disadvantage of such liquids is that they vary considerably in heating value, flash and fire points, viscosity, and moisture content. They are not expected to make a big dent in the oil picture.

Gaseous by-products are more attractive, with refinery gas and coke-oven gas the most suitable. *Refinery gas* is generated in the conversion of crude oil to gasoline and other refinery products. It has a high heating value and is often blended with lower-heating-value gas by-products from the refinery prior to combustion. *Coke-oven gas* results from the manufacture of coke from raw coal in a coke oven where the volatile matter is distilled off and the coke-oven gas separated from liquids and solids in the volatile matter by cooling and extraction. It consists of about half H_2, about a third methane, and the remainder of various other gases. Its heating value ranges from 400 to 600 Btu/ft^3 (14,200 to 21,300 kJ/m^3).

Other less attractive gases include *regenerator gas,* which is produced in refinery catalytic-cracking processes. It includes less than 10 percent CO and high inert gas and solids contents, and is low in heating value. Its one redeeming factor is its availability from the process at high temperatures (more than 1000°F, 540°C), so that its sensible heat is significant when burned on site. Another, *blast-furnace gas,* a product of iron reduction processes, has a higher CO content (about 30 percent) but has a high dust loading that would cause burner plugging and heat-transfer surface fouling if burned directly. It thus requires cleaning prior to burning.

There are many solid wastes that are available as fuels. The ones that have seen most use are wood waste and sugar-cane waste. Both, naturally, are confined in use to those industries where they are available as waste products. *Wood waste* is found in sawmills, where it represents some 50 percent of the mass of the logs, and in pulp, paper, furniture, plywood, and similar industries. It appears as refuse, bark, sawdust, chips, shavings, and slabs. Most woods have about the same chemical composition, having a proximate analysis of about 70 to 75 percent volatile matter, 25 percent fixed carbon, and 0.5 to 5 percent ash on a dry basis, and a heating value ranging between 8300 to slightly over 9000 Btu/lb_m (about 19,300 to 21,000 kJ/kg), but they vary in density and moisture content. To burn wood efficiently, it must be cut down to chip size in a hogger or chipper to permit continuous feeding and should not have a moisture content greater than 60 or 65 percent. Wood-burning furnaces burn the wood in piles, thin beds on traveling-grate spreader stokers or cyclones. Environmentally, wood burns more cleanly than oil or coal. A relatively small 10-MW powerplant that burns wood chips from forest residue has been in operation since 1977 in Burlington, Vermont. The city has plans for a 50-MW wood-burning station. Other plants are in Wisconsin, Montana, and Oregon.

Sugar-cane waste, also called *bagasse,* is that portion of sugar cane that remains after the sugar is extracted. It consists of matted cellulose fibers and fine particles. Its proximate analysis is about 84 percent volatile matter, 12.5 percent fixed carbon, and 3.5 percent ash on a dry basis, but it contains more than 50 percent moisture. To burn it, it is shredded to short fibers and fines. It has a heating value of 3600 to 4200 Btu/lb_m (about 8400 to 9770 kJ/kg). Sugar mills generate enough bagasse to satisfy much of their demand for cogeneration of both process steam and electricity. Bagasse has been burned most successfully in the so-called *Ward furnace,* which is a refractory furnace that contains individual cells. Piles of bagasse are fed via chutes and burn incompletely in the cells, resulting in partially drying the fuel. A secondary furnace above the cells completes the combustion. Spreader-stoker furnaces have been recently introduced but have not yet proved as simple and reliable as the Ward furnace. Numerous bagasse-fired steam- and electric-generating plants have been in operation at sugar mills throughout the world for decades.

Other solid fuels include food industry by-products, solid wastes, and biomass.

Solid Wastes

Solid wastes, or refuse, are generated by industrial and domestic processes (garbage). Industrial wastes include paper, wood, metal scraps, and agricultural waste products. Domestic wastes include paper, containers, tin, aluminum, food scraps, and sewage. In the United States, solid-waste production is at the rate of about 1 metric ton per person per year and growing at about 5 percent per year. Most of this waste is currently disposed of in land-fill sites near industrial and metropolitan areas. A typical composition of municipal waste in the United States is given in Table 4-6. Some of these wastes, such as paper, some metals, and woods, can be recycled for reuse. Much of it can be burned, since about 50 to 60 percent of it is combustible. Thus there is a potential of burning some 100 million tons of refuse each year for heating or generating steam for electric powerplants.

Table 4-6 Typical composition of municipal waste in the United States

Material	%	Material	%
Paper	50.7	Leather, rubber, plastic	3.3
Food	19.1	Wood	2.9
Metal	10.0	Textiles	2.6
Glass	9.7	Miscellaneous	1.7

The heat content of refuse varies widely, sometimes up to 100 percent, but averages about half that of coal on a mass basis. Thus the energy potential of refuse is not too great, being about 1 percent of the total U.S. needs. The attractive feature, however, is getting rid of a good portion of the total refuse by burning it rather than by dumping it in land-fill sites that are becoming scarce. Another advantage of refuse as fuel is its much lower sulfur content as compared with that of coal or oil.

The main problems are the wide assortment of constituents, a high moisture content, dangers of explosions due to careless volatile-fuel dumping and metal sparks during processing, yet-unknown effects on the operation of large powerplants, and of course, the wide variations in heat content. Because of the latter problem, a practice now receiving wide acceptance is to burn a mixture of solid waste and fossil fuel, with the refuse supplying 10 to 20 percent of the required heat input to the boilers.

Refuse must be carefully prepared for burning. First it is discharged from collecting trucks into a raw-refuse receiving building. It is then shredded in a hammermill and conveyed to a storage bin. The head end of the conveyor is equipped with a magnetic separator that removes ferrous material which may be sent for recycling. An air-classification system removes most of the remaining noncombustibles and heavy particles that cause abrasion. The costs here are in collecting, transporting, and processing the refuse and in the dual-fuel system.

Refuse burning in incinerators has been a wide practice in many parts of the world. One of the first successful refuse-burning electric-generating powerplants, in Bern, Switzerland, burns about 200 tons/day and produces heated water, steam, and electricity. Munich, Germany, has a notable installation, and other cities around the world are following suit.

Some attention and research are directed toward the conversion of organic wastes to synthetic fuels, such as those produced from coal and biomass (below). These processes are largely in the developmental stages.

4-11 SYNTHETIC FUELS

Expected future shortages and supply problems of naturally found liquid and gaseous fuels, as well as environmental problems that restrict the burning of coals, are responsible for the large developmental efforts going on around the world toward the production of synthetic fuels. *Synthetic fuels,* also called *synfuels,* are gaseous and liquid fuels produced largely from coal but also from various wastes and biomass (Sec. 4-12). This production must, among other things, be economical and environmentally acceptable. In this section we shall discuss coal conversion to these two fuels by (1) *gasification* and (2) *liquefaction*. As indicated earlier, the modern use of the available vast resources of coal requires the use of low-sulfur coal, leaving the bulk (about 66 percent) of it unused, cleaning it prior to combustion (washing, froth floating, microwave, and magnetic separation), during combustion (fluidized bed), or after combustion (precipitators, scrubbers). Some of these methods are as yet commercially unproven, and some are costly.

Synthetic fuels in the form of coal-converted clean gaseous or liquid fuels are another alternative. Besides use for steam generation in powerplants, they can be used for domestic, industrial, and transportation purposes.

The basic idea of coal gasification is not new. Gas was manufactured from coal around 1800 when wood and charcoal were growing scarce for smelting of iron, and coal was carbonized to coke by removing its volatiles and using them as a by-product gas. It had a heating value of 550 to 600 Btu/scf (*s*tandard *c*ubic *f*oot), was distributed to urban areas, and was called *town gas*. Coal liquefaction is more recent, initial research having been done in Germany in the 1920s and 1930s. During World War II the Germans and the Japanese, denied access to much of their fuel needs, developed several liquefaction processes that ran their entire war machines, including fuel oil, lubricants, and motor and aviation fuels. One of the German processes, called the Fischer-Tropsch process, is in use commercially in oil-short South Africa today.

Of course, these operations were and are the result of necessity, in one case need in a war economy, in the other political and economic independence. Where such circumstances do not exist, the processes involved must be able to compete economically and environmentally against available fuels or must show potential improvements in expected future shortages to be pursued actively. Because prices of natural fossil fuels have moderated in the early 1980s, the United States government has withdrawn its support from many synthetic-fuel projects. However, it is believed that several such fuels will appear in commercial quantities before the end of this century, and certainly in the next.

Coal Gasification

As stated above, town gas was first produced from coal more than a 100 years ago and had a heating value of 550 to 600 Btu/scf. The next step was *water gas,* similar to our present-day synthetic gas in that it is mainly hydrogen and carbon monoxide, made at about 1800°F (1000°C) by the action of water vapor on coal. It has a heating value of 250 to 325 Btu/scf. This value is increased to 500 to 550 Btu/scf by carburetion with oil cracked at a higher temperature. *Producer gas* obtained from the partial combustion of coal, coke, or wood with added water vapor has a heating value of only 100 to 180 Btu/scf. It is used locally and is not suitable for distribution because distribution costs per Btu are inversely proportional to heating value. It was also used in "gasogens" on motor vehicles in some countries during World War II.

Several coal-gasification processes exist, including three that survived World War

II: the German Lurgi, the Koppers-Totzek, and the Wellman-Galusha processes. These have now evolved into processes that differ slightly, depending upon the particular rank of coal used.

The basic process involves several steps. First, the coal feedstock is crushed by usual methods (Sec. 4–6). If a *caking coal,** such as certain bituminous coals, is used, the feedstock will have to be pretreated by oxidizing its surface to avoid plugging the gasifier. The following step occurs in the gasifier where the coal undergoes chemical reaction with a mixture of air and steam or oxygen and steam. The reactions with air or oxygen are partial combustion because a rich mixture is used, i.e., one having a fuel-to-air ratio greater than chemically correct, or stoichiometric, which also means that the oxygen in either case is not sufficient to convert all the carbon to carbon dioxide. Coal gasification results in three gas mixtures, classified according to their heating value. They are called *low-Btu, medium-Btu,* and *high-Btu* gas. These are obtained in the following ways.

Low-Btu gas The feedstock is reacted with a mixture of air and steam. The air may be enriched in oxygen (i.e., oxygen-to-nitrogen ratio greater than atmospheric) but will be less than stoichiometric. The reactions taking place are

In air
$$C + O_2 + 3.76N_2 \rightarrow CO_2 + 3.76N_2 \qquad (4\text{-}9)$$

CO_2 from this reaction reacts further with additional C in the rich mixture to give

$$C + CO_2 + 3.76N_2 \rightarrow 2CO + 3.76N_2 \qquad (4\text{-}10)$$

In steam
$$C + H_2O \rightarrow CO + H_2 \qquad (4\text{-}11)$$

The result is a gas composed principally of CO, H_2, N_2, and some CO_2. The N_2 may be less than shown if the air is oxygen-enriched. The CO_2 appears if the air is increased beyond that shown or because of stratification (imperfect mixing) in the gasifier. It may also contain small amounts of H_2O, CH_4, and C_2H_6. The gas has a heating value range of 180 to 350 Btu/scf, depending upon the composition of reactants and resulting composition of products.

Medium-Btu gas The feedstock is burned with a mixture of oxygen (rather than air) and steam in the same reactions as given by Eqs. (4-9) to (4-11) but with the nitrogen removed. The result is a gas, composed principally of CO and H_2, that has a heating value range of 250 to 500 Btu/scf, again depending upon the composition of reactants and resulting products. The increase in heating value is a result of the absence of the diluting effect of nitrogen.

In the literature, the term *low-Btu gas* is sometimes used to refer to both low- and medium-Btu gases.

The next step in processing low- and medium-Btu gases is *quenching* to condense the tars and heavy oils that come with the feedstock and did not burn. This is followed by a *purification process* in which the hydrogen sulfide in the gas, formed by the

* A caking coal is one which softens and agglomerates as a result of the application of heat.

combination of the sulfur in the coal with hydrogen gas, is converted to elemental sulfur by an absorption process* and, simultaneously, a *cleaning process* in which char (solid coal residue, mainly fixed carbon and ash), dust, and ash are removed. These processes occur at low temperature with aqueous solutions at 100 to 220°F so that energy in the form of sensible heat of the product gases is lost to the environment.

High-Btu gas Purified medium-Btu gas may be converted to a high-Btu gas by two additional steps. The first is *shift conversion,* in which CO from the CO-rich gas is saturated with steam and passed through a catalytic reactor thus producing more hydrogen and carbon dioxide

$$CO + H_2O \rightarrow H_2 + CO_2 \tag{4-12}$$

The ratio of H_2 to CO_2 in the products can be changed by changing the composition of the reactants. The CO_2 is removed in a wash plant.

The next step is that of *methanation,* which is the production of methane, CH_4, from the available mixture of CO and H_2 in a catalytic reactor

$$3H_2 + CO \rightarrow CH_4 + H_2O \tag{4-13}$$

and the H_2O is removed. High-Btu gas is thus largely methane, with a heating value of 950 to 1000 Btu/scf, approaching that of natural gas, which is 950–1100 Btu/scf. It has all the characteristics of natural gas but without the sulfur and other pollutants.

Various methods under development for the production of low- and high-Btu gas may be found in the literature [32]. Because the production of high-Btu gas is more complex and expensive than low-Btu gas, it is intended for use in lieu of natural gas in domestic and industrial applications. Low- and medium-Btu gases are considered for use as utility fuels.

Purified low-Btu gas can be fired directly in a conventional steam generator. It has the advantages of being sulfur and ash free, thus eliminating the need for precipitators and scrubbers (Chap. 17). The lack of SO_2 in the flue gases also permits lower stack temperatures, which results in improved efficiency. In addition, its content of the CO_2 and H_2O inerts reduces combustion temperatures and hence the formation of nitrogen oxides. The disadvantages are the large volumetric rate of flow of the gas for a given heat input to the steam generator, the high costs, and the large demand for water as a source of steam and hydrogen and for cooling in the coal gasification process.

A Combined-Cycle Powerplant

An attractive utilization of low-Btu gas for electric generation is as a fuel for a combined-cycle powerplant. Such cycles are covered more fully in Chap. 8. However, a brief description of such a cycle integrated with a coal gasifier is appropriate at this

* In contrast to removal of SO_2 from the steam-generator flue gases, the removal of H_2S at this stage is easier and is accomplished by any of several commercially available and proven processes.

time. A *combined-cycle* is one which uses a gas turbine at the high-temperature end and a steam turbine at the low-temperature end.

Low-Btu gasifiers operate at different pressures and temperatures, depending upon the process used. Some operate at high pressures, up to 35 bar, and at exit temperatures between 1000 and 2000°F (540 to 1100°C). As indicated above, the exit gas must be cooled for purification and cleaning. Normally this cooling represents a large loss of energy to the environment. A combined cycle (Fig. 4-14) takes advantage of the high gasifier pressure, and recovers much of that heat loss by a gas-to-gas heat exchanger.

In one proposed design [33] the gas leaves the gasifier at 1 at about 1000°F (540°C) and 300 psia (20 bar), gives up some of its heat in a regenerative heat exchanger leaving at 2, where it is further cooled to 3 by an external heat exchanger to temperatures suitable for purification and cleaning from 3 to 4. The gas then recovers the heat it gave up in the regenerative heat exchanger leaving it at 5. It then enters a combustion

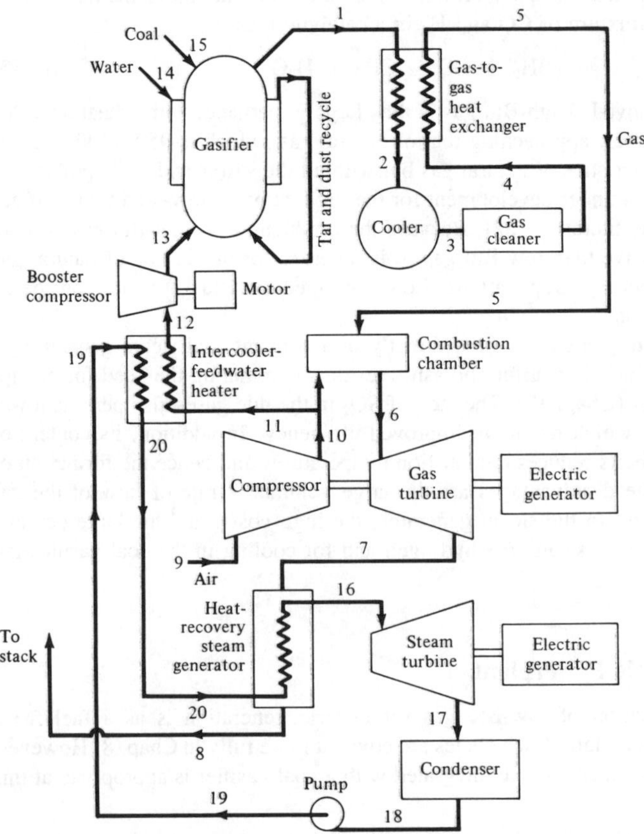

Figure 4-14 Coal gasifier combined-cycle powerplant [33].

chamber where it mixes with compressed air from the gas turbine–driven compressor and leaves at 6 at about 1800°F (980°C). It expands in the gas turbine, leaving it at 7 at about 965°F (520°C). It then enters a heat-recovery steam generator (HRSG) where it generates steam and leaves to the stack at 8 at about 260°F (125°C).

The gas turbine drives one of two electric generators and the compressor. The compressor receives atmospheric air at 9 at about 60°F (15°C) and compresses it to about 600°F (315°C). The compressor has a dual role: it supplies combustion air to the combustion chamber at 10 and gasifier air at 11. The latter is first cooled in a steam-cycle feedwater heater to 12, then boosted by a motor-driven compressor to the gasifier pressure at 13. The gasifier is designed to generate its own steam from feedwater 14. The feedstock 15 reacts there with the steam-air mixture to generate the low-Btu gas at 1.

The steam cycle is fairly standard. Superheated steam at 300 psia (20 bar) and 900°F (480°C) is generated in the HRSG at 16, expands through a steam turbine that drives the second electric generator, and exhausts at 17 to a condenser. The condensate at 18 is pumped, 19, to the feedwater heater which receives its heat from the compressed gasifier air. No bled turbine steam is used in this design, although that form of feedwater heating can be used. The feedwater then enters the HRSG at 20, completing the cycle.

The above cycle (with irreversibilities taken into account) shows an overall efficiency of 34.3 percent, which compares favorably with modern 2400 psi/1000°F/1000°F conventional steam powerplants with scrubbers. An advanced conceptual design with gas turbine inlet at 2800°F (not yet developed) and 2400 psi/1000°F steam shows an efficiency of 43.3 percent [33].

Coal Liquefaction

Coal liquefaction is the conversion of coal into a liquid fuel for direct energy production or a liquid substitute for refinery feedstock from which other liquid fuels may be obtained.

Coal-liquefaction technology, first researched in the 1920s and 1930s, was stimulated in both Germany and Japan by World War II. Japan produced aviation gasoline in a large plant in North Korea which converted coke from coal to calcium carbide in electric furnaces, then to acetylene, acetaldehyde, butyraldehyde, octanol, and finally octane. In a plant in Taiwan, the Japanese also used starch from root vegetables which they fermented to butanol, which was then converted to butyraldehyde, octanol, and octane. The most important German processes were the *Bergius,* which is no longer in use, and the *Fischer-Tropsch* process, which is still used commercially by the Union of South Africa's SASOL Corporation. South Africa has no indigenous oil of its own but has coal. Thanks to coal liquefaction, it is completely independent of foreign oil.

Long, complex hydrocarbon molecular chains have a lower hydrogen-carbon atomic ratio than shorter molecules, like that of octane. In coal liquefaction, the long molecules are shortened by adding hydrogen. The needed hydrogen is generated, and desulfurization is accomplished, in the same manner as for coal gasification. The Fischer-Tropsch process first produces a mixture of CO and H_2 from coal and steam. This is followed by catalytic reactions at about 300°F (150°C) and 150 bar, which yield a range of hydrocarbons from gaseous methane to higher liquid hydrocarbons. These are then separated with methane going as pipeline gas and the rest going to different liquid fuels.

Some half-dozen new processes are currently under development [34] in pilot plants. Scaling these up to commercial sizes is one of the major problems. Another problem, as in coal gasification, is the large demand for water that restricts the use of the large western coal reserves in the United States, which are not located near large supplies of water.

Oil Shale

Shale is a fine-grained rock, formed by the hardening of clay, that splits into thin layers when broken. *Oil shale* is *not* oil-impregnated shale but rather is a shalelike rock impregnated with a waxy organic material called *kerogen,* a substance which originated from vegetation that degraded over millions of years to oil that got absorbed in inorganic matter. Stratification under pressure produced the oil shale. Kerogen can be decomposed to yield synthetic crude oil, called *syncrude* or *shale oil,* by heating in retorts or by underground combustion. One ton of oil shale can produce 25 to 35 gal (95 to 130 L) of low-sulfur oil by retorting at about 950°F (500°C). This oil makes a good refinery feedstock that can be processed further to various useful liquid and gaseous fuels, although one low in fuel oil.

There are vast amounts of oil-shale deposits throughout the world, the largest being in the United States in Colorado, Wyoming, and Utah. It is said that the United States deposits can, by present technology, produce 50 times as much syncrude as all the natural crude produced in the United States to date. The first production was started in France around 1840. The only commercial (publicly owned) facilities nowadays exist in the Soviet Union and China.

Several pilot plants have been successfully operated in the United States, although commercial operations have run into major economic and practical difficulties, resulting in one major cancellation in 1982. Some of the current problems are (1) the large demand for water (3 times as much water is needed as oil produced) in the largely desert areas where oil shale is found in the United States; (2) environmental concerns arising from surface mining; (3) disposal of the spent rock (10 times the mass of the syncrude produced), which has increased in volume due to "puffing" upon heating; and (4) the large amounts of energy required for mining, transportation, processing, and disposal. Doubtless to say, such problems will be less restrictive in future decades when the cost and availability problems of natural crude make oil-shale plants competitive.

Tar Sands

Another potential and very rich source of oil, but one which is even less attractive than oil shale, is found in tar sands in such places as Alberta, Canada, about 10 degrees from the arctic circles. *Tar sand* is a thick, extremely viscous bitumen locked in sands and silt to form a sodden, sticky semiplastic material. It is believed that it contains 2 to 3 times the oil reserves of all the Middle East. Although small pilot plants have operated in the inhospitable terrain, the problems of commercial exploitation seem, at present, to be insurmountable.

One last word regarding synthetic fuels: Coal and other conversion plants, while they are designed to produce clean sulfur- and ash-free fuels, are *not* themselves pollution free. The plants generate enormous amounts of air pollutants such as CO_2, H_2S, SO_2, and NO_x; liquid effluents such as phenols, cresols, xylenols, thiocyanates, and ammonium sulfides; and solid wastes such as ash, slag, and sludge. In addition, carcinogenic compounds in the form of polycyclic aromatic hydrocarbons and amines may be produced from coal tars and coal-derived oils. It is obvious that very careful disposal schemes must be designed at the plant sites.

4-12 BIOMASS

Biomass is organic matter produced by plants, both terrestrial (those grown on land) and aquatic (those grown in water) and their derivatives. It includes forest crops and residues, crops grown especially for their energy content on "energy farms," and animal manure [35]. Unlike coal, oil, and natural gas, which take millions of years to form, biomass can be considered a renewable energy source because plant life renews and adds to itself every year. It can also be considered a form of solar energy as the latter is used indirectly to grow these plants by photosynthesis.

Biomass includes wood waste and bagasse, which have already been covered above, plus other matter. All are highly dispersed and bulky and contain large amounts of water (50 to 90 percent). Thus, it is not economical to transport them over long distances, and conversion into usable energy must take place close to the source, which is limited to particular regions. However, biomass can be converted to liquid or gaseous fuels, thereby increasing its energy density and making feasible transportation over long distances.

Terrestrial crops include (1) sugar crops such as sugar cane and sweet sorghum; (2) herbaceous crops, which are nonwoody plants that are easily converted into liquid or gaseous fuels; and (3) silviculture (forestry) plants such as cultured hybrid poplar, sycamore, sweetgum, alder, eucalyptus, and other hardwoods. Current research focuses on the screening and idenification of species that are suitable for short-rotation growing and on the optimum techniques for planting, fertilization, harvesting, and conversion.

Animal and human waste are an indirect terrestrial crop from which methane for combustion and ethylene (used in the plastics industry) can be produced while retaining the fertilizer value of the manure.

Aquatic crops are grown in fresh, sea, and brackish waters. Both submerged and emergent plants are considered, including seaweeds, marine algae, and of particular interest, the giant California kelp.

Biomass Conversion

Biomass conversion, or simply *bioconversion,* can take many forms: (1) direct combustion, such as wood waste and bagaase (above), (2) thermochemical conversion, and (3) biochemical conversion.

Thermochemical conversion takes two forms: gasification and liquefaction. *Gasification* takes place by heating the biomass with limited oxygen to produce low-Btu gas or by reacting it with steam and oxygen at high pressure and temperature to produce medium-Btu gas. The latter may be used as fuel directly or used in liquefaction by converting it to methanol (methyl alcohol CH_3OH) or ethanol (ethylalcohol CH_3CH_2OH), or it may be converted to high-Btu gas.

Biochemical conversion takes two forms: Anaerobic digestion and fermentation. *Anaerobic digestion* involves the microbial digestion of biomass. (An *anaerobe* is a microorganism that can live and grow without air or oxygen. It gets its oxygen by the decomposition of matter containing it.) It has already been used on animal manure but is also possible with other biomass. The process takes place at low temperatures, up to 65°C, and requires a moisture content of at least 80 percent. It generates a gas consisting mostly of carbon dioxide and methane with minimal impurities such as hydrogen sulfide. The gas can be burned directly or upgraded to synthetic natural gas by removing the CO_2 and the impurities. The residue may consist of protein-rich sludge that can be used as animal feed and liquid effluents that are biologically treated by standard techniques and returned to the soil.

Fermentation is the breakdown of complex molecules in organic compounds under the influence of a ferment such as yeast, bacteria, enzymes, etc. Fermentation is a well-established and widely used technology for the conversion of grains and sugar crops into ethanol. Some 60 million gal were produced in the United States in 1979, with capacity projected to 500 million gal per year by 1985 by the use of surplus grain. It is intended for mixing with gasoline to produce *gasohol* (90 percent gasoline, 10 percent ethanol). By the early 1980s the scheme had not met with great success because of the high cost and the high energy required in the process. One scheme considered for cutting costs of ethanol production by fermentation is in finding less expensive grains or sugars and a process that requires less energy. Glucose produced by hydrolysis of an abundant carbohydrate polymer called *lignocellulose* is being considered for the former.

Biomass energy concepts under study are resulting in the cultivation of large forests in areas not suitable for food production that may yield 10 to 20 tons/acre per year. The energy forest would perhaps be 50 to 200 mi^2. The trees are to be harvested by automated means, then chipped and pulverized for burning in a powerplant that would be located in the middle of the forest. Sycamore is a promising tree that yields up to 16 tons/acre per year. All of it is used except the foliage, which contains the nutrients and is returned to the soil. A harvested sycamore produces a number of sprouts that are themselves ready for harvesting in 2 to 3 years. Thus no replanting and little fertilization are necessary. It is estimated that an energy farm 350 mi^2 in area could produce 400 MW of electricity. The electricity costs are uncertain at this time and the possibilities depend upon the costs from other sources, such as oil or coal.

Other schemes envision aquatic farms growing algae, tropical grasses, floating kelp, water hyacinth (one of the most pernicious weeds in the world, one that shelters disease-carrying organisms, causes floods, disrupts hydroelectric plants, and interferes with traffic on major waterways such as the Nile and the Congo), and others. In controlled environments they could yield several hundred tons/acre year. One interesting idea is to use hot condenser cooling water to grow algae in large quantities or increase the yield of other crops.

Finally, while the efficiency of solar energy use in the growth of crops, the photosynthesis efficiency, is rather low, about 3–5%, means of increasing it to 10–11% are under study.

4-13 THE HEAT OF COMBUSTION

In Chap. 1 we treated the first law of thermodynamics from a mechanical engineering point of view, ignoring such energy terms as chemical, electrical, and magnetic. When dealing with combustion systems, we can no longer ignore the chemical energy in the fluid.

The Open System

The first law equation for the steady-state steady-flow system (SSSF) Eq. (1-1*d*) will now be written but for a chemically reactive system (Fig. 4-15), with changes in kinetic and potential energies ignored, as

$$H_R + \Delta Q = H_p + \Delta W_{sf} \tag{4-14}$$

where H_R and H_p are the enthalpies of the reactants and products, respectively, evaluated for their constituents at their respective pressures and temperatures; ΔQ is the net heat added to the system from the surroundings (which is usually negative because in combustion heat is usually rejected to the surroundings); and ΔW_{sf} is the net work done by the system, if any.

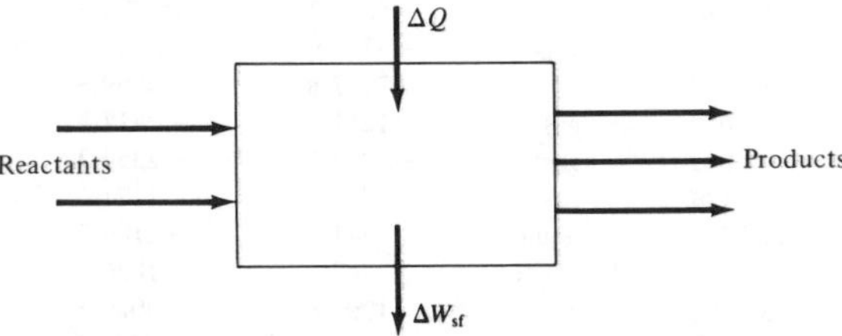

Figure 4-15 A steady-state steady-flow chemically reactive system.

Because the reactants and products are usually composed of several constituents each, Eq. (4-14) is written in the form

$$\sum_R (mh) + \Delta Q = \sum_P (mh) + \Delta W_{sf} \qquad (4\text{-}15)$$

where m is the mass and h the specific enthalpy of each constituent. To define the enthalpy, consider the complete combustion of ethane in oxygen

$$C_2H_6 + 3.5O_2 \rightarrow 2CO_2 + 3H_2O \qquad (4\text{-}16)$$

The enthalpies of the various reactants and products are those that start at the same datum of composition and temperature, in this case the elemental substances and the datum temperature commonly chosen as 25°C (77°F). For example, C_2H_6 is formed from elemental carbon C and hydrogen H_2, CO_2 from C and O_2, and so on. These are exothermic reactions that, when they begin at 25°C and are cooled back to 25°C after the reaction takes place, yield 1211.3 and 3846.7 Btu/lb$_m$ of product, respectively. In other words, in steady flow at 25°C, the formation reactions are

$$2C + 3H_2 \rightarrow C_2H_6 + 1211.38 \text{ Btu/lb}_m\ C_2H_6$$

and

$$C + O_2 \rightarrow CO_2 + 3846.7 \text{ Btu/lb}_m\ CO_2$$

Table 4-7 Enthalpies of formation h_f at 25°C (77°F) and 1 atm pressure*

				h_f	
Substance	Formula	M	State	Btu/lb$_m$	kJ/kg
Carbon	C	12.011	solid	0	0
Oxygen	O_2	32.000	gas	0	0
Hydrogen	H_2	2.016	gas	0	0
Nitrogen	N_2	28.016	gas	0	0
Sulfur	S	32.060	solid	0	0
Carbon monoxide	CO	28.011	gas	−1697.6	−3948.3
Carbon dioxide	CO_2	44.011	gas	−3846.7	−8946.8
Water	H_2O	18.016	liquid	−6825.7	−15,875.5
			vapor	−5774.6	−13,430.8
Methane	CH_4	16.043	gas	−2007.8	−4669.8
Ethane	C_2H_6	30.070	gas	−1211.3	−2817.3
Propane	C_3H_8	44.097	gas	−1013.1	−2356.3
Butane	C_4H_{10}	58.124	gas	−933.7	−2171.6
Octane	C_8H_{18}	114.230	liquid	−941.4	−2189.5
			vapor	−785.1	−1826.0
Nitric oxide	NO	30.008	gas	−1298.8	−3020.8
Nitrogen dioxide	NO_2	46.008	gas	−315.3	−733.3
Sulfur dioxide	SO_2	64.060	gas	−1992	−4632.8

* Based on data from Ref. 36.

The quantities 1211.3 and 3846.7 leave the system and hence are negative. They are given the name *enthalpies of formation*, h_f. Table 4-7 gives values of h_f for various substances at 25°C. Note that the elemental substances, C, O_2, etc., have zero enthalpies of formation. Also note that if the combustion equation involves a liquid, such as octane, or water there will be two enthalpies of formation, depending upon whether they start or end the reaction in a liquid or a vaporized state.

At temperatures other than the datum of 25°C, the enthalpies of formation must include the sensible heat that is a product of the temperature difference and $c_p(T)$, the specific heat at constant pressure which varies with temperature (Sec. 1-2). Table 4-8 gives values of the enthalpies of formation of several substances as a function of temperature.

Because chemical equations, such as Eq. (4-16), are balanced in terms of moles and not masses, and because $m = nM$, Eq. (4-15) is now written in the form

$$\sum_R (nMh_f) + \Delta Q = \sum_P (nMh_f) + \Delta W_{sf} \tag{4-17}$$

where n and M are the number of moles and molecular mass of each constituent, respectively.

Table 4-8 Enthalpies of formation h_f at different temperatures and 1 atm pressure, Btu/lb$_m$*†

Temperature, K	CO_2	CO	H_2O	O_2	H_2	N_2
298	−3846.7	−1697.6	−5774.6	0	0	0
400	−3807.5	−1651.9	−5692.2	40.7	631.4	52.8
500	−3765.4	−1606.5	−5609.4	81.8	1255.5	90.8
600	−3720.4	−1560.2	−5524.0	124.3	1880.4	136.5
700	−3673.1	−1513.0	−5435.9	168.1	2506.9	183.3
800	−3623.7	−1464.5	−5345.0	213.0	3137.4	231.0
900	−3572.6	−1415.0	−5251.1	258.8	3773.3	279.8
1000	−3520.2	−1364.5	−5154.3	305.3	4414.2	329.5
1100	−3466.5	−1313.1	−5054.3	352.5	5062.5	380.2
1200	−3411.9	−1261.0	−4951.4	400.2	5717.8	431.6
1300	−3356.4	−1208.2	−4845.7	448.4	6381.9	483.7
1400	−3300.2	−1154.8	−4737.2	497.0	7055.6	536.5
1500	−3243.4	−1100.9	−4626.2	546.0	7739.1	590.8
1600	−3186.1	−1046.6	−4512.8	595.3	8434.0	643.5
1700	−3128.3	− 992.0	−4397.1	646.0	9136.4	697.6
1800	−3070.1	− 937.0	−4279.6	694.9	9848.2	752.2
1900	−3011.6	− 881.7	−4160.1	745.3	10568	807.0
2000	−2952.8	− 826.1	−4038.9	795.9	11296	862.1
2100	−2893.6	− 770.3	−3916.1	846.8	12031	917.5
2200	−2834.2	− 714.3	−3791.8	929.3	12774	973.1
2300	−2774.6	− 658.2	−3666.2	949.6	13523	1028.9
2400	−2713.1	− 601.8	−3539.4	1001.5	14279	1084.9
2500	−2654.9	− 545.3	−3411.5	1053.7	15043	1141.1

* To convert to kJ/kg multiply by 2.32584.

† Based on data from [36].

Example 4-2 Find the useful heat generated by the combustion of 1 lb_m and 1 scf (standard cubic foot) of ethane in a furnace in a 20 percent deficient air if the reactants are at 25°C and the products at 1500 K. Assume that hydrogen, being more reactive than carbon, satisfies itself first with the oxygen it needs and burns completely to H_2O. Five percent of the heat of combustion is lost to the furnace exterior.

SOLUTION The stoichiometric equation for ethane in air is

$$C_2H_6 + 3.5O_2 + 13.16N_2 \rightarrow 2CO_2 + 3H_2O + 13.16N_2$$

(where there are 3.76 mol N_2/mol O_2 in atmospheric air, thus $13.16 = 3.5 \times 3.76$). With 20 percent deficient air multiply the O_2 and N_2 mol by 0.8. H_2 will burn completely to H_2O and C will burn partially to CO_2 and partially to CO:

$$C_2H_6 + 2.8O_2 + 10.528N_2 \rightarrow aCO_2 + bCO + 3H_2O + 10.528N_2$$

Carbon balance: $$a + b = 2$$

Oxygen balance: $$a + \frac{b}{2} + \frac{3}{2} = 2.8$$

Thus $a = 0.6$, $b = 1.4$, and the combustion equation is

$$C_2H_6 + 2.8O_2 + 10.528N_2 \rightarrow 0.6CO_2 + 1.4CO + 3H_2O + 10.528N_2$$

As there is no work done in a furnace, Eq. (4-17) is written as

$$\Delta Q = \sum_P (nMh_f)_{1500K} - \sum_R (nMh_f)_{25°C}$$

$$\sum_P (nMh_f)_{1500K} = 0.6 \times 44.011 \times -3243.4 + 1.4 \times 28.011 \times -1100.9$$

$$+ 3 \times 18.016 \times -4626.2 + 10.528 \times 28.016 \times +590.8$$

$$= -85647.2 + -43172.2 + -250036.9 + -174257.9$$

$$= -204{,}598.4 \text{ Btu/(lb·mol } C_2H_6)$$

$$\sum_R (nMH_f)_{25°C} = -1211.3 + 0 + 0 = -1211.3 \text{ Btu/(lb·mol } C_2H_6)$$

Thus $$\Delta Q = -204{,}598.4 - -1211.3 = -203387.1 \text{ Btu/(lb·mol } C_2H_6)$$

$$= \frac{-203{,}387.1}{30.070} = -6763.8 \text{ Btu/lb}_m\ C_2H_6$$

$$= -6763.8 \times 2.32584 = -15731.5 \text{ kJ/kg } C_2H_6$$

$$\Delta Q_{useful} = 0.95 \times -6763.8 = -6425.6 \text{ Btu/lb}_m\ C_2H_6$$

$$= -14{,}944.9 \text{kJ/kg } C_2H_6$$

A standard cubic foot is obtained at 1 atm and 60°F. Since C_2H_6 is a gas, its density at these conditions is obtained from $PV = mRT$

$$\text{or} \qquad \rho = \frac{m}{V} = \frac{P}{RT} = \frac{14.696 \times 144}{(1545/30.70) \times (60 + 460)} = 0.0792 \text{ lb}_m/\text{scf}$$

$$\text{Thus} \qquad \Delta Q_{\text{useful}} = -6425.6 \times 0.0792 = -508.9 \text{ Btu/scf}$$

The Closed System

Combustion equations for fuels burning in a closed system, such as a cylinder or bomb, may be obtained by writing the first-law equation for the closed system. Thus, by analogy with Eq. (4-14)

$$U_R + \Delta Q = U_P + \Delta W_{\text{nf}} \tag{4-18}$$

when ΔW_{nf} is the nonflow work. For gases

$$H \equiv U + PV = U + nR_oT$$

where R_o is the universal gas constant. Thus combustion calculations for a closed system may be carried out using the enthalpies in Tables 4-6 and 4-7 by modifying Eq. 4-18 to

$$\sum_R (nMh_f - nR_oT) + \Delta Q = \sum_P (nMh_f - nR_oT) + \Delta W_{\text{nf}} \tag{4-19}$$

4-14 HEATING VALUES

We have repeatedly used the term *heating value* in this chapter to indicate the useful energy content of different fuels. Actually there is more than one heating value for each fuel and these should be carefully differentiated. We have already explained the difference between the *higher heating value* (HHV) and the *lower heating value* (LHV). The former is the heat released when water vapor in the products due to the combustion of hydrogen in the fuel condenses, the latter when it stays in the vapor state (Sec. 4-3). Heating values are commonly tabulated in the literature as the heat released when complete combustion begins at a standard temperature, such as 77°F or 25°C, and the products are cooled to the *same* temperature, in a steady-flow adiabatic system without work. Thus it can be calculated from Eq. (4-17) by putting $\Delta W_{\text{sf}} = 0$ and replacing ΔQ by HV for heating value

$$\text{HV} = \Big[\sum_P (nMh_f) - \big(\sum_R nMh_f\big)\Big]_{T_1} \tag{4-20}$$

where T_1 represents the standard temperature.

The heating value given by Eq. (4-18) is for an open system and is thus sometimes called the *enthalpy of combustion*. It depends upon T_1; upon whether the water vapor formed in combustion condenses (in which case it is the HHV), or not (in which it is the LHV); and upon whether the reactant fuels were in a liquid or vapor state, as it takes a certain amount of energy to vaporize a liquid fuel such as octane.

Example 4-3 Find the higher and lower heating values of ethane at 25°C (77°F).

SOLUTION Using Eqs. (4-16) and (4-20) and Table 4-6, HHV is calculated by assuming that H_2O in the products has condensed.

$$\sum_P (nMh_f)_{25°C} = 2 \times 44.011 \times -3846.7 + 3 \times 18.016 \times -6825.7$$

$$= -707{,}509.6 \text{ Btu/(lb·mol } C_2H_6)$$

$$\sum_R (nMh_f)_{25°C} = 1 \times 30.070 \times -1211.3 + 3.5 \times 32.00 \times 0$$

$$= -36.423.8 \text{ Btu/(lb·mol } C_2H_6)$$

$$\text{Thus HHV} = -707{,}509.6 - -36{,}4238 = -671{,}085.8 \text{ Btu/(lb·mol } C_2H_6)$$

$$= \frac{-671{,}085.5}{30.070} = -22{,}317.5 \text{ Btu/lb}_m\ C_2H_6$$

$$= -22{,}317.5 \times 2.32584 = -51{,}906.8 \text{ kJ/kg } C_2H_6$$

LHV is calculated by using $h_f = -5774.6$ Btu/lb$_m$ for H_2O vapor instead of -6825.7 Btu/lb$_m$ for H_2O liquid, resulting in

$$\text{LHV} = -614{,}276.0 \text{ Btu/(lb·mol } C_2H_6)$$

$$= -20{,}428.2 \text{ Btu/lb}_m = -47{,}512.7 \text{ kJ/kg}$$

Note that in calculating the heating values, writing the complete combustion equation in oxygen alone suffices because the addition of nitrogen on both sides of the equation with h_f the same at T_1 (0 at 25°C) does not alter the results. Note also that the same results are obtained with stoichiometric and lean mixtures (excess oxygen or air), so long as complete combustion (C to CO_2, H_2 to H_2O, etc.) is assured.

Heating values for complex fuels such as coal and gas mixtures are obtained by writing the stoichiometric equation for the fuel and following a similar procedure as above.

Heating values for a closed system are obtained by modifying Eq. (4-19) to

$$\text{HV}_{\text{nf}} = \sum_P (nMh_f - nR_oT) - \sum_R (nMh_f - nR_oT)$$

and since the temperature of the products is brought back to that of the reactants T_1

$$\text{HV}_{\text{nf}} = \text{HV} - (n_P - n_R)R_oT_1 \tag{4-21}$$

where HV_{nf} = nonflow, a closed system heating value

HV = heating value as defined for a steady-flow system

n_p and n_R = total number of moles of products and reactants, respectively

HV_{nf} is sometimes called the *internal energy of combustion*. Needless to say, there are as many HV_{nf}s as there are HVs; that is, they both depend upon T_1, whether the water vapor formed in the combustion process condenses or not, so there are HHV_{nf} and LHV_{nf}, and whether fuels begin the reaction in a liquid or vapor state.

In the literature, heating values of different fuels are usually listed for an open system and for a given T_1, which unfortunately varies from source to source. For example, 60°F and 77°F (25°C) are used, though the numerical differences between them are fairly small.

4-15 COMBUSTION TEMPERATURES

If the heat of combustion or a portion of it is kept in the gases, it will have the effect of increasing their enthalpy (internal energy in a closed system) and hence their temperature. The high temperature is then normally used to effect heat transfer from the gases to the heat-transfer surfaces (water tubes, superheaters, and reheaters in a steam generator) by radiation and/or convection. The same basic first-law equation used to calculate heats of combustion and heating values of fuels can be used to calculate combustion temperatures.

Example 4-4 A low-Btu gas from a coal-gasification plant has the following volumetric analysis: CO 22.9 percent, H_2 10 percent, CO_2 4.4 percent, N_2 62.7 percent. It enters the furnace at 1 atm and 800 K and burns in 100 percent air at 25°C in a steady-flow combustion process during which it loses 10.72 Btu/ft^3 at inlet conditions to the surroundings. Find the temperature of the products.

SOLUTION Recalling that volumetric and molal analyses are equal for gases, the stoichiometric combustion equation (for 100 percent air) is

$$0.229CO + 0.100H_2 + 0.044CO_2 + 0.627N_2 + \left(0.229 + \frac{0.100}{2}\right)O_2$$

$$+ \left(0.229 + \frac{0.100}{2}\right) 3.76N_2 \rightarrow (0.229 + 0.044)CO_2 + 0.100H_2O$$

$$+ \left[\left(0.229 + \frac{0.100}{2}\right)3.76 + 0.627\right]N_2$$

or $$(0.229CO + 0.100H_2 + 0.044CO_2 + 0.627N_2) + 0.279O_2 + 1.049N_2$$
$$\rightarrow 0.273CO_2 + 0.100H_2O + 1.676N_2$$

Using Eq. (4-17)

$$\sum_P (nMh_f)_{T2} = \sum_R (nMh_f) + \Delta Q$$

Using $PV = nR_oT$, 1 mol of gas, between parentheses in the combustion equation above, occupies

$$V = \frac{R_oT}{P} = \frac{1545\ (800 \times 1.8)}{14.696 \times 144} = 1051.3\ \text{ft}^3$$

Thus $\Delta Q = -10.7 \times 1051.3 = 11{,}270$ Btu/mol gas.

$\sum_R (nMh_f)$ are figured at 800 K for the fuel and 25°C for the air. Thus

$$\sum_R (nMh_f) = 0.229 \times 28.011 \times -1464.5 + 0.100 \times 2.016 \times 3137.4$$
$$+ 0.044 \times 44.011 \times -3623.7 + 0.627 \times 28.016 \times 231.0$$
$$+ 0 + 0 = -11{,}721\ \text{Btu/(lb·mol fuel)}$$

Therefore

$$\sum_P (nMh_f) = -11{,}721 - 11{,}270 = -22{,}991\ \text{Btu/(lb·mol fuel)}$$

It is now necessary to solve the problem by finding a value for the combustion temperature that results in a $\sum_P (nMh_f)$ of $-22{,}991$ Btu/(lb·mol fuel). This value is found by trial and error to be 1400 K. To confirm

$$\sum_p (nMh_f)_{1400\ \text{K}} = 0.273 \times 44.011 \times -3300.2 + 0.1 \times 18.016 \times -4737.2$$
$$+ 1.676 \times 28.016 \times 536.5 = -22{,}995\ \text{Btu/(lb·mol fuel)}$$

Thus the combustion temperature is 1400 K = 1127°C = 2520°R = 2060°F.

The Adiabatic-Combustion Temperature

It can be seen from Example 4-4 that a lean mixture, e.g., 120 percent combustion air, would yield a lower combustion temperature because of the dilution effect of the excess air. A rich mixture would also yield a lower temperature because of incomplete combustion.

A fuel burning with no heat exchange with the surroundings and no work done will result in the *adiabatic-combustion* or *adiabatic-flame temperature*. It is greater if the fuel is burned in oxygen than in air because of the dilution effect of nitrogen and greater for a stoichiometric than either a lean or rich mixture because, as above, a lean mixture has a dilution effect, whereas a rich mixture results in incomplete combustion.

At the adiabatic-flame temperature, $\Delta Q = 0$ and

$$\sum_P (nMh) = \sum_R (nMh) \tag{4-22}$$

Example 4-5 A desulfurized coal has a moisture-ash-free ultimate analysis of C 81.1 percent, H_2 4.4 percent, O_2 2.7 percent, N_2 1.8 percent on a mass basis. It burns in 120 percent air. The reactants were at 25°C. Find the adiabatic-flame temperature.

SOLUTION The relative number of moles of each constituent is found by dividing its mass percent by its molecular mass M. Thus the molal composition of the coal would be

$$\frac{0.811}{12.011}C + \frac{0.044}{2.016}H_2 + \frac{0.027}{32}O_2 + \frac{0.018}{28.016}N_2$$

or

$$0.06752C + 0.02183H_2 + 0.00084O_2 + 0.00064N_2$$

Normalizing to 1 mol of coal gives

$$0.7434C + 0.2403H_2 + 0.0092O_2 + 0.0071N_2$$

The stoichiometric combustion equation is

$$\begin{aligned}0.7434C + 0.2403H_2 + 0.0092O_2 + 0.0071N_2 + \left(0.7434 + \frac{0.2403}{2}\right.\\ \left.- 0.0092\right)O_2 + \left(0.7434 + \frac{0.2403}{2} - 0.0092\right)3.76N_2 \rightarrow 0.7434CO_2\\ + 0.2403H_2O + \left(0.7434 + \frac{0.2403}{2} - 0.0092 + 0.0071\right)N_2\end{aligned}$$

or

$$\begin{aligned}0.7434C + 0.2403H_2 + 0.0092O_2 + 0.0071N_2 + 0.85435O_2\\ + 3.21236N_2 \rightarrow 0.7434CO_2 + 0.2403H_2O + 3.21946N_2\end{aligned}$$

120 percent air means that atmospheric air is increased by 20 percent. The combustion equation is therefore

$$\begin{aligned}&0.7434C + 0.2403H_2 + 0.0092O_2 + 0.0071N_2 + (1.20 \times 0.85435)O_2\\ &+ (1.2 \times 3.21236)N_2 \rightarrow 0.734CO_2 + 0.2403H_2O\\ &+ (0.2 \times 0.85435)O_2 + (1.2 \times 3.21236 + 0.0071)N_2\end{aligned}$$

or

$$\begin{aligned}0.7434C + 0.2403H_2 + 0.0092O_2 + 0.0071N_2 + 1.02522O_2\\ + 3.85483N_2 \rightarrow 0.7434CO_2 + 0.2403H_2O + 0.17087O_2 + 3.86193N_2\end{aligned}$$

All reactants in this particular case are made up of elemental substances (not true if the fuel contained complex molecules such as a hydrocarbon), so that at 25°C

$$\sum_R (nMh_f) = 0$$

The temperature of the products that yields $\sum_P (nMh_f) = 0$ is now obtained by trial and error. At 2100 K

$$\begin{aligned}\sum_P (nMh_f) &= 0.7434 \times 44.011 \times -2893.6 \\ &\quad + 0.2403 \times 18.016 \times -3916.1 \\ &\quad + 0.17087 \times 32 \times 846.8 \\ &\quad + 3.86193 \times 28.016 \times 917.5 = -7725\end{aligned}$$

Similarly, at 2200 K

$$\sum_P (nMh_f) = 1221 \text{ Btu/(lb·mol fuel)}$$

By interpolation, the adiabatic-flame temperature = 2186 K = 3935°R = 3475°F.

The combustion temperatures as calculated in the previous two examples are higher than actual because *dissociation* of some of the products takes place at high temperatures. For example, some CO_2 dissociates to CO and $\frac{1}{2}O_2$, which is an endothermic reaction that lowers the temperature. *Chemical equilibrium* occurs at a certain temperature when the reaction rate is the same in both directions, i.e., when

$$CO + \frac{1}{2}O_2 \rightleftarrows CO_2 \qquad (4\text{-}23)$$

and

$$H_2 + \frac{1}{2}O_2 \rightleftarrows H_2O \qquad (4\text{-}24)$$

The effect of dissociation on the theoretically obtained temperatures above is greater the higher the temperatures but is lower for lean mixtures because the excess oxygen tends to drive the reactions toward completion and the higher the pressure. Thus the effect of dissociation is minimal in steam-generator furnace combustion, which occurs with lean mixtures around 3000°F.

Finally, it should be recalled that all combustion calculations can be done if there is heat exchange to or from the environment (as in Example 4-4), work done by or on the system, or even changes in kinetic or potential energy between products and reactants by including the proper terms into Eq. (4-17).

PROBLEMS

4-1 A sample of coal has the following molal analysis C 67.35%, H_2 26.26%, O_2 2.28%, N_2 0.57%, S 1.37%, H_2O 2.17%. Write the complete combustion equation in stoichiometric air and calculate the coal ultimate analysis, mass percent.

4-2 Write the complete combustion equation for the anthracite coal in Table 4-2, assuming stoichiometric air and find the dew point, degrees centigrade, of the combustion products if the total pressure is 1 bar.

4-3 H_2 burns in pure oxygen in a chemically correct (stoichiometric) mixture. Write the combustion equation and calculate (*a*) the mass of products per unit mass of H_2, and (*b*) the lower heating value of H_2 if its higher heating value is 61,100 Btu/lb_m.

4-4 Calculate the higher and lower heating values, in Btus per pound mass, using the Dulong-type formula, of the anthracite coal in Table 4-2, if the total pressure is 1 atm.

4-5 A gaseous fuel that is derived from coal (Sec. 4-11) has the following ultimate volumetric analysis: H_2 47.9%, methane (CH_4) 33.9%, ethylene (C_2H_4) 5.2%, CO 6.1%, CO_2 2.6%, N_2 3.7%, and O_2 0.6%. It burns in 110 percent of theoretical air. Calculate (*a*) the volume flow rate of air required per unit volume flow rate of the gas when both are measured at the same pressure and temperature, and (*b*) the dew point of the combustion products, in degrees fahrenheit, if the total pressure is 2 atm.

4-6 10,000 U.S. gal of a fuel oil are burned per hour in 20 percent excess air. The fuel oil has the following ultimate analysis by mass: C 87%, S 0.9%, H_2 12%, ash 0.1%. Write the combustion equation and find the volume flow rate of air required, in cubic feet per minute, if the fuel has a density of 7.73 lb_m/ U.S. gal and the air is at 1 atm and 60°F.

4-7 A southern California natural gas has the following ultimate analysis by mass: H_2 23.3%, C 74.72%, N_2 0.76%, and O_2 1.22%. The flue gases have the following volumetric analysis: H_2O 15.583%, CO_2 8.387%, O_2 3.225%, N_2 72.805%. Calculate (*a*) the percent theoretical air used in combustion and (*b*) the dew point, in degrees centigrade, if the flue gases are at 2 bars.

4-8 A fuel oil composed only of carbon, hydrogen, and sulfur is used in a steam generator. The volumetric flue gas analysis on a dry basis is: CO_2 11.7%, CO 0.440%, O_2 4.002%, SO_2 0.176, and N_2 83.682. Find (*a*) the fuel mass composition, (*b*) the air-fuel ratio by mass, (*c*) the excess air used, in percent, and (*d*) the dew point, in degrees centigrade, of the flue gases if their pressure is 2 bar.

4-9 A fuel oil burned in a steam generator has a composition which may be represented by $C_{14}H_{30}$. A dry-basis flue-gas analysis shows the following volumetric composition: CO_2 11.226%, O_2 4.145%, CO 0.863%, N_2 83.766%. Write the complete combustion equation for 1 mol of fuel and calculate (*a*) the air-to-fuel ratio by mass, (*b*) the excess air, in percent, and (*c*) the mass of water vapor in the flue gases per unit mass of fuel.

4-10 A crushed bituminous coal to be used in a fluidized-bed combustion chamber varies in size between 1/4 and 3/4 in and has a density of 80 lb_m/ft^3. The coefficient of drag when fluidized is 0.60. Calculate (*a*) the minimum gas velocity that fluidizes all the coal if the gas is at 1600°F and 9-atm pressure, and (*b*) the pressure drop in the bed, psi. Assume that the coal in the collapsed state has a height of 2 ft and a porosity of 0.25 and that the gas density can be approximated by that of pure air.

4-11 10,000 tons of coal are burned in a powerplant per day. The coal has an as-received ultimate analysis of C 75%, H_2 5%, O_2 6.7%, H_2O 2.5%, S 2.3%, N_2 1.5%, ash 7.0%. It burns in excess air in a fluidized-bed combustor. Calculate (*a*) the mass of calcium carbonate to be added, in tons per day, and (*b*) the mass of calcium sulfate to be disposed of, in tons per day. (The molecular mass of calcium = 40.)

4-12 Write the chemical formula and sketch the molecule for the following hydrocarbons: (*a*) ethane, (*b*) ethene or ethylene, (*c*) decane, (*d*) iso-decane (2,2,3,3 tetramethyl hexane), (*e*) pentatriacontane (do not sketch), (*f*) isobutene (2-methyl propene), (*g*) 1,5-heptadiene (the numbers indicate the positions of the carbon atoms that precede double bonds), (*h*) cyclohexane, (*i*) naphthalene, (*j*) 1-methyl napthalene (a methyl radical CH_3 attached to a carbon atom instead of a hydrogen atom), (*k*) tetracontane (do not sketch), and (*l*) dotriacontahectane (do not sketch).

4-13 A coal-oil mixture (COM) is composed of the bituminous medium volatility coal listed in Table 4-2 and a distillate oil no. 2 that has an ultimate analysis on a mass basis of C 87.2%, H_2 12.5%, S 0.3%, N_2

0.02%, and negligible O_2 and ash, and has a higher heating value of 19,430 Btu/lb$_m$. The mixture is 50-50 on a heating-value basis. It burns in 15 percent excess air. Write the complete combustion equation and calculate the air-to-fuel ratio by mass.

4-14 Purified medium-Btu gas at 77°F is burned in 110 percent theoretical air. It leaves the combustion chamber at 1700°F. 5 percent of the heat released is lost to the environment. Calculate the useful heat added to the chamber in Btus per pound mass and Btu per standard cubic foot (at 14.696 psia and 60°F) of the gas.

4-15 A purified low-Btu gas at 77°F burns in stoichiometric air and gives off 52.49 Btu/lb$_m$in the combustion chamber. Calculate the exit temperature, in degrees Fahrenheit.

4-16 Gaseous propane at 77°F is mixed with air at 300°F and ignited. What percentage theoretical air must be used if the temperature of the products is 1700°F? Assume an adiabatic process.

4-17 100,000 standard cubic feet per minute of a purified low-Btu gas enters a gas turbine combustion chamber at 440°F, where it burns adiabatically in 150 percent theoretical air. It then drives a gas turbine and leaves it at 1160°F. Calculate the power input to the turbine in megawatts.

4-18 Ethane at 25°C burns completely in stoichiometric air in a vertical cylinder so that a piston is pushed upward. The cylinder diameter is 0.15 m and the piston mass is 20 kg. The air pressure above the piston is 1 bar. The initial cylinder volume is 0.1 m^3. Calculate the final cylinder volume, in cubic meters, if the products of combustion at the end of expansion were at 1600 K. (The Universal gas constant in SI units is 8314.3 J/kg·mol·K.)

4-19 300 lb$_m$/h of liquid octane is burned in an internal combustion engine with 125 percent theoretical air. The initial temperature is 77°F and the products leave the cylinder at 1160°F. 85% C burns to CO_2; the rest burns to CO. (*a*) Write the complete combustion equation for 1 mol of C_8H_{18}, (*b*) find the heat released in the process, Btu/mol C_8H_{18}, and (*c*) find the horsepower developed if the engine efficiency is 15 percent.

4-20 A chemically correct mixture of gaseous methane and air at 77°F is admitted into a nozzle where it burns completely. Calculate the nozzle exit velocity in feet per second if the exit temperature is 2000 K.

4-21 Calculate the higher and lower heating values in Btus per pound mass at 25°C of fuel oil no. 2 (Table 4-5), assuming that O_2 and N_2 have mass percentages of 0.1 each.

4-22 Gaseous propane is mixed with air at 77°F and burned. The products reached a temperature of 1520°F. There was a loss of 10 percent of the reactants energy by heat transfer. Calculate the percentage theoretical air used.

4-23 Calculate the adiabatic flame temperature, in degrees fahrenheit, for the subbituminous coal in Table 4-2 if it burns from 77°F with 100 percent air. Ignore the effect of sulfur on the energy balance.

CHAPTER

FIVE

TURBINES

5-1 INTRODUCTION

Using steam to provide mechanical work probably owes its birth to the need for pumping water from coal mines. The very first successful attempt at this was a "pumping engine" built by Thomas Savery (1650–1715) in England. In Savery's engine, steam at pressures between 50 and 100 psig (4.5 to 8 bar) acted directly upon the surface of water in a chamber to force it up through a pipe. A crude check valve prevented reverse flow. After the chamber was cleared of water, steam supply was manually cut off and cold water was applied over the chamber, thus condensing the steam inside and creating a vacuum that sucked in more water. The direct contact between steam and water caused large condensation losses, and the lack of safety valves was responsible for many explosions.

At about the same time, Denis Papin (1647–1712), who invented the safety valve, conceived of the idea of separating the steam and water by a piston, and Thomas Newcomen (1663–1729) designed and built an engine with one. In it low-pressure steam was admitted to a vertical cylinder, where it pushed the piston upwards. The steam left in the cylinder was then condensed by a jet of outside cold water, creating a vacuum in the cylinder. The outside atmospheric pressure pushed the piston back on the working stroke, hence the name "atmospheric engine." The piston was attached to one end of a beam that had a fulcrum about midpoint. A piston in a separate pumping cylinder was attached to the other end. The pump piston was smaller in diameter than the steam piston, so a greater water pressure than steam pressure was obtained. The various valves of Newcomen's engine were operated manually at first. Automatic operation, the first on record, was conceived by a small lad who was hired to operate the valves. Being smaller and lazier than the others, as the story goes, he noticed the regular pattern of beam and valve operation and rigged up a string mechanism that allowed the beam to actuate the valves. Newcomen's engine used one-third less coal per hph than Savery's.

It was not until some 60 years later that James Watt* developed the ideas of the "modern" reciprocating steam engine. Working as an instrument maker, he was called upon one day in 1764 to repair a Newcomen engine and noticed the waste of steam condensed in the cylinder. In 1765 he conceived of the idea of a separate condenser, and subsequently ideas such as the working stroke caused by steam expansion, the double-acting cylinder, the flyball throttling governor, the conversion of reciprocating to rotary motion (in 1781), and other important features. His now-famous engine was a major contributor to the industrial revolution. Watt's engine used 60 percent less coal than Newcomen's and 75 percent less than Savery's.

The next major improvement was made by Corliss (1817–1888), who developed the quick-closing intake valves that bear his name which reduce throttling during closing. The Corliss engine used about half as much coal as Watt's but still four or five times as much as modern steam-turbine powerplants. Next came Stumpf (1863–?), who developed the "uniflow engine," which was designed to further reduce condensation losses.

The reciprocating steam engine reached its pinnacle in size when it was called upon to drive the then-giant 5-MW electric generators early in the twentieth century. No larger engines were built then or after, although performance continued to improve mainly with the uniflow engine. But about that time electric generators were capable of getting bigger with no reciprocating engines large enough to drive them. Enter the steam turbine, not a new idea by any means, for the need for it was foreseen by many inventors in the late 1800s. Like many great inventions, it became practical when the world needed it.

Actually, history's first recorded steam turbine was one built by Hero of Alexandria about the first century A.D. It consisted of a hollow sphere that was free to turn about a horizontal axis between two fixed tubes that connected it to a caldron or boiler (Fig. 5-1). Steam, generated in the caldron, enters the sphere and exits to the atmosphere tangentially via two nozzles situated in the plane perpendicular to the axis of rotation and pointing in opposite directions. The steam leaving the nozzles causes the sphere to rotate in a manner similar to that of water ejecting from a lawn sprinkler. Hero's turbine thus operated on the *reaction principle* (Sec. 5-3). Much later, about 1629, a steam turbine used a jet of steam that impinged on blades projecting from a wheel,

* James Watt (1736–1819) was a Scottish self-educated instrument maker and inventor. Besides his famous engine, he is credited with the first meaningful research on the properties of steam. He ran into considerable financial difficulties that necessitated borrowing money from benefactors before his engine became a financial success. In 1769 he took a patent for "A New Invented Method of Lessening the Consumption of Steam and Fuel in Fire Engines." In 1775 he went into partnership with Matthew Boulton, a manufacturer in Birmingham. The partnership lasted 25 years. Unlike many great inventors, his achievements were recognized during his lifetime. He was awarded a doctor of laws degree from Glasgow University in 1806, was made a foreign associate of the French Academy of Sciences in 1814, and was offered a baronetcy, which he declined because of modesty.

Figure 5-1 The aeolipile of Hero of Alexandria (from *Aeolus*, "god of the winds," and *pila*, a "ball") the first recorded steam turbine in history.

thus causing it to rotate. This turbine operated on the *impulse principle* (Sec. 5-2). Later yet, in 1831, William Avery of the United States built the first steam turbines, which were used commercially in sawmills and woodcutting shops, with at least one tried on a locomotive. The Avery turbine had similarities to Hero's in that it used a hollow shaft with two 2.5-ft-long hollow arms attached to it at right angles with a small opening at the end of each and each pointing in opposite directions. Steam supplied through the hollow shaft exited through the openings and caused the shaft to rotate. Avery's, like Hero's, was therefore a reaction turbine. The turbines, though claiming similar efficiencies as contemporary reciprocating steam engines, were abandoned because of high noise level, difficult regulation, and frequent breakdowns.

The turbines that were destined to take over from the reciprocating engine, however, came about late in the nineteenth century as a result of the efforts of a handful of men, the most prominent of whom were Gustav de Laval* of Sweden and Charles Parsons† of England. de Laval first developed a small, high-speed (42,000 r/min)

* Carl Gustav Patrick de Laval (1845–1913), an engineering graduate of the University of Upsala, Sweden, was an inventor whose main income came from a cream separator and was spent on various other unprofitable inventions. The turbine he invented was intended for a cream separator. Also active in public affairs, he became a member of both houses of parliament and was honored repeatedly for his contributions to technology.

† Sir Charles Algernon Parsons (1854–1931), an upper-class Englishman, was motivated by the need to find a steam drive for ships. He is credited with developing the *reaction-stage principle*.

reaction turbine but did not consider it practical and so turned to the development of a reliable single-stage impulse turbine, which bears his name until today. He is also credited with being the first to employ a convergent-divergent nozzle for use in that turbine. The first unit was tested in 1890 and the first commercial unit, 5 hp, went into service in 1891. In 1892 he built a 15-hp turbine with two wheels for ships: one for forward and one for astern propulsion. Parsons developed the multistage reaction turbine, a low-speed machine for use in ships. The first Parsons turbine was built in 1884. The first ship ever to use a turbine was launched in 1895 and, naturally, was called the "Turbinia." It also had two sets of elements, one for forward and one for astern propulsion. Later on some turbine-generator sets were installed in both ships and powerplants.

In addition to de Laval and Parsons, C. E. A. Rateau of France developed the multistage impulse principle (pressure staging), Charles G. Curtis of the United States developed the velocity-compounded impulse stage, and George Westinghouse, also of the United States, secured American rights to the Parson's turbine and installed the first U.S. commercial units of 400-kW capacity at the Westinghouse Air Brake Co., Wilmerding, Pennsylvania.

Shortly after the turn of the century, steam turbines began to replace reciprocating steam engines in powerplants. Rapid development ensued. By 1909 12-MW units were installed in the Fisk powerplant in Chicago. The turbine performance and efficiency exceeded those of the reciprocating engine and allowed the use of superheated steam on an expanding scale. This led to the use of cast steel rather than cast iron in turbines. Capacities rose steadily. A 208-MW unit was installed in New York in 1929. The rise was helped in 1937 by the use of hydrogen-cooled, 3600-r/min electric generators. By the late 1950s capacities reached 450 MW. In the post-World War II era, capacities rose beyond 1000 MW and 3600 r/min became a standard in the United States for 60-Hz current (3000 for 50-Hz current in most of the rest of the world) for high-pressure units and 1800 r/min (1500 non-U.S.) for the low-pressure units used in water-cooled nuclear reactors. The steam turbine today is the mainstay of electric-power generation and promises to continue in that role for the foreseeable future.

Gas turbines are as old as windmills since a windmill is essentially a gas (air) turbine. An early gas device called the *smokejack* was operated by hot gases rising through a fireplace chimney. The smokejack is believed to have been first sketched by Leonardo da Vinci and later described by John Wilkins, an English clergyman, in his book *Mathematical Magick* in 1648. Other attempts followed, including work by John Barber of England, who received a patent in 1871 for a device that compressed air and produced gas in a cylinder and then burned and directed the mixture to a turbine wheel through nozzles. The first significant advance was by F. Stolze of Germany, whose turbine consisted of components similar to those of today's gas turbines: a separately fired combustion chamber and a multistage axial-flow compressor coupled directly to a multistage reaction turbine. The efficiencies of compressor and turbine and the gas temperatures, however, were too low to permit successful operation. The first successful gas turbine was built in 1903 in France. It consisted of a multistage reciprocating compressor, combustion chamber, and a two-row impulse turbine. It had a thermal efficiency of about 3 percent. Further progress was slow.

In more modern times, during World War II developers in Switzerland, a country isolated by the war, developed the technology for power generation by gas turbines. Sir Frank Whittle of England was one of many who recognized the applicability of gas turbines for jet propulsion of aircraft. Such efforts led to the development of the jet fighter and subsequently jet transport in many countries.

The gas turbine is now used in the utility industry mainly as a peaking unit (to deliver excess power during periods of high demand), for powering isolated locations, on oil pipeline routes, and more recently, in combined-cycle (gas and steam) powerplants (Chap. 8).

5-2 THE IMPULSE PRINCIPLE

Before discussing the impulse turbine, a review of the *impulse principle* may be useful. Consider a horizontal fluid jet impinging in the $+x$ direction on a fixed vertical flat plate (Fig. 5-2a). That fluid will spread out along the plate, its velocity in the direction of the jet reduced to zero, and will impart to it a horizontal force F in the $+x$ direction. This force is called an *impulse* and is equal to the change in momentum of the jet in the $+x$ direction

$$F = \frac{\dot{m}}{g_c}(V_s - 0) = \frac{\dot{m}}{g_c}V_s \tag{5-1}$$

where
F = force or impulse, lb_f or N
$\dot{m}$ = mass-flow rate of the jet, lb_m/s or kg/s
V_s = velocity in the horizontal direction, ft/s or m/s
g_c = conversion factor, 32.2 $lb_m \cdot ft/(lb_f \cdot s^2)$ or 1 $kg \cdot m/(N \cdot s^2)$

Now consider that the plate is free to move in the horizontal direction (Fig. 5-2b) with a velocity V_B. $V_s - V_B$ will be the velocity of the jet relative to the plate. Now the force on the plate is

$$F = \frac{\dot{m}}{g_c}(V_s - V_B) \tag{5-2}$$

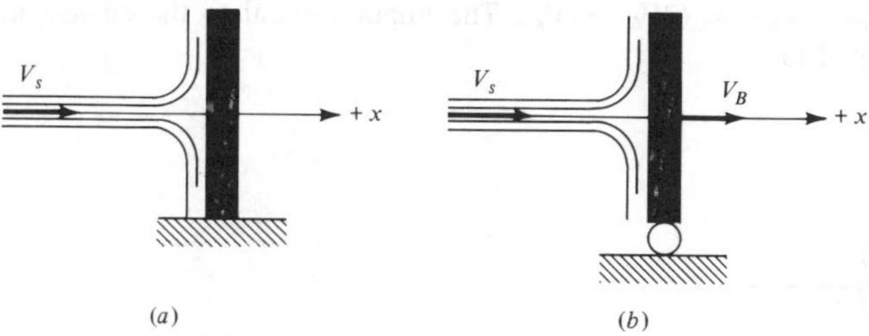

Figure 5-2 The impulse of a fluid jet on (a) a fixed flat plate and (b) a moving flat plate.

and because the plate is moving, work is being done on it equal to the product of force and distance. In a unit time, that distance is V_B, so that the work done per unit time, or the power, in ft · lb$_f$/s or watts, is

$$\dot{W} = FV_B = \frac{\dot{m}}{g_c} V_B(V_s - V_B) \tag{5-3}$$

The efficiency of the flat plate is obtained by dividing the work, Eq. (5-3), by the initial power of the jet $\dot{m}V_s^2/2g_c$ (kinetic energy per unit time). Thus

$$\eta_{\text{plate}} = 2\left[\frac{V_B}{V_s} - \left(\frac{V_B}{V_s}\right)^2\right] \tag{5-4}$$

where $\dot{W}$ = work per unit time, a power, ft · lb$_f$/s or watts. Note the power is zero if V_B is zero (fixed plate) or if $V_B = V_s$, since no force will be exerted by the jet. Thus there is an optimum plate velocity V_B which maximizes the power. It is found by differentiating $\dot{W}$ with respect to V_B and equating the derivative to zero, or

$$\frac{d}{dV_B}\left[\frac{\dot{m}}{g_c}(V_sV_B - V_B^2)\right] = \frac{\dot{m}}{g_c}(V_s - 2V_B) = 0$$

Thus the optimum plate velocity is half the jet velocity

$$V_{B,\text{opt}} = \frac{V_s}{2} \tag{5-5}$$

and the maximum power is

$$\dot{W}_{\text{max}} = \frac{\dot{m}V_s^2}{4g_c} \tag{5-6}$$

which also happens to be half the power or kinetic energy per unit time of the jet. Thus the maximum efficiency of converting the jet energy to the plate is 50 percent, as can be verified from Eq. (5-4).

Now, instead of a flat plate, consider a cylindrical blade that allows the jet to reverse direction (Fig. 5-3). Again the jet enters the blade with a *relative velocity* $V_s - V_B$. Consider further that the blade is frictionless and that there is neither expansion nor contraction of the fluid between blade entry and exit. The fluid relative exit velocity is therefore also $V_s - V_B$. The absolute velocity of the jet at exit in the $+x$ direction will now be $V_B - (V_s - V_B) = (2V_B - V_s)$. The impulse equal to the change in momentum is then equal to

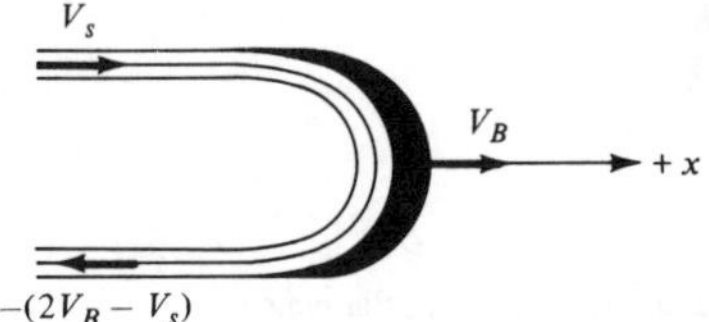

Figure 5-3 The impulse of a fluid jet on a 180° curved blade.

$$F = \frac{\dot{m}}{g_c}[V_s - (2V_B - V_s)] = \frac{2\dot{m}}{g_c}(V_s - V_B) \tag{5-7}$$

and the work per unit time is FV_B or

$$\dot{W} = 2\frac{\dot{m}}{g_c}V_B(V_s - V_B) \tag{5-8}$$

We shall now define a *blade efficiency* η_b as the ratio of the power, Eq. (5-8), to the initial power of the jet, $\dot{m}V_s^2/2g_c$, or

$$\eta_b = 4\left[\frac{V_B}{V_s} - \left(\frac{V_B}{V_s}\right)^2\right] \tag{5-9}$$

and F, $\dot{W}$, and η_b, Eqs. (5-7) to (5-9), for the blade are twice the values for the flat plate, Eqs. (5-2) to (5-4).

To find the optimum blade velocity that results in maximum power, again differentiate $\dot{W}$ with respect to V_B and equate to zero, also giving

$$V_{B,\text{opt}} = \frac{V_s}{2} \tag{5-10}$$

and

$$\dot{W}_{\max} = \frac{\dot{m}V_s^2}{2g_c} \tag{5-11}$$

The optimum blade velocity is half the jet velocity, the same as for the flat plate, but the maximum power is twice that for the flat plate and equal to the total kinetic energy (per unit time) of the jet. In other words, the *maximum blade efficiency* as can be verified from Eq. (5-9) is

$$\eta_{B,\max} = 100\% \tag{5-12}$$

Because the blade moves away from the jet, continuous power can be obtained only if a series of blades were mounted on the circumference of a wheel so that as the wheel rotates they continually face the jet. A high-speed jet needs a nozzle which has physical dimensions so that it is impossible to have the jet impinging on the blades in their direction of motion but at a shallow angle θ (Fig. 5-4). The blade-entrance angle also cannot be zero from horizontal because it should correspond nearly to the relative fluid direction. The blade-exit angle also cannot be zero from horizontal or

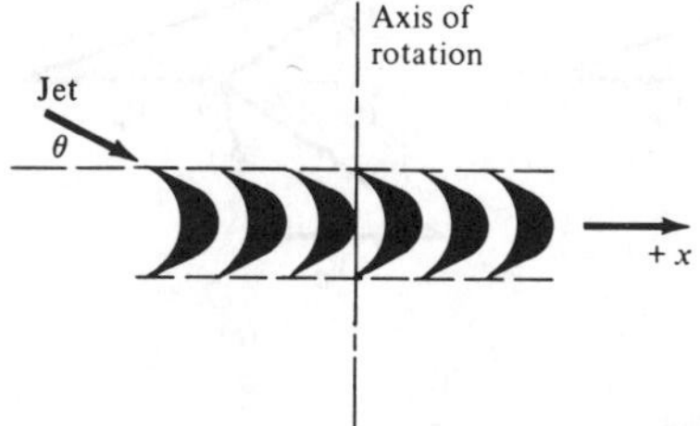

Figure 5-4 Top view of a row of impulse blades on wheel.

else the fluid would not be able to leave the row of successive blades. The practical blade, therefore, is turned around an angle less than 180°.

The Velocity Diagram

To evaluate the work on the blade, which is in the direction of motion, one then needs to construct a velocity vector diagram, shown for a single blade in Fig. 5-5. Figure 5-5*a* shows the velocity diagram in relation to the blade. Figure 5-5*b* and *c* shows simplified versions of it, called "extended" diagrams, with the blade shape removed. In these diagrams

V_{s1} = absolute velocity of fluid leaving nozzle
V_B = blade velocity
V_{r1} = relative velocity of fluid (as seen by an observer riding on the blade)
V_{r2} = relative velocity of fluid leaving blade
V_{s2} = absolute velocity of fluid leaving blade
θ = nozzle angle
ϕ = blade entrance angle
γ = blade exit angle
δ = fluid exit angle

The work on the blade may be obtained from impulse-momentum principles as above or from first-law principles. Both methods yield numerically identical results.

From *impulse momentum,* the force, a vector quantity in the direction of motion of the blade, is equal to the change in momentum of the fluid in the direction of motion, or

$$F = \frac{\dot{m}}{g_c}(V_{s1} \cos \theta - V_{s2} \cos \delta) \tag{5-13a}$$

The component of the steam velocity in the direction of blade motion is called the *velocity of whirl*. Thus $V_{s1} \cos \theta$ is the entrance velocity of whirl V_{w1}, and $V_{s2} \cos \delta$ is the exit velocity of whirl V_{w2} and the force may be written as

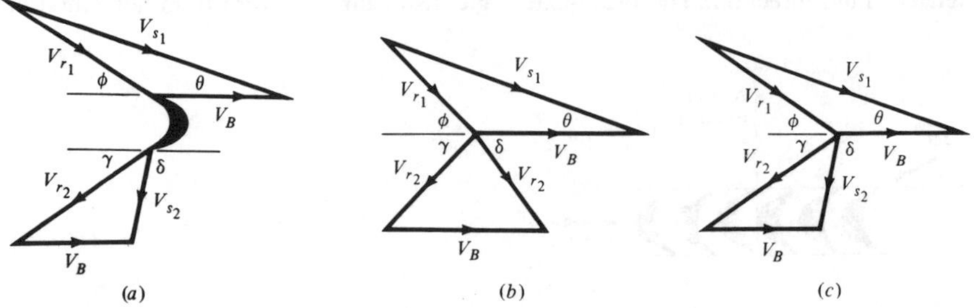

Figure 5-5 Velocity diagrams on a single-stage impulse blade.

$$F = \frac{\dot{m}}{g_c}(V_{w1} - V_{w2}) \tag{5-13b}$$

The work per unit time (power) is equal to the product of force and distance traveled by the blade in unit time or the product of force and velocity. Thus

$$\dot{W} = \frac{\dot{m}V_B}{g_c}(V_{s1}\cos\theta - V_{s2}\cos\delta) \tag{5-14}$$

Note that if both θ and δ are 0, $V_{s2} = V_B - (V_{s1} - V_B)$ in the $+x$ direction (for frictionless, nonexpansion or noncontraction flow), and Eq. (5-14) reverts to Eq. (5-8). Note also that $\cos\delta$ is positive if δ is less than 90° (Fig. 5-5*b*) and negative if δ is greater than 90° (Fig. 5-5*c*), so that the work is greater if δ is greater.

The blade efficiency again is defined as the ratio of the blade work, Eq. (5-14), to the initial energy of the jet $\dot{m}V_{s1}^2/2g_c$ or

$$\eta_B = 2\left[\left(\frac{V_B}{V_{s1}}\right)\cos\theta - \left(\frac{V_B}{V_{s1}}\right)\left(\frac{V_{s2}}{V_{s1}}\right)\cos\delta\right] \tag{5-15}$$

Optimum blade speed. By analogy with the 180° blade, the relative velocity of fluid entering blade in the $+x$ direction is $V_{s1}\cos\theta - V_B$. With no friction, expansion, or contraction, that is also the relative velocity leaving the blade but in the $-x$ direction. The absolute velocity of the fluid leaving the blade in the $+x$ direction, $V_{s2}\cos\delta$, is therefore $V_B - (V_{s1}\cos\theta - V_B) = (2V_B - V_1\cos\theta)$. Equation (5-14) can thus be written in the form

$$\dot{W} = \frac{2\dot{m}V_B}{g_c}(V_{s1}\cos\theta - V_B) \tag{5-16}$$

The optimum blade speed that yields maximum work is again obtained by differentiating $\dot{W}$ with respect to V_B and equating the derivative to zero, giving

$$V_{B,\text{opt}} = \frac{V_{s1}\cos\theta}{2} \tag{5-17}$$

The maximum work is obtained by substituting Eq. (5-17) into Eq. (5-16), giving

$$\dot{W}_{\max} = \frac{\dot{m}}{2g_c}(V_{s1}\cos\theta)^2 = \frac{2\dot{m}}{g_c}V_{B,\text{opt}}^2 \tag{5-18}$$

The maximum blade efficiency is obtained by dividing $\dot{W}_{\max}$ by $\dot{m}V_{s1}^2/2g_c$, giving

$$\eta_{B,\max} = (\cos\theta)^2 \tag{5-19}$$

An examination of the velocity diagram shows that in ideal flow where $V_{r1} = V_{r2}$ and when $\phi = \gamma$, the optimum blade velocity which results in maximum work also results in $\delta = 90°$, or absolute exit velocity straight in the axial direction. In that case $V_{s2}\cos\delta$, the exit velocity of whirl, is zero. Equations (5-17) to (5-19) revert to Eqs. (5-10) to (5-13) for $\theta = 0$.

From the *first-law principles,* with no change in potential energy and no heat transfer, the work is equal to the decrease in enthalpies and absolute kinetic energies of the fluid

$$\dot{W} = (H_1 - H_2) + \dot{m}\left(\frac{V_{s1}^2}{2g_c} - \frac{V_{s2}^2}{2g_c}\right) \tag{5-20}$$

where H_1 and H_2 are the enthalpies of the fluid entering and leaving the blade, respectively. $H_1 - H_2$ is obtained by considering fluid flow relative to the blade (as seen by an observer riding on the blade), where only relative velocities and no work are observed. Thus

$$H_1 - H_2 = \left(\frac{V_{r2}^2}{2g_c} - \frac{V_{r1}^2}{2g_c}\right) \tag{5-21}$$

Combining with Eq. (5-20) gives

$$\dot{W} = \frac{\dot{m}}{2g_c}[(V_{s1}^2 - V_{s2}^2) - (V_{r1}^2 - V_{r2}^2)] \tag{5-22}$$

This is a *general* equation for the work in *any* blade, i.e., including friction, expansion, or contraction of the fluid through the blade passage. In the case of a *pure* impulse blade, where none of these effects is present, $H_1 = H_2$, $V_{r1} = V_{r2}$, and

$$\dot{W}_{\text{pure impulse}} = \frac{\dot{m}}{2g_c}(V_{s1}^2 - V_{s2}^2) \tag{5-23}$$

Values of V_{s1} and V_{s2} are obtained from the nozzle equation, V_B and the various blade angles. In an impulse blade with friction, the usual method of describing the effect of friction is by the use of a *velocity coefficient* k_v, less than one and given by

$$k_v = \frac{V_{r2}}{V_{r1}} \tag{5-24}$$

We shall now define *stage efficiency* $\eta_{\Delta H}$. This is the work of the blade divided by the total enthalpy drop of the fluid for the whole stage, i.e., including nozzle and blade, if expansion of the fluid were reversible and adiabatic, ΔH_s

$$\eta_{\Delta H} = \frac{\dot{W}}{\Delta \dot{H}_s} = \frac{\dot{W}}{\dot{m}\,\Delta h_s} \tag{5-25}$$

Example 5-1 100 lb_m/s of steam enter and leave the nozzle of an impulse turbine stage at 400 psia and 500°F, and 200 psia, respectively. The nozzle efficiency is 90 percent. The nozzle angle is 20°. The blade is symmetrical, travels at optimum velocity, and has a velocity coefficient of .97. Calculate the blade angle, stage power in horsepower and megawatts, and the blade and stage efficiencies.

SOLUTION From the steam tables, App. A, steam conditions entering nozzle (Fig. 5-6) are $h_n = 1307.4$ Btu/lb_m, $s_n = 1.5901$ Btu/($lb_m \cdot$ °R). Steam conditions

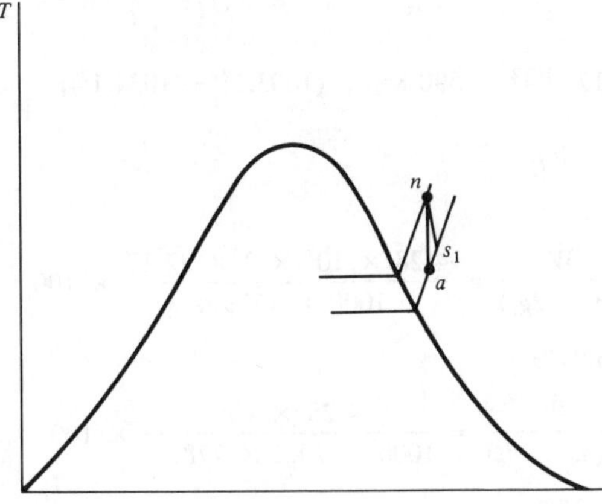

Figure 5-6 *T-s* diagram for nozzle of Example 5-1.

leaving nozzle if it were adiabatic reversible at a are $s_a = 1.5901$ Btu/(lb$_m$ · °R) and by interpolation at 200 psia, $h_a = 1237.2$ Btu/lb$_m$.

$$\text{Nozzle isentropic enthalpy drop} = 1307.4 - 1237.2 = 70.2 \text{ Btu/lb}_m$$

$$\text{Nozzle actual enthalpy drop} = 0.9 \times 70.2 = 63.18 \text{ Btu/lb}_m$$

$$V_{s1} = \sqrt{2 \times 32.2 \times 778.16 \times 63.18} = \underline{1779.4 \text{ ft/s}}$$

Refer to Fig. 5-5

where $\theta = 20°$, $V_{s1} \cos \theta = 1672.1$ ft/s

$$V_{B,\text{opt}} = V_{s1} \cos \theta/2 = 836.05 \text{ ft/s}$$

$$V_{r1} \sin \phi = V_{s1} \sin \theta = 608.6 \text{ ft/s}$$

$$V_{r1} \cos \phi = V_{s1} \cos \theta - V_B = 836.05 \text{ ft/s}$$

from which $V_{r1} = 1034.1$ ft/s and $\phi = 36.05°$

$$V_{r2} = k_v V_{r1} = 1034.1 \times 0.97 = 1003.1 \text{ ft/s}$$

$$V_{r2} \sin \gamma = V_{s2} \sin \delta, \; \gamma = \phi$$

$$V_{r2} \cos \gamma + V_{s2} \cos \delta = V_B$$

from which $V_{s2} = 590.8$ ft/s, $\delta = 87.57°$

Refer to Eq. (5-14)

$$\dot{W} = \frac{1000 \times 836.1}{32.17}(1672.1 - 25.1) = 4.28 \times 10^7 \text{ ft} \cdot \text{lb}_f\text{/s}$$

$$= 77{,}818 \text{ hp} = 58.03 \text{ MW}$$

Refer to Eq. (5-22)

$$\dot{W} = \frac{1000}{2 \times 32.17} [(1779.4^2 - 590.8^2) + (1003.1^2 - 1034.1^2)]$$

$$= 4.28 \times 10^7 \text{ ft} \cdot \text{lb}_f/\text{s}$$

which confirms Eq. (5-14).

$$\text{Blade efficiency} = \frac{\dot{W}}{(\dot{m}V_{s1}^2/2g_c)} = \frac{4.28 \times 10^7 \times 2 \times 32.17}{1000 \times 1779.4^2} \times 100$$

$$= 86.97\%$$

$$\text{Stage efficiency} = \frac{\dot{W}}{\dot{m}(h_n - h_a)} = \frac{4.28 \times 10^7}{1000 \times 70.2 \times 778.16} \times 100$$

$$= 78.35\%$$

5-3 IMPULSE TURBINES

Impulse turbines or turbine stages are simple, single-rotor or multirotor (compounded) turbines to which impulse blades are attached. Impulse blades can be recognized by their shape. They are usually symmetrical and have entrance and exit angles, ϕ and γ respectively, around 20°. Because they are usually used in the entrance high-pressure stages of a steam turbine, when the specific volume of steam is low and requires much smaller flow areas than at lower pressures, the impulse blades are short and have constant cross sections.

Impulse turbines are also characterized by the fact that most or all of the enthalpy, and hence the pressure, drop occurs in the nozzles (or fixed blades that act as nozzles) and little or none in the moving blades. What pressure drop occurs in the moving blade is a result of friction that gives rise to the velocity coefficient k_v discussed above. Single-rotor and compounded impulse steam turbines will now be discussed.

The Single-Stage Impulse Turbine

The *single-stage impulse turbine,* also called the *de Laval turbine* after its inventor (see Introduction, Sec. 5-1), consists of a single rotor to which impulse blades are attached. The steam is fed through one or several convergent-divergent nozzles which do not extend completely around the circumference of the rotor, so that only part of the blades are impinged upon by the steam at any one time. The nozzles also allow governing of the turbine by shutting off one or more of them.

The velocity diagram for a single-stage impulse turbine has been shown in Fig. 5-5. Figure 5-7 shows the overall pressure and absolute steam-velocity changes in the nozzle and blade passages for an ideal turbine. As can be seen the pressure drop occurs in the nozzle and not in the blades. Maximum velocity, and hence kinetic energy, of

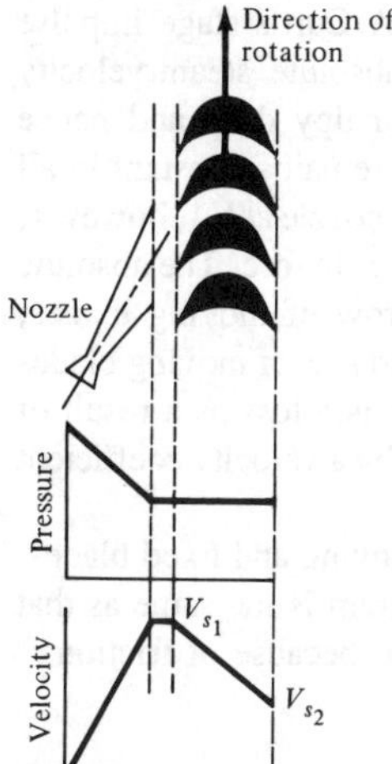

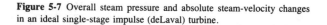

Figure 5-7 Overall steam pressure and absolute steam-velocity changes in an ideal single-stage impulse (deLaval) turbine.

the steam occurs at nozzle exit and decreases from V_{s1} to V_{s2} in the blades. The linear changes in pressure and velocity shown are only schematic and do not represent the actual processes.

Compounded-Impulse Turbines

It has been shown that the optimum blade speed in a single-stage impulse turbine is 0.5 V_{s1} cos θ or roughly one-half of the incoming absolute steam velocity, θ being small. Steam expanding from modern boiler conditions, say 2400 psia and 1000°F to the condenser pressure of 1 psia (or even to atmospheric pressure) in a single nozzle stage, will have velocities of about 5400 ft/s (1645 m/s), meaning a blade speed of 2700 ft/s (820 m/s). Such a speed is far beyond the maximum allowable safety limits because of centrifugal stresses on the rotor material. In addition, the large steam velocities result in large friction losses (proportional to the square of the velocity) and a reduction in turbine efficiency. The high rotor speeds would also necessitate large and bulky reduction gearing to the electric generator. To overcome these difficulties, two methods have been utilized, both called *compounding* or *staging*. One is the *velocity-compounded* turbine, and the other the *pressure-compounded* turbine.

The Velocity-Compounded Impulse Turbine

The velocity-compounded turbine was first proposed by C. G. Curtis (see Introduction) to solve the problems of a single-stage impulse turbine for use with high pressure and temperature steam. The *Curtis stage* turbine, as it came to be called, is composed of one stage of nozzles as the single-stage turbine, followed by two rows of moving blades instead of one. These two rows are separated by one row of fixed blades attached to the turbine stator, which has the function of redirecting the steam leaving the first row of moving blades to the second row of moving blades. A Curtis stage impulse turbine is shown in Fig. 5-8 with schematic pressure and absolute steam-velocity changes through the stage. In the Curtis stage, the total enthalpy drop and hence pressure drop occur ideally in the nozzles so that the pressure remains constant in all three rows of blades. The kinetic energy of the steam leaving the nozzle at V_{s1}, however, is utilized in both rows of moving blades instead of a single row as before. The absolute velocity of the steam decreases from V_{s1} to V_{s2} in the first row of moving blades, remains essentially constant in the fixed blades, enters the second row of moving blades at V_{s3}, and leaves at V_{s4}. Ideally $V_{s2} = V_{s3}$, but actually there is a loss as a result of friction in the fixed blades so that $V_{s3} < V_{s2}$ and they are related by a velocity coefficient k_v similar to that of Eq. (5-24).

The velocity diagram for a Curtis stage, with friction in moving and fixed blades, is shown in Fig. 5-9. The procedure for constructing that diagram is the same as that for the single-stage impulse turbine (Fig. 5-5). In the diagram, because of friction

$$\begin{aligned} V_{r2} &< V_{r1} & \frac{V_{r2}}{V_{r1}} &= k_{v1} \\ V_{s3} &< V_{s2} & \frac{V_{s3}}{V_{s2}} &= k_{v2} \\ V_{r4} &< V_{r3} & \frac{V_{r4}}{V_{r3}} &= k_{v3} \end{aligned} \tag{5-26}$$

Using an analysis similar to that used for the single-stage impulse turbine, it is easy to write expressions for the work of the Curtis stage using either a momentum-impulse or first law analysis. The latter yields

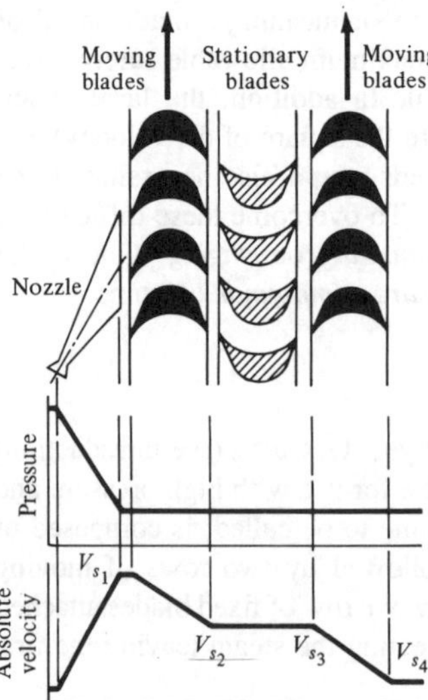

Figure 5-8 Overall steam pressure and absolute steam velocity changes in an ideal velocity-compounded impulse turbine (a Curtis stage).

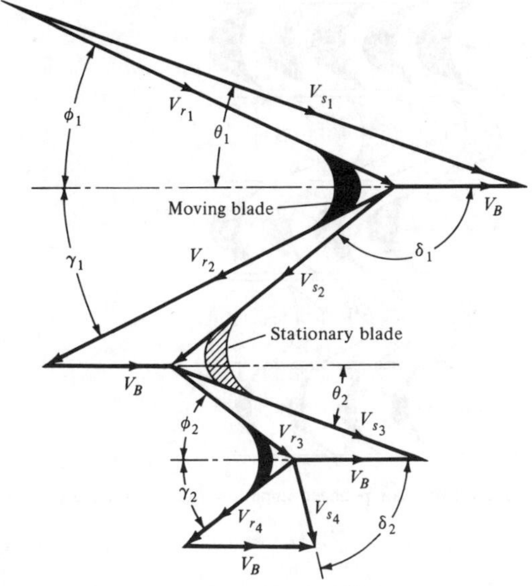

Figure 5-9 Velocity diagram for a Curtis stage, drawn with V_B the same in both moving blades and with friction taken into account. Note the effect of velocity change on blade angles.

$$\dot{W} = \frac{\dot{m}}{2g_c}\{[(V_{s1}^2 - V_{s2}^2) + (V_{r2}^2 + V_{r1}^2)] + [(V_{s3}^2 - V_{s4}^2) + (V_{r4}^2 - V_{r3}^2)]\} \quad (5\text{-}27)$$

where the quantities between the two sets of brackets are due to the work of the first and second rows of moving blades, respectively. The blade efficiency of the Curtis stage turbine is obtained, as usual, by dividing $\dot{W}$ from Eq. (5-17) by $\dot{m}(V_{s1}^2/2g_c)$, and the efficiency of the stage is obtained by dividing $\dot{W}$ by the adiabatic reversible enthalpy drop for the stage.

Although the Curtis stage is composed of two rows of moving blades, a velocity-compounded turbine can be composed of any number of such rows, all sharing in the kinetic energy of the incoming steam at V_{s1}. Such staging usually is built with successively increasing blade angles, which results in flatter and thinner blades toward the last row (Fig. 5-10), constructed for three rows of moving blades. An expression for the work of a three-stage turbine may be easily obtained by extending the terms in brackets in Eq. (5-27) to three.

An expression for the optimum velocity may be derived for an ideal frictionless turbine. It shows

$$V_{B,\text{opt}} = \frac{V_{s1} \cos \theta_1}{2n} \quad (5\text{-}28)$$

where θ_1 is the nozzle angle and n the number of stages (rows of moving blades). The exit velocity of whirl (due to V_{s6}) is zero. Notice that the optimum blade velocity is $\frac{1}{n}$th that of a single-stage impulse. In the actual turbine with friction, $V_{B,\text{opt}}$ is slightly less than that given by Eq. (5-28) [37].

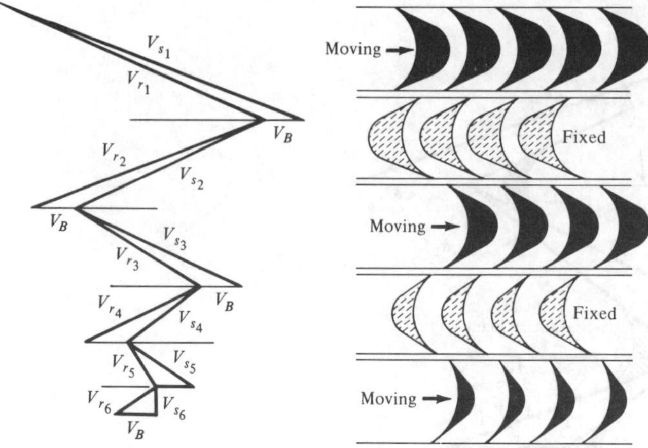

Figure 5-10 velocity diagram and blades for a velocity-compounded impulse turbine with three rows of moving blades.

The work ratio of the highest-to-lowest-pressure stages, in an ideal turbine, can be found to have the ratio 3:1 for a two-stage turbine (Curtis), 5:3:1 for a three-stage turbine, 7:5:3:1 for a four-stage turbine, and so on. This points to one of the major drawbacks of velocity compounding, namely that the lower-pressure stages produce such little work that staging beyond two stages (Curtis) is uneconomical. Another drawback is the still-high steam velocities that result in large friction, especially in the high-pressure stages.

If blade speeds must be reduced below that afforded by two stages, the second method of compounding is resorted to.

The Pressure-Compounded Impulse Turbine

To alleviate the problem of high blade velocity in the single-stage impulse turbine, the total enthalpy drop through the nozzles of that turbine are simply divided up, essentially equally, among many single-stage impulse turbines in series. Such a turbine is called a *Rateau turbine,* after its inventor. Thus the inlet steam velocities to each stage are essentially equal and due to a reduced Δh. From the nozzle equation, ignoring inlet velocities to the nozzles

$$V_{s1} = V_{s2} = \cdots = \sqrt{2g_c \frac{\Delta h_{\text{tot}}}{n}} \tag{5-29}$$

where Δh_{tot} is the total specific enthalpy drop of the turbine and n the number of stages.

A two-stage pressure-compounded turbine is shown in Fig. 5-11. Note that although the enthalpy drops per stage are the same, the pressure drops are not. An examination of a Mollier steam chart shows, for example, that if we were to divide

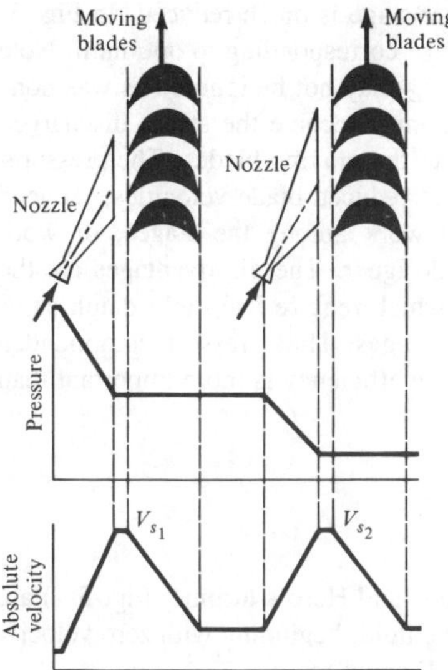

Figure 5-11 A two-stage pressure-compounded impulse turbine (Rateau).

the total enthalpy drop from 1000 psia and 1000°F to 1 psia in isentropic expansion, approximately 580 Btu/lb$_m$, into four equal parts, approximately 145 Btu/lb$_m$, the pressure drops in the first to fourth stages would roughly be 650, 260, 75, and 15 psi, respectively.

Figure 5-12 shows a velocity diagram of a three-stage pressure-compounded impulse turbine with friction so that $V_{r2} < V_{r1}$, etc. The individual triangles are constructed in the identical manner of the single-stage impulse and the equations for that turbine

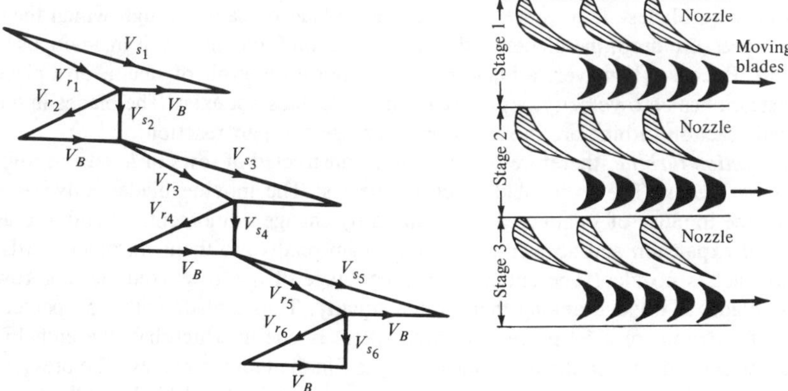

Figure 5-12 Velocity diagrams and nozzles and blades for a three-stage pressure-compounded impulse (Rateau) turbine.

are applicable here, with the exception that Δh per stage is much reduced. In Fig. 5-12 the exit velocities of whirl for all stages are zero, corresponding to optimum. Note also that in calculating V_{s3}, the kinetic energy in V_{s2} may not be ignored as was done in Eq. (5-29), because the nozzles of each stage must receive the steam discharged by the preceding stage, expand it, and redirect it to the moving blades. The pressure-compounded impulse turbine has the advantages of reduced blade velocities, reduced steam velocities (and hence friction) and equal work among the stages, or work distribution among the stages as desired by the designer. The disadvantages are the pressure drop across the fixed rows of nozzles, which require leak-tight diaphragms to avoid steam leakage, and the large number of stages. Thus pressure-compounded turbines are ordinarily used for large turbines where efficiency is more important than capital cost.

5-4 THE REACTION PRINCIPLE

A fixed nozzle, a rocket, a whirling lawn sprinkler, and Hero's turbine (Fig. 5-1) are devices that cause a fluid to exit at high speeds. The fluid, beginning with zero velocity inside, creates a force in the direction of motion F equal to

$$F = \dot{m}\frac{V}{g_c} \tag{5-30}$$

and there is a corresponding and equal force tending to move the devices in the opposite direction. This force is called *reaction*. The devices mentioned above are, therefore, reaction machines that may have to hold stationary (the nozzle), move in a straight line (the rocket), or in a rotary fashion (the sprinkler and Hero's turbine). In all these, the pressure drop (caused by enthalpy drop) that causes the high velocities occurs inside the devices.

In the impulse turbine, no pressure drop (except that caused by friction) occurs in the moving blades. If now we can imagine a blade passage through which there will be a pressure drop, the blade would have a reaction force moving it in an opposite direction. Because, however, a blade passage is not a reservoir of steam, or a place where steam has zero velocity, a pure reaction blade does not exist. The blades in the so-called reaction turbine are in reality part impulse and part reaction.

A *reaction turbine,* therefore, is one that is constructed of rows of fixed and rows of moving blades. The fixed blades act as nozzles. The moving blades move as a result of the impulse of steam received (caused by change in momentum) and also as a result of expansion and acceleration of the steam relative to them. In other words, they also act as nozzles. The enthalpy drop per stage of one row fixed and one row moving blades is divided among them, often equally. Thus a blade with a 50 percent degree of reaction, or a 50 percent reaction stage, is one in which half the enthalpy drop of the stage occurs in the fixed blades and half in the moving blades. The pressure drops will not be equal, however. They are greater for the fixed blades and greater for the high-pressure than the low-pressure stages, as explained for the pressure-compounded impulse turbine.

The term reaction turbine is used despite the fact that pure reaction turbines are not built. (Some European writers prefer the terms *equal-pressure* and *unequal-pressure turbines* for the impulse and reaction turbines, respectively.)

5-5 REACTION TURBINES

Reaction turbines, originally invented by C. A. Parsons, are illustrated by Fig. 5-13 with three stages, each composed of a row of fixed blades and a row of moving blades. The stationary blades are designed in such a fashion that the passages between them form the flow areas of nozzles. They are therefore nozzles with *full steam admission* around the rotor periphery.

The moving blades of a reaction turbine are easily distinguishable from those of an impulse turbine in that they are not symmetrical and, because they act partly as nozzles, have a shape similar to that of the fixed blades, although curved in the opposite direction.

The schematic pressure line (Fig. 5-13) shows that pressure continually drops through all rows of blades, fixed and moving, though the pressure change is greater the greater the pressure. The absolute steam velocity changes within each stage as shown and repeats from stage to stage.

Figure 5-14 shows a typical velocity diagram for two-reaction stages. To construct such a diagram, the Δh per stage is determined, say by dividing the total enthalpy drop of the turbine by the number of stages. This Δh is further divided among the fixed Δh_f and moving Δh_m rows of each stage, each typically half Δh. V_{s1} is calculated

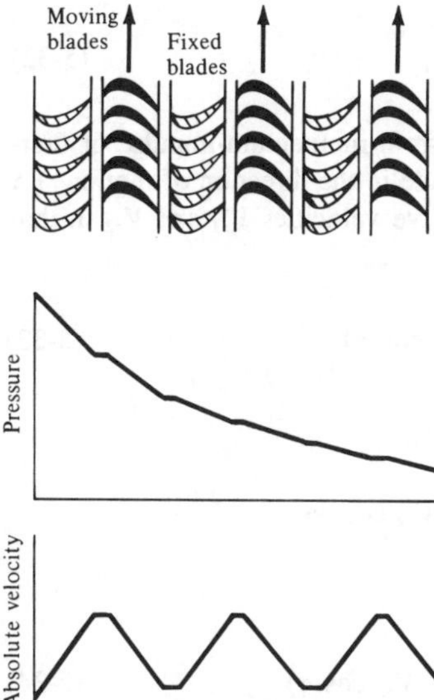

Figure 5-13 Three stages of reaction turbine with overall steam pressures and absolute velocities.

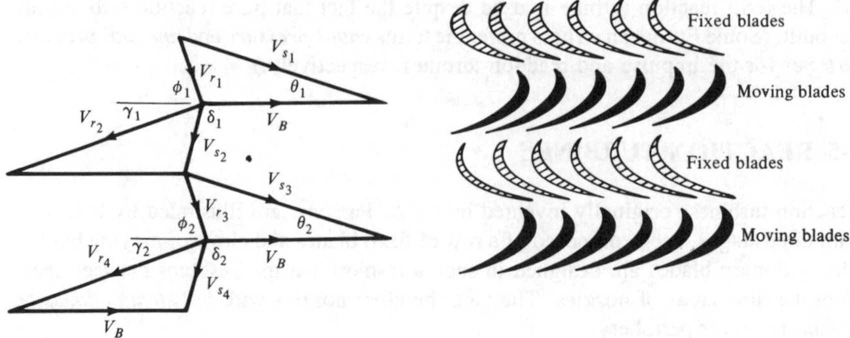

Figure 5-14 Velocity diagram for a two-stage reaction turbine.

from Δh_f. A blade speed V_B is chosen, say, to correspond to optimum conditions equal to $V_{s1} \cos \theta$ (compared with $V_{s1} \cos \theta/2$ for the impulse turbine). V_{r1} is then found. Note that γ is nearly equal to θ but is much less than ϕ here.

The second half of the enthalpy drop Δh_m occurs in the moving blade. This results in increasing the velocities *relative* to the blades. In other words, Δh_m in the moving blades increases its relative velocity. Thus V_{r1} is increased to V_{r2} (in the impulse turbine V_{r2} was equal to or less than V_{r1} because of friction). For the same V_B, get the new V_{s2}, which enters the next row of fixed blades to be increased to V_{s3}, and so on. Thus

$$\frac{V_{s3}^2 - V_{s2}^2}{2g_c} = \Delta h_f \tag{5-31}$$

and

$$\frac{V_{r4}^2 - V_{r3}^2}{2g_c} = \Delta h_m \tag{5-32}$$

The work of a reaction stage can also be obtained from momentum-impulse or first-law principles. The change in momentum on the blade in the direction of motion $+x$ is due to the change in the components of the relative velocities V_{r1} and V_{r2} in that direction. For one reaction stage in general

$$F = \frac{\dot{m}}{g_c}(V_{r1} \cos \phi + V_{r2} \cos \gamma) \tag{5-33}$$

but as $V_{r1} \cos \phi = V_{s1} \cos \theta - V_B$

$$F = \frac{\dot{m}}{g_c}(V_{s1} \cos \theta - V_B + V_{r2} \cos \gamma)$$

The rate of work, or power, $\dot{W} = FV_B$

$$\dot{W} = \dot{m}\frac{V_B}{g_c}(V_{s1} \cos \theta - V_B + V_{r2} \cos \gamma) \tag{5-34}$$

From first-law principles, Eq. (5-22), repeated here, applies.

$$\dot{W} = \frac{\dot{m}}{2g_c}[(V_{s1}^2 - V_{s2}^2) - (V_{r1}^2 - V_{r2}^2)] \tag{5-22}$$

Optimum Blade Speed

The optimum blade speed can be easily obtained for the case where the fixed moving blades are similar, so that $\theta = \gamma$, and Eq. (5-34) is written in the form

$$\dot{W} = \dot{m}\frac{V_B}{g_c}(2V_{s1}\cos\theta - V_B)$$

Again differentiating $\dot{W}$ with respect to V_B and equating to zero

$$\frac{d\dot{W}}{dV_B} = 2V_{s1}\cos\theta - 2V_B = 0$$

or

$$V_{B,\text{opt}} = V_{s1}\cos\theta \tag{5-35}$$

and

$$\dot{W}_{\max} = \frac{\dot{m}}{g_c}(V_{s1}\cos\theta)^2 = \frac{\dot{m}}{g_c}(V_B)^2_{\text{opt}} \tag{5-36}$$

The *efficiency* of a reaction stage is dependent upon the efficiency of the fixed blades (nozzles) and the efficiency of the moving blades. These are explained with the help of Figs. 5-14 and 5-15, which is a Mollier chart representation of Fig. 5-14.

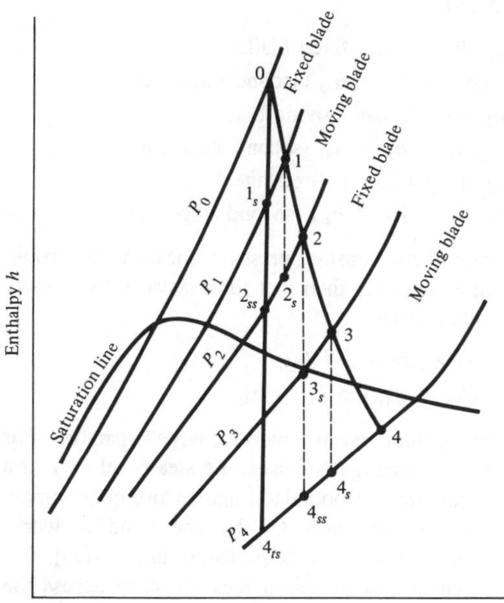

Figure 5-15 Condition curve for a two-stage reaction turbine, drawn on the Mollier (enthalpy-entropy) chart.

Lines P_0, P_1, etc., represent constant pressure lines, which diverge to the right on the Mollier chart. The actual expansion line 0-1-2-3-4 represents the actual condition of the steam in the two stages and is called a *condition curve*.

The *fixed-blade*, or *nozzle*, *efficiency* η_N is the ratio of the kinetic energy change to the adiabatic reversible (isentropic) energy change across the fixed blade. For the first fixed blade

$$\eta_N = \frac{(1/2g_c)(V_{s1}^2 - V_{s0}^2)}{\Delta h_{f,s}} = \frac{h_0 - h_1}{h_0 - h_{1s}} \tag{5-37}$$

The *moving-blade efficiency* η_B is the work of the blade, Eqs. (5-22) or (5-34), divided by the total energy available to that blade, which consists of the kinetic energy of the incoming steam at V_{s1} plus adiabatic reversible (isentropic) enthalpy drop across it. Note that the latter is greater than Δh_m because of the friction (irreversibility) in the blade, which causes an increase in entropy. Thus

$$\eta_B = \frac{\dot{W}}{\dot{m}(V_{s1}/2g_c) + \Delta h_{\text{ms}}} = \frac{\dot{W}}{\dot{m}[(V_{s1}^2/2g_s) + (h_1 - h_{2s})]} \tag{5-38}$$

The *stage efficiency* η_{stage} of a reaction stage is the work of the moving blade in the stage divided by the adiabatic reversible (isentropic) enthalpy drop for the entire stage, including fixed and moving blades. Thus

$$\eta_{\text{stage}} = \frac{\dot{W}}{\dot{m}\,\Delta h_s} = \frac{\dot{W}}{\dot{m}(h_0 - h_{2ss})} \tag{5-39}$$

where, as shown in Eqs. (5-37) to (5-39)

Δh_{fs} = isentropic enthalpy drop across fixed blade
= $h_0 - h_{1s}$ (first stage), = $h_2 - h_{3s}$ (second stage), etc.

Δh_{ms} = isentropic enthalpy drop across moving blade
= $h_1 - h_{2s}$ (first stage), = $h_3 - h_{4s}$ (second stage), etc.

Δh_s = isentropic enthalpy drop across entire stage
= $h_0 - h_{2s}$ (first stage), = $h_2 - h_{4ss}$ (second stage), etc.

Note that because of the divergence of the constant pressure lines, the isentropic enthalpy drop charged to a row or stage is greater than that for a succession of rows or stages or for an entire turbine. In other words

$$h_1 - h_{2s} > h_{1s} - h_{2ss}$$
$$(h_0 - h_{2ss}) + (h_2 - h_{4ss}) > h_0 - h_{4ts}$$

The reaction turbine is an efficient machine that is suited for large capacities. For a given blade speed, limited by material centrifugal stresses, the steam velocity in a reaction turbine is about half that in a pressure-compounded impulse turbine [compare Eqs. (5-35) and (5-17)], resulting in low-friction losses. On the other hand, its work, for the same V_B, is about half that of an impulse stage [Eqs. (5-36) and (5-18)].

Contrary to an impulse stage, a reaction stage has a pressure drop across the moving blades. This makes it less suitable for work in the high-pressure stages where ΔP per unit enthalpy drop is high, which results in steam leakage around the tips of the blades, which in turn leads to throttling and a loss of availability. It follows then that impulse staging is preferable in the entrance stages of a turbine, when the pressures are high, steam specific volumes are low, and the blade height is small so that steam velocities would be correspondingly low. In the low-pressure stages, reaction stages are preferred because the ΔP across the moving blades is less; the blades become progressively longer so that the tip clearance becomes smaller relative to the blade height, i.e., relative to the steam volume. With large reaction blading, V_B is larger, negating the disadvantage of lower power per stage than an impulse stage of the same V_B.

Axial Thrust

Turbine rotors are subjected to an axial thrust as a result of pressure drops across the moving blades and changes in axial momentum of the steam between entrance and exit. This axial thrust must be counteracted to keep the rotor in place.

In *impulse turbines,* there is no pressure drop across the moving blades if the turbine is ideal and little pressure drop caused by friction in a real turbine. In addition there is an axial force on the row because of the change in the axial component of momentum of the steam from entrance to exit. This is given by (see Fig. 5-5)

$$F_{\text{axial}} = \frac{\dot{m}}{g_c}(V_{r1} \sin \phi - V_{r2} \sin \gamma) \tag{5-40}$$

This axial thrust results in no work. In the case of pure symmetrical impulse blades, $V_{r1} = V_{r2}$, $\phi = \gamma$ and that thrust is zero. The total axial thrust on an impulse turbine rotor is, in any case, small and poses no severe problems.

The case of the *reaction turbine* is different. The axial components of the steam entering and leaving a reaction turbine are nearly equal (see Fig. 5-14), so that the axial thrust due to the change in axial momentum of the steam is, like an impulse turbine, essentially zero. There is, however, a large and continual pressure drop across the moving blades. Although that pressure drop decreases in the low-pressure stages, the effect is counterbalanced by an increasing blade height and hence area. The resulting axial thrust is quite large and must be coped with. In small turbines, this is done by a thrust bearing on the rotor shaft or by one or more dummy pistons (discs) inside the casing of sufficient area with high-pressure steam on one face only, the other face sealed by a labyrinth packing.

In modern large utility steam turbines, the common solution is to have double-flow turbines or turbine sections in which steam enters in the center, expands both left and right, and leaves to the next lower-pressure section or to the condenser at opposite ends. This gives the turbine an X shape (Fig. 5-16) with each side's axial thrust canceling the other. It also results in dividing up the blade heights and hence areas and axial thrusts in two and reduces blade tip speeds.

Figure 5-16 A double-flow low-pressure turbine section rotor.

Twisted Blades

Reaction blades are high, especially in the latter stages. Their height, often one-third of the mean blade diameter, reaches 43 in. (about 1.1 m) in some cases (refer to Table 5-2). The velocity diagrams constructed so far in this chapter assumed constant blade speeds V_B, given by

$$V_B = \pi DN \tag{5-41}$$

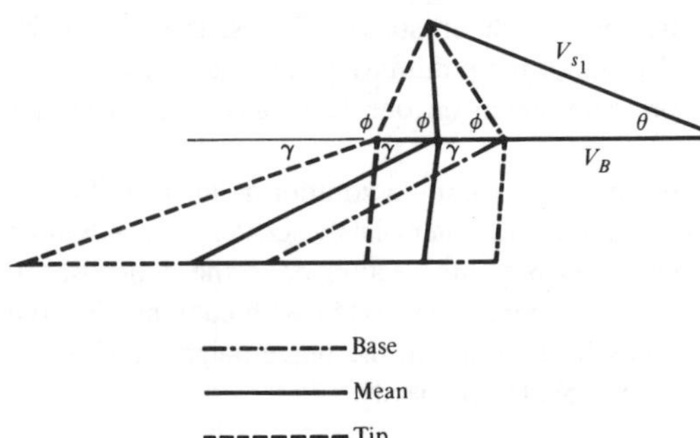

Figure 5-17 Effect of reaction blade height on entrance and exit angles, necessitating a warped radial shape. Drawn for same V_{s_1}, θ, and same exit whirl.

Figure 5-18 33 1/2-in. reaction blading showing twisted construction [38].

where D is the diameter of the blade and N the number of revolutions per unit time. Although N is constant, D for a high blade obviously is not (high-pressure impulse blades, if used, are so short compared with the rotor shaft diameter that V_B for them can be considered constant).

Thus V_B increases with radius from base to tip of the blade, resulting in changes in the shapes of the velocity diagrams along the blade length, as shown in Fig. 5-17, which is drawn for optimum conditions at midpoint. It can be seen that, assuming V_{s1} and θ do not vary in the radial direction, the blade entrance angle ϕ increases, and exit angle γ decreases, from base to tip, which necessitates giving the blade a twisted shape. It can also be seen that the degree of reaction varies from base to tip with the blade somewhat resembling an impulse blade at the base and having maximum reaction at the tip. Such blades are called *twisted, warped,* or *vortex blades*. Figure 5-18 shows a turbine wheel with $33\frac{1}{2}$-in-long reaction blades with this characteristic [38]. See also Fig. 5-21.

5-6 TURBINE LOSSES

Supersaturation

When steam expands *rapidly* from a superheated state across the saturated vapor line (point 1, Fig. 5-19), a condition of *metastable equilibrium* exists in which the steam does not *immediately* condense upon crossing point 1. Instead there is no change in the character of the steam, which continues to follow the laws governing superheated steam for some distance past point 1 until a certain lower pressure is reached. At that point condensation suddenly takes place, and the condition of the system is once again in *thermodynamic equilibrium,* which has a quality dictated by the pressure and specific

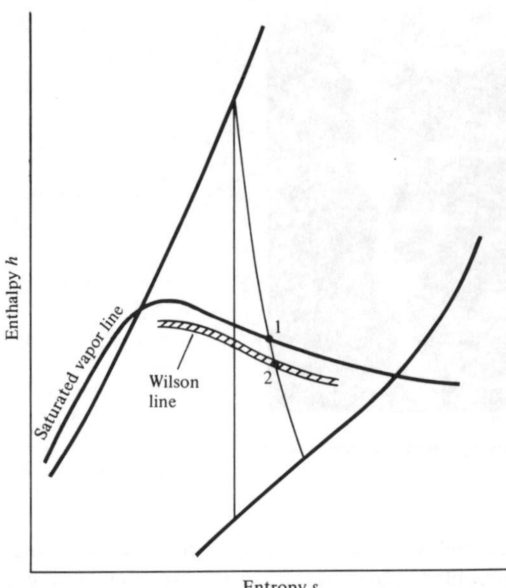

Figure 5-19 Supersaturation condition and the Wilson line, shown on the Mollier (h-s) diagram.

volume (or entropy) at point 2. The phenomenon occurs in both turbines and nozzles, where rapid expansion occurs.

Steam in the region 1-2 is called *supersaturated,* or *undercooled,* steam. The locus of points 2 at various pressures, really a band or a zone, is called the *Wilson line* (Fig. 1-19). It is about 60 Btu below the saturated-vapor line on the Mollier chart.

Initial condensation results in liquid droplets of very small diameters and thus large curvature (inversely proportional to diameter). The vapor pressure of a highly curved surface is greater than that of a flat or less-curved surface at the same temperature because a molecule on a highly curved surface is freer to leave that surface as it is restrained by fewer adjacent molecules. Droplet diameters below which this effect is pronounced are believed to be around 10 Å (1 angstrom Å = 10^{-10} m). Conversely, for the same vapor pressure a small drop will be at a lower temperature than a larger one or than the saturation temperature corresponding to that pressure. Thus when expansion occurs rapidly to a given pressure and no condensation takes place, a lower temperature will be reached before the first droplets form. Once they form and grow, thermodynamic equilibrium returns to the system.

This phenomenon is further illustrated by the use of a modified Mollier chart (Fig. 5-20) that represents the region of most interest in steam-turbine supersaturation. In thermodynamic equilibrium, a two-phase mixture at a given temperature has one and only one corresponding saturation pressure (e.g., 4.74 psia and 160°F). In supersaturation, the steam behaves somewhat like a gas and the temperature lines in the superheat region extend into the two-phase region, as shown by the dashed lines in the figure. At any given pressure then a supersaturated fluid such as at b has a lower temperature, about 105°F, than if it were in thermodynamic equilibrium (160°F). In other words, the steam is undercooled. The ratio of actual pressure to the pressure corresponding to the lower temperature is called the *degree of supersaturation*, or *degree of undercooling*.

Example 5-2 Compare the final conditions and the steady-flow work when superheated steam at 11.5 psia and 240°F expands (1) isentropically to 4.74 psia when expansion occurs to a supersaturated state or (2) slowly and thermodynamic equilibrium is maintained. Assume supersaturated vapor obeys $PV^{1.32}$ = constant.

SOLUTION The initial conditions (point a, Fig. 5-20) from the steam tables are $h_a = 1165.0$ Btu/lb$_m$, $v_a = 35.88$ ft^3/lb$_m$, and $s_a = 1.8047$ Btu/(lb$_m$ · °R).

This steam expands isentropically to point b. If it does so rapidly and becomes supersaturated, it will behave as a gas and obey $Pv^{1.32}$ = constant. Thus

$$v_b = v_a \left(\frac{P_a}{P_b}\right)^{1/1.32} = 35.88\left(\frac{11.5}{4.74}\right)^{0.7575} = 70.22 \text{ ft}^3/\text{lb}_m$$

$$T_b = T_a \left(\frac{P_b}{P_a}\right)^{(1.32 - 1)/1.32} = (240 - 460)\left(\frac{4.74}{11.5}\right)^{0.2424} = 564.6°\text{R} = 104.6°\text{F}$$

From Table 1-3

$$\text{Steady-flow work } w_{sf} = \frac{n}{1-n}(P_b v_b - P_a v_a)$$

$$= \frac{1.32 \times 144}{1 - 1.32}\left(\frac{4.74 \times 70.22 - 11.5 \times 35.88}{778.16}\right)$$

$$= 60.1 \text{ Btu/lb}_m$$

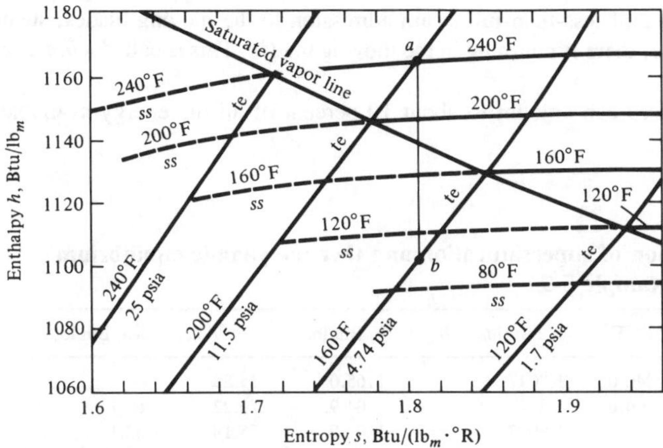

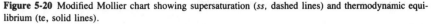

Figure 5-20 Modified Mollier chart showing supersaturation (*ss*, dashed lines) and thermodynamic equilibrium (te, solid lines).

Therefore

$$h_b = h_a - w_{sf} = 1165.0 - 60.1 = 1104.9 \text{ Btu/lb}_m$$

If the steam expands slowly and maintains thermodynamic equilibrium, the conditions at b are obtained from the steam tables at 4.74 psia and $s_b = s_a =$ 1.8047 Btu/lb$_m$, giving $x_b = 0.9728$, $h_b = 1102.9$ Btu/lb$_m$, and $v_b = 75.18$ ft^3/lb$_m$. The steady-flow work $w_{sf} = h_a - h_b = 1165.0 - 1102.9 = 62.1$ Btu/lb$_m$. The solution is summarized in Table 5-1.

It can be seen that supersaturation results in a lower temperature, justifying the dual name "undercooling," lower volume, and reduced work.

When expansion crosses the Wilson line, it reverts to thermodynamic equilibrium by sudden condensation. This releases the enthalpy of vaporization of the condensed vapor and results in a sudden pressure rise and reduction in specific volume and velocity. The phenomenon is called *condensation shock,* which is similar, though not identical, to normal shocks that occur in supersonic nozzles. It is an irreversible process that results in further loss in availability.

Fluid Friction

Fluid friction is the biggest cause of all turbine losses. It occurs throughout the turbine. There is, to begin with, friction in the steam nozzles. Next there is blade friction, which we tried to minimize by decreasing steam velocities by compounding, etc. Also there is turbulence in the blades when the blade shape does not possess the proper entrance angle for steam at other than design loads. There is also friction between the steam and the rotor discs that carry the blades. Here rotor design is important (Sec. 5-8). In addition, the rotor and blade rotation impart a centrifugal action on the steam, thus causing part of it to flow radially to the casing and be dragged along by the moving blades. In case of less-than-full steam admission to the moving blades, such as for an impulse stage, there is churning in the moving blades. This is called a *fanning* loss.

Fluid friction losses can amount to about 10 percent of all the energy available to the turbine.

Table 5-1 Comparison of supersaturation and thermodynamic equilibrium for the data from Example 5-2

Properties	P, psia	T, °F	s, Btu/(lb$_m$° · R)	h, Btu/lb$_m$	v, ft^3/lb$_m$	w_{sf}, Btu/lb$_m$
Initial	11.5	240.0	1.8047	1165.0	35.88	—
Final,*ss**	4.74	104.6	1.8047	1104.9	70.22	60.1
Final,te*	4.74	160.0	1.8047	1102.9	75.18	62.1

* *ss* = supersaturation, te = thermodynamic equilibrium.

Leakage

Steam leakage can occur within and to the outside of a turbine. Within the turbine steam can leak between the tips of moving blades and the casing when there is a pressure drop across them, such as in a reaction turbine. This leakage is greater the greater the pressure drop, i.e., in the higher-pressure stages, and the greater the ratio of tip clearance to blade height. The leaking steam is throttled and represents a loss of available energy. In a pressure-compounded (Rateau) impulse turbine, leakage occurs between the base of the stationary diaphragms that carry the nozzles and the shaft.

Leakage can also occur to the outside of the turbine at the various shaft bearings. This is minimized by the use of proper seals or packings, such as a *labyrinth packing*.

Leakage loss can account for about 1 percent of the total energy available to a turbine.

Moisture Loss

Besides the losses encountered as a result of supersaturation in the two-phase region (above), the presence of liquid droplets causes further losses. These droplets have both a size and velocity distribution, not unlike those in a liquid nozzle spray. Some low-speed droplets splash against the moving blades, i.e., strike them at off-design angles, and thus reduce the mechanical work of the rotor. Others are accelerated by the steam and remove some of its energy through momentum exchange. The result is that turbine sections that operate in the two-phase region are substantially less efficient than those that operate in the superheat region.

Turbines are usually designed to operate with exit-moisture content of no more than about 12 percent (88 percent minimum quality). Higher moisture content (often coupled with high oxygen content in boiling-water reactors) cause blade erosion as a result of the impingement of water droplets on the blades, surface washing, and the so-called *wire drawing* caused by high-velocity water leaking through narrow passages. Oxygen, in addition, causes corrosion. If steam expansion causes higher moisture content than 12 percent, moisture extraction at certain stages in the turbine is resorted to keep the moisture content within this limit. This is the practice in boiling-water reactor turbines, for example. The moisture extracted represents a mass-flow, and hence work, loss to the turbine, though this effect can be minimized by combining it with bled steam for feedwater heating. Moisture extraction can be accomplished by constructing the moving blades with grooves on the back side, where the drops are known to collect (Fig. 5-21). The drops are then thrown radially by the centrifugal force of the rotating blades into a collecting chamber in the casing where they may be bled to a feedwater heater or the condenser [39].

From an operational point of view, contaminants in the steam besides oxygen, such as particulate matter and chemicals such as sodium and chlorine from water treatment operations, can cause stress-corrosion cracking (caused by otherwise tolerable stressing, but in a corrosive atmosphere) and erosion. This calls for better water-chemistry control, monitoring, and maintenance [40].

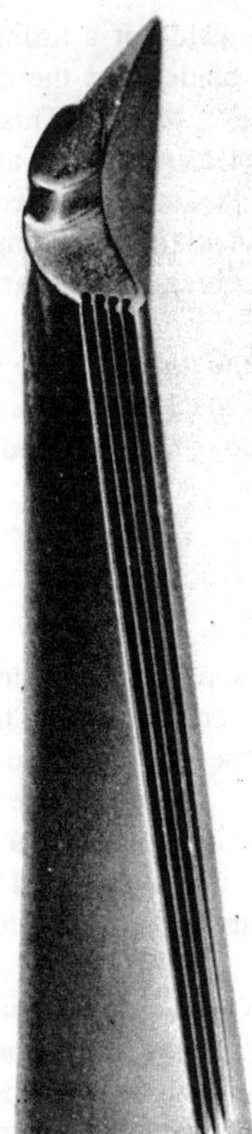

Figure 5-21 A grooved moisture-extracting turbine blade [39].

Leaving Loss

We have noted a velocity residual in individual turbine stages, both impulse and reaction. The corresponding kinetic energy is usually recovered in subsequent stages, except for that due to the last row in the turbine. The velocity leaving this row, as a result of low-pressure steam of maximum specific volume, and the corresponding kinetic energy represent a loss to the turbine. This velocity is approximately normal to the plane of rotation near rated load but has a large forward component at lighter loads. The magnitude of this velocity can be changed by the designer by the proper combination of last-blade height, speeds, and area of the exhaust ducts to the condenser. (We already noted that low-pressure turbine sections are double-flow and have two exhaust ducts each.) Whereas high leaving velocities result in energy loss, velocities that are too low result in inordinately high blades, large exhaust ducts, and increased capital costs. In modern large turbines, leaving velocities around 900 to 1000 ft/s (270 to 300 m/s) are typical. They result in a 2 to 3 percent loss to the turbine.

Exhaust hoods to the condenser that gradually increase in area like a diffuser further decrease the exhaust-steam velocity and increase its pressure as it enters the condenser. Such hoods allow the turbine to operate down to a slightly lower pressure than that required by the condenser (which is dictated by the available cooling-water temperature), thus increasing the turbine work. This practice is more common with 3000- and 3600-r/min than 1500- and 1800-r/min turbines because the latter have larger blades and exhaust ducts to begin with.

Heat-Transfer Losses

Heat-transfer losses from turbines, as usual, are caused by conduction, convection, and radiation. Conduction occurs internally through metal between stages and is aided by convection, which is largely the result of the high steam velocities. Conduction also takes place between the turbine casing and the foundation. Convection and radiation losses occur from the turbine casings to the surroundings in the turbine hall and are more pronounced for the high-pressure section, where steam temperatures are the highest. That section, however, is smallest in diameter and is usually well insulated. The larger low-pressure sections operate with steam not much above room temperature, and are usually uninsulated.

Although a turbine hall usually feels warm, the total heat losses per unit mass flow of steam of large turbines are so small that they are negligible. This is not the case, however, with small turbines, such as those used for mechanical drives, for which the heat-transfer losses usually amount to several percent of the turbine energy.

Mechanical and Electrical Losses

Now that the turbine has extracted work from the system, it must deliver it to the electric generator. In so doing, it encounters frictional losses in bearings, governor mechanism, and reduction gearing, if present. It must also supply mechanical work to accessories such as oil pumps, etc.

Mechanical losses are practically constant and independent of load and thus increase in percentage as load decreases. On the other hand, the percentage is also smaller the larger the turbine. In general, mechanical losses are fairly small, amounting to 1 percent or less of the turbine energy.

As turbines are usually rated by the electric generator output, a knowledge of the extent of generator losses is essential. Modern large electric generators are hydrogen-cooled, well designed, and very efficient. Efficiencies around 98 to 99 percent are common, increasing slightly with load, and are somewhat higher for 1500- and 1800-r/min than 3000- and 3600-r/min generators.

5-7 TURBINE EFFICIENCIES

We have already seen that, because constant pressure lines diverge on a Mollier chart (true also for gases), the isentropic enthalpy drops charged to a turbine stage are greater than those for multiple stages and for the entire turbine. It follows that the efficiency of a stage is less than that of a turbine section, etc. This can be seen with the help of Fig. 5-15 for a reaction turbine, though it applies to all turbines

$$\text{Efficiency of 2 stages} = \frac{h_0 - h_4}{h_0 - h_{4ts}} > \text{efficiency of 1 stage } \frac{h_2 - h_4}{h_2 - h_{4ss}}$$

The ratio of the total individual isentropic enthalpy drops to the enthalpy drop of a turbine section or whole turbine is called the *reheat factor* R_h. R_h is obviously greater than 1.0, with values ranging between just above 1.0 to perhaps 1.065, depending upon the pressure range. For the two stages of Fig. 5-15

$$R_h = \frac{(h_0 - h_{2ss}) + (h_2 - h_{4ss})}{h_0 - h_{4ts}} \tag{5-42}$$

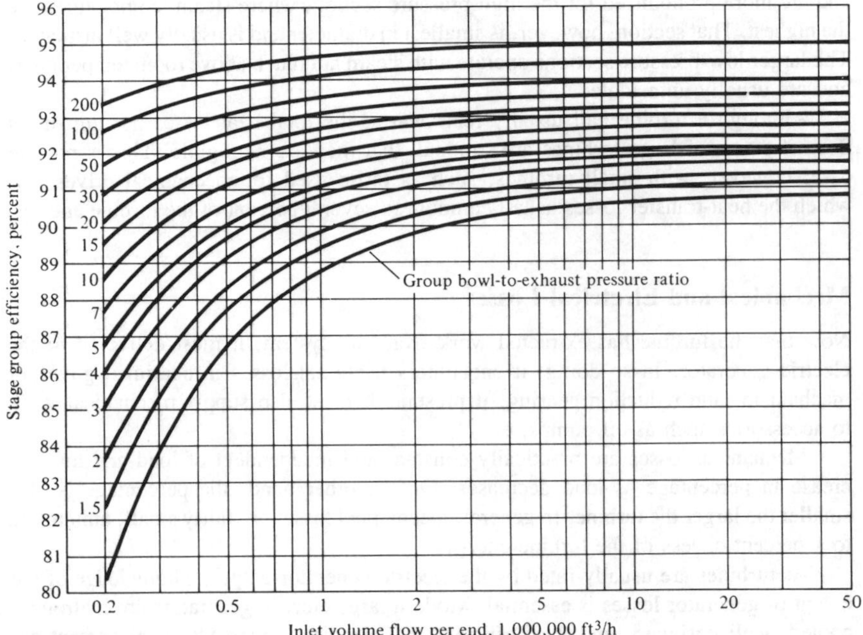

Figure 5-22 Typical stage group efficiencies for turbine section operation in superheat region as a function of steam inlet volume flow and ratio of bowl (inlet) pressure to group exhaust pressure. Leaving loss not included [41].

If the designer wishes to have equal work from the stages, dividing the isentropic enthalpy drop for the whole turbine in equal parts will not, because of the divergence of the pressure lines, result in equal actual work in each stage; that is, if $h_0 - h_{2ss} = h_{2ss} - h_{4ts}$, it does *not* follow that $h_0 - h_2 = h_2 - h_4$. To get the same actual work, the designer must take into account this divergence.

We have also seen that turbine stages operating in superheated steam are more efficient than those operating in the two-phase region. The performance and efficiency of stages or of a whole steam turbine are undoubtedly a complex function of many variables. Accurate knowledge of the condition curve of a turbine, which is affected by the individual stage rather than the whole turbine efficiency, is necessary, for example, to do the cycle analysis in reheat and moisture-extraction turbines.

Methods for predicting the performances and efficiencies of various steam turbines are often manufacturer's proprietary information, but some may be found in the literature [41, 42]. One such method [41] predicts the performance of large turbines used in modern nuclear powerplants operating with low-superheat or saturated steam in Figs. 5-22 and 5-23 for the superheat and two-phase regions, respectively. These figures give base efficiencies for a volumetric flow (10^6 ft^3/h, 7.867 m^3/s) and other data. Corrections for the departure from these data are available in the original paper.

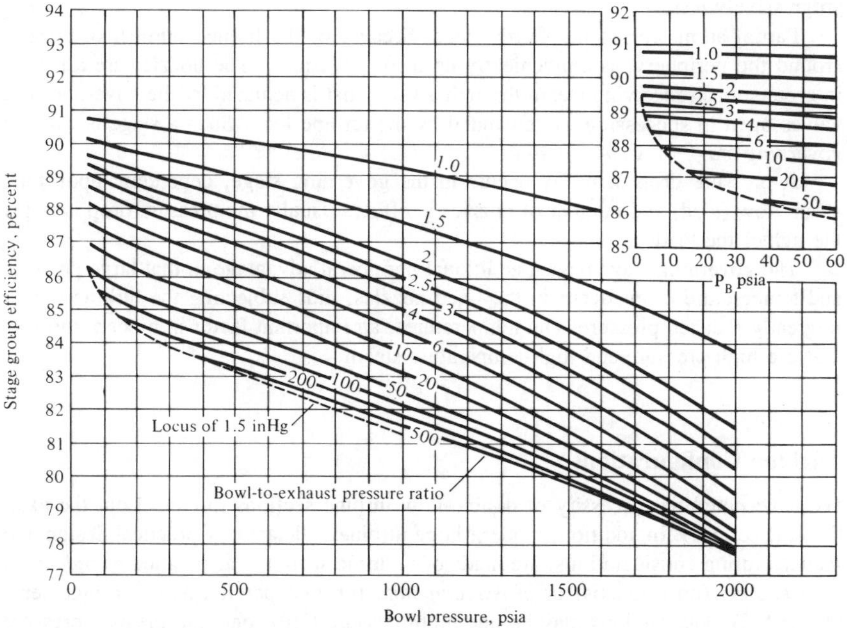

Figure 5-23 Typical stage group efficiencies for turbine section operating in two-phase region as a function of bowl (inlet) pressure and ratio of bowl pressure to group exhaust pressure. Leaving loss not included [41].

5-8 TURBINE ARRANGEMENTS

Combination Turbines

In earlier times, turbines used to be built as pure impulse, of one type or another, or pure reaction. With the expiration of patents, manufacturers were free to use combinations, especially in medium and large sizes. One popular arrangement used to be a Curtis stage (two-row velocity-compounded impulse) followed by a series of Rateau (pressure-compounded impulse) stages. Another was a de Laval (single-stage impulse) followed by a Rateau or reaction turbine.

A more common arrangement is a combination of a Curtis stage followed by a large number of reaction stages. There are certain advantages of this arrangement. The impulse stage is more suited to the high pressure of admission than the reaction stage because there is virtually no pressure drop in the moving blades. Recall that for the same enthalpy drop, a much larger pressure drop occurs at high pressure. Also the clearance between the blade tip and the casing is greater for the shorter high-pressure blades, which aggrevates the leakage problem should there be a pressure drop across the moving blades. After the impulse stage, the pressure is sufficiently low that the more efficient reaction stages can now be used. They become progressively longer, the clearance proportionately less, and the pressure drop across their moving blades progressively less.

Partial admission to the Curtis stage, because of the limited number of nozzles around the periphery, is conveniently used for governing. The nozzles are arranged in groups, each receiving steam through a valve that is actuated by the governor. The valves open in succession as demanded by the turbine load. Such a stage is called a *governing stage* or a *control stage*.

A pressure drop naturally occurs in the governing stage, depending upon total steam flow (load) and number of nozzles in effect. Usually the pressure drop is larger the lighter the load.

The governing stage has an additional peripheral advantage in that large pressure and temperature drops occur in the fixed nozzles, thus subjecting the turbine proper to greatly reduced pressures and temperatures, an important factor in modern turbines that use high-pressure and high-temperature steam.

Turbine Configurations

We have noted the necessity of double-flow turbine sections to cancel out the axial thrust (Sec. 5-5). In addition, modern large turbines, dictated by practical design and manufacturing considerations, are made of multiple *sections,* also called *cylinders,* in both *tandem* (on one axis) or *cross-compound* (on two parallel axes) arrangements (Table 5-2). The sections may be one high-pressure (HP), one intermediate-pressure (IP), and two low-pressure (LP) sections, all in tandem, but with the two LP sections operating in parallel as far as steam flow is concerned. They may be one HP, three

Table 5-2 Turbine-generator configurations*

Fossil	Fossil	Nuclear
TC-2F LSB 26, 30 and 33.5 in Two casings 3600 r/min HI-IP LP G 125–400 MW	TC-6F LSB 26, 30 and 33.5 in Five casings 3600 r/min HP IP LP LP LP G 550–1000 MW	TC-4F LSB 38 and 43 in Three casings 1800 r/min HP LP LP G 450–1000 MW
TC-4F LSB 26, 30 and 33.5 in Three casings 3600 r/min HP-IP LP LP G 250–650 MW	TC-6F LSB 30 and 33.5 in Five casings 3600 r/min (double reheat) HP-IP IP LP LP LP G 450–725 MW	TC-6F LSB 38 and 43 in Four casings 1800 r/min HP LP LP LP G 600–1100 MW
TC-4F LSB 26, 30 and 33.5 in Four casings 3600 r/min HP IP LP LP G 550–850 MW LP LP G	CC-4F LSB 38 and 43 in Four casings 3600/1800 r/min HP IP G 3600 r/min LP LP G 1800 r/min 600–1250 MW	

* Data provided by the General Electric Company. TC = tandem compound, CC = cross compound, F = number of flow ducts to condenser, LSB = last-stage blade.

LP, etc. The multiciplicity of the LP sections further reduces the blade lengths, giving for example, *last-stage-blade* or *-bucket* (LSB) lengths of 43 in for a 1000-MW, 1800-r/min (water-cooled nuclear reactor) turbine; 33.5 in for a 1000-MW, 3600-r/min (fossil and high-temperature nuclear) turbine; etc.

Configurations are also affected by admission requirements. Figure 5-24*a* shows a number of *straight-through* turbine sections (for simplicity). Figure 5-24*b* shows a turbine with *single reheat* (steam expanding part way, reheated in the steam generator, and readmitted to the turbine). Figure 5-24*c* shows an *extraction* turbine in which steam is bled for feedwater, for process steam uses (cogeneration, Sec. 2-15), or both. Figure 5-24*d* shows an *induction* turbine, the reverse of an extraction turbine, in which low-pressure steam is injected at a low-pressure stage. This low-pressure steam comes from a process, in special-design utility boilers as Dresden I nuclear powerplant or early British gas-cooled reactor systems [3], or in some combined gas-turbine–steam-turbine cycles (Sec. 8-11).

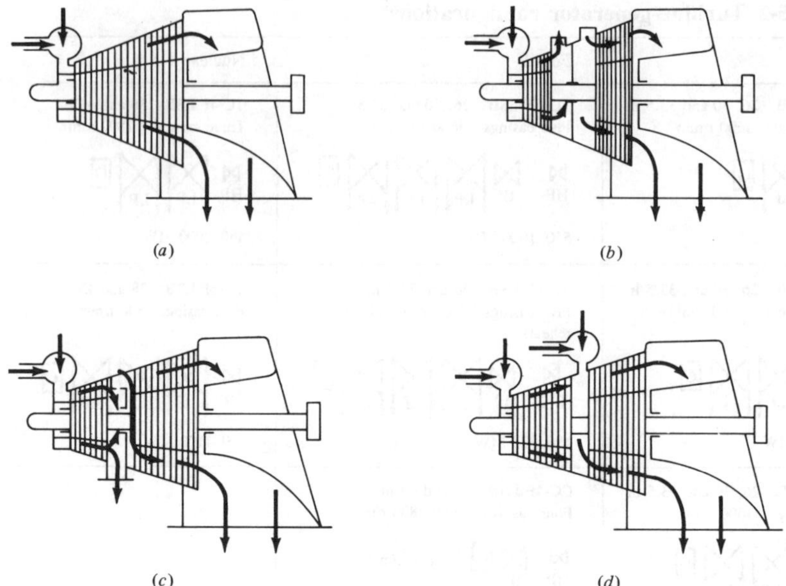

Figure 5-24 Turbine arrangements as affected by different steam paths: (*a*) straight through; (*b*) single reheat; (*c*) extraction; (*d*) induction.

Figures 5-25 and 5-26 show cross sections for a 3600-r/min fossil-fueled turbine and a 1800-r/min nuclear-fueled turbine, respectively.

Turbine Rotors

The rotor is the heart of the turbine. Current designs are shown in Fig. 5-27. Figure 5-27*a* and *b* shows two versions of a rotor produced from a *single forging*. Figure 5-27*c* shows a *composite construction* produced by shrinking rotor discs on a central shaft. Figure 5-27*d* shows a *drum-type rotor* composed of separate rotor discs that are welded together. This last design is receiving acceptance for the very large units being built today, 500 to 1000 MW and larger, which would otherwise be extremely heavy and uneconomical and would pose severe mechanical problems..

Materials for such rotors are carefully chosen to yield lasting resistance to softening and creep, receive uniform heat treatment, and have lasting ductility and good resistance to scale. Besides materials, attention must be paid to manufacturing methods and operating stresses [43].

Water-cooled nuclear-reactor turbines pose fewer problems than fossil-fueled turbines because they operate with lower steam temperatures, around 540°F versus 1000°F (285°C versus 540°C).

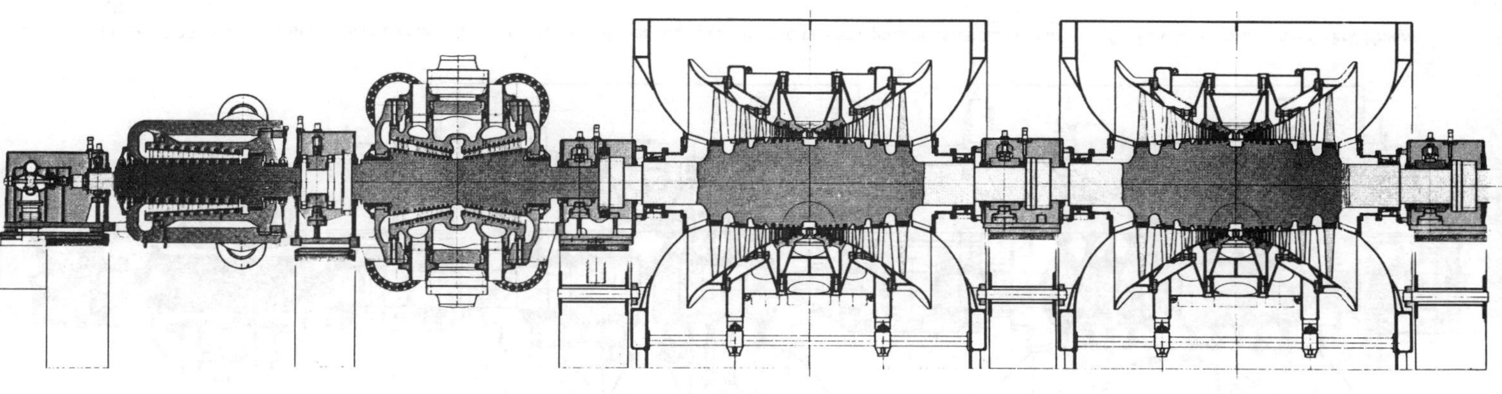

Figure 5-25 A 3600-r/min tandem compound steam turbine with one high-pressure, one intermediate-pressure, and two low-pressure sections. (*Courtesy of Allis Chalmers and Kraftwerk Union KWU.*)

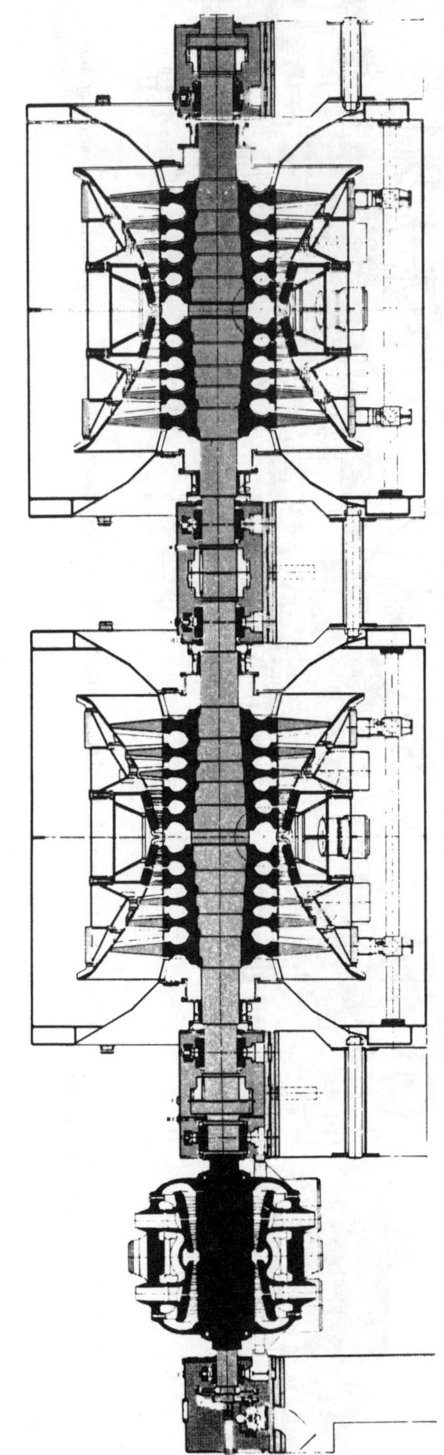

Figure 5-26 A 1800-r/min tandem compound steam turbine with one high-pressure and two low-pressure sections. (*Courtesy of Allis Chalmers and Kraftwerk Union KWU.*)

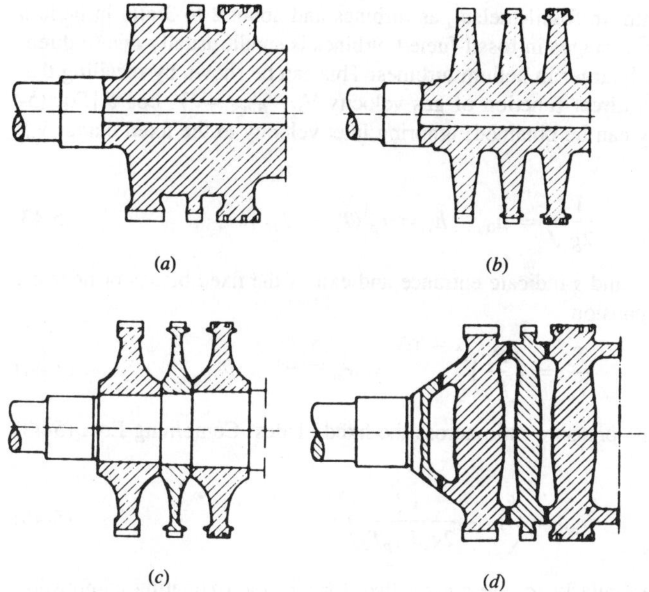

Figure 5-27 Different turbine rotor designs.

5-9 GAS TURBINES

Gas turbines for utility service are normally used for peak power production but sometimes also for intermediate and base-load duties when called upon during a major plant outage. Gas-turbine cycles will be covered in Chap. 8.

There are two basic types of gas turbines: radial-flow and axial-flow. The *radial-flow gas turbine* is similar in appearance to a centrifugal compressor, with the exception, of course, that gas flow is radially inward instead of radially outward. Radial-flow turbines are widely used in small sizes. They form a compact rigid rotor when combined with centrifugal compressors. A common use of such a combination is for *turbochargers* on stationary and marine diesel engines and, more recently, on both diesel and gasoline motor vehicle engines. Radial gas turbines, however, are not as suitable to the high-temperature gases necessary for good thermal efficiency (Chap. 8), and except for small sizes, are not as efficient as axial gas turbines. *Axial-flow gas turbines* resemble the steam turbines discussed in this chapter. Because we are concerned with larger sizes, the discussion that follows pertains to axial-flow gas turbines.

Gas-turbine stages are similar to those of steam turbines, except that the fluid is either a pure gas, such as helium, which is proposed for use with high-temperature gas-cooled nuclear reactors, or air-and-combustion products in a fossil-fueled gas turbine. There are, obviously, no steam-condensation problems to worry about.

The inlet pressures to gas turbines are much lower than those for steam turbines, being about 6 to 10 atm in fossil-fueled gas turbines and about 2 to 3 atm in helium turbines. The number of stages in fossil-fueled turbines is small, usually one to three, but the number is much larger in helium turbines. This can be shown by recalling that blade velocity V_B is a direct function of gas velocity V_s: $V_{B,\text{opt}} = V_s \cos \theta$ [Eq. (5-35)]. The gas velocity can be obtained, ignoring inlet velocity to the fixed blades V_o, from

$$\frac{V_s^2}{2g_cJ} = h_0 - h_s = c_p\,(T_o - T_s) \tag{5-43}$$

where the subscripts o and s indicate entrance and exit of the fixed blades or nozzles. For a gas in ideal expansion

$$\frac{T_s}{T_o} = \left(\frac{P_s}{P_o}\right)^{(k-1)/k} = r_{pf}^{(1-k)/k} \tag{5-44}$$

where $r_{pf} = P_o/P_s$, the pressure ratio across the fixed blades. Combining Eqs. (5-43) and (5-44)

$$r_{pf} = \left(1 - \frac{V_s^2}{2g_cJ\;c_pT_o}\right)^{k/(1-k)} \tag{5-45}$$

For helium k and, particularly, c_p are greater than for air (approximating combustion gases). Equation (5-45) shows then that the pressure ratio across a single stage (related to that across the fixed blades by the degree of reaction) is much lower for helium than for air. Thus while the overall pressure ratio of a helium turbine is less than that of a combustion turbine, the pressure ratio per stage is far less and the number of stages is greater. Note that the overall pressure ratio is equal to the pressure ratio per stage to a power equal to the number of stages. A cross section of a 36-MW air-combustion turbine with a 16-stage compressor and a 3-stage turbine is shown in Fig.

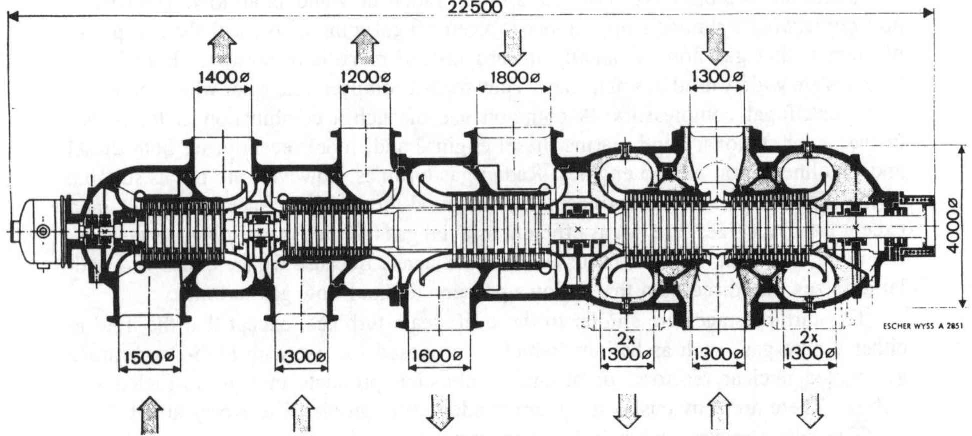

Figure 5-28 A proposed design for a helium-driver gas turbine, dimensions in millimeters [44].

8-12. Figure 5-28 shows a cross section of a proposed design for a single-shaft, 600-MW helium turbine showing, from left to right, 9-stage low-pressure, 10-stage medium-pressure, and 12-stage high-pressure axial compressors, and a double-flow turbine with a total of 20 stages [44].

Gas-Turbine Blading

In steam-turbine practice, relatively inexpensive straight blading, i.e., untwisted, is used and designed on conditions at mean diameter, except in the long-bladed low-pressure stages where the large change in blade velocity with radius necessitates the use of twisted or vortex blading. The gas-turbine (Brayton) cycle is not as efficient as the Rankine cycle, and the gas-turbine designer, striving to squeeze improvements in efficiency from every stage, has used twisted blading throughout. Gas-turbine blading is invariably of the reaction type, meaning, as in steam-turbine blading, part reaction and part impulse, but the degree of reaction increases from blade root to tip. Hence it has not been the practice of the gas-turbine designer to designate a degree of reaction, or even impulse and reaction blading, but rather to use the so-called vortex theory [45, 46].

A detailed discussion of vortex theory is beyond the scope of this book. We shall assume for simplicity that the degree of reaction is constant along the blade and hence draw the velocity diagrams for gas-turbine blades in the same manner* as steam-turbine reaction blades (Sec. 5-5). The velocities, however, are calculated from gas relationships. Helium, being a monatomic gas, has constant specific heats and is relatively easy to do. Combustion-gas properties can either be approximated by variable specific heat air (the air-to-fuel ratios are usually high) or obtained by the use of gas tables that take into account variable specific heats, the fuel-air mixture, and dissociation, App. I [48].

For the case of helium, or other gas with assumed constant specific heats, the enthalpy drop across the turbine Δh_T per unit mass-flow is given with reference to Fig. 5-29 by

$$\Delta h_T = c_p(T_o - T_e) \tag{5-46}$$

$$\frac{T_{e,s}}{T_o} = r_{p,T}^{(k-1)/k} \tag{5-47}$$

and

$$\frac{T_0 - T_e}{T_o - T_{e,s}} = \eta_T \tag{5-48}$$

where T_o and T_e = the inlet and exit temperatures, °R or K, respectively

$T_{e,s}$ = exit temperature if turbine were adiabatic reversible, °R or K

* We may add here that, in gas-turbine practice, blade and velocity angles are measured from the axial direction instead of the tangential direction as in steam practice and as we have done throughout this chapter. This practice we will ignore here and continue the same procedure as for steam blading.

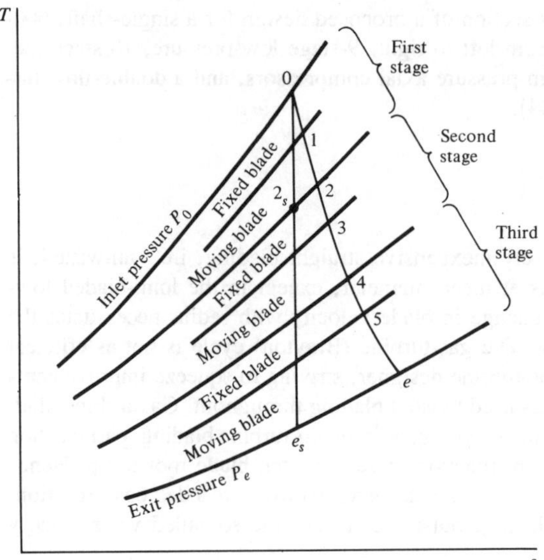

Figure 5-29 *T-s* diagram for gas turbine expansion.

c_p = specific heat at constant pressure = 1.250 Btu/(lb_m · °R) or 5.233 kJ/(kg · K) for helium and 0.240 Btu/(lb_m · °R) or 1.005 kJ/(kg · K) for air at low temperatures

k = ratio of specific heats = 1.659 for helium and 1.40 for air at low temperatures

$r_{p,T}$ = pressure ratio across turbine = ratio of inlet pressure P_o to exit pressure P_e

η_T = turbine polytropic (adiabatic, or isentropic) turbine efficiency

For equal work by the stages, the total temperature difference $T_o - T_e$ is divided equally (for constant specific heat), and the stage exit temperatures, and pressures, are found. For example, T_2 and T_4 are the exit temperatures of the first and second stages in a three-stage turbine (Fig. 5-29). The stage is now divided according to the degree of reaction. For a 60 percent reaction in the second stage for example, T_3, the fixed-blade exit temperature, is found from $(T_2 - T_3)/(T_2 - T_4) = 0.40$. The fixed-blade exit velocity, and the moving blade inlet velocity V_{s3}, is obtained from the nozzle equation for gases

$$V_{s3}^2 - V_{s2}^2 = 2g_c J c_p (T_2 - T_3) \tag{5-49}$$

where V_{s2}, the inlet velocity to the fixed blades, is obtained from the velocity diagram of the previous stage, and $J = 778.16$ ft · lb_f/Btu, if English units are used. $V_{s,o}$ may be negligible, however.

The velocity diagrams are now constructed in the manner of Figure 5-14, using Eqs. (5-31) and (5-32) where here $\Delta h_f = c_p(T_2 - T_3)$ and $\Delta h_m = c_p(T_3 - T_4)$.

For the case of combustion gases with variable specific heats, products of combustion and dissociation, the gas tables [8] (see App. I) are used. This is best illustrated by an example.

Example 5-3 A gas turbine using 200 percent theoretical air receives combustion gases at 2460°R. The first stage has a pressure ratio of 2.0, an efficiency of 0.9, and a 60 percent reaction (assumed constant along the blades). Referring to Figs. 5-14 and 5-29, take $\theta = 20°$, V_B corresponding to optimum, and calculate (1) the stage exit temperature T_2, (2) the fixed-blade exit temperature T_1 and velocity V_{s1}, (3) the moving-blade inlet and exit angles, and (4) the exit velocity for zero exit whirl. The molecular weight of the gases is 28.88.

SOLUTION Using the gas tables for 200 percent theoretical air, App. I:

$T_o = 2460°\text{R}$, $h_o = \dfrac{19168.6}{28.88} = 663.7\ \text{Btu/lb}_m$, and $P_{ro} = 521.1$.

$$P_{r2} = \frac{521.1}{2} = 260.55$$

Thus

$$T_{2s} = 2099.7°\text{R} \quad h_{2s} = 555.4\ \text{Btu/lb}_m$$

$$\eta_{\text{stage}} = 0.9 = \frac{h_0 - h_2}{h_0 - h_{2s}}$$

Therefore

$$h_2 = 566.2\ \text{Btu/lb}_m$$

Thus

$$T_2 = 2135.8°\text{R}$$

For the stage $\Delta h = h_0 - h_2 = 97.5\ \text{Btu/lb}_m$. For 60 percent reaction $\Delta h_f = 0.4 \times 97.5 = 39.00\ \text{Btu/lb}_m$. $h_1 = h_0 - 39.00 = 624.7\ \text{Btu/lb}_m$. Thus

$$T_1 = 2331.3°\text{R}$$

Therefore

$$V_{s1} = \sqrt{2 \times 32.2 \times 778.16 \times 39.00} = 1398.0\ \text{ft/s}$$

$$V_B = V_{b,\text{opt}} = V_{s1} \cos\theta = 1398.0 \cos 20 = 1313.7\ \text{ft/s}$$

$$\phi_1 = 90°$$

$$V_{r1} = V_{s1} \sin \theta = 478.1 \text{ ft/s}$$

$$\Delta h_m = 0.6 \times 97.5 = 58.50 \text{ Btu/lb}_m$$

$$\frac{V_{r2}^2 - V_{r1}^2}{2g_c} = \Delta h_m = 58.50 \times 778.16 = 45524.0 \text{ ft·lb}_f$$

Therefore

$$V_{r2} = \sqrt{2 \times 32.2 \times 45524.0 + 478.1^2} = 1777.7 \text{ ft/s}$$

For zero exit whirl $\delta = 90°$. Thus

$$\gamma = \cos^{-1} \frac{V_B}{V_{r2}} = \cos^{-1} \frac{1313.7}{1777.7} = 42.3°$$

$$V_{s2} = V_{r2} \sin \gamma = 1777.7 \sin 42.3 = 1197.7 \text{ ft/s}$$

PROBLEMS

5-1 A flat plate mounted on wheels, Fig. 5-2*a*, receives a perpendicular jet of water from a 5-cm² nozzle at a velocity of 20 m/s. calculate the maximum power, in watts, imparted to the plate and the velocity of the plate, in meters per second, corresponding to that maximum power. Take water density as 1000 kg/m³.

5-2 A large movable cylindrical blade, Fig. 5-2*b*, receives a jet of air of 1-in² cross-sectional area. The air is at 2-atm pressure and 1000°F. Find the necessary air velocity, in feet per second, to produce a maximum power to the blade of 45 kW.

5-3 Steam enters a single-stage impulse (DeLaval) turbine at 900 psia and 900°F and leaves at 300 psia. Flow is adiabatic and reversible. The nozzle angle is 20°. The blade speed corresponds to maximum blade efficiency. The moving blade is symmetric. Determine the velocity diagram and find (*a*) the velocity of the steam leaving the nozzle, in feet per second, (*b*) the blade entrance angle, (*c*) the horsepower developed for a steam flow of 1 lb_m/s, and (*d*) the blade efficiency.

5-4 A single-stage impulse turbine is required to develop 50 MW of power. Steam enters the nozzles saturated at 70 bars and leaves at 50 bars. The blades are symmetrical and have a velocity coefficient of 0.96. Calculate (*a*) the minimum steam flow, in kilograms per second, that would result in the required power, (*b*) the blade efficiency, and (*c*) the stage efficiency.

5-5 Steam expands ideally in a turbine from 2500 psia and 1000°F to 1 psia. Compare the maximum steam velocities and the number of stages required by (*a*) a velocity-compounded impulse turbine, (*b*) a pressure-compounded impulse turbine, and (*c*) a 50 percent reaction turbine if the optimum blade velocity may not exceed 885 ft/s in any of them. Take all nozzle angles to be 25°.

5-6 1.08×10^6 lb_m/h of steam enter a Curtis stage with an absolute velocity of 4000 ft/s. The nozzle angle and discharge angle of stationary blades are both 20°. The moving blades are symmetric and rotate at 600 ft/s. Assuming ideal steam flow in the nozzle and blades, determine the velocity diagram and find (*a*) the power, in horsepower and megawatts, developed in the stage, and (*b*) the blade efficiency.

5-7 A Curtis stage receives 3.6×10^6 lb_m/h of steam at 2380 ft/s and 20° angle. The blade speed is 550 ft/s. The velocity coefficients in moving and stationary blades are 0.905 and 0.932, respectively. Determine the velocity diagram and find (*a*) the total stage power in feet-pound force per second, horsepower, and kilowatts and (*b*) the blade efficiency of the stage.

5-8 Steam enters a Curtis impulse stage at 1000 psia and 1000°F, and exits at 1 atm. The nozzle angle is 20° and its efficiency is 87 percent. The fixed blade exit angle is 25°. The moving blades are symmetrical.

All velocity coefficients are 0.97. For optimum work, calculate (*a*) the blade velocity, in feet per second, (*b*) the work done by each stage, in Btus per pound mass of steam, (*c*) the stage efficiency.

5-9 A Rateau turbine operating between 1000 psia and 1000°F and 1 atm has symmetrical blades, nozzle angles of 20°, nozzle efficiencies of 97 percent, and velocity coefficients of 0.97. (The same data as that for the Curtis stage in Prob. 5-8). Calculate (*a*) the number of stages necessary to limit the optimum blade velocity to that of the Curtis stage (1009.6 ft/s), (*b*) the work of each stage, in Btus per pound mass, (*c*) the turbine efficiency, and (*d*) the percent error in steam inlet velocity to the second stage due to ignoring the absolute steam velocity leaving the first stage.

5-10 A 50 percent reaction turbine operates between 1000 psia and 1000°F and 1 psia (the same conditions as for the velocity-compound and pressure-compound impulse turbines of Probs. 5-8 and 5-9). All steam expansion efficiencies in fixed and moving blades are 87 percent. All steam absolute angles (θ and γ) are 20°. The turbine has the same optimum blade speed as the impulse turbines of Probs. 5-8 and 5-9 (1009.6 ft/s), assumed constant for all stages for simplicity. (*a*) Find the number of stages, (*b*) determine the steam velocity diagram, (*c*) calculate the work done by each stage, in Btus per pound mass, and (*d*) calculate the first stage efficiency.

5-11 A two-nozzle aeolipile similar to that of Hero of Alexandria contains saturated steam at 6 bars and exhausts to 1 atm. The nozzles are 60 percent efficient, have exit areas of 20 cm^2 each and their areas are 2 m apart. Calculate the torque, in joules, on the turbine shaft.

5-12 Consider one stage in a 50 percent reaction turbine. 1 lb$_m$/s steam enters the stage at 100 psia and 400°F and leaves at 40 psia. The adiabatic efficiency of the stage is 0.90. The blades have exit angles of 20°. The blade-speed ratio (blade velocity to incoming steam velocity) is 0.8. Determine the velocity diagram and find (*a*) the pressure at the exit of the fixed blades, in psia, (*b*) the blade speed, in feet per second, and (*c*) the horsepower developed in the stage.

5-13 A 50 percent rection stage in a steam turbine undergoes a total of 20 Btu/lb$_m$ enthalpy drop. The nozzle efficiency and angle are 88 percent and 25°, respectively. The blades move at 420 ft/s and have V_{r1} = 332 ft/s, V_{s2} = 386 ft/s, and γ = 22°. The steam flow is 1.08 × 10^6 lb$_m$/h. Find (*a*) the work done by the stage in horsepower and megawatts, (*b*) the blade efficiency, (*c*) the stage (nozzle and blade) efficiency, and (*d*) the blade velocity corresponding to maximum efficiency, in feet per second.

5-14 A reaction turbine has 33-in-long blades that receive a constant steam velocity 800 ft/s along their entire lengths. The blades are designed for optimum conditions at midlength. They are attached to a 60-in-diameter rotor. Assuming ideal frictionless flow, calculate the blade entrance and exit angles (ϕ and γ) at the base, midpoint, and tip of the blades, respectively. Assume a constant exit whirl along the length of the blades.

5-15 A steam turbine having N stages develops 20,000 hp when receiving 110,000 lb$_m$/h, 800 psia, 900°F steam, and exhausting at 1 psia. All stages have equal enthalpy drops and equal efficiencies. For a reheat factor r_h = 1.07, find (*a*) the individual stage efficiency and (*b*) the turbine efficiency.

5-16 10^6 lb$_m$/h of steam enters an ideal steam turbine stage at 100 psia and 350°F and leaves it at the Wilson line, assumed in this case to occur at 50 psia. Calculate the stage exit temperature, in degrees Fahrenheit, and the power produced in the stage, in megawatts, in the cases of (*a*) supersaturation, assuming that supersaturated steam expansion may be represented by perfect gas laws with a polytropic exponent n = 1.32, and (*b*) thermodynamic equilibrium.

5-17 Consider a simple combination turbine with one Delaval impulse stage and one 60 percent reaction stage. The nozzle of impulse stage receives steam at 900 psia and 900°F and leaves it at 300 psia. The nozzle efficiency is 97 percent and its angle is 20°. The blade speed is optimum and its velocity coefficient is 0.95. The impulse stage is followed on the same shaft by the reaction stage which exhausts to 100 psia. The steam entrance angles for that stage are also 20°. The efficiency of the fixed blades (nozzles) is 90 percent. Because of different diameters of the impulse and reaction moving blades, the velocity of the reaction blade is 1.5 that of the impulse blade. (*a*) Determine all steam velocities and draw the velocity diagram of the combined turbine, (*b*) calculate the work of each stage, in Btus per pound mass, and (*c*) calculate the individual stage and the turbine efficiencies.

5-18 It is required to compare the design of an ideal helium to an ideal air gas turbine. Both have the same inlet gas temperature of 2000°R and maximum gas velocity of 1250 ft/s. Find the number of stages of each if the overall pressure ratios are 2 for the helium turbine and 6 for the air turbine. Take c_p = 1.25 Btu/lb$_m$ · °R for helium and 0.243 Btu/lb$_m$ · °R for air.

5-19 A gas turbine composed of two reaction stages receives 1 lb$_m$/s of combustion gases (assumed to be pure air) at 80 psia and 1800°F. It exhausts at 15 psia. All stages are 50 percent reaction. Flow is considered adiabatic and reversible. The blade exit angles are 20°. The blade speeds correspond to maximum efficiency. Both stages produce equal power. Determine the velocity diagram and find the turbine power in horsepower and megawatts. For simplicity, assume V_B = constant, and c_p = 0.24 Btu/lb$_m$ · °R and k = 1.4.

5-20 An ideal helium gas turbine has six rows of 50 percent reaction blades and an overall pressure ratio of 2.5. All stages have the same enthalpy drop. The maximum helium temperature is 1000°F. The blade speed corresponds to optimum work. All blade entrance angles are 20°. Considering the high-pressure stage, determine the velocity diagram, and calculate (*a*) the helium exit temperature and (*b*) the horsepower and megawatts developed in the stage for a helium flow of 1 lb$_m$/s.

5-21 An ideal gas turbine receives combustion gases at 6 atm and 2000°F, and exhausts to 1 atm. It has two stages of 50 percent reaction blading producing equal work per stage. $\gamma = \theta = 20°$, $V_{r2} = V_1$. Consider the turbine to be adiabatic and reversible. For the high-pressure stage (*a*) draw the velocity diagram, assuming optimum blade speed, (*b*) find the horsepower for 1 lb$_m$/s air flow, and (*c*) the blade efficiency. For simplicity assume the gases to have a constant c_p = 0.24 Btu/lb$_m$ · °R k = 1.4.

5-22 An ideal-gas turbine composed of two reaction stages receives 1 lb$_m$/s of combustion gases with 200 percent theoretical air at 80 psia and 1800°F. It exhausts at 15 psia. All stages are 50 percent reaction. The blade exit angles are 20°. The blade speeds correspond to maximum efficiency. Both stages produce equal power. Determine the velocity diagram and find the turbine power in kilowatts. For simplicity, assume V_B = constant. The molecular weight of the gases is 28.88. Use the gas tables, App. I.

5-23 A reaction gas turbine stage with 59.16 percent degree of reaction (based on an isentropic enthalpy drop) receives 10^6 lb$_m$/s of combustion gasses with 200 percent theoretical air at 2560°R. The pressure ratio across the stage is 1.987. The fixed blades (nozzles) exit angle is 22° and their efficiency is 86.72 percent. The stage efficiency is 83.94 percent. The moving blade speeds are optimum. The molecular weight of the gases is 28.88. Using the gas tables determine the velocity diagram and calculate (*a*) the gas velocity, in feet per second, and temperature, in degrees Rankine, entering the moving blades, (*b*) the gas velocity, in feet per second, and temperature, in degrees Rankine, leaving the stage, and the stage power, in megawatts.

5-24 A reaction gas turbine stage with 59.16 percent degree of reaction receives combustion gases with 200 percent theoretical air at 2560°R. The pressure ratio across the stage is 1.987. The fixed blade (nozzle) efficiency is 86.72 percent. The stage efficiency is 83.94 percent. Calculate (*a*) the gas velocity entering the moving blades, in feet per second, (*b*) the stage efficiency, and (*c*) the power developed, in megawatts. The molecular weight of the gases is 28.88.

CHAPTER

SIX

THE CONDENSATE-FEEDWATER SYSTEM

6-1 INTRODUCTION

We will now continue following the working fluid around the powerplant cycle. We started with steam generation and combustion in Chaps. 3 and 4 and then continued on with turbines in Chap. 5. In this chapter we will cover the major components that make up the condensate-feedwater system, which takes us back to the steam generator. These are primarily heat-transfer equipment, such as the condenser and feedwater heaters, but they also include some important items that are necessary for the efficient operation of the cycle, such as boiler water makeup and treatment.

The plant heat-rejection system, which deals with the circulating cooling water of the condenser, requires separate and special attention and will be covered in the next chapter.

The primary purpose of the condenser is to condense the exhaust steam from the turbine and thus recover the high-quality feedwater for reuse in the cycle. In so doing, it actually performs an even more useful function. If the circulating cooling-water temperature is low enough, as is usually the case, it creates a low back pressure (vacuum) for the turbine to exhaust to. This pressure is equal to the saturation pressure that corresponds to the condensing steam temperature, which in turn is a function of the cooling-water temperature. As is now known, the enthalpy drop, and hence turbine work, per unit pressure drop is much greater at the low-pressure than the high-pressure end of a turbine. A condenser, by lowering the back pressure by only a few psi, increases the work of the turbine, increases the plant efficiency, and reduces the steam flow for a given plant output. The lower the pressure, the greater the effects; hence, thermodynamically, it is important to use cooling-water temperatures that are the lowest available. Condensing powerplants are therefore much more efficient than noncondensing ones. All modern powerplants are of the condensing type. A condenser is a major and very important piece of equipment in a powerplant.

There are primarily two types of condensers: *direct-contact* and *surface condensers*. The latter are used in the majority of powerplants.

The main purpose of feedwater heaters (Chap. 2) is to improve cycle efficiency by heating the condensate and feedwater before returning it to the steam generator. The heating could be as high as 400 to 500°F (200 to 260°C) in a fossil-fueled powerplant but is lower in a water-cooled nuclear-reactor powerplant. There are also two basic types of feedwater heaters: the closed, *surface* or shell-and-tube type, and the *open,* direct-contact or *deaerating* type.

6-2 DIRECT-CONTACT CONDENSERS

Direct-contact, or *open, condensers* are used in special cases, such as when dry-cooling towers are used (Sec. 7-6), in geothermal powerplants (Chap. 12), and in powerplants that use temperature differences in ocean waters (OTEC) (Chap. 15). Modern direct-contact condensers are of the spray type. Early designs were of the barometric or jet types.

The Spray Condenser

Direct-contact condensers, as the name implies, condense the steam by mixing it directly with the cooling water. In the *spray condenser* this is done by spraying the water into the steam. Thus turbine exhaust steam at point 2, Figs. 6-1 and 6-2, mixes

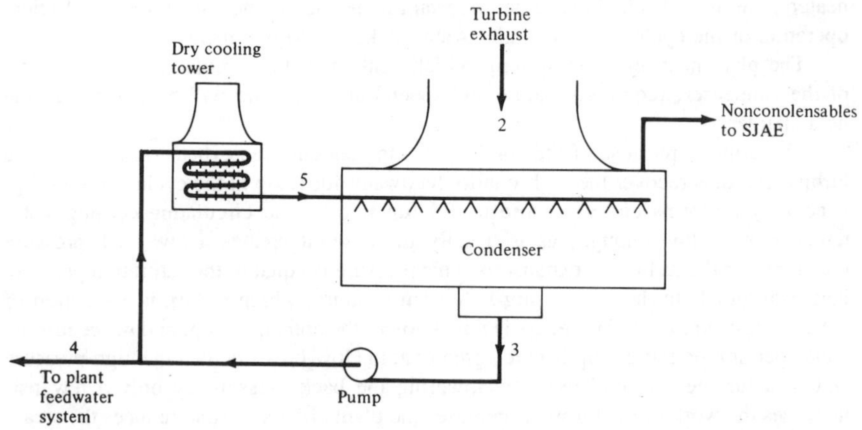

Figure 6-1 Schematic flow diagram of a direct-contact condenser of the spray type. SJAE = steam-jet air ejector.

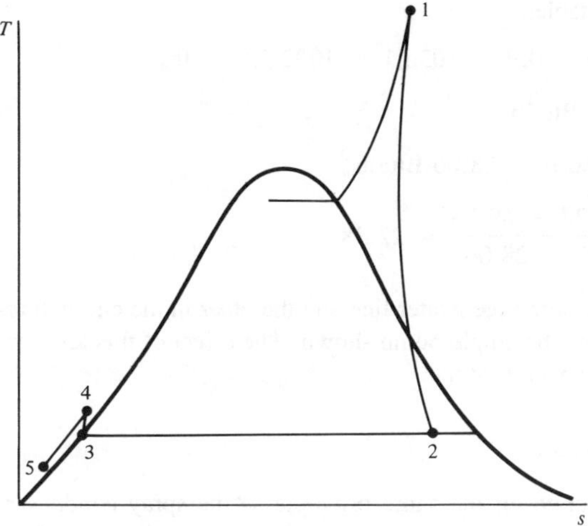

Figure 6-2 *T-s* diagram of condensate and cooling water in a direct-contact condenser system. (Difference between line 4-5 and saturated-liquid line is exaggerated).

with cooling water at 5 to produce nearly saturated condensate at 3, which is pumped to 4. Part of the condensate, equal to the turbine exhaust flow, is sent back to the plant as feedwater. The rest is cooled, usually in a dry- (closed-) cooling tower to point 5. The cooled water at 5 is sprayed into the turbine exhaust, and the process is repeated. Thus cooling water continually circulates. Its purity must be maintained because it mixes with the steam, hence its use with closed-type dry-cooling tower. Otherwise the mixture at 4 is discarded, as in geothermal or OTEC plants.

A mass balance on the system, where $\dot{m}$ denotes mass-flow rates, gives

$$\dot{m}_2 = \dot{m}_4 \tag{6-1}$$

and

$$\dot{m}_3 = \dot{m}_2 + \dot{m}_5$$

An energy balance, where h denotes specific enthalpies, gives

$$\dot{m}_2 h_2 + \dot{m}_5 h_5 = \dot{m}_3 h_3 \tag{6-2}$$

and the ratio of circulating cooling water to steam flow is given by

$$\frac{\dot{m}_5}{\dot{m}_2} = \frac{h_2 - h_3}{h_3 - h_5} \tag{6-3}$$

Thus circulating-water flow is much greater than steam flow because $h_2 - h_3$ represents a large fraction of the large latent heat of vaporization at the reduced pressure, whereas $h_3 - h_5$ represents the much smaller sensible heat of the liquid.

Example 6-1 Find the ratio of circulating water to steam flows if the condenser pressure is 1 psia and the cooling tower cools the water to 60°F. Assume turbine exhaust at 90 percent quality.

SOLUTION From the steam tables

$$h_2 = 69.73 + 0.9 \times 1036.1 = 1002.22 \text{ Btu/lb}_m$$

$$h_3 = 69.73 \text{ Btu/lb}_m$$

$$h_5 \approx h_f \text{ at } 60°\text{F} = 28.06 \text{ Btu/lb}_m$$

$$\frac{\dot{m}_5}{\dot{m}_2} = \frac{1002.22 - 69.73}{69.73 - 28.06} = 22.38$$

In Fig. 6-1, two pumps, one in the feedwater line and the other in the circulating-water line, may be used instead of the single pump shown. The effect of this arrangement on mass and energy balances is minimal.

Barometric and Jet Condensers

These early type condensers operate on the same principle as the spray condenser, except that no pump is required. The vacuum in the condenser is obtained by virtue of a static head, as in the barometric condenser (Fig. 6-3*a*), or a diffuser, as in the jet condenser (Fig. 6-3*b*).

In the *barometric condenser,* the cooling water is made to cascade down a series of baffles in the form of water curtains or sheets of high surface-to-volume ratio to

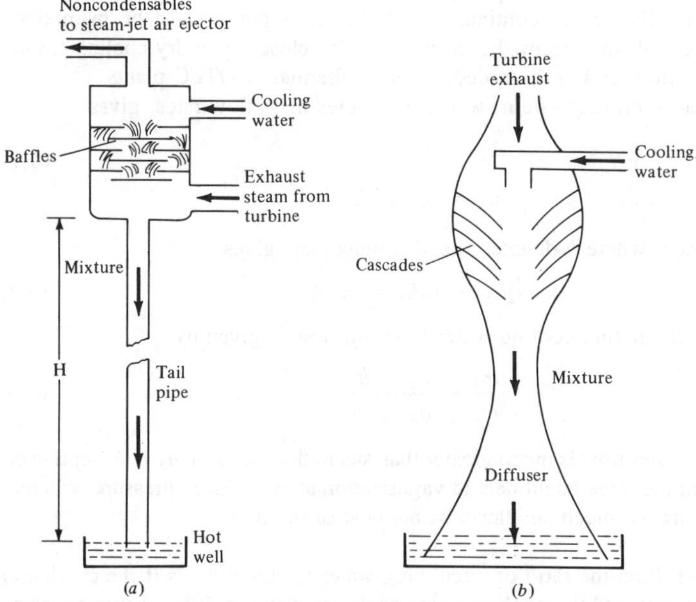

Figure 6-3 Schematics of early direct-contact condensers: (*a*) barometric, (*b*) diffuser or jet.

mix thoroughly with the turbine exhaust that is trying to rise from a lower inlet. The steam condenses and the mixture goes down a *tail pipe* to the hot well.

The tail pipe compresses the mixture to atmospheric pressure at the hot well by virtue of its static head and thus replaces the pump used in the spray-type condenser. The pressure differential created by the tail pipe must overcome the pressure difference between the atmosphere P_{atm} and the condenser proper P_{cond}, plus the friction pressure drop caused by the mixture flow ΔP_f in the tail pipe itself. Thus

$$\rho H \frac{g}{g_c} = P_{atm} - P_{cond} + \Delta P_f \tag{6-4}$$

where ρ = density of mixture, lb_m/ft^3 or kg/m^3

H = height of tail pipe, ft or m

g = gravitational acceleration, ft/s^2 or m/s^2

g_c = conversion factor, 32.2 $lb_m \cdot ft/(lb_f \cdot s^2)$ or 1 $kg \cdot m/(N \cdot s^2)$

It can be seen that for low values of the friction pressure drop the tail pipe height H is around 32 ft (9.6 m) and is higher the larger the friction. Friction is reduced by increasing tail-pipe diameter, which results in a tall, heavy system.

In the *jet-type condenser* (Fig. 6-3*b*), the height of the tail pipe is reduced by replacing it with a diffuser. The diffuser acts on the same principle as the diverging section of a convergent-divergent nozzle in subsonic flow. It thus helps raise the pressure in a shorter distance than the tail pipe. Even though the height is reduced, the mass and cost of the system are probably increased, however.

In both barometric and diffuser-type condensers, the mixture is split and cooled in the same manner as in the spray-type condenser. In all direct-contact condensers, as in surface-type condensers (below), noncondensable gases must be removed, usually with a steam-jet air ejector (SJAE).

6-3 SURFACE CONDENSERS: GENERAL

Surface condensers are the most common type used in powerplants. They are essentially shell-and-tube heat exchangers, in which the primary heat-transfer mechanisms are the condensing of the saturated steam on the outside of the tubes and the forced-convection heating of the circulating water inside the tubes. Figure 6-4 is a schematic of a surface condenser with two passes on the water side. It is composed of a steel shell with water boxes on each side, the right one divided to allow for the two water passes. The water tubes are rolled at each end into tube sheets, and there are steel support plates at intermediate points between the tube sheets to prevent tube vibration. The hot well that receives the condensate acts as a reservoir, with a capacity equal to the total condensate flow during a prescribed time, e.g., 1 min.

Surface condensers have grown in size since late in the last century to present-day sizes that exceed 1 million ft^2 ($\sim$93,000 m^2) of heat-transfer surface area. A

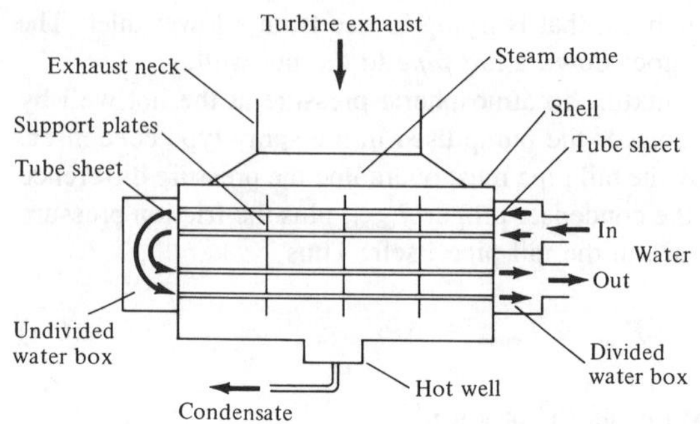

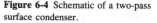

Figure 6-4 Schematic of a two-pass surface condenser.

landmark of development was two 80,000 ft^2 (7432 m^2) units built in 1929 by the Foster Wheeler Corporation for the predecessor of the Commonwealth Edison Co. of Chicago, Illinois. These were and remained through the late 1940s the largest units built. They were constructed of sections of cast-iron shells that were bolted together and were single-pass with full reverse backwashing provided by valving built integrally with the water boxes.

Besides size, philosophies of design have also changed over the years. The early surface condensers used simple circular tube sheets that supported as many tubes that could be tightly packed between them. They were 3 or 4 ft in diameter, and the tubes were no more than 10 or 12 ft in length. As powerplants and their condensers grew in size, this simple design resulted in heat-transfer problems because the upper tubes shielded the steam from effective condensing and in high-steam-pressure drop problems because of the long tortuous path of the steam through the bundle. (We would like to see the minimum pressure at turbine exhaust, if possible.) The heat-transfer problem was solved by using larger spacings between tubes (called ligaments) and placing them in rows that provided lanes or steam paths to allow the steam to penetrate deeply into the lower tubes.

The next improvement tackled the high-pressure drops by cutting the bundle in half, thus in effect forming two smaller bundles side by side. This gave the condenser a kidney cross-sectional shape that was popular with most manufacturers through the 1940s. Even this solution was not sufficient for the larger units that were coming on line because the bundles were still too deep for effective steam penetration. Four tube bundles were then used. This also helped reduce condenser height, an increasingly frequent requirement because of the low available head room in the plant.

The current design philosophy is to have a tube layout in the shape of a funnel with most tubes, and the largest tube passage area, where the steam enters the condenser from the turbine. As the steam condenses and its volume decreases, there are fewer tubes and smaller passage areas. Steam is made to enter the tube bundle, or bundles, from all sides toward a central air cooler for deaeration (below). In effect a low and balanced (to avoid cross-flow) steam-pressure drop should be ensured.

The tubes are rolled into the tube sheets at both ends to prevent leakage of the circulating water into the steam. An *expansion joint* allows for the different rates of expansion between the tubes and the shell. The tube sheets are usually made of Muntz metal, which is similar to brass.

A problem of steam distribution, other than vertical penetration, is end-to-end or horizontal distribution that arises with present-day long-tube units. Tube lengths of 30 to 50 ft (~9 to 15 m) are commonplace. Multipressure condensers (below) may have 70- to 90-ft (~21- to 27-m) tubes. Long tubes result in larger changes in water temperature inside them and hence greater changes in condensing ability. Thus, the tubes would be too close at the cold end, where condensing is good, and too open at the hot end. A design compromise is, of course, necessary. This results in some short-circuiting that may be counteracted by cross baffles.

Another distribution problem is the result of the unavoidably unequal steam flow from the turbine exhaust duct to the condenser tubes. Thus special attention must be paid to the design of the connection between turbine and condenser (called the *exhaust neck*), such as adding a well-tapered *steam dome* above the tube bundle to minimize this problem. An expansion joint is usually provided between the turbine exhaust and the condenser steam inlet. This permits the condenser to be rigidly mounted on the floor. Another, less common, arrangement is to bolt the condenser directly to the turbine exhaust duct and support it on springs that allow a certain vertical movement and reduce the strain on the turbine casing.

Number of Passes and Divisions

Condensers are designed with one, two, or four cooling-water passes. The number of passes determines the size and effectiveness of a condenser. Four passes are seldom used in utility installations. A *single-pass condenser* is one in which the cooling water flows through all the condenser tubes once, from one end to the other. In a *two-pass condenser* the water enters half the tubes at one end of a divided inlet water box, passes through these tubes to an undivided water box at the other end, reverses direction, and passes through the other half of the tubes back to the other side of the divided water box. A single-pass condenser with the same total number and size of tubes, i.e., the same heat-transfer area, and with the same water velocity, requires twice as much water flow but results in half the water temperature rise and thus lower condenser pressure. Thus such a single-pass condenser is good for plant thermal efficiency and reduces thermal pollution, but requires more than twice the water and hence four times the pumping power.

Water boxes are often divided beyond the divisions required by the number of passes. A divided water box single-pass condenser, for example, may have a partition in both the inlet and outlet water boxes at opposite ends of the condenser. This allows half the condenser to operate while the other half is being cleaned or repaired. In the case of a divided two-pass condenser, the water boxes are divided into four quarters. Divided water boxes have duplicate inlet and outlet connections, each with its own circulating-water circuit. Valves in the division plates permit backwashing by reversing water flow for cleaning purposes.

Single- and Multipressure Condensers

As is now known, large powerplants usually have two or more low-pressure turbine sections in tandem. The condenser may be divided into corresponding sections or shells, situated below the low-pressure turbine sections.

If the turbine exhaust pressure in all sections is the same, i.e., when the exhaust ducts are not isolated from each other, we would have a *single-pressure condenser*. If the exhaust ducts are isolated from each other, these individual condenser shell pressures will increase because the circulating-water temperature will increase as it flows from shell to shell. We would then have a *multipressure condenser*.

A multipressure condenser results in efficiency improvement because the average turbine back pressure is less compared with that of a single-pressure condenser (which is determined by the highest circulating-water temperature). Multipressure condensers are more commonly used in nuclear powerplants. They are usually single-pass units

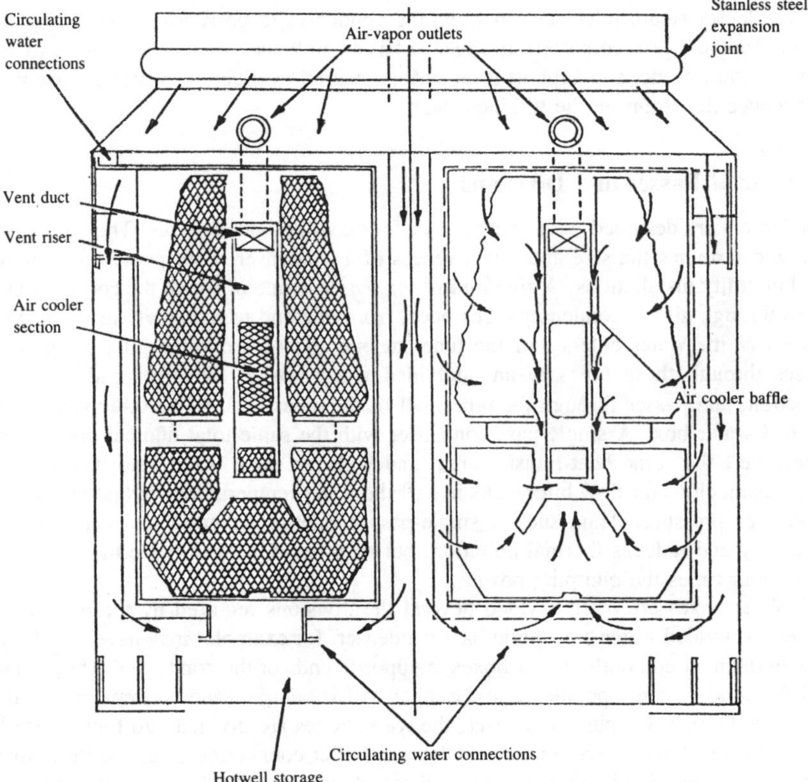

Figure 6-5 cross section of a single-pass, divided-box surface condenser. Note the radial pattern of steam flow.

arranged with their tubes parallel to the turbine shaft. They are roughly as long as the low-pressure turbine sections combined, often 70 to 90 ft (~21.3 to 27.4 m). Single-pressure condensers, on the other hand are usually 30 to 50 ft (~9.1 to 15.2 m) long and are often arranged with their tubes perpendicular to the turbine shaft.

In essence, condensers are almost custom-designed to suit individual requirements of steam flow, available cooling-water flow and temperature, available space, and other variables.

Figure 6-5 shows a cross section of a typical modern large condenser. It is a single-pass, single-pressure, radial-flow type condenser in which the steam enters the bundles from top, sides, and bottom and flows toward the center of the tube nest. At that point most of it has condensed, leaving only air and other noncondensable gases that are cooled before being removed by the deaeration system (below). Figure 6-6 shows a two-pass divided-box surface condenser.

Table 6-1 lists typical condenser dimensions for plants up to 500-MW capacity.

Tube Sizes and Materials

Tube sizes and gauges (Birmingham Wire Gauge, BWG) are listed for condenser and feedwater heater tubes in App. K. Note that the higher the gauge number, the thinner

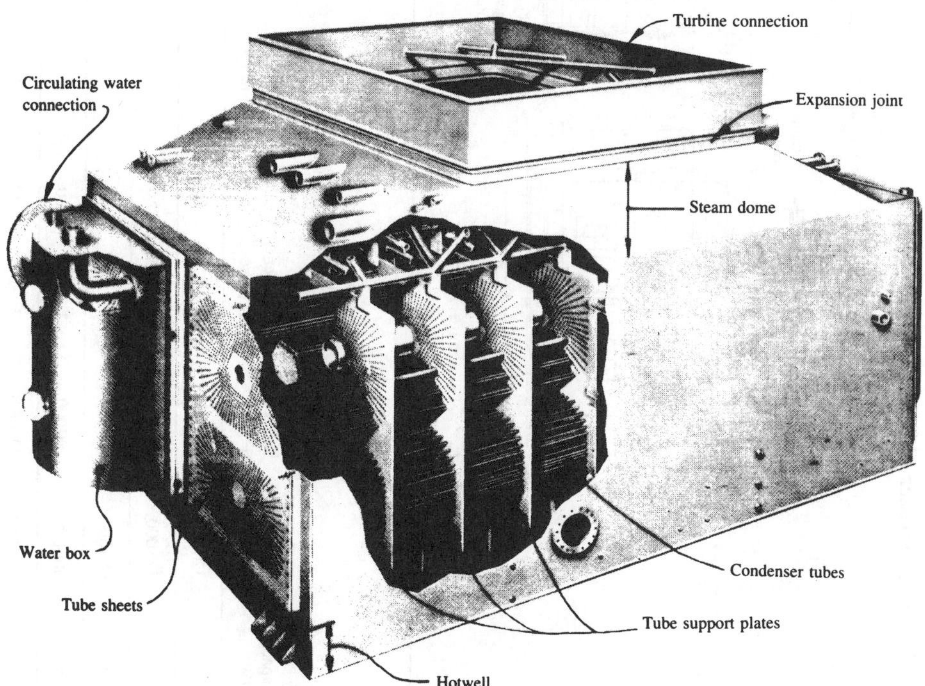

Figure 6-6 A two-pass divided-box surface condenser.

Table 6-1 Condenser dimension sheet*

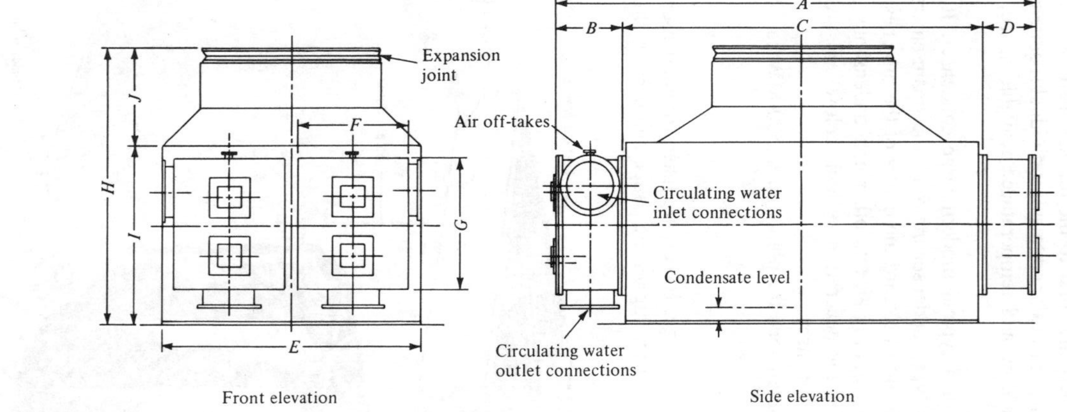

Capacity, MW	Steam flow, lb_m/h	Surface area, ft^2	Number of passes	Circulating water inlet temp., °F	Vacuum, in Hg	Hotwell storage, min	Condenser tubes		Connections			
									Circulating water		Turbine exhaust	
							Length	OD in.	Inlet, in.	Outlet, in.	Width	Length
50	320000	42500	2	80	2	5	26	7/8	30	30	9′–5″	12′–5″
75	376000	55000	2	80	2	5	26	7/8	36	36	9′–5″	12′–5″
100	486000	75000	2	80	2	5	28	7/8	42	42	14′–0″	14′–0″
125	600000	90000	2	80	2	5	28	7/8	48	48	14′–0″	14′–0″
150	710000	110000	2	80	2	5	28	7/8	54	54	14′–0″	14′–0″

200	910000	150000	2	80	2	5	30	7/8	60	60	17′–4″	19′–4″
250	1100000	180000	2	80	2	5	30	7/8	66	66	18′–0″	20′–0″
300	1310000	210000	2	80	2	5	30	7/8	72	72	18′–0″	20′–0″
400	1740000	280000	2	80	2	5	30	7/8	78	78	18′–0″	20′–0″
500	2140000	340000	2	80	2	5	30	7/8	84	84	18′–0″	20′–0″

Table 6-1 Condenser dimension sheet (continued)

	Dimensions									
Capacity MW	*A*	*B*	*C*	*D*	*E*	*F*	*G*	*H*	*I*	*J*
50	31′–6″	3′–6″	26′–0″	2′–0″	19′–3″	7′–1″	9′–0″	19′–4″	12′–4″	7′–0″
75	32′–8″	4′–0″	26′–0″	2′–6″	21′–2″	8′–0″	10′–1″	20′–5″	13′–5″	7′–0″
100	35′–4″	4′–6″	28′–0″	3′–0″	22′–10″	8′–10″	11′–3″	22′–7″	14′–7″	8′–0″
125	36′–6″	5′–0″	28′–0″	3′–6″	24′–4″	9′–7″	12′–3″	23′–7″	15′–7″	8′–0″
150	37′–6″	5′–6″	28′–0″	4′–0″	26′–2″	10′–6″	13′–5″	25′–9″	16′–9″	9′–0″
200	40′–6″	6′–0″	30′–0″	4′–6″	29′–0″	11′–9″	15′–0″	28′–0″	19′–0″	9′–0″
250	41′–8″	7′–6″	30′–0″	5′–0″	31′–0″	12′–9″	16′–4″	30′–4″	20′–4″	10′–0″
300	42′–6″	7′–0″	30′–0″	5′–6″	34′–0″	13′–9″	17′–7″	31′–7″	21′–7″	10′–0″
400	43′–6″	7′–6″	30′–0″	6′–0″	38′–0″	15′–9″	20′–3″	34′–3″	24′–3″	10′–0″
500	44′–6″	8′–0″	30′–0″	6′–6″	41′–0″	17′–3″	22′–1″	36′–1″	26′–1″	10′–0″

* Courtesy Westinghouse Electric Corporation.

Note: Based on using admiralty tubes.

and lighter the tubes. 5/8-in. tubes (all sizes refer to OD) are easily clogged and are used only for small and special applications. Modern condensers commonly use 7/8 or 1.0 in. of 18 gauge, which is adequate for the water pressures encountered in condensers.

Tube materials in common use are listed in Table 6-2. Admiralty metal,* the popular choice for a long time, although still occasionally specified, is being supplanted by type 304 stainless steel. Type 304 stainless, now readily available at reasonable cost, has excellent erosion and corrosion resistance in fresh water and immunity to ammonia and sulfide attack. It also eliminates the risk of introducing copper ions into the feedwater, a potential possibility with other materials. Its disadvantages are a low thermal conductivity and low resistance to chloride attack and biofouling. 90-10 copper-nickel is another choice for fresh water.

In the case of seawater and brakish water, 90-10 copper-nickel is the primary choice whether these waters are clean or polluted. 70-30 copper-nickel is preferred in the case of clean waters where shell-side ammonia is a problem. Copper-nickel has excellent corrosion resistance in salt and brakish water and good immunity to stress-corrosion cracking. Aluminum-brass is another tube material, though at present a remote possibility with polluted water.

Other materials that have seen use are arsenical copper and aluminum-bronze. It is expected that, for the foreseeable future, stainless steel and copper-nickel will dominate the market for condenser tubes. Other materials of promise include titanium and AL6X, which are expected to see increasing service under severe conditions. In any case no one material can function perfectly without periodic cleaning, and there is a growing interest in devising methods of on-line cleaning.

Deaeration

In steam and other vapor cycles, it is important to remove the noncondensable gases that otherwise accumulate in the system. The noncondensables are mostly air that leaks from the atmosphere into those portions of the cycle that operate below atmospheric pressure, such as the condenser, but also include other gases caused by the decomposition of water into oxygen and hydrogen by thermal or radiolytic (under the influence of nuclear radiation) action and by chemical reactions between water and materials of construction. The presence of noncondensable gases in large quantities has undesirable effects on equipment operation for several reasons.

1. They raise the total pressure of the system because that total pressure is the sum of the partial pressures of the constituents. Thus in a condenser the pressure will be the sum of the saturation pressure of the steam, determined by its temperature, and the partial pressure of the noncondensables. An increase in condenser pressure lowers plant efficiency.

* Admiralty metal: 70 to 73 percent copper (Cu), 0.9. to 1.2 percent tin (Sn), 0.07 percent maximum iron (Fe), the rest zinc (Zn).

2. They blanket the heat-transfer surfaces such as the outside surface of the condenser tubes, thus resulting in a severe decrease in the condensing heat-transfer coefficient and hence in condenser effectiveness.
3. The presence of some noncondensables results in various chemical activities. Oxygen causes corrosion, most severely in the steam generator. Hydrogen, which is capable of diffusing through some solids, causes hydriding (e.g., uranium hydride, which swells and ruptures nuclear-fuel elements). Hydrogen, methane, and ammonia are also combustible.

The process of removing noncondensables is called *deaeration*. Most fossil powerplants have a deaerating feedwater heater (Sec. 6-6), but whether or not a plant has such a heater, or other separate deaerator, it is essential that the condenser itself be the place of good deaeration. Manufacturers usually guarantee a maximum oxygen concentration in the condensate leaving the condenser. For some time, this maximum was set at 0.03 cm^3/L (0.003 percent by volume), but this has been reduced recently to 0.01 cm^3/L, required, and 0.005 cm^3/L frequently guaranteed.

Good deaeration within a condenser requires time, turbulence, and good venting equipment. The cold condensate falling from the lower tubes must have sufficient falling height and scrubbing steam for reheat and deaeration. The scrubbing steam is provided by allowing some of the incoming steam to pass through an open flow area directly to the bottom tubes to reheat the condensate. The reason is that noncondensables are more easily released from a hotter than a colder liquid.

Once the noncondensables are released, they are cooled to reduce their volume before being pumped out of the condenser. For this a number of water tubes, about 6 to 8 percent in the center of the tube bundle, are set aside for this function (Figs. 6-5 and 6-7). This, called an *air-cooler section,* is baffled to separate the noncondensables from the main steam flow. The noncondensables flow toward the cold end of the condenser, where they connect to a vent duct that leads to the venting equipment.

The venting equipment, as other components, went through several stages of development, including reciprocating compressors, called *dry-vacuum pumps,* which were used for some time. These were, however, superseded by *jet pumps,* which have now found almost universal acceptance because of their simplicity and lack of moving parts and, hence, low maintenance and good reliability.

The jet pumps used on condensers have come to be known as *steam-jet air ejectors (SJAE)* because they use a steam jet as their motive or driving flow. They are usually multistage units, usually two or three. Figure 6-8 shows a two-stage SJAE. It uses main steam at a reduced pressure that enters a driving-flow nozzle in the first-stage ejector, from which it exits with high velocity and momentum and reduced pressure. This reduced pressure draws in the noncondensables from the condenser. By a process of momentum exchange, the gases are entrained by the steam jet. The combined flow of steam and gas is now compressed in the diffuser of the first-stage ejector and discharged into a small intercondenser, where the steam is condensed by passing across cooling pipes in much the same manner as the main condenser. Cooling here, however, is accomplished by the main condenser condensate and is part of the feedwater heating system, resulting in improvement in efficiency of the plant.

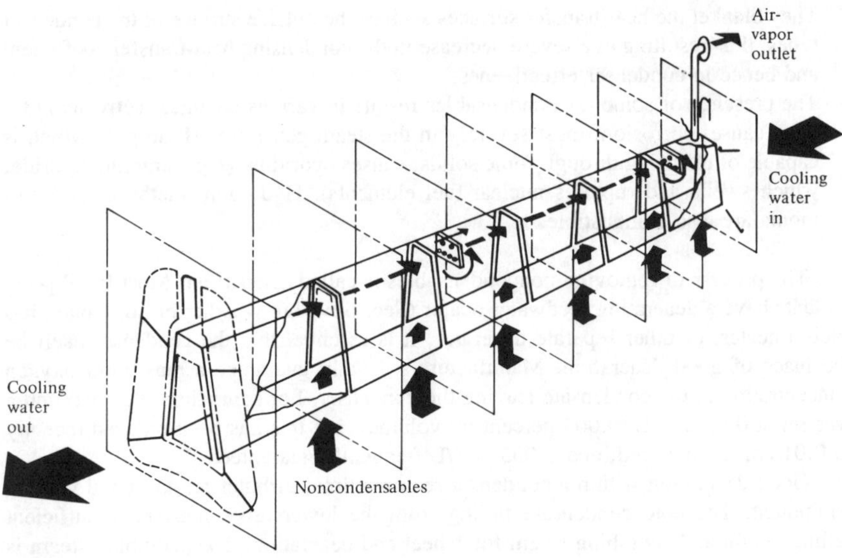

Figure 6-7 Condenser air-removal system. *(Courtesy Foster Wheeler Energy Corporation.)*

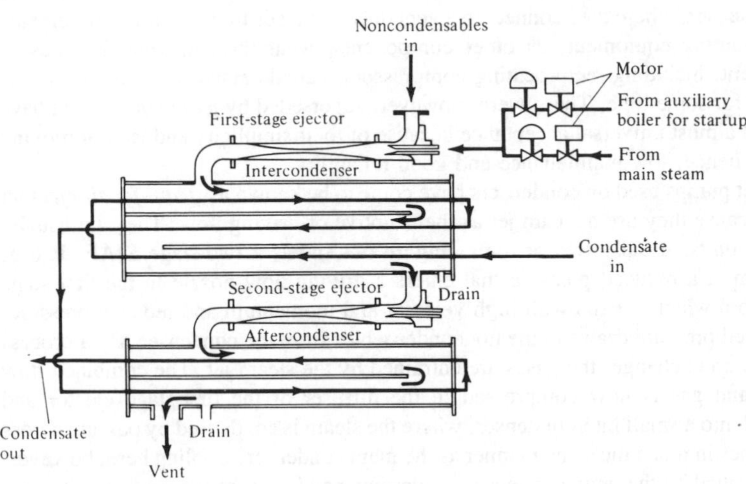

Figure 6-8 A two-stage steam-jet air ejector (SJAE).

The condensed steam is drained and returned to a low-pressure part of the cycle. The noncondensables and any remaining steam are then passed to the second-stage ejector, where they are compressed further and passed to an aftercondenser. A third-stage ejector may or may not be necessary to bring the system to the discharge pressure or, in nuclear powerplants, to the off-gas system.

6-4 SURFACE-CONDENSER CALCULATIONS

Heat-Transfer Surface Area

The heat-transfer calculations for determining the number of tubes and total surface area required by a surface condenser are rather complex. They require a knowledge of the total heat load on the condenser, i.e., the heat removed by the turbine exhaust steam, low-pressure feedwater heater drains, steam-jet air ejector (SJAE) drains, small drive turbine exhausts, etc. They also require a knowledge of the heat-transfer mechanisms and coefficients in various parts of the condenser. The heat-transfer mechanisms are condensation of the steam over the colder but varying-temperature tube surfaces, conduction through the tube walls, forced convection of the circulating water inside the tubes, and forced convection of the noncondensables in the air-removal section.

The outer tube surfaces are usually clean when the condenser is new but quickly develop an oily film that changes condensation from dropwise to filmwise condensation [2,47]. This is expected in most condensing equipment, and heat-transfer coefficients are conservatively based on the lower filmwise condensation mechanism. The heat-transfer coefficient here depends upon the difference between the steam-saturation temperature and tube-wall temperature (being inversely proportional to that difference to the power 0.25), on the relative positions of tubes, steam velocity and turbulence, the extent of noncondensables, and the existence of superheated steam, if any. The circulating-water heat-transfer coefficient depends upon its velocity and, hence, temperature and the cleanliness of the inside surface. Because this circulating water may be obtained from a natural body of water, algae and other deposits accumulate on that surface, thus affecting heat transfer and necessitating frequent cleaning.

Because all the above are variables with many uncertainties, manufacturers have usually based their designs in general accordance with a method proposed by the Heat Exchange Institute Standards for Steam Surface Condensers [48]. The method is based on the usual heat transfer equation

$$Q = UA\, \Delta T_m \tag{6-5}$$

where Q = heat load on condenser, Btu/h or J/s

U = overall condenser heat-transfer coefficient, based on outside tube area, Btu/(h · ft^2 · °F) or J/(s · m^2 · K) or W/(m^2 · K)

A = total outside tube surface area, ft^2 or m^2

ΔT_m = log mean temperature difference in the condenser, °F or °C, given by

Table 6-2 Constants in Eq. (6-7)

Tube outer diameter, in	3/4	7/8	1.0
C_1 [V in ft/s, U in Btu/(h · ft² · °F)]	270	263	251
C_1 [V in m/s, U in W/(m² · K)]	2777	2705	2582

Water temperature, °F	35	40	45	50	55	60	70	80	90	100
C_2	0.57	0.64	0.72	0.79	0.86	0.92	1.00	1.04	1.08	1.10

Tube material	304 stainless steel	Admiralty, arsenic-copper	Aluminum-brass, Muntz metal	Aluminum-bronze, 90-10 Cu-Ni	70-30 Cu-Ni
C_3 18 gauge	0.58	1.00	0.96	0.90	0.83
17 gauge	0.56	0.98	0.94	0.87	0.80
16 gauge	0.54	0.96	0.91	0.84	0.76

C_4 0.85 for clean tubes, less for algae covered or sludged tubes

$$\Delta T_m = \frac{\Delta T_i - \Delta T_o}{\ln(\Delta T_i/T_o)} \tag{6-6}$$

ΔT_i = difference between saturation-steam temperature and inlet circulating water temperature (Fig. 6-9), °F or °C

ΔT_o = difference between saturation-steam temperature and outlet circulating-water temperature (Fig. 6-9), °F or °C, also called *terminal temperature difference,* TTD

The overall heat-transfer coefficient U is given empirically by

$$U = C_1C_2C_3C_4\sqrt{V} \tag{6-7}$$

where V = circulating-water velocity in the tubes at inlet (cold) conditions, ft/s or m/s

C_1 = dimensional factor depending upon tube outer diameter

C_2 = dimensionless correction factor for circulating-water inlet temperature

C_3 = dimensionless correction factor for tube material and gauge

C_4 = dimensionless cleanliness factor

Refer to Table 6-2 for the factors in Eq. (6-7).

In using Eqs. (6-5) to (6-7), it is necessary to know the steam-saturation temperature and the circulating-water inlet temperature, hence ΔT_i, and to select a value for the *terminal temperature difference* (TTD) of the condenser, which in this case is ΔT_o. For a given ΔT_i, calculated U, and selected ΔT_o, the tube surface area is calculated and the condenser design is fixed. For a given ΔT_i, a large TTD results in a large ΔT_m and small condenser (small A) but an increased water flow because the water-temperature rise ($T_2 - T_1$, Fig. 6-9) is reduced. A small TTD results in a larger condenser,

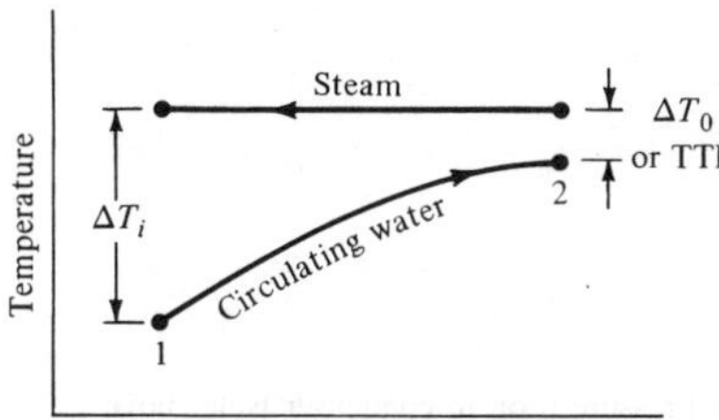

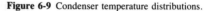

Figure 6-9 Condenser temperature distributions.

reduced water flow, and higher exit-water temperature. An oversized condenser, bigger than design, will increase capital cost but reduce operating costs because it would decrease ΔT_m and lower condenser pressure. Proper design, therefore, depends upon many factors, such as capital costs, operating costs, water availability, and environmental concerns.

The circulating-water inlet temperature should be sufficiently lower than the steam-saturation temperature to result in reasonable values of ΔT_o. It is usually recommended that ΔT_i should be between about 20 and 30°F (11 to 17°C) and that ΔT_o, the TTD, should not be less than 5°F (2.8°C).

Circulating-Water Flow and Pressure Drop

It is important to determine the necessary circulating-water flow and the pressure drop through the condenser because this, along with other parts of the circulating-water system (Chap. 7), determine the pump horsepower necessary.

The water mass-flow rate $\dot{m}_w$ is simply given by

$$\dot{m}_w = \frac{Q}{c_p(T_2 - T_1)} \tag{6-8}$$

where c_p is the specific heat of the water and T_1 and T_2 are the inlet and exit temperatures, respectively.

The pressure drop in the condenser is composed of (1) the pressure drop in the water boxes and (2) the friction pressure drop in the tubes. Again these depend upon many factors, such as the flow pattern in and the size of the water boxes, the inlet and exit of the tubes at the tube sheets, the size and length of the tubes, and the water temperatures and velocities. The Heat Exchange Institute recommends the values given in Figs. 6-10 and 6-11. The pressure drops are given in terms of *head H*, which is related to the pressure loss ΔP by

$$\Delta P = \rho H \frac{g}{g_c} \tag{6-9}$$

where ρ is the density, g the gravitational acceleration, and g_c the conversion factor 32.2 $\text{lb}_m \cdot \text{ft}/(\text{lb}_f \cdot \text{s}^2)$ or 1.0 $\text{N} \cdot \text{m}/(\text{kg} \cdot \text{s}^2)$

Water inlet velocities in condenser tubes are usually limited to a maximum 8

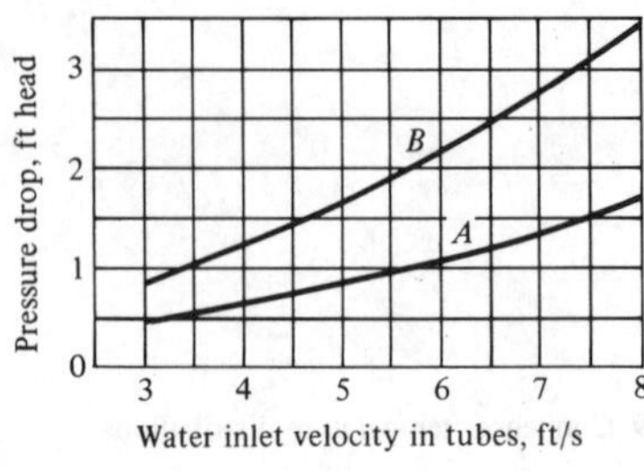

Figure 6-10 Pressure drop in condenser water boxes, expressed as head in feet: (*a*) one-pass, (*b*) two-pass.

ft/s (~ 2.5 m/s) to minimize erosion, and a minimum of 5 or 6 ft/s (1.5 to 1.8 m/s) for good heat transfer. Values between 7 and 8 ft/s (~ 2.1 to 2.5 m/s) are most common.

Example 6-2 Design a condenser that would handle 3 × 10^6 lb_m/h of 90 percent quality steam at 1 psia, as well as 360,000 lb_m/h of 112°F drain water from the low-pressure feedwater heater, and 1875 lb_m/h of 440°F drains from the steam-jet air ejector. Fresh cooling water is available at 70°F.

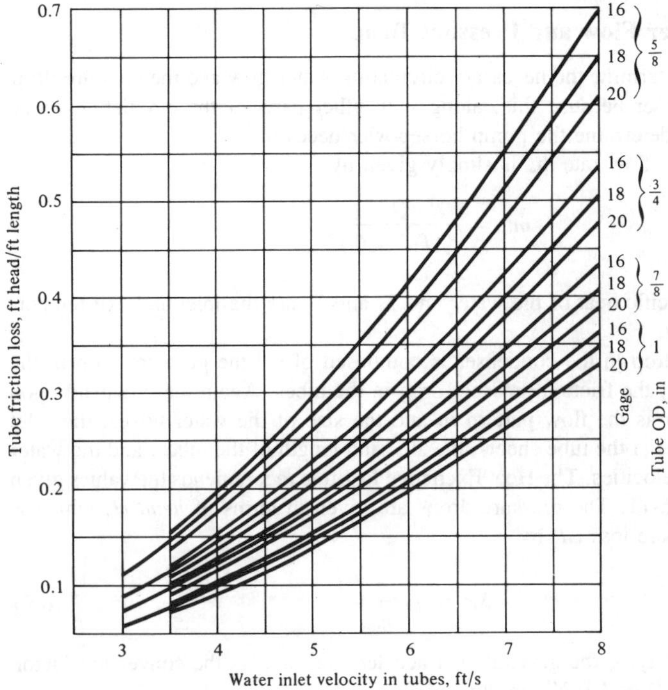

Figure 6-11 Pressure drop in condenser tubes, expressed as head in feet per foot length of tubes.

SOLUTION

Heat-transfer calculations:

Select:

1. A two-pass condenser
2. Type 304 stainless steel tubing
3. Tubes: 50 ft in length, 7/8 in OD, 18 BWG
4. TTD = 6°F
5. Inlet water velocity = 7 ft/s

$$\begin{aligned}\text{Heat load } Q &= \text{turbine exhaust} + \text{low-pressure drain} + \text{SJAE drain}\\ &= 2 \times 10^6 \times xh_{fg} + 240{,}000(h_{112^\circ F} - h_f) + 1250(h_{440^\circ F} - h_f)\\ &= 3 \times 10^6 \times 0.9 \times 1036.1 + 360{,}000(79.98 - 69.7)\\ &= 2.798 \times 10^9 + 3.701 \times 10^6 + 0.656 \times 10^6\\ &= 2.802 \times 10^9 \text{ Btu/h}\end{aligned}$$

$$\Delta T_i = t_{sat} - 75 = 101.74 - 70 = 31.74°\text{F}$$

$$\Delta T_o = 6°\text{F}$$

$$\Delta T_m = \frac{31.74 - 6}{\ln(31.74/6)} = 15.45°\text{F}$$

$$\begin{aligned}U &= 263 \times 1.00 \times 0.58 \times 0.85\sqrt{7}\\ &= 343.0 \text{ Btu/(h} \cdot \text{ft}^2 \cdot {}^\circ\text{F)}\end{aligned}$$

Thus

$$\text{Total tube surface area} = \frac{2.802 \times 10^9}{343.0 \times 15.45} = 528{,}675 \text{ ft}^2$$

For 7/8-in tubes, surface area per foot is 0.2291/ft² (see App. K). Therefore

$$\text{Total length of tubes} = \frac{528{,}675}{0.2291} = 2{,}307{,}615 \text{ ft}$$

and

$$\text{Number of tubes} = \frac{2{,}307{,}615}{50} = 46{,}150$$

Water calculations:

$$T_2 - T_1 = \Delta T_i - \Delta T_o = 31.74 - 6 = 25.74°\text{F}$$

For $c_p = 0.99$ Btu/(lb$_m \cdot$ °F)

$$\dot{m}_w = \frac{2.802 \times 10^9}{0.99 \times 25.74} = 1.0996 \times 10^8 \text{ lb}_m\text{/h}$$

Note: A check on $\dot{m}_w$, obtained above from Eq. 6-8, and that obtained by multiplying the water velocity by the density by the total cross-sectional area of flow (in this case of half the tubes) should be made at this point. If not equal, a correction for the TTD and/or the tube length would have to be made.

$$\text{Volumetric flow rate} = \dot{m}v_{70°\text{F}} = \frac{1.0996 \times 10^8 \times 0.01605}{60}$$

$$= 29{,}414 \text{ ft}^3/\text{min} = 29{,}414 \times 7.481 = 220{,}043 \text{ gal/min}$$

Pressure drop in water boxes = 2.7 ft head (Fig. 6-10)

$$= 62.3 \times 2.7 \times \frac{32.2}{32.2} = 168.3 \text{ lb}_f/\text{ft}^2 = 1.17 \text{ psi}$$

Pressure drop in tubes = 0.32 ft head/ft (Fig. 6-11)

$$= 62.3 \times 0.32 \times \frac{32.2}{32.2} = 19.936 \text{ lb}_f/\text{ft}^2 \text{ per ft} = 0.1384 \text{ psi/ft}$$

Allow for 1.2 in thick tube sheets, and thus each pass would have a length of $50 + 2 \times 1.2/12 = 50.2$ ft

$$\text{Total pressure drop in tubes} = 0.1384 \times 2 \times 50.2 = 13.90 \text{ psi}$$

$$\text{Total pressure drop in condenser} = 1.17 + 13.90 = 15.07 \text{ psi}$$

The power attributed to the water flow in the condenser only would be

$$\frac{\dot{m}\,\Delta P}{\rho} = \dot{m}v\,\Delta p = 3.7916 \times 10^9 \text{ ft} \cdot \text{lb}_f/\text{h} = 1915 \text{ hp}$$

6-5 CLOSED FEEDWATER HEATERS: GENERAL

It has been demonstrated (Chap. 2) that regenerative feedwater heaters are indispensable in Rankine-cycle type powerplants if improved cycle performance is to be expected. They raise the temperature of the feedwater before it enters the economizer or steam drum. Both open- and closed-type feedwater heaters are used. In small industrial systems, only one open-type feedwater heater may be used. Utility and large industrial plants use a multiple of feedwater heaters, typically five to seven closed and one open, which doubles up as a deaerator. Nuclear powerplants of the pressurized- or boiling-water reactor types do not use open-type feedwater heaters, but gas-cooled and fast-breeder reactor powerplants do. Open-type feedwater heaters will be described in the next section.

Closed-type feedwater heaters are shell-and-tube heat exchangers. In essence they are small condensers that operate at higher pressures than the main condenser because bled steam is condensed on the shell side, whereas the feedwater, acting like circulating condenser water, is heated on the tube side.

Closed feedwater heaters are placed within the cycle to receive bled steam from the turbines at pressures determined roughly by equal temperature increments from the condenser to the boiler saturation temperatures (Sec. 2-13). They are therefore classified as low-pressure (LP) and high-pressure (HP) heaters, depending upon their location in the cycle. The LP heaters are usually located between the condensate pump and the open heater, which is followed by the main boiler-feed pump. The HP heaters are located between that pump and the economizer. Occasionally a boiler-feed booster pump is located up-stream of the main boiler feed pump, in which case the feedwater heaters are classified as LP, IP (intermediate pressure), and HP. The shell-side (bled-steam) pressure extends from vacuum to several hundred psia in LP heaters and may exceed 1200 psia in HP heaters. Tube-side (feedwater) pressures after the boiler feed pump are higher than the maximum steam pressure because of the pressure drop through the feedwater system and may exceed 5000 psia in supercritical-pressure cycles (Sec. 2-14).

When bled steam entering a feedwater heater is superheated, as is often the case in fossil-fueled high-pressure and some low-pressure heaters, the heater includes a *desuperheating zone* where the steam is cooled to its saturation temperature (Fig. 6-12). This is followed in all closed feedwater heaters by a *condensing zone* where the latent heat of vaporization is removed and the steam is condensed to a saturated liquid. This liquid, now called the *heater drain,* is, in all heaters, except sometimes the one or two lowest-pressure heaters, cooled below its saturation temperature in a *subcooling zone* or a *drain-cooling zone* before the drain is cascaded backwards or pumped forward (Chap. 2).

The above is a *three-zone* closed-type feedwater heater (Fig. 6-12). There are, however, *two-zone* heaters that include a desuperheating and a condensing zone or a condensing and a subcooling zone. And there are also *single-zone* heaters that include only a condensing zone. A drain-cooling zone, instead of being integral with the shell, may be located external to it. Details of construction and standards of feedwater heaters are given in Ref. 49.

Closed feedwater heaters could be either horizontal or vertical, depending upon space availability. Vertical heaters are designed with their head (water box) down or

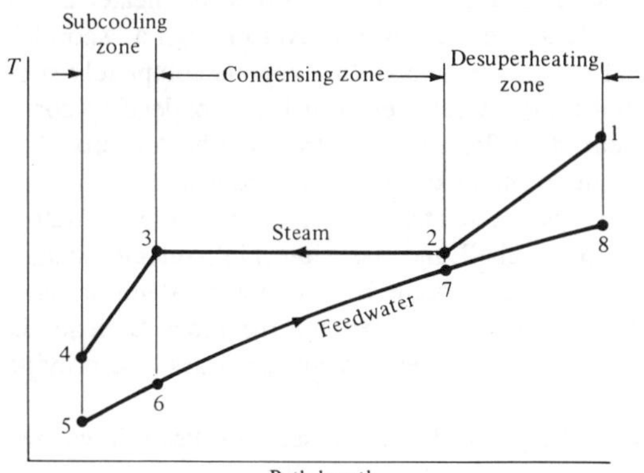

Figure 6-12 Temperature-path length diagram for a three-zone closed feedwater heater.

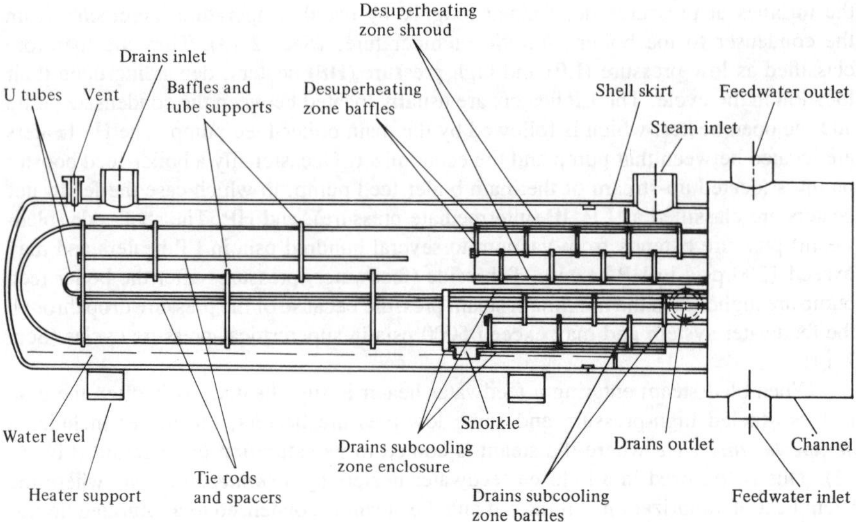

Figure 6-13 A three-zone horizontal closed-type feedwater heater [49].

up, the latter only in special circumstances. Figure 6-13 shows a typical horizontal three-zone closed feedwater heater. The feedwater tubes are usually in the form of U-tube bundles, although occasionally they are straight, like those of a condenser. The feedwater enters via a divided water box through the subcooling zone, flows through the U tubes in a generally parallel but opposite direction to the steam flow, and leaves to the water box through the desuperheating zone.

The bled steam flows first through the desuperheating zone, which is usually separated from the rest of the heater by a shroud and which is fitted with a series of vertically cut baffles that provide both a good heat-transfer path and proper tube support. The condensing zone is the major portion of the heater and also provides a baffled steam-flow path. The subcooling zone is separated from the rest of the heater by an end plate and an enclosure but connects to the main water level through a "snorkle" entrance. While the main water level is low, the subcooling zone is completely submerged with liquid by the differential pressure between it and the condensing zone. The liquid flows through properly spaced baffles and exits from the heater through a nozzle next to the tube sheet to cascade to the next lower-pressure heater.

Cascaded drain from the higher-pressure heater or from a moisture-separator-reheater are throttled upon entrance to the shell and flash into high-velocity steam. The tubes are protected from erosion by this steam by the use of stainless steel impingement plates. The drain inlet is usually at or beyond the tubes U bend, in horizontal heaters, in order to keep the flashing steam from the tubes and provide sufficient volume for it.

Figure 6-14 shows a vertical closed-type feedwater heater with head down, the usual arrangement.

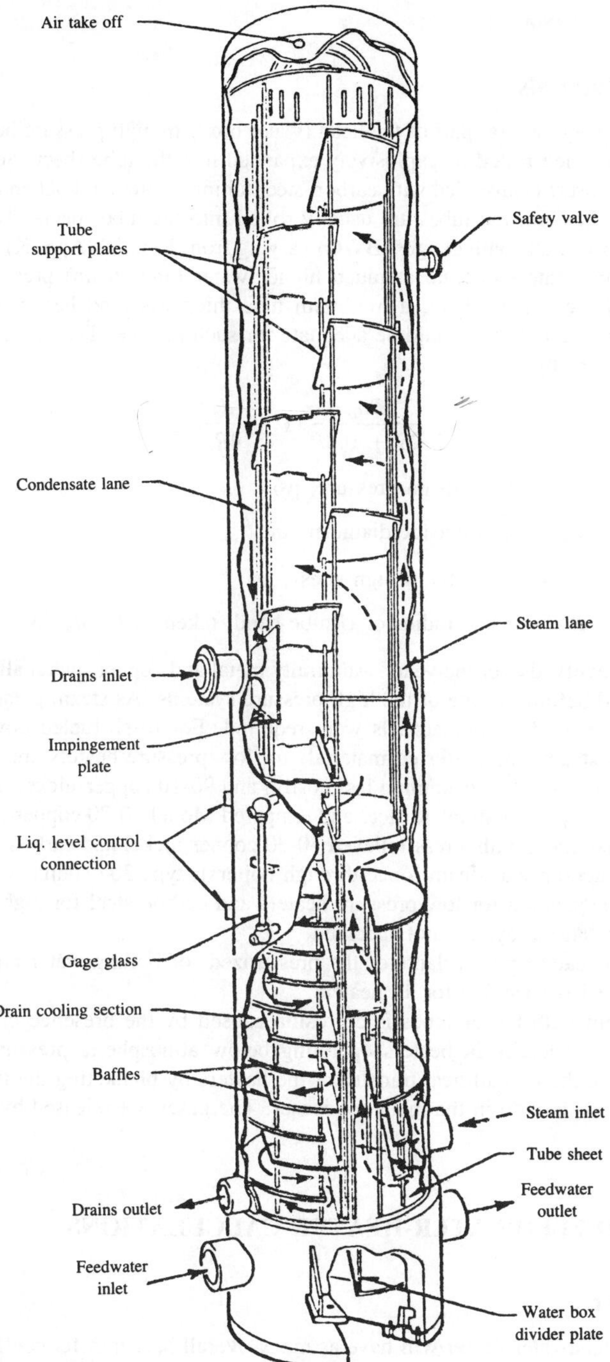

Figure 6-14 A three-zone vertical closed-type feedwater heater [49].

Tubes and Materials

The tube sheet is an integral part of the head (water box). In high-pressure heaters the tubes are welded then rolled or explosively expanded into the tube sheet. Sometimes a stainless steel insert is provided with carbon-steel tubing to prevent inlet end erosion. In low-pressure heaters, the tubes are usually rolled into the tube sheets. Tube sizes vary from 5/8 to 1 in OD with gauges (BWG) varying from 10 to 20 (App. K). Because closed feedwater heaters operate at much higher water (and steam) pressures than condensers and are usually bent, a minimum tube thickness (and hence maximum gauge) must be selected that would be adequate for such service. The minimum tube thickness t is given by

$$t = \frac{Pd_o}{2S + 0.8P}\left(1 + \frac{d_o}{4R_u}\right) \tag{6-10}$$

where P = tube design pressure, psi

d_o = tube outside diameter, in

S = allowable design stress, psi

R_u = inside radius of U-tube bend, taken as 1.5 d_o, in

Tube materials evolved over the years. Admiralty metal and copper-nickel alloys were extensively used before the age of the high-pressure systems. As steam pressures and temperatures increased, new materials were required. For fossil-fueled powerplants with subcritical steam, the preferred materials for low-pressure heaters are now type 304 stainless steel (which is becoming less costly) and 90-10 copper-nickel. For high-pressure heaters, type 304 stainless steel and tempered Monel 30-70 copper-nickel are the preferred materials, with stress-relieved 70-30 copper-nickel and carbon steel also specified. For supercritical steam (once-through boilers), type 304 stainless steel and carbon steel are specified for low-pressure heaters and carbon steel for high-pressure heaters. Copper-base alloys are not specified.

For nuclear-reactor powerplants of the pressurized- or boiling-water types, type 304 stainless steel is specified for all heaters.

One problem with heater tubes is corrosion caused by the presence of noncondensable gases, especially in heaters operating below atmospheric pressure. These gases also reduce the overall heat transfer of the heaters by blanketing the tube outer surfaces, much as they do in the main condenser. The gases are released by a proper vent mechanism.

6-6 CLOSED-FEEDWATER-HEATER CALCULATIONS

Heat Transfer

A closed-type feedwater heater will have as many overall heat-transfer coefficients as there are zones. For example, a three-zone heater will have one coefficient each for the desuperheating, condensing, and subcooling zones. These may be combined into one overall heat-transfer coefficient, if desired, but they must be evaluated independently. Once they are calculated, the total number of tubes for the heater and the zone lengths are calculated.

The overall heat-transfer coefficient for any zone is given by the usual relationship

$$U = \frac{1}{(1/h_o) + (tA_o/RA_m) + (A_o/h_iA_i)} \tag{6-11}$$

where U = overall heat-transfer coefficient of the zone in question, based on the outside area of the tubes, Btu/(h · ft^2 · °F) or W/(m^2 · s)

h_o = heat-transfer coefficient of fluid outside tubes, Btu/(h · ft^2 · °F) or W/(m^2 · K)

k = thermal conductivity of tube material, Btu/(h · ft · °F) or W/(m · K)

h_i = heat-transfer coefficient of water inside tubes, Btu/(h · ft^2 · °F) or W/(m^2 · K)

A_o = outside surface area of tubes per unit length, ft^2/ft or m^2/m

A_i = inside surface area of tubes ft^2 per unit length, ft^2/ft or m^2/m

A_m = log mean area of tubes per unit length

$= (A_o - A_i)/\ln \frac{A_o}{A_i} \approx \frac{A_o + A_i}{2}$ for thin tubes, ft^2/ft or m^2/m

t = thickness of tubes, ft or m

In the *desuperheating zone*, h_o is evaluated for forced convection of the superheated steam, between points 1 and 2 in Fig. 6-12, and h_i is evaluated for forced convection of feedwater inside the tubes between points 7 and 8.

In the *condensing zone*, h_o is the result of steam condensation between points 2 and 3 and h_i of feedwater forced convection between 6 and 7.

In the *subcooling zone*, h_o and h_i are due to forced convection of water outside the tubes between 3 and 4 and inside the tubes between 5 and 6, respectively.

For single- or two-zone heaters, the unapplicable zones are simply deleted.

The outer tube surface area A_o, and hence the zone lengths, is then obtained from

$$A_o = \frac{Q}{U\,\Delta T_m} \tag{6-12}$$

where Q is the heat load and ΔT_m the logarithmic mean temperature difference for the zone in question. (Note that an overall ΔT_m for the heater based on temperatures at 1, 4, 5, and 8 cannot be obtained because of the discontinuities in the steam temperature line 1-2-3-4.)

The values of h_o for each zone can be obtained from standard heat-transfer calculations.

As in the main condenser, there are many flow and other variables and uncertainties. Condensation heat-transfer coefficients are usually high, a few thousand

Btu/(h · ft² · °F), and the tube thicknesses are small, so the main heat-transfer resistance is due to the boundary layer of the feedwater inside the tubes. An empirical calculation for U in the *condensing zone*, shown in the form of a graph (Fig. 6-15), thus uses the average film (boundary-layer) temperature as well as water velocity as parameters. The average film temperature is defined arbitrarily as

$$T_f = T_s - 0.8\,\Delta T_m \tag{6-13}$$

where T_f = average film temperature, °F

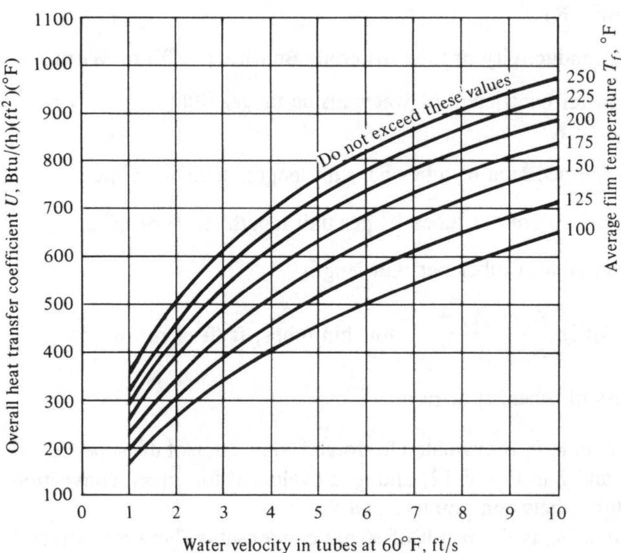

Gauge	Adms.	90/10 Copper-nickle	80/20 Copper-nickle	70/30 Copper-nickle	Monel	302,304 Stainless steel
18	1.00	0.97	0.95	0.92	0.89	0.85
17	1.00	0.94	0.91	0.87	0.85	0.80
16	1.00	0.91	0.88	0.84	0.82	0.77
15	0.99	0.89	0.86	0.82	0.79	0.74
14	0.96	0.85	0.82	0.77	0.75	0.70
13	0.93	0.81	0.78	0.73	0.70	0.65
12	0.90	0.77	0.73	0.68	0.65	0.60
11	0.87	0.74	0.70	0.65	0.62	0.57
10	0.83	0.69	0.66	0.60	0.58	0.52
9	0.80	0.65	0.62	0.56	0.54	0.48

Multipliers of base heat-transfer rates for various tube materials and gauges (for tube OD $\frac{5}{8}$ to 1 in. inclusive)

Figure 6-15 Overall heat-transfer coefficient for condensing zone only as a function of average film temperature as defined by Eq. (6-13) [49].

T_s = saturated-steam temperature = $T_{2\text{-}3}$, °F

ΔT_m = log mean temperature difference in condensing zone, °F

$$\Delta T_m = \frac{(T_3 - T_6) - (T_2 - T_7)}{\ln[(T_3 - T_6)/(T_2 - T_7)]} \tag{6-14}$$

There is an upper limit for U for values of ΔT_m 250°F or higher. The variation of water velocity with temperature is taken care of in Fig. 6-15 by specifying a velocity based on water density at 60°F, 62.4 lb_m/ft^3. Correction factors for various tube materials and gauges are included in Fig. 6-15.

Pressure Drop

Pressure drops of the feedwater in feedwater heaters are usually large because of the flow friction in long small-diameter tubes in several heaters. Calculations for such pressure drops are necessary for the design of condensate and boiler feed pumps.

An empirical correlation for the pressure drop in closed feedwater heaters, also applicable for external drain coolers, is given by the following equation [49]

$$\Delta P = \frac{F_1 F_2 (L + 5.5 d_i) N}{d_i^{1.24}} \tag{6-15}$$

where ΔP = total tube-side pressure drop, psi

F_1 = factor depending upon water velocity, also corrected for density at 60°F (Fig. 6-16)

F_2 = factor depending upon average water temperature T_{av} (Fig. 6-16), given by $T_{av} = T_s - \Delta T_m$, °F

d_i = inside tube diameter, in (App. K)

N = number of tube passes

L = length of tubes in one pass

The usual water velocities in closed feedwater heaters, corrected to 60°F, are 6 to 8 ft/s.

Example 6-3 Find the number and length of tubes for a three-zone feedwater heater that is used after the reheater (no. 2 high-pressure heater) in a subcritical fossil-fueled powerplant, for the following data:

Feedwater: 3.0×10^6 lb_m/h, 3000 psia, 370°F in, 398°F out
Bled steam: 140,000 lb_m/h, 240 psia, 800°F
Drain in from no. 1 HP heater: 300,000 lb_m/h, 520 psia, 410°F
Drain out, total: 440,000 lb_m/h, 380°F
Overall heat-transfer coefficients: desuperheating zone = 125 Btu/(h · ft² · °F), drain-cooling zone = 300 Btu/(h · ft² · °F)

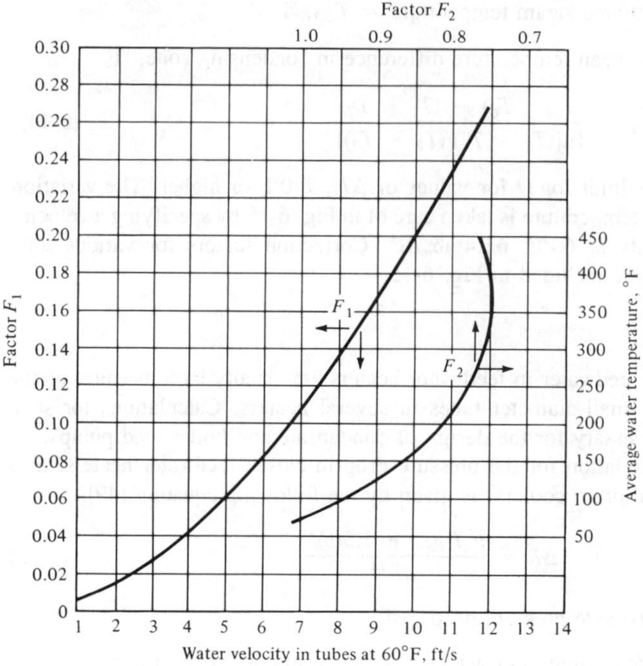

Figure 6-16 Factors in Eq. (6-15) for pressure drop in tubes of closed feedwater heaters and drain coolers [49].

SOLUTION

Tubes:

Choose a U-tube design, tubes 3/4 in OD, type 304 stainless steel. To find gauge, use maximum allowable stress = 20,000 psi. Eq. (6-10) gives minimum tube wall thickness

$$t = \frac{3000 \times 0.75}{2 \times 20{,}000 + 0.8 \times 3000}\left(1 + \frac{0.75}{4 \times 1.5 \times 0.75}\right) = 0.062 \text{ in}$$

From App. K, 16 BWG with 0.065 in thickness is selected.

Number of tubes:

Volumetric water flow corrected to 60°F = $3 \times 10^6\ v_{60°\text{F}} = 3 \times 10^6 \times$ 0.016033 (from steam tables) = 48,099 ft^3/h. Cross-sectional area of 16 BWG, 3/4-in tubes = 0.302 in^2 (App. K). Choose water velocity at 60°F = 8 ft/s. Therefore

$$\text{Number of tubes} = \frac{48{,}099}{3600} \times \frac{144}{0.302 \times 8} = 796 \text{ tubes}$$

which will be rounded to 800 tubes, (meaning 1600 tubes in a cross section of the U-tube bundle).

Desuperheating zone:
Referring to Fig. 6-12: $h_1 = 1423.8$ Btu/lb$_m$, $h_2 = 1200.6$ Btu/lb$_m$, and heat load $Q_s = 140{,}000(h_1 - h_2) = 31{,}248{,}000$ Btu/h. But $Q_s = 3 \times 10^6(h_8 - h_7)$. h_8 at 398°F and 3000 psia (assuming pressure drop relatively small) = 376.37 Btu/lb$_m$. Thus $h_7 = 365.95$ Btu/lb$_m$, corresponding to $T_7 = 388.1$°F. The log mean temperature difference in desuperheating zone is

$$\Delta T_{m,s} = \frac{(800 - 398) - (397.39 - 388.1)}{\ln[(800 - 398)/(397.39 - 388.1)]} = \frac{392.71}{3.7675} = 104.24°\text{F}$$

Outside tube area in desuperheating zone $A_{o,s}$ is

$$A_{o,s} = \frac{Q_s}{U_s\,\Delta T_{m,s}} = \frac{31{,}248{,}000}{125 \times 104.24} = 2398.16\ \text{ft}^2$$

From App. K, outside tube surface area per foot length = 0.1963 ft^2. Therefore the length of desuperheating zone tubes L_s is

$$L_s = \frac{2398.16}{0.1963 \times 800} = 15.27\ \text{ft} = 4.655\ \text{m}$$

Drain-cooler zone:
$h_3 = 372.27$ Btu/lb$_m$, h_4 (at 380°F and 240 psia) = 353.63 Btu/lb$_m$, h_5 (at 370°F and 3000 psia) = 347.06 Btu/lb$_m$. Heat load $Q_{dc} = (140{,}000 + 300{,}000)(h_3 - h_4) = 8{,}201{,}600$ Btu/h. But $Q_{dc} = 3 \times 10^6(h_6 - h_5)$. Therefore, $h_6 = 349.79$ Btu/lb$_m$, corresponding to $T_6 = 372.6$°F. Log mean temperature difference in drain-cooler zone

$$T_{m,\text{dc}} = \frac{(397.39 - 372.6) - (380 - 370)}{\ln[(397.39 - 372.6)/(380 - 370)]} = \frac{14.79}{0.9079}$$
$$= 16.29°\text{F}$$

Outside tube area in drain-cooler zone is

$$A_{o,\text{dc}} = \frac{8{,}201{,}600}{300 \times 16.29} = 1678.25\ \text{ft}^2$$

Therefore length of drain-cooler zone tubes L_{dc} is

$$L_{\text{dc}} = \frac{1678.25}{0.1963 \times 800} = 10.69\ \text{ft} = 3.26\ \text{m}$$

Condenser zone:
h_2 (at 410°F and 520 psia) = 386.21 Btu/lb$_m$. Heat load in condenser zone Q_c is due to both bled steam and drain from the higher-pressure heater, called 2′. $Q_c = 140{,}000(h_2 - h_3) + 300{,}000(h_{2'} - h_3) = 120.149 \times 10^6$ Btu/h. Log mean temperature difference in condensing zone is

$$\Delta T_{m,c} = \frac{(397.39 - 372.6) - (397.39 - 388.1)}{\ln[(397.39 - 372.6)/(397.39 - 388.1)]}$$

$$= \frac{15.5}{0.9815} = 15.79°\text{F}$$

Average film temperature = 397.39 − 0.8 × 15.79 > 250°F. Therefore U from Fig. 6-15, at 8 ft/s = 910 Btu/(h · ft² · °F). Correction factor for 16 gauge stainless steel = 0.770. Therefore overall heat-transfer coefficient in condensing zone is

$$U_c = 910 \times 0.770 = 700.7 \text{ Btu/(h} \cdot \text{ft}^2 \cdot °\text{F)}$$

Outside tube area in condensing zone $A_{o,c}$ is

$$A_{o,c} = \frac{120.149 \times 10^6}{700.7 \times 15.79} = 10{,}857.94 \text{ ft}^2$$

Length of condenser zone tubes L_c

$$L_c = \frac{10{,}857.94}{0.1963 \times 800} = 69.14 \text{ ft} = 21.07 \text{ m}$$

Total tube length:

Allowing for a 3-in-thick tube sheet, total heater tube length L_{tot} is

$$L_{tot} = 15.27 + 10.69 + 69.14 + 2 \times \frac{3}{12}$$

$$= 95.6 \text{ ft} = 29.14 \text{ m}$$

Thus the average length of the U-tube bundle is 47.8 ft, which may be rounded to 50 ft, or 15 m.

Pressure drop in tubes:

From Eq. (6-15)

$$\text{Average water temperature in heater} = \frac{370 + 398}{2} = 384°\text{F}$$

The average water temperature, obtained for the condenser zone (the largest zone), is 397.39 − 15.79 = 381.6°F, which is adequate for Fig. 6-16. Thus F_1 = 0.1275, F_2 = 0.75, d_i (App. K) = 0.620 in.

$$\Delta P = \frac{0.1375 \times 0.75(100 + 5.5 \times 0.620)}{0.620^{1.24}} = 19.29 \text{ psi} = 1.330 \text{ bar}$$

6-7 OPEN FEEDWATER HEATERS

An *open feedwater heater*, also called *direct-contact* and *deaerating (DA) heater*, is one that heats the feedwater by directly mixing it with bled steam from the turbine (Sec. 2-8). Usually only one such heater is used in fossil- and high-temperature nuclear-

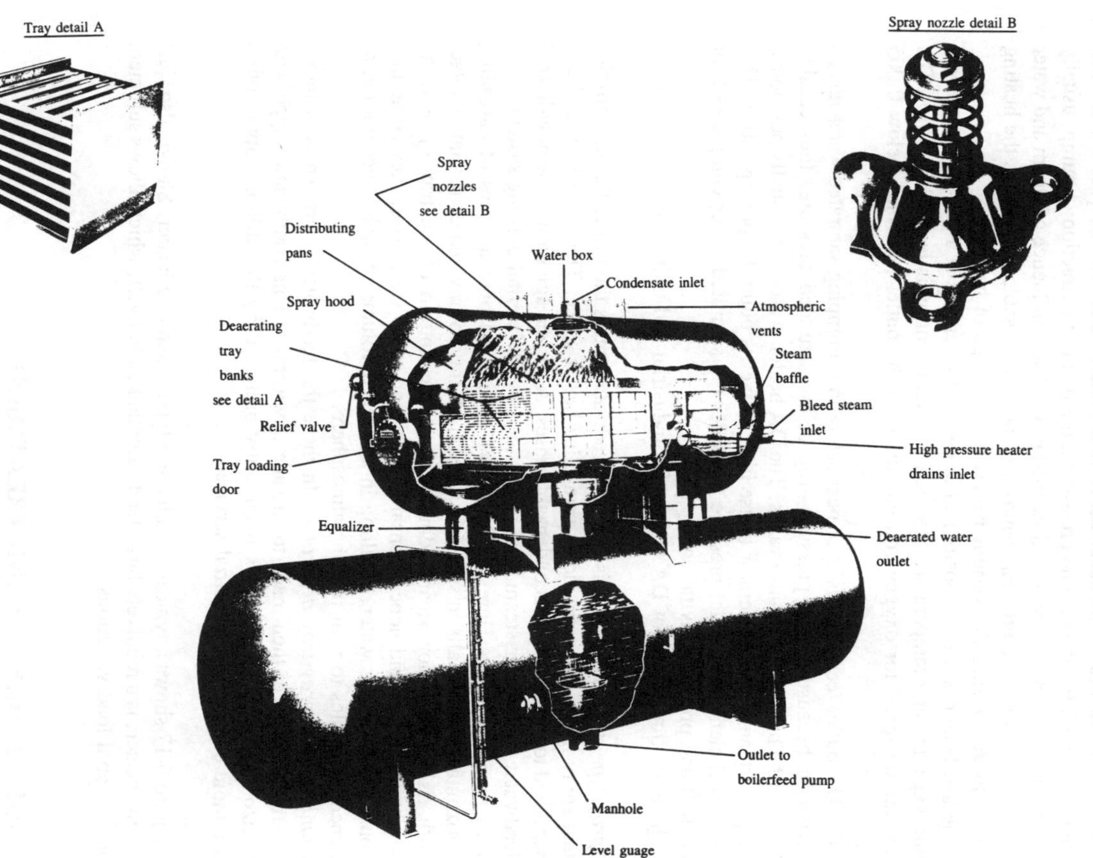

Figure 6-17 A typical combination open-type deaerating feedwater heater.

reactor powerplants. None are used in water-cooled nuclear-reactor powerplants, which rely on a more elaborate condenser deaeration system. Because the pressure in such a heater cannot exceed the turbine pressure at the point of extraction, a pump, usually the main boiler feed pump, must follow the heater. The confluence of steam and water flows makes possible the efficient removal of noncondensables as well as the heating of the feedwater. Hence the various names for this type of heater.

The DA heater is usually positioned in the feedwater line at a pressure to prevent air inleakage and at a temperature at which oxygen retention is least likely. Most DA heaters are designed for oxygen concentrations in the outlet feedwater below 0.005 cm^3/L.

The DA outlet feedwater is at or near saturation. Pumping saturated water results in cavitation because of the pressure drop below saturated pressure, thus causing flashing on the back side of pump vanes. The DA heater is therefore usually positioned in the powerplant steam-generator house high above its pump by perhaps 60 ft. This provides sufficient pump inlet pressure to render the saturated water compressed (or subcooled) and prevents cavitation.

There are three types of DA heaters for industrial and utility use.

1. *Spray-type deaerators* In this type the feedwater enters the heater through nozzles that spray it into the extraction-steam-filled heater space. The water is heated and scrubbed to release the noncondensable gases. A second agitation of the now-heated feedwater by another steam flow is provided by an internal baffling system.
2. *Tray-type deaerators* Here the feedwater is directed onto a series of cascading horizontal trays. It falls in sheets or tubes from tray to tray and comes into contact with rising extraction steam admitted from the bottom of the tray system. As scrubbing occurs and noncondensable gases and some steam rise, they come into contact with colder water, resulting in a reduced volume of high concentration of noncondensables to vent into the atmosphere.
3. *Combination spray-tray deaerators* In these the feedwater is first sprayed into a steam-filled space, then made to cascade down trays. This combination type with horizontal stainless steel trays is currently preferred by the utilities. Spray types are more common in industrial service.

Figure 6-17 shows a typical combination-type deaerating heater. Shown also, just below the heater, is a relatively large feedwater tank, a hotwell, which allows sufficient water for rapid load variations.

6-8 BOILER MAKEUP AND TREATMENT

Steam powerplants lose a fraction of their water-steam circuit because of leakage from fittings and bearings, escape with noncondensable gases in deaeration processes, boiler blowdown, and other causes. The fraction is 0.5 to 1.5 percent of the flow rate, depending upon design and age of the plant, with nuclear powerplants in the low end of the range. This fraction should be made up. The water added must be well treated to maintain water and steam purity in order to prevent deposition of suspended solids and scale on boiler surfaces and also silica deposition and corrosion damage to turbine blades and the condensate-feedwater system.

A water makeup system begins by *pretreating* the raw water. This is followed by a *demineralizing system,* which is essential to all powerplants. For plants with stringent water-quality requirements, such as those using once-through boilers and boiling-water nuclear reactors, a *condensate-polishing system* is used to further "polish" the water. Raw water usually has different concentrations of a variety of impurities present in it. They may include suspended solids and turbidity, organics, hardness (calcium and magnesium), alkalinity (bicarbonates, carbonates, hydrates), other dissolved ions (sodium, sulfate, chloride, etc.), silica, and dissolved gases (O_2, CO_2).

Pretreatment

The extent of pretreatment depends upon the source of the raw water used. Well water usually requires simple filtration. Raw water from a surface source, such as a river or lake, on the other hand, requires more elaborate pretreatment. The first step is *clarification,* in which the water is chlorinated to prevent biofouling of the equipment. The suspended solids and turbidity are then made to coagulate by special chemicals and by being brought together by a slow agitation in the middle of the clarifier vessel (chlorination that oxidizes organic matter also helps them coagulate). The coagulated matter then settles by gravity in the clarifier and is removed.

In the next step, the clarified water, depending upon its hardness and alkalinity, undergoes *softening*. Hardness, the chief source of scale in heat exchangers, boilers, and pipelines, is caused by the presence of calcium and magnesium salts containing Ca^{2+} and Mg^{2+}. Alkalinity is mostly bicarbonate HCO_3^- but is also carbonate and hydrate. Softening is usually done in a cold process using lime/soda ash. Lime, calcium hydroxide $Ca(OH)_2$, precipitates calcium bicarbonate as calcium carbonate $CaCO_3$, and magnesium salts as magnesium hydroxide Mg $(OH)_2$ according to

$$Ca(OH)_2 + Ca^{2+}(HCO_3^-)_2 \rightarrow 2CaCO_3 \downarrow + 2H_2O \quad (6\text{-}16)$$

$$2Ca(OH)_2 + Mg^{2+}(HCO_3^-)_2 \rightarrow 2CaCO_3 \downarrow + Mg(OH)_2 \downarrow + 2H_2O \quad (6\text{-}17)$$

$$(CaOH)_2 + Mg^{2+} (SO_4^-) \rightarrow Mg(OH)_2 \downarrow + CaSO_4 \quad (6\text{-}18)$$

The soda ash, sodium carbonate Na_2CO_3, is added to react with calcium chloride and calcium sulfate to form calcium carbonate

$$Na_2CO_3 + CaSO_4 \rightarrow CaCO_3 \downarrow + Na_2SO_4 \quad (6\text{-}19)$$

The products, calcium carbonate and magnesium hydroxide, are insoluble in water and settle to the bottom of the vessel. In softening, calcium, magnesium, and carbonate alkalinity are reduced to a few 10 ppm each. A problem of sludge removal, however, arises. Environmental regulations do not permit the sludge to be discharged. Instead, it is "dewatered," either in a settling basin or by thickeners and centrifuges. The water is then discharged to its source or recycled. Other softening processes use hot-process phosphate and zeolite softening.

The next step in pretreatment is *filtration,* which further removes residual suspended solids and turbidity. Filtration can be done under gravity or pressure, although the latter is preferred. Various filter media are used, sand being the most common. The pressure difference across the filtering medium is an indication of solid accumulation. When it reaches a given limit, the solids are removed from the bed by backwashing and are discharged to waste. Further filtration by activated charcoal may be necessary to absorb organics and remove residual chlorine from the chlorination process.

Demineralization

Demineralization is the process of removing dissolved solids by *ion exchange*. Two types of resins are used: cation and anion resins. *Cation resin* has a positively charged hydrogen ion attached to a negatively charged polymer. The hydrogen ion is exchanged for the cations calcium, magnesium, and sodium. *Anion resin* is similar to the cation except that it has a negatively charged hydroxide ion (OH^-), attached to a positively charged polymer structure. The hydroxide ion is exchanged for the anions sulfates, chlorides, and bicarbonates (alkalinity). Both ion-exchange processes are reversible, and the resins are restored to their original form by regeneration. Regeneration is called for when the resins are about to be exhausted and traces of dissolved solids begin showing up in the demineralizer exit. It is accomplished by passing a concentrated acid through the cation resin and caustic or sodium hydroxide through the anion resin.

A typical demineralizing "train" (Fig. 6-18) consists of a series of demineralizers that contains weak-acid cation, strong-acid cation, weak-base anion, strong-base anion, and mixed-bed units, which contain both cation and anion resins used for polishing to produce very high-quality water. Sometimes a decarbonator is added to the train to help the anion resin in reducing alkalinity. This reduces the amount of strong-base anion used. This is a mechanical dealkalizer that operates by blowing air up through downward-flowing water to drive CO_2 from the water, thus removing the alkalinity.

The demineralizing and regeneration equipment, including tanks, pumps, etc., is programmed to operate automatically. The result is high-quality water, at least equal to the best from evaporators (below), that can be used directly in the steam plant without scale formation or corrosion.

A method of demineralization that is gaining acceptance for boiler makeup is *reverse osmosis*. Osmosis is the diffusion of a solvent, in this case water, through a semipermeable membrane from a region of no or low solid concentration to a region of high solid concentration. The membrane blocks the passage of the solutes, in this case the dissolved solids. The motion is in the direction of the high partial pressure of the purer water to the low partial pressure of the less pure water. It is a response to an osmotic pressure, which can be relatively large even for dilute solutions. Osmosis plays a vital role in many biological processes such as the passage of nutrients and waste material through the cell wall of animal tissue. The diffusion can be prevented by applying to the region of high concentrations an external pressure equal to the osmotic pressure. If a greater pressure is applied, the flow is reversed. In our case,

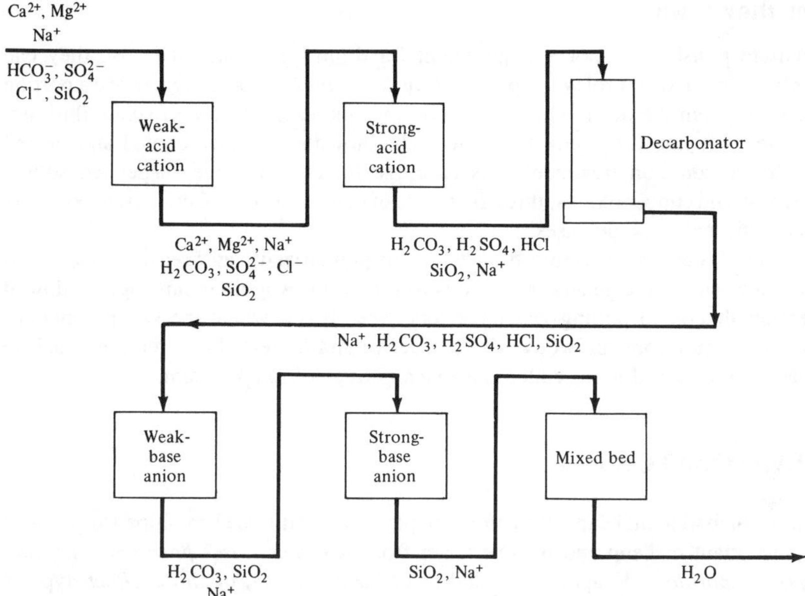

Figure 6-18 A typical demineralizing train.

pure water flows from the region of high concentrations to the region of pure water. This is reverse osmosis. The membranes used are expensive, and coagulation, settling, and filtration are used first to protect them.

Condensate Polishing

Although the above systems produce high-quality makeup water for the plant, the water in its journey through the cycle can pick up metallic ions, such as iron and copper, from pipelines, etc., as well as impurities due to condenser inleakage from the circulating water. As mentioned above, powerplants such as those using once-through boilers and boiling-water nuclear reactors (Sec. 10-7) require continual high-quality water and use a condensate polishing system. *Condensate polishing* is accomplished by passing the condensate through large demineralizing vessels that contain both cation and anion resins. The resins remove dissolved solids in the above manner as well as act as filters for impurities or suspended solids.

Condensate-polishing systems are similar for once-through boilers and boiling-water reactors, except that they are shielded in the latter case because the water goes through the reactor and the solids become radioactive. Some pressurized-water reactor systems (Sec. 10-2) require polishing to meet changing water-quality standards.

Boiler Blowdown

Condensate polishing is not a requirement for drum-type boilers because they can utilize blowdown to control water purity. *Blowdown lines* periodically remove a portion of the water from the drum where concentrations increase because of steam flashing. This water is replaced by pure feedwater. The blowdown is then cooled and treated for reuse. A common treatment uses cartridge filters to remove suspended solids, notably iron and copper oxides, through the use of demineralizers. Condensate polishers may be indicated in some cases.

Nuclear boilers also feature blowdown. In pressurized-water-reactor steam generators (Sec. 10-3), the general practice is to use throwaway filter cartridges followed by demineralizers. In boiling-water reactors (Sec. 10-7), where the steam generator is also the reactor core, blowdown is radioactive and is treated and returned back to the reactor in a closed loop, called the *reactor-water cleanup system*.

6-9 EVAPORATORS

One form of boiler makeup, used in older plants and still used in ships (to produce both powerplant makeup and potable water from seawater), uses *flash-type distilling units,* or *evaporators*. Evaporators could be of the *one-stage,* or *single-effect,* type or the *multistage,* or *multieffect,* type. Usually two and sometimes three effects are used.

Older units had water sprayed on a bank of tubes heated by steam from the plant. This design has been largely replaced by the submerged type, which consists of a shell containing the steam tubes, which in turn are completely submerged by the water.

Figure 6-19 is a schematic of a single-effect evaporator. *Raw water,* pretreated (but not demineralized) as above, is introduced at point w. Steam from the plant, called *motive steam,* extracted at a relatively low pressure, is introduced at m, goes into a steam chest, flows through the tubes, and is discharged as condensate at c. The condensate is returned to the plant. The tubes slope slightly toward discharge to allow proper drainage. The pretreated water boils and pure saturated vapor is extracted at v to go to the plant as makeup. Care must be taken to avoid carryover of raw water with the vapor because that would defeat the purpose of the evaporator. This is done by proper separation and baffling for the receiving tube (much as in a boiler steam

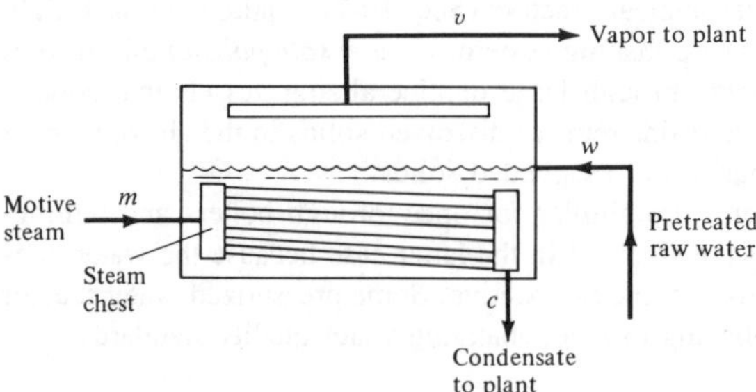

Figure 6-19 Schematic of a single-effect submerged evaporator.

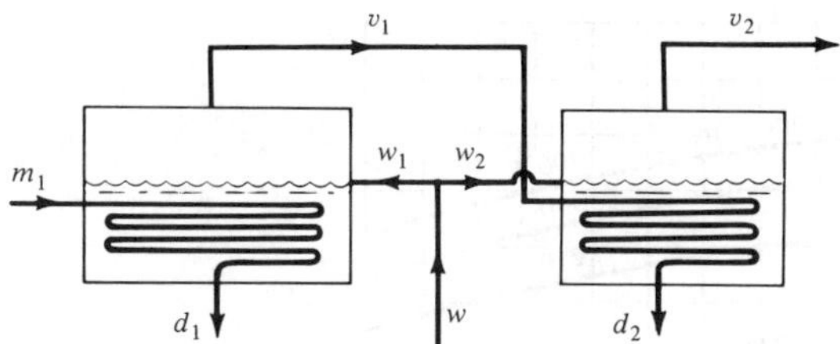

Figure 6-20 Schematic of a double-effect evaporator.

drum) and by keeping the rate of vapor formation per unit water surface area at a low enough level to reduce violent boiling at that surface. The raw-water surface is kept at approximately the centerline of the shell, where it would have maximum area. The level is maintained by a float-controlled valve on the raw-water line.

Scale that collects on the outside tube surfaces should be periodically cleaned to maintain heat-transfer rates. This is done during shutdown either manually or by subjecting the tubes to alternate high and low temperatures. Periodical blowdown is necessary to remove the raw-water sediment that accumulates in the shell.

Figure 6-20 shows a double-effect evaporator. Here pretreated raw water is introduced into two shells at w_1 and w_2, respectively. The motive steam m_1 condenses to d_1 and is returned to the plant. The vapor produced in the first-effect evaporator v_1 becomes the motive steam for the second-effect evaporator. Its condensate d_2 is returned to the plant. Both the vapor v_2 produced in the second effect and the condensate d_2 constitute the boiler makeup.

A temperature-enthalpy (or heat-transfer length) diagram is shown in Fig. 6-21. It is now necessary to define *heat head* as the difference between the saturation temperature of the motive steam and the saturation temperature of the vapor.

The overall heat-transfer coefficient U of an evaporator is a function of both motive-steam saturation temperature and the heat head and is given empirically by the curves in Fig. 6-22. Heat heads below 20°F are not effective. Those above 100°F result in film boiling and a reduction in heat transfer. The most effective heat heads are in the range 35 to 55°F.

Single-effect evaporators, which can be built in parallel multiples, average approximately 0.8 lb_m of vapor produced per lb_m of motive steam. Double-effect evaporators produce almost double, 1.5 lb_m/lb_m, while triple-effect evaporators improve

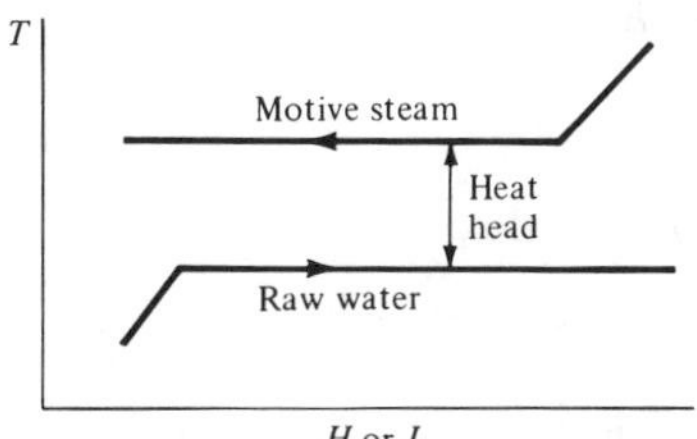

Figure 6-21 Temperature vs. enthalpy (or path-length) diagram of a single-effect evaporator.

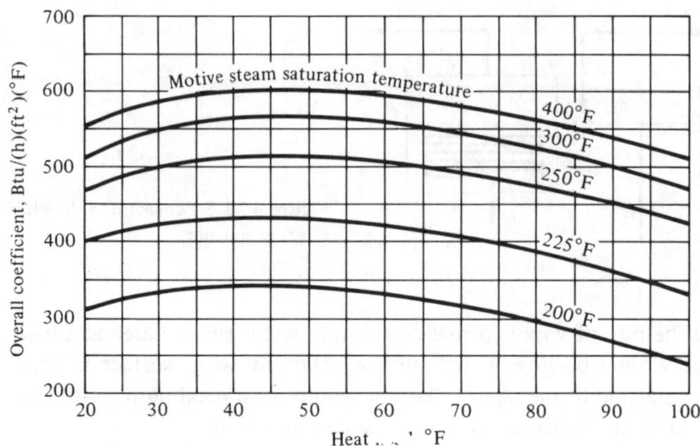

Figure 6-22 Overall heat-transfer coefficient for evaporators.

this further to about 2.0 lb_m/lb_m, but at the cost of increased complexity and capital cost.

Heat heads are usually divided up equally among the effects used. If the heat head and the overall heat-transfer coefficient for each heat head are known, the heat-transfer areas can then be easily obtained.

Example 6-5 Calculate the mass-flow rate of motive steam required and the heat-transfer areas of a double-effect evaporator to produce 100,000 lb_m/h of vapor. Pretreated water is available at 60°F. Motive steam is available at 70 psia, saturated. The vapor leaves the evaporator to a feedwater heater at 15 psia.

SOLUTION Referring to Fig. 6-20 and the steam tables: T_{m1} = 302.93°F, h_{m1} = 1180.6 Btu/lb_m, and h_{d1} = 272.7 Btu/lb_m. T_{v2} = 213.03°F, h_{v2} = 1150.9 Btu/lb_m. $T_{v1} = \frac{1}{2}(302.93 + 213.03) = 258$°F, P_{v1} = 34.24 psia, h_{v1} = 1166.7 Btu/lb_m, h_{d2} = 226.72 Btu/lb_m. h_w = 28.06 Btu/lb_m.

Assume raw water flows to the first and second effects to be w_1 and w_2 lb_m/lb_m motive steam at m_1, respectively.

Energy balance on the first effect:

$$w_1(h_{v1} - h_w) = h_{m1} - h_{d1}$$

Therefore

$$w_1 = \frac{1180.6 - 272.7}{1166.7 - 28.06} = 0.797\ lb_m/lb_m$$

Second effect:

$$w_2(h_{v2} - h_w) = w_1(h_{v1} - h_{d2})$$

Therefore

$$w_2 = 0.797 \left(\frac{1166.7 - 226.72}{1150.9 - 28.06} \right) = 0.797 \times 0.837$$

$$= 0.667 \text{ lb}_m/\text{lb}_m$$

$$\text{Total purified water} = w_1 + w_2 = 0.797 + 0.667 = 1.464 \text{ lb}_m/\text{lb}_m$$

$$\text{Motive steam required} = \frac{100{,}000}{1.464} = 68{,}306 \text{ lb}_m/\text{h}$$

First-effect heat-transfer area:

$$Q = \dot{m}_{w1}(h_{v1} - h_w) = \frac{0.797}{1.464} \times 100{,}000(1166.7 - 28.06)$$

$$= 62 \times 10^6 \text{ Btu/h}$$

$$\text{Heat head} = T_{m1} - T_{v1} = 302.93 - 258 = 45°\text{F}$$

From Fig. 6-22 at T_{m1} and heat head, $U = 565$ Btu/(h · ft^2 · °F)

$$\text{Area} = \frac{Q}{U \times \text{heat head}} = \frac{62 \times 10^6}{565 \times 45} = 2438 \text{ ft}^2$$

Second-effect heat-transfer area:

$$Q = \dot{m}_{w2}(h_{v2} - h_w) = \frac{0.667}{1.464} \times 100{,}000(1150.9 - 28.06)$$

$$= 51.16 \times 10^6 \text{ Btu/h}$$

$$\text{Heat head} = T_{v2} - T_{v1} = 45°\text{F}$$

At T_{v1} and heat head, $U = 520$ Btu/(h · ft^2 · °F)

$$\text{Area} = \frac{51.16 \times 10^6}{520 \times 45} = 2186 \text{ ft}^2$$

PROBLEMS

6-1 A direct-contact condenser of the spray type receives 60 kg/s of 92% quality steam at 0.06 bar. A dry cooling tower system cools the feedwater to 15°C. Calculate the condensate pump volume flow rate, in cubic meters per second, and the pump power input, in kilowatts, if the pressure losses in the dry cooling tower system are 2 bar and the pump efficiency is 0.7.

6-2 A barometric condenser is used to condense 100,000 lb$_m$/h of 90 percent quality steam at 1 psia using cooling water at 60°F. The tail pipe inside diameter is 1 ft. The friction factor in it is 0.020. Find the necessary height of the tail pipe, in feet.

6-4 It is required to compare the cooling water requirements and pumping powers of single-pass and two-pass surface condensers having the same total heat-transfer surface area and inlet water conditions. Assume

at 15°C. Calculate the height of the tail pipe, in meters, if the frictional losses in bar are given by $40V^2$, where V is the water velocity, in meters per second.

6-4 It is required to compare the cooling water requirements and pumping powers of single-pass and two-pass surface condensers having the same total heat-transfer surface area and inlet water conditions. Assume that both have 20,000 40-ft-long 7/8-in-OD 18-BWG tubes. Cooling water enters the tubes at 64°F and 6.5 ft/s. The pressure drops are 0.4 psi/ft in the tubes, 0.6 psi in each water box and $5 \times 10^{-15}\ \dot{m}^2$ in the intake, ducting, and discharge system, where $\dot{m}$ is the mass flow rate, in pound mass per hour. Calculate for each (*a*) the cooling water requirements, in pound mass per hour and cubic feet per minute, and (*b*) the pumping power, in kilowatts, if the pumps are 70 percent efficient.

6-5 It is required to compare the effects of single-pass and two-pass surface condensers on the powerplant for the same total heat-transfer surface area and inlet water conditions. Both condensers have 20,000 40-ft-long 7/8-in-OD, 18-BWG tubes. Cooling water enters the tubes at 64°F and 6.5 ft/s (the same conditions as Prob. 6-4). Assume for simplicity that both condensers have the same overall heat-transfer coefficient, the same steam inlet of 2×10^6 lb_m/h and 92.6 percent quality, but that the single-pass condenser operates at 100°F. Calculate for each (*a*) the condenser operating temperature, in degrees Fahrenheit, and pressure, in pound per square inch absolute, (*b*) the cooling water exit temperature, in degrees Fahrenheit, and (*c*) the heat removed, in Btus per hour and megawatts.

6-6 Consider a powerplant operating on an ideal Rankine cycle without feedwater heating. Steam enters the turbine saturated at 1000°F at the rate of 5×10^6 lb_m/h. The turbine exhausts to a surface condenser at 1 psia. The condenser has an overall heat-transfer coefficient of 435 Btu/h·ft^2·°F, a heat-transfer surface area of 402,467 ft^2 and cooling water inlet temperature of 60°F. Calculate (*a*) the plant net power, in megawatts, (*b*) the plant efficiency, in percent, (*c*) the cooling water exit temperature, in degrees Fahrenheit, and (*d*) the water flow rate, in pound mass per hour and in cubic feet per minute.

6-7 It is required to evaluate the effects of changing surface condenser size on powerplant performance. Consider for simplicity an ideal Rankine cycle without feedwater heaters. The turbine inlet steam is saturated at 1000 psia and has a mass flow rate of 5×10^6 lb_m/h. Condenser cooling water is available at 60°F. For a condenser heat-transfer area of 402,467 ft^2 and overall heat transfer coefficient of 435 Btu/h·ft^2·°F, the condenser operates at 1 psia (the same conditions as Prob. 6-6). Now consider that the same plant operates with a condenser that has 15 percent more tubes (and surface area) and 15 percent more cooling water flow. Calculate (*a*) the new exit water temperature, in degrees Fahrenheit, (*b*) the new condenser pressure, in pounds per square inch absolute, and (*c*) the new plant power, in megawatts, and efficiency. Assume that the overall heat-transfer coefficient is unchanged.

6-8 A two-pass surface condenser contains 30,000 10-m-long 3/4-in 18-BWG 90-10 Cu-Ni Tubes. Cooling water enters the tubes at 10°C and 2 meters per second. The condensing temperature is 38°C. Calculate (*a*) the capacity of the condenser (heat rejected) in megawatts, and (*b*) the exit water temperature, in degrees centigrade. Take for water $c_p = 4184$ J/kg · K and density $= 1000$ kg/m^3.

6-9 Consider for simplicity an ideal Rankine cycle with turbine inlet and exit steam at 2500 psia and 1000°F, and 1 psia, respectively, and no feedwater heating. The steam mass flow rate is 2×10^6 lb_m/h. The plant has a two-pass surface condenser with 45-ft-long 7/8-in 18-BWG-type 304 stainless steel tubes. Cooling water enters the tubes at 70°F and 7 ft/s. With no flows into the condenser other than turbine steam, find (*a*) the plant power, in megawatts, and efficiency and (*b*) the total number of tubes if the condenser terminal temperature difference is 8°F.

6-10 It is required to evaluate the effects of changing surface condenser inlet water temperature on powerplant performance. Consider the ideal Rankine cycle and condenser of Prob. 6-9 (with 25,696 tubes) but the cooling water inlet temperature is changed from 70° to 60°F. Calculate (*a*) the new exit cooling water temperature, in degrees Fahrenheit, (*b*) the new condenser pressure, in pounds per square inch absolute, and (*c*) the new plant power, in megawatts.

6-11 A 1100-MW two-pass surface condenser has 50-ft-long 18-BWG tubes. It uses seawater with tube inlet at 7 feet per second and 80°F, and 100°F exit. The condenser is situated 18 ft above sea level. The pressure drops in the circulating water system, expressed in head are: 4 ft in the inlet tunnel, 6 ft in the inlet pipe, 3 ft in the outlet pipe, and 2 ft in the outlet tunnel. Calculate (*a*) the water flow rate in pound mass per hour and gallons per minute, and (*b*) the pumping in horsepower and megawatts required if the pumps efficiency is 80 percent. Use for seawater c_p = 0.962 Btu/lb_m · °R, and density = 64 lb_m/ft^3.

6-12 A condensing only feedwater heater uses 7/8-in-OD 90-10 copper-nickel tubes. It receives 84,000 lb_m/h of 95 percent quality bled steam at 20 psia, and 160,000 lb_m/h of drain from the next higher pressure heater at 240°F. 3.9 × 10^6 lb_m/h of feedwater goes through the heater at 7 ft/s, 2000 psia, and 195°F. The terminal temperature difference is 5°F. Determine the size, length, and number of tubes based on a U-tube design. Take a maximum allowable stress in the tubes of 15,000 psi.

6-13 A closed type two-pass feedwater heater receives 100,000 lb_m/h steam at 300 psia and 450°F and feedwater at 380°F. The terminal temperature differences is −2.65°F. No drain cooler is used. Find (*a*) the mass flow rate of feedwater heated, and (*b*) the water pressure drop, in pounds per square inch, if the adjusted cold water velocity is 8 ft/s, the tube length is 20 ft, and the tube inside diameter is 0.652 in.

6-14 A one-pass condensing only closed feedwater heater receives saturated extraction steam at 80 psia, and 10^7 lb_m/h of feedwater at 240°F. The terminal temperature difference is 4°F. Calculate (*a*) the total tube heat-transfer area if they are made of 16 gage admiralty metal and the velocity adjusted to 62.4 lb_m/ft^3, is 8 ft/s, (*b*) the amount of extraction steam required, in pound mass per hour, and (*c*) the new log mean temperature difference and the new bled steam temperature and pressure if the water velocity was changed to 6 ft/s at the same inlet temperature.

6-15 A condensing only closed type feedwater heater receives saturated bled steam at 80 psia. 0.9 × 10^6 lb_m/h of feedwater enter the tubes at 260°F and 6 ft/s. The tubes are 45.5-ft-long U-shaped 3/4-in-OD 18-BWG admiralty metal. The terminal temperature difference is 5°F. Calculate (*a*) the mass flow rate of the bled steam, in pound mass per hour, (*b*) the number of tubes, and (*c*) the pressure drop in the heater, in pounds per square inch.

6-16 A closed-type two-pass feedwater heater with desuperheating zone but no drain cooler uses 20-ft-long 3/4-in 18-BWG admiralty tubes. 10^6 lb_m/h of feedwater enter the tube at 260°F and 8.527 ft/s. Bled steam is at 80 psia and 350°F. The terminal temperature difference is 4°F. The overall heat-transfer coefficient in the desuperheating zone is 120 Btu/h · ft^2 · °F. Calculate (*a*) the mass flow rate of bled steam, in pound mass per hour, (*b*) the number of tubes, (*c*) the pressure drop in the heater, in pounds per square inch.

6-17 A sample of water upon analysis showed that each liter contained 0.0018 g · mol calcium bicarbonate, 0.0005 g · mol magnesium bicarbonate, 0.0008 g · mol magnesium sulfate, and 0.0006 g · mol sodium sulfate, Calculate (*a*) the total hardness in parts per million, (*b*) the masses of lime and soda ash, kilograms per day needed to soften 100,000 L of the water per day and (*c*) the mass of sludge, in kilograms, that must be disposed of per day. Take the density of water as 1000 kg/m^3.

6-18 A single-effect evaporator receives motive steam at 7 bars and 180°C and pretreated raw water at 9 bars and 16°C. It has 123 m^2 of heat-transfer area. Calculate the mass flow rates of motive steam and vapor produced in kilograms per second. Ignore the pressure drop in the evaporator.

6-19 A 35,000 lb_m/h double-effect evaporator receives motive steam at 40 psia and 300°F. Pretreated raw water enters at 60°F. The pressure in the second effect is 5 psia. Calculate, for equal heat heads, (*a*) the amount of motive steam required, in pound mass per hour, and (*b*) the surface areas for each evaporator, in square feet.

6-20 A double-effect evaporator is to evaporate 15,000 lb_m/h of 60°F raw water using 25 psia motive steam. The vapor from the second effect is at 3.0 psia. Find (*a*) the heat head for each effect, (*b*) the mass flow rate of motive steam, in pound mass per hour, (*c*) surface area of each effect, in square feet, and (*d*) total purified water, in pound mass per hour, if the pressure of the motive steam was changed to 17 psia during part load operation.

CHAPTER

SEVEN

THE CIRCULATING-WATER SYSTEM

7-1 INTRODUCTION

The circulating-water system supplies cooling water to the turbine condensers and thus acts as the vehicle by which heat is rejected from the steam cycle to the environment. The system also supplies lesser amounts of auxiliary cooling water for turbine and steam-generator buildings, for the fire protection system, and for general station yard use. In the case of nuclear powerplants, it supplies, in addition, auxiliary cooling water to the reactor building (for cooling a closed-loop cooling-water circuit to limit radioactivity release back to the environment), water for dilution and dispersion of radioactive wastes released from the plant, and water for decay heat removal when necessary. The total auxiliary requirements are usually about 5 percent of condenser flow requirements.

The circulating-water system is called upon to reject heat to the environment in an efficient manner but one that also conforms to thermal-discharge regulations. Its performance is vital to the efficiency of the powerplant itself because a condenser operating at the lowest temperature possible results in maximum turbine work and cycle efficiency and in minimum heat rejection. Hence, a good heat-rejection system makes its own job easier; i.e., it is called upon to reject less heat and is smaller and requires less cooling water.

The heat rejected by the circulating-water system is greater than that converted to useful work by the steam cycle. In currently operating powerplants, new and old, the heat rejected varies from 1.5 to 3.0 times the useful work output of these plants. This is given by the equation

$$\dot{Q}_R = \left(\frac{1}{\eta} - 1\right)\dot{W} \tag{7-1}$$

Table 7-1 Effect of cycle efficiency on heat rejection by 1000-MW powerplants

$\dot{W}$	η	$\dot{Q}_A$	$\dot{Q}_R$	$\dot{Q}_R/\dot{W}$
1000	0.20	5000	4000	4.0
1000	0.25	4000	3000	3.0
1000	0.33	3000	2000	2.0
1000	0.40	2500	1500	1.5
1000	0.50	2000	1000	1.0

where $\dot{Q}_R$ is the heat rejection rate, $\dot{W}$ the power, and η the cycle efficiency. To evaluate the effect of η on $\dot{Q}_R$ one should compare plants with the same $\dot{W}$, say 1000 MW (Table 7-1). The first and last efficiencies roughly represent old small industrial units and hoped-for advanced future units. $\eta = 0.25$ is for older but still-operating powerplants. $\eta = 0.33$ represents current nuclear powerplants of the pressurized- and boiling-water-reactor type. $\eta = 0.40$ represents modern high-temperature fossil-fueled and gas-cooled and fast-breeder nuclear-reactor powerplants. It can be seen that an improvement of 7 percent in efficiency, from 25 to 33 percent, results in 33 percent savings in heat rejection (not 8 percent as one is inclined to guess), whereas an efficiency improvement of 7 percent from 33 to 40 percent, results in 25 percent savings.

7-2 SYSTEM CLASSIFICATION

Circulating-water systems are broadly classified as (1) once-through, (2) closed-loop, and (3) combination systems.

Once-through systems In once-through systems, water is taken from a natural body of water such as a lake, river, or ocean and pumped through the condenser, where it is heated and then discharged back to the source (Fig. 7-1). There are generally three methods of discharge:

1. *Surface discharge,* by which the condenser water is released in a relatively thin layer on the surface of the original body of water. The resulting plume is cooled by evaporation of some of the water to the atmosphere and by mixing with the cooler water below.
2. *Submerged discharge,* by which the water is released as a buoyant jet below the surface of the body. The jet mixes with the cooler water, and heat is eventually dissipated by evaporation from the mixture.
3. *Diffuser discharge,* by which the water is let out through a number of nozzles from a long pipe submerged across the flowing system, such as across a river. The jets may point upstream to promote more rapid mixing. Again heat rejection is accomplished by evaporation from the mixture.

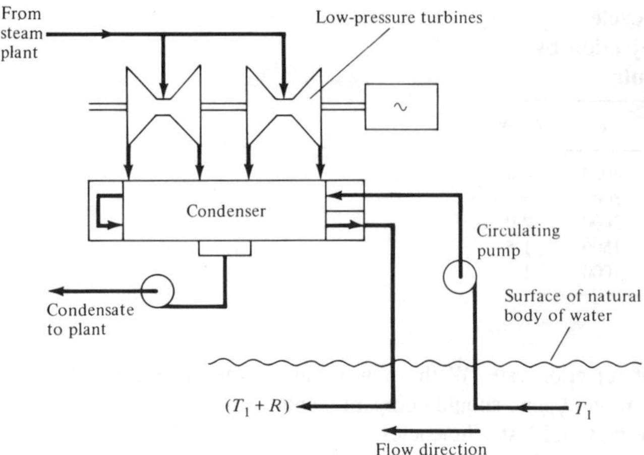

Figure 7-1 Schematic of a once-through circulating-water system.

Once-through cooling is, thermodynamically, the most efficient means of heat rejection. It uses the lowest-temperature heat sink available to the powerplant. There are, however, instances when water is scarce or when environmental regulations limit the utilization of surface waters or the amount of heating they may be subjected to. In such cases, the less efficient closed-loop systems are used.

Closed-loop systems In closed-loop systems, water is taken from the condenser, passed through a cooling device, and returned to the condenser. Often a reservoir is placed between the cooling device and the condenser. A nearby natural body of water is still necessary to supply makeup water to replace that lost by evaporation during the cooling process and to receive blowdown from it. There are several types of cooling devices available for closed-loop systems. These are:

1. *Cooling towers* may be of the wet type, (Sec. 7-4), dry type (Sec. 7-6), or combination wet-dry type (Sec. 7-8). The dry-type cooling tower is the least efficient of all heat rejection methods but requires no makeup water and hence is suited for desert installations or where the use of natural waters is absolutely prohibited. Cooling towers are also classifed as *natural draft* or *mechanical draft*. A cooling tower operating in a closed loop is said to be operating in the *closed mode* (Fig. 7-2).
2. *Spray ponds* rely upon winds that blow across the ponds and cool fine sprays of water by evaporation.
3. *Spray canals* are like spray ponds except the water is sprayed into large droplets so that drift loss (water carryover with the wind) is reduced, but at a lower mass-transfer rate and hence lower cooling efficiency.

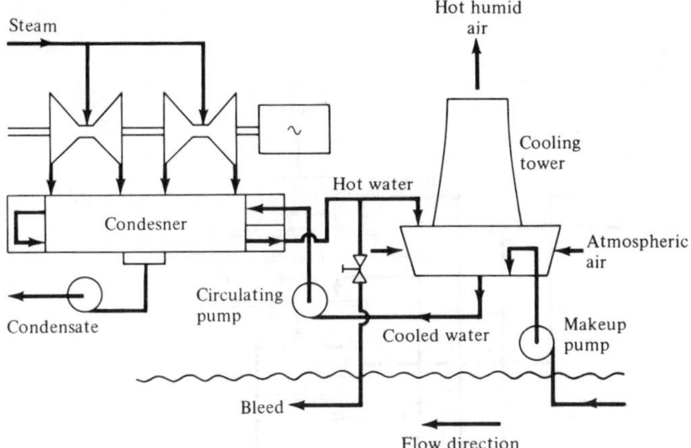

Figure 7-2 Schematic of a wet-cooling tower operating in the closed mode.

4. *Cooling lakes* are areas in which the water is cooled naturally by evaporation and radiation. They are the least efficient of the three artificial bodies of water and therefore require the greatest acreage, but also the least mechanical equipment.

Combination systems Combination systems combine once-through systems with a cooling device, usually a cooling tower, that cools the water before returning it to the natural body of water. A cooling tower operating this way is said to be operative in the *open mode* (Fig. 7-3), also called the *terminal-difference* mode. A cooling tower

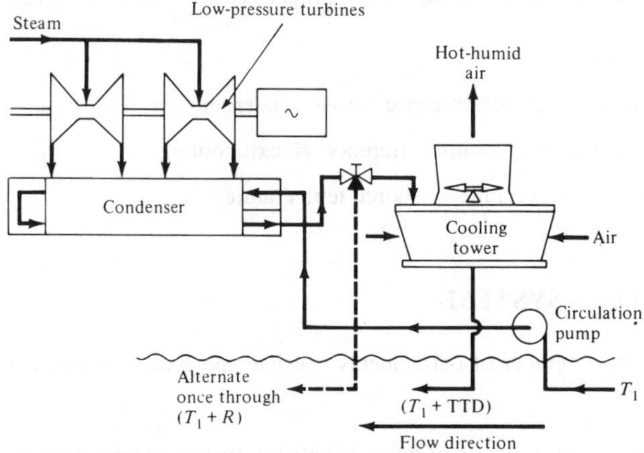

Figure 7-3 A wet-cooling tower operating in the open mode. An alternate once-through system is shown by the dashed line.

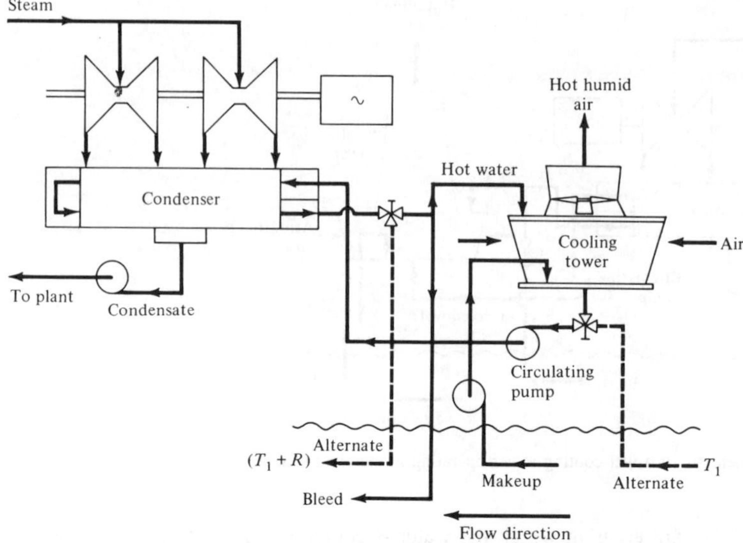

Figure 7-4 A wet-cooling tower operating in the helper mode.

may also operate in the *helper mode* (Fig. 7-4), in which both once-through and cooling-tower closed mode share cooling duties. This is the case when water supplies may be unreliable, as during drought periods, upstream intermittent use, or river temperature limitations during certain periods. The cooling tower may carry anywhere from 0 to 100 percent of the cooling load. A cooling-tower installation may also be operated in the closed, open, or helper modes by the proper use of a variety of gates and valves.

In Figs. 7-1 through 7-4

$$R = \text{range} = \text{temperature rise across condenser}$$

$$\text{TTD} = \text{terminal temperature difference} = \text{exit cooling tower temperature} - \text{source temperature}$$

7-3 THE CIRCULATION SYSTEM

The system is composed of a number of components. Besides the condenser and the cooling device, they are:

1. An intake structure from a natural body of water in an open system or from cooling towers, spray ponds, etc., in a closed or helper mode
2. A circulating-water pumping station

3. Circulating-water conduits
4. Flow gates
5. Vacuum breaking system
6. Cold- and warm-water channels

The *intake system* begins with a *skimmer wall* that assists in obtaining cooler bottom water from the natural source. The intake structure, located at the land end of an intake channel, is usually constructed of reinforced concrete. It houses the necessary equipment for screening debris from the water (Fig. 7-5). The water passes through *bar racks* that are mounted across each inlet bay to intercept large debris. The bar-rack grill sections extend from the inlet-bay floor to the top of the intake structure with an incline, usually 15° from vertical.

The debris is removed from the upstream face of the bar racks by a *traversing trash rake,* usually operator actuated, that travels across the top of the intake structure on two parallel rails. Its main component is a *rake basket,* typically 10 ft wide, equipped with teeth. As the basket is lowered, it is opened and the teeth are turned away from the bar rack to clear the debris from the rack. As the basket is raised, it is closed so that the teeth penetrate between the rack bars to clean out the embedded debris. When it reaches the discharge position at the top, it is opened to dump the trash into the trash cart. When full, the latter can be removed from the end of the trash-rake frame and towed to a disposal area.

The main circulating-water pumps are further protected by *traveling screens* at their intakes (Fig. 7-6). These screens are composed of a series of screen panels, which typically have 3/8-in^2 openings and are connected in a continuous loop across rotating drive sprockets. As water flows through the initially stationary screens, debris is deposited on them and held by the force of the moving water. When sufficient debris accumulates, the pressure drop across the screens reaches a point at which the screen drive is automatically actuated. The screen panels rotate at top and are washed by a spray, and the debris collects in a large tray. Manual operation of the screens is also provided from the control room. There are typically six such traveling screens per circulating pump.

The cooling water now enters a number of large *circulating-water pumps* operating in parallel. The number of pumps depends upon the plant and pump sizes, but a usual arrangement is three pumps per pumping station or unit. In nuclear powerplants, the size of the circulating pumps is such that any one is sufficient to dissipate reactor-shutdown decay heat. In emergency conditions that pump can be started by the two standby diesel generators that are part of these powerplants. Each pump discharge to the condensers is equipped with a motor-operated butterfly valve, which is interlocked with its pump motor so that both will start together to minimize surges in the condenser.

The combined flow goes through a tunnel to the condenser inlet conduit. The conduit, which could be 6.5 ft (2 m) in diameter, has a motor-operated butterfly valve in a vertical run of it. It directs the water to each side of the condenser divided inlet water box. A similar arrangement on the condenser discharge side, also with a motor-operated butterfly valve, permits throttling for testing or equalization of flows in case there is more than one condenser. The discharge from the condensers now goes through

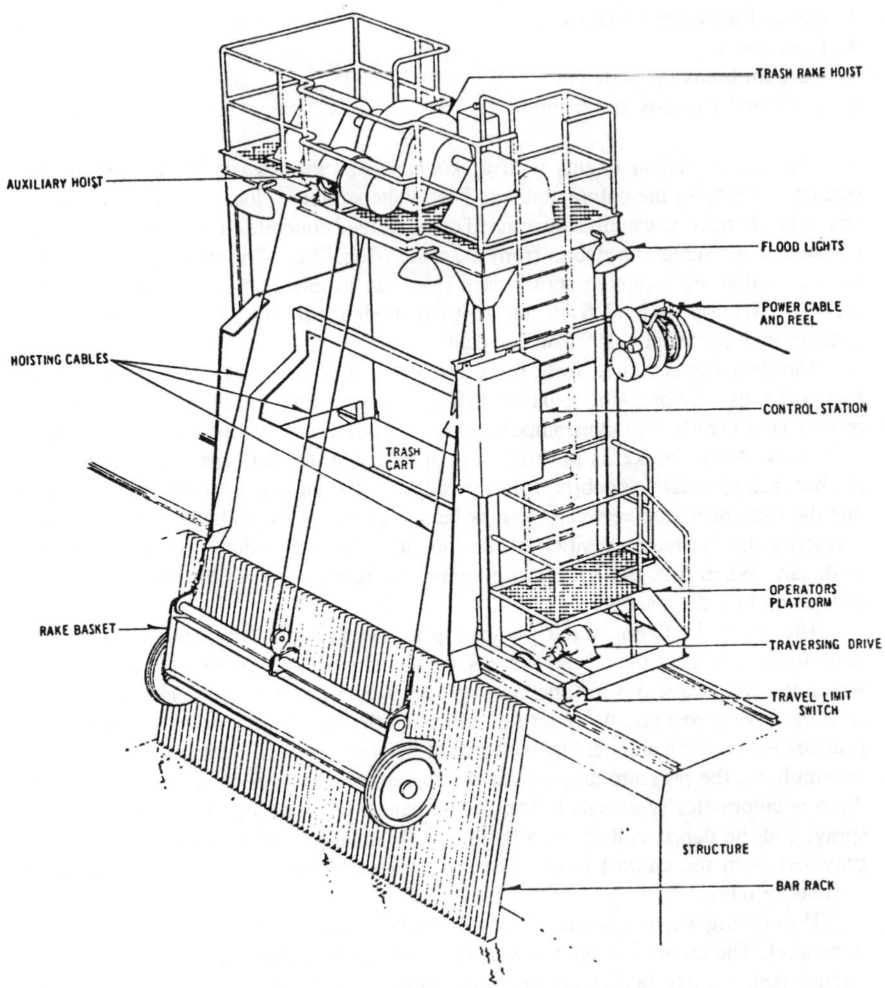

Figure 7-5 A typical intake system bar rack and traversing trash rake.

a discharge tunnel to a warm-water channel that feeds the cooling tower. Alternatively, when conditions permit, the warm water is returned to the pump reservoir by means of diffuser pipes. These are partially perforated, corrugated, galvanized steel pipes laid side by side across the bottom of the reservoir. They provide thermal mixing with the reservoir water.

When the circulating-water system is operated in the closed or helper modes, the warm water from the condenser is directed to the cooling towers by pumping stations, one per tower, typically composed of two pumps and located in separate reinforced

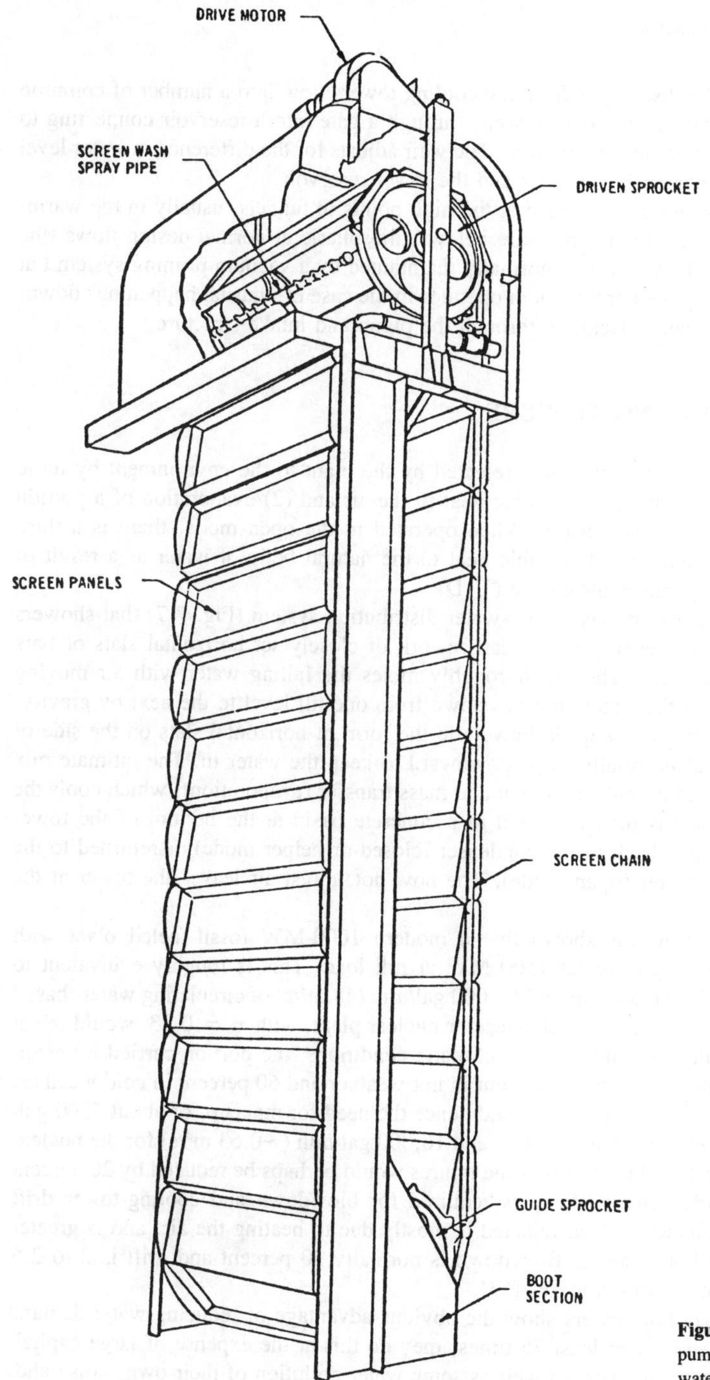

Figure 7-6 A typical pump intake traveling water screen.

concrete pits. The discharges from the cooling towers flow into a number of common open cold-water channels, over a weir, through a gate into a reservoir connecting to the circulating pump intake structure. The weir adjusts for the difference in water level between the higher cooling towers and the lower reservoir.

Vacuum systems are provided at the high points in tunnels, usually in the warm-water channel, that are at a pressure below atmospheric at normal design flows (the syphon effect). The vacuum is normally maintained by a vacuum priming system but must be broken by the control-room operator in the case of unusual happenings downstream that may cause backflow through the pump and intake structure.

7-4 WET-COOLING TOWERS

Wet-cooling towers dissipate heat rejected by the plant to the environment by these mechanisms: (1) addition of sensible heat to the air and (2) evaporation of a portion of the recirculation water itself. When operated in the open mode, there is a third mechanism: (3) addition of sensible heat to the natural body of water as a result of the terminal temperature difference (TTD).

Wet cooling towers have a hot-water distribution system (Fig. 7-7) that showers or sprays the water evenly over a latticework of closely set horizontal slats or bars called *fill*, or *packing*. The fill thoroughly mixes the falling water with air moving through the fill as the water splashes down from one fill level to the next by gravity. Outside air enters the tower via louvers in the form of horizontal slats on the side of the tower. The slats usually slope downward to keep the water in. The intimate mix between water and air enhances heat and mass transfer (evaporation), which cools the water. Cold water is then collected in a concrete basin at the bottom of the tower where it is pumped back to the condenser (closed or helper mode) or returned to the natural body of water (open mode). The now hot, moist air leaves the tower at the top.

In Table 7-1 it was shown that a modern 1000-MW fossil-fueled plant with $\eta = 0.40$ would reject about 1500 MW at full load. This is roughly equivalent to 512×10^6 Btu/h and uses about 760,000 gal/min (48 m^3/s) of circulating water, based on an 18°F (10°C) range. A water-reactor nuclear plant, with $\eta = 0.33$, would reject 683×10^6 Btu/h. Depending upon climatic conditions, the portion carried by evaporative mechanism is about 75 percent in hot weather and 60 percent in cold weather. It would result in the evaporation, and hence the need for makeup, of about 7500 gal/min ($\sim$0.47 m^3/s) for the fossil plant and 10,000 gal/min ($\sim$0.63 m^3/s) for the nuclear plant in hot weather. In cold times the figures would perhaps be reduced by 20 percent (Table 7-2). Additional makeup is required for blowdown and cooling-tower drift (below). The balance of heat rejected is mostly due to heating the air, and is greater in cold than in hot weather. Blowdown is normally 20 percent and drift is 2 to 2.5 percent of the evaporation losses [53].

Although cooling towers show the obvious advantage of reducing water demand for available water by at least 75 times, they do this at the expense of large capital, land, and operational costs, as well as some water pollution of their own, noise and, it is often said, sight pollution. Nevertheless, environmental regulations and thermal pollution of once-through systems, and the increasing scarcity of dependable natural water supplies in many parts of the world, are forcing utilities into building more and more cooling towers (and some other systems such as cooling ponds). Once-through systems, it should be recalled, are the most efficient cooling systems and should be preferred if there is an abundance of natural water.

Wet-cooling towers are classified as either (1) *mechanical-draft* or (2) *natural-draft* cooling towers. Each of these types is further classified as (1) *counterflow* or (2) *cross-flow* cooling towers (Fig. 7-7).

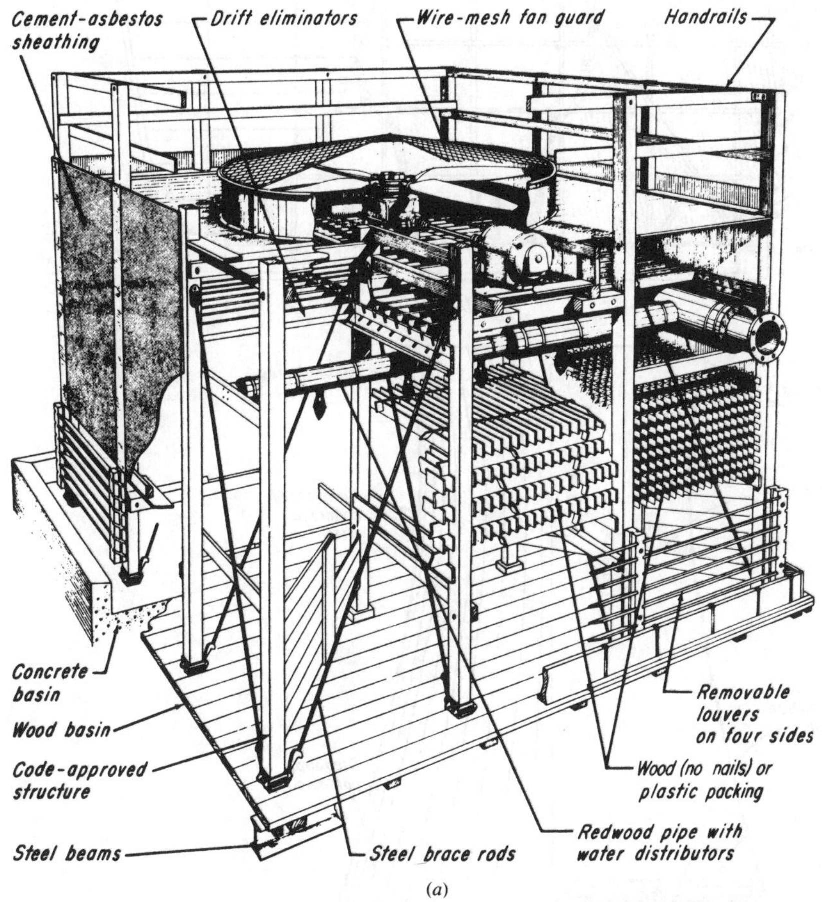

(*a*)

Figure 7-7 The four types of wet-cooling towers: (*a*) mechanical-draft counterflow, (*b*) mechanical-draft cross-flow, (*c*) natural-draft counterflow, (*d*) natural-draft cross-flow [51,52]. (Not drawn to same scale.) (Continued on next two pages.)

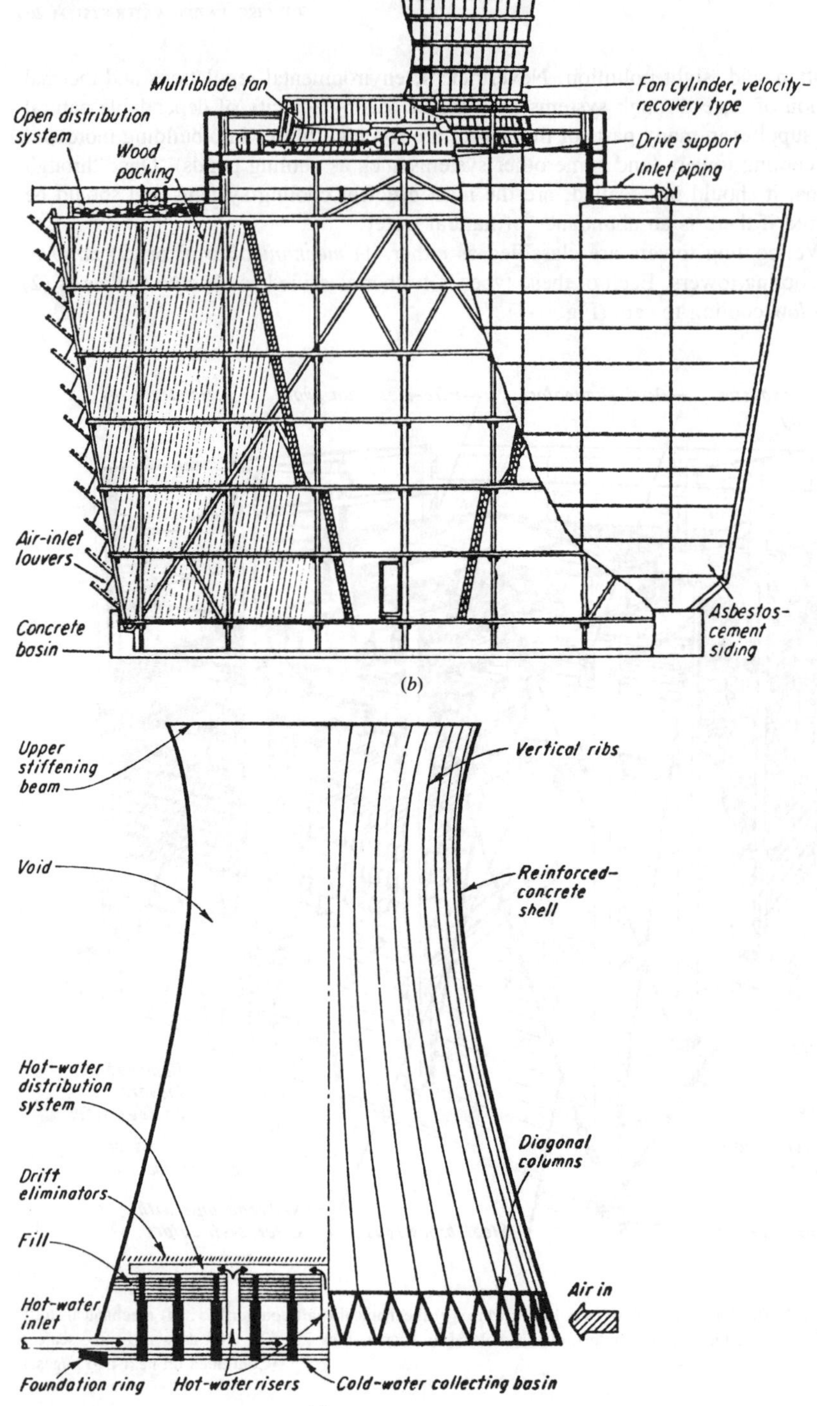

Multiblade fan
Fan cylinder, velocity-recovery type
Open distribution system
Wood packing
Drive support
Inlet piping
Air-inlet louvers
Concrete basin
Asbestos-cement siding

(b)

Upper stiffening beam
Vertical ribs
Void
Reinforced-concrete shell
Hot-water distribution system
Diagonal columns
Drift eliminators
Fill
Air in
Hot-water inlet
Foundation ring
Hot-water risers
Cold-water collecting basin

(c)

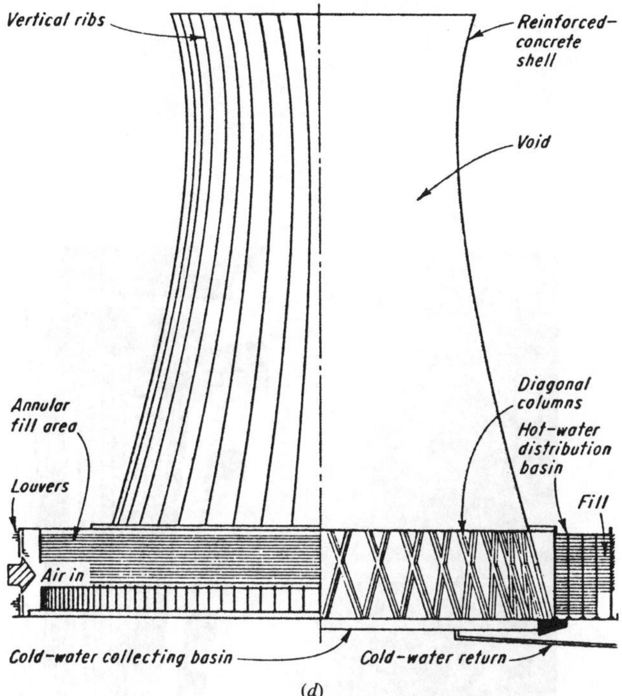

(*d*)

Figure 7-7 (continued)

Table 7-2 Typical evaporation from wet-cooling towers serving a 1000-MW plant

η	Q_R, MW	Circulating-water flow, °R = 18°F = 10°C gal/min	m³/s	Climate	Evaporation gal/min	m³/s	%
0.33	2000	758,000	47.8	Hot	10,000	0.63	1.3
				Cold	8000	0.50	1.0
0.40	1500	568,000	35.8	Hot	7500	0.47	1.3
				Cold	6000	0.38	1.0

Figure 7-8 Multicell induced-draft cooling towers showing cross-flow in-line, cross-flow round, and counterflow octagonal arrangements. (*Courtesy the Marley Cooling Tower Company, Mission, Kansas.*)

Mechanical-Draft Cooling Towers

In *mechanical-draft* cooling towers, the air is moved by one or more mechanically driven fans. As in steam generators, the fans could be of the *forced-draft* (FD) type, which would be mounted on the lower sides to force air into the tower. This type would theoretically be preferred because the fans would operate on cooler air and hence consume less power. However, experience with this type of fan has shown some disadvantages because of air distribution problems, leakage, recirculation of the hot and moist exit air back to the tower, and frost accumulation at the fan inlets during winter operation.

The majority of mechanical-draft cooling towers for utility application are therefore of the *induced-draft* (ID) type. With this type, air enters the sides of the tower through large openings at low velocity and passes through the fill. The fan is located at the top of the tower, where it exhausts the hot, humid air to the atmosphere. Induced-draft cooling towers are usually multicell with a number of fan stacks on top and come in various arrangements: rectangular, octagonal, circular, etc. (Fig. 7-8).

The fans are usually multibladed and large, ranging from 2 to 33 ft (0.6 to 10 m) in diameter (Fig. 7-9). They are driven by electric motors, as large as 250 hp, at relatively low speeds through reduction gearing. They are of the propeller type, which move large volumetric flow rates at relatively low static pressures. They have adjustable-pitch blades for minimum power consumption, depending upon load and climatic conditions. The fans and their drives are designed to function satisfactorily in the hot, humid atmosphere they are in. The blades are usually made of cast aluminum, stainless steel, or fiberglass, although plastic and laminated wood have been used.

The airflow into the tower is roughly horizontal. The flow in the fill however is either horizontal, in which case it is a *cross-flow* cooling tower, or vertical, in which case it is a *counterflow* cooling tower (Fig. 7-7*a* and 7-7*b*).

The main advantages of mechanical-draft cooling towers are the assurance of moving the required quantity of air at all loads and climatic conditions, low initial capital and construction costs, and a low physical profile. Their main disadvantages are power consumption, operating and maintenance costs, and greater noise [54].

Natural-Draft Cooling Towers

Natural-draft cooling towers were developed in Europe. The first ones, erected in Holland early in this century, were made of wood. They evolved to wood on steel and finally to the reinforced concrete used today. The early shapes were nearly cylindrical, followed by an inverted truncated cone on top of another, and finally the hyperbolic shape of today. Their most extensive use has been in England. In the United States, the first unit was built in 1972, and the number being built has risen dramatically since. One hundred units were expected to be constructed by 1980 [51].

Natural-draft cooling towers use no fans. They depend for airflow upon the natural driving pressure caused by the difference in density between the cool outside air and the hot, humid air inside. The driving pressure ΔP_d is given by

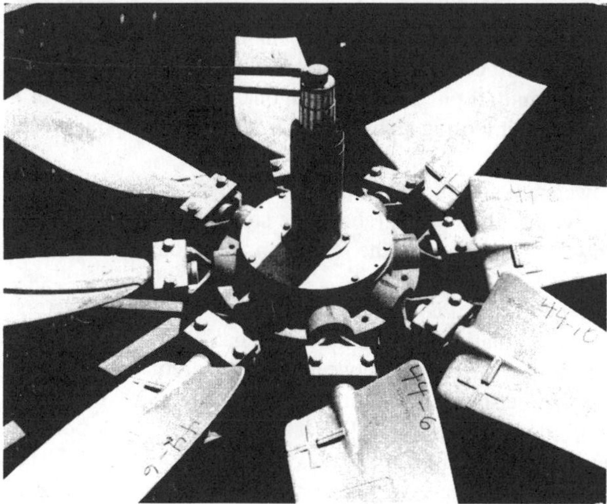

Figure 7-9 Typical fans used in induced-draft cooling towers. On the bottom is one with a large diameter. On the top is one with automatic variable-pitch control. *(Courtesy the Marley Cooling Tower Company, Mission, Kansas.)*

$$\Delta P_d = (\rho_o - \rho_i)H\frac{g}{g_c} \tag{7-2}$$

where ρ_o = density of outside air, lb_m/ft^3 or kg/m^3

ρ_i = density of inside air, taken at exit of the fill, lb_m/ft^3 or kg/m^3

H = height of tower above the fill, ft or m

g = gravitational acceleration, ft/s^2 or m/s^2

g_c = conversion factor = 32.2 $lb_m \cdot ft/(lb_f \cdot s^2)$ or 1.0 $kg \cdot m/(N \cdot s^2)$

This driving pressure must balance the air-pressure losses through the tower.

Because $\rho_o - \rho_i$ is relatively small, H must be large to result in the desired ΔP_d. Natural-draft cooling towers are therefore very tall, often a few hundred feet. The tower body, above the water distribution system and the fill, is an empty shell of circular cross section but with a hyperbolic vertical profile. Natural-draft cooling towers are therefore often referred to as *hyperbolic towers*. The hyperbolic profile has been found to offer superior strength and the greatest resistance to outside wind loading (pressure due to strong winds) compared with other forms, so that substantially less material is needed and thicknesses can be as low as 6 or 7 in at the waist. It has little to do with inside airflow. Natural-draft cooling towers are made of reinforced concrete, very large volumes of concrete, and sit on "stilts" or diagonal support columns in a shallow basin of water. They are an imposing sight from afar, dwarfing the powerplant itself (Fig. 7-10).

Natural-draft cooling towers are also of the counterflow or cross-flow types (Figs. 7-7*c* and 7-7*d*). In counterflow, the fill is inside (above the stilts), spread over a large area, and is therefore shallower. In cross-flow, the fill sits in a ring outside the tower outside the stilts.

The choice between different types of cooling towers depends upon many factors, the most important of which are climatic and economic. Mechanical-draft towers are the choice when the approach (the difference between the temperature of the cold water leaving the tower and the wet-bulb temperature of the outside air) is low and when a broad range of water flow is expected. The latter factor is made possible because they are usually built as multicell units with a variable-airflow fan, a design which offers versatility and good response to changes in cooling parameters and demands.

Natural-draft towers are selected most often: (1) in cool, humid climates (low wet-bulb temperatures and high relative humidity); (2) when there is a combination of low wet-bulb temperatures and high condenser-water inlet and outlet temperature, i.e., a broad range and long approach; or (3) in cases of heavy winter loads. Economic factors favor them when, because of their large capital costs, a long amortization period can be arranged. In general, they are favored for very large powerplants where fewer

Figure 7-10 A natural-draft cooling tower next to a nuclear-reactor containment building. *(Courtesy Research-Cottrell, Bound Brook, New Jersey.)*

and larger towers can be built. In the early 1980s mechanical-draft towers still commanded the lion's share in utility and large industrial installations, although natural-draft towers are expected to gradually increase their share in the future.

The choice between cross-flow and counterflow of either type is less clear. Cross-flow towers offer less resistance to airflow, thus allowing higher air velocities than counterflow towers. They are, however, less efficient. Both designs, therefore, have counterbalancing advantages and limitations, and the choice depends upon the particular application. Natural-draft towers, however, recently ordered have been in the counterflow variety only.

The Water-Distribution System

The *water-distribution system* dispenses hot condenser water evenly over the fill. There are several types, among them are: (1) *Gravity distribution,* used mainly on cross-flow towers, consists of vertical hot-water risers that feed into an open concrete basin, from which the water flows by gravity through orifices to the fill below (Fig. 7-11*a*). (2) *Spray distribution,* used mainly on counterflow towers, has cross piping with spray-downward nozzles (Fig. 7-11*b*). (3) *Rotary distribution* consists of two slotted distributor arms that rotate about a central hub through which water comes in under pressure. The slots are aimed downward but slightly to one side, resulting in a curtain of water at an angle and a reaction force that rotates the arms at 25 to 30 r/min. The speed of rotation can be varied by adjusting the slot angle (Fig. 7-11*c*).

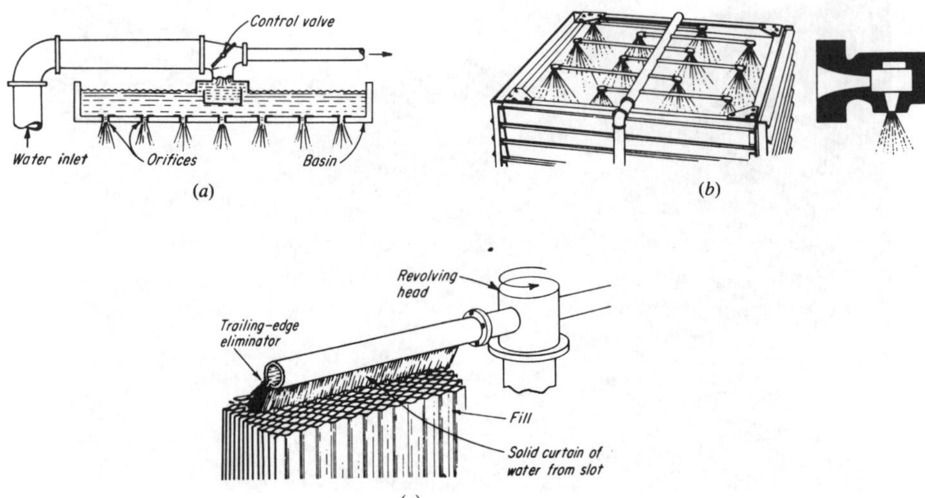

Figure 7-11 Water-distribution systems: (*a*) gravity, (*b*) spray, (*c*) rotary [51].

The Fill

The *fill*, or *packing*, is the heart of the cooling tower. It should provide both good water-air contact for high rates of heat transfer and mass transfer and low resistance to airflow. It should also be strong, light, and deterioration-resistant. There are basically two types of fill: (1) *splash type* and (2) *film*, or *nonsplash*, *type*.

Splash packing is made of bars stacked in decks (Fig. 7-12*a*) that break the water into drops as it falls from deck to deck. The bars come in different shapes [52]—narrow, square, or grid—are smooth or rough, and are made of different materials—redwood, high-impact polystyrene, or polyethylene. Splash fill provides excellent heat and mass transfer between water and air.

Film fill is usually made of vertical sheets that have a rough adsorbent surface that wets well and allows the water to fall as film that adheres to the vertical surfaces (Fig. 7-12*b*). This exposes the maximum water surface to the air without breaking it into drops or small rivulets. Film fill also comes in different shapes and materials: redwood battens, cellulose corrugated sheets, asbestos-cement sheets, and waveform

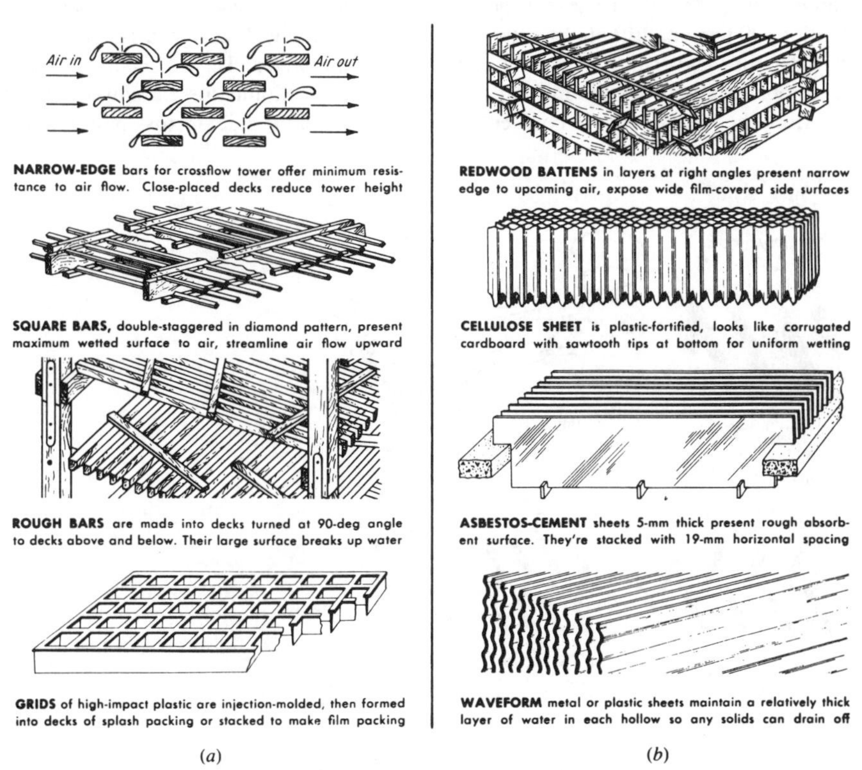

Air in
Air out
NARROW-EDGE bars for crossflow tower offer minimum resistance to air flow. Close-placed decks reduce tower height
SQUARE BARS, double-staggered in diamond pattern, present maximum wetted surface to air, streamline air flow upward
ROUGH BARS are made into decks turned at 90-deg angle to decks above and below. Their large surface breaks up water
GRIDS of high-impact plastic are injection-molded, then formed into decks of splash packing or stacked to make film packing
(*a*)
REDWOOD BATTENS in layers at right angles present narrow edge to upcoming air, expose wide film-covered side surfaces
CELLULOSE SHEET is plastic-fortified, looks like corrugated cardboard with sawtooth tips at bottom for uniform wetting
ASBESTOS-CEMENT sheets 5-mm thick present rough absorbent surface. They're stacked with 19-mm horizontal spacing
WAVEFORM metal or plastic sheets maintain a relatively thick layer of water in each hollow so any solids can drain off
(*b*)

Figure 7-12 Types of fill: (*a*) splash (*b*) film [52].

metal or plastic. Film fill presents less resistance to airflow and requires less total height than splash fill.

The present trend in materials for wet-cooling towers favors concrete structures with plastic fill, drift eliminators (below), fan stacks, fan blades, and manifolds, valves, and nozzles. The concrete-plastic combination results in longer life and less maintenance.

Drift and Drift Eliminators

Drift is water entrained by and carried with the air as unevaporated drizzle or fine droplets. This water is thus lost to the circulating-water system and does not contribute to heat removal by evaporation. (The problem is somewhat similar to the carryover of droplets by the steam in a boiler drum.) Drift is minimized by *drift eliminators* (Fig. 7-13), which are baffles that come in one, two, or three rows. The baffles force the air to make a sudden change in direction. The momentum of the heavier drops separates them from the air and impinges them against the baffles, thus forming a thin film of liquid that falls back into the tower. The baffles are made of wood, metal, or plastic. 100 percent drift elimination is not possible, but a well-designed system results in water loss less than 0.2 percent of the circulating water. This, of course, should be compensated for by the makeup system. The drift eliminators are situated at air exit from the fill, above it in counterflow towers, and to the side in cross-flow towers.

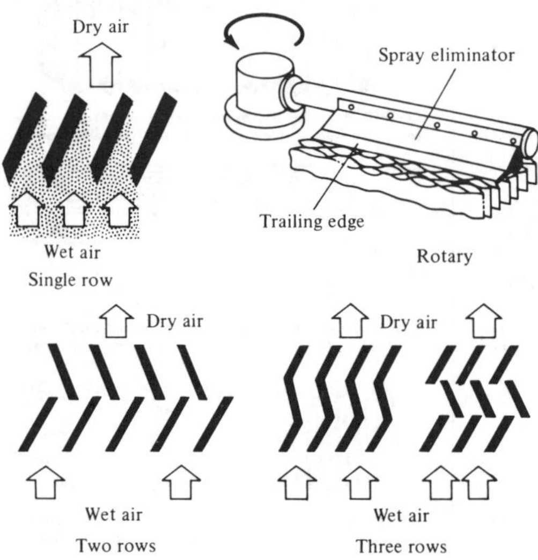

Figure 7-13 Types of drift eliminators [52].

The Basin

The *cold-water basin,* situated beneath the tower, collects and strains the water before it is pumped back to the condenser (in the closed mode). It also receives the circulating-water makeup. It has makeup inlet and outlet, quickfill, overflow, drain, and bleedoff connections. Large utility tower basins are usually made of concrete. They are sized to permit tower operation for several hours without adding makeup. The drain is used to remove silt deposits and control water level in case of flow surges. Water leaves the basin via a sloped canal at the bottom and through screens that prevent debris from entering the pumps.

Makeup

The total makeup required by a cooling tower is the sum of that which would compensate for evaporation, drift, and bleeding or blowdown. The latter is water intentionally discharged from the circulating-water system and replaced with fresh water. This water, in addition to makeup for evaporation and drift, keeps the concentration of salts and other impurities down. Otherwise these concentrations would continuously build up as the water continues to evaporate. As indicated earlier (Table 7-2), the evaporation loss rate is 1 to 1.5 percent of the total circulating-water flow rate. The drift loss is much less, perhaps 0.03 percent. Large quantities of drift cannot be tolerated because, in addition to the plume [55], they can cause water and ice deposition problems at and near the plant site. Bleeding accounts for an amount comparable to the evaporation loss if the solids concentrations are to be kept low but may be reduced if higher concentrations can be tolerated. Table 7-3 gives a breakdown of water makeup requirements for typical 1000-MW fossil-fired and water-reactor nuclear powerplants for different cooling ranges [53]. It shows that the total makeup is independent of range but that the percentage of total flow is not.

Circulating-Water Quality and Blowdown Pollution

The circulating water is warm and fully aerated, contains suspended solids, is relatively high in conductivity, and is a good biological nutrient.

It is almost always corrosive, requiring a corrosion inhibitor such as chromates. To prevent scale and deposits that foul heat-transfer surfaces, it often needs scale inhibitors. Silt washed off the air by the cooling tower usually is in the form of small colloidal matter and is difficult to remove, but can be treated by a family of chemicals called *polyelectrolytes*. These keep the particles suspended in flowing water but allow them to precipitate in basins where they can be removed as mud. Microbiological growth (algae, slime, bacteria), besides fouling, contributes to corrosion by shielding metal surfaces and thus producing oxygen-concentration cells. In addition, the decomposition of the organisms produces H_2S, CO_2, and other products, themselves corrosive. Chlorination, alternated with biocides, is used (organisms can build up tolerance to chlorine, if used alone).

Table 7-3 Typical makeup demand for 1000-MW powerplants

Range, °F	Tower water flow, gal/min	Drift, gal/min	Evap., gal/min	Bleed required to maintain concentration,* gal/min: 2	4	6	8
			Fossil-fueled plant				
15	600,000	180	6750	6,570	2070	1170	784
25	360,000	108	6750	6,642	3142	1242	856
35	257,153	77.2	6750	6,673	2173	1273	887
45	200,000	60	6750	6,690	2190	1290	904
Total makeup required, gal/min				13,500	9100	8100	7714
			Nuclear-fueled plant				
15	900,000	270	10,125	9,855	3,105	1,755	1,176
25	540,000	162	10,125	9,963	3,213	1,863	1,284
35	385,715	115.7	10,125	10,009	3,259	1,909	1,331
45	300,000	90	10,125	10,035	4,285	1,935	1,356
Total makeup required, gal/min				20,250	13,500	12,150	11,571

* Concentration = (evap. + drift + bleed)/(drift + bleed).

Note: Based on heat-rejection rates of 4500 and 6750 Btu/kWh, respectively. Data based on (1) evaporation 0.075 percent of recirculated gal/min per °F range and (2) 0.03 percent drift loss. Data from Ref. 53.

Because of all the above additives, cooling-tower bleed or blowdown can be an unacceptable source of pollution to the natural body of water. (Boiler blowdown is another source of pollution. It is mainly thermal but also contains small quantities of phosphates and organics.) Bleed may contain chemicals and various minerals contained in or added to the circulating water, including chromate inhibitors, various phosphates, organic and inorganic compounds, combined with some heavy metals.

Bleed, therefore, has to be handled with care. Depending upon the size of plant and the extent of contaminants, it may be discharged to the body of water, treated before being returned, or allowed to mix with other plant wastes, such as boiler blowdown, etc., and treated all in one installation.

An example of treatment is the removal of chromates by reducing them from hexavalent form to trivalent chrome with $FeSO_4$, then precipitated as $Cr(OH)_3$ by elevating the pH with lime (or an alkaline stream from somewhere else in the plant). The resulting sludge can be disposed of in various ways or reused. Another alternative is to use nonpolluting inhibitors, but these are, at least for now, of questionable ability and economics [56].

A Hybrid Wet Tower

Natural-drift cooling towers are huge in material, height, and land requirements but consume little power for operation. Mechanical-draft cooling towers are smaller and

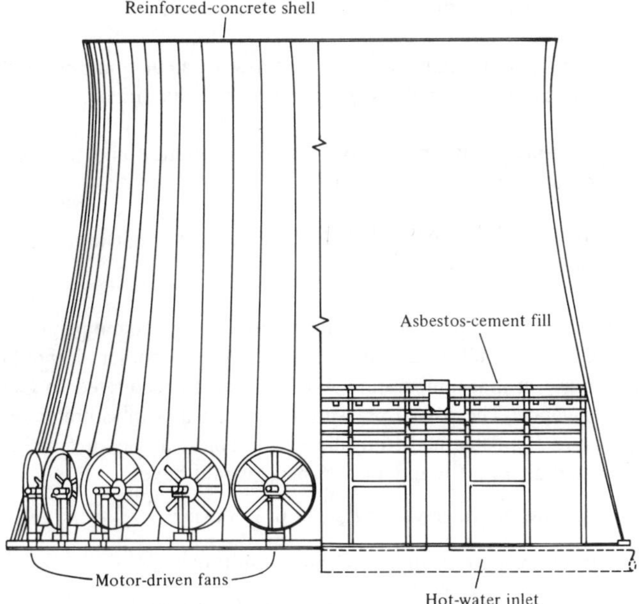

Figure 7-14 A hybrid or fan-assisted hyperbolic tower [51].

cheaper but consume power and suffer from air recirculation. As often happens, someone has come up with a compromise. This is the *hybird* or *fan-assisted hyperbolic* tower, which combines the best features of natural- and mechanical-draft towers.

The hybrid tower (Fig. 7-14) has a reinforced concrete hyperbolic shell, similar but smaller than that of the natural-draft tower. In addition, it has a number of large electrically driven forced-draft fans situated around the periphery of the base. The hybrid tower will have roughly two-thirds the base diameter (45 percent of the land area) and half the height of a natural-draft tower designed for the same performance, and hence will have less of a visual impact on the landscape. The fans will give better airflow control than natural-draft and will consume less power than mechanical draft. The high velocity and height of the exit air will eliminate the hot-air recirculation problems associated with mechanical-draft towers. Hybrid towers can also be operated during the cold part of the year without fans as natural-draft units. Construction costs will be intermediate between natural- and mechanical-draft towers.

7-5 WET-COOLING TOWER CALCULATIONS

Wet-cooling tower calculations involve energy and mass balances. The energy balances here will be based on the first-law steady-state steady-flow (SSSF) equation. There are, however, three fluids entering and leaving the system: the cooling water, the dry air, and the water vapor associated with it. The mass balance should also take into account these three fluids.

Because air humidity is an important factor in wet-tower calculations, a short review of tower and psychrometric terminology is appropriate at this time.

Approach This is the difference between the cold-water temperature and the wet-bulb temperature of the outside air.

Range (or cooling range) The range is the difference between the hot-water temperature and the cold-water temperature.

Saturated air This is air that can accept no more water vapor at its given temperature. A drop in that temperature would result in condensation and the new cooler air would also be saturated. An increase in that temperature would make it unsaturated so that it could accept more water vapor. The partial pressure of water vapor in saturated air equals the saturation pressure P_{sat} (obtained from the steam tables) at the air temperature. For example, saturated air at 60°F and 14.696 psia, would have

$$\text{Partial pressure of water vapor } P_{sat} = 0.256 \text{ psia}$$

$$\text{Partial pressure of dry air } P_a = 14.440 \text{ psia}$$

$$\text{Total pressure } P = 14.696 \text{ psia}$$

Recall from thermodynamics of gas mixtures that the partial pressure ratios, the volume ratios, and the mole fractions of constituents are all equal. Water vapor in air is at such low pressure (its partial pressure) that it is treated as a gas with little error.

Relative humidity This is equal to the partial pressure of water vapor in air P_v, divided by the partial pressure of water vapor that would saturate the air at its temperature P_{sat}. Relative humidity is given the symbol ϕ.

$$\phi = \frac{P_v}{P_{sat}} \tag{7-3}$$

Thus, for air at 60°F, 14.696 psia, and 50 percent relative humidity

$$P_v = \phi P_{sat} = 0.5 \times 0.256 = 0.128 \text{ psia}$$

$$P_a = P - P_v = 14.696 - 0.128 = 14.568 \text{ psia}$$

$$P = 14.696 \text{ psia}$$

and $\phi = 100$ percent refers to saturated air.

Absolute humidity (also called humidity ratio) This is the air per unit mass of dry air (da). Absolute humidity is given the symbol ω. Using $PV = mRT$ for both water vapor and dry air

$$\omega = \frac{m_v}{m_a} = \frac{53.3P_v}{85.7P_a} = \frac{0.622P_v}{P - P_v} \tag{7-4}$$

where 53.3 and 85.7 are the gas constants for dry air and water, respectively. The absolute humidity of saturated air is then given by

$$\omega_{sat} = \frac{0.622P_{sat}}{P - P_{sat}} \tag{7-5}$$

In the above example, air at 60°F, 14.696 psia, and 50 percent ϕ would have $\omega = 0.005465$ lb$_m$ water vapor/lb$_m$ da and $\omega_{sat} = 0.01103$.

Because saturation pressure increases rapidly with temperature (as all vapor pressures do), it can be deduced that warm air can hold much more moisture than cool air.

Dry-bulb temperature This is the temperature of the air as commonly measured and used. In psychrometric work, it is called dry-bulb to distinguish it from the wet-bulb temperature (below). It is the temperature as measured by a thermometer with a dry mercury bulb, a thermocouple, etc., and is given the symbol T or T_{db}.

Wet-bulb temperature This is the temperature of the air as measured by a psychrometer, in effect a thermometer with a wet gauze on its bulb, hence the name. Air is made to flow past the gauze. If the air is relatively dry, water would evaporate from the gauze at a rapid rate, cooling the bulb and resulting in a much lower reading than if the bulb were dry. If the air is humid, the evaporation rate is slow and the wet-bulb temperature approaches the dry-bulb temperature. If the air is saturated, i.e., $\phi = 100$ percent, the wet-bulb temperature equals the dry-bulb temperature. Thus, for a given T, the wet-bulb temperature is lower the drier the air. The wet-bulb temperature is given the symbol T_{wb}.

Dew point The temperature below which water vapor in a given sample of air begins to condense is called the dew point. It is equal to the saturation temperature corresponding to the partial pressure of the water vapor in the sample. Thus for the above air at 60°F and $\phi = 50$ percent, the dew point equals the saturation temperature at 0.128 psia, which is 41.3°F. For saturated air T, T_{wb}, and the dew point are all equal.

Psychrometric chart This is a chart that relates ϕ, ω, T_{wb}, and T but which may also contain additional information such as enthalpy and specific volume (App. M). Psychrometric charts are calculated for 1 standard atmosphere pressure and are based on a *unit mass of dry air plus associated water vapor*, that is, on $1 + \omega$. Thus all data in App. M are for a fixed 1.0 lb$_m$ of dry air (da) plus a variable ω lb$_m$ water vapor/lb$_m$ da. Check the above examples on the chart.

Energy balance The first-law SSSF equation with three fluids will now be written for the tower fill as a system (Fig. 7-15). It applies to all types of wet towers. Changes

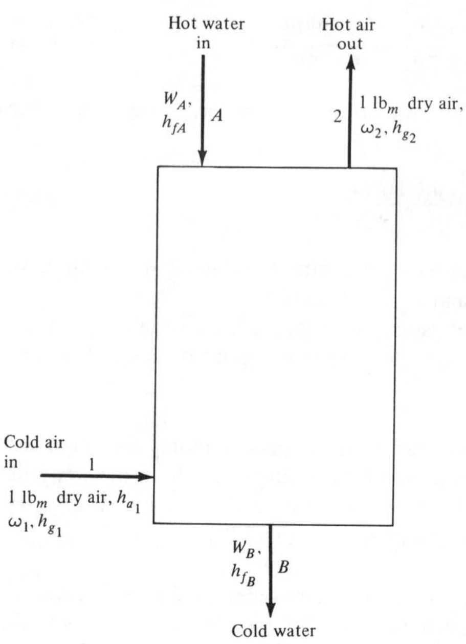

Figure 7-15 The tower fill as a steady-state steady-flow system.

in potential and kinetic energies and heat transfer are all negligible. No mechanical work is done. Thus only enthalpies of the three fluids appear. Following psychrometric practice, the equation is written for a unit mass of dry air

$$h_{a1} + \omega_1 h_{v1} + W_A h_{WA} = h_{a2} + \omega_2 h_{v2} + W_B h_{WB} \tag{7-6}$$

where

h_a = enthalpy of dry air, Btu/lb_m or J/kg

ω = *mass of water vapor per unit mass of dry air*

= *absolute humidity, dimensionless*

h_v = enthalpy of water vapor. Btu/lb_m or J/kg

W = mass of circulating water per unit mass of dry air, dimensionless

h_W = enthalpy of circulating water, Btu/lb_m or J/kg

The subscripts 1 and 2 refer to air inlet and exit, and the subscripts A and B refer to circulating-water inlet and exit, respectively.

Because of the low pressures and temperatures commonly encountered in towers, the above equation can be simplified with little error by the following approximations.

$$h_{a2} - h_{a1} = c_p\,(T_2 - T_1) \qquad c_p = 0.24\ \text{Btu/(lb}_m \cdot {}^\circ\text{F)}$$

$$h_v = h_g \text{ from the steam tables}$$

$$h_W = h_f \text{ from the steam tables}$$

The air leaving the system at 2 is often saturated.

Mass balance The dry air goes through the tower unchanged. The circulating water loses mass by evaporation. The water vapor in the air gains mass due to the evaporated water. Thus, based on a unit mass of dry air

$$\omega_2 - \omega_1 = W_A - W_B \tag{7-7}$$

Equation (7-6) can now be written in the form

$$\omega_1 h_{g1} + W_A h_{f,A} = c_p(T_2 - T_1) + \omega_2 h_{g2} + [W_A - (\omega_2 - \omega_1)]h_{f,B} \tag{7-8}$$

At pressures P other than 1 standard atmosphere, where the psychrometric chart would not apply, and when T and T_{wb} are known, the value of P_v, and hence ω, from Eq. (7-4), is obtained from the empirical Carrier equation:

$$P_v \ \text{(psia)} = P_{v,\text{wb}} - \frac{(P - P_{v,\text{wb}})(T - T_{\text{wb}})}{2800 - 1.3T_{\text{wb}}} \tag{7-9}$$

where $P_{v,\text{wb}}$ is the vapor pressure corresponding to T_{wb}, psia, and T and T_{wb} are in °F.

Air densities We will now evaluate ρ_o and ρ_i in Eq. (7-2), repeated below.

$$\Delta P_d = (\rho_o - \rho_i)H\frac{g}{g_c} \tag{7-2}$$

The driving pressure ΔP_d, which should equal the air-pressure losses in the tower, is used to calculate H in natural-draft cooling towers. Here ρ_o is the outside air density and ρ_i the inside air density at the air exit of the fill, and H the height of the tower above the fill. The density of any gas mixture is equal to the sum of the densities of its constituents as they all occupy the same volume. Thus, using $\rho = m/V = P/RT$,

$$\rho_o = \frac{P - P_{v1}}{R_a T_1} + \frac{P_{v1}}{R_v T_1} \tag{7-10}$$

and

$$\rho_i = \frac{P - P_{v2}}{R_a T_2} + \frac{P_{v2}}{R_v T_2} \tag{7-11}$$

where the subscripts 1 and 2 again refer to air inlet and outlet, R_a and R_v are the gas constants of dry air and water, respectively, and the temperatures are in degrees absolute.

The densities can also be obtained from the specific volume v lines of the psychrometric chart if $P = 1$ standard atmosphere. Note, however that these are given in ft^3/lb_m da, so the densities of the moist-air mixture would be $1 + \omega$ divided by v as given in the chart. Note also that the v lines are rather widely spaced, which leads to less accurate evaluations. Small inaccuracies in ρ_o and ρ_i could lead to large errors in the difference between them.

Example 7-1 A natural-draft cooling tower with a range of 20°F receives 360,000 gal/min of 90°F circulating water. The outside air is at 60°F, 14.696 psia, and 50 percent relative humidity. The exit air is at 80°F saturated. Calculate (1) the amount of makeup water to compensate for evaporation, (2) the outside air required in ft^3/min, and (3) the height of the cooling tower if the total pressure losses are 0.0105 psi.

SOLUTION Referring to Fig. 7-15,

$$\omega_1 = 0.005465 \text{ (calculated earlier)}$$

$$\omega_2 = \frac{0.622 \times 0.50683}{14.696 - 0.50683} = 0.022217$$

$$h_{g1} = 1087.7 \text{ Btu/lb}_m \qquad h_{g2} = 1096.4 \text{ Btu/lb}_m$$

$$T_A = 90°\text{F} \qquad T_B = 90 - 20 = 70°\text{F} \qquad v_{f,A} = 0.016050 \text{ ft}^3/\text{lb}_m$$

$$\dot{W}_A = \text{mass flow rate of water entering tower} = \frac{360{,}000 \times 0.1337}{0.01605}$$

$$= 3 \times 10^6 \text{ lb}_m/\text{min}$$

$$h_{f,A} = 58.018 \text{ Btu/lb}_m \qquad h_{f,B} = 38.052 \text{ Btu/lb}_m$$

Referring to Eq. (7-8)

$$0.005465 \times 1087.7 + W_A \times 58.018 = 0.24\,(80 - 60) + 0.022217 \times 1096.4 + [W_A - (0.022217 - 0.005465)] \times 38.052$$

Therefore

$$W_A = 1.1308 \text{ lb}_m \text{ water/lb}_m \text{ da}$$

$$\text{Dry air required} = \frac{\dot{W}_A}{W_A} = \frac{3 \times 10^6}{1.308} = 2.653 \times 10^6 \text{ lb}_m/\text{min}$$

$$\text{Makeup due to evaporation} = 2.653 \times 10^6 \,(\omega_2 - \omega_1)$$

$$= 0.0444 \times 10^6 \text{ lb}_m/\text{min}$$

$$= \frac{0.0444 \times 10^6 \times 0.01605}{0.1337} = 5335 \text{ gal/min}$$

Referring to Eqs. (7-10) and (7-11)

$$\rho_o = \frac{(14.696 - 0.128) \times 144}{53.34(60 + 460)} + \frac{0.128 \times 144}{85.76(60 + 460)}$$

$$= 0.07563 + 0.00041$$

$$= 0.07604 \text{ lb}_m/\text{ft}^3$$

$$\rho_i = \frac{(14.696 - 0.50683) \times 144}{53.34(80 + 460)} + \frac{0.50683 \times 144}{85.76(80 + 460)}$$

$$= 0.07094 + 0.00158$$

$$= 0.07251 \text{ lb}_m/\text{ft}^3$$

$$\text{Total outside air required} = 2.653 \times 10^6 (1 + \omega_1)$$

$$= 2.653 \times 10^6 (1 + 0.005465)$$

$$= 2.667 \times 10^6 \text{ lb}_m/\text{min}$$

$$\text{Actual volume flow rate of outside air} = \frac{2.667 \times 10^6}{0.07604}$$

$$= 35.072 \times 10^6 \text{ ft}^3/\text{min}$$

Referring to Eq. (7-2)

$$\text{Height of tower above fill} = \frac{\Delta P_d}{(\rho_o - \rho_i)g/g_c}$$

$$= \frac{0.0105 \times 144}{0.07604 - 0.07251}$$

$$= 428.33 \text{ ft}$$

7-6 DRY-COOLING TOWERS

A *dry-cooling tower* is one in which the circulating water is passed through finned tubes over which the cooling air is passed. All the heat rejected from the circulating water is thus in the form of sensible heat to the cooling air. A dry-cooling tower can be either mechanical-draft or natural-draft.

Dry-cooling towers have attracted much attention in recent years. They permit plant siting without regard for large supplies of cooling water. Typical sites are at or near sources of abundant fuel, which cuts down fuel transportation costs; at or near the utility load-distribution center, which cuts down transmission costs; and at existing plants that need to be expanded but do not have sufficient water for the addition. Other advantages of dry-cooling towers are that they are less expensive to maintain than wet towers and do not require large amounts of chemical additives and periodic cleaning as do wet towers. Their main disadvantage is that they are not as efficient as evaporative cooling, and the result is higher turbine back pressure, lower plant cycle efficiency, and increased heat rejection. (The situation worsens at high atmospheric air temperatures.) Small dry-cooling towers have seen extensive service in such installations as industrial-process cooling, air conditioning, and atmospheric-air-cooled heat exchangers. Large utility dry-cooling towers have seen more usage in Europe where they have been developed, with a number of installations in successful operation. The recent attraction in the United States is certain to grow as powerplants get bigger and available water supplies dwindle so that even the makeup water needed by a wet tower will be burdensome, not to speak of once-through cooling.

Another important plus for dry-cooling towers is the increasingly restrictive environmental legislation on thermal pollution of once-through systems, blowdown pollution, and fogging and icing of wet towers, which are a real menace in certain localities.

Because of the above-mentioned advantages, dry-cooling towers are intended to operate only in the closed mode. There are two basic dry-cooling tower types: *direct* and *indirect*.

Direct Dry-Cooling Towers

This system combines the condenser with the tower (Fig. 7-16). Turbine exhaust steam is admitted to a steam header through large ducts to minimize pressure drop and is condensed as it flows downward through a large number of finned tubes or coils in

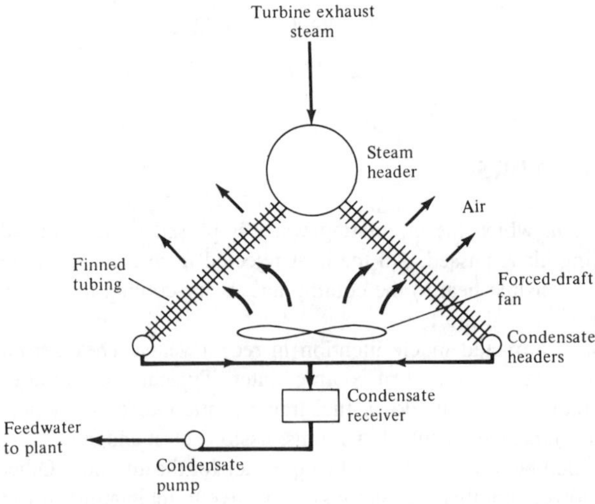

Figure 7-16 Schematic cross section of a direct dry-cooling tower.

parallel (only two are shown). The latter are cooled by atmospheric air flowing in a natural-draft cooling-tower setup or, as shown, by a forced-draft fan. As with surface condensers (Sec. 6-3), there is a system for removing noncondensables and air. There also is a system for the prevention of freezing in cold weather (below). The condensate flows by gravity to condensate receivers and is pumped back to the plant feedwater system by the condensate pump.

Direct systems operate at the disadvantage of high vacuum in the cooling coils and the need for large steam ducts. They are limited to small-size powerplants. The largest direct installation in the United States is the 330-MW minemouth Wyodak powerplant near Gillette, Wyoming, built by Pacific Power and Light Co. and the Black Hills Power and Light Co. Turbine exhaust steam is admitted to the coils via two 13-ft diameter ducts.

Indirect Dry-Cooling Towers

These are of three general designs. The first uses a conventional surface condenser (Fig. 7-17). The circulating water leaving it goes through finned tubing cooled by atmospheric air in the tower. The latter could be natural-draft or, as is shown, induced-draft. In this design there are two heat exchangers in series and two temperature drops, one between steam and water and one between water and air. This double irreversibility imposes a severe penalty on turbine back pressure, thus necessitating operating at condenser pressures of about 2.5 to 4.0 psia (0.17 to 0.27 bar) compared with 0.5 to

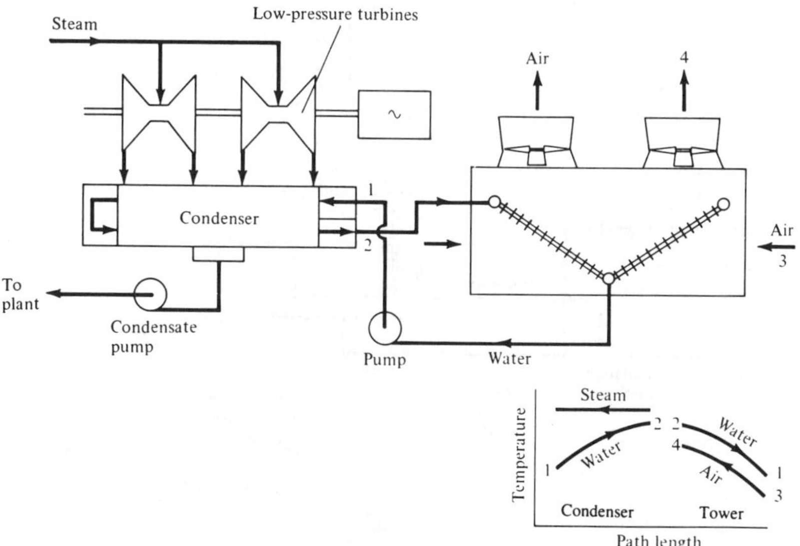

Figure 7-17 Schematic of an indirect dry-cooling tower with a conventional surface condenser.

1.0 psia (0.034 to 0.069 bar) for once-through systems. The results are loss in cycle efficiency and increased heat rejection.

The second design of indirect dry-cooling towers eliminates the intermediate water loop and uses an open- or direct-contact condenser (also called a jet or spray condenser, Sec. 6-2). As operation (of all dry-cooling towers) is in the closed mode and no atmospheric or surface water impurities enter the system through makeup, the circulating water can be mixed with the steam from the plant, hence the open-type condenser. The turbine exhaust steam enters the open condenser where the cold circulating water is sprayed into the steam for intimate mixing (Fig. 7-18). The condensate falls to the bottom of the condenser, from which most of it is pumped by a recirculation pump under positive pressure to finned tubing or coils in the tower. This part, cooled, returns to the condenser sprays. The balance of the condensate, equal to the steam mass flow, is pumped to the plant feedwater system by the condensate pump. Again the tower may be natural-draft or forced-draft. The ratio of circulation to feedwater is large (Sec. 6-2). Alternately only one pump may be used on the condensate from the condenser and flows adjusted by proper valving. Condensate polishers (Sec. 6-8) may be used to maintain the circulation water at condensate quality. Another optional component is a water-recovery turbine, between tower exit and condenser inlet, that is connected to the drive shaft of the circulating-water pump to recover some of the work of that pump. This indirect system is expected to be more efficient, more economical, and more feasible for large plants.

The third indirect dry-cooling tower design uses a circulating vaporizing coolant

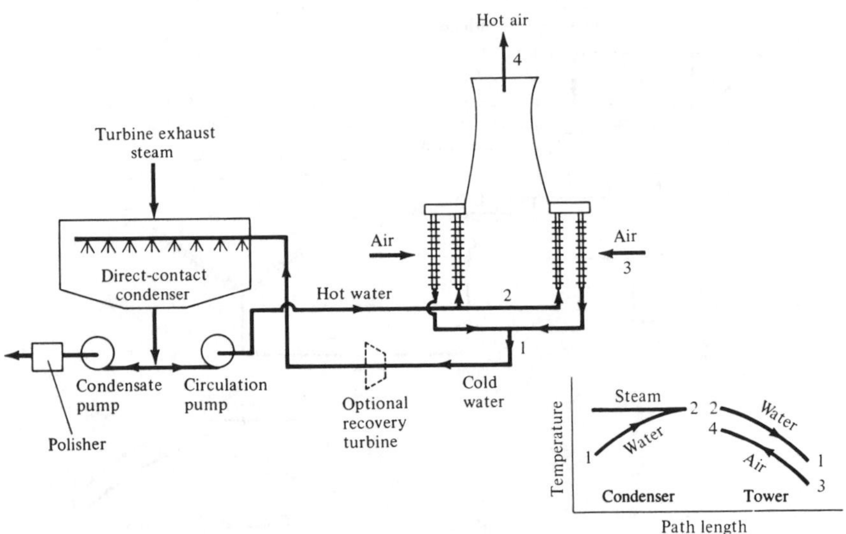

Figure 7-18 Schematic of an indirect dry-cooling tower with an open-type condenser.

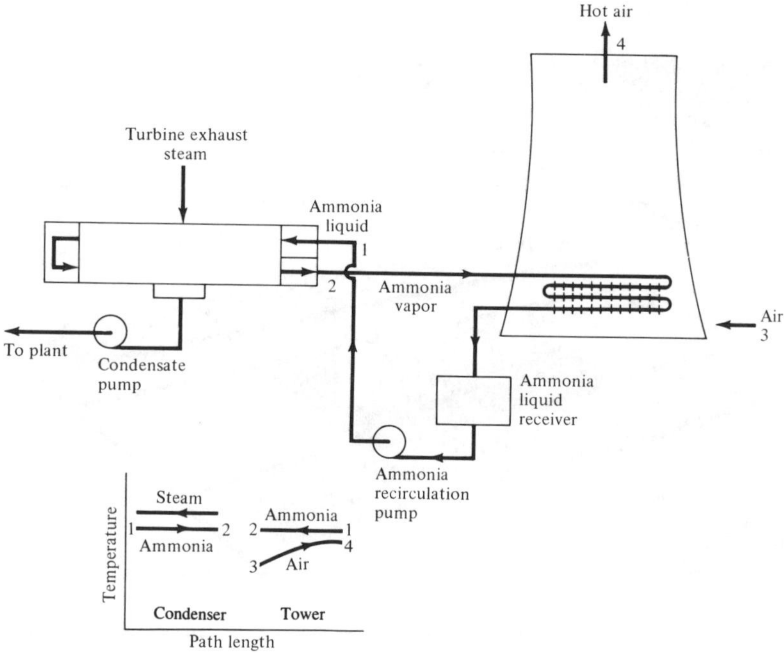

Figure 7-19 Schematic of an indirect dry-cooling tower with a surface condenser and a two-phase recirculating coolant (ammonia).

instead of water. The one that has been developed uses ammonia as the heat-transfer medium between the steam and air (Fig. 7-19). The use of ammonia enables phase-change heat-transfer boiling in the condenser tubes and condensation in the tower tubes. Nearly saturated liquid ammonia enters the surface condenser and is vaporized to saturated vapor. The vapor flows to the lower finned coils and is condensed to saturated liquid. The latter is pumped back to the condenser. The boiling and condensation modes have much higher heat-transfer coefficients, on the tube side, than forced convection of a single-phase fluid (as water in Fig. 7-17). This results in (1) a lower temperature difference between steam and ammonia, and between ammonia and air, and (2) reduced size and power requirements of the equipment. An optional addition is a compressor on the ammonia vapor to raise its temperature sufficiently above that of the air during particularly hot days, which would result in enhanced heat transfer in the tower. This makes the system resemble that of a vapor-compression refrigeration system. The work of the compressor can be partly recovered by the placement of an expander (turbine) in the liquid line.

A 6-MW demonstration plant using an ammonia system is undergoing tests near Bakersfield, California. The plant is part of the 150-MW Kern powerplant of the

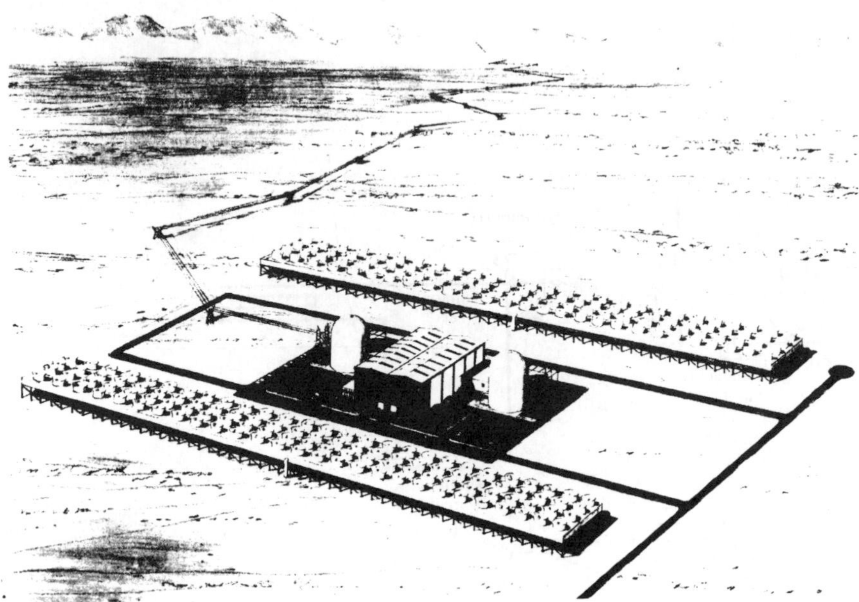

Figure 7-20 An artist's conception of a dry-cooling-tower system serving a twin-reactor nuclear powerplant in an arid area. *(Courtesy the Marley Cooling Tower Company, Mission, Kansas.)*

Pacific Gas and Electric Co. The tests are sponsored by the Electric Power Research Institute, the Department of Energy, and a consortium of utilities [57].

Thermodynamic data of amomnia may be found in App. C.

Because of their lower heat-transfer capabilities, dry-cooling towers in general are larger and require more land area than wet towers. Figure 7-20 is an artist's conception of the dry-cooling tower system needed for a twin-reactor nuclear powerplant.

7-7 DRY-COOLING TOWERS AND PLANT EFFICIENCY AND ECONOMICS

A parameter of importance to dry cooling towers is the *initial temperature difference* (ITD), given by

$$\text{ITD} = \text{temperature of hot water or other fluid entering the tower} - \text{temperature of outside cooling air}$$

The temperature of the outside air here is the dry-bulb temperature. Wet-bulb temperatures do not apply to dry-cooling towers. The most economical ITD range is 50 to 60°F (10 to 16°C).

The heat dissipated by a dry-cooling tower is proportional to the product of the size of cooling surface in the tower and the ITD, other things being equal. Thus for a given tower, an increase in load causes an increase in ITD, and hence an increase in the temperature of the circulating water entering the condenser and the back pressure on the turbine. For example, if the outside air is at 74°F, ITD = 60°F, the tower inlet circulating water temperature is 134°F. With a 6°F condenser terminal temperature difference, the steam condensing temperature would be 140°F, corresponding to 2.89 psia (0.2 bar).

If the plant heat rejection increases by 10 percent, ITD becomes 66°F and the condensing temperature and pressure would climb to 146°F and 3.37 psia (0.23 bar).

On the other hand, if the heat rejection and ITD remain the same, but the outside air temperature increases from 74°F to 90°F, the condensing temperature and pressure would be 156°F and 4.31 psia (0.3 bar).

This dependency of tower performance on load and outside temperature is much more pronounced for dry towers than for wet ones because the latter depend upon evaporation for much of their heat-rejection mechanism. For a wet tower operating with the same air at 90°F dry-bulb, and 76°F wet-bulb, and with 18°F approach, 24°F range, and 6°F TTD, the corresponding values would be 124°F and 1.890 psia (0.13 bar).

In many parts of the United States, 90°F air is exceeded during a good part of the summer and condensing temperatures of dry-cooling towers would range from 150°F to 160°F, corresponding to 4.31 and 4.74 psia (0.3 to 0.33 bar). During extremely hot weather condensing pressures may reach 6 or 7 psia (0.4 to 0.48 bar).

These turbine back pressures, which compare with 0.5 to 1.0 psia for once-through cooling systems, result in the loss of turbine work and cycle efficiency and an increase in heat rejection, thus necessitating a bigger tower yet. So, although dry-cooling towers save water, thermal pollution, fogging, etc., they are not as effective as wet towers, a natural consequence of heat transfer to air being lower than evaporative heat transfer. Figure 7-21 shows the operating turbine back pressures when the ambient temperature is changed from a base of 100°F. For example, if the tower is designed to yield a turbine back pressure of 6 inHg absolute (2.95 psia, 0.2 bar) at 100°F, an ambient temperature of 60°F would reduce the turbine back pressure to 2 inHg abs (1 psia, 0.068 bar).

The turbine work losses for the 90°F air example above would be around 6.5 percent, compared with 0.4 percent for a wet tower at the same conditions. Recall (Chap. 2) that a small change in pressure at the low-pressure end of turbine expansion results in a much larger change in work than a corresponding change at the high-pressure end.

The effect of using dry-cooling towers on electricity unit production costs versus using once-through and wet-cooling towers is given in Table 7-4 for three nuclear plants under construction (1, 2, and 3) and a hypothetical nuclear plant (4) in various parts of the United States. The study, conducted by R. W. Beck and Associates, was based on 1975 completion date prices and the figures should now be used only in a relative sense. Unit power costs are in mill/kWh (1 mill = 1/1000 of a U.S. dollar = 0.1¢). They are the sum of fixed charges on the capital costs, the fuel cycle

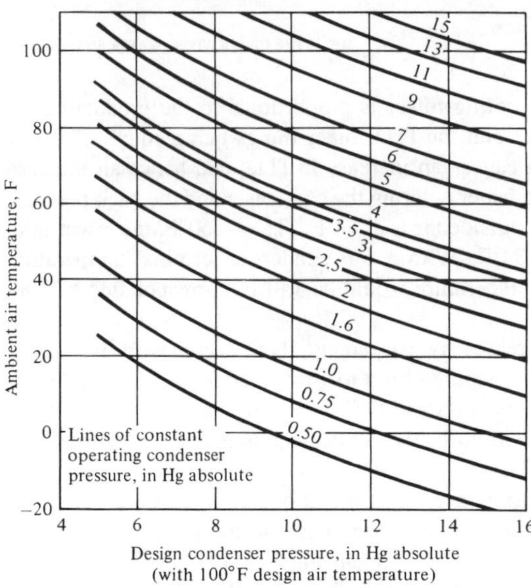

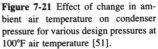

Figure 7-21 Effect of change in ambient air temperature on condenser pressure for various design pressures at 100°F air temperature [51].

Table 7-4 Increase in busbar cost as a result of using dry-cooling rather than wet-cooling towers, mill/kWh*

Plant cooling system†	Fixed charges 12%			Fixed charges 15%			Fixed charges 18%		
	Fuel costs, ¢/10⁶ Btu 15	18	21	15	18	21	15	18	21
				Mechanical-draft-dry-cooling system					
1	0.99	1.02	1.05	1.18	1.21	1.24	1.36	1.39	1.42
2	0.81	0.82	0.83	0.94	0.95	0.97	1.07	1.09	1.10
3	0.71	0.73	0.74	0.85	0.87	0.87	0.98	0.99	1.01
4	0.72	0.73	0.74	0.86	0.87	0.89	0.99	1.01	1.03
				Natural-draft dry-cooling system					
1	1.04	1.07	1.10	1.25	1.27	1.30	1.44	1.47	1.50
2	0.88	0.89	0.89	1.03	1.30	1.05	1.17	1.18	1.19
3	0.81	0.82	0.82	0.96	0.97	0.98	1.11	1.11	1.13
4	0.76	0.77	0.72	0.91	0.91	0.92	1.05	1.06	1.07

* Data taken from Ref. 51.

† 1 = Once-through, northeastern plant, 906.6 MW
2 = Mechanical-draft wet tower, southeastern, 860.6 MW
3 = Natural-draft wet tower, western, 928.5 MW
4 = Mechanical-draft wet tower, eastern, 860.6 MW

costs, and operation and maintenance costs and hence depend on the initial plant costs and plant efficiency.

The above dry-cooling tower "effects" must be coped with. Large turbines capable of operating at high back pressures need to be used and optimized with dry cooling. Units capable of back pressures up to 15 inHg abs (7.4 psia, 0.5 bar) can be obtained by strengthening or eliminating the last row of blades of existing turbines or by designing a new line of turbines.

The loss of effectiveness during periods of excessively high ambient temperatures and/or high electric demand can be countered by the use of water. Several schemes are considered:

1. The deluge or spraying with water of the outside tube surfaces. (The tower may be designed to have the tubes surrounded by a wet-tower-type fill for that purpose.)
2. Cooling the inlet air by presaturation (by water evaporation) or by sensible-heat removal [58].
3. Using an additional standby wet-cooling tower.
4. Using an additional standby circulating-water system, a scheme considered with ammonia towers (above).
5. Using wet-dry cooling towers (below).

7-8 WET-DRY-COOLING TOWERS

We have seen that wet-cooling towers consume some water by evaporation, drift, and bleed and that they suffer from plume problems. We have also seen that dry-cooling towers impose a penalty on powerplant operation, especially during periods of hot weather. As often happens in such cases, a compromise wet-dry cooling tower has been developed that reduces these disadvantages.

As the name implies, a *wet-dry-cooling tower* operates by a combination of wet and dry cooling. It has two air paths in parallel and two water paths in series. Figure 7-22 shows a mechanical- (induced-)draft wet-dry cooling tower. The top part of the tower, below the fan, is the dry section that contains finned tubing. The lower, wider part is the wet section that contains the fill. The hot circulating water is admitted through a header at midsection. It first flows up and down through the dry section finned tubing. It then leaves the dry section and gravitates through the fill of the wet section towards the cold-water basin.

Ambient air is drawn in two streams through the dry and wet sections. The two streams converge and mix inside the tower before leaving it. Because the first stream is dry heated and comes out even drier (lower relative humidity) than ambient air, whereas the second stream normally is saturated, the mixture leaving the tower is subsaturated (Fig. 7-23).

There are, therefore, two primary advantages of the wet-dry-cooling tower: (1) The subsaturated exit air provides a less visible plume than a completely wet tower and, in favorable weather conditions, completely eliminates it. (2) Because the water is precooled in the dry section, evaporative losses and hence makeup are substantially reduced.

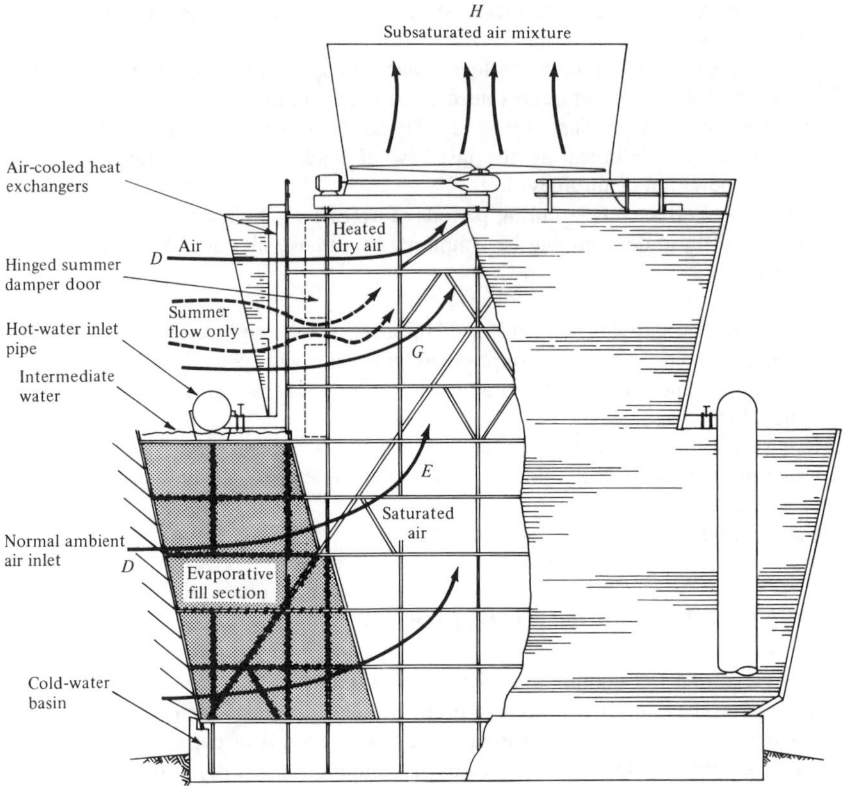

Figure 7-22 A wet-dry cooling tower with an induced-draft fan [51].

The ratio of heat removed in the dry and wet sections is adjusted both by design and operation. For example, if water consumption is the main problem, the tower is built with a larger dry section than wet section. In addition, evaporation can be varied by a control device that would regulate airflow through the wet section. In cold winter weather, the tower can operate nearly dry by blocking the air through the wet section. On the other hand, a doorlike damper, located in the heated air stream between dry heat exchanger and fan, reduces airflow through the dry section, thus allowing airflow to increase in the wet section. This damper is used in hot weather where the dry section (like a completely dry-cooling tower) would pose a penalty on powerplant performance. The damper is not intended to be completely closed, however, because of the necessity for plume control.

Wet-dry-cooling towers are seeing increasing service in those in-between cases of moderate water availability and the need for good though not complete plume reduction.

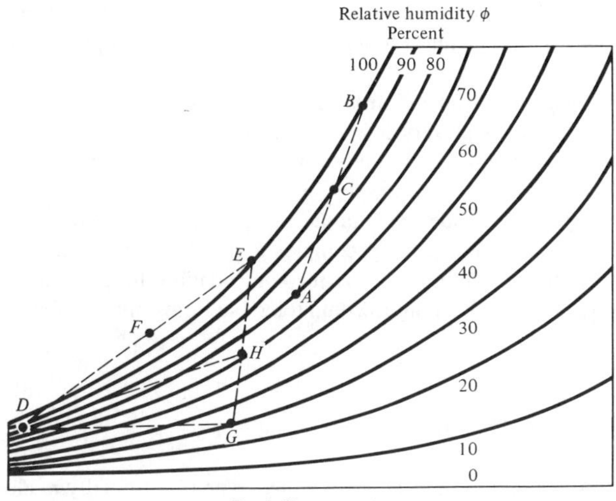

Figure 7-23 Psychrometric chart representation of wet- and wet-dry cooling towers. A-B = wet-cooling tower in summer. Exit air at B mixes with outside air at A outside tower resulting in C and vanishing plume visibility. D-E wet cooling-tower operation in winter. Exit at E mixes with outside air at D resulting in condensation ($\phi > 100$ percent) and foggy plume at F. With dry portion of wet-dry cooling tower DG, exit air would be subsaturated at H, resulting in no visible plume.

7-9 COOLING-TOWER ICING

Icing is a problem in both wet and dry towers in cold weather. In wet towers, ice may form where the cold air entering the tower first comes into contact with the water. It usually forms on air-intake louvers and on fill nearest to them. If allowed to build up, it may restrict airflow and/or cause structural damage. In natural-draft wet towers this condition is controlled by increasing warm-water flow to the outer periphery near the air intake, thus melting the ice and preventing new ice formation. In mechanical-draft wet towers, where the amount of ice formed is directly proportional to airflow, icing is controlled by reducing fan speed or shutting it down completely or by using fully adjustable air-intake towers. This reduces cooling and allows the warmer water to melt the ice. In extreme cases, such as extended periods of extremely cold weather, the fan may be reversed so that both warm air and water help in melting the ice. In multicell mechanical-draft towers, fans are usually shut off alternately for 12-h periods during ice-forming periods.

In dry-cooling towers, the problem is that of freezing of condenser cooling water inside the heat-exchanger tubes during times of extremely cold outside air. Water has a minimum specific volume (or maximum density) at about 4°C, or 39°F. When it freezes to ice at 0°C, its specific volume increases about 10 percent. If the expansion is unhindered, no problems arise, as for example when a crust of ice forms on the inner surface of a tube with water flowing through it. However, if plugs of ice form or the flow is stopped by valves, the increase in volume of the ice and trapped water can cause severe stresses in the tube walls.

Normal full-load (full water-flow) operation usually poses no problems even in freezing weather. The problem arises at very low ambient temperatures, at reduced load, and/or if the tower or a section of it is shutdown and not properly drained of water.

Theoretical analyses of freezing inside tubes in the static case (water at rest), in steady flow, and in transient conditions are fairly complex and will not be attempted here. A review and analysis of the problem, as well as recommendations for preventing ice formation inside dry tower tubes or keeping it within tolerable limits, may be found in Ref. 58.

7-10 COOLING LAKES OR PONDS

Cooling lakes, also called *ponds,* are the oldest and simplest type of artificially made heat-rejection systems. Hot circulating water from the condenser is simply dumped into an artificially made lake (or a modified natural lake, regulations permitting) and left to cool. Cool water from the lake is returned to the circulating-water system. Cooling is accomplished naturally by evaporation, by thermal radiation to the sky, especially during cloudless nights, and by convection to the wind. The cooling rates, unaided by any mechanical means such as sprays (below), are slow, and very large land areas are required for such lakes. The water residence time has to be long, thus requiring a large separation distance between intake and outlet. The effectiveness of a cooling lake may be enhanced by making the hot water circulate around a series of baffle dikes or natural barriers so that most of the lake area is utilized for cooling. Some grill work, partitions, and troughs may also be added.

Cooling lakes may be formed by enclosing a relatively flat area with a dike and filling it with water. The result is a *pershed lake*. Another method is damming a natural watershed, which results in a *natural contour lake*.

The main disadvantages of a cooling lake, as mentioned above, are the very low cooling effectiveness and therefore the very large acreage required. Others are the costs of added structures and the relative lack of freedom in choosing its shape. Its advantages are simplicity, low maintenance, the ability to operate for extended periods without makeup water, and the low power requirements, as the only mechanical equipment needed are pumps for occasional addition of makeup water. Circulating-water pumps are of course needed by cooling lakes, as by all other systems. Another advantage is that the appearance of cooling lakes is less objectionable than cooling towers and that they can be turned into recreational areas.

The heat transfer to and from a large body of water such as a cooling lake is a complex combination of many processes [59]. The processes by which heat is *added* are:

Heat addition to the lake from the powerplant
Absorption of short-wave radiation from the sky

Absorption of long-wave radiation from the atmosphere
Convection through the bottom from the earth
Transformation of kinetic energy to sensible energy
Heat addition due to chemical processes
Condensation of water vapor onto the lake

The processes by which heat is *rejected* are:

Reflection of short-wave solar radiation by the water
Reflection of long-wave atmospheric radiation
Long-wave radiation emitted by the water
Conduction of heat to the atmosphere
Evaporation from the lake into the atmosphere.

The net effect, cooling of the lake, is not constant over the total lake area because it is a function of the local water temperature as well as climatic conditions, such as dry- and wet-bulb temperatures of the air above the water, wind conditions, solar radiation, clouds, day and night variations, and others.

The lower limit of cooling is the *equilibrium temperature*. This is the temperature that the water asymptotically approaches if all parameters are held constant. The equilibrium temperature is given by [60]

$$\frac{T_C - T_E}{T_H - T_E} = e^{-\frac{UA}{\dot{m}c}} \tag{7-12}$$

where T_E = equilibrium temperature, °F or °C

T_C = cold-water temperature, °F or °C

T_H = hot-water temperature, °F or °C

U = overall heat-transfer coefficient, Btu/(h · ft^2 · °F) or W/(m^2 · °C)

A = lake surface area, ft^2 or m^2

$\dot{m}$ = water mass-flow rate, lb$_m$/h or kg/s

c = water specific heat, Btu/(lb$_m$ · °F) or J/(kg · K)

and the heat dissipation rate $\dot{Q}_R$, Btu/h or W, is given by

$$\dot{Q}_R = \dot{m}c(T_H - T_C) \tag{7-13}$$

Values of T_E and U depend upon the weather conditions, including solar and sky radiation, dry- and wet-bulb temperatures, and wind speed. They can be evaluated by procedures outlined in Refs. 61 and 62. The quantity $A/\dot{m}c$ is then obtained and used to size the lake for design conditions of U, T_E, T_H, and T_C. The same quantity can also be used to analyze off-design performance. In a numerical example [59], the quantity $UA/\dot{m}c$ was 1.173 and U was 11.0 Btu/(h · ft^2 · °F) [62.4 W/(m^2 · °C)].

The foregoing is a simplified model that neglects stratification; i.e., it assumes the entire surface area is equally effective and that there are constant and uniform weather conditions. Further uncertainties include such effects as buoyancy, turbulence, and the effects of makeup water and extraneous streams. A rigorous treatment of the problem requires a computerized hydrodynamic model, which is beyond the scope of this book.

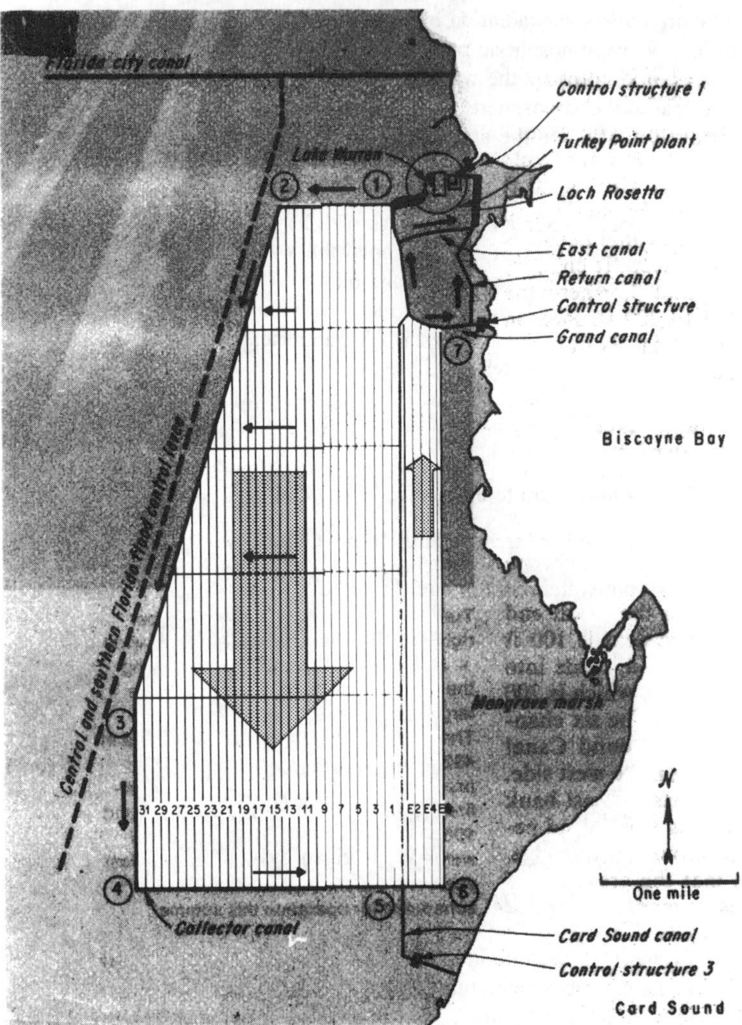

Figure 7-24 The cooling-canal system of the Turkey Point generating station at Biscayne Bay, Florida.

Long-term tests on actual cooling lakes [63] have shown that the average heat dissipation from the surface was about 3.5 Btu/d(h · ft^2 · °F), or 20 W/(m^2 · °C), and that an area between 1 and 2 acres (4000 to 8000 m^2) is required per megawatt of plant output.

Figure 7-24 shows a plan for a closed cooling-canal system at the Turkey Point generating station of the Florida Power and Light Company at Biscayne Bay which has two 432-MW fossil-fuel plants and two 666-MW nuclear plants. The system is composed of 38 channels, with a total length of 168 mi. Water enters a feeder canal at 1 and branches into 32 channels traveling south at 0.25 ft/s (0.076 m/s) to a collector canal at 4, then north through 6 channels back to the plant. Control structures around the circuit adjust flow, blowdown, and makeup. Regulations restrict the salinity, temperature, and flow of purge water into Card Sound at 5.

7-11 SPRAY PONDS AND CANALS

Spray ponds and canals, in a sense, operate like wet-cooling towers in that they depend primarily upon evaporation for cooling the circulating water. A *spray pond* is similar in water-flow pattern to a cooling lake or pond, only much smaller. Like a cooling lake, a spray pond can be turned into a recreational area. A *spray canal* usually consists of a long open canal or series of canals. In both, water is sprayed into the air above the surface. The spray in the canal is usually coarser than in the pond in order to reduce the drift loss (carryover with the air outside the canal boundaries) at the expense of lower heat-transfer rates because of the coarser drops.

The water is sprayed by modules that are self-contained units consisting of electrically driven propeller-type pumps that distribute the water through a diffuser system to spray nozzles, usually four per module. The spray pattern form a single diffuser can range up to 50 ft (~15 m) in diameter and between 10 and 20 ft (~3 to 6 m) high (Fig. 7-25). The modules are properly spaced within the pond (Fig. 7-26). They may be fixed or floating. In the latter, floatation is provided by polyurethane-filled fiberglass or stainless steel floats, which can easily be moored into place and attached to anchors.

The spray, consisting of a very large number of small drops, greatly increases the total water surface area in contact with the air. Cooling occurs as the spray is propelled upwards and then falls down to the surface. Additional cooling is to be expected from the surface as is the case of a cooling lake. Heat transfer is by evaporation, conduction to the air, and some radiation. In warm weather evaporation is the predominant mode of heat transfer, whereas conduction is predominant in cold weather. Evaporation, as usual, is affected by weather conditions: dry- and wet-bulb temperatures and wind speed. The water temperature after spraying approaches the lower wet-bulb rather than the dry-bulb temperature of the air, an advantage of spray cooling.

A simplified governing equation for a spray canal is [60]

$$\frac{T_C - T_{\text{wb}}}{T_H - T_{\text{wb}}} = exp\left(-\frac{N(1 - f)\, r\, ntu\, b}{c}\right) \tag{7-14}$$

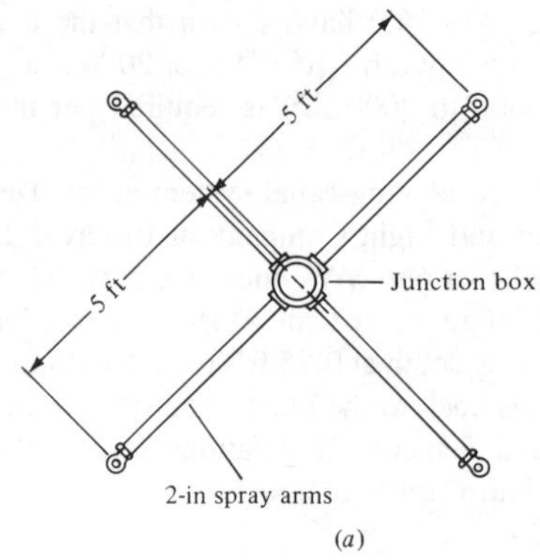

(a)

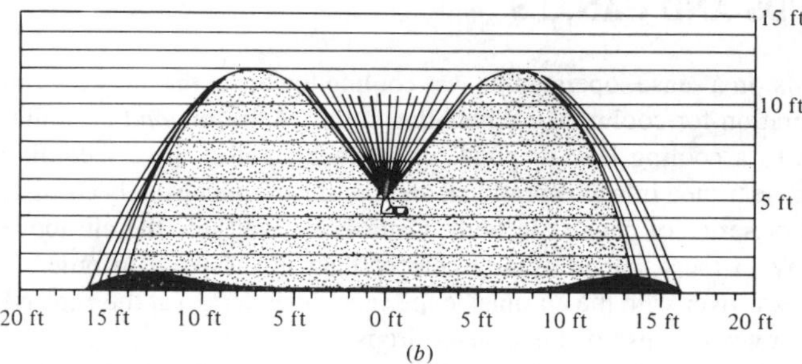

(b)

Figure 7-25 A typical spray module and pattern from a diffuser system consisting of four nozzles [63]. (*a*) Plan view of spray module. (*b*) Spray pattern. Spraco #1751 nozzle; pressure, 7 lb/in² gage; flow rate, 53 gal/min.

where T_{wb} = wet-bulb temperature of atmospheric air, °F or °C

N = number of nozzles

f = interference allowance to correct for local wet-bulb temperature elevation

r = ratio of nozzle flow rate to recirculation water flow rate

b = rate of change of enthalpy with temperature dh/dT, Btu/(lb$_m$ · °F) or J/(kg · K), evaluated at a film temperature $T_f = 0.5\,(T_{wb} + T_H)$, given in the table below

ntu = a characteristic number of the spray system

T_f, °F	32	40	60	80	100
b, Btu/(lb$_m$ · °F)	0.38	0.48	0.64	0.99	1.6

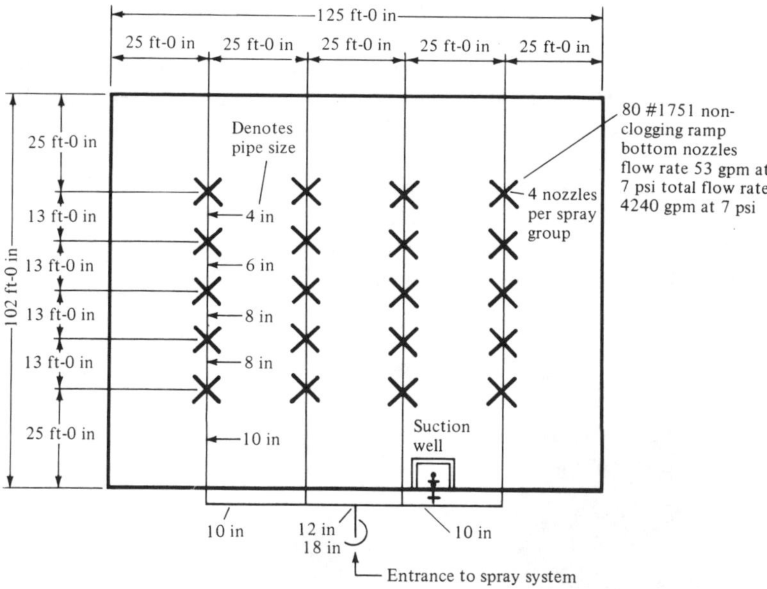

Figure 7-26 A typical layout of a spray pond [63].

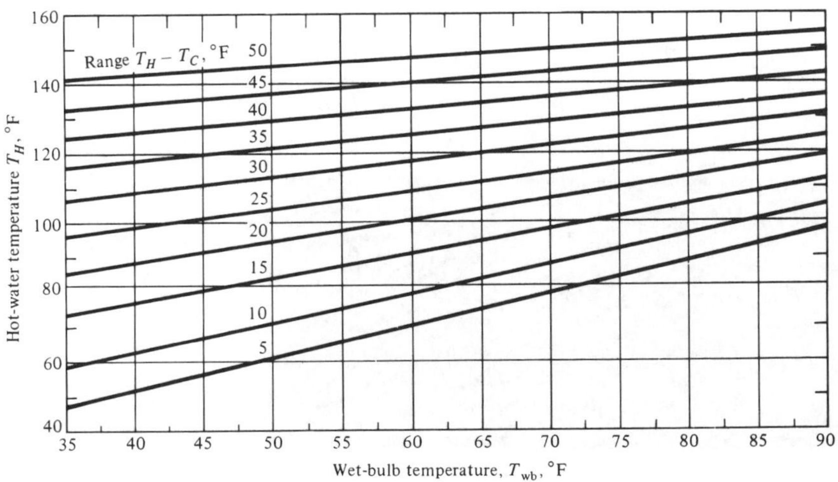

Figure 7-27 Spray-pond cooling range [63].

In a numerical example [60], the quantity $N(1 - f)r$ was taken as 3.45 and *ntu* was taken as 0.15, both dimensionless.

The many variables in a spray-cooling system, such as climatic conditions, winds, spray nozzle design, and pond layout, make the use of theoretical analyses useful for comparative evaluations only and make empirical and test results more useful for specific cases. Figure 7-27 shows a chart prepared by Spray Engineering Company of Burlington, Massachusetts [63], based on numerous tests over several years under many climatic conditions and in several locales in the United States and abroad. The chart gives the range $T_H - T_C$ for given values of T_H and T_{wb}.

The same tests have shown that heat dissipation by a spray pond averaged 127

Figure 7-28 Spray pond at Rancho Seco nuclear powerplant.

Btu/(h · ft^2 · °F), or 720 W/(m^2 · °C). Comparing this with 3.5 Btu/(h · ft^2 · °F) for a cooling lake shows that a spray pond would be, on the average, 4 percent as large in area as a cooling lake. Others estimate the ratio to be closer to 10 percent. The difference is, nonetheless, impressive. The areas are designed from known nozzle-module flow rate and the diameter of the spray pattern obtained. For example [63] a four-nozzle module under 7 psig pressure delivers 212 gal/min and sprays it over a 32-ft diameter area (Fig. 7-25). It is important to provide a sufficiently free airspace between sprays so that they will not interfere with one another and thus reduce the efficiency of cooling. The number of modules is easily obtained from the total water flow. (The recommended spacing is as shown in Fig. 7-26.) Figure 7-28 is a photograph of a typical spray-pond installation at Rancho Seco nuclear powerplant, near Sacramento, California.

Finally, cooling lakes and spray ponds and canals can be combined with cooling towers, once-through or other circulating-water cooling systems, both open and closed, to supplement cooling requirements both for new plant construction and existing installations.

PROBLEMS

7-1 A 900-MW powerplant of the pressurized-water nuclear-reactor type uses once-through cooling from a river. Regulations do not permit cooling water discharge back to the river hotter than 5°C above river ambient. Estimate (*a*) the necessary water flow, in kilograms per second, and (*b*) the reduction, in percent, in this water flow if the plant efficiency is improved by 1 percentage point. Take c_p for water = 4.184 kJ/kg · K.

7-2 A 900-MW powerplant of the pressurized-water nuclear-reactor type uses a wet cooling tower in the closed mode. The condenser range is 10°C. Estimate, for hot-weather conditions, (*a*) the amount of makeup water flow in kilograms per second, (*b*) the percent makeup flow of the total condenser flow, and (*c*) the reduction, in percent, in the makeup if the plant efficiency is improved by 1 percentage point.

7-3 A natural-draft cooling tower is 450 ft high. Air enters the tower at 14.696 psia, 50°F, and 50 percent relative humidity and leaves in a saturated condition. The pressure drop in the tower is 0.015 psi. Calculate (*a*) the air exit temperature, in degrees Fahrenheit, and (*b*) the makeup due to evaporation, in pound mass per pound mass of dry air.

7-4 A natural-draft cooling tower receives 250,000 ft^3/min of air at standard atmospheric pressure, 70°F, and 45 percent relative humidity. The air leaves saturated at 100°F. 3500 gallons per minute of water at 104°F is cooled. Calculate (*a*) the temperature of the water at tower exit, and (*b*) the necessary height of the tower if the total pressure losses are 2.00 lb_f/ft^2.

7-5 An induced-draft cooling tower cools 90,000 gallons per minute of water from 84 to 68°F. Air at 29.75 inHg absolute pressure, 70°F dry bulb and 60°F wet bulb, enters the tower and leaves saturated at 80°F. Find (*a*) the volume flow rate of air, in cubic feet per minute, and (*b*) the makeup water required, in pound mass per hour.

7-6 A wet cooling tower receives 1.5×10^6 lb_m/min of condenser water at 96°F, and 1.25×10^6 lb_m/min of air at 1 standard atmosphere, 62°F, and 50 percent relative humidity. The air leaves saturated at 82°F. Calculate (*a*) the exit water temperature, in degrees Fahrenheit, and (*b*) the percent condenser cooling water makeup due to evaporation.

7-7 A cooling tower is situated above sea level where the pressure is 14.5 psia. Air at 50°F and 50 percent relative humidity enters the tower at the rate of 2×10^6 lb_m/min and leaves saturated at 80°F. The condenser cooling water enters at 96°F. The condenser range is 18°F. Calculate (*a*) the mass flow rate of the condenser cooling water, and (*b*) the makeup due to evaporation, in pound mass per minute.

7-8 A mechanical-draft cooling tower receives 3.3 × 10^6 lb_m/min of condenser water at 92°F. The condenser range is 20°F. The outside air is at 60°F dry bulb, 46°F wet bulb, and 14.696 psia. The exit air is saturated at 82°F. The pressure drop in the tower is 0.0125 psi. Find the fan horsepower requirements in the case of (*a*) an induced-draft fan, and (*b*) a forced-draft fan. Assume that both fans have an efficiency of 65 percent.

7-9 A cooling tower receives 3.3 × 10^6 lb_m/min of condenser cooling water at 92°F. The condenser range is 20°F. The outside air is at 60°F dry bulb, 46°F wet bulb, and 14.696 psia. The exit air is saturated at 82°F. The pressure drop in the tower is 0.0125 psi. (The conditions are the same as those in Prob. 7-8.) Calculate (*a*) the height of the tower if it is of the natural-draft type, and (*b*) the forced-draft horsepower requirements if a hybrid wet tower half the height of the natural draft tower is used. Because the pressure drop is mainly in the fill, assume that both have the same pressure drop. The fan efficiencies are 65 percent.

7-10 A hybrid tower is designed to cool 2 × 10^8 lb_m/h of condenser water from 94 to 72°F. The outside air is at one standard atmosphere, 66°F dry bulb, and 60°F wet bulb. It leaves the tower saturated at 86°F. The pressure drop in the tower is given by 5.75 × 10^{-15} $\dot{m}_a^2$ psia, where $\dot{m}_a$ is the mass flow rate of air in pound mass per hour. Calculate the height of the hybrid tower if the total fan power should not exceed 3000 kW. The fan efficiency is 0.65.

7-11 An induced-draft wet cooling tower situated on a natural body of water at 60°F operates in the helper mode. The total condenser water flow is 3 × 10^6 lb_m/min. The condenser exit water is at 96°F. The tower receives half the condenser water, and 1 × 10^6 lb_m/min of air at 1 standard atmosphere, at 62°F, and 50 percent relative humidity. The air leaves the tower saturated at 82°F. Calculate (*a*) the water supply from the lake in pound mass per minute, assuming losses due to evaporation only, (*b*) the tower water exit temperature, and (*c*) the condenser inlet temperature, in degrees Fahrenheit.

7-12 It is desired to compare the effect of two types of dry cooling tower systems on powerplant performance. The towers are of the indirect cooling type. One operates with a surface condenser with an 8°F terminal temperature difference, and the other with a direct-contact condenser with 0°F terminal temperature difference. Consider for simplicity a simple ideal Rankine cycle with inlet saturated steam at 1000 psia. Further consider that both have the same condenser cooling water mass flow rate of 7.21 × 10^7 lb_m/h, the same inlet temperature of 70°F, the same dry tower air temperature range of 60°F in and 90°F out, and the same air mass flow rate of 5 × 10^8 lb_m/h. Calculate for each case (*a*) the condenser temperature, in degrees Fahrenheit, and pressure, in psia, (*b*) the steam mass flow rate, in pound mass per hour, (*c*) the cycle efficiency, and (*d*) the cycle work, in megawatts, ignoring the pump work.

7-13 A wet-dry cooling tower receives 20 × 10^6 lb_m/h of condenser cooling water at 110°F. The water is cooled in the dry section to 90°F and in the wet section to 70°F. The outside air is at 1 standard atmosphere, 60°F, and 40 percent relative humidity. The air leaves both dry and wet sections at 81°F. Calculate (*a*) for the dry section the outside cooling air required, in pound mass per hour, and the relative humidity leaving it, (*b*) for the wet section the outside cooling air required, in pound mass per hour, (*c*) the condition (temperature and relative humidity) of the air leaving the tower, (*d*) the required makeup of condenser cooling water due to evaporation, in pound mass per hour, and (*e*) the percentage savings in that makeup because of the use of a wet-dry instead of an all-wet tower.

7-14 A cooling lake is used to cool a 200-MW powerplant that has a 39 percent efficiency. The condenser cooling water inlet and exit temperatures are 75° and 95°F, respectively. The lake equilibrium temperature and overall heat-transfer coefficient based on the available climatic conditions are 65°F and 3.5 Btu/h · ft^2 · °F. Determine the surface area of the lake, in acres.

7-15 A spray pond is used to cool a 200-MW powerplant that has a 39 percent efficiency. The condenser cooling water outlet temperature is 90°F. The atmosphere is at 70°F and 60 percent relative humidity. The spray nozzles have ntu = 0.15, r = 0.015 and f = 0.25. Determine (*a*) the numer of spray modules required, (*b*) the flow per module, in gallons per minute, and (*c*) the approximate pond area, in acres.

CHAPTER

EIGHT

GAS-TURBINE AND COMBINED CYCLES

8-1 INTRODUCTION

Gas turbines are used by themselves in a very wide range of services, most notably for powering aircraft of all types but also in industrial plants for driving mechanical equipment such as pumps, compressors, and small electric generators in electrical utilities and for producing electric power for peak loads as well as for intermediate and some base-load duties.

There is also growing interest in using gas turbines in combined-cycle plants. These plants use combinations of gas and steam turbines in a variety of configurations of turbines, heat-recovery boilers, and regenerators.

Gas turbines for industrial and utility applications have many advantages. Compared with steam plants they, and their total systems, are small in size, mass, and initial cost per unit output. They are available with relatively short delivery times and are quick to install and put to use. They are quick-starting, often by remote control, and are smooth-running. They offer flexibility in supplying process needs, such as compressed air, in addition to electric power and in using a range of liquid and gaseous fuels, including the new synthetic fuels like low-Btu gas (Sec. 4-11). They are also subject to fewer environmental restrictions than other prime movers.

Gas turbines have one major disadvantage that prevents utilities from using them as major base-load prime movers: their present low cycle efficiency. Another disadvantage is their incompatibility with solid fuels. The combinations of low capital cost and low efficiency have determined their use primarily as power-peaking units where they are not expected to be on line for more than 1000 or 2000 h/year and where a large steam plant designed to meet peak loads would operate at an uneconomical load factor during most of the year.

Improvement in gas-turbine cycle efficiencies can be effected by a boost in the inlet combustion gas temperatures to the turbine from the present 2000 to 2300°F (1100 to 1260°C). Manufacturers are engaged in expensive research and development work to raise this to close to 2800°F (1540°C), with an eye on 3000°F (1650°C) for the future.

The use of gas turbines in combined cycles is one scheme to overcome their present low cycle efficiency for utility base-load use, while at the same time offering the utilities the gas-turbine advantages of quick starting and flexible operation over a wide range of loads.

Gas turbines are available in one- or two-shaft models (Fig. 8-1). The latter has two shafts that rotate at different speeds. One shaft has the compressor and a turbine that drives it, the other has the power turbine connected to the external load. Or one shaft might have high-pressure sections of the compressor and the turbine, while the other has the low-pressure compressor, turbine, and external load. In either case, the portion of the system containing the compressor, combustion chamber, and high-pressure turbine is sometimes called the *gas generator*. The two-shaft configuration allows the load to be driven at variable speed, which is well suited to many industrial applications. Gas turbines designed for aircraft propulsion are sometimes modified and used for industrial service [64]. Single-shaft turbines have the compressor, turbine, and load on one shaft running at constant speed. This configuration is used to drive small generators as well as large generators for utility use.

Gas-Turbine Cycles

The hot gas emerging from a combustor or a gas-cooled reactor can be used directly as the primary working fluid, i.e., by expanding through a gas turbine, or indirectly, by heating a secondary fluid acting as the working fluid. For each of these two cases, i.e., the direct or the indirect cycle, we may also have an open or a closed cycle. Following are the possible combinations.

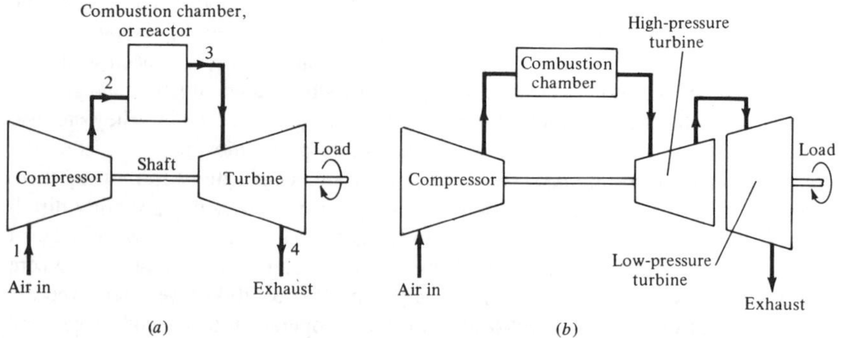

Figure 8-1 Schematic of a direct open gas-turbine cycle: (*a*) single shaft and (*b*) two shaft.

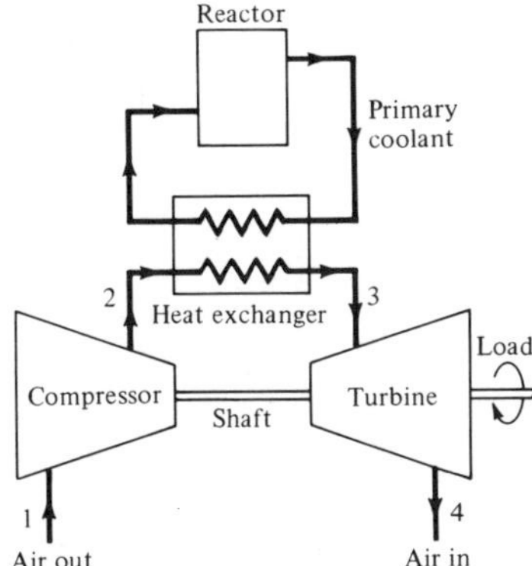

Figure 8-2 Schematic of an indirect open gas-turbine cycle.

Direct open cycle The direct open cycle is shown schematically in Fig. 8-1. The gas enters a compressor at point 1, where it is compressed to point 2. The gas then enters the combustion chamber or reactor, where it receives heat at constant pressure (ideally) and emerges hot at point 3. From there it expands through the turbine to point 4. The hot exhaust mixes with the atmosphere outside the cycle, and a fresh cool supply is taken in at point 1. The turbine supplies the compressor power. Useful power may be supplied by the turbine (or by the gas expanding further in a nozzle that supplies propulsion to the vehicle carrying the power plant, such as a jet aircraft). Because this is an open cycle, air is the only feasible working fluid (on earth).

Indirect open cycle The elements of the indirect open cycle (Fig. 8-2) are similar to those in the direct open cycle except that here the air is a secondary fluid that receives its heat from a primary coolant in a heat exchanger. This cycle is suitable for uses where environmental concerns prevent the air from receiving heat directly, such as from a nuclear reactor where radioactivity releases may spread to the atmosphere. Nuclear-reactor use, however, is best served by a closed cycle (see below).

Direct closed cycle In the direct closed cycle (Fig. 8-3) the gas coolant is heated in the reactor, expanded through the turbine, cooled in a heat exchanger, and compressed back to the reactor. In this cycle a gas other than air may be used. No effluent of radioactive gases passes into the atmosphere under normal operating conditions. Closed cycles permit pressurization of the working fluid with consequent reduction in the size of rotating machinery. The most suitable working fluid in this case is helium.

Indirect closed cycle The indirect closed cycle combines the indirect open cycle and the direct closed cycle in that the reactor is separated from the working fluid by a heat exchanger, whereas the working gas rejects heat to the atmosphere via a heat exchanger (Fig. 8-4). The primary coolant may be water, a liquid metal, or a gas such as helium.

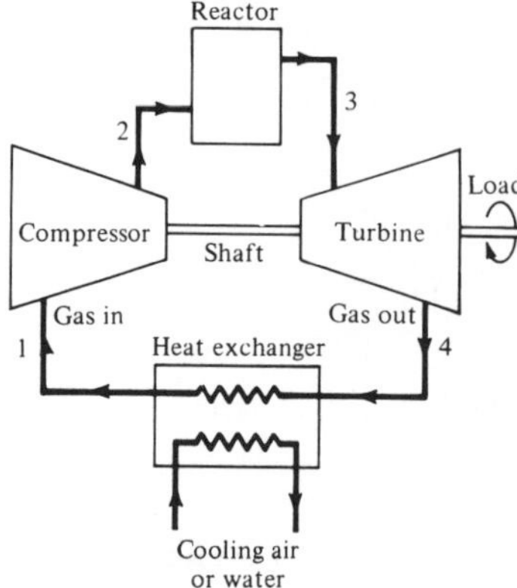

Figure 8-3 Schematic of a direct closed gas-turbine cycle.

8-2 THE IDEAL BRAYTON CYCLE

The ideal cycle for gas turbine work is the *Brayton cycle*. It is composed of two adiabatic-reversible (and hence isentropic) and two constant-pressure processes (Fig. 8-5). The gas is compressed isentropically from point 1 to 2, heated at constant pressure from 2 to 3, and then expanded isentropically through the turbine from point 3 to 4. Cooling occurs from point 4 to point 1, either in a heat exchanger (closed cycle) or in the open atmosphere (open cycle).

The work done in the turbine (a steady-flow machine) per unit time (power), with

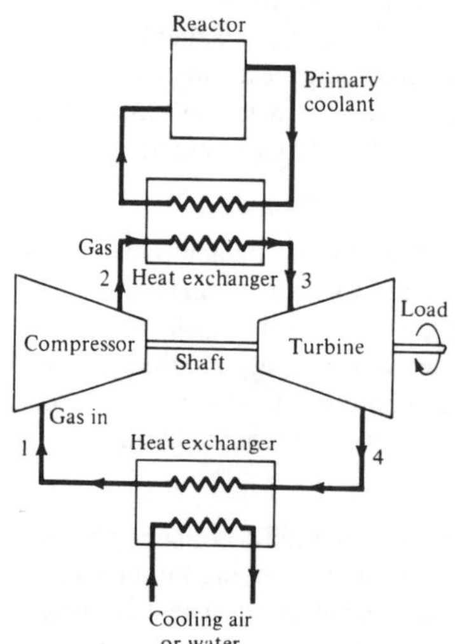

Figure 8-4 Schematic of an indirect closed gas-turbine cycle.

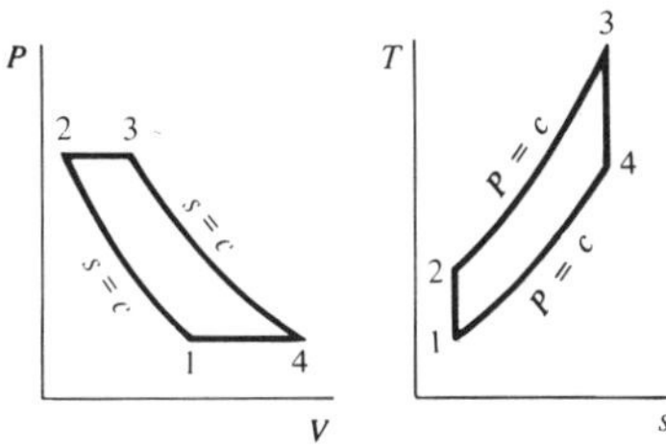

Figure 8-5 P-V and T-s diagrams of an ideal Brayton cycle.

relatively negligible change in the kinetic energy of the gas, $\dot{W}_T$ in Btu/h or W, is equal to the rate of change in its enthalpy. Thus

$$\dot{W}_T = \dot{H}_3 - \dot{H}_4 = \dot{m}(h_3 - h_4) \tag{8-1}$$

where $\dot{H}$ = total enthalpy of flowing gas, Btu/h or W

h = specific enthalpy, Btu/lb$_m$ or J/kg

$\dot{m}$ = mass rate of flow of gas, lb$_m$/h or kg/s

For a gas, Eq. (8-1) may be rewritten in the form

$$\dot{W}_T = \dot{m} \int_{T_3}^{T_4} c_p(T)\, dT \tag{8-2}$$

where $c_p(T)$ is the specific heat at constant pressure of the gas, which is a function of temperature T.

We will now assume constant specific heats for simplicity, a procedure sometimes referred to as analysis for the air-standard cycle (when c_p is a constant for air). However, it should be noted that specific heats for monatomic gases such as helium and argon are essentially constant and independent of temperature (except when highly compressed, at very low temperatures and high pressures). Specific heats increase with temperature for diatomic gases such as air and N_2 and increase even faster with temperature for triatomic gases such as CO_2 (Fig. 8-6 and Table 8-1). The following analysis, therefore, is *exact* for monatomic gases and only approximate for others. For a constant c_p, Eq. (8-2) becomes

$$\dot{W}_T = \dot{m}c_p(T_4 - T_3) \tag{8-3}$$

Using the perfect-gas laws (Table 1-2), we can write Eq. (8-3) in terms of the pressure ratio across the turbine r_{p_T} given by

$$r_{p_T} = \frac{P_3}{P_4}$$

which is related to the absolute temperature ratio across the turbine by

$$\frac{T_3}{T_4} = r_{p_T}^{(k-1)/k} \tag{8-4}$$

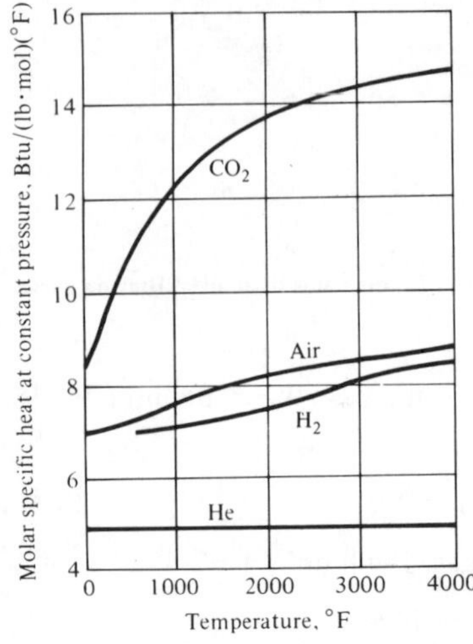

Figure 8-6 Variation of molar c_p with temperature for various gases.

where k is the ratio of specific heats at constant pressure and constant volume

$$k = \frac{c_p}{c_v}$$

and since

$$c_p - c_v = R \tag{1-14}$$

where R is the gas constant, a constant c_p means a constant c_v and a constant k. Equations (8-3) and (8-4) can now be combined to give

$$\dot{W}_T = \dot{m}c_pT_3\left(1 - \frac{1}{r_{PT}^{(k-1)/k}}\right) \tag{8-5}$$

Table 8-1 c_p and k for gases at low pressure*

		c_p, Btu/(lbm · °F)†			k, dimensionless		
Gas	M	Low	1000°F	2000°F	Low	1000°F	2000°F
H_2	2.016	3.421			1.405		
He	4.003	1.250	1.250	1.250	1.659	1.659	1.659
Air	28.97	0.240	0.264	0.289	1.400	1.353	1.314
N_2	28.02	0.249	0.269	0.293	1.401	1.357	1.321
A	39.95	0.125	0.125	0.125	1.668	1.668	1.668
CO_2	44.01	0.202	0.280	0.313	1.293	1.192	1.169

* Data from Ref. 65.
† To convert to kJ/(kg · °C), multiply by 4.1868.

Using a pressure ratio across the compressor r_{p_c} where

$$r_{p_c} = \frac{P_2}{P_1}$$

and

$$\frac{T_2}{T_1} = r_{pc}^{(k-1)/k} \tag{8-6}$$

The absolute magnitude of the rate of work of the compressor would also be given by

$$|\dot{W}_c| = \dot{m}(h_2 - h_1)$$

$$= \dot{m}c_p T_2\left(1 - \frac{1}{r_{p_c}^{(k-1)/k}}\right) \tag{8-7}$$

Assuming $r_{p_T} = r_{p_c} = r_p$, that is, no pressure losses in the cycle, a common assumption in the ideal case, the net work rate of the cycle $\dot{W}_n$ is given by

$$\dot{W}_n = \dot{W}_T - |\dot{W}_c| = [\dot{m}c_p(T_3 - T_2)]\left(1 - \frac{1}{r_p^{(k-1)/k}}\right) \tag{8-8}$$

The expression within the brackets on the right-hand side of this equation is obviously the heat added $\dot{Q}_A$ in Btu/h or W

$$\dot{Q}_A = \dot{m}c_p(T_3 - T_2) \tag{8-9}$$

The second expression must then be the cycle thermal efficiency η_{th}, a function of both r_p and k

$$\eta_{th} = 1 - \frac{1}{r_p^{(k-1)/k}} \tag{8-10}$$

Although the above equations pertain to constant specific-heat gases, the trends they predict apply to all gases. The thermal efficiency of the cycle for any one gas (same k) is a sole function of r_p, increases asymptotically with it, and is independent of initial or maximum cycle temperatures T_1 and T_3. (As we shall see later, this is not true for nonideal cycles.) However, while η_{th} increases indefinitely with r_p, the specific power, the power per unit mass-flow rate (and the specific work or work per unit mass of working fluid) does not and reaches a maximum at an optimum r_p. This can be seen by rewriting Eq. (8-8) in terms of T_1 and T_3, using Eq. (8-6). Again for $r_{p_T} = r_{p_c} = r_p$

$$\frac{\dot{W}_n}{\dot{m}} = c_p\,(T_3 - T_1 r_p^{(k-1)/k})\left(1 - \frac{1}{r_p^{(k-1)/1}}\right) \tag{8-11a}$$

or

$$\frac{\dot{W}_n}{\dot{m}} = c_p\left[T_1(1 - r_p^{(k-1)/k}) + T_3\left(1 - \frac{1}{r_p^{(k-1)/k}}\right)\right] \tag{8-11b}$$

Examination of Eqs (8-8) and (8-11) shows the following:

1. Other things being equal, that is, for the same T_1, T_3, r_p, and k, the work per unit mass of gas is a direct function of c_p. Hence, helium can produce more than five times $\dot{W}_n/\dot{m}$ than air (at low temperatures).
2. Other things being equal, gases with higher values of k, that is, higher $(k - 1)/k$, produce more work per unit mass of gas than gas with lower values of k. Again, this shows an advantage for He over air (k for air decreases with temperature).
3. For any one gas, an increase in r_p from its lowest value of 1.0 (where the work is zero) decreases one part of Eq. (8-11) and increases the other. The net work thus goes through a maximum at an optimum value of r_p. This state of affairs can be shown graphically by the three ideal cycles of Fig. 8-7. These operate between the same temperatures T_1 and T_3 and have the same inlet and exhaust pressures but different values of r_p. The net work in each case is represented by the enclosed area of the cycle.

The optimum pressure ratio can be evaluated for ideal cycles by differentiating the net work in Eq. (8-11) with respect to r_p and equating the derivative to zero. This gives a value of T_2 expressed by

$$T_2 = (T_1 T_3)^{1/2} \tag{8-12}$$

and since $T_2/T_1 = T_3/T_4 = r_p^{(k - 1/k)}$ (for same pressure ratio), then

$$(T_2 = T_4)_{\text{opt}} \tag{8-13}$$

and

$$r_{p_{\text{opt}}} = \left(\frac{T_2}{T_1}\right)^{k/(k - 1)} = \left(\frac{T_3}{T_1}\right)^{k/2(k - 1)} \tag{8-14}$$

Note that the quantity $k/2(k - 1)$ decreases as k increases. Thus, for fixed initial and maximum cycle temperatures, the optimum pressure ratio for monatomic gases (He) is, in general, lower than for diatomic gases (air, N_2). These in turn have lower ratios than the triatomic gases (CO_2). It follows that a monatomic gas, for example, may

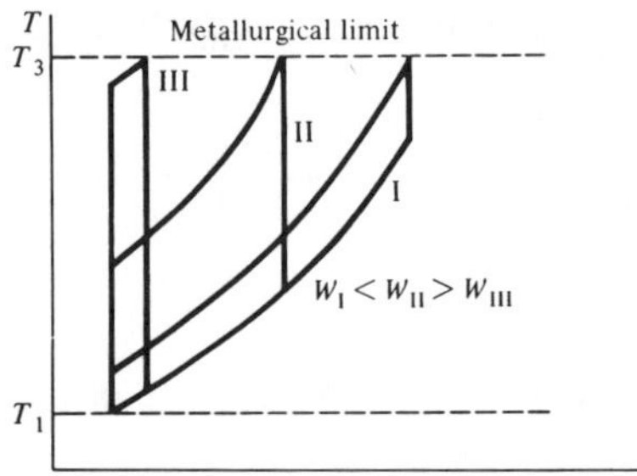

Figure 8-7 Effect of pressure ratio on ideal Brayton cycle work given by enclosed areas on T-s diagram. T_1 and T_3 are fixed for all cycles.

operate at lower maximum pressures or, if the pressure in the low-pressure section of the cycle is increased (in closed cycle), may operate with a larger average density. This is accompanied by a reduction in plant size and mass.

Example 8-1 Find the pressure ratio required by an ideal Brayton cycle to produce a net work of 600 Btu/lb$_m$ of (1) helium and (2) air (with c_p at low temperatures). The cycle has initial and maximum temperatures of 500°R and 2500°R, respectively. Also calculate the optimum pressure ratio for both gases.

SOLUTION

(1) *Helium:*

$$c_p = 1.250 \qquad k = 1.659 \qquad (k - 1)/k = 0.3972$$

From Eq. (8-11):

$$600 = 1.25(2500 - 500r_p^{0.3972})\left(1 - \frac{1}{r_p^{0.3972}}\right)$$

or

$$(r_p^{0.3972})^2 - 5.04(r_p)^{0.3972} + 5 = 0$$

which yields two values of $r_p = 2.16$ and 26.62

$$r_{p_{\text{opt}}} = \left(\frac{2500}{500}\right)^{[1.659/2(1.659 - 1)]} = 7.58$$

From Eq. (8-11), this optimum pressure ratio yields a maximum work of 954.8 Btu/lb$_m$. Note that $r_p = 1.0$ results in zero work. There is also a maximum value of r_p that results in zero work and beyond which the work becomes negative. This is the r_p that makes $T_2 = T_3$. For He it is given by

$$r_p^{0.3972} = \frac{T_2}{T_1} = \frac{T_3}{T_1} = \frac{2500}{500} = 5$$

which yields a maximum r_p of 57.5.

(2) *Air:*

$$c_p = 0.24 \qquad k = 1.4 \qquad (k - 1)/k = 0.2857$$

$$600 = 0.24(2500 - 500r_p^{0.2857})\left(1 - \frac{1}{r_p^{0.2857}}\right)$$

$$r_p^{0.2857} - r_p^{0.2857} + 5 = 0$$

which yields imaginary values of r_p indicating that air is incapable of producing 600 Btu/lb$_m$

$$r_{p_{\text{opt}}} = \left(\frac{2500}{500}\right)^{[1.4/2(1.4 - 1)]} = 16.72$$

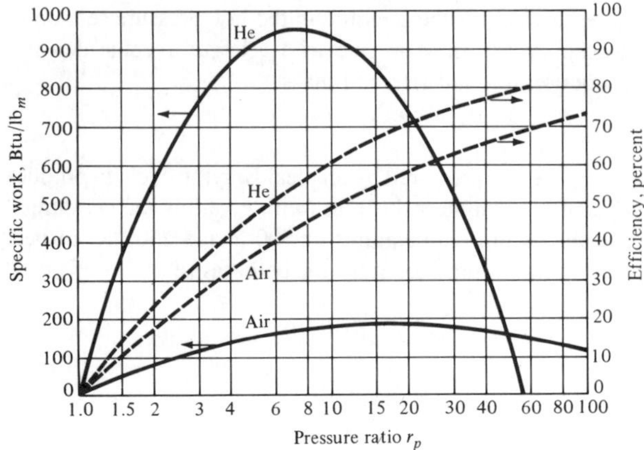

Figure 8-8 Specific power and efficiency versus pressure ratio for ideal Brayton cycles operating with helium and air.

Compared with 7.58 for He, the maximum work at $r_p = 16.72$ from Eq. (8-11) is 183.3 Btu/lb$_m$, less than the 600 Btu/lb$_m$ asked for and the 954.8 Btu/lb$_m$ He is capable of. The maximum r_p for air, obtained as for He, is 279.6.

Figure 8-8 shows results of calculations for $\dot{W}/\dot{m}$ and η_{th} for ideal Brayton cycles using He and air as working fluids, where the initial and maximum temperatures T_1 and T_3 are the same as in Example 8-1, i.e., 500°R and 2500°R, respectively. Note that the specific work Btu/lb$_m$ of He is generally much higher than that of air, and in the practical range of pressure ratios, occurs at much lower pressure ratios. Recall, however, that while helium turbines operate with lower overall pressure ratios than air, they need many more stages (Sec. 5-9). To obtain the specific work based on a unit mole (or volume at same P and T), multiply the ordinates of Fig. 8-8 by the molecular mass of each gas.

8-3 THE NONIDEAL BRAYTON CYCLE

The Brayton cycle with fluid friction is represented on the P-V and T-s diagrams of Fig. 8-9 by 1-2-3-4. Both the compression process with fluid friction 1-2 and the expansion process with fluid friction 3-4 show an increase in entropy as compared with the corresponding ideal processes 1-2$_s$ and 3-4$_s$. Drops in pressure during heat addition (process 2-3) and heat rejection (process 4-1) are neglected in this analysis, so the turbine pressure ratio equals the compressor pressure ratio as before.

The compression and expansion processes with fluid friction can be assigned *polytropic,* also called *adiabatic* or *isentropic efficiencies* (Sec. 1-8), as follows.

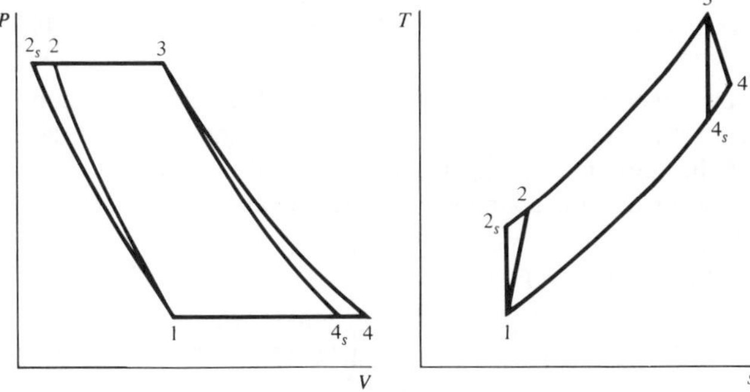

Figure 8-9 *P-V* and *T-s* diagrams of ideal and nonideal Brayton cycle.

$$\eta_c = \text{compressor polytropic efficiency} = \frac{\text{ideal work}}{\text{actual work}}$$

$$= \frac{h_{2s} - h_1}{h_2 - h_1} \tag{8-15}$$

If we assume constant specific heats

$$\eta_c = \frac{T_{2s} - T_1}{T_2 - T_1} \tag{8-16}$$

and

$$\eta_T = \text{turbine polytropic efficiency} = \frac{\text{actual work}}{\text{ideal work}}$$

$$= \frac{h_3 - h_4}{h_3 - h_{4s}} \tag{8-17}$$

and for constant specific heats

$$\eta_T = \frac{T_3 - T_4}{T_3 - T_{4s}} \tag{8-18}$$

where in each case the smaller work is always in the numerator.

The net power of the cycle is $\dot{W}_n$ = power of turbine − |power of compressor|. For constant specific heats

$$\dot{W}_n = \dot{m}c_p[(T_3 - T_4) - (T_2 - T_1)] \tag{8-19a}$$

or

$$\dot{W}_n = \dot{m}c_p\left[(T_3 - T_{4s})\eta_T - \frac{T_{2s} - T_1}{\eta_c}\right] \tag{8-19b}$$

This equation can be written in terms of the initial temperature T_1, a chosen metallurgical limit T_3, and the compressor and turbine efficiencies (above) to give

$$\dot{W}_n = \dot{m}c_pT_1\left[\left(\eta_T\frac{T_3}{T_1} - \frac{r_p^{(k-1)/k}}{\eta_c}\right)\left(1 - \frac{1}{r_p^{(k-1)/k}}\right)\right] \tag{8-19c}$$

The second quantity in parentheses can be recognized as the efficiency of the corresponding ideal cycle, i.e., one having the same pressure ratio and using the same fluid. As in the case of the ideal cycle, the specific power of the nonideal cycle, $\dot{W}_n/\dot{m}$, attains a maximum value at some optimum pressure ratio and is a direct function of the specific heat of the gas used.

The heat added in the cycle, Q_A, is given by

$$\dot{Q}_A = \dot{m}c_p(T_3 - T_2) = \dot{m}c_p\left[(T_3 - T_1) - \left(T_1\frac{r_p^{(k-1)/k} - 1}{\eta_c}\right)\right] \tag{8-20}$$

The efficiency of the nonideal cycle can then be obtained by dividing Eq. (8-19c) by Eq. (8-20). Although the efficiency of the ideal cycle is independent of cycle temperatures, except as they may affect k, and increases asymptotically with r_p, the efficiency of the nonideal cycle is very much a function of the cycle temperatures. It also assumes a maximum value at an optimum pressure ratio for each set of temperatures T_1 and T_3. The two optimum pressure ratios, for specific power and for efficiency, are not the same, and this necessitates a compromise in design.

Another effect of nonideality is fluid friction in heat exchangers, piping, etc. This results in a pressure drop between 2 and 3 (Fig. 8-9) and a pressure at 4 greater than at 1. In other words the pressure ratio across the compressor r_{p_c} would be greater than the pressure ratio across the turbine r_{p_T}. General equations that would take these and other effects on the Brayton cycle will be presented next. Further nonidealities result from mechanical losses in bearings friction and auxiliaries, heat losses form combustion chambers, and air bypass to cool the turbine blades (Sec. 8-7).

Figures 8-10 and 8-11 show results of calculations [66] for η and $\dot{W}/\dot{m}$ of a simple air-combustion Brayton cycle (solid lines) and of one with a regenerator (dashed lines). Regeneration is explained below. For the simple cycle, the following data were assumed.

T_1 = 15°C = 59°F = constant
P_1 = 1.013 bar = 1 atm = constant
η_c = 90%; η_T = 87%
Mechanical losses = 1%
Combustion chamber losses = 2%
Air bypass = 3%
Pressure losses: at inlet = 1%
in combustion chamber = 3%
at outlet = 2%
in regeneration = 4%

Actual, variable properties of air and combustion gases were used.

It can be seen that both η and $\dot{W}/\dot{m}$ are strongly dependent on T_3, which necessitates operating at as high a T_3 as metallurgically possible. They are also strong functions

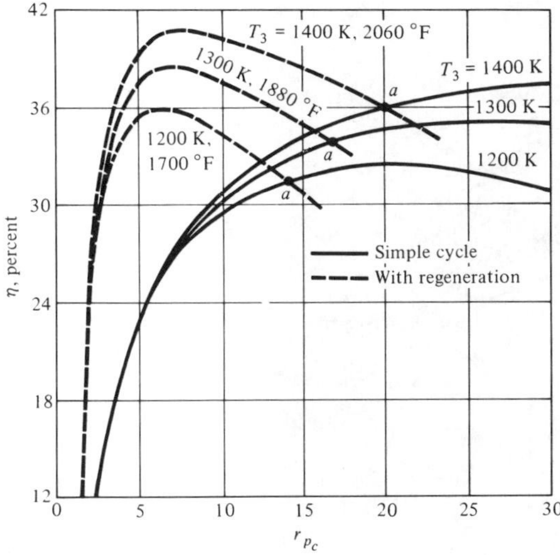

Figure 8-10 Efficiency versus compressor pressure ratio of a nonideal Brayton cycle, showing effects of maximum temperature and regeneration [66].

of r_{p_c}, with optimum r_{p_c} increasing with T_3 for both efficiency and specific power. It can also be seen that the optimum r_{p_c} is greater for η than for power.

Figure 8-12 is a photograph of a single-shaft, direct-cycle, open air combustion gas-turbine package. It shows a 16-stage axial compressor, one of ten combustion chambers, and a three-stage turbine. A diesel engine for starting is shown on the left.

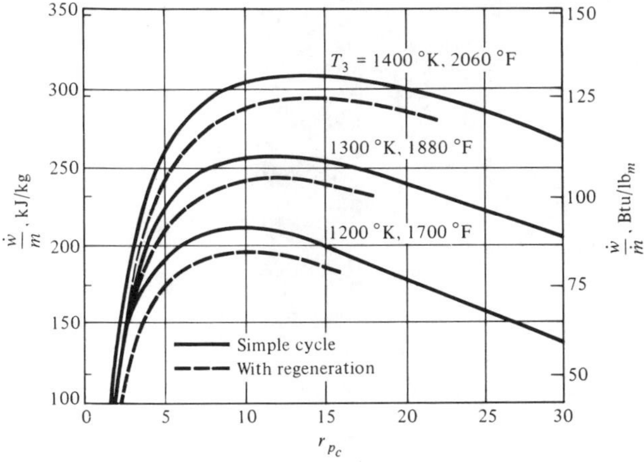

Figure 8-11 Specific power versus compressor pressure ratio of a nonideal Brayton cycle, showing effects of maximum temperature and regeneration [66].

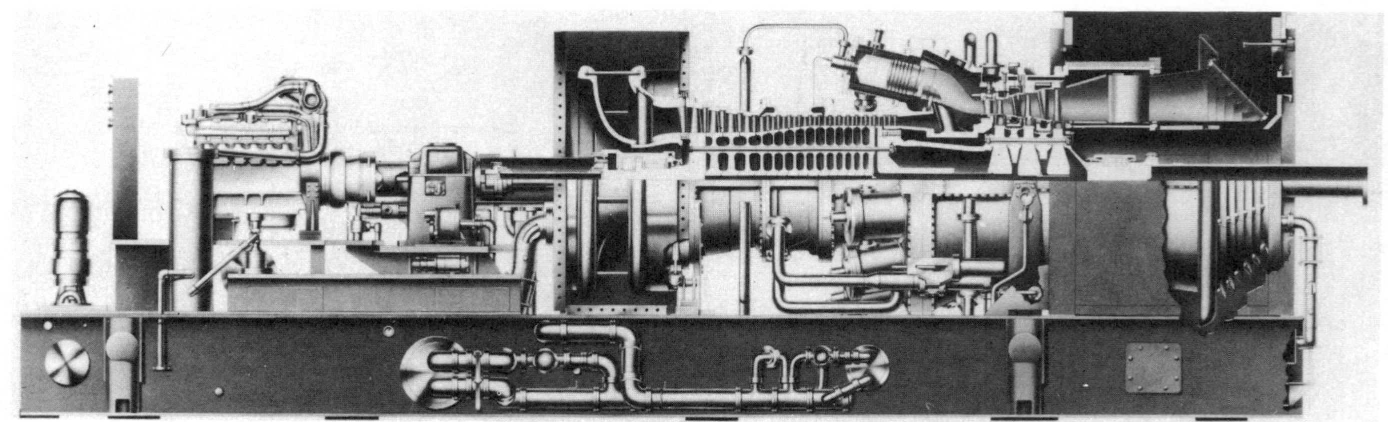

Figure 8-12 35.75 MW direct-cycle gas-turbine powerplant. *(Courtesy Gas Turbine Division, General Electric Company, Schenectady, New York.)*

The powerplant, General Electric model MS6001, produces 35.75 MW, is 30.50 percent efficient, runs at 5100 r/min, and has overall dimensions, including electric generator (not shown), 38 m (122 ft) long, 11 m (36 ft) high, and 8 m (26 ft) wide.

8-4 MODIFICATIONS OF THE BRAYTON CYCLE

Although the above calculations show reasonable efficiencies, they nonetheless do not take into account some of the complications of a real powerplant and are hence on the optimistic side. While a simple gas-turbine cycle is economically adequate for many purposes, such as peaking units and jet transports, cyclic and base-loaded units require modifications to improve the output and efficiency (and hence the heat rate). Besides increasing T_3, the modifications are:

1. Regeneration
2. Compressor intercooling
3. Turbine reheat
4. Water injection

Regeneration

Regeneration, as in steam cycles, is the internal exchange of heat within the cycle. In the Brayton cycle, T_4 is often higher than T_2 and heat addition is from 2 to 3 (Fig. 8-9). Regeneration, therefore, is used to preheat the compressed gas at 2 by the exhaust gases at 4 in a surface-type heat exchanger called the *regenerator* or, sometimes, the *recuperator*. Figure 8-13 shows such an arrangement for a closed cycle, suitable for He, but also used equally effectively for open cycles with air.

If the regenerator were 100 percent effective, the temperature of the gas entering the combustion chamber or nuclear reactor would be raised from T_2 to $T_{2''}$. The net work of the cycle would be maintained, except for the effect of the added pressure loss in the regenerator, but the heat added would be materially reduced from $H_3 - H_2$ to $H_3 - H_{2''}$, with corresponding increase in cycle efficiency. Actually, the regenerator effectiveness is never 100 percent, and the compressed gases are heated instead to a lower temperature such as $T_{2'}$. Regenerator *effectiveness,* ε_R, is defined as the ratio of the actual to maximum possible temperature change. In other words

$$\varepsilon_R = \frac{T_{2'} - T_2}{T_4 - T_{4''}} \tag{8-21a}$$

and since point 4″ is at the same temperature as point 2

$$\varepsilon_R = \frac{T_{2'} - T_2}{T_4 - T_2} \tag{8-21b}$$

Figures 8-10 and 8-11 include the effects of adding a regenerator with $\varepsilon_R = 0.75$, shown by the dashed lines. It can be seen that the effect of adding a regenerator on efficiency is remarkable and shifts the optimum pressure ratio for efficiency to lower

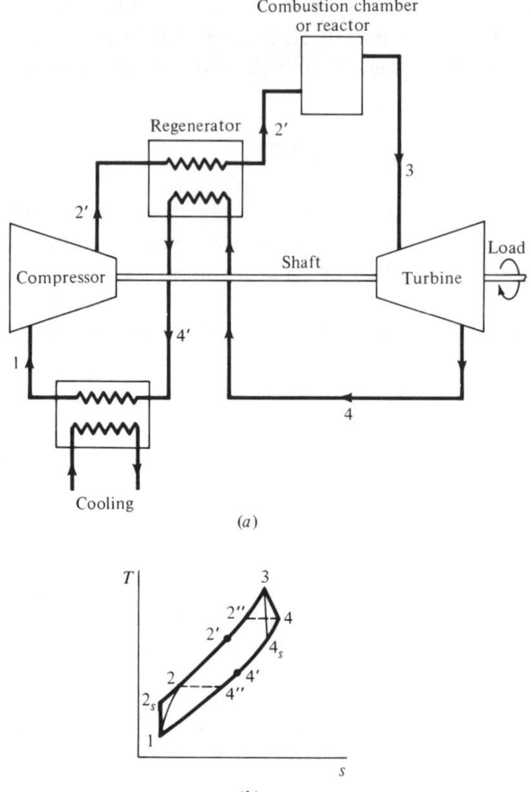

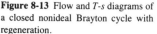

Figure 8-13 Flow and *T-s* diagrams of a closed nonideal Brayton cycle with regeneration.

values. This is because the lower the pressure ratio, the greater the difference between T_4 and T_2 and the greater the reduction in cycle heat input. At very low value of r_p, the effect of reduced cycle work predominates and the curves drop, though still higher than those for the simple cycle. The efficiency curves for a cycle with regenerator cross those for the simple cycle at points such as *a*, beyond which the effect of a regenerator on efficiency is negative. These points represent pressure ratios at which the exhaust gases are cooler than those after compression.

The effect of the regenerator on the specific power curves is only to reduce them somewhat because of the added pressure losses in the regenerator.

Because regenerative gas-turbine cycles are more efficient than simple gas-turbine cycles, thus reducing fuel consumption by 30 percent or more, they are now used by utilities for meeting cycling duty as well as base-load assistance in driving pumps, compressors, and the like.

Compressor Intercooling

The work in a flow system, such as a compressor or turbine, is given by Eq. (1-9), here repeated

$$W = -\int_1^2 V\,dP \tag{1-9}$$

For a perfect gas where $PV = mRT$, this equation can be written as

$$W = -\int_1^2 mRT\,\frac{dP}{P} \tag{8-22}$$

For a given dP/P, therefore, the work is directly proportional to temperature. A compressor working between points 1 and 2 (Fig. 8-13), therefore, would expend more and more work as the gas approaches point 2. Since compressor work is negative and a drain on the net cycle work, it is advantageous to keep T low while reaching the desired pressure P_2. This can theoretically be done by continuous cooling of the compressed gas to keep it at T_1, as shown by the lower horizontal dashed line of Fig. 8-14. However, this is not physically possible, and cooling, instead, is done in stages. Figure 8-13, drawn for simplicity for ideal (isentropic) compression and expansion, shows two stages of intercooling where gas is partially compressed from 1 to 2, cooled back to 1′ at constant pressure (ideally), compressed again to 2′, intercooled to 1″, then finally compressed to 2″. Ideally $T_1 = T_{1'} = T_{1''}$ and $T_2 = T_{2'} = T_{2''}$. In that case we have three compressor sections operating in tandem with equal work because for any one compressor section (from Table 1-2)

$$W = m\,\frac{nR(T_2 - T_1)}{1 - n}$$

where n is the polytropic exponent for compression (equal to k for ideal compression). When the temperature rises are equal, the pressure ratios are equal because

$$r_p = \left(\frac{T_2}{T_1}\right)^{n/(n-1)}$$

and the pressure ratio per stage is given by

$$r_{p,\text{stage},c} = \sqrt[Nc]{r_{p,\text{tot},c}} \tag{8-23}$$

where N_c is the number of compressor sections. Thus for an overall compressor pressure ratio of 10 and 3 sections, the pressure ratio per stage is $\sqrt[3]{10} = 2.154$ (not $10/3 = 3.33$). The improvement in the cycle is in increased work and efficiency. The increase in work is the result of the reduction in total compressor work since

$$(H_2 - H_1) + (H_{2'} - H_{1'}) + (H_{2''} - H_{1''}) < H_x - H_1$$

because of operation at lower temperatures. This can also be easily seen from the T-s diagram where the work has increased by area 2-1′-2′-1″-2″-x-2. The heat added has also increased by $H_x - H_{2''}$. However, the work increase of the cycle more than offsets the increase in heat addition, resulting in an improvement in efficiency.

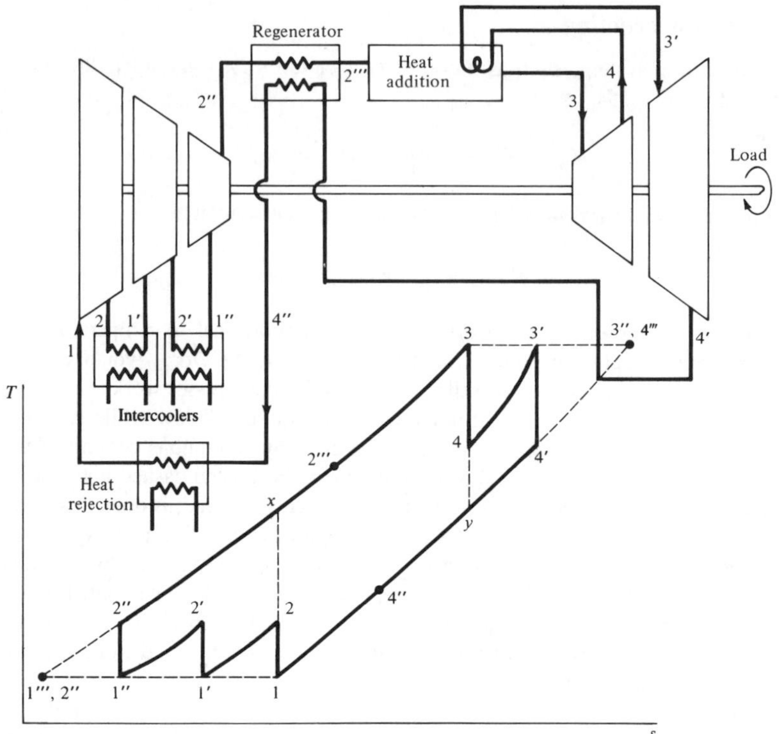

Figure 8-14 Flow and T-s diagrams of a closed ideal Brayton cycle with two stages of intercooling, one stage of reheat, and regeneration.

Intercoolers can be air-cooled heat exchangers but are more commonly water-cooled.

Turbine Reheat

Equation (8-22) has shown that compressor work can be decreased by keeping the gas temperatures in the compressor low. It also shows that turbine work can be increased by keeping the gas temperatures in the turbine high. This can also be done theoretically by continuous heating of the gas as it expands through the turbine, as shown by the upper horizontal dashed line of Fig. 8-14. Note that if cooling and heating were at constant temperatures, and if the rest of the cycle were ideal, we would have an ideal Ericsson cycle, which has the same efficiency as a Carnot cycle operating between the same temperature limits T_1 and T_3.

Again, continuous heating is not practical and reheat is done in steps or stages. Figure 8-14 shows two turbine sections and one stage of reheat. The gas expands in the high-pressure section of the turbine from 3 to 4, is then reheated at constant pressure (ideally) to 3′, and finally expands in the low-pressure section of the turbine to 4′. For $T_3 = T_{3'}$ and $T_4 = T_{4'}$, the pressure ratio per turbine stage is

$$r_{p,\text{stage},T} = \sqrt[N_T]{r_{p,\text{tot},T}} \tag{8-24}$$

The increase in cycle work is shown by the area 4-3′-4′-y, whereas the heat added is increased by $H_{3'} - H_4$. The net effect is an increase in both work and efficiency.

Intercooling, reheat, and regeneration can all be combined in one cycle as shown in Fig. 8-14.

General equations for the specific power and heat added for a composite cycle as the one discussed above, for the case of constant specific heat, but with nonidealities taken into account, are

$$\frac{\dot{W}_n}{\dot{m}c_p} = T_3\eta_T(n_T + 1)\left(1 - \frac{1}{r_{p_T}^{(k-1)/k}}\right) - T_1\frac{n_c + 1}{\eta_c}\left(r_{p_c}^{(k-1)/k} - 1\right) \tag{8-25}$$

$$\frac{\dot{Q}_A}{\dot{m}c_p} = T_3\left\{(n_T + 1) - (n_T + \varepsilon_R)\left[1 - \eta_e\left(1 - \frac{1}{r_{p_c}^{(k-1)/k}}\right)\right]\right\} - T_1(1 - \varepsilon_R)\left[1 + \frac{1}{\eta_c}\left(r_{p_c}^{(k-1)/k} - 1\right)\right] \tag{8-26}$$

where
$\dot{W}_n$ = net power = $\dot{W}_T - |\dot{W}_c|$

η_T = turbine adiabatic efficiency, assumed same for all turbine sections

η_c = compressor adiabatic efficiency, assumed same for all compressor sections

r_{p_T} = overall turbine pressure ratio

r_{p_c} = overall compressor pressure ratio

ε_R = regenerator effectiveness

n_T = number of reheat stages (e.g., 1 in Fig. 8-14)

n_c = number of intercooling stages (e.g., 2 in Fig. 8-14)

The efficiency of the cycle may now be obtained by dividing Eq. (8-25) by Eq. (8-26). The greater the number of reheat and intercooling stages there are, the higher the efficiency. However, this is attained at the cost of the capital investment and size of the plant. The design of the plant should be optimized, with consideration given to capital versus operating (fuel, etc.) expenses and to size.

Water Injection

Water injection is a method by which the power output of a gas-turbine cycle is materially increased and the efficiency is only marginally increased. In some aircraft-propulsion units and some stationary units, water is injected into the compressor and evaporates as the air temperature rises through the compression process. The heat of vaporization thus reduces the compressed air temperature, reducing the compressor work, an effect similar to that of intercooling (above).

In gas-turbine cycles that have regenerators, water injection is more beneficial if it is injected between the compressor and regenerator [67, 68]. The method can be used on both single- and two-shaft units. Figure 8-15 shows a schematic of a two-shaft unit with water injection between compressor and regenerator. On the *T-s* diagram, 1-2-4-5-7-9′-1 represents the cycle without water injection, in which 4 and 9′ are the compressed air and exhaust exits of the regenerator, respectively. With water injection the compressed air at 2 is cooled at nearly constant pressure by the evaporating water to 3. (A small increase in pressure does take place from 2 to 3.) The cooled compressed air at 3 is then preheated in the regenerator to a temperature almost the

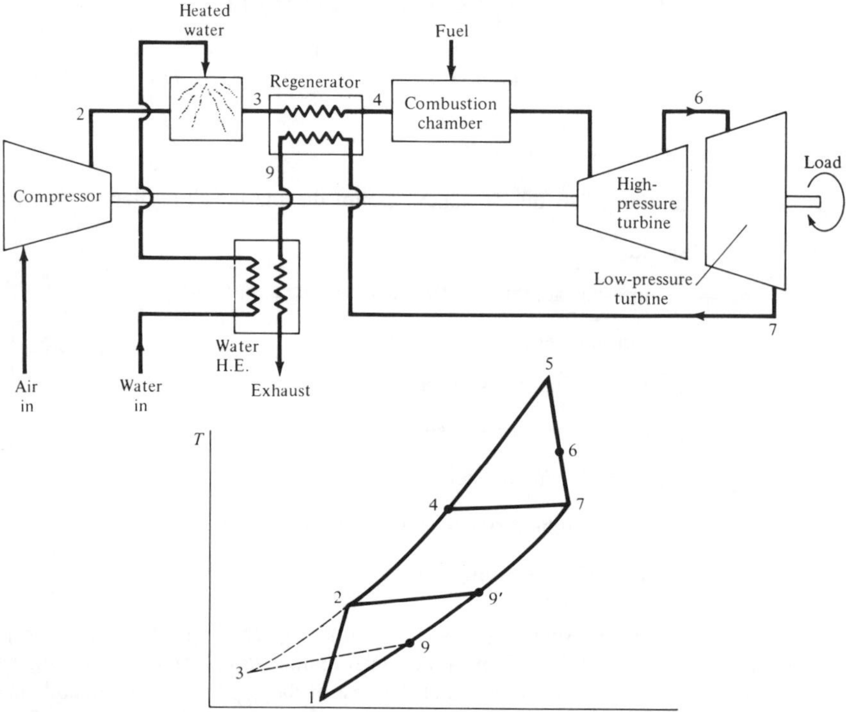

Figure 8-15 Flow and *T-s* diagrams of a two-shaft gas-turbine cycle with water injection and regeneration.

same as (actually very slightly below) 4. The added heat required to heat the moist air back from 3 to 2 is obtained from the exhaust gases by heat between 9′ and 9, which would have been lost to the cycle in any case. 9, then represents the new exhaust temperature. The inlet water may (as shown) or may not be preheated by the gases at 9 before injection.

The quantity of water vapor to be injected is that which would saturate the compressed air at T_3. A greater amount of water results in liquid carrythrough which, although it results in somewhat increased work, also results in reduced efficiency compared with that of saturated air and in fouling of the regenerator, local severe temperature differences, and associated thermal stresses.

The increase in work of a turbine plant with water injection is, in part, a result of increased turbine work due to the increased mass-flow rate of air and water vapor without a corresponding increase in compressor work. The increased mass stems from the saturated vapor at point 3 (Fig. 8-15) minus the water vapor originally in the air at point 1. Using Eq. (7-4), this is given by

$$\omega_3 - \omega_1 = 0.622\left(\frac{P_{v,3}}{P_3 - P_{v,3}} - \frac{P_{v,1}}{P_1 - P_{v,1}}\right) \tag{8-27}$$

where ω_3 and ω_1 = mass of water vapor per unit mass of dry air at points 3 and 1, respectively

$P_{v,3}$ and $P_{v,1}$ = partial pressure of water vapor, saturated at point 3, and function of relative humidity (Sec. 7-5) at point 1

P_3 and P_1 = pressure of the air-vapor mixtures at points 3 and 1, respectively; P_3 is very nearly the same as the compressor exit pressure P_2

The temperature at point 3 can be obtained by an energy balance on the dry air and water vapor

$$h_{a,3} - h_{a,2} = (\omega_3 - \omega_1)(h_{w,} - h_{v,3}) \tag{8-28}$$

where $h_{a,3} - h_{a,2}$ = enthalpy change of dry air

$h_{v,3}$ = enthalpy of saturated vapor of T_3

h_w = enthalpy of injected water

Example 8-2 Air at 60°F, 14.696 psia, and 60 percent relative humidity is compressed by a compressor with a pressure ratio of 10 and 89.63 percent polytropic efficiency. The air is then saturated by water at 60°F. Find the mass of added water per unit mass of original air and the temperature of the saturated compressed air. For simplicity use $c_p = 0.24$ Btu/(lb$_m$ · °F) and $k = 1.40$ for air.

SOLUTION Refer to Fig. 8-14 and use the steam tables (App. A). For $T_1 =$ 60 °F: $P_{\text{sat},1} = 0.25611$ psia, and $P_{v,1} = 0.6 \times 0.25611 = 0.1537$ psia.

Thus
$$\omega_1 = \frac{0.622 \times 0.1537}{14.696 - 0.1537} = 0.00657$$

$$\omega_2 = \omega_1 = 0.00657$$

$$P_2 = 14.696 \times 10 = 146.96 \text{ psia}$$

$$T_{2,s} \text{ (isentropic compression)} = (60 + 460)(10)^{(1.4 - 1)/1.4} = 1004°\text{R}$$

$$h_w = \text{enthalpy of injected water at 60°F} = 28.06 \text{ Btu/lb}_m$$

From Eq. (8-16)

$$\frac{1004 - 520}{T_2 - 520} = 0.8963$$

Therefore

$$T_2 = 1060°\text{R} = 600°\text{F}$$

$$h_{a,3} - h_{a,2} = c_p(T_3 - T_2) = 0.24(T_3 - 600)$$

The problem requires a trial-and-error solution. Assume values of T_3 that would satisfy Eq. (8-28):

T_3,°F	$P_{v,3}$	ω_3 (Eq. 7-4)	$\omega_3 - \omega_1$	$h_{v,3}$	$(\omega_3 - \omega_1)(h_w - h_{v,3})$	$h_{a,3} - h_{a,2}$
220	17.186	0.08237	0.0758	1153.4	−85.30	−91.20
222	17.860	0.08605	0.0795	1154.2	−89.50	−90.72
224	18.556	0.08989	0.0833	1154.9	−93.87	−90.24

By interpolation

$$T_3 = 222.5°\text{F} \qquad \omega_3 - \omega_1 = 0.0804$$

Thus the mass increases by 8.04 percent of dry air, or $0.0804/(1 + \omega_1) = 0.08$, or 8 percent of original air.

Another reason for the increase in work of a cycle with water injection is that the optimum pressure ratio for work increases. The cycle efficiency, too, is shown to increase, and its optimum pressure ratio also increases, though to a lesser degree than the work [67]. It would thus be advantageous to select a higher pressure ratio to increase the work, provided the cycle efficiency is not disadvantaged.

Another advantage for water injection is that partial load operation could be affected by reduction of the water injection rate while the turbine inlet temperature T_5 is kept constant, which would maintain high efficiency during that portion of the load. When the load drops below that requiring water injection, the turbine inlet temperature is reduced.

The exhaust emissions are also favorably affected by water injection. Emissions of CO and unburned hydrocarbons in gas-turbine powerplants are not significant because of the high air-to-fuel ratios used in them. They become of concern only at very high loads when the air-to-fuel ratios are reduced. The oxides of nitrogen, NO_x, however, are becoming a problem in gas-turbine combustion because of the steadily increasing combustion temperatures in modern units. It has been found that water injection reduces NO_x by at least half [69].

8-5 CYCLE ANALYSIS WITH VARIABLE PROPERTIES

As indicated above, cycle analysis with constant properties—in effect c_p and k, which are needed to calculate enthalpies and P, V, and T relationships—are accurate (except at extremely low temperatures and high pressure) for monatomic gases such as helium. For diatomic gases (air, N_2, CO), and more so for triatomic gases (CO_2, H_2O in gas form), and larger molecules (NH_3), the use of constant c_p and k yields results that are useful only to predict trends of variables for the particular gas considered. For such gases, a knowledge of the variation of c_p with temperature is necessary. For air, for example, c_p in Btu/1b$_m$°R is given by [70]

$$c_p = 0.2317 + 9.0083 \times 10^{-6}T + 2.1998 \times 10^{-8}T^2 - 9.0067 \times 10^{-12}T^3 \tag{8-29}$$

The value of the specific heat at constant volume c_v is obtained by subtracting the gas constant for air, i.e., 53.34/778.16 = 0.0685 from Eq. (8-29). The value of k as a function of T is then obtained as $c_p(T)/c_v(T)$. The change in enthalpy is obtained by

$$\Delta h = \int_{T_1}^{T_2} c_p(T)\, dT \tag{8-30}$$

or by

$$\Delta h = \bar{c}_p(T_2 - T_1) \tag{8-31}$$

where $\bar{c}_p$ is an average specific heat given by

$$\bar{c}_p = \frac{\int_{T_1}^{T_2} c_p(T)\, dT}{T_2 - T_1} \tag{8-32}$$

To obtain P, V, and T relationships for polytropic reversible processes an average value of k should also be found.

Such procedure suffers from two drawbacks. (1) It is complicated, especially when the final temperature of a given process is unknown, and (2) it does not take into account the effects of fuel and combustion-products composition and dissociation. In analyzing gas-turbine cycles, it is more convenient to rely on tabulated properties of air and products of combustion and component gases as given in the gas tables by Keenan and Kaye [8]. It is probably best to illustrate the procedure by an example.

Example 8-3 A single-shaft gas-turbine 25-MW plant has one stage of intercooling, no reheat, and a regenerator. Air enters the compressor at 1 atm and 520°R. The compressor has an overall pressure ratio of 10, and each of its two sections has a polytropic efficiency of 90 percent. The intercooler cools the air back to 520°R. The regenerator has an effectiveness of 75 percent. The air-fuel ratio corresponds to 200 percent of theoretical air. The fuel may be represented by $CH_{2.145}$. Because of pressure losses through regenerator and combustion chamber and also at exit, the turbine pressure ratio is 9.2. The turbine inlet temperature is 2500°R. It has a polytropic efficiency of 87 percent. The mechanical efficiency of the system is 95 percent. The electric-generator efficiency is 98 percent. Calculate the various pressures and temperatures around the cycle, the plant efficiency, and the necessary airflow. Estimate the plant efficiency if no regenerator were used.

SOLUTION Refer to Fig. 8-16 and the gas tables or App. I.

(1) *Compressor:*

$$r_p \text{ per section} = \sqrt{10} = 3.1623$$

$$T_1 = 520°\text{R},\ P_{r,1} = 1.2147,\ h_1 = 124.27$$

$$P_{r,2} = 1.2147 \times 3.1623 = 3.8412$$

Therefore

$$T_{2,s} = 722°\text{R} \qquad h_{2,s} = 172.88$$

$$\eta_c = 0.9 = \frac{h_{2,s} - h_1}{h_2 - h_1} = \frac{172.88 - 124.27}{h_2 - 124.27}$$

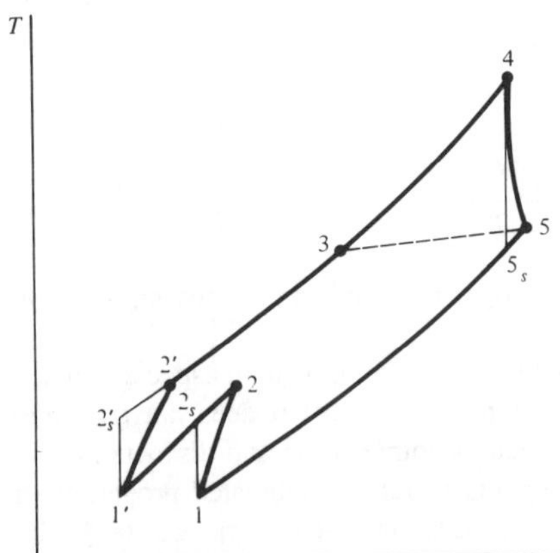

Figure 8-16 *T-s* diagram of gas-turbine powerplant of Example 8-3.

Therefore

$$h_2 = 178.28 \qquad \text{and} \qquad T_2 = 744°\text{R} = 284°\text{F}$$

For $T_{1'} = 520°\text{R}$, calculations for the high-pressure section of the compressor 1′-2′ are identical to those of the low pressure section 1-2. Thus

$$T_{2'} = 744°\text{R} \qquad h_{2'} = 178.28$$

The total work of the compressor is

$$|2(h_2 - h_1)| = 2(178.28 - 124.27) = 2 \times 54.01 = 108.02 \text{ Btu/(lb}_m \text{ air)}$$

(2) *Turbine:*

$$T_4 = 2500°\text{R} \qquad p_{r,4} = 559.8 \qquad \bar{h}_4 = 19520.7$$

$$p_{r,5} = \frac{55948}{9.2} = 50.75$$

Therefore

$$T_{5,s} = 1482°\text{R} \qquad \bar{h}_{5,s} = 10905.5$$

$$\eta_T = 0.87 = \frac{h_4 - h_5}{h_4 - h_{5,s}} = \frac{19520.7 - \bar{h}_5}{19520.7 - 10905.5}$$

Therefore

$$\bar{h}_5 = 12025.5 \qquad \text{and} \qquad T_5 = 1620°\text{R}$$

Turbine work is

$$\bar{h}_4 - \bar{h}_5 = 19520.7 - 12025.5 = 7495.2 \text{ Btu/(lb} \cdot \text{mol)}$$

(3) *Regenerator:*

$$\text{Effectiveness } \varepsilon_R = \frac{T_3 - T_{2'}}{T_5 - T_{2'}} = 0.75 = \frac{T_3 - 744}{1620 - 744}$$

Therefore

$$T_3 = 1401°\text{R}$$

From the air tables (App. I)

$$h_3 = 343.16 \text{ Btu/(lb}_m \text{ air)}$$

It is now important to base all calculations on the same basis, say 1 lb_m of air. The enthalpies of the combustion gases are based on 1 lb · mol of components. At 200 percent theoretical air, the molecular mass of the products = 28.880; therefore

$$h_4 = \frac{14520.7}{28.88} = 675.9 \text{ Btu/lb}_m \text{ products}$$

and

$$\text{Turbine work} = \frac{7495.2}{28.88} = 259.5 \text{ Btu/lb}_m \text{ products}$$

The air-fuel ratio for 100 percent theoretical air is obtained by writing the chemical equation

$$CH_{2.145} + \left(1 + \frac{2.145}{4}\right)O_2 + 3.76\left(1 + \frac{2.145}{4}\right)N_2$$

$$\rightarrow CO_2 + \frac{2.145}{2}H_2O + 3.76\left(1 + \frac{2.145}{4}\right)N_2$$

or

$$CH_{2.145} + 1.53625O_2 + 5.7763N_2 \rightarrow CO_2 + 1.0725H_2O + 5.7763N_2$$

The air-fuel ratio is

$$\frac{1.53625 \times 32 + 5.7763 \times 28}{12 + 2.145} = 14.91$$

For 200 percent theoretical air, air-fuel ratio = 29.82. Thus there are 1 + 1/29.82 = 1.0335 lb_m products of combustion per lb_m air. Therefore

$$\text{Heat added} = h_4 - h_3 = 675.9 \times 1.0335 - 343.16 = 355.38 \text{ Btu/lb}_m \text{ air}$$

$$\text{Net cycle work} = \text{turbine work} - |\text{compressor work}|$$

$$= 259.5 \times 1.0335 - 108.02 = 160.17 \text{ Btu/lb}_m \text{ air}$$

$$\text{Net plant work} = 160.17 \times 0.95 \times 0.98 = 149.12 \text{ Btu/lb}_m \text{ air}$$

$$\text{Plant efficiency} = \frac{149.12}{355.38} = 0.42 = 42\%$$

Since 1 MW = 3.412×10^6 Btu/h

$$\text{Airflow necessary} = \frac{25 \times 3.412 \times 10^6}{149.12 \times 3600} = 158.9 \text{ lb}_m\text{/s}$$

Without a regenerator the plant efficiency would be approximately $W_{net}/(h_4 - h_2')$ or

$$\frac{149.12}{675.9 \times 1.0335 - 178.28} \times 100 = 28.66\%$$

The approximation is because, without a regenerator, the turbine pressure ratio would be slightly higher than 9.2 and the plant work and efficiency slightly higher. However, the difference between the thermal efficiencies as calculated is indicative of the large effect of regeneration.

8-6 DESIGN FOR HIGH TEMPERATURE

It should be evident by now that it is becoming more necessary to operate gas-turbine plants with higher and higher turbine inlet temperatures to achieve higher efficiencies and outputs. This also means higher pressure ratios because optimum pressures increase with increasing turbine inlet temperatures for both efficiency and power. High-pressure ratio units have higher capital costs than lower-pressure ones, but the decrease in fuel consumption rapidly pays back for this capital cost differential. Another concern that goes with higher temperatures is increased potential for corrosion, which has to be dealt with. As indicated earlier, research and development is underway to raise turbine inlet temperatures from the present 2000 to 2300°F (1090 to 1260°C) to near 2800°F (1540°C). Such temperatures are well above those that modern steam turbines have to cope with, which are around 1000 to 1200°F (540 to 650°C). The present range is suitable for peaking service, and with regeneration, for cyclic and some base-load service. It is also competitive with steam plants when used in a combined cycle (Sec. 8-7). Future ranges would make them competitive on their own.

There are several approaches to the problems associated with high gas temperatures. In general they can be categorized as developing suitable (1) materials, (2) cooling, and (3) fuels.

Materials

The components that suffer most from a combination of high temperatures, high stresses, and chemical attack are those of the turbine first-stage fixed blades (nozzles) and moving blades. They must be weldable and castable and must resist corrosion, oxidation, and thermal fatigue. Heat resistant materials and precision casting are two recent advances, largely attributable to aircraft engine developments. Cobalt-based alloys have been used for the first-stage fixed blades (which are subjected to the highest temperatures but not the high stress of the moving blades). These alloys are now being supplemented by vacuum-cast nickel-base alloys that are strengthened through solution- and precipitation-hardened heat treatment. For the moving blades, cobalt-based alloys with high chromium content are now used.

Ceramic materials are also being developed, especially for the turbine inlet fixed blades. Developmental problems here are inherent brittleness, which causes fabrication problems and raises uncertainties about the mechanical properties of ceramic materials.

Cooling

Early turbines operated uncooled, as do many present-day ones. The increases in temperatures we are witnessing require cooling, however.

The thermal stresses in high-temperature turbine moving blades are caused by the high rotational speeds, uneven temperature distributions in the different blade cross sections, and static and pulsating gas forces that may give rise to dangerous vibrational stresses. Other thermal stresses occur during start-up, shutdown, and load changes. Thermal stresses are thus caused by steady-state as well as transient operation. The latter give rise to low-cycle fatigue, which reduces blade life. In addition there are problems of creep rupture, high-temperature corrosion, and oxidation. It is generally agreed that blade surfaces should be kept below about 1650°F (900°C) to reduce corrosion to a tolerable degree.

A blade is cooled by being made hollow so that a coolant can circulate through it. A hollow blade is lighter than a solid blade and has a much lower Biot number* and hence a fairly uniform temperature distribution.

The coolants that have been used and/or are under consideration are air and water (and steam). The ranges for these are air for gas temperatures up to about 2100°F (1150°C), water for gas temperatures above 2400°F (1315°C), and a hybrid system for the intermediate range. In the hybrid system, water cooling is used for the highest temperature components, mainly the inlet fixed blades, and air for the remaining blades and rotor [71].

Air Cooling

Air cooling is of three kinds: convection cooling, film or transpirational cooling, and impingement cooling. Cooling air is obtained directly from the compressor, thus bypassing the combustion chamber. In *convection cooling* the air is made to flow inside the hollow blade, entering at the leading edge, reversing direction a few times, and leaving at the trailing edge to enter the main gas stream. *Film cooling* is used in conjunction with convection cooling and never alone. Based on aerospace technology, it involves air flowing through holes or slots from the inside of the blade to the outside boundary layer to form a protective insulating film between the blade and the hot gases. Besides the cooling effect, it also helps prevent corrosion of the blades. This transpirational air must be very clean for proper effectiveness.

Figure 8-17 shows an example of air cooling on inlet fixed blades [72]. The upper vertical cross section shows air entering at the top from the stator. It then flows downward by the leading edge in two parallel paths, changes direction three or four times, and leaves at the trailing edge. The middle path includes longitudinal ribs or fins to enhance heat transfer. Other designs may include roughened internal surfaces and transverse ribs and webs. A good design should allow good heat transfer with little pressure drop.

The middle horizontal cross section through the blade (Fig. 8-17*b*) shows the internal paths in pure convection cooling. The lower horizontal cross section (*c*) shows two rows of holes A and B on the suction side of the blade for film cooling.

For the moving blades (Fig. 8-18), the cooling air enters the blade root from the rotor, flows radially through ducts in the hollow blade, changes direction, and leaves through slots from the blade trailing edge. Figure 8-18 shows three such ducts, one change of direction, and several exit slots.

The third method, *impingement cooling,* is one in which the cooling air is made to impinge on the internal surface of the blade, which provides particularly intensive cooling.

* In transient heat transfer between a solid body and a fluid, a low Biot number (Bi) indicates a more uniform temperature distribution within the body than a high Biot number. Bi is given by hL/k, where h is the heat-transfer coefficient, L is a characteristic length of the body, in this case the blade wall thickness, and k the thermal conductivity of the material of the body. A Biot number less than 0.1 means that all parts of the body are at the same temperature at any one instant of time and that the body can be treated as a lumped capacity in transient heat-transfer analysis.

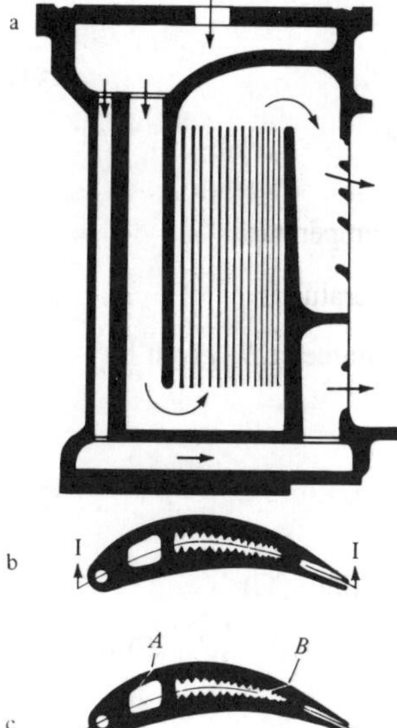

Figure 8-17 Air-cooled gas-turbine fixed blade [72].

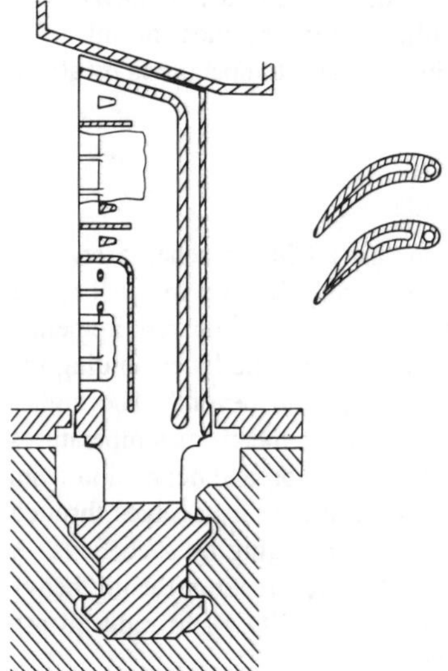

Figure 8-18 Air-cooled gas-turbine moving blade [72].

A *blade-cooling effectiveness* ε_{bc} is defined as

$$\varepsilon_{bc} = \frac{T_g - T_b}{T_g - T_{a,i}} \tag{8-33}$$

where T_g = gas temperature

T_b = mean blade surface temperature

$T_{a,i}$ = inlet cooling air temperature

ε_{bc} has been found to be a function of a dimensionless parameter B, given by

$$B = \frac{\dot{m}_a c_{p,a}}{h_g A_b}$$

where m_a = mass-flow rate of air, lb_m/h or kg/s

$c_{p,a}$ = specific heat or air, Btu/(lb_m · °F) or J/(kg · °C)

h_g = average heat-transfer coefficient of gas, Btu/(h · ft^2 · °F) or W/m^2 · °C

A_b = blade surface area, ft^2 or m^2

It is to be noted that calculation of h_g is not a simple matter, especially when film or impingement cooling is involved; the calculation requires elaborate computer programs and needs to be supplemented by experimental data. Local values between 200 and 500 Btu/(h · ft^2 · °F) [1.1 to 2.8 kJ/(m^2 · °C)] have been reported [72]. Figure 8-19 shows a plot of ε_{bc} versus B. It shows that combined convection and film cooling is superior, followed by impingement and then convection cooling. It is believed that combined convection and film (transpirational) cooling offers the most promise for air-cooling schemes and permits an increase in turbine inlet-gas temperatures by about 120°F (67°C).

Water Cooling

As gas temperatures exceed 2100°F (1150°C) or so, air cooling reaches a state of rapidly diminishing returns because of the quickly increasing demand for cooling air that bypasses the combusion chamber. Despite their additional equipment requirements, the regions of hybrid and pure water cooling are reached, with the latter holding the greatest promise for gas temperatures of 2400°F (1315°C) and beyond. The higher heat capacity and heat-transfer capability of water permit lower metal gas temperatures (for the same gas temperatures) and hence reduced hot corrosion and deposition from contaminated fuels. Water cooling also eliminates the need for air passages through the blades, as in film cooling, which would be subject to plugging by such fuels. A test program sponsored by EPRI (the Electric Power Research Institute) using heavy ash-bearing fuels showed metal temperatures below 850°F (450°C) and reduced ash accumulation on the blades with water cooling [71].

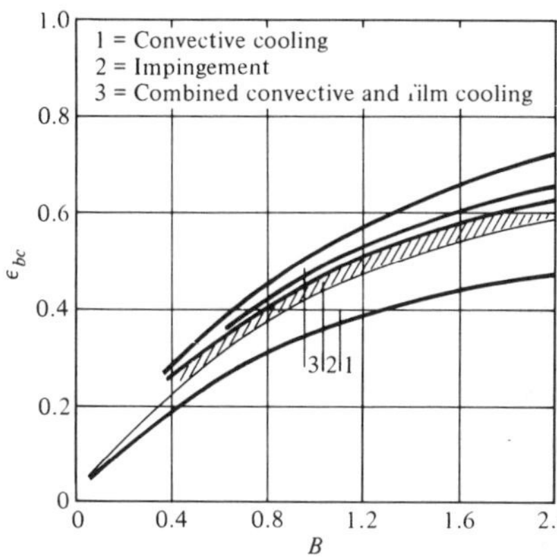

Figure 8-19 Comparison of different processes of blade cooling [72].

In designing a gas turbine for water cooling, standard aerodynamic design is followed but the hot gas paths are kept as short as possible to minimize surface areas that require cooling. The fixed blades or nozzles are hollow and contain series and parallel water-flow paths (not unlike air-cooled blades). In one turbine design [71], cooling water circulates in, through, and out of these paths in a closed loop. Heat removed from these blades is recovered in a heat exchanger for use in the steam portion of a combined-cycle plant (below). The inlet water is hot enough to prevent thermal shock and is at a pressure high enough (about 1250 psia, 86 bar) to prevent boiling and keep the water in single phase. In that design, the first-stage fixed blades are of composite construction with a core of high-strength material (Nitronic 50), a surrounding copper matrix in which water-cooling tubes are imbedded, and a covering skin, all bonded by hot isostatic pressing. Second- and third-stage fixed blades are cast IN-718 with drilled water paths. Such blades have been manufactured and successfully tested at operating conditions.

Moving blades are cooled by an open-loop water system. Water enters the blades at lower pressures and is allowed to boil, with steam ejecting from the blade tips to mix with the hot gas stream. Unvaporized water moves radially by centrifugal force and may be collected in a circumpherential cavity in the casing (Fig. 8-20) [73]. Closed-loop systems are being considered but pose some design difficulties. The moving blades in the above design are forged from IN-718, with the cooling paths drilled after forging.

Fuels

The advantages of increased gas-turbine combustion temperatures—that is, increased efficiency, power, and reduced fuel consumption—are partly negated by the increasing cost of fuels normally used by gas turbines. So far gas turbines have relied primarily on natural gas and clean liquid fuels. Natural gas now is being conserved for domestic use, and liquid hydrocarbons have increased in cost by an order of magnitude during

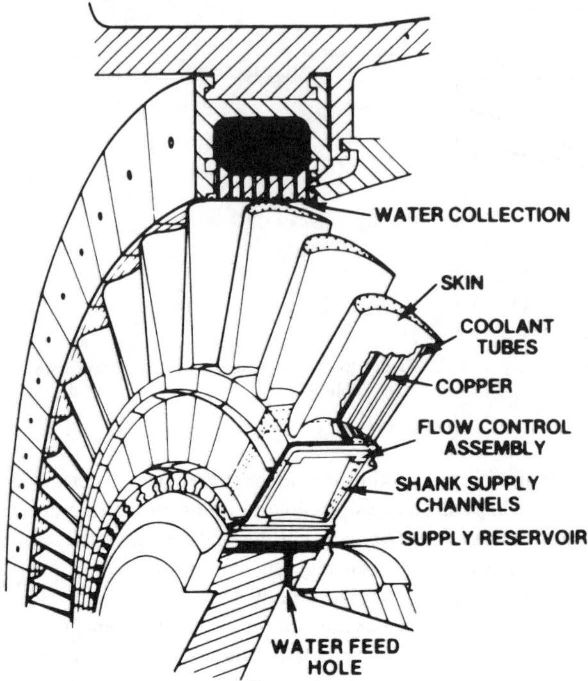

Figure 8-20 A water-cooled gas-turbine moving blade [73].

the decade of the 1970s alone. By contrast steam powerplants are making increasing use of various abundant coal-combustion systems (Chap. 5) and cheap nuclear fuels.

Residual liquid fuels, the residue left after the profitable light fractions have been extracted from the crude, have been used in gas turbines to some extent [74]. They are (1) viscous and (2) tend to polymerize (form sludge or tar) when overheated. (3) Their high carbon content leads to excessive carbon deposits in the combustion chamber. (4) Their contents of alkali metals, such as sodium, combine with sulfur to form sulfates that are corrosive. (5) They have other metals like vanadium with compounds that form during combustion also being corrosive. (6) They have a relatively high ash content that deposits mostly on the inlet fixed blades, thus reducing gas flow and power output.

The rate of corrosion increases with increasing gas temperatures. Early turbines designed for residual fuel use operated at temperatures below 1650°F (900 K) to avoid the problem. Ash deposition is not a problem with intermittent operation because of successive expansions and contractions, but it is a serious problem with steady operation.

Fortunately progress is being made at keeping gas turbines competitive by the development of systems that prepare cheaper lower-grade fuels for gas-turbine use

[75]. Washing the fuel with water and separating the mixture with centrifugal or electrostatic separators is found to remove the alkalis. Fuel additives such as magnesium have been found to neutralize vanadium. Additives and protective coatings are also used to reduce corrosion [76].

A promising scheme is pressurized-fluidized-bed combustion (PFBC) (Sec. 4-8), which makes cheap, abundant coal readily available as a gas-turbine fuel. In the PFBC, the addition of limestone will remove enough sulfur to meet environmental regulations. However, much development work remains to be done in order to reduce particulate matter from the fluidized-bed gaseous product, which can rapidly destroy turbine blading. Other alternatives (for coal use) are the use of synthetic fuels from coal gasification and liquefaction. A low-Btu gas use in a combined cycle has already been discussed in Sec. 4-11.

8-7 COMBINED CYCLES: GENERAL

Combined-cycle powerplants are those which have both gas and steam turbines supplying power to the network. The idea of combined cycles has grown out of the need to improve the simple Brayton-cycle efficiency by utilizing the waste heat in the turbine exhaust gases. This, as we have seen, can be done by regeneration (Sec. 8-4). Regeneration reduces the heat lost up the stack down from some 70 to 60 percent of the energy input. The sole purpose of a regenerator is to improve efficiency; it does not increase the power output. In fact, because of additional pressure losses it imposes on the plant, a regenerator reduces the turbine pressure ratio and hence the net plant output by a few percent. Aside from this small reduction in power output, regenerators with their large heat-exchange surfaces and large gas and air piping make the plant more costly. Another effect is that the optimum pressure ratio for maximum efficiency moves sharply to lower values with regeneration, resulting in reduced power as can be seen from Figs. 8-10 and 8-11.

Simple cycles operate near maximum power because they are not used in service where efficiency is the prime concern. Regenerative cycles, however, are meaningful only if they are operated near maximum efficiency. Thus they would have their output reduced from a simple cycle by a much larger percentage, perhaps 10 to 14 percent. In certain applications, an economic compromise between capital and operating costs would have to be found.

It can be seen that raising the efficiency of a gas-turbine plant by regeneration, while used for stationary applications, is costly. A means, therefore, was sought whereby *both* efficiency and power are increased. The solution was found in using the large quantity of energy leaving with the turbine exhaust to generate steam for a steam-turbine powerplant. This is a natural solution as the gas turbine is a relatively high-temperature machine (2000 to 3000°F, 1100 to 1650°C), whereas the steam turbine is a relatively low-temperature machine (1000 to 1200°F, 540 to 650°C). This joint operation of the gas turbine at the "hot end" and the steam turbine at the "cold end," is called a combined-cycle powerplant.

Besides both high efficiency and high power outputs, combined cycles are characterized by flexibility, quick part-load starting, suitability for both base-load and cyclic operation, and a high efficiency over a wide range of loads. They have the potential of using coal [77] as well as synthetic and other fuels. Their obvious disadvantage is in their complexity, as they in essence combine two technologies in one powerplant complex.

The idea of combined cycles is not new, having been proposed as early as the beginning of this century. It was not, however, until 1950 that the first plant was installed. This was followed by a rapid rise in the number of installations, especially in the 1970s. An estimated 100 plants, with a total of 150,000-MW output, had been installed by the end of the 1970s throughout the world [78].

There have been many suggested types of combined cycles, the most important of which comprise:

1. A heat-recovery boiler with or without supplementary firing
2. A heat-recovery boiler with regeneration and/or feedwater heating
3. A heat-recovery boiler with multipressure steam cycle
4. A closed-cycle gas turbine with steam-cycle feedwater heating

Some examples of these are presented next.

8-8 COMBINED CYCLES WITH HEAT-RECOVERY BOILER

Figure 8-21 shows a schematic flow diagram of such a combined cycle. A simple gas-turbine cycle, consisting of air compressor (AC), combustion chamber (CC), and gas turbine (GT) is used with the turbine exhaust gas going to a *heat-recovery boiler* (HRB) to generate superheated steam. That steam is used in a standard steam cycle, which consists of turbine (ST), condenser (C), pump (CP), closed feedwater heaters (FWH), and deaerating heater (DA). The HRB consists of an economizer (EC), boiler (B), steam drum (SD), and superheater (SU). The gas leaves the HRB to the stack. Both gas and steam turbines drive electric generators (G).

The gas turbine is usually operated with a high air-fuel ratio, approximately 400 percent theoretical air, to make sufficient air available in the gas-turbine exhaust for further combustion.

For low-powered combined cycles the steam-turbine output is less than the gas-turbine output, by as much as 50 percent, and the number of feedwater heaters is small, often one deaerator and one closed-type. To increase the output for short periods during load peaks, supplementary fuel burners may be fitted to the HRB to increase the steam mass-flow rate. This can also be done on a continuous basis, but limits on the amount of fuel added are posed by the design of the HRB, which usually lacks refractory lining, water-cooled walls, etc. A limit of about 1400°F (760°C) on the gas temperatures in the HRB is common. This however, is usually sufficient to increase the steam-turbine output by perhaps 100 percent and the total cycle output by 30 percent.

In large combined-cycle plants used for base-load operation, where efficiency is

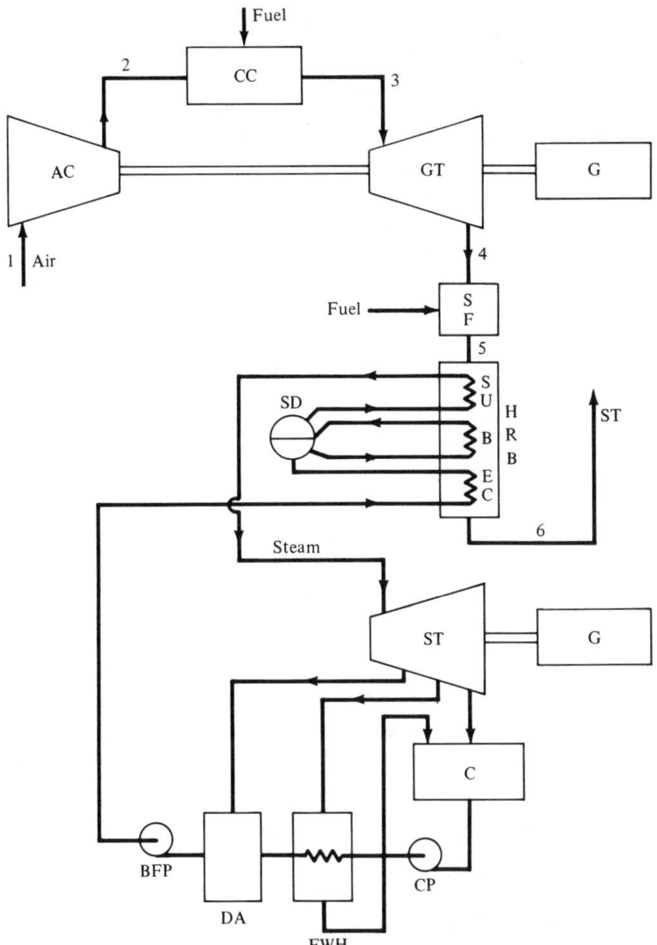

Figure 8-21 Schematic flow diagram of a combined cycle with heat-recovery boiler (HRB).

of prime importance, separate supplementary firing equipment (SF) is interposed between the gas turbine and the HRB. The steam-turbine output is usually greater than the gas-turbine output by up to 8:1. The steam cycle is therefore designed for high efficiency with reheat and a full complement of feedwater heaters. A forced-draft fan may be installed ahead of the SF to operate the steam cycle on its own when the gas turbine is cut off.

The fuel used in supplementary firing may be the same high-grade fuel used in the gas turbine. This is the simplest solution as such fuel causes few problems in the

SF and HRB. However, cheaper lower-grade fuels such as heavy oil or coal, can also be used in the SF.

Example 8-4 A combined cycle as shown in Fig. 8-20 with supplementary firing has gas-turbine cycle fuel represented by $CH_{2.145}$, air-fuel ratio corresponding to 400 percent of theoretical air, gas-turbine inlet temperature of 2600°R, pressure ratio of 8, and a polytropic efficiency of 0.90. Supplementary firing using the same fuel raises the gas temperature to 2000°R before entering the heat-recovery boiler. The gas leaves to the stack at 800°R. Steam is generated at 1200 psia and 1560°R from feedwater at 780°R. Calculate the heat added in supplementary firing, Btu/lb_m of air, and the mass ratio of airflow to steam flow.

SOLUTION Use the gas tables for 400 percent theoretical air (App. I) and refer to Fig. 8-22.

$$T_3 = 2600°\text{R} \qquad p_{r,3} = 586.4 \qquad \bar{h}_3 = 19979.7 \text{ Btu/(lb} \cdot \text{mol gas)}$$

$$p_{r,4} = \frac{p_{r,3}}{8} = 73.3 \qquad T_{4,s} = 1579°\text{R} \qquad \bar{h}_3 = 11499.0$$

$$\eta_T = 0.9 = \frac{\bar{h}_3 - \bar{h}_4}{\bar{h}_3 - \bar{h}_{4,s}} = \frac{19979.7 - \bar{h}_4}{19979.7 - 11499.0}$$

Therefore

$$\bar{h}_4 = 12347.1 \text{ Btu/(lb} \cdot \text{mol gas)} \qquad T_4 = 1685°\text{R}$$

Assume that supplementary firing changes the mixture from 400 percent at 4 to 200 percent of theoretical air at 5 (use the gas tables for 200 percent theoretical air).

$$T_5 = 2000°\text{R} \qquad \bar{h}_5 = 15189.3 \text{ Btu/(lb} \cdot \text{mol gas)}$$

$$T_6 = 80\text{C}°\text{R} \qquad \bar{h}_6 = 5676.3 \text{ Btu/(lb} \cdot \text{mol gas)}$$

From Example 8-3, the air-fuel ratio (A-F) for stoichiometric mixture of $CH_{2.145}$ fuel is 14.91. The products-to-air mass ratio is 1/(1 + 1/A-F). This is 1.0335 for

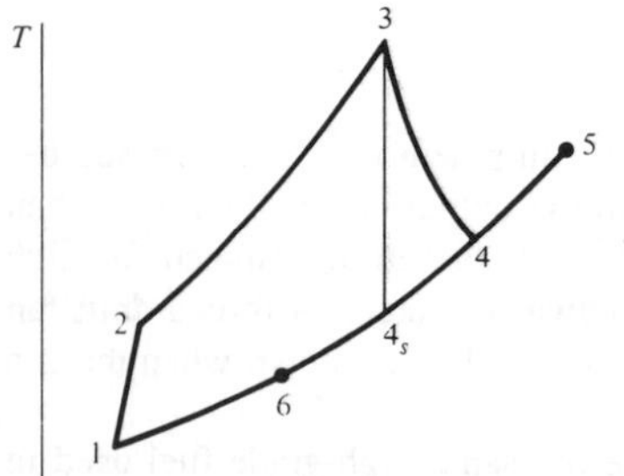

Figure 8-22 *T-s* diagram for the gas in Example 8-4.

200 percent and 1.0168 for 400 percent theoretical air. Also, the molecular masses for these products are 28.880 and 28.925, respectively. Based on 1 lb_m of air

$$h_4 = \frac{12{,}347.1}{28.925} \times 1.0168 = 434.04 \text{ Btu/lb}_m \text{ air}$$

$$h_5 = \frac{15{,}189.3}{28.880} \times 1.0335 = 543.56 \text{ Btu/lb}_m \text{ air}$$

and

$$h_6 = \frac{5676.3}{28.880} \times 1.0335 = 203.13 \text{ Btu/lb}_m \text{ air}$$

Therefore, heat added in SF $= h_5 - h_4 = 543.56 - 434.04 = 109.52$ Btu/lb_m air and, heat added in HRB $= h_5 - h_6 = 543.56 - 203.13 = 340.43$ Btu/lb_m air.

For steam entering turbine, $h = 1556.9$ Btu/lb_m steam. For feedwater entering the HRB, $h = 290.4$ Btu/lb_m steam. Therefore

$$\Delta h_{\text{steam}} = 1556.9 - 290.4 = 1266.5 \text{ Btu/lb}_m \text{ steam}$$

So

$$\text{Mass ratio of air to stream} = \frac{340.43}{1266.5} = 0.269$$

Variations of the cycle shown in Fig. 8-21 are used to extract the maximum amount of energy from the gas leaving the heat-recovery boiler before exhausting it to the stack. Depending upon the temperature of that gas, it may be used for (1) partial heating (regeneration) of the compressed air leaving the compressor, (2) feedwater heating of the steam cycle in a closed-type feedwater heater, or (3) generating steam in a dual- or multipressure steam cycle. This last variation is described in Sec. 8-10.

Because gas turbines are not yet built in sizes as large as steam turbines, combined cycles are often built in combinations of more than one gas turbine plus one steam turbine. Such combinations show certain advantages, not only in higher total plant output but also in higher availability, flexibility in service, and part-load efficiency. An example is the STAG powerplant, described next.

8-9 THE STAG COMBINED-CYCLE POWERPLANT

The STAG (for steam and gas) 330-MW combined-cycle powerplant [79] is a cyclic plant built for the Jersey Central Power and Light Company. It is located at the Gilbert Generating Station on the Delaware River, south of Phillipsburg, N.J.

The plant was designed by the General Electric Company and is composed of four GE Model-7000 gas turbines exhausting to supplementary firing in the form of auxiliary burner sections within four heat-recovery boilers. The HRBs generate su-

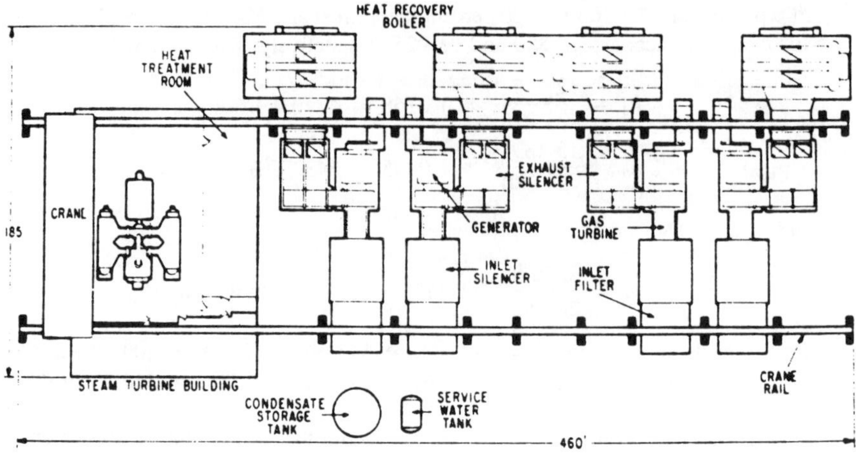

Figure 8-23 Layout of the STAG combined-cycle plant [72].

perheated steam for one steam turbine. The steam plant feedwater system includes one low-pressure closed-type feedwater heater with drain cascaded to the condenser and one open-type deaerating feedwater heater. The plant flow diagram can be schematically represented by Fig. 8-21, except that there are four gas turbines. Figure 8-23 shows the STAG plant layout. The plant data follow.

Gas turbines:	Four GE Model 7000, each rated at 49.5 MW base and 54.9 MW peak at 80°F (27°C) inlet.
Turbine exhaust:	970°F (521°C). Dampers used to bypass gas to atmosphere when operating alone, or to direct gas to the HRB when operating in combined-cycle mode. Silencers are located ahead of bypass stack and HRB.
HRB:	Four, single-pressure, burner-and-steam-generator sections are factory-assembled modules [80] for site erection. Forced-recirculation in boiler section.
Feedwater:	267°F (130°C) at economizer inlet.
Steam:	1250 psig (87 bar), 950°F (510°C), 995,200 lb_m/h (125 kg/s).
Steam turbine:	One high-pressure and one double-flow low-pressure, tandem-compound sections, nonreheat, rated at 129.6 MW with 3.5 inHg (0.12 bar) back pressure.
Fuel:	No. 2 distillate oil initially. Corrosion-resistant first-stage gas-turbine materials allow future use of heavier fuel.

STAG, as most combined cycles, has operational flexibility. Each of the four gas turbines and the steam turbine may be started, controlled, and loaded independently from a centralized control room. Either one or more gas turbines may be operated, each with or without its HRB and with its HRB supplementary fired or unfired. Gas-turbine start-ups are staggered by 30 s. Interlocks will prevent steam-turbine starting if no HRBs are operational. Steam pressures of 600, 800, 1000, and 1250 psig are used with 1, 2, 3, and 4 gas turbines and HRBs, respectively. The total gas-turbine output of 198 MW is available within 30 min and the total plant output of 330 MW within 1 hour after an overnight shutdown.

The efficiency and operational flexibility of the plant are illustrated by the heat-rate curves of Fig. 8-24. The top four short curves are indicative of operation on 1, 2, 3, 4 gas turbines, exhausting to the atmosphere, respectively. They have a heat-rate range of 13,000 to 13,500 Btu/kWh, which corresponds to thermal efficiencies of 26.3 to 25.3 percent. The curves labeled *F* are for part load combinations with the steam turbine and minimum HRB firing. Other notations are as explained in the table. It can be seen that best heat rate, at point A_1, for peak loads and full HRB firing, is 8700 Btu/kWh corresponding to 39.2 percent efficiency.

Because STAG is partly a gas-turbine plant, the cooling-tower and circulating-water makeup requirements are about half those of a conventional steam-cyclic plant of the same output. STAG had a short lead time from order to operation and was budgeted at \$53,300,000, corresponding to about \$160/kW capital cost.

Point	Mode: Gas Turbine	Mode: Steam Turbine	HRB Firing	GT Comp Flow
A_1	Peak	VWO	Full	Full
A	Peak	PL	Full	Full
B	Base	VWO	Full	Full
C	Base	PL	Full	Full
D	Base	PL	Min	Full
E	PL	PL	Min	Min
F	PL	PL	Min	Min

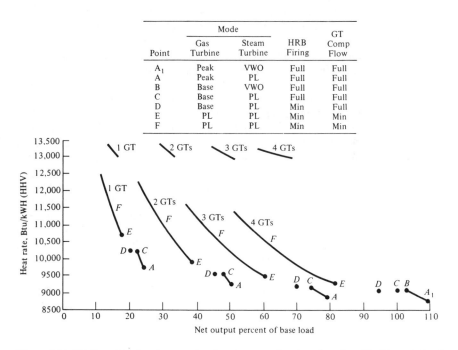

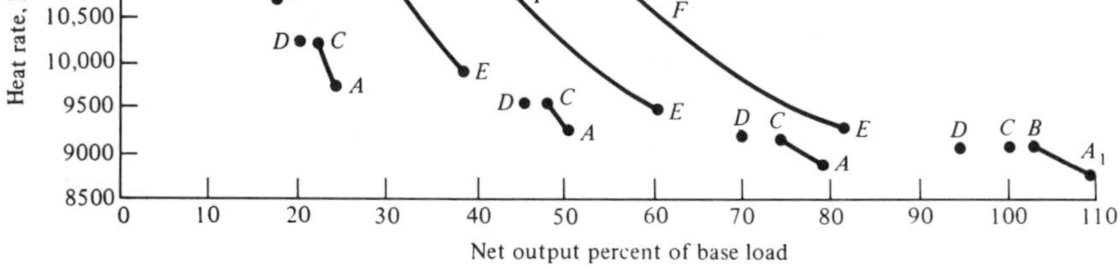

Figure 8-24 STAG combined-cycle powerplant heat rates as a function of number of turbines, mode of operation and plant load. Data for 80°F air and 3.5 inHg condenser pressure [72]. VWO = very-wide open throttle, PL = part load.

Another U.S. built combined-cycle plant is PACE, built for the St. Joseph Power and Light Company of northwestern Missouri [81].

8-10 COMBINED CYCLES WITH MULTIPRESSURE STEAM

A combined cycle with multipressure steam reduces the temperature of the gas leaving the heat-recovery boiler and hence results in increased efficiency of the plant as a whole. With steam cycles operating around 1300 psia (90 bar), saturation temperatures around 575°F (300°C), and feedwater into the HRB at about 265°F (130°C), the gas temperature leaving the HRB to the stack is still at about 300 to 400°F (150 to 200°C). Some of the energy leaving with that gas can be utilized in a multipressure steam cycle. The simplest such cycle is a dual-pressure* one, although triple-pressure cycles have been considered.

A *dual-pressure cycle* (Fig. 8-25) shows a heat-recovery boiler with two steam circuits in it. One, a high-pressure circuit, feeds steam to the steam turbine at its inlet; the other, a low-pressure circuit, feeds steam to the same turbine at a lower-pressure stage. A corresponding temperature-enthalpy diagram of both gas and steam circuits in the HRB is shown in Fig. 8-26.

Exhaust gas leaving the gas turbine enters supplementary firing (SF) at 4 and the heat-recovery boiler (HRB) at 5, leaving it to the stack (ST) at 6. Condensate leaves the steam condenser (C) at 8 and enters the condensate pump (CP) and two closed-type feedwater heaters (FWH) and one open-type deaerating heater (DA). It then enters the boiler feed pump (BFP) at 9, where it is pumped to 10 to a lower pressure than that of steam maximum. Process 10-11 is feedwater heating in a low-pressure economizer, followed by evaporation to 12 and superheat to 13. Superheated low-pressure steam at 13 enters the steam turbine at a low-pressure stage.

Water from the low-pressure steam drum at 11 is pumped by a booster pump (BP) to 14 and goes to the high-pressure economizer. Evaporation occurs from 15 to 16 and superheat to 17. High-pressure superheated steam at 17 enters the steam turbine first stage.

It can be seen from the *T-H* diagram (Fig. 8-26) that low-pressure steam boils at a temperature (12) below that of high-pressure steam (16), and hence there are two pinch points between the gas line and the saturated steam lines. It can also be seen that a single high-pressure steam circuit would be represented by 10′-15-16-17 with gas leaving to the stack at 6′. Adding the low-pressure circuit allows the gas to leave at a lower temperature (6), thus extracting more energy from it and increasing the overall cycle efficiency.

An example of a dual-pressure combined cycle is the Donge-Geertruidenberg plant of PNEM in Holland [78]. It has a gas-turbine output of 76.7 MW and a steam-turbine

* The idea is not unique to combined cycles. Low-temperature gas-cooled-reactor powerplants of the British Magnox type (Sec. 10-11) have used dual-pressure cycles for a similar purpose. An early boiling-water-reactor powerplant (Dresden I) used dual pressure, but for a different purpose [3].

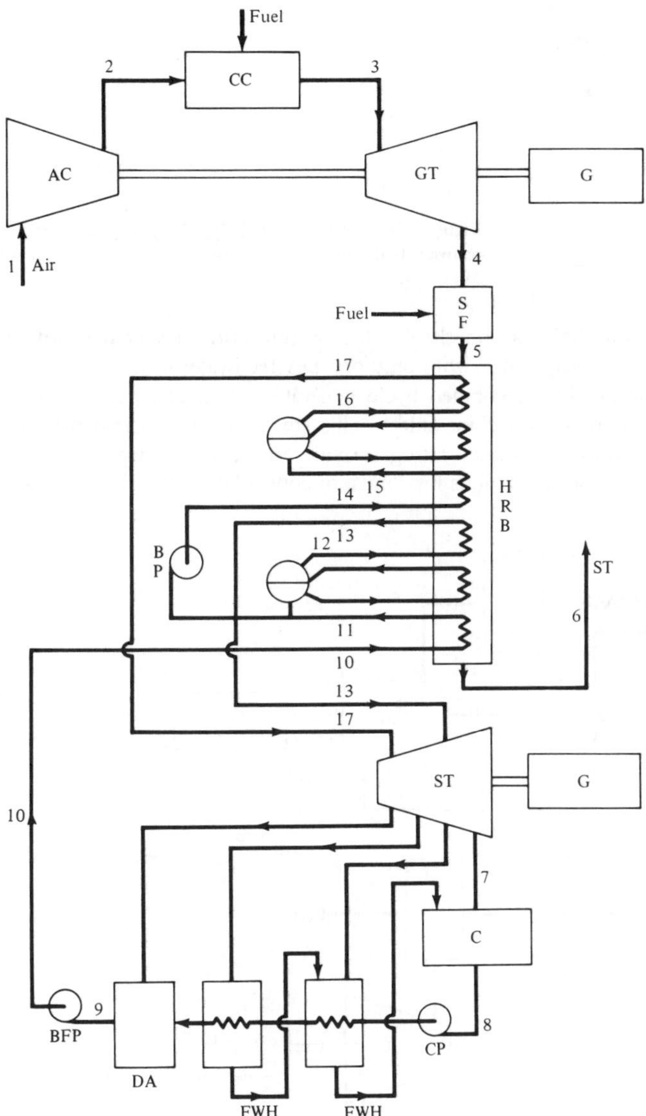

Figure 8-25 A schematic diagram for a dual-pressure combined cycle.

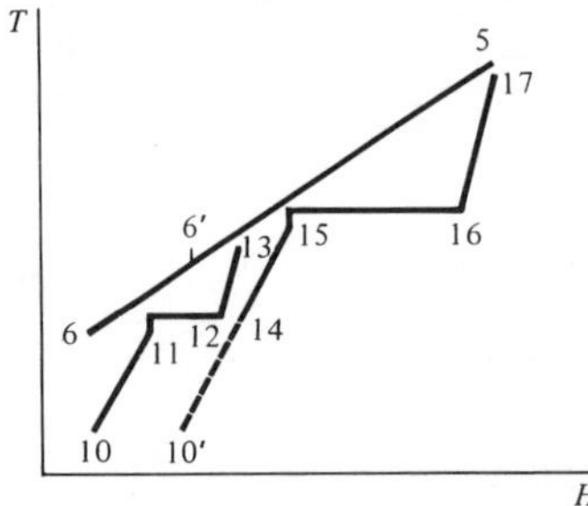

Figure 8-26 Temperature-enthalpy (T-H) diagram of the heat recovery boiler of the dual-pressure combined cycle shown in Fig. 8-25.

output of 47.3 MW and attains a remarkable 46.1 percent efficiency at standard air conditions of 15°C (59°F) and 1 atm, with only one DA feedwater heater.

A proposed triple-pressure combined cycle generates steam at an intermediate pressure between the two steam-turbine inlets. This steam is injected into the gas-turbine combustion chamber to reduce nitrogen oxide emissions to meet NO_x rigid standards, if applicable. The water, thus lost, must of course be continually made up.

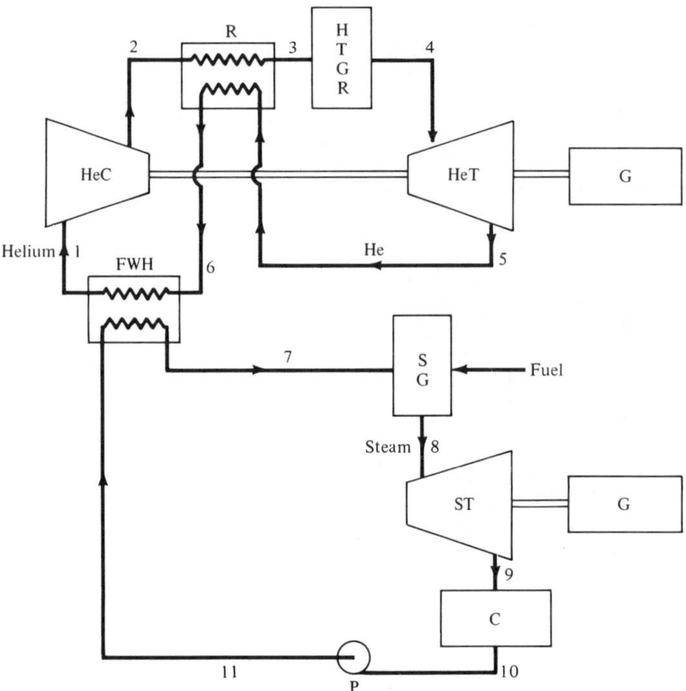

Figure 8-27 A schematic diagram of a combined cycle with a nuclear gas turbine and fossil-fuel-fired steam turbine [75].

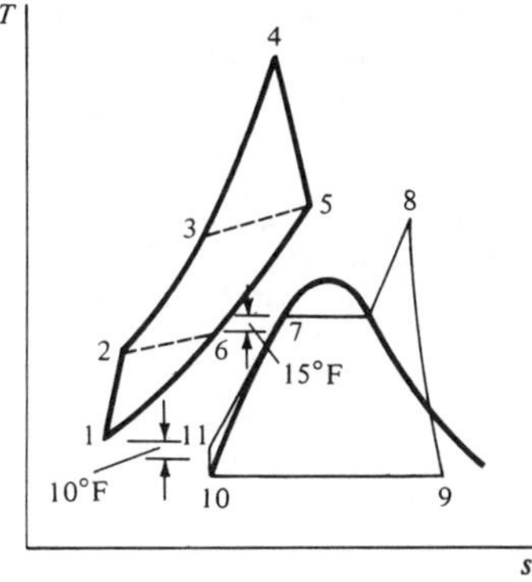

Figure 8-28 *T-s* diagrams of helium and steam cycles shown in Fig. 8-27.

8-11 A COMBINED CYCLE FOR NUCLEAR POWERPLANTS

A combined cycle for nuclear plants [82] presupposes that a high-temperature gas-cooled nuclear reactor (HTGR) (Sec. 10-12) is the heat source for the gas-turbine cycle. Such a reactor-turbine combination uses helium gas as the reactor coolant and gas-turbine cycle working fluid in a closed cycle.

Helium at 1 (Figs. 8-27 and 8-28) is compressed in a helium compressor (HeC) to 2. It is then preheated in a regenerator (R) to 3, enters the HTGR, and leaves at 4 at 1435 to 1470°F (780 to 800°C). It expands in the helium turbine (HeT) to 5 and enters the regenerator. The energy left in the gas at 6 is finally transferred to the steam cycle in a closed-type steam feedwater heater (FWH). The helium, back at 1, reenters the helium compressor.

The steam cycle is fairly standard. Feedwater leaves the FWH at 7, enters the fossil-fueled steam generator (SG), and leaves it as superheated steam at 8. This expands in the steam turbine (ST) and enters the condenser at 9. Condensate at 10 is pumped by pump (P), after which it enters the FWH to repeat the cycle.

It can be seen that the gas and steam cycles are coupled only by the FWH, that the heat generated in the HTGR is completely utilized in both cycles, and that heat is rejected only in the steam condenser.

PROBLEMS

8-1 Show that the efficiency of an ideal Brayton cycle at the optimum pressure ratio is a function of minimum and maximum temperatures only, and independent of the working fluid.

8-2 It is required to compare the optimum pressure ratios and corresponding efficiencies of three ideal Brayton cycles using air, helium, and carbon dioxide as working fluids. The minimum and maximum temperatures are 10°C and 1115°C, respectively. Use properties at low temperatures.

8-3 A gas turbine cycle operates with air only with a constant specific heat of 0.24 Btu/lb$_m$ · °R. The inlet air is at 14.696 psia and 60°F. The maximum cycle temperature is 1800°F. The compressor and turbine have the same pressure ratio of 8 and polytropic efficiencies of 0.85. A regenerator with 75 percent effectiveness is considered. Calculate (*a*) all temperatures around the cycle, in degrees Fahrenheit, (*b*) the efficiency of the cycle, in percent, (*c*) the heat added, in Btus per hour, to produce 10-MW power, (*d*) the efficiency of the cycle if no regenerator is used, (*e*) the efficiency of a comparable ideal cycle with a perfect regenerator.

8-4 A Brayton cycle operating on air only with a constant $c_p = 0.24$ Btu/lb$_m$ · °R, has compressor and turbine polytropic efficiencies of 0.8 and 0.9, respectively, and the same pressure ratio of 6.0. The inlet and maximum temperatures are 40 and 1940°F, respectively. A regenerator with 85 percent effectiveness is used. Calculate (*a*) the net work, in Btu/lb$_m$ (*b*) the cycle efficiency, and (*c*) the air mass flow rate for a 20-MW output.

8-5 A Brayton cycle uses helium as a working fluid with a mass flow rate of 200 lb$_m$/s. The compressor and turbine polytropic efficiencies are 0.8 and 0.9 and their pressure ratios are 2.5 and 2.4, respectively. The inlet and maximum temperatures are 100 and 2000°F, respectively. A regenerator with 85 percent effectiveness is used. Calculate (*a*) the cycle efficiency, (*b*) the percent improvement in cycle efficiency due to regeneration, and (*c*) the cycle power, in megawatts.

8-6 A helium gas turbine cycle has compressor and turbine polytropic efficiencies of 0.8 and 0.9 and pressure ratios of 2.5 and 2.4, respectively. The compressor has one stage of intercooling and the turbine has one stage of reheat. The cycle has a regenerator with 85 percent effectiveness. The cycle minimum and maximum temperatures are 100 and 2000°F, respectively. Calculate (*a*) the cycle efficiency, and (*b*) the cycle power for a helium mass flow rate of 200 lb$_m$/s.

8-7 Air at 14.696 psia, 40°F, and with 65 percent relative humidity enters the compressor of a gas turbine cycle. The compressor and turbine have the same pressure ratio of 6 and polytropic efficiencies of 0.80 and 0.90, respectively. Water at 60°F is injected into the compressor exit air, saturating it. Calculate (*a*) the air temperature after water injection (*b*) the percent increase in mass flow rate due to water injection, (*c*) the compressor work in Btus per pound mass of original air, and (*d*) the compressor work if water is injected during the compression process to the same temperature as (*a*), in Btus per pound mass of original air. Use a constant $c_p = 0.24$ Btu/lb$_m$ · °R.

8-8 Air at 14.696, 40°F, and with 65 percent relative humidity enters the compressor of a gas turbine cycle. The compressor and turbine have the same pressure ratio of 6 and polytropic efficiencies of 0.80 and 0.90, respectively. Water at 60°F is injected into the compressor exit air, saturating it. The air then goes through an 85 percent effective regenerator. The turbine inlet temperature is 1940°F. Calculate (*a*) the heat added, in Btus per pound mass of original air, (*b*) the cycle efficiency, (*c*) the original air mass flow rate, in pound mass per hour, for a 20-MW net cycle output, and (*d*) the necessary water injection rate, in pound mass per hour. For simplicity, use a constant $c_p = 0.24$ Btu/lb$_m$ · °R. Compare these results with those of Prob. 8-4 that have the same data but no water injection.

8-9 A simple Brayton cycle without regeneration, intercooling or reheat has compressor and turbine pressure ratios of 9.286 and 8.276, polytropic efficiencies of 0.862 and 0.904, and minimum and maximum temperatures of 40 and 2040°F. Air enters the compressor at the rate of 6×10^6 lb$_m$/h. The combustion gases correspond to 400 percent theoretical air. Find the net cycle power, in megawatts, and cycle efficiency using (*a*) a constant specific heat of 0.24 Btu/lb$_m$ · °R and constant $k = 1.4$, and (*b*) the gas tables. The molecular weight of the combustion gases = 28.925. The ratio of products to air by mass = 1.0168.

8-10 A 50-MW combustion gas turbine cycle using 200 percent of theoretical air has one stage of intercooling no reheat and a 79.8 percent effective regenerator. The compressor and turbine pressure ratios are 8.63 and 8.231 and have polytropic efficiencies of 90 and 90.52 percent, respectively. The inlet and maximum temperatures are 40 and 2140°F. Using the gas tables, calculate (*a*) the net cycle work, in Btus per pound mass air, (*b*) the cycle efficiency, and (*c*) the air mass flow rate, in pound mass per hour, if the mechanical and electrical efficiencies are 95 percent each.

8-11 A combustion Brayton cycle uses 200 percent of theoretical air. The cycle has an inlet temperature of 500°R and a turbine inlet temperature of 2000°R. The compression ratios in the compressor and turbine are assumed equal to 9. One stage of intercooling and one stage of reheat are used. Assume that all rotary machines have efficiencies of 0.85. A 0.85 effective regenerator is used. Calculate the net work of the cycle and the overall efficiency, using the gas tables.

8-12 It is required to compare the three methods of air cooling of the blades of high-temperature combustion gas turbines: (1) convective, (2) impingement, and (3) combined convection and film (transpiration). Assume for all cases that the blades have a mean surface area of 100 cm^2 and that the blade surface temperature is not to exceed 900°C, with a cooling air inlet temperature to the blades of 300°C. Assume also that the cooling air mass flow rate is 0.02 kg/s per blade and that the heat transfer coefficient is the same in all cases, 2.0 $kW/m^2 \cdot C$. Estimate *(a)* the permissible maximum turbine gas temperature for each of three methods and *(b)* the net work, in kilojoules per kilogram of air, and the cycle efficiency for methods 1 and 3 only. Take $k = 1.4.$ and 1.314 for compressor and turbine, and $c_p = 1.005$, 1.2267 and 1.1057 kJ/kg · K for compressor, turbine, and heat addition, respectively.

8-13 Consider a combined gas-steam turbine cycle with both machines ideal. Air enters the compressor at 1 atm and 500°R. 200 percent theoretical air gases enter the gas turbine at 2400°R. The pressure ratio for both compressor and turbine is 5.624. No intercooling, reheat, or regeneration are used in the gas-turbine cycle. The turbine exhaust is used directly without supplementary firing and leaves to the stack at 1000°R. Steam is produced at 1000 psia and 1000°F. The condenser pressure is 1 psia. No feedwater heating is used and the pump work may be ignored, Draw the flow and *T-s* diagrams labeling points correspondingly, and calculate *(a)* the heat added, per pound mass of air, *(b)* the steam flow per pound mass of air, *(c)* the combined work, in Btus per pound mass of air, *(d)* the combined plant efficiency, and *(e)* the plant efficiency if the steam cycle is inoperative.

8-14 A combined gas-steam-turbine powerplant is designed with four 50-MW gas turbines and one 120-MW steam turbine. Each gas turbine operates with compressor inlet temperature 505°R, turbine inlet temperature 2450°R, pressure ratio for both compressor and turbine 5, and compressor and turbine polytropic and mechanical efficiencies 0.87 and 0.96. The gases leaving the turbines go to a heat-recovery boiler then to a regenerator with an effectiveness of 0.87. The turbine gases correspond to 200 percent of theoretical air. The steam cycle has a turbine steam inlet at 1200 psia and 1460°R, one open-type feedwater heater (not optimally placed) with feedwater temperature to heat-recovery boiler at 920°R, condenser pressure 1 psia, and turbine polytropic and mechanical efficiencies 0.87 and 0.96. All electric generator efficiencies are 0.96. Supplementary firing at full load raises the gas temperature to 2000°R. Draw the cycle flow and *T-s* diagrams and find *(a)* the required steam mass flow rate in the steam turbine in pound mass per hour, *(b)* the required air mass flow rate in each gas turbine in pound mass per hour, *(c)* the heat added in the gas cycle and in supplementary firing at full load, *(d)* the stack gas temperature in degrees Fahrenheit, *(e)* the cycle efficiency at full load, and *(f)* the efficiency at startup when only one gas turbine is used at its full load with no supplementary firing or regeneration. (Ignore the steam cycle pump work.)

8-15 Consider the steam generator in a combined cycle of the dual-pressure variety (Fig. 8-25). The combustion gases, corresponding to 200 percent theoretical air, leave supplementary firing at 1500°R and the steam generator at 810°R. Steam is generated at 1000 psia and 1000°F, and 200 psia and 500°F, from feedwater at 300°F. The high- and low-pressure steam mass flow rates are equal. Using the gas tables, calculate *(a)* the gas mass flow rate, in pound mass per pound mass of total steam, and *(b)* the gas exit temperature, in degrees Fahrenheit, if only high-pressure steam is generated, for the same gas mass flow rate. The gas molecular weight = 28.880.

8-16 A combined cycle of the type shown in Fig. 8-27 uses helium in the gas turbine portion. The compressor and turbine have the same pressure ratio of 2.0, the same polytropic efficiencies of 0.85, and inlets at 140 and 2040°F, respectively. The regenerator effectiveness is 0.85. The steam cycle operates with 1000 psia and 1000°F steam and condenses at 1 psia. The steam turbine polytropic efficiency is 0.85. The feedwater heater has a terminal temperature difference of 28°F. Calculate *(a)* the helium mass flow rate, in pound mass per hour, for a combined output of 200 MW, and *(b)* the gas, steam, and combined cycle efficiencies. Ignore the steam cycle pump work.

CHAPTER

NINE

PRINCIPLES OF NUCLEAR ENERGY

9-1 INTRODUCTION

There is strategic as well as economic necessity for nuclear power in the United States and indeed most of the world. The strategic importance lies primarily in the fact that one large nuclear powerplant saves more than 50,000 barrels of oil per day. At $30 to $40 per barrel (1982), such a powerplant would pay for its capital cost in a few short years. For those countries that now rely on but do not have oil, or must reduce the importation of foreign oil, these strategic and economic advantages are obvious. For those countries that are oil exporters, nuclear power represents an insurance against the day when oil is depleted. A modest start now will assure that they would not be left behind when the time comes to have to use nuclear technology.

The unit costs per kilowatthour for nuclear energy are now comparable to or lower than the unit costs for coal in most parts of the world. Other advantages are the lack of environmental problems that are associated with coal- or oil-fired powerplants and the near absence of issues of mine safety, labor problems, and transportation bottlenecks. Natural gas is a good, relatively clean-burning fuel, but it has some availability problems in many countries and should, in any case, be conserved for small-scale industrial and domestic uses. Thus nuclear power is bound to become the social choice relative to other societal risks and overall health and safety risks.

Other sources include hydroelectric generation, which is nearly fully developed with only a few sites left around the world with significant hydroelectric potential. Solar power, although useful in outer space and domestic space and water heating in some parts of the world, is not and will not become an economic primary source of electric power.

Yet the nuclear industry is facing many difficulties, particularly in the United States, primarily as a result of the negative impact of the issues of nuclear safety, waste disposal, weapons proliferation, and economics on the public and government. The impact on the public is complicated by delays in licensing proceedings, court, and ballot box challenges. These posed severe obstacles to electric utilities planning nuclear powerplants, the result being scheduling problems, escalating and unpredictable costs, and economic risks even before a construction permit is issued. Utilities had to delay or cancel nuclear projects so that in the early 1980s there was a de facto moratorium on new nuclear plant commitments in the United States.

It is, however, the opinion of many, including this author, that despite these difficulties the future of large electric-energy generation includes nuclear energy as a primary, if not the main, source. The signs are already evident in many European and Asian countries such as France, the United Kingdom, Japan, and the U.S.S.R.

In a powerplant technology course, it is therefore necessary to study nuclear energy systems. We shall begin in this chapter by covering the energy-generation processes in nuclear reactors by starting with the structure of the atom and its nucleus and the reactions that give rise to such energy generation. These include fission, fusion, and different types of neutron-nucleus interactions and radioactivity. The following two chapters will cover the two main classes of nuclear powerplants in use or under active development in the world today, the thermal and the fast-breeder reactor powerplants.

9-2 THE ATOMIC STRUCTURE

In 1803 John Dalton, attempting to explain the laws of chemical combination, proposed his simple but incomplete *atomic hypothesis*. He postulated that all elements consisted of indivisible minute particles of matter, *atoms,** that were different for different elements and preserved their indentity in chemical reactions. In 1811 Amadeo Avogadro introduced the molecular theory based on the *molecule,* a particle of matter composed of a finite number of atoms. It is now known that the atoms are themselves composed of subparticles, common among atoms of all elements.

An atom consists of a relatively heavy, positively charged *nucleus* and a number of much lighter negatively charged *electrons* that exist in various orbits around the nucleus. The nucleus, in turn, consists of subparticles, called *nucleons*. Nucleons are primarily of two kinds: the *neutrons,* which are electrically neutral, and the *protons,* which are positively charged. The electric charge on the proton is equal in magnitude but opposite in sign to that on the electron. The atom as a whole is electrically neutral: the number of protons equals the number of electrons in orbit.

One atom may be transformed into another by losing or acquiring some of the above subparticles. Such reactions result in a change in mass Δm and therefore release (or absorb) large quantities of energy ΔE, according to Einstein's law

* Much earlier, in the fifth century B.C., the Greek Democritus declared that the simplest thing out of which everything is made is an atom. (The Greek word *atomos* means "uncut".)

$$\Delta E = \frac{1}{g_c} \Delta mc^2 \qquad (9\text{-}1)$$

where c is the speed of light in vacuum and g_c is the familiar engineering conversion factor (Sec. 9-4). Equation (9-1) applies to *all* processes, physical, chemical, or nuclear, in which energy is released or absorbed. Energy is, however, classified as *nuclear* if it is associated with changes in the atomic nucleus.

Figure 9-1 shows three atoms. Hydrogen has a nucleus composed of one proton, no neutrons, and one orbital electron. It is the only atom that has no neutrons. Deuterium has one proton and one neutron in its nucleus and one orbital electron. Helium contains two protons, two neutrons, and two electrons. The electrons exist in orbits, and each is quantitized as a lumped unit charge as shown.

Most of the mass of the atom is in the nucleus. The masses of the three primary atomic subparticles are

$$\text{Neutron mass } m_n = 1.008665 \text{ amu}$$

$$\text{Proton mass } m_p = 1.007277 \text{ amu}$$

$$\text{Electron mass } m_e = 0.0005486 \text{ amu}$$

The abbreviation amu, for *atomic mass unit,* is a unit of mass approximately equal to 1.66×10^{-27} kg, or 3.66×10^{-27} lb_m. These three particles are the primary building blocks of all atoms. Atoms differ in their mass because they contain varying numbers of them.

Atoms with nuclei that have the same number of protons have similar chemical and physical characteristics and differ mainly in their masses. They are called *isotopes*. For example, deuterium, frequently called *heavy hydrogen,* is an isotope of hydrogen. It exists as one part in about 6660 in naturally occurring hydrogen. When combined with oxygen, ordinary hydrogen and deuterium form *ordinary water* (or simply water) and *heavy water,* respectively.

The number of protons in the nucleus is called the *atomic number Z*. The total number of nucleons in the nucleus is called the *mass number A*. As the mass of a

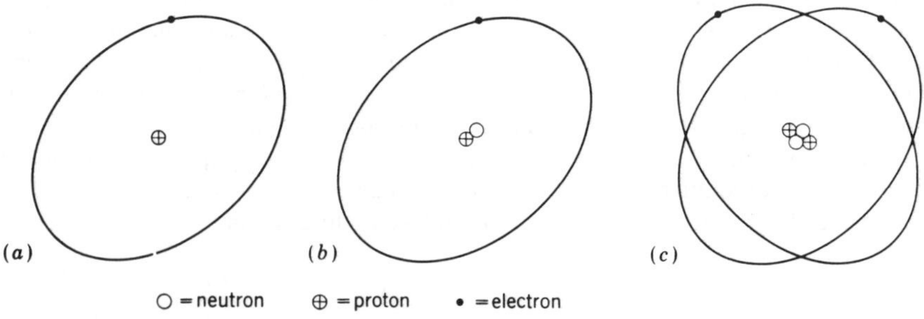

Figure 9-1 Structure of some light atoms: (*a*) hydrogen; (*b*) deuterium or heavy hydrogen, and (*c*) helium.

neutron or a proton is nearly 1 amu, A is the integer nearest the mass of the nucleus, which in turn is approximately equal to the atomic mass of the atom. Isotopes of the same element thus have the same atomic number but differ in mass number. Nuclear symbols are written conventionally as

$$_{Z}X^{A}$$

where X is the usual chemical symbol. Thus the hydrogen nucleus is $_1H^1$, deuterium is $_1H^2$ (and sometimes D), and ordinary helium is $_2He^4$. For particles containing no protons, the subscript indicates the magnitude and sign of the electric charge. Thus an electron is $_{-1}e^0$ (sometimes e^- or β^-) and a neutron is $_0n^1$. Symbols are also often written in the form He-4, helium-4, etc. Another system of notation, written as $^{A}_{Z}X$, will not be used in this text.

Many elements (such as hydrogen, above) appear in nature as mixtures of isotopes of varying abundances. For example, naturally occurring uranium, called *natural uranium,* is composed of 99.282 mass percent U^{238}, 0.712 mass percent U^{235}, and 0.006 mass percent U^{234}, where the atomic number is deleted. It is 92 in all cases. Many isotopes that do not appear in nature are synthesized in the laboratory or in nuclear reactors. For example, uranium is known to have a total of 14 isotopes that range in mass numbers from 227 to 240.

The known elements, their chemical symbols, and their atomic numbers are listed alphabetically in App. J. Figure 9-2 shows, schematically, the structure of H^1, He^4, and some heavier atoms and the distribution of their electrons in various orbits.

Two other particles of importance are the positron and the neutrino. The *positron* is a positively charged electron having the symbols $_{+1}e^0$, e^+, or β^+. The *neutrino* (little neutron) is a tiny electrically neutral particle that is difficult to observe experimentally. Initial evidence of its existence was based on theoretical considerations. In nuclear reactions where a β particle of either kind is emitted or captured, the resulting energy (corresponding to the lost mass) was not all accounted for by the energy of the emitted β particle and the recoiling nucleus. It was first suggested by Wolfgang Pauli in 1934 that the neutrino was simultaneously ejected in these reactions and that it carried the balance of the energy, often larger than that carried by the β particle itself. The importance of neutrinos is that they carry some 5 percent of the total energy produced in fission. This energy is completely lost because neutrinos do not react and are not stopped by any practical structural material. The neutrino is given the symbol ν.

There are many other atomic subparticles. An example is the *mesons,* unstable positive, negative, or neutral particles that have masses intermediate between an electron and a proton. They are exchanged between nucleons and are thought to account for the forces between them. A discussion of these and other subparticles is, however, beyond the scope of this book.

Electrons that orbit in the outermost shell of an atom are called *valence electrons*. The outermost shell is called the *valence shell*. Thus, hydrogen has one valence electron and its K shell is the valence shell, etc. Chemical properties of an element are a function of the number of valence electrons. The electrons play little or not part in nuclear interactions.

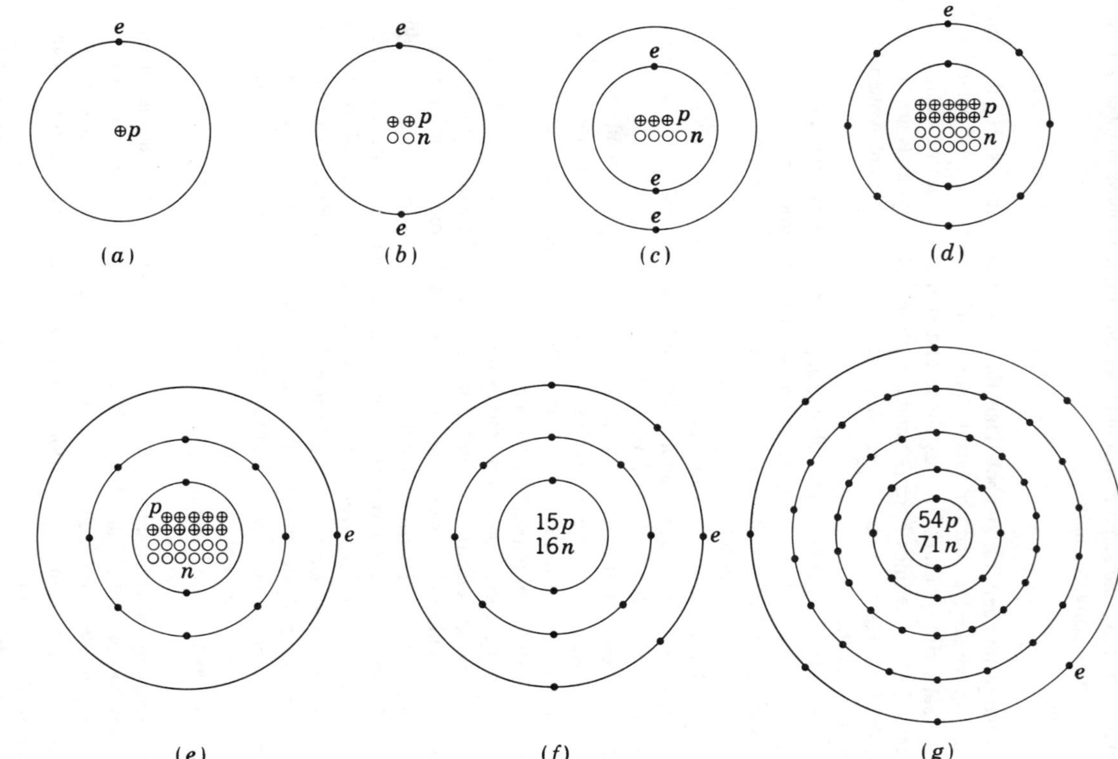

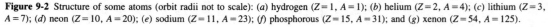

Figure 9-2 Structure of some atoms (orbit radii not to scale): (*a*) hydrogen ($Z=1$, $A=1$); (*b*) helium ($Z=2$, $A=4$); (*c*) lithium ($Z=3$, $A=7$); (*d*) neon ($Z=10$, $A=20$); (*e*) sodium ($Z=11$, $A=23$); (*f*) phosphorous ($Z=15$, $A=31$); and (*g*) xenon ($Z=54$, $A=125$).

9-3 CHEMICAL AND NUCLEAR EQUATIONS

Chemical reactions involve the combination or separation of whole atoms. For example

$$C + O_2 \rightarrow CO_2 \tag{9-2}$$

This reaction is accompanied by the release of about 4 electron volts (eV). An *electron volt* is a unit of energy in common use in nuclear engineering. 1 eV = 1.6021 × 10^{-19} joules (J) = 1.519 × 10^{-22} Btu = 4.44 × 10^{-26} kWh. 1 *million electron volts* (1 MeV) = 10^6 eV.

In chemical reactions, each atom participates as a whole and retains its identity. The molecules change. The only effect is a sharing or exchanging of valence electrons. The nuclei are unaffected. In chemical equations there are as many atoms of each participating element in the products (the right-hand side) as in the reactants (the left-hand side). Another example is one in which uranium dioxide (UO_2) is converted into uranium tetrafluoride (UF_4), called green salt, by heating it in an atmosphere of highly corrosive anhydrous (without water) hydrogen fluoride (HF), with water vapor (H_2O) appearing in the products

$$UO_2 + 4HF \rightarrow 2H_2O + UF_4 \tag{9-3}$$

Water vapor is driven off and UF_4 is used to prepare gaseous uranium hexafloride (UF_6), which is used in the separation of the U^{235} and U^{238} isotopes of uranium by the gaseous diffusion method. (Fluorine has only one isotope, F^{19}, and thus combinations of molecules of uranium and fluorine have molecular masses depending only on the uranium isotope.)

Both chemical and nuclear reactions are either *exothermic* or *endothermic,* that is, they either release or absorb energy. Because energy and mass are convertible, Eq. (9-1), chemical reactions involving energy *do* undergo a mass decrease in exothermic reactions and a mass increase in endothermic ones. However, the quantities of energy associated with a chemical reaction are very small compared with those of a nuclear reaction, and the mass that is lost or gained is minutely small. This is why we assume a preservation of mass in chemical reactions, undoubtedly an incorrect assumption but one that is sufficiently accurate for usual engineering calculations.

In nuclear reactions, the reactant nuclei do not show up in the products, instead we may find either isotopes of the reactants or other nuclei. In balancing nuclear equations it is necessary to see that the same, or equivalent, nucleons show up in the products as entered the reaction. For example, if K, L, M, and N were chemical symbols, the corresponding nuclear equation might look like

$$_{Z1}K^{A1} + {}_{Z2}L^{A2} \rightarrow {}_{Z3}M^{A3} + {}_{Z4}N^{A4} \tag{9-4}$$

To balance Eq. (9-4), the following relationships must be satisfied.

$$Z_1 + Z_2 = Z_3 + Z_4 \tag{9-5a}$$

and

$$A_1 + A_2 = A_3 + A_4 \tag{9-5b}$$

Sometimes the symbols γ or ν are added to the products to indicate the emission of electromagnetic radiation or a neutrino, respectively. They have no effect on equation balance because both have zero Z and A, but they often carry large portions of the resulting energy.

Although the mass numbers are preserved in a nuclear reaction, the masses of the isotopes on both sides of the equation do not balance. Exothermic or endothermic energy is obtained when there is a reduction or an increase in mass from reactants to products, respectively.

Example 9-1 One exothermic reaction occurs when common aluminum is bombarded with high-energy α particles (helium-4 nuclei), resulting in Si^{30} (a heavy isotope of silicon whose most abundant isotope has mass number 28). In the reaction, a small particle is emitted. Write the complete reaction and calculate the change in mass.

SOLUTION The reaction is

$$_{13}Al^{27} + {}_2He^4 \rightarrow {}_{14}Si^{30} + {}_{Z,4}X^{A,4}$$

where X is a symbol of a yet unknown particle. Balancing gives

$$Z_4 = 13 + 2 - 14 = 1 \qquad \text{and} \qquad A_4 = 27 + 4 - 30 = 1$$

The only particle satisfying these is a proton. Thus the complete reaction is

$$_{13}Al^{27} + {}_2He^4 \rightarrow {}_{14}Si^{30} + {}_1H^1 \tag{9-6}$$

The isotope masses of the nuclei showing up in this reaction are:

Reactants		Products	
Al^{27}	26.98153 amu	Si^{30}	29.97376 amu
He^4	4.00260 amu	H^1	1.00783 amu
Total	30.98413 amu	Total	30.98159 amu

Thus there is a *decrease* in mass, as $\Delta m = 30.98159 - 30.98143 = -0.00254$ amu. The corresponding energy is negative; i.e., energy is released or is exothermic. In nuclear reactions, the results depend on a small difference between large numbers, which makes it necessary to carry the isotope masses to the fourth or fifth decimal places.

An example of an endothermic nuclear reaction is

$$_7N^{14} + {}_2He^4 \rightarrow {}_8O^{17} + {}_1H^1 \tag{9-7}$$

The sum of the masses of these reactants and products are 14.00307 + 4.00260 = 18.00567 amu and 16.99914 + 1.00783 = 18.00697 amu, respectively. Thus there is a net gain in mass of 0.00130 amu, which means that energy is absorbed and the reaction is endothermic.

In the above two reactions, the positively charged α particles must be accelerated to high kinetic energies to overcome electrical repulsion and bombard the positively charged aluminum or nitrogen nuclei. The reactants possess initial kinetic energy equal to the kinetic energy of the α particle plus the kinetic energy of the nucleus, though the latter is usually negligible. (This process is analogous to raising a fuel-air mixture to its ignition temperature by adding activation energy before combustion can take place.) When the reactions are completed, the energy released will be equal to the initial energy of the reactants plus the energy corresponding to the lost mass (or minus the energy corresponding to the gained mass).

This energy shows up in the form of kinetic energy of the resultant particles, in the form of γ energy, and sometimes as *excitation energy* of the product nucleus, if any become so excited. The total kinetic energy of the products is divided among the nuclei and particles in such a manner that the lighter particles have higher kinetic energies than the heavier ones.

The isotope masses used above included the masses of the orbital electrons. The nuclear masses can be computed by subtracting the sum of the masses of Z orbital electrons. For example, the mass of the Al^{27} nucleus $= 26.98153 - 13 \times 0.0005486 = 26.97440$ amu, and so on. Such corrections are unnecessary in most cases because the same number of electrons show up on both sides of the equation. For example, in Eq. (9-6), the energy produced corresponds to the change in masses of the nuclei as given by

$$\Delta m = [(M_{\text{Si}} - 14m_e) + (M_{\text{H}} - m_e)] - [(M_{\text{Al}} - 13m_e) + (M_{\text{He}} - 2m_e)]$$

where M is the isotope atomic mass and m_e the mass of the electron. It can be seen that the number of electrons balance and that

$$\Delta m = (\text{M}_{\text{Si}} + M_{\text{H}}) - (M_{\text{Al}} + M_{\text{He}})$$

The priniciple holds even if neutrons (whose mass, 1.008665 amu, does not include any electrons) are involved. In general then

$$\Delta m = \Sigma M_{\text{products}} - \Sigma M_{\text{reactants}} \tag{9-8}$$

and the electron masses are neglected. This rule applies even if an electron appears on either side of the equation. An example is

$${}_{16}\text{S}^{35} \rightarrow {}_{17}\text{Cl}^{35} + {}_{-1}e^{0} \tag{9-9}$$

In this case

$$\Delta m = [(M_{\text{Cl}} - 17m_e) + m_e] - (M_{\text{S}} - 16m_e) = M_{\text{Cl}} - M_{\text{S}}$$

An *exception,* however, is in reactions involving positrons

$${}_{6}\text{C}^{11} \rightarrow {}_{5}\text{B}^{11} + {}_{+1}e^{0} \tag{9-10}$$

In this case

$$\Delta m = [(M_{\text{B}} - 5m_e) + m_e] - (M_{\text{C}} - 6m_e) = M_{\text{B}} - M_{\text{C}} + 2m_e$$

Two electron masses are added if the positron is on the right-hand side of the equation and subtracted if it is on the left-hand side.

9-4 ENERGY FROM NUCLEAR REACTIONS

The energy corresponding to the change in mass in a nuclear reaction is calculated from Einsteins law, Eq. (9-1), here repeated

$$\Delta E = \frac{1}{g_c} \Delta mc^2 \tag{9-1}$$

where g_c is a conversion factor* that has the following values

1.0	$kg \cdot m/(N \cdot s^2)$
1.0	$g \cdot cm^2/(erg \cdot s^2)$
32.2	$lb_m \cdot ft/(lb_f \cdot s^2)$
4.17×10^8	$lb_m \cdot ft/(lb_f \cdot hr^2)$
0.965×10^{18}	$amu \cdot cm^2/(MeV \cdot s^2)$

Thus if Δm is in kilograms and c in meters per second, ΔE will be in joules. Since $c = 3 \times 10^8$ m/s, Eq. (9-1) can be written in the form

$$\Delta E \text{ (in J)} = 9 \times 10^{16}\ \Delta m \text{ (in kg)} \tag{9-11}$$

But as it is convenient to express the masses of nuclei in amu $= 1.66 \times 10^{-27}$ kg and the energy in joules (J) and MeV, Eq. (9-11) becomes

$$\Delta E \text{ (in J)} = 1.49 \times 10^{-10}\ \Delta m \text{ (in amu)} \tag{9-12}$$

and

$$\Delta E \text{ (in MeV)} = 931\ \Delta m \text{ (in amu)} \tag{9-13}$$

a useful relationship to remember. The reaction in Example 9-1 thus produces $-0.00254 \times 931 = -2.365$ MeV of energy. Mass-energy conversion factors are given in Table 9-1.

9-5 NUCLEAR FUSION AND FISSION

Nuclear reactions of importance in energy production are *fusion, fission,* and *radioactivity*. Radioactivity will be discussed in Sec. 9-8. In *fusion,* two or more light nuclei fuse to form a heavier nucleus. In *fission,* a heavy nucleus is split into two or more lighter nuclei. In both, there is a decrease in mass resulting in exothermic energy.

* The same as in force $= 1/g_c \times$ mass $\times$ acceleration.

Table 9-1 Mass-energy conversion factors

	Energy				
Mass	MeV	J	Btu	kWh	MW · day
amu	931.478	1.4924×10^{-10}	1.4145×10^{-13}	4.1456×10^{-17}	9.9494×10^{-13}
kg	5.6094×10^{29}	8.9873×10^{16}	8.5184×10^{13}	2.4965×10^{10}	5.9916×10^{14}
lb_m	2.5444×10^{29}	4.0766×10^{16}	3.8639×10^{13}	1.1324×10^{10}	2.7177×10^{14}

Fusion

Energy is produced in the sun and stars by continuous fusion reactions in which four nuclei of hydrogen fuse in a series of reactions involving other particles that continually appear and disappear in the course of the reactions, such as He^3, nitrogen, carbon, and other nuclei, but culminating in one nucleus of helium and two positrons

$$4\,{}_1H^1 \rightarrow {}_2He^4 + 2\,{}_{+1}e^0 \tag{9-14}$$

resulting in a decrease in mass of about 0.0276 amu, corresponding to 25.7 MeV. The heat produced in these reactions maintains temperatures of the order of several million degrees in their cores and serves to trigger and sustain succeeding reactions.

On earth, although fission preceded fusion in both weapons and power generation, the basic fusion reaction was discovered first, in the 1920s, during research on particle accelerators. Artificially produced fusion may be accomplished when two light atoms fuse into a larger one as there is a much greater probability of two particles colliding than of four. The 4-hydrogen reaction requires, on an average, billions of years for completion, whereas the deuterium-deuterium reaction requires a fraction of a second.

To cause fusion, it is necessary to accelerate the positively charged nuclei to high kinetic energies, in order to overcome electrical repulsive forces, by raising their temperature to hundreds of millions of degrees resulting in a plasma. The plasma must be prevented from contacting the walls of the container, and must be confined for a period of time (of the order of a second) at a minimum density. Fusion reactions are called *thermonuclear* because very high temperatures are required to trigger and sustain them [3,83]. Table 9-2 lists the possible fusion reactions and the energies produced

Table 9-2

	Fusion reaction		
Number	Reactants	Products	Energy per reaction, MeV
1	D + D	T + p	4
2	D + D	$He^3 + n$	3.2
3	T + D	$He^4 + n$	17.6
4	He^3 + D	$He^4 + p$	18.3

by them. n, p, D, and T are the symbols for the neutron, proton, deuterium (H^2), and tritium (H^3), respectively.

Many problems have to be solved before an artificially made fusion reactor becomes a reality [3,83]. The most important of these are the difficulty in generating and maintaining high temperatures and the instabilities in the medium (plasma), the conversion of fusion energy to electricity, and many other problems of an operational nature. Fusion powerplants will not be covered in this text.

Fission

Unlike fusion, which involves nuclei of similar electric charge and therefore requires high kinetic energies, *fission* can be caused by the neutron, which, being electrically neutral, can strike and fission the positively charged nucleus at high, moderate, or low speeds without being repulsed. Fission can be caused by other particles, but neutrons are the only practical ones that result in a sustained reaction because two or three neutrons are usually released for each one absorbed in fission. These keep the reaction going. There are only a few fissionable isotopes. U^{235}, Pu^{239}, and U^{233} are fissionable by neutrons of all energies. U^{238}, Th^{232}, and Pu^{240} are fissionable by high-energy neutrons only. An example, shown schematically in Fig. 9-3, is

$$_{92}U^{235} + {_0n^1} \rightarrow {_{54}Xe^{140}} + {_{38}Sr^{94}} + 2{_0n^1} \tag{9-15}$$

The immediate (prompt) products of a fission reaction, such as Xe^{140} and Sr^{94} above, are called *fission fragments*. They, and their decay products (Sec. 9-7), are called *fission products*. Figure 9-4 shows fission product data for U^{235} by thermal and fast neutrons (Secs. 9-10 and 9-11) and for U^{233} and Pu^{239} by thermal neutrons [84]. The products are represented by their mass numbers.

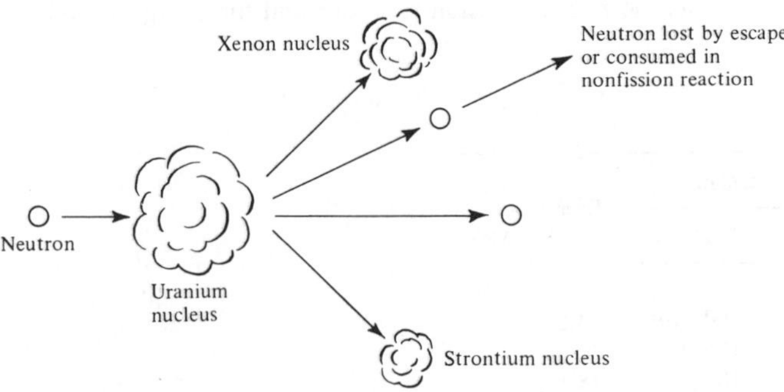

Figure 9-3 A typical fission reaction.

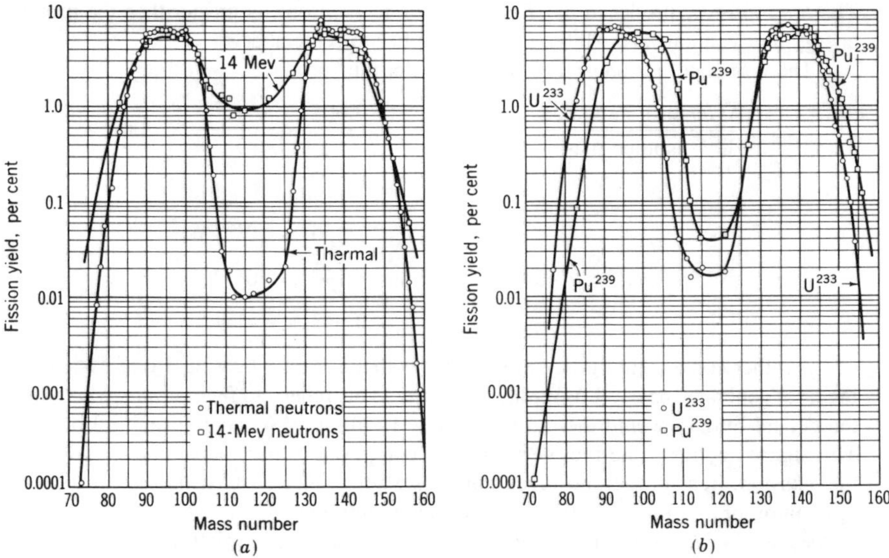

Figure 9-4 Fission product yield data for (*a*) U^{235} by thermal and 14 MeV neutrons and (*b*) U^{233} and PU^{239} by thermal neutrons [84].

9-6 ENERGY FROM FISSION AND FUEL BURNUP

There are many fission reactions that release different energy values. The one in Eq. (9-15), for example, yields 196 MeV. Another

$$_{92}U^{235} + {}_0n^1 \rightarrow {}_{56}Ba^{137} + {}_{36}Kr^{97} + 2{}_0n^1 \tag{9-16}$$

has the mass balance

$$235.0439 + 1.00867 \rightarrow 136.9061 + 96.9212 + 2 \times 1.00867$$

$$236.0526 \rightarrow 235.8446$$

$$\Delta m = 235.8446 - 236.0526 = -0.2080 \text{ amu}$$

Thus

$$\Delta E = 931 \times -0.2080 = -193.6 \text{ MeV} = -3.1 \times 10^{-11} \text{ J}$$
$$= -2.937 \times 10^{-11} \text{ Btu}$$

On the *average* the fission of a U^{235} nucleus yields about 193 MeV. The same figure roughly applies to U^{233} and Pu^{239}. This amount of energy is *prompt,* i.e., released at the time of fission. More energy, however, is produced because of (1) the slow decay of the fission fragments into fission products and (2) the nonfission capture of excess neutrons in reactions that produce energy, though much less than that of fission.

The *total energy,* produced *per* fission reaction, therefore, is greater than the prompt energy and is about 200 MeV, a useful number to remember.

The complete fission of 1 g of U^{235} nuclei thus produces

$$\frac{\text{Avogadro's number}}{U^{235}\text{ isotope mass}} \times 200\text{ MeV} = \frac{0.60225 \times 10^{24}}{235.0439} \times 200$$

$$= 0.513 \times 10^{24}\text{ MeV}$$

$$= 2.276 \times 10^{24}\text{ kWh}$$

$$= 8.190 \times 10^{10}\text{ J}$$

$$= 0.948\text{ MW-day}$$

Another convenient figure to remember is that a reactor burning 1 g of fissionable material generates nearly 1 MW-day of energy. This relates to fuel *burnup*. Maximum theoretical burnup would therefore be about a million MW-day/ton (metric) of fuel. This figure applies if the fuel were entirely composed of fissionable nuclei and all of them fission. Reactor fuel, however, contains other nonfissionable isotopes of uranium, plutonium, or thorium. *Fuel* is defined as all uranium, plutonium, and thorium isotopes. It does not include alloying or other chemical compounds or mixtures. The term *fuel material* is used to refer to fuel plus such other materials.

Even the fissionable isotopes cannot be all fissioned because of the accumulation of fission products that absorb neutrons and eventually stop the chain reaction. Because of this—and owing to metallurgical reasons such as the inability of the fuel material to operate at high temperatures or to retain gaseous fission products [such as Xe and Kr, Eqs. (9-15) and (9-16)] in its structure except for limited periods of time—burnup values are much lower than this figure. They are, however, increased somewhat by the fissioning of some fissionable nuclei, such as Pu^{239}, which are newly converted from fertile nuclei, such as U^{238} (Sec. 11-2). Depending upon fuel type and *enrichment* (*mass* percent of fissionable fuel in all fuel), burnups may vary from about 1000 to 100,000 MW · day/ton and higher.

9-7 RADIOACTIVITY

Radioactivity is an important source of energy for small power devices and a source of radiation for use in research, industry, medicine, and a wide variety of applications, as well as an environmental concern.

Most of the naturally occurring isotopes are *stable*. Those that are not stable, i.e., *radioactive,* are some isotopes of the heavy elements thallium ($Z = 81$), lead ($Z = 82$), and bismuth ($Z = 83$) and all the isotopes of the heavier elements beginning with polonium ($Z = 84$). A few lower-mass naturally occurring isotopes are radioactive, such as K^{40}, Rb^{87}, and In^{115}. In addition, several thousand artificially produced isotopes of all masses are radioactive. Natural and artificial radioactive isotopes, also called *radioisotopes,* have similar disintegration rate mechanisms. Figure 9-5 shows a *Z-N* chart of the known isotopes.

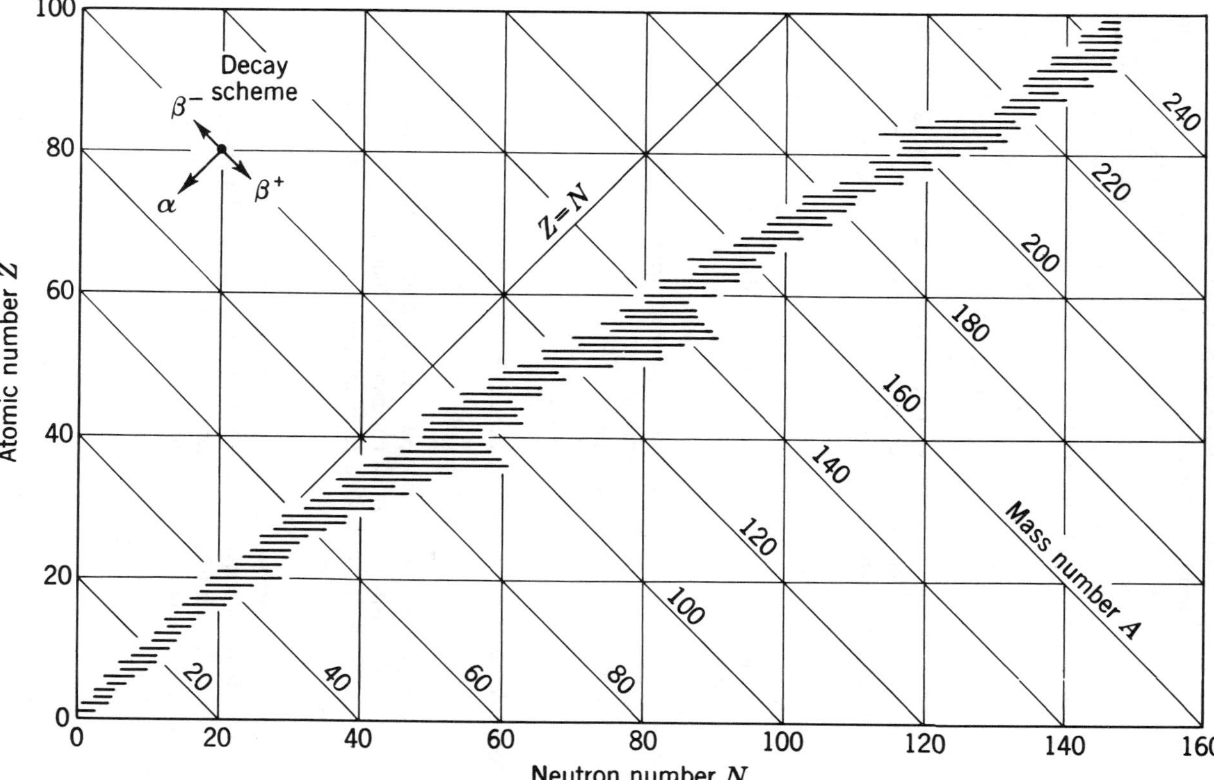

Figure 9-5 *Z-N* chart of the unknown isotopes.

Radioactivity means that a radioactive isotope continuously undergoes spontaneous (i.e., without outside help) disintegration, usually with the emission of one or more smaller particles from the *parent* nucleus, changing it into another, or *daughter*, nucleus. The parent nucleus is said to *decay* into the daughter nucleus. The daughter may or may not be stable, and several successive decays may occur until a stable isotope is formed. An example of radioactivity is

$$_{49}In^{115} \rightarrow {}_{50}Sn^{115} + {}_{-1}e^0 \qquad (9\text{-}17)$$

where the parent, In^{115}, is a naturally occurring radioisotope and its daughter, Sn^{115}, is stable.

Radioactivity is *always* accompanied by a *decrease* in mass and is thus always exothermic. The energy liberated shows up as kinetic energy of the emitted particles and as γ radiation. The light particle is ejected at high speed, whereas the heavy one recoils at a much slower pace in an opposite direction.

Naturally occurring radioisotopes emit α, β, or γ particles or radiations. The artificial isotopes, in addition to the above, emit or undergo the following particles or reactions: positrons; orbital electron absorption, called K capture; and neutrons. In addition, neutrino emission accompanies β emission (of either sign).

Alpha decay Alpha particles are helium nuclei, each consisting of two protons and two neutrons. They are commonly emitted by the heavier radioactive nuclei. An example is the decay of Pu^{239} into fissionable U^{235}

$$_{94}Pu^{239} \rightarrow {}_{92}U^{235} + {}_{2}He^4 \qquad (9\text{-}18)$$

Beta decay An example of β decay, besides Eq. (9-17), is

$$_{82}Pb^{214} \rightarrow {}_{83}Bi^{214} + {}_{-1}e^0 + \nu \qquad (9\text{-}19)$$

where ν, the symbol for the neutrino, is often dropped from the equation. The penetrating power of β particles is small compared with that of γ-rays but is larger than that of α particles. β- and α-particle decay are usually accompanied by the emission of γ radiation.

Gamma radiation This is electromagnetic radiation of extremely short wavelength and very high frequency and therefore high energy. γ-rays and x-rays are physically similar but differ in their origin and energy: γ-rays from the nucleus, and x-rays from the atom because of orbital electrons changing orbits or energy levels. Gamma wavelengths are, on an average, about one-tenth those of x-rays, although the energy ranges overlap somewhat. Gamma decay does not alter either the atomic or mass numbers.

Positron decay Positron decay occurs when the radioactive nucleus contains an excess of protons. It effectively converts a proton into a neutron. An example is

$$_{7}Ni^{13} \rightarrow {}_{6}C^{13} + {}_{+1}e^0 \qquad (9\text{-}20)$$

Because the daughter has one less proton than the parent, one of the orbital electrons is released to maintain atom neutrality. It combines with an emitted positron according to

$$_{+1}e^0 + _{-1}e^0 \rightarrow 2\gamma \tag{9-21}$$

The two particles therefore undergo an *annihilation* process, which produces γ energy equivalent to the sum of their rest masses $2m_ec^2$, or $-(2 \times 0.0005486)931 = -1.02$ MeV.

The reverse of the annihilation process is called *pair production*. In this, a γ photon of at least 1.02-MeV energy forms a positron-electron pair. This is an endothermic process that converts energy to mass. It occurs in the presence of matter and never in a vacuum.

K capture K capture also takes place when a nucleus possesses an excess of protons but not the threshold of 1.02 MeV necessary to emit a positron. Instead it captures an orbital electron from the orbit or shell nearest to the nucleus, called the K shell; hence the name K capture. The vacancy in the K shell is filled by another electron falling from a higher orbit. Thus K capture is accompanied by x-ray emission from the atom. K capture also effectively changes a proton into a neutron. The process is shown in Fig. 9-6. An example of K capture is

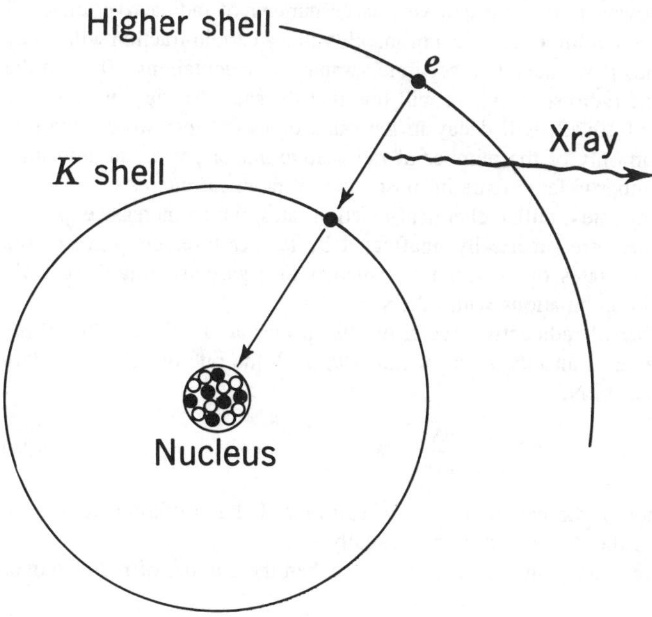

Figure 9-6 K capture.

$$_{29}\text{Cu}^{64} + {}_{-1}e^{0} \rightarrow {}_{28}\text{Ni}^{64} \tag{9-22}$$

in which Ni^{64} is stable. Because the parent acquires an electron, the electron symbol is on the left-hand side.

Neutron emission This occurs when a nucleus possesses an extremely high excitation energy. The *binding energy* of a neutron in a nucleus (the energy that would have to be added to a nucleus to expel a neutron) varies with mass number but averages about 8 MeV. Thus, if the excitation energy of a nucleus were at least 8 MeV, it could decay by the emission of a neutron. An example is

$$_{54}\text{Xe}^{137} \rightarrow {}_{54}\text{Xe}^{136} + {}_{0}n^{1} \tag{9-23}$$

The parent Xe^{137} is a fission product resulting from the β decay of the fission fragment I^{137} (called a *precursor*). In neutron decay the daughter is an isotope of the parent. It is a rare occurrence except in nuclear reactors where it is the source of *delayed fission neutrons*, which are of utmost importance in reactor control.

9-8 DECAY RATES AND HALF-LIVES

There can be no indication of the time that it takes any one particular radioactive nucleus to decay. However, if there is a very large number of radioactive nuclei of the same kind, there is a definite statistical probability that a certain fraction will decay in a certain time. Thus if we have two separate samples, one containing 10^{20} and the other 10^{30} of the same radioisotopes, we will find that the same fraction in each, say one-half or $10^{20}/2$ and $10^{30}/2$, will decay in the same time. In other words, the rate of decay is a function only of the number of radioactive nuclei present at any time, provided that the number is large (true in most cases of practical interest).

Radioactive-decay rates, unlike chemical-reaction rates, which increase exponentially with temperature, are practically unaffected by temperature, pressure, or the physical and chemical states of matter, i.e., whether in a gaseous, liquid, or solid phase or in chemical combinations with others.

If N is the number of radioactive nuclei of one species at any time θ, and if dN is the number decaying in an increment of time $d\theta$, at θ, the rate of decay $-dN/d\theta$ is directly proportional to N.

$$-\frac{dN}{d\theta} = \lambda N \tag{9-24}$$

λ is a proportionality factor called the *decay constant*. It has different values for different isotopes and the dimension time^{-1}, usually s^{-1}.

Integrating between an arbitrary time, $\theta = 0$, when the number of radioisotopes was N_0, gives

$$-\int_{N_0}^{N} \frac{dN}{N} = \lambda \int_{0}^{\theta} d\theta$$

Thus

$$N = N_0 e^{-\lambda\theta} \tag{9-25}$$

The rate of decay λN is also called the *activity* A and commonly has the dimension disintegrations per second (dis/s) or s^{-1}. The initial activity A_0 is equal to λN_0. Thus

$$A = A_0 e^{-\lambda\theta} \tag{9-26}$$

A common way of representing decay rates is by the use of the *half-life,* $\theta_{1/2}$. This is the time during which one-half of a number of radioactive species decays or one-half of their activity ceases. Thus

$$\frac{N}{N_0} = \frac{A}{A_0} = \frac{1}{2} = e^{-\lambda\theta_{1/2}}$$

and

$$\theta_{1/2} = \frac{\ln 2}{\lambda} = \frac{0.6931}{\lambda} \tag{9-27}$$

and the half-life is inversely proportional to the decay constant. Starting at $\theta = 0$ when $N = N_0$, one-half of N_0 decay after one half-life; one-half of the remaining atoms, or one-quarter of N_0, decay during the second half-life; one-eighth of N_0 during the third, and so on (Table 9-3 and Fig. 9-7). The fraction of the initial number of parent nuclei or activity remaining after n half-lives is equal to $(1/2)^n$.

Theoretically, it takes an infinite time for the activity to cease. However, about 10 half-lives reduce the activity to less than one-tenth of 1 percent of the original—negligible in many cases. Half-lives of radioisotopes vary from fractions of a microsecond to billions of years, and no two have the same half-lives. They are "fingerprints" by which a particular radioactive species may be identified. This is done by measuring

Table 9-3 Activity and half-life

Number of half-lives	N/N_0 or A/A_0	
0	1	1.00000
1	1/2	0.50000
2	1/4	0.25000
3	1/8	0.12500
4	1/16	0.06250
5	1/32	0.031250
6	1/64	0.015625
7	1/128	0.007813
8	1/256	0.003906
9	1/512	0.001953
10	1/1024	0.000977
11	1/2048	0.000488

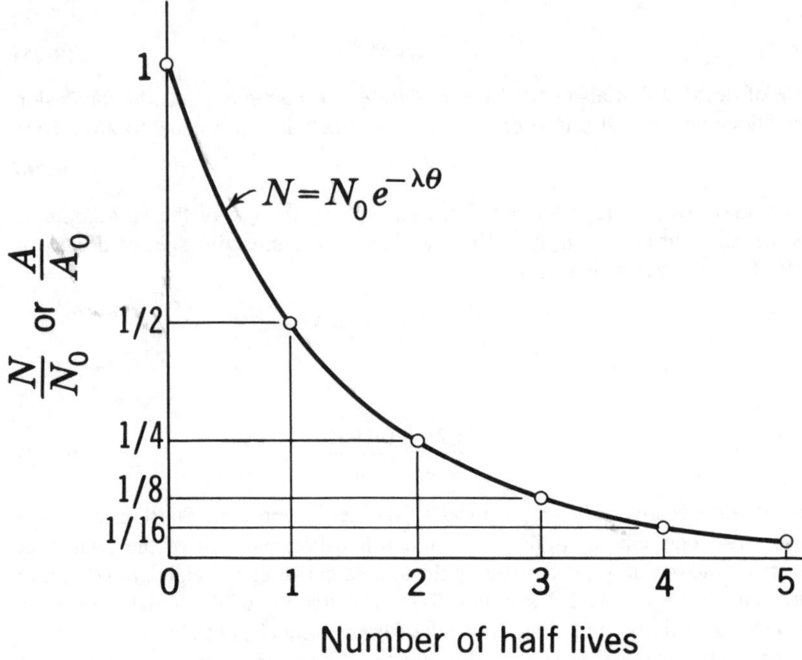

Figure 9-7 Radioactive-decay rates as a function of half-life.

the change in activity with time and computing λ from the slope of the activity history on a semilog plot (Fig. 9-8), from which $\theta_{1/2}$ and the unknown specie are identified.

There are cases that involve two transitions from the parent isotope with two decay rates and two half-lives. In some cases two different half-lives represent one transition. Table 9-4 gives the half-lives and type of activity of some important radioisotopes. Note that the readily fissionable isotopes U^{233}, U^{235}, and Pu^{239} have extremely long half-lives, so they can be stored practically indefinitely. U^{233} and Pu^{239} are artificially produced (from Th^{232} and U^{238}, respectively, themselves very long-lived), whereas U^{235} is found in nature.

The energy generated by the decaying fission products results in continuing, though much-reduced, energy generation in a reactor after shutdown and must be removed by an adequate coolant system.

Example 9-2 Radium 226 decays into radon gas. Compute (1) the decay constant and (2) the initial activity of 1 g of radium 226. The atomic mass is 226.0245 amu.

SOLUTION

(1) Half-life of Ra^{226} = 1600 yr = 5.049×10^{10} s (from Table 9-4)

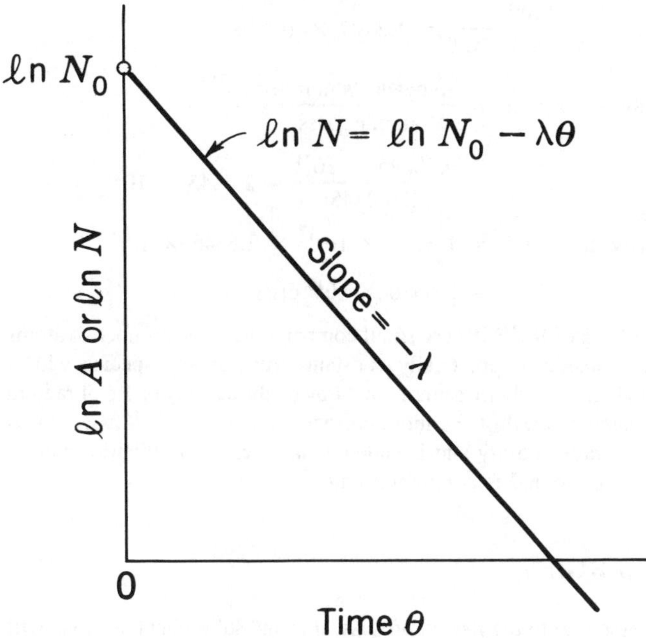

Figure 9-8 Radioactive-decay curve on a semilog plot.

Table 9-4 Half-lives of some isotopes

Isotope	$\theta_{1/2}$	Activity
Tritium (H^3)	12.26 yr	β
Carbon 14	5730 yr	β
Krypton 87	76 min	β
Strontium 90	28.1 yr	β
Xenon 135	9.2 h	β and γ
Barium 139	82.9 min	β and γ
Radium 223	11.43 days	α and γ
Radium 226	1600 yr	α and γ
Thorium 232	1.41×10^{10} yr	α and γ
Thorium 233	22.1 min	β
Protactinium 233	27.0 days	β and γ
Uranium 233	1.65×10^5 yr	α and γ
Uranium 235	7.1×10^8 yr	α and γ
Uranium 238	4.51×10^9 yr	α and γ
Neptunium 239	2.35 days	β and γ
Plutonium 239	2.44×10^4 yr	α and γ

$$\lambda = \frac{0.6931}{5.049 \times 10^{10}} = 1.3727 \times 10^{-11} \text{ s}^{-1}$$

(2) Number of atoms per gram $= \dfrac{\text{Avogadro's number}}{\text{atomic mass}}$

$$= \frac{0.60225 \times 10^{24}}{226.0245} = 2.6645 \times 10^{21}$$

$$\text{Initial activity } A_0 = \lambda N_0 = 1.3727 \times 10^{-11} \times 2.6645 \times 10^{21}$$

$$= 3.6576 \times 10^{10} \text{ dis/s}$$

Thus the activity of 1 g of Ra^{226} is very small compared with the number of atoms in it and may be considered practically constant, true for any species with a sufficiently long half-life. Early measurements showed the activity of 1 g of radium to be 3.7×10^{10} dis/s instead of the more correct value above. 3.7×10^{10} was adopted as a unit of radioactivity and is called a *curie* (ci). A millicurie (mci) is one-thousandth of a curie and is a common unit.

9-9 NEUTRON ENERGIES

Because neutrons are essential to the fission process, this and subsequent sections will deal with them and their interactions. As any other body, the kinetic energy of a neutron KE_n is given by

$$KE_n = \frac{1}{2g_c} m_n V^2 \tag{9-28}$$

where m_n = mass of neutron

V = speed of neutron

g_c = conversion factor (the same as in Sec. 9-6)

The term *kinetic* is occasionally dropped and the symbol KE_n simplified to E_n, so that neutron energy means neutron kinetic energy. E_n is commonly expressed in eV or MeV. Since m_n = 1.008665 amu, then

$$E_n = \frac{1}{2 \times 0.965 \times 10^{18}} \times 1.008665 V^2$$

or

$$E_n = 5.227 \times 10^{-19} V^2 \text{ MeV} = 5.227 \times 10^{-13} V^2 \text{ eV} \tag{9-29}$$

where V is the centimeters per second.

The newly born fission neutrons have energies ranging between less than 0.075 to about 17 MeV. When they travel through matter, they collide with nuclei and are decelerated, mainly by the lighter nuclei, thus giving up some of their energy with each successive collision. This process is called *scattering*.

Neutrons are classified into three categories according to energy: *fast* (greater than 10^5 eV), *intermediate*, and *slow* (less than 1 eV). One of main reactor classifications is the energy range of the neutrons causing fission. A *fast reactor* is one dependent primarily on fast neutrons for fission. A *thermal reactor* is one utilizing mostly *thermal neutrons* (Sec. 9-10).

Newly born fission neutrons carry, on an average, about 2 percent of a reactor

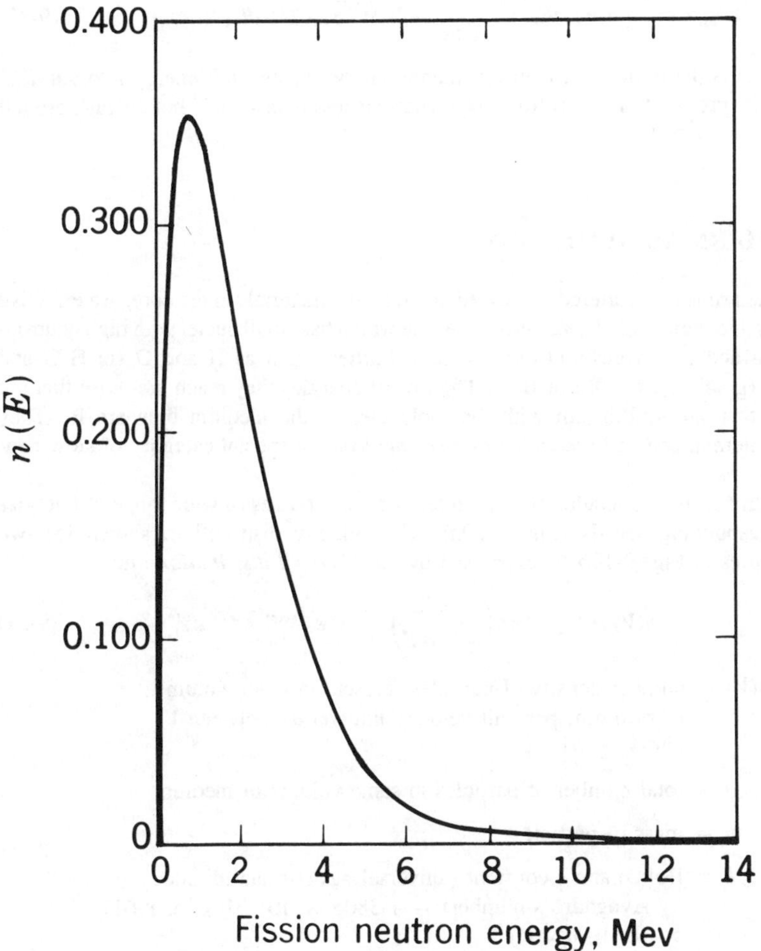

Figure 9-9 Energy spectrum of fission neutrons.

fission energy. They are either *prompt* or *delayed*. Prompt neutrons are released at the time of fission, within about 10^{-14} s (from fission fragments with a neutron-proton ratio the same as the original nucleus but greater than that corresponding to their mass number). Delayed neutrons, produced in radioactive decay of some fission products (Sec. 9-7), constitute only 0.645 percent of the total fission neutrons in U^{235} fission (less for Pu^{239} and U^{233}). Their energies are small compared with those of prompt neutrons but they play a major role in reactor control.

Prompt neutrons have an energy distribution shown in Fig. 9-9 and given (for U^{235} and Pu^{239} fission) by [85]

$$n(E)\, dE_n = \sqrt{\frac{2}{\pi e}} \sinh \sqrt{2E_n}\, e^{-E_n}\, dE_n \tag{9-30}$$

where $n(E)$ is the number of neutrons having energy E_n per unit energy interval dE_n. Most of the prompt fission neutrons have energies less than 1 MeV but average around 2 MeV.

9-10 THERMAL NEUTRONS

Fission neutrons are scattered or slowed down by the materials in the core. An effective scattering medium, called a *moderator,* is one which has small nuclei with high neutron-scattering and low neutron-absorption probabilities, such as H and D (in H_2O and D_2O), C (graphite), and Be or BeO. The lowest energies they reach are those that put them in thermal equilibrium with the molecules of the medium they are in. They become thermalized and are called *thermal neutrons,* a special category of slow neutrons.

Neutrons, like molecules at a given temperature, possess a wide range of energies and corresponding speeds (Fig. 9-10*a*). The velocity distribution, shown for two temperatures in Fig. 9-10*b*, is expressed by the *Maxwell distribution law*

$$n(V)\, dV = 4\pi n\left(\frac{m}{g_c 2\pi kT}\right)^{1.5} V^2 e^{-g_c(mV^2/2kT)}\, dV \tag{9-31}$$

where $n(V)$ = number density of particles, present in given volume of medium, per unit velocity interval dV between V and $V + dV$

n = total number of particles in same volume of medium

m = mass of particle

k = Boltzmann's constant (universal gas constant divided by Avogadro's number) = 1.3805×10^{-23} J/K, or 8.617×10^{-11} MeV/K)

T = absolute temperature

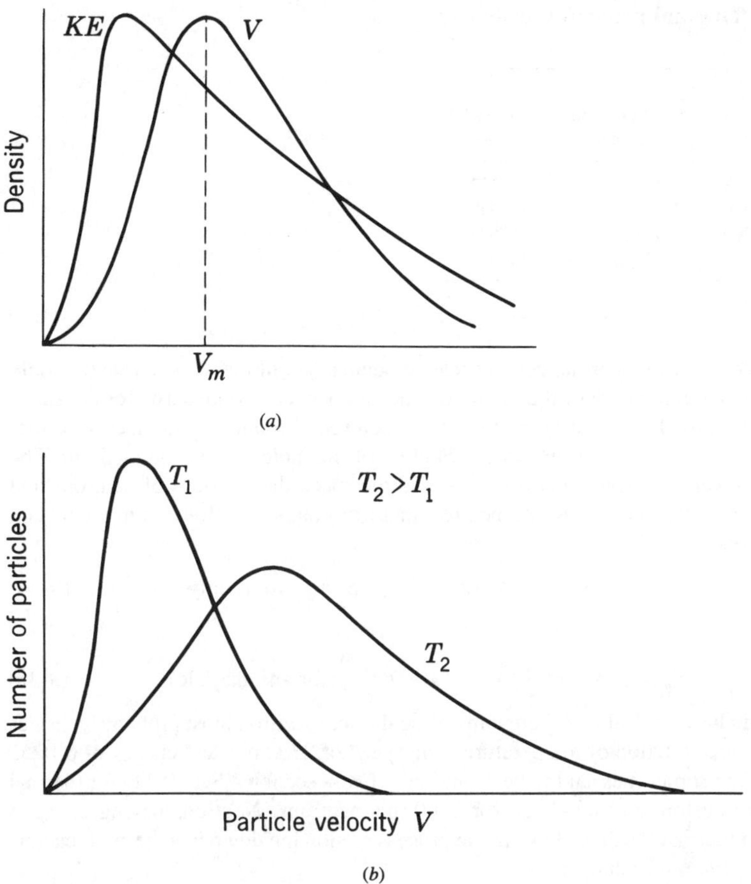

Figure 9-10 (*a*) Kinetic energy and velocity distributions of thermal neutrons at a given temperature. (*b*) Velocity distribution at two temperatures.

The *most probable speed* V_m is the one that corresponds to the maximum number density evaluated by differentiating the right-hand side of Eq. (9-31) with respect to V and equating the derivative to zero, resulting in

$$V_m = \left(\frac{g_c 2kT}{m}\right)^{0.5} \tag{9-32}$$

The *energy corresponding to the most probable speed* (which is not the same as the most probable energy) is

$$E_m = \frac{1}{2g_c} m V_m^2 = kT \tag{9-33}$$

Table 9-5 Thermal-neutron speeds and energies

Temperature		Most probable	Corresponding
°C	°F	speed, m/s	energy, eV
20	68	2,200	0.0252
260	500	2,964	0.0459
537.8	1000	3,656	0.0699
1000	1832	4,580	0.1097

Thus the energy of the thermalized particle is independent of mass and a function only of the absolute temperature of the medium. The independence is also true for the shape of the energy-distribution curve so that when neutrons become thermalized in a medium, they possess the same energy distribution of the molecules of the medium. The speeds, however, are dependent on mass, and the speed distributions of neutrons and molecules are different. Using the neutron mass in grams and Boltzmann's constant in eV/K gives

$$V_m \text{ (in m/s)} = 128.39T^{0.5} \qquad \text{(for a neutron only)} \qquad (9\text{-}34)$$

and

$$E_m \text{ (in eV)} = 8.617 \times 10^{-5}T \qquad \text{(for any particle)} \qquad (9\text{-}35)$$

where T is in kelvins. Table 9-5 contains some thermal-neutron most probable energies and speeds as a function of temperature. The speed of 2200 m/s and energy of 0.0252 eV at 20°C are sometimes said to be "standard." Cross sections (Sec. 9-11) for thermal neutrons are customarily tabulated for 2200-m/s neutrons. Neutrons having energies greater than thermal, such as those in the process of slowing down in a thermal reactor, are called *epithermal* neutrons.

9-11 NUCLEAR CROSS SECTIONS

Assume a beam of monoenergetic neutrons of intensity I_0 neutrons/s impinging on a body having a target area A cm^2 and a nuclear density N nuclei/cm^3 (Fig. 9-11). The nuclei have radii roughly 1/1000 those of atoms and, therefore, have a cross-sectional area, facing the neutron beam, that is very small compared with the total target area. Taking one nucleus into consideration, we may use the analogy of a large number of peas (neutrons) being shot at a basketball (nucleus) in the center of a window (target area). The number of peas that will collide with the basketball is proportional to its cross-sectional area. However, the fraction colliding with the basketball, or the *probability* of collision, is equal to the cross-sectional area of the basketball divided by the area of the window. The actual cross-sectional area of a nucleus is obtained from its radius r_c, which is given by [86]

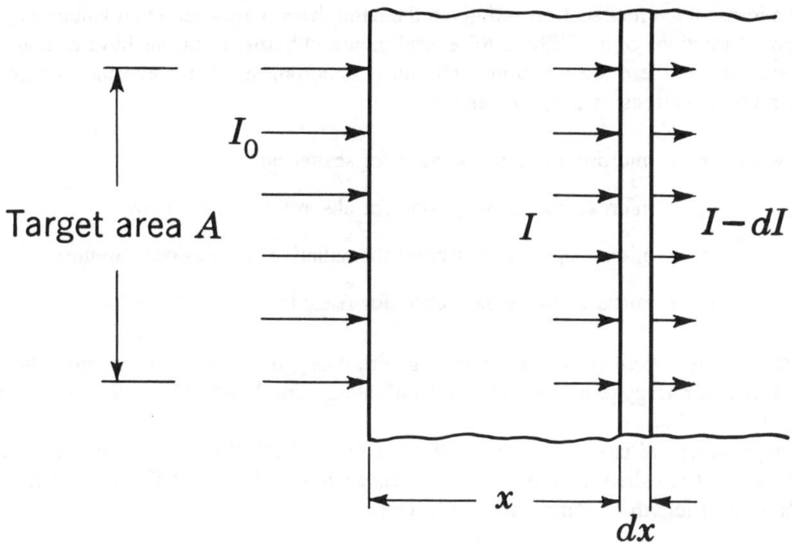

Figure 9-11 Neutron beam striking target area A.

$$r_c = r_0 A^{1/3} \tag{9-36}$$

where r_0 is a constant varying for different nuclei with an average of 1.4×10^{-13} cm and A is the mass number. The average cross-secitonal areas of nuclei therefore is about 10^{-24} cm^2.

The probability of neutrons colliding or interacting with nuclei is proportional to an *effective,* rather than actual, cross-sectional area of the nuclei in question. This is called the *microscopic cross section,* or simply the *cross section,* of the reaction and is given the symbol σ. It varies with the nucleus, type of reaction, and neutron energy.

In Fig. 9-11 the number of nuclei in volume $A\ \Delta x$ is $(NA\ \Delta x)$. As the neutron beam passes through Δx, some of the neutrons are removed (by absorption or scatter) from the beam. The fraction removed is equal to the ratio of *effective* cross-sectional areas of the nuclei, $\sigma(NA\ \Delta x)$, to the total area A. Thus, if at x and $x + \Delta x$ the beam intensities become I and $I - \Delta I$, respectively, it follows that, in the limit

$$\frac{-dI}{I} = \frac{\sigma(NA\ dx)}{A} = \sigma N\ dx$$

Integrating

$$-\int_{I_0}^{I} \frac{dI}{I} = \sigma N \int_0^x dx$$

from which

$$I = I_0 e^{-\sigma N x} \tag{9-37a}$$

σ has the units of area. Because nuclei are small, cm^2 is too large a unit. Instead, the actual cross-sectional area of an average nucleus, 10^{-24} cm^2, was taken as the unit of the microscopic cross section and given the name *barn*. Cross-sectional values vary from small fractions of a millibarn to several thousand barns. Neutrons have as many cross sections as there are reactions. The most important are the scattering and absorption cross sections: σ_s, σ_a, σ_c, and σ_f

where σ_s = microscopic cross section for scattering*

σ_a = microscopic cross section for absorption $= \sigma_c + \sigma_f$

σ_c = microscopic cross section for radiative (nonfission) capture

σ_f = microscopic cross section for fission

Sometimes only a *total* cross section σ_t is given, where $\sigma_t = \sigma_s + \sigma_a +$ any other. Cross-sectional energy plots for some nuclei of interest are shown in Figs. 9-12 through 9-14 [87].

The product σN is equal to the total cross sections of all the nuclei present in a unit volume. It is called the *macroscopic cross section* and is given the symbol Σ. It has the unit of length^{-1}, commonly cm^{-1}. Thus

$$\Sigma = N\sigma \tag{9-38a}$$

and Eq. (9-37*a*) can be written in the form

$$I = I_0 e^{-\Sigma x} \tag{9-37b}$$

Macroscopic cross sections are also designated according to the reaction they represent. Thus $\Sigma_f = N\sigma_f$, $\Sigma_s = N\sigma_s$, etc. The reciprocal of macroscopic cross section for any reaction is the *mean free path* for that reaction. It has the symbol λ, not to be confused with the decay constant in radioactivity (Sec. 9-8).

For an element of atomic mass A_t and density ρ (g/cm^3), N (nuclei/cm^3) can be calculated from

$$N = \rho \frac{\text{Avogadro's number}}{A_t} \tag{9-39}$$

9-12 NEUTRON FLUX AND REACTION RATES

The number of neutrons crossing a unit area per unit time in one direction is called the *neutron current* and is proportional to the gradient of *neutron density*. In a reactor core, however, the neutrons travel in all directions. If n is the neutron density (neutrons/

* Scattering is of two kinds, *inelastic* and *elastic*. *Inelastic scattering* occurs with high-energy neutrons, and the reduction in neutron kinetic energy shows up partly as kinetic energy and partly as excitation energy of the struck nucleus. *Elastic scattering* occurs with low-energy neutrons when the neutrons have slowed down and no longer possess sufficient energy to excite the nucleus. The struck neutron is not excited, and the kinetic-energy loss of the neutron is equal to the kinetic-energy gain of the nucleus.

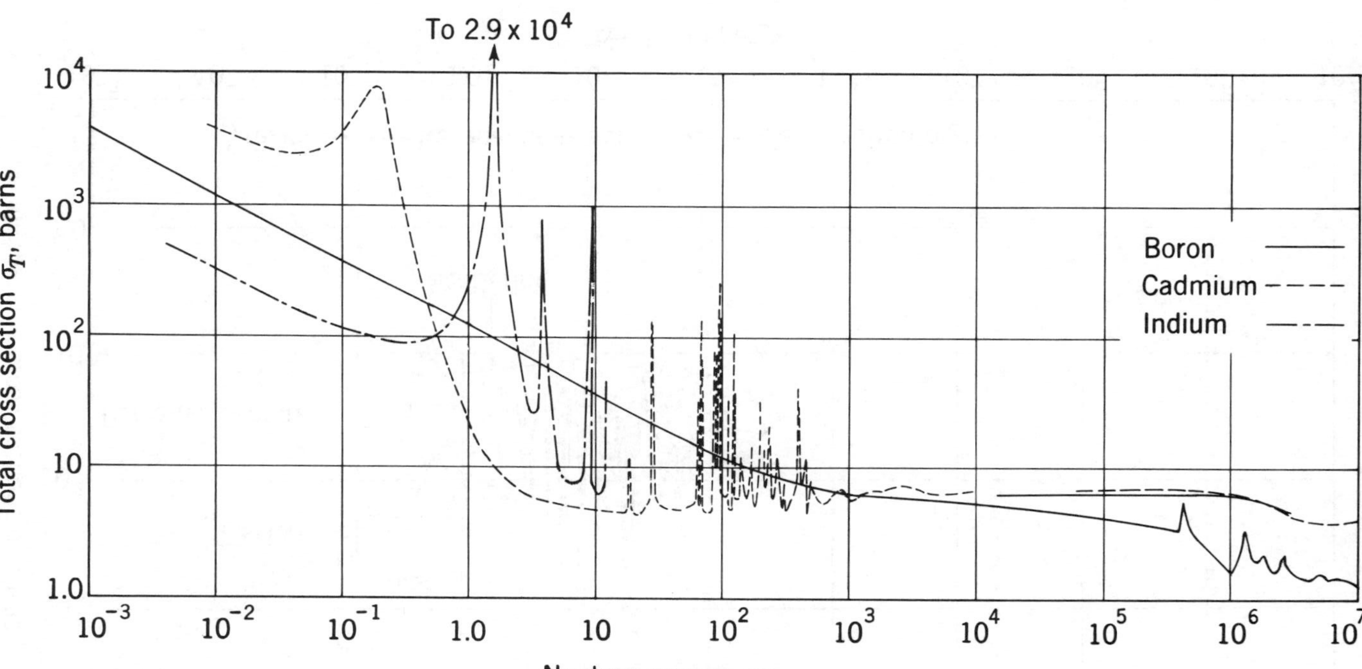

Figure 9-12 Neutron cross sections for cadmium, indium, and boron.

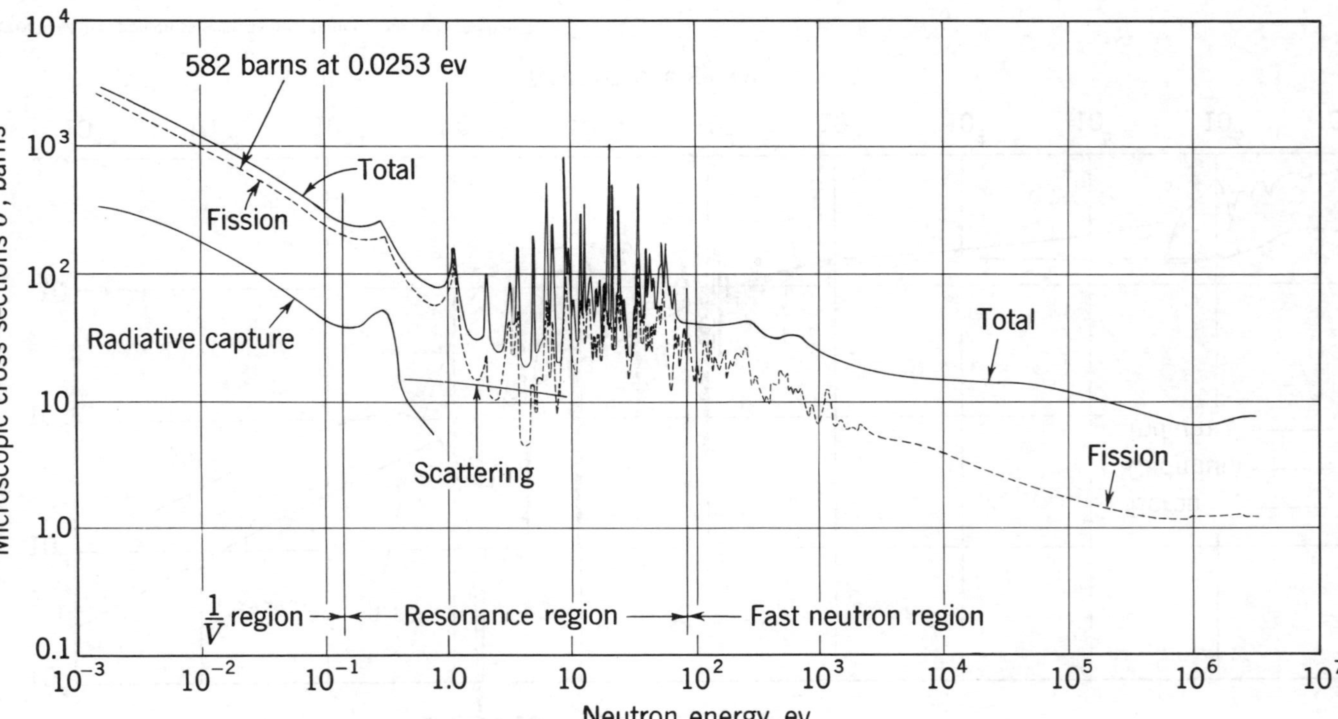

Figure 9-13 Neutron cross sections for U^{235}.

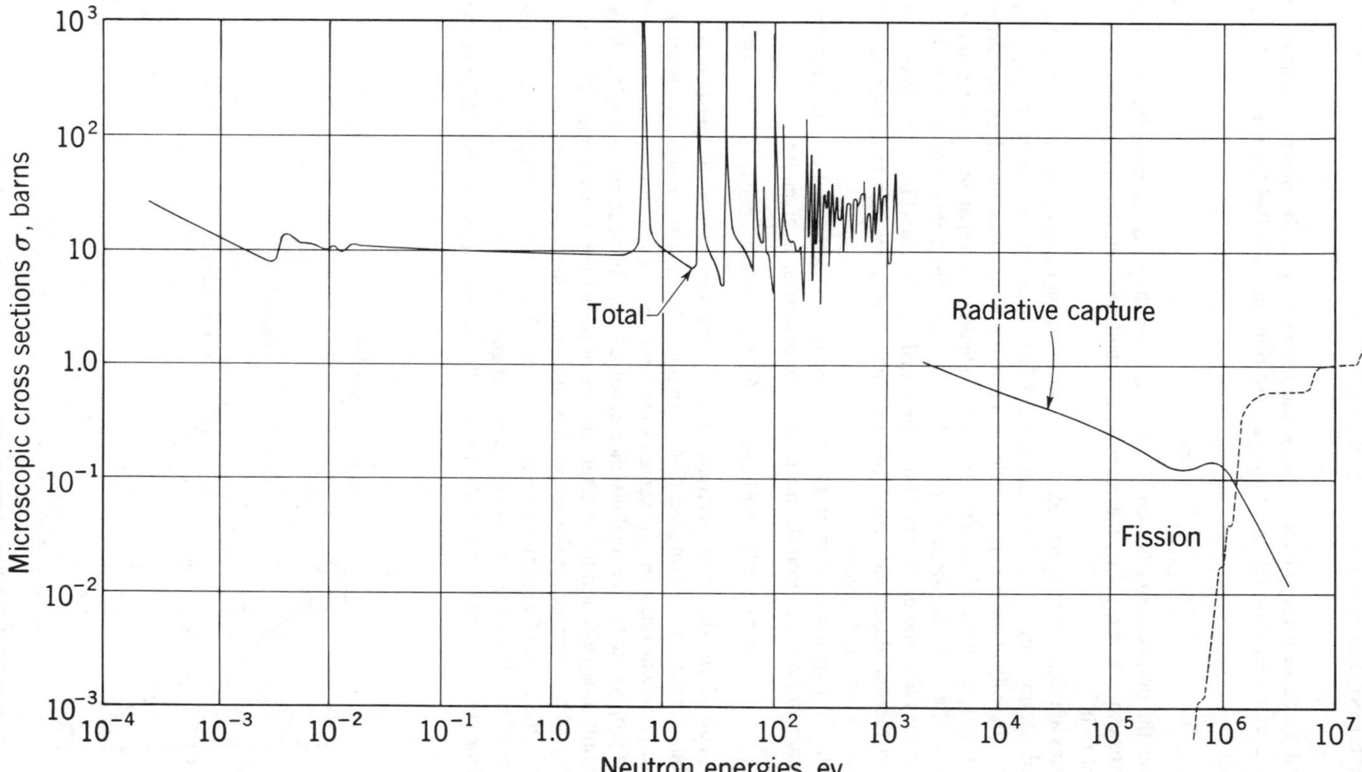

Figure 9-14 Neutron cross sections for U^{238}.

cm^3) and V the neutron velocity (cm/s), the product nV is the number of neutrons crossing a unit area from all directions per unit time and is called the *neutron flux* ϕ. Thus

$$\phi = nV \tag{9-40}$$

ϕ has the unit neutrons/(s · cm^2), which is often dropped. Because flux involves all neutrons at a given point, the reaction rate between neutrons and nuclei is proportional to it.

Fluxes are dependent upon velocity V, that is, upon energy, but are often quoted for broad energy ranges, such as thermal and fast. In a reactor core they vary from maximum, usually at the core geometric center, to minimum, near the edges. In thermal heterogeneous reactors, where the fuel and moderator are separate, fission neutrons are born in the fuel and thermalized in the moderator. Fast fluxes thus peak above average in the fuel, whereas thermal fluxes peak in the moderator (Fig. 9-15). Maximum full-power thermal fluxes vary from 10^6 for small training reactors to as high as 10^{15} for power and research reactors.

Now, if a medium containing nuclei of density N is subjected to a neutron flux ϕ, the *reaction rate,* between the nuclei and the neutrons, is given by

$$\text{Reaction rate} = nV\,N\sigma = \phi\Sigma \text{ reactions/(s} \cdot \text{cm}^3) \tag{9-41}$$

where σ and Σ are the cross sections of the particular reaction in question (i.e., absorption, scatter, etc.). Equation (9-41) simply states that the number of neutrons entering a particular reaction (the same as the number of reactions) per unit time and volume is proportional to the total distance traveled by all the neutrons in a unit volume during a unit time *(nV)* and to the total number of nuclei per unit volume (N); σ, the probability of the reaction, is the proportionality factor. Since in general N is fixed in a medium, the rates of a particular reaction (fixed σ) are directly proportional to the neutron flux. It will suffice here to state that heat generation by fission at a given point in a reactor core is proportional to the neutron flux at that point. A knowledge of the

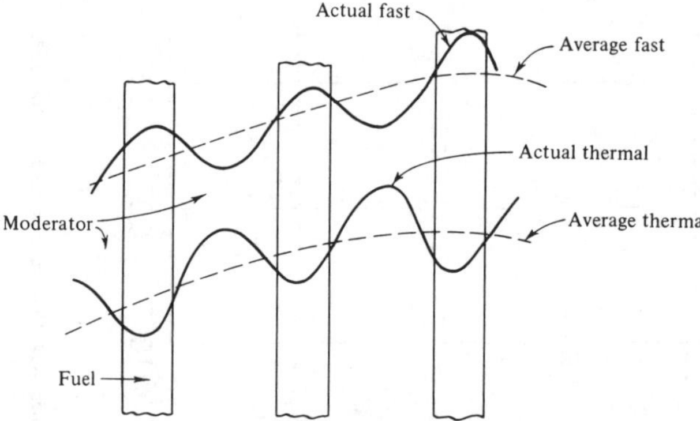

Figure 9-15 Neutron-flux distributions in fuel and moderator in a heterogeneous thermal reactor.

neutron-flux distribution in a reactor core is therefore necessary for the study of heat generation and removal in that core [2].

9-13 THE VARIATION OF NEUTRON CROSS SECTIONS WITH NEUTRON ENERGY

$\sigma - E_n$ plots are usually made on log-log coordinates. In many but not all cases, scattering cross sections are so small compared with absorption cross sections that the total cross sections shown are very nearly equal to the absorption cross sections. Also, for many nuclei, scattering cross sections vary little with neutron energy.

Variations of absorption cross sections with neutron energy, such as those in Figs. 9-12 through 9-14, are represented by three regions which, beginning with low neutron energies, are (1) the *1/V region,* (2) the *resonance region,* and (3) the *fast-neutron region*.

1/V region In the low-energy range, the absorption cross sections for many, but not all, nuclei are inversely proportional to the square root of the neutron energy E_n

$$\sigma_a = C_1\left(\frac{1}{E_n}\right)^{0.5} \tag{9-42a}$$

Thus

$$\sigma_a = C_1\left[\frac{1}{m_n(V^2/2g_c)}\right]^{0.5} = C_2\frac{1}{V} \tag{9-42b}$$

where C_1 and C_2 are constants, m_n is the neutron mass, and V is the neutron velocity.

This relationship, known as the *1/V law,* indicates that the neutron has a higher probability of absorption by a nucleus if it is moving at a lower velocity and is thus spending a longer time in the vicinity of that nucleus. The $1/V$ law may also be written in the form

$$\frac{\sigma_{a,1}}{\sigma_{a,2}} = \frac{V_2}{V_1} = \left(\frac{E_{n,2}}{E_{n,1}}\right)^{0.5} \tag{9-42c}$$

where the subscripts 1 and 2 refer to two different neutron energies within the $1/V$ range. Absorption cross sections for monoenergetic neutrons, within the $1/V$ region, may thus be calculated at any energy from tabulated values at 2200 m/s.

On the log-log plots of Figs. 9-12 to 9-14, the $1/V$ region is a straight line with a slope of -0.5. The upper limit of the $1/V$ region is different for different nuclei, being around 0.3 eV for indium, 0.05 eV for cadmium, 0.2 eV for U^{235}, 150 eV for boron, etc.

Resonance region Following the $1/V$ region, most neutron absorbers exhibit one or more peaks occurring at definite neutron energies, called *resonance peaks*. They affect neutrons in the process of slowing down. Note that indium has but one peak, whereas

U^{235} and U^{238} have many. Uranium 238 has very high resonance-absorption cross sections, with the highest peak, about 4000 barns, occurring at about 7 eV. This fact affects the design of thermal reactors because U^{238} absorbs many of the neutrons passing through the region and affects the reactor neutron balance. Many elements, especially those of low mass numbers, do not exhibit resonance absorption and thus can be used as reactor construction materials, especially if their absorption cross sections are low.

Fast-neutron region Following the resonance region, cross sections usually undergo a gradual decrease as neutron energies increase. At very high energies, the sum of absorption and scattering cross sections approaches twice the actual cross-sectional area of the target nucleus, that is, $2\pi r_c^2$. Combining with Eq.(9-36)

$$\sigma_t = 2\pi r_0^2 A^{2/3}$$

Using $r_0 = 1.40 \times 10^{-13}$ cm and 1 barn $= 10^{-24}$ cm^2

$$\sigma_t = 0.125 A^{2/3} \text{ barns} \tag{9-43}$$

In the very high neutron energy range, therefore, total cross sections are rather low, usually less than 5 barns each for the largest nuclei. Some nuclei, such as boron, carbon, and beryllium, exhibit some resonance in the high energy range (Fig. 9-12), but the peaks are rather low, and the phenomenon is of little importance.

9-14 FISSION REACTOR TYPES

With the information on nuclear-fuel cross sections and their variation with neutron energies in hand, we are now able to put together a qualitative picture of the effects of fuel enrichment on reactor core design and configuration.

We already know that in fission one neutron is absorbed by the fissioning nucleus but between two and three fast fission neutrons are produced. The exact number is given the symbol ν, given in Table 9-6 for the three fissionable nuclei.

In order for a reactor using U^{235} as fuel to operate at a steady rate, no more than 1.47 neutrons should be lost to the fission process. Such losses occur in two ways: (1) nonfission absorption in reactor-core materials, which include structural materials, coolant channel walls, fuel cladding, coolant, moderator, and the fuel material itself;

Table 9-6 Fuel constants

Nucleus	Number of fission neutrons per fission, ν	Number of fission neutrons per thermal neutron absorbed, η
U^{233}	2.51 ± 0.03	2.28 ± 0.02
U^{235}	2.47 ± 0.03	2.07 ± 0.02
Pu^{239}	2.90 ± 0.04	2.10 ± 0.02

and (2) leakage of the core, a function of both core materials and size. More leakage occurs if the neutron mean free path in the core materials is large and if the reactor size is small, i.e., its surface-volume ratio is large. For the purpose of discussion, we will fix the fractions of neutrons leaking and engaging in nonfission absorption in reactor materials other than the fuel itself. The only variables, then, will be nonfission and fission absorptions in the fuel.

The simplest fuel to use in a reactor is natural uranium, composed of about 0.7 percent U^{235}, about 99.3 percent U^{238}, and a trace of U^{234}. Let us now try to build a reactor core from a solid mass of this relatively cheap and plentiful fuel with possible small holes for cooling (Fig. 9-16). Let us start with 110 newly born fast neutrons within this fuel and assume that 10 will leak out of the core during their lifetime. The remaining 100 neutrons will be subjected only to scatter, radiative capture, and fission in the fuel. Because the cross-sectional curves of Figs. 9-13 and 9-14 do not contain all reactions, we will use Table 9-7, which contains the average cross sections of the above reactions for U^{235}, U^{238}, and natural uranium.

The average energy of the newly born fast neutrons is 2 MeV. At this energy the

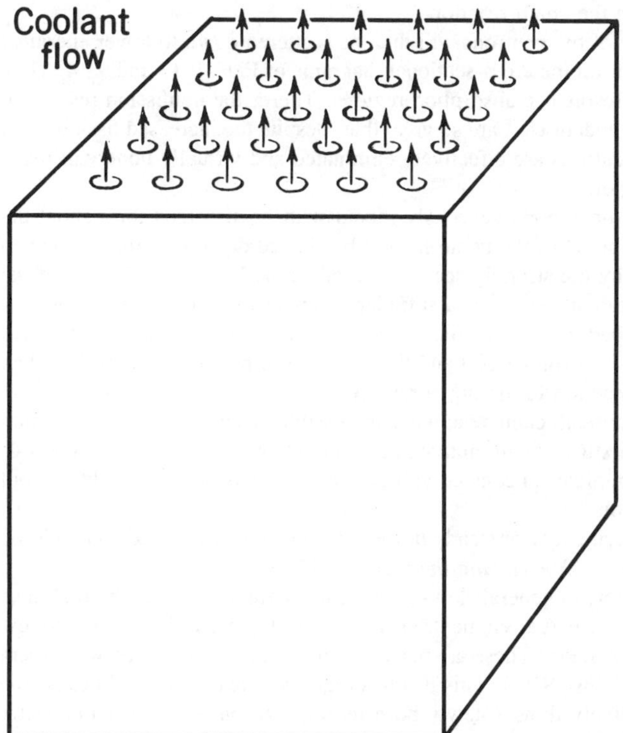

Figure 9-16 A homogeneous mass of natural uranium metal with coolant flow channels.

Table 9-7 Microscopic cross sections for uranium fuels

Neutron	Microscopic cross sections, barns	Nucleus U^{235}	Nucleus U^{238}	Natural uranium
Fast, 1 MeV	σ_s	5.30	6.6	6.6
	σ_c	0.093	0.14	0.14
	σ_f	1.20	0.018	0.026
Thermal, 0.0253 eV	σ_s	6.00	8.00	8.00
	σ_c	112.00	2.73	3.47
	σ_f	577.10	0	4.16

fission cross section of U^{238} is about 0.53 barn. The majority of these neutrons, however, will possess the most probable energy, just below 1.0 MeV (Fig. 9-9), where the fission cross section in natural uranium is so low compared with the scattering cross sections that only very few of the 100 original neutrons will engage in fission, producing 2.47 neutrons each. This production is so far below the original 100 that, alone, it cannot sustain the chain reaction.

The great majority of the neutrons will then be scattered down to lower energies, from the right to the left on the cross-sectional diagrams of Figs. 9-13 and 9-14. They first have to cross the resonance-absorption regions. There, the nonfission resonance cross sections of the abundant U^{238} are so great that, despite the increased fission cross sections of U^{235}, the neutrons are effectively eliminated and virtually none will reach the thermal energy region.

Now, if a moderator is homogeneously mixed with or dispersed throughout the natural-uranium lump, some of the neutrons will be slowed down past the resonance-energy region of U^{238} by the strongly scattering moderator. However, because of the presence of U^{238} throughout the core, a sufficient number of neutrons of resonance-energy range are absorbed in U^{238} that the few neutrons reaching thermal energies will be less than 1/2.47 of the original 100 and the new fission neutrons will be less than 100. This causes the reaction to die down rapidly.

Thus a sustained (critical) chain reaction is impossible in a mass of natural uranium or in a homogeneous mixture of natural uranium and moderator. Actually, it is possible to store natural uranium plates in contact with each other to any desired height without fear of a critical reaction.

In order to obtain criticality or steady power, three methods are used: (1) building a *heterogeneous reactor,* (2) *enriching the fuel,* or (3) both.

If natural uranium or, in general, low-enriched fuels are to be used, the fuel must be subdivided into separate fuel elements in the form of pins, rods, plates, hollow cylinders, pellets, spheres, etc. These are placed in the core, with space between them filled with a moderator (Fig. 9-17). This is a heterogeneous reactor core. Because the fuel elements are relatively thin, a newly born fission neutron, even near the center of the elements, has a good chance of escape before attaining resonance energies because of the low fuel-scatter cross sections and the short distance it has to travel to

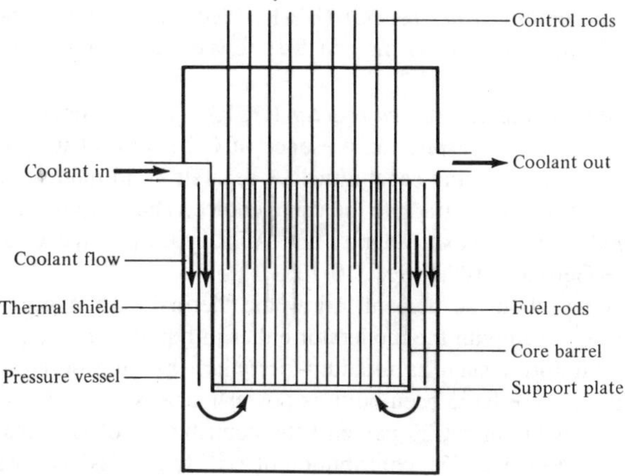

Figure 9-17 Schematic of a heterogeneous reactor.

get out of the element—a distance that is short compared with the neutron mean free path for scatter in the fuel. Thus this neutron spends the resonance-energy period outside the fuel element and escapes resonance capture. Once in the moderator, the neutron becomes thermalized in a few collisions. When it reenters the fuel, at thermal energies, the probability of fission by U^{235} far exceeds that of nonfission absorption by U^{235} and U^{238}, and a chain reaction is possible.

Actually, some resonance absorption is unavoidable, but not enough to adversely affect neutron economy. Also some fission of U^{238} by high-energy neutrons occurs in a heterogeneous reactor. Some nonfission absorption by U^{238} results in the production of Pu^{239}, a fissionable isotope, a process similar to that which occurs in fast-breeder reactors (Sec. 11-2).

Because neutron economy is difficult to come by with natural uranium, the moderator has to have a very low neutron-absorption cross section. This eliminates ordinary water and other hydrogenous materials, such as the organics, as moderators with natural uranium because they have relatively high neutron-absorption cross sections and consequently low moderating-to-absorption ratios. All water- and organic-moderated and cooled reactors must use fuels slightly enriched in U^{235}. A heterogeneous natural-uranium reactor can be built, however, with moderators of low neutron-absorption cross sections, such as graphite or heavy water, D_2O. The latter, called the *heavy-water reactor* (HWR), is the basis of the Candu reactor (Sec. 10-14).

The second method of attaining criticality is to enrich the fuel by artificially increasing the percentage of U^{235} (or other fissionable material) in it. In this case, the effect of U^{238} is less pronounced. Low-enriched fuels (of the order of a few percent), however, still have to be of the heterogeneous type for reasons similar to those given for natural uranium. Ordinary water, often incorrectly called *light water,* may be used as a moderator, resulting again in thermal heterogenous reactors.

Light-water reactors are of two main types: (1) the *pressurized-water reactor*

(PWR) (Sec. 10-6) and (2) the *boiling-water reactor* (BWR) (Sec. 10-7). Both constitute the largest number of reactors built during the first few decades of nuclear power.

A third thermal-reactor type is the *gas-cooled reactor* (GCR). The gas coolant is either CO_2, which is a poor moderator despite the presence of C because of its low density, or helium, which has no moderating capabilities. A separate solid moderator such as graphite, Be, or BeO is used, with graphite the usual choice. There are several types of such reactors, including the British Magnox and AGR types [3], the U.S. HTGR (Sec. 10-12), and the German THTR (Sec. 10-13).

Slightly higher enrichments (about 20 percent) allow the thermal homogeneous reactors to be built. The fuel is mixed with the moderator either in liquid form, called a *fluid-fueled reactor,* or solid form, such as uranium-zirconium hydride or UO_2-polyethylene mixtures. None of these have been built as power reactors.

In highly enriched fuels (beyond about 20 percent) the contribution of U^{238} resonance is no longer of prime importance. The contribution of U^{235} to the fission cross sections of the mixture outweighs U^{238} nonfission-capture effects. In this case, no moderator is necessary, and we may have a *fast reactor,* i.e., one relying primarily on high-energy (fast) neutrons for fission. Fast reactors can be homogeneous but are usually heterogeneous.

The fission energy generated in the fuel in all reactor types must be removed by a *coolant,* which leaves the reactor at a higher temperature than when it entered it. The rate of heat removal must be such that the fuel operates within safe temperature or boiling limits [2]. The energy carried out by the coolant is used in a thermodynamic cycle to generate electricity. It is fortunate that in most thermal reactors the coolant, ordinary or heavy water, can double as a moderator. Gas-cooled thermal reactors use either CO_2 or He for cooling and need a separate moderator. Fast reactors use no moderator and need a coolant that does not moderate the neutrons and that has a high heat-transfer coefficient because of the large power densities in fast-reactor cores. Liquid metals, particularly molten sodium (Na) are the most common. Such reactors are called *liquid-metal fast-breeder reactors* (LMFBR) (Secs. 11-3 to 11-5). Gases, particularly helium, have also been considered as fast-reactor coolants. Such reactors are called *gas-cooled fast reactors* (GCFR) (Sec. 11-6).

Table 9-8 contains the most common fission-power reactor types in use commercially, or under serious development, in the world today.

Table 9-8 The most-common fission-power reactor types

Neutron energy	Reactor type	Coolant	Moderator	Fuel, enrichment	Examples
Thermal	PWR	H_2O	H_2O	UO_2, low	Sec. 10-2
	BWR	H_2O	H_2O	UO_2, low	Sec. 10-7
	GCR	CO_2, He	Graphite	UC, low	Sec. 10-11
	HWR	D_2O	D_2O	UO_2, natural	Sec. 10-14
Fast	LMFBR	Na	None	$Pu^{239}O_2 + U^{238}O_2$	Sec. 11-3 to 11-5
	GCFR	He	None	$Pu^{239}O_2 + U^{238}O_2$	Sec. 11-6

9-15 REACTOR CONTROL

There are several methods to control a nuclear reactor, i.e., to start, increase, decrease, and turn off its power. The most common method is the use of *control rods*. These rods may be shaped like the fuel rods themselves and are interspersed throughout the core (Fig. 9-17). Instead of containing fuel, they contain a *neutron absorber,* also called poison, such as boron, cadmium, or indium. They have high neutron-absorption cross sections (Fig. 9-12) and do not contribute to neutron multiplication.

Such control rods are designed with sufficient absorber to change the neutron balance in the core so that less than one neutron is left for fission for each neutron engaging in fission, a situation that would lead to a power decrease and eventual shutdown. The control rods are operated by control-rod drives that can move them in and out of the core around a power equilibrium position which is usually a partially inserted position. The rods are moved out to increase power at a prescribed safe rate or are moved in to decrease power. In either case, when the required power is attained the rods are returned to the equilibrium position.

The rods can also be used to adjust power levels selectively within the core to help even out the radial power distribution. *Fuel zoning,* i.e., the use of variable fuel enrichment within the core (usually three roughly concentric zones are used) is also used to attain the same purpose.

Control rods can be suddenly and completely inserted in the core to shut it down in case of an emergency. Such an operation is called a *scram*. It can be done automatically or manually by a visible colored button on the reactor console in the control room. A number of rods may be built into the system which are usually fully withdrawn from the core during normal operation and which have the sole function of shutdown by becoming fully inserted upon demand.

The equilibrium position varies with the life of the core. As more fuel is depleted, i.e., is fissioned with time, fewer neutrons are produced and more neutrons are absorbed in nonfission reactions. The rods are then slowly moved to less-inserted positions to compensate for this loss of neutrons and to keep the core in equilibrium.

The rods are said to possess a *reactivity worth* that should be adequate to cover operational control during the life of the core, fuel depletion, safety shutdown, and such other effects as Doppler, samarium, and xenon poisoning [2].

Another method of control in pressurized-water reactors (Sec. 10-5), called *chemical shim,* is used in addition to, not in lieu of, control rods. Chemical shim is the use of a soluble absorber, usually boric acid, in the moderator coolant. The concentration of this absorber in the moderator coolant is decreased slowly during the core lifetime to overcome the effect of fuel depletion. The concentrations are sufficient to permit operating the core almost unrodded, i.e., with all control rods nearly fully withdrawn.

It should be mentioned here that boiling-water reactors do not use chemical shim, and that the core of a BWR can be in equilibrium at several positions of the control rods because of the strong effect of the steam voids on reactivity (Sec. 10-8).

Another control system is the use of *reflectors*. These are mechanically operated devices, situated just outside the core, that contain material that reflects some of the neutrons escaping the core back into it. The reflectors are swung away or toward or are axially moved with respect to the core to increase or decrease power. This method is used only for small-power reactors and in special cases, such as the SNAP 10A reactor which was operated in space [3].

Another method of control of lesser use yet is the use of some movable fuel rods. These are withdrawn from the core to decrease power and inserted to increase it, the opposite of the poison control rods.

PROBLEMS

9-1 Einstein's law (Eq. 9-1) applies to all processes: physical, chemical, and nuclear. Find the percent change in mass for the following physical and chemical processes: (*a*) when copper is heated from 100 to 1000°C, if the specific heat of copper is taken as 0.40 kJ/kgK, and (*b*) when carbon and oxygen burn to carbon dioxide, releasing 30,435 kJ/kg of carbon.

9-2 Find the percent change in mass for the following nuclear process: (*a*) when radioactive polonium-210 undergoes alpha decay for 365 days. Po^{210} has a half-life of 138.4 d and an atomic mass of 209.9829 amu. The products Pb^{206} and He^4 have atomic masses of 205.9745 and 4.00260 amu, respectively.

9-3 Find the percent change in mass for the following nuclear processes (*a*) when uranium-235 undergoes complete fission, and (*b*) when hydrogen-1 undergoes complete fusion.

9-4 A nuclear reactor powerplant operated continually for one year producing 500 MW. The powerplant efficiency is 33 percent. The reactor contained 75 metric tons of 3 percent enriched uranium dioxide fuel. Calculate (*a*) the mass of U^{235} consumed in kilograms, and (*b*) the fuel burnup, in MWd/T.

9-5 Write the complete decay reactions and calculate the energy generated, in million electron volts and joules per reaction of the following radioisotopes: (*a*) Po^{211}, an alpha emitter, (*b*) Zr^{95}, a beta emitter, (*c*) P^{30}, a position emitter, (*d*) Cu^{64}, in a K-capture reaction, and (*e*) Kr^{87}, a neutron emitter. The necessary atomic masses in amu are: Si^{30} = 29.97376, P^{30} = 29.97863, Ni^{64} = 63.9280, Cu^{64} = 63.9288, Kr^{86} = 85.9109, Kr^{87} = 86.9136, Zr^{95} = 94.9072, Nb^{95} = 94.9060, Pb^{207} = 206.9759, and Po^{211} = 210.9866.

9-6 Rutherford once postulated that when the earth was formed, it contained an equal number of atoms of U^{235} and U^{238}. From this he was able to determine the age of the earth. His answer was not far from that obtained by astronomical investigations. What is the Rutherford age of the earth?

9-7 Radioactivity exists almost everywhere, even inside our own bodies. The human body contains about 0.35 percent by mass of natural potassium, which contains 0.0118 atomic percent of radioactive potassium-40. K^{40} has an atomic mass of 39.9740 amu and a half-life of 1.28×10^9 years. Calculate the radioactivity, microcuries, in a 175-lb_m person.

9-8 When pure ordinary water is passed through a reactor core as a coolant-moderator, it becomes slightly radioactive. The most important of the radioactivities is due to the absorption of a neutron by an oxygen-16 nucleus. This absorption reaction results in a proton and a radioactive nucleus that has a 7.2-s half-life, as products. (*a*) Identify the nucleus, and (*b*) calculate the percent radioactivity left in the water 28.8 s after the above reactions.

9-9 *Carbon dating* is used to determine the age of materials of organic origin. The process involves the determination of the amount of *radiocarbon* (carbon-14) in them. When a nitrogen-14 nucleus in the atmosphere is bombarded by slow neutrons emitted in cosmic radiation, they result in carbon-14 and a proton. Carbon-14, like ordinary carbon-12, converts to $C^{14}O_2$, which constitutes 0.1 percent of all CO_2 in the atmosphere. Both are absorbed by living organisms. When absorption ceases due to the death of the organisms, the fraction of carbon-14 begins to decrease by radioactive decay. Estimate the age of an old manuscript if the amount of carbon-14 was determined by analysis to be 0.030 percent of all carbon in it.

9-10 Tritium decays by emitting low-energy β particles. This radiation acts on a phosphor producing illumination. Illuminators can thus be made by adding tritium to a phosphor in the form of paint which are sealed in a plastic container that is transparent to illumination but that blocks the β particles so that no hazard is encountered. The illuminators are used for such devices as locks, timepieces, aircraft safety markers, exit signs, etc. Regulations limit the amount of original radioactivity in such devices, depending upon service. Assuming that 4 mci are permitted for an aircraft safety device, calculate (a) the maximum mass of tritium that can be used, in grams, and (b) the percent decrease in luminosity (which is proportional to radioactivity) after 10 years of service.

9-11 Promethium-147 is another β emitter that is used in the manufacture of illuminators when combined with a phosphor. (It has a half-life of 2.5 years. It produces more luminosity than tritium at a lower cost.) Calculate (*a*) the activity, in millicuries, of 1×10^{-4} g of Pm^{147}, and (*b*) the percent decrease in illumination (proportional to the radioactivity) after 1 year of operation.

9-12 In fast-breeder reactors, plutonium-239 is the primary fuel. A relatively stationary Pu^{239} nucleus (atomic mass 239.0522 amu) is fissioned by a 1.0-million electron volts neutron resulting in two fission fragments: krypton-93 and cerium-144. Kr^{93} undergoes five stages of β decay and Ce^{144} two stages of β decay, both to stable products which have atomic masses of 92.060 and 143.9099 amu. (a) Identify all the fission products, and (*b*) calculate the total energy produced in million electron volts per Pu^{239} nucleus, and kilowatt-hours per gram of Pu^{239}.

9-13 When a spent fuel rod is removed from a reactor core, it is stored in an on-site storage pool of water so that the most intense, short-lived radioactive fission products decay, and the rod is safe for further handling and possible shipment to a reprocessing plant. Consider for simplicity only the radioisotope xenon-133 which β decays into a stable isotope with a half-life of 5.27 days. If a 30-kg fuel rod contains 0.1 percent by mass of Xe^{133} when removed from the core, what is the minimum time it should be stored in the pool so that the activity may not exceed 300 mci?

9-14 Boron-10 is used in reactor cores as a control-rod material. Natural boron has an atomic mass of 10.8110 a density of 2.3 g/cm^3 and contains 19.78 atomic percent of B^{10} which has an atomic mass of 10.01294 amu and a microscopic absorption cross section for 2200 m/s thermal neutrons of 3837 barn. Calculate the number of such neutrons absorbed per second by a 1-kg piece of natural boron.

9-15 Consider that 1 kg of pure ordinary water is subjected to an instantaneous thermal neutron pulse of 10^{10} per centimeters squared. Calculate (*a*) the radioactivity, in millicuries, generated immediately after the pulse in which oxygen-16 is converted to radioactive nitrogen-16 which has a half-life of 7.2 s, and (b) the radioactivity, in millicuries, one hour after the event. Ordinary water has a molecular mass of 18.01534 amu. O^{16} constitutes 99.759 atomic percent of all oxygen in ordinary water and has a microscopic cross section for thermal neutrons of 0.000178 barn.

9-16 The earth rotates around the sun with a mean radius of 149.5×10^6 km. The solar energy flux as measured just outside the earth's atmosphere (called the solar constant S, Sec. 13-3) is 1.353 kW/m^2. The reactions in the sun are of the hydrogen-1 fusion type, Eq. 9-14. Estimate (*a*) the total power generated by the sun, in megawatts, (b) the energy per fusion reaction, in million electron volts and joules, and (*c*) the mass of hydrogen-1 burned, in the sun in metric tons per day.

9-17 SNAPS (systems for nuclear auxiliary power) are devices that generate electric power directly from the heat generated by radioisotopic "fuels," in which case they are given odd numbers; or by fission nuclear reactors, in which case they are given even numbers. Direct generation is usually accomplished by thermoelectric energy conversion. An example is the Apollo lunar surface experiment package (ALSEP), called SNAP-27, which was placed on the lunar surface by the Apollo astronauts during their lunar landings in the late 1960s and early 1970s. SNAP-27 used plutonium-238 as "fuel" in the form of plutonium carbide PuC. Pu^{238} is an alpha emitter with an 86-year half-life. Assuming that the fuel deployed has a mass of 1 kg and that the efficiency of the thermoelectric conversion device is 8 percent, calculate (*a*) the power generated upon deployment, in watts, and (b) the power generated 5 years later. Atomic masses in amu: Pu^{238} = 238.0495, U^{234} = 234.0409, C = 12.0112. Density of PuC = 12.5 g/cm^3.

9-18 Calculate the power generated per unit volume in million electron volts per cubic centimeter, Btus per hour per cubic foot, and kilowatts per meter squared of a 3.5 percent enriched uranium dioxide fuel element in a thermal reactor if the effective fission cross section is 350 barns and the neutron flux is 10^{14}. The density of UO_2 is 10.5 g/cm^3.

9-19 A 12-ft-high fuel element has a 3.5 percent enriched uranium dioxide fuel with a U^{235} density of 8.3 $\times 10^{20}$ nuclei/cm^3. The fuel element diameter is 0.9 cm. The neutron flux in the fuel has a maximum value at the center plane of the element $\phi_c = 10^{14}$. It varies in the axial direction sinusoidally according to ϕ $(z) = \phi_c \pi z/H$, where z is the axial distance from the center plane and H is the height of the element. The effective fission cross section is 350 barn. Calculate the power generated by the fuel element, in kilowatts, and the average linear power, in kilowatts per foot.

9-20 A reactor core contains 43,120 vertical fuel elements of the type described in Prob. 9-19. The core may, for simplicity, be considered cylindrical with a diameter of 14 ft. The element in Prob. 9-19 is situated in the center of the core. The neutron flux varies radially according to $\phi(r) = J_0(2.4048\, r/R)$ where J_0 is the Bessel function of the first kind, zero order, r is the radius measured from the core center line, and R is the radius of the core. Consider an approximation where each fuel element generates power evenly over an area $\pi R^2/n$, where n is the number of fuel elements. Calculate the total power generated by the core in megawatts. Note: $\int_0^R r\, J_0(2.4048\, r/R) = (R/2.4048)\, [r\, J_1(2.4048\, r/R)]_0^R$, where J_1 is the Bessel function of the first kind, first order, and $J_1(0) = 0$ and $J_1(2.4048) = 0.519$.

9-21 The neutron fission cross for uranium-235 for 2200 m/s neutrons is 577.1 barn. A uranium dioxide fuel pellet 0.9 cm diameter and 1.5 cm high is subjected to a monenergetic neutron flux of 10^{14}, which was thermalized to a temperature of 260°C. The U^{235} density in the pellet is 8.3×10^{20} nuclei/cm³. The cross sections are within the $1/V$ range. (The energy of neutrons, like that of a perfect gas is given by kT, where k is the Boltzmann constant and T the absolute temperature.) Calculate the power generated by the pellet, in kilowatts.

9-22 Lithium is considered for use as a blanket material surrounding fusion reactors of the D-T type. It would receive high-energy neutrons from the D-T reaction, moderates (slows them down), absorbs them, and converts their kinetic energy to heat. That heat is then to be used in a thermodynamic cycle for power generation. Lithium also acts as a breeder of tritium for use with new deuterium fuel to keep the reactor going. Naturally occurring lithium is composed of 7.42 percent Li^6 and 92.58 Li^7. Upon neutron absorption each produces one helium nucleus and one tritium nucleus. (*a*) Write the complete nuclear equation for each. (*b*) Calculate the energy of each reaction, in million electron volts, stating whether exothermic or endothermic. (*c*) What needs to be done to naturally occurring lithium to make it a net energy generator assuming, for simplicity, that both reactions are equally probable. Atomic masses in amu: $Li^6 = 6.01512$, $Li^7 = 7.01600$, $H^3 = 3.01605$.

CHAPTER

TEN

THERMAL-FISSION REACTORS AND POWERPLANTS

10-1 INTRODUCTION

Nuclear fission was first discovered in Germany by Otto Hahn and Fritz Strassmann in 1938. Since then, great strides have led to its utilization in first, unfortunately, destructive uses, then in peaceful uses for meeting the increasing demand for abundant and reliable electric power. The time span between discovery and utilization was dizzyingly short when compared with other technologies. Because of the pace of development and the still lingering destructive specter, public acceptance problems have arisen.

As with all complex technologies, the first generation of nuclear-fission powerplants needed improvements in design, construction, and operation. The present efforts now are concentrating on improving the design, safety, and operation of proven systems, such as, the pressurized-water, boiling-water, and gas-cooled thermal reactors, and on building demonstration plants of the fast-breeder reactor.

The Hahn and Strassmann experiments that led to the discovery of fission came about, as with many great scientific discoveries, almost by accident. It showed that uranium was split, or fissioned, into smaller elements. This was the opposite of what was expected, namely the formation of larger and heavier elements than uranium, the now called *transuranium elements*. Within a few short weeks, worldwide interpretations led to significant and far-reaching effects on the technological, economic, and political future of the world.

Ten days after publication of the Hahn and Strassmann experiments, on 16 January 1939, Lise Meitner and Otto Robert Frisch published notes in *Nature* in which they made theoretical interpretations of these experiments. On 7 April 1939, Frederic Joliet, Hans von Halban, and Lew Kowarski published the paper "Liberation of Neutrons in the Nuclear Explosion of Uranium," which dealt with the possibility of a nuclear chain reaction.

Earlier, in 1905, Albert Einstein, then a young physicist and an assistant at the patent office in Bern, published in the German journal *Annals of Physics* a $2\frac{1}{2}$-page supplement to his theory of relativity entitled "Is the Inertia of a Body Dependent on Its Energy Content?" In it, Einstein arrived at his famous theory of the convertability of mass m and energy E, expressed as

$$E = mc^2$$

where c is the velocity of light (300,000 km/s). Einstein himself calculated that if mass is reduced by 1 g, an amount of energy equal to 9×10^{20} ergs is produced. He wrote at the time:

> The mass of a body is a measure of its energy content. If the energy is changed, then the mass will change in the same way.

At a later date, he wrote:

> Is it not impossible that substances whose energy content can be varied to a high degree (for example, the radium salts) will make it possible to test the theory? If the theory is in accordance with the facts, then radiation transmits inertia between the emitting and the absorbing body.

On 2 August 1939, Einstein, then living in the United States, wrote a historic letter to President Franklin D. Roosevelt that drew attention to the possibility of an atomic bomb and, considering the possibility of a German lead, urgently advised the president to make preparations for the production of nuclear weapons in the United States. This dramatic event was made possible by Hahn and Strassmann's discovery, the final necessary link in a chain of scientific discoveries that made the whole thing feasible.

Unfortunately, then, the first use of fission was for destructive purposes, a birth from which the nuclear industry continues to suffer today. However, for the sake of completeness, a word on weaponry is appropriate. The first fission bomb exploded at Hiroshima, Japan, had a uranium content of approximately 50 kg, and had the equivalent destructive power of 20,000 tons of trinitrotoluene (TNT). Of the 50-kg content, only 1 kg actually fissioned, and of that only 1 g of mass was converted to energy and disappeared. The second bomb, which exploded at Nagasaki, Japan, used plutonium as fuel. The largest known nuclear explosion was detonated by the USSR in 1961. It was a hydrogen bomb (fission plus fusion) and had the equivalent destructive power of 60 million tons of TNT. In all such explosions considerable amounts of fission products are formed, thus producing large amounts of lethal radioactive radiations as a by-product. Although this may have been "desirable" from a military standpoint, it is definitely unacceptable for peaceful, commercial uses of nuclear energy. Such radiations must be minimized and contained.

The first known thoughts regarding harnessing the tremendous explosive powers of fission for the production of energy were voiced by a 38-year-old German nuclear physicist named Werner Heisenberg (who had previously received a Nobel Prize at age 31) in a 1939 paper entitled "The Possibility of Large-Scale Energy Production Using Uranium Fission." In it, he wrote:

> The data available at present indicate that the uranium fission processes discovered by Hahn and Strassmann can also be used for large-scale energy production. The most reliable method for developing a suitable machine is the enrichment of the uranium-235 isotope. The greater the degree of enrichment, the smaller the size of the machine needed. The enrichment of uranium-235 is the only method that allows the volume of the machine to stay small, that is about 1 cubic metre. Moreover, it is the only method of producing explosive substances that exceed by several decimal powers the explosive force of the strongest explosive known to date. It is, however, also possible to use normal uranium without uranium-235 enrichment, if the uranium is combined with another substance that slows down the neutrons of the uranium without absorbing them. Water is not suitable for this purpose, but present data indicate that heavy water and very pure carbon fulfill this purpose.

In a February 1940 paper, Heisenberg described the construction and operation of a nuclear "reactor." The theoretical concepts presented in that paper do not differ greatly from those currently used in present-day reactors.

Practical work on peaceful energy production required the use of enriched fuel. The enrichment process posed almost insurmountable difficulties. Several methods were considered, including the ultracentrifuge, the diffusion process, and others. In Germany the ultracentrifuge process was pursued but did not meet with much success, and the use of natural uranium and heavy water to produce plutonium was pursued. Only small amounts of heavy water were available. The large amounts needed were sought from Norsk Hydro, the Norwegian hydroelectric utility located in Vermork near Rjukan in southern Norway. This company was engaged in the production of ordinary hydrogen for ammonia synthesis, and heavy water was produced as a waste product. The plant was destroyed in a daring raid by allied forces.

In the United States, huge diffusion installations erected at considerable cost succeeded in separating the chemically identical, but nuclearly different, uranium isotopes. Also a team lead by Enrico Fermi worked with great intensity on a natural uranium reactor moderated by graphite instead of heavy water, a process tried by the Germans but not pursued further by them because their graphite was not pure enough and absorbed too many neutrons. The Germans slowed down in any case because of the indifference of the Third Reich. The United States, on the other hand, gave high priority to nuclear research. It was interested in both uranium-235, as fuel for both peaceful and military purposes, and plutonium, a transuranium element discovered in 1940 by Edwin McMillan and Glenn Seaborg (a discovery that earned them both a Nobel Prize in 1951).

On 2 December 1942, a coded message was sent to Washington. It read "The Italian sailor has arrived in the new world." This signalled that the world's first nuclear reactor went critical. It was situated under the stands of the University of Chicago football stadium. The "Italian sailor" referred to was Enrico Fermi, who had come to the new world only on 2 January 1937. That reactor, called the Chicago Pile-1 (CP1), was 9 m wide, 9.5 m long, and 6 m high. It contained about 52 tons of natural uranium and about 1350 tons of graphite. Cadmium rods were used as control devices. The experiment produced an output of 0.5 W and lasted only a few minutes. However, it was definite proof that a continuous chain reaction was possible, a feat that had eluded scientists previously. The Fermi chain reaction was the event that signalled the dawn of the nuclear age.

Parallel efforts of isotope separation at Oak Ridge, large plutonium production reactors at Hanford, and research at Los Alamos (a town constructed for atomic research in 1943 with people like Niels Bohr, James Chadwick, Enrico Fermi, Hans Bethe, and J. Robert Oppenheimer as the leader) paved the way to making the United States the leading atomic power at the time.

Many events took place after this, including the construction of hydrogen weapons by both the United States and the USSR. But the event of most concern to us here is the famous and dramatic Atoms for Peace address by President Dwight Eisenhower to the United Nations General Assembly on 8 December 1953. In part, he said:

> The United States knows that peaceful power from atomic energy is no dream of the future. That capability, already proved, is here now—today. Who can doubt, if the entire body of the world's scientists and engineers had adequate amounts of fissionable material with which to test and develop their ideas, that this capability would rapidly be transformed into universal, efficient and economic usage?
>
> To hasten the day when fear of the atom will begin to disappear from the minds of people and the governments of the East and West, there are certain steps that can be taken now . . . without irritations and mutual suspicions incident to any attempt to set up a completely acceptable system of a worldwide inspection and control. The atomic energy agency could be made responsible for the impounding, storage and protection of the contributed fissionable and other materials.
>
> The more important responsibility of this atomic energy agency would be to devise methods whereby this fissionable material would be allocated to serve the peaceful pursuits of mankind. Experts would be mobilized to apply atomic energy to the needs of agriculture, medicine and other peaceful activities. A special purpose would be to provide abundant electrical energy in the power-starved areas of the world. Thus, the contributing powers would be dedicating some of their strength to serve the needs rather than the fears of mankind.

The first reactor to produce electricity was a small 5-MW unit built near Moscow, USSR. But perhaps the most significant contribution to commercial nuclear power was the development of the nuclear submarine. The *U.S.S. Nautilus,* launched in 1954, signaled the age of controlled nuclear power. Another milestone was the commissioning of the world's first full-scale powerplant at Calder Hall, England, a 180-MW(e) gas-cooled, graphite-moderated, natural-metallic-uranium-fueled reactor that can be considered the true descendent of the Fermi pile.

In the United States, and then most of the world, development proceeded with water-cooled-and-moderated reactors that use slightly enriched fuel. The first plant had a 60-MW(e) pressurized-water reactor (PWR) that began operation in 1956 in Shippingport, Penn. It was followed by a 184-MW(e) boiling-water-reactor (BWR) plant that began operation in 1960 in Dresden, Ill. Capacities reached 500 MW(e) with the San Onofre, Calif., PWR in 1968 and the Oyster Creek BWR in 1969 and inched upward to the present 1000 to 1250 MW(e) in several plants around the world.

In the previous chapter we learned that there are two kinds of fission reactors, thermal and fast. Thermal-reactor powerplants, i.e., those using thermal reactors as a heat source, will be discussed in this chapter. Fast-breeder-reactor powerplants will be discussed in the next. Thermal reactors are those in which fission is primarily caused by thermal neutrons. They, therefore, need a moderator to thermalize the neutrons as well as a coolant to remove the heat generated by the fission process. The moderator and coolant can be one and the same, such as light water or heavy water, or different, such as a graphite moderator and a gas coolant such as helium or carbon dioxide.

Although there have been many concepts for thermal reactors, we will limit our discussion in this chapter to four powerplant types that have been built commercially. These are

1. Pressurized-water-reactor (PWR) powerplants
2. Boiling-water-reactor (BWR) powerplants
3. Gas-cooled-reactor (GCR) powerplants
4. Heavy-water-reactor (PHWR) powerplants

There are 511 thermal reactor powerplants that are operable, under construction, or on order in the world as of early 1983. Of these there are 284 PWRs, 132 BWRs, 53 GCRs, and 39 PHWRs; the rest not yet decided. They represent about 391,600 MW of power. In addition, there are 7 fast-breeder reactor powerplants representing about 3280 MW. In the United States the number is 144 thermal and one fast breeder, representing about 135,000 MW [88]. In 1982 the United States had 76 operating plants that generated 280 billion kWh, about one eighth of the country's consumption.

10-2 THE PRESSURIZED-WATER REACTOR (PWR)

In a PWR, the coolant pressure is higher than the saturation pressure corresponding to the maximum coolant temperature in the reactor, so that no coolant boiling takes place. A PWR powerplant is composed of two loops in series, the coolant loop, called the *primary* loop, and the water-steam or *working-fluid* loop (Fig. 10-1), the coolant picks up reactor heat and transfers it to the working fluid in the steam generator. The steam is then used in a Rankine-type cycle to generate electricity.

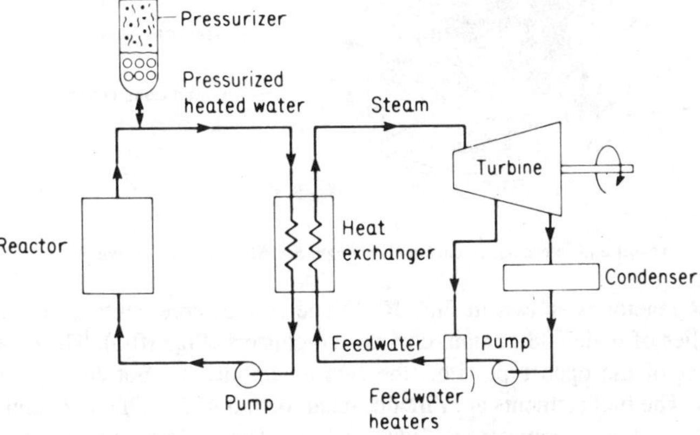

Figure 10-1 Schematic arrangement of a PWR powerplant.

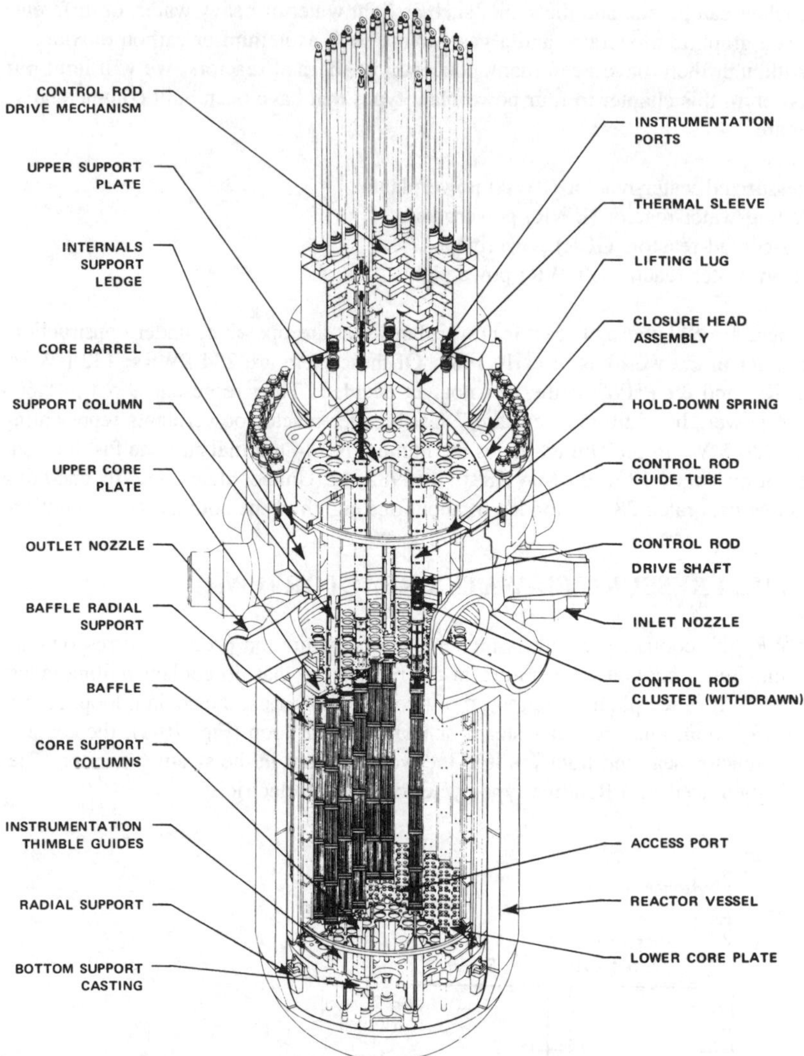

Figure 10-2 PWR reactor vessel and internals. *(Courtesy Westinghouse Electric Corporation.)*

A typical PWR reactor is shown in Fig. 10-2. The reactor core contains a total of 121 fuel assemblies of which 33 contain control-rod clusters (Fig. 10-3). The core, unlike in a BWR, is of the open type, i.e., the fuel assemblies are not enclosed in individual channels. The fuel elements are Zircaloy-clad rods, 0.422 in OD, containing UO_2 pellets, each 0.3669 in in diameter and 0.600 in long. The cladding tube is sealed at both ends by a plug welded to it. Sufficient void is left at the top to accommodate

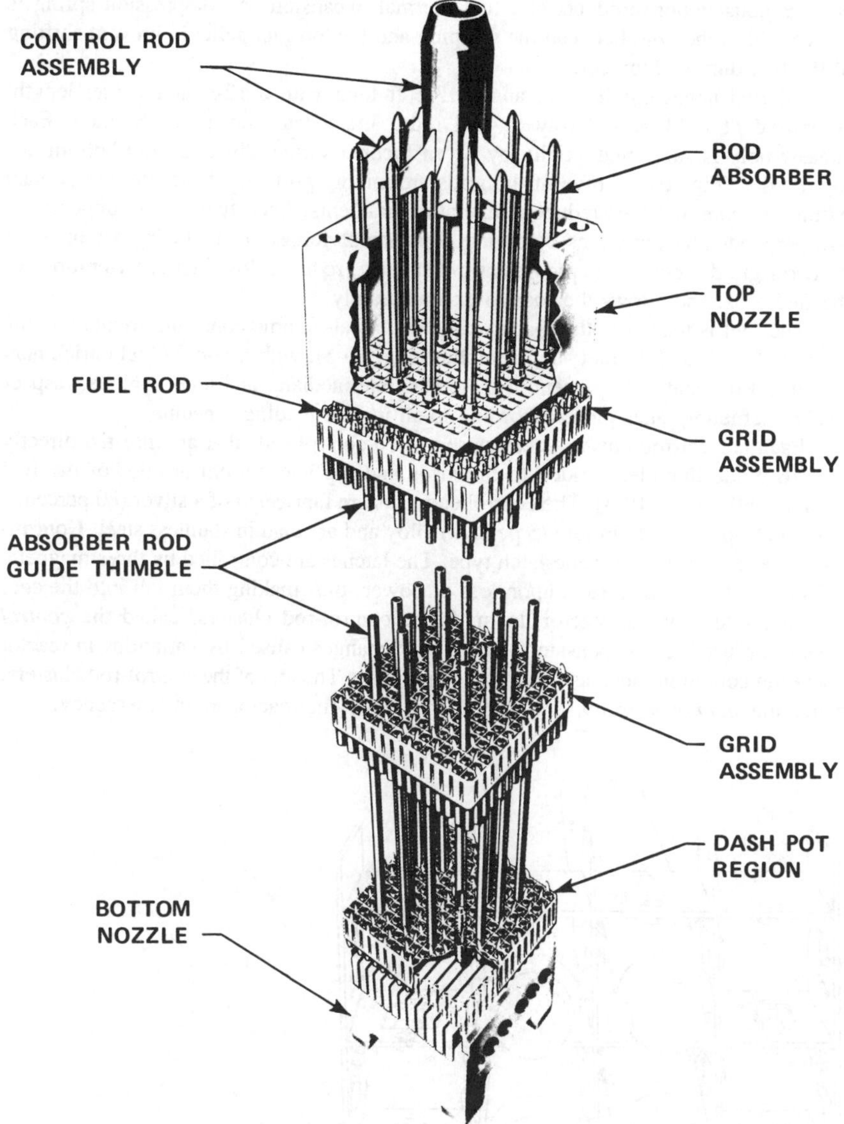

Figure 10-3 Typical control-rod cluster fuel assembly. *(Courtesy Westinghouse Electric Corporation.)*

both gaseous fission products and fuel thermal expansion. A compression spring is placed within the void between the top plug and the top fuel pellet to prevent shifting of the fuel during shipment.

All fuel subassemblies are about 13.5 ft long with a 12-ft active fuel length, composed of a 14 × 14 array of fuel rods, and located on a square pitch. Each subassembly is supported axially by seven Inconel spring clip grids and bottom and top nozzles (Fig. 10-4). Five of the grids are mixing grids that help intermix coolant within the core and thus reduce temperature gradients. Each fuel rod is supported in two perpendicular directions by spring clips whose forces (11 to 14 lb_f) are opposed by two rigid dimples. This provides rigid support, reduces flow-induced vibrations of the fuel rods, and allows the rods to expand axially.

The fuel is loaded in three approximately equal-volume concentric regions in the core of 40, 40, and 41 fuel subassemblies (Fig. 10-5), with first-core fuel enrichment of 3.40, 3.03, and 2.27 percent in the outer, intermediate, and inner regions, respectively. Refueling takes place according to an inward loading schedule.

Each control-rod cluster is composed of 16 control rods that are inserted directly into 16 guide thimbles welded to the grids and top and bottom nozzles of the fuel subassemblies (Fig. 10-3). The control elements are fabricated of a silver (80 percent)-indium (15 percent)-cadmium (5 percent) alloy and are clad in stainless steel. Control-rod drives are of the magnetic-latch type. The latches are controlled by three magnetic coils that release the clusters upon loss of power, thus making them fall into the core by gravity to shut the reactor down. Some control-rod clusters, called the *control group,* are used to compensate for reactivity changes caused by variations in reactor operating conditions such as power or temperature. The rest of the control-rod clusters, called the *shutdown group,* are used to shut down the reactor in an emergency.

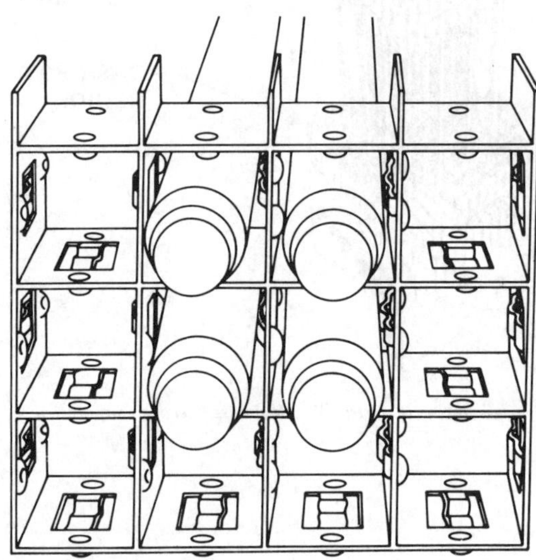

Figure 10-4 Fuel element spring clip grid detail. *(Courtesy Westinghouse Electric Corporation.)*

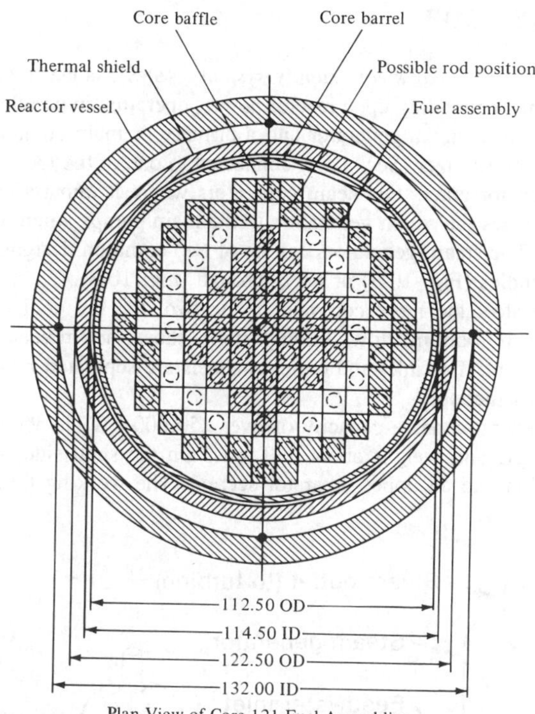

Figure 10-5 Typical PWR reactor cross section at the core.

The core is surrounded by a form-fitting baffle (Fig. 10-5) that restricts the bulk of upward coolant flow to the fuel. The baffle is in turn surrounded by the core barrel. A small amount of coolant is allowed to flow between baffle and barrel. The coolant is diffused uniformly into the core by a perforated flow-mixture plate situated between the core support plate and the lower core plate. A thermal shield, supported by the core barrel, is provided to intercept core radiations and protect the pressure vessel.

The primary coolant enters the reactor vessel at about 552°F (289°C) via a number of inlet nozzles (two for 500 MW, four for 1000 MW and larger, three for intermediate) and flows downward through the annulus between the core barrel and reactor vessel wall (Fig. 10-5), thus cooling the thermal shield on both sides. It then enters a plenum at the bottom of the vessel, reverses direction, goes upward through the core where it picks up fission heat, and leaves through an equal number of exit nozzles at about 605°F (318°C). The maximum coolant temperature at the exit of the center fuel assemblies is about 650°F (343°C). The reactor coolant pressure is 2235 psig (155 bar), greater than the saturation pressure at 650°F.

10-3 THE PWR PRIMARY LOOP

The primary loop, also called the *nuclear steam supply system* (NSSS), consists of the reactor and a number of loops, depending upon reactor power, operating in parallel (Fig. 10-6). Each loop consists of a steam generator and a primary or main coolant pump. In addition, there is one pressurizer (Sec. 10-4) connected to one of the loops.

The coolant leaving the reactor enters the steam generators where it imparts its heat to the working fluid and leaves the steam generators to the main pumps where it is pumped back to the reactor. The steam generators can be of two common designs, shell-and-tube, with U-tube bundles (Fig. 10-7) or once-through (Fig. 10-8).

In the U-tube steam generator, the more common of the two, the hot coolant enters an inlet channel head at the bottom, flows through the U tubes, and reverses direction to an outlet at the bottom. The inlet and outlet channels are separated by a partition. The tubes are made of Inconel.

A typical U-tube steam generator has a capacity of over 250,000 kW, is about 67 ft high and 14 ft in diameter, and weighs about 330 tons. On the shell side, it consists of an evaporator section and an upper separator section. The working fluid

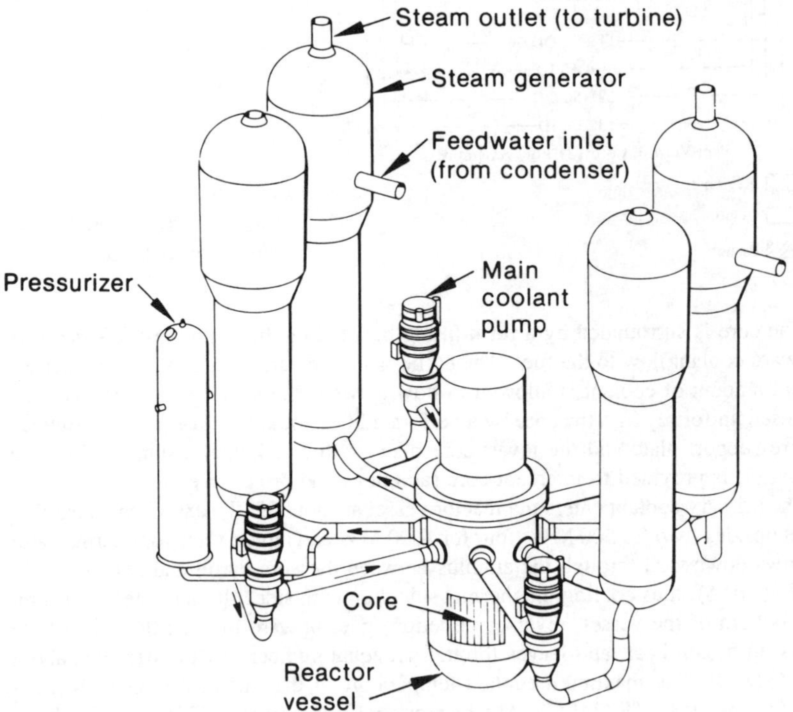

Figure 10-6 PWR nuclear steam supply system (NSSS).

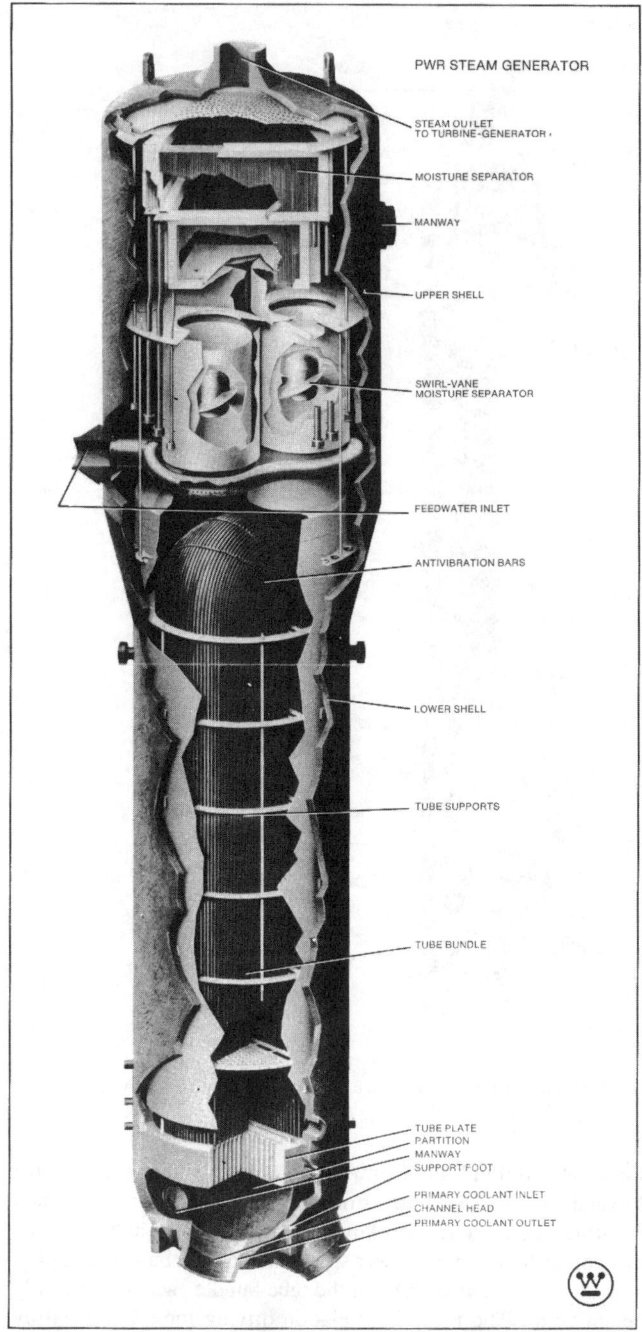

Figure 10-7 U-tube PWR steam generator. *(Courtesy Westinghouse Electric Corporation.)*

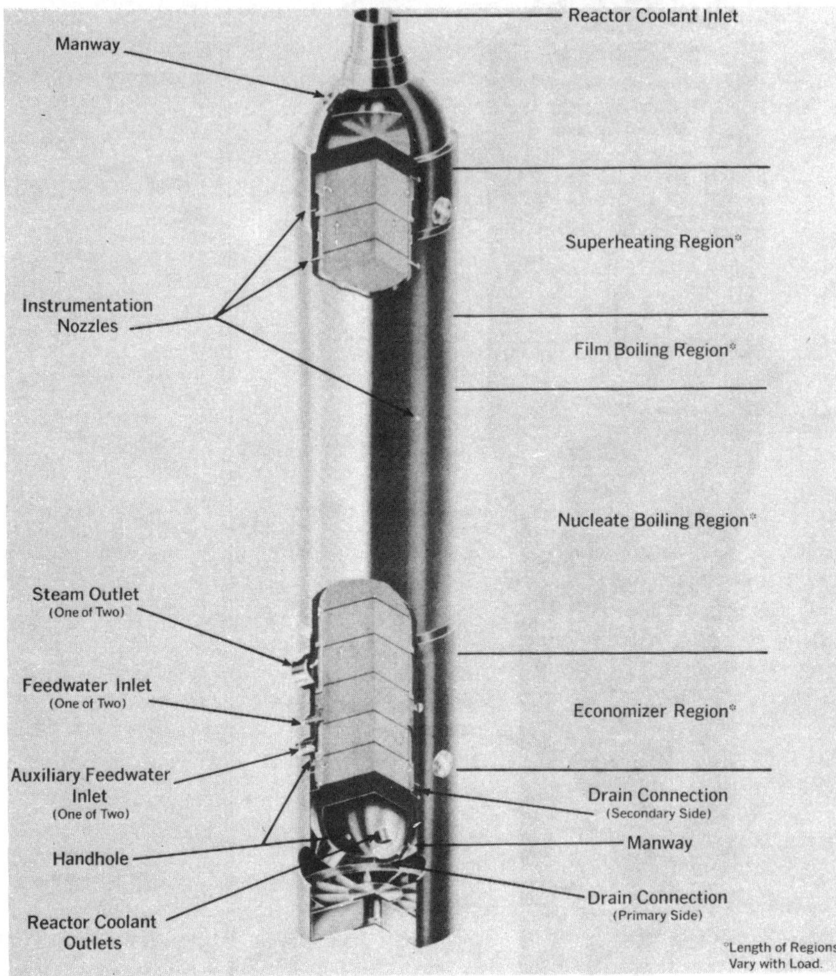

Figure 10-8 Once-through PWR steam generator. *(Courtesy Babcock and Wilcox Company.)*

(feedwater) enters the generator through the feedwater inlet nozzle above the U-tube bundle and mixes with water that is recirculated from the moisture separators located near the top of the generator. The mixture then flows downward to the bottom of the shell via an annular downcomer between the lower shell and the tube bundle wrapper. It then flows upward by natural circulation through the tube bundle, where it partially boils into a steam-water mixture. The natural circulation driving force is caused by the density difference between the water in the downcomer and the two-phase mixture in the evaporator section.

The steam-water mixture is separated in the upper shell, first by swirl vane separators and finally by vane-type separators. Dry saturated steam discharges through the steam outlet nozzle at the top. The saturated water leaves the separators and mixes with feedwater before entering the downcomer.

Because boiling occurs in the same compartment with water, this type of steam generator can produce only saturated steam (or steam with very low moisture, about $\frac{1}{4}$ percent).

In the once-through steam generator (Fig. 10-8), the primary coolant enters at top, flows downward through the tubes, and exits at bottom to the main pumps. Feedwater is on the shell side moving in a general counterflow fashion to the primary water. Because of the once-through feature of this type, a dry or a low degree of superheat steam is possible and no separators are required.

Although the once-through generator results in somewhat better turbine efficiency, it has the disadvantage of containing a low water volume and, therefore, a reduced reserve in case of a reactor accident.

Steam is usually produced at about 1020 psia (70.3 bar). The main pumps are large vertical single-stage centrifugal shaft-seal pumps designed to handle large volumes of water at high pressure and temperature.

10-4 THE PRESSURIZER

It has been shown that, in PWR primary loops, the coolant is maintained at a pressure (around 2250 psia, 155 bar) greater than the saturation pressure corresponding to the maximum coolant temperature in the reactor. This avoids bulk boiling of the coolant and keeps it in the liquid phase throughout the loop. Because liquids are practically incompressible, small changes of volume—caused by changes in coolant temperatures becaues of normal load changes or accidental nuclear reactivity insertions or caused by unforeseen expansions or contractions in the loop components—cause severe or oscillatory pressure changes. These may be quite unsafe when they are positive, that is, when the pressures increase. They cause flashing into steam and consequent disruption of the reactor nuclear characteristics and possible burnout of reactor fuel elements. They cause cavitation when they are negative, that is when the pressures decrease. For these reasons, it is necessary to provide a surge chamber that will accommodate coolant volume changes while maintaining pressure within acceptable limits. Such a chamber is called a *pressurizer*. There are two types of pressurizers in common use: vapor pressurizers and gas pressurizers. Pressurized-water reactors conveniently use vapor pressurizers because their coolant, water, can be vaporized, which results in a more compact pressurizer. Gas-type pressurizers are used in liquid-metal-cooled fast-breeder reactors.

A *vapor pressurizer* is essentially a small boiler (Fig. 10-9) in which liquid, the same as the primary coolant, is maintained by controlled electrical heating at a constant temperature and consequently a constant vapor pressure above its full surface. This pressure is the same as that of the primary coolant at the junction between the pressurizer and the hot leg of the primary loop. Thus the pressurizer temperature is higher than

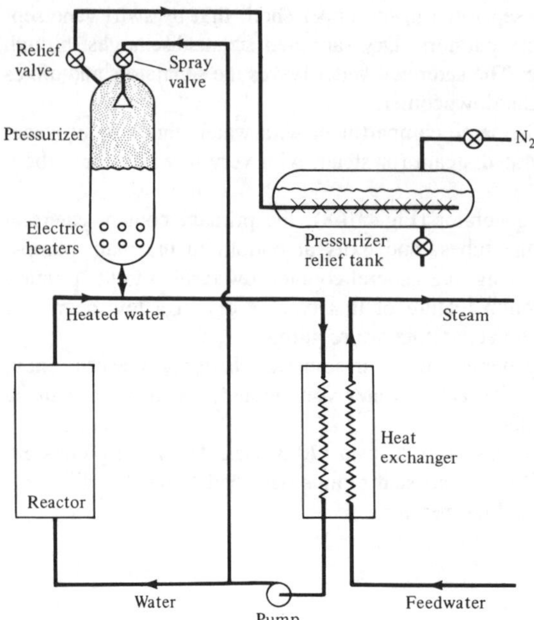

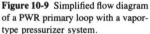

Figure 10-9 Simplified flow diagram of a PWR primary loop with a vapor-type pressurizer system.

the primary-coolant temperature because the latter is subcooled. For example, if the primary-loop pressure and temperature at the junction are 2250 psia and 605°F, the pressurizer temperature would be 653°F.

The heaters are of the electric immersion type, located in the lower section of the pressurizer vessel. These heaters are also used to heat the pressurizer and its contents at the desired rate during plant startup.

The bottom of the pressurizer is connected to the hot leg of the primary coolant system (Fig. 10-9). A spray nozzle located at the top of the pressurizer is connected, via control valves, to the cold leg of the primary coolant system after the pump. Under normal full-power operation, the pressurizer is about half full of water. The top half is full of vapor.

During a positive surge, the volume of the primary coolant increases and the vapor in the top half is compressed. Entry of the cooler primary coolant into the pressurizer condenses some of the vapor, thus limiting the pressure rise. In addition, the spray valves are power-actuated, and a cool spray (under pump pressure) enters the top, which helps condense vapor at a rapid rate and limits pressure rise. (The spray valves may also be manually operated.) A small continuous spray is usually provided to prevent excessive cooling of the spray piping and to maintain equal boron concentrations in the primary coolant and pressurizer water. Boron is used for chemical shim (Sec. 10-5).

A negative surge decreases the primary-coolant volume and expands the vapor in the pressurizer, thus causing a momentary reduction in pressurizer pressure. The liquid in the pressurizer then partially flashes into vapor, and, assisted by further steam generation because of the automatic actuation of the electric heaters, the pressure is maintained above a minimum allowable limit.

A power-operated relief valve is attached to the top of the pressurizer to protect against pressure surges that are beyond the capacity of the pressurizer. The relief valve, in such a case, discharges steam into a pressurizer relief tank that is partly filled with water under a nitrogen blanket at near-room temperature and in which the vapor condenses. The condensate then goes to a waste-disposal system.

In a typical design of a 500-MW PWR, both the pressurizer and pressurizer relief tank had volumes of 800 ft^3 each, compared with 3700-ft^3 volume of the reactor pressure vessel. Figure 10-10 shows a photograph of a pressurizer for a PWR plant built by Westinghouse Electric Corporation.

The following is a treatment of the pressure change in an insurge (when a liquid expansion in the primary circuit causes a rush of primary coolant into the pressurizer) or, conversely, an outsurge. The situation is a nonsteady flow case. The general energy equation [Eq. (1-1)] in that case is written as

$$\begin{aligned} \text{PE}_1 + \text{KE}_1 + \Delta m_i h_i + m_{f,1} u_{f,1} + m_{g,1} u_{g,1} + \Delta Q \\ = \text{PE}_2 + \text{KE}_2 + \Delta m_e h_e + m_{f,2} u_{f,2} + m_{g,2} u_{g,2} + \Delta W \end{aligned} \tag{10-1}$$

where the subscripts 1 and 2 refer to conditions within the pressurizer before and after the insurge (or outsurge) and i and e refer to inlet and exit fluids. The changes in potential energy PE and kinetic energy KE as well as the work ΔW are zero. Equation (10-1) then becomes

$$\Delta m_i h_i + m_{f,1} u_{f,1} + m_{g,1} u_{g,1} + \Delta Q = \Delta m_e h_e + m_{f,2} u_{f,2} + m_{g,2} u_{g,2} \tag{10-2}$$

The above is an energy balance. A mass balance gives

$$\Delta m_i + m_{f,1} + m_{g,1} = \Delta m_e + m_{f,2} + m_{g,2} \tag{10-3}$$

A volume balance gives

$$m_{f,1} v_{f,1} + m_{g,1} v_{g,1} = m_{f,2} v_{f,2} + m_{g,2} v_{g,2} = V \tag{10-4}$$

where Δm = mass of water entering or leaving pressurizer

m_f, m_g = masses of water and steam within pressurizer

h = enthalpy of water entering or leaving pressurizer

u_f, u_g = internal energies of water and steam within pressurizer

v_f, v_g = specific volumes of water and steam within pressurizer

ΔQ = heat added from electric heaters, less heat lost to the ambient (relatively small)

V = total volume of pressurizer

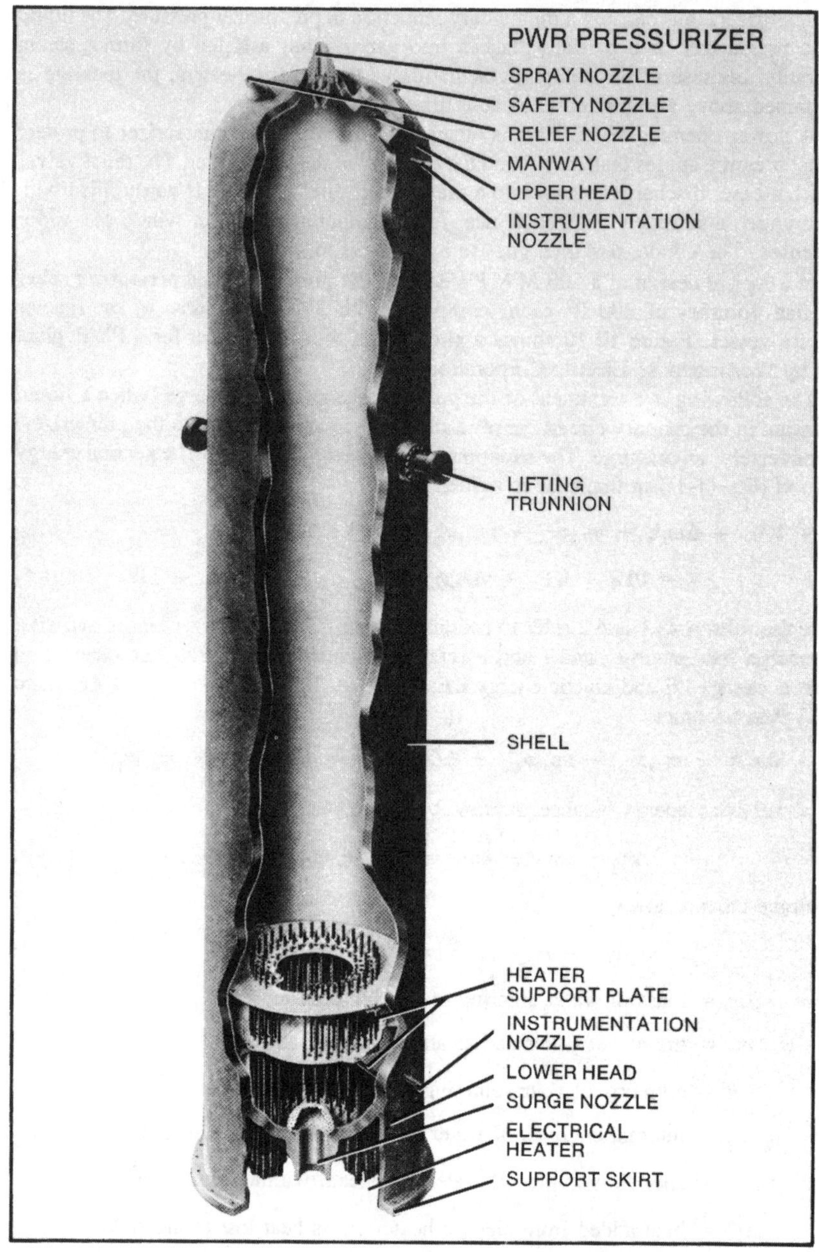

Figure 10-10 A PWR pressurizer. *(Courtesy Westinghouse Electric Corporation.)*

Noting that the pressurizer is large enough so that there will always be both water and steam in it,* the values of u and v above are for the saturated fluids at the corresponding pressures.

The above three equations are solved for the case of an insurge, where $\Delta m_c h_i$ is the sum of the incoming fluids from the hot leg and the spray from the cold leg, each times its specific enthalpy, and

$$\Delta m_e = 0 \tag{10-5}$$

or an outsurge, where

$$\Delta m_i = 0 \tag{10-6}$$

Solutions, for given primary-system conditions and permissible pressure fluctuations result in required total pressurizer volume. For given pressurizer volume and primary-system temperature fluctuations, the solutions (usually by trial and error) give resulting pressure fluctuations. As expected, the larger the pressurizer, the smaller the pressure surges.

Basic equations, for vapor-type pressurizers, that predict system transient pressure-time relationships during changes in reactor power level have been formulated [89] from energy, mass, and volume balances.

A *gas pressurizer* is simply a large volume of gas situated above the primary coolant that compresses or expands whenever the primary coolant expands or contracts, respectively. The gas, not miscible with the coolant, thus acts as a cushion to limit pressure changes in the primary system. Because of the absence of condensing or vaporizing of the gas, as occurs with the steam in a vapor pressurizer, a gas pressurizer is usually large in volume. It is limited in use to low-pressure systems such as the liquid metal fast breeder reactors. It is discussed here merely for comprehensiveness of coverage.

The pressure rise (or decrease) in a gas pressurizer is easily obtained from the gas laws $PV = mRT$ [Eq. (1-30*a*)] and $PV^n = C$ (Table 1-3) giving

$$\frac{\Delta P}{P_1} = \left(\frac{V_1}{V_1 + \Delta V}\right)^n - 1 \tag{10-7}$$

where ΔP = system pressure rise or decrease, lb_f/ft^2 or Pa

P_1 = initial system pressure, lb_f/ft^2 or Pa

V_1 = initial volume of gas, ft^3 or m^3

ΔV = volume change of gas

= negative of volume change of primary coolant, ft^3 or m^3

n = coefficient of polytropic exponent during gas compression or expansion

= 1 for an isothermal process

= ratio of specific heats of gas k for an adiabatic reversible process.

* The pressurizer must never be allowed to be completely filled with water, a situation referred to as operating "solid," as it would lose all control over system pressure.

10-5 CHEMICAL-SHIM CONTROL

The term *chemical shim* refers to the use of a soluble neutron absorber, such as boric acid, that is dissolved in the primary reactor coolant. Control is then accomplished by varying the concentration of this absorber in the coolant. This, of course, is a slow process and is used only to control slowly varying reactivity effects in addition to conventional control rods.

Boric acid has good water solubility and has been used experimentally in both pressurized- and boiling-water reactors. For commercial power reactors, however, it is restricted to the former use. Since boron has no radioactive isotopes, no coolant radioactivity problems arise from it. (Boric acid has also been used as a shutdown device in many reactors.) The concentration of boric acid in the coolant is changed at startup and during the lifetime of a core to compensate for (1) changes in reactivity resulting from changes in moderator temperature from a cold shutdown condition to a hot operating, zero-power condition; (2) changes in reactivity caused by the buildup of neutron-absorbing xenon 135 and samarium 149 concentrations in the core; and (3) reactivity losses resulting from fuel depletion and the buildup of long-lived fission products other than zenon and samarium. Rapid reactor transients are handled by the usual control rods.

Boron concentration in the coolant may be adjusted by the *feed-and-bleed method*. By this method boron content is increased by feeding into the core some of a more concentrated boron solution than is in the core and decreased by feeding in some pure water or a less concentrated solution. Some coolant must necessarily be bled off to make room for the feed. This may be processed by distillation and the resulting concentrate and distillate reused for subsequent adjustments.

Since chemical shim permits a reduction in the amount of reactivity controlled by control rods, the number and/or size of rods may be reduced, the result being simplified design and reduced costs. Also, the rod blackness may be reduced; i.e., rod materials of lower neutron cross sections may be used. Note that because of strong water moderation a pressurized-water reactor normally needs a large number of control rods.

A chemical absorber does not by itself materially affect the spatial power distribution since it is uniformly distributed throughout the core except for a minor effect due to increased density in the lower half of the core because of cooler water temperatures there. The use of chemical shim results, however, in improvements in spatial power distribution and therefore increased average-to-maximum power density in the core because of reduced blackness, size, or degree of insertion of control rods.

The control rods that remain inserted in a chemically shimmed core will, as in any core, distort the axial power distribution (which is normally sinusoidal) so that peak power density is pushed from the core center to a point farther from the rod entrance (Fig. 10-11). The axial maximum-to-average power ratio is low when the rod is fully withdrawn because of the uniformity of the channel, increases with insertion up to a maximum, and then decreases as the rod approaches full insertion. Near full insertion the ratio increases again to the original value because the rod channel approaches uniform composition again. A hypothetical rod of zero reactivity worth (zero

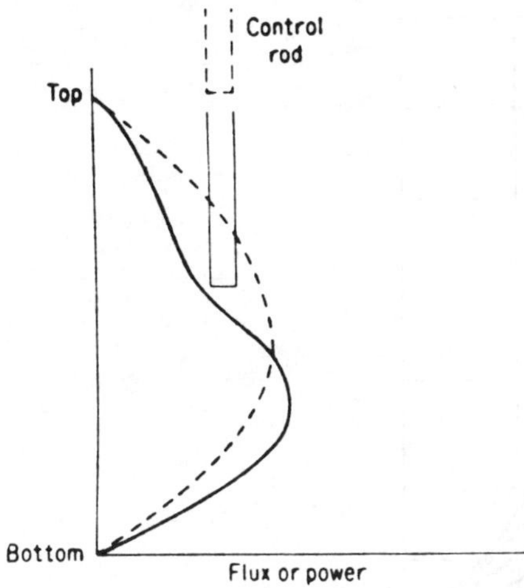

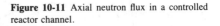

Figure 10-11 Axial neutron flux in a controlled reactor channel.

neutron absorption) would not perturb the channel at all. The degree of skewing is of course severer the higher the reactivity worth of the rod. These effects, which apply whether or not chemical shim is used, are shown by Fig. 10-12, which is representative of large pressurized-water reactors. It contains lines of different rod worths. It can be seen that the effects of partially inserted rods on peaking become severer the larger the reactivity and that, for a given rod reactivity, a more favorable axial distribution (lower axial maximum-to-average ratio) is obtained by the use of relatively low worths (and deeper insertion).

It can also be seen that rods of lower worths, obtainable when chemical shim supplements rods, are conducive to a more favorable axial power distribution.

At the beginning of core life, control rods are usually used to flatten the radial power distribution. Because of burnup, these rods are later moved out and the power distribution becomes less flat. When chemical shim is used, however, power can be flattened at the beginning of core life by spatial variations in fuel enrichment or core composition. This favorable power distribution is then maintained to a great degree by simply varying the concentration of the chemical absorber, while the rod positions are maintained. Large PWRs with chemical shim are now operated practically unrodded (i.e., rods almost completely withdrawn).

With favorable power distributions, the use of chemical shim results in increased fuel burnup for a given number of control rods. Note that without a chemical absorber, fuel burnup is limited by the number of control rods.

On the debit side, although chemical shim results in a decrease in number, size, or blackness of control rods, the effect is not linear. There actually is a slight decrease in rod worth with increasing concentrations of chemical absorber. The boron concentration necessary to ensure a minimum shutdown margin at room temperature is estimated at 3200 ppm. Figure 10-13 shows the boron worth as a function of boron

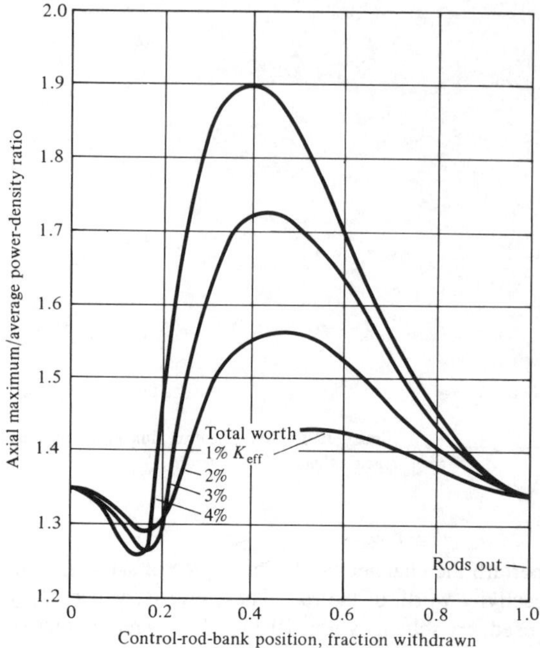

Figure 10-12 Effect of reactivity worth and depth of insertion of control rods on axial power density [90].

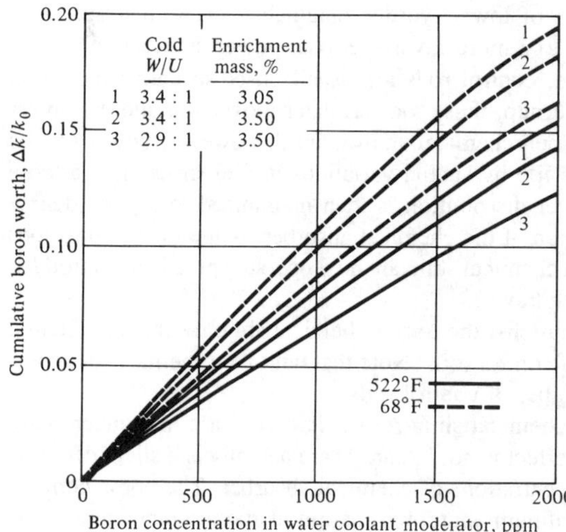

Figure 10-13 Calculated reactivity worth of dissolved boron [91].

Table 10-1 Typical reactivity requirements in a chemically shimmed PWR*

Reactivity requirement	Reactivity, %	
	Rods	Boron
Safety shutdown, cold to operating temperature change	3.0	
Doppler effect	2.2	2.0
Samarium poisoning	—	0.8
Xenon poisoning	—	2.2
Operating control	0.8	
Core lifetime (fuel depletion)	—	9.0

* Data from Ref 92.

concentration for two moderator-fuel ratios, two fuel enrichments, and two temperatures.

One problem with chemical shim is referred to as *hideout* and *plateout*. *Hideout* is defined as the precipitation of boron from solution onto solid surfaces or in deposits of corrosion products adhering to surfaces of the core and coolant system. It may later reenter the system as a result of changes in operation or water conditions. This is *plateout*. Hideout and plateout are of course undesirable because they would result in positive and negative reactivity drifts. But they represent the only significant safety problem resulting from the use of chemical shim. Tests have shown that boron deposition occurs to a limited extent onto corrosion products adhering to surfaces rather than on clean surfaces. Indications are also that it occurs on fouled surfaces when boiling occurs when the thickness of fouling is at least 0.3 to 0.4 mil. In this case deposition was proportional to the rate of evaporation. No rapid release of boron from a surface into the coolant has been observed. However, reactivity insertion from such rapid release is assumed in order to safely estimate the control rod requirements in a chemically shimmed reaction.

Typical reactivity requirements in a pressurized-water, chemically shimmed reactor are given in Table 10-1.

10-6 A PWR POWERPLANT

The secondary or working-fluid system of a twin-reactor, 1000-MW (total) powerplant [93] is shown in Fig. 10-14. It is composed on the shell side of the steam generators, a turbine-generator, condenser, two condensate pumps, five stages of feedwater heating, two feedwater pumps, and auxiliary equipment.

The turbine-generator system is designed to produce a guaranteed maximum of 502.841-MW gross and an expected maximum of about 523.7-MW gross. The 1800-

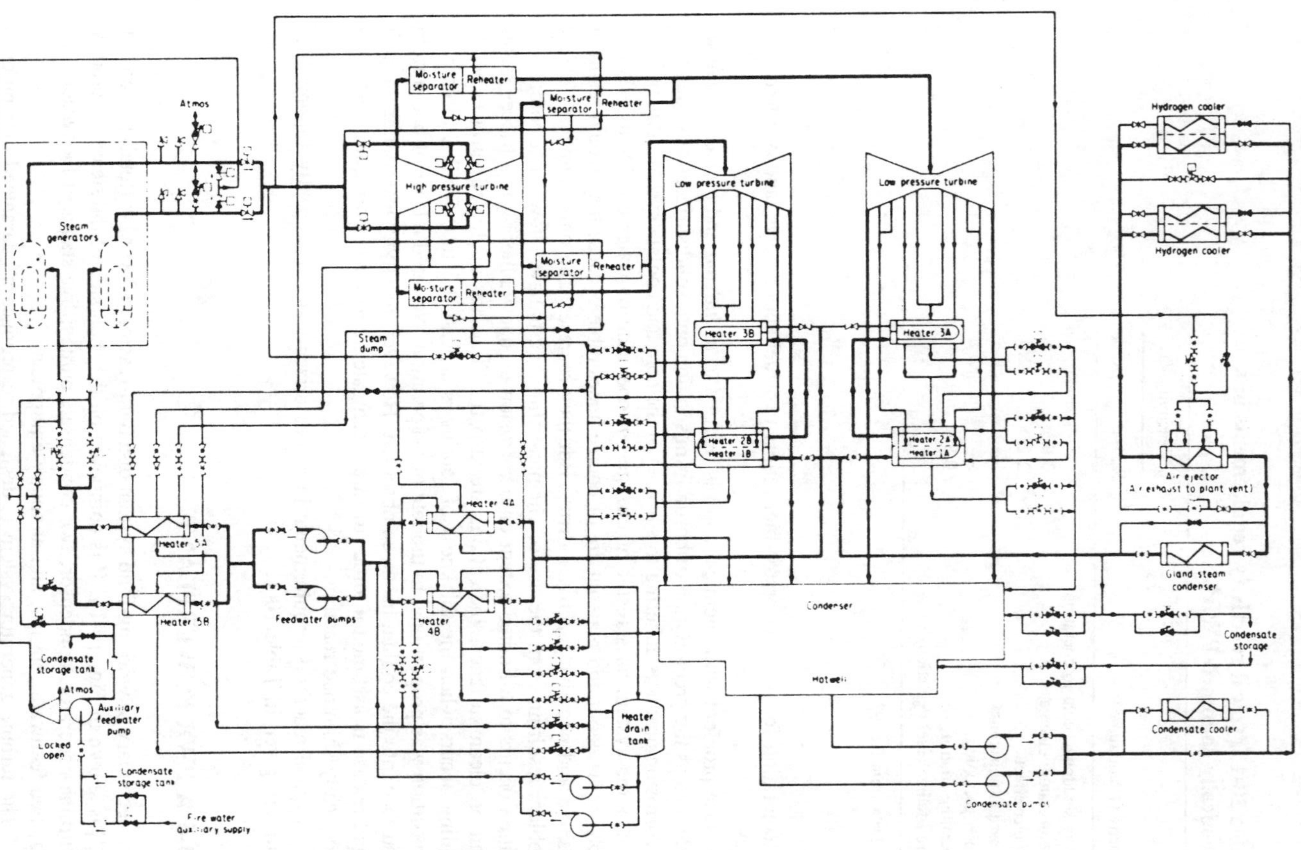

Figure 10-14 Flow diagram of the working fluid in a PWR powerplant [93].

r/min steam turbine is composed of one double-flow, high-pressure element in tandem with two double-flow, low-pressure elements.

There are four combination moisture separator-reheater assemblies between the high- and low-pressure units. Wet steam from the exhaust of the high-pressure elements enters each assembly at one end, is distributed by internal manifolds, and rises through a wire mesh where the moisture is removed. It then flows over the tubes in the reheater where it is heated by high-pressure steam from the steam generators. This enters the other end of each assembly, passes through the tubes and leaves as condensate to the high-pressure feedwater heater. The reheated steam flows back to the low-pressure turbines.

The AC generator, with rotating rectifier-exciter, is mounted on the turbine shaft. The generator is rated at 582,000 kVA and is hydrogen-cooled.

Turbine exhaust steam from four manifolds enters a radial-flow-type condenser. The condenser has a deaerating hotwell with sufficient storage for 3 min operation at full throttle and an equal free volume for surge flow. A steam-jet air ejector is provided. It has four first-stage elements and two second-stage elements mounted on shells of intermediate and after condensers. The ejector is driven with high-pressure steam from the steam-generator outlet.

Condensate from the condenser hotwell is pumped by two condensate pumps then normally passed through hydrogen coolers, air ejectors, a gland steam condenser, then through the first four feedwater-heater stages. It is then pumped by two feedwater pumps through the fifth and last feedwater-heater stage back to the steam generators. An auxiliary feedwater pump driven by a steam turbine is provided for decay-heat removal in case of loss of power. Steam for this turbine is produced from reactor decay heat. The normally closed steam valves to this turbine open automatically on loss of power, or manually. All feedwater heaters are of the closed type and are twin units operating in parallel. The lowest two stages are joined in a duplex arrangement. Steam for the heaters is obtained from five extraction points, one from the high-pressure turbine, one from high-pressure exhaust, and three from the low-pressure turbines. Drains from the high-pressure heater cascade to the second-stage heater and then to a drain tank. Heater drain pumps force water from the three lowest-pressure heaters to cascade to the condenser.

10-7 THE BOILING-WATER REACTOR (BWR)

The BWR has a function closely resembling that of the boiler in a conventional fossil-fuel steam powerplant and is basically simpler than it. In the boiler, heat is transmitted from the furnace to the water indirectly, partly by radiation, partly by convection, and partly by conduction, with combustion gases used as an intermediate agent or coolant. In the boiling-water reactor, the coolant is in direct contact with the heat-producing nuclear fuel and boils in the same compartment in which the fuel is located. It boils because the reactor pressure is maintained at about 1000 psia (about 70 bar), less than half that in a PWR, with the fuel temperatures roughly comparable. Because water and vapor coexist in the core, a BWR produces saturated steam at about 545°F (285°C). The coolant thus serves the triple function of coolant, moderator, and working fluid.

In its simplest form (Fig. 10-15), a boiling-water-reactor powerplant consists of a reactor, a turbine generator, a condenser and associated equipment (air ejector, cooling system, etc.), and a feed pump. Slightly subcooled liquid enters the reactor core at the bottom, where it receives sensible heat to saturation plus some latent heat of vaporization. When it reaches the top of the core, it has been converted into a very wet mixture of liquid and vapor. The vapor separates from the liquid, flows to the turbine, does work, is condensed by the condenser, and is then pumped back to the reactor by the feedwater pump.

The saturated liquid that separates from the vapor at the top of the reactor or in a steam separator flows downward via downcomers within or outside the reactor and mixes with the return condensate. This recirculating coolant flows either naturally, by the density differential between the liquid in the downcomer and the two-phase mixture in the core, or by recirculating pumps in the downcomer (not shown in the figure). This is similar to what happens in modern large fossil-fueled steam generators. Modern large boiling-water reactors are of the internal, forced recirculation type.

The saturated liquid that separates from the vapor at the top of the reactor or in a steam separator flows downward via downcomers within or outside the reactor and mixes with the return condensate. This recirculating coolant flows either naturally, by the density differential between the liquid in the downcomer and the two-phase mixture in the core, or by recirculating pumps in the downcomer (not shown in the figure). This is similar to what happens in modern large fossil-fueled steam generators. Modern large boiling-water reactors are of the internal, forced recirculation type.

The ratio of the recirculation liquid to the saturated vapor produced is called the *recirculation ratio*. It is a function of the core average exit quality [Eq. (10-9), below]. Boiling-water core exit qualities are low, between 10 and 14 percent, so that recir-

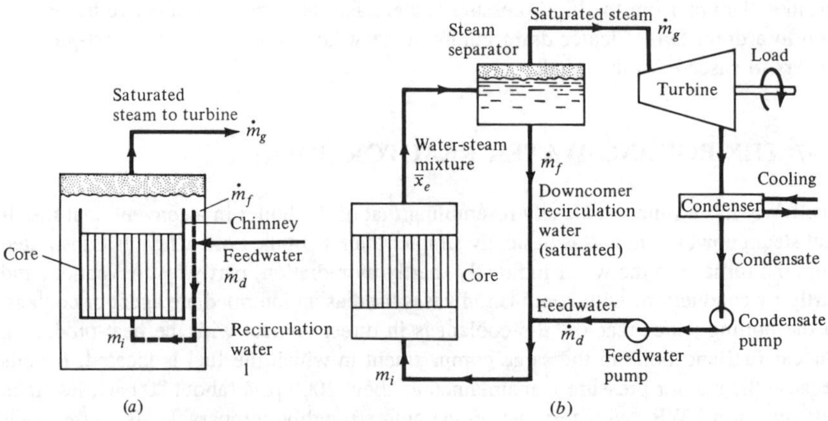

Figure 10-15 Schematic of a BWR system: (*a*) internal and (*b*) external recirculation.

culation ratios in the range of 6 to 10 are common. This is necessary to avoid large void fractions in the core, which would materially lower the moderating powers of the coolant and possibly result in low heat-transfer coefficients or vapor blanketing and burnout [2].

In either kind, a slightly subcooled liquid enters the core bottom at a rate of $\dot{m}_i$, mass per unit time. This liquid rises through the core and chimney, if any. The chimney is an unheated section above the core that helps to increase the driving pressure in natural circulation. The resulting vapor separates and proceeds to the powerplant at a mass flow rate of $\dot{m}_g$. The saturated recirculation liquid flows via the downcomer at mass flow rate of $\dot{m}_f$. There it mixes with the relatively cold return feedwater $\dot{m}_d$ from the power plant to form the slightly subcooled inlet liquid $\dot{m}_i$.

An overall mass balance in the rector core is given by

$$\dot{m}_d = \dot{m}_g \tag{10-8a}$$

$$\dot{m}_g + \dot{m}_f = \dot{m}_i \tag{10-8b}$$

The *average* exit quality of the entire core $\bar{x}_e$, that is, the quality of all the vapor-liquid mixture at the core exit, is given by

$$\bar{x}_e = \frac{\dot{m}_g}{\dot{m}_g + \dot{m}_f} = \frac{\dot{m}_d}{\dot{m}_d + \dot{m}_f} = \frac{\dot{m}_d}{\dot{m}_i} \tag{10-9}$$

The *recirculation ratio* R is the ratio of recirculation liquid to vapor produced. It is given by modifying Eq. (10-9) as follows

$$R = \frac{\dot{m}_f}{\dot{m}_g} = \frac{1 - \bar{x}_e}{\bar{x}_e} \tag{10-10}$$

Now, if the incoming feedwater has a specific enthalpy h_d Btu/lb$_m$ or kJ/kg and the recirculated liquid has a specific enthalpy h_f (at the system pressure), a heat balance is obtained, if we assume no heat losses to the outside (a good assumption) and neglect changes in kinetic and potential energies, as follows

$$\dot{m}_i h_i = \dot{m}_f h_f + \dot{m}_d h_d \tag{10-11}$$

where h_i is the specific enthalpy of the liquid at the reactor-core inlet. Equation (10-11) can be modified to

$$h_i = (1 - \bar{x}_e)h_f + \bar{x}_e h_d \tag{10-12}$$

Rearranging gives the following expression for $\bar{x}_e$

$$\bar{x}_e = \frac{h_f - h_i}{h_f - h_d} \tag{10-13}$$

The condition of the liquid entering the bottom of the core is given by the *enthalpy of subcooling*

$$\Delta h_{\text{sub}} = h_f - h_i = \bar{x}_e(h_f - h_d) \tag{10-14}$$

or by the *degree of subcooling*

$$\Delta t_{sub} = t_f - t_i \tag{10-15}$$

where t_i is the core-inlet liquid temperature, corresponding to h_i.

The *total heat generation* Q_t can be obtained from a heat balance on the core as a system or on the reactor as a system. The two relationships, which yield identical results, are

$$Q_t = \dot{m}_i[(h_f + \bar{x}_e h_{fg}) - h_i] \tag{10-16}$$

$$Q_t = \dot{m}_g(h_g - h_d) \tag{10-17}$$

Shown diagramatically in its simplest form in Fig. 10-16, a boiling-water-reactor system supplies saturated (or 0.4 percent moisture) vapor directly to the turbine. After expansion through the turbine, the exhaust wet vapor is condensed and the condensate is pumped back to the reactor. The direct cycle has the advantages of simplicity and of relatively low capital costs. Because of the direct loop arrangement, there are no heat exchangers. Because of the large steam volume above the water, no pressurizer is needed.

A major drawback of the direct cycle is that the reactor is not load-following. To illustrate this, let us assume that the reactor is operating at some power level determined by its flow and control-rod setting. Let us also assume that a larger load is applied to the turbine, thus causing the turbine governor to increase the throttle valve opening, i.e., call for more steam. This reduces the flow resistance in the steam passages between the core and turbine. This in turn reduces the reactor pressure. A reduction in pressure causes flashing, i.e., increases steam voids in the core. This decreases moderation and decreases reactor power, which results in less steam generation,* opposite to the desired effect. A similar argument applies to a demand for reduced load.

* It can be shown that only above 500 psia (34.5 bar) and at very high qualities, far beyond those found in BWRs, the opposite effect takes place; i.e., a reduction in pressure causes a reduction in voids [3].

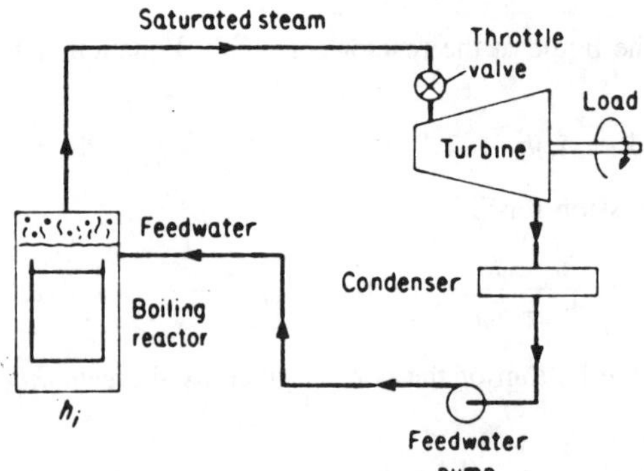

Figure 10-16 Schematic of a direct-cycle BWR plant.

This state of affairs has been corrected by one of several methods: by-pass control, a dual-pressure cycle, an overmoderated reactor in which voiding results in a reduced loss of neutrons due to water absorption and hence increased power, and recirculation control [3]. The last method is the one adopted in current designs. It is described in the next section.

10-8 BWR LOAD FOLLOWING CONTROL

The *recirculation control* method is based on a direct cycle but with variable recirculation flow in the downcomer. It is shown schematically in Fig. 10-17. Equation (10-17) is rewritten with the help of Eqs. (10-8*a*) and (10-9) as

$$Q_t = \bar{x}_e \dot{m}_i (h_g - h_d) \tag{10-18}$$

h_g, the saturated steam enthalpy at the system pressure, and h_d, the feedwater enthalpy, are both weak functions of load. Thus the plant load Q_t is therefore proportional to the product of $\bar{x}_e$ and m_i, the flow in the downcomer.

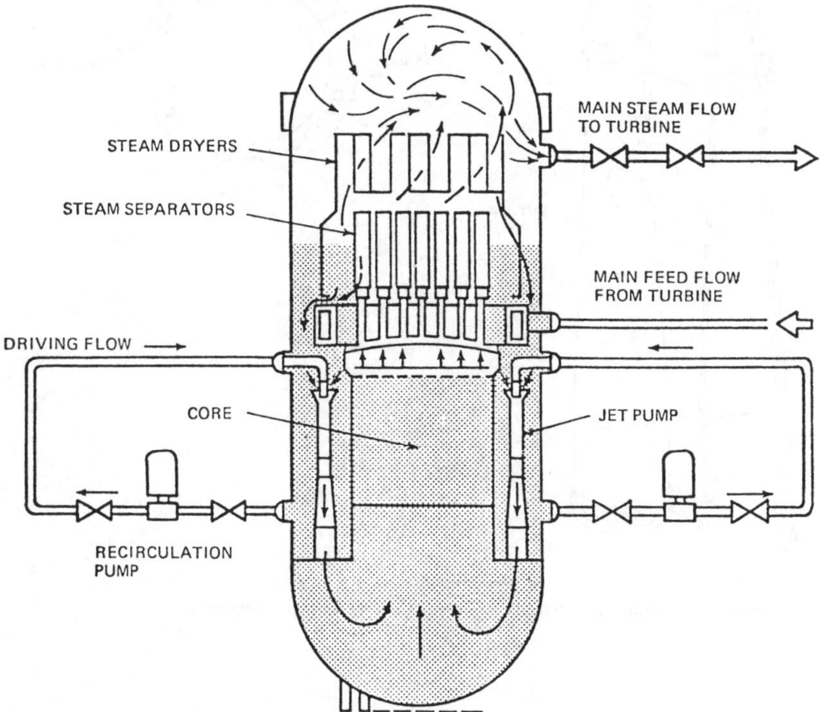

Figure 10-17 BWR reactor vessel internal flow paths.

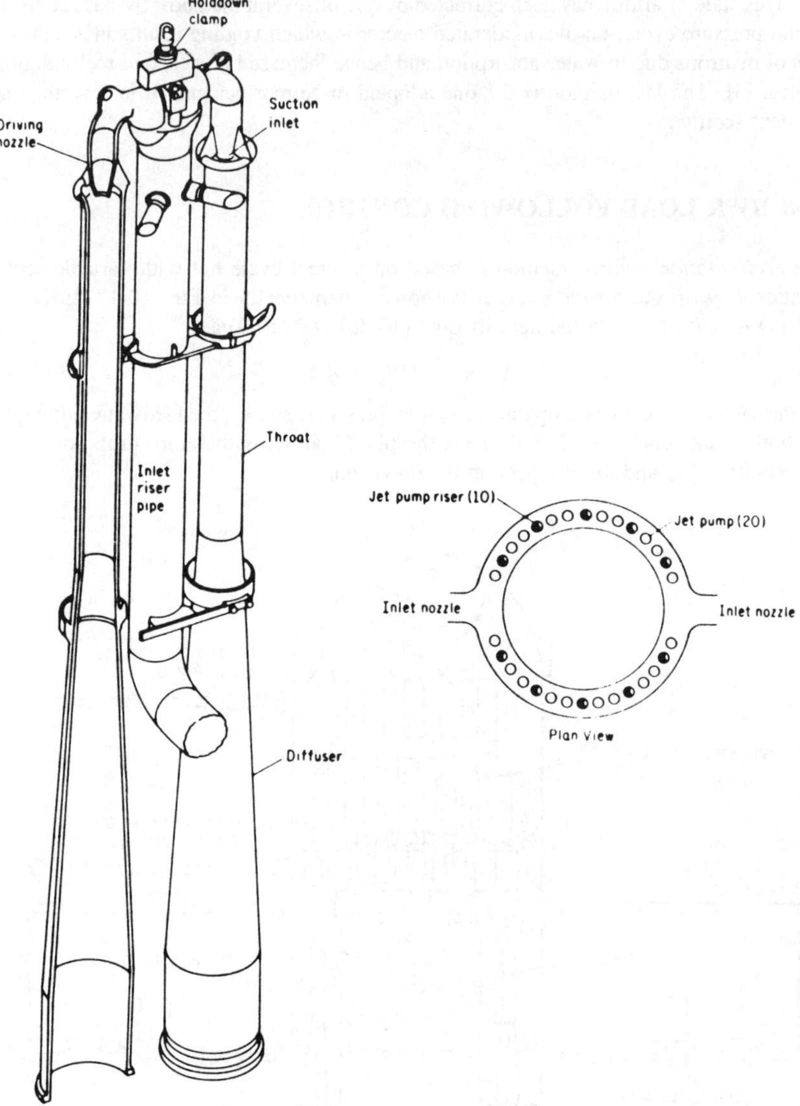

Figure 10-18 Jet-pump arrangement. *(Courtesy General Electric Company.)*

If the reactor power is to be increased by a certain percentage, $\dot{m}_i$ is increased by the same percentage. This momentarily decreases $\bar{x}_e$, as Q_t has not yet changed. $\bar{x}_e$ is a measure of the qualities in the core as a whole. When it decreases, the reactor power increases (due to improved moderation), in turn increasing $\bar{x}_e$ to the original value where nuclear equilibrium returns. We are now at the new $\dot{m}_i$, the original $\bar{x}_e$; thus Q_t is at the new desired value. The same holds true if a reduction in load is desired. In general then, except at very low loads, Q_t is essentially directly proportional to $\dot{m}_i$, hence the name recirculation control.

The recirculation system consists of 2 external recirculation pump loops and 20 internal jet pumps located inside the reactor vessel (Figs. 10-17 and 10-18). The two external recirculation pumps are single-stage vertical centrifugal units, with mechanical shaft seals, driven by variable-frequency motor-generator sets. Changing the pump speed changes recirculation flow, which in turn changes reactor power. One third of the total reactor recirculation water is pumped through the two external pumps. This flow leaves the reactor vessel through two outlet nozzles and returns to the jet-pump inlet-riser pipes to become the *driving flow* for the jet pumps. The pressure to which the driving fluid is raised in the external pumps is, however, higher than required in an all-pumped recirculation system (Fig. 10-19). This driving flow goes through the jet-pump nozzles and acquires high velocity and momentum. By a process of mo-

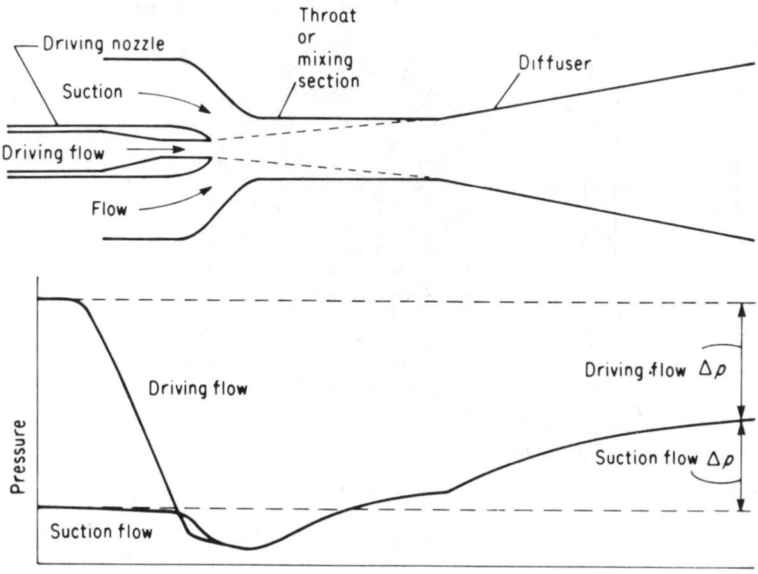

Figure 10-19 Jet-pump flows and pressure diagram (94).

mentum exchange, it entrains the remainder of the recirculation flow, called the *suction flow,* which is at a lower inlet pressure. The combined flow then enters the throat or mixing section of the jet pump where the momentum decreases and the pressure rises. Additional pressure recovery to the exit pressure occurs in the diffuser. The resultant suction flow Δp is sufficient to overcome losses through the reactor.

The jet-pump design has the advantages of fewer moving parts, lower probabilities of major line ruptures, capability to reflood the vessel in case of such rupture (loss-of-coolant accident, LOCA), improved natural circulation (lower pressure losses) in the event of power loss to the circulation pumps, and relative freedom from cavitation, a problem that arises in pumps where the liquid is near saturation conditions.

Figure 10-20 is a control "operation map." It shows the effect recirculation flow ratio $\dot{m}_i/\dot{m}_{io}$ on power ratio Q_t/Q_{to}, where the subscript *o* indicates conditions at full load. To understand this map it is important to note one distinguishing feature of BWRs. Unlike PWRs, which can be critical at one control-rod position and one

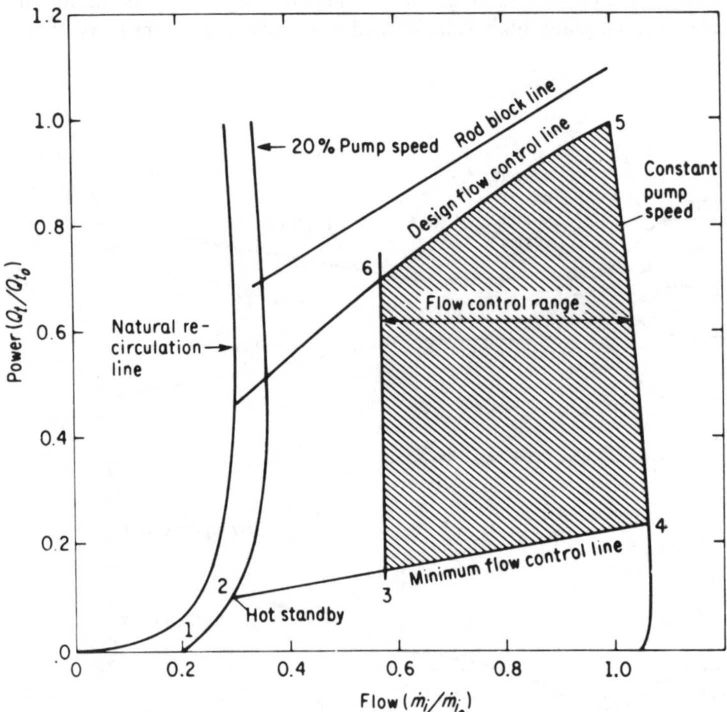

Figure 10-20 Operation map of BWR recirculation control [94].

chemical-shim concentration at a given time during the core lifetime and in which changes in these cause an increase or decrease in power until they are returned to original, BWRs can be critical at several control-rod positions because the resulting voids compensate for the change in reactivity due to the change in the rod positions.

Lines 3-4 and 5-6, and similar somewhat parallel lines in between, indicate recirculation control at various control-rod positions. They are nearly linear, as indicated by Eq. (10-18), and each indicates a control range of approximately 25 percent, without control-rod motion. Zero pump speed results in the natural circulation line at the left. Lines 3-6 and 4-5, and similar parallel lines in between, represent constant recirculation-pump speed lines.

The recirculation pumps are driven by motor-generator sets with adjustable-speed couplings that vary the frequency supply to the pump motors and hence their speed. To change reactor power, a demand signal from the operator, or a load-frequency error signal from the governor, is supplied to a master controller and compared with the actual generator speed. An error signal is used to adjust the speed of the recirculation pumps.

10-9 THE CURRENT BWR SYSTEM

Current BWR designs are of the direct-cycle, forced-internal-recirculation-type described above. The reactor (Fig. 10-21) is about 21 ft ID and 73 ft high for a 1000-MW plant. The fuel rods are similar to those of a PWR. They are Zircaloy-clad, 12 ft high in active length, and contain enriched UO_2 pellets. The fuel assemblies, however, are either 7 × 7 or 8 × 8 arrays and are enclosed in a Zircaloy-4 channel (Figs. 10-22 and 10-23). The control rods are cruciform in shape and occupy the space between four channels. They enter the reactor vessel from the bottom (vs. the top in a PWR). This is because the voids, mainly in the top part of the core, cause neutron peaking in the lower part where the control rods would be most effective. The blades of the cruciform contain sealed stainless steel tubes fitted with compacted boron carbide powder (Fig. 10-24). The reactor also contains temporary control curtains of borated stainless steel to supplement the control rods for the initial core.

The water-steam mixture leaves the top of the core and passes through steam separators and dryers located within the pressure vessel (Figs. 10-25 and 10-26). The separators are standpipes, each of which contains a centrifugal-type steam separator located at top. In each, the mixture impinges on vanes that impart a spin to separate water from steam. The separated water enters a pool surrounding the standpipes, from which it enters the downcomer annulus. Steam, still slightly wet, flows upward and outward through the dryers, where moisture is removed through troughs and tubes to a pool and back to the downcomer. The dry (or 0.4 percent wet) steam, separated from the wet steam by shrouds, finally leaves the reactor vessel through outlet nozzles and goes directly to the turbine.

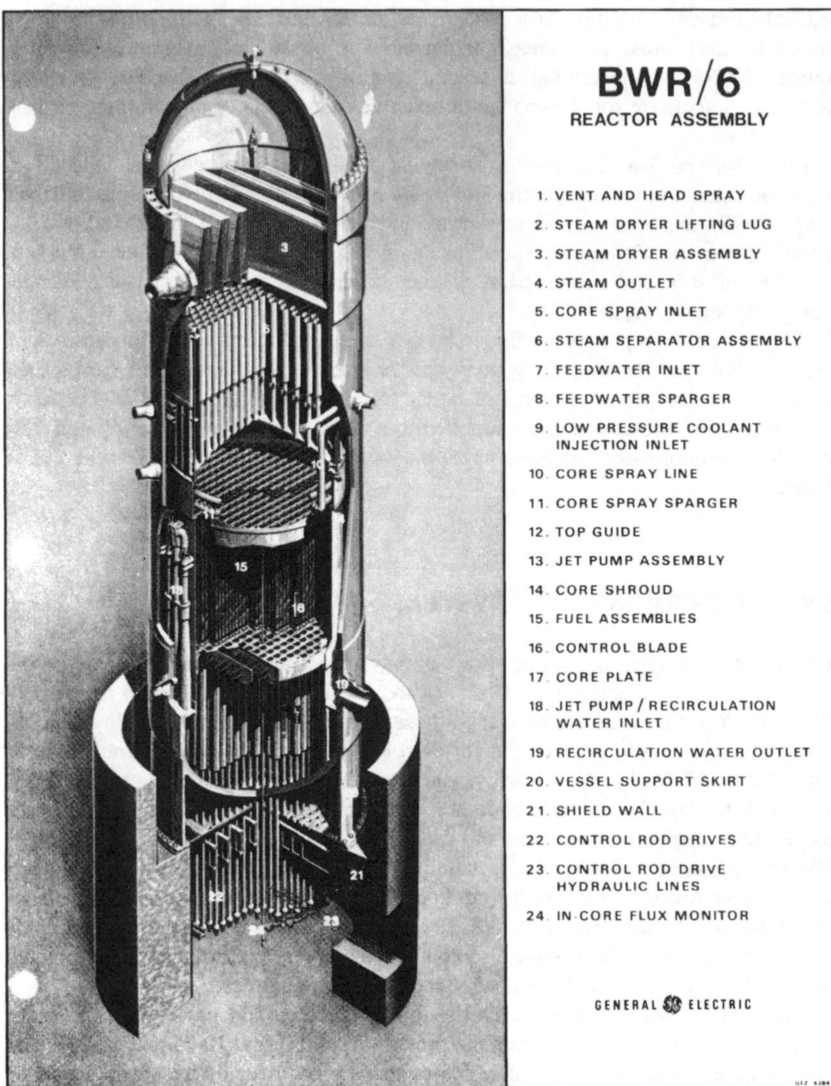

Figure 10-21 BWR reactor vessel and internals. *(Courtesy General Electric Company.)*

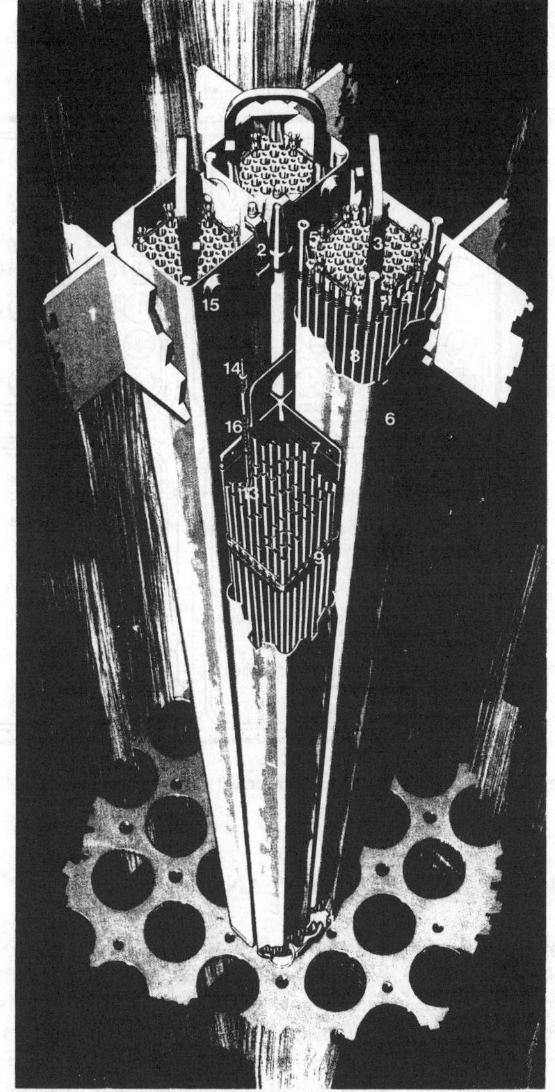

Figure 10-22 BWR fuel assemblies and control-rod arrangement. *(Courtesy General Electric Company.)*

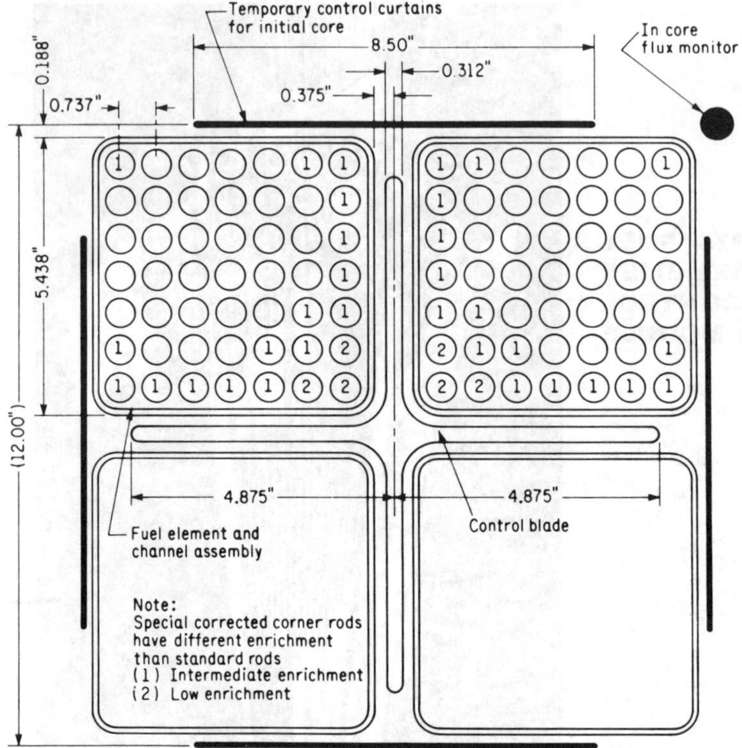

Figure 10-23 BWR core lattice arrangement, cross section of Fig. 10-22.

10-10 A BWR POWERPLANT

Figure 10-27 shows a typical flow diagram of a BWR powerplant [95]. It produces 780 MW from 0.4 percent moisture steam at 965 psia (66.7 bar) at turbine inlet, condensing to 2 inHg absolute (0.067 bar). The turbine is a tandem-compound (with one high-pressure section and two low-pressure sections), four-flow (ducts to the condenser), 1800-r/min, nonreheat unit. The cycle has a moisture separator between the high-pressure and low-pressure turbines and five feedwater heaters, all with 5°F terminal temperature difference and 10°F drain-cooler temperature difference. The feedwater leaving the high-pressure feedwater heater to the reactor is at 381.7°F (194.3°C). The throttle steam flow is 9,610,000 lb_m/h (1210 kg/s). The flow to the condenser after moisture separation and bleeding steam for feedwater heating is 5,711,820 lb_m/h (719 kg/s). It has a "best point" heat rate of 10,287 Btu/kWh, corresponding to 33.18 percent thermal efficiency.

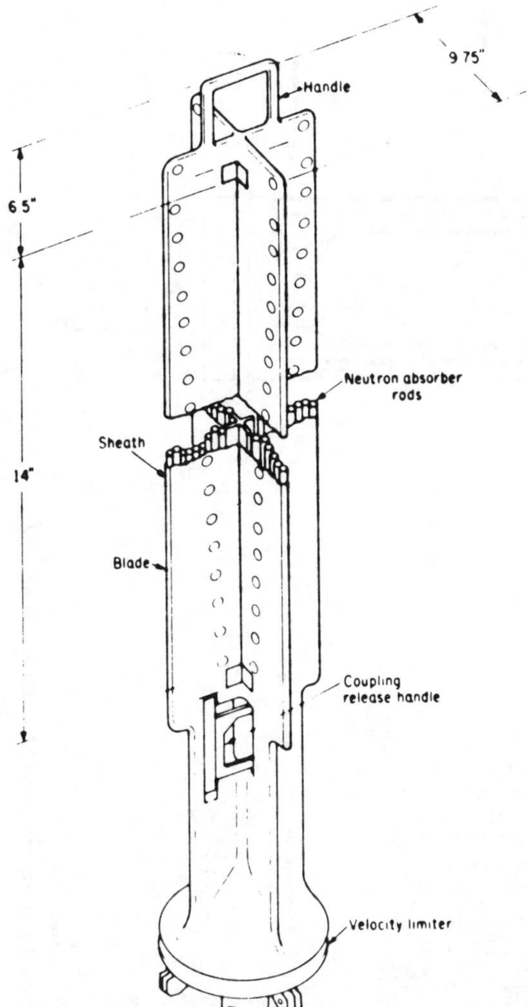

Figure 10-24 BWR control rod. *(Courtesy General Electric Company.)*

One final observation concerns radioactivity in the steam, since it is produced in the reactor and used in the balance of the plant. If mineral content of the coolant in water-cooled reactor systems is kept low enough (below about 1 ppm), the main radioactivity will be that due to the neutron capture by the oxygen in the water. Neutron capture by hydrogen converts it into nonradioactive deuterium. The most important of the oxygen reactions is the $O^{16}(n,p)N^{16}$ reaction

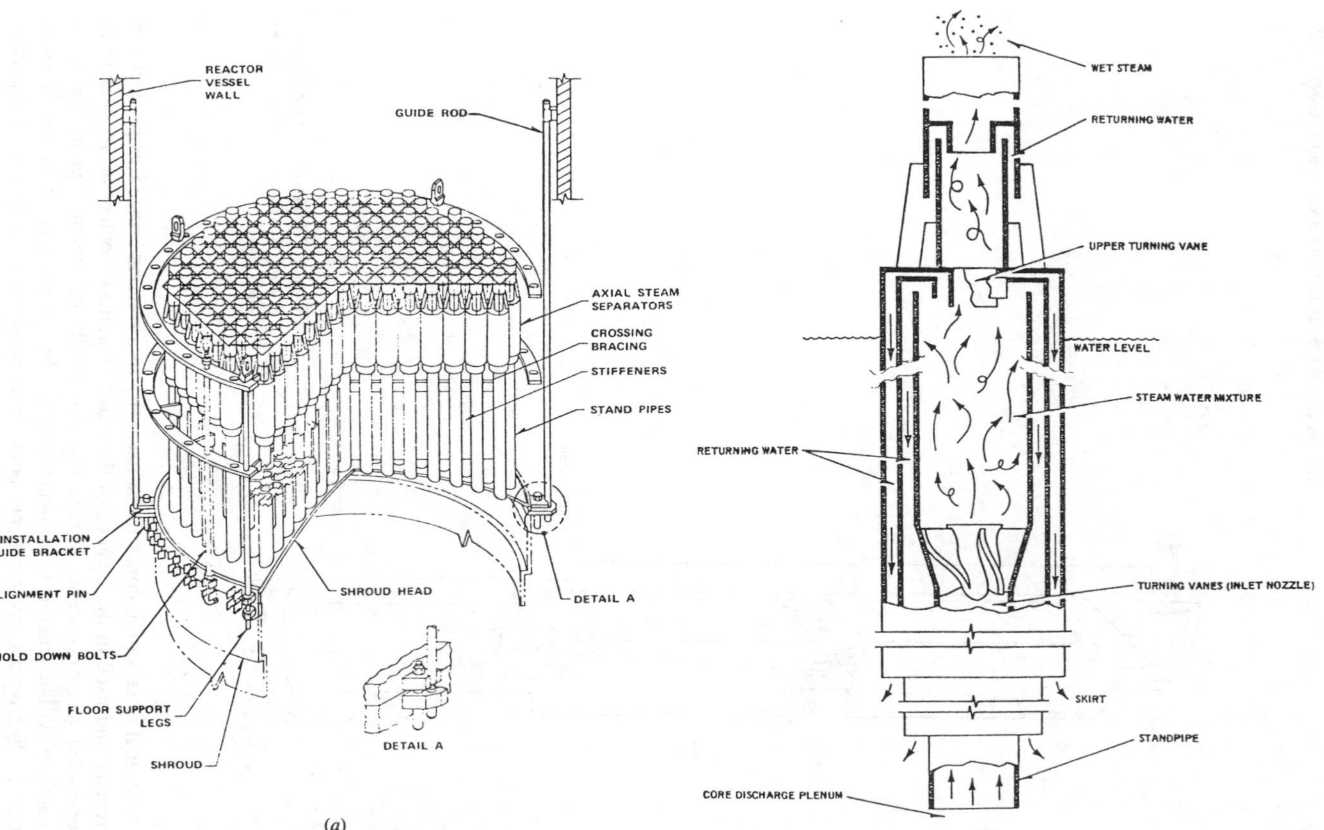

Figure 10-25 (*a*) BWR shroud head and steam separator assembly. (*b*) Steam separator detail.

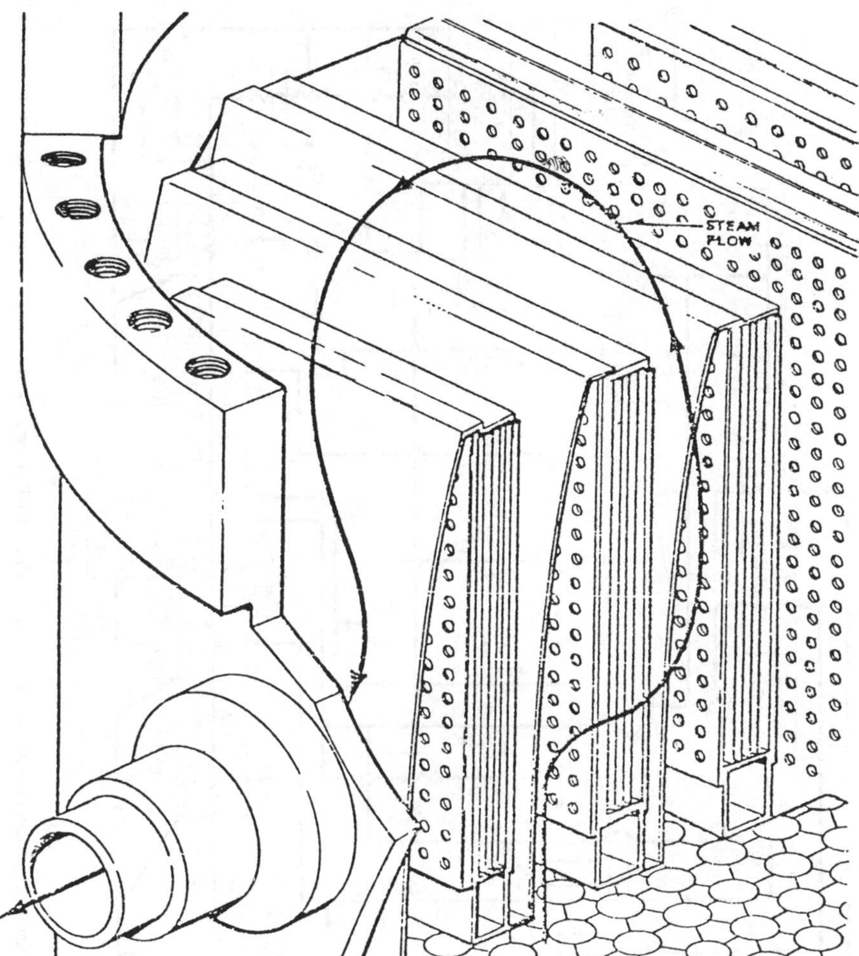

Figure 10-26 BWR steam dryer assembly.

$$_8O^{16} + {}_0n^1 \rightarrow {}_7N^{16} + {}_1H^1 \tag{10-19}$$

which has a microscopic cross section of 1.4×10^{-5} barn. Nitrogen 16 is a radioactive β and γ emitter (reverting to O^{16}) with a half-life of 7.2 s. The β rays are mainly of 3.8, 4.3, and 10.5 MeV energy. The γ rays are mainly of 6.13 and 7.10 MeV energy. The half-life of N^{16} is short enough so that only a small fraction of radioactivity remains when the steam reaches the turbine. Thus moisture remaining in contact with turbines and other equipment after shutdown is not dangerously radioactive, and maintenance work can usually be undertaken a short time after shutdown.

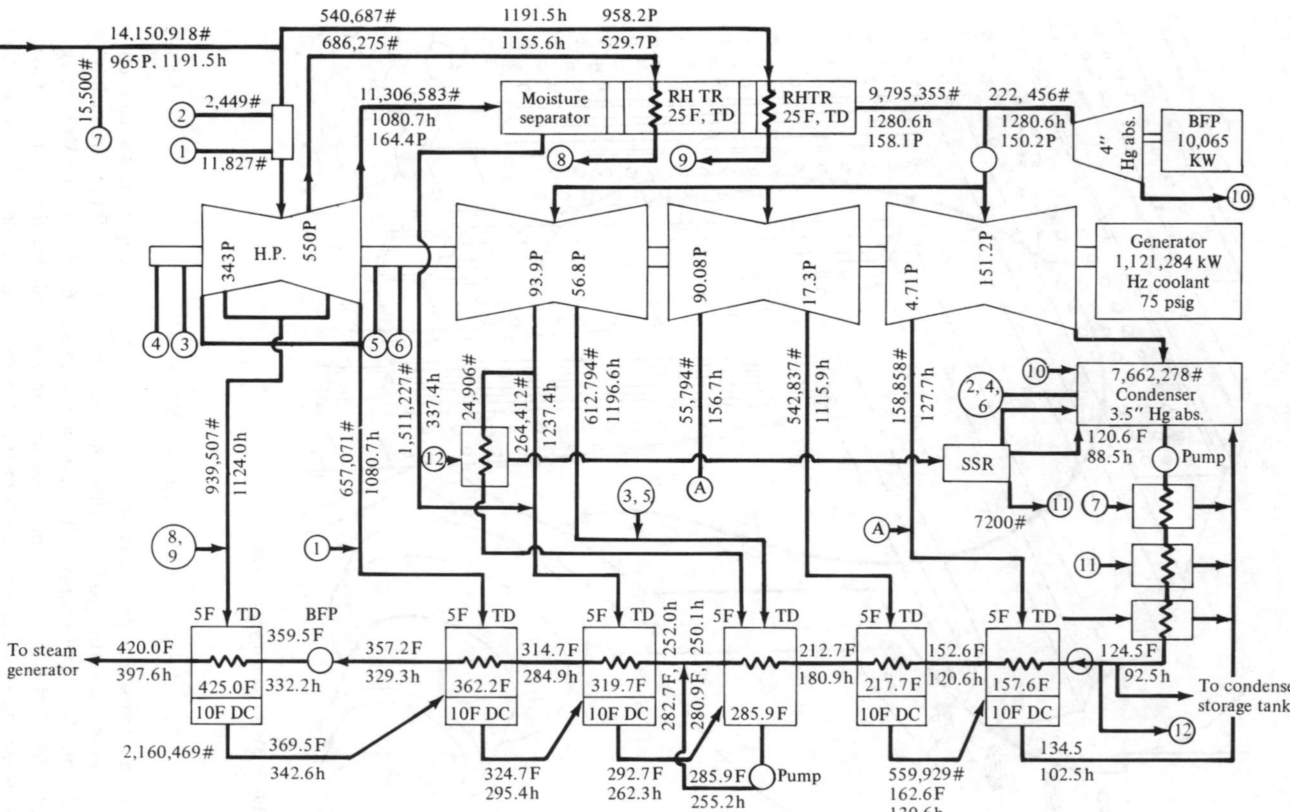

Figure 10-27 Flow diagram and mass and energy balances at rate load of twin-unit LaSalle BWR powerplant at Seneca, Ill. Capacity of each unit = 1121 MW gross, 1078 MW net.

Another reaction of somewhat less importance than the above is $O^{17}(n,p)N^{17}$. O^{17} is present to the extent of 0.037 percent of all oxygen. Nitrogen 17 is another β emitter of 4.16-s half-life. The microscopic cross section of this reaction is higher than the one above by a factor of about 10^3. This is outweighed, of course, by the smaller concentration of O^{17}. A third reaction of some importance is $O^{18}(n,\gamma)O^{19}$. Oxygen 19 is also a β emitter converting to stable fluorine 19, with a half-life of 29.0 s. This and other reactions that are not very important produce only a few weak radiations.

If the mineral content of the coolant is high, long-lived and strong radiations result. The radioactive particles may embed themselves in component parts, thus making maintenance difficult. Coolant treatment is thus an important feature in nuclear plants. Also, care should be taken in designing components to avoid pockets and crevices that may collect and retain radioactive particles.

10-11 THE GAS-COOLED REACTOR (GCR)

Gas-cooled reactors have received great attention, particularly in the United Kingdom and, to a lesser extent, in France, Germany, the United States, and the USSR. Both natural- and enriched-uranium fuels with CO_2 as coolant and graphite as moderator are used in the U.K. and France. In the United States and Germany, enriched fuels and helium coolant are used. Heavy-water-moderated, gas-cooled reactors have received some attention [96,97].

The attractiveness of gas cooling lies in the fact that, in general, gases are safe, are relatively easy to handle, have low macroscopic neutron cross sections, are plentiful and cheap (except helium), and may be operated at high temperatures without high pressurization.

The main disadvantages are the lower heat-transfer and heat-transport characteristics of gases, which require large contact surfaces and flow passages within the reactor and heat exchangers, and their high pumping requirements (between 8 to 20 percent of plant gross power), which necessitate careful attention to the problems of fluid flow, pressure drops, etc.

To partially overcome the inherent disadvantages of gas coolants and at the same time obtain attractive thermodynamic efficiencies, it is necessary to operate the fuel elements at as high temperatures as possible (commensurate with metallurgy) and permit a high gas-temperature rise in the reactor by reducing the gas mass-flow rate and pressurizing the gas. Because the fuel operates at high temperatures, fuel-element and cladding-material choice and fabrication in gas-cooled reactors present major problems, and the trend is toward using oxide and carbide fuel elements in such reactors. Also, because gas-cooled reactors are inherently large, they are particularly suited to large-capacity power plants. The reactor itself may impose structural and foundation problems. The size of the units can, of course, be reduced to a certain extent by fuel enrichment.

The British Program

One of the largest single programs per capita for civilian nuclear power is the British development and construction of a series of graphite-moderated, CO_2-cooled-reactor powerplants. This effort started in October 1956, when the first powerplant, using natural-metallic-uranium, magnesium-clad fuel, was put in operation at Calder Hall, England [98]. This powerplant type, originally known as "Calder Hall" but subsequently as "Magnox" because of magnesium cladding, was devised because of economic and military (production of Pu^{239}) necessity. It had the advantages of using familiar and economical materials and fuels that do not require enrichment. The coolant is cheap and plentiful. The steam temperatures and pressures are, however, rather low, being a few hundred degrees Fahrenheit and a few hundred psi, well below current, highly efficient fossil-fueled powerplants. Because of their low temperatures, the Magnox stations used a special dual-pressure cycle [3]. The Magnox program is composed of nine twin-reactor stations for a total of 4845 MW.

A second British program called the AGR (advanced gas-cooled reactor), based on the prototype at Windscale, was started in the late 1960s. Its objectives were to construct a nuclear steam supply system with steam conditions comparable to those in modern fossil-fueled power stations and with a degree of integrity that would permit siting nearer centers of population. AGR powerplants use enriched ceramic fuels clad in stainless steel but otherwise still use graphite moderators and CO_2 gas coolant. They operate at clad surface temperatures of about 1520°F (~825°C) with gas outlet temperatures of up to 1230°F (~665°C), resulting in steam temperatures of up to 1050°F (~565°C). The AGR stations are characterized by prestressed-concrete pressure vessels, double containment of all access penetrations, and provisions for refueling on load for high availability. The AGR program is composed of four twin-reactor stations and one four-reactor station for a total of 4750 MW.

An example of the AGR program is Hinkley Point B (Hinkley Point A was a Magnox), a twin-reactor station with a total output of 1250 MW net [99]. Figure 10-28 shows a cross section of one of the reactors in a building it shares with the other. The core is a 16-sided stack of graphite blocks, of which 308 form fuel channels and 81 are for the control rods. The vertical stacks are composed of 12 graphite blocks each, 10 of which house the fuel and the top and bottom ones act as neutron reflectors (Fig. 10-29). The core is surrounded by an annular graphite reflector, two blocks wide, and a top graphite shield. Each fuel channel contains eight fuel elements. Each is 40.8 in (1039 mm) long and consists of a 36-pin cluster of stainless steel–clad 0.57-in (14.5-mm)-diameter UO_2 pellets. The pins are 0.6 in (15.25 mm) in diameter. The total fuel is 122.5 tons of uranium and is enriched up to 2.6 percent U^{235}.

The CO_2 coolant enters at bottom at 590°F (310°C) and 617 psia (42.6 bar) and leaves at top at 1210°F (655°C) and 586 psia (40.4 bar). It then flows down 12 superheat-reheat boilers surrounding the core, within the cavity of a prestressed concrete reactor vessel (PCRV), leaving at 529°F (276°C) and 583 psia (40.2 bar). It is then pumped back to the reactor by eight constant-speed, electrically driven (28 MW) centrifugal circulators with variable-guide-vane flow control.

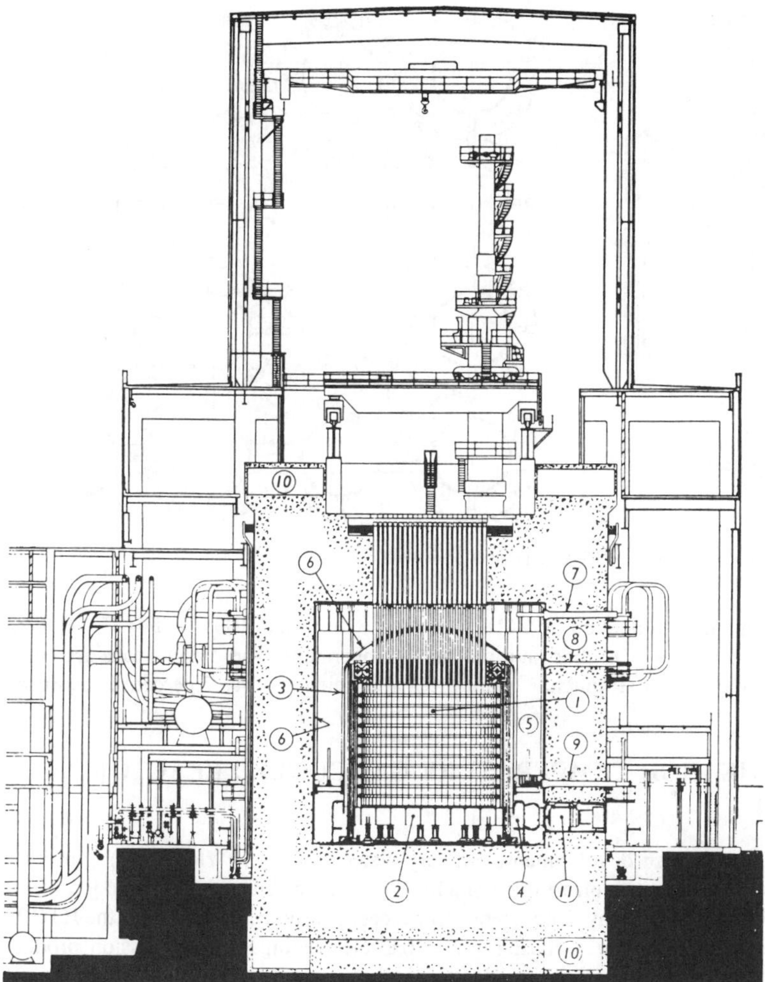

Figure 10-28 Hinkley Point B AGR building: (1) reactor core, (2) supporting grid, (3) gas baffle, (4) circulator outlet gas duct, (5) boiler, (6) thermal insulation, (7) reheat steam penetrations, (8) main stream penetrations, (9) boiler feed penetrations, (10) cable stressing gallery, (11) gas circulators. *(Courtesy The Nuclear Power Group, Limited.)*

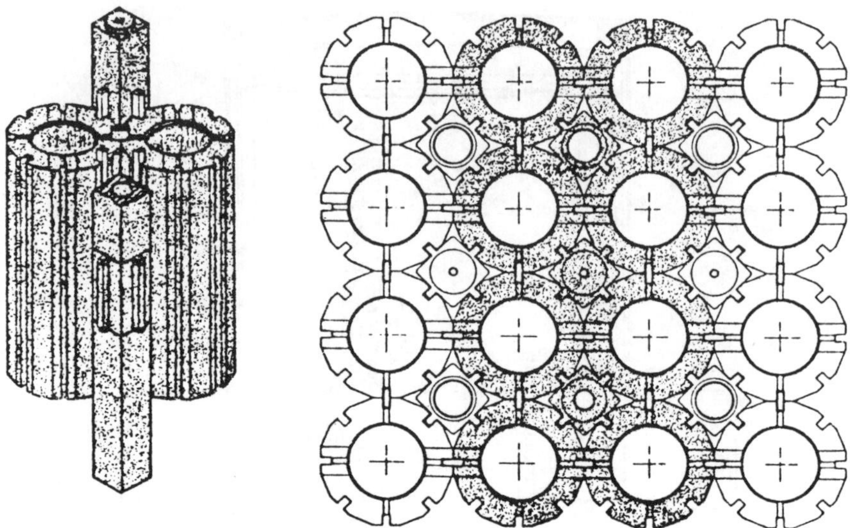

Figure 10-29 Cross section of Hinkley Point B core [99].

The steam is generated at 2418 psia (166.8 bar) and 1005°F (541°C). Reheat steam is at 590 psia (40.7 bar) and the same temperature. These conditions are comparable to modern fossil units. A plant thermal efficiency of 41.7 percent is claimed.

10-12 THE HIGH-TEMPERATURE GAS-COOLED REACTOR (HTGR)

In contrast to British development, ordinary-water reactors (PWR, BWR) constitute the bulk of the American effort in commercial nuclear power, and only a modest effort in gas cooling is under way. The American concept is usually referred to as the HTGR. In that concept, graphite is used for fuel particle coating (primary fission product barrier), fuel structural material, moderator, and coolant channel walls. Helium is used as coolant. The use of an all-ceramic fuel element results in low parasitic neutron capture in the core and therefore high conversion ratios (Sec. 11-2) and good fuel-cycle economics. The coated-fuel-particle design results in high specific powers and high fuel burnups.

Figure 10-30 shows coated fuel particles and the rod and graphite block in which they are contained [100]. The block is hexagonal, 14 in (35.6 cm) across sides, and 2.5 ft (76.2 cm) high. The coated particles are contained in over 200 fuel holes interspersed with a lesser number of coolant holes through which helium flows. The coated fuel particles are miniature spherical fuel elements containing UC_2 fuel and ThC_2 fertile material, with diameters of 200 $\pm$50 micrometers (μm) for the fuel and 400 $\pm$100 μm for the fertile material. They are clad with a three-layer coating of

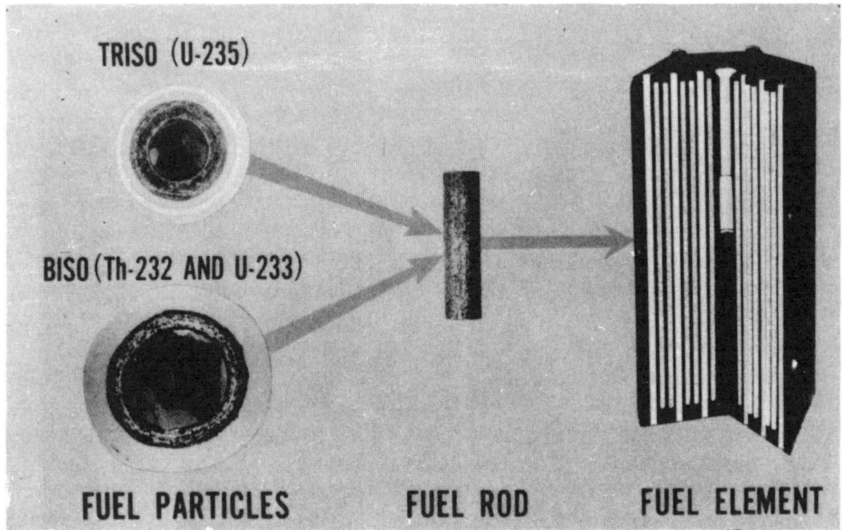

Figure 10-30 Coated particles, fuel element, and hexagonal block of HTGR. *(Courtesy GA Technologies, Inc.)*

pyrolytic carbon 100 μm thick for the fuel and 125 μm thick for the fertile material. This cladding here, as elsewhere, isolates the fuel and prevents chemical reactions and minimizes the release of fission products. The two sizes for fuel and fertile materials make it easy to separate the two for fuel recycle purposes. The Th^{232} is converted to fissionable U^{233}, which partly fissions in place and is partly processed for use in other reactors. With thorium plentiful in many parts of the world, this system can materially extend our fuel resources.

The first U.S. efforts in HTGR development were a 40-MW plant at Peach Bottom near Philadelphia (now decommissioned) and the 330-MW Fort St. Vrain plant near Platteville, Colorado. The latter plant had many engineering problems that discouraged many utilities that had once opted for the HTGR from proceeding with it despite its many favorable features. Research and development, however, are still continuing on large HTGRs in the hope that it will be revived in the future. The following is a description of a large HTGR.

The program envisions the building of units of 770- and 1160-MW capacity. The conceptual design for the latter is based on a helium-cooled, 3000-MW(t) reactor (Figs. 10-31 and 10-32). The reactor is housed in a vertical prestressed concrete reactor vessel (PCRV), which is prestressed vertically by individual cables and circumferentially by

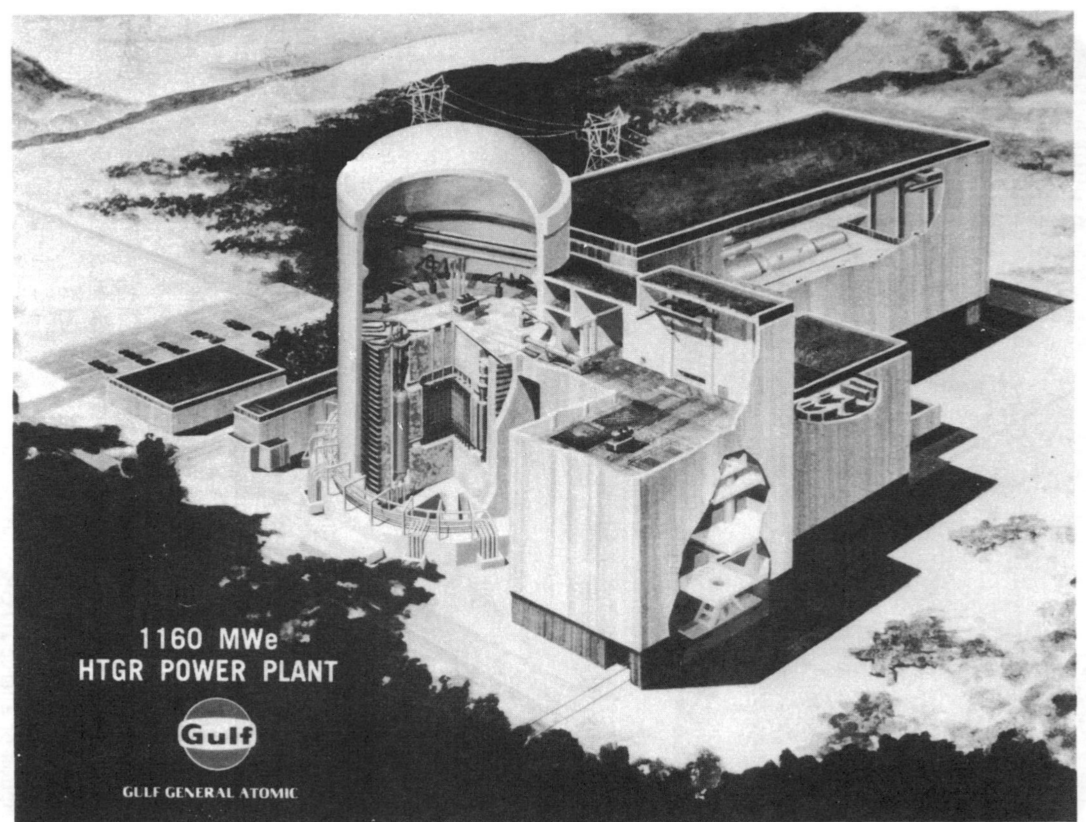

Figure 10-31 A proposed 1160-MW HTGR plant layout. *(Courtesy GA Technologies, Inc.)*

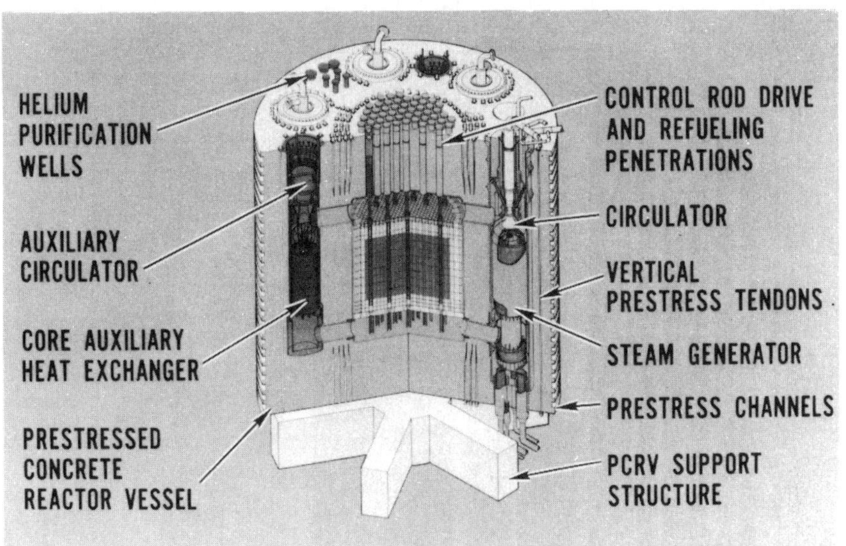

Figure 10-32 HTGR PCRV and nuclear steam supply system. *(Courtesy GA Technologies, Inc.)*

wrapped wires under tension. It is 29.4 m in diameter, 27.6 m high, and includes the reactor core and reflector, helium and nuclear steam supply systems, decay-heat removal system, control rods and their drives, and facilities. The reactor core and reflector are located in a vertical cylindrical center cavity that is 11 m in diameter and 14.5 m high. That cavity is surrounded by nine other cavities, six of which house steam generators and steam-turbine-driven circulators. Three other cavities, arranged between the six main cavities, house auxiliary cooling systems. This gives the reactor block a "telephone dial" appearance when viewed from the top. All cavities are provided with ceramic thermal insulation, gas-tight steel liners, and liner cooling systems. The PCRV penetrations are steel-lined and sealed by single covers with double gaskets preceded by flow resistors.

The reactor core is made up of the hexagonal graphite blocks with holes that contain the coated fuel particles (Figs. 10-30 and 10-33). In addition, each block has a central pickup hole, coolant channels, and three dowel pins for proper positioning with respect to each other. One block in seven has two additional large holes to accommodate control rods and one large hole to receive absorber granules (below). The blocks are stacked up vertically 14 to a column. Each column is composed of three upper graphite reflector blocks, eight fuel blocks, and three lower reflector blocks. There are 493 such adjacent columns for a total of 6902 blocks (1479 upper reflector,

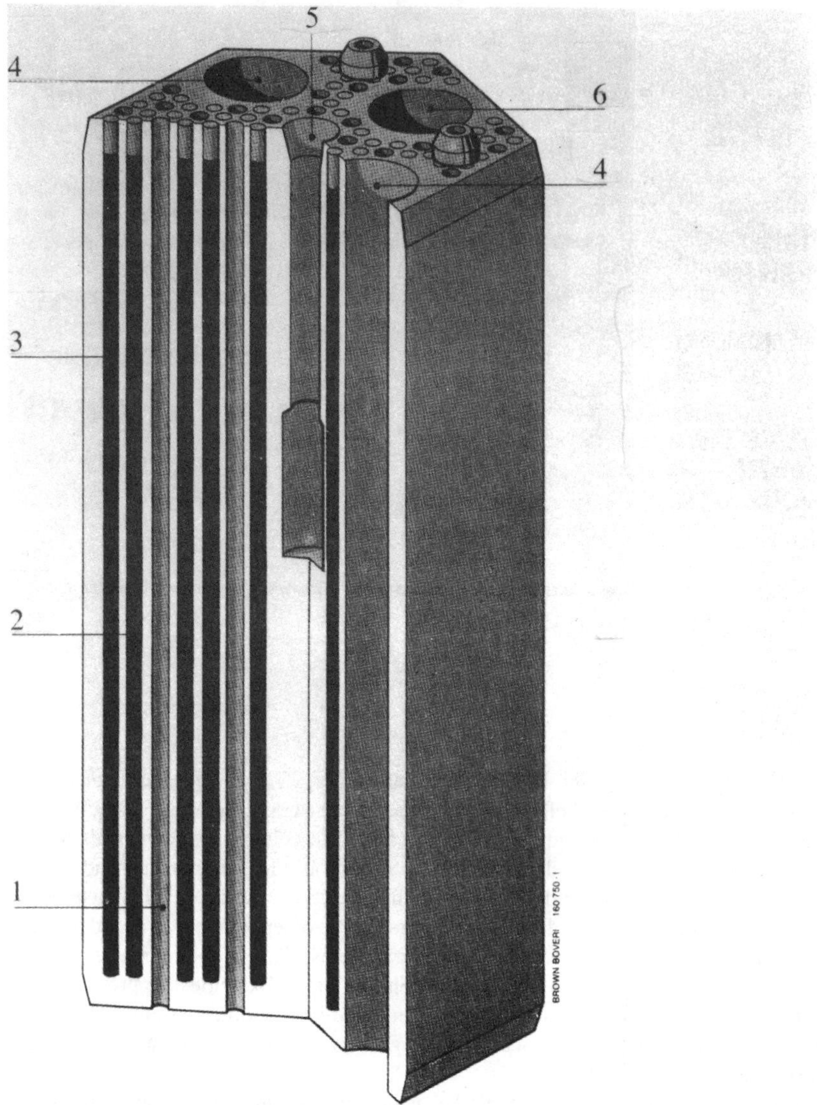

Figure 10-33 HTGR hexagonal graphite fuel block. (1) coolant channel, (2) fuel rod, (3) combustible poison rod, (4) control rod channel, (5) fuel handling pickup hole, (6) hole for absorber granules (reserve shutdown). *(Courtesy GA Technologies, Inc.)*

3944 fuel, and 1479 lower reflector) (Fig. 10-34). The fuel blocks are also surrounded by side reflectors. Groups of seven fuel columns make up a fuel region that is loaded and unloaded as a unit through a common refueling penetration in the top of the PCRV. The whole core fits within a cylindrical steel liner, the only nonceramic part of the core.

The control system consists of 146 control rods operated in pairs. The neutron absorber is boron carbide. Each pair travels in channels (item 4, Fig. 10-33) in the centermost fuel column of each of the seven-column regions. The rods are actuated by electric-motor drives that are located in standpipes in the top of the PCRV but can be made to drop under gravity for quick shutdown. A backup shutdown system consists

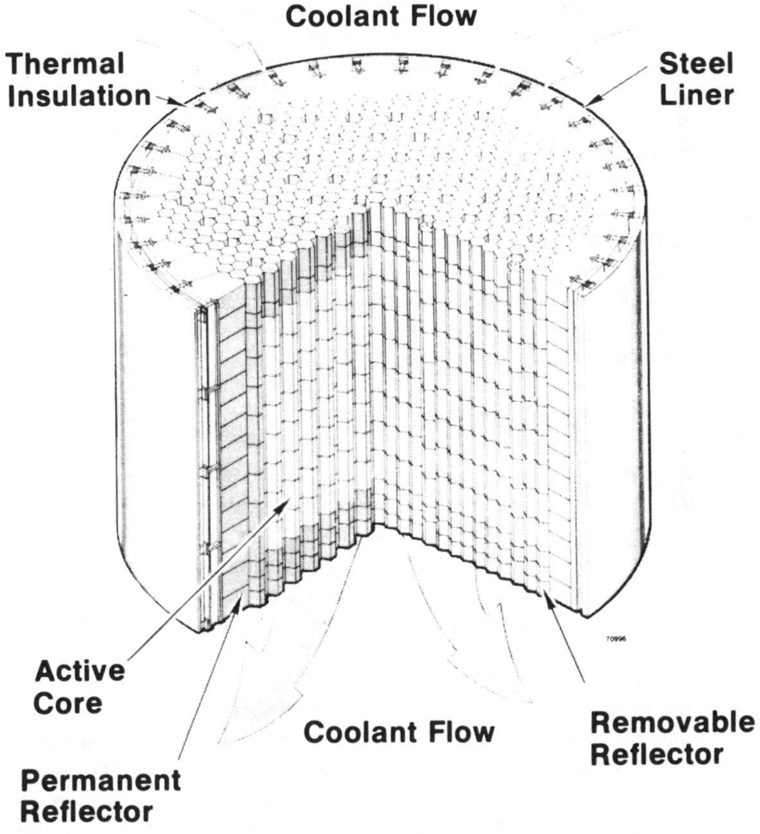

Figure 10-34 HTGR core arrangement. *(Courtesy GA Technologies, Inc.)*

of a neutron-absorbing granular material that drops into holes in the same fuel columns (item 6, Fig. 10-33).

The Primary Heat-Transfer System

Helium is circulated at an average pressure of about 48 bar (700 psi). It enters the reactor core at the top at about 240°C (644°F), flows downward through the coolant holes in the graphite blocks (item 1, Fig. 10-33), and exits at 760°C (1,400°F) to a hot gas plenum at bottom. From there it goes radially to the bottom entrance of the six steam-generator cavities, enters the six circulators at the top of the cavities at its lowest temperature, and returns radially to a "cold" gas plenum at the top of the core.

The steam generators are shell-and-tube heat exchangers in which water and steam flow in tube coils and helium flows on the shell side. They are composed of a reheater at the bottom and an economizer-boiler-superheater at the top. The hot helium enters the steam generator above the reheater, flows downward through it, reverses direction and goes up a central pipe to the top of the steam generator, reverses direction again and flows downward through the superheater, boiler, and economizer, and then upward again through an outside annular space between the generator and its cavity to the helium circulator. The water-steam flow in the generator tubes is upward. The steam flow in the reheater tubes is downward.

The helium circulators are single-stage axial blowers and are driven by 14,500-hp, single-stage steam turbines at 6755 r/min. The circulator impeller and turbine wheel of each unit are mounted on the same shaft, which has water-lubricated bearings. A labyrinth and sealing system prevent water from entering the helium coolant and helium from entering the steam system.

The Steam Plant

Superheated steam from the six steam generators, a total of about 1,000 kg/s ($\sim 8 \times 10^6$ lb_m/h), combines into a single header and enters the high-pressure section of the turbine at 166 bar (2,400 psig) and 510°C (950°F) (Fig. 10-35). The full steam flow leaves that section, divides up to drive the six single-stage circulator turbines, then goes to the reheater in the steam generator. It leaves the reheater at 38 bar (550 psia) and 538°C (1000°F), recombines, and enters the intermediate section of the turbine and three double-flow low-pressure sections, exhausting through six ducts to a common condenser at 0.09 bar (1.3 psia).

Because steam in the HTGR is comparable in pressure and temperature to that in a fossil powerplant, the turbine is a standard 3600 r/min unit (PWR and BWR plant turbines run at 1800 r/min). All turbine sections are double-flow, resulting in six exhaust pipes to the common condenser. The condensed steam is pumped by three 50 percent flow condensate pumps (one on standby). It passes through three low-pressure feedwater heater stages, each consisting of two heaters in parallel, and a direct contact deaerating heater that has a 500-m^3 storage tank. The feedwater is then pumped by three feedwater pumps (one on standby) and enters the economizer section of the steam

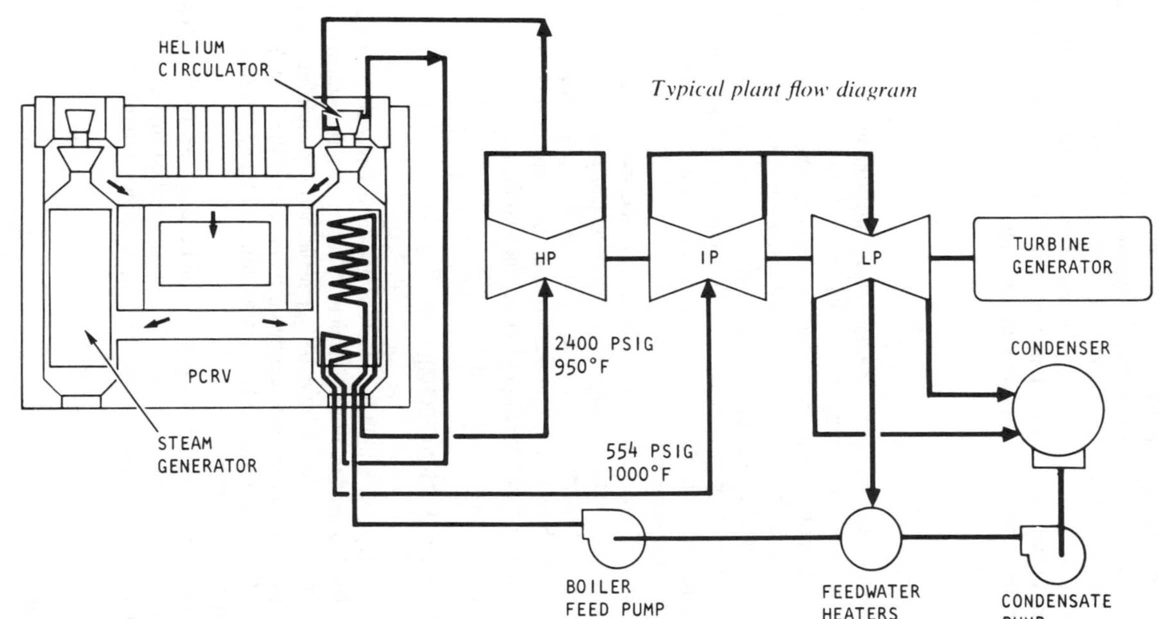

Figure 10-35 Schematic of HTGR plant flow diagram.

118°C (370°F). The feedwater pumps are driven by 16-MW condensing steam turbines whose condensate is fed back to the feedwater line.

The electric generator is a two-pole, three-phase machine, rated at 27 kV and 1170 MW. Both stator and rotor are water-cooled. The plant is started up and shut down with steam generated in auxiliary boilers or from other units on the plant site and fed into the exhaust line of the high-pressure turbine. From there it flows to drive the helium circulator turbines and, via reducing valves, to drive the feedwater pump turbines.

Decay-Heat Removal

Under normal shutdown conditions, the core decay heat (the heat that continues to be generated due to the radioactivity of the fission products) is initially sufficient to generate enough steam to operate the helium circulator turbines at the required reduced load. When decay heat is no longer sufficient to do this, the same auxiliary boilers used for plant startup take over and supply steam to these turbines. In addition, there are the three auxiliary heat exchangers located within the PCRV. These are helium-water heat exchangers with electrically driven helium circulators. The cooling water is in turn cooled in an air-blast heat exchanger.

It is to be noted that, because of the large mass and specific heat of graphite, the HTGR core has a very large heat capacity. It is estimated that it can absorb 20 times as much heat per unit temperature rise as other reactor types of comparable power. This results in a much slower temperature rise in the event of loss of coolant, thus allowing ample time for the initiation of emergency cooling.

10-13 PEBBLE-BED REACTORS

An interesting concept of a high-temperature gas-cooled reactor is the *pebble-bed reactor* under development in West Germany. The first effort resulted in a 15-MW reactor powerplant built at Julich by Arbeitsgemeinschaft Brown/Boveri/Krupp-Reaktorbau and called AVR. A second-generation plant, called the Thorium High-Temperature Reactor (THTR-300), is under construction near Hamm-Uentrop; commercial operation is expected in 1986.

A pebble-bed reactor is one in which the fuel and moderator are mixed together in the form of spherical pebbles that are randomly packed into a relatively simple vessel to form the reactor core. The coolant, helium, flows through the voids between the spheres.

The THTR-300

In the THTR-300 (Fig. 10-36), the core contains 675,000 60-mm-diameter fuel elements (Fig. 10-37). Each has a 5-mm-thick outer graphite shell surrounding a graphite matrix that contains (12.1 percent by volume) 40,000 400-μm-diameter biso-coated

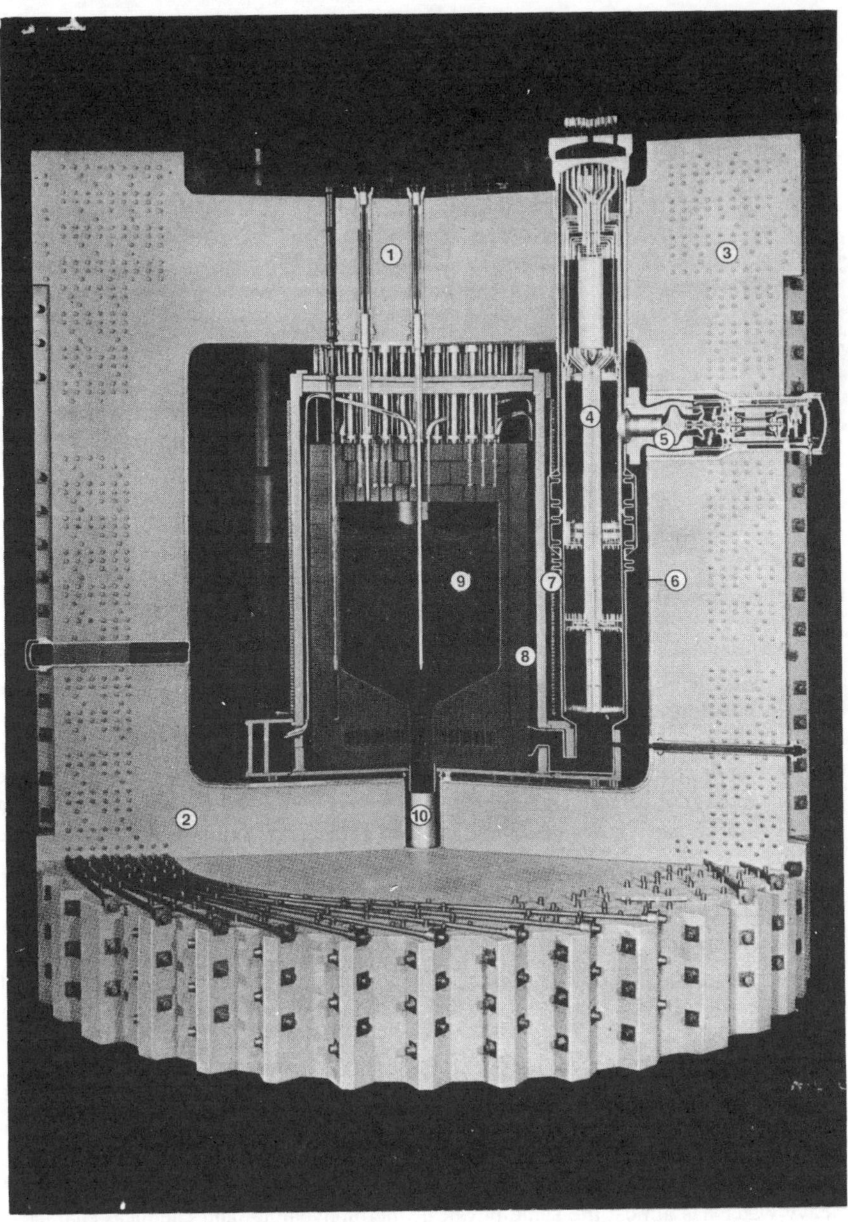

Figure 10-36 The THTR-300 pebble-bed reactor. (1) shutdown rod, (2) prestressed concrete vessel, (3) tendon rings, (4) steam generators, (5) helium circulators, (6) thermal insulation, (7) thermal shield, (8) side reflector, (9) core, (10) pebble-removal tube. *(Courtesy Hochtemperatur-Kernkraftwerk Gmbh, West Germany.)*

Diameter	60 mm
Fuel-free zone	5 mm
Volume loading of particles	12·1%
Number of particles per fuel element	40000
Matrix density	1·70 g cm^{-3}
Operation time	1100 full power days
Maximum surface temperature	1000°C
Maximum fuel temperature	1150°C
Maximum fast-neutron fluence ($E > 0 \cdot 1$ MeV)	$6 \cdot 4 \times 10^{21}$ cm^{-2}
Maximum burn-up	14·1% FIMA

Figure 10-37 Cross section of the THTR-300 spherical fuel element.

fuel particles, similar to those shown in Fig. 10-30. The particles in the THTR are not segregated by fuel but contain mixed UO_2-ThO_2.

One advantage of a pebble-bed reactor is the relative ease with which it continually charges and discharges fuel on load. Done pneumatically, the spherical elements leave the core through a central pipe, after which they are checked for breakage and sent to a burnup measuring device. They are then either recycled or sent to a discharge facility.

Helium at 39 bar (566 psia) enters the core at top at 270°C (518°F) and flows downward and leaves it at 750°C (1382°F). Surrounding the core, and within a PCRV (similar to HTGR), are six vertical steam generators and six helium circulators. The latter (unlike in the HTGR) are horizontally mounted through the PCRV. Superheated steam at 550°C (1022°F) is produced. The condenser water is cooled by a dry-cooling tower.

10-14 THE PRESSURIZED HEAVY-WATER REACTOR (PHWR)

Heavy water has almost the same physical, thermodynamic, and chemical characteristics as ordinary water (214.6°F versus 212°F normal boiling point, 38.9°F versus 32°F melting point, and 1.10 versus 1.0 g/cm^3 density at room temperature). Power reactors that use heavy water at high temperature must, as those using ordinary water, operate at high pressure. Heavy water, however, has markedly different nuclear characteristics. As a moderator, it requires a neutron to travel about twice as far as ordinary water in order to lose the same energy fraction. On the other hand, heavy water absorbs practically no neutrons, which is not true for ordinary water.*

Because of near zero neutron absorption, heavy water can be used as coolant-moderator with natural-uranium fuels. Heavy water also results in good fuel burnups and conversion ratios. A further improvement is obtained by dividing the heavy water into a cooled moderator region and a hot coolant region in a *pressure-tube* design (below).

Other economic advantages of the PHWR are low-cost natural fuel because of the absence of the very costly enriching process and the absence of criticality hazards during fuel fabrication. The use of natural fuels can also be particularly attractive to those countries that do not possess enriching facilities. On the debit side, the production of heavy water is, in itself, a costly operation.

The economic incentives to develop heavy-water reactors thus vary in different countries and the economics depend on different methods of estimations and projections. In Canada, for example, it has been shown that natural-uranium heavy-water-moderated reactors can produce power at a lower cost than enriched-fuel reactors, and Canada has concentrated on developing this type of reactor. This view, plus other advantages already cited, has been shared elsewhere, and Canadian-type reactors (Candu) are now being built in such countries as Argentina, India, Korea, and Pakistan. In the United States, however, most studies have indicated cheaper power from enriched-fuel light-water reactors (BWR, PWR) and that, even if heavy water is used, power would be cheaper if enriched fuels are used. The overriding factors in this case are the higher capital costs and the large expensive heavy-water inventory, coupled with high fixed-charge rates. Hence, the United States effort in D_2O reactors is rather meager.

Because of the long neutron paths for moderation, heavy-water-moderated reactors require large moderator-to-fuel-volume ratios. Such reactors, therefore, require large-diameter pressure vessels. Large power reactors operating at high temperature and pressure, therefore, require larger, thicker, and costlier pressure vessels than ordinary-water reactors of comparable output. Both pressure vessel and pressure-tube designs have been used. The latter design [101] allows the use of a lower-pressure, less costly vessel but with the added expense of constructing a leak-tight *calandria* vessel free of differential expansion. It also results in the separation of coolant and moderator, and, as already mentioned, a cool-moderator design can then be easily incorporated.

Some other problems associated with D_2O reactors are the loss by leakage of the expensive D_2O and the high activity associated with the decay of tritium formed in the reactor.

We will next present the Candu reactor of the pressurized heavy-water-moderated and cooled, pressure-tube variety. Other reactor types of interest that have seen various degrees of development are the D_2O-moderated, gas-cooled, organic-cooled, and ordinary-water-cooled reactors, all made possible by the pressure-tube concept. In the

* The ratio of slowing down to absorption is called the *moderating ratio*. It is about 90 times greater for heavy water than ordinary water.

latter concept the ordinary water may be allowed to boil, which results in the generation of steam within the pressure tubes.

The Candu Reactor Powerplant

The *Candu*-type reactors are designed by Atomic Energy of Canada, Limited. Their main features are a horizontal pressure-tube calandria reactor vessel, fuel elements

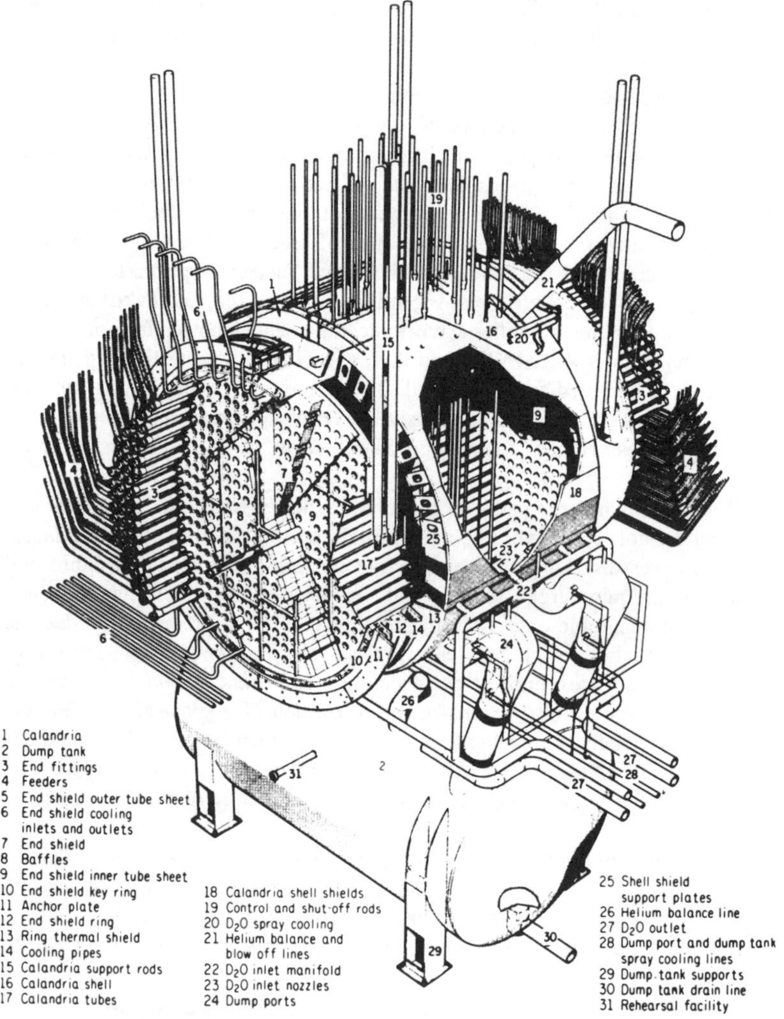

Figure 10-38 The Pickering heavy-water (Candu) reactor assembly. *(Courtesy Atomic Energy of Canada, Limited.)*

containing natural-uranium dioxide and housed in horizontal presure tubes, which also carry the pressurized heavy-water coolant, and a relatively cool heavy-water moderator on the shell side.

One such reactor (Fig. 10-38) has a stainless steel calandria shell (1) that has integral-steel end shields (7) cooled by ordinary water (6) and stainless steel circumferential shell thermal shields (13). The calandria contains 390 horizontal pressure tubes (17). Each pressure tube in turn contains a coolant tube that is supported in sliding bearings at the end shields of the calandria. The calandria and coolant tubes are separated by a sealed annulus containing CO_2 or nitrogen. There is a total of 276 tons of heavy water in the reactor. Below the calandria is a dump tank (2) with a heavy aggregate concrete vault to provide shielding from reactor radiation. The concrete is cooled by one layer of cooling coils embedded in it. The dump tank contains a helium atmosphere.

The fuel elements are hermetically sealed Zircaloy-2 rods that contain compacted and sintered naturally enriched UO_2 pellets. Twenty-eight such elements are attached mechanically at their ends to form a cylindrical fuel assembly (Fig. 10-39) 4.03 in in diameter and 19.5 in long. Spacers attached to the cladding maintain 0.05 in space between the elements.

The heavy-water coolant enters the fuel channels at 480°F (249°C) and leaves at 560°F (293°C) and 1300 psig (90.7 bar) via feeder pipes (4). Twelve U-shaped tube-in-shell heat exchangers generate a total of 6.46×10^6 lb_m/h (814 kg/s) of steam at 585 psia (40.3 bar) and 483.5°F (250.8°C) from the ordinary-water working fluid.

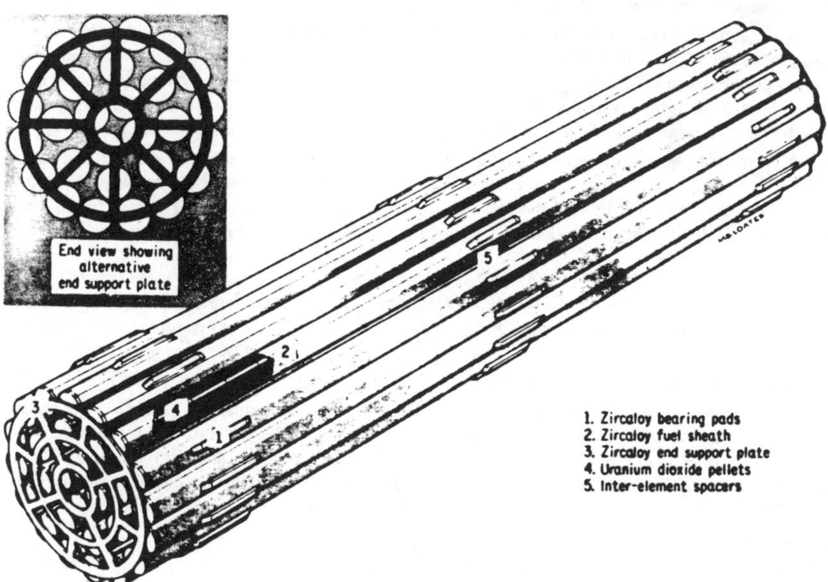

Figure 10-39 Candu 28-element fuel bundle.

The reactor is protected by 11 shutoff rods and by the moderator dump. The latter is made possible in this design by the fact that dumping the moderator would not starve the fuel from the coolant, a situation which would result in fuel meltdown. It is, however, resorted to only in case the shutoff rods fail to shut down the reactor.

The fuel is loaded and removed in the reactor *on power* by two coordinated fueling machines located at opposite ends of the reactor, each on the underside of a bridge. The bridges move vertically, whereas the machines move horizontally. In an on-power refueling operation, both machines locate, by remote control, onto an end fitting (3) of the selected channel. The machines pressurize themselves, then remove and store end closures and shield plugs. One machine then advances a fuel carrier containing the new fuel subassembly into the channel end fitting. Simultaneously the other machine extends an empty fuel carrier at the other end of the channel. The charge machine then rams the new fuel through a fuel latch into the channel, thus probing the spent fuel to enter the carrier in the discharge machine. The carriers are then withdrawn, the shield plugs and end closures are replaced, and the machines are detached from the end fittings. After the required number of bundles have thus been loaded, the discharge machine traverses to a spent-fuel port, where it jettisons the spent fuel to an underwater storage canal via a transfer mechanism and a conveyor.

PROBLEMS

10-1 A PWR powerplant pressurizer operates at a steady state at 2200 psia with a constant spray flow of 8 lb_m/min from the cold leg at 552°F. Calculate the amount of heat added by the electric heaters, in kilowatts, if the pressurizer heat losses to the ambient are 6144 Btu/h.

10-2 A pressurized water reactor has inlet and exit water at 290 and 320°C, respectively. It has a 30 m^3 vapor pressurizer which is normally 60 percent full of water at a pressure of 140 bar. A case of an insurge occurred during which 0.25 m^3 of water entered the pressurizer from the primary circuit hot leg, 0.05 m^3 entered through the spray, and 50 kWh was added by the electric heaters. Determine the internal energy of the pressurizer contents before and after the event, in kilojoules. Ignore heat losses to the ambient.

10-3 A pressurized-water reactor operating at 2000 psia has primary water entering at 550°F and leaving at 610°F. A 1000-ft^3 pressurizer is normally half full of water. During a transient the pressure rose to 2100 psia, 200 lb_m of spray water entered the pressurizer, and the pressurizer became 60 percent full of water. Ignoring heat losses to ambient, calculate the amount of heat added by the electric heaters, in kilowatt hours.

10-4 A PWR operates at 2000 psia and 600°F hot leg. It has a 1000 ft^3 pressurizer which is normally half full of water. The primary loop is 10,000 ft^3 in volume. The hot leg temperature suddenly rose to 605°F. Calculate (*a*) the final composition of the pressurizer, and (*b*) the final system pressure. Ignore spray flow and heat.

10-5 Repeat Prob. 10-4, except for the case when the hot-leg temperature dropped to 595°F.

10-6 It is desired to design a vapor pressurizer for a PWR that operates normally at 2200 psia. The primary loop has a volume of 8000 ft^3 and a hot-leg temperature of 610°F. The pressurizer is to be normally 60 percent full of water and designed to prevent a system pressure rise of more than 1 percent for a 10°F hot-leg temperature rise. Ignoring the spray and heat added or lost to the ambient, calculate the necessary pressurizer volume.

10-7 Derive Eq. (10-7) for gas pressurizers.

10-8 Repeat Prob. 10-6, but design a gas pressurizer. Assume two gases (*a*) argon and (*b*) helium, and calculate the pressurizer volume for each with the polytropic exponent corresponding to isothermal and to adiabatic reversible processes. Compare these volumes to that of the vapor pressurizer of Prob. 10-6.

10-9 A sodium-cooled reactor has 5000 ft^3 of primary coolant at an average temperature of 1400°R. A cover gas of argon acts as a pressurizer. Calculate the volume of argon necessary to limit the pressure rise to 1 percent for a coolant temperature rise of 100°F. Use a polytropic exponent corresponding to an (*a*) isothermal and (*b*) adiabatic reversible process.

10-10 A sodium-cooled reactor has 20,000 ft^3 of primary coolant at 600 psia and 1300°R average, and 8000 ft^3 of argon cover gas. Find the pressure rise, in pounds per square inch, if the coolant average temperature rose to 1400°R. Assume that the gas is compressed adiabatically and reversibly.

10-11 A PWR primary loop is 8000 ft^3 in volume and operates at an average temperature of 580°F. It has a 1000-ft^3 vapor pressurizer which normally contains 60 percent water by volume at 2200 psia. An accident occurred in which the relief valve suddenly stuck in an open position and fluid discharged to the relief tank. The system pressure steadily dropped to 1600 psia, during which time the electric heaters were fully activated to help slow down the rate of pressure drop. At 1600 psia, the primary loop average temperature was 550°F, the pressurizer was 95 percent full of steam, the heaters were turned off to protect them against overheating, and the emergency core cooling system (ECCS) was activated. This replenished the primary loop with water to prevent uncovering and damage to the fuel elements. The relief tank is assumed to remain at nearly atmospheric pressure, but there is a 15.3-psi pressure drop in the line connecting it to the pressurizer relief valve. Ignoring the effect of spray and heat losses to ambient, calculate (*a*) the initial mass composition of the pressurizer, in pounds mass, (*b*) the condition of the fluid leaving the relief valve at the instant it opened (pressure, temperature, and quality or degree of superheat), (*c*) the total loss of fluid from the primary loop (before ECCS) assuming for simplicity that its temperatures remained the same, in pounds mass, and (*d*) the condition of the fluid leaving the relief valve at the instant the ECCS came on the line.

10-12 A PWR powerplant producing 10^7 lb_m/h of 1100 psia saturated steam uses reheat from live steam (similar to that in Fig. 10-14). The high- and low-pressure turbines exhaust at 250 and 1 psia, respectively. For simplicity, assume that both turbines and pump are adiabatic reversible, that there are no feedwater heaters, and that the reheater drain is returned to the steam generator. Calculate (*a*) the fraction of live steam that is diverted to the reheater, if steam is reheated to 550°F, (*b*) the cycle net power, in megawatts, and (*c*) the cycle efficiency.

10-13 Calculate the gross station heat rate, in Btus per kilowatt-hour, and the thermal efficiency of the PWR powerplant shown in Fig. 10-14.

10-14 A chemically shimmed PWR uses 3.50 percent enriched fuel with a water-to-fuel ratio of 2.9. It has 4 percent worth control rods which are 30 percent inserted in the core. The boron concentration is 1600 ppm, hot conditions. Find (*a*) the axial to average power density ratio, (*b*) the total core reactivity, percent, and (*c*) the rod insertion and approximate boron concentration that would result in a 5 percent reduction in axial to average power density for the same total core reactivity as in *b*.

10-15 A chemically shimmed PWR uses 3.05 percent enriched fuel and a water-to-fuel ratio of 3.4. It is required to vary the reactivity of the core by 1 percent with rod insertions of 30 percent minimum and 60 percent maximum, without causing the maximum-to-average power density to exceed 1.72. Find (*a*) the total rod worth, percent, and (*b*) the boron concentration that would be required for a total core reactivity of 16 percent (hot conditions).

10-16 A boiling-water reactor operating at a pressure of 70 bar produces 1200 kg/s of saturated steam from feedwater at 200°C. The average core exit quality is 10 percent. Calculate (*a*) the recirculation ratio, (*b*) the core inlet enthalpy, in kilojoules per kilogram, and temperature, in degrees Celsius, (*c*) the degree of subcooling, in degrees Celsius, and (*d*) the heat generated in the reactor, in megawatts.

10-17 A 1000 MW boiling-water reactor powerplant with 33 percent efficiency was operating at a 75 percent of rated load with a steam mass flow rate of 1150 kg/s, a reactor core pressure of 70 bar, and an average exit quality of 13.6 percent. The plant uses recirculation control. Find (*a*) the feedwater temperature, in degrees Celsius, (*b*) the core degree of subcooling, in degrees Celsius, (*c*) the downcomer flow at 75 percent load, (*d*) the average exit quality immediately after initiation of load change to 80 percent, and when load has changed to 80 percent, and (*e*) the steam and downcomer flows, in kilograms per second after load change.

10-18 A forced-recirculation BWR core operating at 1000 psia generates 2500 MW. The core average exit quality is 10 percent. The average core and downcomer densities are 40 and 50 lb_m/ft^3, respectively. The core is 12 ft high. Feedwater is added at top of downcomer. The total flow pressure losses are 6 psi. Find: (*a*) the core inlet temperature and enthalpy, (*b*) the steam generated, in pounds mass per hour, (*c*) the natural driving pressure, in pounds force per square inch, and (*d*) the pump power in kilowatts required to supplement the natural driving pressure if its efficiency is 0.75. (Note power = $\dot{m}\ \Delta P/\rho\eta$.)

10-19 Calculate the gross station heat rate, in Btus per kilowatt hour, and the thermal efficiency of the BWR powerplant shown in Fig. 10-27.

10-20 A pebble-bed reactor has a 14 ft diameter, 15-ft-high core containing 2.4-in-diameter fuel elements. The helium coolant flow is 1.36×10^6 lb_m/h at a pressure of 40 atm and core inlet and outlet temperatures of 520 and 1380°F, respectively. The blowers are 70 percent efficient and the pressure drop in the primary system is 50 percent greater than in the core alone. Find (*a*) the blower work, in megawatts, (*b*) the core heat generation, in megawatts, and (*c*) the steam generator capacity, in megawatts. (The pressure drop in a randomly packed pebble reactor in pounds force per square foot is given by

$$1.56 \times 10^{-6}\ (H/d^{1.27})\ (\mu^{0.27}/\rho)\ G^{1.73}$$

where H = core height, ft
d = pebble diameter, ft
μ = gas viscosity, lb_m/ft h
ρ = gas density, lb_m/ft^3
G = apparent gas mass velocity, $lb_m/h \cdot ft^2$, based on the total bed cross-sectional area)

CHAPTER

ELEVEN

FAST-BREEDER REACTORS AND POWERPLANTS

11-1 INTRODUCTION

The growth of nuclear power in the world's electric utility industry to date has primarily relied upon water-cooled and gas-cooled thermal neutron reactors. These are burners that in effect consume fissionable fuel with low conversion ratios (Sec. 11-2). The expected continued growth in the use of nuclear power and consequent consumption of available nuclear fuels have prompted the reevaluation of long-term goals. Ways of conserving uranium reserves and of keeping fuel-cycle costs down as fuel costs go up are therefore being sought. The fast-breeder reactor is the logical step in that direction.

Fast reactors are those whose neutrons are not slowed down to thermal energies by a moderator. Coolant and other reactor materials, however, moderate the neutrons to a certain extent in a fast reactor so that the neutron spectrum extends from fission energies that may be as high as 17 MeV but average about 2 MeV down to about 0.05 or 0.1 MeV. A *hard-spectrum* fast reactor is one in which the neutron density distribution extends over a narrow range nearer the high end of the energy spectrum. A *soft-spectrum* fast reactor, on the other hand, is one in which more moderation occurs, such as by liquid sodium and oxide fuels; hence the distribution extends to lower energies. A breeder reactor is one in which more fissionable fuel is produced than is consumed, or one that has a conversion, or breeding, ratio greater than 1.

Present-day burner reactors produce some Pu^{239} from the U^{238} in their fuel. Some of this plutonium is burned during the life of the core, but some remains. The remainder can be used to fuel initial fast-breeder reactors in a reactor "mix" until a self-supporting fast-breeder-reactor system is at hand. Such a system would convert the abundant U^{238}

(99.3 percent of all natural uranium) to fissionable Pu^{239} and hence extend our fuel reserves for centuries to come.

Table 11-1 lists the U.S. energy reserves as compiled by the Departments of Commerce and Energy in the late 1970s. The figures are best estimates and could be off by a factor of 2. This does not materially alter the conclusions, which are equally valid for the rest of the world.

A quad (Q) is a quadrillion, or 10^{15}, Btu, or roughly 10^{18} J. At present, U.S. energy consumption is about 60 Q/year, of which about 20 Q are consumed in the generation of about 8 Q of electricity. The balance is used for domestic services, industrial processing, and transportation. The "Years" column in Table 11-1 shows the time before depletion at present rates of consumption, if the particular fuel cited is used alone for *all* energy. "Present reactors" in the "Remarks" column refers to water-cooled thermal reactors (PWR and BWR). With breeding, our fuel resources are extended about 55 times from 50 to 2750 years. If an economical way is found to mine low-grade ores containing 1/55 as much uranium, we would have another factor of 10 to 12 to over 30,000 years. If the 0.003 ppm known to exist in sea water were extracted (the means are not yet available), the uranium reserves would be sufficient to supply the world for 1 or 2 million years. (Here would be a good use of solar energy to evaporate water from the sea.) These figures may seem overly optimistic, but they do dramatize the need to develop the fast-breeder reactor as a necessary power producer for the next century and probably beyond. Fusion has not yet been scientifically demonstrated and is not expected to make a big dent in that picture for some time to come.

Fast reactors have been built in the United States, the United Kingdom, the Soviet Union, France, Japan, and Germany, and Italian and Indian efforts are underway (Table 11-2). Some early reactors were cooled by NaK or mercury, but the majority are cooled by Na. Fuels varied from uranium and plutonium in the form of metal alloys, oxides, or carbides. The majority now favor the mixed-oxide fuel PuO_2 - UO_2.

The first fast reactor was built in the United States in 1946. Called Clementine,*

* Named after the ballad "In a cavern, in a canyon . . . Excavating for a mine . . . dwelt a miner, forty niner . . . and his daughter Clementine. . . ." 49 was the code number for plutonium in World War II, and the reactor was built in a cavern.

Table 11-1 Energy reserves in the United States

Fuel	Quads	Years	Remarks
Coal	5000	60	
Oil	500	6	
Gas	500	6	
U_3O_8	1600	20	No breeding
U_3O_8	4000	50	Present reactors
U_3O_8	220,000	2750	Breeding; high-grade ore
U_3O_8	>2,500,000	>30,000	Breeding; all economic ore

Table 11-2 The world's fast-breeder reactors

Country	Reactor	Power MW(t)	Power MW(e)	Coolant	Design	Initial fuel	Status*
USA	Clementine	0.025	—	Hg		Pu metal	1946,D
	LAMPRE-I	1	—	Na		Molten Pu	D
	EBR-I	1.4	0.2	NaK		U metal	1951,D
	EBR-II	62.5	16.0	Na	Pool	U metal	1963,O
	Enrico Fermi	200	62	Na	Loop	U metal	1966,D
	SEFOR	20	—	Na	Loop	PuO_2-UO_2	D
	FFTF	400	—	Na	Loop	PuO_2-UO_2	1980,O
	CRBR	975	350	Na	Loop	PuO_2-UO_2	1987,C
	CDS	2550	1000	Na	Loop	PuO_2-UO_2	1992,P
USSR	BR-1	0	—			PuU metal	D
	BR-2	0.1	—	Hg		Pu metal	D
	BR-5/10	10	—	Na	Loop	Pu metal	1959,O
	BOR-60	60	12	Na	Loop	PuO_2-UO_2	1969,O
	BN-350	1000	350	Na	Loop	PuO_2-UO_2	1972,O
	BN-600	1470	600	Na	Pool	PuO_2-UO_2	1980,O
	BN-1600-1	4200	1600	Na	Pool		1990,P
FRANCE	Rapsodie	40	—	Na	Loop	U metal	1967,O
	Phénix	563	250	Na	Pool	PuO_2-UO_2	1973,O
	Super Phénix I	3000	1200	Na	Pool	PuO_2-UO_2	1983,C
	Super Phénix 2&3	3750	1500	Na	Pool	PuO_2-UO_2	1990-91,P
	Super Phénix 3&4	3750	1500	Na	Pool	PuO_2-UO_2	1992-93,P
UK	Dounreay	60	13.5	NaK	Pool	U metal	1962,D
	PFR	600	250	Na	Pool	PuO_2-UO_2	1974,O
	CDFR-1	3250	1300	Na	Pool	PuU metal	1992,P
GERMANY (FRG)	KNK-2	58	20	Na	Loop	UO_2	1977,O
	SNR-300	762	327	Na	Loop	PuU metal	1985,C
	SNR-2	3250	1300	Na	Loop	PuU metal	1993,P
JAPAN	JOYO	75–100	—	Na	Loop	PuO_2-UO_2	1977,O
	MONJU	714	300	Na	Loop	PuO_2-UO_2	1987,C
	DFBR	2800	1000	Na		PuO_2-UO_2	1993,P
ITALY	PEC	135	—	Na	Loop	PuO_2-UO_2	1983,C
INDIA	Madras FBTR	42	17	Na	Loop	PuO_2-UO_2	1983,C

* Dates of initial or expected operation; C-under construction; D-decommissioned, O-operating, P-in the planning-design stage.

it was plutonium-fueled, produced 25 kW(t), was mercury-cooled (one of only two to date), and has long been shut down. The main initial effort was the Experimental Breeder Reactors EBR-I and EBR-II [102]. The 200 kW(e) EBR-1 was the first reactor of any kind to produce electricity in the United States. Construction began in November 1947; it achieved criticality in August 1951 and produced electric power in December 1951. It was NaK-cooled and used a stainless steel–clad fully enriched 2 percent zirconium-uranium alloy for fuel and a stainless steel–clad natural-uranium external blanket. It was discontinued after demonstrating the feasibility of fast reactors and breeding. EBR-II is a 62.5-MW(t), 20-MW(e), sodium-cooled reactor fueled with stainless steel–clad 49.5 percent enriched-uranium-alloy. The blanket is stainless steel–clad depleted uranium. EBR-II has an integrated fuel manufacturing and reprocessing facility and will operate on a fuel-recycle basis, ultimately with plutonium. These reactors were followed by Enrico Fermi [3] and the experimental reactors SEFOR [3] and FFTF. A 350-MW(e) demonstration plant, the Clinch River Breeder Reactor (CRBR), was under construction in the early 1980s (Sec. 11-4).

The British effort is centered around Dounreay [103] and PFR. The first is NaK-cooled with tubular fuel clad in vanadium on the inside and partly finned niobium on the outside. Natural uranium is used in external breeder blankets surrounding the core. Control is achieved by axially moving 12 fuel assemblies (of 10 fuel rods each), situated around the periphery of the core. These are divided into two safety, six control, and four shutoff assemblies. A large 1300-MW(e) reactor, CDFR-1, is in the planning-design stage.

The USSR effort includes three experimental reactors built since 1955. BR-1 is PuU-fueled and U-reflected, designed to carry out cross-sectional measurements at various neutron energies. BR-2, Pu-fueled and mercury-cooled, was used to test nuclear parameters, breeding ratios, and liquid metals. BR-5 (upgraded to BR-10) is UC-fueled, sodium-cooled, with a uranium-nickel blanket. The program includes three operating reactors: BOR-60; BN-350, a dual-purpose reactor that produces 150 MW(e) and process steam for a 120,000-m^3/day water desalination plant; and BN-600, a 600-MW(e) power reactor. A 1600-MW(e) plant, BN-1600-1, is in the planning-design stage.

France has what must be considered the most ambitious program. It started with Rapsodie, a 20-MW(t) sodium-cooled and plutonium-uranium-oxide-fueled reactor. A 600-MW(t), 250-MW(e) reactor, the Phénix,* has been in operation since 1973 and has given excellent service. Under construction and in the planning stages are the Super Phénix class of reactors (Sec. 11-5).

Gas-cooled fast-breeder reactors are recieving some attention in the United States and elsewhere (Sec. 11-6).

11-2 NUCLEAR REACTIONS, CONVERSION, AND BREEDING

A typical neutron-flux spectrum in a liquid-metal-cooled fast reactor compared with that in a water-cooled thermal reactor is shown in Fig. 11-1. The objective in the latter is to maintain a chain reaction with thermal neutrons having energies below 1 eV. In a fast-breeder reactor, the objective is to maintain a chain reaction with fast neutrons that have an average energy of about 1 MeV by fission in U^{235} and Pu^{239}. It also must provide additional fast neutrons sufficient to convert U^{238} to Pu^{239}.

* Named after the mythical, beautiful, lone Egyptian bird which lived in the desert for 500 or 600 years and consumed itself in fire, then rose from the ashes to start another long life. It has become a symbol of immortality.

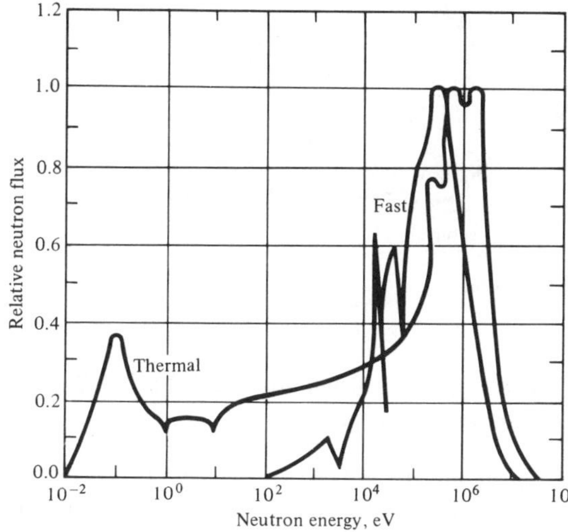

Figure 11-1 Typical neutron flux spectra for fast (LMFBR) and thermal (PWR) reactors.

Typical fission reactions in fast reactors are the same as those in thermal ones, as for example

$$_{92}U^{235} + {}_0n^1 \rightarrow {}_{56}Ba^{137} + {}_{36}Kr^{97} + 2{}_0n^1 \tag{11-1}$$

$$_{92}U^{233} + {}_0n^1 \rightarrow {}_{56}Ba^{136} + {}_{36}Kr^{96} + 2{}_0n^1 \tag{11-2}$$

and

$$_{94}Pu^{239} + {}_0n^1 \rightarrow {}_{56}Ba^{137} + {}_{38}Sr^{100} + 3{}_0n^1 \tag{11-3}$$

A typical nonfission reaction is the one that occurs in the sodium coolant, which is composed of 100 percent Na^{23}

$$\left.\begin{aligned} &{}_{11}Na^{23} + {}_0n^1 \rightarrow {}_{11}Na^{24} \\ &{}_{11}Na^{24} \xrightarrow[\beta]{14.8\ h} {}_{12}Mg^{24} + {}_{-1}e^0 \end{aligned}\right\} \tag{11-4}$$

where Na^{24} is a highly radioactive isotope that emits 2.76 MeV γ radiation and 1.3 MeV β decays with a 14.8-h half-life to stable Mg^{24}.

A fast neutron reaction with U^{238} results in a series of reactions that culminate in Pu^{239}, shown in Fig. 11-2 and given by

$$\left.\begin{aligned} &{}_{92}U^{238} + {}_0n^1 \rightarrow {}_{92}U^{239} + \gamma \\ &{}_{92}U^{239} \xrightarrow[\beta]{24\ min} {}_{93}Np^{239} + {}_{-1}e^0 \\ &{}_{93}Np^{239} \xrightarrow[\beta]{2.3\ days} {}_{94}Pu^{239} + {}_{-1}e^0 \end{aligned}\right\} \tag{11-5}$$

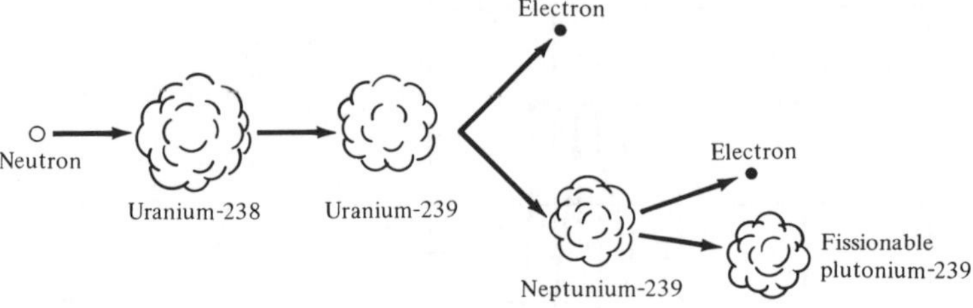

Figure 11-2 Schematic of the Pu^{239} breeding chain.

This is a *breeding* reaction that converts *fertile* U^{238} into *fissionable* Pu^{239}.* We will now define two important fuel parameters. These are:

ν = the average number of neutrons emitted per *fission,* as in Eqs. (11-1) to (11-3). ν depends primarily on the fissile isotope, and to a lesser degree on neutron energy. It is highest for fast fission in plutonium, being about 3.0.

η = the *fission factor,* equal to the average number of neutrons emitted per neutron *absorbed* (and not necessarily causing fission). η is a strong function of neutron energy (Fig. 11-3).

The Conversion and Breeding Ratios

When a neutron is *absorbed* in the fuel, it produces η neutrons. One of these must be reserved for further absorption to keep the reaction going. There will also be losses by parasitic capture in reactor coolant and materials of construction and by leakage. These losses we will designate L, neutrons lost per neutron absorbed. The rest of the neutrons per neutron absorbed will be available for the breeding reaction, Eq. (11-5), and are called the *conversion* or *breeding* ratio C, or

$$C = \eta - 1 - L \tag{11-6}$$

The maximum possible C, $C_{\max}$, is obtained if L were zero, or

$$C_{\max} = \eta - 1 \tag{11-7}$$

Depending upon η and L, C can be much less than unity, and the reactor is called a *burner*. A reactor with a low C is generally called a *converter*. One with high C but less than 1.0 is called an *advanced converter*. For C less than 1.0, it can be shown [104] that there is a maximum theoretical limit to the amount of fertile nuclei that can

* Some two billion years ago in what is now Gabon in Africa, large deposits of natural uranium ore, more enriched than now (about 3 percent), mixed with geologic water causing thermal fission. The low-power-density but huge natural reactors (at least six were identified) stayed critical for many centuries. Some U^{238} was converted to Pu^{239}, showing that not all plutonium is manufactured.

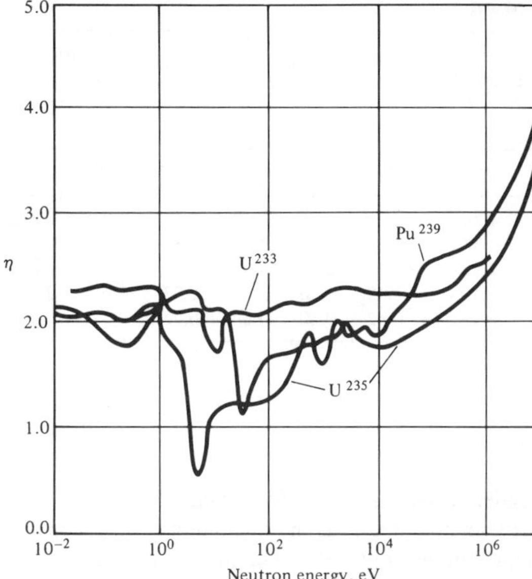

Figure 11-3 The variation of the fission factor η with neutron energy for three fissionable nuclei.

be converted to fissionable nuclei. This maximum depends upon both C and the initial number of fissionable nuclei.

$C = 1$ means that the reactor is producing a number of fissionable nuclei equal to what it consumes; $C > 1$ means that there is no limit to conversion. In both these cases it is theoretically possible to consume all fissionable and fertile nuclei present. In practice, however, as with burner reactors, the fuel elements must be reprocessed and replaced periodically because fission products absorb neutrons and reduce reactivity or because of metallurgical considerations. A reactor with $C \geqslant 1$ is a *breeder*.

The term *breeding ratio* has the same meaning as conversion ratio. *Breeding* (or *conversion*) *gain G* is the gain in fissionable nuclei per fissionable nucleus consumed. Thus

$$G = C - 1 = \eta - 2 - L \tag{11-8}$$

and a maximum gain, G_{max}, is

$$G_{max} = C_{max} - 1 = \eta - 2 \tag{11-9}$$

Average values of ν, η, and G_{max} are given in Table 11-3 for fissionable fuels for broad thermal and fast neutron energy ranges and for fertile materials for fast neutrons.

Allowing for the fact that L is not zero, it can be seen that best breeding can occur in fast reactors fueled with Pu^{239} (and Pu^{241}, which is not readily available), followed by U^{233} and U^{235}. In thermal reactors, good conversion or breeding can be expected only from U^{233}, hence the use of Th^{232} as a fertile fuel in gas-cooled thermal reactors (Sec. 10-11). Th^{232} breeds U^{233} in reactions similar to those of U^{238} breeding Pu^{239}.

Table 11-3 Fuel production constants

	Fissionable fuels								Fertile Materials	
	Thermal				Fast				Fast	
Constants	U^{233}	U^{235}	Pu^{239}	Pu^{241}	U^{233}	U^{235}	Pu^{239}	Pu^{241}	Th^{232}	U^{238}
σ_f*	527	577	790	1000	2.2	1.4	1.78	2.54	0.025	0.112
σ_a*	580	675	1185	1400	2.35	1.61	2.05	2.83	—	0.273
ν	2.51	2.40	2.90	2.98	2.59	2.50	3.0	3.04	2.04	2.6
η	2.28	2.06	2.10	2.13	2.42	2.20	2.6	2.73	2.0	2.27
G_{max}	0.28	0.06	0.10	0.13	0.42	0.20	0.6	0.73	0	0.17

* In barn.

The Doubling Time

Of economic importance to a breeder reactor is its *doubling time*. This is the time required to produce as many new fissionable nuclei as the total number of fissionable nuclei that are both normally contained in the core and tied up in the reactor fuel cycle (i.e., in fabrication, reprocessing, etc.). In general, doubling times range between 10 and 20 years, the shorter the better.

The number density of new fissionable nuclei produced in a breeding reactor during time θ, ΔN_b nuclei/cm^3, may be given by

$$\Delta N_b = \Delta N_{ff} G = \Delta N_{ff}(\eta - 2 - L) \tag{11-10}$$

but

$$\Delta N_{ff} = F_c (N_0)_{ff} (\sigma_a)_{ff} \overline{\phi}\, \theta \tag{11-11}$$

where

ΔN_{ff} = number of original fissionable fuel nuclei consumed (by neutron absorption) during time θ per cm^3

$(N_0)_{ff}$ = number of fissionable fuel nuclei present in the core and tied up in the fuel cycle at arbitrary time 0, nuclei/cm^3

F_c = fraction of $(N_0)_{ff}$ that is in the core, dimensionless

$(\sigma_a)_{ff}$ = microscopic absorption cross section of fissionable fuel, cm^2

$\overline{\phi}$ = average reactor neutron flux, neutrons/s · cm^2

θ_d, called the *simple doubling time*, by definition, occurs when $\Delta N_b = (N_0)_{ff}$. Thus

$$\theta_d = \frac{1}{F_c (\sigma_a)_{ff} \overline{\phi}\, (\eta - 2 - L)} \tag{11-12}$$

Note that θ_d is shortened by operating at high power levels (high $\overline{\phi}$) and that it is also inversely proportional to the breeding gain ($\eta - 2 - L$). In practice θ_d slightly increases with reactor life because L increases with fission product buildup, as a result of the finite time required after start-up to build up the fissile inventory in a breeder-reactor blanket (a region surrounding the core that contains fertile material), the time taken for fuel and blanket element reprocessing, the economics of reprocessing, etc. It is economically desirable to have doubling times short enough that new Pu^{239} inventories are continually provided for new breeders. In other words, the plutonium is thought to be invested and the dividends are compounded. This gives rise to a *compound doubling time*, θ_{dc}, related to θ_d by

$$\theta_{dc} = \frac{0.6931}{\ln\,[1 + (1/\theta_d)]} \tag{11-13}$$

θ_d and θ_{dc} may be based on core inventory only (excluding fuel cycle inventory). Thus $F_c = 1.0$ and their values would be shortened.

11-3 LMFBR PLANT ARRANGEMENTS

Because sodium and other liquid metals suffer from high induced radioactivities, Eq. (11-4), and are generally chemically active, intermediate coolant loops are used between the primary radioactive coolant and the steam cycle (Fig. 11-4). The intermediate coolant is usually also a liquid metal, often Na or NaK. The intermediate loop guards against reactions between the radioactive primary coolant and water. Such reactions result, among other things, in the radiolytic decomposition of steam-generator water

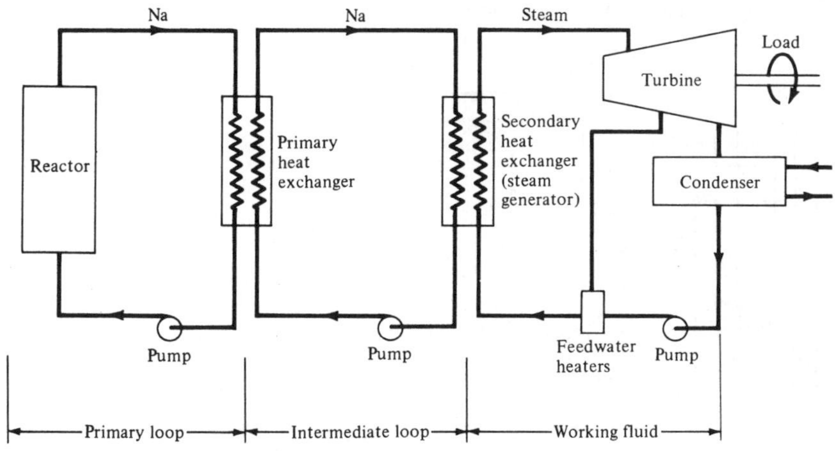

Figure 11-4 Schematic arrangement of a liquid-metal fast-breeder reactor (LMFBR) powerplant.

by the strong γ radiations emitted by Na^{24}. The intermediate loop also ensures against the high pressure water or hydrogen entering the reactor.

There are two primary-loop designs that are being considered. These are (1) the *loop,* or *pipe,* type and (2) the *pool, tank,* or *pot* type.

The *loop* type, represented schematically in Fig. 11-5, is the more conventional of the two, being the design used in all U.S. operating sodium-cooled plants to date, with the exception of EBR-II. In it, the reactor vessel, heat exchangers, liquid metal pumps, and other components of the primary system and their interconnecting piping are separated within a large building containing an inert atmosphere to preclude sodium fires in case of a sodium leak. The major advantages of this design are the accumulated experience with it and the separation or decoupling of the components of the primary system. It has the disadvantage of large and multiple shielding of pipeways, equipment cells, and of the large and complex structure resulting from the spread of the components. The design of the interconnecting piping is complex and requires expansion loops to accommodate thermal expansion. Stress concentrations at the pipe-reactor vessel joints pose critical problems. Breaks or leaks in the piping system may seriously affect reactor-core cooling. Leaks also would necessitate extended shutdowns for repairs.

In the *pool* system, represented schematically in Fig. 11-6, the entire primary system, including reactor, primary heat exchangers, and pumps, is submerged in a large tank filled with molten sodium. That tank is part of the primary coolant loops. The heat exchangers discharge coolant directly into the tank, and the pumps receive coolant directly from it. The main advantages of the pool design are the relative insensitivity to sodium leaks in the primary system and a more compact primary-system arrangement. The disadvantages result from the close coupling of the various components, which leads to accentuated mechanical and thermal interactions, and the rather complex structure of the pool closure that must serve the multiple functions of shield, inert gas closure, and support of equipment above and must contain all the necessary penetrations to the components.

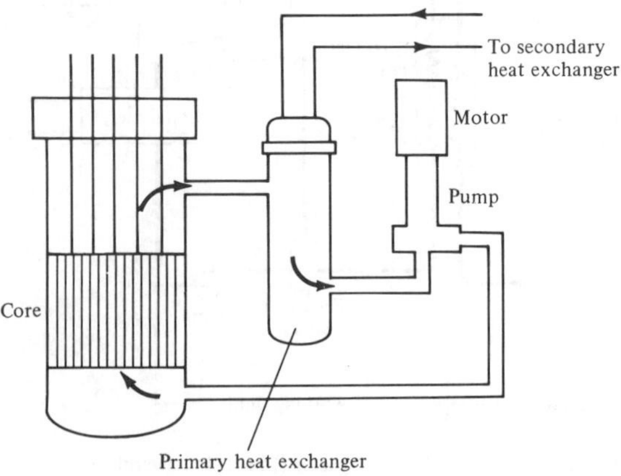

Figure 11-5 Schematic of a loop-type LMFBR.

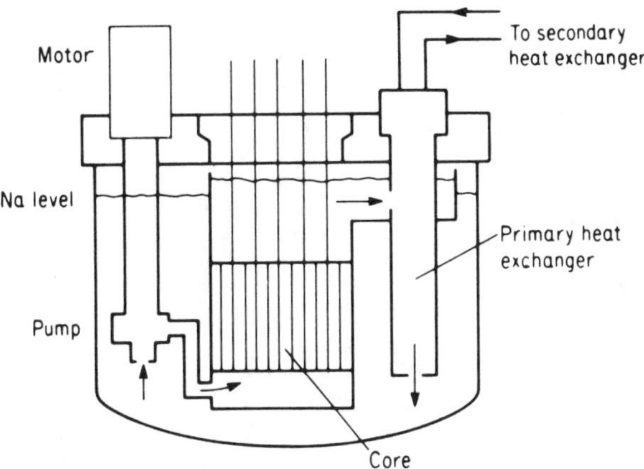

Figure 11-6 Schematic of a pool-type LMFBR.

In general, it is now believed that the pool system has the edge in safety and economy while the loop system has the edge in that it is a straightforward mechanical design. Both types are being used in current designs. The pool is used in French, British, and recent USSR designs. The loop is used in U.S., early USSR, German, and Japanese designs (Table 11-2).

The next two sections cover the Clinch River Breeder Reactor Project, an American demonstration plant of the loop design, and Super Phénix, a French commercial-size plant of the pool design.

11-4 THE CLINCH RIVER BREEDER REACTOR PROJECT

The Clinch River Breeder Reactor (CRBR) project is a demonstration liquid-metal fast-breeder reactor plant. It is owned by DOE with Westinghouse as primary contractor with responsibility for reactor and primary heat-transfer system manufacturing. General Electric is responsible for the intermediate heat-transport and steam-generation systems and turbine-generator manufacturing, and Rockwell International is responsible for refueling and auxiliary maintenance systems. Several subcontractors are entrusted with various component design and manufacturing. Burns and Roe is the architect-engineer, and Stone and Webster is the constructor. CRBR is situated on 100 acres of a 1364-acre TVA site on the Clinch River near Oak Ridge, Tenn. Figure 11-7 shows a general layout of the plant [105].

CRBR has been facing economic and political delays because of cost overruns and governmental caution against the universal production of plutonium and its implications regarding nuclear weapons proliferation. This argument may be answered by the fact that plutonium production can be accomplished easier, faster, and cheaper by means other than a complex LMFBR. Because of the need for fast-breeder reactors

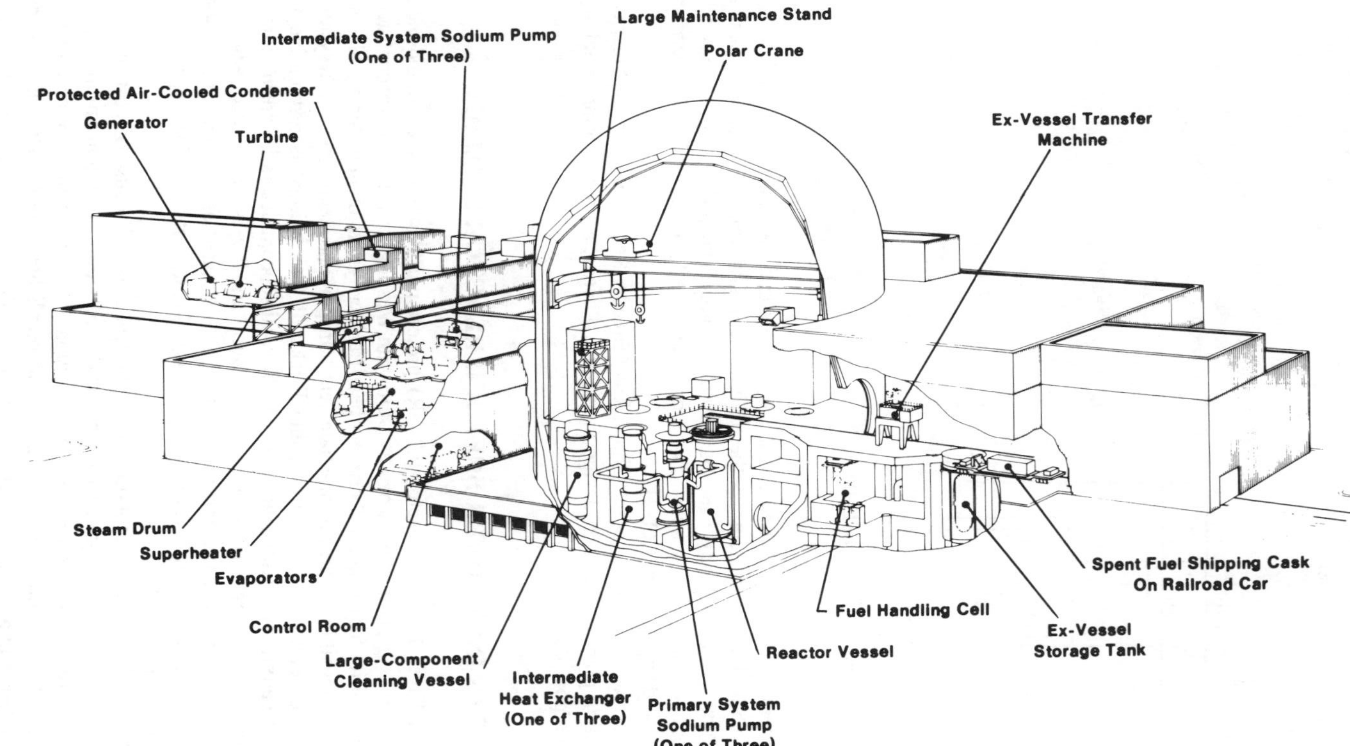

Figure 11-7 Layout of the Clinch River Breeder Reactor (CPBR) powerplant. *(Figures 11-7 through 11-18 courtesy Project Management Corporation.)*

to expand our dwindling energy supplies, there is optimism that CRBR will eventually be completed. Initial startup is now planned for the late 1980s.

CRBR is a loop design, so chosen to utilize experience gained by Westinghouse from the Fast Flux Test Facility (FFTF), also a loop design. It is designed to produce 375 MW(e) gross from 975-MW(t) reactor thermal power. Table 11-4 lists some design data of CRBR.

Reactor core Figure 11-8 shows a cross section of the CRBR reactor core. It is composed of 156 fuel assemblies containing a mixed oxide fuel (PuO_2-UO_2), interspersed with 76 inner blanket assemblies, 6 assemblies that are used alternately for fuel or inner blanket, and 15 control assemblies. This mix is surrounded by 132 radial blanket assemblies which is turn are surrounded by 306 removable radial shield assemblies. All assemblies are contained within hexagonal ducts having the same external dimensions. Refueling is planned annually.

Table 11-4 Design data of the Clinch River breeder-reactor plant

Power	375 MW(e) gross, 975 MW(t)
Fuel	PuO_2-UO_2 mixed oxide, 1511 kg fissile Pu at beginning of life
Fuel elements	0.23-in (5.8-mm)-diameter rods, SS-316 clad, pitch-diameter ratio 1.26 spaced on a triangular array by spiral wire wraps, 217 rods per assembly, 156 assemblies
Blankets	Depleted UO_2, 76 inner blanket and 132 radial blanket assemblies
Control rods	Nine primary worth ~\$22, six secondary worth ~\$13
Core	6.2-ft (1.89-m) diameter, 3.0 ft (0.91 m) high
Maximum neutron flux	5.5×10^{15}
Power peaking factor	1.28 axial, 1.18 radial at beginning of cycle
Linear power rating	16 kW/ft (52.5 kW/m) peak, 14.3 kW/ft (46.9 kW/m) average
Fuel burnup	80,000 MW · day/ton initial core, 110,000 peak
Breeding ratio	1.24, 30-year doubling time
Core temperatures	Fuel cladding wall, max. 1215°F (657°C); sodium 730°F (388°C) inlet, 995°F (535°C) outlet
Primary pumps (3)	33,700 gal/min (2.13 m^3/s) each, 458 ft (139.6 m) Na head, core pressure drop 123 psi
Intermediate systems (3)	Cold leg 651°F (344°C); hot leg 936°F (502°C); pumps 29,500 gal/min (1.86 m^3/s) each; 330 ft (100.6 m) Na head
Steam systems (3)	Two evaporators and one superheater per loop, 3.33×10^6 lb_m/h (420 kg/s) each, superheater outlet 906°F (486°C), 1535 psig (105.8 bar), feedwater 468°F (242°C), to steam separator drum, 548°F (287°C) to evaporators
Turbine	3600 r/min tandem-compound, nonreheat, six extractions to feedwater heaters, total extracted steam 1.15×10^6 lb_m/h (145 kg/s)
Generator	Synchronous 3600 r/min, 485 MVA self-excited three-phase wye connected 60 Hz (6.17 m), 22,000 V, 0.9 power factor
Dimensions	Reactor vessel 20.25-ft (6.17-m) diameter, 54.67 ft (16.66 m) high; intermediate heat exchangers, 8.8-ft (2.68-m) diameter, 52.1 ft (15.88 m) high; steam generators 4.33-ft (1.32-m) diameter, 65.0 ft (19.81 m) high; containment 186 ft (56.7 m) diameter
Gross thermal efficiency	39%

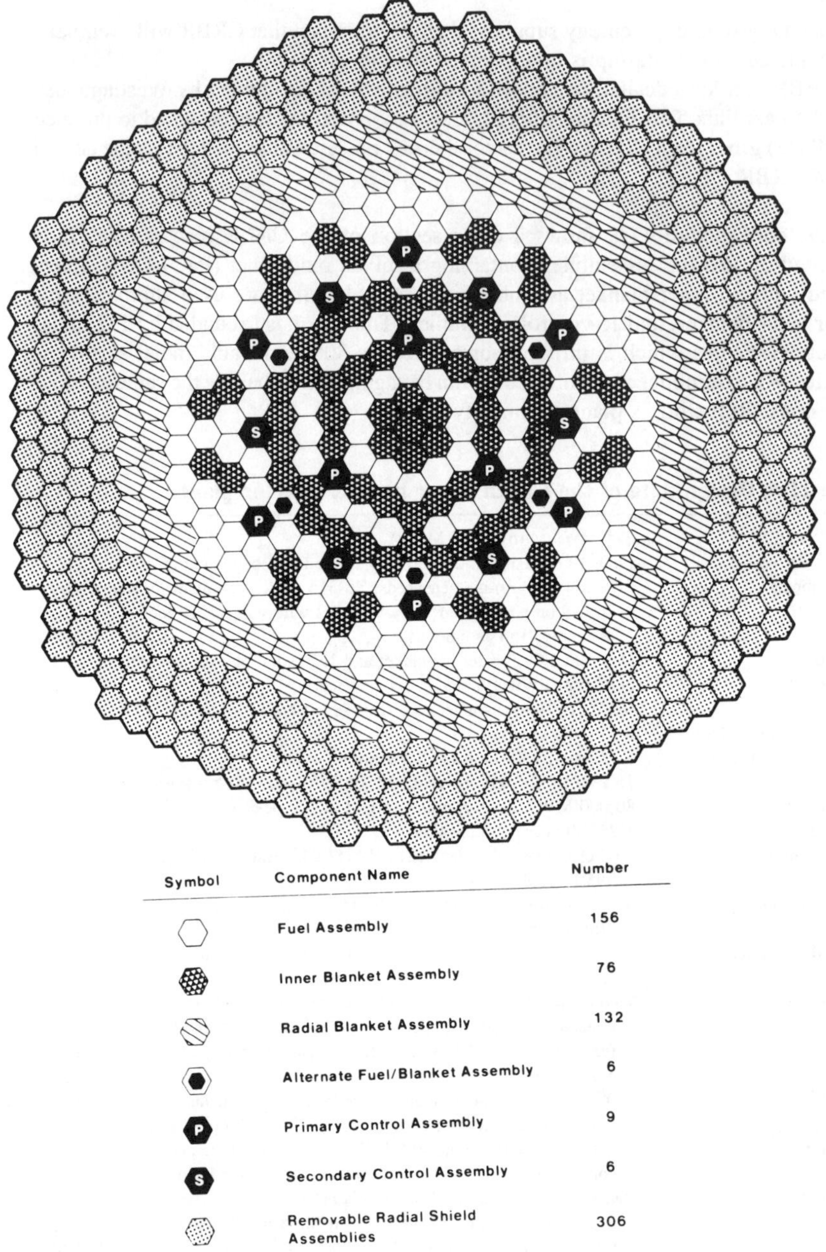

Figure 11-8 Cross section of the CRBR core.

The fuel-rod design is based on that of the FFTF. Each fuel assembly (Fig. 11-9) contains 217 0.23-in OD rods arranged in a triangular array and spaced by wire wraps of 12-in spiral pitch along the rod length, giving a rod pitch-to-diameter ratio of 1.26. Each fuel rod contains a 36-in-long stack of fuel pellets and two 14-in-long stacks of depleted UO_2 blanket pellets above and below the fuel pellets. The rods are 316 stainless steel and are capped at top and bottom, with a fission gas plenum at the top. An axial spring prevents pellet movement during handling, preirradiation, and shipping.

The inner and radial blanket assemblies (Fig. 11-10) contain 61 rods each, also on a triangular pitch and spaced by wire wraps. Each rod contains a 64-in-long stack of 0.470-in-diameter depleted UO_2 pellets. This length is equal to the total of fuel and upper and lower blankets in the fuel rods. The cladding is 0.506 in OD and 0.015 in thick. All blanket pellets breed plutonium from the depleted UO_2, generate some of the power, and provide some of the shielding.

The removable radial shield assemblies contain 19 rods each. Their principal function is to provide neutron shielding and to limit radiation damage to such reactor structures as the core barrel and pressure vessel. In addition there is a fixed stainless steel segmented-ring radial shield located between the removable assemblies and the core barrel. All shields are designed to ensure a 10 percent residual ductility for a 30-year design life.

The control rod system is composed of nine primary and six secondary hexagonal assemblies. The primary assemblies (Fig. 11-11) contain 37 pins each. The neutron absorber is boron carbide, B_4C, 92 percent enriched in B^{10} and contained in a 36-in stack of pellets in each pin. The primary control rods are used for primary shutdown (scram) as commanded by a plant protection system and for reactivity control of the reactor. Their drive mechanisms are mounted on the intermediate rotating plug of the reactor closure head. The secondary control rods have the sole function of providing an independent backup shutdown for the primary rods. The two systems are designed to be independent and are not subject to the same mode failure. Either one can achieve reactor shutdown with the other completely inoperable and with one of its own control rods inoperative.

The core is supported by a 24-in-thick type 304 stainless steel plate that is welded to the core barrel and through which 61 lower sodium inlet modules are positioned (Fig. 11-12). Each module supports seven of the fuel, control, inner and radial blanket, and some of the removable radial shield assemblies, for a total of 427. The modules have seven different configurations to control the sodium flow distribution to the different assemblies. In addition, flow-control orifice cartridges provide intermediate-pressure sodium to the remaining shielding components and low-pressure sodium to the annulus between reactor vessel and core barrel.

Located directly above the core is a welded internal structure made of 316 stainless steel with local Alloy 718 protection. Its principal functions are as backup core hold-down, for positioning instruments, mixing the core exit sodium and directing it to the upper region of the reactor vessel outlet plenum, and providing guidance and cross-flow protection for control-rod drivelines. A core restraint system, among other things,

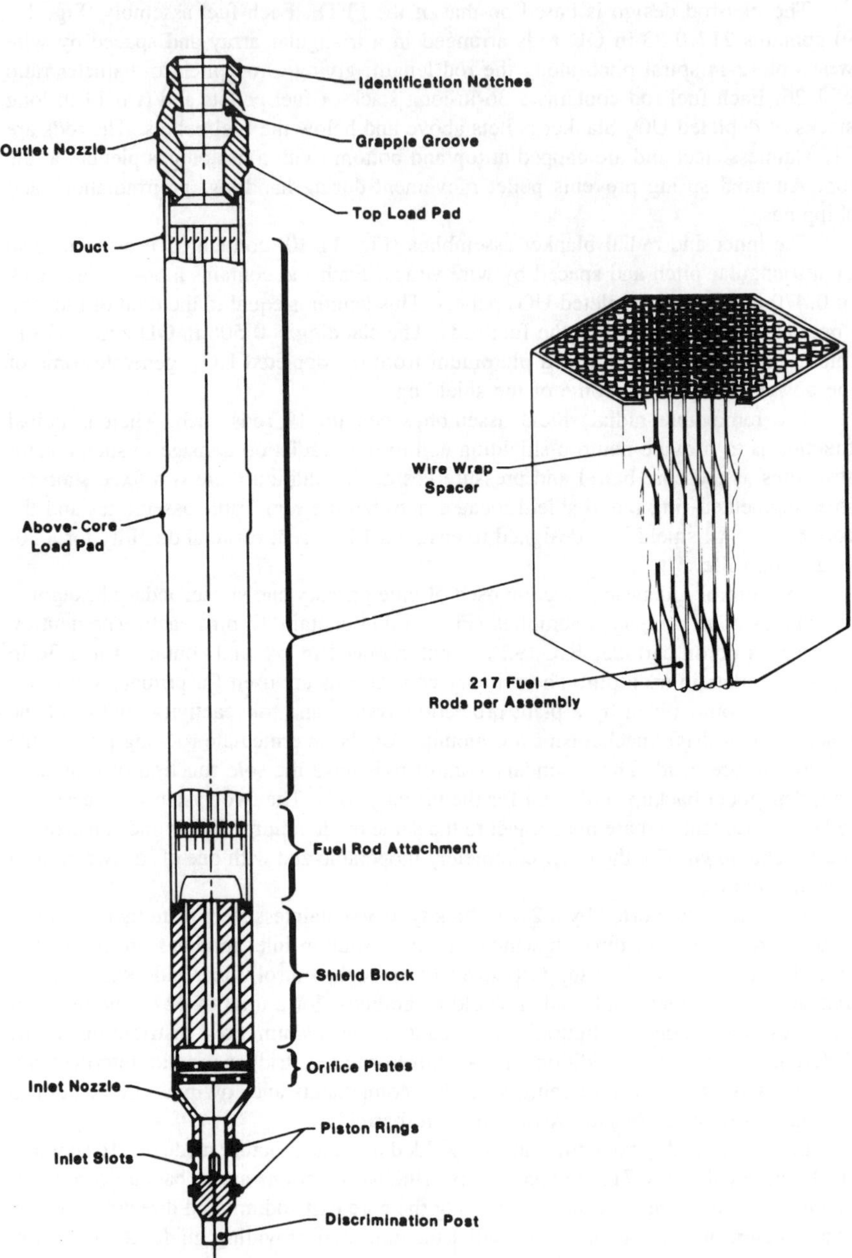

Figure 11-9 CRBR fuel rod and assembly.

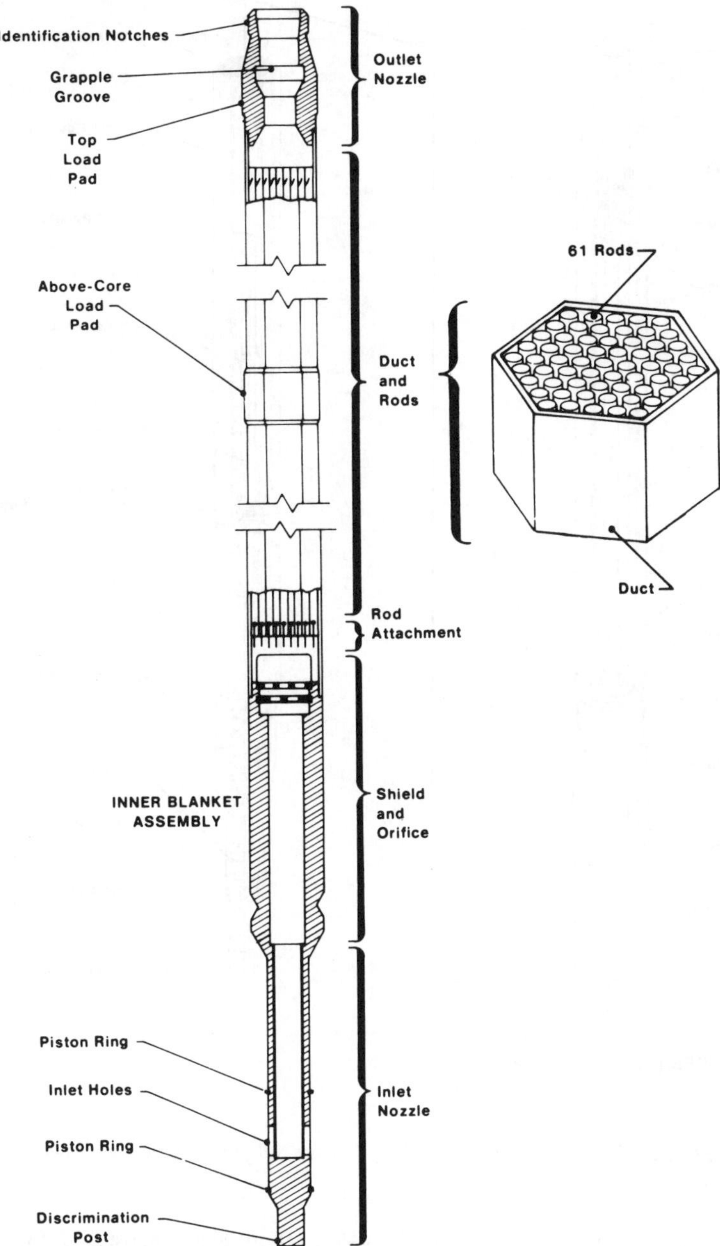

Figure 11-10 CRBR inner blanket assembly.

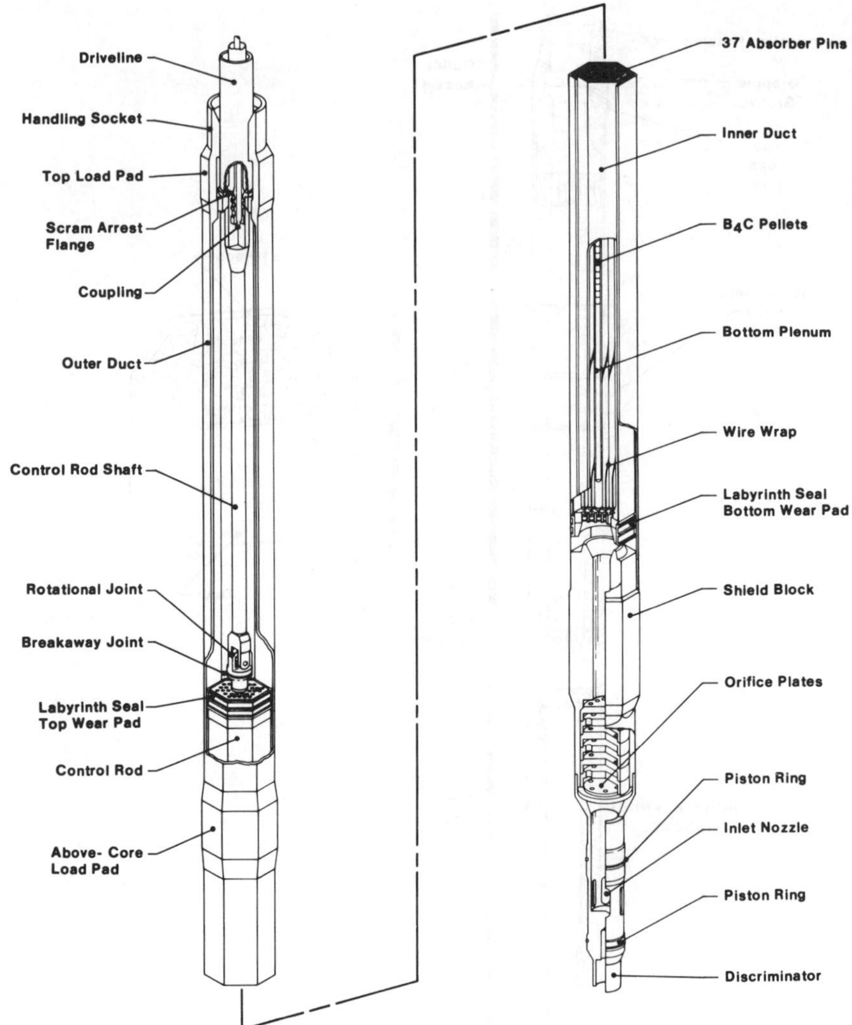

Figure 11-11 CRBR primary control-rod assembly.

maintains clearances between core assemblies that are subject to creep and swelling caused by fast neutron irradiation.

The reactor The reactor vessel (Fig. 11-12) is 59 ft (18 m) high, almost 27 ft (8.23 m) in diameter at its widest, and weighs 505 tons. Sodium enters through three 2-ft-diameter pipes and exits through three 3-ft-diameter pipes. Four smaller nozzles in

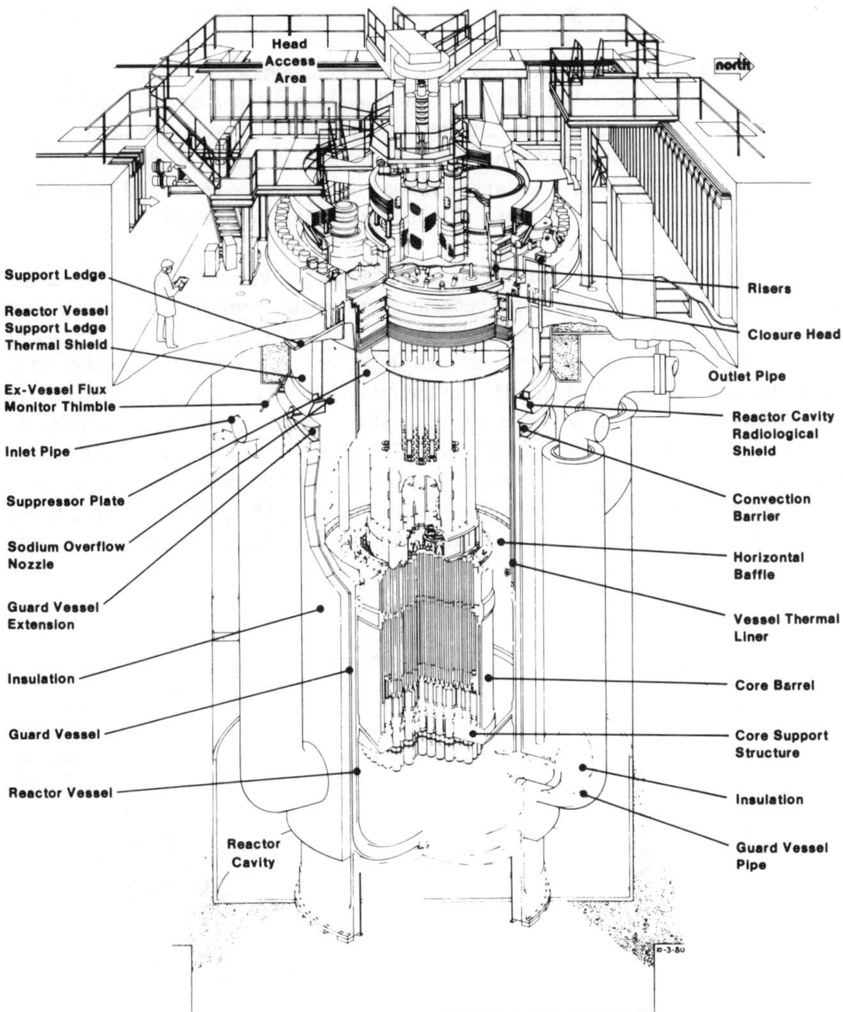

Figure 11-12 The CRBR reactor system.

the upper regions are for entry and exit of cover gas, sodium overflow, and makeup flow. The vessel wall is kept below 900°F by directing 2 percent of sodium flow into the annulus between the vessel and a 20-ft-diameter, 25-ft-long stainless steel thermal liner. A 22-ft-diameter, 47-ft-high guard vessel surrounds the liner and provides secondary containment, which ensures that the core is always covered by sodium and that undamaged pipes can remove heat from it in the unlikely event of a leak from the reactor vessel or piping. The whole reactor system is situated in a 40-ft-diameter reactor cavity.

The upper closure head assembly (Fig. 11-13) contains various penetrations such as those for the control rod assemblies, as well as three independently rotatable plugs for positioning components of the refueling system. The smallest plug is nested within and eccentric to the intermediate plug, which, in turn, is nested inside and eccentric to the large plug. The plugs rotations in respect to the reactor vessel and one another provide straight pull access to fuel and other removable components.

The heat-transfer system The reactor is connected to three primary loops, each in turn connected to one of three independent intermediate loops (Fig. 11-14). In each primary loop, sodium enters the reactor vessel at 730°F (388°C) and exits at 995°F (535°C). A primary sodium pump, located in the hot leg, pumps the hot sodium to an intermediate heat exchanger (IHX) at the rate of 33,700 gal/min (2.126 m^3/s) and a pressure of 175 psi (12 bar). The primary sodium leaves the IHX and returns to the reactor via a check valve and a flow meter. The primary piping is 1/2-in-thick welded stainless steel and has a 3-ft diameter between reactor and pump and 2-ft diameter between pump and IHX and between IHX and reactor. The primary pumps, IHXs, and some associated piping are surrounded with guard vessels for the same reasons

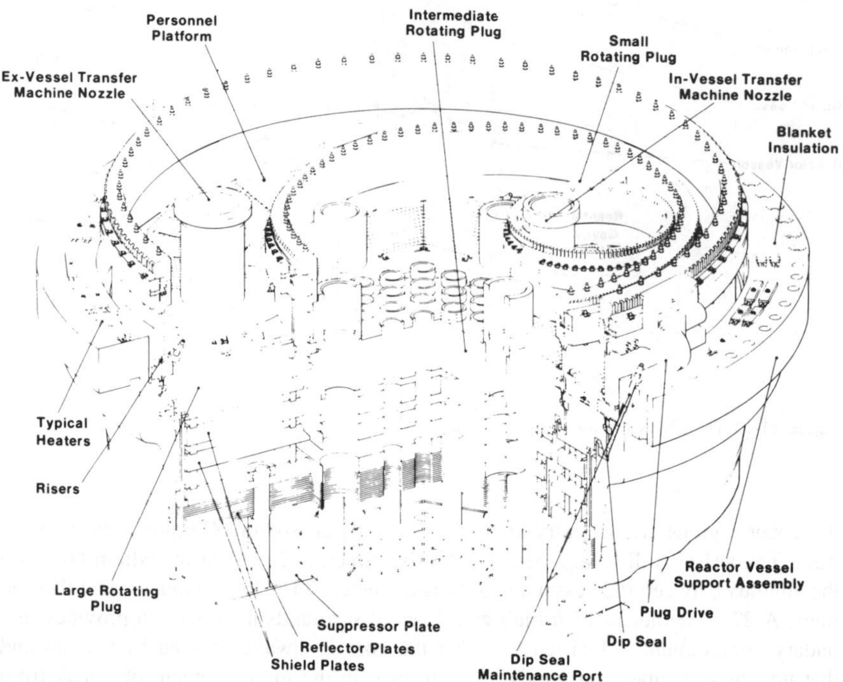

Figure 11-13 The CRBR reactor closure head assembly.

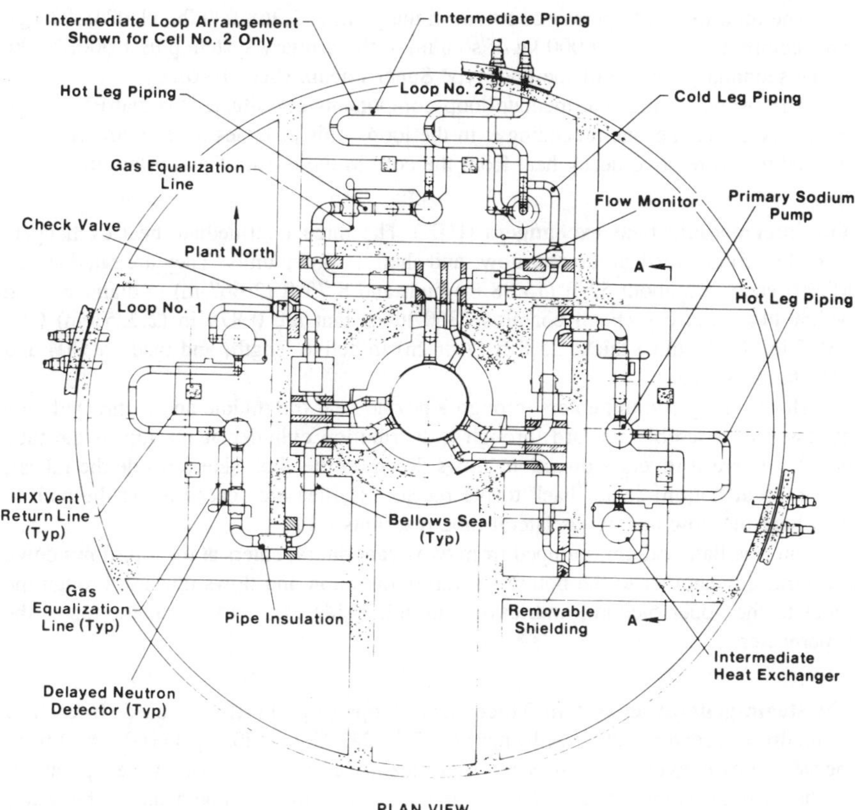

Figure 11-14 Plan view of the CRBR primary heat-transport system.

as the reactor vessel. The whole reactor system is situated in a 40-ft-diameter reactor cavity. Argon is used as cover gas for sodium in the reactor and other components. (Argon is also used as cover gas for NaK, fuel handling cell atmosphere and other places, while nitrogen is provided for inerted cell atmospheres and other functions. An elaborate gas receiving, processing, and decontamination system is provided.)

The sodium in each intermediate loop enters the IHX at 651°F (344°C) and exits at 936°F (502°C) and 180 psi (12.4 bar) to go to the superheater, which it leaves at 885°F (474°C). It then divides into two paths, each of which enters an evaporator. Sodium leaving both evaporators is combined and pumped by an intermediate cold-leg pump back to the IHX via a flow meter at 29,500 gal/min (1.86 m^3/s) and 195 psi (14.5 bar). Thus there are two pumps, one superheater, two evaporators, and a steam drum (below) for each of the three loops. The higher pressure of the intermediate sodium in the IHX ensures that any leakage would be from the intermediate to the primary side, thus preventing contamination by the highly radioactive primary sodium. All intermediate-loop piping is 1/2-in-thick welded stainless steel 2 ft in diameter.

The total mass of operating sodium in the plant is 2,368,000 lb_m (1,074,107 kg). This compares with 3,500,000 kg of sodium in the primary system plus 1,500,000 kg in the secondary system of the pool-type Super Phénix (Sec. 10-6).

The primary- and intermediate-loop components are situated so that each component is above the one preceding it in the loop. This provides natural circulation of the sodium to remove decay heat from the core in the event of pump failure.

The intermediate heat exchangers (IHX) The three intermediate heat exchangers (Fig. 11-15) are vertical, counterflow, and shell-and-tube units. They are rated at 325 MW(t) each, are about 52 ft (15.88 m) high and 8.75 ft (2.667 m) in diameter, and weigh 115 tons, dry. Each contains 2850 tubes that are 0.875 in (2.225 cm) OD, 0.045 in (1.143 mm) thick, 25.8 ft (7.86 m) in active length, and made of 304 and 316 stainless steel.

Hot primary sodium enters through a nozzle near the middle and is directed by a bypass seal assembly upward through a distribution cylinder at the top of the tube bundle. There it reverses direction, flows down on the shell side (outside the tubes), and exits at bottom center back to the reactor. Baffles are spaced along the path to distribute the flow and act as lateral tube restraints.

Intermediate sodium, pumped from two evaporators, enters at the top, flows down a central downcomer to the bottom, reverses direction and flows upward through the tubes to the upper plenum, and leaves through a side exit nozzle near the top to the superheater.

The steam-generating system Three steam loops (Fig. 11-16) are used. In each, a steam drum receives heated feedwater at 468°F (242°C) and 2032 psi (140.1 bar) from the steam power cycle, as well as saturated steam-water mixture from two evaporators. In the drum, steam separates from water and goes to the superheater. The water remaining is force-recirculated to the two evaporators by a recirculating pump (not unlike a forced external recirculation BWR, Sec. 10-7) and enters them at 548°F (287°C) and 2034 psi (140.2 bar). There, it receives heat from the intermediate sodium and leaves the drum as a mixture of steam and water at 628°F (331°C) and 1896 psi (130.7 bar) of 50 percent quality at full load, corresponding to a recirculation ratio of 2:1.

The separated saturated steam enters the superheater at 625°F (329°C) and 1851 psi (127.6 bar) where it receives heat from the hot intermediate sodium and leaves at 906°F (486°C) and 1450 psi (100 bar).

The evaporators and superheaters All evaporators and superheaters are essentially identical in design to minimize cost and spare parts (Fig. 11-17). They are shell-and-tube counterflow "hockey stick" design heat exchangers that are vertical except for a 90° bend at the top. This provides for differential thermal expansion between the tube bundle and shell. Each is 65 ft (19.8 m) high and 4.33 ft (1.32 m) in diameter and weighs 115 tons, dry. Each contains 757 tubes that are 5/8 in (1.5875 cm) OD, 0.109 in (2.7686 mm) thick, 46 ft (14.02 m) in active length, and made of Cr-Mo steel.

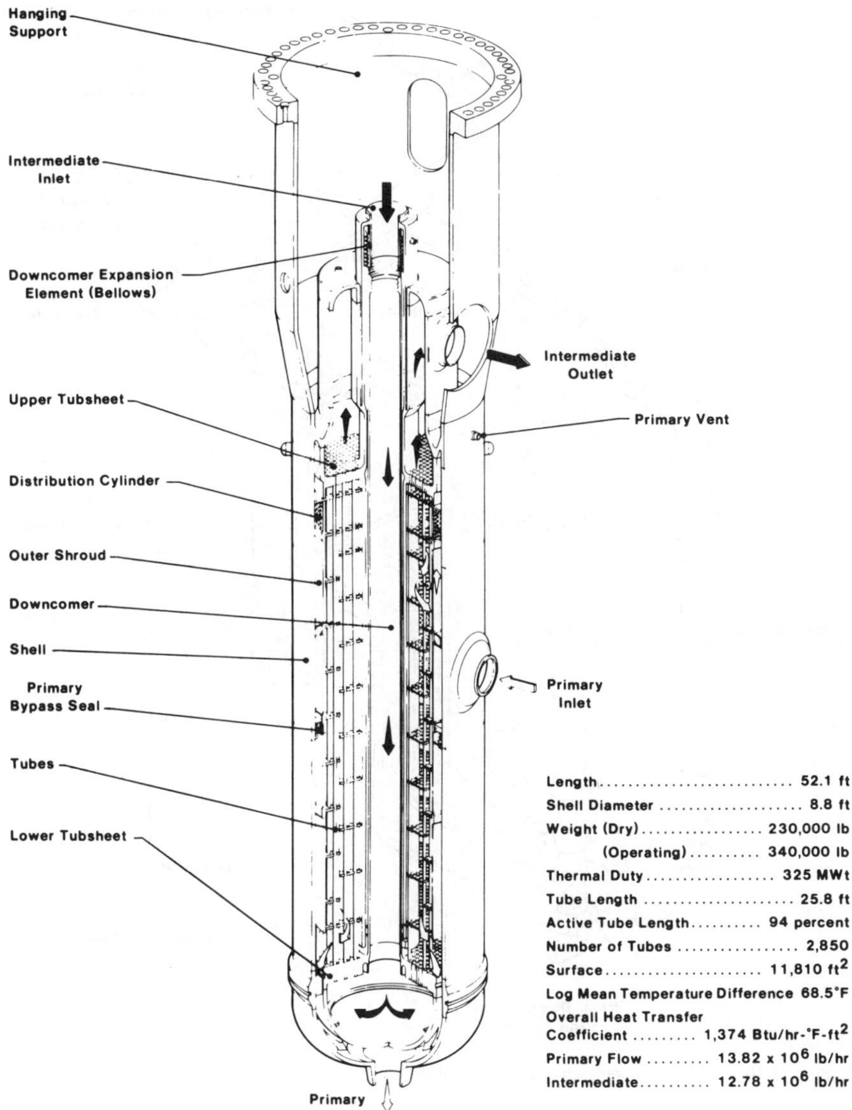
Hanging Support
Intermediate Inlet
Downcomer Expansion Element (Bellows)
Intermediate Outlet
Upper Tubsheet
Primary Vent
Distribution Cylinder
Outer Shroud
Downcomer
Shell
Primary Inlet
Primary Bypass Seal
Tubes
Lower Tubsheet
Primary Outlet
Length .... 52.1 ft
Shell Diameter .... 8.8 ft
Weight (Dry) .... 230,000 lb
(Operating) .... 340,000 lb
Thermal Duty .... 325 MWt
Tube Length .... 25.8 ft
Active Tube Length .... 94 percent
Number of Tubes 2,850
Surface .... 11,810 ft²
Log Mean Temperature Difference 68.5°F
Overall Heat Transfer Coefficient .... 1,374 Btu/hr-°F-ft²
Primary Flow .... 13.82 x 10⁶ lb/hr
Intermediate .... 12.78 x 10⁶ lb/hr

Figure 11-15 The CRBR intermediate heat exchanger (IHX).

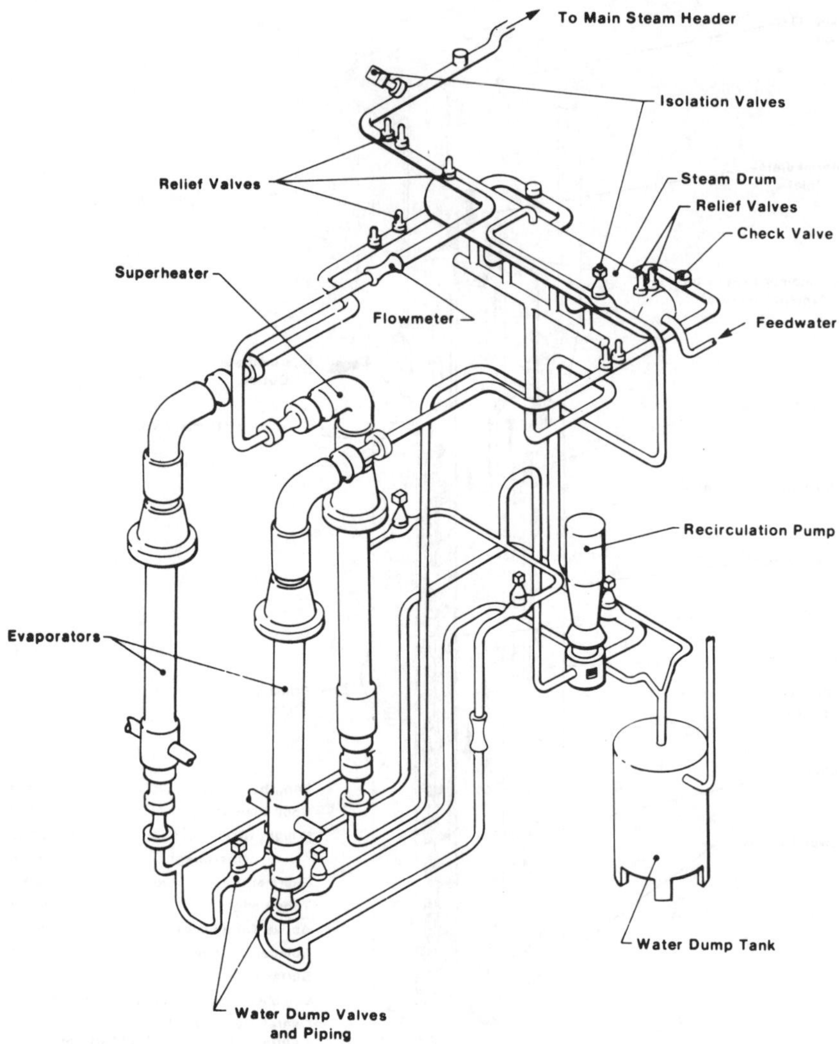

Figure 11-16 CRBR steam-generation loop (one of three).

Intermediate sodium from the IHX first enters the superheater, and sodium leaving the superheater enters each of the two evaporators, through a nozzle just below the bend, flows down on the shell side, and exits from two nozzles near the bottom. As in the IHX, baffles are spaced in that path to distribute the sodium and support the tubes.

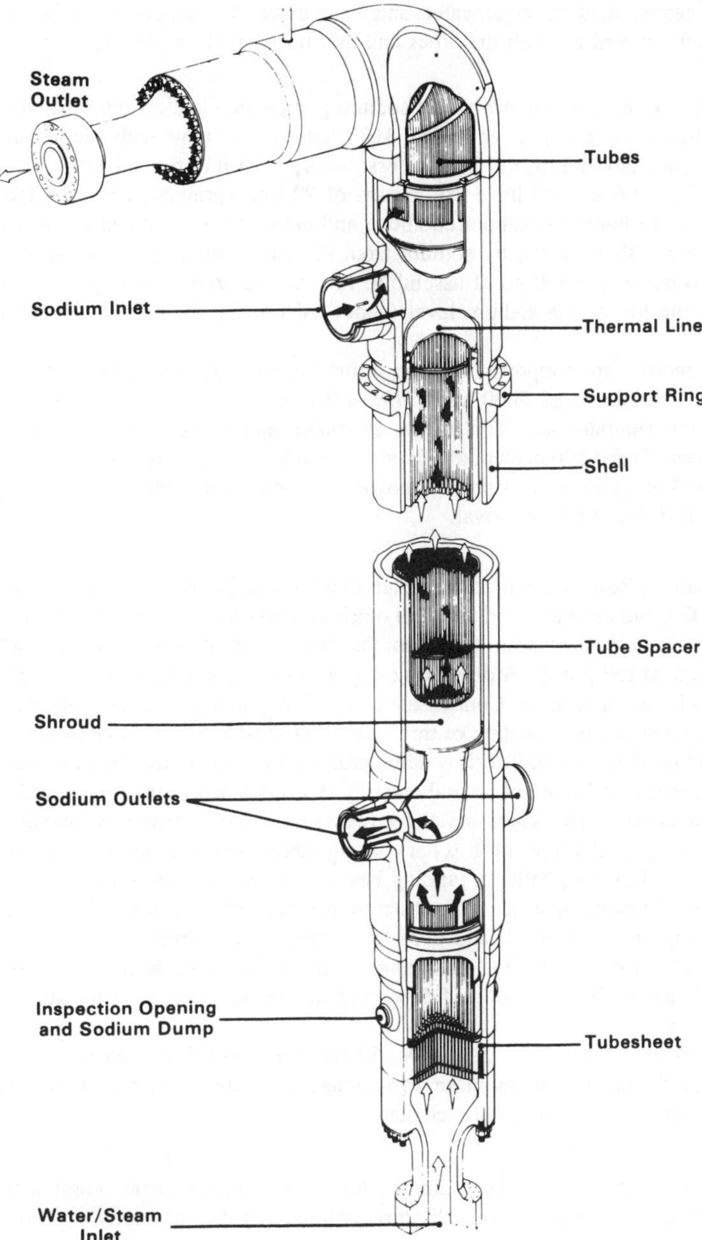

Figure 11-17 CRBR module for both steam evaporators and superheaters.

Saturated steam enters the superheater, and water enters the evaporators, at bottom center; they flow upward through the tubes and exit horizontally at the top.

The pumps The primary and intermediate sodium pumps are identical motor-driven, vertically mounted, single-stage, centrifugal 316 stainless steel units with double suction impellers. They have an overall height, less motor, of 23 ft (7 m) and a maximum diameter of 8.5 ft (2.6 m) and have dry masses of 82 tons (primary) and 64.5 tons (intermediate). The impeller is supported above and below by two sodium-lubricated hydrostatic bearings that are fed by sodium from the pump discharge. Both suction and discharge occur at the bottom of assembly. Pressurized argon cover gas fills the space between the top of the sodium level (about halfway up the driveshaft) and a thermal shield.

The drive motors are supported by one thrust bearing and two radial bearings each. They are variable-speed 5000-hp (3773.5-kW) units powered by variable-frequency AC motor-generator sets. This allows a variable speed range that permits load following between 40 and 100 percent of full power. Independently-powered, constant-speed 75-hp (60-kW) pony motors provide power for sodium circulation during start-up, shutdown, and decay-heat removal.

The steam plant A flow diagram of the steam plant at rated power is shown in Fig. 11-18. The 3600-r/min turbine is composed of one high-pressure and three low-pressure sections. The total turbine inlet steam flow from the three superheaters is about 3,320,000 lb_m/h (418.3 kg/s) at full power. A deaerating single-pass condenser is located directly below and parallel to the turbine. Condensate at 101°F is pumped from the condenser hot well to a demineralizer consisting of three parallel mixed-bed ion exchangers (one always on standby) that provide full-flow demineralization. It then flows through three low-pressure closed feedwater heaters and one tray-type open deaerating heater, which also serves as a storage tank. Three boiler feedwater pumps (one always on standby) then pump the water to 2160 psi (148.9 bar) through three high-pressure closed feedwater heaters. The last one, called a tapping heater, uses blowdown from the steam drums to heat the feedwater instead of bleed steam from the turbine as usual. Feedwater leaves the tapping heater at 468°F and goes to the three steam drums via regulating valves. A total of 1,150,000 lb_m/h (145 kg/s) of steam is bled from the turbine for the six feedwater heaters. This is about 34.6 percent of the steam entering the turbine high-pressure stage.

The turbine drives a 416-MW, 22,000-V, 60-Hz, 0.9 power-factor generator. The generator is a totally enclosed three-phase wye-connected single winding machine with a hydrogen-cooled rotor and water-cooled stator.

Heat sink Heat rejection from the main condenser and miscellaneous component cooling water systems is accomplished by three pumps that deliver 185,200 gal/min (11.68 m^3/s) of cooling water. This water is in turn cooled by a forced-draft, counterflow wet-cooling tower capable of handling full turbine and component cooling

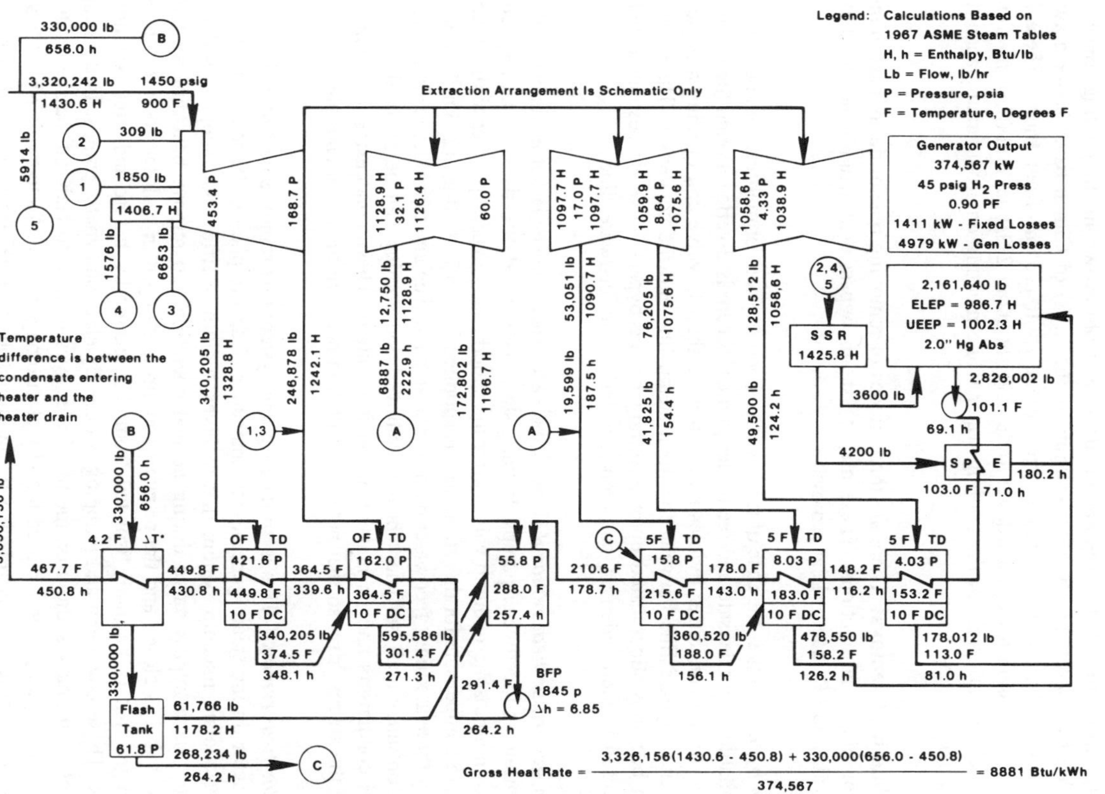

Figure 11-18 CRBR steam powerplant flow diagram, data at rated power.

loads with 76°F wet-bulb temperature air yielding a turbine back pressure of less than 5 inHg. Makeup due to evaporation, drift, and blowdown in the cooling tower is provided by approximately 5790 gal/min (0.365 m^3/s) of Clinch River water drawn in via two submerged perforated pipes about 75 ft (23 m) from shore. There is continuous cooling-tower blowdown to the river at about 2210 gal/min (0.14 m^3/s) to control chemical levels in the cooling water while meeting environmental standards on discharge to the river.

Residual heat removal There are three modes of removing reactor decay heat in the case of plant malfunctions. These are (1) steam bypass, (2) steam-generator heat removal, and (3) reactor heat removal.

Steam bypass This is the case of turbine trip, nonoperation, or malfunction, but power is available to condensate and feedwater pumps. Steam is diverted around the turbine directly to the condenser by closing the turbine throttle valve and opening a bypass valve in a bypass line between the main steam line and the condenser. The steam, which becomes superheated upon throttling by the bypass valve, is desuperheated, condensed in the condenser, and then recycled to the feedwater line.

Steam generator heat removal This is the case where power is not available to the condensate and feedwater pumps. Initially steam is dumped directly to the atmosphere via steam atmospheric dump valves after the superheaters. Makeup water to the steam drums to compensate for this steam loss is supplied from a 72,000-gal (273 m^3) storage tank by three auxiliary feedwater pumps, one steam driven and two electrically driven. This continues until sufficient heat is removed and cooling can be handled by air-cooled condensers and no makeup water is required. These condensers are situated above the steam drum so that natural circulation between them can take place.

Reactor heat removal This is an independent system that removes heat directly from the reactor during steam-generator maintenance or severe plant malfunction. It uses a reactor-coolant makeup system that receives sodium overflow and supplies sodium makeup via a bypass valve during normal operation. To remove heat directly this bypass valve is closed and the primary sodium is pumped in a closed loop to a heat exchanger that is normally isolated from the reactor by before-and-after isolation valves. These now open and pumping is done by electromagnetic pumps. In the heat exchanger, primary sodium on the tube side gives its heat to NaK on the shell side, which is also pumped in a closed loop by electromagnetic pumps to two air-blast heat exchangers where heat is rejected to the atmosphere. NaK remains liquid at room temperatures and thus requires no auxiliary heating to remain fluid. The electromagnetic pumps and other components in this system are powered by the plant emergency diesel generators.

The last two systems are situated in hardened buildings and expected to operate under all postulated accidents.

11-5 THE SUPER PHÉNIX LMFBR

Among the industrialized countries, France is one of the poorest in energy resources. In the 1970s she was importing more than three quarters of her energy. She has, therefore, embarked on an aggressive nuclear program that is expected to reduce energy imports to 50 percent by the year 2000. The program calls for more than 40,000 MW by 1985, 60,000 MW by 1990, and 100,000 MW by 2000.

The proven uranium reserve inside France, and where it has controlling interest, is estimated at 160,000 tons. If used solely for light-water reactors (PWRs and BWRs), it would suffice for only 32,000 MW for 30 years, or a total of 960,000 MW-years. If used in fast-breeder reactors, however, it could produce 50,000,000 MW-years, hence the urgent need in France to develop a breeder program [106].

Development of the liquid-metal fast-breeder reactor (LMFBR) in France started in the early 1960s and can be represented so far by three main reactor types. The first is Rapsodie, a 20- to 40-MW(t) experimental reactor that went into operation in 1967. It produced no electricity and was used to check technical features and operations for use in the design of succeeding LMFBRs. The next is Phénix, a 250-MW demonstration plant whose construction started in 1964 and which went critical in 1973 and began on-line operation in July 1974. Phénix is a pool-type reactor with three secondary loops and modular steam generators. Early in its life it had some minor fuel problems and needed some redesign of its intermediate heat exchangers but otherwise has had an excellent operational record since 1978.

The third and latest major step in the French breeder program is the Super Phénix, a class of reactors which is an extrapolation of Phénix. The first of this class, a near-commercial plant called Super Phénix Mark I, is under construction at Creys-Malville on the Rhône River, east of Lyons, in cooperation with four other European countries. It is a 1200-MW plant that retains the pool design that was originally chosen for safety but also proved stable and manageable with Phénix.

The future, while somewhat clouded by a reassessment of energy future needs in 1981, is bright. A study of Super Phénix Mark II is already underway with contracts expected to be awarded in 1983 for possibly a twin-reactor station with operation in 1990. Following this, several identical twin-reactor plants are considered on a reasonable time schedule. Based on the use of French-produced plutonium in various reactors, forecasts call for 16- to 23-GW breeders in operation by the year 2000.

General arrangement Figures 11-19 and 20 show the general layout and Table 11-5 lists some design data of Super Phénix Mark I [107] (with data in parentheses pertaining to Phénix.) The containment structure is a 64-m ID, 80-m-high reinforced concrete circular building that houses the reactor and its auxiliary circuits, fuel, and other active handling equipment; the primary heat-transfer loop; part of the intermediate heat-transfer loop; and temporary storage of radioactive wastes.

There are four buildings situated symmetrically around the containment structure, each of which houses a 750-MW steam generator and auxiliary equipment. This design maximizes physical separation between the intermediate loops. Interspersed between

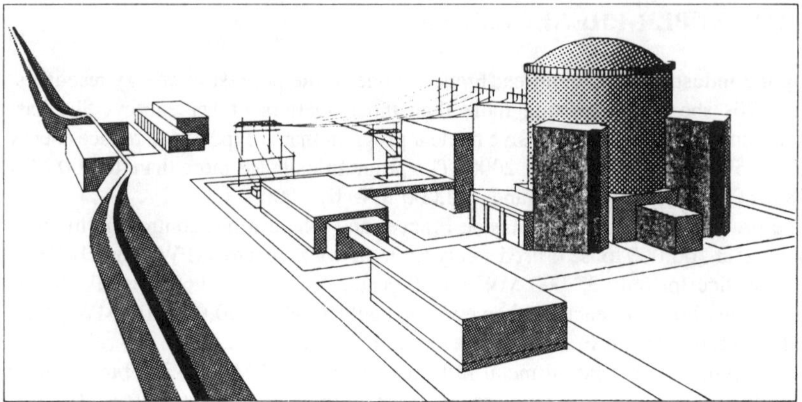

Figure 11-19 General layout of Super Phénix powerplant at Creys-Malville [107].

these are four shorter buildings, one for liquid-waste processing and three nuclear-service buildings that accommodate cooling-water circuits, sets powering the primary pumps, ventilation equipment, and subsidiary circuits for steam, organic liquids, nitrogen, argon, etc. Upstream of these buildings, and separated from them, are two turbine halls, each housing one 600-MW turboalternator. Closer to the river, and connected to the reactor and turbine buildings by underground and sky passages, is the electrical sections building. It houses the control room in its center and duplicate auxiliary electrical units in the two outer wings, as well as the diesel generators. Other buildings on the premises include one to the west for housing sodium storage tanks; one to the north used as an engineering and assembly shop for large components; one to the east for condenser cooling water intake, filtration, and pumping from the Rhône; and one, downstream of the latter, for cooling-water discharge to the river.

Reactor and fuel Figure 11-21 shows the reactor block that houses the nuclear steam supply system (NSSS) and other components. The main reactor vessel contains the sodium pool and the entire primary sodium system consisting of the core with its fuel and fertile and shielding assemblies at the center, surrounded by four primary sodium pumps and eight intermediate heat exchangers (IHX). In addition it contains handling machines for loading and unloading assemblies and control rods. The main vessel is made of 316L stainless steel and is a 21-m-diameter and 18-m-high cylinder with a torispherical base. Its thickness varies between 2.5 and 6.0 cm. An argon blanket is provided over the sodium at a pressure of about 100 millibars, gage.

The main vessel is surrounded at a distance of 70 cm by a 304 stainless steel safety vessel which is 22.5 m in diameter and 19 m deep and has a torispherical base. The safety vessel serves to recover sodium in the unlikely event of a leak in the main vessel and thus keeps the core flooded and the sodium free to circulate normally to remove decay heat after reactor shutdown. It is provided with a recovery tank and a

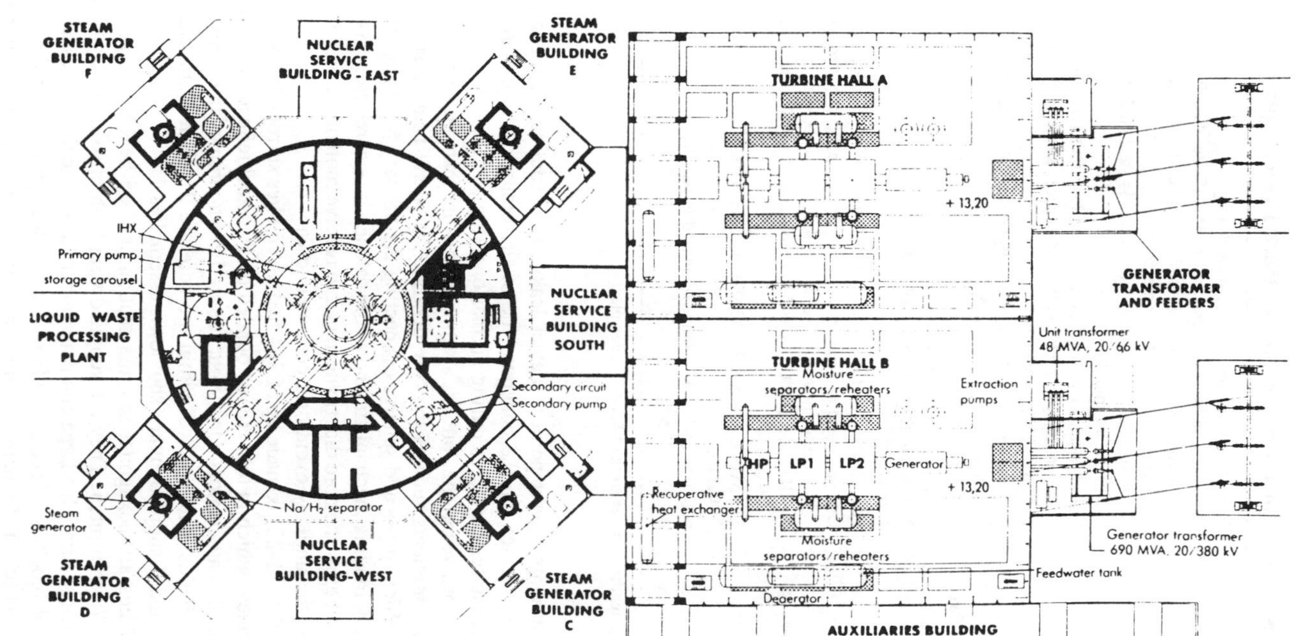

Figure 11-20 A plan view of the Super Phénix powerplant [107].

Table 11-5 Design data for the Super Phénix Mark I reactor plant

Power	3000 (590)* MW(t), 1240 (264) MW(e) gross
Fuel	PuO_2 UO_2 mixed oxide, enrichment 15.12% Pu^{239} equivalent
Fuel elements	364 assemblies, 5.4-m-long stainless steel clad rods, 8.65 (6.6) mm OD, 2.7 m long, 271 rods per assembly
Radial blanket	233 assemblies, 5.4 m (4.3) long, stainless steel clad, 1.950 m long; 91 rods per assembly
Control rods	Main system: 21 assemblies, 31 1.3-m-long rods each; back up shutdown system: three assemblies, three rods each; all stainless steel clad
Core	10.820 (1.227) m^3 volume
Reactor vessel	Stainless steel cylindrical with torispherical bottom, 21 m ID, 19.5 m high, contains 3500 tons Na
Maximum neutron flux	6.2×10^{15} (7.2×10^{15})
Linear power rating	450 (430) W/cm maximum
Fuel burnup	70,000–100,000 (50,000) MW · day/ton
Breeding ratio	1.24 (1.12)
Core temperatures	Fuel cladding wall max. 620 (650)°C; sodium 395°C inlet, 545°C outlet
Primary systems (4)	One pump and two intermediate heat exchangers per system arranged symmetrically around the core within the reactor vessel
Intermediate systems (4)	Na IHX inlet 345°C, outlet 525°C, nominal flow 3.27 ton/s each; total Na 1500 tons
Steam generators (4)	750 MW each, inlet 235°C, 210 bar, outlet 487°C, 177 bar, nominal flow 340 kg/s each
Turboalternators (2)	600 MW each, 3000 r/min, connected in parallel
Gross thermal efficiency	41.5% (44.75%)

*Numbers in parentheses refer to the Phénix reactor plant.

sodium detector at its lowest point to detect leaks. Two cooling circuits are placed outside the safety vessel, on the concrete side of the biological shield, to maintain acceptable concrete temperatures during normal operation.

Within the main vessel the core has a support structure that supports and positions various inner components such as the core diagrid, the lateral neutron shielding, the inner vessel, baffles, pipes and spheres connecting the primary pumps to the diagrid, the loading and unloading mechanism, and a core catcher. The core catcher is a stepped conical structure below the core diagrid designed to receive molten fuel in a noncritical configuration in the unlikely event of a core meltdown.

The core (Fig. 11-22) is made up of 364 fuel assemblies surrounded by 233 fertile blanket assemblies, which in turn are surrounded by 197 steel reflector assemblies that serve as neutron-flux attenuators. On the outside are 1076 nonremovable steel protective lateral neutron shields. Interspersed within the core are 21 main regulating control-rod assemblies and 3 supplementary shutdown assemblies.

All assemblies are hexagonal in cross section and are held to the core support structure on a 179-mm pitch. They are supplied with high-pressure (5 bar) sodium in the support structure through a large enough number of ports in their bases to minimize the possibility of blockage. Feed is made radially so that pressure on the foot of the assembly prevents liftoff. Orificing to adjust flow to power is situated within the base of each assembly.

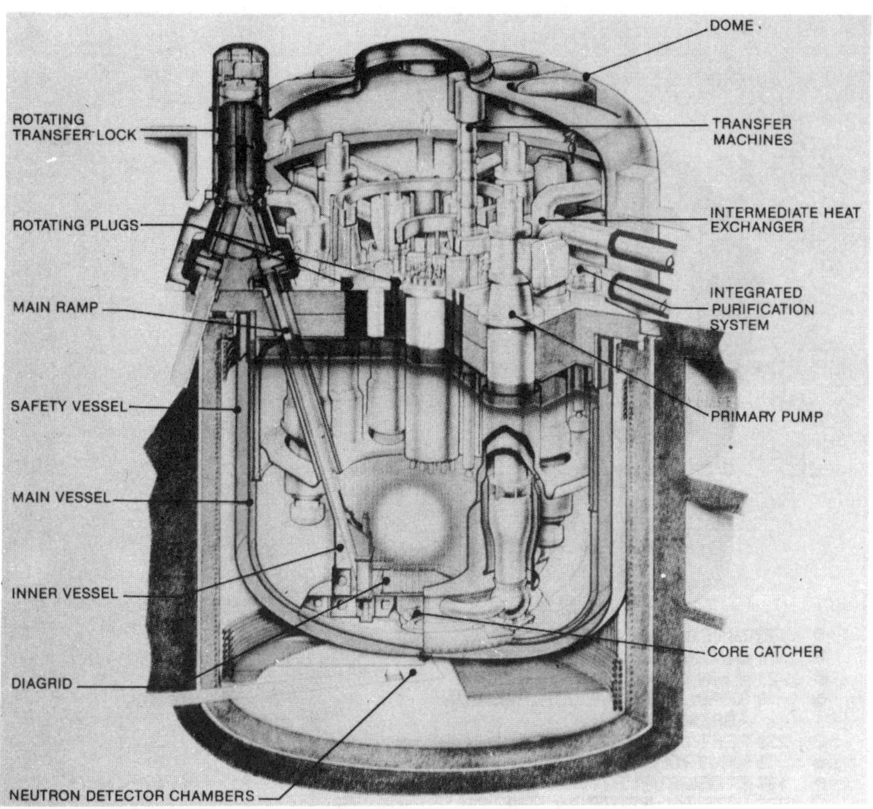

Figure 11-21 Super Phénix reactor block [107].

The fuel assemblies (Fig. 11-23), are cold-worked 316 stainless steel, are 5.4 m long and 173 mm across the flats, and contain 271 fuel rods each. There are 193 assemblies in an inner and 171 in an outer zone, with $PuO_2/(PuO_2 + UO_2)$ enrichments of 15 and 18.8 percent, respectively. Orificing of the fuel assemblies results in six flow rates: 44.5, 42.0, and 39.0 kg/s for the inner zone and 40.3, 36.9, and 30.0 kg/s for the outer zone.

Each fuel assembly has a hollow hexagonal steel block at the top with a centrally drilled hole for sodium exit. It acts as an integral upper shield that, while lengthening the assemblies, makes it unnecessary to have special shielding for nearby components within the pool, such as the IHXs, core cover, pumps, etc.

The fuel rods are 8.5 mm OD and 2.7 m long. The cladding is cold-worked 316 stainless steel that contains a 0.162-m-long retaining spring at the top, followed by a 0.3 m zone of fertile UO_2 pellets, 1.0 m of the mixed-oxide fuel pellets, another 0.3 m of fertile UO_2 pellets and, at the bottom, a 0.85-m fission gas expansion chamber.

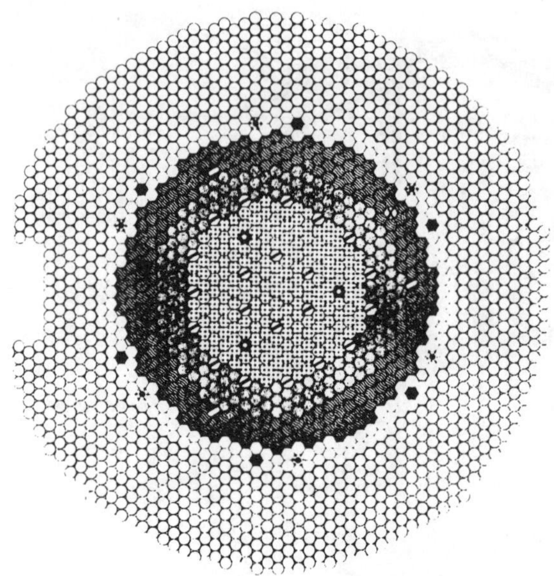

193 INNER FUEL ASSEMBLIES
171 OUTER FUEL ASSEMBLIES
21 MAIN CONTROL ASSEMBLIES
3 SUPPLEMENTARY SHUTDOWN ASSEMBLIES
233 FERTILES ASSEMBLIES
3 NEUTRON GUIDES
197 STEEL REFLECTOR ASSEMBLIES
1076 LATERAL NEUTRON SHIELD ASSEMBLIES
6 CLEAN-UP POSITIONS FOR INNER FUEL ASSEMBLIES
6 CLEAN-UP POSITIONS FOR OUTER FUEL ASSEMBLIES

Figure 11-22 Cross section of the Super Phénix reactor core.

Helically wound around the rods are 1.20-mm wires that, besides correctly spacing the pins, minimize pin vibration, increase sodium flow turbulence, and permit some fuel swelling in relation to the hexagonal channel.

Nominal fuel pellet linear power is 450 W/cm and nominal cladding temperature is 620°C. Fuel burnup is planned at 70,000 MW · day/ton, which results in refueling every 2 years, after which fuel is sent for reprocessing. (Some Phénix assemblies have reached 65,000 MW · day/ton, and some experimental rods in Rapsodie have exceeded 160,000 MW · day/ton.) The limit on burnup is more due to steel irradiation damage, in particular swelling, than to oxide fuel damage.

The fertile assemblies In addition to the fertile material above and below the fuel in the fuel rods, the fertile assemblies contain fertile material, also of depleted UO_2. They are similar in design to the fuel assemblies but contain 91 rods of 15.8 mm OD each. Each rod is 1.944 m long, 1.6 m of which is the fertile section, equal to total

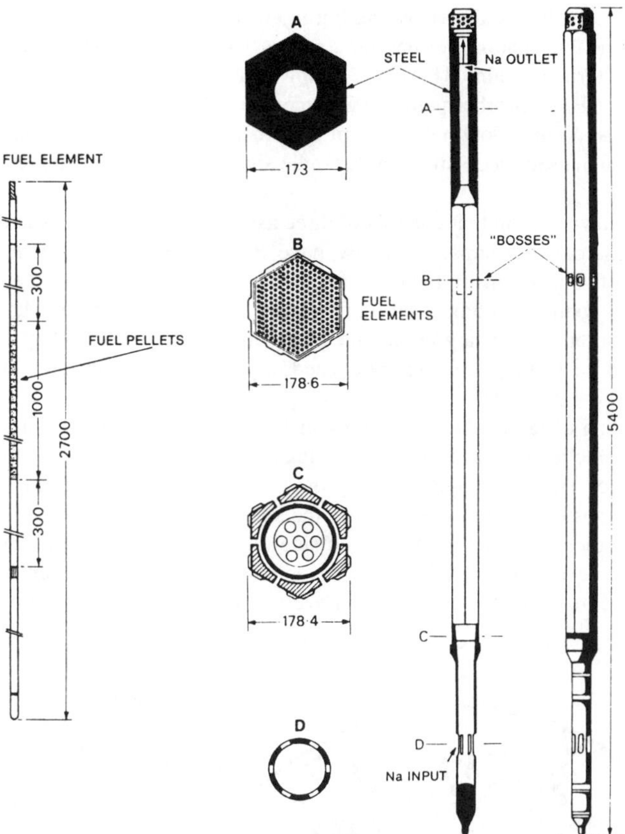

Figure 11-23 Super Phénix fuel rod and assembly.

length of fuel and fertile pellets in the fuel rods. The rest is a gas expansion chamber. The pellets here are thicker than the fuel pellets and the expansion chamber is smaller because the specific power is lower. As with the fuel rods, the fertile rods have 0.9-mm wires helically wound around them and hollow hexagonal steel upper neutron shields.

Orificing allows three flow rates in the fertile assemblies, depending upon their position and residence time in the reactor: 9.3 kg/s (3 years), 3.5 kg/s (4 years), and 1.2 kg/s (5 years). These rates are designed for cooling at the end of life when plutonium loading and hence power are greatest.

The control and shutdown rods There are 21 main control assemblies that consist of one group of five situated on an inner circle and one group of six situated on an outer circle within the core (Fig. 11-22). They have the multiple functions of shutdown, load, and temperature compensation and control. They are always partially inserted in the core and are progressively withdrawn as fuel burnup increases. The outer sleeve of each assembly is identical to that of the fuel but contains a 149-mm diameter, 1.3-m-long cylinder that, in turn, contains 31 rods, 21 mm in diameter each. The rods contain a 1.1-m section of B_4C absorber pellets, 90 percent enriched in B^{10}, enclosed in 316 stainless steel cans. Vents allow helium from the n,α reaction to escape to an expansion chamber, a design used successfully in Rapsodie and Phénix. The assemblies are sodium-cooled.

There is also a shutdown system that consists of three assemblies situated between the inner and outer main control groups. They also have B_4C pellets but a different interior. They have only two positions in the core, fully withdrawn and fully inserted, and therefore perform no control or compensation functions. Actuated by electromagnets, they are inserted on an emergency command to the main control system by a self-actuated system or if the sodium temperature exceeds a preset value, between 650 and 700°C.

Loading and unloading of removable assemblies such as the fuel, control, fertile, and lateral shielding assemblies is to a loading and unloading chamber that is fixed to the core support structure. This is accomplished by the two off-center rotating plugs (Fig. 11-24) using two transfer machines attached to the small rotating plug (Fig. 11-21). The assemblies are then removed to storage outside the reactor by means of an inclined ram and a rotating transfer lock. A core cover plug is housed in the small rotating plug above the core with its bottom just above the assembly heads but sub-

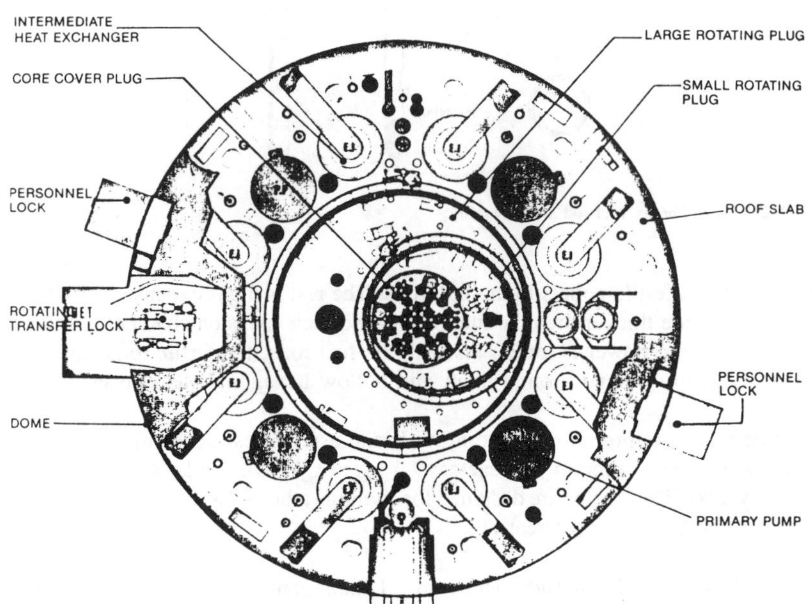

Figure 11-24 Plan view of Super Phénix reactor block.

merged in sodium. Its functions are to deflect sodium flow from the assemblies and to properly position the rod mechanisms and core instrumentation.

The heat-transfer systems As with all sodium-cooled reactor plants, a primary sodium system transfers heat from the reactor to an intermediate sodium loop, which in turn transfers heat to the steam cycle. In this pool design, the entire primary system is inside the main vessel. It consists of four pumps and eight intermediate heat exchangers (IHX) arranged symmetrically in groups of three (one pump and two exchangers) on a 16.2-m-diameter circle slightly off-center to the core (Fig. 11-24).

Unlike CRBR, Super Phénix uses "cold leg" primary pumps. Each draws 4100 kg/s of cool (392°C) sodium from the IHX outlet region through a skirt, which ensures uniform feed to its inlet diffuser (Fig. 11-25). It then delivers sodium axially to a

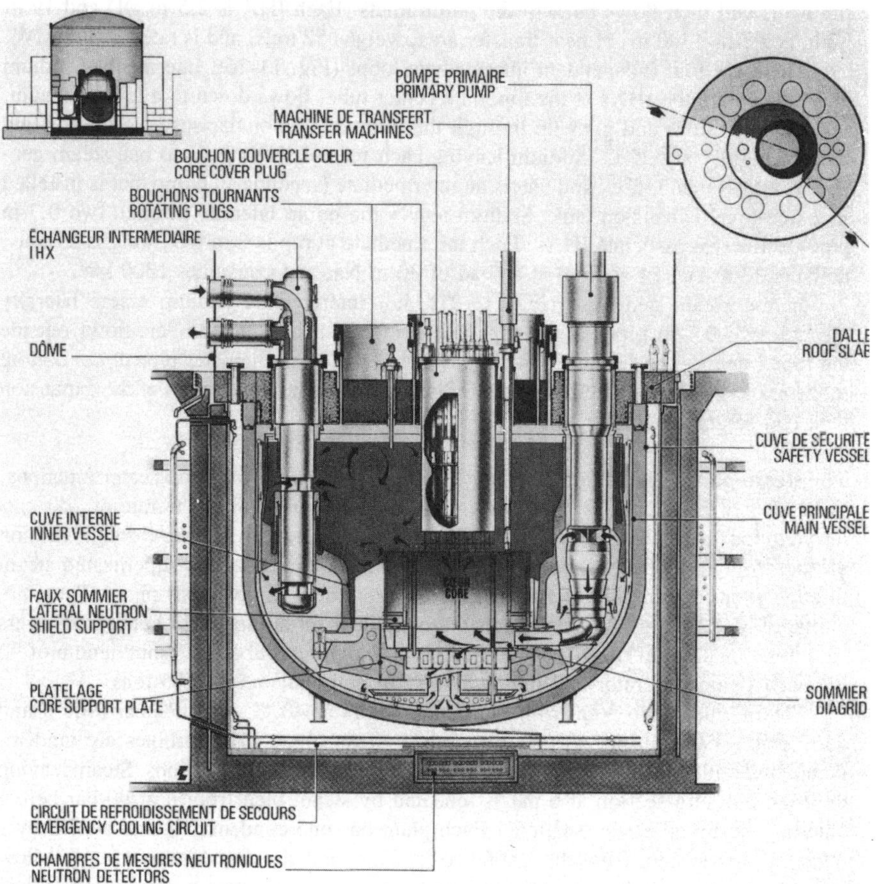

Figure 11-25 Cross section of Super Phénix reactor block showing primary coolant circuit. *(Courtesy Novatome, Le Plessis-Robinson, France.)*

plenum below the core diagrid with a head of 65-m Na. A one-way valve in the delivery pipe prevents flowback in case of individual pump shutdowns. The pumps are 2.5 m OD and 12 m long and weigh 125 tons each. They are driven by 3300-kW asynchronous motors that are powered by variable voltage and variable frequency current. This current is provided by a variable-speed alternator that is connected to a constant-speed asynchronous motor by a hydraulic coupler. This allows a pump speed variation from 15 to 100 percent of its 460 r/min nominal speed. In addition, each pump can be powered by a pony motor that runs at about 15 percent of nominal speed, and that receives its power from the emergency diesels.

Primary sodium enters and leaves the core at 395 and 545°C, respectively. It then enters the shell-and-tube IHXs at 542°C via inlet ports at top, flows outside the tubes and leaves at 392°C through outlet ports at bottom to the "cold" IHX outlet region in the pool, and then flows back to the pump inlets. Each IHX is 2.5 m OD and 19 m high, contains 1300 m^2 of heat-transfer area, weighs 52 tons, and is rated at 375 MW.

There are four independent intermediate loops (Fig. 11-26). Intermediate sodium at 345°C enters the IHXs at the top via a center tube, flows down to a lower plenum, reverses direction, and goes up through the tubes to an annular space at the top and leaves laterally at 525°C. Sodium leaving each pair of IHXs goes to one steam generator, leaves it at 345°C, and enters an intermediate (secondary) pump that is installed in a free-level expansion tank. Sodium leaves the pump laterally through two 0.7-m pipes to the two associate IHXs. Each intermediate pump is 2 m OD, weighs 35 tons, delivers 3300 kg/s of sodium at a head of 30 m Na, and consumes 1300 kW.

In the steam generator (Fig. 11-27), hot intermediate sodium enters laterally through two 0.7-m pipes at the top, mixes in a distribution chamber, flows outside the tube bundle, and leaves at 345°C via one central 1-m-diameter pipe at the bottom to the expansion tank. There are free sodium levels in the upper parts of the expansion tank and steam generator.

The steam plant The steam generator combines both boiler and superheater functions. Feedwater at 235°C and 210 bar enters each steam generator at the rate of 340 kg/s through four water chambers at the lower end. It flows in a once-through fashion through four sets of tubes from the water chambers and leaves as superheated steam through four corresponding steam chests that combine into two exit pipes. There are a total of 357 20-mm ID tubes wrapped around the central pipe in 17 concentric layers for a total of 2700-m^2 heat-transfer area. Each individual tube has a total length of 92 m. Each steam generator is 3 m OD and 25 m high and weighs 175 tons.

Steam enters two 3000 r/min turboalternators at 487°C and 177 bar (909°F and 2567 psia). Each turbine has its own reheat steam plant. The turbines are tandem-compound with one high-pressure section and two low-pressure sections. Steam leaving the high-pressure section at 6 bar is reheated by steam bled from it at 90 bar before entering the low-pressure sections. Each plant has one condenser that is cooled by a once-through system from the Rhône river. The condensate goes through a full-flow water treatment unit (to avoid fouling and corrosion in the steam generator). It then goes to two twin low-pressure feedwater heaters, a deaerating heater with storage tank,

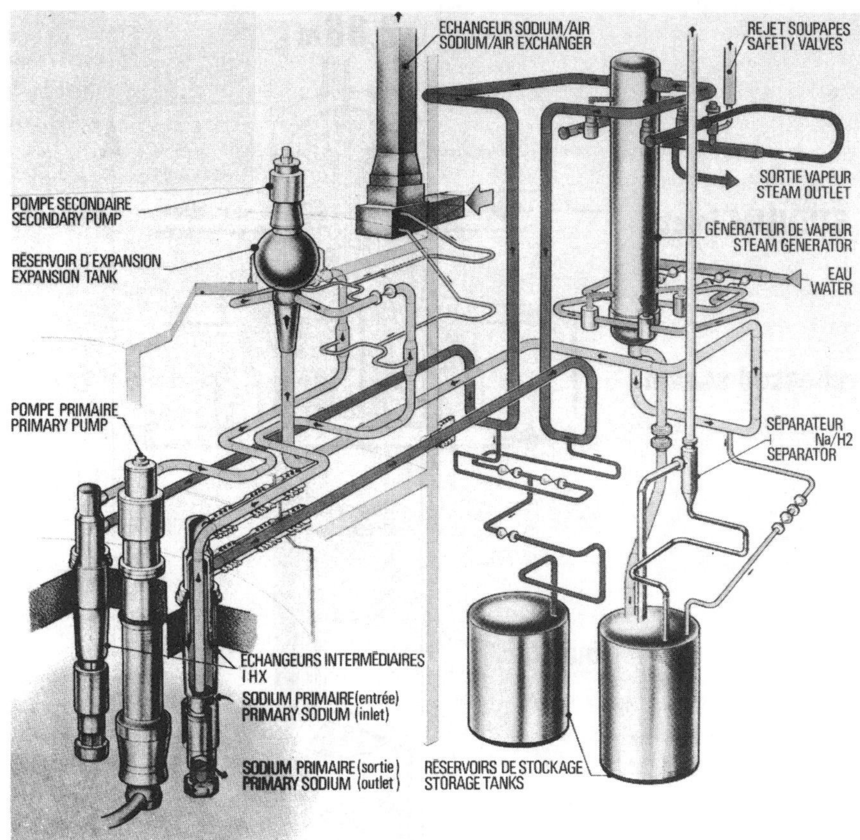

Figure 11-26 Super Phénix intermediate sodium loop (one of four). *(Courtesy Novatome.)*

three boiler feed pumps, one steam-driven and full-flow, the other two electrically driven for start-ups and other special operations. These are followed by two twin high-pressure feedwater heaters.

To avoid sodium freezing (about 100°C) in steam generators during shutdown or transients, water in the storage tank is maintained at 150°C by a variety of steam inlets.

Decay-heat removal In normal plant shutdown the reactor decay heat is channeled as usual by the primary and intermediate sodium loops and the steam generators. The resulting steam is shunted to the two turbine condensers and a special water loop called the shutdown–start-up circuit. In the event of the steam loop being inoperative, a backup system of sodium-air heat exchangers connected to each of the four intermediate sodium loops is used. These exchangers are cooled by forced convection with air

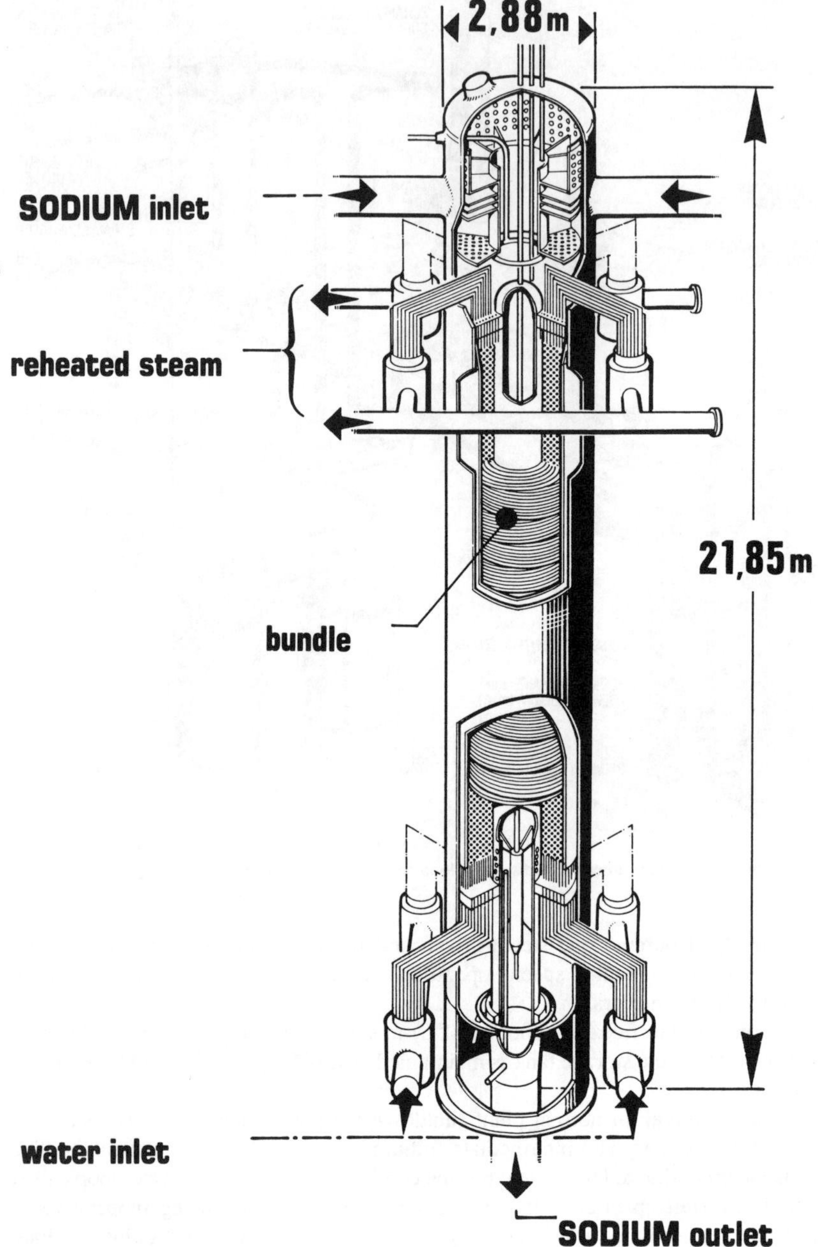

Figure 11-27 Super Phénix steam generator. *(Courtesy Novatome.)*

blowers but are capable of removing the decay heat by natural convection alone in case of power failure. Additional backup systems operate in the event of failure of all four intermediate loops. They are composed of two independent water circuits in the reactor cavity and four sodium-to-sodium heat exchangers in the reactor vessel which connect to sodium in heat exchangers situated on top of the steam-generator building.

11-6 THE GAS-COOLED FAST-BREEDER REACTOR (GCFBR)

Research and development for the gas-cooled fast-breeder reactor (GCFBR) started in 1962 by Gulf General Atomic Corporation with some utility and, somewhat later, European participation. Studies were made on fuel development, systems, safety, physics, heat transfer, and fluid flow. Preliminary designs were made for a reactor experiment, a 300-MW demonstration plant, and a 1000-MW commercial plant. The studies also included the possible future uses of the GCFBR with a direct gas-turbine cycle for power production and with a high-temperature gas generator for industrial applications.

Figures 11-28 and 11-29 show the proposed 300-MW demonstration plant and its nuclear steam supply system. As were the thermal neutron AGR and HTGR gas-cooled reactors (Chap. 10), it is contained in a cylindrical prestressed-concrete reactor vessel

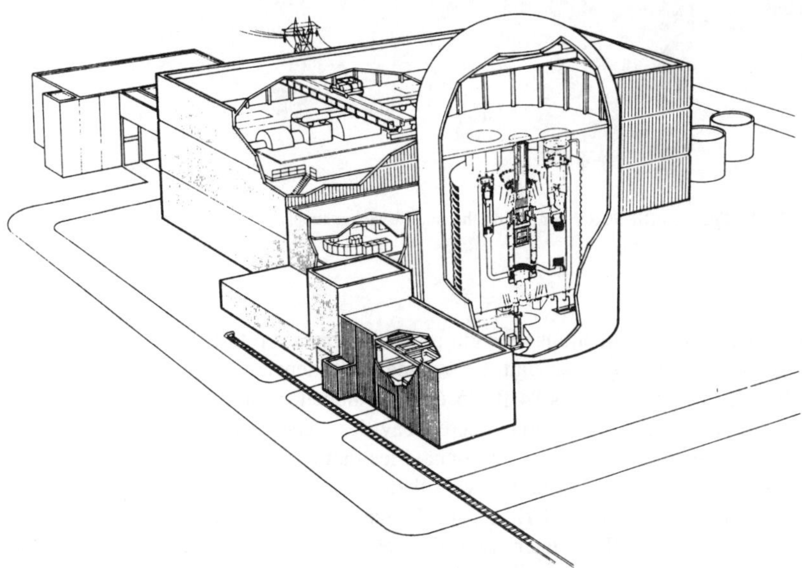

Figure 11-28 Layout of the 300-MW gas-cooled fast-breeder reactor (GCFBR) demonstration plant. *(Courtesy GA Technologies, Inc.)*

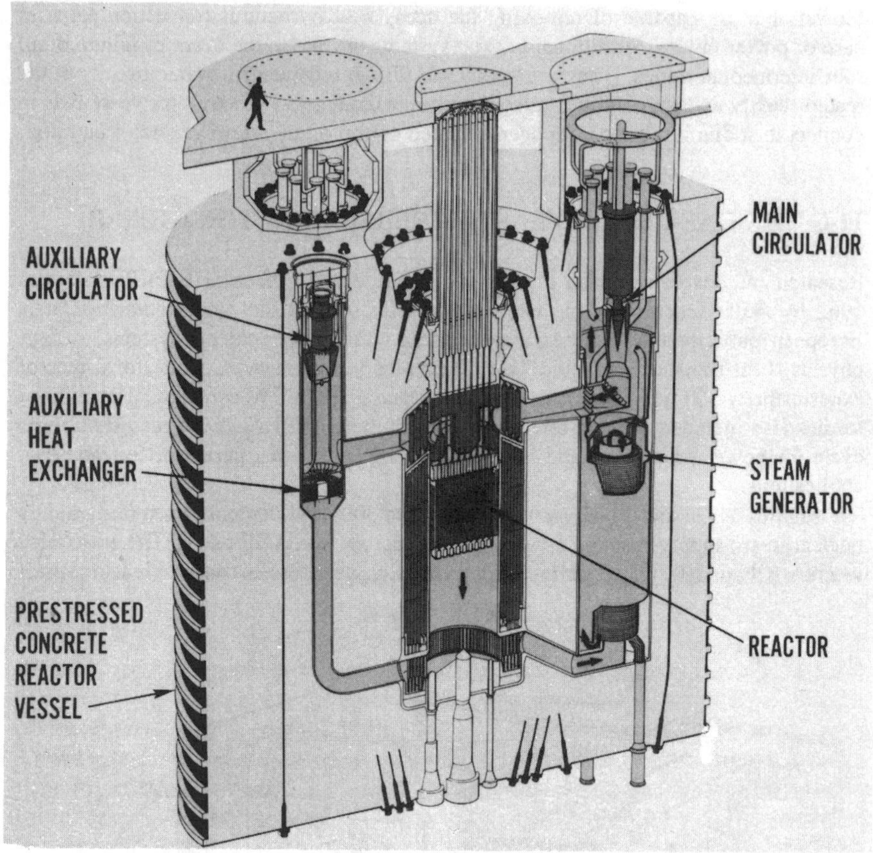

Figure 11-29 The GCFBR PCRV containing the nuclear steam supply system.

(PCRV) that has linear tendons extending through from top to bottom and circumferential steel cables wrapped around the outside under tension.

The PCRV in this case is 84 ft (25.6 m) in diameter and 71 ft (21.6 m) high and contains seven steel-lined interconnected cavities. The central cavity contains the reactor and is surrounded by three main cavities that contain the steam generators and helium circulators and, alternating with them, three that contain auxiliary cooling equipment. Helium coolant, at 1250 psia (86 bar), passes downward through the core (as in the HTGR) then flows laterally to the bottom of the three main cavities. In each it goes up through a steam generator, through a 22,300-hp single-stage, axial, steam-driven main helium circulator, and then back laterally to the top of the core. Although these three heat-transfer loops are in interconnected cavities, they are designed to operate independently of one another, thus increasing the reliability of the system.

The PCRV is housed in a secondary containment building. The space between the two contains atmospheric air and is accessible during plant operation for inspection and maintenance of equipment outside the PCRV.

The reactor core and fuel The reactor core is composed of 211 hexagonal fuel and radial blanket assemblies. The assemblies, 10 ft (3.05 m) high and 6.5 in (16.5 cm) across the flats, are supported by clamping to a top grid plate where they are at their coolest. The grid plate, 11 ft (3.35 m) in diameter and 2 in (5.08 cm) thick, is in turn supported by a cylindrical structure connected to the liner of the central penetration in the PCRV. It contains closely spaced 6-in (15.24-cm) diameter holes that accommodate the circular top extensions of the fuel assemblies.

The assemblies are spaced about 1/4 in (6.3 mm) apart and are to be rotated during reloading to accommodate and reduce swelling (Sec. 9-13). Reloading is done by changing one-third of the core approximately annually and requires shutdown and depressurization.

The fuel assemblies (Fig. 11-30) contain 271 fuel rods except 27 that are used as control assemblies and contain 234 rods. The rods are the same as those used in LMFBRs except that their cladding surface is roughened to increase the gas-coolant heat-transfer coefficient. They are composed of annular pellets of mixed-oxide fuel (PuO_2-UO_2) stacked in 0.25-in (0.63-cm) OD, 20-mil (0.5-mm) thick 316 stainless steel cladding. An axial blanket of depleted UO_2 pellets is provided above and below the fuel within each rod.

The radial blanket assemblies occupy two rows outside the fuel assemblies and are of the same external dimensions. They, however, contain 127 larger rods containing depleted UO_2 pellets.

Each fuel rod has a fission product trap at the top. Fission gases are allowed to vent to an annular trap in the assembly and to a helium purification system. This system equalizes pressures of the fission gases inside and the coolant outside the rods, thus relieving mechanical stresses on the cladding. It also limits the release of radioactivity from failed elements into the coolant. Radioactivity monitors on the vent lines of separate groups of assemblies are used to detect and locate failed cladding.

The power plant Figure 11-31 shows a flow diagram of the plant. Helium at 1250 psia (86 bar) enters the reactor top at 595°F (313°C) and leaves at 1010°F (543°C). It then enters the three steam generators and steam-driven helium circulators (one shown) and goes back to the reactor. There are no intermediate loops as in a LMFBR as there are no problems of primary-coolant radioactivity or coolant-water chemical reactions as with sodium.

The steam generators are once-through helical-coiled units with reheaters. Feedwater leaving the last feedwater heater divides into three paths. Each enters a steam generator and leaves as superheated steam at 2900 psia (200 bar) and 875°F (468°C). That steam partially expands as it is made to drive the helium circulator turbine, after

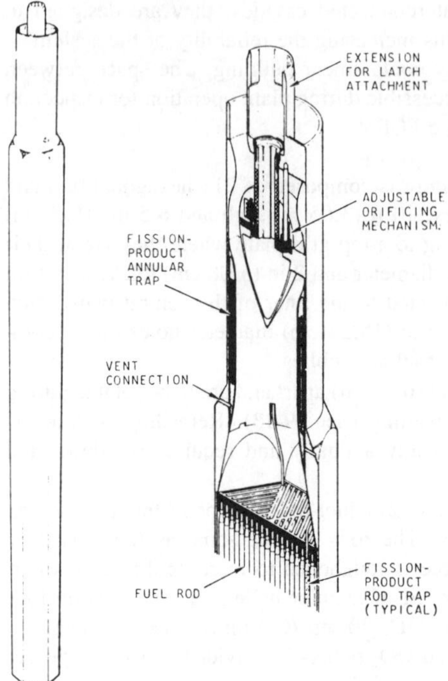

Figure 11-30 The GCFBR fuel assembly.

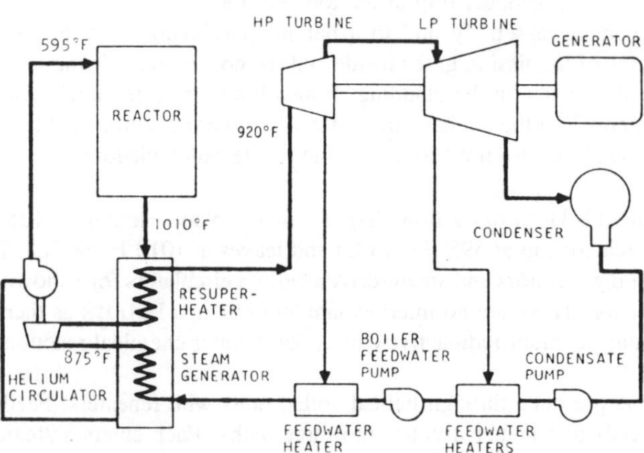

Figure 11-31 Simplified flow diagram of the GCFBR demonstration plant.

Table 11-6 Design parameters of the GCFBR

GCFBR	Demonstration	Commercial
Power	310 MW(e)	1000 MW(e)
Core: height	40 in (101 cm)	54 in (1.36 m)
diameter	80 in (2.01 cm)	107 in (2.71 m)
Fuel	Mixed oxide	Mixed oxide, 16% fissile
Fuel pins	0.25 in (6.3 mm) OD	0.25 in (6.3 mm) OD
Cladding	316 stainless steel, 0.48 mm thick	316 stainless steel, 0.286 mm thick
Fuel fissile rating	0.6 MW(t)/kg	1.1 MW(t)/kg
Max fuel linear rating	12.5 kW/ft (410 W/cm)	15 kW/ft (490 W/cm)
Cladding surface hot spot	1275°F (690°C)	1382°F (594°C) outlet
Coolant	Helium	Helium
Reactor coolant conditions	1250 psia (86.2 bar), 595°F (313°C) inlet, 1010°F (543°C) outlet	1250 psia (86.2 bar), 602°F (317°C) inlet, 1100°F (594°C) outlet
Throttle steam conditions	1225 psia (84.5 bar), 920°F (493°C)	1250 psia (86.2 bar), 925°F (496°C)
Plant thermal efficiency	37.6%	38%
Breeding ratio, doubling time	1.33	1.5 (~8 years)
Maximum fuel burnup	—	100,000 MW · day/ton

which it is admitted to the reheater. It then combines with the other two paths to enter the high-pressure turbine at 1225 psia (84.5 bar) and 920°F (492°C). The balance of the steam plant is essentially identical to a fossil powerplant because it uses superheated steam of similar properties. Note the difference between this steam cycle and that of the HTGR (Sec. 10-12). In the latter, the steam leaving the main steam generator expands in the high-pressure turbine and then drives the helium-circulator turbine before being reheated and sent back to the lower-pressure section of the main turbine.

The three main cooling loops are also used to remove decay heat from the core after shutdown. As in the HTGR, decay heat initially generates enough steam to drive the circulator turbines to help remove that decay heat, an advantage of the steam cycle design that uses the main steam flow into the circulator turbines. A half hour after shutdown, small auxiliary boilers, also used for plant start-up, come on line to supply steam to the circulator turbines. In addition, there are the three separate electric motor-driven auxiliary cooling loops (Fig. 11-30).

Table 11-6 includes data for the GCFBR demonstration plant as well as for a commercial-size extrapolation of GCFBR. The large plant is expected to have thinner cladding, higher cladding temperature, higher fuel linear heat rating, and a higher breeding ratio. These improved parameters are the results of the increased freedom of selecting fuel-fertile and fuel-coolant ratios in a large GCFBR.

PROBLEMS

11-1 A large commercial gas-cooled fast-breeder reactor using a plutonium-uranium mixed oxide fuel has a breeding ratio of 1.5 and a compound doubling time of 6 years. Estimate the average neutron flux in the reactor. Assume that the fuel in the core represents 80 percent of the reactor fuel inventory.

11-2 A large sodium-cooled fast-breeder reactor uses a plutonium-uranium oxide fuel mix. It has an average neutron flux of 3×10^{15}. The number of neutrons lost by leakage and parasitic absorption is 0.13 per neutron absorbed. Calculate (*a*) the reactor simple and doubling times, in years, based on the fuel in the core only, and (*b*) the neutron losses per neutron born in fission that would render the reactor a nonbreeder.

11-3 A fast-breeder reactor generates 3000 MW of heat. The fuel is composed of 20% $Pu^{239}O_2$, 80% $U^{238}O_2$ by mass. The average neutron flux is 10^{16}. Estimate the total mass of the fuel material in the core. Ignore fast fission in U^{238} and take neutron losses by leakage and parasitic absorption as 0.25 per neutron absorbed.

11-4 Estimate the maximum fraction (zero losses) of all neutrons that is available for breeding for the three fuels U^{233}, U^{235}, and Pu^{239} if they existed in monoenergetic neutron fluxes at 1, 10^2, 10^4, and 10^6 eV.

11-5 A fast-breeder reactor core that generates 2800 MW is fueled with $Pu^{239}O_2$-$U^{238}O_2$. The mass of $Pu^{239}O_2$ is 4 tons (metric). Ignoring fast fission in U^{238}, estimate (*a*) the maximum theoretical breeding ratio and gain, and (*b*) the corresponding minimum simple and compound doubling times, in years. Assume that the fuel in the core represents 75 percent of the reactor fuel cycle inventory.

11-6 A fast breeder reactor powerplant generates 1000 MW with a 39 percent efficiency. The core is fueled with $Pu^{239}O_2$-$U^{238}O_2$ with Pu-U nuclear number ratio of 1 : 4. The core average neutron flux is 10^{16}. Estimate (*a*) the number and mass of Pu^{239} nuclei originally in the core, (*b*) the number and mass of the Pu^{239} nuclei consumed per day, and (*c*) the number and mass of Pu^{239} nuclei bred and gained per day. Assume neutron losses by leakage and parasitic absorption to be O.25 per neutron absorbed in the fuel.

11-7 A thermal-neutron breeder reactor using a U^{233}-Th^{232} breeding cycle is considered. The fuel is composed of 15 percent $U^{233}O_2$ by mass. The average neutron flux is 10^{14}. The reactor generates 2000 MW. Find (*a*) the total mass of fuel in the core, (*b*) the breeding ratio and gain if the neutron losses due to leakage and parasitic capture are 0.2 per neutron absorbed, (*c*) the simple and compound doubling times, in years, if the fuel in the core is 80 percent of the total reactor fuel inventory, and (*d*) the mass of $U^{233}O_2$ bred per day.

11-8 From the data of the Clinch River Breeder Reactor calculate the overall heat transfer coefficients, in Btus per hour per square foot per degree Fahrenheit for (*a*) the intermediate heat exchangers, (*b*) the evaporators, and (*c*) the superheaters.

11-9 From the data of Super Phénix, calculate the overall heat transfer coefficients in watts per square meter per kelvin and Btus per hour per square foot per degree Fahrenheit of (*a*) the intermediate heat exchanger, and (*b*) the economizer, evaporator, and superheater of the steam generator. Use for the specific heat of sodium 1.256 kJ/kg · K.

11-10 Assume that a gas-cooled fast-breeder reactor powerplant operates on a similar cycle and has the same power and efficiency and helium and steam pressures and temperatures as the demonstration GCFBR (Sec. 11-6). Assume further that the feedwater enters the steam generator at 468°F and that 35 MW of steam power is consumed in driving the helium circulators. Draw a temperature-path-length diagram of the steam generator and calculate (*a*) the helium mass flow rate, in pounds mass per hour, (*b*) the steam mass flow rate, in pounds mass per hour, and (*c*) the heat transferred, in Btus per hour in the economizer, evaporator, and first and second superheaters, respectively.

CHAPTER

TWELVE

GEOTHERMAL ENERGY

12-1 INTRODUCTION

With this chapter we begin a series of four chapters on the so-called *renewable energy resources*,* which are defined as those resources that draw on the natural energy flows of the earth. In this book, the ones that we are concerned with are those that arise from the earth's interior heat (this chapter), the sun (Chap. 13), the wind (Chap. 14), and the oceans (Chap. 15). Another, biomass, was covered briefly in Sec. 4-12.

Renewable energy resources are so named because they recur, are seemingly inexhaustible, and are free for the taking. The recurrence is often periodic, ranging from daily (the sun) to a few short years (biomass). Their main disadvantages are in their intermittency, lack of dependability, and their usually extreme low energy densities. Despite this, they were the predominant energy forms used by humankind during its early developmental millenia. When the earth's populations grew and nature alone could no longer support life on earth, human beings, as often is the case, discovered solutions when they were needed. The solutions came in the form of fossil fuels that were concentrated in certain pockets of the earth and that delivered much higher energy densities; they eventually brought on the industrial revolution and thus the modern way of life as we know it today. To be sure, fossil fuels are also renewable, but on

* Another term that is often used interchangeably with renewable energy is *alternative energy*. This term seems to mean different things to different people. Some include nuclear fuel used in fission, especially when its use is extended by breeder reactions (Sec. 11-2), and/or fusion. To others, the term precludes all established forms of energy. Others limit it to solar and solar-induced energies such as the wind and the rise and fall of waters. Still others make no distinction between the two terms.

a geologic time scale of hundreds of millions of years. They, however, are consumed much faster than they are renewed, so for all practical purposes, they have to be considered as finite. The same conclusion could be applied to nuclear fuels, which were formed with the earth. Their use, however, is believed to extend far beyond fossil fuels because of breeding, a reality today, and fusion, when it becomes a workable system.

With high energy density come high temperatures and therefore higher efficiencies. Advances in metallugy had the multiplier effect of further increasing temperatures and efficiency. Together with abundant and cheap fuels, these effects relegated renewable energy systems to doing odds and ends here and there forever, or so it seemed. Then came the 1970s and the great economic squeeze of the oil producers and resultant soaring prices; the era of cheap energy came to an end. This, and the environmental concerns of the same decade, reopened the era of the renewables (or alternatives). Goaded by public fears and pressures, a rush was on for broad technological solutions of the problems of the renewable systems. Before they were technologically and economically demonstrated, many people were inclined to overstate the case for them, which led to further public confusion about their true nature and their possible contributions to the total electric-energy picture. (Most renewable resources are aimed at electric production as their main contribution.) With time, however, and probably inevitably, cooler heads prevailed and a more sensible reexamination began taking place. This is what we will do in these four chapters: reexamine the renewables and follow up on modern solutions to their old problems.

We will begin our examination with geothermal energy, the one renewable resource that has practically no intermittency, has the highest energy density, and is economically not far removed from conventional technologies. Geothermal energy is classified as renewable because the earth's interior is and will continue in the process of cooling for the indefinite future. Hence, geothermal energy from the earth's interior is almost as inexhaustible as solar or wind energy, so long as its sources are actively sought and economically tapped.

12-2 PAST, PRESENT, AND FUTURE

Geothermal energy is primarily energy from the earth's own interior. The natural heat in the earth has manifested itself for thousands of years in the form of volcanoes, lava flows, hot springs, and geysers. These were mostly picturesque, often awesome proofs that vast heat stores lie beneath the earth's crust. In earlier times, natural steam that spouted from the earth was used only therapeutically. Roman documents, more than 2000-years-old, tell of a steam field that is now Larderello, south of Florence, a site that was to become history's first geothermal electric-generating station.

In the United States, geothermal fields were first discovered in 1847 by William Bell Elliot, an explorer-surveyor who was hiking in the mountains between Cloverdale and Calistoga, California, in search of grizzly bears. He discovered steam seeping out of the ground along a quarter of a mile on the steep slope of a canyon near Colb Mountain, an extinct volcano, now known as the Geysers. Telling friends that he came upon the gates of hell, the word spread and the area became something of a tourist attraction.

The Geysers is really a misnamed field, as a geyser, like Old Faithful in Yellowstone National Park, periodically and dramatically spews jets of water and steam. In the Geysers, however, steam is continuously vented through fissures in the ground. These vents are called *fumaroles*.

Historically, the first applications of geothermal energy were for space heating, cooking, and medicinal purposes. The earliest record of space heating dates back to 1300 in Iceland. In the early 1800s, geothermal energy was used on what was then a large scale by the Conte Francesco de Laderel to recover boric acid. The first mechanical conversion was in 1897 when the steam of the field at Larderello, Italy, was used to heat a boiler producing steam which drove a small steam engine. The first attempt to produce electricity (our main concern in this book) also took place at Larderello in 1904 with an electric generator that powered four light bulbs. This was followed in 1912 by a condensing turbine; and by 1914, 8.5 MW of electricity was being produced. By 1944 Larderello was producing 127 MW. The plant was destroyed near the end of World War II, but was fortunately rebuilt and expanded and eventually reached 360 MW in 1981.

In the United States, the first attempt at developing the Geysers field was made in 1922. Steam was successfully tapped, but the pipes and turbines of the time were unable to cope with the corrosive and abrasive steam. The effort was not revived until 1956 when two companies, Magma Power and Thermal Power, tapped the area for steam and sold it to Pacific Gas and Electric Company. By that time stainless steel alloys were developed that could withstand the corrosive steam, and the first electric-generating unit of 11-MW capacity began operation in 1960. Since then 13 generally progressively larger units have been added to the system. The latest, No. 17 (Fig. 12-1), is a 109-MW unit that began operation in September 1982 and which brought the Geysers total capacity to 909 MW. Two more units are under construction and four more are planned, which will bring the total capacity to 1514 MW by the late 1980s.

Other electric-generating fields of note are in New Zealand (where the main activity at Wairakei dates back to 1958), Japan, Mexico (at Cerro Prieto), the Phillipines, the Soviet Union, and Iceland (a large space-heating program). These and other electric-generating fields are listed in Table 12-1.

Future world projections for geothermal electric production, based on the decade of the 1970s, are 7 percent per year. In the last four years of that decade, however, the growth rate was 19 percent per year (Fig. 12-2). In the United States, the projections are for growth between 13.5 and 22 percent per year through the 1980s, which is 2.5 to 4 times the 5.3 percent per year growth rate of the total electric-generating capacity. This includes the steam field at the Geysers and other fields of different types (Sec. 12-3).

The U.S. Geological Survey [109] predicts a U.S. potential from currently identified sources to be around 23,000 MW of electric power and around 40×10^{15} Btu (about 42×10^{15} kJ) of space and process heat for 30 years with existing technology, and 72,000 to 127,000 MW of electricity and 144 to 294×10^{15} Btu of heat from unidentified sources. Areas of geothermal potential in the North American continent

Figure 12-1 The 109-MW unit No. 14 of the Geysers. *(Courtesy Pacific Gas and Electric Co.)*

Table 12-1* World geothermal-energy utilization as of December 1979

Country	Electricity, MW		Space and process heat, MW
	Installed	Under construction	
USA	773	641	30
Italy	421	—	1
New Zealand	203	150	70
Japan	166	100	37
Mexico	150	30	
El Salvador	60	35	
Iceland	33	30	475
USSR	6	58	300
Phillipines	60	605	
Turkey	0.5	—	
Hungary	—	—	363
France	—	—	5
Total	1872	1649	1281

*From Ref. 108.

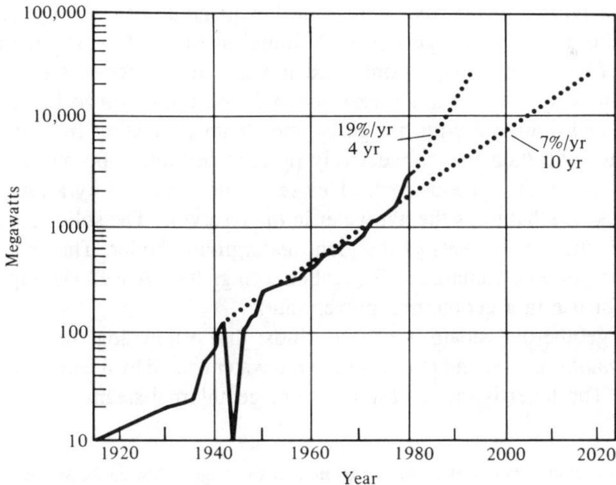

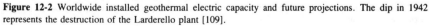

Figure 12-2 Worldwide installed geothermal electric capacity and future projections. The dip in 1942 represents the destruction of the Larderello plant [109].

are mainly west of the Great Plains from Canada to Mexico, with a geopressured zone (described in the next section) extending along the Gulf Coast and a low-temperature zone extending down the eastern seaboard. These include about 1.8 million acres of land of known sources in the western states and additional 96 million acres of prospective value. Between 800 and 1000 acres are needed for 100 MW of production for 30 years.

It can be seen that while geothermal energy is not the sought-after sole and long-range solution to our energy problems (the U.S. total installed electric capacity in 1982 is nearly 500 million MW), it nevertheless represents a not insignificant factor if its resources are developed in a careful and efficient manner.

12-3 ORIGIN AND TYPES OF GEOTHERMAL ENERGY

As indicated earlier, geothermal energy is heat transported from the interior of the earth. It is recoverable in some form such as steam or hot water.

The earth is said to have been created as a mass of liquids and gases, 5 to 10 percent of which was steam. As the fluids cooled, by losing heat at the surface, an outer solid crust formed and the steam condensed to form oceans and lakes in depressions of that crust. The crust now averages about 20 mi (32 km) in thickness. Below that crust, the molten mass, called *magma,* is still in the process of cooling.

Earth tremors in the early Cenozoic period* caused the magma to come close to the earth's surface in certain places and crust fissures to open up. The hot magma near the surface thus causes active volcanoes and hot springs and geysers where water exists. It also causes steam to vent through the fissures (fumaroles).

Figure 12-3 shows a typical geothermal field. The hot magma near the surface (A) solidifies into igneous rock† (B). (Igneous rock found at the surface is called volcanic rock.) The heat of the magma is conducted upward to this igneous rock. Ground water that finds its way down to this rock through fissures in it will be heated by the heat of the rock or by mixing with hot gases and steam emanating from the magma. The heated water will then rise convectively upward and into a porous and permeable reservoir (C) above the igneous rock. This reservoir is capped by a layer of impermeable solid rock (D) that traps the hot water in the reservoir. The solid rock, however, has fissures (E) that act as vents of the giant underground boiler. The vents show up at the surface as geysers, fumaroles (F), or hot springs (G). A well (H) taps steam from the fissure for use in a geothermal powerplant.

It can be seen that geothermal steam is of two kinds: that originating from the magma itself, called *magmatic steam,* and that from ground water heated by the magma, called *meteoritic steam*. The latter is the largest source of geothermal steam.

* The Cenozoic period is a geologic era that started some 60 million years ago following the Mesozoic period and includes the present. It is characterized by the appearance and development of mammals.

† From the Latin *igneus* meaning "of fire" or "fiery," from *ignus*, "fire"; specifically formed by volcanic action or great heat.

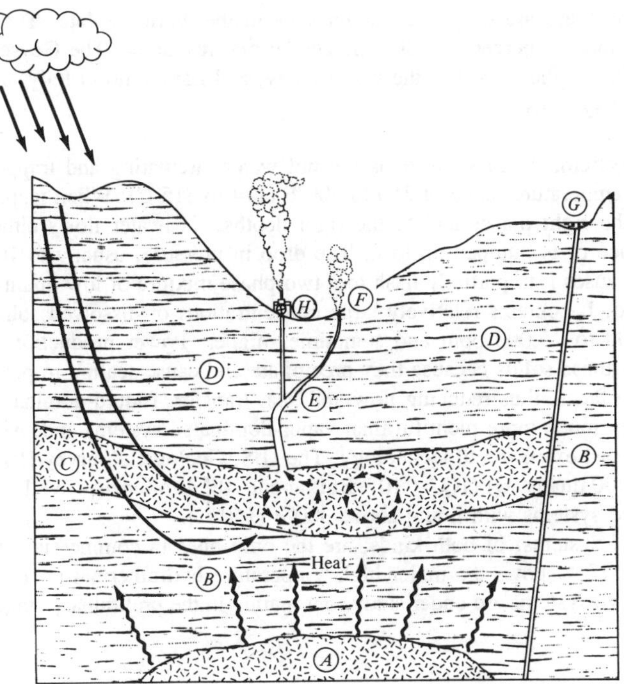

Figure 12-3 A typical geothermal field.

Not all geothermal sources produce steam as described above. Some are lower in temperature so that there is only hot water. Some receive no ground water at all and contain only hot rock. Geothermal sources are therefore of three basic kinds: (1) *hydrothermal,* (2) *geopressured,* and (3) *petrothermal*. These are explained below.

Hydrothermal Systems

Hydrothermal systems are those in which water is heated by contact with the hot rock, as explained above. Hydrothermal systems are in turn subdivided into (1) *vapor-dominated* and (2) *liquid-dominated* systems.

Vapor-dominated systems In these systems the water is vaporized into steam that reaches the surface in a relatively dry condition at about 400°F (205°C) and rarely above 100 psig (8 bar). This steam is the most suitable for use in turboelectric powerplants, with the least cost. It does, however, suffer problems similar to those encountered by all geothermal systems, namely, the presence of corrosive gases and erosive material and environmental problems (see below). Vapor-dominated systems, however, are a rarity; there are only five known sites in the world to date. These systems account for about 5 percent of all U.S. geothermal resources. The Geysers plant in the United States, the largest in the world today, and Larderello in Italy, arc both vapor-dominated systems.

Liquid-dominated systems In these systems the hot water circulating and trapped underground is at a temperature range of 350 to 600°F (174 to 315°C). When tapped by wells drilled in the right places and to the right depths, the water flows either naturally to the surface or is pumped up to it. The drop in pressure, usually to 100 psig (8 bar) or less, causes it to partially flash to a two-phase mixture of low quality, i.e., liquid-dominated. It contains relatively large concentrations of dissolved solids ranging between 3000 to 25,000 ppm and sometimes higher. Power production is adversely affected by these solids because they precipitate and cause scaling in pipes and heat-exchange surfaces, thus reducing flow and heat transfer. Liquid-dominated systems, however, are much more plentiful than vapor-dominated systems and, next to them, require the least extension of technology. The U.S. Geological Survey [109] shows from 900 to 1400 quads (Q) (1Q = 10^{15} Btu, about 10^{18} J) of energy available from liquid-dominated systems with liquid above 300°F (150°C).

The hydrothermal systems, of both kinds, are the only ones in commercial operation today. Figure 12-4 shows the major high-temperature hydrothermal areas of the world. The next two systems are under study, but mainly in the preliminary stages at this time (1982).

Geopressured Systems

Geopressured systems are sources of water, or brine, that has been heated in a manner similar to hydrothermal water, except that geopressured water is trapped in much deeper underground acquifers,* at depths between 8000 to 30,000 ft. (about 2400 to 9100 m). This water is thought to be at the relatively low temperature of about 325°F (160°C) and is under very high pressure, from the overlying formations above, of about 15,000 psia (more than 1000 bar). It has a relatively high salinity of 4 to 10 percent and is often referred to as brine. In addition, it is saturated with natural gas, mostly methane CH_4, thought to be the result of decomposition of organic matter.

Such water is thought to have thermal and mechanical potential to generate electricity. The temperature, however, is not high enough and the depth so great that there is little economic justification of drilling for this water for its thermal potential alone. What is drawing attention, however, is the amount of recoverable methane in solution that can be used for electric generation. The U.S. Geological Survey estimates 100 Q of electricity from the thermal content of geopressured water and 500 Q of energy in the gas. Studies have been under way to determine the economic feasibility of generating electricity by a combined cycle, one that involves the combustion of the methane as well as heat from the thermal content of the water.

* An *aquifer* is a water-bearing stratum of permeable rock, gravel, or sand.

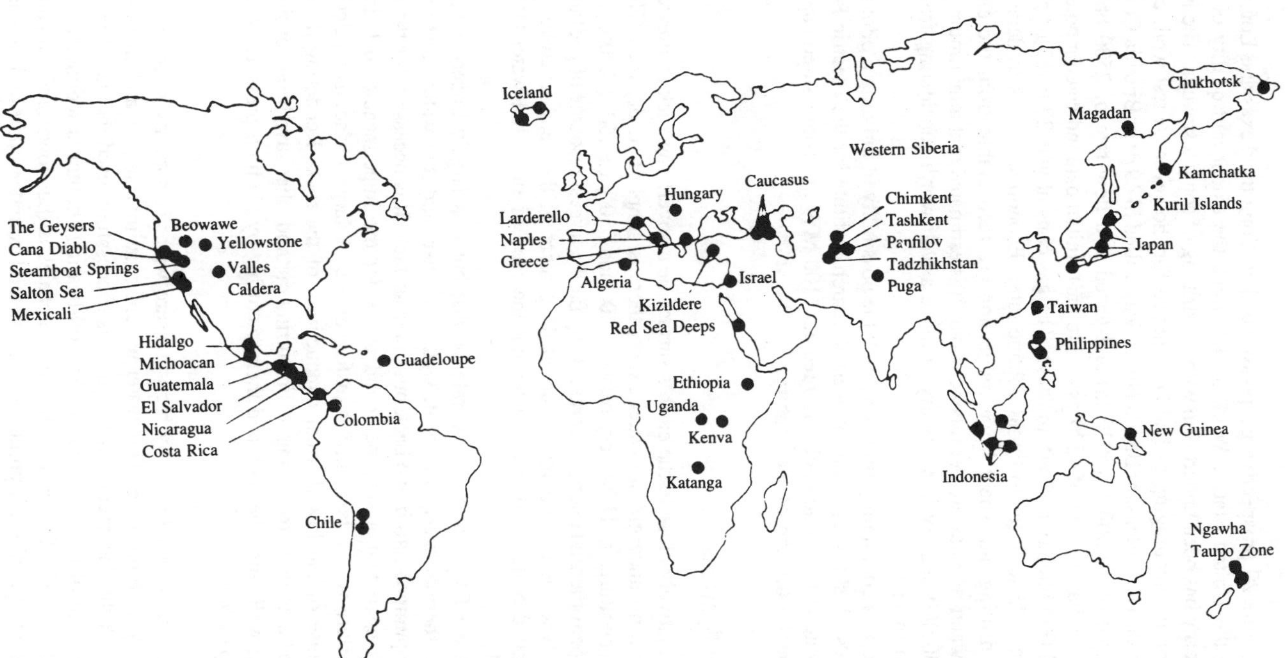

Figure 12-4 Major high-temperature hydrothermal areas of the world [110].

There are some 20 prospective geopressured sites along the Texas and Louisiana Gulf coasts in the United States. Work to determine the extent and quality of geopressurized energy has been undertaken with the drilling of some test wells. The initial results were not as encouraging as had been hoped for, however. A test well, called the General Crude-DOE Pleasant Bayou No. 2, was drilled in 1979 in Brazilia County, Texas, to a depth of 16,500 ft (5030 m) and tested at a flow rate of 2500 bbl/day (~300 m^3/day) and a pressure of 4570 psig (316 bar). Initial data indicated a potential flow rate of 30,000 bbl/day (~3575 m^3/day). The gas content was 20 to 25 ft^3/bbl of water, or about 4.75 to 6 gas-to-liquid volume ratio. Economic studies to determine if the cost of drilling and spent brine reinjection are recoverable from the energy content in the water and gas are yet inconclusive. It is estimated that a minimum yield of 40,000 to 50,000 bbl/day is necessary to make a well worth considering from an economic point of view.

Further work is continuing, however, with the possibility of building a pilot plant in the late 1980s. A study by the Southwest Research Institute for the Electric Power Research Institute (EPRI) optimistically predicts 1100 MW of geopressured capacity could be on line by the end of the century.

Petrothermal Systems

Magma lying relatively close to the earth's surface heats overlying rock as previously explained. When no underground water exists, there is simply hot, dry rock (HDR). The known temperatures of HDR vary between 300 and 550°F (~150 to 290°C). This energy, called petrothermal energy, represents by far the largest resource of geothermal energy of any type, as it accounts for about 85 percent of the geothermal resource base of the United States. Other estimates put the ratio of steam:hot water:HDR at 1:10:1000 [111].

Much of the HDR occurs at moderate depths, but it is largely impermeable. In order to extract thermal energy out of it, water (or other fluid, but water most likely) will have to be pumped into it and back out to the surface. It is necessary for the heat-transport mechanism that a way be found to render the impermeable rock into a permeable structure with a large heat-transfer surface. A large surface is particularly necessary because of the low thermal conductivity of the rock. Rendering the rock permeable is to be done by fracturing it. Fracturing methods that have been considered involve drilling wells into the rock and then fracturing by (1) high-pressure water or (2) nuclear explosives.

High-pressure water Fracturing by high-pressure water is done by injecting water into HDR at very high pressure. This water widens existing fractures and creates new ones through rock displacement. This method is successfully used by the oil industry to facilitate the path of underground oil. The oil-bearing stratum is sedimentary rock that is softer than HDR. The cost to the oil companies is thus lower and, in addition justified by the additional oil it produces. The method is under study by Los Alamos Scientific Laboratory (LASL) with support from the Department of Energy (DOE), Japan, and West Germany.

Nuclear explosives Fracturing by nuclear explosives is a scheme that has been considered as part of a program for using such explosives for peaceful uses, such as natural gas and oil stimulation, creating cavities for gas storage, canal, and harbor construction, and many other applications [112]. In the United States the program is called *Plowshare*.* Fracturing by this method would require digging in shafts suitable for introducing and sealing nuclear explosives and the detonation of several such devices for each 200-MW plant operating for 30 years. Initial studies revealed that large explosions, about 5 megatons (TNT equivalent), at substantial depths would be required before the scheme became economic. The principal hazards associated with this are the ground shocks, the danger of radioactivity releases to the environment, and the radioactive material that would surface with the heated water and steam.

A variation of the above concept would be to generate heat by the nuclear explosions themselves in deep salt formations. This would create an underground pool of molten salt that may be exploited for many years. Both schemes have many problems that are difficult to assess without actual experimentation. Not much progress has been made beyond the study stage.

12-4 OPERATIONAL AND ENVIRONMENTAL PROBLEMS

Steam and water from both hydrothermal systems contain, besides the dissolved solids in the water, entrained solid particles and noncondensable gases. The entrained solids must be removed as much as possible, usually by centrifugal separators at the well head, before they enter plant equipment, and by strainers, usually before turbine entry.

The noncondensable gas content varies from 0.2 to 4.0 percent, depending upon the particular well and its age. The younger the well, the higher the percentage, as the noncondensables tend to vent out a bit faster than the H_2O. The noncondensables themselves are mostly CO_2 (about 80 percent) plus varying amounts of methane CH_4, hydrogen H_2, nitrogen N_2, ammonia NH_3, and hydrogen sulfide H_2S. Besides finding their way with the fluid into the plant equipment, the noncondensables also partly escape to the atmosphere via the particle centrifugal separators, the condensor ejectors, and in some cases the cooling towers.

The presence of the noncondensable gases has several effects. First, the large quantity of these gases, relative to noncondensables in conventional steam systems, necessitates the careful design of adequate gas ejectors to maintain vacuum in the condenser. Second, although the presence of acid-forming gases causes no particular problems in dry steam lines that are made of ordinary carbon steels, their corrosive effect in wet conditions necessitates the use of stainless steel in all equipment exposed to wet steam or condensate. Such equipment includes turbine erosion shields and shaft seals, exhaust duct lining, condenser lining, condensate lines and pumps, and metal parts in cooling towers. (Condensers in geothermal plants may be of the direct-contact

* "And he shall judge among the nations, and shall rebuke many people, and they shall beat their swords into plowshares and their spears into pruning hooks. Nation shall not lift sword against nation, neither shall they learn war anymore." (Isiah 2:4)

type, and hence cooling towers are exposed to geothermal condensate.) In the turbine, steam nozzles and blades subjected to dry or high-quality steam are usually made of 11 to 13 percent chrome steel. The nozzles are usually designed with large throat areas on a wide pitch to minimize scaling. Because nickel is particularly sensitive to H_2S corrosion, it is not recommended for use in the rotor. The cooling towers are usually designed with plastic fill and concrete shells, the latter coated with coal-tar epoxy. Aluminum is recommended for condensate pipes and valves that are made large enough to allow low velocities and hence erosion. Aluminum is also recommended for switchyard structures that are in the open but in a generally corrosive atmosphere.

Another effect of H_2S is that it is corrosive to bare copper, particularly in the humid atmosphere around geothermal plants. Unprotected copper is to be avoided in plant electrical equipment that requires special attention. Electrical relays, motor control equipment, excitation gear, switchgear, and others are often kept in "clean rooms" under positive gauge pressure to isolate them from the corrosive atmosphere outside. Static-type exciters, instead of copper-commutator, motor-driven exciters, are used. Auxiliaries are usually motor-driven to avoid the additional corrosion of steam that occurs with turbine drives.

A further effect of the noncondensables is that they are environmentally undesirable because they partly escape into the atmosphere. Most are corrosive in the normally damp atmosphere of the plant site and are noxious and toxic and hence major air pollutants. The most objectionable are H_2S and, to a lesser extent, NH_3.

Another environmental problem caused by geothermal plants is land *surface subsidence*. This occurs because of the extraction of large quantities of underground fluids, though this is partly alleviated by reinjecting the spent brine or condensate into the ground, a procedure widely used in the oil industry. Reinjection also minimizes surface pollution. Large extractions and reinjections also pose the possibility of seismic disturbances.

Noise pollution is another problem. Exhausts, blowdowns, and centrifugal separation are some of the sources of noise that necessitate the installation of silencers on some equipment.

Geopressured water, in addition to the above problems, is thought to carry large quantities of sand, especially at the high flows required. The result is increased erosion and scaling problems.

12-5 VAPOR-DOMINATED SYSTEMS

As indicated previously, vapor-dominated systems are the rarest form of geothermal energy but the most suitable for electricity generation and the most developed of all geothermal systems. They have the lowest cost and the least number of serious problems.

Figures 12-5 and 12-6 show a schematic and *T-s* diagram of a vapor-dominated power system. Dry steam from the well (1) at perhaps 400°F (200°C) is used. It is nearly saturated at the bottom of the well and may have a shut-off pressure up to 500 psia (~35 bar). Pressure drops through the well causes it to slightly superheat at the

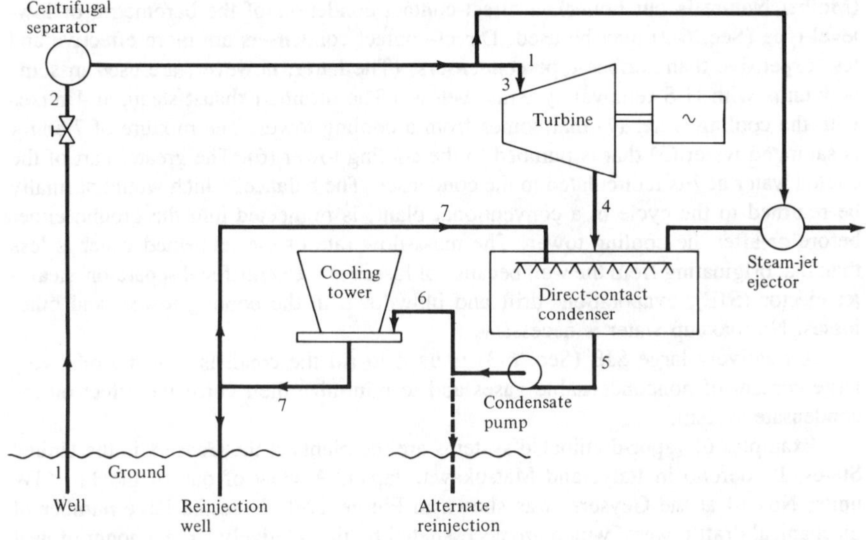

Figure 12-5 Schematic of a vapor-dominated powerplant.

well head (2). The pressure there rarely exceeds 100 psia (~7 bar). It then goes through a centrifugal separation to remove particulate matter and then enters the turbine after an additional pressure drop (3). Processes 1-2 and 2-3 are essentially throttling processes with constant enthalpy. The steam expands through the turbine and enters the condenser at 4.

Because turbine flow is not returned to the cycle but reinjected back into the earth

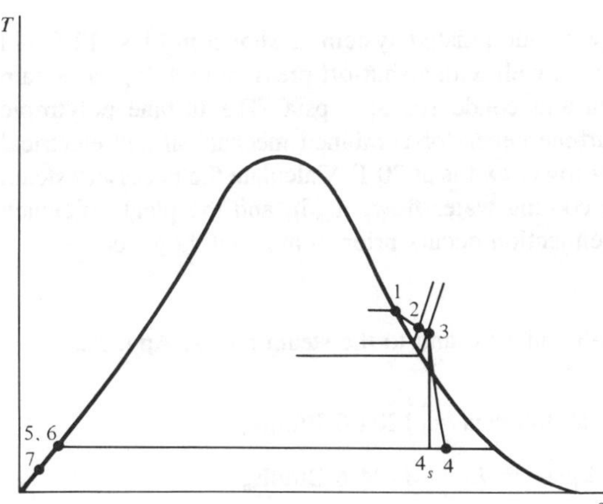

Figure 12-6 *T-s* diagram of the cycle shown in Fig. 12-5.

(Mother Nature is our boiler), a direct-contact condenser of the barometric or low-level type (Sec. 6-2) may be used. Direct-contact condensers are more effective and less expensive than surface-type condensers. (The latter, however, are used in some new units with H_2S removal systems, below.) The turbine exhaust steam at 4 mixes with the cooling water (7) that comes from a cooling tower. The mixture of 7 and 4 is saturated water (5) that is pumped to the cooling tower (6). The greater part of the cooled water at 7 is recirculated to the condenser. The balance, which would normally be returned to the cycle in a conventional plant, is reinjected into the ground either before or after the cooling tower. The mass-flow rate of the reinjected water is less than that originating from the well because of losses in the centrifugal separator, steam-jet ejector (SJE), evaporation, drift and blowdown in the cooling tower, and other losses. No makeup water is necessary.

A relatively large SJE (Sec. 6-3) is used to rid the condenser of the relatively large content of noncondensable gases and to minimize their corrosive effect on the condensate system.

Examples of vapor-dominated systems are the plants at the Geysers in the United States, Larderello in Italy, and Matsukawa, Japan. A view of one of the 110-MW units, No. 14 at the Geysers, was shown in Figure 12-1. Note the large number of mechanical draft towers, which are necessitated by the relatively large amount of heat rejected. Geothermal plants use much lower temperature and pressure steam and higher condenser pressures than conventional plants, and hence they are much lower in efficiency, having heat rates some 2 to 3 times those of the best fossil-fueled plants. Other differences are the large-diameter steam piping as a result of the large specific volume of the low-pressure steam, the large SJE, and because of the low efficiency, large turbines, condensers, and plant auxiliaries. In order to reduce the amount of cooling water needed, and therefore cooling-tower flow, the turbines are usually operated at relatively high back pressures, about 4 inHg absolute (~2 psia, 0.135 bar) or higher.

Example 12-1 A 100-MW vapor-dominated system as shown in Figs. 12-5 and 12-6 uses saturated steam from a well with a shut-off pressure of 400 psia. Steam enters the turbine at 80 psia and condenses at 2 psia. The turbine polytropic efficiency is 0.82 and the turbine-generator combined mechanical and electrical efficiency is 0.9. The cooling-tower exit is at 70°F. Calculate the necessary steam flow, lb_m/h and ft^3/min; the cooling-water flow, lb_m/h; and the plant efficiency and heat rate, Btu/kWh, if reinjection occurs prior to the cooling tower.

SOLUTION Refer to Figs. 12-5 and 12-6 and to the steam tables, App. A.

$$h_1 = h_g \text{ at 400 psia} = 1204.6 \text{ Btu/lb}_m$$

and

$$h_3 \text{ at 80 psia} = h_1 = 1204.6 \text{ Btu/lb}_m$$

Thus

$$T_3 = 350°\text{F (38°F superheat)}$$

$$s_3 = 1.6473 \text{ Btu/(lb}_m \cdot °\text{R)}$$

$$v_3 = 5.801 \text{ ft}^3/\text{lb}_m$$

$$s_{4,s} \text{ at 2 psia} = s_3 = 1.6473 = 0.1750 + x_{4,s}(1.7450)$$

Therefore

$$x_{4,s} = 0.8437$$

$$h_{4,s} = 94.03 + 0.8437(1022.1) = 956.4 \text{ Btu/lb}_m$$

$$\text{Isentropic turbine work} = h_3 - h_{4,s} = 1204.6 - 956.4$$

$$= 248.2 \text{ Btu/lb}_m$$

$$\text{Actual turbine work} = 0.82 \times 248.2 = 203.5 \text{ Btu/lb}_m$$

$$h_4 = 1204.6 - 203.5 = 1001.1 \text{ Btu/lb}_m$$

$$h_{5,6} \text{ (ignoring pump work)} = 94.03 \text{ Btu/lb}_m$$

$$h_7 = h_f \text{ at } 70°\text{F} = 38.05 \text{ Btu/lb}_m$$

$$\text{Turbine steam flow} = \frac{100 \times 3.412 \times 10^6}{203.5 \times 0.9} = 1.863 \times 10^6 \text{ lb}_m/\text{h}$$

$$\text{Turbine volume flow} = \frac{1.863 \times 10^6 \times v_3}{60} = 1.8 \times 10^5 \text{ ft}^3/\text{min}$$

$$\text{Cooling-water flow } \dot{m}_7\text{:} \quad \dot{m}_7(h_5 - h_7) = \dot{m}_4(h_4 - h_5)$$

Therefore

$$\dot{m}_7 = \frac{1001.1 - 94.03}{94.03 - 38.05}\dot{m}_4 = 16.2\dot{m}_4 = 16.2 \times 1.863 \times 10^6$$

$$= 30.187 \times 10^6 \text{ lb}_m/\text{h}$$

$$\text{Heat added} = h_1 - h_6 = 1204.6 - 94.03 = 1110.57 \text{ Btu/lb}_m$$

$$\text{Plant efficiency} = \frac{203.5 \times 0.9}{1110.57} = 0.1649 = 16.49\%$$

$$\text{Plant heat rate} = \frac{3412}{0.1649} = 20{,}690 \text{ Btu/kWh}$$

H_2S Removal

H_2S is found in the Geysers steam at concentrations around 200 ppm. It is toxic, noxious, and poses major air quality problems. Because of recent environmental regulations on its release to the atmosphere, the latest Geysers units (13 through 17) use conventional shell-and-tube surface-type condensers so that the cooling water does not mix with the turbine exhaust until after the noncondensable gases have been removed. This reduces plant efficiency somewhat because surface condensers are less effective than direct-contact ones. In addition, a process called the *Stretford process* is used to remove H_2S. This process was originally developed for the coal industry and is said to achieve more than 90 percent H_2S abatement when used in conjunction with a surface condenser, and commercial-grade sulfur is produced as a by-product [111]. Figure 12-7 shows unit No. 15 of the Geysers, a 59-MW unit that uses the Stretford process (shown to the left in the picture). The system has not always proven satisfactory in operation, however.

Pacific Gas and Electric Co., DOE, and EPRI are testing methods of upstream H_2S abatement. One approach that has proven technically feasible in small-scale experiments operates on steam upstream of the turbine. That steam is cooled and condensed in a vessel at a temperature where H_2S and other undesirable gases do not condense and are removed. The purer water is then reevaporated and sent to the turbine. The loss of heat in the condensing process is reduced by a regenerative-type heat exchanger that is placed so that the incoming steam reevaporates the condensed steam.

Figure 12-7 59-MW unit No. 15 of the Geysers, with H_2S removal by the Stretford method shown to the left. *(Courtesy Pacific Gas and Electric Co.)*

Some loss of availability, however, does occur, thus posing a further penalty on cycle efficiency. Studies have shown this to be of the order of a few percent. The concept is simple, economical, and has the advantage of removing the gases before they reach the turbine. The experiments showed a 94 percent H_2S abatement. A scaled-up pilot unit is planned for the mid-1980s.

12-6 LIQUID-DOMINATED SYSTEMS: FLASHED STEAM

Although the largest geothermal power generation to date (1982) comes from vapor-dominated systems, these systems are rare, and the natural expansion of generation must come from liquid-dominated systems, which are much more abundant, though not so much as geopressured or petrothermal systems. However, as indicated earlier, liquid-dominated systems require the least extension of technology. The known resources show that water is available above 300°F (150°C), with some up to 600°F (315°C). When tapped, the water can flow naturally under its own pressure or be pumped to the surface. The drop in pressure causes it to partially flash into steam and arrive at the well head as a low-quality, i.e., liquid-dominated, two-phase mixture.

The water comes with various degrees of salinity, ranging from 3000 to 280,000 ppm of dissolved solids, and at various temperatures. There are, therefore, various systems for converting liquid-dominated systems into useful work that depend upon these variables. Two methods stand out: (1) the *flashed-steam system,* suitable for water in the higher-temperature range, and covered in this section, and (2) the *binary-cycle system,* suitable for water at moderate temperatures (Sec. 12-7). A third method, called the *total-flow system,* awaits further development (Sec. 12-8).

The Flashed-Steam System

This system, reserved for water in the higher-temperature range, is illustrated by the flow and *T-s* diagrams of Figs. 12-8 and 12-9. Water from the underground reservoir at 1 reaches the well head at 2 at a lower pressure. Process 1-2 is essentially a constant enthalpy throttling process that results in a two-phase mixture of low quality at 2. This is throttled further in a flash separator resulting in a still low but slightly higher quality at 3. This mixture is now separated into dry saturated steam at 4 and saturated brine at 5. The latter is reinjected into the ground.

The dry steam, a small fraction of the total well discharge (because of the low quality at 3), and usually at pressures below 100 psig (8 bar), is expanded in a turbine to 6 and mixed with cooling water in a direct-contact condenser with the mixture at 7 going to a cooling tower in the same fashion as the vapor-dominated system. The balance of the condensate after the cooling water is recirculated to the condenser is reinjected into the ground.

Example 2-2 A flashed-steam system such as that shown in Fig. 12-8 uses a hot-water reservoir that contains water at 460°F and 160 psia. The separator pressure

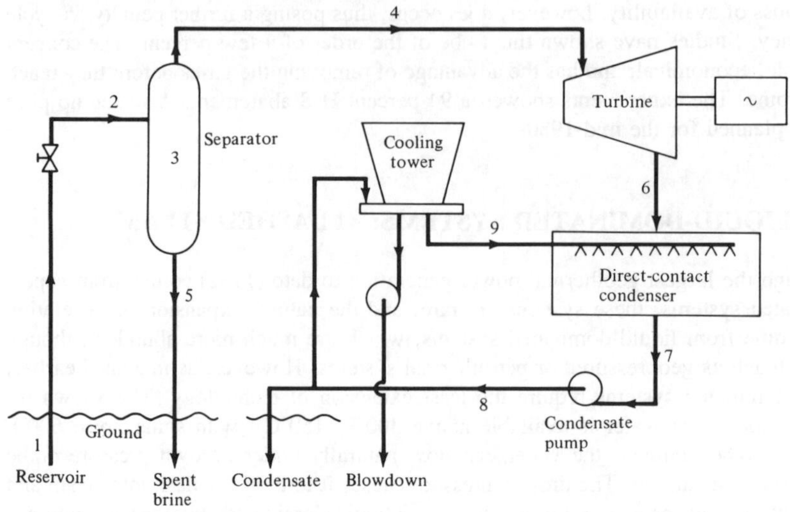

Figure 12-8 Schematic of a liquid-dominated single-flash steam system.

is 100 psia. Find (1) the mass-flow rate of water from the well and of reinjected brine per unit mass-flow rate of steam into the turbine and (2) the ratio of enthalpies of spent brine to steam.

SOLUTION

$$h_1 \approx h_f \text{ at } 460°\text{F} = 441.5 \text{ Btu/lb}_m$$

$$h_3 = h_1 = (h_f + x_3 h_{fg})_{100 \text{ psia}}$$

$$441.5 = 298.5 + x_3(888.6)$$

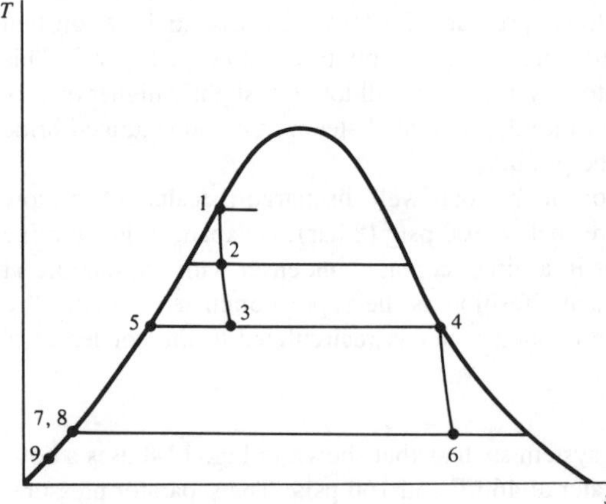

Figure 12-9 *T-s* diagram of the cycle shown in Fig. 12-8.

Therefore

$$x_3 = 0.161$$

1. Mass of water from well per unit mass of steam $= 1/x_3 = 6.21$.
   Mass of reinjected brine per unit mass of steam $= 6.21 - 1 = 5.21$.
2. Ratio of enthalpy at 5 to enthalpy at 4 $= 5.21(h_5/h_4) = 5.21 \times (298.54/1187.2) = 1.31$.

The flashed-steam system is a more difficult proposition than the vapor-dominated system for several reasons: (1) much larger total mass-flow rates through the well, as shown by the above example; (2) a greater degree of ground surface subsidence as a result of such large flows; (3) a greater degree of precipitation of minerals from the brine, resulting in the necessity for design of valves, pumps, separator internals, and other equipment for operation under scaling conditions; and (4) greater corrosion of piping, well casing, and other conduits.

Flashed-steam systems have been widely used in Japan, New Zealand, Italy, Mexico, and elsewhere. An example is the 10-MW Onuma plant in Akita Prefecture in northern Honshu, Japan, which has been in operation since 1973. In this plant steam enters the turbine at 127°C (661°F) and 2.45 bar (35.6 psia). Another is the 75-MW Cerro Prieto plant located in the Mexicali-Imperial Rift Valley in Mexico, 35 km south of the U.S. border. It has been in operation since 1973. Additional units are being built there with a potential between 400 and 1000 MW. In the United States, development has lagged because of the availability of lower-cost energy sources. With the energy crisis, however, activity in this field began with a 10-MW pilot plant built by the Southern California Edison Company and Union Oil Company of California that went into operation in 1981 in Brawley, California. Several plants, in the 20- to 50-MW range are being planned in California, Nevada, New Mexico, and Utah.

Improvements in the Flashed-Steam System

The spent brine leaving the separator at 5 (Fig. 12-8) has a large mass-flow rate and a large total energy compared with that in the steam used to drive the turbine at 4. In Example 12-2, the ratio of the brine enthalpy to the steam enthalpy was found to be 1.31:1. Improvements in the cycle would therefore use some of this otherwise lost energy in the cycle. Two methods are being developed:

1. *Double flash*. Depending upon the original water conditions, the brine at 5 is admitted to a second, lower-pressure separator, where it flashes to a lower-pressure steam that would be admitted to a low-pressure stage in the turbine. The new lower-pressure brine carries less energy with it and represents a reduced energy loss to the cycle. Figures 12-10 and 12-11 show a schematic flow and T-s diagram of a double-flash steam system. The saturated brine from the first-stage flash separator at 5 is reflashed in a second-stage separator at lower pressure to 6. The lower-pressure steam from that separator is admitted to the admission turbine at a lower-pressure stage. The remaining spent brine at 8 is reinjected into the ground. An example of the double-flash system is the 50-MW Hatchobaru plant built on the island of Kyushu in Japan.

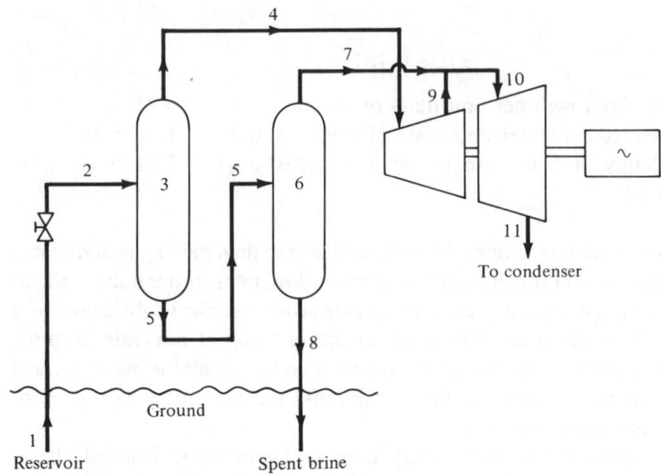

Figure 12-10 Schematic of a liquid-dominated, double-flash steam system. The cycle below the turbine is the same as in Fig. 12-8.

It uses an innovative steam condenser and gas extraction system and a dual-admission, double-flow steam turbine.

2. *Water turbine*. Here the spent brine at 5, still at high pressure, is used instead to drive a water turbine and an additional electric generator operating in parallel with the steam-turbine generator. A variation of this principle, being development under an EPRI contract, uses a so-called *rotary separator turbine* (RST). In this system, the

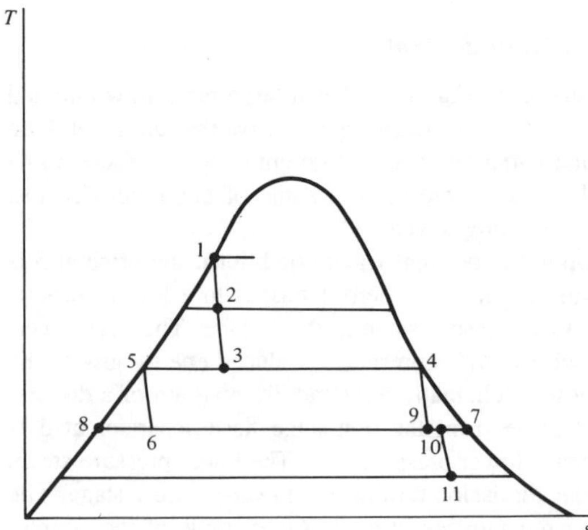

Figure 12-11 *T-s* diagram of the cycle shown in Fig. 12-10.

fluid leaving the well head as a two-phase mixture of steam and water is partially expanded in a nozzle. This increases the steam fraction (as was done in the separator of Fig. 12-8) but also imparts a higher kinetic energy to the denser water, thus facilitating separation of the two phases. This separation takes place in a rotating drum by centrifugal acceleration. The steam is admitted to a steam turbine in the usual fashion. The water at high kinetic energy is used in a special liquid turbine, after which it is reinjected into the ground. A 20-kW unit based on this system is undergoing tests at Roosevelt Hot Springs, Utah. Initial results show a 15 to 20 percent improvement in utilization of the original water energy over a single flashed-steam system. Larger RST units are being developed for possible commercial use in the late 1980s [113].

12-7 LIQUID-DOMINATED SYSTEMS: BINARY CYCLE

About 50 percent of hydrothermal water is in the moderate temperature range of 300 to 400°F (~150 to 205°C).* If used in a flashed-steam system, it would have to be throttled down to such low pressures that result in excessively large specific volume flows as well as even poorer cycle efficiencies. Instead this water is used as a heat source for a closed cycle that uses another working fluid that has suitable pressure-temperature-volume characteristics. This is likely to be an organic with a low boiling point, such as isobutane (2-methyl propane) C_4H_{10} (normal boiling point at one atm pressure: 14°F, −10°C), Freon-12 (normal boiling point: −21.6°F, −29.8°C) (App. B), ammonia (App. C), or propane (App. D). The working fluid would operate at higher pressures, corresponding to the source-water and heat-sink temperatures.

Figure 12-12 shows a schematic flow diagram of a binary-cycle system. Hot water or brine from the underground reservoir circulates through a heat exchanger and is pumped back to the ground. In the heat exchanger it transfers its heat to the organic fluid thus converting it to a superheated vapor that is used in a standard closed Rankine cycle. The vapor drives the turbine and is condensed in a surface condenser; the condensate is pumped back to the heat exchanger. The condenser is cooled by water from a natural source, if available, or a cooling-tower circulation system. The blowdown from the tower may be reinjected to the ground with the cooled brine. Makeup of the cooling-tower water must be provided, however.

In the binary cycle there are no problems of corrosion or scaling in the working cycle components, such as the turbine and condenser. Such problems are confined only to the well casing and the heat exchanger. The heat exchanger is a shell-and-tube unit so that no contact between brine and working fluid takes place.

The first binary cycle was installed in the Soviet Union on the Kamchatka peninsula in 1967. It had a gross output of 680 kW using a low-temperature water reservoir at 80°C (176°F) and Freon-12 as working fluid. The in-plant power consumption (pump-

* Water in lower temperature ranges is unsuitable for power production. It is however suitable for direct utilization for domestic and industrial process heating.

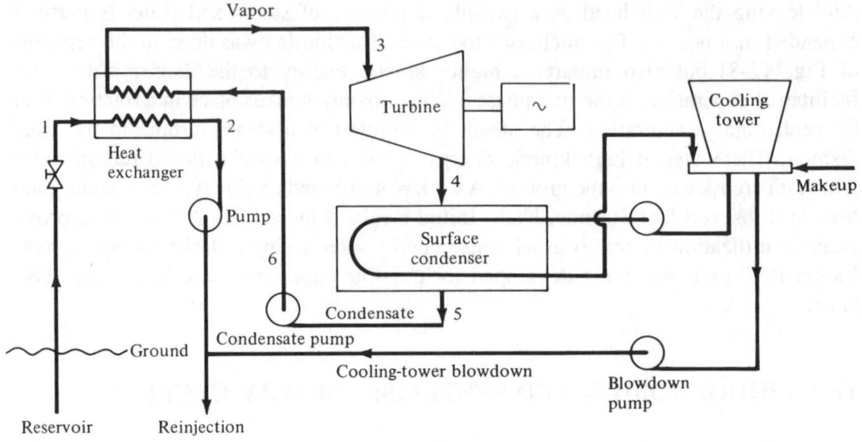

Figure 12-12 Schematic of liquid-dominated binary-cycle system.

ing, etc.) was 35 percent; the net output was 440 kW. The first binary cycle to be built in the United States is an 11-MW plant built by the Magma Company at East Mesa, in the Imperial Valley site in California. The site has a potential of 10,000 MW, but a more conservative projection is about 1500 MW, including a 260-MW unit at East Mesa as well as others at Heber, Westmoreland, Brawley, and Salton Sea.

The $14 million unit at East Mesa has production wells at depths of 5200 to 7500 ft (1585 to 2290 m) which receive 360°F (182°C) pressurized brine to 11 shell-and-tube heat exchangers. The heat exchangers are of the one-pass, counterflow type with very long tubes and no baffles to minimize scaling. The brine leaves the heat exchangers at 180°F and is reinjected into the ground. The brine has 9000 to 10,000 ppm of dissolved solids (mainly calcium chloride and salt), another reason for using a binary cycle. The working fluid is isobutane. A parallel loop using propane is provided. Isobutane was chosen in preference to Freon-12, primarily because of its lower cost. The Magma Company contracted with San Diego Gas and Electric Company to sell it East Mesa power at 25 mill/kWh.

The circulating-water system used in the isobutane condenser consists of two cooling ponds, one of which contains sprays used only at night to take advantage of the cool night air to cool the warm water. The cooled water goes to the other pond, which has sufficient capacity for cooling the plant during daytime.

A second U.S. binary-cycle plant is being built at Raft River, Idaho. It is a 10-MW plant that uses water at 300°F (~150°C). A full-scale $122 million, 45-MW demonstration plant is planned for construction at Heber in the Imperial Valley. It is co-sponsored by EPRI, DOE, San Diego Gas and Electric Company, and various other agencies.

12-8 LIQUID-DOMINATED SYSTEMS: TOTAL-FLOW CONCEPT

In the flashed-steam system, some useful energy is discarded with the separated brine regardless of how many stages of separation are used. Thermodynamically, therefore, direct expansion of the fluid from the well head to the condenser has the potential of converting the greatest fraction of available energy in the fluid to mechanical work. This means that the total well-head flow is to be expanded to the condenser pressure, hence the name *total-flow concept* [114]. In principle this concept is simple, as can be seen from the flow and *T-s* diagrams of Figs. 12-13 and 12-14. Again, hot brine from the well at 1 is throttled to 2, where it becomes a two-phase mixture of low quality. Instead of separating the two phases at this point, the full flow is expanded to 3, condensed to 4, and reinjected into the ground at 5. Comparing the *T-s* diagrams of Figs. 12-9 and 12-14 shows that the throttling process 2-3 that occurs in the flash separator in the former is no longer necessary and that, considering equal pressures at 2, the full available energy at 2 is used in the latter, while part of it is destroyed in the former as a result of throttling. In addition the flow in the flashed-steam turbine and hence the work per unit flow from the well head is only a small fraction, equal to the quality in the flash separator, x_3, of the total flow and work that would occur in the latter.

Flashed-steam systems, and for that matter vapor-dominated systems, rely on axial-flow multistage steam turbines similar to those used in conventional powerplants except that they are designed for much lower pressures. These turbines use relatively clean high-quality or even superheated steam. The total-flow concept, on the other

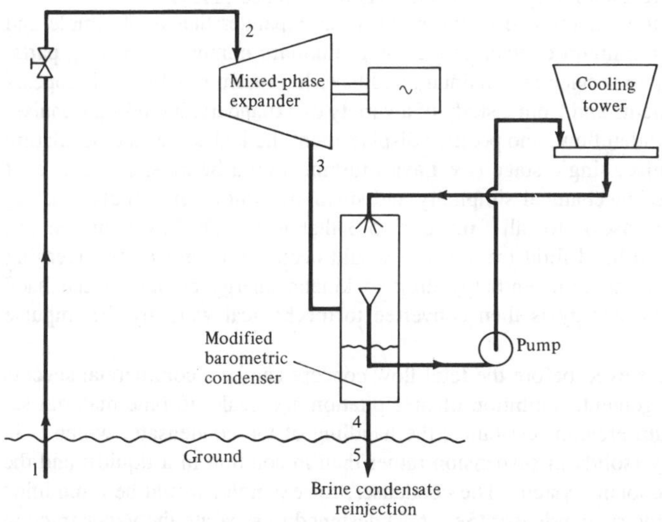

Figure 12-13 Schematic of a liquid-dominated total-flow concept.

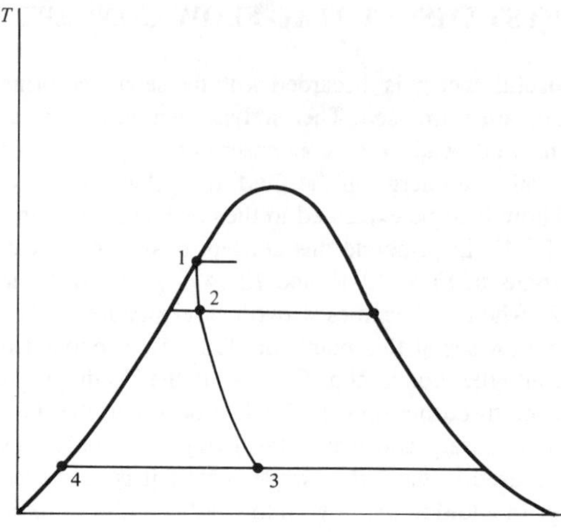

Figure 12-14 *T-s* diagram of cycle in Fig. 12-13.

hand, requires the use of a *mixed-phase expander* powered by a two-phase mixture of low quality (2, Fig. 12-14). Such expanders must be able to overcome the losses associated with the impingement of liquid droplets on blades (turbines operate less efficiently as the quality decreases). They must also be able to withstand the corrosive and erosive effects of the significant quantities of dissolved solids in the brine. Such expanders have not yet been developed, although experimental and analytical work at Lawrence Livermore Laboratory, California, has taken place [113].

The considerations mentioned above point to an expander that is of simple and clean design, with minimum contact surface and minimum number of moving parts. It must also be easy to service and maintain and have long-term reliability. It appears from these requirements and from a study of a variety of expanders, including impulse, reaction, axial and radial flow, and positive displacement (helical screw and oscillating vane), that an impulse, single-stage (De Laval) turbine would be most suitable, as it has the advantage of mechanical simplicity and small size (although impulse staging may work in some cases). Recall from Sec. 5-3 that in the De Laval turbine full expansion of the well-head fluid from 2 to 3 would occur in a converging-diverging nozzle that converts the fluid enthalpy drop to kinetic energy at the turbine back pressure. The kinetic energy is then converted to mechanical work by the impulse blading.

Problems to be solved before the total-flow concept can be a commercial success include brine management; inhibition of precipitation and scale; turbine material selection for maximum erosion resistance; the handling of the condensate, which is in the form of a slurry (solids in suspension rather than in solution in a liquid); and the carryover into the coolant system. The condenser, for example, would be a modified direct-contact barometric condenser (Sec. 6-2) designed to separate the vapor fraction from the brine fraction to prevent fouling of the circulating-water system. No water

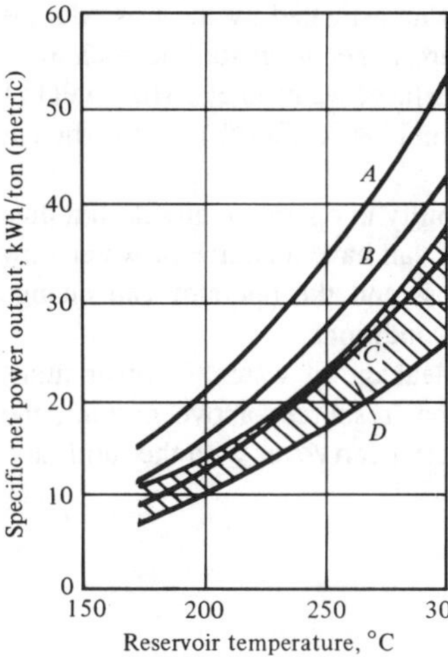

Figure 12-15 Comparison of various liquid-dominated systems: (*A*) total-flow concept; (*B*) two-stage flash; (*C*) single-stage flash; (*D*) binary-cycle range [114].

makeup in the cooling tower is required as that can be supplied from the condensed steam before reinjection into the ground.

Figure 12-15 shows the results of an analytical comparison of various liquid-dominated systems. The curves indicate that the total-flow concept produces the highest specific power per unit mass-flow rate at the well head. The calculations were based on the same condenser temperatures of 120°F, the same turbine efficiencies of 85 percent, and the same in-plant power requirements of 30 percent of the gross. Differences in these would naturally change the differences between the curves.

12-9 PETROTHERMAL SYSTEMS

As indicated earlier (Sec. 12-3), petrothermal systems are those that are composed of hot dry rock (HDR) but no underground water. They represent by far the largest geothermal resource available. The rock, occurring at moderate depths, has very low permeability and needs to be fractured to increase its heat-transfer surface.

The thermal energy of the HDR is extracted by pumping water (or other fluid) through a well that has been drilled to the lower part of the fractured rock. The water moves through the fractures, picking up heat. It then travels up a second well that has been drilled to the upper part of the rock and finally back to the surface. There, it is used in a powerplant to produce electricity.

A feature of this scheme is that, as the reservoir heat is depleted with time, temperature differences within the rock result in stresses that cause the original fractures to propagate, thereby unlocking more HDR surface to the water and resulting in a pancake-shaped fracture zone.

Figure 12-16 shows the petrothermal concept as investigated by the Los Alamos Scientific Laboratory (LASL). It envisages a fracture zone in crystalline rock with bottom pipe at about 7500 ft (~2300 m) below the surface, water in at 1610 psia (111 bar) and 65°C (150°F) and out at 2000 psia (138 bar) and 280°C (540°F), and sufficient flow for a 150-MW heat exchanger.

It is believed that HDR systems offer more flexibility in operation and design than other geothermal systems. For example, the designer can have a choice of water flow rates and temperatures by drilling to various depths, and the operator can change pumping pressure and hence flow rates to suit load conditions.

Problems that are faced by developers include leakage of water (or other fluid) underground and the necessity of makeup for it from resources above ground, the effect of the water or fluid on rock composition, material carryover with the fluid, and

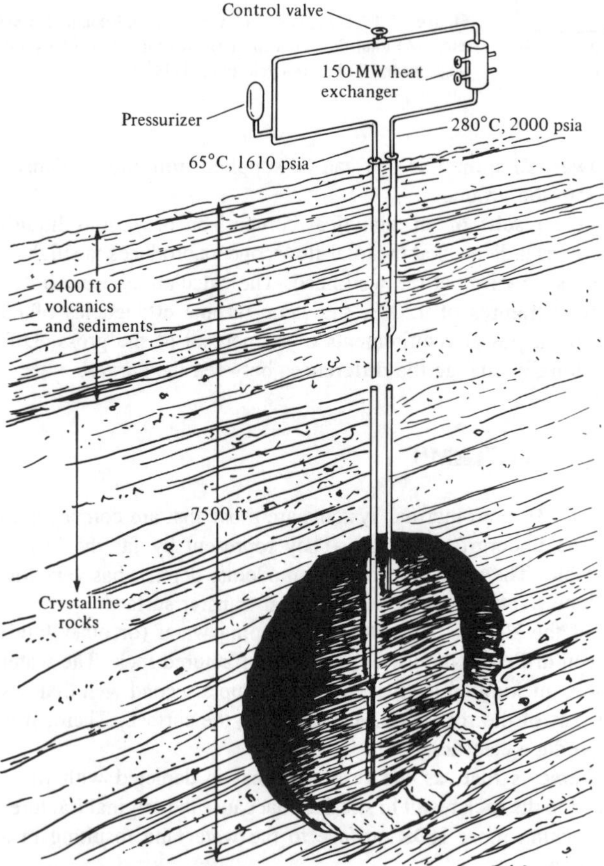

Figure 12-16 Petrothermal energy extraction (LASL).

cost. It should be noted that two wells are to be drilled instead of the one for hydrothermal energy and that these wells are drilled deeper and in much harder rock. This is expected to make petrothermal exploitation very costly, unless the underground rock being developed is very hot.

Many more studies on the mechanical, thermodynamic, and economic aspects of petrothermal systems are necessary before commercial exploitation becomes feasible. So far (1982), LASL has drilled several test wells at a facility near Fenton Hill, New Mexico. Based on these tests, consideration is being given to constructing a 40- to 50-MW plant for start-up, tentatively, in the late 1980s.

12-10 HYBRID GEOTHERMAL-FOSSIL SYSTEMS

The concept of hybrid geothermal-fossil-fuel systems utilizes the relatively low-temperature heat of geothermal sources in the low-temperature end of a conventional cycle and the high-temperature heat from fossil-fuel combustion in the high-temperature end of that cycle. The concept thus combines the high-efficiency of a high-temperature cycle with a natural source of heat for part of the heat addition, thus reducing the consumption of the expensive and nonrenewable fossil fuel.

There are two possible arrangements for hybrid plants [115]. These are (1) *geothermal preheat,* suitable for low-temperature liquid-dominated systems, and (2) *fossil superheat,* suitable for vapor-dominated and high-temperature liquid-dominated systems.

Geothermal-Preheat Hydrid Systems

In these systems the low-temperature geothermal energy is used for feedwater heating of an otherwise conventional fossil-fueled steam plant. Geothermal heat replaces some, or all, of the feedwater heaters, depending upon its temperature. A cycle operating on this principle is illustrated in Fig. 12-17. As shown, geothermal heat heats the feedwater throughout the low-temperature end prior to an open-type deaerating heater (DA) (Sec. 2-8). The DA is followed by a boiler feed pump and three closed-type feedwater heaters with drains cascaded backward (Sec. 2-9). These receive heat from steam bled from higher-pressure stages of the turbine. No steam is bled from the lower-pressure stages because geothermal brine fulfills this function.

Fossil-Superheat Hybrid Systems

In these systems, the vapor-dominated steam, or the vapor obtained from a flash separator in a high-temperature liquid-dominated system, is superheated in a fossil-fired superheater.

Figures 12-18 and 12-19 show schematic flow and *T-s* diagrams of a system proposed in [115]. It comprises a double-flash geothermal steam system. Steam produced at 4 in the first-stage flash separator is preheated from 4 to 5 in a regenerator by exhaust steam from the high-pressure turbine at 7. It is then superheated by a fossil

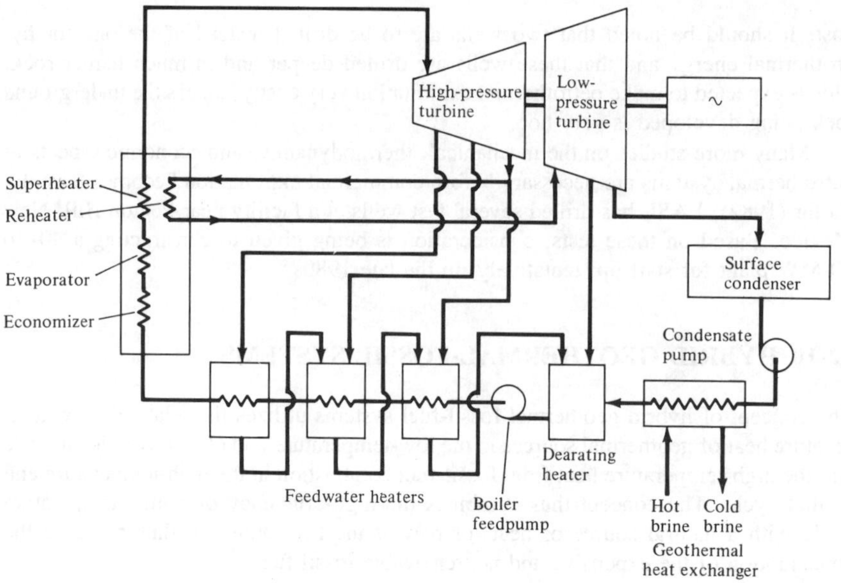

Figure 12-17 Schematic of a geothermal-preheat hybrid system.

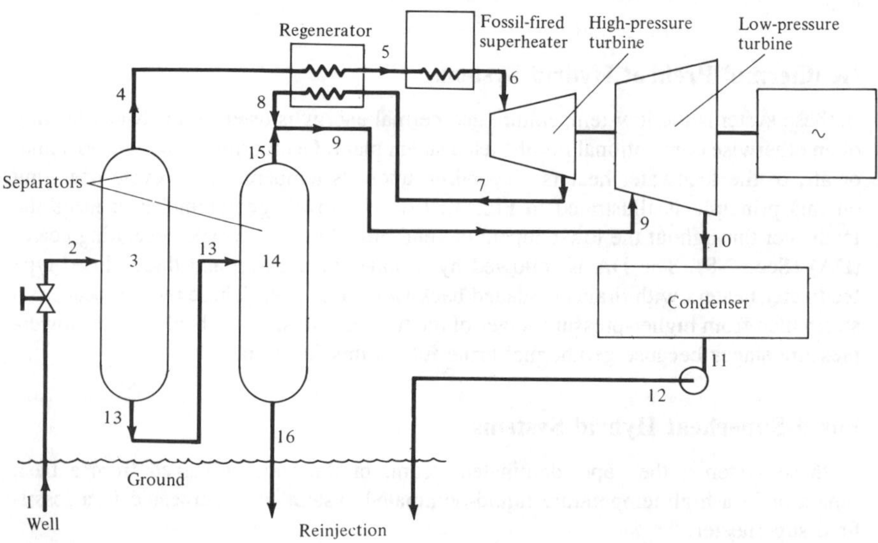

Figure 12-18 Schematic of a fossil-superheat hybrid system with two-stage flash evaporation, regenerator, and fossil-fired superheater.

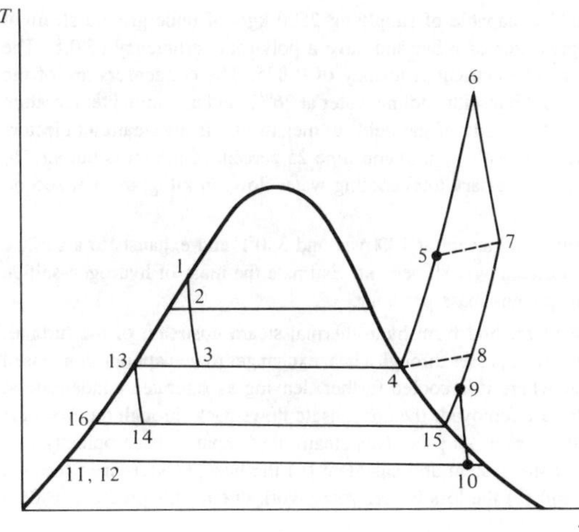

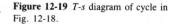

Figure 12-19 T-s diagram of cycle in Fig. 12-18.

fuel–fired superheater to 6 and expands in the high-pressure turbine to 7 at a pressure near that of the second-stage steam separator. It then enters the regenerator, leaves it at 8, where it mixes with the lower-pressure steam produced in the second-stage flash separator at 15, and produces steam at 9, which expands in the lower-pressure turbine to 10. The condensate at 11 is pumped and reinjected into the ground at 12. The spent brine from the second-stage evaporator is also reinjected into the ground at 16.

PROBLEMS

12-1 Using a T-s diagram, (*a*) compare the range of temperatures and/or steam qualities or superheats of liquid-dominated and vapor-dominated hydrothermal systems at ground level, and (*b*) calculate the range of well mass flow rates necessary to produce a unit dry-steam mass flow rate at turbine inlet.

12-2 A geopressured well produces water at the surface at a temperature of 240°F, a salinity of 6 mass percent of pure water, and a methane gas content of 25 ft^3 per barrel (U.S. liquid) of water. The water flow rate is 50,000 bbl/day. Calculate (*a*) the mass of solids to be removed, in pounds mass per day, (*b*) the available thermal power in pure water, in Btus per hour and megawatts, and (*c*) the available chemical power, in Btus per hour and megawatts. Take water specific heat = 1 Btu/lb$_m$ · °R, and methane higher heating value = 23,560 Btu/lb$_m$.

12-3 A 250-MW vapor-dominated hydrothermal powerplant uses well steam that is saturated at 450 psia at shutoff. The steam is throttled to a turbine inlet pressure of 140 psia. A direct-contact condenser operates at a pressure of 5 psia with a cooling-water inlet temperature of 48°F. The turbine polytropic efficiency is 0.80 and the turbine-generator combined mechanical-electrical efficiency is 0.90. Calculate (*a*) the steam mass flow rate, in pounds mass per hour, (*b*) the condenser cooling water mass flow rate, in pounds mass per hour, (*c*) the cycle thermal efficiency and heat rate, in Btus per kilowatt hour, and (*d*) the difference between well flow and disposal flow, in pounds mass per hour, if the cooling tower losses by evaporation and drift are 2 percent.

12-4 A vapor-dominated hydrothermal field is capable of supplying 2500 kg/s of underground steam at 215°C. The turbines operate with throttle pressures of 6 bar and have a polytropic efficiency of 0.8. The turbine generators have a combined mechanical-electrical efficiency of 0.875. The condensers are of the surface type (for H_2S removal) and operate at 0.4 bar with cooling water at 26°C, and a terminal temperature difference of 10°C. Determine (*a*) the total net capacity of the field, in megawatts, if the steam jet ejectors use 1 percent of the steam, and the auxiliaries are estimated to consume 25 percent of the gross output, (*b*) the gross and net plant efficiencies and (*c*) the necessary total cooling water flow, in kilograms per second and cubic meters per second.

12-5 A = 50 MW turbine receives hydrothermal steam at 100 psia and 350°F and exhausts to a surface condenser at 5 psia. The turbine polytropic efficiency is 80 percent. Estimate the mass of hydrogen sulfide that can be removed from the condenser, in pounds mass per hour.

12-6 Consider a hydrogen sulfide abatement method from hydrothermal steam upstream of the turbine. Steam comes into the system at 100 psia and 350°F, passes through a heat exchanger to reevaporate condensed steam, and then to a second heat exchanger where it is cooled further, leaving as saturated condensate at 324°F. After H_2S and other noncondensables are removed, the condensate flows back through the first heat exchanger, where it evaporates to saturated steam at 90 psia. This steam now expands isentropically in a turbine to 5 psia. Draw the flow diagram of the system and calculate (*a*) the heat transferred in the two heat exchangers, in Btus per pound mass, and (*b*) the loss in isentropic work due to this process, in Btus per pound mass.

12-7 A liquid-dominated geothermal plant with a single flash separator receives water at 400°F. The separator pressure is 150 psia. A direct-contact condenser operates at 5 psia. The turbine has a polytropic efficiency of 0.75. For a cycle output of 10 MW, find (*a*) the steam mass flow rate, in pounds mass per hour, (*b*) the well water mass flow rate, in pounds mass per hour, (*c*) the reinjected brine mass flow rate, in pounds mass per hour, (*d*) the cycle efficiency, and (*e*) the cooling water mass flow rate if such water is available at 80°F. Ignore the pump work.

12-8 A liquid-dominated geothermal plant with double flash separators receives water at 440°F. The pressures in the separators are 150 and 80 psia. A direct contact condenser operates at 5 psia. The two turbine sections have polytropic efficiencies of 0.75. For a cycle output of 10 MW calculate, (*a*) the high- and low-pressure steam mass flow rates, in pounds mass per hour, (*b*) the well water mass flow rate, in pounds mass per hour, (*c*) the reinjected brine mass flow rate, in pounds mass per hour, and (*d*) the cycle efficiency. Compare these answers with those of Prob. 12-7, which has the same data but a single flash system. Ignore the pump work.

12-9 A geothermal plant uses a liquid-dominated heat source and a binary cycle. 5×10^6 lb_m/h of underground brine enter a shell-and-tube heat exchanger at 280°F and leave at 110°F to be reinjected into the ground. The working fluid, Freon-12, leaves the heat exchanger as saturated vapor at 230°F. It expands in a turbine that has a polytropic efficiency of 0.70 to a surface condenser at 128.24 psia. The mechanical-electrical efficiency of the turbine-generator is 0.9. The condensate pump has a polytropic efficiency of 0.65. Calculate (*a*) the Freon-12 mass flow rate, in pounds mass per hour, (*b*) the plant power, in megawatts, and (*c*) the plant efficiency and heat rate, in Btus per kilowatt-hour.

12-10 A 10-MW binary cycle geothermal powerplant uses ammonia as working fluid. Brine from the ground enters and leaves the vapor generator at 320°F and 120°F, respectively. Ammonia vapor is generated at 200 psia and 300°F and expands in the turbine to 120 psia with a polytropic efficiency of 0.72. Calculate (*a*) the turbine and net cycle work, in Btus per pounds mass of NH_3, (*b*) the mass flow rate of ammonia, in pounds mass per hour, if the combined mechanical-electrical efficiency of the turbine generator is 0.80, (*c*) the mass flow rate of underground water, in pounds mass per hour, and (*d*) the cycle efficiency.

12-11 A hot-water geothermal plant of the total-flow type receives water at 440°F. The pressure at turbine inlet is 150 psia. The plant uses a direct contact condenser that operates at 5.0 psia. The turbine has a polytropic efficiency of 0.65. For a cycle net output of 10 MW, calculate (*a*) the hot water flow, in pounds mass per hour, (*b*) the condenser cooling water flow, in pounds mass per hour, if such water is available at 80°F, and (c) the cycle efficiency. Ignore the pump work.

12-12 A hybrid geothermal-fossil powerplant of the type shown schematically in Fig. 12-18 receives underground water at 440°F. The pressures in the separators are 150 and 80 psia. A direct-contact condenser operates at 5 psia. The regenerator is 80 percent effective. The high pressure steam is superheated in the fossil-fired superheater to 500°F. Because of operations with superheated steam, the high- and low-pressure turbine sections have polytropic efficiencies of 0.88 and 0.85, respectively. For a cycle output of 10 MW, calculate (*a*) the well water mass flow rate, in pounds mass per hour, (*b*) the high- and low-pressure steam mass flow rates, in pounds mass per hour, (*c*) the heat added in the superheater, in Btus per hour, and (*d*) the cycle efficiency. Ignore the pump work.

CHAPTER

THIRTEEN

SOLAR ENERGY

13-1 INTRODUCTION

Of all the renewable energy sources, solar energy received the greatest attention in the decade of the 1970s and has been the hub of much emotion and pressure. Many regarded it as *the* solution for reducing the use of fossil and nuclear fuels and for a cleaner environment. Solar energy, as a result, has been the object of usually inflated, overly optimistic predictions ranging from largely supplementing to eventually replacing all the current means of production of both electric-power and thermal-energy requirements.

Solar energy, in sheer size, does have the potential to supply all energy needs: electric, thermal, process, and chemical, and even transportation fuels. It is, however, very diffuse, cyclic, and often undependable. It, therefore, needs systems and components that can gather and concentrate it efficiently for conversion to any of these uses and that can do the conversion as efficiently as possible. Much of the construction materials used for collecting and converting solar radiation, whether thermal or photovoltaic, are themselves very energy-intensive. For example, large quantities of aluminum, steel, copper, concrete, glass, and plastic are needed, all of which require large quantities of energy for conversion from ore to finished products. Direct conversion of solar radiation to electricity (photovoltaic) requires the extremely energy-intensive processing of sand to crystalline silicon. Both systems of converting the collected radiation to usable energy (solar thermal and photovoltaic) are rather inefficient, making them large per unit energy output and hence exacerbating the need for the materials mentioned above.

A rational assessment of the role of solar energy in the total energy picture would have to take into account both the costs and the real, rather than the imaginary, environmental impact of that energy.

The costs are of two kinds: (1) the monetary cost of the finished products and of assembling them into workable systems and (2) the quantities of conventional energy sources such as coal, oil, natural gas, and nuclear fuel that have to be consumed in the processes of producing the above finished products and in constructing the systems. It is logical that one should expect the energy thus consumed to be a small fraction of the net energy generated during the life of the plant (a truism for any power system). It seems that this fraction is larger for solar systems than fossil or nuclear systems, often bordering on the unacceptable, except in special cases where solar energy is uniquely suited.

On the environmental question, solar-energy systems are not pollution-free as many of its proponents claim. (No power system, and for that matter, no human endeavor to improve the quality of life, is. The real criterion for success is whether the gains sufficiently outweigh the losses.) The mining of large quantities of mineral ores, the processing of these ores to the finished material, the manufacturing of the equipment, and the construction of the plants are processes that generate their own pollution and health and occupational safety risks. In addition, the low efficiency of electric-generating plants, whether thermal or photovoltaic, results in large heat-rejection rates in a concentrated manner (even though the energy input is diffuse) from the plants, hence thermal pollution.

Although solar energy may be used in many markets, such as in active and passive space heating and cooling, industrial process heating, desalination and water heating [116], and in electric generation, it is the latter we are concerned with in this book. After investments by federal and other agencies amounting to several hundred million dollars in the 1970s on all these technologies, only one is in commercial use today (1982), namely flat-plate collectors for space and water heating. All the others remain in various stages of research and development. It now appears that solar-electric systems are not expected to make engineering and economic sense as central electric-generating plants of hundreds of megawatts capacities (the economical sizes needed now by the world) in the foreseeable future. They do, however, make sense in cases where the power requirements are relatively small and in remote areas, both on earth and in outer space, that lack proximity to an efficient electricity-distribution network. On earth, they are most particularly suited to remote desert areas where the land is available, relatively inexpensive, and adequate for the very large collector surfaces required, and where the sunshine, although cyclic, is abundant and dependable.

13-2 HISTORY

Solar energy's place in the total needs of the human race is unquestionable. It grows all our crops and creates the fresh water for these crops and for our survival. It has, in addition, been used for millenia for passive heating of dwellings, as can be ascertained by examination of archaeological sites around the world. Serious experiments to use it to generate energy for specialized uses began, however, in the eighteenth century. In 1774 Joseph Priestly concentrated it on mercuric oxide. The result was the generation of a gas that was found "to cause a candle to burn brightly and a mouse to live longer" which he called "air in a much greater perfection." This experiment, in reality, had led to the discovery of oxygen. It also led the French chemist, Lavoisier, to explain combustion as the chemical combination with oxygen.

A century later, a relatively large solar distilling plant was installed in a desert in northern Chile. It used 4800 m^2 of land and contained slanted roofs of glass plates that transmitted the sun's rays to troughs of water below them. The vaporized water condensed on the cooler underside of the glass, ran down, and was collected in channels. The plant produced some 25 m^3 of fresh water per day, which was used in a nitrate mine. It ran for 40 years until the mine was exhausted.

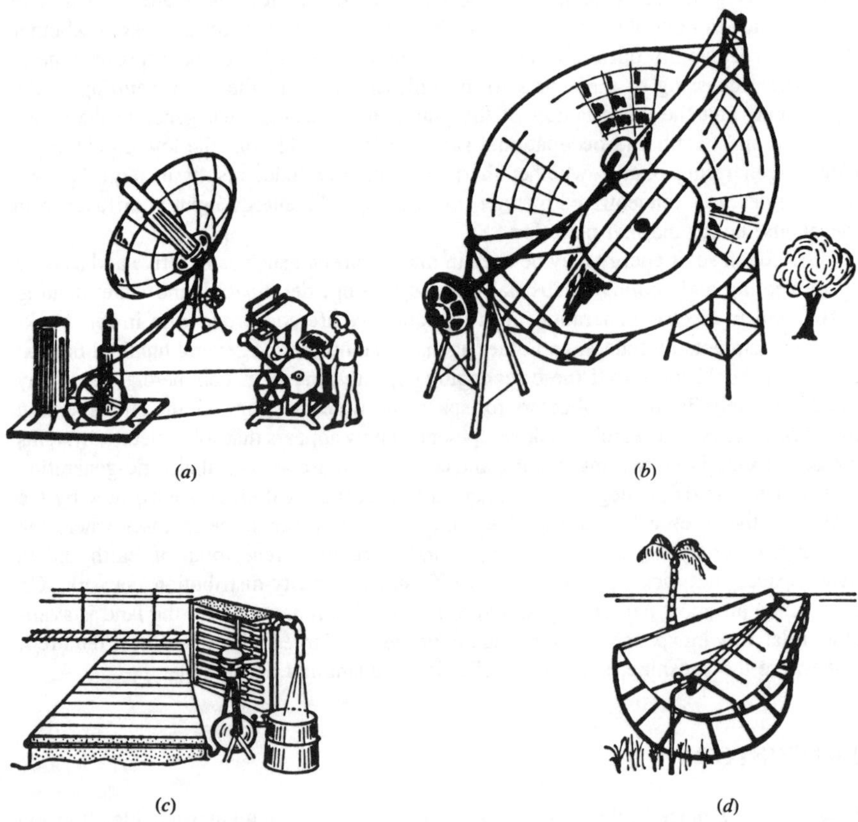

Figure 13-1 Some early examples of solar-power conversion to mechanical power. (*a*) Solar steam engine, Paris, 1878. (*b*) Solar steam engine, Pasadena, California, 1901. (*c*) Solar steam engine with flat-plate collector, Philadelphia, Pennsylvania, 1907, 1911. (*d*) Solar steam engine, Cairo, Egypt, 1913.

Solar-to-mechanical conversion was first demonstrated at an exhibition in Paris in 1878 when sunlight was concentrated by a focusing collector on a steam boiler that ran a small steam engine that in turn ran a small printing press. In 1901, a larger (6 m^2) focusing collector in the form of a truncated cone generated steam for a 4.5-hp engine. In 1907 and 1911, near Philadelphia, F. Shuman built solar steam engines of several horsepower that were used for pumping water. In 1913, in collaboration with C. V. Boys, Shuman built the then large 50-hp solar steam engine, which used a long parabolic collector that focused solar radiation onto a central pipe (with a concentration ratio of 4.5). The engine pumped irrigation water from the Nile, near Cairo. (The four systems mentioned above are shown in Fig. 13-1.)

An early attempt at solar-electric conversion, as well as energy storage, was made by J. A. Harrington in New Mexico around 1915. Sunlight was focused on a boiler that ran a steam engine that pumped water into a 19-m^3 tank 6 m above. The water thus stored was made to run down into a water turbine that powered an electric generator that lighted small electric bulbs inside a small mine.

These and other activities could not, however, survive the competition of cheap fossil fuels, and little activity in solar development took place for 30 years or so. Renewed interest began in the 1950s as a result of the efforts of a few such as Farrington Daniels of Wisconsin, an interest that was accelerated by the energy crisis of the 1970s.

13-3 EXTRATERRESTRIAL SOLAR RADIATION

The total quantity of solar energy incident upon the earth is immense, but the energy is very diffuse and, because of the earth's rotation and orbit around the sun, cyclic both daily and seasonally. It also suffers from atmospheric interference from clouds, particulate matter, gases, etc. In this section we will deal with the energy incident on the earth outside of its atmosphere. This is called *extraterrestrial radiation.*

The earth rotates around the sun on a slightly elliptical orbit with major and minor axes differing by 1.7 percent. The earth is closest to the sun on December 21 at a distance of about 1.45×10^{11} m, and farthest on June 22 at about 1.54×10^{11} m; the average distance is 1.49×10^{11} m (Fig. 13-2). The sun has a diameter of about 1.39×10^9 m and subtends an angle of only 32 minutes at the earth. For all practical purposes, therefore, the sun's rays may be considered parallel when they reach the earth. The sun has an effective black-body temperature, as seen from the earth, of 5762 K.* The spectral distribution of solar energy at the mean sun-earth distance outside the earth's atmosphere is shown in Fig. 13-3 by curve *A* as beam irradiance flux per unit wavelength width in $W/(s^2 \cdot \mu m)$ versus wavelength in μm (micrometers). As seen, it does not follow a smooth black-body spectral distribution curve. This spatial relationship between sun and earth and the sun's effective temperature result

* This is the temperature of a black body (a perfect radiator with both emissivity and absorptivity equal to 1.0) that radiates the same amount of energy as the sun. The sun's interior is much hotter and denser than its surface. At its center, the temperature is estimated at 8×10^6 to 40×10^6 K and the density at about 10^5 kg/m^3.

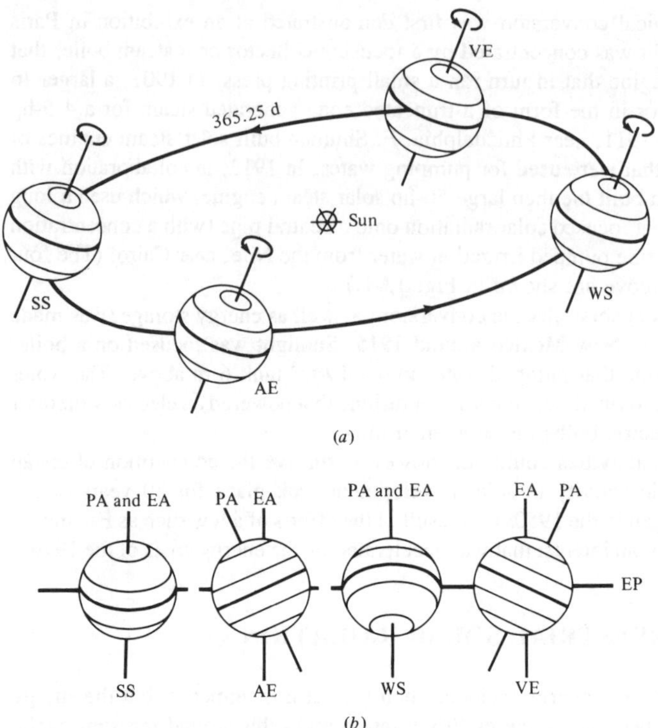

Figure 13-2 Orientation of earth and sun. (*a*) Earth's orbit around sun; SS = summer soltice, June 22; AE = autumnal equinox, September 23; WS = winter soltice, December 22; VE = Vernal (spring) equinox, March 21. (*b*) The earth as seen from the sun; PA = polar axis; EA = ecliptic axis; EP = ecliptic plane. Circles from north to south: Arctic circle (66.5° N. lat.), tropic of capricorn (23.5° N. lat.), equator, tropic of cancer (23.5° S. lat.), antarctic circle (66.5° S. lat.).

in extraterrestrial solar radiation intensity that is nearly constant and is called the *solar constant S*. It is equal to the area under curve *A* in Fig. 13-3 and has the values

$$S = 1353 \text{ W/m}^2 = 1.353 \text{ kW/m}^2$$

$$S = 1.940 \text{ langley*/min}$$

$$S = 428 \text{ Btu/h} \cdot \text{ft}^2$$

$$S = 4871 \text{ kJ/(h} \cdot \text{m}^2)$$

The slightly elliptical orbit of the earth around the sun causes the actual extraterrestrial intensity to deviate only slightly from the solar constant with a range roughly ±3

* The *langley* is a unit frequently used in solar engineering. It is equal to 1.0 cal/cm^2.

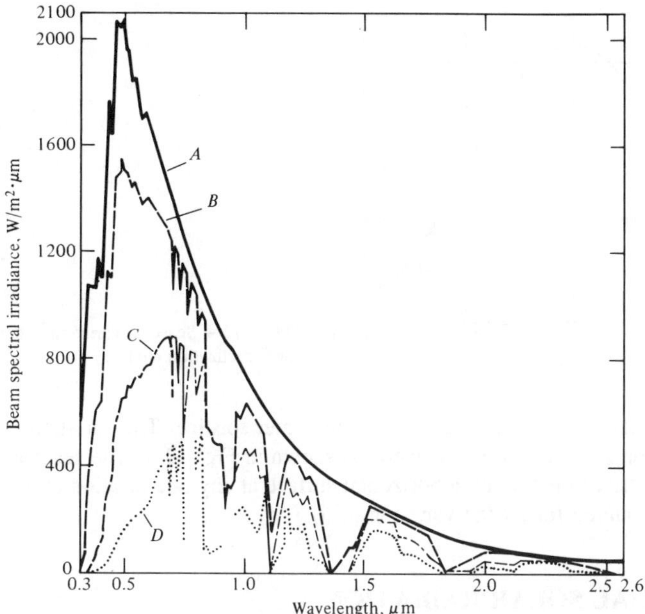

Figure 13-3 Solar-energy beam radiation as a function of wavelength (*A*) extraterrestrial with air mass m_a = 0 irradiance, I = 1353 W/m^2; (*B*) terrestrial, m_a = 1.0, I = 924.9 W/m^2; (*C*) terrestrial, m_a = 4, I = 528.9 W/m^2; (*D*) terrestrial, m_a = 10, I = 234.5 W/m^2. Curve *A* according to NASA/ASTM standards. Curves *B*, *C*, *D* and computed for 20-mm precipitate water vapor, 3.4 mm ozone, and clear atmosphere [117].

percent. Other small variations occur with different periodicities and as a result of sunspots.

Extraterrestrial solar radiation is all of the *beam-radiation* type, also called *direct radiation*. This is radiation received from the sun in essentially straight rays or beams that are unscattered by the atmosphere.

The earth's *polar axis* (the axis of rotation of the earth) is inclined a permanent 23.45° from a normal to the *ecliptic plane* (EP), the plane of the earth's orbit around the sun (Fig. 13-2*b*). The *ecliptic axis* (EA) of the earth is the one that goes through its center but which is always perpendicular to the ecliptic plane. The polar and ecliptic axes are in a plane normal to the sun's rays at the two equinoxes and in a plane parallel to them at the two solstices. Thus, although the earth's inclination is fixed in space, it appears to wobble around the sun with its northern hemisphere inclined toward it in the summer and away from it in the winter, with both hemispheres "seeing" it equally only at the equinoxes.

The angle between the sun's rays and the earth's equatorial plane (normal to the polar axis) is called the sun's *declination angle* θ_d. In the northern hemisphere θ_d is zero at both equinoxes and has a maximum value that corresponds to the tilt angle

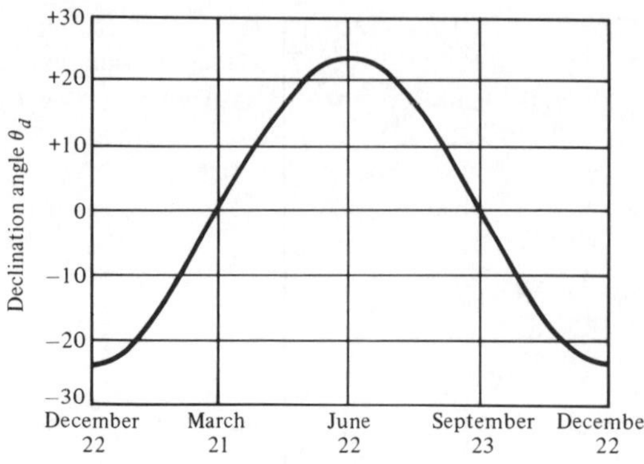

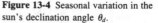

Figure 13-4 Seasonal variation in the sun's declination angle θ_d.

+23.45° at the summer solstice and −23.45° at the winter solstice. The variation of θ_d during the year, for the northern hemisphere, is shown in Fig. 13-4. The seasonal variation of the terrestrial radiation on a horizontal surface at any one location on the earth's surface is accounted for by the variation in θ_d.

13-4 TERRESTRIAL SOLAR RADIATION

The solar energy falling on the earth's surface is called *terrestrial radiation*. The rate of terrestrial energy falling on a unit surface area in W/m^2, $J/(s \cdot m^2)$, or $Btu/(h \cdot ft^2)$ is variably referred to as radiation, irradiation, irradiance, insolation, or energy flux. Terrestrial radiation varies significantly, both daily because of the earth's rotation and seasonally because of the change in the sun's declination angle. During both the vernal and autumnal equinox, the sun's rays are perpendicular to the earth's surface at the equator and cause equal radiation intensities in the northern and southern hemispheres. At the summer solstice the earth's axis is tilted towards the sun so that its northern hemisphere receives more radiation than the southern hemisphere. At the winter solstice, the reverse is true, with local solar intensities strongly dependent upon latitude. Besides variations caused by the spatial changes between the sun and earth, terrestrial radiation is subject to reductions and changes due to the presence of various gases, vapors, and particulate matter in the earth's atmosphere. The extraterrestrial radiation is said to be *attenuated* by two mechanisms: scattering and absorption.

Scattering is a mechanism by which part of a radiation beam is scattered laterally and is, therefore, attenuated by the air molecules, water vapor, and the dust in the atmosphere. The mechanism is dependent upon the type of scattering medium and the wavelength of radiation and is a rather complex phenomenon. It is known, however, that scattered, or diffuse, radiation is mostly of shorter wavelengths, which is the reason the sky appears blue.

Absorption of solar radiation in the atmosphere is mainly by ozone O_3, water vapor H_2O, and carbon dioxide CO_2. X-ray and other very short wavelength radiations of the sun are absorbed high in the ionosphere by N_2, O_2, and other components. The main effects are caused by: (1) Ozone in the ultraviolet (short wavelength) rays with absorption complete below 0.29 μm and decreasing between 0.29 and 0.35 μm, where it ceases except for a weak absorption band near 0.6 μm. (2) Water vapor with absorption strong in wavelength bands in the infrared (long wavelength) part of the spectrum, at 1.0, 1.4, and 1.8 μm. (3) Carbon dioxide with no absorption bands in the short or visible parts of the spectrum but several between 2.36 to 3.02 μm, 4.01 to 4.08 μm, and 12.5 to 16.5 μm. The extraterrestrial radiation is very low in that region (less than 5 percent of the total spectrum) so that absorption by both H_2O and CO_2 causes the energy received at wavelengths greater than 2.3 μm at the earth's surface to be insignificant.* For terrestrial applications, therefore, only the wavelength range 0.29 to 2.5 μm is significant.

The *terrestrial solar radiation*, that incident on the earth's surface, is composed of two parts: (1) *beam radiation*, defined above, and (2) *diffuse radiation*. The latter is radiation that had its direction changed by atmospheric scattering. It is sometimes called *sky radiation* because it is the radiation one feels if standing in the shade or under a small cloud on an otherwise sunny day. (It should not, however, be confused with radiation emitted by the atmosphere itself).

The incident radiation on the earth's surface is usually presented in terms of dimensionless *air mass* m_a, defined as

m_a = air mass = ratio of optical thickness of the atmosphere through which beam radiation passes to the surface to its optical thickness if the sun were at the zenith, i.e., directly above (dimensionless)

Thus, $m_a = 0$ means extraterrestrial, $m_a = 1$ indicates sea level on the earth when the sun is at the zenith, $m_a = 2$ when the sun is at a zenith angle θ_z equal to 60°. θ_z is the angle subtended by the solar beam and the zenith. The air mass m_a is related to zenith angles from $\theta_z = 0°$ to $\theta_z = 70°$, at sea level, by

$$m_a = (\cos \theta_z)^{-1} \tag{13-1}$$

For values of θ_z greater than 70°, the curvature of the earth becomes significant and m_a becomes progressively smaller than that given by Eq. (13-1).

Figure 13-3 shows the spectral distribution of extraterrestrial *beam radiation* ($m_a = 0$) as well as terrestrial beam radiations for air masses $m_a = 1$, 4, and 10. The spectral distribution of the *total radiation* is the sum of *both* the beam and diffuse components. Measurements show that the diffuse component distribution is similar to that of the total, though shifted somewhat toward the short wavelengths. (Recall that scattering occurs most at short wavelengths.) This suggests that the spectral-energy distribution from an overcast sky is similar to that from a clear sky.

Considering a horizontal surface on a clear day, both beam and diffuse radiation depend upon the hour of the day. Both are maximum around noon and both decrease

* The CO_2 infrared absorptions are, however, significant when one considers low-temperature radiation from the earth back to space, which typically covers a spectral range of 3 to 50 μm with a peak near 10 μm. An increase in CO_2 concentrations in the atmosphere are cause for concern because of the "greenhouse effect" (Sec. 17-6).

toward dusk, but beam radiation decreases at a faster rate. Thus the ratio of diffuse-to-beam radiation also depends upon the hour of the day, as it is very low around noon and increases slowly toward unity and slightly exceeds it at dusk. Table 13-1 shows calculated beam and diffuse radiations for a clear August day in the Midwest (Madison, Wisconsin), including beam radiation on a surface normal to the radiation, i.e., one that tracks the sun, $I_{b,n}$; beam radiation on a horizontal surface $I_{b,h}$; diffuse radiation I_d; and cumulative radiation on a horizontal surface $I_{c,h}$. All are in MJ/m^2 for 1-h periods as shown. The totals for the day would be twice the sums of the columns. Thus the day total beam radiation on a normal surface is $2 \times 15.38 = 30.76$ MJ/m^2 per day. Likewise the day total beam radiation, diffuse, and cumulative on a horizontal surface would be 18.82, 3.74, and 22.56 MJ/m^2 per day, respectively. The ratio of diffuse to beam radiation and their total are given only for a horizontal surface. The reason is that for a surface normal to the beam radiation, diffuse radiation is affected by the decrease of the "total" sky seen by the surface as well as reflectance effects from the ground.

The computations given in Table 13-1 dealt with a clear day. The available terrestrial solar energy at a given time and place is influenced not only by time of day or year, location, and scattering as mentioned above but also by cloudiness. All effects may be combined in one parameter called the *clearness index* C_i, defined as

C_i = clearness index = ratio of the average radiation on a horizontal surface for a given period to the average extraterrestrial radiation for the same period

The averaging could be monthly, daily, or hourly, in which case C_i would be a monthly, daily, or hourly clearness index. C_i varies widely from near 30 to as high as 70 percent in some localities on earth, with its value going down to zero in some locations because of bad weather even in the daytime.

Table 13-2 shows measured cumulative daily radiations on a horizontal surface in selected locations in the United States.

The total radiation received by a surface however is undoubtedly greater from a

Table 13-1 Beam and diffuse radiations for a clear August day in Madison, Wisconsin*

Hour, p.m.	$I_{b,n}$	$I_{b,h}$	I_d	$I_d/I_{b,h}$	$I_{c,h}$
12–1	2.96	2.50	0.36	0.14	2.86
1–2	2.90	2.31	0.35	0.05	2.66
2–3	2.77	1.95	0.34	0.17	2.29
3–4	2.53	1.44	0.31	0.22	1.75
4–5	2.12	0.87	0.27	0.31	1.14
5–6	1.38	0.31	0.20	0.65	0.51
6–7	0.72	0.03	0.04	1.33	0.07

*From Ref. 116.

Table 13-2 Average cumulative daily solar radiation on a horizontal surface*

Location	Latitude	Radiation, MJ/m² per day											
		Jan.	Feb.	Mar.	Apr.	May	June	July	Aug.	Sept.	Oct.	Nov.	Dec.
Albuquerque, N. Mex.	33.93	11.5	15.2	20.1	25.3	28.8	30.4	28.2	26.0	22.4	17.6	12.9	10.9
Chicago, Ill.	41.98	5.8	8.6	12.6	16.6	20.3	22.8	22.1	19.5	15.4	11.0	6.4	4.6
Fairbanks, Alaska	64.82	0.8	3.2	9.8	16.1	20.0	22.1	18.6	15.2	7.7	3.6	1.1	0.3
Los Angeles, Calif.	33.93	10.5	13.8	18.4	22.1	23.4	24.0	26.2	23.6	19.1	14.9	11.4	9.6
Phoenix, Ariz.	33.43	11.6	11.5	20.6	26.7	30.4	31.1	28.2	26.0	22.9	17.9	13.1	10.6
Madison, Wis.	43.13	5.8	5.7	12.9	15.9	19.8	22.1	22.0	19.4	14.7	10.3	5.7	4.4
San Francisco, Calif.	37.78	11.5	15.6	16.5	21.8	21.8		27.1	24.0	19.8	13.9	9.3	7.3
Seattle, Wash.	47.45	5.7	9.1	11.1	16.6	21.0	21.8	23.7	19.1	13.7	7.9	4.4	2.7

*From Ref. 116.

surface that is normal to beam radiations, as can be seen when comparing columns 2 and 6 of Table 13-1, even though diffuse radiation is not included in the former. This is the reason why a radiation "collecting" surface would be more effective if it were made to track the sun, i.e., to change angle, so that it would always remain perpendicular to the rays emanating from it. We will see later, however, that in a central-receiver system, the tracking is such that the collecting surface should alter its angle so that it reflects solar radiation to a fixed in-place central receiver.

The total amount of solar radiation received by the earth is, as indicated earlier, immense. Because the sun's rays are essentially parallel and hence perpendicular to the earth's projected area, the extraterrestrial power P_e received is given by the solar constant S times the projected area of the earth, or

$$P_e = S\pi R^2 \tag{13-2}$$

The radius of the earth R is 6.378×10^6 m and $S = 1.353$ kW/m^2, so that

$$P_e = 1.353 \times \pi(6.378 \times 10^6)^2 = 1.73 \times 10^{14} \text{ kW}$$

$$\text{Energy per year} = 1.73 \times 10^{14} \times 8766 \text{ h/year} = 1.516 \times 10^{18} \text{ kWh/year}$$

$$= 1.516 \times 10^{18} \times 3.6 = 5.457 \times 10^{18} \text{ MJ/year}$$

The continental United States land area is about 2.885×10^6 mi^2, about 7.5×10^{12} m^2. It thus receives extraterrestrial radiation at the rate of $7.5 \times 10^{12} \times 1.353 = 10^{13}$

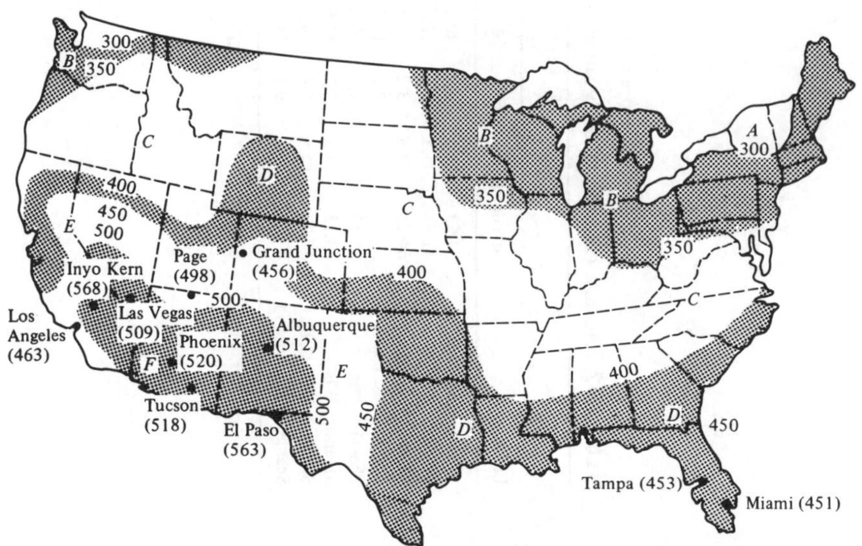

Figure 13-5 Annual mean daily solar radiation in the United States, areas $A < 300$, $B > 300$, $C > 350$, $D > 400$, $E > 450$, $F > 500$ langley/day. (1 langley = 1 cal/cm^2. 1 langley/day = 4.846×10^{-4} kW/m^2 = 0.01163 kWh/m^2 per day.)

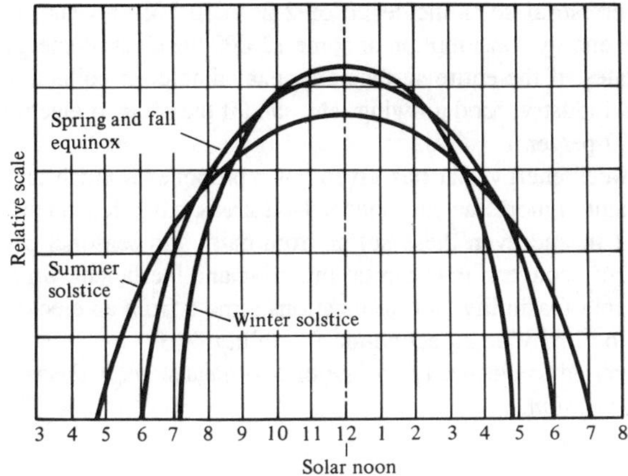

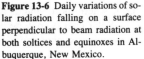

Figure 13-6 Daily variations of solar radiation falling on a surface perpendicular to beam radiation at both soltices and equinoxes in Albuquerque, New Mexico.

kW when the sun is shining. For a yearly average sunshine of 12 h/day, the total yearly extraterrestrial radiation would be $10^{13} \times 8766/2 = 4.43 \times 10^{16}$ kWh/year. Assuming an average clearness index C_i of 50 percent, the terrestrial radiation received by the continental United States would be about 2.2×10^{16} kWh/year. Figure 13-5 shows the annual terrestrial mean daily solar radiation falling on the United States in langleys (1 langley $= 1$ cal/cm^2 $= 41.87$ kJ/m^2 $= 0.0116$ kWh/m^2 $= 3.69$ Btu/ft^2). Figure 13-6 shows an example of solar radiation falling on a surface normal to the sun's rays in Albuquerque, New Mexico, versus time of day based on the mean sun time* at the two solstices and either equinox. It is interesting to note that during the winter solstice, the maximum radiation is highest because the sky is clearest though it does not last as long during the day.

13-5 SOLAR-ELECTRIC CONVERSION SYSTEMS

The yearly consumption of all forms of energy in the continental United States in the early 1980s was 78×10^{15} Btu, or 2.286×10^{13} kWh, of which about 25 percent or 20×10^{15} Btu, or 5.86×10^{12} kWh, was used to generate electricity. At an average conversion efficiency of about 30 to 32 percent, the total electric-power consumption was about 1.8×10^{12} kWh/year.

* The *mean sun time* (MST) on earth is calculated from the *local standard time* (LST) for each longitude. The 360° of the earth are divided into twenty-four 15°-wide geographical time zones (artificially modified by political boundaries). An imaginary longitudinal line runs through the center of each time zone. It is called the *standard meridian* of that zone. (In the United States the standard meridians are located at longitudes 75° for eastern standard time, 90° for central standard time, 105° for mountain standard time, and 120° for Pacific standard time.) At each meridian the local standard time and the mean sun time are identical. At points within a zone east or west of its meridian, the mean sun time is later or earlier, respectively, than the local standard time. Because the earth rotates 360° in 24 h or $24 \times 60 = 1440$ min, each degree of rotation corresponds to $1440/360 = 4$ min. Thus

It can be seen that the terrestrial solar incidence of 2.2×10^{16} kWh/year can supply some 1000 times all the energy consumption or some 12,000 the electric-energy consumption in the United States, if the entire surface area is available for collection and nothing is left for housing, industry, food growing, etc., and if the solar-to-electric conversion efficiency were 100 percent.

It is this abundance of "free" energy that has given solar proponents much ammunition, though their arguments ignore the fact that not all areas have dependable solar radiation (high average C_i), and even these suffer from daily and seasonal periodicity, and that solar-to-electric conversion is capital-intensive and inefficient. Still, such conversion is an engineering feasibility and, in locations remote from an electric grid, where the demands are low, may be an economic feasibility.

Solar energy may be converted to electricity by one of two means: *solar thermal conversion* or *photovoltaic conversion*.

Solar-Thermal Conversion

By this method solar radiation is converted to heat that in turn is added to a thermodynamic cycle to produce mechanical work and electricity. For this to be efficient, and hence economical, it is necessary to collect and concentrate the diffuse solar radiation in an efficient manner to arrive at a reasonably high-temperature heat source. The *collectors* gather the sun's energy and direct it onto *receivers* that contain the working fluid of the thermodynamic cycle. The receivers that are the subject of the most serious considerations are:

1. Central receivers.
2. Dispersed or distributed receivers. These could be of (a) the point-focus type or (b) the line-focus type.
3. Ponds.

$$\text{MST} = \text{LST}\begin{pmatrix} +\text{ degrees east} \\ \text{or} \\ -\text{ degrees west} \end{pmatrix} \text{of standard meridian} \times 4 \text{ min}$$

During periods of daylight saving time, LST is obtained by subtracting 1 h from the local daylight time (LDT).

The true or local sun time, called the *apparent solar time* (AST), and the mean sun time (MST) would be the same if the earth rotated around the sun in a circular orbit and thus at constant velocity. The somewhat elliptical orbit of the earth causes its velocity to vary and, therefore, the sun to appear on earth to be earlier or later than the mean sun time. The difference between AST and MST is called the *equation of time* (ET), which is not an equation at all but a variable correction factor that depends upon the time of the year as shown:

21st day of	Jan.	Feb.	Mar.	Apr.	May	June	July	Aug.	Sept.	Oct.	Nov.	Dec.
ET, min	−11.2	−13.9	−7.5	+1.1	+3.3	−1.4	−6.1	−2.4	+7.5	+15.4	+13.8	+1.6

Thus AST = MST + equation of time.

The conversion systems used in solar-thermal-electric conversion are of many types. These include:

1. Rankine cycle, using steam or other working fluid
2. Brayton cycle, using helium or air as working fluid
3. Hybrid systems
4. Repowering systems

In addition, storage systems may be necessary. Systems producing process heat only are also receiving some attention.

Photovoltaic Conversion

Photovoltaic systems consist of direct-conversion devices in the form of cells that convert the solar radiant-energy photons to electricity without benefit of a thermodynamic cycle or working fluid. They can be their own collectors or can use concentrating collectors that focus the solar input on them.

The cells produce low currents and voltages and are therefore usually combined into modules that in turn are combined in panels and then arrays to meet specific power requirements. The cells are made of

1. Single-crystal silicon
2. Silicon with many crystals
3. Thin films with a wide range of single chemical compounds or combinations of them

These various devices will be covered in the next few sections in this chapter.

13-6 SOLAR-THERMAL CENTRAL-RECEIVER SYSTEMS

The central-receiver approach to solar-thermal-electric systems uses a large field of reflecting mirrors called *heliostats* that redirect the sun's energy and concentrate it on a central receiver mounted on top of a tower (Fig. 13-7). The heliostats are individually guided, since they cover a large field, so that each focuses the sun's energy it receives on the central receiver at all hours of sunlight. In the receiver the concentrated solar energy is absorbed by a circulating fluid. The fluid could be water, which vaporizes into steam that is used to drive a turbogenerator in a Rankine cycle, or an intermediate fluid that transports the heat to the steam cycle.

The system should incorporate storage for nighttime and cloudy periods, as shown in Fig. 13-7. The receiver output is made greater than that required by the steam cycle, and the excess output during periods of greatest solar incidence is bypassed to a thermal storage system. During periods of low or no solar incidence, the feedwater is shunted

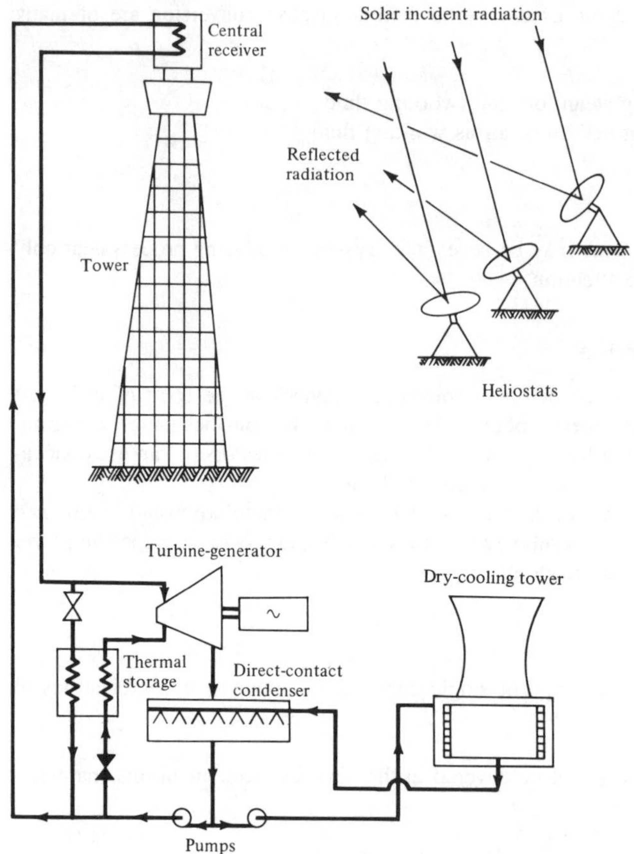

Figure 13-7 Schematic of a solar-thermal central-receiver system powerplant.

to the storage system, instead of to the receiver, where it vaporizes for use in the turbine. Proper valving in the system allows operation in either mode.

Because solar-thermal electric plants are most likely to be located in hot arid areas where land is plentiful (for the large heliostat field) and where the sun's energy is plentiful and dependable, but where cooling water is scarce, the condenser water is most probably cooled by a dry-cooling tower. Such towers are less effective and cause a reduction in Rankine cycle efficiency but require practically no makeup water (Sec. 7-7).

In the next five sections, the major subsystems of the central-receiver concept will be presented, with many of the design features and data obtained from the Solar One plant experience. Solar One is a 10-MW(e) (peak) pilot plant located in the Mojave Desert in California that went into operation and testing in mid-1982 (Sec. 13-11).

13-7 THE HELIOSTATS

The *heliostats** are reflecting mirrors that are steerable so that they can reflect the sun's rays on the central receiver at almost all times during the daylight hours. In essence, they keep the sun stationary as far as the receiver is concerned. They are made to track the sun, not by being perpendicular to its rays (as in distributed systems, Sec. 13-12) but by being at such an angle to these rays that they reflect them to the stationary receiver. This angle depends upon both the time of day and the position of the individual heliostat with respect to the receiver.

A heliostat is composed of a reflective surface or mirror, mirror support structure, pedestal, foundation, and control and drive mechanisms. Current designs have total reflective areas between 40 and 70 m^2. Ideally, the surface should be slightly parabolic with the focal length equal to the distance from the surface to the receiver but, because that distance is long, spherical or even flat surfaces offer good performance. There are two types of reflecting surfaces: *glass* and *plastic*.

A *glass heliostat* is typically divided into 10 to 14 panels rather than a single large surface for ease of manufacture and transportation (Fig. 13-8). A glass panel is typically a second-surface mirror (similar to household mirrors). The panels are usually rectangular, 1.2 × 3.6 m (4 × 12 ft) being typical. They are made of thin (1.5 to 3 mm) low-iron glass sheets to minimize absorption. The panels, when assembled, are canted slightly with respect to each other, with the canting varying throughout the field, to focus the sun's rays on the receiver.

Heliostats that use glass are much further developed at this time (1982) than plastic ones mainly because of their higher reflectance and strength. Current heliostats, therefore are constructed of glass and silver for the reflecting surface plus steel and aluminum or copper for the supporting structure. They must be constructed in a massive, sturdy manner to withstand strong wind loads and other severe weather conditions. They range widely in mass, from 5 to 60 kg/m^2 of mirror area, excluding foundation. The 10-MW(e) Solar One pilot plant uses 1818 glass heliostats of the type shown in Fig. 13-8, each containing 12 separate panels.

The heliostats currently account for about 45 percent of the total capital cost of a solar-thermal central-receiver powerplant (the tower and receiver account for about 11 percent each, the thermal storage system for about 17 percent, and the rest is divided among the land and the balance of the plant). A reduction in heliostat costs is therefore a major goal in reducing plant capital costs, which are currently estimated at 2 to 5 times the cost of conventional and nuclear powerplants, with a heliostat cost of about \$250/m^2. DOE second-generation heliostat studies indicate the cost would go down to \$110/m^2 to \$150/m^2 (in 1980 dollars) if ordered in quantity.

The *plastic heliostat,* although lower in reflectance and strength than the glass heliostat, promises lower costs, lower mass of the reflector surface, and hence lower mass and costs of the support structure and drive mechanism. A typical plastic heliostat design by Boeing (Fig. 13-9) has a reflecting, stretched plastic film disc of 16.7-m^2 area that is protected from wind loads by an air-supported plastic bubble.

* From the Greek *helios,* "the sun." In Greek mythology: the sun god, son of Hyperion.

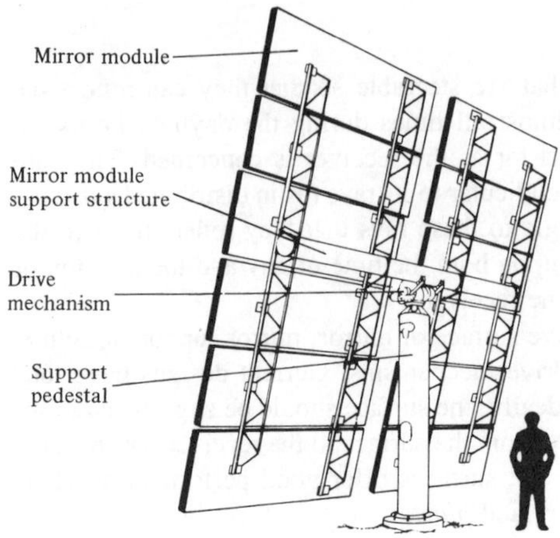

Figure 13-8 A typical glass-heliostat system, rear view *(McDonnell Douglas.)*

The Heliostat Field

The *heliostat field* supplying a central receiver, also called the *collector subsystem,* has a shape that must be optimized to suit the topography of the area and the power level of the plant. The field may be on a flat terrain, on the side of a hill, etc. In the northern hemisphere, the noontime sun is always south of the central-receiver tower, so a north field is usually most cost-effective because its cosine loss (below) is least.

For small plants, of less than 100 MW of thermal-energy input, a totally north field is optimum (Fig. 13-10*a*). As plant size increases, the field becomes larger and many heliostats are farther from the tower. The atmosphere around the plant attenuates the reflected radiation from the most distant north heliostats. The receiver input can then be improved by relocating the distant heliostats to the east and west of the tower and, as plant size increases further, to the south of it (Fig. 13-10*b*). In such cases, the additional cosine loss is less than the atmospheric attenuation loss from the distant north heliostats.

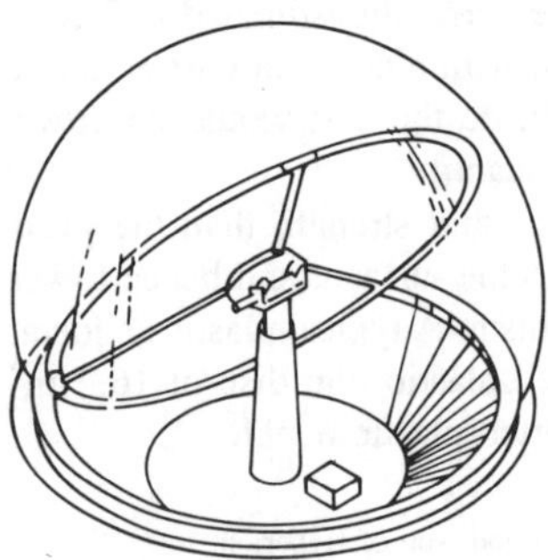

Figure 13-9 A typical plastic heliostat system *(Boeing.)*

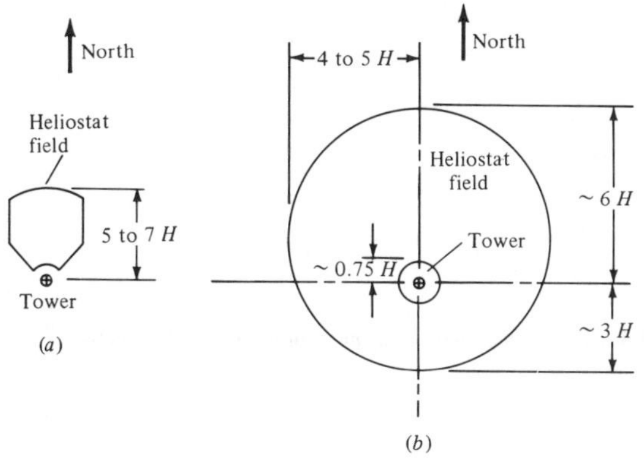

Figure 13-10 Optimum heliostat field shapes oriented for the northern hemisphere: (*a*) small plant < 100 MW(t) and (*b*) large plant > 5000 MW(t). Dimensions in multiples of receiver-tower height *H* [118].

The energy losses betwen the extraterrestrial (solar constant) and terrestrial insolations have already been discussed. There are, in addition, energy losses between the incident energy on the heliostat field and the receiver. These may be summarized by:

1. *Shadowing*. This is caused by one heliostat casting a shadow on the reflective surface of another at certain times of the day. (The tower shadow on the heliostats has a negligible effect.)
2. *Cosine loss*. Since the reflective surface of a heliostat is not perpendicular to the beam radiation from the sun, but assumes an angle that would reflect that beam to the central receiver, the area of solar flux intercepted by the heliostat is less than its reflective surface by the cosine of the angle between the surface and the perpendicular to the beam.
3. *Blocking*. This is the result when the reflected light from one heliostat is partially blocked by the back side of another.
4. *Reflective losses*. These result from absorption of the incident radiation by the glass and silvering and by scattering by dirt on the reflective surface. They prevent some of the light from being specularly reflected to the receiver.
5. *Attenuation*. This represents losses by absorption and scatter by water vapor, haze, fog, smoke, and particulates in the atmosphere between the heliostat and receiver.

The collector field, and the heliostat layout within that field, should therefore be optimized to transport the required thermal energy to the receiver at the lowest cost. As always, optimization implies trade-offs between usually conflicting requirements. For example, reducing shadowing and blocking by increasing spacing between heliostats results in greater attenuation, greater land use for a given energy input, and a larger image at the receiver, resulting in greater spillage (below).

Studies have shown that the most cost-effective layout of heliostats within their field is a radial stagger layout (Fig. 13-11). It minimizes losses by blocking and attenuation and results in minimum land use.

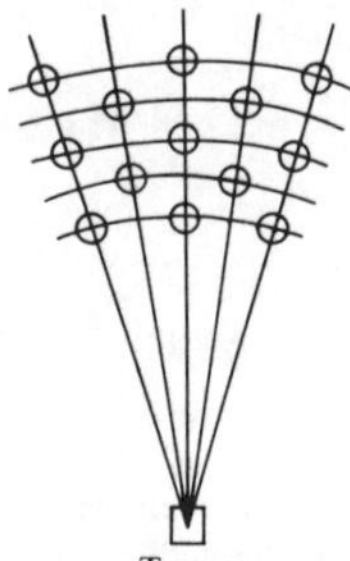

Figure 13-11 Radial stagger heliostat layout with respect to the receiver.

Heliostat Control

The insolation to the field varies from hour to hour, as well as seasonally (Fig. 13-6). At sunrise most of the heliostats are shadowed except in the extreme western sector. A similar situation occurs at sunset. A minimum practical amount of energy, usually considered at 10 percent of peak power, can only be collected when the sun is at an elevation of 15° or greater from horizontal. The reflected power and the receiver losses (below) are shown compared to the normal incident terrestrial insolation for the summer solstice, equinox, and winter solstice in Fig. 13-12, with a rapid rise and fall at 15° from sunrise and sunset. The receiver loss is nearly constant and independent of time of day as it operates at a fixed design temperature. Hence, that loss is a greater fraction of the input during off noon operation.

The daily operating cycle of solar-energy collection therefore includes morning start-up, operation during the power periods shown in Fig. 13-12, and evening shut-down. During nonoperation, the heliostats are placed in the *stow*, or storage, position. This is usually a horizontal face down or a vertical position. For morning start-up, they are maneuvered from the stow position to a standby position with the reflected sun rays aimed at a point adjacent to but not at the receiver itself. This maneuver may take as much as 15 min and may consume as much as 0.1 kWh per heliostat. The receiver is then started up (coolant pumped through), after which the heliostats are sequentially moved from the standby position to the operating position, i.e., with reflected sun rays aimed at the receiver surface. This prevents overheating of the receiver. Receiver flow controls also help prevent overheating.

During normal operation, the heliostats are in a *sun-tracking mode,* in which each is adjusted by a controller to keep the reflected sun rays at the aim point. Different heliostats can have slightly different aim points in order to spread the energy flux uniformly over the receiver surface. Sun tracking of the heliostats is accomplished by one of two methods. The first uses active reflected-beam sensors that control drive motors on the heliostat to orient the reflective surfaces so that the beam is continuously reflected to its aim point. The second method uses a preprogrammed computer control that orients the reflective surfaces according to their position with respect to the receiver,

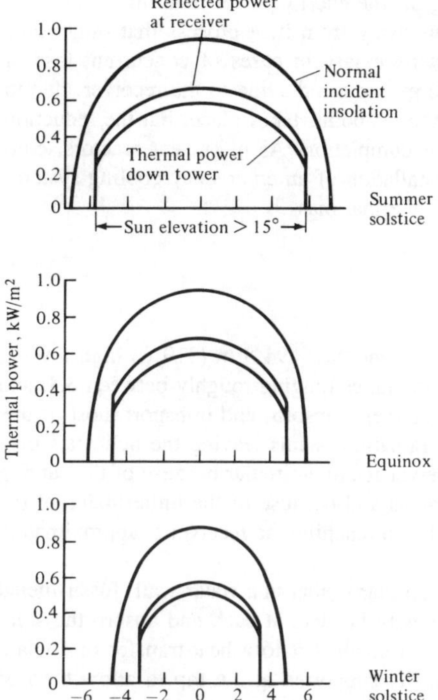

Figure 13-12 Incident normal insolation, reflected power reaching the receiver, and thermal power converted by the receiver at the two solstices and equinoxes, showing effects of start-up and shutdown [118].

time of day, and day of year. Tracking in either method is done by adjusting two axes of rotation, azimuth and elevation, of the reflective surface.

The shutdown procedure is the reverse of the start-up procedure. The heliostats are moved from the sun-tracking mode positions to the standby position and then to the stow position. The stow position is used during the night but also during plant outages and periods of extreme weather (vertical stow in hail; horizontal for high winds; vertical slightly slanted upwards, for rain to take advantage of its cleaning action).

The control systems must be capable of extremely accurate reflective surface orientation (within a few milliradians) because even a tiny angle error would result in the reflected beam widely missing its aim point as a result of the long distance between heliostat and receiver. This accurate operation must also be maintained in strong winds by the rigidity of the heliostat support end structure.

Loss-of-coolant accidents (LOCA) require great attention to avoid damage to the receiver (a similar concern to that in nuclear reactors where a loss-of-coolant accident can result in overheating and damage to the reactor core and possible core fuel meltdown). A LOCA must be met by shutting off the energy power input to the receiver by rapidly moving the heliostat aim points away from it, a process that might take 30 s. A back-up power-supply system is necessary in cases of concurrent loss of electric power. In that case, a passive-energy input reduction to the receiver, due to defocusing as the sun travels away from the set point, takes place, but the reduction is gradual and slow, taking 2 to 3 min for completion. As in nuclear reactors, consideration may have to be given to the installation of an emergency cooling system, especially for heavy receivers with a large thermal mass.

13-8 THE RECEIVER

Central receivers sit atop tall towers. [Solar One has a 94.5 m (310 ft) high tower.] They are subjected to peak radiation energy fluxes ranging roughly between 300 and 700 kW/m^2. The receiver is designed to intercept, absorb, and transport most of this energy to a heat-transfer fluid. Although radiation beams leaving the heliostats may be rectangular, the beam reaching the receiver is more circular because of the canting and focusing of the heliostat reflecting panels and because of the finite image of the sun. The flux distribution of the reflected beam reaching the receiver is approximately gaussian.

The heat-absorbing surface is usually similar to that in a water-wall, fossil-fueled boiler (Sec. 3-4): Panels of parallel tubes with headers at each end absorb the solar energy incident on their outside surface and conduct it to a heat-transfer or coolant fluid flowing inside. The panels are typically supported at the top to allow thermal expansion to occur downward.

There are two basic types of receivers under consideration: cavity and external (Fig. 13-13). A cavity receiver has coolant-tube panels lining the inner walls of the cavity. The tube-panel arrangement within the cavity is concave toward the heliostat field, and the panel area is 2 to 3 times the aperture area. The aperture area is sized about the same as the sun's image from the farthest heliostats. The aperture-to-panel area ratio is sized to minimize the sum of thermal losses and spillage (below). A cavity receiver may be designed with one or more cavities and apertures, each with a collecting sector with an angle between 60 and 120°. Figure 13-13*a* shows a four-aperture cavity receiver. The panels of adjacent cavities may form a common wall, heated from both sides, to reduce thermal stresses that result from one-sided heating (as in an external receiver).

The external receiver (Fig. 13-13*b*) has the coolant-tube panels lining the outside of the receiver. The panels may either be flat, for small plants, or slightly convex toward the heliostat field, for large plants. The external receiver shown is a multipanel polyhedron approximating a cylinder with a height-to-diameter ratio ranging between 1:1 to 2:1.

The coolant tubes range in diameter from 0.75 to 2.2 in (20 to 56 mm) and in thickness from 0.049 to 0.25 in (1.2 to 6.4 mm), depending upon pressure, material, and other engineering design considerations.

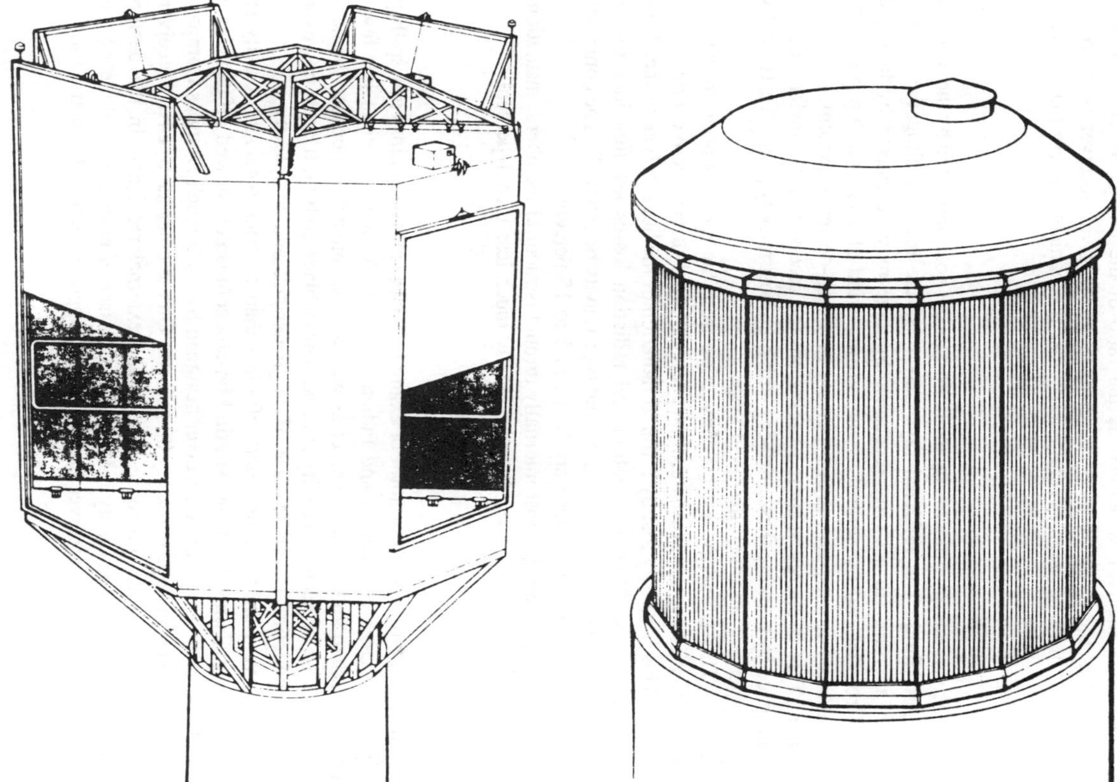

Figure 13-13 Typical central receivers: (*a*) four-aperture cavity type and (*b*) external type [118].

Receiver Efficiency

The efficiency of a receiver is defined as the ratio of the energy absorbed to that intercepted by it at the design point. The efficiency of a given receiver is the result of design compromises between several energy-loss mechanisms. These may be summarized as:

1. *Spillage*. This is energy reflected by the heliostats but not intercepted by the receiver heat-transport fluid. The reflected rays may miss the receiver altogether or fall outside an aperture (in a cavity receiver). Spillage may be caused by heliostat tracking errors caused by control system errors, wind effects, steering backlash, etc. Spillage is normally less than 5 percent in a well-designed system.
2. *Reflection*. This is energy scattered back from the receiver heat-transfer surface. It is minimized by painting these surfaces with high-absorptivity paint. It is typically less than 5 percent in a well-designed system.
3. *Convection*. This is energy lost by convection from the receiver body to the surrounding air. It is the sum of natural and forced (wind-driven) convection.
4. *Radiation*. This is energy lost by infrared radiation of the hot receiver surface back to the environment. Both convection and radiation losses are functions of the receiver temperature, its configuration, and type (cavity or external). The combined convection and radiation losses vary between 5 to 15 percent.
5. *Conduction*. This is energy lost internally through structural members, insulation, etc. It is the least of the losses, being typically much less than 1 percent.

Design optimization is important. For example, a large receiver enjoys low spillage but suffers from large convection and radiation losses. A cavity receiver has fewer reflection losses than an external receiver as well as less convection because the heat-transfer surfaces are protected (a cavity approximates a black body*). It does, however, have greater conduction losses because of its greater size and complexity.

Cavity receivers are therefore more efficient than external receivers. Cavity receivers designed by Honeywell and Martin Marietta companies showed efficiencies in excess of 90 percent. An external receiver designed by McDonnell Douglas Company showed an efficiency of approximately 80 percent. On the other hand, cavity receivers are much larger, heavier, and more costly than external receivers. In the designs proposed for Solar One, the cavity receivers had masses of more than 250 metric tons each, whereas the external receiver had a mass of approximately 136 metric tons.

* A black body can be closely approximated by a small aperture to a larger cavity. The radiation entering through the aperture is repeatedly reflected within the cavity with very little of it finding its way back out. The radiation is thus trapped inside, and the result is an absorbtivity nearly equal to unity.

13-9 THE HEAT-TRANSPORT SYSTEM

The heat-transport system is composed of the heat-transfer fluid (primary coolant), receiver piping, piping between receiver and power-generating equipment, and pumps. The primary coolant may or may not be the same as the powerplant working fluid. Five primary coolants have been extensively studied:

1. *Water-steam.* Water receives heat in the receiver, where it is converted to steam, which doubles as powerplant working fluid. The phase change in the receiver occurs as in conventional steam generators, either in a drum boiler with separate boiler and superheater sections, or in a once-through boiler without drum (Sec. 3-8). Steam may be generated at 1000 to 1100°F (540 to 600°C) and 1000 to 2000 psia (70 to 140 bar). The use of an all water-steam system has the advantages of least extension of technology and hence lower developmental costs. It is the one used in Solar One. Its performance, however, is considerd somewhat inferior to the other systems still under development.

2. *Liquid metals.* Liquid metals, particularly sodium Na, are under intensive development as heat-transport fluids for fast-breeder nuclear reactors (Secs. 11-3 to 11-5). Their main advantage is the high heat-transfer coefficients they are capable of, which result in a more-compact and lighter receiver. Sodium freezes at 208°F (98°C) and thus requires freeze protection during plant shutdown, either by heating the pipes or by draining them. Sodium is also chemically active, oxidizing rapidly in air or water. A cover gas such as argon, which is heavier than air, is used (as in sodium-cooled nuclear reactors) to protect the coolant and prevent oxidation and fires. Cold and hot traps need to be used to remove oxides and carbon respectively, from the sodium. Sodium may be operated up to 1000°F (540°C) as a single-phase liquid coolant [it has a normal boiling point of 1621°F, (883°C)] and must transport the solar heat to a steam generator at the plant site [119]. Liquid metals have low vapor pressures at the operating temperatures of interest and hence do not require high pressurization in the heat-transfer loop.

3. *Molten salts.* Molten salts are used as heat-transport agents in the chemical industry and have been used experimentally as nuclear-reactor fluid fuels. They, as liquid metals, do not require high pressurization at operating temperatures and, in addition, have high volumetric heat capacities. Nitrate salt mixtures have been proposed for solar central-receiver systems [120]. The salt can also operate at high temperature and transfer its heat to a steam generator at the plant site. It does, however, suffer from an even higher freezing point, about 290 to 430°F (140 to 220°C), which requires precautions against solidification during shutdown.

4. *Gases.* Gases can be operated at any temperatures desired that are compatible with component materials, often in excess of 1550°F (840°C). Pressurization is necessary only to overcome pressure losses in the coolant loop and to increase their mass-flow rates. The low heat capacities of gases require large volume flow rates and hence large flow velocities. Pressurization increases density and thus increases the volumetric heat capacity and decreases velocities, and hence pressure losses, for a given heat-transport rate. The gases considered are air and helium. They can be used as heat-

transport agents that generate steam in a steam generator in a Rankine-cycle powerplant, or they can be used directly in a Brayton-cycle powerplant gas-turbine of the open type (with air) [121], in a closed cycle (with helium) [122], or in a combined cycle [123] (Sec. 8-8).

5. *Heat-transfer oil*. Oils have the advantages of low-corrosion characteristics with most materials and can be selected for low vapor pressures. They are, however, flammable, can cause flow problems at low temperatures, and usually suffer pyrolitic damage, i.e., they decompose under high-temperature conditions, called *pyrolitic decomposition* (in a nuclear-radiation environment it is called radiolytic decomposition). The decomposition results in low boilers (gases and low-molecular-weight liquids) that cause an increase in system pressure and in high boilers (thick, gummy residue-type oils) that foul heat-transfer surfaces. They can only therefore be operated over a relatively narrow temperature range. Two suggested oils for solar central-receiver systems are Therminol 66 and Carolina HT-43 [124], which have an operating range of 20 to 600°F (−7 to 315°C). Oils are also used to transport receiver heat to a steam generator at the plant site.

The piping that takes the primary fluid from the receiver, down the tower, and across the heliostat field to the plant site is a major cost item of the total powerplant. Pipe sizes for a 400-MW(t) receiver range in ID from 1.0 to 1.5 ft (0.3 to 0.5 m) for molten salt to 8 ft (2.4 m) for air at 175 psia (12 bar) and in length up to 2000 ft (600 m) for both the hot and cold runs.

Because of the differences in their densities and heat-transport properties, the heat-transport fluids influence the mass of the receiver. For a 400-MW(t) receiver, the mass may range between 550,000 lb_m (2.5 metric tons) for a sodium-cooled external receiver to 5.5 million lb_m (2500 metric tons) for an air-cooled cavity receiver.

13-10 THE THERMAL-STORAGE SYSTEM

Although several energy-storage systems are possible, such as chemical, electrochemical (batteries), mechanical (compressed air, pumped hydro, flywheel), and single- or multiphase thermal energy (Chap. 16), the probable location of most solar powerplants precludes most such systems as being cost-ineffective. The one exception seems at present to be thermal single-phase energy storage. Such a system stores as sensible heat, i.e., without change in phase of the storage medium, some of the energy of the primary coolant after it has passed through the receiver. The energy thus stored is extracted when needed for use in the electric-generating station (or industrial process).

Energy storage is needed in solar-thermal-electric powerplants (and solar process-heat applications) because of normally variable solar insolation, nonsolar periods, and abrupt insolation changes in inclement weather. Conventional (fossil-fired) backup systems may, of course, be used during such periods, as in hybrid systems (Sec. 13-13), but thermal storage is another option that must be used in pure solar systems. There are two types of thermal storage that are considered with solar systems. These are (1) *single-tank*, or *thermocline*, and (2) *dual-tank*, or *hot-cold*, systems.

The *single-tank,* or *thermocline, storage system* is shown schematically in Fig. 13-7. Storage takes place by circulating some of the hot primary coolant through the storage medium and returning the cooled primary coolant from the bottom of the storage tank back to the receiver for reheating. Heat extraction during times of need is accomplished by reversing the process: cold primary coolant from the powerplant is heated by the storage medium, drawn from the top of the tank to the power system, and then returned to the bottom of the tank. The thermal gradient of hot at top, cold at bottom maintains stratification, allowing the hot fluid to remain afloat on the top, and gives the tank thermal stability. In addition, a solid storage medium of low thermal conductivity and high volumetric heat capacity, such as rock, is used to help impede mixing of hot and cold fluids. A solid storage medium is a necessity in case the primary coolant is a gas (such as air or helium) because of the low heat capacity of gases. A porous solid makes a good storage medium for gases.

Figure 13-14 shows temperature-time characteristics of a thermocline thermal-storage system. Note that the temperature gradient is never completely flat at any one time, a factor that affects the performance of such systems (below).

A variation of the thermocline system is one in which a separate storage fluid, other than the primary coolant, may be used. This would be a liquid of low thermal conductivity and high volumetric heat capacity, usually an oil. It may be used alone as a single medium or be assisted by a solid storage medium, such as rock in a dual-

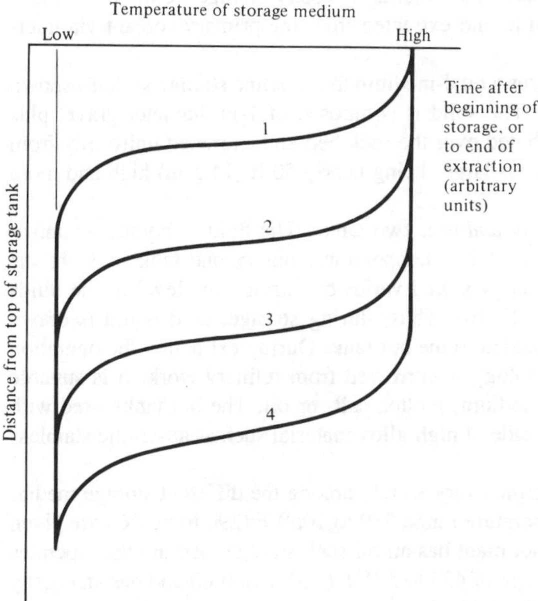

Figure 13-14 Temperature-time characteristics of a thermocline storage system.

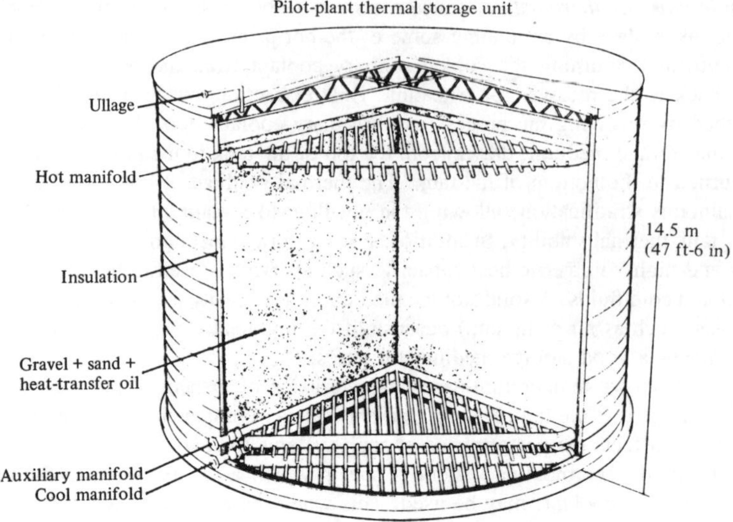

Figure 13-15 Thermocline storage system for the Solar One pilot powerplant.

medium system as above, to reduce the volume of costly storage liquid and impede its mixing. Heat would be added to and extracted from the primary coolant via intermediate heat exchangers.

The Solar One powerplant uses a dual-medium thermocline storage system, shown in Fig. 13-15. The liquid is oil. The solid is composed of 1-in diameter gravel plus sand. The oil is uniformly distributed over the rock bed and extracted uniformly from it by diffuser manifolds. The tank is large, being nearly 50 ft (14.5 m) high and more than that in diameter.

The *dual-tank,* or *hot-cold, system* uses two tanks. The fluid, a liquid, is, unlike the stratified thermocline system, at one temperature, hot in one tank, cold in the other. The amount of sensible energy stored varies by varying the level of the fluids in the well-insulated tanks (Fig. 13-16). Thus, during storage, cold liquid is drawn from the cold tank, heated, and added to the hot tank. During extraction the operation is reversed. This dual-tank technology is borrowed from refinery work. It is suitable only for liquid coolants such as sodium, molten salt, or oil. The hot tanks used with sodium and molten salt need be made of high-alloy material such as austenitic stainless steel.

Thermal-energy-storage densities vary widely among the different storage media. Some typical values for the temperature range 550 to 1050°F (290 to 565°C) are given in Table 13-3. The Solar One pilot plant has an oil-rock storage medium that operates at the much lower temperature range of 425 to 575°F (220 to 300°C) and has an energy storage density of 5300 Btu/ft^3 (0.05 MWh/m^3).

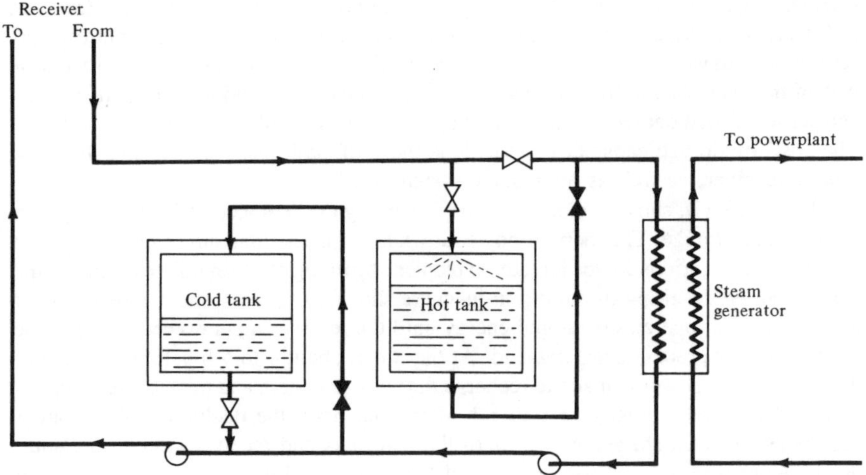

Figure 13-16 Schematic of a dual-tank, or hot-cold, thermal-storage system. Valves shown in normal plant operating and storing mode.

Insulation is an important consideration in storage tanks to minimize loss of energy during holding periods. Another important consideration is the fraction of stored energy that is usually extracted. This fraction is called the *storage utilization factor*. In dual-tank or hot-cold storage systems it is usually in excess of 0.98. In single-tank or thermocline systems the factor is lower because of the temperature gradient that exists at any one time during charging or discharging (Fig. 13-14). During discharge, there comes a time when the temperature drops below maximum, thus necessitating an operational limit on the total energy extracted. For thermocline systems the storage utilization factor is nearly 0.8 for liquid-solid systems but less than 0.5 for gas-solid systems. Note that a low utilization factor does not mean a loss of energy because in a well-insulated system energy losses are small. It simply means that a large and costly tank(s) is required.

Storage sizing can be made to supply enough thermal energy, together with direct

Table 13-3 Typical thermal-storage densities

	Energy storage density	
Medium	Btu/ft³	MWh/m³
Sodium	8000	0.08
Rock, 25% void fraction	14,000	0.15
Molten salt (nitrate)	21,000	0.22

thermal power from the receiver, to operate the powerplant for 24 h/day. However, if the sizing is based on the long days of the year, it will not have sufficient capacity for the short days of the year. The opposite results in an excess storage capacity during part of the year. Sizing thus depends upon many factors, including variations in daily and seasonal load demands, cloud conditions, availability of backup power, and cost. The costs of storage capacity must include those of additional heliostats to fill it and a large receiver, as well as the storage system itself.

Figure 13-17*a* shows a cloudless daily operating cycle of a central-receiver system with thermal storage. The thermal-power down tower is less than direct solar insolation because of the various losses already explained. Morning start-up utilizes some of the stored energy, resulting in the dip in its curve shown between 4 and 6 h before solar noon. The plant operates on stored energy until the receiver input exceeds the end use (powerplant or process heat) demand. At that time, about 4 to 5 h before solar noon, the excess solar power is made to recharge the storage system and the receiver supplies both end use and storage. At 4 to 5 h after solar noon the declining solar input is supplemented from storage energy until the heliostats and receiver systems are shut-down. After that, all end use is supplied from storage. Figure 13-17*a* assumes, for

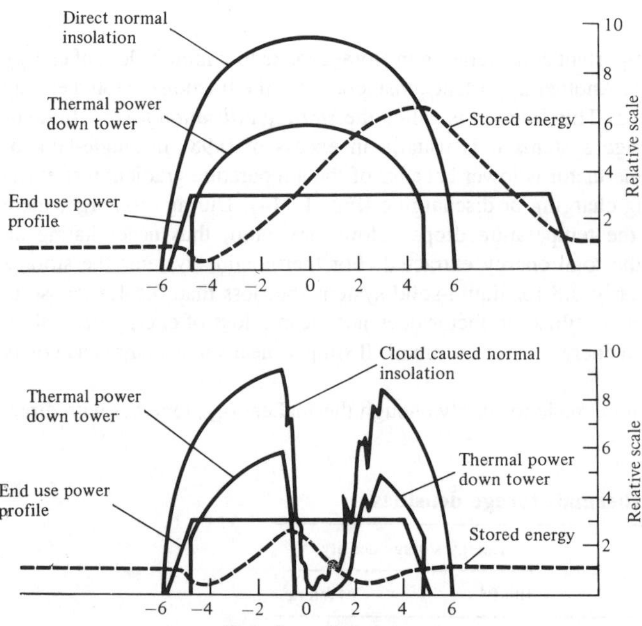

Figure 13-17 Typical daily operating cycle of a central-receiver system with thermal storage and constant power (end use): (*a*) cloudless day and (*b*) cloudy day [118].

simplicity, a constant end use, which is not usually the case. Figure 13-17*b* shows the effect of a cloudy period during the day. Note that the scales for all curves, except stored energy should be in power units, e.g., kilowatts. For the latter, the scale should be in energy units, e.g., kilowatt hours.

13-11 WORLD EXPERIENCE

Central-receiver systems are receiving attention in many parts of the world. In the United States, with 68 solar-thermal projects of various kinds taking place, the major activity in the late 1970s to early 1980s was the development, construction, and operation of Solar One, the 10-MW(e) (peak) pilot plant referred to earlier. The plant cost $140 million and is situated in the Mojave Desert at Dagget, about 12 mi southeast of Barstow, California. It is owned by Southern California Edison Company and is financed jointly by the Department of Energy (DOE) and a group of utilities. Figure 13-18 shows an overall view of Solar One.

Figure 13-18 Overall view of Solar One, a 10-MW(e) (peak) central-receiver pilot powerplant near Barstow, California.

McDonnell Douglas Corporation and Martin Marietta Corporation were the major contractors. The plant uses water-steam as the heat-transport and working fluid with thermal storage using oil as the storage heat-transport fluid. It started operation in the summer of 1982. It reached a 10.4-MW(e) net output on 10 October 1982. The storage system, rated at 7 MW(e) maximum, had not met that goal at that time. The first 2 years of operation are devoted to experimental testing of the major subsystems and evaluating requirements and costs of operation and maintenance. This period will then be followed by full power tests to obtain long-term operating, maintenance, and reliability data.

In Almeria, Spain, the International Energy Agency, several European countries and the United States have funded a 500-kW(e) plant that uses molten sodium as the heat-transport and thermal-storage fluid. The sodium-receiver exit temperature is 975°F (525°C). The plant has 93 heliostats of 430 ft^2 (40 m^2) each that are supplied by the Martin Marietta Corporation. It has a cavity receiver atop a 140-ft (43-m) tower. The plant started operation in September 1981. An interesting feature of this plant is the installation of a parallel distributed system of 500 kW(e) using line-focus receivers (Sec. 13-12) to compare its performance with that of the central-receiver system. Another plant in Almeria, Spain, scheduled for completion in late 1982, is a 1.2-MW(e) central-receiver system called CESA-1. It has 300 heliostats, 430 ft^2 (40 m^2) each, a cavity receiver, water-steam at 975°F (525°C) as the heat-transport and working fluid, and a 3-MWh(th) thermal storage using Hytec as the storage fluid. It is funded by Spain and the United States, with heliostats built in France and Germany.

In Adrano, Sicily, Italy, a 1-MW(e) central-receiver plant, called Eurelios, began operation in May 1981. It has a cavity receiver atop a 180-ft (55-m) tower that produces steam at 950°F (510°C), and a 30-min thermal-storage subsystem to even out cloud transients that uses Hytec as the storage fluid. The project was developed by a consortium from Italy, France, and Germany.

In France, a 2-MW(e) plant called Themis, has 200 heliostats, 560 ft^2 (52 m^2) each, a cavity receiver with molten salt heated to 975°F (525°C) as the heat-transport fluid, and a 4-MWh(th) thermal-storage system using Hytec as the storage fluid. Start-up was scheduled for early 1982.

Japan has its "Project Sunshine" in the form of a 1-MW(e) plant in Nio Town, Nagawa Prefecture. It has 807 small heliostats of 172 ft^2 (16 m^2) area each, a semicavity receiver atop a 197 ft^2 (60-m) tower producing steam at 480°F (250°C), and a 3-MWh(th) thermal storage system using pressurized water as the storage medium.

13-12 DISTRIBUTED SOLAR-THERMAL SYSTEMS

Another approach to solar-thermal-energy conversion, besides the central-receiver-tower concept, is the *distributed* or *dispersed solar-thermal system*. Rather than hundreds of heliostats focusing solar energy on a single distant receiver atop a tall tower, this system is characterized by the use of a large number of collectors, called *concentrators*, each focusing the solar energy it receives directly on its own receiver to heat locally a heat-transport fluid. The fluid is combined with those from other concentrators for thermal-electric conversion. Distributed systems are of two kinds: *point focus* and *line focus*.

The *point-focus system* generally uses concentrators each in the form of a mirrored parabolic dish that tracks the sun but focuses the captured energy on a receiver mounted at the focal point of the parabola, a few feet from it. Figure 13-19 shows two advanced concentrator concepts by Boeing, Inc., and E-Systems Corp., which are believed to offer the potential for low-cost production. The Boeing design uses a reflecting surface made of thin plastic films on sheet steel substrates. The E-Systems design uses a domed Fresnel-lens concept with the lenses focusing the solar energy at a common focal point.

The Jet Propulsion Laboratory (JPL) operates a facility for testing dish-concentrator systems and components at Edwards Air Force Base in California. The U.S. Department of Energy is in the process of building (1982) an array of parabolic dish concentrators to be used for electric production, heating and cooling, and low-temperature process steam for a knitwear plant nearby. Figure 13-20 shows a number of point-focus concentrators undergoing testing at the Sandia Laboratories, Albuquerque, New Mexico.

The *line-focus system*, also called the *trough system,* uses concentrators in the form of long troughs of cylindrical or parabolic cross sections, which are lined with mirrors to collect and concentrate the sun's radiation onto a focal linear conduit through which the primary coolant flows (Fig. 13-21). Because of their geometry such troughs are usually made to track the sun in only one plane, by being rotated about their focal line. Thus, other than solar noon, they receive sun's rays that get more inclined with respect to their projected surface as the sun deviates from solar noon. They, therefore, usually operate in the lower-temperature ranges of 200 to 600°F (about 90 to 315°C). Line-focus systems are thus believed suitable only for small-sized electric-generating systems for which thermal efficiency is not of prime importance and for other applications such as driving irrigation pumps (one of their earliest uses, Fig. 13-1*d*), providing industrial process heat, space heating and cooling, and other industrial applications, but not for large-scale electric production.

A variation of the line-focus system, called the *line-focus bowl,* which is being tested under DOE sponsorship, uses a stationary mirror-lined spherical collector that concentrates solar energy on a conduit receiver that moves during the day. Heat received from a number of such bowls can be used for small-scale electric production, space heating and cooling, the production of process steam, etc.

Of all the distributed concepts, the line-focus systems are the beneficiaries of the most experience in industrial and commercial process-heat installations where the temperature requirements are lower than necessary for efficient large-scale utility electric production. Point-focus systems, on the other hand, are least developed. It is believed, however, that they have the potential for achieving the temperatures necessary for utility electric production. Being modular, they have the additional advantages of meeting the demands of a wide range of load requirements and of adding to existing capacities by incremental installation.

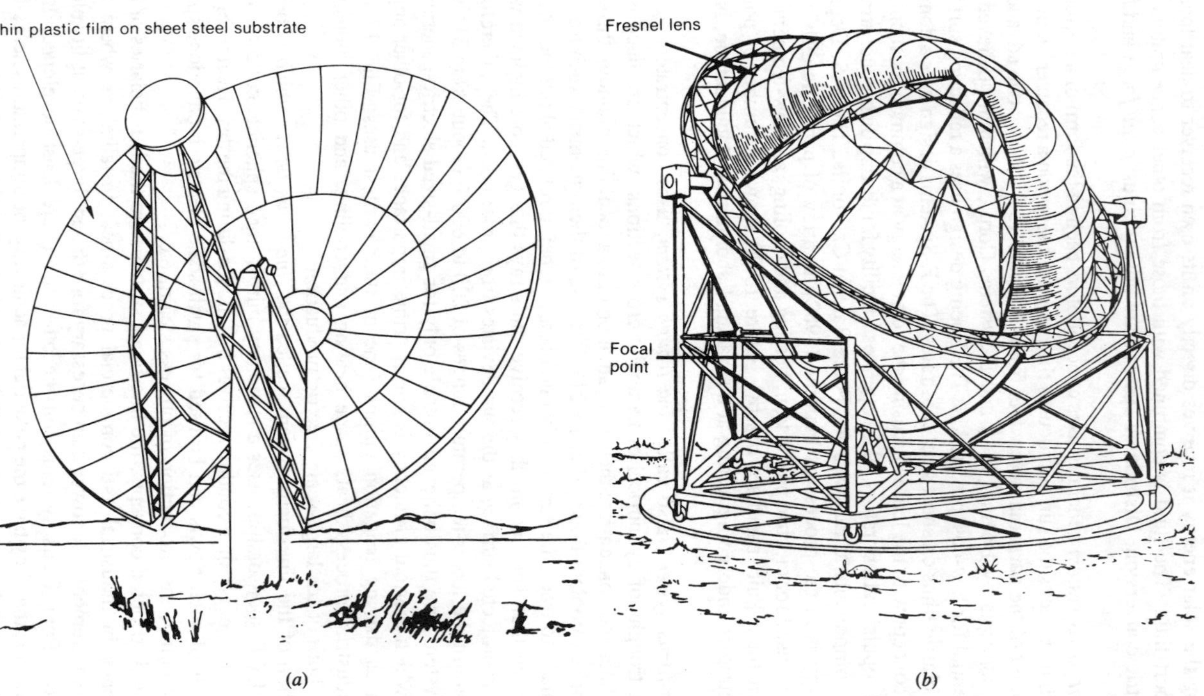

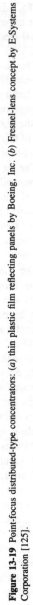

Figure 13-19 Point-focus distributed-type concentrators: (*a*) thin plastic film reflecting panels by Boeing, Inc. (*b*) Fresnel-lens concept by E-Systems Corporation [125].

Figure 13-20 Point-focus dish concentrators. *(Courtesy Sandia Laboratories.)*

Figure 13-21 Line-focus trough concentrators. *(Courtesy Sandia Laboratories.)*

Studies by DOE and EPRI show that "of all direct solar options under consideration by electric utilities (solar-thermal, photovoltaic, solar ponds, etc.), the prospects for solar-thermal conversion look the best," and that of all the solar-thermal conversion concepts, the central-receiver concept holds the greatest potential for electric utility applications [126]. The studies optimistically predict that, if carried successfully through the test and demonstration phases, solar-thermal central-receiver powerplants may become competitive with gas- and oil-fired powerplants in certain locations, such as in the southwestern United States.

13-13 OTHER SOLAR-THERMAL POWER SYSTEMS

Repowering

It is believed by some that the next step that will pave the way toward commercialization of solar-thermal systems, after the pilot plants such as Solar One and others around the world are tested, would most probably be the *repowering plant*. The repowering concept implies the adding of a solar-thermal-conversion plant to an existing fossil-fueled plant in sunny locations to replace some of the fossil fuel used in the latter, perhaps 20 percent. It is believed that the concept of repowering might be economically attractive in case the fossil fuels being replaced are gas or oil, which are expensive, but it is doubtful that would be the case with coal or nuclear fuels.

Hybrid Plants

A *hybrid system* is conceived to compensate for the intermittency of solar energy, not by storage but by using the sun as the source of heat when available and sufficient and a fossil fuel, such as oil or gas, at other times. DOE and EPRI studies of the hybrid concept have concentrated on the Brayton cycle as the main prime mover. The primary coolant, which would double up as the receiver heat-transport fluid and gas-turbine working fluid, could be (1) air, (2) helium, or (3) other gas mixtures. (An interesting variation is the use of gases that undergo reversible chemical reaction with temperature, such as SO_3, to give proper specific heat and enthalpy properties.) Although a Brayton cycle is simpler in concept than a Rankine cycle, it requires much higher temperatures in order to attain comparable efficiencies. In general a Rankine-cycle powerplant, solar or fossil, uses steam at temperatures of 1000 to 1100°F (about 540 to 590°C) and pressures of 1000 psia (about 70 bar) and beyond. Solar Brayton-cycle plants operate with gases at much lower pressures but, to be competitive, at much higher temperatures. This poses severe design and materials problems on the receiver. DOE and EPRI have sponsored studies aimed at the design and operation of receivers containing high-temperature metallic alloy and ceramic components. Figure 13-22 shows a ceramic honeycomb receiver developed by Sanders Associates and tested at the JPL Parabolic Disc Test Site, Edwards, California. The receiver is mounted on an 11-m-diameter test concentrator so that concentrated solar energy passes through

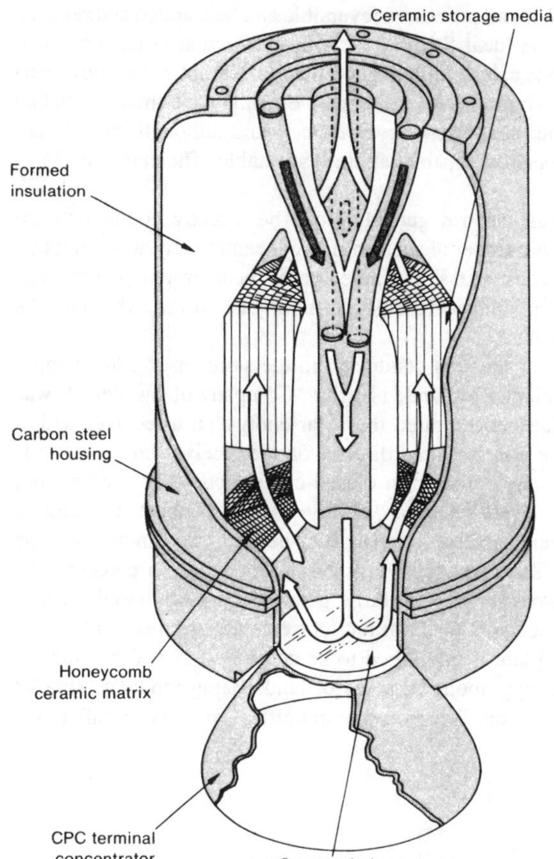

Figure 13-22 High-temperature honeycomb ceramic matrix receiver for Brayton-cycle applications [125].

the quartz window and is absorbed by the ceramic honeycomb. Air is pumped downward, reverses direction, and goes up the honeycomb, leaving at top center. Tests have shown air outlet temperatures ranging between 1470 and 2600°F (800 to 1430°C), which are useable in present-day gas turbines or those with cooled blades currently under development (Sec. 8-7).

Stirling-Cycle Conversion

Another hybrid concept uses the Stirling cycle. The ideal Stirling cycle is a regenerative cycle composed of four reversible processes: two constant volume and two constant temperature (Fig. 13-23). Heat is added and rejected during the constant temperature processes 2-3 and 4-1 respectively. Heat is exchanged internally between the constant-volume processes 3-2 and 1-2. With all processes reversible and heat added and rejected only at constant temperatures, the ideal Stirling cycle has a thermal efficiency equal to that of Carnot. No actual engine can be built that simulates a Carnot cycle. However, actual Stirling engines have been built based on the ideal Stirling cycle (much as actual Diesel engines are based on the ideal Diesel cycle, etc.), and although they do not have efficiencies equal to the ideal cycle, they attain respectable efficiencies of 35 to 40 percent.

A Stirling engine driving an electric generator can be directly coupled to the receiver of a solar concentrator. A number of such engine-concentrator systems, hooked in parallel electrically, can then constitute a distributed point-focus-system powerplant. In addition the engines can have an integral fossil-fuel (oil or gas) burner, which results in a distributed hybrid powerplant.

An experiment at the JPL test site was conducted in the summer of 1981 using a model P-40 Stirling engine developed by United Stirling Company of Sweden. It was mounted on a parabolic dish that concentrated the solar energy on a receiver with a fused silica aperture. The engine was mounted directly on the receiver body (Fig. 13-24). The engine working fluid, air, operated in closed-cycle mode with a maximum temperature of 1500°F (816°C). Heat rejection was by a water coolant. The engine was equipped with fossil burners, and no storage was needed. The induction-type alternator produced 25 kW(e). The conversion efficiency was about 35 percent. The experiment was designed to pave the way for large generating plants based on dish Stirling modules with dish diameters 9 to 22 m and corresponding power outputs of 15 to 100 kW(e). A 10-MW(e) plant is envisaged to be composed of 150 modules of 65 to 75 kW(e) each and to occupy about 50 acres of land. Being modular, smaller systems can be erected to supply small remote communities, military installations, etc.

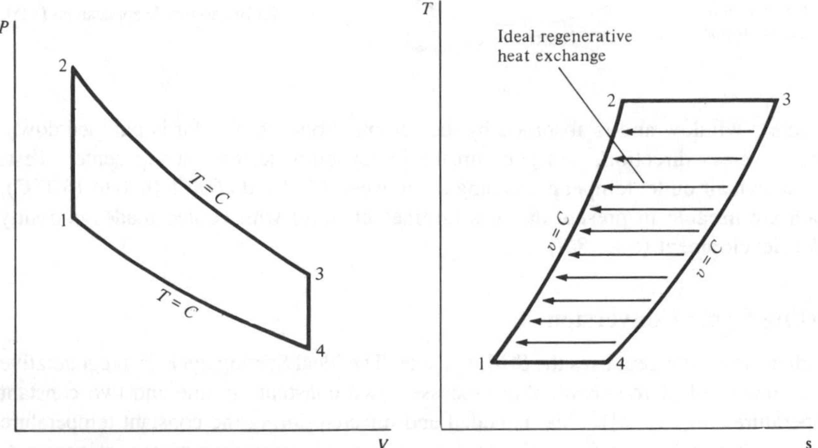

Figure 13-23 P-V and T-s diagrams of an ideal Stirling cycle.

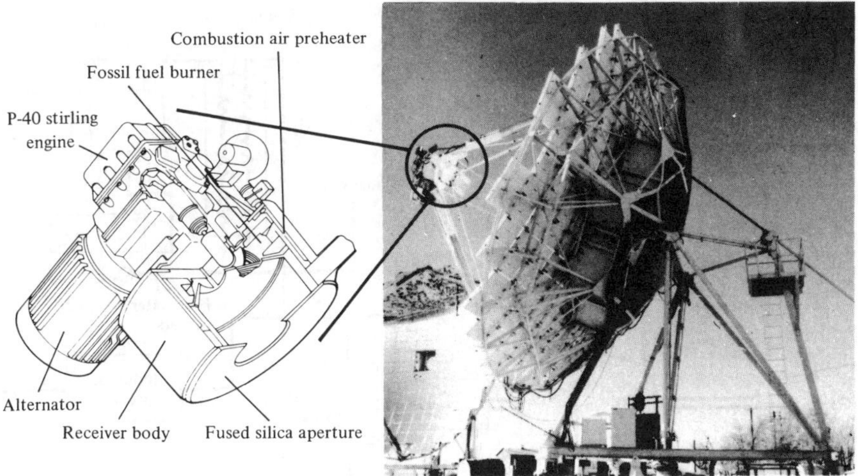

Figure 13-24 JPL test bed with parabolic bed concentrator and Stirling engine. *(Courtesy Jet Propulsion Laboratory.)*

Combined-Cycle Systems

Combined cycles are those using a combination of Brayton- and Rankine-cycle-type powerplants with the gas turbine of the Brayton cycle occupying the high-temperature end and exhausting to the steam generator of the Rankine cycle (Sec. 8-8).

Figure 13-25 shows a combined-cycle with a two-shaft gas turbine and a solar central-receiver system. Atmospheric air is compressed by the compressor into a high-temperature receiver where it is heated to perhaps 1500°F (815°C). It then expands through the compressor turbine and through the power turbine, which drives an electric generator. The exhaust air is then ducted to thermal storage and/or to the steam generator of the Rankine cycle, depending upon solar generation and load demand. During periods of no solar insolation the steam generator is heated by air driven through the thermal-storage system by a separate compressor. Supplemental auxiliary firing for the steam generator may be used.

An alternative scheme uses the gas-turbine exhaust air for industrial process heating.

13-14 SOLAR PONDS

Large bodies of water, such as lakes, receive vast quantities of solar radiation because of their large areas. Such bodies usually have water that contains solids in solution with concentrations that vary little with depth. When solar energy is received and absorbed below the surface (water at the surface is partially transparent to solar ra-

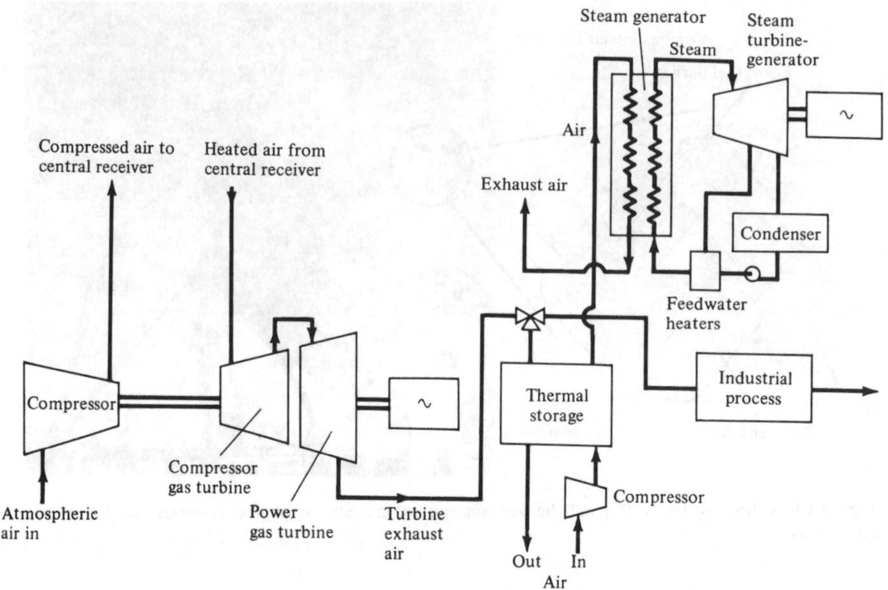

Figure 13-25 Schematic of a combined gas-steam cycle adapted to a solar central receiver.

diation), the warmer water becomes less dense and rises to the top. The result is a convective circulation that keeps cooler water at the bottom and retransports the solar energy back to the surface. There it is dissipated partly by conduction and convection to the atmosphere but mainly by evaporation since the vapor pressure at the surface is usually higher than the partial pressure of water vapor in the atmosphere. The latent heat of vaporization of the diffusing vapor is extracted from the water itself, and energy balances result in the water temperature remaining below ambient.

Some natural saltwater lakes, however, have exhibited the opposite behavior, namely, that the temperature gradient is reversed with warmer water at the bottom and cooler water at the top. That temperature gradient was found to exist because these lakes contained a nonuniform vertical concentration of salts with greater concentration at the bottom than at the top. This causes the bottom water to have a greater density and to remain at the bottom, even though it may be hotter. The solar energy absorbed in the deep layers is effectively trapped there because the effect of salt on density offsets the effect of thermal expansion. The temperature gradient is now reversed and temperature is higher at the bottom than at the top.

Solar ponds, also called *solar salt ponds,* take advantage of this phenomenon in two ways: (1) the conversion of solar energy to useful work as a result of the temperature difference between bottom and top and (2) the use of the pond as a thermal-storage medium so that the conversion can take place at a rate that is largely unaffected by daily and weekly fluctuations in solar input. A solar pond is, therefore, in effect, a combined solar collector and storage medium.

Figure 13-26 shows a schematic of an artificial solar pond and conversion system. The bottom warm water is used as the heat source for the powerplant, the top cool water as the heat sink. Because these layers flow out and in to their respective heat exchangers, they are subjected to convection currents. The pond is therefore divided into three layers. The central layer is nonconvective and contains the required salinity (and density) gradient with salinity greatest at the bottom. This central layer isolates a low-salinity (and density) convective layer at top, which is the heat sink, and a high-salinity (and density) convective layer at the bottom, which is both heat source and storage medium and which is insulated from the atmosphere above.

In constructing solar ponds, about one-third of a ton of salt is added per square meter of pond area. Provisions must be made to prevent leakage of the solution into the ground and subsequent heat transfer to it. In a well-designed and constructed solar pond, the bottom layer can reach about 200°F (93°C).

The hot bottom water is pumped through an evaporator and back to the bottom. The working-fluid vapor drives a turbogenerator. The turbine exhaust is condensed in the condenser, and the condensate is fed back to the evaporator. The condenser cooling water is obtained from the top cool layer of the pond. Because the temperature difference between the heat source and heat sink is not great enough, a turbine cycle working fluid other than water-steam, such as an organic fluid, is used.

The salt in an artificial solar pond tends to spread throughout the height of the pond by diffusion, a process that is slow but one that if allowed to continue would result in uniform salinity, convection currents, and an ineffective pond. The required

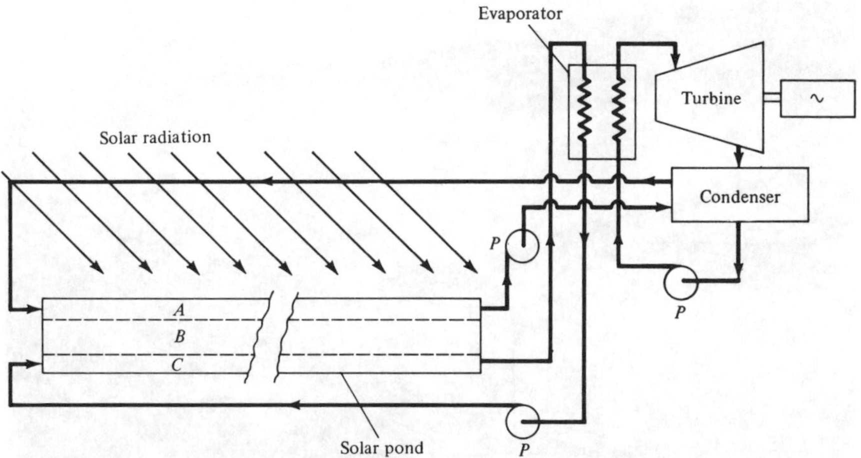

Figure 13-26 Schematic of a solar pond powerplant. A = low salinity-density convective layer; B = salinity-density gradient nonconvective layer; C = high salinity-density convective storage layer.

gradient is maintained by methods that include injecting salt to the bottom layer and flushing the top layer with fresh water.

Because of the relatively low temperature differences encountered in solar ponds, the thermal efficiency of the electric powerplants is nowhere near a fossil, nuclear, or even solar-thermal power system. They do, however, have the potential for trapping vast amounts of solar energy in regions where both solar energy and land are abundant. Besides electric generation, they may be more suitable for industrial uses such as space heating and cooling, crop drying, desalination, and other process heat. The necessity of having both large land mass and sunny climates, and also available water and salt, make solar ponds unlikely to be large contributors to the total energy picture, although they may have appeal in certain locations where there is land, sun, and salt or brakish water. Several test solar ponds have operated near the Dead Sea in Israel, for example. A 5-MW facility is under construction there (1982). In the United States studies are underway by the NUS Corporation to investigate the feasibility of a demonstration 300-kW solar pond facility in eastern California. Figure 13-27 shows an artist's conception of such a facility. The studies are aimed at determining the optimum pond areas, depths, salinity, and clarity and at evaluating the seasonal performance of ponds and conversion systems, the effects of atmospheric and ground and soil conditions, as well as costs of construction, operations, and maintenance.

Figure 13-27 Artist's rendering of a solar pond powerplant for the Los Angeles Water and Power. *(Courtesy NUS Corporation.)*

13-15 PHOTOVOLTAIC-ENERGY CONVERSION

Photovoltaic energy conversion is a direct conversion technology that produces electricity directly from sunlight without the use of a working fluid such as steam or gas and a mechanical cycle such as Rankine or Brayton. Photovoltaic systems, therefore, appear simple, convenient and, lacking moving parts, dependable. In addition they are modular, so that arrays of identical modules can be assembled to meet various power needs ranging from small residential systems installed on rooftops to relatively large central systems.

The basic unit of a photovoltaic system is the *solar cell*. The most common solar cells are made of highly refined silicon that is solidified as a single crystal in ingots (by the slow withdrawal of a seed from molten silicon), from which thin wafers are cut and polished (with the loss of almost half the original material). A doping material is introduced into the wafers to convert them into semiconductors (below) that are used to make photovoltaic cells. When sunlight strikes the cells, an electric potential is produced and current flows when the cell is connected to an external load.

Silicon solar cells are typically circular wafers, about 3 in (7.6 cm) in diameter and 300 μm thick, although square or rectangular wafers are being developed to increase the useful fraction of the total area exposed to sunlight when the cells are assembled side by side. A single cell typically produces a power of 1 W at a voltage of 0.5 V. They are then connected electrically in series-parallel arrangement, called a *module*, to produce the required current and voltage. A module is typically 4 × 4 ft (1.2 × 1.2 m). Several modules make up a *panel*. A panel is the design unit for assembling large photovoltaic *arrays* to meet the required power generation. Figure 13-28 shows a photovoltaic array field covering three-fourths of an acre and consisting of 2366 modules. It supplies 60 kW to a U.S. Air Force base at Mt. Laguna, California.

Photovoltaic electricity was first generated for a meaningful purpose in 1958 to power a radio transmitter on the Vanguard space satellite. It has since been used successfully to generate electricity in remote areas, such as in space, for navigation aids, in telecommunications on offshore installations, for isolated microwave relay stations, and similar uses. Coupled with batteries for storage, solar cells generate electricity for such uses more economically than any other method.

The costs, however, are very high when compared with other means where fossil or nuclear fuels are available and where the demands for electricity are very large. The unit costs of photovoltaic electricity started around $200 per peak watt for electricity from satellite modules in 1959. The price came down to $22 by 1976. Currently (1982) with annual manufacture of cells nearly 4 MW of peak electric capacity for remote needs, the price came down further to around $10, still a very high cost that can be afforded only by the space and defense industries. Government planners optimistically estimated a goal of 500 MW of annual photovoltaic electricity by 1986, a volume production that would cut the cost to $0.70 per peak watt (in 1980 dollars). This market goal does not now look likely to be reached, and the price is expected to remain relatively high.

It is to be noted further that the cost of end-use electricity is not dependent only

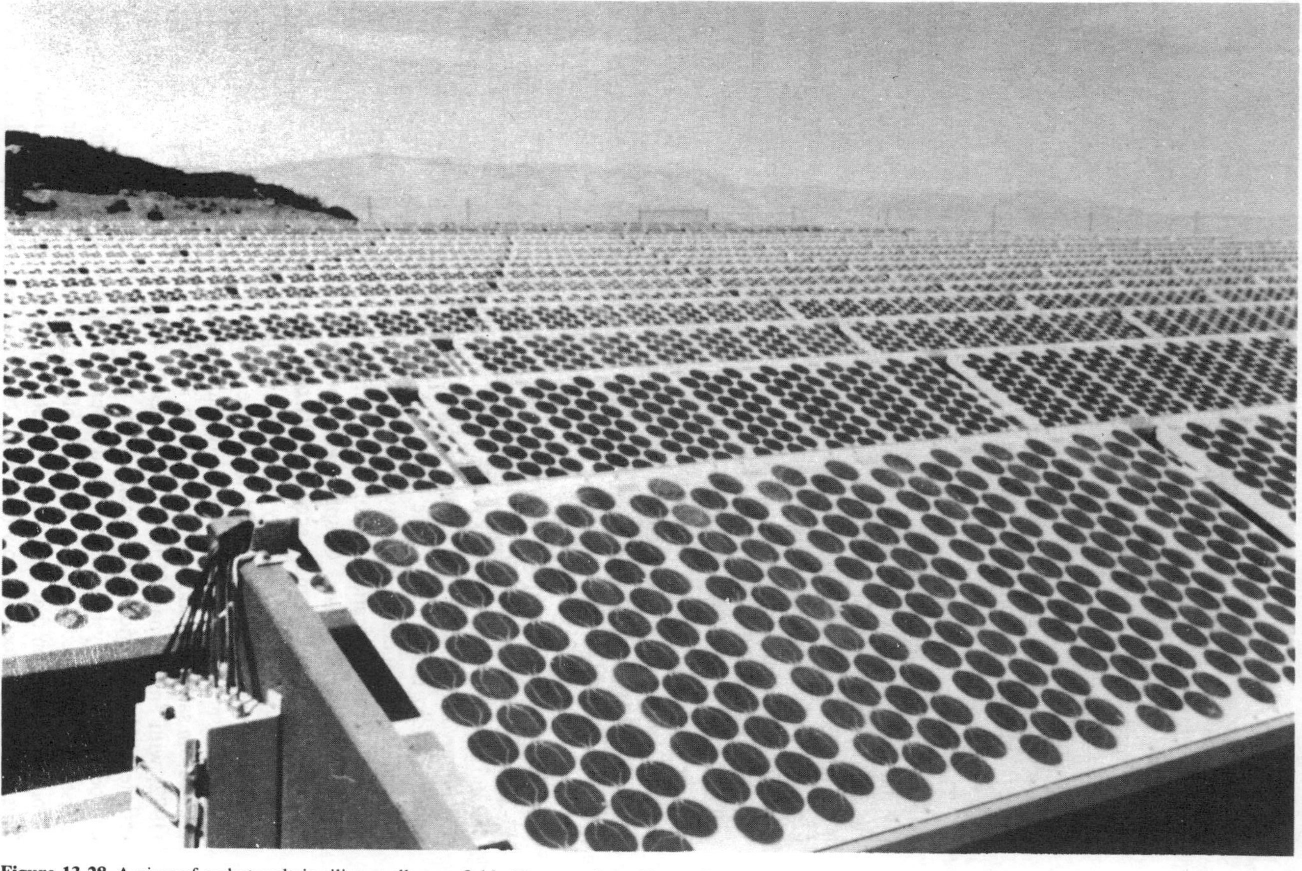

Figure 13-28 A view of a photovoltaic silicon-cell array field. *(Courtesy Solar Energy Research Institute, Golden, Colorado.)*

upon the cells alone but also upon the cost of the balance of the plant of which the cells are only a part. The nonphotovoltaic part of the plant, such as land, foundations, structures, wiring, dc-to-ac power conversion, connections to the utility grid, and cooling, currently represent two-thirds of the total capital cost. Thus, at $10 per peak watt the total plant capital cost would be about $30,000 per peak kW. This is some 15 to 20 times the cost of fossil or nuclear plants. Considering that a photovoltaic plant operates at peak only a small part of the time, the real ratio is much higher.

Efforts at reducing all costs include a search for lower-cost cell-manufacturing techniques [127] and for materials other than silicon. Besides single-crystal silicon, materials being investigated include multicrystal silicon, thin films of cadmium and copper sulfide, and amorphous (noncrystalline) silicon.

13-16 SOLID-STATE PRINCIPLES

In this and the next section we will cover some of the basic solid-state phenomena that are important in the study of the mechanism of direct conversion of solar radiation to electricity. These two sections may be bypassed without breaking the continuity of the chapter.

Solid Phenomena

A metallic crystalline solid contains atoms that have nuclei surrounded by electrons that are tightly bound to them (the shell model) and by outer electrons that are weakly bound to them. The latter are called *valence,* or *conduction,* electrons. They are free to migrate in the *interior* of the metal because there are no *net forces* on them because of the other free electrons or the ionized nuclei and their bound electrons. They thus move in an essentially equipotential field, each having a constant electrostatic *potential energy* E_i that is independent of its location inside the crystal, and they are considered to belong to the entire crystalline solid rather than to any one particular atom. These valence electrons constitute the primary mechanism of electric and heat conduction in a metal. At the surface, however, there are no positive ions on one side to give them equal attractive forces. Thus, while they are easily moved by electric fields within the metal, they encounter an *energy barrier* at the surface and require considerably more energy to get them out of the metal.

One therefore has an *electron gas* that is *confined* within the metal. It is not, however, equivalent to an ordinary gas whose energy distribution is given by the Maxwell-Boltzmann law. Electrons, instead, exist in states restricted by the so-called *Pauli exclusion principle* (which specifies that no two electrons in the same atom can exist in the same state at the same time), and their energy distribution is given by the *Fermi-Dirac distribution law,* given for temperatures that are not too high (not greater than 3000 K) by [128]

$$n(E)\,dE = \left[\frac{4\pi}{h^3}(2m_e)^{3/2}\right]\frac{E^{1/2}}{1 + e^{(E - E_F)/kT}}\,dE \qquad (13\text{-}3)$$

where $n(E)\, dE$ is the number of electrons per unit volume in the energy range dE. The quantities between brackets are constants: h is Planck's constant (6.625×10^{-34} J · s), and m_e is the mass of the electron (9.13×10^{-28} g). The balance of the righthand side is energy dependent. k is the Boltzmann constant [1.38×10^{-23} J/(molecule · K)]. The quantity E_F is called the *Fermi energy*. It is a constant for many cases of interest, being nearly independent of temperature.

A plot of the energy distribution of an electron gas, Eq. (13-3), is shown in Fig. 13-29 at different temperatures. The quantity $E^{1/2}$ contributes the parabolic rise of the curves from $E = 0$. The quantity

$$P(E) = \frac{1}{1 + e^{(E - E_F)/kT}} \qquad (13\text{-}4)$$

is called the *Fermi-Dirac probability distribution function*. Note that, unlike a classical gas, free electrons do *not* all have zero energy at absolute zero but rather finite energies *up to a maximum* given by E_F, the Fermi energy. At $T = 0$, the probability distribution function is 1 from $E = 0$ to $E = E_F$, meaning that the probability that any state between energies 0 and E_F is occupied by an electron is 1 and is zero for $E > E_F$.

At higher temperatures, the high-energy portion of the distribution is different from that at $T = 0$. The probability that states much less than E_F are occupied is still 1, the extent depending upon the value of T. For $E = E_F$ the probability is exactly 1/2 (independent of temperature). For energies much greater than E_F, the exponential term is much greater than the 1 in the denominator of Eq. (13-4), and the probability reduces to the Maxwell-Boltzmann distribution probability (given by $Ae^{-E/kT}$, where A is a constant).

The Fermi energy can be computed by evaluating the total number of free electrons per unit volume n as

$$n = \int_0^{E_F} n(E)\, dE = \frac{4\pi}{h^3}(2m_e)^{3/2} \int_0^{E_F} E^{1/2}\, dE$$

$$= \frac{8\pi}{3h^3}(2m_e)^{3/2} E_F^{\;3/2}$$

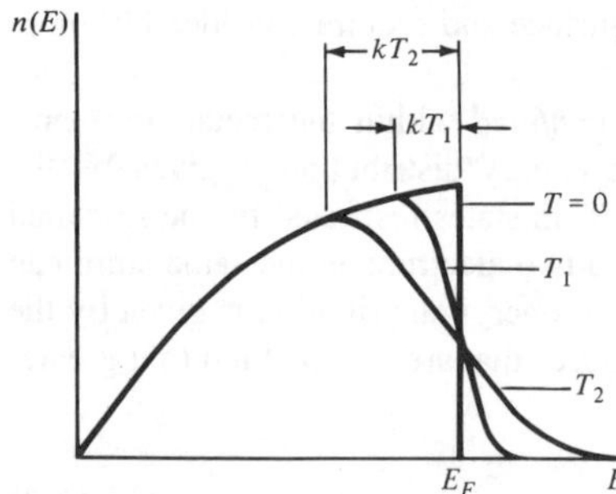

Figure 13-29 Energy distribution of an electron gas at different temperatures, $T_2 > T_1 > 0$.

from which

$$E_F = \frac{h^2}{2m_e}\left(\frac{3n}{8\pi}\right)^{2/3} \tag{13-5}$$

The values of E_F for metals are typically of the order of a few electron volts (eV),* about 7 eV for copper and 3.1 eV for sodium. It can also be easily shown that the *average energy* of a free electron at $T = 0$ is equal to 3/5 E_F, or a few electron volts, much higher than the average kinetic energy of a particle in a classical gas even at high temperatures, which is given by 3/2 kT, thus being zero at $T = 0$, about 0.025 eV at room temperature, and 0.1 eV at 1160 K.

The behavior of free electrons at temperatures higher than 0 is important. At moderate temperatures, the corners of the zero temperature distribution (Fig. 13-29) are only slightly rounded. The difference increases as the temperature increases and becomes significant only within the energy range

$$|E - E_F| \approx kT \tag{13-6}$$

and, as indicated previously, the probability that the state $E = E_F$ is occupied is 1/2. The free electrons whose energies are much less than the Fermi energy remain locked in the same energy states they occupied at $T = 0$ irrespective of temperature. A fraction of the most energetic electrons, those having energies within kT of the Fermi energy, occupy higher energy states than the Fermi energy. Such electrons can be elevated to these energies by collision with energetic particles such as high-energy photons or by thermal excitation (the basis for thermionics).

The Band Theory

Photovoltaic devices, like thermoelectric devices, require a knowledge of the distinction between conductors, insulators, and semiconductors. Electrical materials differ in their electrical conductivities by factors as high as 10^{30}. The *band theory* of solids helps us understand this distinction as well as the phenomena of electrical and thermal conduction. A quantitative treatment of the band theory involves detailed wave mechanics and is beyond the scope of this book. The following is an adequate (for our purposes) qualitative presentation of some aspects of that theory.

In a system of noninteractive (or isolated) atoms of the same species, as in a gas, the valence electrons occupy a set of single-energy levels, corresponding to permitted states of that species (Fig. 13-30*a*). When these atoms are brought together, such as in a crystalline solid, they interact strongly and thus cause the energy levels to spread throughout a set of *energy bands* (Fig. 13-30*b*). The regions between the energy bands of the interacting atoms, like those between the levels of noninteracting atoms, cannot be occupied by electrons and are called *forbidden bands*. The widths and spacings of

* An electron volt (eV) is a unit of energy equal to the energy acquired by a single electron charge when accelerated through a potential of 1.0 V. 1 eV = 1.6021×10^{-19} J = 4.44×10^{-26} kWh = 1.519×10^{-22} Btu. 1 million electron volts (MeV) = 1.6021×10^{-13} J.

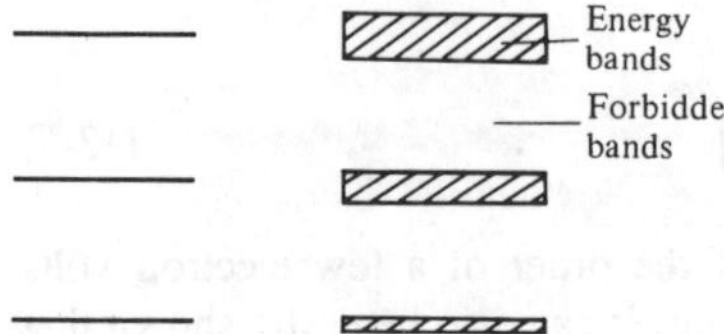

Figure 13-30 (*a*) Electron energy levels in a noninteractive atom. (*b*) Electron energy bands in an interactive atom.

the energy and forbidden bands depend upon the crystalline material in question. Energy bands may be occupied, only partially filled, or completely unoccupied. The shading in Figs. 13-30 to 13-32 indicates which bands are occupied.

High conductivity in a metallic crystal is caused by its uppermost band *not* being completely filled with electrons. An external electric field causes the electrons in this band to gain small amounts of energy that are sufficient to promote them to the continuum of available states immediately above this band. Distribution of electrons among the available states may be varied slightly by energizing them. The shift is controlled by the Fermi-Dirac statistics (above), and therefore significant shift occurs only for those electrons that are within a range $\pm kT$ of energy about the uppermost filled level at $T = 0°$.

Figure 13-31 shows energy bands of a conductor (sodium) and an insulator (diamond). For sodium, the 3*s* band is partially filled, the 3*p* band is unoccupied, and the two bands overlap. The number of unoccupied levels readily available for the 3*s* band electrons is therefore large, which results in high electrical conductivity. Diamond, on the other hand, has two 2*p* energy bands, one filled, one unfilled, separated by a forbidden band that is 6 eV wide. This *gap* is much larger than *kT* at room or high temperatures (above) so thermal excitation, or weak electrical or other fields, will not impart sufficient energy to promote the electrons to the unoccupied 2*p* band and create electron flow or an electric current. The same applies to photons of visible light that do not lose energy to the electrons, which is the reason why diamond is transparent to visible light.

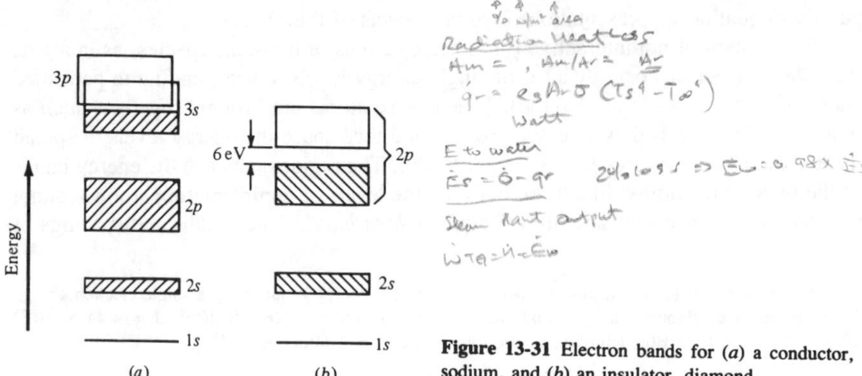

Figure 13-31 Electron bands for (*a*) a conductor, sodium, and (*b*) an insulator, diamond

The uppermost filled, or partially filled, energy band is called the *valence band,* and the next empty but available energy band is called the *conduction band*. A substance is a conductor if these two bands are separated by a very narrow forbidden gap or if they overlap. A substance is an insulator if they are separated by a large gap.

13-17 SEMICONDUCTORS

There is a class of crystalline solids where the forbidden band between the valence and conduction bands is relatively small, about 1 eV. Silicon and germanium are examples, with gaps of 1.1 and 0.7 eV, respectively. Such solids behave as insulators at low temperatures. At very high temperatures, an appreciable number of electrons receive sufficient thermal energy to be promoted into the conduction band, the number being a function of both the temperature and the gap width—and thus the material becomes a conductor. Such a material (a pure material) is called an *intrinsic semiconductor*.

Semiconductors are manufactured by adding controlled amounts of impurity to the pure material, 10^{17} atoms per cubic centimeter being typical. These are called *extrinsic,* or *doped, semiconductors*. The impurity adds an allowed energy level (not band) in the forbidden band between the valence and conduction bands. There are two types: *n-type* semiconductors and *p-type* semiconductors (Fig. 13-32).

In *n-type semiconductors* the allowed energy level is near the bottom of the conduction band; this is called the *donor level*. It has an abundance of electrons that, at room temperatures, are easily excited into the conduction band (since the gap is very narrow). By contrast only a few electrons (one shown) are intrinsically (due to the pure material) elevated from the valence to the conduction bands.

In *p-type semiconductors* the allowed energy level is near the top of the valence band; this is called the *acceptor level*. The impurity has an electron deficiency. At room temperatures, electrons are easily excited from the valence band to the acceptor level, thus leaving a deficiency of electrons in the valence band. The absence of an electron is called a *hole* and may be considered a positively charged particle. Again

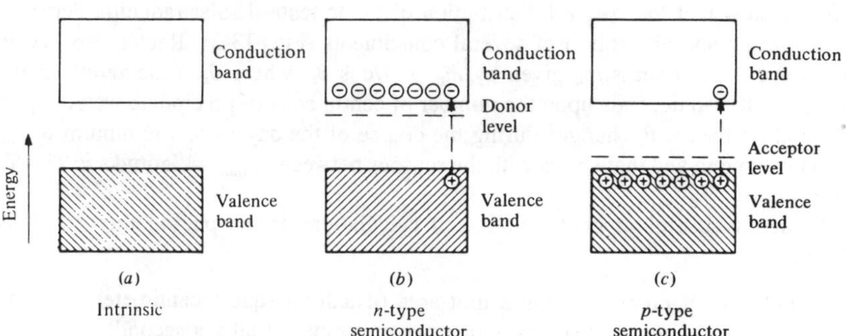

Figure 13-32 Electron energy bands for three types of semiconductors.

intrinsic motion of electrons is small by comparison and conduction in a *p*-type semiconductor, therefore, is largely the result of the movement of holes to the valence band. In both types of semiconductors, the conductivity is largely determined by the amount and form of the impurities.

Photons

Energy, on a microscopic scale, is transferred not continuously but in discrete pieces, the smallest of which is the *quantum*. Light is radiant energy, whose quantum is called a *photon*. Max Planck was the first to suggest that the energy of a photon is proportional to the frequency of radiation, or

$$E_p = h\nu = h\frac{c}{\lambda} \tag{13-7}$$

where E_p = photon energy

h = Planck's constant, 6.6256×10^{-34} J · s, 4.13576×10^{-15} eV · s

ν = frequency of radiation, hertz

c = speed of light, 2.997925×10^8 m/s

λ = wavelength of radiation, m

At any given frequency ν, light energy is thus a whole or multiple of $h\nu$, never a fraction of it. Light thus has a dual personality. It is characterized both as energy transferred in discrete amounts, the photons, and in waves having a frequency and, therefore, wavelength. Radiation from the sun, or any other radiant source, is therefore composed of energy chunks, the photons, each one carrying a quantity of energy exactly equal to its frequency times Planck's constant.

Thus a monoenergetic radiation beam having a wavelength of half a micrometer, 0.5 μm = 0.5×10^{-6} m, the energy of a single photon is

$$E_p = 4.13576 \times 10^{-15} \frac{3 \times 10^8}{0.5 \times 10^{-6}} = 2.48 \text{ eV}$$

Recall now that the spectral distribution of the terrestrial solar radiation depends upon scattering and absorption of several constituents (Fig. 13-3). Recall also that it depends upon an air mass m_a given by $m_a = 1/\cos \theta_z$, where θ_z is the zenith angle (Sec. 13-4). It also depends upon the number of centimeters of precipitate water vapor ω in the atmosphere. θ_z changes during the course of the day from a minimum $\theta_{z,\min}$ that occurs at noon and that varies with the seasons between $\theta_{z,\min}$ = latitude $\pm 23.45°$ and 90°.

A quantity useful in photovoltaic-cell calculations is the photon flux ϕ_p. It is defined as

ϕ_p = number of photons crossing a unit area (usually a square centimeter) perpendicular to beam radiation per unit time (usually a second)

The solar energy flux E'' is related to the photon flux simply by

$$E'' = \Sigma_i \phi_{p,i}\, h\nu_i \tag{13-8a}$$

where the subscript i indicates a frequency or energy range. Equation (13-8a) may be simplified by assuming an average frequency ν_{av} and corresponding wavelength λ_{av} and obtaining a total photon flux ϕ_p. Thus

$$E'' = \phi_p h\nu_{av} = \phi_p h(c/\lambda_{av}) \tag{13-8b}$$

Outside the earth's atmosphere, the solar energy flux is equal to the solar constant $S = 1359$ W/m^2, or 0.1359 W/cm^2, $m = 0$, $\omega = 0$, and the average photon energy $h\nu_{av} = 1.48$ eV.

Thus
$$\phi_p \text{ (extraterrestrial)} = \frac{0.1359}{1.48} \times \frac{1}{1.6021 \times 10^{-19}}$$
$$= 5.8 \times 10^{17} \text{ photons/(s} \cdot \text{cm}^2)$$

Both terrestrial insolation and the average photon energy decrease as m increases and as ω increases. The photon flux thus decreases from $5.8 \times 10,^{17}$ above, to about half that amount at $m = 3$ and $\omega = 5$ [129].

When the sun's rays, i.e., the sun's photon flux, strike a pn semiconductor junction, they help generate electron-hole pairs; i.e., they cause electrons to be raised to the conduction band in the n material and holes to be moved to the valence band in the p material. When connected to a load, electrons will thus diffuse from n to p across the junction, thus creating an electric current through the load and hence electric power, which is a function of the photon flux.

13-18 THE SOLAR CELL

Figure 13-33 shows a schematic representation of a solar cell composed of pn semiconductor junctions. For single-crystal silicon, p is obtained by doping silicon with boron and is typically 1 μm thick; n is obtained by doping silicon with arsenic and is typically 800 μm thick. Thin film cells are composed of copper sulfide for p, typically 012 μm thick, and cadmium sulfide for n, typically 20 μm thick.

The sun's photons strike the cell on the microthin p side and penetrate to the junction. There they generate electron-hole pairs. When the cell is connected to a load as shown, the electrons will diffuse from n to p. The direction of current I is conventionally in the opposite direction of the electrons.

Typical voltage-current characteristics are shown in Fig. 13-34 at two different solar radiation levels. For each

V_o = open-circuit voltage

I_o = short-circuit current

P_m = point of maximum power $(VI)_{max}$

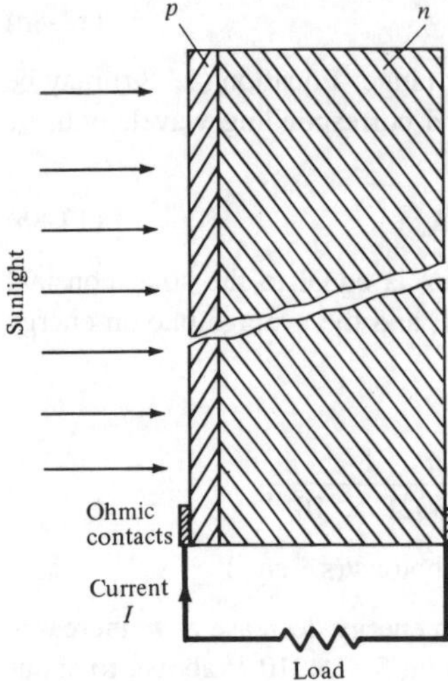

Figure 13-33 A schematic cross section of a solar cell.

Solar cells do not convert all solar radiation falling upon them to electricity. Weak, low-frequency (long-wavelength) photons do not possess sufficient energy to dislodge electrons. Strong, high-frequency (short-wavelength) photons are too energetic, and although they dislodge electrons, some of their energy is left over unused. Table 13-4 shows a breakdown of solar energy wavelength distribution, the fraction utilized by a typical cell and the percent solar energy converted, ideally, to electricity.

There, thus, is a maximum theoretical efficiency of solar cells, around 48 percent. *Efcciency* is defined as the ratio of electric power output of the cell, module, or array to the power content of sunlight over its total exposed area. Efficiencies of modules

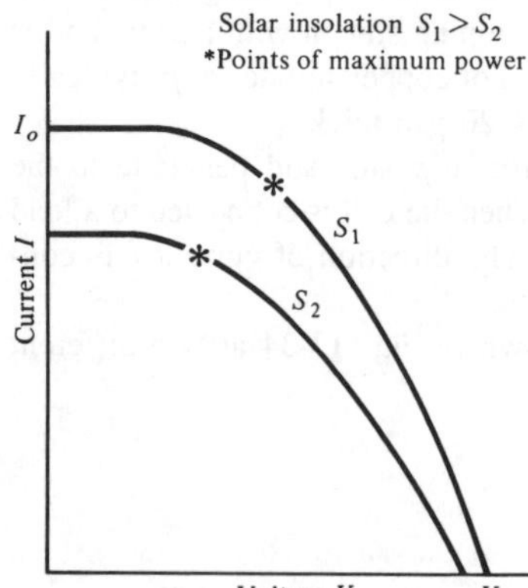

Figure 13-34 Typical performance characteristics of a silicon solar cell at two solar radiations.

Table 13-4 Ideal spectral solar energy utilized by silicon cells

Wavelength range, μm	Solar energy, %	Fraction converted, by cell	Solar energy converted, %
<0.3	0	0	0
0.3–0.5	17	0.36	5
0.5–0.7	28	0.55	15
0.7–0.9	20	0.73	15
0.9–1.1	13	0.91	12
>1.1	22	0	0
		Total	48

or arrays are therefore lower than those of the cells because of the areas between the individual cells. The fraction of cell to total areas is called the *packing factor*.

Actual efficiencies are much lower, however, because part of the solar energy is reflected back to the sky, absorbed by nonphotovoltaic surfaces, or converted to heat; and because of various electrical losses and the recombination of the electron-hole pairs, a process that is encouraged by increased temperatures. Cells are usually laboratory rated at 1000 W/m^2 and 28°C but normally operate at 50 to 60°C. This reduces the efficiency by 1 or 2 percent. With modules and arrays, there are additional losses that result from the mismatch between individual cells in a module and between modules in an array. Thus the best single-crystal cells yield efficiencies of about 16 to 17 percent. Mass-produced modules yield efficiencies seldom exceeding 10 percent. Table 13-5 gives a typical energy balance of a nonconcentrating-silicon photovoltaic array, showing an array efficiency of about 8 percent. Methods of increasing cell efficiency besides a search for other materials include *concentration, thermophotovoltaic,* and *cascade* systems.

Concentrators improve cell efficiency. They subject the cells to concentrated sunlight by putting them at the foci of parabolic or trough concentrators. EPRI-sponsored research indicates cell efficiencies of about 25 percent can be achieved with a concentration of about 500 suns.

Concentration has other advantages: With improvements in efficiency, fewer cells as well as fewer arrays, structures, and tracking equipment are needed per unit output, and the high cost becomes a bit more bearable. Concentrators, however, are more expensive than flat-plate photovoltaic arrays, are effective only with direct solar radiation, i.e., excluding most of the diffuse radiation, and lose about 15 percent of that radiation in various optical losses. In addition the high cell temperatures and high currents result in higher electrical losses. Concentrators, therefore, are cost-effective only if the improvement in cell efficiency results in performance far exceeding those obtained from flat-plate arrays.

Thermophotovoltaic systems, considered seriously at one time, involved the use of highly concentrated light that is absorbed by a refractory material that becomes hot and reradiates the solar energy to silicon cells at longer wavelengths. Such wavelengths are more efficiently converted to electricity by the silicon cells (column 3, Table 13-

Table 13-5 Typical energy balance of a nonconcentrating silicon photovoltaic conversion array, arbitrary units

<table>
<tr><th>Input on array</th><th colspan="6">Energy distribution</th></tr>
<tr><td rowspan="7">100</td><td rowspan="2">In nonphotovoltaic material</td><td>12</td><td colspan="4">Reflection by and absorption in cover glass</td></tr>
<tr><td>13</td><td colspan="4">Absorption by frames, structures, earth</td></tr>
<tr><td rowspan="5">In photovoltaic material</td><td rowspan="5">75</td><td>Nonelectric</td><td>64</td><td colspan="2">Dissipation as heat in silicon</td></tr>
<tr><td rowspan="4">Electric</td><td rowspan="4">11</td><td>1.5</td><td>Losses due to cell temperature above 28°C</td></tr>
<tr><td>0.5</td><td>Losses due to cell and module mismatch</td></tr>
<tr><td>1.0</td><td>Losses in wiring and dc-to-ac conversion</td></tr>
<tr><td>8.0</td><td>Delivered as ac power</td></tr>
</table>

4). The temperatures required, however, are in the neighborhood of 3400 to 3500°F (1870 to 1925°C), which results in severe materials problems. It is now believed that thermophotovoltaic systems do not attain efficiencies higher than concentrated systems to warrant these high temperatures.

Cascade systems are in essence multijunction systems in which different cells are subjected to different regions of the solar spectrum at which they operate most efficiently. Such systems are believed to yield conversion efficiencies even higher than 25 percent, but they are still in the developmental stage.

13-19 PHOTOVOLTAIC-ENERGY STORAGE

Solar-thermal systems, as has been noted, usually use thermal storage. Photovoltaic systems, on the other hand, must share conventional powerplant grids or must use electrical storage if their output is to last longer than sunlight. Several schemes are being considered. Some are:

1. *Electrochemical storage*. This is storage of electric energy by conversion to chemical energy in batteries. The most common and most highly developed is the lead-acid battery. Other battery systems, still in the developmental stages, have higher energy-to-mass ratios than the lead-acid battery. Large electric energy storage, on a utility scale, in lead-acid or other batteries, however, is not economically feasible.

2. *Pumped-hydro storage*. This method is more suitable to large powerplants. It involves the use of surplus electric energy to pump water into high reservoirs during sunny periods or periods of low demand and the extraction of power during evening or cloudy periods or periods of high demand by running the same down through waterturbines. The storage and regeneration are done via reversible pump-turbine motor-generator sets (Sec. 16-3). The energy extracted is less than that stored because of losses in both pumping and generation. Another problem is the necessity of finding sites with suitable topography near solar powerplants, which tend to be located in desertlike flat terrain. The system is in use with conventional plants as a load leveler but on a somewhat limited scale.

Variations of hydropump storage would be underground pumping and compressed-air storage (Sec. 16-4).

3. *Cryogenic storage*. This is a system in which electric energy is directly stored in large underground electrical coils at liquid-helium temperatures, about 4 K. At that temperature the electrical resistivity of the coils is nearly zero. The system is currently under development, and the costs are still an unknown factor (Sec. 16-7).

13-20 SATELLITE SOLAR-POWER SYSTEMS

The earth *satellite solar-power system* (SSPS) is based on technological advances stemming from the space program. First proposed by Glaser of Arthur D. Little, Inc., in the late 1960s [130], the concept has since received wide attention and serious consideration.

The concept involves the placement of earth satellites that would function as solar-energy collecting stations in geostationary or synchronous orbits around the earth. Such orbits would be at an altitude of about 22,300 mi (36,000 km) and would be equational, i.e., parallel to the earth's equational plane. A satellite traveling from east to west in that plane would have an angular velocity equal to that of the earth and would then appear stationary with respect to any point on earth. (Geostationary civilian and military communication satellites are in common use around the earth.)

Figure 13-35 shows a schematic of the concept. The satellites would have large collectors of photovoltaic arrays. They would also have conversion systems that would convert the electric power generated by the arrays into power at microwave frequencies. A large transmitting antenna on each satellite would beam the microwave energy from its fixed position relative to the earth to a receiving station on the surface of the earth. That station would have a large receiving antenna that would reconvert the microwave power into ac electric power and feed it into a conventional power transmission grid. The satellites, being so high above the earth, would be in sunlight *most* of the day, and no electric-energy storage would be needed.

The attitude controls of an SSPS, possibly through use of laser technology, must see to it that the collector areas are constantly facing the sun and that the transmitting antenna is constantly facing the receiving antenna on earth. Still, the SSPS would have to pass through the earth's shadow once a day (Fig. 13-36), so that a complete

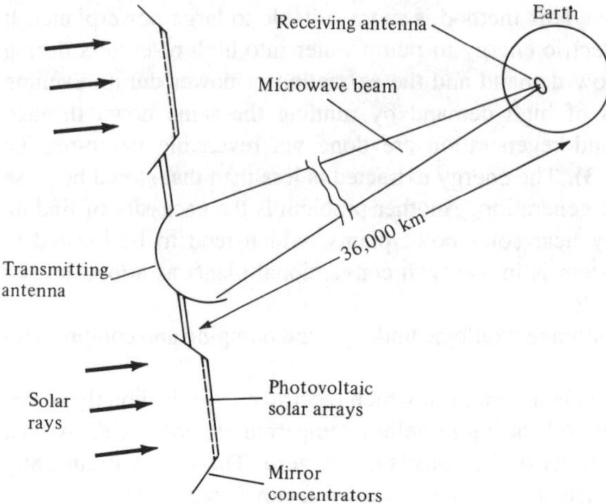

Figure 13-35 A schematic of an earth satellite solar-power system (SSPS).

cutoff of power from any one satellite is experienced about 5 percent of the time. A possible solution to this would be a system that would consist of two geostationary satellites separated by about 7900 mi (12,700 km) and thus about 20° out of phase, both having a direct line of sight to the same receiving antenna on earth. Such a system would ensure that one would be illuminated during the time the other is in the earth's shadow. This woulu mean a 50 percent power cutoff during roughly 10 percent of the time, instead of a 100 percent cutoff during 5 percent of the time, and a possibly

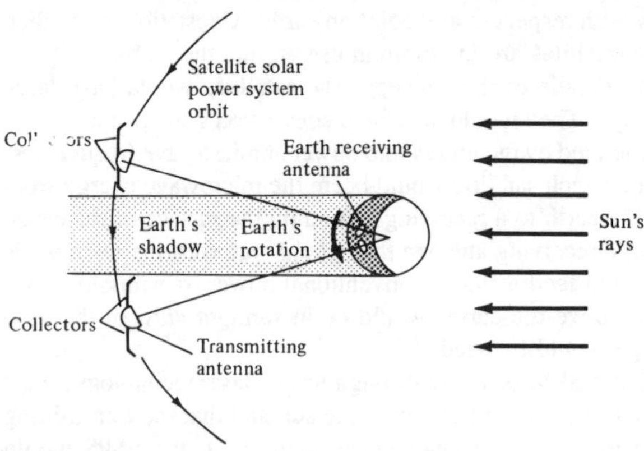

Figure 13-36 Effect of the earth's shadow on SSPS operation, showing satellite and receiving antenna at two different times (not to scale).

better match to load demands. Additional satellites would even the power output further.

Solar insolation in space is 40 percent larger than insolation on earth. It is available to the satellite for nearly 0.95×24 h/day, whereas on earth it is available for only 4 h/day figured at maximum intensity. Thus, the output of an SSPS would be about 8 times that of a photovoltaic plant located on the surface of the earth, with a comparable solar collector area.

Energy analyses on SSPS systems [130, 131] show them to be adaptable to various power ranges. A plant resulting in 10 GW on earth is believed to require two square photovoltaic arrays, each 4 km on the side, a 1-km-diameter transmitting antenna, and a 7-km-diameter receiving antenna on earth. A much higher density system envisages the transmission of 20 GW to an earth receiving antenna only 3 km in diameter. The microwave power density on that antenna would be less than 10 kW/m,^{2} about an order of magnitude greater than terrestrial solar radiation density. Highly efficient solid-state rectifiers would be required to absorb this power, and the converted power would be distributed through superconducting transmission lines operating at cryogenic temperatures.

It is estimated (Arthur D. Little, Inc.) that a number of SSPSs with a total solar-collector area of about 4900 km^2 of arrays of single-crystal silicon cells operating at their upper theoretical maximum efficiency of 24 percent would supply the total electric-energy demands of the United States in the year 2000. Such arrays would have a total mass of about 1460 million metric tons, excluding structures, etc. This is considered impractical, and other more efficient cell materials are being sought. Thin-film organic cells, still early in their development stage, are believed to attain efficiencies of 80 percent and require about 920 km^2 of collector area and a mass of about 5.9 million metric tons [130].

The launching, deployment, and assembly of the large structures of an SSPS, such as the collectors, transmitting antenna, attitude-control systems, and operator life-support systems in outer space, now seem like problems of gigantic proportions. However, it is not unreasonable to believe that if and when SSPSs are really ready to go into space, space technology will have also advanced to a point where such undertakings have become feasible. Space transportation is currently estimated to account for about 45 percent of the total capital cost of the system.

The transmitted energy should have frequencies in a spectral region having the least absorption and scattering in the earth's atmosphere. Most of the absorption occurs at the 22 GHz line of the water vapor molecule and the 60 GHz line of the oxygen molecule [131]. Below 10 GHz the attenuation caused by molecular absorption is approximately 10 dB or less. Attenuation by rain, cloud droplets, snow, and hail depends upon their size distribution and composition. The most serious is the result of rain clouds and may reach 4 percent at 3 GHz, and a 1-km path through wet hail may cause 13 percent attenuation at the same frequency.

Scattering, which causes a broadening of the main microwave beam, is not expected to be a significant problem. A 5-GW beam operating at 3 GHz is believed to scatter only 3 mW isotropically in a 1-km-high storm; the result is a scattered beam density of only about 2×10^{-4} mW/cm^2 at a 10-km range.

Environmental Effects

Studies have been made to ascertain expected various environmental and biological effects of SSPS microwave beams [132]. Some of these are:

1. *Stratospheric pollution*. This is caused by the space vehicles that deploy the satellites and transport personnel and equipment. This is already a problem of concern due to present military and civilian supersonic aircraft and space shuttle exhaust. Although not fully evaluated, the injection of water vapor and NO_x is believed to lower the ozone concentrations and requires further investigation.
2. *Thermal pollution*. This is a result of waste heat at the receiving antenna. It is believed to be in the neighborhood of 15 percent, which is about a fourth of the waste heat generated by the most efficient conventional powerplants and hence of little concern.
3. *Land exclusion*. A 5-GW plant is believed to require a land area of about 105 mi^2 (270 km^2) from which the public would be excluded (of which one-third would be covered by the receiving antenna.) This compares favorably to the land required by conventional plants of the same output. Offshore locations may also be an attractive alternative.
4. *Radio-frequency interference*. The most desirable SSPS frequencies are in heavy use for worldwide communications and are based on internationally assigned frequencies. Thus, there is a high probability of interference with existing communications systems. Such radio-frequency interference can be controlled by careful selection of frequencies, narrow-band operation, and the use of filters. Detailed and specific effects will have to be carefully studied, however.
5. *Biological effects*. Standards for microwave exposure are already in effect. In the United States, the maximum, based on microwave heating of body tissue, is 10 mW/cm.^{2} In the USSR, the maximum is based on possible effects on the central nervous system and is a much more conservative 0.01 mW/cm.^{2} The induced biological effects of the SSPS thus need careful study. It is important that adequate, fail-safe controls of the microwave-beam pointing apparatus be incorporated in the development program.

   Microwave effects on birds in flight through the microwave beam need evaluation. Preliminary evidence suggests that birds are affected at 25 mW/cm^2 exposures in the X-band. Effects on aircraft flying through the beam are also of concern, although the shielding effect of the metal fuselage and the short exposure time at aircraft speeds make it unlikely to result in significant human exposure. A more worrisome effect, perhaps, is that of microwaves on aircraft operations, such as fuel tank susceptibility to electrical discharges (now a standard protective feature of aircraft design) and interference with aircraft communications and radar equipment.

PROBLEMS

13-1 From the solar constant and the average distance between the sun and the earth, estimate (*a*) the total power emitted by the sun, in megawatts, and (*b*) the mass loss of the sun, in tons per day (Sec. 9-4).

13-2 For the city of Albuquerque, New Mexico, estimate the average yearly and maximum and minimum monthly clearing indices on a horizontal surface, and the corresponding possible power generation, in kilowatthours per square meter per day, if the average solar-electric conversion efficiency is 0.08.

13-3 A central-receiver solar thermal powerplant uses 1000 heliostats that have 50 m^2 of reflecting surface each. The steam plant has an overall efficiency of 27 percent and generates 3.0-MW peak power. Using a reasonable solar energy insolation on the heliostats on a clear sunny day, estimate (*a*) the overall efficiency of the solar powerplant and (*b*) the efficiency of energy transmission to the steam in the central receiver, both at peak conditions.

13-4 A central receiver solar thermal powerplant uses 1000 heliostats that have 60 m^2 of reflective surface each. The overall efficiency of the plant is 5 percent. The efficiency of the steam powerplant is 30 percent. A constant 20 percent of the incident energy on the receiver is assumed to go to storage during operation. Estimate the powerplant output, in megawatts, at peak conditions and at shutdown. Assume all efficiencies are constant during the day.

13-5 A 300-ft-long 6-ft-wide parabolic trough concentrator receives normal solar radiation at 905 W/m^2. A pipe at the focal line receives 450 lb_m/h of water at 200 psia and at 100°F and exits at 180 psia. Calculate the pipe exit conditions. Assume concentrator reflective losses of 5 percent.

13-6 A central receiver thermal solar powerplant has 2670 heliostats, each composed of 12 1.2 × 3.6 m reflecting glass panels. The average solar insolation during 10 h of operation is 635 W/m^2. The steam cycle has a 10-MW average output and an overall efficiency of 30 percent, assumed constant. The average power lost in radiation transmission from solar insolation on the heliostats to that received by the steam generator is 35 percent. Calculate (*a*) the sizes, in cubic meters, of storage with 25 percent void rock and with sodium, and (*b*) the number of hours the storage system can run the plant at average output.

13-7 A central receiver solar thermal powerplant with storage has 1000 heliostats that have 60 m^2 of reflective surface each. The transmission efficiency between incidence and the receiver is 25 percent. The steam plant efficiency is 30 percent, both considered constant during the day. Assume that the solar incidence in June follows a sine curve beginning with zero at −6 h before solar noon, and ending with zero at +6 h after solar noon, with a peak of 1 kW/m^2. The plant operates between −5.5 h and +5.5 h around solar noon. During that period the end use power profile is constant at 2.5 MW. Calculate (*a*) the peak powerplant output, in megawatts, (*b*) the powerplant output at startup, in megawatts, (*c*) the total energy output during operation, in megawatt hours, (*d*) the total equivalent electric energy going into storage, in megawatt hours, and (*e*) the time end use stops withdrawing from storage and energy input to storage begins, hours before solar noon.

13-8 A parabolic dish solar concentrator has 60 m^2 of reflective surface. A Stirling engine is mounted at the focal point. It is assumed to have a constant efficiency of 38 percent and is connected to an electric generator that has an efficiency of 95 percent. The concentrator reflective losses are 4 percent. The peak solar insolation is 1 kW/m^2. Assuming for simplicity that the concentrator shape can be approximated by a spherical surface subtended by a 60° solid angle from its center, determine (*a*) the peak power of the generator, in kilowatts, (*b*) the estimated power when the sun's elevation is 15° (summer solstice), and (*c*) the fuel oil, in grams per second, necessary to produce peak power at night. Assume fuel oil with a heating value of 42,000 kJ/kg is used.

13-9 A solar thermal central receiver system uses 2000 heliostats each 40 m^2 in reflective surface area. The receiver converts feedwater from the condenser at 1 psia to steam at 500 psia and 600°F. No feedwater heaters are used. Assuming absorption by the steam in the receiver to be 60 percent of all radiation incident on the heliostats, that 20 percent of all receiver energy goes to storage at peak radiation, and that the combined turbine-generator polytropic-mechanical-electrical efficiency to be 80 percent, find for the conditions at peak radiation (*a*) the heat added to the cycle, in Btus per hour, (*b*) the turbine steam flow rate, in pounds mass per hour, and (*c*) the generator output, in megawatts.

13-10 A combined thermal solar powerplant operates on a flow diagram similar to but not identical with that of Fig. 13-25. At peak solar insolation, 10^7 lb_m/h of atmospheric air at 520°R is compressed to 9.76 atm, after which it is heated in the central receiver, entering the high-pressure gas turbine at 9.6 atm at 1960°R. It leaves the low-pressure turbine at 20 psia to the steam generator, leaving it at 600°R. Superheated steam is generated at 1000 psia and 780°F and expands in the steam turbine to 1 psia. The compressor and both gas turbines have polytropic efficiencies of 0.82 and 0.88, respectively. The steam turbine has a polytropic efficiency of 0.90. The compressor and all turbines have mechanical efficiencies of 0.95. The electric generator has a combined mechanical-electrical efficiency of 0.96. For simplicity ignore storage, feedwater heating and condensate pump work. Calculate (*a*) the steam mass flow rate, in pounds mass per hour, (*b*) the net power output of the plant, in megawatts, if 20 percent of the gross power is used internally, (*c*) the combined plant net efficiency, and (*d*) the number of heliostats needed if their reflective areas are 100 m^2 each and the radiation transmission losses to the air in the tower are 40 percent of solar insolation.

13-11 A 7.5-cm-diameter circular photovoltaic solar cell is subjected to a solar energy flux of 2.5×10^{17} photons/s $\cdot$ cm^2 at an average photon wavelength of 0.838 μm. Calculate (*a*) the solar insolation on the cell, in watts per square meter, and (*b*) the maximum theoretical power that can be produced by the cell, in watts.

13-12 A 0.08-m-diameter circular photovoltaic cell receives the following solar fluxes, in photons per second per square centimeter, in the wavelength ranges between the parentheses: 0.5×10^{17} (0.3–0.5 μm), 0.85×10^{17} (0.5–0.7 μm), 0.5×10^{17} (0.7–0.9 μm), 0.45×10^{17} (0.9–1.1 μm), and zero below 0.3 and above 1.1 μm. Estimate (*a*) the number of such cells necessary to produce 10 kW of ac power, and (*b*) the array area on which they are mounted.

13-13 Calculate the overall efficiency of a satellite solar-power system that produces 10 GW on earth from two square photovoltaic arrays, each 5 km on the side. Assume a 0.95 packing fraction of the cells.

CHAPTER

FOURTEEN

WIND ENERGY

14-1 INTRODUCTION

Wind energy is rightfully an indirect form of solar energy since wind is induced chiefly by the uneven heating of the earth's crust by the sun. The topic would have fit in Chap. 13 except that that chapter was already quite lengthy and a separate chapter devoted to wind energy seemed appropriate.

Winds can be broadly classified as *planetary* and *local*. *Planetary winds* are caused by greater solar heating of the earth's surface near the equator than near the northern or southern poles. This causes warm tropical air to rise and flow through the upper atmosphere toward the poles and cold air from the poles to flow back to the equator nearer to the earth's surface. The direction of motion of planetary winds with respect to the earth is affected by the rotation of the earth. The warm air moving toward the poles in the upper atmosphere assumes an easterly direction (in both the northern and southern hemispheres) that results in the *prevailing westerlies* (Fig. 14-1). (Winds are named according to the direction they come *from*.) At the same time, the inertia of the cool air moving toward the equator nearer the earth's surface causes it to turn west, resulting in the *northeast trade winds* in the northern hemisphere and the *southeast trade winds* in the southern hemisphere.

The western motion toward the equator and the eastern motion toward the poles result in large counterclockwise circulation of air around low-pressure areas in the northern hemisphere and clockwise circulation in the southern hemisphere. The westerlies control events between the 30° and 60° latitudes (where a majority of the earth's population lives). Because the earth's axis is inclined to its orbital plane around the sun (Fig. 13-2), seasonal variations in the heat received from the sun result in seasonal

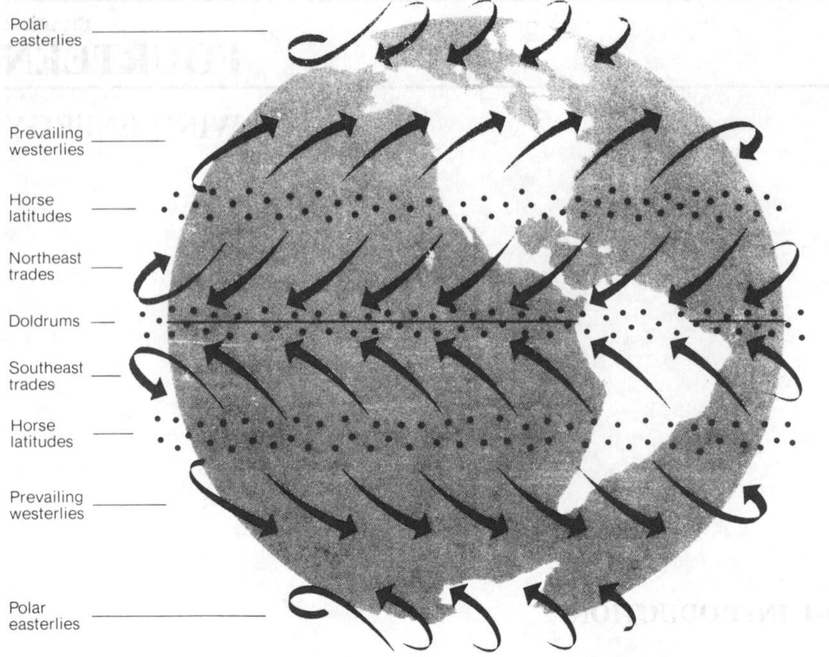

Figure 14-1 Planetary winds in the earth's atmosphere *(EPRI.)*

variations in the velocity and direction of the wind from the general flow pattern described above.

Local winds are caused by two mechanisms. The first is *differential heating* of land and water. Solar insolation during the day is readily converted to sensible energy of the land surface but is partly absorbed in layers below the water surface and partly consumed in evaporating some of that water. The land mass becomes hotter than the water, which causes the air above land to heat up and become warmer than the air above water. The warmer lighter air above the land rises, and the cooler heavier air above the water moves in to replace it. This is the mechanism of shore breezes. At night, the direction of the breezes is reversed because the land mass cools to the sky more rapidly than the water, assuming a clear sky. The second mechanism of local winds is caused by *hills and mountain sides*. The air above the slopes heats up during the day and cools down at night, more rapidly than the air above the low lands. This causes heated air during the day to rise along the slopes and relatively cool heavy air to flow down at night.

It has been estimated [133,134] that about 2 percent of all solar radiation falling on the face of the earth is converted to kinetic energy in the atmosphere and that 30 percent of this kinetic energy occurs in the lowest 1000 m (3280 ft) of elevation. It is thus said that the total kinetic energy of the wind in this lowest kilometer, if harnessed, can satisfy more than 3 times the energy demands of the United States at the early 1980s' rates. It is also claimed that wind power is pollution free and that its source of energy is free. Such are the seemingly compelling arguments for wind-power, not unlike those for solar-power. Although solar energy is cyclic and predictable, and even dependable in some parts of the globe, wind energy, however, is erratic, unsteady, and often treacherous except in very few areas. It does, however, have a place in the total energy picture, particularly for those areas with more or less steady winds, especially those that are far removed from central power grids, and for small, remote domestic and farm needs.

14-2 HISTORY OF WIND POWER

Human beings have always dreamt of converting wind power to mechanical and, more recently, electric power. Wind, more than any other renewable energy source, has intrigued serious and amateur inventors over the ages. It is said that more patents for wind systems have been applied for than almost any other device to date.

In ancient times the kinetic energy of the wind was used to propel ships by sails. Windmills, however, are more recent, having been used for a little over a thousand years. The earliest reference to windmills appeared in Arab writings from the ninth century A.D. that described mills that operated on the borders of Persia and Afganistan some two centuries earlier.

Vertical-Axis Windmills

These early machines, sometimes referred to as the *Persian windmills,* were vertical-axis machines. They evolved from ships. Sails, firs nade of canvas and then of wood, were attached to a large horizontal wheel. The wind pressure against the sails caused the wheel to turn. A vertical axle attached to the wheel usually turned a grindstone to grind grain into flour, hence the name windmill. Similar mills were known to have been used in the thirteenth century A.D. in China to evaporate seawater for the production of salt, and later in the Crimea, Europe, and the United States, though few of them remain today.

One of the most successful early forms of the vertical-axis mill is the one named after Savonius of Finland. The *Savonius windmill* had single or multiple S-shaped sails and a vertical axis (Fig. 14-2). Vertical-axis machines that are still being investigated are the Madaras (Magnus effect) and Darrieus machines (Secs. 14-9 and 14-10).

One advantage of vertical-axis machines is that they operate in all wind directions and thus need no yaw adjustments.

Horizontal-Axis Windmills

The vertical-axis windmill was changed, after the idea of a windmill reached Europe, into a vertical-wheel horizontal-axis configuration. The first designs had sails built on a post that could be made to face into any wind direction. The vertical wheel drove

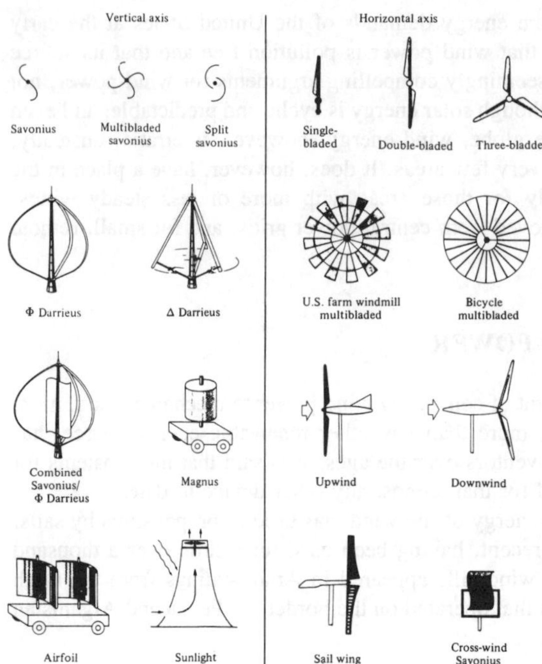

Figure 14-2 Some of the more workable (though not necessarily economical) windmills of the vertical- and horizontal-axis types [135].

a vertical axle through gears. Such machines first appeared in France and England in the late twelfth century and were called *post mills*. Various modifications of these mills evolved in Europe and America throughout the middle ages and were used for grinding grain, drainage, pumping, sawmilling, and other purposes. In the Netherlands, where very large land areas are below sea level and constantly threatened with flooding, windmills were widely used to pump water out of the fields and into canals which took it back to the sea. Nowadays, only a few romantic mills remain and pumping relies mostly on electric drives.

Electrical Generation

The first windmill to drive an electrical generator was built by P. La Cour of Denmark late in the nineteenth century. After World War I, sails with airfoil cross sections (like the blades of an airplane propeller) were developed for use in windmills; the result was what is now called the *propeller-type windmill,* or *wind turbine*. In 1931, such a windmill was built in the Crimea and produced low-voltage electricity that was fed directly into the local grid.

Experiments on twin-bladed mills were conducted in the United States, resulting, notably in 1940, in a large mill called the Smith-Putnam machine, which was designed by Palmer Putnam with the assistance of Theodore von Karman, the famed aerodynamicist, and others. The plant, rated at 1.25 MW of ac power [135], was constructed by the S. Morgan Smith Company of York, Pennsylvania, atop a 2000-ft (610-m) high mountain called Grandpa's Knob, near Rutland, Vermont. The mill had a twin-bladed, 175-ft (55-m) diameter propeller-type rotor that weighed 16 tons, was mounted on top of a 110-ft (34-m) tower, and rotated at 28 r/min. After intermittent operations for a number of years, one of the blades broke off in March 1945, near the hub where a previously discovered weakness was not corrected because of wartime shortages. The project was abandoned when it was decided that, even if repaired, the plant could not compete economically with fossil or hydro powerplants.

The tranditional nonelectrical windmill still survives around the world, however (1000 mills were still operating in Portugal in 1965), mainly in areas not easily accessible to central power grids and in areas of mainly peasant economy.

14-3 PRINCIPLES OF WIND POWER

Total Power

The total power of a wind stream is equal to the rate of the incoming kinetic energy of that stream KE_i, or

$$P_{tot} = \dot{m} KE_i = \dot{m} \frac{V_i^2}{2g_c} \tag{14-1}$$

where

P_{tot} = total power, W or ft · lb_f

$\dot{m}$ = mass-flow rate, kg/s or lb_m/h

V_i = incoming velocity, m/s or ft/h

g_c = conversion factor = 1.0 kg/(N · s^2) or 4.17×10^8 lb_m · ft/(lb_f · h^2)

The mass-flow rate is given by the continuity equation

$$\dot{m} = \rho A V_i \tag{14-2}$$

where

ρ = incoming wind density, kg/m^3 or lb_m/ft^3

A = cross-sectional area of stream, m^2 or ft^2

Thus

$$P_{tot} = \frac{1}{2g_c} \rho A V_i^3 \tag{14-3}$$

Thus the total power of a wind stream is directly proportional to its density, area, and the cube of its velocity. Figure 14-3 is a plot of P_{tot} as a function of A and V_i for wind at 1 standard atmosphere pressure and 15°C (59°F).

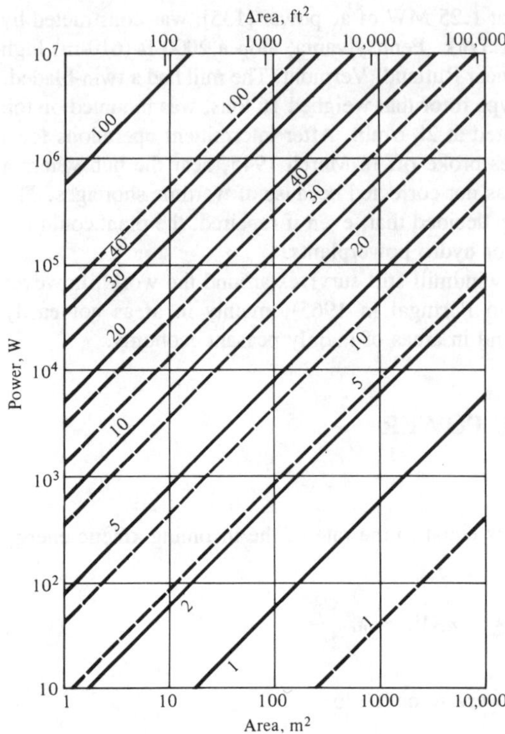

Figure 14-3 Total power available in wind streams of incoming velocities V_i and cross-sectional areas A. P = 1 atm., T = 15°C = 59°F. Solid lines, V_i in m/s. Dashed lines, V_i in mi/h.

Maximum Power

It will shortly be apparent that the total power discussed above cannot all be converted to mechanical power. Consider a horizontal-axis, propeller-type windmill, henceforth to be called a wind turbine, which is the most common type used today. Assume that the wheel of such a turbine has thickness *a-b* (Fig. 14-4); that the incoming wind pressure and velocity, far upstream of the turbine, are P_i and V_i, and that the exit wind pressure and velocity, far downstream of the turbine, are P_e and V_e, respectively. V_e is less than V_i because kinetic energy is extracted by the turbine.

Considering the incoming air between i and a as a thermodynamic system, and assuming that the air density remains constant (a good assumption since the pressure and temperature changes are very small compared to ambient), that the change in potential energy is zero, and no heat or work are added or removed between i and a, the general energy equation (Eq. 1-1c) reduces to the kinetic and flow energy terms only.

Thus

$$P_i v + \frac{V_i^2}{2g_c} = P_a v + \frac{V_a^2}{2g_c} \tag{14-4a}$$

or
$$P_i + \rho\frac{V_i^2}{2g_c} = P_a + \rho\frac{V_a^2}{2g_c} \tag{14-4b}$$

where v and ρ are the specific volume and its reciprocal, the density, respectively, both considered constant. Equation (14-4b) is the familiar Bernouilli equation.

Similarly, for the exit region *b-e*

$$P_e + \rho\frac{V_e^2}{2g_c} = P_b + \rho\frac{V_b^2}{2g_c} \tag{14-5}$$

The wind velocity across the turbine decreases from *a* to *b* since kinetic energy is converted to mechanical work there. The incoming velocity V_i does not decrease abruptly but gradually as it approaches the turbine to V_a and as it leaves it to V_e. Thus $V_i > V_a$ and $V_b > V_e$, and therefore, from Eqs. (14-4) and (14-5), $P_a > P_i$ and $P_b < P_e$; that is, the wind pressure rises as it approaches, then as it leaves the wheel.

Combining Eqs. (14-4) and (14-5) gives

$$P_a - P_b = \left(P_i + \rho\frac{V_i^2 - V_a^2}{2g_c}\right) - \left(P_e + \rho\frac{V_e^2 - V_b^2}{2g_c}\right) \tag{14-6}$$

It is reasonable to assume that, far from the turbine at *e*, the wind pressure returns to ambient, or

$$P_e = P_i \tag{14-7}$$

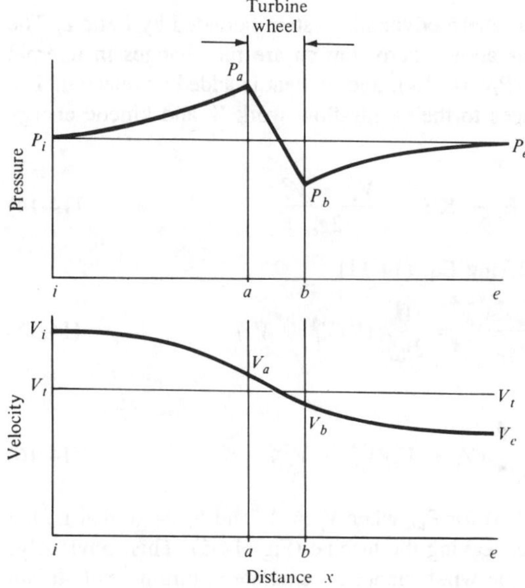

Figure 14-4 Pressure and velocity profiles of a wind moving through a horizontal-axis propeller-type wind turbine.

and that the velocity within the turbine, V_t, does not change because the blade width *a-b* is thin compared with the total distance considered, so that

$$V_t \approx V_a \approx V_b \tag{14-8}$$

Combining Eqs. (14-6) to (14-8) gives

$$P_a - P_b = \rho\left(\frac{V_i^2 - V_e^2}{2g_c}\right) \tag{14-9}$$

The axial force F_x, in the direction of the wind stream, on a turbine wheel with projected area, perpendicular to the stream A, is given by

$$F_x = (P_a - P_b)A = \rho A\left(\frac{V_i^2 - V_e^2}{2g_c}\right) \tag{14-10}$$

This force is also equal to the change in momentum of the wind $\Delta(\dot{m}V)/g_c$ where $\dot{m}$ is the mass-flow rate given by

$$\dot{m} = \rho A V_t \tag{14-11}$$

Thus

$$F_x = \frac{1}{g_c}\rho\, A V_t(V_i - V_e) \tag{14-12}$$

Equating Eqs. (14-10) and (14-12) gives

$$V_t = \frac{1}{2}(V_i + V_e) \tag{14-13}$$

We shall now consider the total thermodynamic system bounded by i and e. The changes in potential energy are, as above, zero, but so are the changes in internal energy ($T_i = T_e$) and flow energy ($P_i v = P_e v$), and no heat is added or rejected. The general energy equation now reduces to the steady-flow work W and kinetic energy terms

$$W = \text{KE}_i - \text{KE}_e = \frac{V_i^2 - V_e^2}{2g_c} \tag{14-14}$$

The power P is the rate of work. Using Eq. (14-11)

$$P = \dot{m}\frac{V_i^2 - V_e^2}{2g_c} = \frac{1}{2g_c}\rho A V_t(V_i^2 - V_e^2) \tag{14-15}$$

Combining with Eq. (14-13)

$$P = \frac{1}{4g_c}\rho A(V_i + V_e)(V_i^2 - V_e^2) \tag{14-16}$$

Equation (14-15) reverts to Eq. (14-3) for P_{tot} when $V_t = V_i$ and $V_e = 0$; that is, the wind comes to a complete rest after leaving the turbine (Fig. 14-5). This, obviously, is an impossible situation because the wind cannot accumulate at turbine exit. It can

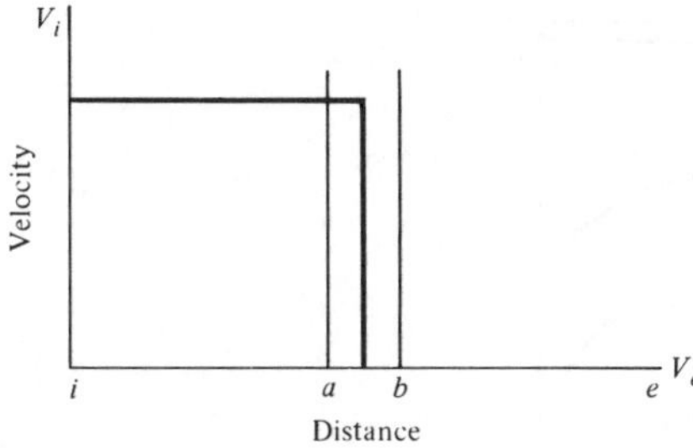

Figure 14-5 Total conversion of incoming wind kinetic energy to work.

be seen from Eq. (14-16), where V_e is positive in one term and negative in the other, that too low or too high a value for V_e results in reduced power. There thus is an *optimum exit velocity* $V_{e,\text{opt}}$ that results in maximum power $P_{\max}$, which is obtained by differentiating P in Eq. (14-16) with respect to V_e for a given V_i and equating the derivative to zero, i.e., $dP/dV_e = 0$, which gives

$$3V_e^2 + 2V_iV_e - V_i^2 = 0$$

This is solved for a positive V_e to give $V_{e,\text{opt}}$

$$V_{e,\text{opt}} = \frac{1}{3}V_i \tag{14-17}$$

Combining with Eq. (14-16) gives $P_{\max}$

$$P_{\max} = \frac{8}{27g_c}\rho A V_i^3 \tag{14-18}$$

The *ideal,* or *maximum, theoretical efficiency* $\eta_{\max}$ (also called the *power coefficient*) of a wind turbine is the ratio of the maximum power obtained from the wind, Eq. (14-18), to the total power of the wind, Eq. (14-3), or

$$\eta_{\max} = \frac{P_{\max}}{P_{\text{tot}}} = \frac{8}{27g_c} \times 2g_c = \frac{16}{27} = 0.5926 \tag{14-19}$$

In other words, a wind turbine is capable of converting no more than 60 percent of the total power of a wind to useful power.

Actual Power

Like steam- and gas-turbine blades (Sec. 5-5), wind-turbine blades experience changes in velocity dependent upon the blade inlet angle and the blade velocity. Because the blades are long, the blade velocity varies with the radius to a greater degree than steam- or gas-turbine blades and the blades are therefore twisted. The maximum efficiency (or power coefficient) given by Eq. (14-19) assumes ideal conditions along the entire blade. A rigorous treatment of the power extracted from the wind by a propeller-type wind turbine shows that the power coefficient is strongly dependent on

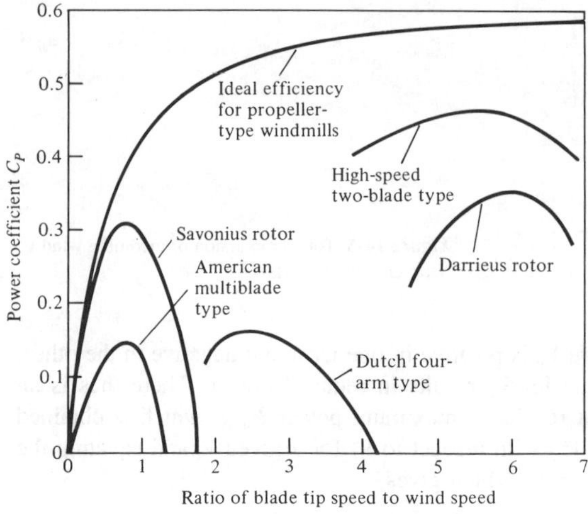

Figure 14-6 Power coefficient of various windmills vs. ratio of blade tip to wind speeds [135].

blade-to-wind speed ratio, that it reaches its maximum value of about 0.6 only when the maximum blade speed, i.e., the blade speed at the tip, is some 6 or 7 times the wind speed, and that it drops rapidly at blade tip-to-wind speed ratios below about 2.0. Figure 14-6 shows the power coefficient for an ideal propeller-type wind turbine and various other wind turbines.

Because a wind-turbine wheel cannot be completely closed, and because of spillage and other effects, practical turbines achieve some 50 to 70 percent of the ideal efficiency. The *real efficiency* η is the product of this and $\eta_{\max}$ and is the ratio of actual to total power

$$P = \eta P_{\text{tot}} = \eta \frac{1}{2g_c} A V_i^3 \tag{14-20}$$

where η varies between 30 and 40 percent for real turbines.

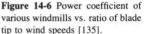

Forces on the Blades

There are two types of forces operating on the blades of a propeller-type wind turbine. They are the *circumferential forces* in the direction of wheel rotation that provide the torque and the *axial forces* in the direction of the wind stream that provide an *axial thrust* that must be counteracted by proper mechanical design.

The *circumferential force,* or *torque,* T is obtained from

$$T = \frac{P}{\omega} = \frac{P}{\pi D N} \tag{14-21}$$

where T = torque, N or lb_f

ω = angular velocity of turbine wheel, m/s or ft/s

D = diameter of turbine wheel = $\sqrt{4A/\pi}$, m or ft

N = wheel revolutions per unit time, s^{-1}

For a turbine operating at power P, Eq. (14-20), the torque is given by

$$T = \eta \frac{1}{8g_c} \frac{\rho D V_i^3}{N} \tag{14-22}$$

For a turbine operating at maximum efficiency $\eta_{max} = 16/27$, the torque is given by T_{max}

$$T_{max} = \frac{2}{27g_c} \frac{PDV_i^3}{N} \tag{14-23}$$

The *axial force*, or axial thrust, given by Eq. (14-10), here repeated, is

$$F_x = \frac{1}{2g_c}\rho A(V_i^2 - V_e^2) = \frac{\pi}{8g_c}\rho D^2(V_i^2 - V_e^2) \tag{14-10}$$

The axial force on a turbine wheel operating at maximum efficiency where $V_e = 1/3\, V_i$ is given by

$$F_{x,max} = \frac{4}{9g_c}\,\rho A V_i^2 = \frac{\pi}{9g_c}\,\rho D^2 V_i^2 \tag{14-24}$$

The axial forces are proportional to the square of the diameter of the turbine wheel, which makes them difficult to cope with in extremely large-diameter machines. There is thus an upper limit of diameter that must be determined by design and economical considerations.

Example 14-1 A 10-m/s wind is at 1 standard atm pressure and 15°C temperature. Calculate (1) the total power density in the wind stream, (2) the maximum obtainable power density, (3) a reasonably obtainable power density, all in W/m², (4) the total power (in kW) produced if the turbine diameter is 120 m, and (5) the torque and axial thrust N if the turbine were operating at 40 r/min and maximum efficiency.

SOLUTION For air, the gas constant R = 287 J/(kg · K). 1 atm = 1.01325×10^5 Pa.

$$\text{Air density } \rho = \frac{P}{RT} = \frac{1.01325 \times 10^5}{287(15 + 273.15)} = 1.226 \text{ kg/m}^3$$

(1) Eq. (14-3):

$$\frac{P_{\text{tot}}}{A} = \frac{1}{2g_c}\rho V_i^3 = \frac{1}{2 \times 1}1.226 \times 10^3 = 613 \text{ W/m}^2$$

(2) Eq. (14-18):

$$\frac{P_{\max}}{A} = \frac{8}{27g_c}\rho V_i^3 = \frac{8}{27 \times 1}1.226 \times 10^3 = 363 \text{ W/m}^2$$

(3) Assuming $\eta = 40\%$:

$$\frac{P}{A} = 0.4\left(\frac{P_{\text{tot}}}{A}\right) = 0.4 \times 613 = 245 \text{ W/m}^2$$

(In English units this corresponds to 22.76 W/ft² at 22.37 mi/h.)

(4) $$P = 0.245 \times \frac{\pi D^2}{4} = 0.245 \times \frac{\pi 120^2}{4} = 2770 \text{ kW}$$

(5) Eq. (14-23):

$$T_{\max} = \frac{2}{27g_c}\frac{\rho D V_i^3}{N} = \frac{2}{27 \times 1} \times \frac{1.20 \times 1.226 \times 10^3}{40/60}$$

$$= 16{,}347 \text{ N } (=3675 \text{ lb}_f)$$

Eq. (14-24):

$$F_{x,\max} = \frac{\pi}{9g_c}\rho D^2 V_i^2 = \frac{\pi}{9 \times 1}(1.226 \times 120^2 \times 10^2)$$

$$= 616{,}255 \text{ N } (=138{,}540 \text{ lb}_f)$$

14-4 WIND TURBINE OPERATION

Equations (14-18) and (14-20) demonstrate an inherent weakness of all wind machines, namely, the strong dependence of the power produced on wheel diameter and wind speed, being proportional to turbine wheel area, i.e., to the square of its diameter, and to the cube of the wind velocity. The latter dependence means that even small fluctuations in wind velocity, which almost always occur, would mean large fluctuations in power. For example, a drop in wind velocity by only 20 percent would result in the loss of a little less than half the power, whereas a drop in velocity to approximately one-half (0.464) would result in a drop in power to one-tenth, assuming constant efficiency (Fig. 14-7). Actually, a greater loss in power occurs because many of the machine losses are independent of wind velocity; the results are an increase in machine fractional losses and a decrease in efficiency as wind velocity and power decrease. The maximum machine efficiency occurs over a relatively narrow power range.

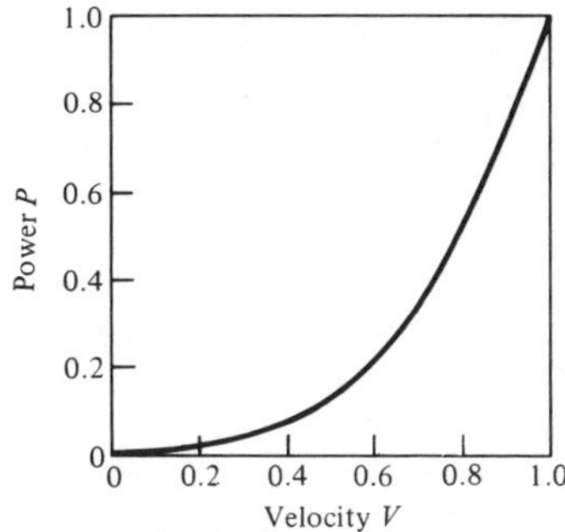

Figure 14-7 Power-velocity characteristics of a wind turbine with constant efficiency, arbitrary units.

Flat Rating

Severe fluctuations in power are, of course, undesirable. They pose power-oscillation problems on the grid and severe strains on the windmill hardware. From an economic point of view, a windmill designed to produce a rated power output corresponding to the maximum, or near maximum, prevailing wind velocity at a given site would generate low powers, with the full capacity of the turbine and electric generator unused much of the time.

It is, therefore, more cost-effective to design a windmill to produce rated power at less than the maximum prevailing wind velocity, using a smaller turbine and generator, and to maintain a constant output at all wind speeds above rating. This is called *flat rating* (Fig. 14-8).

Because of the severe loss in efficiency and power at low wind velocities, a wind turbine is also designed to come into operation at a minimum wind speed, called the *cut-in velocity*. To protect the turbine wheel against damage at very high wind velocities, it is designed to stop operation (such as by feathering the blades) at a *cut-out velocity*. Thus the wind turbine operates with variable load over a narrow range between the cut-in and the rated velocities and at constant power between the rated and the cut-out velocities and ceases operation above the cut-out velocity.

An example of flat rating is that of the MOD-2, a second-generation class of wind turbines, each of which produces 2.5 MW (peak) (Sec. 14-8). The cut-in, rated, and cut-out wind velocities are 9.0, 20.0, and 36.0 mi/h, respectively.

Wind-turbine ratings are usually given for a wind velocity occuring at a reference height, usually 30 ft (9.1 m) above grade, and with an availability factor, usually 90 percent. The *availability factor* is defined as the fraction of time, during a given period, that the turbine is actually on line (Fig. 14-9). The actual wind velocity, at the propeller hub, that determines the turbine power is usually higher [Eq. (14-28), Sec. 14-5].

Because they do not always operate at rated power because of changing wind velocities, the overall load factor is much lower than the availability factor. The *overall load factor*, also called the *plant operating factor* and the *plant capacity factor*, is the ratio of the total energy generated during a given period of time to the total rated

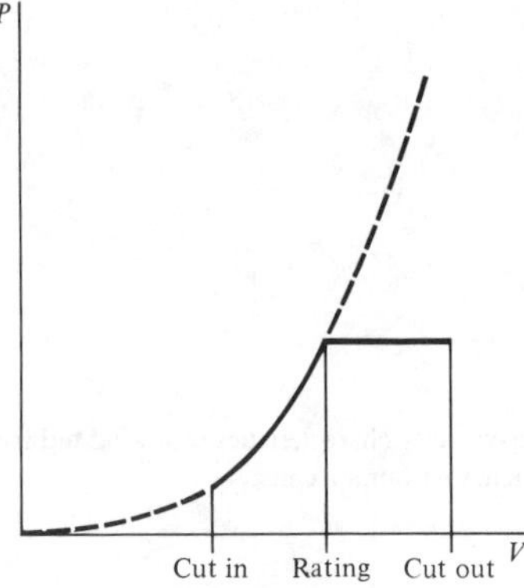

Figure 14-8 Power-velocity characteristics of a flat-rated wind turbine.

generation capacity during the same period (Fig. 14-9). This factor thus takes into account operations at less than rated wind velocity, nonoperation below cut-in and above cut-out velocities, and power outages caused by various situations, such as repairs, maintenance, etc.

The overall load factor is typically 30 to 40 percent (the MOD-0A 200-kW machine in Oahu, Hawaii, accumulated an impressive 45 percent factor in 1981 because of favorable wind patterns there). Considering an average load factor of 1/3, a wind powerplant of a particular rating would have to be nearly 2.5 times as large as a conventional powerplant of the same rating operating at a load factor of about 80 percent. The unit capital cost, \$/kW rating, for a wind powerplant would have to be multiplied by about 2.5 to yield a more realistic cost to a utility. This is one of the economic burdens of wind power.

14-5 SITE CHARACTERISTICS

Small Machines

Wind turbines are broadly classified as small or large. *Small wind turbines* are those of less than 100 kW rated capacity. They are usually used for generating power for

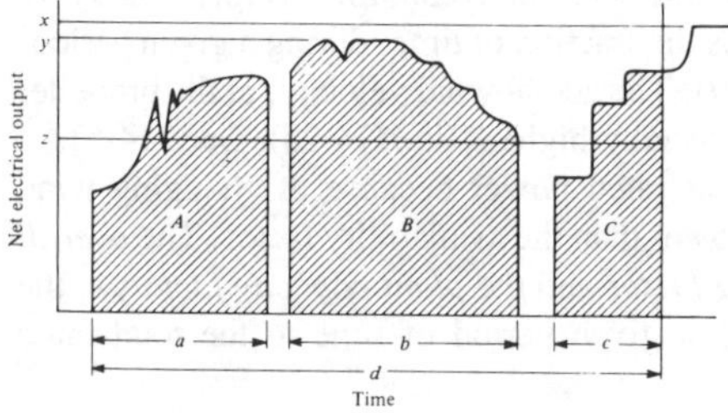

Figure 14-9 Hypothetical diagram of operation of a powerplant: x = rated net capacity of plant, y = peak load during period d, z = average load during period d. *Plant overall load factor*, or *plant operating factor*, $= (A + B + C)/xd$. *Plant load factor* $= z/y$. *Plant availability factor* $= (a + b + c)/d$. d is usually taken as 1 year.

local use in a particular location. Siting procedures for small machines involve accurately evaluating the wind characteristics and finding the best acceptable site within that location. Guidelines for siting small machines may be found in the *Handbook for Siting Small Wind Energy Conversion Systems* [136].

Large Machines

Large wind turbines are those of 100 kW rated capacity or greater. They are used to generate power for distribution in central power grids. Siting procedures for such machines are approached differently. The geographic area in which they may be located may be very large, and there usually is a great deal of freedom in site selection. The main objective is to minimize the cost of power production for the entire grid, and this requires rather extensive analysis. A handbook for siting large machines was in preparation at the time of this writing, but various techniques have already been developed as tools in the different stages of siting. These include flow modeling, measurements, and biological, geological, topographical, social, and cultural indicators of wind. Some are included in the proceedings of a 1979 Conference and Workshop on Wind Energy Characteristics and Siting [137].

Suitable sites for large windmills around the world depend upon favorable wind activity and, in the United States, have been consolidated under a program area called Wind Energy Prospecting. They result in maps showing annual average wind-power density, W/m^2, such as that shown in Fig. 14-10. The map is general in the sense that good sites may occasionally be found in low power density areas and vice versa.

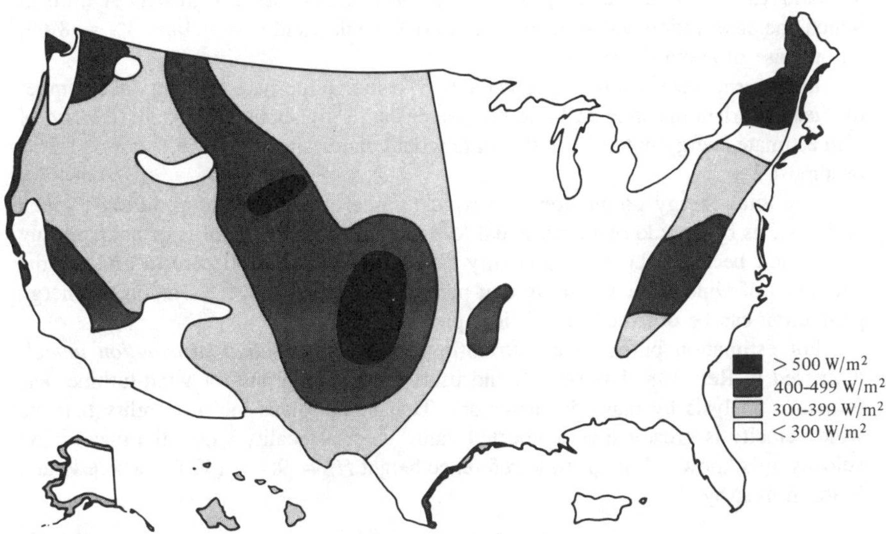

Figure 14-10 Annual distribution of average wind power in the United States *(EPRI)*.

Mean Wind and Energy Velocities

Wind-power potential can be estimated from the *mean wind velocity* $\overline{V}$, which is based on measurements over a period of time. It is given by

$$\overline{V} = \frac{\sum_{i}^{n} V_i}{n} \tag{14-25}$$

where $\sum_{i}^{n} V_i$ is the sum of all velocity observations and n is the number of these observations.

Powerplant sizing would, however, be grossly underestimated if it were rated at the mean wind velocity. A more representative figure would be based on the *mean energy velocity* $\overline{V}_E$, which because of the power dependence on the cube of the wind velocity, is given by

$$\overline{V}_E = \left(\frac{\sum_{i}^{n} V_i^3}{n} \right)^{1/3} \tag{14-26}$$

The ratio of $\overline{V}_E$ to $\overline{V}$ depends upon the distributions of velocity in the observations. Taking a hypothetical case in which 10 observations yielded velocities of 0 to 9 (in any units) with increments of of 1, $\overline{V}$ would be 4.5 while $\overline{V}_E$ would be 5.872, greater by a little over 30 percent. Another in which nine observations are zero and one was 9 would yield $\overline{V} = 0.9$ and $\overline{V}_E = 4.177$, which is 364 percent greater. A third in which one observation was zero and nine were 9 would yield $\overline{V} = 8.1$ and $\overline{V}_E = 8.69$, an increase of about 7 percent.

Long-term observations taken by the U.S. National Storms Laboratory at a number of sites in Oklahoma showed $\overline{V}_E$ to be greater than $\overline{V}$ by about 25 percent. The use of $\overline{V}$ to estimate energy potential in that case would underestimate it by a factor of 1.25^3, or almost 2.

The total energy production of a wind turbine of a given rating during a given period is, as by now known, not equal to its rating times the number of hours during that period because (1) the availability factor of less than 100 percent and (2) the variations of wind velocity during that period. The effect of this variation on energy production can be obtained only if it is precisely known.

For estimation purposes a distribution, called the *Weibull distribution model*, discussed in Ref. 138, has been found useful and appropriate for wind-turbine performance analysis by many investigators. That model gives the probability that the wind velocity is greater than a selected value V for a locality where the mean wind velocity $\overline{V}$ is known. It is given at reference height $H_r = 9.1$ m (30 ft), where V_i and V are in m/s, by

$$P(V_i > V) = e^{-(V/C_r)^{K_r}} \tag{14-27}$$

where $P(V_i > V)$ = probability that incoming wind velocity exceeds a value V at reference height H_r

$$K_r = 1.09 + 0.20V$$

$$C_r = \frac{\overline{V}}{\Gamma[1 + (1/K_r)]} \text{ m/s}$$

Γ = gamma function

Figure 14-11 shows *wind-distribution curves* (also called *exceedance* and *endurance curves*) for three values of $\overline{V}$ [139]. The abscissa shows the number of hours per year during which the wind velocity exceeds that given by the ordinate. For example, for $\overline{V}$ = 15 mi/h, the wind velocity exceeds 20 mi/h about 2000 h/year, or about 22 percent of the time. [Some sites in Oklahoma, one of the high power density states, have reported winds at 8 knots (about 9.2 mi/h) or greater about 70 percent of the time.]

A height above ground of 30 ft (9.1 m) is usually used as a reference elevation for which wind velocities are tabulated or listed (as in Fig. 14-11). The velocity to be used in determining the power of a wind turbine, however, is that at the hub of the turbine wheel, which is 100 to 260 ft (~30 to 80 m) for current large wind turbines. The ratio of wind velocity V at a height H to the reference velocity V_r at reference height H_r is given by a wind-shear model [137] as

$$\frac{V}{V_r} = \left(\frac{H}{H_r}\right)^{\alpha} \tag{14-28}$$

where
$$\alpha = \alpha_O\left(1 - \frac{\log V_r}{\log V_O}\right)$$

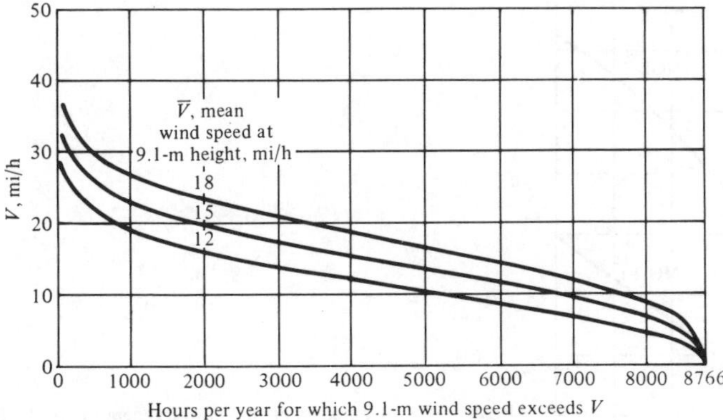

Figure 14-11 Wind-speed distribution (exceedance) curves for three values of the mean wind speed $\overline{V}$ at a height of 30 ft (9.1 m) [139].

$$\alpha_O = \left(\frac{Z_O}{H_r}\right)^{0.2}$$

Z_O = surface roughness length, varying widely but taken as 0.4 m for evaluation purposes to conform with the empirical values used in the model

V_O = a fixed velocity = 67.1 m/s

At elevations H, other than the reference H_r (30 ft, 9.1 m), the Weibull distribution model is used, but the parameters K_r and C_r are modified to K_H and C_H, given by

$$K_H = \frac{K_r}{1 - \alpha_O\left[\left(\log \frac{H}{H_r}\right) \Big/ \log V_O\right]} \tag{14-29}$$

and

$$C_H = C_r\left(\frac{H}{H_r}\right)^{\alpha_H} \tag{14-30}$$

where

$$\alpha_H = \alpha_O\left(1 - \frac{\log C_r}{\log V_O}\right) \tag{14-31}$$

The annual energy output for any horizontal-axis wind turbine can now be estimated for a given mean wind velocity $\overline{V}$ using the wind-distribution curves by calculating the power output at each wind velocity and integrating it over the time duration for that wind velocity.

Specifications for wind turbines usually include estimated annual power output,

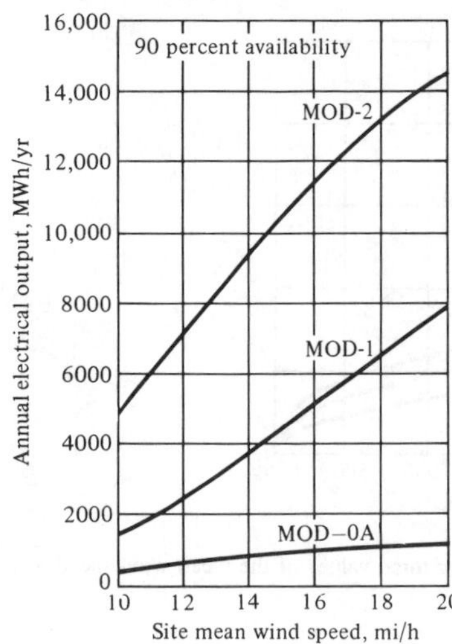

Figure 14-12 Computed annual power generation of three large horizontal-axis wind turbines, assuming 90 percent availabiity factor, Weibull wind-speed distribution at 30-ft elevation, and various aerodynamic, mechanical, and electrical losses [139].

in MWh/year, computed for the reference elevation of 9.1 m (30 ft), for a machine availability factor of 90 percent, for a Weibull wind-velocity distribution, and including the effects of aerodynamic, mechanical, and electrical losses, up to the busbar (beyond the step-up transformer). Such calculations have been made for three large machines, the MOD-0A, MOD-1, and MOD-2 (Sec. 14-6) at various values of the mean wind speed $\overline{V}$ (Fig. 14-12).

14-6 NEW DEVELOPMENT: SMALL MACHINES

As indicated earlier, small machines are classified as those less than 100 kW rated capacity. They are designed primarily for local use at or near their intended load for farm, home, or rural use. In the United States, the Department of Energy has a "Small Wind Energy Conversion Systems (SWECS)" test program that is managed and operated by Rockwell International at a test center at Rocky Flats, Colorado (Fig. 14-13). The program is designed to develop small advanced machines with the goal of their earliest possible commercialization by attempting to make their costs competitive with the costs of power obtained from conventional systems and by dealing with the problems of small consumer and institutional acceptance. The program is producing

Figure 14-13 The Rocky Flats, Colorado, Small Wind Machine Test Center, SWECS. *(Courtesy DOE.)*

technical data on machine costs, construction, output, durability, and behavior under varying as well as extreme weather conditions.

The first cycle of the SWECS program included nine contracts to build machines in the 1-, 8-, and 40-kW ranges. In the 40-kW range, McDonnell Aircraft Company is building a three-bladed, 65-ft-diameter rotor geared to a vertical axis, while Kaman Aerospace Corporation is building two prototypes—each a twin-bladed, 64-ft-diameter rotor, horizontal-axis machine—one to generate electricity and the other to drive equipment, such as irrigation pumps, mechanically.

14-7 NEW DEVELOPMENTS: LARGE MACHINES

The large horizontal-axis wind turbine is currently receiving the lion's share of research and development funds and is the beneficiary of the most advanced technology for harnessing wind power. In the United States the development effort is managed by the National Aeronautics and Space Administration (NASA) for the Department of Energy and by the U.S. Bureau of Reclamation. The program has gone through three generations of machines (Fig. 14-14).

The *first generation* started with the MOD-0 machine, which was rated at 100 kW in 18-mi/h wind (at the reference height of 30 ft) and first tested in September 1975 at a NASA experimental station near Sandusky, Ohio. MOD-0, not shown in Fig. 14-14, has a power coefficient of 0.375, a power train efficiency of 0.75, a cut-in wind velocity of 8 mi/h (13 km/h), and a rotor speed of 40 r/min in winds over 6 mi/h (10 km/h). The blades are fully feathered in 60-mi/h winds, and the system was designed to withstand 150-mi/h (240-km/h) winds. The rotor, transmission train, generator, and controls were mounted on a bed plate atop a 100-ft (30-m) tower of pinned-truss design. The controls included a blade pitch change mechanism and a yaw control that rotated the bed plate at 1/6 r/min to keep the rotor aligned to the wind. The gear box had a speed ratio of 45 : 1. The alternator weighed 1425 lb_m (645 kg), rotated at 1800 r/min, and was rated at 125 kVA. The installation had a capital cost of $500,000, or a unit capital cost of $5000 per rated kW, 4 to 5 times that of conventional systems. The real cost is of course higher because it must take into account the overall load factor.

Four updated versions of the MOD-0 machine, called MOD-0A and rated at 200 kW, were subsequently built between 1977 and 1980. Their design data are shown in Fig. 14-14. The most improved and successful of the MOD-0A machines is the one built in Oahu, Hawaii. It generated 1.14 GWh between July 1980 and November 1981 for an overall load factor of about 0.45. The MOD-0A machines were followed by a single MOD-1 machine built at Howard's Knob, Boone, North Carolina, in 1979. Figures 14-15 and 14-16 show an overall view and a cross section of the nacelle of MOD-1.

The first-generation machines described above were research prototypes used for field experimentation, for the assessment of structural and aerodynamic performance, and to gain operating experience. Data gathered have been used in the design of the second-generation machines.

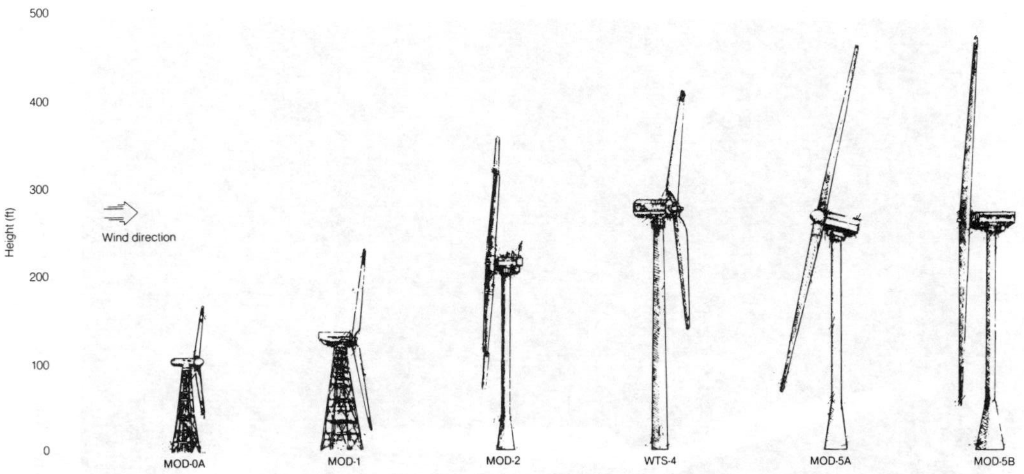

Height (ft): 500, 400, 300, 200, 100, 0

Wind direction

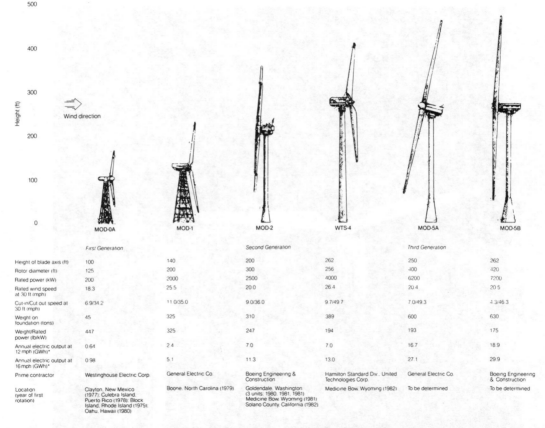

	MOD-0A	MOD-1	MOD-2	WTS-4	MOD-5A	MOD-5B
	First Generation		*Second Generation*		*Third Generation*	
Height of blade axis (ft)	100	140	200	262	250	262
Rotor diameter (ft)	125	200	300	256	400	420
Rated power (kW)	200	2000	2500	4000	6200	7200
Rated wind speed at 30 ft (mph)	18.3	25.5	20.0	26.4	20.4	20.5
Cut-in/Cut out speed at 30 ft (mph)	6.9/34.2	11.0/35.0	9.0/36.0	9.7/49.7	7.0/49.3	4.3/46.3
Weight on foundation (tons)	45	325	310	389	600	630
Weight/Rated power (lb/kW)	447	325	247	194	193	175
Annual electric output at 12 mph (GWh)*	0.64	2.4	7.0	7.0	16.7	18.9
Annual electric output at 16 mph (GWh)*	0.98	5.1	11.3	13.0	27.1	29.9
Prime contractor	Westinghouse Electric Corp.	General Electric Co.	Boeing Engineering & Construction	Hamilton Standard Div., United Technologies Corp.	General Electric Co.	Boeing Engineering & Construction
Location (year of first rotation)	Clayton, New Mexico (1977); Culebra Island, Puerto Rico (1978); Block Island, Rhode Island (1979); Oahu, Hawaii (1980)	Boone, North Carolina (1979)	Goldendale, Washington (3 units: 1980, 1981, 1981) Medicine Bow, Wyoming (1981) Solano County, California (1982)	Medicine Bow, Wyoming (1982)	To be determined	To be determined

Figure 14-14 Stages in the development of large horizontal-axis wind turbines in the United States. *(Courtesy Electric Power Research Institute.)*

Figure 14-15 The MOD-1 wind turbine at Howard's Knob, Boone, North Carolina *(Courtesy DOE.)*

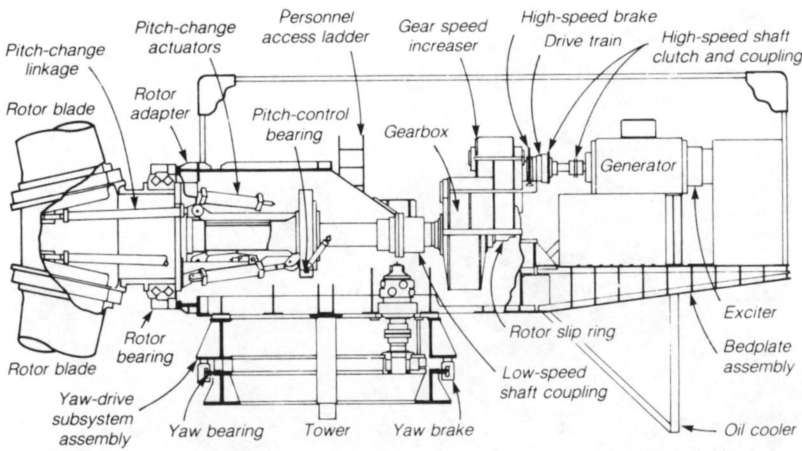

Figure 14-16 The MOD-1 nacelle showing rotor and blades (in part) on the left, electrical generator on the right, and yaw drive on the bottom.

The *second-generation machines,* called MOD-2 (Fig. 14-14), have two-bladed upwind turbines and are rated at 2.5 MW and situated atop 200-ft (61-m) cylindrical towers. The rotor is 300 ft (91 m) in diameter (which is the length of a football field) and is constructed as a continuous blade without a protruding hub. Speed control is accomplished by pitching the tips of the rotor. The cut-in, rated, and cut-out wind velocities (at tower top) are 14, 27.5, and 45 mi/h (6.3, 12.3 and 20.1 m/s), respectively. Five such machines have been built, three in Goodnoe Hills, Goldendale, Washington (in 1980, 1981, and 1981), and one each in Medicine Bow, Wyoming (in 1981), and Solano County, California (in 1982). Figure 14-17 is an artist's rendering of one of the Goodnoe Hills MOD-2 machines, and Fig. 14-18 is a cross section of the MOD-2 nacelle. Some design parameters of the MOD-2 machines are given in Table 14-1.

The *third-generation machines,* called MOD-5, are currently (1982) in the design stage. They are designed as advanced multimegawatt systems with the goal of reducing unit capital costs approximately 30 percent below the second-generation machines. There are two MOD-5 designs (Fig. 14-14), both upwind and twin-bladed. MOD-5A is rated at 6.2 MW with a rotor diameter of 400 ft (122 m). MOD-5B is rated at 7.2 MW with a rotor diameter of 420 ft (128 m). They are being designed with new features, such as two-speed and variable-speed operation, induction generators, and blades made of laminated wood.

Much of the early work on blades and their controls benefited from the experience of the helicopter rotor industry. The advanced wind turbines, however, have rotor diameters much greater than those common in helicopters, though their control systems are somewhat less complicated.

Figure 14-17 Artist's conception of the MOD-2 wind turbine at Goodnoe Hills, Washington. *(Courtesy DOE.)*

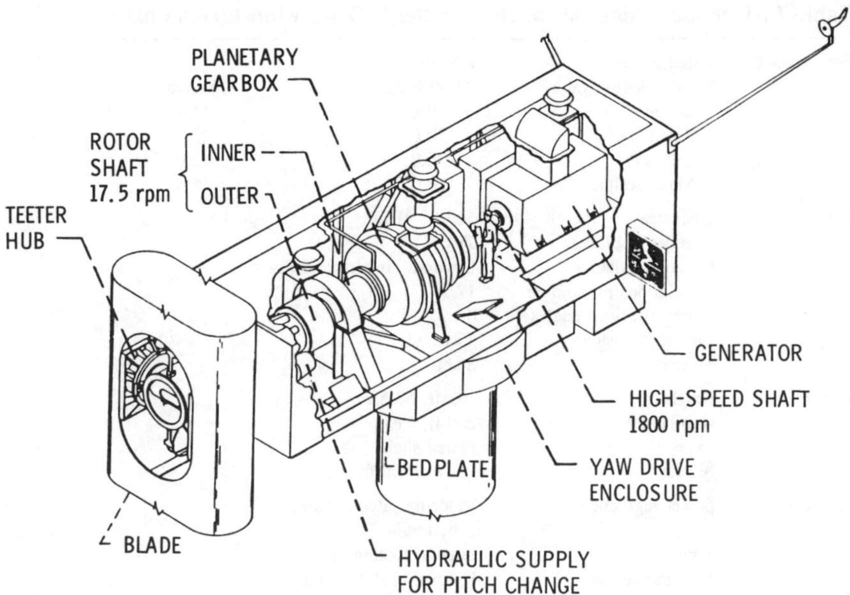

Figure 14-18 The MOD-2 nacelle [139].

Materials for the blades have varied over the years, including steel, aluminum, fiberglass, laminated wood, and combinations such as steel spar with foam trailing edge (MOD-1).

Novel Wind Turbines

Although research and development funds have gone largely to propeller-type wind turbines, some have been expended to investigate other designs that may offer future technical and economic advantages. Two of these will be covered in the next sections, the *Madaras concept* and the *Darrieus rotor*. While they are referred to here as "novel," the principles behind them are not new.

14-8 THE MAGNUS EFFECT

The Madaras concept is based on the *Magnus effect,* which was first demonstrated experimentally by Magnus in 1852. The effect showed that if a horizontal cylinder is rotated about its axis and moved through a still wind, a lift force is produced. Similarly, if a stationary horizontal cylinder is rotated about its axis in a cross wind, it will experience a lift force. The effect is equally applicable to a vertical cylinder being

Table 14-1 Some design parameters of the MOD-2 wind-turbine plant

Performance	Rated power	2.5 MW	
	Wind velocity, mi/h	At 30-ft height	At hub
	Cut-in	9.0	14.0
	Rated	20.0	27.5
	Cut-out	36.0	45.0
	Max. design	120.0	125.0
Rotor	Diameter	300 ft, 91 m	
	Number of blades	Two	
	Location, rotation	Upwind, counterclockwise	
	Revolutions per minute	17.5	
	Cone, tilt, twist angles	0°, 2°, 8°	
	Tip length, each	45 ft, 13.7 m	
	Material	Steel	
Tower	Height	192 ft, 58.5 m	
	Hub height	200 ft, 61 m	
	Type	Flared shell	
	Access	Power manlift	
Controls	Power regulation	Rotor-tip pitch control, hydraulic	
	Yaw	Internal toothing gear	
	Yaw motor, rate	Hydraulic, 0.25 deg/s	
	Supervisory	Microprocessor	
Generator	Rating, power factor	3125 kVA, 0.8	
	Voltage, frequency	4160 (three-phase), 60 Hz	
	Revolutions per minute	1800	
	Gearbox	Three-stage planetary	
	Gear step-up ratio	103	
Mass	Rotor	180,000 lb_m, 81,670 kg	
	Rotor and nacelle	364,000 lb_m, 165,150 kg	
	Tower	255,000 lb_m, 115,700 kg	

rotated about its axis in a cross wind. It will experience a force perpendicular to its axis, which will cause it to move in a direction essentially perpendicular to that of the wind.

The magnus effect can be explained by first considering a cylinder in nonviscous flow (Fig. 14-19). The cylinder has a length much greater than its diameter, so end effects are unimportant and flows around it can be considered two dimensional.

In Fig. 14-19*a* the cylinder is not rotating and a uniform stream is flowing past it, which gives rise to streamlines as shown. The velocity at any point on the surface of the cylinder is given [140] by

$$V_\theta = 2V_i \sin \theta \tag{14-32}$$

where V_θ = air velocity on the cylinder surface at angle θ

V_i = incoming uniform air velocity, far from the cylinder

θ = polar angle, measured from the stagnation point

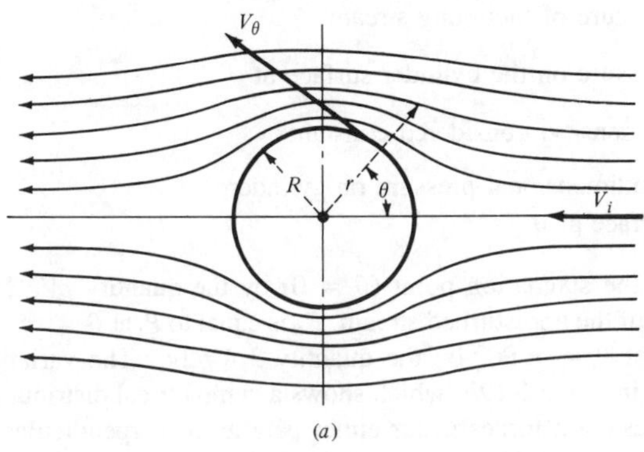

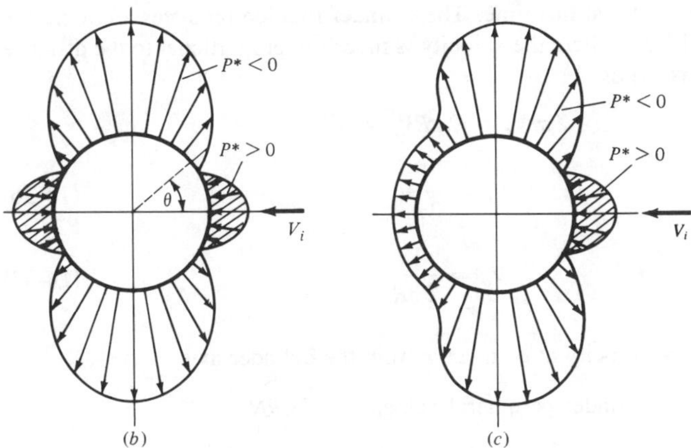

Figure 14-19 Streamlines and pressure distributions for two-dimensional flow past a nonrotating cylinder. $P^* = (P_O - P_i)/(\rho V_i^2/2g_c)$. *a* and *b*, nonviscous flow; *c*, viscous flow.

Applying Bernoulli's equation to any position θ on the surface of the cylinder gives

$$\frac{P_i}{\rho} + \frac{V_i^2}{2g_c} = \frac{P_\theta}{\rho} + \frac{V_\theta^2}{2g_c} \tag{14-33}$$

Combining with Eq. (14-32) gives

$$P_\theta = P_i + \frac{1}{2g_c}\rho V_i^2(1 - 4\sin^2\theta) \tag{14-34}$$

or

$$P_\theta^* = \frac{P_\theta - P_i}{\rho V_i^2/2g_c} = 1 - 4\sin^2\theta \tag{14-35}$$

where P_i = pressure of incoming stream

P_θ = pressure on the cylinder surface at θ

ρ = air density, considered constant

P_θ^* = nondimensional pressure on cylinder surface at θ

Thus P_θ is greater than P_i at the stagnation point ($\theta = 0$) by the quantity $\rho V_i^2/2g_c$, which is the *dynamic pressure* of the undisturbed stream. P_θ is equal to P_i at $\theta = \pm 30°$ and $\pm 150°$ and less than P_i at $\theta = \pm 90°$ by the quantity $3\rho V_i^2/2g_c$. The variation with θ is shown by the P_θ^* plot in Fig. 14-19*b*, which shows a symmetrical distribution about the cylinder. Thus no resultant forces occur either parallel or perpendicular to the undisturbed stream; i.e., there is neither lift nor drag on the cylinder for the case of ideal nonviscous flow.

Now we will consider that the cylinder is rotated about its axis but that the air far from the cylinder is at rest this time. The cylinder rotation produces a circulatory flow around it (Fig. 14-20) where the velocity is inversely proportional to the distance from the cylinder axis. Thus

$$2\pi r V_r = 2\pi R V_p = \Gamma$$

or

$$V_r = \frac{\Gamma}{2\pi r} \tag{14-36}$$

and

$$V_p = \frac{\Gamma}{2\pi R} \tag{14-37}$$

where V_r = velocity at distance r from the cylinder axis

V_p = cylinder peripheral velocity = $2\pi RN$

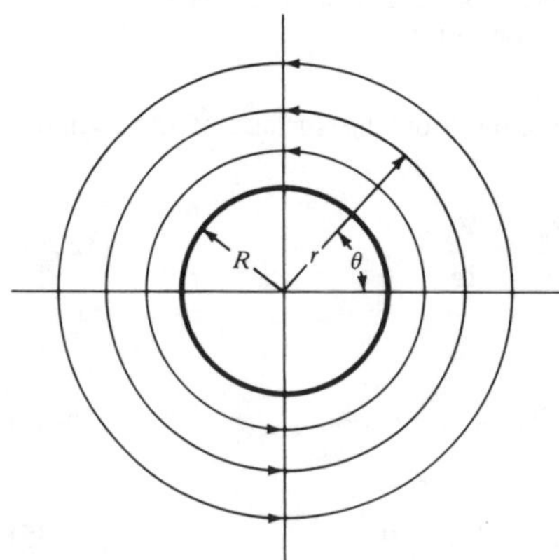

Figure 14-20 Circulatory flow around a rotating cylinder in an otherwise still air.

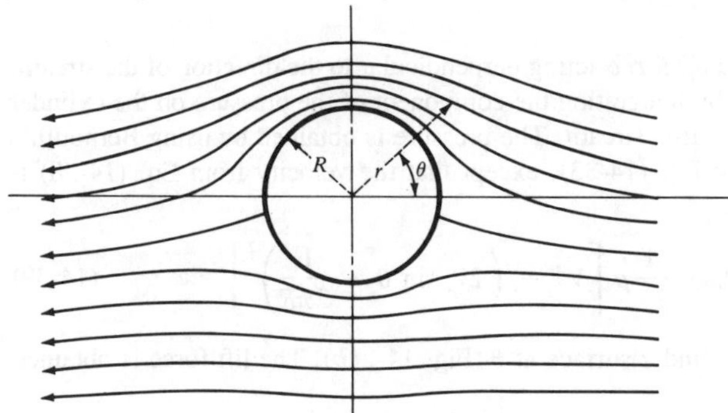

Figure 14-21 Velocity streamlines around a rotating cylinder in a stream of air.

r = radial distance from cylinder axis

R = cylinder radius

Γ = a constant = $2\pi RV_p$ = $(2\pi R)^2 N$ having the dimension length²/time and called the *circulation constant*

N = number of revolutions per unit time

Now we will combine the two cases from above, i.e., a rotating cylinder in a stream of air. When Figs. 14-19*a* and 14-20 are superimposed, it is obvious that the two velocity patterns will reinforce each other when in the same direction, i.e., at the cylinder top, and oppose each other at the bottom. Thus there will be a velocity increase at the top and a velocity decrease at the bottom (Fig. 14-21). The velocity V on the cylinder surface is given by the sum of V_θ and V_p from Eqs. (14-32) and (14-37), or

$$V = 2V_i \sin\theta + \frac{\Gamma}{2\pi R} \tag{14-38}$$

where $\sin\theta$ is positive between $\theta = 0$ and 180° and negative between 0 and $-180°$. The pressures on the cylinder will therefore be higher at the bottom (where the velocities are lower) and lower at the top (where the velocities are higher) than in the case of no cylinder rotation. Note the stagnation points where $V = 0$ are no longer at $\theta = 0$ and 180° but are slightly on the lower side of the cylinder.*

* The two stagnation points approach each other as V_p increases until only one stagnation point occurs at $\theta = -90°$ when $V_p = 2V_i$ and $\Gamma = 4\pi RV_i$.

The Lift Force

There will, therefore, be a *lift force* acting perpendicular to the direction of the stream. The lift force is obtained by integrating the component of the pressure on the cylinder that is perpendicular to the free stream. The pressure is obtained by using Bernoulli's equation, as was done for Eq. (14-33), except that the velocity from Eq. (14-38) is used in place of P_θ. Thus

$$P = P_i + \frac{1}{2g_c}\rho\left[V_i^2 - \left(2V_i \sin\theta + \frac{\Gamma}{2\pi R}\right)^2\right] \tag{14-39}$$

where P = pressure on cylinder surface at θ (Fig. 14-21*b*). The lift force is obtained from

$$F_L = \int_{-\pi}^{+\pi} - P \sin\theta \, R \, d\theta \, H \tag{14-40}$$

Combining with Eq. (14-39) and integrating results in

$$F_L = \frac{2\pi}{g_c}\rho RHV_pV_i = \frac{4\pi^2}{g_c}\rho R^2HNV_i \tag{14-41}$$

where F_L = lift force on cylinder

H = length (or height) of cylinder

It is convenient to give the lift force in terms of a dimensionless *lift coefficient* C_L, the dynamic pressure of the free stream $\rho V_i^2/2g_c$, and the projected area of the cylinder $A = 2RH$, or

$$F_L = C_L A\rho\frac{V_i^2}{2g_c} \tag{14-42}$$

Comparing this with Eq. (14-41) gives, for nonviscous flow

$$C_L = 2\pi\frac{V_p}{V_i} \tag{14-43}$$

or the value of the lift coefficient in nonviscous flow is directly proportional to the ratio of the cylinder peripheral to the free stream velocities.

As indicated, the above development of the Magnus effect was based on ideal nonviscous flow. For a real fluid, the effects of viscosity and boundary layer separation result in a pressure distribution for nonrotating cylinders as shown in Fig. 14-19*c*. Experimental measurement on rotating cylinders in real fluids show values of C_L that are much lower than those given by Eq. (14-43), reaching a maximum between 9 and 10 beyond a speed ratio V_p/V_i of 4 or 5 (Fig. 14-22). Equation 14-42 should be used with a value of C_L obtained from Fig. 14-22.

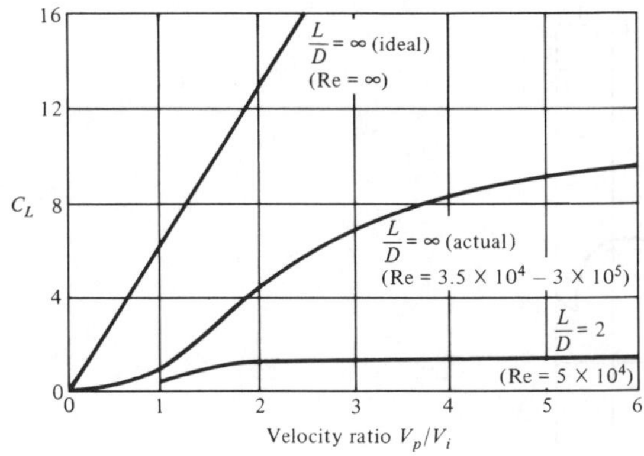

Figure 14-22 Lift coefficient of a rotating cylinder in an air stream as a function of the ratio of the cylinder peripheral and free stream velocities.

The Drag Force

The *drag force* for an ideal fluid was found above to be zero. For a real fluid the imbalance in pressures (Fig. 14-19*c*), results in a drag force in the direction of the free stream, given, similarly to Eq. (14-42), by

$$F_D = C_D A \rho \frac{V_i^2}{2g_c} \tag{14-44}$$

where C_D = drag coefficient, dimensionless. It is a function of the Reynolds number

$$\text{Re} = \frac{D V_i \rho}{\mu} \tag{14-45}$$

where D = diameter of cylinder = $2R$

μ = viscosity of fluid

The relationship between C_D and Re for long smooth cylinders is shown in Fig. 14-23.

The force acting on the cylinder is the resultant force between the lift and drag forces.

Example 14-2 Calculate the lift, drag, and resultant forces on a 6-ft-diameter, 90-ft-long smooth cylinder rotating at 140 r/min in a 30 mi/h wind. The wind is at 1 standard atm pressure and 60°F.

SOLUTION For air at 1 atm and 60°F

$$\rho = \frac{P}{RT} = \frac{14.69 \times 144}{53.34 \times (60 + 460)} = 0.0763 \text{ lb}_m/\text{ft}^3$$

$$\mu = 0.046 \text{ lb}_m/(\text{h} \cdot \text{ft})$$

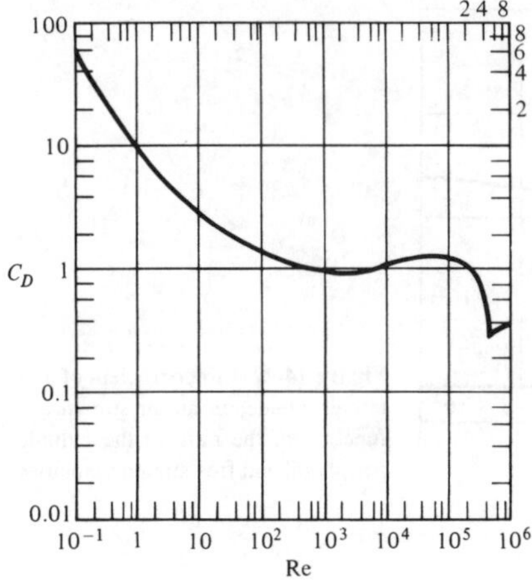

Figure 14-23 Drag coefficient as a function of Reynolds number for long smooth cylinders [140].

$$V_p = \pi DN = \pi \times 6 \times \frac{140}{60} = 44 \text{ ft/s}$$

$$V_i = 30 \text{ mi/h} = \frac{30 \times 5280}{3600} = 44 \text{ ft/s}$$

$$V_p/V_i = 1.0$$

$$\text{Re} = \frac{DV\rho}{\mu} = \frac{6 \times 44 \times 0.0763}{0.046/3600} = 1{,}576{,}400$$

Therefore

$$C_L = 1.0 \qquad \text{(Fig. 14-22)}$$

and

$$C_D = 0.4 \qquad \text{(by extrapolation from Fig. 14-23)}$$

$$\text{Dynamic pressure } \rho V_i^2/2g_c = 0.0763 \times 44^2/(2 \times 32.2) = 2.294 \text{ lb}_f/\text{ft}^2$$

$$\text{Lift force} = 1.0 \times 6 \times 90 \times 2.294 = 1239 \text{ lb}_f$$

$$\text{Drag force} = 0.4 \times 6 \times 90 \times 2.294 = 495 \text{ lb}_f$$

$$\text{Resultant force} = \sqrt{1239^2 + 495^2} = 1334 \text{ lb}_f$$

$$\text{Ideal lift coefficient} = 2\pi \times 1 = 2\pi$$

$$\text{Ideal lift force} = 2\pi \times 60 \times 90 \times 2.294 = 7785 \text{ lb}_f$$

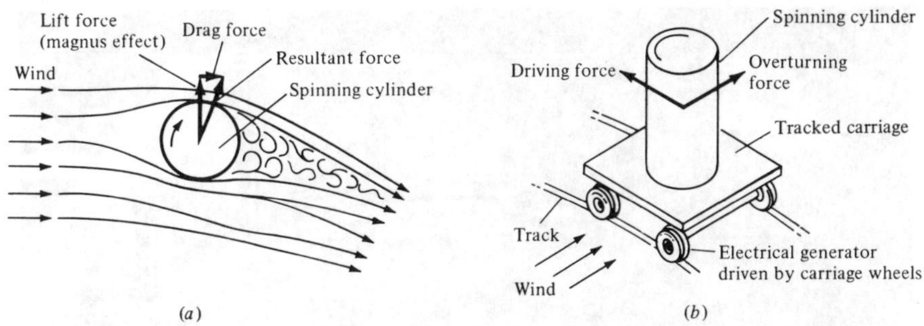

Figure 14-24 Resultant force of lift and drag propelling a spinning vertical cylinder along a track [135].

14-9 THE MADARAS ROTOR WIND MACHINE

Based on the Magnus effect, the Madaras rotor powerplant was invented by Julius D. Madaras in 1912 and was soon sponsored by the Detroit Edison Company, which actually built a pilot model of it at Burlington, New Jersey, in 1933. It generated much interest because of the novel idea behind it and because it was sponsored by an important utility.

The concept consists of several tall vertical cylinders that are rotated about their axes in the presence of a wind. The resultant of the lift and drag forces, more in the direction of lift because of its larger value (Fig. 14-24*a*), propels the cylinders horizontally* along a track (Fig. 14-24*b*).

The powerplant concept as proposed by Madaras consists of a circular train track with a continuous train of four-wheel flat cars on it (Fig. 14-25). Each flat car has a 90-ft-high cylinder that is rotated about its vertical axis at 120 r/min by a small electric motor. The resultant force along each cylinder moves flat cars around the circular track. To keep the force in the same circumferential direction along the track, the direction of rotation of the cylinders is reversed twice each complete turn around the track. The carriage wheels drive electrical generators, and the electricity generated is transmitted to the mains by a trolley.

The Madaras project generated a lot of excitement and attracted a number of highly qualified engineers who agreed that it was a sound, economical, and safe means of electric-power generation. It was relatively insensitive to wind velocity changes, above about 6 mi/h, and was believed able to withstand winds of 100 to 120 mi/h. Its capital cost was estimated in 1928 at $40/kW for a 10-MW plant. The pilot plant erected at Burlington, New Jersey, in 1933, unfortunately, blew down in a high wind before it was even tested with the loss of several million dollars and much time and effort.

It was the failure of both the Madaras and Smith Putnam (Sec. 14-2) machines that caused commercial utilities in the United States to quietly abandon efforts at

* The Magnus effect was used to propel a ship across the Atlantic Ocean in the early 1930s.

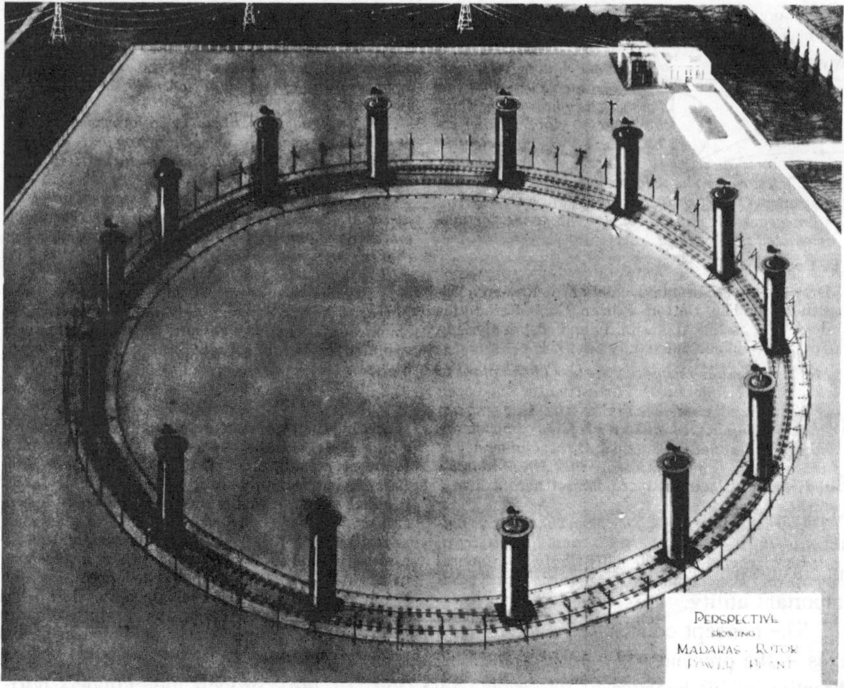

Figure 14-25 The Madaras concept in perspective [135].

generating power from the wind back in the 1930s and 1940s. It was thus up to the federal government to sponsor wind-power research and development when interest was rekindled with the energy crisis of the 1970s.

14-10 THE DARRIEUS MACHINE

The *Darrieus* wind turbine, shown in two forms, the ϕ and the Δ in Fig. 14-2 is an invention of Georges Darrieus of France, who patented it in 1920. The ϕ-Darrieus has received renewed interest as a machine that may efficiently fill the gap between the small domestic and large utility wind turbines. Unlike the large modern machines, but like the ancient Persian ones, it is a vertical-axis machine, but here the similarity with the ancient machines ends. A Darrieus wind turbine is invariably described as looking like an egg beater, with either two or three slender wings or blades that loop over the top and that are joined at top and bottom to the vertical axis (Fig. 14-26).

Figure 14-26 A Darrieus windmill under testing at West Texas State University *(Courtesy DOE.)*

The Darrieus machine has the advantages of spinning in any wind direction, thus requiring no yaw control, and a vertical axis that allows the power-conversion equipment, such as an electrical generator or pump, to be located in a stationary ground installation.

The Darrieus concept is under investigation by Sandia Laboratories for the DOE. Several machines have been constructed and were undergoing tests in the early 1980s, the largest being used for a deep-well irrigation system in Bushland, Texas. It has two blades 37 ft in diameter and 55 ft high, produces 56 hp (42 kW) in 30-mi/h (13.4-m/s) wind, and is used to pump 300 gal/min of water from a 280-ft-deep well. Pump speed is maintained in this experimental unit by connecting the pump in parallel to an electrical motor or a governed Diesel engine.

14-11 OTHER WIND TURBINE DESIGNS

Besides the propeller, Madaras, and Darrieus, there has been a plethora of designs for wind machines. Wind, more than any other renewable energy source, has intrigued professional and amateur inventors over the ages. In recent years DOE has been evaluating literally dozens of designs for wind systems, some have already been shown in Fig. 14-2. The governing consideration, as always, is the economic one: what size and cost are needed per unit power output.

One intriguing powerplant design, called an *aeroelectric plant,* uses the flow up a tower that looks much like a cooling tower. Its walls are heated by solar radiation. Because the walls are circular, the sun's rays need not be tracked as it changes position in the sky during the day. The heated walls in turn heat the inside air and a flow up the tower is established.

This air flow is made to drive a number of air turbines located near the top of the tower. The turbines, in turn, drive electrical generators. The driving pressure causing airflow is given by the now familiar expression

$$\Delta P_d = (\rho_o - \rho_i)H\frac{g}{g_c} \tag{14-46a}$$

and, since air can be treated as a perfect gas where $\rho = P/RT$

$$\Delta P_d = \frac{1}{R}\left(\frac{P_o}{T_o} - \frac{P_i}{T_i}\right)H\frac{g}{g_c} \tag{14-46b}$$

where ΔP_d = driving pressure needed to overcome the pressure losses in the tower and to drive the air turbines

P_o = outside air pressure

P_i = average inside air pressure

R = air gas constant

T_o = outside air absolute temperature

T_i = average inside air absolute temperature

H = height of tower

g = acceleration of gravity

g_c = conversion factor

A variation of the above system is one proposed by P. Carlson of California. The interior air in a very tall tower would be cooled by pumping water to the top. The water evaporates in the low-pressure air there, causing a downward flow of the cooled air. The driving pressure is obtained from Eq. (14-46*a*) but with P_i/T_i and P_o/T_o in reversed positions. They are to be calculated in a manner similar to that for wet cooling

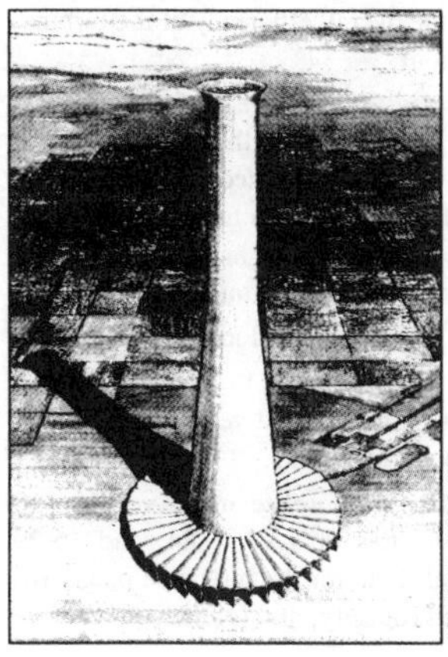

Figure 14-27 The Carlson aeroelectric tower *(ASME.)*

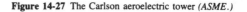

towers (Sec. 7-5). A conceptual design of such a plant called for a 1.5-mi-high, 900-ft-diameter tower located in a hot desert and 10 wind turbines surrounding the tower periphery at the bottom producing 2500 MW (Fig. 14-27).

PROBLEMS

14-1 A 27 mi/h wind at 14.65 psia and 70°F enters a turbine wheel that has a 1000-ft^2 cross-sectional area. Calculate (*a*) the power of the incoming wind, (*b*) the theoretical maximum attainable turbine power, (*c*) a reasonably attainable turbine power, all in horsepower and kilowatts, (*d*) the torque, and (*e*) the axial thrust, in pound force, if the turbine wheel rotates at 30 r/min.

14-2 Consider two cases: (*i*) a constant wind velocity twice the mean velocity and operating half the time, and (*ii*) a constant wind velocity three times the same mean velocity operating one third of the time. At all other times the wind velocity is zero. Determine for each case (*a*) the ratio of total energy in the wind to that if it operated continuously at the mean velocity, and (*b*) the mean energy velocity as a function of the mean velocity.

14-3 A 15-ft-diameter wind turbine operates in 25 ft/s wind at 1 atm and 60°F. The turbine is used to pump 60°F water from a 30-ft-deep well. How much water is pumped, in cubic feet per day, if the overall efficiency of the wind-turbine-pump system is 0.25?

14-4 A wind turbine with a diameter of 100 ft operates during a 24-h period in which the wind velocity may be approximated by half a sine wave as $V(\theta) = V_m \sin(\pi\theta/24)$, where θ is time in hours and $V_m = 10^5$ ft/h. The wind density is 0.076 lb$_m$/ft^3. Calculate (*a*) the theoretical maximum turbine energy during that 24-h period, in foot-pound force and kilowatt hours, and (*b*) the wind mean energy velocity, in feet per hour.

14-5 Consider a wind velocity pattern during a 24-h period in which the wind velocity increased steadily and linearly from zero to 48 ft/s. The wind density is 0.0768 lb$_m$/ft^3. A 200-ft-diameter wind turbine operates in that wind with cut-in, rating, and cut-out wind velocities of 12, 24, and 42 ft/s, respectively. The turbine-generator overall efficiency is assumed, for simplicity, constant at 0.25. Calculate (*a*) the power generated during the day, in kilowatt hours, and (*b*) the wind mean energy velocity, in feet per second.

14-6 A wind at 1 atm and 62.4°F had a velocity pattern during one day that could be approximated by half a sine wave with an amplitude of 1.1×10^5 ft/h. A 100-ft-diameter wind turbine with a cut-in, rated, and cut-out velocities of 4.0×10^4, 8.0×10^4, and 1.2×10^5 ft/h, respectively, operated in that wind.

Calculate (*a*) the energy generated, in kilowatt hours during that day, if the overall efficiency of the turbine-generator is 0.34, and (*b*) the wind mean energy velocity, in feet per hour.

14-7 A wind at 14.5 psia and 55°F had a velocity pattern during one day that followed a full sinusoidal wave with a mean velocity of 20 mi/h and an amplitude of 10 mi/h. A wind turbine with a 30,000-ft^2 area operated in that wind. Calculate for that day (*a*) the total energy of the wind, (*b*) the theoretical maximum energy developed by the turbine, (*c*) the theoretical maximum energy developed by the turbine if the wind had a constant velocity at the mean value, (*d*) a reasonably attainable turbine energy in the actual wind, all in foot-pound force and kilowatt hours, and (*e*) the wind mean energy velocity, in miles per hour.

14-8 Repeat Prob. 14-7, but assume that the wind turbine had cut-in and cut-out velocities of 8 and 20 mi/h, respectively, and was not flat-rated.

14-9 Two wind patterns follow the Weibull distribution model. One has a mean wind velocity of 12 m/s, the other 8 m/s. Calculate the percent of time that each exceeds a velocity of 10 m/s.

14-10 A wind pattern that follows the Weibull distribution model has a reference mean velocity 10 m/s. Consider a wind turbine that has cut-in, rated, and cut-out velocities of 4.97, 8.84, and 16.29 m/s, respectively, all at the reference height above ground of 9.1 m. Calculate for the reference height (*a*) the turbine availability factor, (*b*) the percent time it operates below rated capacity, and (*c*) the percent time it operates at rated capacity.

14-11 Repeat Prob. 14-10 but for a wind turbine with a hub 50 m above ground.

14-12 Wind at 1 bar and 20°C has a velocity of 15 m/s, measured at the reference height above ground of 9.1 m (30 ft). Calculate (*a*) the power density in the wind at the reference height. Estimate (*b*) the power density at the hub of a large 60-m-diameter wind turbine, 60 m above ground, (*c*) the maximum theoretical power delivered by the turbine, and (*d*) a reasonably attainable power by the turbine generator if the latter has an efficiency of 94.2 percent, all in watts per square meter.

14-13 A 30-m-long 2-m-diameter cylinder rotates at 120 r/min in a 10 m/s wind that is at 1 atm and 15°C. Calculate (*a*) the ideal and actual lift forces, in newtons, (*b*) the ideal and actual drag forces, in newtons, and (c) the positions of the stagnation points.

14-14 A Madaras rotor wind machine consists of 14 vertical cylinders such as the one described in Prob. 14-13. Assume for simplicity that the wind at 1 atm and 15°C is at standstill and that the cylinders move around a 220-m-diameter track at 10 m/s circumferential velocity. Calculate (*a*) the magnitude, in newtons, and direction of the resultant force on each cylinder with respect to the direction of motion, (*b*) the maximum power generated by one cylinder, in kilowatts, (*c*) the power of the entire machine if it takes each cylinder 5 percent of the travel distance to change its direction of rotation (ignore the power produced during that change), and (*d*) the energy produced by the machine during one complete circle, in kilowatt hours.

14-15 An aeroelectric wind powerplant with turbines on top has a 1.5-mi-high tower. The outside air is at 1 atm and 60°F. The inside air is at an average temperature of 160°F. The tower diameter is very large so that only 2.5 percent of the driving pressure goes to overcome frictional and other losses in it. The balance drives the turbines. Calculate (*a*) the total pressure drop in the tower, in pounds per square inch, and (*b*) the power produced by the turbines, in watts, per unit air volume flow rate, in cubic feet per hour, if they have an overall efficiency of 0.75.

14-16 A Carlson aeroelectric wind powerplant has a 1.5-mi-high tower. The outside air is at 1 atm, 60°F, and 20 percent relative humidity. Water is injected at top so that the air becomes saturated. The process is one of evaporative cooling which is a constant wet-bulb temperature process. Calculate (*a*) the driving pressure on the turbines, in pounds per square inch, and (*b*) the power produced by the turbines, in watts, per unit air volume flow rate, in cubic feet per hour, if they have an overall efficiency of 0.75. Ignore pressure losses in tower.

CHAPTER

FIFTEEN

ENERGY FROM THE OCEANS

15-1 INTRODUCTION

Solar energy, which may be used directly (Chap. 13), creates other forms of energy that can also be harnessed to generate power. One, the wind, is caused by the uneven solar heating and cooling of the earth's crust combined with the rotation of the earth (Chap. 14). Another is the result of the absorption by the seas and oceans of solar radiation, which causes, like the wind, ocean currents and moderate temperature gradients from the water surface downward, especially in tropical waters. The oceans and seas constitute some 70 percent of the earth's surface area, so they represent a rather large storage reservoir of the solar input.

The temperature gradient can be utilized in a heat engine to generate power. This is called *ocean temperature energy conversion* (OTEC). OTEC may be considered solar energy once removed. Because the temperature difference is small, even in the tropics, OTEC systems have very low efficiencies and consequently have very high capital costs.

The wind, whose energy is also solar energy once removed, generates large *ocean waves* with energies that can be used to generate power. Ocean-wave energy thus is said to be solar energy twice removed. Ocean waves, unfortunately, vary widely with time and place in amplitude and frequency, and hence in their energies, much like the wind that causes them. There are, however, areas in the world where energetic waves persist a good deal of the time. Such waves have been considered for power generation by a wide variety of ingenious means.

A third form of energy that emanates from the sun-ocean system stems from the mechanism of surface water evaporation by solar heating. This forms clouds that condense into rains. Part of that rain falls over land, which causes river flows, which may be trapped behind dams to even out the variations in these river flows and thus become the source of either low-head (river) or high-head (dam) *hydroelectric energy*. Hydroelectricity accounts for some 25 percent of the world's total electricity capacity. Some countries, such as Norway, Switzerland, and Canada, because of favorable topographies and rainfall, far exceed this average. In the United States, the percentage was 15 percent in the 1970s, going down to some 10 percent in the 1980s. Hydroelectricity used to account for much more than these percentages in the nineteenth century, but the advent of cheap coal and steam powerplants, and the lack of additional favorable hydro sites, caused a decline in its use. Many small river hydroelectric plants were abandoned. With the coming of the energy crisis of the 1970s, interest was renewed in these small plants to supplement input to power grids.

A fourth major, though different, source of energy in the oceans that can be exploited for power generation is the *tides*. Tides are primarily caused by lunar, and only secondarily by solar, gravitational forces acting together with those of the earth on the ocean waters to create tidal flows. These manifest themselves in the rise and fall of waters with ranges (height differences) that vary daily and seasonally and come at different times from day to day. They also vary widely from place to place, being as low as a few centimeters but may exceed 8 to 10 m (25 to 30 ft) in some parts of the world. The potential energy of the tides can be trapped to generate power, but at extremely high capital costs.

In this chapter, power production from the energies associated with the ocean temperature difference (OTEC), the waves, and the tides will be discussed. Hydroelectric power is a major topic in its own right and deserves a separate course of study. The interested reader is advised to consult the many books that are devoted to it. See, for example, the *Handbook of Applied Hydraulics* [141].

15-2 OCEAN TEMPERATURE DIFFERENCES

As the seas and oceans of the earth constitute about 70 percent of its surface area, the total terrestrial solar-energy incidence on them is immense, being equal to the total extraterrestrial solar energy received by the earth, which is about 1.516×10^{18} kWh/year, or about 5.457×10^{18} MJ/year, times an average clearness index of 0.5 (Sec. 13-14), times the fraction of the area 0.7 or about 0.53×10^{18} kWh/year, or 1.9×10^{18} MJ/year. This corresponds to an average terrestrial incidence on the waters of the solar constant $S = 1353 \text{ W/m}^2 \times 0.5 = 676 \text{ W/m}^2$. This energy is not totally absorbed by the water because some of it is reflected back to the sky. A good estimate of the amount absorbed is obtained from the annual evaporation, which averages 120 cm (or 120 cm^3/cm^2, or 1.20 m^3/m^2).* At an average water surface temperature of 20°C (68°F), the latent heat of vaporization is 2454 kJ/kg and the sea water density is a little over 1000 kg/m^3. The annual energy absorbed would therefore be $1.20 \times 1000 \times 2454$, or about 3×10^6kJ/m^2 per year, which is equivalent to about

* This amount is, of course, replenished by rainfall back on the water and by runoff from land.

95 W/m^2 or about 14 percent of the incidence. This figure varies, being a little higher than 100 W/m^2 in the tropics to much less in arctic waters.

Solar-energy absorption by the water takes place according to *Lambert's law of absorption,* which states that each layer of equal thickness absorbs the same fraction of light that goes through it. In other words

$$-\frac{dI(y)}{dy} = \mu I$$

or (15-1)

$$I(y) = I_o e^{-\mu y}$$

where I_o and $I(y)$ are the intensities of radiation at the surface ($y = 0$) and at a distance y below the surface. μ is an *extinction coefficient* (also called *absorption coefficient*) that has the unit length^{-1}. μ has values of 0.05 m^{-1} for very clear fresh water, 0.27 m^{-1} for turbid fresh water, and 0.50 m^{-1} for very salty water. Thus the intensity falls exponentially with depth and, depending upon μ, almost all of the absorption occurs very close to the surface of deep waters. Because of heat and mass transfer at the surface itself, the maximum temperatures occur just below the surface.

Considering deep waters in general, the high temperatures are at the surface, whereas deep water remains cool. In the tropics, the ocean surface temperature often exceeds 25°C (77°F), while 1 km below the temperature is usually no higher than 10°C (50°F). Water density decreases with an increase in temperature (above 3.98°C, where pure water's density is maximum, decreasing again below this temperature, the reason ice floats). Thus there will be no thermal convection currents between the warmer, lighter water at the top and the deep cooler, heavier water. Thermal conduction heat transfer between them, across the large depths, is too low to alter this picture, and thus mixing is retarded, so the warm water stays at the top and the cool water stays at the bottom.

It is said, therefore, that in tropical waters there are two essentially infinite heat reservoirs, a heat source at the surface at about 27°C (81°F) and a heat sink, some 1 km directly below, at about 4°C (39°F); both reservoirs are maintained annually by solar incidence.

The concept of ocean temperature energy conversion (OTEC) is based on the utilization of this temperature difference in a heat engine to generate power, a concept first recognized by the Frenchman d'Arsonval in 1881. The maximum temperature difference on earth is in the tropics and is about 15°C (59°F). Ocean currents carry the 27 to 28°C warm tropical surface waters on a journey to the arctic circles during which they are gradually cooled to 4°C and maximum density (Fig. 15-1). In the arctic circle they then settle below the surface, and a surface–deep water siphon is created that keeps cold water below the surface.

The surface temperatures (and temperature differences) vary both with latitude and season (Fig. 15-2), both being maximum in tropical, subtropical, and equatorial waters, i.e., between the two tropics, making these waters the most suitable for OTEC systems [143].

The claims for OTEC systems are just as grandiose as those for most other renewable energy systems. For example, it is said that the Gulf stream is known to

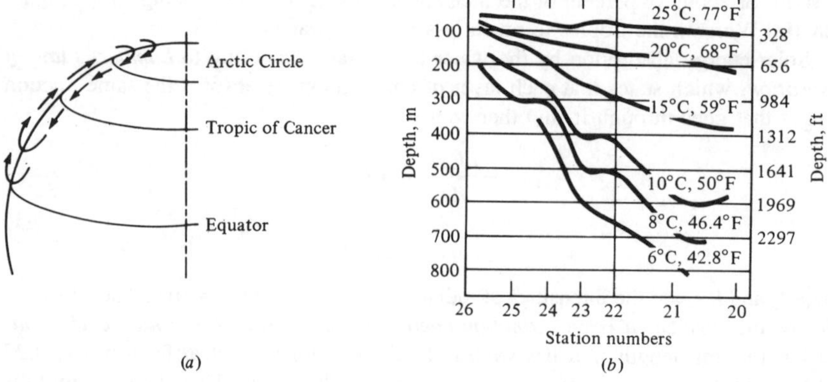

Figure 15-1 (*a*) Ocean-water movement in the northern hemisphere. (*b*) Measured water temperatures at various stations and depths in the Straits of Florida [142].

carry some 30 to 40 million m³/s of near tropical sea water through the Gulf of Florida, flowing in a path about 32-km (20-mi) wide. Within 800 km (500 mi) of that path, the temperature differences between surface and deep waters varies between 22°C (40°F) and 15°C (27°F). Assuming a practical conversion efficiency of 2 percent (below), the Gulf stream represents an annual power potential of 700×10^{12} kWh. An array of conversion plants moored on 1-mi (1.6-km) spacings along the length and breadth of that path would be capable of an annual 26×10^{12} kWh. Such are the claims for OTEC, but as we will soon learn, practical and financial problems effectively preclude such dreams.

The maximum possible efficiency of a heat engine operating between two temperature limits cannot exceed that of a Carnot cycle operating between the same temperature limits. For heat source and sink temperatures, T_1 and T_2, the Carnot efficiency η_C is

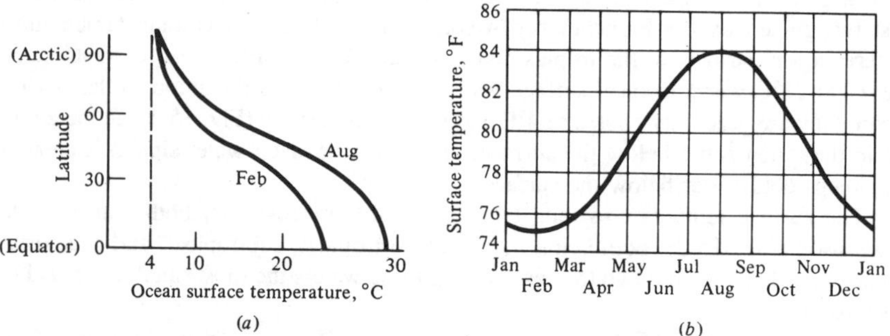

Figure 15-2 Ocean surface temperature as a function of (*a*) latitude and (*b*) the seasons, in tropical waters.

$$\eta_C = \frac{T_1 - T_2}{T_1} \tag{15-2}$$

It is thus important to have a large temperature difference $T_1 - T_2$. In OTEC systems this may average 20°C (68°F) compared with about 500°C (932°F) for modern fossil powerplants. Taking the temperature difference of 20°C and a surface temperature of 27°C, the Carnot efficiency would be

$$\eta_C = \frac{20}{27 + 273} = 0.0667 = 6.67\%$$

Thus for 27°C (300 K) surface temperature an OTEC system can have no efficiency greater than $\Delta T/300$ or 1/3 ΔT percent, where ΔT is the difference between surface and deep waters. Because of temperature drops in steam or other vapor generator and condenser in an actual system (external irreversibilities, Chap. 1), inefficiencies in turbine and pumps, pressure drops of the large volumes of water and working fluid flows (internal irreversibilities), and inefficiencies of other components, the efficiency of a real OTEC powerplant seldom exceeds 2 percent.

The extremely low efficiency of an OTEC system implies extremely large powerplant heat exchangers and components. At 2 percent efficiency the heat exchangers must handle 50 times the net output of the plant. Although there are no fuel costs, the capital costs are extremely high and so are the unit capital costs, \$/kW. In addition to the large size per unit power generation, the developmental problems and the uncertainties of market penetration make the financial risks associated with the development of large OTEC technologies so high as to effectively preclude most utilities. (The same can be said for other ocean energy systems, as well as most solar-electric systems, especially the solar satellite system, Sec. 13-20).

There are two basic designs for OTEC systems: the *open cycle,* also known as the *Claude cycle,* and the *closed cycle,* also known as the *Anderson cycle*. These are covered in the next two sections.

15-3 THE OPEN, OR CLAUDE, CYCLE

The first OTEC plant to be constructed was built by the Frenchman Georges Claude in 1929 on the Mantanzas Bay in Cuba [144]. It used the warm waters of the Gulf Stream as a heat source and a submarine cliff adjacent to the bay that descends nearly vertically to depths of 100 to 200 m as the heat sink. The warm surface water was at 25°C (77°F). The cold water, at 11°C (51.8°F), was tapped by a 2-km-long, 2-m-diameter pipe that weighed 400 tons. The laying of the cold-water pipe was the most difficult part of the construction: two such pipes were lost before a third was successfully installed.

The Claude plant used an *open cycle* (also called the Claude cycle) in which seawater itself plays the multiple role of heat source, working fluid, coolant, and heat sink. Schematic flow and corresponding *T-s* diagrams are shown in Figs. 15-3 and 15-4, respectively.

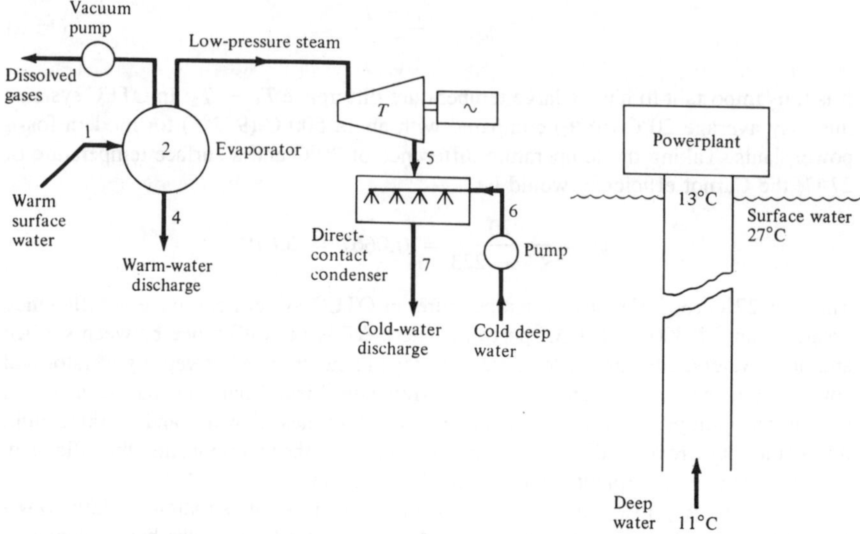

Figure 15-3 Flow diagram and schematic of a Claude (open-cycle) OTEC powerplant.

In the cycle shown warm surface water at 27°C (80.6°F) is admitted into an evaporator in which the pressure is maintained at a value slightly below the saturation pressure corresponding to that water temperature. Water entering the evaporator, therefore, finds itself "superheated" at the new pressure. For example, in Fig. 15-4, the warm water at 27°C has a saturation pressure of 0.0356 bar (0.517 psia), point 1. The

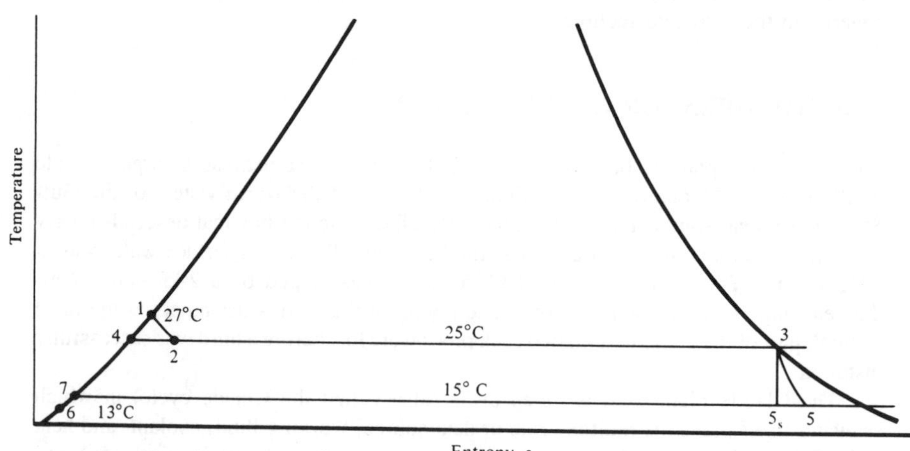

Figure 15-4 *T-s* diagram corresponding to Fig. 15-3.

evaporator pressure is 0.0317 bar (0.459 psia), which corresponds to 25°C (77°F) saturation temperature. This temporarily superheated water undergoes *volume boiling* (as opposed to *pool boiling*, which takes place in conventional boilers due to an immersed heating surface), causing that water to partially flash to steam to an equilibrium two-phase condition at the new pressure and temperature of 0.0317 bar and 25°C, point 2. Process 1-2 is a throttling and hence constant enthalpy process. The low pressure in the evaporator is maintained by a vacuum pump that also removes the dissolved noncondensable gases from the evaporator.

The evaporator now contains a mixture of water and steam of very low quality at 2. The steam is separated from the water as saturated vapor at 3. The remaining water is saturated at 4 and is discharged as brine back to the ocean. The steam at 3 is, by conventional powerplant standards, a very low-pressure, very high specific-volume working fluid (0.0317 bar, 43.40 m^3/kg, compared to about 160 bar, 0.021 m^3/kg for modern fossil powerplants). It expands in a specially designed turbine that can handle such conditions to 5. In Fig. 15-4, the condenser pressure and temperature at 5 are 0.017 bar (0.247 psia) and 15°C (59°F). Since the turbine exhaust system will be discharged back to the ocean in the open cycle, a direct-contact condenser (Sec. 6-2) is used, in which the exhaust at 5 is mixed with cold water from the deep cold-water pipe at 6, which results in a near-saturated water at 7. That water is now discharged to the ocean.

The cooling water reaching the condenser at 13°C is obtained from deep water at 11°C (51.8°F). This rise in temperature is caused by heat transfer between the progressively warmer outside water and the cooling water inside the pipe as it ascends the cold-water pipe. In Claude's first plant, that pipe was oversized to minimize this heat transfer and carried some 4000 m^3/h of water of which only 10 percent was needed for actual cooling.

There are thus three temperature differences, all about 2°C: one between warm surface water and working steam, one between exhaust steam and cooling water, and one between cooling water reaching the condenser and deep water. These represent external irreversibilities that reduce the overall temperature difference between heat source and sink from $27 - 11 = 16$°C (28.8°F) to $25 - 15 = 10$°C (18°F) as the temperature difference available for cycle work. It is obvious that because of the very low temperature differences available to produce work, the external differences must be kept to absolute minimum to realize as high an efficiency as possible. Such a necessary approach, unfortunately, also results in very large warm and cold water flows and hence pumping power, as well as large heavy cold water pipes.

Example 15-1 A Claude cycle producing 100 kW (gross) operates on the conditions of Fig. 14-4. The turbine has a polytropic efficiency of 0.80 and the turbine-generator has a combined mechanical-electrical efficiency of 0.90. Calculate the surface and deep-water flow rates in kg/s and m^3/s, and the gross cycle and plant efficiencies.

SOLUTION Use the low-temperature steam data (in SI units) shown in Table 15-1 (from App. A).

Table 15-1 Saturated steam data at low temperatures (SI units)

Temp.,	Pressure		Specific volume, m^3/kg		Enthalpy, kJ/kg			Entropy, kJ/(kg · K)		
°C	bar	psia	v_f	v_g	h_f	h_{fg}	h_g	s_f	s_{fg}	s_g
13	0.01497	0.2171	0.0010007	88.18	54.60	2470	2525	0.1952	8.6324	8.8276
15	0.01704	0.2471	0.0010010	77.97	62.97	2465	2528	0.2244	8.5562	8.7806
25	0.03166	0.4592	0.0010030	43.40	104.8	2442	2547	0.3672	8.1898	8.5570
27	0.03564	0.5169	0.0010036	38.82	113.2	2437	2550	0.3951	8.1196	8.5147

The evaporator

$$h_1 = h_f \text{ at } 27°\text{C} = 113.2 \text{ kJ/kg} \qquad v_1 = 0.0010036 \text{ m}^3/\text{kg}$$

$$h_2 = h_1 = 113.2 = (h_f + x_2 h_{fg}) \text{ at } 25°\text{C}$$

$$= 104.8 + x_2 \times 2442$$

Therefore

$$x_2 = 0.00344, \text{ or } 0.344\%$$

Warm water mass-flow rate per unit turbine mass-flow rate

$$\dot{m}_w = \frac{m_1}{m_3} = \frac{m_2}{m_3} = \frac{1}{x_2} = \frac{1}{0.00344} = 290.7$$

The turbine

$$h_3 = h_g \text{ at } 25°\text{C} = 2550 \text{ kJ/kg}, \ s_3 = 8.5570 \text{ kJ/(kg·K)}$$

For an adiabatic reversible turbine, the expansion is to 5_s

$$s_{5,s} = s_3 = 8.5570 = (s_f + x_{5,s} s_{fg}) \text{ at } 15°\text{C}$$

$$= 0.2244 + x_{5,s} \times 8.5562$$

Therefore

$$x_{5,s} = 0.9739, \text{ or } 97.39\%$$

$$h_{5,s} = (h_f + x_{5,s} h_{fg}) \text{ at } 15°\text{C} = 62.97 + 0.9739 \times 2465 = 2463.6 \text{ kJ/kg}$$

$$\text{Adiabatic reversible turbine work} = h_3 - h_{5,s} = 2550 - 2463.6$$

$$= 86.4 \text{ kJ/kg}$$

$$\text{Actual turbine work } w_T = 86.4 \times \text{polytropic efficiency}$$

$$= 86.4 \times 0.8$$

$$= 69.1 \text{ kJ/kg}$$

$$h_5 = h_3 - \text{actual work} = 2550 - 69.1 = 2480.9 \text{ kJ/kg}$$

at which $x_5 = 0.9809$ or 98.09% and $v_5 = 76.48$ m^3/kg

$$\text{Turbogenerator output} = 69.1 \times 0.88 = 60.8 \text{ kJ/kg}$$

The condenser

$$h_6 \approx h_f \text{ at } 13°\text{C} = 54.60 \text{ kJ/kg} \qquad v_6 = 0.0010007 \text{ m}^3/\text{kg}$$

$$h_7 = h_f \text{ at } 15°\text{C} = 62.97 \text{ kJ/kg}$$

Cold water mass-flow rate per unit turbine mass-flow rate

$$\dot{m}_c = \frac{h_5 - h_7}{h_7 - h_6} = \frac{2480.9 - 62.97}{62.97 - 54.60} = 288.9$$

The cycle

$$\text{Turbine mass-flow rate } \dot{M}_T = \frac{\text{turbine work}}{w_T} = \frac{100}{69.1}$$

$$= 1.447 \text{ kg/s}$$

$$\text{Turbine volume flow rate at throttle} = \dot{M}_T v_3 = 1.447 \times 43.40$$

$$= 62.8 \text{ m}^3/\text{s}$$

$$\text{Turbine volume flow rate at exhaust} = \dot{M}_T v_5 = 1.447 \times 76.48$$

$$= 110.7 \text{ m}^3/\text{s}$$

$$\text{Warm-water mass-flow rate } \dot{M}_w = \dot{M}_T \dot{m}_w = 1.447 \times 290.7$$

$$= 420.6 \text{ kg/s}$$

$$\text{Warm-water volume flow rate } \dot{V}_w = \dot{M}_w v_1 = 420.6 \times 0.0010036$$

$$= 0.422 \text{ m}^3/\text{s}$$

$$\text{Cold-water mass-flow rate } \dot{M}_c = \dot{M}_T \dot{m}_c = 1.447 \times 288.9$$

$$= 418.0 \text{ kg/s}$$

$$\text{Cold-water volume flow rate } \dot{V}_c = \dot{M}_c v_6 = 418.0 \times 0.0010007$$

$$= 0.418 \text{ m}^3/\text{s}$$

$$\text{Gross cycle efficiency} = \frac{w_T}{q_A} = \frac{h_3 - h_5}{h_3 - h_7} = \frac{69.1}{2487}$$

$$= 0.0278 = 2.78\%$$

$$\text{Gross plant efficiency} = 0.0278 \times 0.9 = 0.0250$$
$$= 2.5\%$$

Note: The gross plant power, 100 kW, and the gross plant efficiency, 2.5 percent, do not take into account pumping and other auxiliary power inputs to the plant.

It can be seen that very large ocean-water mass and volume flow rates are used in open OTEC systems and that the turbine is a very low-pressure unit that receives steam with specific volumes more than 2000 times that in a modern fossil powerplant. Thus the turbine resembles the few last exhaust stages of a conventional turbine and is thus physically large.

Another attempt at building a Claude-type powerplant was undertaken by the French corporation Energie Electrique de la Cote d'Ivoire at Abidjan in the Ivory Coast, Africa, in the 1950s. The plans were for a 7000-kW plant operating on a temperature difference of 20°C (36°F). The cold-water pipe was 2.4 m in diameter and extended to a depth of about 4.8 km at about 5 km from the shore. In actual operation about a quarter of the plant's gross output was used to drive the various accessories. Full output was never realized because of maintenance difficulties on the cold-water pipeline.

15-4 MODIFICATIONS OF THE OPEN OTEC CYCLE

Improvements and modifications of the Claude cycle have been proposed in an attempt to change it into an economically viable system. The attempts focused mainly on two areas: a different, more efficient type of evaporator, called the *controlled flash evaporator,* and the use of the plant for the *cogeneration* of electricity and fresh water.

The Controlled Flash Evaporator

The principle of controlled flash evaporator (CFE) [145] has been successfully used for the production of pure boiler feedwater from heated condenser cooling seawater in a conventional powerplant, as well as for the production of fresh water from hot tropical seawater.

In the volume-boiling flash evaporator used in the early cycle, the water violently flashes upon depressurization. Bubbles of vapor emanate from the turbulent surface carrying with them some entrained saltwater, which increases the corrosion problem. This may be alleviated by adding demisters (which entrap the brine droplets) but at the expense of further pressure and temperature drop.

A controlled flash evaporator, by contrast, has the warm seawater (called brine) descend by gravity in thin films, about 1 mm thick, in a quiescent manner down chutes. A typical CFE chamber is a vertical structure, 2.5 m (~8 ft) or more in height, that contains a large number of such vertical parallel chutes. The water film, on both sides of the chutes, vaporizes without the violent processes of bubble formation and bursting that occur in a conventional evaporator. Pure vapor, devoid of entrained solids, emanates from the films and increases in mass and volume flows as it progresses downward. Because of energy transfer caused by the evaporative process, both brine and vapor progressively cool down as they descend down the chutes. The pure low-pressure steam is then admitted to the turbine, where it expands to the condenser (Fig 15-5).

A minimum of deaeration is required in a CFE because no violent release of noncondensable gases takes place. Deaeration is accomplished by an "ingestor" that operates on a principle similar to that of a steam-jet air ejector SJAE (Sec. 6-3) except that low pressure is obtained by a liquid jet aspirator that uses cooled brine from the chute exit for its liquid. The noncondensables and the liquid leave the ingestor via a barometric leg (diffuser), and the gases come out of solution upon discharge to atmospheric pressure.

The high quality of the vapor going through the turbine from the CFE is well suited to electrical and fresh-water cogeneration (described below).

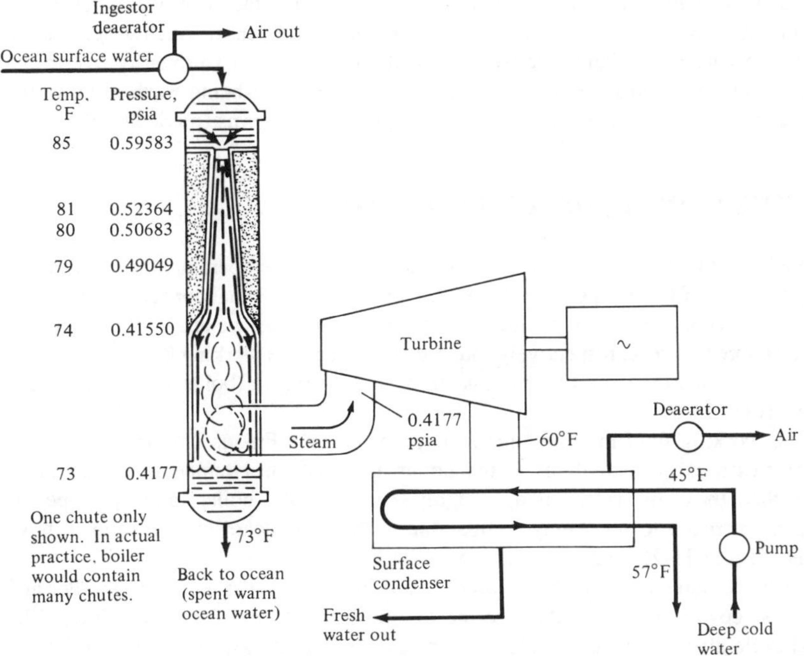

Figure 15-5 Schematic of an open cycle with both controlled-flash evaporator and a surface condenser for the cogeneration of electricity and fresh water.

Electrical and Fresh-Water Cogeneration

Here a shell-and-tube or plate-type surface condenser is used in the plant instead of the direct-contact condenser used in the Claude cycle. The deep cold water is pumped through the tubes or on one side of the plates of the surface condenser and is discharged back to the ocean without mixing with the turbine exhaust steam (Fig. 15-5). The turbine exhaust becomes a fresh-water condensate. It can be pumped to atmospheric pressure and used as fresh water for various uses. When combined with a controlled flash evaporator, the quality of the condensate is so high as to meet standards for potable water with a salt content between 1 and 5 ppm.

A disadvantage of the surface condenser, however, is that it is less effective than a direct-contact condenser and thus operates at a slightly higher turbine exhaust temperature (and therefore pressure), which results in a slight reduction in net cycle work and efficiency.

The plant shown in Fig. 15-5 shows turbine inlet at 0.4177 psia (0.0288 bar) and 73°F (22.8°C) and exhaust at 0.25611 psia (0.0177 bar) and 60°F (15.6°C). If fresh water were not desired, a direct-contact condenser could have been used with turbine exhaust at perhaps 0.2330 psia (0.0161 bar) and 57°F (13.9°C), and a higher plant efficiency and output.

Other ideas for use with the open cycle involve using the deep water after it has been brought up to the surface and passed through the condenser. Such water, rich in nutrients because it contains the remains of the organisms that primarily live in the sunny water layers near the surface, can then be used after being warmed up in the condenser as a food source for marine life in surface sea farms (mariculture).

15-5 THE CLOSED, OR ANDERSON, OTEC CYCLE

Although the first attempt at producing power from ocean temperature differences was the open cycle of Georges Claude in 1929, d'Arsonval's original concept in 1881 was that of a closed cycle that also utilizes the ocean's warm surface and cool deep waters as heat source and sink, respectively, but requires a separate working fluid that receives and rejects heat to the source and sink via heat exchangers (boiler and surface condenser) (Fig. 15-6).

The working fluid may be ammonia, propane, or a Freon. The operating (saturation) pressures of such fluids at the boiler and condenser temperatures are much higher than those of water, being roughly 10 bar at the boiler, and their specific volumes are much lower, being comparable to those of steam in conventional powerplants (Table 15-2). (See also the Appendix.)

Such pressures and specific volumes result in turbines that are much smaller and hence less costly than those that use the low-pressure steam of the open cycle. The closed cycle also avoids the problems of the evaporator. It, however, requires the use of very large heat exchangers (boiler and condenser) because, for an efficiency of about 2 percent, the amounts of heat added and rejected are 50 times the output of the plant. In addition, the temperature differences in the boiler and condenser must

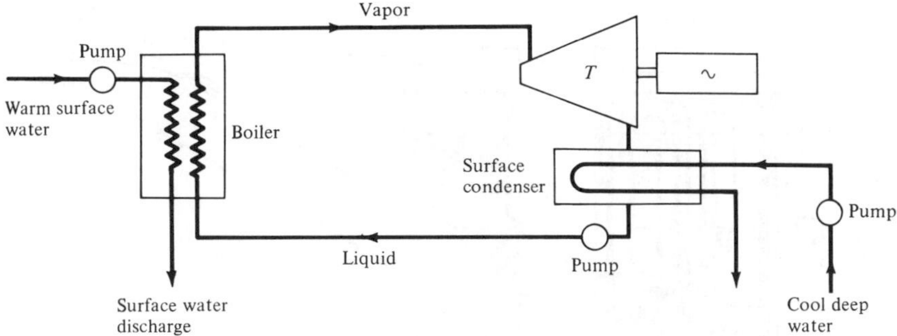

Figure 15-6 Schematic of a closed-cycle OTEC powerplant.

be kept as low as possible to allow for the maximum possible temperature difference across the turbine, which also contributes to the large surfaces of these units.

The closed-cycle approach was first proposed by Barjot in 1926, but the most recent design was by Anderson and Anderson in the 1960s [146]. The closed cycle is sometimes referred to as the *Anderson cycle*. The Andersons chose propane as the working fluid with a 20°C (36°F) temperature difference between warm surface and cool water, the latter some 600 m (~2000 ft) deep. Propane is vaporized in the boiler at 10 bar (145 psia) or more and exhausted in the condenser at about 5 bar.

In order to minimize the mass and the amount of material (and hence cost) used to manufacture the immensely large heat exchangers, the Anderson OTEC system employs thin plate-type heat exchangers instead of the usual heavier and more expensive shell-and-tube heat exchangers. To help reduce the thickness of the plates, the heat exchangers are placed at depths where the static pressure of the water in either exchanger roughly equals the pressure of the working fluid. Thus if propane is the working fluid in the boiler at 80°F (26.7°C) and 143.6 psia (9.9 bar) (Table 15-2), the boiler

Table 15-2 Comparison of saturation pressures and specific volumes of four fluids at low temperatures

	Saturation pressure, psia		Specific volume, ft³/lb$_m$			
			v_f		v_g	
Fluid	40°F	80°F	40°F	80°F	40°F	80°F
Steam	0.1216	0.5068	0.01602	0.01607	2.4458	633.3
Ammonia	73.32	153.0	0.02533	0.0267	3.971	1.955
Propane	77.80	143.6	0.03055	0.0327	1.330	0.745
Freon-12	51.68	98.76	0.0116	0.0123	0.792	0.425
Steam	Superheated at 2400 psia and 1000°F, v = 0.321					

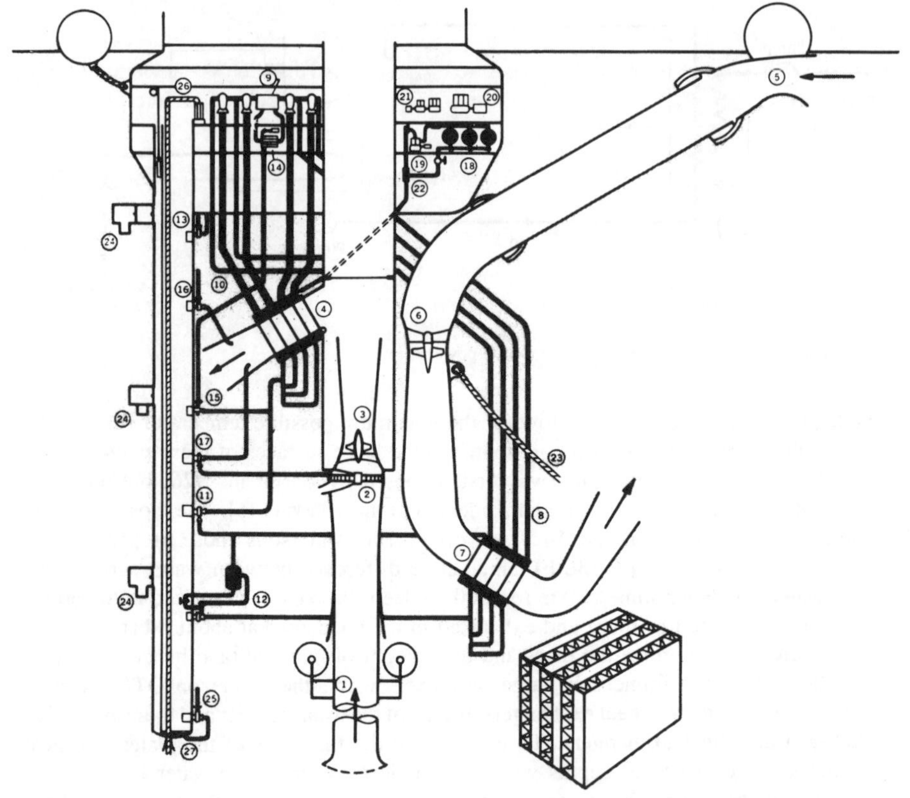

Figure 15-7 Closed-cycle floating powerplant [142].

would be placed at a depth of (g/g_c) 143.6 × 144/64 ≈ 325 ft (~100 m),* where 64 is the average density of seawater in $lb_m/ft.^3$ The condenser, on the other hand, if operating at 40°F (14.4°C) and 77.8 psia (5.36 bar), would be placed at (g/g_c) 77.8 × 144/64 ≈ 175 ft (≈53 m). In other words the powerplant would be designed so that it would be mostly submerged and anchored so that the condenser would actually be higher than the boiler. The turbine, pumps, storage tanks, compressors, and other plant equipment would be placed at intermediate decks. Figure 15-7 shows such a floating plant. Note the arrangement by which the condenser (4) receives cool water from the cold-water pipe (1). That pipe extends 2000 ft (~610 m) below the surface. The condenser is placed above the boiler (7), which receives warm surface water from a buoyantly supported surface inlet pipe (5). Note also the decompression chambers

* 1 bar is roughly equivalent to 10 m of seawater head.

required for operating and maintenance personnel. The inset in the figure shows the proposed design of the plate-type heat exchangers. The plant design was based on a 20°C (36°F) warm-cold-water temperature difference and propane as the working fluid. Roughly 14 percent of the gross power is expected to be consumed internally. The size of the plant is gigantic indeed.

15-6 RECENT OTEC DEVELOPMENTS

Problems facing commercial development of OTEC systems are legion. Some, like the low temperature differences and evaporator problems, have already been discussed. In addition, for open systems, turbines capable of generating 10 MW or more using low-pressure steam are yet to be developed. For closed systems, heat exchangers much larger than any that have been built must be designed and built. For all systems, pumps capable of handling larger amounts of water than any developed so far have to be developed. It is said that a relatively small-output (by modern powerplant standards) OTEC plant producing 100 MW would have a water flow comparable to that going through Boulder Dam. Long ($\sim$1 km), large-diameter (up to 30 m) cold-water pipes, larger than ever made, have to be designed, fabricated, and installed. The whole plant has to be stationed and moored at depths at which no vessel has ever been before. The deployment of the electrical cable required to carry the power to shore must be carefully done because it would be subjected to severe stresses from its own weight as well as from ocean currents and eddies.

In addition, an OTEC powerplant will have to be capable of withstanding severe ocean storms over its lifetime, corrosion by seawater salts, erosion due to large volume flows, biofouling due to algae growth, and encrustation by various marine life such as barnacles.

Because commercial-size OTEC systems are very large and suffer from various problems when placed in a stationary position, recent OTEC demonstration plants have instead concentrated on small-scale plants and used an *OTEC plant ship* concept. In this concept, the plant is built on or as part of a specially designed ship. A so-called *grazing plant ship* would move at low speed (about 0.5 knots) in search of the warmest surface water. The electricity generated must then be used on board to power manufacturing of energy-intensive products like aluminum, magnesium, nickel, various alloys, semiconductors, etc., for deep sea bed mining of such minerals as manganese, cobalt, and nickel, or for chemical processing of products such as ammonia.

The continental United States has a relatively small ratio of coastal to inland areas and relatively limited warm-water resources. Few commercial electric utilities have therefore shown interest in OTEC, and it fell to the federal government to undertake research and development in this field. The case for OTEC is more promising for islands that have low population densities, that are dependent upon imported fossil fuels for generating electricity, and that have relatively large warm-water resources in relation to their energy needs. The U.S. Department of Energy (DOE) has therefore aimed its first efforts at such islands as Hawaii, Puerto Rico, the Virgin Islands, Guam,

American Samoa, and others. The federal OTEC program started in the early 1970s as one of six options selected for investigation in an effort to reduce the United States' dependence on imported oil.

The first result of the program was a test facility called OTEC-1, which was installed on a 26,000-ton converted tanker and operated in July 1980. The effort was funded by DOE and built by TRW, Inc., and Global Marine Development Company. The ship was first moored 29 km (18 mi) off Kawaihae Harbor, Hawaii in 1220-m (4000-ft) deep waters. OTEC-1 was a closed-cycle "heat exchanger" facility without turbine or electrical generator. Its aims were to test heat-exchanger design and performance and evaluate the extents of corrosion and biofouling. It contained a 1-MW titanium shell-and-tube heat exchanger designed to simulate a larger 10-MW unit. It had three cold-water pipes 853 m (2800 ft) long and 1.2 m (4 ft) diameter each. OTEC-1, now retired, was to have been succeeded by a complete electric-generating system for operation in the mid-1980s. Federal funding, however, was withheld.

The second U.S. demonstration, called Mini-OTEC, was a small $3 million electric-generating plant funded cooperatively by the State of Hawaii, Lockheed Missile and Space Company, Inc., Alfa-Laval Thermal, Inc., and the Dillington Corporation. This plant was also housed in a ship, a converted U.S. Navy scow, and was operated between August and November 1979 off Keahole Point, Hawaii. It included a titanium plate-type heat exchanger and one cold-water pipe 660 m (2170 ft) long and 0.6 m (2 ft) in diameter. It produced 50 kW of gross electric power but only between 12 and 15 kW net. This performance was found to be close to design predictions.

For the contiguous United States, only the Gulf of Mexico and the lower eastern coast of Florida are near waters that have sufficient temperature differences and that are close enough to land for ease of power transmission. Because of these limitations, the future of OTEC in the United States is uncertain at best.

In the international arena, a number of countries have recently been active in OTEC research and development. These include Japan, France (which, of course, had been an early pioneer in the field via the Claude system), Sweden, West Germany, and the Netherlands. An interesting effort that may prove to be successful is a 100-kW land-based plant that the Pacific Ocean equatorial island nation of Nauru has contracted for with the Japanese. The hope is that with this and future larger units, Nauru might develop into a center for energy-intensive industries.

15-7 OCEAN WAVES

Like the wind (Chap. 14) and OTEC, ocean and sea waves are caused indirectly by solar energy. Waves are caused by the wind, which in turn is caused by the uneven solar heating and subsequent cooling of the earth's crust and the rotation of the earth. Wave energy at its most active, however, can (like wind energy) be much more concentrated than incident solar energy even at the latter's peak. Devices that convert energy from waves can therefore produce much higher power densities than solar devices.

This "harvesting" of energy from waves has been the subject of some of humankind's wildest dreams for ages. However, it was the recent energy crisis that prompted serious attempts at harnessing the waves for the production of electricity. During the decade of the 1970s dozens of patents were filed to do this, though most are complicated and rather fragile in the face of the gigantic power of ocean storms.

The main advantage of power from waves, like most of the so-called alternative energy sources, is a free and renewable energy source. In addition, wave-power devices, unlike solar or wind devices, do not use up large land masses, are relatively pollution free and, because they remove energy from the waves, leave the water in a relatively placid state in their wakes. Thus a string of devices situated, as they must, where there is a large amount of wave activity, can in addition to producing electricity protect coastlines from the destructive action of such waves, minimize erosion, and even help create artificial harbors. A concept of staggered array of devices has been proposed to do this.

There are several disadvantages. Like most of the so-called alternative energy sources, waves lack dependability, and there is relative scarcity of accessible sites of large wave activity. In addition most devices that have been proposed are relatively complicated and lack the necessary mechanical strength to withstand the enormous power of stormy seas. Economic factors such as the capital investment, costs of maintenance, repair and replacement, as well as problems of biological growth of marine organisms, are all relatively unknown and seem to be on the large side.

World sites that may be suitable include the Molakai and Alenuihaha channels in the Hawaiian islands, where 6 to 10-ft-high (crest to trough) waves are typical during normal trade wind periods, the Pacific coast of North America, the Arabian Sea off Pakistan and India, the North Atlantic coast of Scotland, the coast of New England, and others [147]. Figure 15-8 is a spectrum of wave heights and periods for the New England coast. It shows that waves 6 ft high or more with periods 6 s long or less occur about half the time.

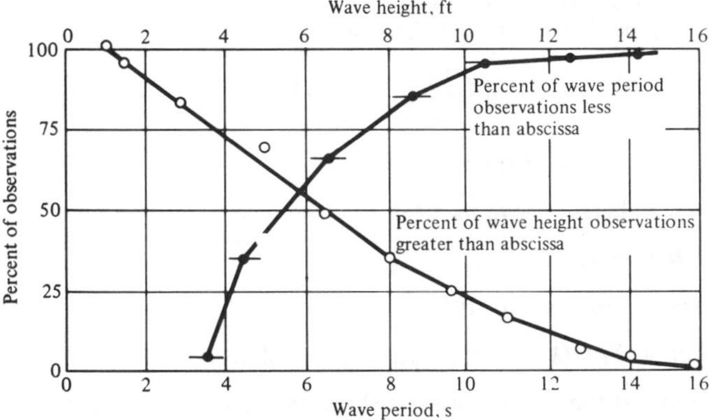

Figure 15-8 Spectrum of wave heights and periods for the New England coast [147].

The basic mechanism of wave motion will be presented in the next section. The theory of energy conversion from waves will be presented in the following section. These will be followed by sections describing a few selected wave-power devices.

15-8 WAVE MOTION

A two-dimensional progressive wave that has a free surface and is acted upon by gravity (Fig. 15-9) is characterized by the following parameters:

λ = wavelength = $c\tau$, m or ft

a = amplitude, m or ft

$2a$ = height (from crest to trough), m or ft

τ = period, s

f = frequency = $1/\tau$, s^{-1}

c = wave propagation velocity λ/τ, m/s or ft/s

n = phase rate = $2\pi/\tau$, s^{-1}

The period τ and wave velocity c depend upon the wavelength and the depth of water (Table 15-3 and Fig. 15-10). The range below the dashed line in the table represents the most common waves.

The relationship between wavelength and period can therefore be well approximated by

$$\lambda = 1.56\tau^2 \qquad (\lambda \text{ in m, } \tau \text{ in s}) \tag{15-3a}$$

or

$$\lambda = 5.12\tau^2 \qquad (\lambda \text{ in ft, } \tau \text{ in s}) \tag{15-3b}$$

Figure 15-9 shows an isometric of a two-dimensional progressive wave, represented by the sinusoidal simple harmonic wave shown at time 0. Cross sections of the wave are also shown at time 0 and at time θ. That wave is expressed by

$$y = a \sin\left(\frac{2\pi}{\lambda}x - \frac{2\pi}{\tau}\theta\right) \tag{15-4a}$$

or

$$y = a \sin (mx - n\theta) \tag{15-4b}$$

where

y = height above its mean level, m or ft

θ = time, s

m = $2\pi/\lambda$, m^{-1} or ft^{-1}

$(mx - n\theta) = 2\pi(x/\lambda - \theta/\tau)$ = phase angle, dimensionless

Note that the wave profile at time θ has the same shape as that at time 0, except that it is displaced from it by a distance $x = \theta/\tau = \theta\,(n/m)$. When $\theta = \tau$, $x = \lambda$ and the wave profile assumes its original position.

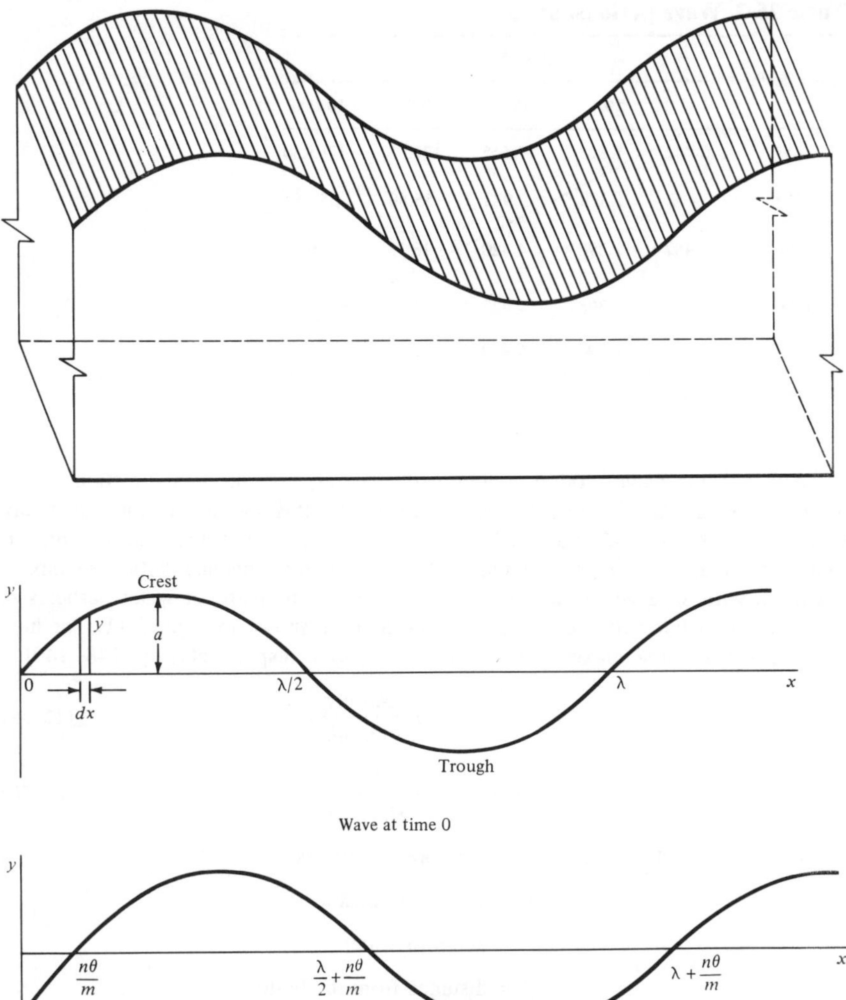

Figure 15-9 A typical progressive wave, a = amplitude, λ = wavelength, showing two-dimensinal wave and amplitudes at time 0 and at time θ.

Table 15-3 Wave periods, s*

Water depth h, ft	Wavelength, ft 1	10	100	1000	10,000
1	0.442	1.873	17.645	176.33	1763.30
10	0.442	1.398	5.923	55.80	557.62
100	0.442	1.398	4.420	18.73	176.45
1000	0.442	1.398	4.420	13.98	59.23
10000	0.442	1.398	4.420	13.98	44.20

* From Ref. 148.

Although the wave motion is continually lateral, i.e., in the x direction with a speed of propagation $c = \lambda/\tau$, the motion of water itself is not, although it deceptively appears to be so. In reality a given particle of water rotates in place in an elliptical path in the plane of wave propagation, with specified horizontal and vertical semiaxes, as can be witnessed when placing a cork on water. The paths of water particles at different depths but with the same mean position are shown in Fig. 15-11. The horizontal and vertical semiaxes of the ellipses are given, respectively, by [148, 149].

$$\alpha = a\frac{\cosh m\eta}{\sinh mh} \tag{15-5a}$$

$$\text{and } \beta = a\frac{\sinh m\eta}{\sinh mh} \tag{15-5b}$$

where α = horizontal semiaxis

β = vertical semiaxis

h = depth of water

η = distance from the bottom

Equations (15-5) show that in general $\alpha > \beta$, that β varies from 0 at the bottom where $\eta = 0$ to a, at the surface where $\eta = h$, and that for large depths $\alpha \approx \beta \approx a$ and the motion is essentially circular at the surface.

A wave therefore possesses both potential and kinetic energies. These are evaluated in the next section.

15-9 ENERGY AND POWER FROM WAVES

The total energy of a wave is the sum of its potential and kinetic energies.

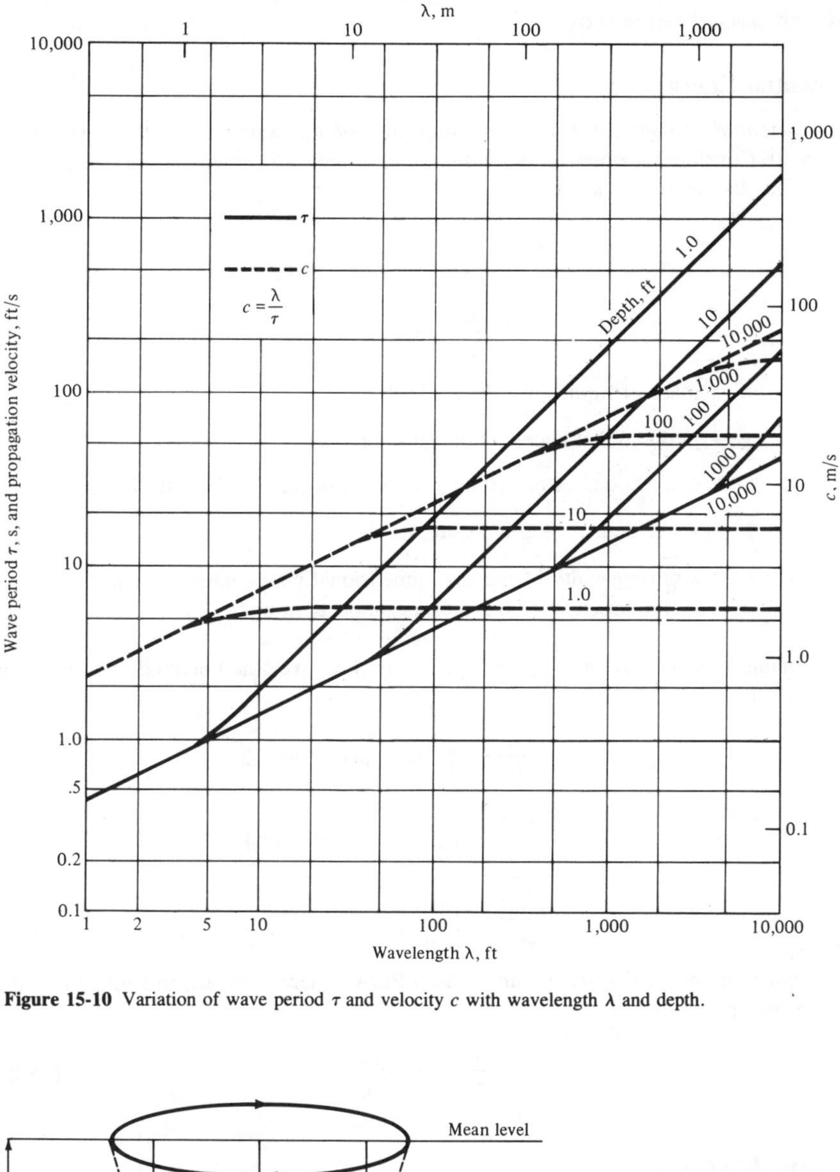

λ, m

1 10 100 1,000

10,000 1,000 100 10 1.0 .5 0.2 0.1

Wave period τ, s, and propagation velocity, ft/s

—— τ

- - - - c

$c = \frac{\lambda}{\tau}$

Depth, ft

1.0 10 100 1,000 10,000

10,000 1,000 100 1000 10,000

100 10 1.0

1,000 100 10 1.0 0.1

c, m/s

1 2 5 10 100 1,000 10,000

Wavelength λ, ft

Figure 15-10 Variation of wave period τ and velocity *c* with wavelength λ and depth.

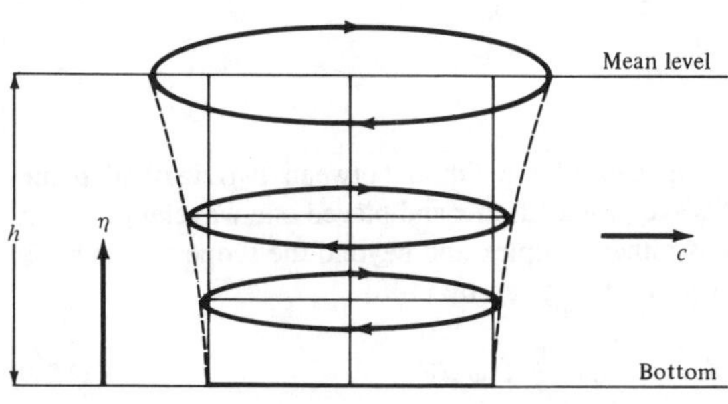

Figure 15-11 Elliptical paths of water particles at different heights. *c* = wave propagation velocity, *h* = water depth, η = distance from bottom.

Potential Energy

The *potential energy* arises from the elevation of the water above the mean level ($y = 0$). Considering a differential volume $y\,dx$ (Fig. 15-9), it will have a mean height $y/2$. Thus its potential energy is

$$dPE = m\frac{yg}{2g_c} = (\rho y\,dx\,L)\frac{yg}{2g_c}$$

$$= \frac{\rho L}{2}y^2\,dx\,\frac{g}{g_c} \tag{15-6}$$

where m = mass of liquid in $y\,dx$, kg or lb_m

g = gravitational acceleration, m/s^2 or ft/s^2

g_c = conversion facto: 1.0 kg · m/(N·s^2) or 32.174 lb_m · ft/(lb_f · s^2)

ρ = water density, kg/m^3 or lb_m/ft^3

L = arbitrary width of the two-dimensional wave, perpendicular to the direction or wave propagation x, m or ft

Combining Eqs. (15-6) and (15-4*b*) and integrating gives the potential energy PE in J or ft · lb_f

$$PE = \frac{\rho La^2}{2}\frac{g}{g_c}\int_0^\lambda \sin^2(mx - n\theta)\,dx$$

$$= \frac{\rho La^2}{2m}\frac{g}{g_c}\left(\frac{1}{2}mx - \frac{1}{4}\sin 2mx\right)_0^\lambda$$

$$= \frac{\rho La^2}{2m}\frac{g}{g_c}\left[\frac{m\lambda}{2}\right] = \frac{1}{4}\rho a^2\lambda L\frac{g}{g_c} \tag{15-7}$$

The *potential energy density* per unit area is PE/A, where $A = \lambda L$, in J/m^2 or ft · lb_f/ft^2, is then given by

$$\frac{PE}{A} = \frac{1}{4}\rho\,a^2\frac{g}{g_c} \tag{15-8}$$

Kinetic Energy

The *kinetic energy* of the wave is that of the liquid between two vertical planes perpendicular to the direction of wave propagation x and placed one wavelength apart. The derivation of kinetic energy is rather complex and beyond the scope of this book. However, from hydrodynamic theory, it is given by [148]

$$KE = \frac{1}{4}i\rho L\frac{g}{g_c}\int \omega\,d\varpi \tag{15-9}$$

where ω is a complex potential given by

$$\omega = \frac{ac}{\sinh (mh)} \cos(mz - n\theta)$$

and z is distance measured from an arbitrary reference point. The integral in Eq. (15-9) is performed over the cross-sectional area bounded between two vertical planes. The result is

$$\text{KE} = \frac{1}{4}\rho\, a^2 \lambda L \frac{g}{g_c} \tag{15-10}$$

and the *kinetic energy density* is

$$\frac{\text{KE}}{A} = \frac{1}{4}\rho\, a^2 \frac{g}{g_c} \tag{15-11}$$

Total Energy and Power

It can be seen that the potential and kinetic energies of a progressive sine wave are identical, so that the total energy E is half potential and half kinetic. The *total energy density* is thus given by

$$\frac{E}{A} = \frac{1}{2}\rho\, a^2 \frac{g}{g_c} \tag{15-12}$$

The power P energy per unit time is given for a wave by energy times frequency. Thus the *power density*, W/m^2 or ft · lb$_f$/(s · ft^2), is given by

$$\frac{P}{A} = \frac{1}{2}\rho\, a^2 f \frac{g}{g_c} \tag{15-13}$$

Example 15-2 A 2-m wave has a 6-s period and occurs at the surface of water 100 m deep. Find the wavelength, the wave velocity, the horizontal and vertical semiaxes for water motion at the surface, and the energy and power densities of the wave. Water density = 1025 kg/m^3.

SOLUTION

Wavelength $\lambda = 1.56 \times 6^2 = 56.16$ m $= 184.25$ ft (Eq. 15-3*a*)

Wave velocity $c = \lambda/\tau = 9.36$ m/s $= 30.71$ ft/s

Wave height $2a = 2$ m Amplitude $a = 1$ m $= 3.28$ ft

$m = 2\pi/\lambda = 2\pi/56.16 = 0.1119$ m^{-1}

At the surface $\eta = h = 100$ m

$$\text{Horizontal semiaxis } \alpha = 1 \times \frac{\cosh 11.19}{\sinh 11.19} = 1 \text{ m} \qquad \text{(Eq. 15-5}a\text{)}$$

$$\text{Vertical semiaxis } \beta = 1 \times \frac{\sinh 11.19}{\sinh 11.19} = 1 \text{ m} \qquad \text{(Eq. 15-5}b\text{)}$$

$$\text{Wave frequency } f = 1/\tau = 1/6 \text{ s}$$

$$\text{Energy density } \frac{E}{A} = \frac{1}{2} \times 1025 \times 1^2 \times \frac{9.81}{1} = 5027.6 \text{ J/m}^2$$

$$= 344.5 \text{ ft} \cdot \text{lb}_f/\text{ft}^2 \qquad \text{(Eq. 15-12)}$$

$$\text{Power density } \frac{P}{A} = \frac{E}{A} f = 5027.6 \times \frac{1}{6} = 837.9 \text{ W/m}^2$$

$$= 0.0778 \text{ kW/ft}^2 \qquad \text{(Eq. 15-13)}$$

Note that, because of the large depth, the semiaxes are equal, so the motion is circular. Note also that they are small compared with the wavelength, so the water motion is primarily vertical.

Two-meter waves, of course, do not occur all the time. However, in regions of high wave activity, 2 m is a median with heavier and calmer seas occurring about 50 percent of the time (Fig. 15-8). The total energy and power densities over a period of time should take this spectrum into account. With these densities proportional to a^2, the average densities would be greater than the values obtained in Example 15-2.

It is instructive to compare these values with the *average daily solar incidence* where, in the southwestern United States, a value of 240 W/m^2, or 0.0223 kW/ft^2, is often used. Thus, wave power density is much higher. A complete comparison should take into account the efficiency of conversion to electric energy as well as other factors, like capital costs for land and equipment, operational costs, costs of energy storage, and other factors. In the next sections some selected devices that convert wave energy to mechanical energy, and hence to electric energy, will be described.

15-10 WAVE-ENERGY CONVERSION BY FLOATS

As seen above, wave motion is primarily horizontal, but the motion of water is primarily vertical. This latter motion is made use of by floats to obtain mechanical power. The concept envisages a large float that is driven up and down by the water within relatively stationary guides. This reciprocating motion is converted to mechanical and then electric power.

A system proposed by Martin [150] is shown in Fig. 15-12. A square float moves up and down with the water, guided by four vertical manifolds that are part of a platform. The platform is stabilized within the water by four large underwater floatation tanks so that it is supported by buoyancy forces and no significant vertical or horizontal

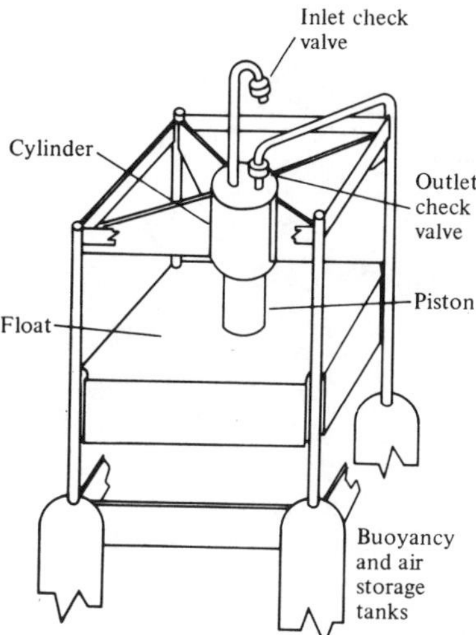

Figure 15-12 Schematic of a float wave-power machine [150].

displacement of the platform due to wave action occurs. Damping fins may be used to further reduce motion if necessary. The platform is therefore expected to be relatively stationary in space, even in heavy seas. An alternative design uses piles to support the platform, if water depth permits. It is proposed that the platform be made of molded plastic with a foamed plastic core to arrive at the required density and strength.

Attached to the float is a piston that moves up and down inside a cylinder that is attached to the platform and is therefore relatively stationary. This piston-cylinder arrangement is used as a reciprocating air compressor. The downward motion of the piston draws air into the cylinder via an inlet check valve. The upward motion compresses the air and sends it through an outlet check valve to the four underwater floatation tanks via the four manifolds. The four floatation tanks thus serve the dual purpose of bouyancy and air storage, and the four vertical manifolds serve the dual purpose of manifolds and float guides.

The compressed air in the buoyancy-storage tanks is in turn used to drive an air turbine that drives an electrical generator. The electric current is transmitted to the shore via an underwater cable.

Assuming an idealized reciprocating compressor cycle, (Fig. 15-13), the work, mass-flow rates, and other relationships may be obtained by common thermodynamic analysis as a function of stroke volume and clearance, wave height, and inlet conditions. The float height is usually subtracted from the wave height $2a$ to obtain the piston stroke.

The air enters the compressor at atmospheric pressure, volume, and temperature P_0, V_0, and T_0. It leaves it as P_1, V_1, and T_1. It, however, cools in the storage tanks back to T_0 (as if it were isothermally compressed along the dotted line). The new volume $V_2 = V_0 P_0/P_1$, and pressure $P_2 \approx P_1$.

This air now expands in the turbine. Assuming ideal (adiabatic reversible) ex-

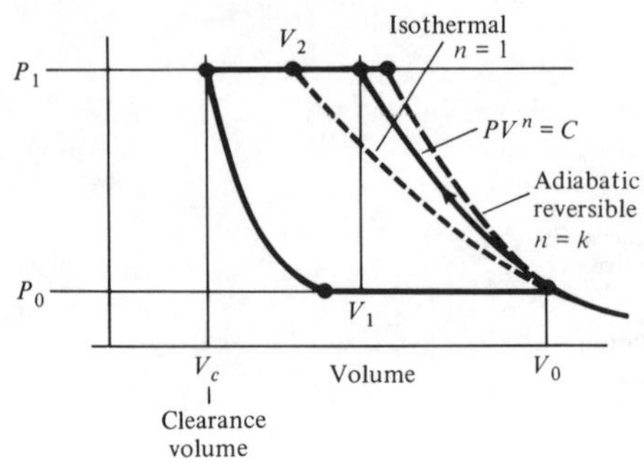

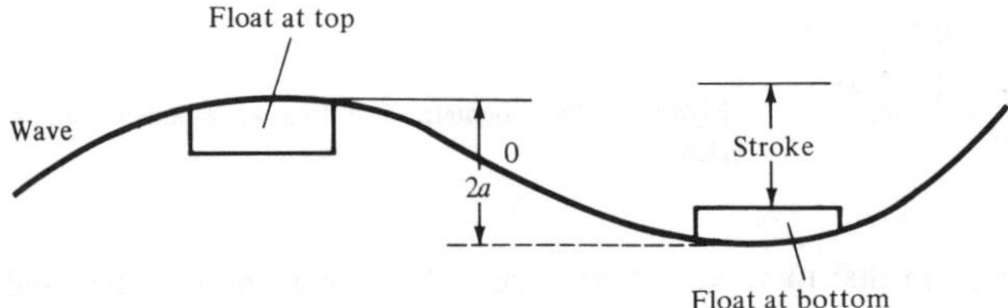

Figure 15-13 Idealized reciprocating compressor cycle and piston stroke.

pansion, the work, in J/kg or Btu/lb$_m$, obtained from that turbine per unit mass of air would be given by

$$\frac{W_{\text{turbine}}}{m} = c_p T_0 \left[1 - \left(\frac{r_c}{r_e}\right)^{(k-1)/k} \cdot r_c^{\,(1-k)/k} \right] \tag{15-14}$$

where

m = mass of air, kg or lb$_m$

c_p = specific heat of air, J/(kg · K) or Btu/(lb$_m$ · °R)

r_c = Pressure ratio in compressor = P_1/P_0

r_e = expansion ratio in turbine = $P_2/P_3 = P_1/P_3$

P_3 = exhaust pressure in turbine, Pa or lb$_f$/ft^2

k = ratio of specific heats for air

r_c/r_e may be taken as 1.1, when allowing for reasonable mechanical clearances

Optimization of a system as above [150] has shown that the power density is roughly given by

$$0.05 \text{ kW/ft}^2$$

This compares with 0.0223 kW/ft^2 incident solar energy in hot regions, which is yet to be multiplied by the efficiency of solar conversion, usually less than 10 percent.

A linear array of such modules perpendicular to wave motion is recommended because the modules attenuate the wave amplitude in the direction of wave motion and would affect other modules in that direction. It would thus take miles of linear arrays to produce 100 MW or more. Other problems of such a scheme, some of them shared by other schemes, are:

1. Waves not perfectly sinusoidal
2. Aspiration of water into intake and even submersion by large waves
3. Water entering turbine
4. Materials problems, such as cost and corrosion (which may be partly overcome by the use of molded plastic)
5. Design to withstand storms
6. Marine growth
7. Power transmission to shore

15-11 HIGH-PRESSURE ACCUMULATOR WAVE MACHINES

In these machines, instead of compressing air the water itself is pressurized and stored in a high-pressure accumulator or pumped to a high-level reservoir, from which it flows through a water-turbine-electrical generator. This is done by transforming large volumes of low-pressure water at wave crest into small volumes of high-pressure water by the use of a composite piston. This piston is composed of a large-diameter main piston and a small-diameter piston at its center.

In one design, that of the *hydraulic accumulator* (Fig. 15-14), the main large piston moves inside a submerged cylindrical generator while the small piston moves inside a power cylinder. Wave water enters and leaves through openings at the bottom, thus causing the main piston to move up and down. A closed water loop exists above the small piston. On the upstroke, the pressure on the main piston is magnified on the small piston by the inverse ratio of the square of their diameters. Thus if the piston diameters are 100 and 20 cm, respectively, a wave head of 2 m would be magnified 25 times to a head H of 50 m. With seawater density averaging 1025 kg/m^3 (64 lb$_m$/ft^3), this head corresponds to a pressure of $\rho H_g/g_c$, which is about 5 bar (~72.5 psi).

The high-pressure water is conducted through a one-way up valve to a hydraulic accumulator at the top of the generator. Two air (or other gas) volumes counterbalance and act as cushions in a chamber above the main piston and in a sealed compartment in the hydraulic accumulator. The latter also maintains the high water pressure. Part of the high-pressure water flows through a Pelton wheel or Francis hydraulic turbine that drives an electrical generator and is then discharged to a storage chamber below the turbine.

On the trough of the wave, the composite piston is pushed downward by the gas pressure above the main piston, which thus acts also as a spring. The turbine exhaust

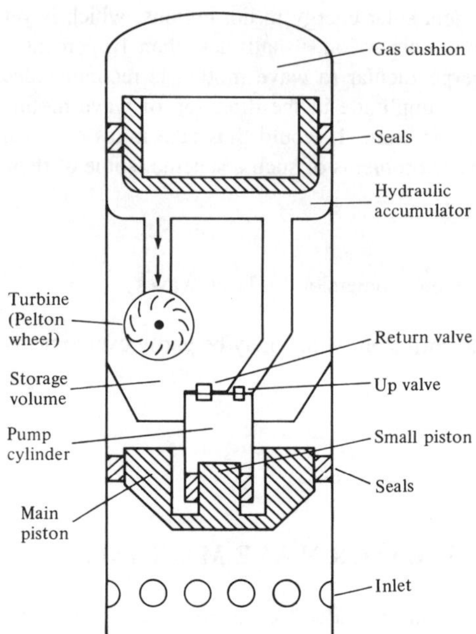

Figure 15-14 Schematic of a hydraulic-accumulator wave machine [151].

water in the storage volume is now sucked into the pump cylinder via a one-way return valve while the up valve is closed, and the cycle is repeated. The hydraulic accumulator is large enough to permit continuous turbine operation even though the waves are cyclic.

A 500-W prototype hydraulic-accumulator generator was constructed in Germany by Harold Kayser in 1975 for the purpose of powering a navigation buoy [151]. A design for a 1-kW generator has piston diameters of 100 and 25 cm and a 12-cm-diameter Pelton-wheel turbine that has 7-mm nozzles and that runs at 2300 r/min. The design calculations for that unit, based on 70 percent system efficiency and 2.5-m waves of 7-s period, show a piston stroke of 0.64 m, a high-pressure water flow rate per pump cycle of 20 dm^3/s, and a continuous Pelton-turbine flow rate of 3 dm^3/s under a head of 50 m.

A liquid, other than water, may be chosen for the closed loop to eliminate corrosion and freezing. Alcohol is one candidate for such a liquid. Particular attention must be paid to the seals between pistons and cylinders that must be gas tight and have low friction. A rolling diaphram seal has been selected. The piston and cylinders are to be constructed of glass fiber–reinforced plastic, ferro-concrete, aluminum, or steel.

An alternative to the high-pressure accumulator concept is the *high-level reservoir concept* (Fig. 15-15). A similar pressure magnification piston is used, but the pressurized water is elevated to a natural reservoir above the wave generator, which would have to be near a shoreline, or to an artificial water reservoir. The water in the reservoir

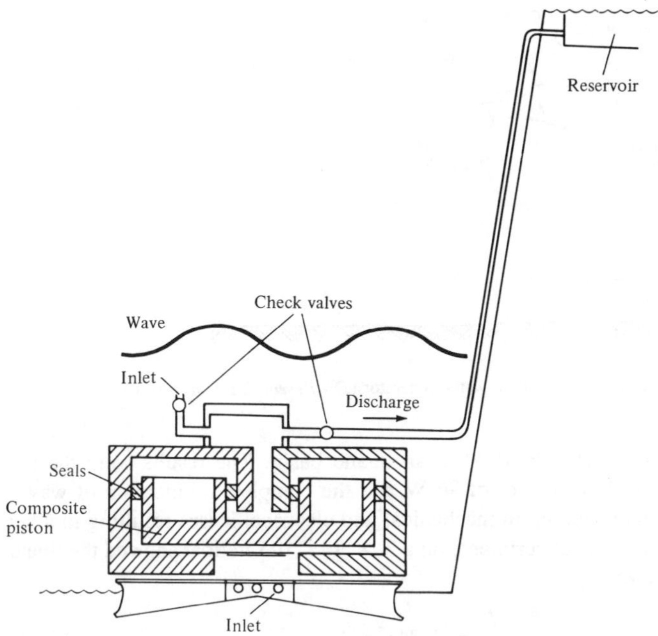

Figure 15-15 Schematic of a high-level reservoir wave machine.

is made to flow through a turbine back to sea level. Calculations show that a 20-m-diameter generator of this type can produce 1 MW.

Although the hydraulic accumulator is a free-floating device, the high-level reservoir machine is stationary. Problems needing solutions are motion and vibration damping in the former and tide compensation in the latter.

15-12 OTHER WAVE MACHINES

The Dolphin-Type Wave-Power Machine

The basic design of a dolphin-type wave-power generator has been worked out by Tsu Research Laboratories in Japan. The major components of the system (Fig. 15-16) are a dolphin, a float, a connecting rod, and two electrical generators.

The float has two motions. The first is a rolling motion about its own fulcrum with the connecting rod. It causes relative revolving movements between the float and the connecting rod. The other is a nearly vertical or heaving motion about the connecting rod fulcrum. It causes relative revolving movements between the connecting rod and the stationary dolphin. In both cases, the movements are amplified and converted by gears into continuous rotary motions that drive the two electrical generators.

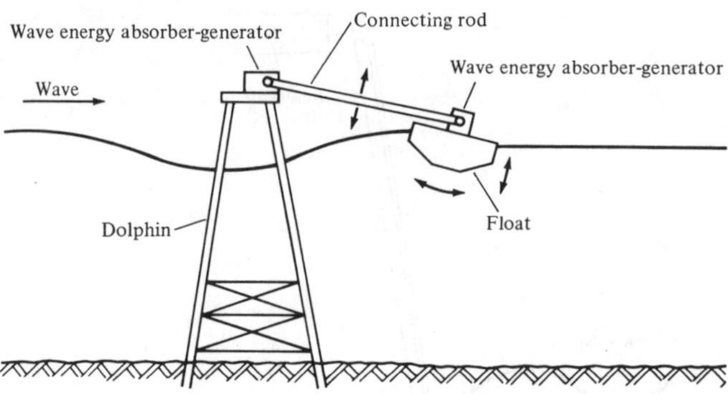

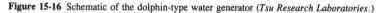

Figure 15-16 Schematic of the dolphin-type water generator (*Tsu Research Laboratories.*)

A scale-model was tested in 1980 in an inland basin. The results were the absorption of the applied wave power of 90 W and the complete elimination of waves downstream. 20 W were used up in mechanical and electrical losses resulting in a net 70 W of electric output. The experiments on a 3 × 1 × 0.5 m float showed the linear power output to be given by

$$\frac{P}{L} = 1.74a^2\tau \tag{15-15}$$

where P/L = power per unit length perpendicular to the wave, kW/m

a = wave amplitude (half the height), m

τ = wave period, s

Thus if the wave height is 2 m and the period is 6 s, the linear-power output is about 10 kW/m. A linear array of units totaling 1 km would thus generate some 10 MW.

Offshore experiments are planned for the future. The system is envisaged to be used for electric-power generation, pumping for desalination equipment, or for uranium extraction from the sea. Because it completely eliminates waves, it can provide suitable sites for fish farming, port facilities, etc.

The Dam-Atoll Concept

The *dam-atoll* is a wave-power conversion device (Fig. 15-17) designed by Wirt and Morrow of the Lockheed Corporation, Burbank, California. It is a massive and robust device that appears to overcome some of the disadvantages of many other devices, namely, complexity and fragility in heavy seas. It is said to be strong enough to survive any ocean storm.

The dam-atoll derives its name from the fact that it incorporates some of the characteristics of both dams and atolls. The principle of operation is based on the

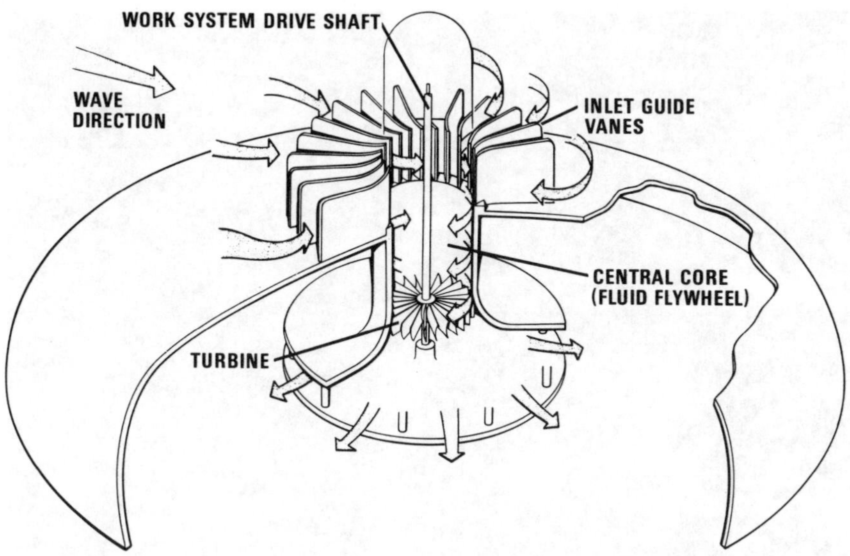

Figure 15-17 Cross section of a dam-atoll wave machine showing water flow (*Lockheed.*)

observed action of waves as they approach atolls (small volcanic islands) in an ocean. The waves wrap themselves around the atolls from all sides, ending in a spiral in the center, driving a turbine before discharging laterally outward. A module, 80 m in maximum diameter and 20 m high, is said to be capable of generating 1 to 1.5 MW in 7- to 10-s period waves. Figure 15-18 shows an artist's conception of a dam-atoll wave energy "farm."

The Multiple-Pontoon Raft

This is a system developed with British government support by Wavepower, Ltd., of Southampton. The concept was designed by Sir Christofer Cockerell, inventor of the hovercraft. A 1-kW scale model underwent sea trials in the solent of the Isle of Wight in 1978.

Taking advantage of favorable wave activity around the British Isles, future full-scale versions may be located off the west coast of Scotland or the western approaches to the English Channel. They would generate 2 MW per raft. A series of rafts 8 to 16 km (5 to 10 mi) offshore and stretching some 24 km (15 mi) parallel to the coast could generate 500 MW. It is believed that for the United Kingdom wave power offers greater potential than any other renewable energy source and that a 1000-km (630-mi) line of wave-power machines might generate half of the United Kindgom's electrical demand.

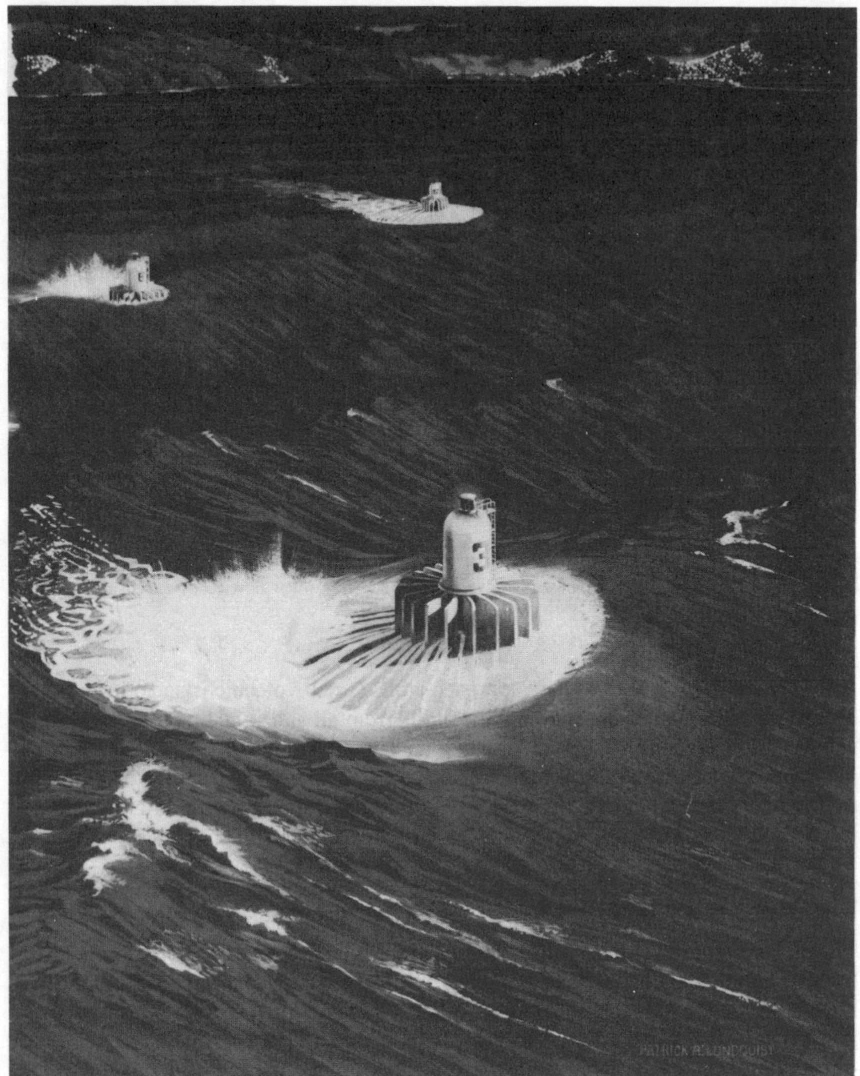

Figure 15-18 An artist's conception of a dam-atoll wave machine farm. *(Courtesy the Lockheed Corporation, Burbank, Calif.)*

15-13 THE TIDES

The tides are yet another source of energy from the oceans. This energy can be tapped from coastal waters by building dams that entrap the water at high tide and release it at low tide back to the sea. Power can then be obtained by turbines from both in-out flows of the water. The amount of energy available is very large but only in a few parts of the world.

Tidal energy is somewhat similar to hydro energy in that it uses the potential energy of water.* Both have been in use for centuries. Tidal "mills" were used in England and Europe. An early example is that of a miller in Woodbridge, Suffolk, who used the tide to mill grain in 1170. Another is a tidal waterwheel that was constructed in the sixteenth century under the London Bridge and that supplied water to London up to the nineteenth century. Other tidal mills were common for hundreds of years. They were used, particularly in the seventeenth and eighteenth centuries, for grinding grain, sawing wood, pumping water, etc. Mills were to be found in Britain, New England, Russia, and other places. One of the earliest scientific works on the tides was written by George Darwin, son of the great biologist Charles Darwin, and published in 1898 [152].

The tides, as we will see, although free, were inconvenient because they come at varying times from day to day, have varying ranges (heads) and, for large outputs, required large capital expenditures. Their early use declined and eventually came to a halt with the coming of the age of steam and cheap coal. With the beginning of the energy crisis in the 1970s, tidal energy, like other renewable energy sources, received renewed attention.

The tides are rhythmic but not constant, nor do they occur on a regular daily schedule. Their occurrence is due to a balance of forces, mainly the gravitational force of the moon but also that of the sun, both acting together with that of the earth to balance the centrifugal force on the water due to the earth's rotation. The result is the rhythmic rise and fall of water. The tides are characterized by their *schedule* and *range* R.

The *tidal schedules* vary from day to day because the orbit of the moon does not occur on a regular, 24-h, daily schedule. Instead, the moon rotates around the earth every 24 h, 50 min. During this time the tide rises and falls twice, resulting in a *tidal cycle* that lasts 12 h 25 min.

The *tidal range* R is defined as

$$R = \text{water elevation at high tide} - \text{water elevation at low tide} \qquad (15\text{-}16)$$

* Hydro energy, not covered in this text, converts the potential energy of water (trapped at relatively high elevation or behind an artificial dam) to mechanical work by a water turbine. Before the introduction of electricity in the nineteenth century, hydro energy was used to power industrial machinery directly. In the mid-nineteenth century, wind and hydro power accounted for some two-thirds of all mechanical power used in the United States. This percentage rapidly dropped, however, with the introduction of steam in the latter part of the nineteenth century. Hydro energy is now used almost exclusively to generate electricity.

The range is not constant. It varies during the 29.5-day lunar month (Fig. 15-19), being maximum at the time of new and full moons, called the *spring tides,* and minimum at the time of the first and third quarter moons, called the *neap tides*. The spring-neap tidal cycle lasts one-half of a lunar month. A typical *mean* range is roughly one-third of the spring range. The actual variations in range are somewhat complicated by seasonal variations caused by the ellipticity of the earth's orbit around the sun.

The variations in daily periodicity and monthly and seasonal ranges must, of course, be taken into account in the design and operation of tidal powerplants. The tides, however, are usually predictable, and fairly accurate tide tables are usually available.

Tidal ranges vary from one earth location to another. They are influenced by such conditions as the profile of the local shoreline and water depth. When these are favorable, a resonancelike effect causes very large tidal ranges. Ranges have to be very large to justify the huge costs of building dams and associated hydroelectric powerplants. Such tides occur only in a few locations in the world. One of the most suitable is the Bay of Fundy between Maine, USA, and New Brunswick, Canada, where the range can be as high as 20 m (~66 ft). Other potential sites are the estuary of the River Severn in Britain, the English Channel, the Patagonian coast of Argentina, the Kislaya inlet on the Barents Sea in the USSR near the Norwegian border, the Rance estuary on the Brittany coast of France, the coast along the Sea of Okhotsk in Japan, and several others with ranges equal to or exceeding 10 m (~33 ft).

The total tidal power that is dissipated throughout the world is estimated at 2.4×10^6 MW, which is about one-third of world consumption in the early 1970s. Of these, some 10^6 MW are dissipated in shallow seas and coastal areas and are not recoverable. Because of the very high capital costs of dams and other structures associated with tide energy-conversion systems, only a small fraction of the rest, and a smaller fraction of increasing world energy needs, is expected to be satisfied by tidal energy.

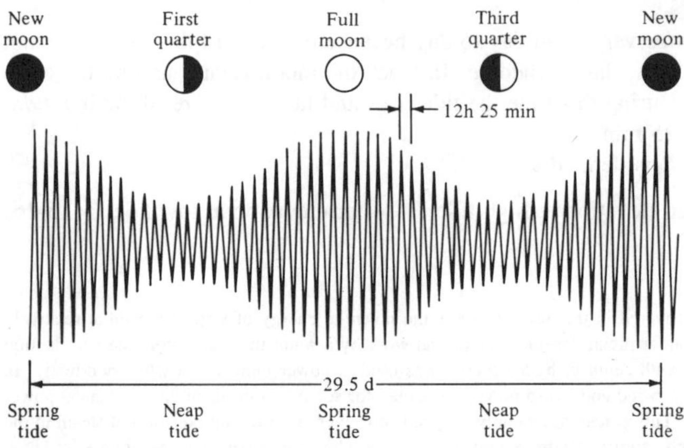

Figure 15-19 Relative high and low tides showing variation in range during lunar month.

We will next discuss three schemes of tidal-energy conversion for the production of electric energy: two depend upon a single-pool or basin, the third on a two-pool design. A discussion of recent developments in tidal electric powerplants may be found in the proceedings of a conference held in Nova Scotia in 1970 [153].

15-14 THE SIMPLE SINGLE-POOL TIDAL SYSTEM

The *simple single-pool tidal system* has one pool or basin behind a dam that is filled from the ocean at high tide and emptied to it at low tide. Both filling and emptying processes take place during short periods of time: the filling when the ocean is at high tide while the water in the pool is at low-tide level, the emptying when the ocean is at low tide and the pool at high-tide level (Fig. 15-20). The flow of water in both directions is used to drive a number of reversible water turbines, each driving an electrical generator. Electric power would thus be generated during two short periods during each tidal period of 12 h, 25 min, or once every 6 h, 12.5 min.

The maximum *energy* that can be generated during one generation period can be evaluated with the help of Fig. 15-21, which shows the case of the pool beginning at high-tide level, emptying through the turbine to the ocean, which is at low tide. (The reverse process results in identical energy).

For a tidal range R, and an intermediate head h at a given time during the emptying process, the differential work done by the water is equal to its potential energy at the time, or

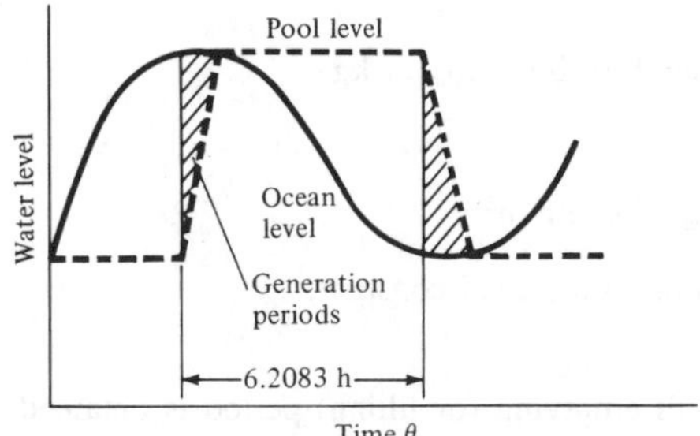

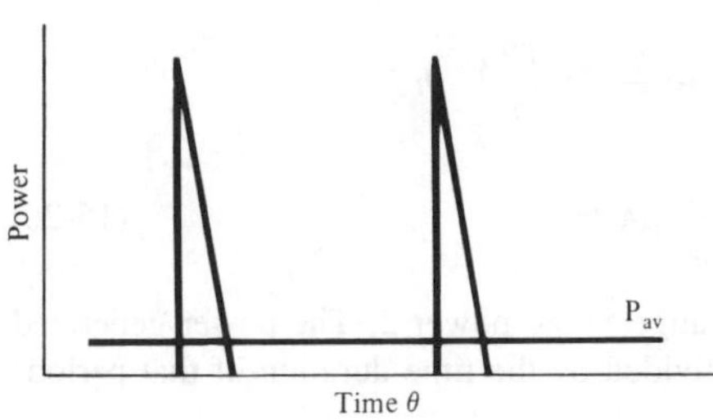

Figure 15-20 Ocean and pool levels and power generated in a simple single-pool tidal system.

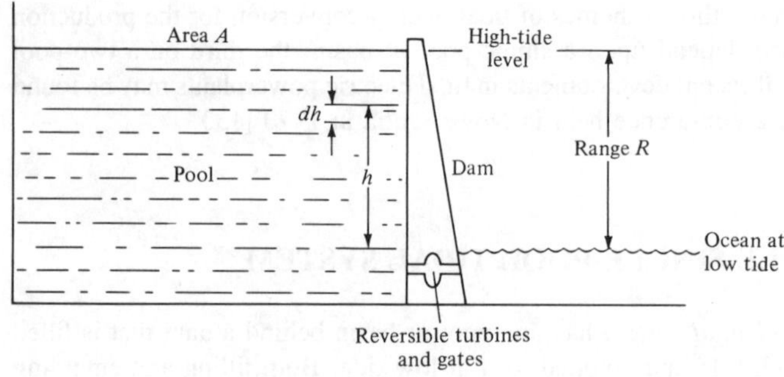

Figure 15-21 Level changes during power production in a single-pool tidal system.

$$dW = \frac{g}{g_c}\, dm\, h \tag{15-17}$$

but

$$dm = -\rho A\, dh \tag{15-18}$$

so that

$$dW = -\frac{g}{g_c}\rho A h\, dh \tag{15-19}$$

where

W = work done by the water, ft · lb_f or J

g = gravitational acceleration, 32.2 ft/s^2 or 9.81 m/s^2

g_c = conversion factor, 32.2 lb_m · ft/(lb_f·s^2) or 1.0 kg/(N · s^2)

m = mass flowing through turbine, lb_m or kg

h = head, ft or m

ρ = water density, lb_m/ft^3 or kg/m^3

A = surface area of pool, considered constant, ft^2 or m^2

The total theoretical work during a full emptying (or filling) period is obtained by integrating Eq. (15-19) as

$$W = \int_R^0 dW = -\frac{g}{g_c}\rho A \int_R^0 h\, dh$$

or

$$W = \frac{1}{2}\frac{g}{g_c}\rho A R^2 \tag{15-20}$$

Thus the work is proportional to the range to the power 2. The power generated during each of the periods is equal to W divided by the time duration of that period.

Zero power is generated during the rest of the time (Fig. 15-20). The *average theoretical power* delivered by the water is W divided by the total time it takes each period to repeat itself, or 6 h, 12.5 m, or 22,350 s. Thus

$$P_{av} = \frac{1}{44,700} \frac{g}{g_c} \rho A R^2 \tag{15-21}$$

where P_{av} = average theoretical power, ft · lb_f/s or W

Assuming an average seawater density of 64 lb_m/ft^3, or 1025 kg/m^3, the average theoretical power per unit pool area would be given by

$$\frac{P_{av}}{A} = 1.43 \times 10^{-3} R^2 \quad \text{ft} \cdot \text{lb}_f/(\text{s}\cdot\text{ft}^2) \tag{15-22a}$$

$$= 0.225 R^2 \quad \text{W/m}^2 \text{ or MW/km}^2 \tag{15-22b}$$

The actual power generated by a real tidal system would be less than the above because of frictional losses and inefficiencies in the turbines and electric generators and might only be 25 to 30 percent of the above.

The power generated, however, could be immense. The Bay of Fundy, for example, has an area of 13,000 km^2 and an average range of 8 m. If we assume an efficiency of 27.5 percent, it has a potential of generating more than 50,000 MW, or 50 GW, which is about twice the electric-power consumption of Canada in 1980.

15-15 THE MODULATED SINGLE-POOL TIDAL SYSTEM

In the simple single-pool system (above), two high-peak, short-duration power outputs occur every tidal period. Such peaks necessitate large turbine-generators that remain idle much of the time. The power peaks also occur at different times every day (50 min later each successive day), at times of high and low tides that almost surely will not always correspond to times of peak power demand, and pose a burden on the electric-power grid they are connected to.

The *modulated single-pool tidal system* partially corrects for these deficiencies by generating power more uniformly at a lower average head, though still with some periods of no generation. Because the average head h is lower and work and power are proportional to h^2, the turbine-generators are much smaller and operate over much longer periods. The resulting total work is reduced, however.

In the system, shown by the ocean and pool level and power diagrams of Fig. 15-22, the reversible turbines are allowed to operate *during* periods of pool filling and emptying instead of at high and low levels only. They cease to operate when the head is too low for efficient operation. Period C_1 begins with both pool and ocean at low-tide level (1), the ocean at the beginning of tide rise, and all gates closed. When the head is sufficient (2), gates to the turbines are opened and water from the ocean is allowed through. Power is generated during period G_1 as both ocean and pool levels rise. The ocean level reaches its peak and begins to decrease but the pool level is still

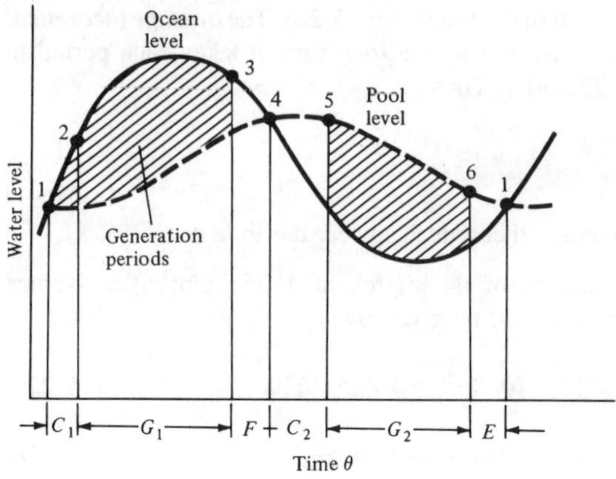

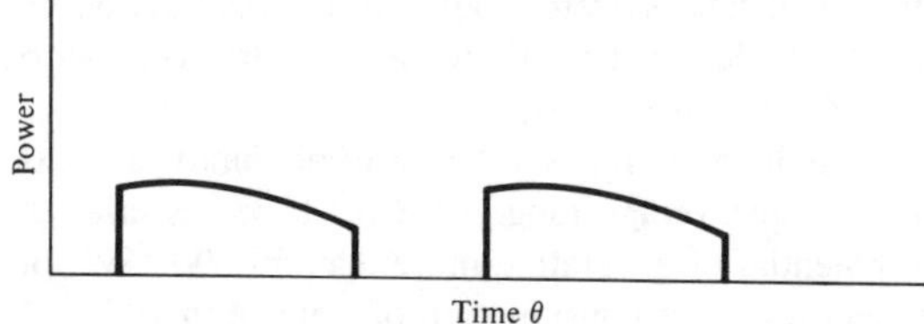

Figure 15-22 Ocean and pool levels in a modulated single-pool tidal system. C = gates closed, G = generation, F = pool filling, E = pool emptying.

increasing until, at 3, the head is too low for efficient generation. The gates to the turbines are closed and bypass gates are opened so that the pool is allowed to fill up during period F to 4. At 4 all gates are once again closed and the pool level remains constant while the ocean level decreases during period C_2. At 5 the head is once again sufficient to allow for turbine water flow in the opposite direction and a second generation period G_2 begins. At 6 generation ceases but the pool is allowed to empty during period E and the system goes back to point 1, repeating the cycle. The power generation shown is certainly not uniform but much more so than in the case of the simple system.

The evaluation of the total work is obtained by assessing the characteristics of the system, i.e., the variation of water mass flow and head with time θ, and integrating their product over the time span during which generation is taking place. When we consider the first-generation period (Fig. 15-23), the general relationships take the form

$$H = f_1(\theta) \tag{15-23}$$

$$y = f_2(\theta) \tag{15-24}$$

$$dW = \frac{g}{g_c}\, dm\, h = \frac{g}{g_c}\, dm\, (H - y) \tag{15-25}$$

but

$$dm = \rho A\, dy \tag{15-26}$$

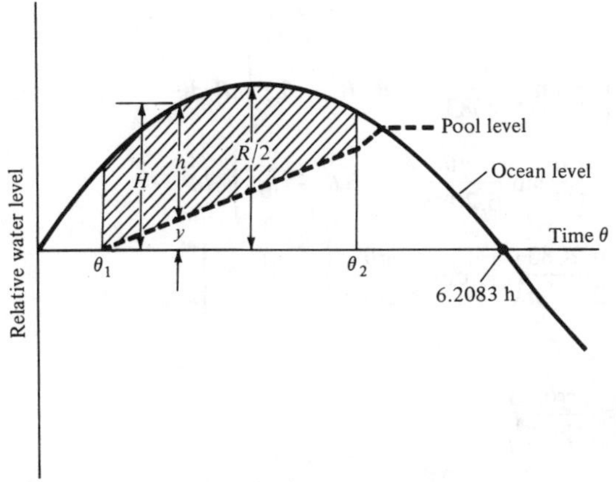

Figure 15-23 Relative ocean and pool levels in the first-generation period of a modulated single-pool tidal system.

Thus

$$dW = \frac{g}{g_c}\,\rho A\left[f_1(\theta) - f_2(\theta)\right] df_2(\theta) \tag{15-27}$$

and

$$W = \frac{g}{g_c}\,\rho A\int_{\theta 1}^{\theta_2}\left[f_1(\theta) - f_2(\theta)\right] df_2(\theta) \tag{15-28}$$

where H = ocean level above mean or other appropriate datum

y = pool level above mean or datum

θ = time

and the other symbols have already been defined. H may be closely approximated by a sinusoidal function of θ such as

$$H = f_1(\theta) = \frac{R}{2}\sin\frac{\pi\theta}{6.2083} \tag{15-29}$$

where θ is in hours and 6.2083 in hours is one-half of a tidal period. y may be approximated by a linear function of θ, starting at 0 at θ_1 for a constant mass-flow rate such as

$$y = f_2(\theta) = aR(\theta - \theta_1) \tag{15-30}$$

where a is a constant having the dimension time^{-1}, e.g., h^{-1}, or y could be a function of $h = H - y$ for a constant flow resistance or some other function determined from operational data. Using the relationships for H and y of Eqs. (15-29) and (15-30), the work during period G_1 (or G_2) would be evaluated from

$$dW = \frac{g}{g_c}\,\rho A\left[\frac{R}{2}\sin\frac{\pi\theta}{6.2083} - aR(\theta - \theta_1)\right]d\left[aR(\theta - \theta_1)\right] \tag{15-31}$$

and

$$W = \frac{g}{g_c}\rho A \int_{\theta_1}^{\theta_2}\left[\frac{R}{2}\sin\frac{\pi\theta}{6.2083} - aR(\theta - \theta_1)\right] aR\, d\theta$$

$$= \frac{g}{g_c}\rho A R^2 \int_{\theta_1}^{\theta_2}\left[\frac{a}{2}\sin\frac{\pi\theta}{6.2083} - a^2(\theta - \theta_1)\right] d\theta$$

$$= \frac{g}{g_c}\rho A R^2 \left[a\frac{6.2083}{2\times\pi}\left(-\cos\frac{\pi\theta}{6.2083}\right) - \frac{a^2}{2}\theta^2\right]_{\theta_1}^{\theta_2}$$

Thus

$$W = \frac{g}{g_c}\rho A R^2\left[0.988a\left(-\cos\frac{\pi\theta_2}{6.2083} + \cos\frac{\pi\theta_1}{6.2083}\right) - \frac{a^2}{2}\left(\theta_2^2 - \theta_1^2\right)\right]_{\theta_1}^{\theta_2} \quad (15\text{-}32)$$

This shows that the work, like in the simple single-pool system, is also a function of R^2.

Example 15-3 Calculate the total energy and average power of a modulated single-pool tidal system using Eqs. (15-29) and (15-30) for H and y and the values $R = 12$ m, $a = 0.0625\ \text{h}^{-1}$, $\theta_1 = 1$ h, $\theta_2 = 4$ h, $A = 10{,}000\ \text{km}^2$, and $\rho = 1025\ \text{kg/m}^3$. Compare the results with those for a simple single-pool system.

SOLUTION Using Eq. (15 -32):

$$W = \frac{9.81}{1}\times 1025 \times 10^{10} \times 12^2\left[0.988 \times 0.0625(0.43795 + 0.87468) - \frac{(0.0625)^2}{2}(16 - 1)\right]$$

$$= 1.448 \times 10^{16}(0.08105 - 0.02930)$$

$$= 7.493 \times 10^{14}\ \text{J}$$

The average power during the generation period of 4 h is

$$P_{\text{av,gen}} = \frac{7.493 \times 10^{14}}{4 \times 3600} = 5.2 \times 10^{10}\ \text{W} = 5200\ \text{MW}$$

The average power during the total period of 6.2083 h is

$$P_{\text{av}} = \frac{7.493 \times 10^{14}}{6.2083 \times 3600} = 3.35 \times 10^{10}\ \text{W} = 33{,}500\ \text{MW}$$

In the simple single-pool system, the corresponding values from Eqs. (15-20) and (15-21) are

$$W = \frac{1}{2}\frac{g}{g_c}\rho AR^2 = \frac{1}{2} \times \frac{9.81}{1} \times 1025 \times 10^{10} \times 12^2 = 7.24 \times 10^{15} \text{ J}$$

and $$P_{av} = 3.24 \times 10^{11} \text{ W or } 324{,}000 \text{ MW}$$

Thus the simple single-pool system produces some 10 times the work and average power of the modulated single-pool system. However, the former does so almost in a "spike," which is very hard on the power grid and requires very large turbines that remain idle most of the time. The latter produces its work over several hours and hence avoids these problems.

The actual work and power above must be multiplied by the efficiency of the system, which is probably in the 25 to 30 percent range.

Only two tidal powerplants have been built in the world to date, one in France, the other in the USSR. The French plant (Fig. 15-24) was built in the Bay of Rance, across the River Rance, near St. Malo in Brittany, France, in 1966. The Bay of Rance has a basin of 22×10^6 m^2 area and a maximum tidal range in excess of 13 m. The plant consists of twenty-four 10-MW reversible turbine generators that operate on the modulated single-basin system described above for a peak power of 240 MW. The average power, however, is 160 MW. An added feature of the Rance plant is that the tidal basin behind the dam is also used for pumped storage from the main power grid. At periods of low power demand and high tide, excess energy from the power grid is used to pump water into that basin raising its level almost 0.5 m above high tide level. This pumping energy is more than recovered during the normal tidal-basin discharge to the ocean when at low tide because the head at discharge is greater than at pumping. Since the energy is proportional to h^2, the energy gain in this operation is estimated at 12:1 maximum. Operating and maintenance difficulties have resulted in a low plant availability factor (fraction of total time plant is on line) of about 25 percent, with the plant operating only about 2000 h/year for a yearly average of about 62 MW. The unit plant capital cost, based on 1962 economics, was $300/kW.

The USSR plant, a small 2-MW demonstration unit, also of the modulated single-basin type, started producing power in 1970 at Kislaya Guba on the Kislaya inlet on the Barents Sea, about 1000 km north of Murmansk, near the Norwegian border. It is believed that the USSR has been studying proposals for larger systems.

15-16 THE TWO-POOL TIDAL SYSTEM

The *two-pool tidal system* is one that is much less dependent on tidal fluctuation but at the expense of more complex and hence more costly dam construction. An inland basin (Fig. 15-25) is enclosed by dam *A* and divided into a high pool and a low pool by dam *B*. By proper gating in dam *A*, the high pool gets periodically filled at high tide from the ocean and the low pool gets periodically emptied at low tide. Water flows from the high to the low pool through the turbines that are situated in dam *B*. The capacities of these two pools are large enough in relation to the water flow between

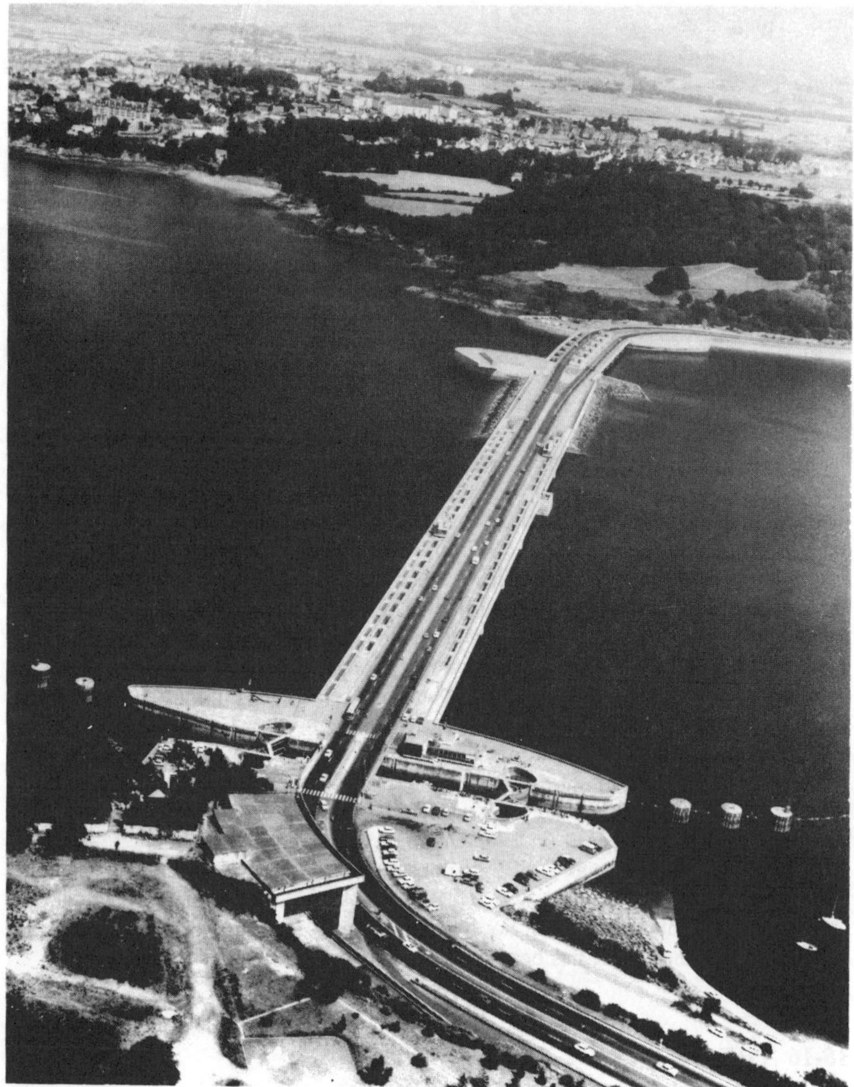

Figure 15-24 A view of the 240-MW (peak) tidal powerplant built on the River Rance, Brittany, France. *(Courtesy Phototeque, Electricité de France, Paris.)*

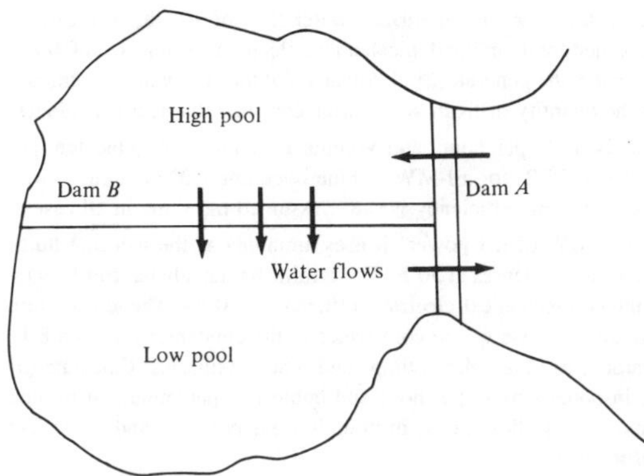

Figure 15-25 Schematic of a two-pool tidal system.

them that the fluctuations in the head are minimized, which results in continuous and much more uniform power generation.

The two-pool system has been considered for the Passamaquoddy Bay tidal project. Passamaquoddy Bay is an arm of the Bay of Fundy. (Enclosing the whole Bay of Fundy is economically prohibitive and not technically feasible, and it is more feasible to enclose its basins and estuaries.) Passamaquoddy Bay itself would constitute the high pool. The adjacent Cobscook Bay would be the low pool. A modest 300-MW plant was under consideration as a joint venture of the U.S. and Canadian governments, but the project was abandoned in 1961 because it was considered uneconomical. Another larger plant of 1-GW capacity was again considered from 1963 to 1974, but no agreement was reached because of similar economic considerations.

PROBLEMS

15-1 Ocean water, just below the topmost layer (where reflection back to the sky and a complex mechanism of heat transfer and evaporative mass transfer occurs) receives 65 W/m^2 of solar energy in one locality. The extinction coefficient is 0.4 m^{-1}. Calculate (*a*) the volumetric heat-generation rate q''', in watts per cubic meter, near the surface and 1 m below the surface, and (*b*) the distance below the surface, in meters, where 99.99 percent of solar energy has been absorbed.

15-2 An OTEC powerplant of the Claude type has an ideal turbine operating between 25 and 15°C. For a turbine output of 1 MW, calculate (*a*) the mass flow rate of steam, in kilograms per second, (*b*) the mass flow rate of warm water, in kilograms per second, if the evaporator temperature drop is 2°C, (*c*) the mass flow rate of cold water, in kilograms per second, if the condenser temperature drop is 2°C, and (*d*) the cold water pipe diameter, in meters, if the water velocity in it is not to exceed 1 m/s.

15-3 An OTEC powerplant of the Claude type uses 10^7 lb$_m$/h of warm surface water at 70°F. The evaporator temperature is 66°F. The condensor pressure is 0.25 psia. The turbine polytropic efficiency is 0.84. The combined turbine-generator mechanical and electrical effeciency is 0.87. Calculate (*a*) the steam mass flow rate into the turbine, in pounds mass per hour, (*b*) the gross generator output, in kilowatts, (*c*) the gross cycle efficiency, and (*d*) the cold-water pipe diameter if the cold-water temperature and velocity in the pipe are 56°F and 1.0 ft/s, respectively.

15-4 Consider the powerplant in Fig. 15-5. Assume an ocean surface water flow of 10^7 lb_m/h, a turbine polytropic efficiency of 0.85, and turbine-generator combined mechanical-electrical efficiency of 0.80. Calculate (*a*) the gross power generated by the turbine generator, in kilowatts, (*b*) the cold-water volumetric flow rate, in cubic feet per minute, and (*c*) the quantity of fresh water produced, in cubic feet per minute.

15-5 Compare the mass flow rates, in pounds mass per hour, and volume flow rates, in cubic feet per minute, at turbine inlet of all four fluids in Table 15-2, for a 1-MW turbine operating with saturated vapor at 70°F and exhausting at 50°F. The turbine polytropic efficiency is 0.88, assumed the same in all cases.

15-6 An Anderson OTEC cycle generates 150 MW of net power. It uses ammonia as the working fluid. Ammonia vapor is generated at 70°F saturated and condenses at 50°F. The turbine has an adiabatic efficiency of 0.80 and the turbine-generator has a combined mechanical-electrical efficiency of 0.88. The temperature drop of the ocean warm water in the evaporator and rise of the cold water in the condenser are both 8°F. 14 percent of the gross output of the generator is used for plant pumps and other auxiliaries. Calculate (*a*) the ammonia mass and volume flow rates, in pounds mass per hour and cubic feet per minute at turbine inlet, (*b*) the warm and cold water mass and volume flow rates, in pounds mass per hour and cubic feet per minute, and (*c*) the plant net thermal efficiency.

15-7 An Anderson OTEC powerplant uses propane as the working fluid. Saturated propane vapor enters the turbine at 70°F and the condenser at 50°F. The turbine polytropic efficiency is 0.88 and the turbine-generator combined mechanical-electrical efficiency is 0.80. 10^9 lb_m/h of hot water enters the evaporator at 82°F and leaves at 74°F. Pumps and other auxiliaries consume 14 percent of the generator gross output. Calculate (*a*) the ammonia mass and volume flow rates, in pounds mass per hour and cubic feet per minute, at the turbine inlet, (*b*) the plant net output, in megawatts, (*c*) the cold-water flow, in pounds mass per hour, if its temperature increases 8°F in the condenser, and (*d*) the plant net efficiency.

15-8 A 2.5-m ocean wave has a 60-m wavelength in 10-m deep water. Calculate (*a*) the probable period, in seconds, (*b*) the major and minor axes, in meters, of the water motion at the surface and at middepth, and (*c*) the wave power kilowatts per meter width of the wave front. Take water density as 1025 kg/m^3.

15-9 A wave-energy generator system is to produce 1 MW in the ocean where the waves have a steady 4-ft amplitude and 227.1-ft wavelength. The floats measure 12 × 12 × 1 ft. Calculate the number of floats necessary if the overall efficiency of the system is 9.6 percent. Sea water density is 64 lb_m/ft^3.

15-10 A wave-energy generator float measures 40 × 40 ft. For simplicity, assume that it experiences waves on a given day that are 7 ft high half the time and 4 ft high the other half. All waves have a wavelength of 184 ft. The overall conversion efficiency of the plant is 5 percent. The water density is 64 lb_m/ft^3. Calculate (*a*) the total energy output during the day in foot pounds force and kilowatt hours and (*b*) the mean energy wave height, and (*c*) the percent error obtained if the "arithmetic mean" wave height were used for the whole day.

15-11 A single square float measuring 10 × 10 × 1 ft is placed in the ocean where the waves have a steady 3.5-ft amplitude and 200-ft wavelength. The float is connected to a piston of 1-ft diameter that operates within a cylinder. The compressed air is expanded in a turbine. Both compression and expansion ratios are 6. The air temperature is 40°F and the water density is 64 lb_m/ft^3. Calculate (*a*) the power, in foot pounds force per second and kilowatts, imparted to the float by the waves, (*b*) the air flow to the turbine, in pounds mass per second, and (*c*) the ideal turbine work, in kilowatts.

15-12 An ocean wave power system of the float type uses a 25-cm-thick float in 2-m waves that have a 60-m wavelength. The cylinder has a diameter of 1 m. The compressor pressure ratio is 10. The air is at 1 atm and 10°C. Calculate the power produced by the turbine if the compressor-turbine system has an efficiency of 76 percent of ideal.

15-13 An ocean wave power device of the high-pressure accumulator type operates in 2.5-m waves that have a wavelength of 60 m. The composite piston has diameters of 20 and 3 m. The water has a density of 1025 kg/m^3. Calculate (*a*) the height of the reservoir above the accumulator if frictional losses in the discharge pipe average 1 bar, (*b*) the average water volumetric flow rate to the reservoir, in cubic meters per second, (*c*) the power generated, in kilowatts, by a turbine-generator of 70 percent efficiency that is situated 2 m above average ocean water level, assuming water level in the reservoir is constant, and (*d*) the length, in meters, of dolphin arrays (Fig. 15-16) that would produce the same power.

15-14 A tidal powerplant of the simple single pool type, has a pool area of 300×10^6 ft^2. The tide has a range of 36 ft. The turbine, however, stops operating when the head on it falls below 9 ft. Calculate the energy generated in one filling (or emptying) process, in foot pounds force and kilowatt hours if the turbine-generator efficiency is 0.73.

15-15 A 300×10^6 ft^2 tidal pool level follows a sinusoidal curve during the tidal cycle of 6.2083 h. Assume for simplicity that it discharges into a reservoir with constant level. The maximum head is 36 ft. Calculate the power generated during the cycle in foot pounds force and kilowatt hours. Assume for simplicity that the water mass flow rate is constant. Take density of water to be 64 lb$_m$/ft^3.

15-16 Calculate the total energy and average power of a modulated single-pool tidal plant operating with a range of 12 m. The tidal pattern is assumed to be sinusoidal. The water level in the pool may be approximated by the relation $0.0625R(\theta_2 - \theta_1)$ where θ is time, in hours. Power generation occurs between $\theta_1 = 1$ and $\theta_2 = 4$ h. The pool has a constant area of 10,000 km^2. The water density is 1025 kg/m^3.

15-17 A tidal powerplant of the two-pool type has upper pool area A_1 and lower pool area A_2. (a) Derive expressions for (*a*) the energy that can be detained if the upper pool is emptied into the lower pool when the initial head difference between them is R and when all gates to the ocean are closed, and (*b*) the ratio of this energy to that obtained from a single pool of area A. Determine the numerical value of this ratio if $A_1 = A_2 = 0.5\,A$.

15-18 Consider for simplicity that the two-pool tidal powerplant proposed for the Bay of Fundy operates with a constant head of 26 ft. Calculate the necessary ocean flow rate into the system, in cubic feet per day, if the powerplant is to produce 200 GW with a 60 percent efficiency. Take water density as 64 lb$_m$/ft^3.

CHAPTER

SIXTEEN

ENERGY STORAGE

16-1 INTRODUCTION

The need for energy storage arises because the demand for electric energy in a utility system is characterized by hourly, daily, and seasonal variations, whereas the supply from that system, in the majority of cases, has a fixed capacity. That capacity must be selected to correspond to the maximum demand plus a reasonable excess to take care of scheduled plant shutdowns for maintenance and unscheduled shutdowns due to abnormal occurrences. The result of this is large, expensive plants that operate much of the time below capacity, thus causing high operating and capital costs.

An example of electric-energy fluctuations in consumption during a typical week in the life of a largely university town is shown in Figs. 16-1 and 16-2 [154]. They demonstrate the differences between daytime and nighttime consumption, between weekdays and weekends, and between summer and winter. More severe fluctuations occur in an industrial or commercial region, where demands drop on weekends. Seasonal fluctuations in all regions also occur. The picture is more clouded, and the need for energy storage is greater, if plants using renewable forms of energy such as solar and wind are used to generate electricity. It is the output of these plants that fluctuates severely because of their input energy intermittency. Their conversion systems are also much more expensive than those of conventional plants.

The objective of energy storage, therefore, is to counteract the disadvantages that result from the fluctuations in demand for electric energy by assuring a steady high output from existing powerplants. When the demand is lower than capacity, energy is stored. When the demand is higher than that capacity, the stored energy is released. The result then is to be able to supply electricity reliably, efficiently and economically,

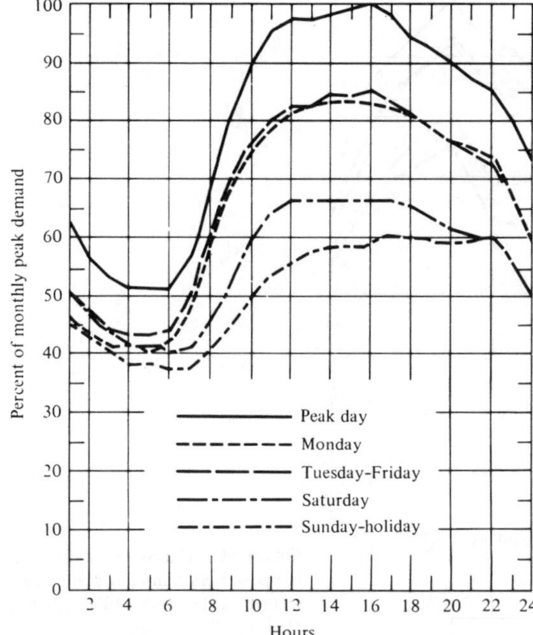

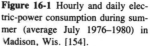

Figure 16-1 Hourly and daily electric-power consumption during summer (average July 1976–1980) in Madison, Wis. [154].

while being able to provide peak electrical demands on short notice during certain times of the day or week.

The need for energy storage was not acute when the generating plants were cheap and the fuel supply plentiful. Indeed, energy storage has, in a sense, been historically done in the form of the latent energy stored by nature in the fuels themselves. The energy density in this "natural" storage is large. Fossil fuels, coal, oil, and gas (at 1000 psia), all contain about a million Btu/ft^3 ($\sim 37 \times 10^6$ kJ/m^3). Natural uranium metal (0.071 percent U^{235}) contains about 3×10^{12} Btu/ft^3 ($\sim 10^{14}$ kJ/m^3). With increasing supply problems of all fuels, the need to conserve resources, and the increasing unit capital costs ($/KW) and production costs (mills*/kWh) of electric-generating plants, suitable methods of *energy management* become necessary. Examples of these are:

1. Supply power peaks by interconnecting power networks that might have different power demands on them.
2. Use newer and more efficient powerplants for base-load generation and use older less efficient plants for peak-power generation.

* 1 mill = one thousands of a dollar = 0.1¢.

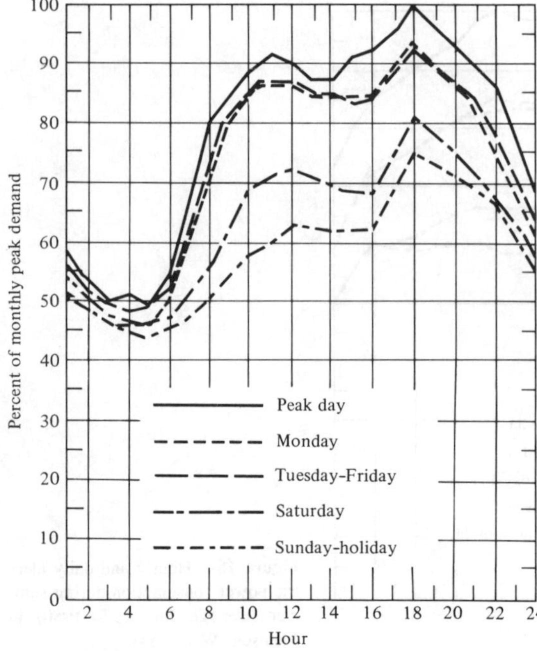

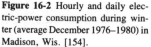

Figure 16-2 Hourly and daily electric-power consumption during winter (average December 1976–1980) in Madison, Wis. [154].

3. Construct smaller, low capital cost, though not so efficient powerplants, as power-peaking units. Examples are special steam plants, small hydroelectric plants, or gas-turbine peaking plants (Chap. 8).
4. Add energy-storage systems.

In general, reliability and economy of electrical supply can best be achieved by having a mix of three types of powerplants: a base-load plant, a cycling plant, and a peaking plant.

Base-load plants are used to provide a base electrical load to the grid. Such plants are usually large, efficient, steam-generating, Rankine-cycle type stations powered by fossil or nuclear fuels. They operate continuously except for scheduled maintenance or forced outages. They have a power operating factor (POF) (Fig. 14-9) between 60 and 70 percent. This relatively high POF results in a comparatively low unit cost of power (mills/kWh).

Cycling plants, also called *intermediate plants,* usually are older, less efficient steam plants, or new ones specifically designed for cyclic operation, such as combined cycles (Sec. 8-8). They operate primarily during hours of high load demand and have an annual POF between 25 and 50 percent. This rather wide range is primarily the result of seasonal variations, such as those due to periods of high industrial output, air-conditioning loads in the summertime, etc.

Peaking plants are specifically designed to provide relatively inexpensive power during peak-demand periods, such as due to abnormal air-conditioning loads and peak-hour domestic demands in the evenings. They operate at a low annual POF of 5 to 15 percent. They also operate at a low availability factor (Fig. 14-9) i.e., intermittently according to system requirements. Their operation may be for as little as 2 or as much as 12 h/day, for as many as 5 days/week.

The last of the courses of action, energy storage, is the one discussed in this chapter. We are here concerned with large-scale storage suitable for incorporation with utility electrical powerplants. Energy storage would allow the plants to be designed for nearly constant load operation below peak demand, a process called *peak shaving*, which would thus reduce the high capital cost of the initial plants. Energy storage, of course, becomes attractive only when the capital and operating costs of the storage system are more than offset by the reduction in the corresponding costs of the original system.

One drawback to all energy-storage systems is that their energy densities are much lower than those mentioned above for fossil and nuclear fuels.

16-2 ENERGY-STORAGE SYSTEMS

There are basically two generic approaches to energy storage in utility systems (Fig. 16-3). These are (1) electrical storage and (2) thermal storage.

Electrical Storage

The primary electric-generating plant is continuously operated in a base-load mode, which results in excess electricity production during the off-peak periods *ab* + *cd* (Fig. 16-4). Electrical storage is then used to hold this excess electricity for use during peak demand, period *bc*. Note that the total energy stored is greater than the total energy supplied because of conversion losses to and from storage. Storage schemes in this category that are in use or under investigation are:

1. Electrical-mechanical energy storage
   a. Potential, pumped hydro
   b. Potential, compressed air
   c. Potential, springs, torsion bars, mass elevation
   d. Kinetic, flywheels
2. Direct electrical energy storage
   a. Batteries
   b. Superconducting coils

Thermal Storage

In thermal storage, all schemes deal with storing energy in a thermal form in a material during periods of low power demand and releasing it back during periods of high demand. The primary electric-generating plant is operated to meet the real-time elec-

Primary heat source
Electrical storage
Thermal storage
Base load input
Load input
Excess heat
Primary electric generating station
Primary electric generating station
Thermal storage
Sensible heat
Latent heat
Chemical reaction
Alternate heat use
Load
Excess electricity
Electrical storage
Electromechanical
Direct
Potential energy
Kinetic energy
Pumped hydro
Compressed air
Springs and torsion bars elevation
Flywheels
Storage batteries
Superconductive magnets
Load
Load
Peak-load generating plant
Peak load
Peak load

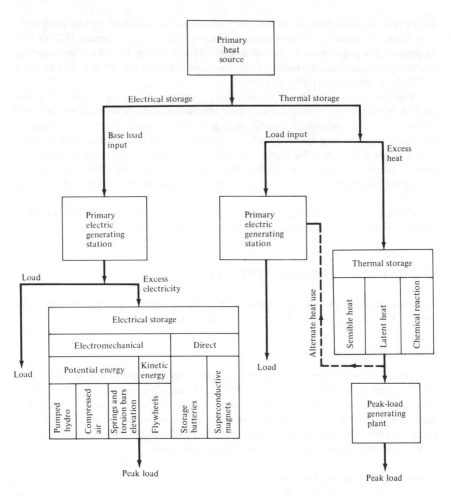

Figure 16-3 Energy storage systems.

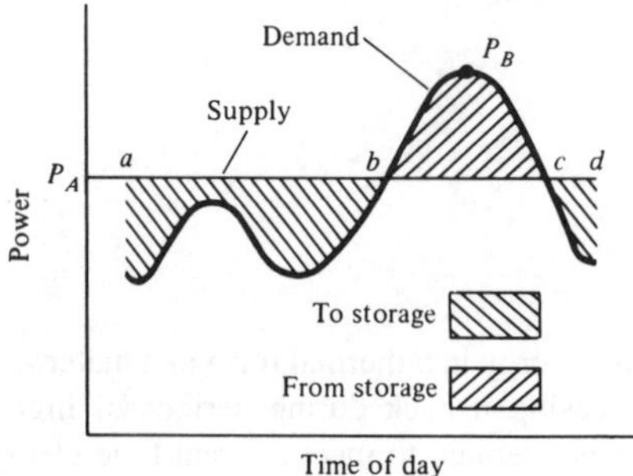

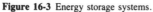

Figure 16-4 Thermal-energy storage with constant thermal input as from fossil or nuclear fuel.

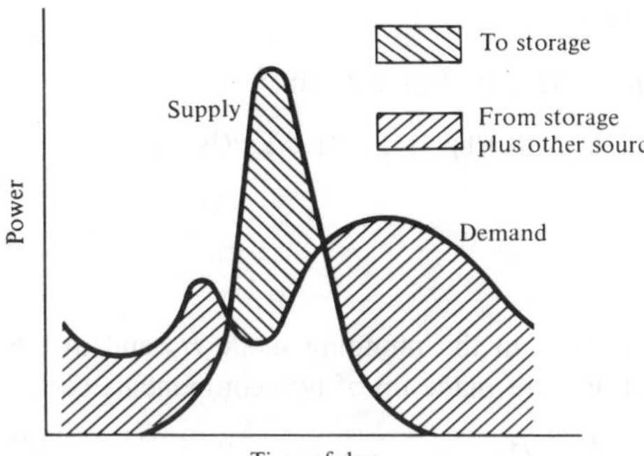

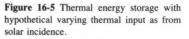

Figure 16-5 Thermal energy storage with hypothetical varying thermal input as from solar incidence.

trical demands during off-peak hours. The available thermal energy input to the plant may be essentially constant, as is that from fossil or nuclear fuel (Fig. 16-4), or varying, as from solar incidence (Fig. 16-5). The excess thermal energy is stored as such and withdrawn to be converted to meet peak electrical demands. Conversion could occur in the primary plant itself or in a separate peaking plant (Fig. 16-4). Again, note that because of storage losses and conversion inefficiencies the total stored energy is greater than that supplied.

Thermal storage schemes include:

1. Sensible heat
2. Latent heat
3. Chemical reaction

Not all the various electrical and thermal energy-storage schemes are suitable for large utility energy storage. Some, like springs, torsion bars, and mass elevation, are very low-capacity systems that are used to power such small devices as watches, clocks, toys, and instruments. They will not be covered in this book. Others, like flywheels and batteries, are in the developmental stages and will probably be suitable for intermediate storage. A few, like pumped hydro, compressed air, and superconductivity are, or will be, suitable for large utility energy storage. These and the various other schemes are covered in the following sections.

16-3 PUMPED HYDRO

Pumped hydro, like compressed air (Sec. 16-4), is a potential-energy storage system suitable for large utility energy storage. It is the most developed and used of all storage systems. The principle behind pumped hydro is simple and follows the law of potential energy PE that is, the raising of mass to an elevation, height, or head H. It is given by

$$PE = \frac{g}{g_c} mH \tag{16-1}$$

where PE = potential energy, ft · lb_f or J

g = gravitational acceleration = 32.2 ft/s^2 or 9.81 m/s^2

g_c = conversion factor = 32.2 lb_m · ft/(lb_f · s^2) or 1.0 kg/(N · s^2)

m = mass, lb_m or kg

H = head, ft or m

The operating heads on the pump turbine in the pumping mode H_p and in the turbine-generating mode H_T are different and are made up of two components each

$$H_p = H + H_\ell \tag{16-2}$$

and

$$H_T = H - H_\ell \tag{16-3}$$

where H is the static head or height and H_ℓ represents the losses during flow conditions (which are different because of the different flow rates).

The pumping and generation powers are given by replacing the mass in Eq. (16-1) with the mass-flow rate $\dot{m}$ lb_m/s or kg/s and using the proper head, or with

$$P_p = \frac{g}{g_c}\rho \dot{Q}_p H_p \tag{16-4}$$

and

$$P_T = \frac{g}{g_c}\rho \dot{Q}_T H_T \tag{16-5}$$

where P_p and P_T = pumping and turbine powers, respectively, ft · lb_f/s or W

ρ = density of water, lb_m/ft^3 or kg/m^3

$\dot{Q}_p$ and $\dot{Q}_T$ = volumetric flow rates in pumping and generation, respectively, ft^3/s or m^3/s

Equation (16-1) shows that 1000 kg raised 100 m will store 9.81 × 10^5 J or 0.2725 kWh. Thus large masses must be elevated to sufficiently large heights to store large quantities of energy. Fortunately large masses are available in pumped-hydro systems by the elevation of large quantities of water from a lower to an upper reservoir. One or both of these reservoirs may be artificially excavated or may be a natural river or lake.

Pumped-hydro systems, unfortunately, require a suitable topography that will allow the design and construction or selection of two reservoirs with sufficient capacities, maximum available elevation difference H, and minimum horizontal distance L between them to reduce flow losses. Values of $L/H < 2$ are considered very favorable, although most existing plants average L/H between 4 and 6 with some nearly as high as 10. Good topographies are obviously not available everywhere.

Although high heads are desirable, some topographies do not allow them, and thus pumped-hydro systems are often classified as *above-ground,* which includes the preferred *high head* and *medium head,* and *underground.*

Above-Ground Pumped Hydro

In *high-head* installations, the upper reservoir may have originally been a stream descending a steep slope which has been dammed to form the reservoir. From that reservoir, water is diverted into a horizontal pressure tunnel driven through the rock to the valve house from which the main steel pipeline slopes down to the powerhouse. At the head of the steel pipeline there usually is a surge tank and a valve house. It contains the main sluice valves, which are automatic isolating valves that come into operation in the case of pipeline burst. Automatic air valves, also, may be used. These contain buoyancy floats that fall when sufficient air separates from the water. These floats are attached to a spindle which then opens the valve to vent the air to the atmosphere. Other automatic air valves allow air to enter the pipeline in case the pipeline is drained. They safeguard the pipeline against internal collapse when thus emptied. A surge tank or surge chamber is built near the mouth of the pressure tunnel to relieve the pipes of undue inertia pressure set up in the tunnel when the flow is checked following a reduction of load. Should this pressure exceed a predetermined amount, water merely spills over the lip of the surge tank. The surge tank also provides a reservoir of water that can be drawn upon when the load on the turbine suddenly increases. At every point of deviation of the pipelines, either in the horizontal or the vertical plane, anchorages are constructed with expansion joints provided immediately below. The powerhouse itself is located as close as possible to the lower reservoir into which the tail race discharges.

In *medium-head* installations, a nearly horizontal open canal or conduit may be carried along the side of the valley as far as the powerhouse site. A relatively short pressure pipe, often called a *penstock,* leads to the turbine.

A typical pumped-hydro system of a conventional design is shown in Fig. 16-6.

Underground Pumped Hydro

To overcome the requirement of a suitable topography, *underground pumped hydro* is being considered in this system. The upper reservoir may be at or near ground level. The lower reservoir is placed underground in natural caverns, old mines, or other underground cavities. Such a system is shown in Fig. 16-7.

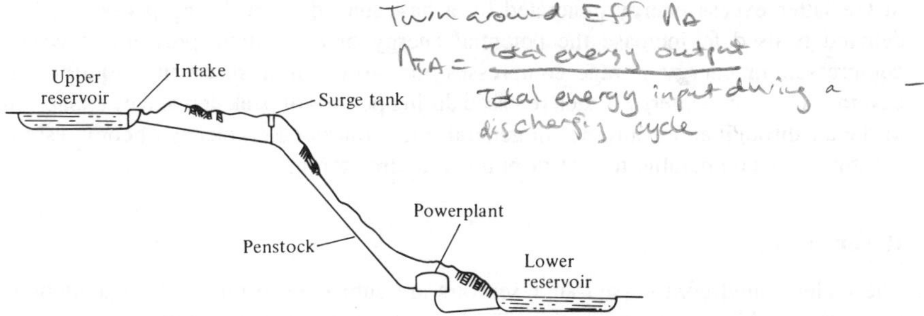

Figure 16-6 Schematic of a conventional above-ground pumped-hydro storage system.

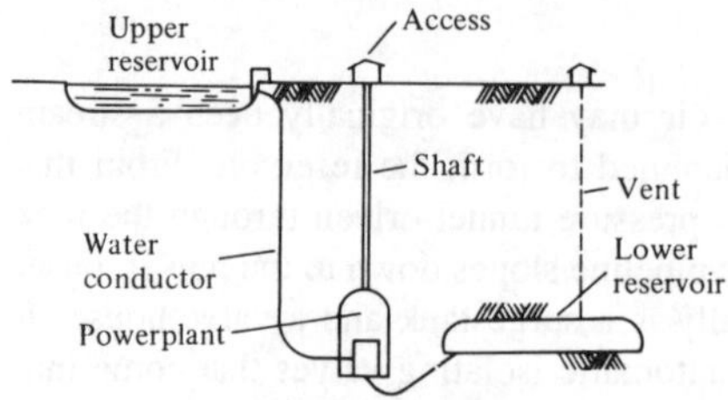

Figure 16-7 Schematic of an underground pumped-hydro storage system.

In all systems, a principal piece of equipment is a reversible pump-turbine or motor-generator set or sets. The excess electric energy supplied by the primary power station during off-peak hours is used to drive it in the motor-pump mode to pump water from the lower to the upper reservoir. During periods of peak demand, the system reverses to the turbine-generator mode to generate the excess electricity needed. (Some old installations use separate conventional pumps and turbines rather than reversible machines.)

The losses in pumped-hydro systems include motor and pump losses and flow losses during upflow; seepage into ground, leakage from pipes and equipment, and evaporation during storage; and turbine and generator losses and flow losses during downflow. The combined efficiency of a pumped-hydro system, called the *turnaround efficiency,* is defined as the total energy output divided by the total energy input during a charge-discharge cycle. In most plants, the turnaround efficiency is in the neighborhood of 65 percent. Pumped-hydro systems are rated according to their power output, usually in megawatts (MW). The maximum power output in the turbine-generator mode is usually greater than the maximum power input in the motor-pump mode, but operation in the latter lasts longer than the former, so the input energy is greater. (Recall that energy = power × time.)

16-4 COMPRESSED-AIR STORAGE

Compressed-air energy storage is analogous to pumped-hydro energy storage. Whereas in the latter excess energy generated by a base-loaded plant during periods of low demand is used to increase the potential energy or hydrostatic pressure of water, compressed-air energy storage compresses and stores air in reservoirs, aquifers, or caverns. The stored energy is then released during periods of peak demand by expansion of the air through an air turbine. In general, the turnaround efficiency of compressed-air storage is comparable to that of pumped-hydro storage.

Reservoirs

The underground compressed-air reservoirs are subjected to repeated fluctuations in pressure, humidity, and temperature. The long-range effects of such fluctuations remain to be determined. Usually multiple reservoirs, operating in parallel, serve one storage system. Three types of reservoirs show the most promise [155]: (1) salt caverns, (2) aquifiers, and (3) hard-rock caverns.

Salt caverns These have been used in the past to store petroleum products. Research so far indicates that they are stable under compressed-air storage loadings for the duration of plant life. The major concerns are cavern geometry, size and spacings, long-term creep and creep-rupture of rock salt, and air leakage.

Aquifers These are naturally occurring porous-rock formations. They have been used for natural gas storage for over 50 years but with annual rather than daily cycling. The effects of the different physical properties of air and its oxygen and the high temperatures of storing remains to be evaluated. Among other concerns are cyclic fatigue of the porous rock, air-water interface movement (water is usually present in aquifers), and the generation and transport of fine particulate matter.

Hard-rock caverns Because of their size these require water-compensating surface reservoirs to maintain air pressure and therefore are more costly than the two reservoir types above. However, they are believed to be most stable in the absence of severe temperature fluctuations (50°C). The major concerns here include the effervescence of air in the water shaft (called the champagne effect), hard-rock properties under cyclic conditions, and the residual strength of hard rock after an initial failure.

Adiabatic and Hybrid Systems

When air is compressed for storage, its temperature will rise (since it is a compressible gas) according to the relationship

$$T_2 = T_1 \left(\frac{P_2}{P_1}\right)^{(n-1)/n} \tag{16-6}$$

where T and P are the absolute temperature and pressure and the subscripts 1 and 2 refer to before and after compression, respectively. n is the polytropic exponent for the nonreversible compression process.

The heat of compression may be retained in the compressed air or in another heat-storage medium and then restored to the air before expanding through the turbine. This is called *adiabatic storage* and results in high storage efficiency. Recall that at a given pressure ratio, turbine work is directly proportional to the inlet absolute temperature (Chap. 8). Recall also that constant-pressure lines on a temperature-entropy diagram for gases diverge at high temperatures so that isentropic work, equal to the vertical distance between any two constant-pressure lines, increases with temperature. Restoring the heat to the air also prevents the turbine parts from freezing if low-temperature air is allowed to expand through it. If the heat of compression is allowed to dissipate, additional heat could be added by fuel combustion to retain the high storage efficiency, but the results would be extra expense and maintenance problems. This is called a *hybrid system*.

Figure 16-8 shows a simple adiabatic compressed-air energy-storage system. The

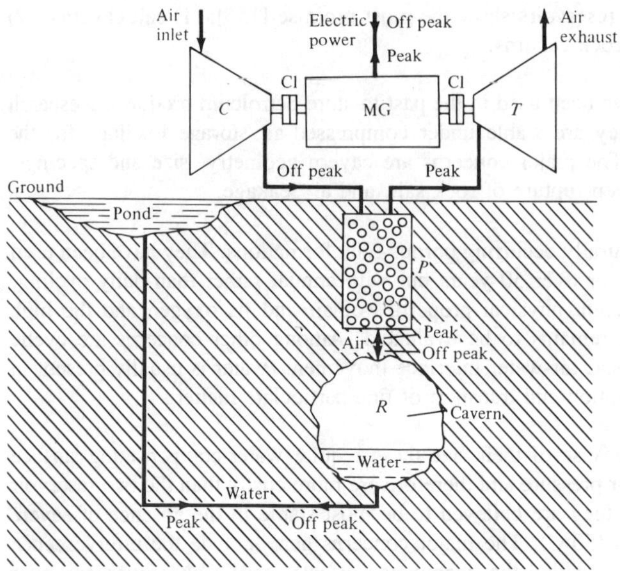

Figure 16-8 Schematic of a simple single-stage adiabatic compressed-air energy storage system with pressure-compensation pond. C = compressor, T = turbine, MG = motor-generator set, P = packed-bed thermal-energy storage, R = air-storage reservoir.

main plant is not shown. During off-peak hours, electric energy from the main plant generator is used by the motor-generator set (MG) operating in the motor mode to drive the compressor (C). The compressed air passes first through a packed bed (P) for sensible thermal-energy storage (Sec. 16-7), then to a constant-pressure underground reservoir (R). The constant pressure is obtained by displacing water to a pressure-compensation pond that has a nearly constant head above the reservoir. During peak hours, air from the reservoir flows through the packed bed picking back sensible heat, then through the air turbine that now drives the motor-generator set in the generator mode. Clutches (Cl) separate the compressor during peak (generation) periods and the turbine during off-peak (storage) periods.

As expected, the air-reservoir (cavern) volume is a strong function of the storage pressure. For a peak unit capacity of 1500 MWh that volume is estimated at nearly 2,000,000 m^3 for 10 bar, or 64,000 m^3 for 100 bar storage pressures. The packed bed thermal-energy storage volume is about a tenth of that of the storage reservoir in most cases. Thus to reduce storage volume and hence cost, operation at high pressure is necessary.

Example 16-1 Calculate the airflow, compressed-air temperature, and storage volume for a 1500-MWh peaking unit charging for 7.5 h. Assume compressor inlet is at 1 bar and 20°C, compressor exit at 100 bar, a compressor polytropic efficiency of 70 percent, a peaking turbine efficiency of 60 percent, and a constant specific heat for air $c_p = 1.05$ kJ/(kg · °C). The air-gas constant $R = 284.75$ kJ/(kg · K).

SOLUTION For a compressor polytropic efficiency of 70 percent and constant specific heat

$$0.7 = \frac{h_{2s} - h_1}{h_2 - h_1} = \frac{T_{2s} - T_1}{T_2 - T_1}$$

where the subscripts 1, 2, and 2*s* are for compressor inlet, exit, and isentropic exit conditions, respectively.

$$T_{2s} = T_1\left(\frac{P_2}{P_1}\right)^{(k-1)/k} = (20 + 273)\left(\frac{100}{1}\right)^{(1.4-1)/1.4} = 1092 \text{ K} = 819°\text{C}$$

$$T_2 = \frac{819 - 20}{0.7} + 20 = 1162°\text{C}$$

(This corresponds to a polytropic exponent $n = 1.5266$.)

For a turbine output of 1500 MWh

$$\text{Storage capacity} = \frac{1500}{0.60} = 2500 \text{ MWh}$$

$$\text{Mass of air required} = \frac{2500 \times 3.6 \times 10^6}{1.05(1162 - 20)} = 7.5 \times 10^6 \text{ kg}$$

Assuming air is stored in the cavern at 100 bar (10^7 Pa) and 20°C

$$\text{Total volume needed} = \frac{7.5 \times 10^6 \times 284.75(20 + 273)}{10^7} = 62{,}575 \text{ m}^3$$

$$\text{Average air flow to cavern during 7.5 h of charging} = 8343 \text{ m}^3/\text{h}$$

It can be seen that a system such as that described above requires a very large compressor with an inlet airflow at 1 bar of about 834,000 m^3/h (~490,000 ft^3/m), capable of an exit pressure of about 100 bar (1450 psia), and an exit temperature of more than 1100°C (~2000°F). Such compressors or combination of compressors need to be developed before a large utility compressed-air energy-storage system can be a reality.

A method of reducing the above temperature is the so-called two-stage compressed-air energy-storage system. It employs two compressor-motor-generator-turbine sets.

The Huntorf Compressed-Air Storage System

The first compressed-air storage system to be built is a 290-MW plant designed by Brown Boveri and built at Huntorf, West Germany [156]. It provides storage for the Nordwestdeutsche Kraftwerke (NWK) utility of Hamburg. The plant (Fig. 16-9) has

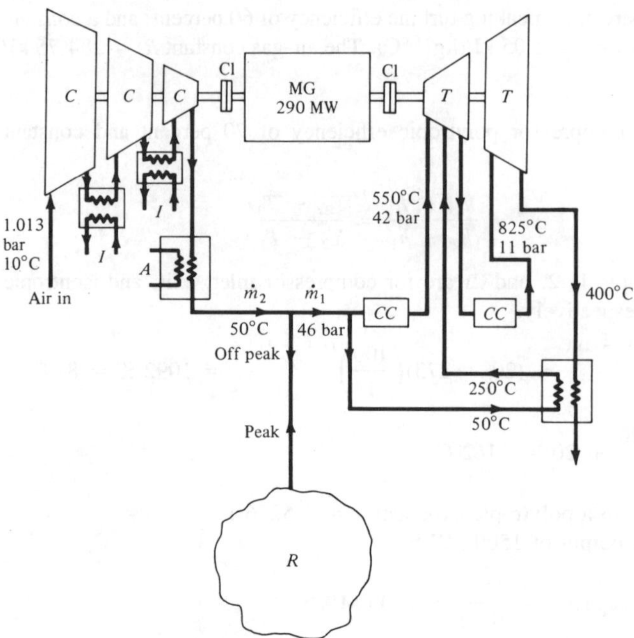

Figure 16-9 The Huntorf hybrid air-storage gas-turbine powerplant. I = intercooler, A = aftercooler, CC = combustion chamber, mass-flow rates: $\dot{m}_2$ = 0.9 to $0.25\dot{m}_1$ [155].

been in operation since 1978. It uses two 150-m-high salt caverns with a total volume of 300,000 m^3 created by leaching a salt dome at a depth between 650 and 800 m below ground. The system is composed of a motor-generator set connected by clutches to a three-stage compressor with intercoolers and a two-stage gas turbine with reheat. It is of the hybrid variety that requires heat addition prior to the gas turbine.

In the storing mode, the compressor pumps atmospheric air into the caverns, where it is stored at 50 to 70 bar (725 to 1015 psia). In the generation mode, the stored air, reduced in pressure to 46 bar (667 psia), enters a natural gas–fueled combustion chamber before the high-pressure section of the gas turbine. Reheat is accomplished by a low-pressure natural-gas burner. Storage occurs daily for about 8 h, generation for about 2 h. The compressor and turbine are each sized independently to suit the power requirements during these periods, an advantage over the usual gas-turbine cycle in which the compressor absorbs more than two-thirds of the turbine output.

Huntorf has shown good availability exceeding 98 percent at times and good reliability. The caverns have shown no detectable creep or stability problems.

In the United States, a 220-MW compressed-air energy-storage system is in the planning stages by Soyland Power Cooperative of Decatur, Illinois. It will have hard-rock reservoirs, with water-compensating surface reservoirs. Siting studies have been conducted by Batelle Pacific Northwest Laboratories. Brown Boveri has been awarded the contract for the electrical and mechanical equipment. Soyland hopes to have the plant in operation by 1985, and they estimate a savings due to energy storage of $34 million (in 1981 dollars) over the first 16 years of operation.

16-5 ENERGY STORAGE BY FLYWHEELS

Flywheels store off-peak energy as kinetic energy. They have been used extensively to smooth out power pulses from reciprocating engines. They are physically connected to the engine crankshafts and are larger the smaller the number of cylinders per engine. (They, for example, were very large for old single-cylinder steam engines.) They operate by storing some of the energy given by the cylinders and releasing it during periods of no power pulses so that the speed and power delivery of the crankshaft are steady and continuous. More recently interest in flywheel energy storage has been generated by motor vehicle designers. In the so-called hybrid automobile, for example, the flywheel stores some of the energy of the gasoline engine during periods of low vehicle demands and releases it during periods of high demands, such as during acceleration, hill climbing, etc., and thus operates the engine at a more steady and hence more efficient output.*

The use of flywheel energy storage by utilities was tried only a few years ago. In this the flywheel rotor is physically connected to a motor-generator set. In the charging mode, during off-peak periods, the motor adds energy to the flywheel. In the generation mode, during periods of peak demand, the flywheel rotor coasts driving the generator.

The fluctuations in speed caused by torque variations are reduced to a minimum by the use of flywheels. As kinetic energy is proportional to the mass times velocity squared, the changes in velocity from the addition or subtraction of kinetic energy are reduced by the use of a large mass. Conversely the energy stored in a flywheel can be increased by increasing the velocity. The velocity of a flywheel is defined as $2\pi Rn$. The energy stored in a flywheel is equal to the kinetic energy, given by

$$E = \frac{1}{2g_c} m(2\pi Rn)^2 = \frac{2\pi^2}{g_c} mR^2n^2 \tag{16-7}$$

where E = energy, ft · lb_f or J

m = mass of flywheel, lb_m or kg

g_c = conversion factor = 32.2 $\text{lb}_m \cdot \text{ft}/(\text{lb}_f \cdot \text{s}^2)$ = 1.0 $\text{kg}/(\text{N} \cdot \text{s}^2)$

* Other uses for flywheels include: a bus electrogyro, which is recharged at bus stops; regenerative braking of subway cars that provides acceleration upon start; electrically powered earth movers to limit peak-power demands; elevator drives including regenerative breaking; and plasma physics experiments to provide very high power peaks.

R = radius of gyration,* ft or m

n = revolutions per second = (r/min)/60

The energy E absorbed (or released) by a flywheel between speeds of rotation n_1 and n_2 is thus given by

$$\Delta E = \frac{2\pi^2}{g_c} mR^2(n_2^2 - n_1^2) \tag{16-8}$$

The ratio of the variation in rotational speed to the mean speed n is called the *coefficient of speed fluctuation* k_s, given by

$$k_s = \frac{n_2 - n_1}{n} = \frac{2(n_2 - n_1)}{n_1 + n_2} \tag{16-9}$$

where
$$n = \frac{n_1 + n_2}{2} \tag{16-10}$$

Combining Eqs. (16-8) through (16-10) results in

$$\Delta E = \frac{4\pi^2}{g_c} k_s mR^2n^2 \tag{16-11}$$

The value of the coefficient k_s depends upon the desired closeness of speed regulation. For engines, it may vary from 0.005 for fine to .2 for coarse regulation. Thus for a given energy absorption ΔE, m and/or R^2 must be high for close speed regulation.

Another important consideration in flywheel design is the stress level a flywheel rotating at very high speed is subjected to. The *theoretical maximum specific energy* (energy stored per unit mass) is dependent upon the stress-to-density ratio and is given by [157].

$$\left(\frac{E}{m}\right)_{\max} = 3.77 \times 10^{-7} k_m \frac{\sigma}{\rho} \tag{16-12}$$

where E/m = specific energy, kWh/ lb_m

k_m = *mass-efficiency factor*, dimensionless

σ = allowable stress, lb_f/ft^2

ρ = density, lb_m/ft^3

* The radius at which the total mass is considered to be concentrated. For a disc of uniform density ρ, uniform thickness t, and outer radius R_o, with r as a variable radius between 0 and R_o:

$$\int_0^{R_o} \frac{1}{2g_c}(2\pi r\, dr\, t\rho)(2\pi rn)^2 = \frac{1}{g_c}(\pi R_o^2 t\rho)(2\pi Rn)^2$$

from which $R = R_o/\sqrt{2} = 0.7071R_o$. Another use of the radius of gyration is in calculating the moment of inertia $I = R_m^2$.

k_m expresses how well a particular flywheel design utilizes the material strength, being a maximum if the stress is uniform throughout the flywheel. The maximum values of k_m are 1.0 in optimum designs with isotropic materials, where radial and tangential stresses are uniform and equal, and 0.5 in optimum designs with materials where only one stress direction can be utilized, such as fiber-reinforced composites. These two maximum values are obtained only in very slender designs where the energy absorbed per unit volume is minimal. In realistic designs k_m and the specific energy are reduced in order to reduce space, mass, and the cost of the safety shield and gas or vacuum chamber (below). The *volumetric specific energy,* energy per unit volume, is given by

$$\left(\frac{E}{V}\right)_{max} = 3.77 \times 10^{-7} k_v \sigma \qquad (16\text{-}13)$$

where E/V = volumetric specific energy, kWh/ft^3

k_v = *volume-efficiency ratio,* dimensionless

k_v expresses how well a particular flywheel design utilizes the material strength and fills the cylindrical volume around the flywheel. For a uniform-density material, k_v equals k_m times the fraction of that cylindrical volume occupied by the flywheel.

The principal parameters that determine the suitability of flywheels for energy storage are the two efficiency ratios k_m and k_v as well as the stress and density. The values of k_m and k_v depend upon the type of material (isotropic, uniaxial composite, variable density) as well as flywheel shape (disc, drum, rod). Figure 16-10 shows the relationship between k_m and k_v for high-performance flywheel designs.

The *strength-density ratio* σ/ρ is high for such materials as glass or silica fibers. However, inevitable manufacturing flaws, which also tend to grow because of stress corrosion, cause these high values to be realized only for short periods or at cryogenic temperatures. Cyclic operation expected in energy storage and release also causes fatigue and growth of small flaws and cracks and is expected to be strength-limiting for most materials.

Vibrational frequencies, coupled with high-cycle fatigue, are also expected to be strength-limiting, especially for slender designs such as thin discs or hoops. Thus the suitability of a design for energy storage depends on the design, on the material, and on the extent of manufacturing flaws and the methods for detecting and reinspecting them—in other words, the stringency of the quality control standards.

Materials for energy-storage flywheels must have high strengths, high strength-density ratio, high resistances to cyclic-crack growth, and high strength density-to-cost ratios. Those under consideration include some alloys, such as the so-called maraging steels, and more promising, composites such as fiber-reinforced plastics. One composite that shows particular promise is a 62 volume percent S-glass in epoxy composite. It has a density of 122.7 lb$_m$/ft^3, an estimated working stress of 21 $\times$ 10^6 lb$_f$/ft^2 for 10^4 cycles (16 $\times$ 10^6 for 10^5 cycles), a stress-density ratio of 170,000 (lb$_f$/ft^2)/(lb$_m$/ft^3) for 10^4 cycles (130,000 for 10^5 cycles), and a cost based on volume production of \$0.80/lb$_m$. Other attractive composites are graphite-epoxy and kelvar-epoxy [157]. By contrast maraging steel has a density of 500 lb$_f$/ft^3, a working stress

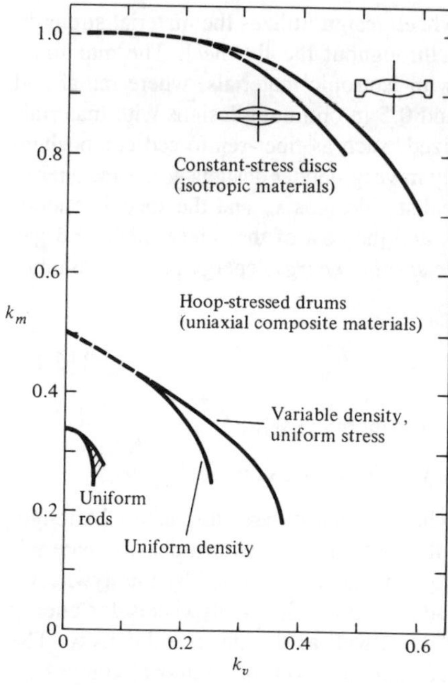

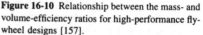

Figure 16-10 Relationship between the mass- and volume-efficiency ratios for high-performance flywheel designs [157].

of 14.5 × $10^6 lb_f/ft^2$ and a stress-density ratio of 29,000, all for 10^4 cycles, and a cost of $3/$lb_m$.

Flywheels for energy storage are systems that include, besides the flywheel itself, a number of subsystems. These are a housing; bearings, with ball bearings believed the most suitable; a vacuum pump to minimize windage losses inside the housing; seals to minimize oil and air leakage into the vacuum chamber; and sometimes a containment ring to protect nearby personnel and equipment from flying fragments in case of flywheel rotor fracture.

Losses in a flywheel energy-storage system include windage, bearing and seal friction, vacuum pump input power, and eddy current (hysteresis) and other inefficiencies in the motor generator (or in transmission systems). In early designs these losses were prohibitively large, and much developmental work still needs to be done to arrive at a technically attractive system.

A NASA conceptual design for a flywheel electric-energy storage system that has nearly zero losses is shown in Fig. 16-11. It is designed for a 10,000-kWh substation. The flywheel rotor, made of a filamentary anisotropic composite material to achieve high strength to mass ratio, has inner and outer diameters and height of 14, 20, and 10 ft (4.27, 6.1, and 3.05 m), respectively. It weighs 735,000 lb_m (333.4 metric tons) and rotates at 1250 r/min. To eliminate rotor-bearing friction, the rotor is magnetically levitated by permanent magnets that are incorporated in the stator with sets of teeth

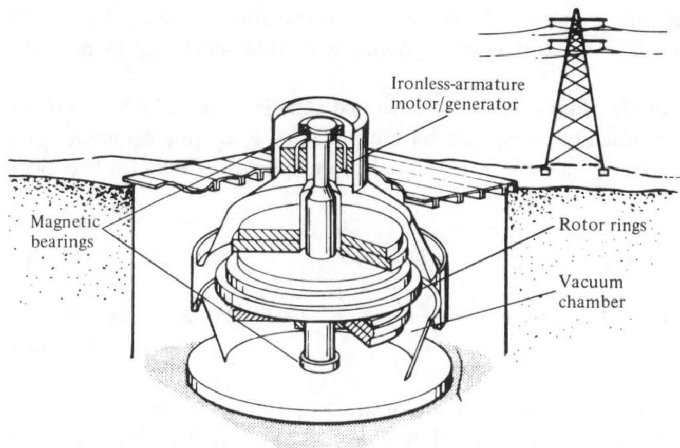

Figure 16-11 A NASA conceptual design of a 10-MWh-capacity flywheel energy storage system.

on both rims facing each other to maintain axial alignment. Alignment in the radial direction is provided by a set of electromagnets in shunt with the permanent magnets. Optical sensors situated along the air gap monitor any radial eccentricity and supply an error signal to the electromagnets to correct the eccentricity.

To eliminate losses from hysteresis and eddy currents, normally present in iron armatures, an ironless motor-generator armature is provided. A contactless electronic commutator is used to control current flow in the motor-generator coils. A permanent magnet in the stator produces a magnetic flux in the gap between rotor and stator. In the motor mode, alternating current is applied to the armature via the electronic commutation that interacts with the flux to produce the necessary force on the rotor. In the generator mode, commutation diodes are connected to the proper phases of the armature coil.

Windage losses are virtually eliminated by operating the flywheel in a sealed evacuated enclosure (vacuum chamber) pumped down to 10^{-3}- to 10^{-5}-torr (1.33×10^{-6} to 1.33×10^{-8} bar) pressure. A safety rotor ring is provided. In addition, the entire system is to be located below ground.

16-6 ELECTRICAL BATTERY STORAGE

The familiar *lead-acid battery* used in motor vehicles is a direct-current battery. It contains a number of voltaic cells (six in a 12-V battery, for example) that are connected in series. Each cell contains several lead plates, connected in parallel, made of grids that are filled with a spongy gray lead, Pb, and which form the anode. Alternating with these are plates of similar design but containing lead oxide, PbO_2, which form the cathode. All plates are immersed in a water solution of sulfuric acid H_2SO_4, which acts as an electrolyte. The electrolyte of each cell is housed separately in its compartment.

In the discharge mode, direct current is generated. The lead in the anode oxidizes to Pb^{2+} ions that immediately precipitate on the plates as lead sulfate $PbSO_4$. The lead oxide in the cathode reduces to Pb^{2+} that also precipitates as $PbSO_4$. The electrochemical reactions are

$$\text{Anode:} \qquad Pb(s) + SO_4^{2-}(aq) \rightarrow PbSO_4(s) + 2e^- \tag{16-14a}$$

$$\text{Cathode:} \qquad PbO_2(s) + 4H^+(aq) + SO_4^{2-}(aq) + 2e^- \rightarrow PbSO_4(s) + 2H_2O(aq) \tag{16-14b}$$

where (s) and (aq) indicate solid and liquid (aqueous) conditions, respectively.

Thus during discharge all plates are slowly covered by $PbSO_4$, which replaces the lead in the anode and the lead oxide in the cathode, and the concentration of sulfuric acid in the solution slowly decreases. For each molecule of Pb, two molecules of H_2SO_4 ($4H^+$ and $2SO_4^{2-}$) are replaced by two molecules of water ($2H_2O$). This is why the extent of charge of a lead-acid storage battery can be checked by measuring the density of the electrolyte; low density indicates low sulfuric acid concentration and a partially discharged cell.

In the charge mode, the battery can be restored to its original condition by reversing the direction of the current (and electron flow). The reactions represented by Eqs. (16-14) are simply reversed. The charging current must be obtained from a direct-current generator (as in older vehicles) or by an alternator equipped with a rectifier (as in modern vehicles).

The lead-acid battery, then, can be charged and discharged over many cycles and has been widely used to satisfy motor vehicle starter, instruments, lights, and other accessory requirements. Their use as the prime movers of all-electrical vehicles, where they would power drive motors and get charged while idle, as during the night, has run into several main drawbacks. These are unacceptable energy-mass and energy-cost ratios and low cycle life, in which they are almost fully discharged and recharged. These same drawbacks have prevented their use as energy-storage systems for large utility purposes.

The lead-acid battery is therefore limited to the small specialized uses it now enjoys for which the relatively low energy-mass ratio and the high cost of chemicals are not crucial factors.

Research and development has been going on for a number of years to develop advanced storage-battery systems that would have greater energy-mass ratios, lower costs, and greater cycle life. One of these is the *nickel-cadmium battery,* which uses a nickel hydroxide cathode, a cadmium anode, and a potassium-hydroxide-solution electrolyte. This battery is characterized by low mass and is primarily used in portable equipment such as radios and cordless appliances. Another is the *silver-zinc battery* which uses a solution of potassium hydroxide, saturated with zinc hydroxide, as an electrolyte. It has a high energy-mass ratio but also a high cost. It is primarily used in applications in which low mass is more important than cost. It also suffers from low cycle life, with only 30 to 300 charge-discharge cycles reported for some designs.

Battery systems that are potentially more suitable for utility applications, however, use soluble or liquid reactants and operate at temperatures other than atmospheric. The ones with the most promise at present are:

1. *Sodium-sulfur batteries*. These use molten sodium as one electrode, a sulfur and sodium sulfite mixture as the other, and a solid aluminum oxide electrolyte. They have a high energy-mass ratio and operate at temperatures of about 250°C (480°F). This relatively low temperature allows the use of Teflon seals and aluminum or glass containers. These batteries have a long cycle life because of the lack of solid-solid transformations. The low mass makes the sodium-sulfur battery a contender for the battery of the electrical car of the future.
2. *Lithium-chlorine* and *lithium-telluride batteries*. These are less developed than sodium-sulfur batteries but have similar favorable characteristics.
3. *Zinc-chlorine batteries*. Here a zinc chloride solution is pumped through graphite cells on which the zinc is deposited and the chloride is liberated in gaseous form that is drawn away, cooled in a heat exchanger, and stored in a separate tank. Chlorine, being toxic, is stored as a relatively harmless icy slush. In the generating mode, the chilled chlorine is pumped as a solution back to the cells where it reacts with the zinc to produce electricity. The Gulf and Western Company had originally developed this battery for the purpose of utility-type energy storage but has initially adapted it to a motor vehicle. (In one test, it ran a car for 175 mi at 42 mi/h). This battery has the advantage of constant electric output that does not drop off during discharge as is the case with most other batteries, for as long as there is chlorine in the storage tank, the battery system develops essentially constant power. The system is of course more complex than can be tolerated for personal automobiles because of the need for a complex charger, pumping, and heat-exchanger equipment.

The turnaround (charge-discharge) efficiencies of most batteries are good, about 70 to 80 percent, compared with some 60 percent for pumped hydro. However, it is clear that at this time high-energy, low-mass, long-cycle-life batteries of a type that would meet large utility peak-load demands are still not commercial realities.

16-7 SUPERCONDUCTING MAGNETIC ENERGY STORAGE

In 1911, the dependence of the electrical resistance of metals on temperature was observed by the Dutch physicist Kamerlingh Onnes. In his investigations, he also discovered that the electrical resistance of mercury dropped suddenly to zero when it was cooled to within a few degrees of absolute zero, a phenomenon which Onnes called *superconductivity*.

Other metals exhibit the same phenomenon. The temperature below which they become superconductive is called the *transition* or the *critical temperature*. All superconducting metals have transition temperatures in the *cryogenic* range*. The phenomenon arbitrarily has found many applications, and cryogenic engineering has become a specialized science as well as an industry all its own. Besides superconductivity, cryogenic applications can be found in medicine (cryogenic surgery, cryogenic ophthalmology, etc.), food preservation, pollution control, and other areas.

In 1970, the main application was the construction of superconducting electromagnets. These were experimentally used for magnetohydrodynamic power generation; bubble chambers to cool electrical generators, motors and transformers; and electric-power transmission and distribution. The latter application promises no-loss transmission. In the 1970s it was determined that it can best be accomplished by the use of high-purity aluminum cables operating at liquid hydrogen temperatures 20 K (−253°C, −423°F). Commerical success depends upon whether the metal, the gas, and the refrigeration systems can be acquired and operated economically.

Superconducting magnetic energy storage is a concept that initially received attention for pulsed energy storage in which the charge and discharge times have been short. It subsequently became apparent that such a concept is suitable for large-scale energy storage by an electric utility [159]. The concept is based on the principle that energy can be stored in the magnetic field associated with a coil. If the coil is made of a material in a superconducting state, i.e., maintained at a temperature below its critical temperature, then once it is charged, the current will not decay and the magnetic energy can be stored indefinitely. The stored energy can be released back to the network by discharging the coil.

The energy E stored in a coil in which a current I circulates is given by

$$E = \frac{1}{2} LI^2 \tag{16-15}$$

where E = energy, J (joule = watt × second)

L = inductance, H [henry = (volt × second)/ampere]

I = current, A (ampere)

* The term is derived from the Greek *kryo,* which means "icy cold or frost." Cryogenics is said to have been born in 1877 with the liquefaction of small quantities of oxygen at 90 K (−183°C, −298°F). In 1908, helium was first liquified at 4.2 K, though a commercial method for liquifying helium was not developed until 1947. Goddard used liquid oxygen to propel a rocket in 1926. Germany used it on a large scale to propel V-2 rockets during World War II, an application that brought cryogenics to the attention of the world. It is now widely used in the space program. Modern techniques can produce temperatures to within a minute fraction of a degree of absolute zero. The cryogenic range has been arbitrarily defined as extending from absolute zero, or 0 K (−273°C, −460°F), to an upper limit of 123 K (−150°C, −238°F), although most of the research has been done below the boiling point of oxygen, about 90 K. The range is considerably below normal refrigeration temperatures. Besides electrical resistance, other material, physical, and chemical properties such as strength, ductility, and thermal conductivity are also greatly affected when they reach cryogenic temperatures.

The inductance L of a coil is a function of its dimensions, which are characterized, for a coil with conductors of a rectangular cross section (Fig. 16-12) by

$$\xi = \frac{2R}{\sqrt{ab}} \tag{16-16}$$

$$\delta = \frac{a}{b} \tag{16-17}$$

and

$$V = 2\pi Rab = \frac{8\pi R^3}{\xi^2} \tag{16-18}$$

where R = mean radius of coil, m

a and b = width and depth of conductor, m

V = volume of conductor in one coil turn, m^3

and L is given by

$$L = f(\xi, \delta)RN^2 \tag{16-19}$$

where $f(\xi, \delta)$ = form function having the units (V · s)/(A · m)

N = number of turns of coil

The energy stored in a coil, Eq. (16-15), can now be written as

$$E = \frac{1}{2} f(\xi, \delta)RN^2I^2 \tag{16-20}$$

Using a current density j given by

$$j = \frac{NI}{ab} \tag{16-21}$$

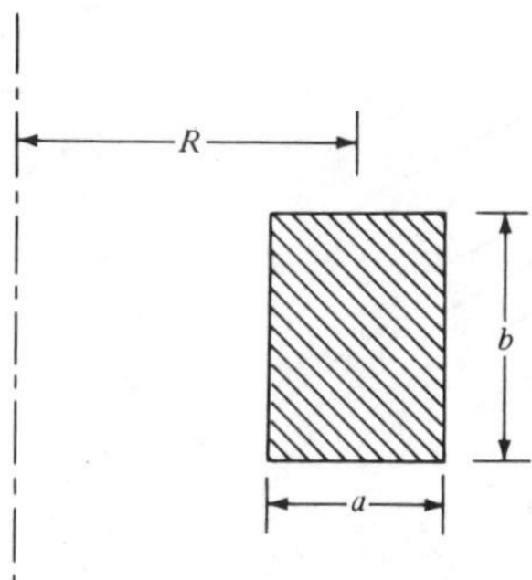

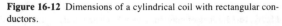

Figure 16-12 Dimensions of a cylindrical coil with rectangular conductors.

gives

$$E = \frac{1}{4}\pi^{-5/3}f(\xi,\delta)\xi^{-2/3}V^{5/3}j^2 \tag{16-22}$$

A coil that gives the maximum value of inductance to volume L/V is called a *Brooks coil* [160]. It has the dimensions

$$a = b \quad \text{and} \quad R = \frac{3}{2}b \tag{16-23}$$

so that for a Brooks coil, $\delta = 1$, $\xi = 3$, and the energy stored in it E_B is given by

$$E_B = 3.028 \times 10^{-8}\ V^{5/3}j^2 \tag{16-24}$$

For a cylindrical coil, other than Brooks, the energy E is given as a fraction E_B, Eq. (16-24), by a factor F less than 1.0

$$F = \frac{E}{E_B} \tag{16-25}$$

F is a function of ξ and δ of the particular coil and is given in Fig. 16-13.

An important parameter is the volume of material per unit energy stored. This is given by

$$\frac{V}{E} = \frac{V}{FE_B} = \frac{0.33 \times 10^8}{FV^{2/3}j^2} \tag{16-26}$$

where for a Brooks coil $F = 1.0$. Equation (16-26) demonstrates the economy of scale of coils as the cost of the coil is proportional to its volume and thus the cost per

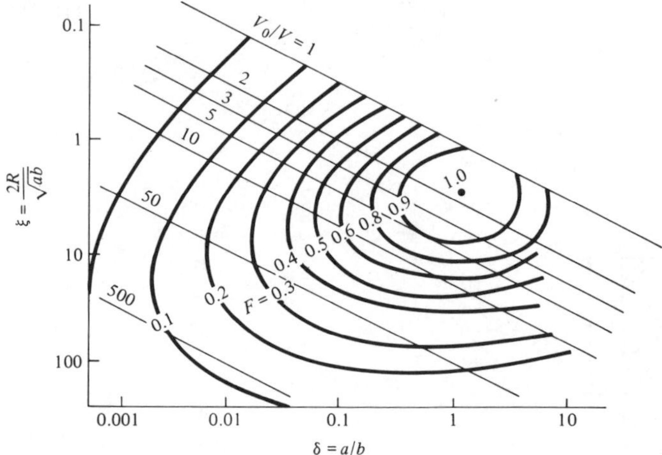

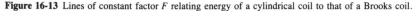

Figure 16-13 Lines of constant factor F relating energy of a cylindrical coil to that of a Brooks coil.

unit energy stored is inversely proportional to its volume to the power 2/3 and to the current density squared. The current density is, however, limited by stability considerations to values between 50×10^6 and 100×10^6 A/m^2.

The main mechanical design problem in magnetic energy storage arises because of the need of a very large structural mass to contain the magnetic field energy. This causes a large radial outward force from the solenoid. The mass is proportional to the material density and the stored energy, and is inversely proportional to the stresses. Such mass, if made of stainless steel, would amount to about 160 kg/kWh and result in unacceptable costs [161]. This consideration led to the selection of bedrock as the structural material with an excavated circular tunnel that would bear the radial outward force and transmit it to the surrounding bedrock.

Figure 16-14 shows an artist's sketch of a section of a proposed 5500-MWh magnetic storage system in which the superconductors are cooled in dewar by superfluid helium (He-II)* at 1 atm pressure and 1.8 K. The total structure is a low-aspect ratio cylinder that has an inner diameter of 1.57 km, a height of 15.7 m, and a thickness of 5 m. The solenoid has 112 turns and carries a 765,000-A current. The conductors are made of aluminum plus NbTi. The struts are made of fiberglass epoxy. The ripples in the conductor and dewar walls are designed to counteract excessive motion on cool down and to reduce magnetic tensile loads. The system has a total radial force of 3.1×10^{10} N, a total axial force of 3.1×10^{11} N, and an average radial pressure of about 4 bar.

The energy transfer between the three-phase ac current from the grid to the dc magnet is accomplished by an ac-dc power converter and inductor using a Graetz thyristor bridge circuitry. The turnaround efficiency (charge-discharge) of this circuitry is said to be greater than 95 percent.

Much of the activity of magnetic energy storage for utility use has taken place in the USA and Japan. A 1-MWh demonstration unit was planned to be built in Japan for the International Science Exposition to be held in Tsukuba Science and Education City in 1985. It would have 12 coils carrying 5000 A, an inductance of 288 H, and major and minor radii of 6.9 and 3.5 m. This plan has now been dropped in favor of a larger 10-MWh unit planned for later construction.

16-8 THERMAL SENSIBLE ENERGY STORAGE

Sensible energy storage is the first of the thermal-energy storage systems to be discussed. In general, thermal-energy storage systems can operate at many desired temperature levels depending upon use and choice of system and material, ranging from refrigeration temperatures to 1250°C. They have found wide use in many industrial applications, such as in the manufacture of cement, iron and steel, glass, aluminum,

* Helium becomes a liquid at 4.2 K and is called helium I. The isotope He^4 (99.99987 percent of all helium) behaves unusually below 2.17 K, the so-called lambda point. It becomes a superfluid, called helium II, that has a thermal conductivity three or four orders of magnitude better than copper at low temperature and surface heat-transfer characteristics about an order of magnitude better than He-I.

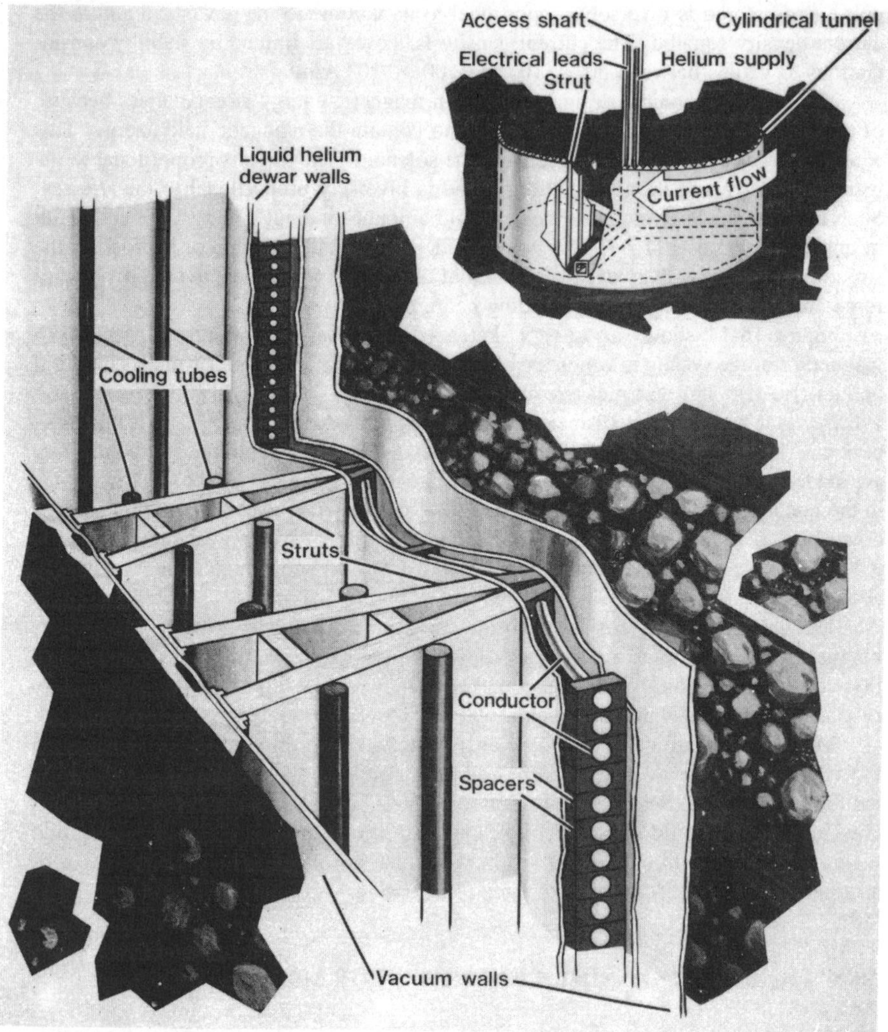

Figure 16-14 Artist's conception of a section of a proposed 5500-MWh magnetic energy storage unit. (*Courtesy of the University of Wisconsin Applied Superconductivity Center.*)

paper, plastics and rubber, and in food processing. Here we will deal with powerplant-related thermal-energy storage, in particular the Rankine-type steam-generator-turbine-condenser cycles.

Sensible energy storage is accomplished by raising the temperature of a material, such as water, an organic liquid, or a solid. The storage density, J/m^3 or Btu/ft^3, is equal to the product of the temperature difference, the specific heat, and the density of the material chosen. This system is simple in concept but has the disadvantages of variable temperature operation and relatively low storage density. Depending upon the coefficient of thermal expansion of the material, large volume changes may be encountered. Sensible energy storage could employ one of the following devices:

1. Pressurized-water storage
2. Organic liquid storage
3. Packed solid beds
4. Fluidized solid beds

Figure 16-15 shows an example of pressurized-water sensible energy storage [162] system in a powerplant in which the primary heat source is either a nuclear reactor or a fossil-fueled furnace. The base-loaded portion of the plant is capable of supplying more steam than needed during periods of low demand. The excess steam is bled at high pressure via turbine extraction (as in feedwater heating) during these periods of low demand. This extracted steam is fed to steel accumulators and mixed with water, thus producing saturated pressurized water. The accumulators are later discharged

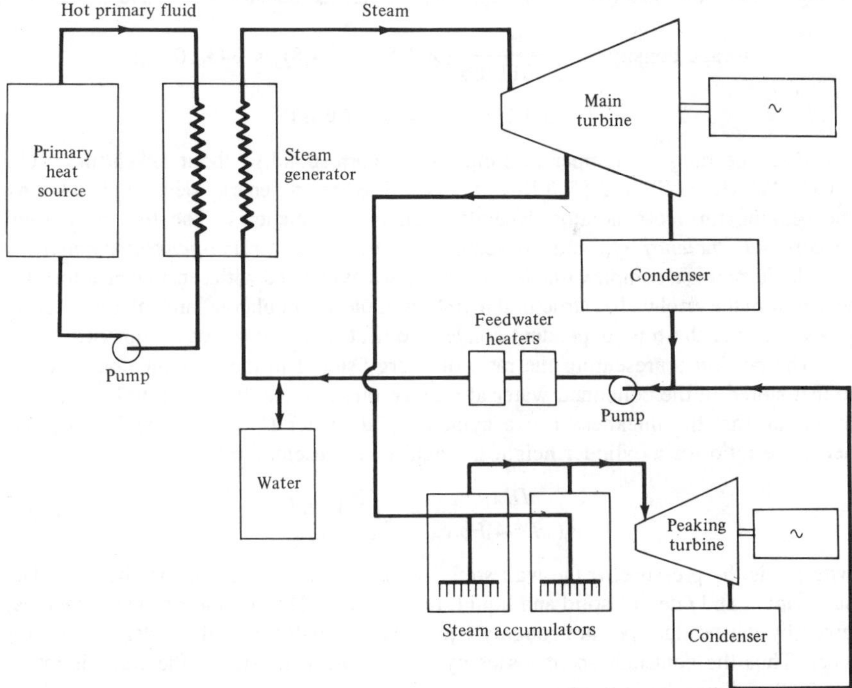

Figure 16-15 Schematic flow diagram of a powerplant with a pressurized-water sensible energy storage system.

through a small peaking turbine during periods of high demand. Discharge continues until a low specified pressure is reached in the accumulators. It has been seen that this results in low and varying steam temperature entering the peaking turbine. Typical values of accumulator high and low pressures are 20 bar, corresponding to a saturation temperature of about 212°C (414°F), and 2 bar, corresponding to a saturation temperature of about 120°C (248°F). Note that while this system involves steam condensing in water during accumulator charge and reevaporating during discharge, the storage medium is the pressurized water in the accumulators and operates over a relatively wide temperature range. Hence it is an example of sensible rather than latent energy storage (Sec. 16-9). The system is sometimes referred to by the misnomer "steam storage."

The storage density of the thermal energy utilized in the peaking turbine per unit volume of the higher-pressure saturated water is given by

$$\text{Storage density} = \frac{1}{v_{f,1}}(h_{f,1} - h_{f,2}) \tag{16-27}$$

where v_f and h_f are the specific volume and enthalpy of saturated water, respectively, and the subscripts 1 and 2 refer to the stored and emptied pressures, respectively. Using the steam tables (SI units), App. A, the storage density for 20 and 2 bars is:

$$\text{Storage density} = \frac{1}{0.0011766}(908.5 - 504.8) = 343{,}107 \text{ kJ/m}^3$$

$$= 95.3 \text{ kWh/m}^3 = 2.7 \text{ kWh/ft}^3$$

Over the range of temperature cited these correspond to about 1kWh/(m^3 · °C), 0.016 kWh/(ft^3 · °F), and 55.6 Btu/(ft^3 · °F). The electric-energy density obtained by the peaking turbine-generator depends upon two efficiencies. The first is *thermal turnaround efficiency* η_{ta}, and the second is the *peaking turbine-generator efficiency* η_p. The former is a complex function of the losses associated with sensible heat transfer to and from the steel walls, structural members of the accumulators, and interconnecting pipework, and the time-dependent convective heat losses to the environment.

The portion representing the ratio of energy stored in the accumulator structure to that stored in the contained water at a given pressure can be shown to be obtained by noting that the thickness t of a cylinder of diameter D is given by $DP/2\sigma$, and hence the ratio for a cylinder height L would be represented by

$$\frac{\pi D L t \rho_s c_s}{(\pi D^2/4) L \rho_f c_f} = 2\left(\frac{P}{\sigma}\right)\frac{\rho_s c_s}{\rho_f c_f} \tag{16-28}$$

where P is the pressure, σ the wall stress, ρ the density, and c the specific heat. The subscripts s and f denote solid and liquid, respectively. The volumetric heat capacities, given by the product ρc, are roughly equal. The ratio P/f is of the order of 0.03 for steel. Thus the contribution of losses by sensible heat transfer to the walls is rather minimal and may be ignored.

The convective heat losses from the water to the environment are therefore the major contributor to the thermal turnaround efficiency. They vary with time and depend upon the water temperature and the overall heat-transfer coefficient U between the water and the outside environment. The time constant τ of the system is given by the product of heat capacity and heat-flow resistance. Thus

$$\tau = \left(\frac{\pi D^2}{4}\,\rho_f c_f L\right)\left(\frac{1}{\pi DLU}\right)$$

$$= \frac{D\rho_f c_f}{4U} \tag{16-29}$$

If the temperature of the liquid at fully charged conditions is T_1, its time-dependent temperature of $T(\theta)$ decreases with time θ following fully charged conditions at (time zero) due to heat losses alone (no energy withdrawal) (Fig. 16-16). Assuming a lumped capacity system, common in transient heat-transfer calculations where the external heat-flow conductance is low compared with the internal heat-flow conductance (low Biot number), $T(\theta)$ is given by the familiar equation

$$\frac{T(\theta) - T_1}{T_\infty - T_1} = 1 - e^{-\theta/\tau} \tag{16-30}$$

where T_∞ is the environmental temperature. Assuming a storage period of θ_s at which the water temperature decreases to T_s

$$\frac{T_s - T_1}{T_\infty - T_1} = 1 - e^{-\theta_s/\tau} \tag{16-31}$$

The thermal turnaround efficiency is given by the energy left in storage at T_s after heat losses divided by the original energy stored. Thus

$$\eta_{ta} = \frac{h_s - h_2}{h_1 - h_2} \tag{16-32}$$

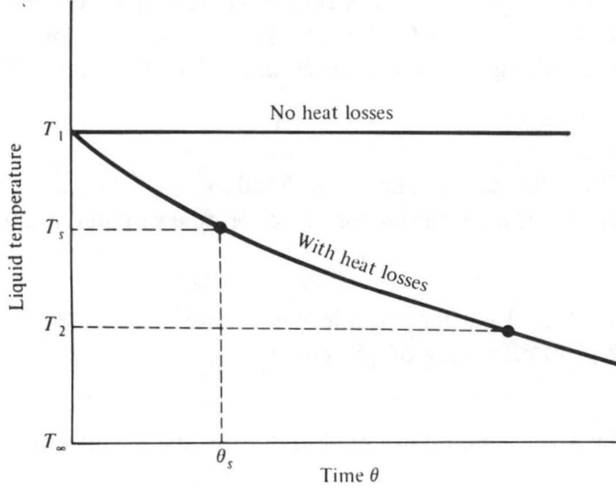

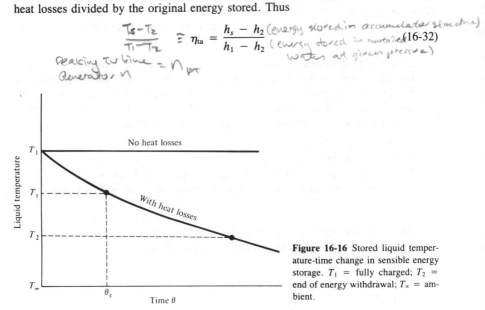

Figure 16-16 Stored liquid temperature-time change in sensible energy storage. T_1 = fully charged; T_2 = end of energy withdrawal; T_∞ = ambient.

where h_1, h_2, and h_s are the compressed liquid enthalpies at T_1, T_2, and T_s, respectively. Assuming, for simplicity, equal specific heats over the temperature ranges in question

$$\eta_{ta} = \frac{T_s - T_2}{T_1 - T_2} \tag{16-33}$$

Combining Eqs. (16-31) and (16-33)

$$\eta_{ta} = 1 - \frac{T_1 - T_\infty}{T_1 - T_2}(1 - e^{-\theta_s/\tau}) \tag{16-34}$$

It can be seen that this efficiency is a strong function of the ratio θ_s/τ. θ_s is several hours in a daily storage system. The only important variable in τ is U, Eq. (16-29), which depends heavily on accumulator insulations, design, and location. Good insulation and design result in low values of U and therefore high values of τ.

Accumulators may be constructed above ground or underground. Underground accumulators are more costly but have higher insulation and generally higher T_∞ and therefore higher values of η_{ta}. Accumulators in general pose safety problems. A pipe or vessel rupture can release large amounts of energy. The choice of above-ground or underground accumulators, therefore, should be based on the careful evaluation of efficiency, cost, operational problems, and safety. Natural underground caverns may be used if such are available near plant sites or if plants can be located near them, and if they can be properly prepared for high pressure hot water storage. They may result in much lower costs.

The efficiency of the peaking turbogenerator is probably low because of variable inlet conditions and the use of low-temperature saturated steam, small size, insufficiency or absence of feedwater heating, and other factors. An efficiency of 20 to 25 percent is probably reasonable, compared with 33 to 40 percent for the base-load plant.

Example 16-2 A base-loaded 1000-MW powerplant is designed with a sensible thermal-energy storage of the type shown in Fig. 16-15. The thermal energy stored is called upon to produce 4000 MWh daily. The accumulators are 4 m in diameter each and are well insulated so that $U = 5$ kJ/(h · m^2 · K). The storage time is 15 h. Maximum and minimum storage pressures are 20 and 2 bar. Calculate:

1. The thermal turnaround efficiency.
2. The total accumulator volume.
3. The storage cost in $/kWh if the accumulators cost $300/m^3.
4. The minimum released energy if an accumulator vessel or pipe ruptured at 20 bar.

The ambient temperature is 20°C. Take the specific heat of water as 4.35 kJ/(kg · K). Assume a peaking plant efficiency of 25 percent.

SOLUTION Following are the water-saturation conditions from the steam tables (App. A)

At 20 bar: $T_1 = 212.37°C \qquad h_{f,1} = 908.5$ kJ/kg $\qquad v_{f,1} = 0.0011766$ m^3/kg

At 2 bar: $T_2 = 120.23°C \qquad h_{f,2} = 504.8$ kJ/kg $\qquad v_{f,2} = 0.0010605$ m^3/kg

At ambient conditions: $T_\infty = 20°C \qquad h_{f,\infty} = 293$ kJ/kg

Take an average density of the water as

$$\frac{1}{2}\left(\frac{1}{v_{f,1}} + \frac{1}{v_{f,2}}\right) = 896.43 \text{ m}^3/\text{kg}$$

then

$$\tau = \frac{D\rho_f c_f}{4U} = \frac{4 \times 896.43 \times 4.38}{4 \times 5} = 785.3 \text{ h}$$

1. Therefore

$$\eta_{\text{ta}} = 1 - \frac{212.37 - 20}{212.37 - 120.23}(1 - e^{-15/785.3})$$

$$= 1 - 2.0878(1 - 0.981) = 0.96$$

For a peaking plant efficiency of 25 percent, the thermal storage needed is given by

$$\frac{4000}{0.25 \times 0.96} = 16{,}667 \text{ MWh}$$

$$16{,}667 \times 3.6 \times 10^6 = 6 \times 10^{10} \text{ kJ}$$

The mass of water needed to be flashed to steam is given by

$$\frac{6 \times 10^{10}}{h_{f,1} - h_{f,2}} = \frac{6 \times 10^{10}}{908.5 - 504.8} = 1.486 \times 10^8 \text{ kg}$$

2. The water volume at 20 bar is

$$1.486 \times 10^8 \times v_{f,1} = 1.486 \times 10^8 \times 0.0011766$$

$$= 174{,}840 \text{ m}^3$$

The volume of the accumulator will be larger to accommodate a steam blanket on top and the water left at end of discharge.

3. The minimum cost of accumulators would therefore be

$$\$300 \times 174{,}840{,}000 \approx \$52.5 \times 10^6$$

4. The energy contained that would be released if a rupture occurred is equal to the mass of the flashed water times the difference between the enthalpy of water at 20 bar at ambient conditions (20°C). The minimum corresponds to the above mass and equals

$$1.486 \times 10^8 \times 908.5 - 293 = 9.146 \times 10^{10} \text{ kJ}$$

16-9 LATENT HEAT ENERGY STORAGE

In this system energy is stored in the form of the latent heat caused by phase change, either by melting a solid or vaporizing a liquid. Energy release is accomplished by reversing the process, i.e., solidifying the liquid or condensing the vapor. The storage density here is equal to the product of the latent heat of fusion (or vaporization) times the density of the storage material. It is greater than that in sensible heat storage because the latent heats are much larger than the specific heats of the single phases of the materials. The system has the additional advantage of operating at essentially constant temperature with low volume changes during phase changes. It also has the advantage of a wide choice of materials with different fusion and evaporation temperatures, which allows a choice of operating temperatures and the ability to generate high-temperature steam for the peaking unit. Some sensible heat storage may be added to latent heat storage by further raising the temperature of the resulting molten solid or vapor.

Latent heat energy storage is not, at this writing, considered a simple, operationally reliable solution to the problem of energy storage in electric-generating powerplants. It is, however, included here as a potential solution along with the problems that must be overcome if such a solution is to become a viable one. Although little work has been done on the application of latent heat energy storage to large powerplants, much work has been done on its use for residential and solar heating applications using fused salts that are available for high- and low-temperature operating ranges [163].

Storage materials must possess, in addition to proper transition temperatures and high latent heat, many other necessary physical and chemical properties. Some of these are good thermal conductivity, containability, stability (considering cyclic operation), nontoxicity, and low cost. No material meets all these requirements but some fluoride salts meet some of them. One of the salts considered most suitable for latent heat storage is the 70% NaF–30% FeF_2 eutectic salt, which has a fusion temperature of about 680°C (2256°F) and potentially possesses the highest storage energy density of any thermal-energy storage material, about 1500 MJ/m^3 (~40,000 Btu/ft^3). $ZnCl_2$ is another with a fusion temperature of about 370°C (~700°F) and a potential storage energy density of about 400 MJ/m^3 (~11,000 Btu/ft^3).

Other materials being suggested are silicon, germanium, and sulfides of germanium. These have high heats of fusion, and like water, they expand upon freezing, so that they tend to float upon solidification, which has advantages in heat transfer. Silicon and germanium, however, have fusion temperatures that are too high for powerplant operation and are very reactive. Germanium sulfides have usable fusion temperatures but tend to solidify to a glassy consistency rather than crystallizing, thus posing an undesirable heat-transfer barrier.

In addition to finding a suitable medium, studies have to evaluate the extents of corrosion, erosion, plant start-up and shutdown, etc. Corrosion problems require that the system be free of oxygen and water vapor, which enter the system because of salt volume changes during heat addition and withdrawal, thus posing interesting engineering problems. It can be seen that there are many design and developmental problems to be overcome before a reliable, economical latent heat energy storage system can be incorporated into an electric-generating powerplant.

A latent heat energy storage conceptual design using the 70% NaF–30% FeF_2 eutectic salt as the storage medium has been proposed by Bundy [164]. The study had the purpose of identifying the technical difficulties and obtaining preliminary cost estimates. It envisaged a high-temperature gas-cooled reactor (HTGR, Sec. 10-12) as the heat source (Fig. 16-17). The helium coolant operates at 48 bar (700 psia), and the various temperatures are indicated on the figure. The storage-system capacity is 7200 MWh, whereas the charge-discharge rates are 600 MW. The peak electric-generating capacity of the plant is 200 MW for 12 h.

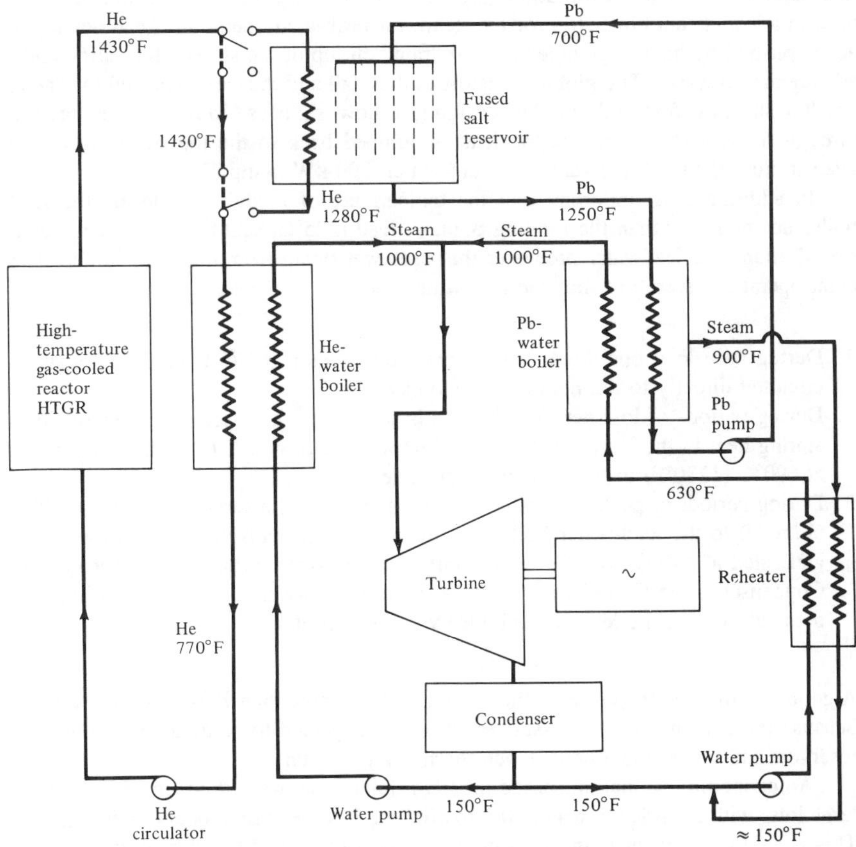

Figure 16-17 A schematic flow diagram of a high-temperature gas-cooled reactor (HTGR) powerplant using a latent heat energy storage system [164].

For the storage capacity of 7200 MWh, 42,000 tons of the eutectic are to be used. In order to give the necessary heat-absorption and heat-release rates, the eutectic is not allowed to freeze completely but instead operates as a slurry. (Recall the undesirability of glassy freezing as with the germanium sulfides.) The total latent heat of fusion is not utilized, and a salt mass of about 80,000 tons is needed. This requires a huge containment vessel about 36 m (120 ft) in diameter and 30 m (100 ft) high.

Another feature of the design is the addition of a secondary heat-transfer loop using molten lead as a heat carrier. Shell-and-tube heat exchangers between helium and the salt slurry, and steam and the salt slurry, have been considered. However, the build-up of solids on the tubes during heat withdrawal would seriously impede the heat-transfer rate and, therefore, the effectiveness of the heat exchanger. This necessitated the addition of the above-mentioned molten-lead loop. During heat withdrawal, therefore, lead is heated from 380°C (700°F) to 675°C (1250°F) in the slurry tank and used in a lead-steam boiler to produce steam for peak-load operation. Lead heating is accomplished by having globules of lead "rain" on top of the slurry, thus sufficiently stirring the reservoir. The globules distort as they fall and thus are expected to "shed" any thin skins of solidified salt. The design lead flow is 2 m^3/s (70 ft^3/s). After passing through the lead-steam boiler, the lead is pumped back to the top of the reservoir against more than 500 psi static pressure via an 800-kW pump.

In addition it is important that the temperature of feedwater entering the lead boiler not be lower than the melting point of lead (325°C, 620°F). This necessitated the addition of a feedwater preheater that uses wet steam from the same boiler. The plant operation, therefore, has the following modes:

1. During base-load operation, helium from the reactor at 775°C (1430°F) is short-circuited directly to the helium-water boiler.
2. During periods of low demand, helium is shunted to the fused salt reservoir, thus storing heat in the slurry at the fusion temperature of 680°C (1260°F) and leaving at 690°C (1280°F) to the helium-water boiler.
3. During periods of peak demand lead is circulated to the reservoir, leaving at 675°C (1250°F) to the lead-water boiler. In all cases steam with the proper flow rate is generated at 540°C (1000°F) and admitted to the same turbine-condenser system. Condensate at 65°C (150°F) is fed back to the helium-water boiler and, during peak demand, also to the lead-water boiler via the preheater.

A good turnaround efficiency for the storage system, more than 90 percent, is expected because the percentage heat losses per cycle are expected to be quite small once the reservoir and other components reach thermal equilibrium.

An economic evaluation, based on 1974 dollars, shows salt costing \$2.15/lb_m in large lots, with the price down to \$0.1 to \$0.5/lb_m if large scale production is attained. This is said to result in a unit energy storage cost of \$3 to \$13/kWh, a unit molten-lead loop cost of \$33/kW, and a unit peak powerplant equipment cost of \$90 to \$110/kW.

16-10 CHEMICAL-REACTION STORAGE

In this mode of energy storage the heat of reaction of reversible chemical reactions is used to store thermal energy during endothermic reactions and to release it during exothermic reactions. Like latent energy storage, this form also offers large energy storage densities and thus has been considered an attractive alternative for some time. (Besides reversible chemical reactions, this category also includes the solution and dissolution of a solid in a liquid and a gas in a solid.) Initial interest focused on low-temperature energy storage for residential heating and cooling. More recent interest contemplates its use to store high-temperature thermal energy suitable for power-generation cycles. In pioneering work by Schulten et al. [165], the following reaction was suggested for long-distance transmission of gas-cooled nuclear-reactor thermal energy

$$CO + 3H_2 \rightleftarrows CH_4 + H_2O \qquad (16\text{-}35)$$

This and other reversible reactions are listed in Table 16-1 with their operating ranges and heats of reaction.

Chemical-energy storage is now explained with the help of Eq. (16-35). Heat is stored by absorbing it in the endothermic direction of the reaction, from right to left. The enthalpy of formation [167] of $CH_4 + H_2O$ (liquid) at 25°C is given by $(-74.9) + (-286) = 360.9$ kJ/(g · mol). The enthalpy of formation of $CO + 3H_2$ at 25°C is $(-110.6) + 0 = -110.6$ kJ/(g · mol). Thus, moving from right to left with both reactants and products maintained at 25°C results in a net energy transfer of $(-110.6) - (-360.9) = +250.3$ kJ/(g · mol). The positive sign indicates energy added to the reaction, i.e., energy is absorbed and the reaction is endothermic. The reverse reaction, from left to right, results in -250.3 kJ/(g · mol), i.e., energy is

Table 16-1 Reversible chemical reactions under consideration for energy storage*

Reaction	Temperature range, K	Heat of reaction at 298 K: Btu/(lb · mol)	Heat of reaction at 298 K: kJ/(g · mol)
$CO + 3H_2 \rightleftarrows CH_4 + H_2O$	700–1200	107,640+	250.3†
$2CO + 2H_2 \rightleftarrows CH_4 + CO_2$	700–1200	106,380	247.4‡
$C_6H_6 + 3H_2 \rightleftarrows C_6H_{12}$	500–750	89,100	207.2
$C_7H_8 + 3H_2 \rightleftarrows C_7H_{14}$	450–700	91,800	213.5
$C_{10}H_8 + 5H_2 \rightleftarrows C_{10}H_{18}$	450–700	135,000	314.0
$C_2H_4 + HCl \rightleftarrows C_2H_5Cl$	420–770	24,120	56.1
$CO + Cl_2 \rightleftarrows COCl_2$		48,420	112.6

* From Ref. [166].

† Higher heating value, including heat of condensation of H_2O.

‡ Heat of reaction per g · mol of CH_4.

released and the reaction is exothermic. The signs have been ignored in Table 16-1. Although the exothermic reaction is common, it will not occur at low temperatures or in the absence of a catalyst or an "igniter," which leads to potentially long storage times. The endothermic reaction is called *reformation** and the exothermic reaction is called *methanation*.†

A schematic of a powerplant with a chemical storage system using the reaction in Eq. (16-35) is shown in Fig. 16-18. During periods of low demand, some heat from the primary heat source is diverted to the reformer (endothermic reactor) to convert the products $CH_4 + H_2O$ to the reactants $CO + 3H_2$, which are stored in a vessel at high pressure, probably 70 bar, but at ambient temperature. During periods of high demand, these reactants are fed to the methanator (exothermic reactor) where heat is generated to run a peak turbine (or generate more steam for the main turbine). In the methanator the reactants are converted to the products $CH_4 + H_2O$, which are stored in a separate vessel for later use in the reformer during periods of low demand.

A thermal turnaround efficiency of this system is estimated (but yet unproven) at 85 to 90 percent [168]. The losses are mainly heat losses to storage vessels and piping and pumping losses of the gases.

The two storage vessels and the two reactors will all have to operate at different pressures. Storage pressures need to be high to minimize vessel size and cost, and the reformer has to operate at low pressures to maximize the rate of the endothermic reaction $CH_4 + H_2O \rightarrow CO + H_2$. There will, therefore, be a large pressure differential between the tube side, which carries the primary heat source fluid (e.g. helium at about 40 bar from a high-temperature gas-cooled reactor), and the gases on the shell side. A similar pressure differential occurs between the steam loop and the gases in the methanator. Such pressure differences require careful design. A complicated scheme to minimize these difficulties envisages the use of compressor-expander sets with a compressor between the reformer and the reactants storage tank and an expander (turbine) between the latter and the methanator. A similar compressor-expander set would be put between the methanator, the products storage vessel, and the reformer, with equal complication. The products expander may supply some of the work required by the reformer compressor because it operates at about the same off-peak time during the cycle. Similarly the reactants expander may supply some of the work required by the methanator compressor during peak hours. Noting the various compressor-expander inefficiencies and the fact that the number of moles of reactants and products are not the same, there will be a net work available during peak operation that may be reabsorbed during off-peak operation. This imbalance in net work of the compressor-expander sets may be remedied by electrical or mechanical storage, and thus we end up with two (or three) forms of energy storage in the same plant.

Some of the problems to be solved before such a storage system can become a

* In general, *reformation* is a process in which low-grade or low-molecular-weight hydrocarbon is catalytically reformed to a higher-grade or higher-molecular-weight hydrocarbon. The term also applies to the endothermic reforming of methane (the process under consideration above) for the production of hydrogen by the reaction of methane and steam in the presence of nickel catalysts.

† *Methanation* is the production of methane from a mixture of carbon monoxide and hydrogen.

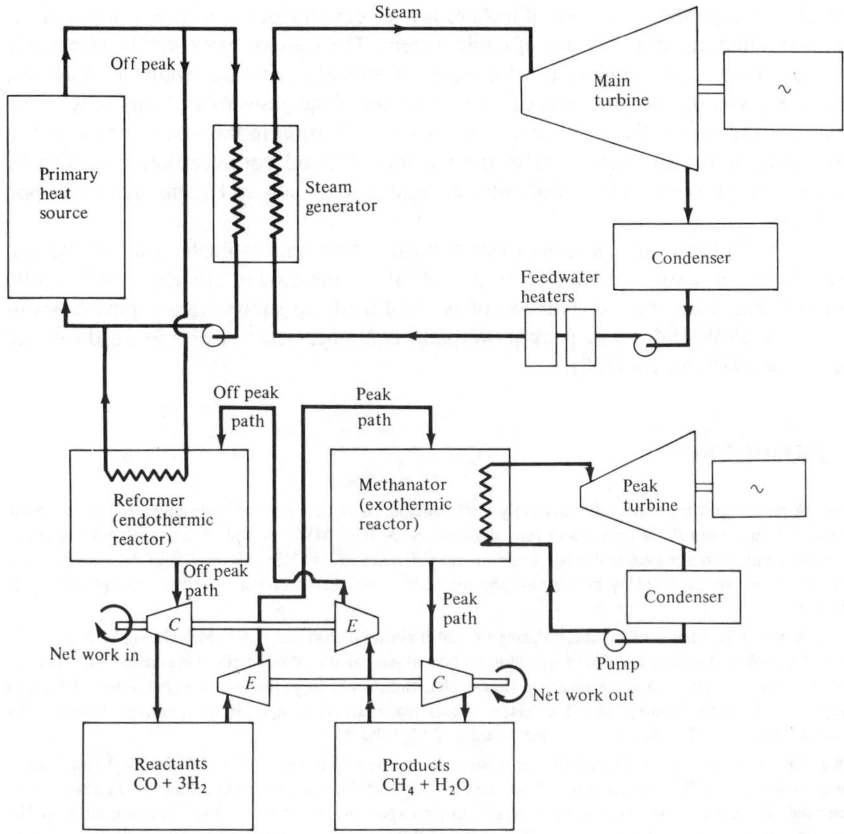

Figure 16-18 Schematic flow diagram of a powerplant with a chemical storage system using the reaction $CO + 3H_2 \rightleftharpoons CH_4 + H_2O$.

serious contender in powerplant use are safety and developmental problems. The major safety problems are those associated with the storage of large volumes of high-pressure flammable and poisonous gases. Developmental problems include the operation of methanators, which in the past were limited to relatively low temperatures to generate steam at high enough temperatures for efficient turbine operation. This entails the availability and use of a suitable catalyst that can operate at high temperatures with a reasonably long lifetime. Other significant problems involve the optimization of the entire cycle to reduce irreversibilities and increase overall efficiency and proper heat-exchanger design in the reformer and methanator. One idea that has been proposed involves the extraction of electrical work from the chemical reaction directly by electrochemical means.

The size of the storage vessels, even with operation near a pressure of 70 bar, which is compatible with current technology for gas storage, is still very large, of the order of hundreds of thousands of cubic meters. The vessels may either be steel tanks or, to reduce costs, underground caverns if suitable sites are available. Caverns, however, would pose problems of contamination of the gases by impurity gases from underground rocks. Such impurities may cause corrosion in the system and result in poisoning of the catalysts used in the reactors. Careful consideration must also be given to the problem of the diffusion of light gases, such as H_2, through rocks and fissures.

A preliminary, and possibly overoptimistic, economic estimate of the above system, based on cavern storage at \$0.5 to \$2/ft^3, a turnaround efficiency of 85 to 90 percent, and 70 percent conversion of original methane shows storage plant costs of \$35 to \$60/kW and a unit peak powerplant equipment cost of \$90 to \$110/kW, all based on 1975 dollars [168].

PROBLEMS

16-1 A power grid has a hypothetical daily load demand curve that may be represented by half a cosine wave with maximun P_m at time 0 and zero at 24 hours. A 1000-MW powerplant is available for baseload operation with a storage system that has a turnaround efficiency of 0.65. Find the peak load P_m, in megawatts, that can be accommodated by the system and the time at which the system switches from generation to storage.

16-2 A power grid has a load pattern during one 24-h period that averages 600 MW during 18 h and 1200 MW during 6 h. A pumped-hydro energy storage system with an elevation of 100 ft is considered. Calculate (*a*) the power output of a powerplant, in megawatts, that would meet the load demand with and without storage, and (*b*) the volume, in cubic feet, of water that must be pumped to meet storage demand. The electric generator efficiency is 0.8. Water density is 62.4 lb$_m$/ft^3.

16-3 An adiabatic compressed-air storage system is required to generate 230 MW during 5 peak hours, from storage during 10 h of operation of the main powerplant. No compressor intercooling or turbine reheat are used. The compressor and turbine have polytropic exponents of 1.5 and 1.3 and pressure ratios of 100 and 80, respectively; and mechanical efficiencies of 0.92 each. The motor-generator set has a combined mechanical-electrical efficiency of 0.97. Atmospheric air is at 1 bar and 20°C. Air enters the heat storage packed bed at 90 bar and 100°C. The packed bed has heat losses of 8 percent and air leakage losses of 1 percent. Calculate (*a*) the turbine air mass flow rate during peak operation, in kilograms per second, (*b*) the minimum volume of the air storage cavern, in cubic meters, (*c*) the energy storage capacity, in megawatt hours, (*d*) the compressor power input, in megawatts, and (*e*) the turnaround efficiency of the system. Assume a constant air specific heat at constant pressure of 1.05 kJ/kg · K.

16-4 A hybrid air-storage energy system is composed of an air compressor with two stages of intercooling, an aftercooler, a storage reservoir, and a gas turbine with one stage of reheat and regeneration. During 10 h of storage, 100 MW are fed to the motor generator, and air enters the compressor at 1 atm and 50°F. The compressor and turbine overall pressure ratios are 64. The compressor sections have polytropic and mechanical efficiencies of 0.7 and 0.8, respectively. Intercooling and aftercooling reduce the temperature back to ambient. During 4 h of peak generation, air leaves storage and enters a 0.85-effective regenerator after which it is heated in the first combustion chamber to 2000°F. The turbine sections have polytropic and mechanical efficiencies of 0.75 and 0.8, respectively. Reheat in the second combustion chamber is back to 2000°F. The motor-generator set has an efficiency of 0.90. Draw a flow diagram for the system and calculate (*a*) the air mass, in pounds mass, and volume, in cubic feet, stored during off peak, and (*b*) the turnaround efficiency of the system taking into account the equivalent electrical energy of the heat added in the turbine if the main plant thermal efficiency is 0.38. Assume that 1 percent of the air in storage is lost by leakage. For simplicity take c_p = 0.24 and 0.26 Btu/lb$_m$ · °R, and k = 1.40 and 1.358 in compressor and turbine, respectively.

16-5 A flywheel in the form of a disc 25 ft in diameter and 10 ft thick runs normally at 3600 r/min. It is made of an anisotropic filament composite material of a uniform density 160 lb$_m$/ft^3. Calculate (*a*) the energy of the flywheel, in kilowatt hours, (*b*) the change in rotational speed and corresponding maximum energy that may be extracted from the flywheel if the coefficient of speed fluctuation may not exceed 0.01.

16-6 A disc-shaped flywheel is made of an isotropic material with a density of 500 lb$_m$/ft^3 and maximum allowable stress of 15 × 10^6 lb$_f$/ft^2. It has a diameter of 15 ft and a thickness of 6 ft. The mass efficiency factor is 0.9. Calculate (*a*) the maximum energy that can be stored in the wheel, based on stress considerations, (*b*) the maximum rotational speed, in revolutions per minute, and (*c*) the coefficient of speed fluctuation if 2 percent of the energy is withdrawn from it.

16-7 A magnetic energy-storage coil is constructed of a conductor of square cross section 40 × 40 cm, and a mean diameter of 32 m. Calculate the number of turns necessary for a stored energy of 5 MWh if the current is 160,000 A.

16-8 Compressed-water thermal energy storage is used in a steam powerplant as shown in Fig. 16-15. Twelve accumulators are used, each 20 m in diameter and 40 m high. They have an overall coefficient of heat transfer of 0.003 kJ/s · m^2 · K. The daily storage time is 16 h. The maximum and minimum storage pressures are 18 and 3 bar. 80 percent of the total accumulator water volume is used to store and release energy to accommodate a steam blanket on top and water residue at bottom during the charge-discharge cycle. The main and peaking steam plant efficiencies are 38 and 23 percent, respectively. The ambient temperature is 10°C. Calculate (*a*) the thermal energy stored that is available for peaking service, in megawatt hours per day, (*b*) the thermal energy input, in megawatt hours per day, (*c*) the peaking plant electrical energy output, in megawatt hours per day, and (*d*) the ratio of peaking electrical energy to the electrical energy that would have been produced by the main plant.

16-9 A latent heat energy storage system is used in conjunction with a high-temperature gas cooled reactor (HTGR) gas turbine powerplant. The storage material is germanium which has a fusion temperature of 1756°F, a latent heat of fusion of 300 Btu/lb$_m$, and a liquid density of 335 lb$_m$/ft^3. The HTGR operates as a heat source for a regenerative helium Brayton cycle. 10^6 lb$_m$/h of helium enter the compressor at 160°F, a 0.85-effective regenerator, and leave the HTGR at 1850°F. During a charge period of 12 h, 50 percent of the helium enters the main turbine and 50 percent is diverted to the latent heat storage heat exchanger. From there, it leaves at 1790°F to enter the main gas turbine at a lower pressure stage. During 4 h of peak operation, the full HTGR flow enters the main turbine and a peaking Brayton cycle operates on the stored heat. Helium enters the peaking compressor at 140°F, then a 0.85-effective regenerator, a heat exchanger in the storage tank where it gets heated to 1720°F, then the peaking gas turbine. Assume for simplicity that all compressors and turbines are ideal with a pressure ratio of 3, and that there are no heat losses from the storage tank. Draw a flow diagram for the powerplant and *T-s* diagrams for the main and peaking cycles, labeling all points correspondingly. Calculate (*a*) the power produced during off peak operation, in megawatts, (*b*) the energy stored, in megawatt hours, (*c*) the minimum germanium mass, in pounds mass, and storage volume, in cubic feet, and (*d*) the power produced during peak operation, in megawatts. (For helium, c_p = 1.25 Btu/lb$_m$ · °R and k = 1.250.)

16-10 A chemical reaction energy storage system of the methanation-reforming type is considered for storing 2550 MWh of electrical energy. The system turnaround efficiency is estimated at 0.85. The main powerplant thermal efficiency is 0.38. Calculate (*a*) the individual masses, in kilograms, of reactant and product gases, (*b*) the necessary storage volumes, in cubic meters, assuming storage is at 70 bar and 500°C, (*c*) the maximum mechanical energy to be stored, in kilojoules, assuming for simplicity that pressure, temperature, and pressure drops are the same in all reactors and tanks, and that all expanders and compressors have an efficiency of 0.60.

CHAPTER

SEVENTEEN

ENVIRONMENTAL ASPECTS OF POWER GENERATION

17-1 INTRODUCTION

It should be recognized at the outset that there is virtually nothing people can do to improve their life standards, or even to maintain these standards in the face of growing populations, not to mention bringing up the standards of the billions of people living in substandard conditions around the world, that does not have adverse side effects, particularly on the environment. We can no longer expect the return of the pristine environment that existed in the early history of humankind. This is true whether people are producing increasing quantities of food and clothing, which have become energy-intensive; building dams and other irrigation schemes; or building powerplants to produce the electricity so necessary for industry and domestic uses and the maintenance of that life standard. The important question is whether or not the benefits of these necessary systems outweigh their adverse effects. Some sentiments have been expressed to the effect that the adverse effects of power generation can be arrested solely by conservation. Conservation means more efficient use of our resources and more efficient electric-energy production and use and is commendable. However, conservation alone can only reduce the rate of increase of environmental degradation. It is thus at least of equal importance to study the causes and effects of the adverse effects on the environment and to minimize them as much as possible.

Another factor to recognize is that nature itself contributes large quantities of contaminants to the environment, mainly due to the natural processes of plant and animal decay and natural background radioactivity.

At this point one needs to distinguish between the often used terms contaminant and pollutant. *Contaminants* are those materials, radiations, or thermal effects that are added to environment beyond what nature itself puts into it. In the 1960s it was estimated that, globally, nature puts into the environment some 10 times the amount of contaminants that people put into it. The contribution of nature is, however, diffuse and thus largely harmless, whereas the contribution by human beings is more localized and concentrated. It follows that *pollutants* are contaminants in concentrations high enough to adversely affect something that people value, such as their environment and health.

Table 17-1 presents a summary of chemical air pollutants from all sources in the United States in 1966. It is to be noted that motor vehicles are the largest contributors, followed by industry and powerplants, and then by small contributions from space heating and refuse. Not included in Table 17-1 is thermal pollution, the warming up of bodies of water and the atmosphere from the above sources as well as nuclear powerplants.

The case of radioactive pollution from nuclear powerplants is even less troublesome than the above sources when compared with the contributions of nature. Table 17-2 presents a summary of average annual exposures per person from all sources. Note that nature is the largest contributor, followed by medical irradiation, whereas releases attributed to the nuclear industry as a whole are relatively miniscule.

Even in abnormal occurrences the picture presented in Table 17-2 holds true. This can be illustrated by two such occurrences, one artificially made, the March 1979 accident at Three Mile Island (TMI-2) powerplant, the other naturally occurring, the May 1980 volcanic eruption of Mount St. Helens in the state of Washington, USA.

The radioactive release from Mount St. Helens was far more significant than that from TMI-2 (Sec. 17-15). It is reasonable to assume that volcanic eruptions and other such natural occurrences have released and will continue to release large amounts of radioactivity into the atmosphere.

Powerplants are, therefore, not the sole or largest contributors to environmental problems. They are, however, a growing concern, as their numbers and sizes will continue to increase in the decades ahead. The powerplant pollutants of most concern are:

Table 17-1 Total United States chemical air pollution (1966)*

	Pollutants	
Source	tons/year	%
Motor vehicles	86	60.6
Industry	23	16.8
Fossil powerplants	20	14.1
Space heating	8	5.6
Refuse disposal	5	3.5

* From Ref. 169.

Table 17-2 Average radioactive exposures per person*

Source	Exposure mrem/yr	Exposure %
Natural background	100.0	67.60
Medical irradiation	45.0	30.70
Fallout	0.9	0.60
Miscellaneous	0.7	0.50
Occupational	0.7	0.45
Nuclear industry	0.2	0.15

* From Ref. 170.

1. From fossil powerplants:
   a. sulfur oxide
   b. nitrogen oxides
   c. carbon oxides
   d. particulate matter
   e. thermal pollution
2. From nuclear powerplants:
   a. radioactivity release
   b. radioactive wastes
   c. thermal pollution

In addition, pollutants such as lead and hydrocarbons are contributed primarily by motor vehicles and will not be covered here.

This chapter is devoted to the discussion of these various aspects of electric-power generation and methods to minimize or alleviate their effects. The first part of the chapter deals with the effects of fossil-fuel powerplants, the latter part with those of nuclear powerplants.

17-2 CONSTITUENTS OF THE ATMOSPHERE

The effects of powerplant pollutants on the environment are primarily to the air and water and, to a lesser extent, the land.

The total mass of the earth's atmosphere is estimated at 5.7×10^{15} tons of various major and trace constituents. Table 17-3 shows approximate masses in tons of the major constituents, N_2 and O_2, as well as other constituents that are of main concern in the earth's atmosphere. The table lists the global (total atmosphere) tonnage as well as the tonnage in a portion of the atmosphere roughly lying between 30 to 60°N latitude and below 20,000 ft altitude. This portion of the atmosphere is less than 20 percent of the total, but it corresponds to the area of much of the world's industry and is therefore the receptacle for much of humankind's additions of contaminants to the

Table 17-3 Major and selected constituents of the atmosphere, tons*

Component	Tonnage: Global	Tonnage: Between 30–60° N latitude	Lifetime
N_2	4.25×10^{15}	0.55×10^{15}	Indefinite
O_2	1.30×10^{15}	0.17×10^{5}	Indefinite
CO_2	2.80×10^{12}	—	Years
CH_4	5.00×10^{9}	—	Years
CO	6.00×10^{8}	—	Years
NO_x	9.00×10^{6}	2.00×10^{6}	Days
SO_2	2.50×10^{6}	5.00×10^{5}	Days
Particulate matter	1.55×10^{8}	3.00×10^{7}	Days to years

* Data from Ref. 169.

global atmosphere. In addition, the rate of atmospheric mixing between the northern and southern hemisphere is slow, so that the region mentioned above may be considered isolated except for extended periods of time.

The atmosphere is recognized as a self-cleansing environment in which contaminants are continually added and removed. Thus, the amount of any one contaminant at any given time is affected by the mass rates of addition and removal of that contaminant. The concentration of that pollutant is affected by these rates as well as the size of the atmosphere itself. The addition of contaminants is due to both natural causes and artificial causes that are of fairly recent origin. The last column in Table 17-3 represents the average lifetime of each constituent. This is the average residence time of the constituent between addition and removal. The figures in this and the tables to follow in this chapter are only ball park figures because of the doubtful accuracy of measurements, so they should be viewed with caution. They are also bound to vary with time. The relative magnitudes are, however, informative.

We will now discuss the most important contaminants, their effects, and methods of coping with them.

17-3 OXIDES OF SULFUR

Sulfur in the atmosphere exists essentially in three forms: sulfur dioxide, SO_2; hydrogen sulfide, H_2S; and various sulfates. H_2S comes primarily from natural sources. The sulfates come from sea spray and from the oxidation of SO_2.

Sulfur dioxide, SO_2, which primarily comes from artificial causes, is therefore our primary concern, even though it is believed to be responsible for something less than 25 percent of all sulfur in the atmosphere. It is produced in the combustion of coal and oil, mainly in powerplants, but also in steel mills, smelters, and similar

Table 17-4 SO_2 flowchart in the 30° to 60°N latitude*

Additions	Natural (volcanoes)	Powerplants	Industry	Space heating	Solid waste	Motor vehicles
tons/year	Negligible	5×10^7	1×10^7	0.6×10^7	0.3×10^7	0.2×10^7

↓ ↓ ↓

Steady state	5×10^5 tons 0.2 ppb, average

↓ ↓ ↓

Removal	Precipitation Gravitation

* Data from Ref. 169.

industries, and space heaters (all also produce some H_2S), with coal responsible for about 70 percent of the total.

Although the mass of SO_2 in the 30 to 60°N latitude region of the atmosphere, which is about 5×10^5 tons (Table 17-3), represents an average of about 0.2 ppb (parts per billion) in that region, the yearly mass added, and removed, is much higher, being of the order of 10^8 tons (Table 17-4). Local SO_2 concentrations vary widely and are usually measured in ppm (parts per million). Urban industrial areas often reach 3.2 ppm, and peaks of 11 ppm have been recorded. SO_2 concentrations below 0.6 ppm produce no ill effects in human beings. Most people, however, become cognizant of sulfur at about 5 ppm and become irritated at about 10 ppm. A 1-h exposure to 10 ppm can cause breathing problems and mucus removal. The effects are further complicated at high air temperatures and humidities, in the presence of aerosols, etc., and they become much more serious when SO_2 is combined with particulates and enters the digestive system.

The effects of SO_2 on bodies of water and on land are discussed in Sec. 17-8.

17-4 OXIDES OF NITROGEN

The oxides of nitrogen come in different forms: nitric oxide, NO; nitrous oxide, N_2O; nitrogen dioxide, NO_2; nitrogen trioxide, NO_3; nitric anhydride, N_2O_5; and nitrous anhydride, N_2O_3. Of these only NO and NO_2 are the significant artificially made oxides; they are commonly referred to as NO_x, pronounced "nox."

Nitric oxide, NO, is formed in the combustion of all fossil fuels. The rate of formation is strongly dependent upon the combustion temperature, being significant only at high temperature. Formation is also dependent upon the oxygen concentration present during combustion and the time allowed for the combustion process. The primary contributors to NO are the motor vehicles, in which combustion does occur at high temperatures, with lesser contributions from the combustion of coal and oil in powerplants, which occurs at lower temperatures. Concentrations of NO in the exhaust of gasoline automobiles depends upon the fuel-air ratio used and vary between about 3000 ppm near stoichiometric conditions to about 6000 ppm for 80 percent lean mixtures.

In the atmosphere NO rapidly oxidizes to NO_2, a process greatly accelerated by photochemical effects (the presence of sunlight) and by the presence of organic material in the air. NO_2 has a more adverse health effect on human beings than NO. It has an affinity for hemoglobin, which carries oxygen to body tissues. It thus deprives them of oxygen. It also forms acid in the lungs and hence is much more toxic than CO (below) for the same concentrations. It also reduces atmospheric visibility. People begin to recognize the existence of NO_2 by its odor when it reaches concentrations of 0.4 ppm and higher. A continuous exposure to 0.06 to 0.1 ppm of NO_2, however, can cause respiratory illness. A few minutes exposure to 150 to 200 ppm causes obliterations of the bronchiols (the smallest divisions of a bronchial tube), and a few minutes exposure to 500 ppm causes acute edema (swelling from the effusion of a watery liquid into cellular tissue).

The atmosphere in the 30 to 60°N latitude region contains 2×10^6 tons of NO_x (Table 17-3), which coresponds to an average of 1 ppb. The additions, and removals, per year are 2×10^8 tons from natural causes, mainly biological, and 5×10^7 tons from artificial causes (Table 17-5). It should be noted here that the two NO_x compounds represent only a small fraction of the total nitrogen compounds entering and leaving the atmosphere. For example ammonia, NH_3, enters and leaves the atmosphere at a rate some 10 times that of NO_x.

While the average in the 30 to 60°N latitude region is about 1 ppb, local urban concentrations of NO_x can be much higher, ranging between 0.02 and 0.9 ppm. Temporary but severely high concentrations up to 3.9 ppm have been recorded in Los Angeles, which has a rate of addition 100 times the world average. (It is instructive to note that tobacco smoke can contain about 250 ppm.)

Although contributions of NO_x by artificial causes on a global scale are far less than those by natural events, local human contributions could be far larger and could

Table 17-5 NO_x flowchart in the 30 to 60°N latitude*

Additions	Natural biological reactions	Motor vehicles	Industry	Powerplants	Space heating	Solid waste
tons/year	20×10^7	2×10^7	2×10^7	0.8×10^7	0.2×10^7	0.2×10^7
		↓		↓	↓	
Steady state	2×10^6 tons, 1.0 ppb, average					
		↓		↓	↓	
Removal	As nitrates by precipitation, vegetation					

* Data from Ref. 169.

thus give rise to severe health and environmental problems. The effects of NO_x on bodies of water and land will also be covered in Sec. 17-8.

17-5 OXIDES OF CARBON

Carbon monoxide, CO, methane, CH_4 and carbon dioxide, CO_2, are the most widely produced of all contaminants, with CO_2 being the largest of all (Table 17-3).

Carbon monoxide is caused in part by natural causes such as marsh gas, coal mines, vegetation, lightning, and forest fires. However, their contribution to production of CO is small compared with that from human-generated causes, of which more than 90 percent is produced by motor vehicles (Table 17-6). Powerplants produce less than 1 percent. The removal of CO is almost entirely due to natural events whose exact nature is not known with certainty.

The CO additions to the entire earth's atmosphere total about 230 million tons. Local additions are often serious. For example, in the Los Angeles basin alone, some 4 million tons are added annually. The basin average CO concentration increased from about 7 ppm to about 11 ppm between 1957 and 1963 compared with the global 0.1 ppm. Emission-control devices are causing a slow decrease in these concentrations. People who smoke voluntarily expose themselves momentarily to concentrations said to reach 42,000 ppm.

The health hazard of CO, like NO, stems from its depriving body tissues of oxygen because of its affinity for hemoglobin, which carries oxygen to body tissues. A CO concentration of 100 ppm causes headache, 500 ppm causes collapse, and 1000 ppm is fatal.

Carbon dioxide, CO_2, is more abundant and, unlike CO, is largely contributed by powerplants and hence of more concern to us here. Natural causes, such as the decay of organic matter, however, contribute much greater amounts of CO_2 than artificial causes (Table 17-7). Although the CO_2 added to the atmosphere by these

Table 17-6 Carbon monoxide flow chart, entire atmosphere*

Additions	Natural events	Motor vehicles	Industry	Space heating	Powerplants	Solid waste
tons/year	0.2×10^8	2.0×10^8	0.04×10^8	0.04×10^8	0.02×10^8	0.02×10^8

↓ ↓ ↓

Steady state	6×10^8 tons, 0.1 ppm, average

↓ ↓ ↓

Removal	Oxidation to CO_2 in atmosphere, dissolution in water, consumption by vegetation, other unknowns

* Data from Ref. 169.

Table 17-7 Carbon dioxide flowchart, entire atmosphere*

	Natural causes			Artificial causes	
Additions	Organic decay	Volcanoes	Respiration	Combustion	Industry
tons/year	2×10^{11}	1×10^{10}	1×10^{9}	2×10^{10}	2×10^{6}
		↓	↓	↓	
Steady state	3×10^{12} tons, 325 ppm average				
		↓	↓	↓	
Removal	Photosynthesis (2×10^{11} tons/year), ocean absorption, rock weathering				

* From Ref. 169.

causes is considered a contaminant, it is not normally considered a pollutant because it is essential to plant and, therefore, human life. The main removal mechanism of CO_2 is *photosynthesis*. This is the process that occurs in green plants in the presence of light by which water, CO_2, and minerals are converted back to oxygen and various organic compounds.

Despite the necessity mentioned above of CO_2 in the atmosphere, there is increasing concern that the growing combustion of fossil fuels for all types of energy will cause the CO_2 concentration in the entire atmosphere to continue to increase with serious effects on the earth's climate as a result of the greenhouse effect.

17-6 THE GREENHOUSE EFFECT

The concentrations of CO_2 in the atmosphere have slowly but continually increased since 1880 with the evergrowing worldwide combustion of fossil fuels. Since the 1950s, the concentrations have risen at an increasing rate (Fig. 17-1). The addition of CO_2 in the latter part of the decade of the 1970s is believed by some to have reached a level that is beyond the capacity of plant life and the oceans to completely remove it, and about half of the CO_2 added is retained in the atmosphere.

The existence of CO_2 in the atmosphere causes a *greenhouse effect*. The atmosphere, analogous to the glass panes of a greenhouse, transmits the radiation from the sun. Because the surface of the sun is at about 6000 K, most of that radiation is in the shortwave and visible portion of the spectrum, and only a small portion of this radiation is absorbed or scattered back to space by the atmosphere. The transmitted radiation is largely absorbed by the surface of the earth, which is warmed by it.

Part of the resulting heat of the earth is transmitted by various modes of heat transfer (conduction, convection, evaporation, and condensation, which has a negative contribution) to the atmosphere and is ultimately reradiated away from the surface. Becaue the surface temperatures are low, this reradiation is mostly in the infrared portion of the spectrum. CO_2 has no emission and absorption bands in the short and

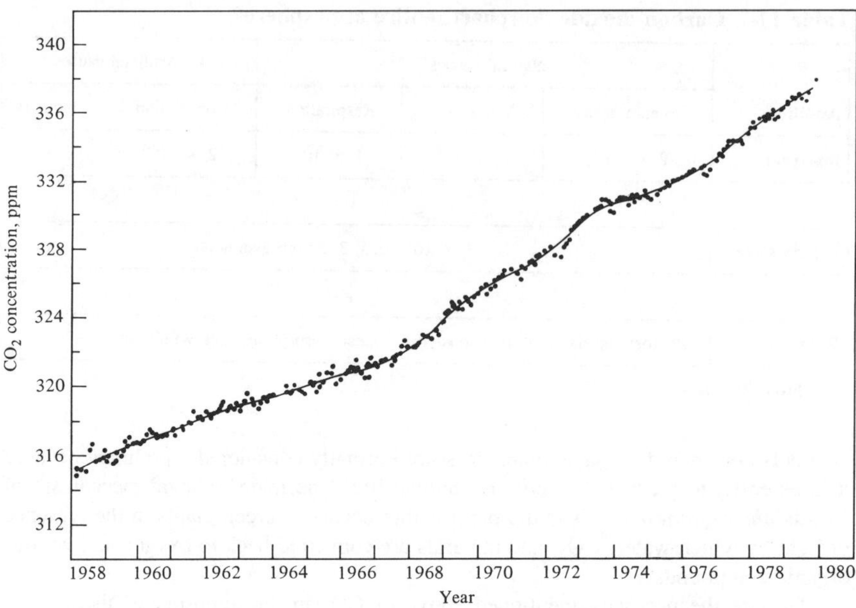

Figure 17-1 Average CO_2 concentrations in the atmosphere up to 1980 with seasonal effects removed. Seasonal fluctuations about average: slightly lower in spring due to absorption by new plants; slightly higher in fall due to release from decaying plants. *(Based on work by C. D. Keeling et al., from Ref. 171.)*

visible wavelengths of the spectrum but does have several infrared bands between the wavelengths 2.36 to 3.02 μm, 4.01 to 4.08 μm, and 12.5 to 16.5 μm. Water vapor, likewise, has only infrared emission and absorption bands in the wavelengths 2.24 to 3.27 μm, 4.8 to 8.5 μm, and 12 to 25 μm. When both CO_2 and H_2O are present, the absorptions are, to a good approximation, additive.

The presence of CO_2 and H_2O in the atmosphere results in absorption of a large portion of the longwave infrared radiations from the surface of the earth and partial radiation of them back to it. In essence, therefore, the atmosphere is not completely transparent to the reradiated energy and, like the panes of a greenhouse, traps much of the energy of the sun. Whereas an equilibrium of sorts existed over the centuries, the growing levels of concentration of CO_2 are expected to cause the earth and its lower atmosphere to warm up to higher temperatures than would otherwise be the case. Indeed, measurements have shown a small warming up already (Fig. 17-2).

It is estimated that if all the fossil-fuel reserves are burned, the current CO_2 concentration in the atmosphere would increase 5- to 10-fold. It is feared, however, that long before this ultimate situation happens, climatic changes might occur that would result in disasterous consequences, such as melting of the polar ice caps, which would result in the raising of sea level and flooding of many coastal areas of the world. Some evidence that this process has already started has been cited.

Somewhat counterbalancing the increasing effect of CO_2 on the atmosphere is the

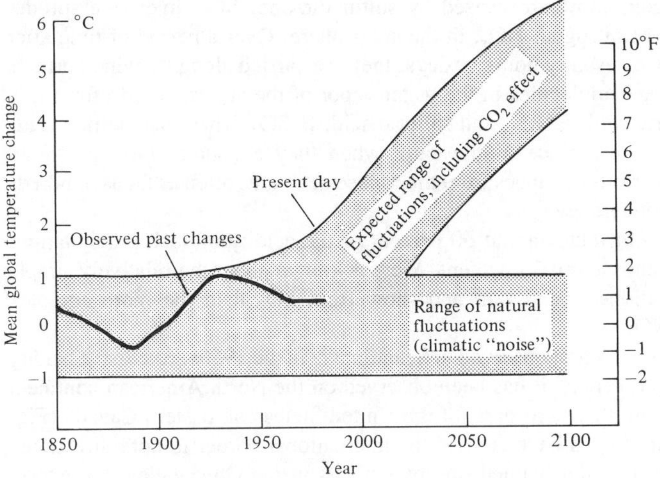

Figure 17-2 Changes in mean global temperature with and without projected CO_2 effects. *(Based on work by J. M. Mitchell, from Ref. 171.)*

increasing content of particulate matter. Particulate matter helps reflect some of the incident radiation from the sun back to space and thus prevents it from reaching the surface of the earth.

17-7 ACID PRECIPITATION

Acid precipitation is the return to earth of the oxides of sulfur and nitrogen in acid form. They take several forms: acid rain and acid snow, dry deposition, and acid fog.

Acid Rain and Acid Snow

Acid fallout in the form of *acid rain* or *acid snow* is one of the more serious environmental hazards of increased concentrations of sulfur and nitrogen oxides in the atmosphere. Pure water has a neutral pH value of 7.0.* The average pH of normal rainfall is a slightly acidic 5.6 because of the formation of mild carbonic acid H_2CO_3 when pure rainwater combines with natural and human-generated carbon dioxide in the atmosphere; it is considered harmless. However, biologists usually consider rainfall with pH values lower than 5.6 to be potentially harmful to plant, animal, and human life. In many parts of the world, however, rainfall has grown increasingly more acidic in recent years. pH values as low as 2.4, about the same acidity as vinegar, have been recorded in England and in widely separated regions of the world.

* *Acidity* and *alkalinity* are measured on a pH scale of 0 to 14. pH = 7.0 denotes neutrality. $0 \leq \text{pH} < 7$ represents acidity. $7 < \text{pH} \leq 14$ represents alkalinity. The strongest acids have pH values below 1.0.

Acid rain and acid snow are caused by sulfur dioxide, SO_2, hydrogen sulfide, H_2S, and the oxides of nitrogen, NO_x, in the atmosphere. Over a period of time after these gases are emitted, usually hours or days, they are carried along by wind currents and combine with water molecules in the water vapor of the atmosphere to form tiny drops, mainly of nitric acid, HNO_3, and sulfuric acid, H_2SO_4. These aerosollike acids are then returned to the surface of the earth when they encounter rain- or snow-producing clouds at various distances from their place of origin, often as far as hundreds or thousands of kilometers away.

In general SO_2 contributes about 60 percent of the acidity of such precipitation, whereas NO_x contributes about 35 percent. More is known about the relatively simple mechanism of formation of acid rain and snow from SO_2 than the more complex mechanism due to NO_x.

Acid precipitation was first observed in Scandinavia in the 1950s. Excessive acidity is a more recent phenomenon. It has been observed on the North American continent in such places as the northeastern parts of the United States and eastern Canada (Fig. 17-3), where the toxic fumes have crossed the international border in both directions, although their source is often blamed on powerplants in the Ohio valley that export them on prevailing winds. Other affected areas in the United States are the northwestern region, the southern Appalacians, and portions of Florida. In Europe, industrialized countries such as Britain, Germany, and France produce millions of tons of these gases that affect many other countries, some as far away as Scandinavia. Acid rain has also been found in parts of Asia and other world spots.

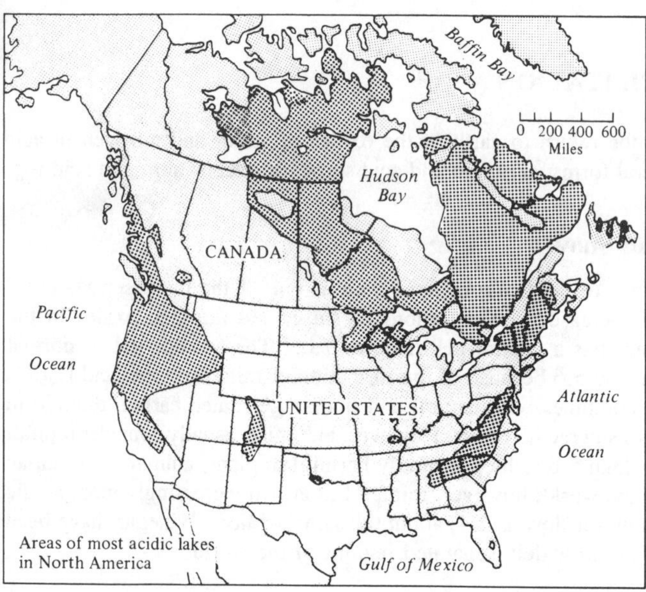

Figure 17-3 Areas of most acidic lakes in North America [172].

This unwelcome traffic across national and international boundaries has prompted multigovernmental conferences to discuss the causes, sources, and extent of acid rain, the outlook for the future, and if possible, to find corrective measures. Treaties between nations defining responsibilities and outlining joint actions may have to be drawn. Expressions of concern were voiced at a 1979 international conference in Toronto by Canadian officials who stated that acid rain is their "most serious environmental problem"; by U.S. officials who announced that "too much damage has already been done by acid rain—too many trout lakes and salmon streams have already been rendered lethal to fish, and valuable wilderness areas are beginning to show signs of acidification"; and by others.

The effect of acid on streams and lakes has resulted in the lowering of pH values of some of them to values betwen 2.8 and 5.0. This acidification causes the decimation or malformation of fish, particularly trout and salmon. It has been estimated that 50,000 lakes in the Adirondack Mountains and Ontario, Canada, have become acidified to the extent that fish population has declined or ceased to exist. High levels of acidity are also said to cause the dissolving of toxic metals such as cadmium, lead, and aluminum out of minerals in the bottom sediments. These toxic metals, like the acids, can destroy fish and also contaminate drinking water.

The effect of acid rain on soil can cause the leaching of essential plant nutrients from the soil and reduce nitrogen fixation by microorganisms, which would make the soil less fertile. It can also dissolve aluminum and cadmium out of soil minerals, as mentioned above, and thus allow them to enter roots and kill trees. A recent (1982) study on the effects of acidity on forests at the University of Vermont showed that on Camel's Hump, a 4100-ft peak in the Green Mountains, half the red spruce trees have died. In West Germany, acid rain is blamed for the death of about 3700 acres of evergreen trees in Bavaria.

Dry Deposition

A particularly lethal form of acidity is called *dry deposition*. This occurs when sulfate particles fall on tree leaves before mixing with water. This dry fallout is said to account for about half the total acid fallout. When these particles mix with surface or rain water, they become concentrated sulfuric acid.

Attempts to alleviate the problem so far (1982) have met with little success. The most serious of these attempts involves dispersal and detoxification of the emissions that cause the acid rain in the first place. Other plans involve the breeding of new plant or animal species that would be acid-resistant. Probably the most effective method is the use of smokestack scrubbers (Sec. 17-9) that remove much of the sulfur from powerplants and industrial furnaces, though they have no effect on nitrogen oxides. They, however, are used only in a few large installations and are expensive to install and operate. Ironically, the dispersal of the emissions by the use of tall smokestacks, thought to be the answer at one time (the solution of pollution is dilution), is effective only in removing particulates and exhaust locally. It has actually encouraged the formation of acid rain by more efficiently sending the emissions to those air currents that transport them to rain clouds, often far away from their point of origin.

Detoxification of lakes by the addition of lime has proven counterproductive because lime has been found to combine with heavy metals in the bottom sediment and to free them, actually increasing the toxicity of water.

The idea of breeding of acid-resistant fish is a proposition that is not taken very seriously by most zoologists.

Acid Fog

A recently noted major acid pollutant is *acid fog*. Its origin is the same as acid rain or snow, i.e., sulfuric and nitric oxides from powerplants and, to a lesser extent, motor vehicles. It forms by the mixing of these pollutants with water vapor near the ground. The acid vapors then begin to condense around very tiny particles of fog or smog, pick up more water vapor from the humid air, and turn into acid fog. When the water in the fog burns off (evaporates) due to the sun or other causes, drops of nearly pure sulfuric acid are left behind. It is these drops that make acid fog so acidic. In Los Angeles and Bakersfield in southern California, the mists have a pH of 3.0 compared with 4 or 4.5 for acid rain. Acid fog 100 times as acidic as acid rain has been detected. Cases have been reported where people had trouble breathing when it was foggy. The problems of fog are now believed by some to be more serious than those of smog in these areas.

Many researchers consider the effects of acid precipitation, especially the changing of soil chemistry, to be irreversible and fear its long-range effects. Monitoring programs of air, soil, and water are being instituted to ascertain these long-range effects. However, the uncertainty about the real extent of the problem adds to the prevailing disquiet regarding it.

17-8 PARTICULATE MATTER

Particulate matter in the atmosphere is composed of smoke, dust, and other solids made of a wide variety of organics and metals. Samples taken in urban areas of the United States between 1960 and 1965 showed average and maximum particulate concentrations in the atmosphere of about 100 and 1250 $\mu g/m^3$, respectively. The Los Angeles basin, by now shown to be a particularly bad case, is estimated to put into the atmosphere some 170,000 tons/year of particulate matter, including that due to SO_2, about 30 times the mass of the atmosphere over that basin. During the London atmospheric pollution crisis of 1962 the particulate concentration rose to 2000 $\mu g/m^3$. There is evidence that air-circulation patterns over industrialized areas cause self-contained dust domes that aggrevate matters over these areas (Fig. 17-4). Nonurban areas in the United States averaged 28 $\mu g/m^3$.

The total particulate matter in the 30 to 60°N latitude region of the atmosphere is about 3×10^7 tons, averaging about 30 $\mu g/m^3$ (Table 17-8). Both natural and human activities seem equally to blame for the presence of particulate matter in the atmosphere. Natural causes include natural dust caused by wind, storms, volcanoes and natural fires, metoritic dust, and fog. Fog, although not strictly a pollutant, con-

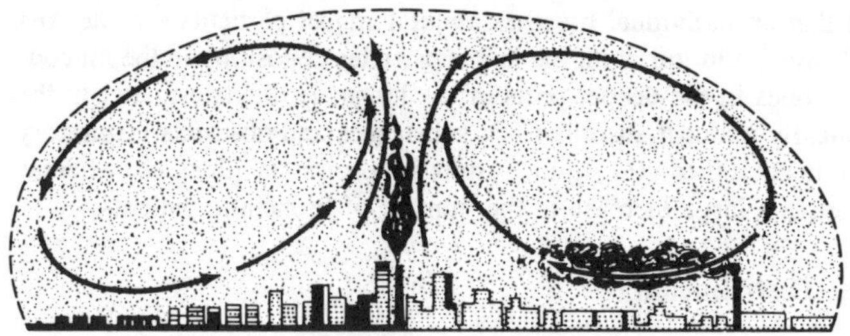

Figure 17-4 Air-circulation pattern over a large industrial urban area, creating a self-contained dust dome.

tributes to undesirable climatic conditions, especially when combined with smoke (smog). Of the human causes, dust and ash that emanate from industrial processes, fossil powerplants, and other combustion processes are the largest contributors, of which powerplants contribute about one-third. Sulfur compounds are larger contributors to particulate matter. This comes about by the SO_2 in the atmosphere (Sec. 17-3) oxidizing to sulfur trioxide, SO_3, which forms H_2SO_4 (sulfuric acid) mist, which in turn reacts with other materials in the air to form, among other things, ammonium and calcium sulfates.

The effects of particulate matter in the atmosphere are many and varied. Besides the obvious effects of decreasing visibility and increasing soiling and corrosion, and the already mentioned effects on climatic conditions, there is a health hazard that is a complex function of concentration and particle size. The size distribution is given by the usual log mean normal distribution curve (below). Numerically, most particles have a diameter below 2 μm, with a numerical average about 1.27 μm. The larger particles, however, although fewer in number, represent a greater mass fraction of the

Table 17-8 Flowchart of particulates in the 30 to 60°N latitude*

	Natural causes			Artificial causes				
Additions	Natural	Meteors	Fog	Industrial	Space heating	Motor vehicles	Solid waste	Sulphates
Tons/year	6×10^7	1×10^3	3×10^7	3×10^7	2×10^6	2×10^6	2×10^6	6×10^7
	↓			↓		↓		
Steady state	3×10^7 tons, 30 μg/m^3 average							
	↓			↓		↓		
Removal	Gravitation, impaction, drop nucleation, washing							

* From Ref. 169.

total. It is estimated that an individual breathes about 1 mg of particulate matter per day during times of heavy pollution, with the larger particles depositing in the mucous lining and the smaller ones in the deeper parts of the lungs. Particulate matter in the atmosphere is intrinsically toxic; it absorbs toxic substances and obstructs respiratory passages. An annual mean 100 to 200 $\mu g/m^3$ results in respiratory illnesses, whereas 300 to 600 $\mu g/m^3$ causes a large increase in the number of bronchitic patients.

Particulate-Matter Distribution

A *Gaussian* or *normal distribution* of a statistical quantity, such as the diameter of particles in a large sample, is one in which the numbers of particles of given diameters fall on a bell-shaped curve, symmetrical about a mean diameter, when plotted against the diameter on a linear scale. It is given by

$$N(d) = \frac{1}{\sqrt{2\pi}S} e^{-\frac{1}{2}\left(\frac{d - d_n}{S}\right)^2} \tag{17-1}$$

where $N(d)$ = number of particles with diameters between d and $d + dd$

d = particle diameter

d_n = mean particle diameter

S = standard deviation of d, having the same dimensions as d and given by

$$S = \left[\frac{\Sigma\left(d_i - \frac{\Sigma d_i}{N}\right)^2}{N - 1}\right]^{0.5} \tag{17-2}$$

where N = total number of particles

Most particle samples from emissions, however, have size distributions that follow a *log normal distribution*. This has the same symmetrical bell-shaped curve as the Gaussian distribution except that the variable is the *logarithm* of the diameter rather than the diameter itself. It is thus given by

$$N(\omega) = \frac{1}{\sqrt{2\pi}S_\omega} e^{-\frac{1}{2}\left(\frac{\omega - \omega_n}{S}\right)^2} \tag{17-3}$$

where $\omega = \ln d$

ω_n = mean value of ω or $\ln d$

$N(\omega)$ = number of particles per unit interval betweeen $\ln d$ and $[\ln d + d(\ln d)]$

S_ω = standard deviation of ω

Figure 17-5a is a plot of the fraction $N(\omega)/N$ showing the usual Gaussian curve but with a logarithmic abscissa scale for d. Figure 17-5b shows a log normal plot of the fraction $N(\omega)/N$ with a linear abscissa scale for d.

The notations on Fig. 17-5b have the following meanings.

Count modal diameter The diameter at which the greatest number of particles occurs, represented by the maximum point on the distribution curve

Count or number median diameter The diameter for which 50 percent of all particles or $N/2$ are larger and 50 percent or $N/2$ are smaller by count

Count mean diameter The arithmetic mean diameter of all particles present, e.g., the sum of diameters of all particles present divided by the total numbers of particles N

Area median diameter The diameter for which the surface area of all particles larger than it constitutes 50 percent of the total surface area

Area mean diameter The diameter of a particle that has a surface area equal to the arithmetic mean of the surface areas of all particles

Mass median diameter The diameter for which the mass of all particles with diameters larger than it constitutes 50 percent of the total mass

Mass mean diameter The diameter of a particle that has a mass equal to the arithmetic mean of the masses of all particles

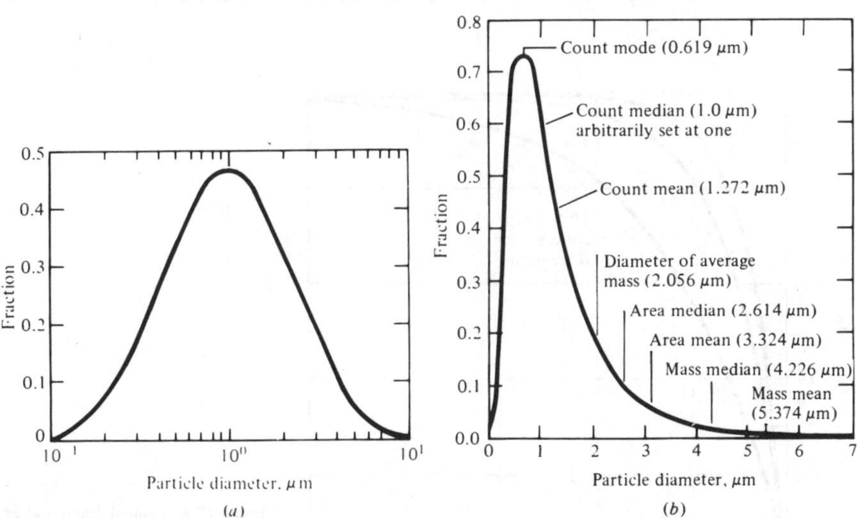

Figure 17-5 Log normal distribution of particle diameters plotted (a) on a semilog scale and (b) on a linear scale.

Devices that are used to clean up gas streams from the particulate matter they carry usually do better with larger particle diameters. They are characterized by (1) a fractional, or grade, efficiency and (2) an overall collection efficiency.

The *fractional,* or *grade, efficiency* is a collection efficiency, fraction, or percent of mass removed, for a given particle diameter. It is rather poor for the smallest sizes, increases rapidly with size, and becomes nearly 100 percent for the largest sizes. Figure 17-6 shows typical fractional efficiency curves for collection equipment. The *overall collection efficiency,* on the other hand, is the fraction or percent of mass removed of all particles or above a specified minimum diameter. It is to be noted that although liquid drops in suspension are usually spherical, solid particles are not. Thus the exact curve depends on what type of "diameter" representations are selected, e.g., the diameter of a spherical particle having the same surface area as the nonspherical particle, or the same volume, the Stoke's diameter, and others [173].

One must then be careful in evaluating the effectiveness of collection devices, which are usually characterized by the overall collection efficiency often above a given minimum diameter. Becaue efficiency is a mass ratio, and given the nature of the size distribution, a good collection efficiency might well mean a good mass removal but a poor numeric removal, as a very large number of small particles escape the collection device.

The majority of particulate matter in the atmosphere usually ranges in size (diameter) from 0.1 to 10 μm, although some are as small as 0.001 μm and as large as 500 μm. Below 0.1 μm, the particles behave like molecules and attain random motions characteristic of collisions with gas molecules. Between 1 and 20 μm the particles tend to be carried along with the air in which they are borne. Above 20 μm the particles

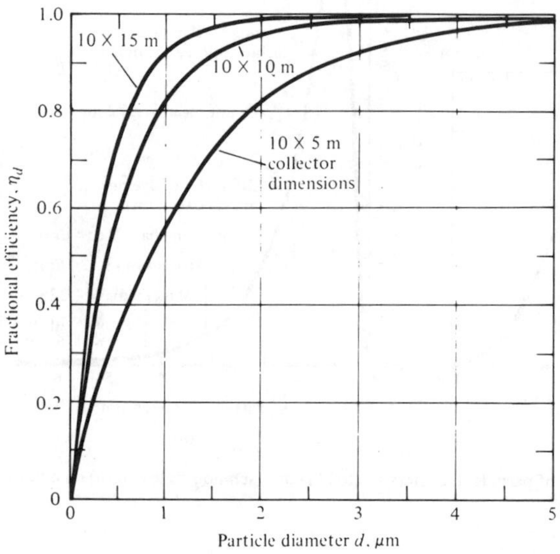

Figure 17-6 Typical fractional efficiency curve for collection equipment. Data plotted using the Deutsch equation for the data of Example 17-1.

attain settling rates that rapidly increase with size (0.3 cm/s for 10 μm, 30 cm/s for 100 μm), and they are, therefore, airborne for relatively short periods of time. The following sizes are of interest:

Particulate matter	*Size,* μm	*Particulate matter*	*Size,* μm
Natural dust	Above 1	Tobacco smoke	0.01–1
Liquid mist	Below 10	Oil smoke	0.03–1
Liquid sprays	Above 10	Coal dust	1–100
Smog	Below 2	Fly ash	1–200
Natural rain	500–10,000	Pulverized coal	3–500

Systems that help solve air quality-control problems associated with fossil-fuel steam-generator exhausts primarily deal with particulate collection and SO_2 emission cleanup. There are four basic types of systems:

1. Flue-gas desulfurization (FGD) (Sec. 17-9):
   a. Wet-type scrubbers
   b. Dry-type scrubbers
2. Particulate collection:
   a. Electrostatic precipitators (Sec. 17-10)
   b. Fabric filters (Sec. 17-11)

17-9 FLUE-GAS DESULFURIZATION (FGD) SYSTEMS

Gas desulfurization can be accomplished by wet, dry, oı alkali scrubbing. These methods are covered in this section.

The Wet Flue-Gas Desulfurization System

The wet FGD system, also called a *wet scrubber,* is commonly based on low-cost lime-limestone* in the form of an aqueous slurry. This slurry, brought into intimate contact with the flue gas by various technique, absorbs the SO_2 in it.

The wet scrubbing process was originally developed in the 1930s by Imperial Chemical Industries (ICI) in England. In the modern version of the process, the flue gas is scrubbed with a slurry that contains lime (CaO) and limestone ($CaCO_3$) as well as the salts calcium sulfite ($CaSO_3 \cdot 2H_2O$) and calcium sulfate (in hydrate form, or natural gypsum, $CaSO_4 \cdot 2H_2O$). The SO_2 in the flue gas reacts with the slurry to form additional sulfite and sulfate salts, which are recycled with the addition of fresh lime or limestone. The chemical reactions are not known with certainty but are thought to be

* Lime is calcium oxide, CaO, a white caustic solid, also known as burnt lime, quicklime, and caustic lime. Limestone is a rock composed almost entirely of calcium carbonate, $CaCO_3$, from which building stones and lime are made. When crystallized by heat and pressure, it becomes marble.

$$\left.\begin{aligned} CaO + H_2O &\rightarrow Ca(OH)_2 \\ Ca(OH)_2 + CO_2 &\rightarrow CaCO_3 + H_2O \\ CaCO_3 + CO_2 + H_2O &\rightarrow Ca(HCO_3)_2 \\ Ca(HCO_3)_2 + SO_2 + H_2O &\rightarrow CaSO_3 \cdot 2H_2O \downarrow + 2CO_2 \\ CaSO_3 \cdot 2H_2O + \frac{1}{2}O_2 &\rightarrow CaSO_4 \cdot 2H_2O \downarrow \end{aligned}\right\} \qquad (17\text{-}4)$$

One technique employs a *spray tower* downstream of the particulate-removal system (electrostatic precipitator or fabric filter) (Fig. 17-7). The flue gas is drawn into the spray tower by the main steam-generator induced-draft fan where it flows in countercurrent fashion to the limestone-slurry spray. A mist eliminator at the upper exit of the tower removes any spray droplets entrained by the gas. The gas may have to be slightly reheated before it enters the stack to improve atmospheric dispersion.

The sprayed limestone slurry collects in the bottom of the tower and is recirculated back to the spray nozzles by a pump. A system of feed and bleed charges a fresh slurry, under pH control, and discharges an equivalent amount from the circulating slurry. The fresh slurry is prepared by mixing the lime-limestone with water in a "slaker-grinder" and stirred in a slurry tank. The bled slurry is sent to a dewatering system, which is in the form of thickeners and filters or centrifuges, where water is removed from the calcium-sulfur salts. The reclaimed water is used to help make fresh slurry.

The wet scrubber has the advantages of high SO_2 removal efficiencies, good reliability, and low flue-gas energy requirements. In addition, it is capable of removing from the flue gases residual particulates that might have escaped the particulate-removal system.

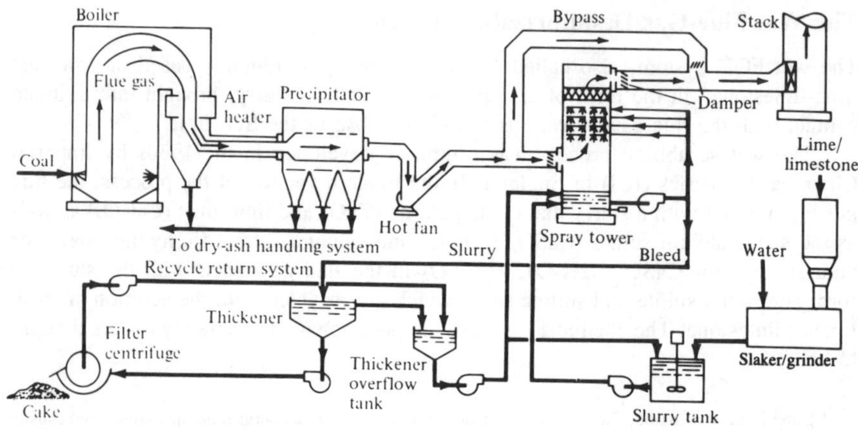

Figure 17-7 Schematic of a wet flue-gas desulfurization (FGD) system with a spray tower (*Electric Forum*.)

A main disadvantage is the buildup of scale in the spray tower and the possibility of plugging. The prevention of such scale is essential to the reliable operation of the tower. Scaling occurs because both calcium sulfite and calcium sulfate have low water solubility, normally around 30 percent, and can therefore form supersaturated water solutions. A minimum liquid-to-gas ratio must therefore be used, its value depending upon the SO_2 content of the flue gas and the expected extent of sulfite oxidation. Precipitation occurs at a finite rate, which necessitates holding the SO_2–absorbing liquor in a delay tank after each pass. An insufficient delay time increases supersaturation and promotes scaling. Another technique for controlling scale is the use of *seed crystals*. These are calcium sulfite and sulfate precipitate crystals, in a supersaturated solution, that are maintained in the SO_2–absorbing liquor. They provide sites around which preferential precipitation takes place and enhance the precipitation rate.

Other disadvantages of the wet scrubber are the reheating of the flue gas, a larger gas pressure drop requiring higher fan power requirements than the dry FGD system (below), and typically higher capital and operating costs.

The waste material from wet scrubbers is a water-logged sludge that poses difficult and costly disposal problems.

The Dry Flue-Gas Desulfurization System

Like the wet scrubber, above, the dry FGD system, also called a *dry scrubber,* utilizes an aqueous slurry of lime, CaO, to capture flue gas SO_2 by forming calcium sulfites and sulfates in spray absorbers (Fig. 17-8). The slurry in this case, however, is atomized, usually by a centrifugal atomizer, into a fine spray that promotes the chemical absorption of SO_2 and, because of the small spray particle size, is quickly dried by the hot flue gases themselves to a particulate suspension that is carried along with the desulfurized gas stream. The reaction particulates as well as those carried by the flue gases (fly ash) are then removed, mainly by a fabric filter (Sec. 17-11), before the gas is drawn by the induced-draft fan to the stack.

A major component of this system is the slurry-generating system. A "slaker" meters lime and water into an agitated tank to prepare a slaked lime slurry which, in turn, is diluted by additional water and processed to remove inert impurities called grits, which are disposed of. The lime slurry is pumped to the spray absorber with the flow controlled by the amount of SO_2 in the flue gas.

Particulates both coming in with the flue gas and generated in the FGD are collected from the absorber and fabric-filter hoppers and sent to a recycling silo for disposal or for recycling of a portion of it with the slurry (depending upon the extent of original utilization of the reactant in the absorber). The recycled slurry is enriched by an alkaline material, such as CaO, MgO, K_2O, or Na_2O.

The main advantages of the dry system are the dry, powdery nature of the waste material, which poses fewer and less costly disposal problems than the wet waste from the wet FGD system (thought these problems are still large), and the mechanical simplicity of the system.

The main disadvantage is that the efficiency of SO_2 removal is lower than that of the wet scrubber. 1979 NSPS (New Source Performance Standards) regulations, which

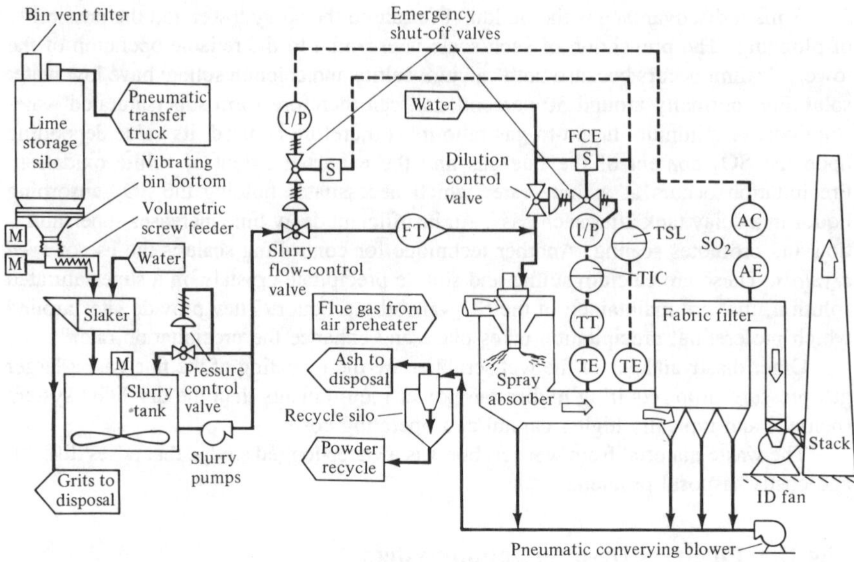

Figure 17-8 Schematic of a dry flue-gas desulfurization (FGD) system (*Electric Forum.*)

specify only 70 percent SO_2 removal in new plants, have encouraged the development of the dry system, however. Other disadvantages are the need for careful design optimization of the spray absorber and the slaker, and the strong dependence of collection efficiency on absorber outlet temperature, which necessitates operating as close as is safe to the saturation temperature that corresponds to the partial pressure of the water vapor in the gas in order to avoid condensation (below the corresponding dew point). This poses problems with filter-bag performance.

Single Alkali Scrubbing

Clear water solutions of either sodium (usually in the form of sodium hydroxide, NaOH, or sodium sulfite, Na_2SO_3) or ammonia (NH_3) are excellent absorbers of SO_2. The advantage of alkali scrubbing is that it avoids the scaling and plugging problems of slurry scrubbing by using alkaline earth. Ammonia scrubbing has the advantage that the scrubber product, ammonium sulfate, can be sold as a fertilizer, but the disadvantage that the process produces troublesome fumes.

A well-developed sodium scrubber is the *Wellman-Lord SO_2 recovery process*, which has found use in powerplants, refineries, sulfuric acid plants, and other industrial installations in the USA and Japan. The process utilizes a water solution of sodium sulfite (Na_2SO_3) for scrubbing and generates a concentrated SO_2 (about 90%), in effect removing the SO_2 gas from other flue gases.

The flue gas from fossil powerplants (or nonferrous smelters) is first pretreated by cooling and removal of particulate matter, such as by electrostatic precipitators (Sec. 17-10), prior to being sent to the absorber (Fig. 17-9). In the absorber the water solution of sodium sulfite absorbs the SO_2 in the pretreated flue gas to produce sodium bisulfite $NaHSO_3$ according to

$$SO_2 + Na_2SO_3 + H_2O \rightarrow 2NaHSO_3 \tag{17-5}$$

The desulfurized gas is reheated before going to the stack in order to improve atmospheric dispersion.

The sodium bisulfite is sent to a forced-circulation evaporator-crystallizer via a surge tank. The evaporator-crystallizer is the heart of the system. The surge tank allows steady flow rates into it despite gas flow and concentration fluctuations. Through the application of low-pressure steam, such as from a turbine exhaust, the sulfite is regenerated in the form of a slurry according to

$$2NaHSO_3 \rightarrow Na_2SO_3 \downarrow + SO_2 \uparrow + H_2O \tag{17-6}$$

The H_2O is separated from the SO_2 in a condenser and recycled to a dissolving tank where the sulfite slurry is redissolved and sent back to the absorber via a solution surge tank that has the same function as the one mentioned above.

A small amount of the circulating solution oxidizes to nonregenerable sodium sulfate crystals that must be disposed of. This necessitates purging a small stream of solution and adding fresh sulfate.

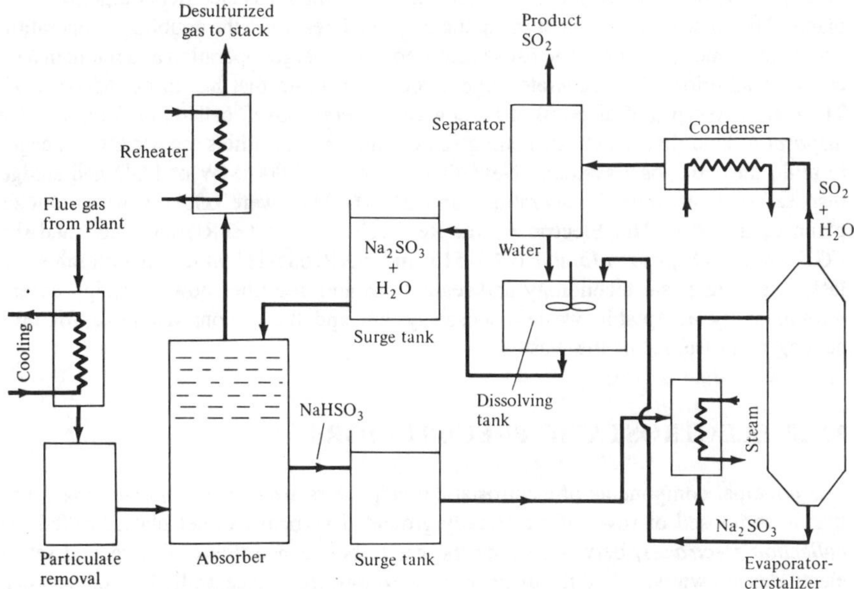

Figure 17-9 Schematic of the Wellman-Lord SO_2 recovery process.

The product SO_2 may be utilized to produce liquid SO_2 or sulfuric acid, on site or in a satellite plant, or to produce elemental sulfur. A well-known process for doing this is called the *Claus process,* which is based on the addition of H_2S according to

$$SO_2 + 2H_2S \rightarrow 3S + 2H_2O \tag{17-7}$$

NO Removal

A process for the removal of NO, also by the addition of H_2S, is proposed. It is given by

$$NO + H_2S \rightarrow S + \frac{1}{2}N_2 + H_2O \tag{17-8}$$

The combined removal of SO_2 and NO is under study. In both reactions, the H_2S must be completely consumed as it is a pollutant itself.

In 1977 the system was estimated to add an additional \$120/kW, or some 12 to 15 percent to the base capital cost of a powerplant. It was said operating costs would increase by about \$60/MBtu.

Most scrubbers in use by 1981 have been of the wet type. There is not sufficient experience with the dry type to establish which of the two may be selected by utilities in the future. Presently all scrubber systems are large and occupy a sizable area of a powerplant, have capital costs that run in the tens of millions of dollars for 500- to 1000-MW plants, and consume a sizable fraction of the gross electrical output of these plants. They also require a lot of maintenance, which results in the doubling of operation and maintenance personnel and causes, consequently, larger operation and maintenance costs. In addition, they generate huge amounts of waste that has to be disposed of. There are two types of disposal of FGD wastes: *wet disposal,* called *ponding,* and *dry disposal* in landfills, which are getting scarce. In general utilities are not always eager to build these disposal systems. Nevertheless, some 19,000 MW of FGD and sludge disposal systems were in operation, and 26,000 MW were under construction or planned, in 1981. The Electric Power Research Institute (EPRI) has published the *FGD Sludge Disposal Manual* (CS-1515 under RP1685-1), which incorporates the latest waste-disposal technology and regulations and describes how to design an environmentally acceptable waste-disposal system and the options available for processing and disposal of the wastes.

17-10 ELECTROSTATIC PRECIPITATORS

The principal components of electrostatic precipitators are *two sets of electrodes*. The first is composed of rows of electrically grounded vertical parallel plates, called the *collection electrodes,* between which the gas to be cleaned flows. The second set of electrodes are wires, called the *discharge electrodes,* that are centrally located between each pair of parallel plates (Fig. 17-10). The wires carry a unidirectional, negatively charged, high-voltage (between 20 and 100 kV, but typically 40 to 50 kV) current

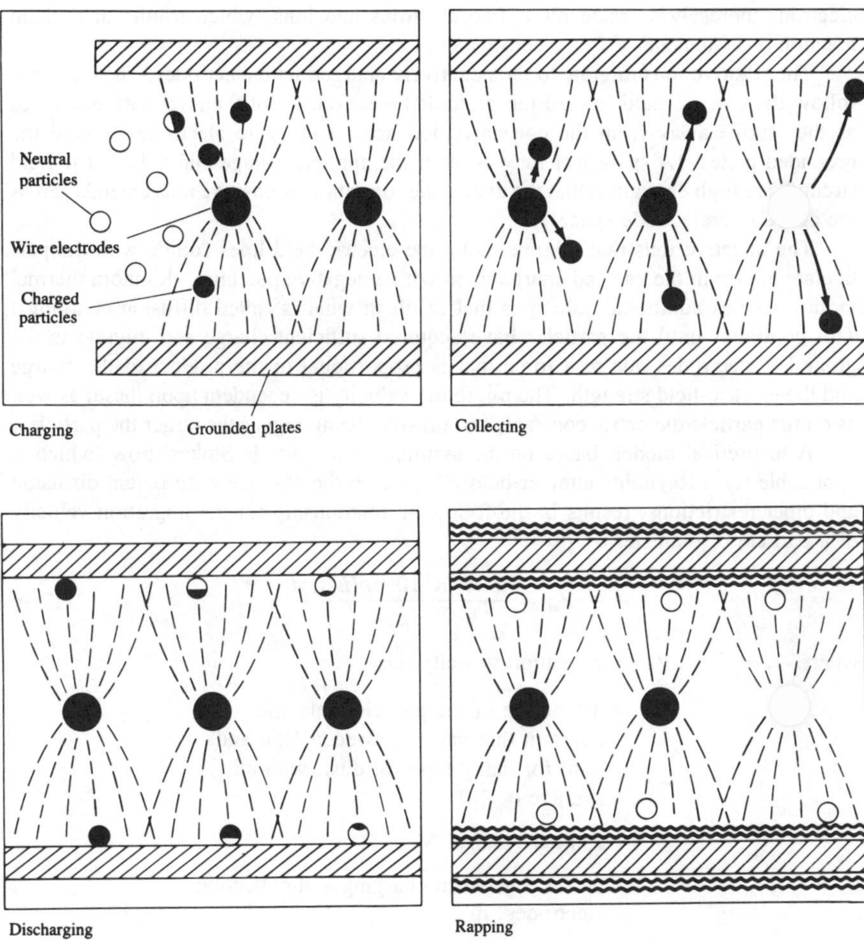

Figure 17-10 Vertical electrodes and grounded plates in an electrostatic precipitator showing four basic operations.

from an external source. The applied high voltage generates a unidirectional, nonuniform electric field whose magnitude is greatest near the discharge electrodes. When that voltage is high enough, a blue luminous glow, called a *corona,* is produced around them. The corona is an indication of the generation of negatively charged gas ions that travel from the wires to the grounded collection electrodes as a result of the strong electric field between them. This process occurs because the electrical forces in the corona accelerate the free electrons present in the gas so that they in turn ionize the gas molecules, thus forming additional electrons and positive gas ions. The new electrons themselves create more free electrons and ions, which results in a chain reaction or avalanche of free electrons and ions.

The positive ions migrate to the negatively charged wire electrodes. The electrons follow the electric field toward the grounded electrodes, but their velocity decreases as they move away from the corona region around the wire electrodes toward the grounded plates. (A sparkover may occur if the applied voltage and the electric-field strength are high enough so that the avalanche formation of positive ions extends across the entire interelectrode space.)

The negative ions that migrate along the electric-field lines collide with the particulate matter in the gas and charge them with a negative potential. (Random thermal motion creates additional collisions that result in what is called diffusion charging.) This continues until the particles have acquired sufficient charge and migrate to the grounded electrodes due to a force that is proportional to the product of the charge and the electric-field strength. The migration velocity is dependent upon these, as well as on the particle dielectric constant, and on size, being higher the larger the particles.

A theoretical model, based on the assumption of particle Stokes' flow, which is applicable for a Reynolds number below 1, and on the absence of turbulent diffusion and other restrictions, results in the following relationship for the migration velocity of a particle of diameter d

$$V_m = \frac{2.95 \times 10^{-12} p(E/s)^2 d}{\mu_g} \tag{17-9}$$

where V_m = migration velocity m/s

p = a function of the particle dielectric constant that varies betweeen 1.50 and 2.40 for many types of dust, with an average of 2.0

E = applied voltage, V

s = distance between charging and collecting electrodes, m

d = particle diameter, m

μ_g = gas viscosity, kg/(m · s)

Equation (17-9) shows that the migration velocity is directly proportional to the particle diameter and the square of the field strength and inversely proportional to the gas viscosity and, hence, is sensitive to changes in gas temperature because the viscosity of gases increases with temperature. Actual migration velocities may deviate considerably from those predicted by Eq. (17-9), and design values are usually based on empirical observations.

When the particles collect on the grounded plates, they are supposed to lose their charge to ground. The electrical resistivity of the particles, however, causes only partial discharging, and the retained charge contributes to forces holding the particles to the plates. Resistivities that are too high or too low pose problems. High restivity causes retention of most of the charge, which increases the forces holding the particles to the plates and makes removal more difficult.* The maximum particle resistivity is known to occur around 300°F. This can then be corrected by operating at high gas temperatures, for instance by installing a "hot side" precipitator, upstream of the steam-generator air preheater where the gas temperature is usually in excess of 600°F. Low resistivity, found in particles with high carbon content such as fly ash from spreader stokers, causes them to lose their charge quickly to ground, and thus the forces that hold them to the plates, and to become reentrained by the flue gas. This can be corrected by increasing the holding forces by greater power input and decreasing reentrainment by designing low gas velocities, low electrode heights, and aspect ratios.

Particle Removal

When dust builds up on the plates, it deposits in a layer of increasing thickness with possible reentry into the gas stream unless it is periodically removed. This removal is done by *rapping* the plates to cause shock vibrations that shake the dust into hoppers at the bottom of the precipitator. Properly controlled and timed removal is critical for precipitator performance. The intensity and frequency of rapping of individual collection plates varies from the inlet to the outlet of the precipitator to suit the changing characteristics and collection rates of the dust. It is also customary to rap sequentially so that only a fraction of the accumulated dust is disturbed at any one time. A rapper device is used. It contains a vertical plunger that strikes the collection electrode support system to deliver the necessary shock wave and then returns back to its top position to prepare for the next strike. In one design the downward impact motion is effected by an electromagnetic coil and the return by a spring. In another the plunger is heavier, the impact is by gravity, and the return is by an electromagnet.

The length of the precipitator passage in the direction of gas flow that is necessary to remove a given particle size is obtained by seeing to it that the time required for the particle to migrate to the collection electrode has to be less than the time it takes it to pass through the precipitator at the same velocity as the gas. Neglecting the charging time, the required length would be given by

$$\frac{s}{V_m} \leqslant \frac{L}{V_g}$$

or

$$L \geqslant s\,\frac{V_g}{V_m} \tag{17-10}$$

where

L = length of passage, m or ft

V_g = gas velocity, m/s or ft/s

* High resistivity may also cause the electric field in the dust layer to accelerate electrons sufficiently to produce positive ions, thus reducing the voltage at which sparkover occurs. It may also generate enough positive ions even when voltage is not high enough for sparkover, the result being a *back corona* that neutralizes the unidirectional field. These effects reduce or completely disrupt precipitator performance.

Collection Efficiencies

An electrostatic precipitator, and all other particle-collection devices, has an *efficiency*, already mentioned in Sec. 17-8. The *overall collection efficiency* η_o is given by

$$\eta_o = \frac{\text{mass or concentration of all particles retained by collector}}{\text{mass or concentration of all particles entering collector}} \tag{17-11}$$

Although the relationships in Eqs. (17-9) and (17-10) are based on highly idealized models, the trends they predict are applicable to the real case. Thus, because the migration velocity V_m is greater for large particles, the length of passage necessary to remove them is smaller. In other words, it is easier to collect larger particles than smaller particles. There is, therefore, a *fractional collection efficiency* η_d (also called the *grade efficiency*) that is given by

$$\eta_d = \frac{\text{mass or concentration of particles of a given size retained by collector}}{\text{mass or concentration of particles of a same size entering collector}} \tag{17-12}$$

From Eq. (17-12) it can be expected that η_d increases with size of diameter d. η_d is also expected to increase with electrode area and decrease with the flue-gas volume flow rate. Several relationships have been proposed for η_d. A well known one is the *Deutsch expression*, which is derived from physical or probability considerations [174,175]

$$\eta_d = 1 - e^{-(AV_m/\dot{Q})} \tag{17-13}$$

where A = area of collector electrodes, m^2 or ft^2

V_m = migration velocity, m/s or ft/s

$\dot{Q}$ = flue-gas volume flow rate for each plate, m^3/s or ft^3/s

Example 17-1 Plot the Deutsch fractional efficiency for an electrostatic precipitator that has plates with height × width of 10 × 10 m, and a spacing of 25 cm. The applied voltage is 50,000 V. The mean flue-gas temperature and velocity between plates are 300°C and 1.5 m/s, respectively. Repeat for plate dimensions of 10 × 15 m and 10 × 5 m.

SOLUTION Refer to Eq. (17-9) and use $p = 2.0$, $s = 0.25/2 = 0.125$ m, and $\mu_g = 2.93 \times 10^{-5}$ kg/(m · s)

$$V_m = \frac{2.95 \times 10^{-12} \times 2 \times (50{,}000/0.125)^2 d}{2.93 \times 10^{-5}} = 3.225 \times 10^4 d$$

Refer to Eq. (17-13) and use

$$\dot{Q} = V_g \times \text{cross-sectional area of flow per plate side}$$

$$= 1.5 \times 10 \times 0.25 \times \frac{1}{2} = 1.875 \text{ m}^3/\text{s}$$

$$A = 10 \times 10 = 100 \text{ m}^2$$

Therefore

$$\eta_d = 1 - e^{-\left(\frac{100 \times 3.255 \times 10^4}{1.875}\right)d} = 1 - e^{-1.72 \times 10^6 d}$$

Similarly, for 10 × 15 m and 10 × 5 m plates

$$\eta_d = 1 - e^{-2.58 \times 10^6 d} \quad \text{and} \quad 1 - e^{-0.86 \times 10^6 d}$$

respectively.

η_d is plotted in Fig. 17-6 as a function of d in micrometers (μm). It can be seen that the plates with the larger path length are more efficient, that η_d is essentially 100 percent for particle sizes beyond 3 μm, and that the efficiency drops drastically for small particles. This is not in accordance with practical experience, from which measurements in many precipitators show a turnaround in fractional efficiency for very fine particles in the submicron range, with minimum efficiencies occurring in the 0.1 to 0.5 μm range and efficiencies going back to 90 to 95 percent below 0.1 μm.

The Deutsch equation is now recommended for use in estimating the operation of a precipitator at off-design conditions. In this case a value V_{mo} is obtained from experimental studies under different operating conditions and used in lieu of V_m, resulting in an overall collection efficiency η_o. V_{mo} is known as the *effective migration velocity*. Thus

$$\eta_o = 1 - e^{-(AV_{mo}/\dot{Q})} \tag{17-14}$$

Equation (17-14) may now be used to estimate the effect of *changes* in volume flow rate $\dot{Q}$ (representing changes in plant load) on η_o, or, for the same V_{mo}, the effect of changes in precipitator design A.

Attempts to arrive at equations more representative of modern high-efficiency precipitator designs [176] are being attempted. Two current ones are

$$\eta_o = 1 - e^{-(AV_{mo}/\dot{Q})x} \tag{17-15}$$

where x is a variable, fitting most data at a value of about 0.5, and

$$\eta_o = 1 - \left(1 + \frac{AV_{mo}}{n\dot{Q}}\right)^{-n} \tag{17-16}$$

where n is a variable, fitting most data in a range between 3 and 5 (but can vary between 2 to 8). Putting $n = \infty$ reverts Eq. (17-16) back to the Deutsch equation. In both equations above, V_{mo} is the effective migration or drift velocity. For utility fly ash its value can vary between 4 and 20 cm/s.

Figure 17-11 shows an electrostatic precipitator used to clean flue gases from an electric-generating powerplant.

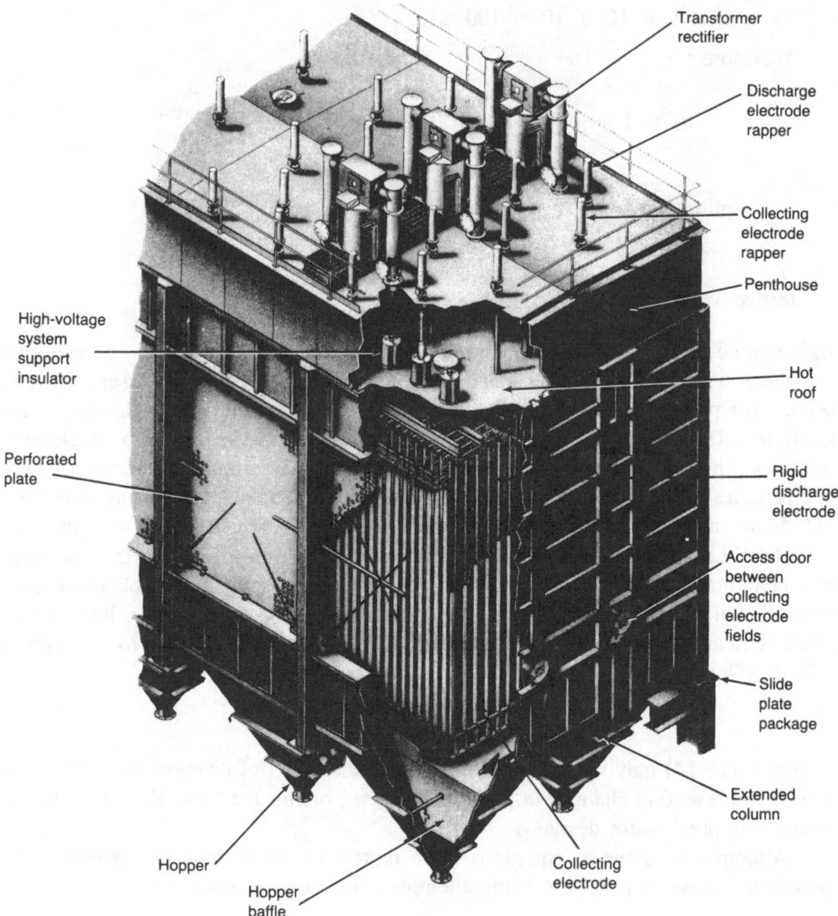

Figure 17-11 An electrostatic precipitator. *(Courtesy Research-Cottrell, Somerville, New Jersey.)*

17-11 FABRIC FILTERS AND BAGHOUSES

Fabric filters are used in powerplants to remove dust particles from a gas stream on a principle similar to that of a household vacuum cleaner except, of course, that the size of fabric filters is far greater. They are made of porous material that retains particulate matter as the carrier gas passes through the voids.

A fabric-filter element is usually made in the form of a long, hollow cylindrical tube that provides a large surface per unit of gas volumetric flow rate. The inverse of this parameter, called the *air-to-cloth,* or *filtering, ratio,* is equal to the superficial gas velocity; i.e., it is based on the surface area rather than the void flow area within the fabric. It ranges typically between 0.5 to 4.0 cm/s.

A fabric-filter system usually contains a large number of vertical cylindrical fabric-filter elements arranged in parallel rows. Such a system is called a *baghouse* (Fig. 17-12). A powerplant baghouse might contain several thousand such cylinders, each ranging in diameter from 5 to 14 in and in height up to 40 ft. The exact number is determined by their size, the required total capacity, plus an additional number to allow for shutdown of portions of the baghouse for cleaning with the plant on load. This points to one disadvantage of baghouses: their size is large in comparison with other types of particulate-removal systems.

In general, the elements have an open bottom and closed top. They rest on a tube sheet above a dirty-air plenum. The sheet distributes the gas evenly to the bags, allowing it to enter the elements at bottom, deposit its particulate matter on the inside of the

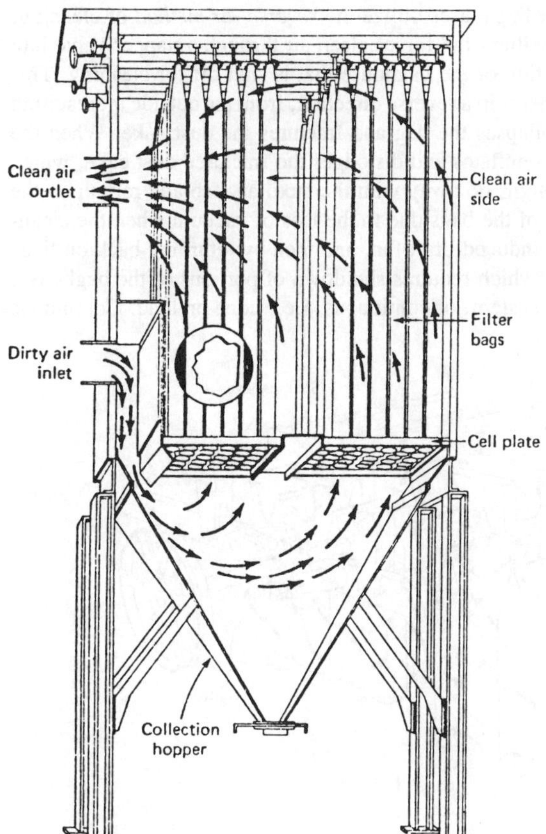

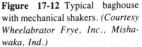

Figure 17-12 Typical baghouse with mechanical shakers. *(Courtesy Wheelabrator Frye, Inc., Mishawaka, Ind.)*

tubes, and pass laterally through the fabric and exit to an outlet manifold where it is drawn out by the plant induced-draft fan.

Collection hoppers, below the tube sheets, are receptacles for collecting the removed particulate matter. This matter usually collects on the inside of the elements in the form of a dust cake that must be made to fall down to the hopper by some means. Such means include mechanical shaking, pulse jets, or reverse airflow.

Mechanical shaking is accomplished by oscillating or vibrating rods attached to the top of the filter elements (Fig. 17-13). The oscillations ripple the bag surfaces, breaking the dust cake and causing it to discharge into the hopper below.

In *pulse-jet* cleaning, the bags are hung from the top, normal flue-gas flow is from the outside in, and dust collects on the outside of the tubes. For cleaning, short pulses of compressed clean air are forced down into the bags through venturis at the top. The pulse fractures and dislodges the dust, which falls to the hopper. The bags in this system are usually shorter than other types (less than 15 ft) to allow for proper cleaning of the tube bottoms.

Most coal-fired powerplant baghouses utilize the *reverse-air method* of cleaning. In reverse-air baghouses, an auxiliary fan forces clean air through a flapper valve into the clean-air plenum of the portion of the baghouse that is shut off for cleaning. This air passes through the filter elements in a reverse direction, from the outside in, resulting in a "backwash" action that collapses the bag and fractures the dust cake. When the bag is brought back on line, it reinflates and dislodges the fractured dust cake, which falls into the hopper. Some designs do away with the auxiliary fan and rely upon the natural collapse and reinflation of the bags due to the loss of vacuum when the clean-air plenum is shut off from the induced-draft fan, and then by returning back on line.

Cleaning may be periodic, which requires shutdown of portions of the baghouse, as explained for the reverse-air system, or continuous, i.e., done on load. Continuous

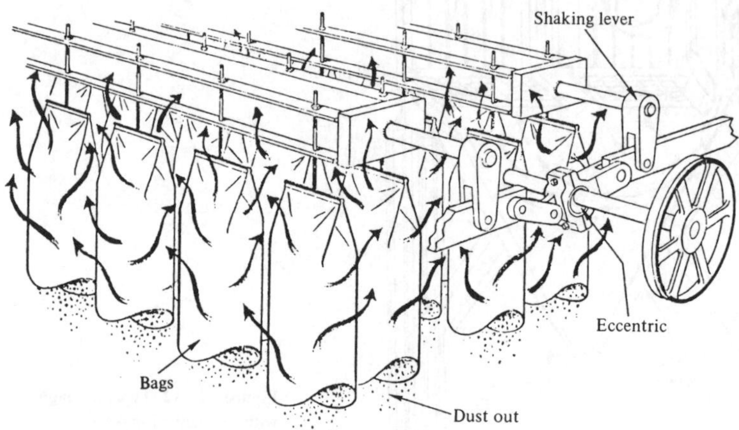

Figure 17-13 Shaker mechanism for the tubular filters of a baghouse. *(Courtesy Western Precipitation Division, Joy Manufacturing Company, Los Angeles, Calif.)*

cleaning is possible with the pulse-jet system because the duration of the pulse is very short.

The bag filter material can fail as a result of high temperature, burning, caking, erosion, chemical attack, and aging. A variety of fabrics have been used, depending upon the type of particulates and flue-gas composition, humidity, and temperature. They include wool, cotton, nylon, glass fiber, polyesters (such as dacron), and aromatic polyamides [177]. Wool and cotton, on one end of the scale, can operate only with low gas temperatures, up to about 175 or 200°F. Glass fibers, at the other end, tolerate high gas temperatures, up to 500 to 550°F. Research is continuing to find fibers that can withstand temperatures beyond 550°F to avoid precooling of the flue gases before they enter the baghouse. A baghouse is usually placed after the air preheater, which has a flue-gas outlet temperature of 300°F or more. High gas temperatures also mean high volume flow rates through the elements. Gas cooling below its dew point, which depends upon the water vapor mole fraction, results in condensation, which is not permissible with fabric filters.

Chemical attack of the fabric is particularly severe with high-sulfur coals. With the shift by many utilities to low-sulfur, high-ash coals, fabric filters have become increasingly attractive in comparison with other particulate-removal systems, such as electrostatic precipitators. Their performance is also less dependent on fuel and flue-gas characteristics, and they have fewer critical design parameters.

Figure 17-14 shows a baghouse of the type used to clean flue gases from fossil-fuel powerplants.

17-12 THERMAL POLLUTION

All thermal powerplants (fossil, nuclear, solar) reject low-availability or low-temperature heat to the environment. A comparison of the amounts of heat rejected by powerplants of different efficiencies should be based on their output, not the heat added. Thus

$$\eta = \frac{W}{Q_A} = \frac{W}{W + Q_R} = \frac{1}{1 + (Q_R/W)} \tag{17-17a}$$

or

$$\frac{Q_R}{W} = \frac{1}{\eta} - 1 \tag{17-17b}$$

where

η = plant net thermal efficiency

W = plant net output

Q_A = heat added

Q_R = heat rejected

For plants of 1000-MW output, the heat added, and the heat rejected, in MW, are given as a function of efficiency in Table 17-9. Thus an increase in thermal efficiency of 10 percent, from 30 to 40, results in a reduction in heat rejected of 35.7 percent,

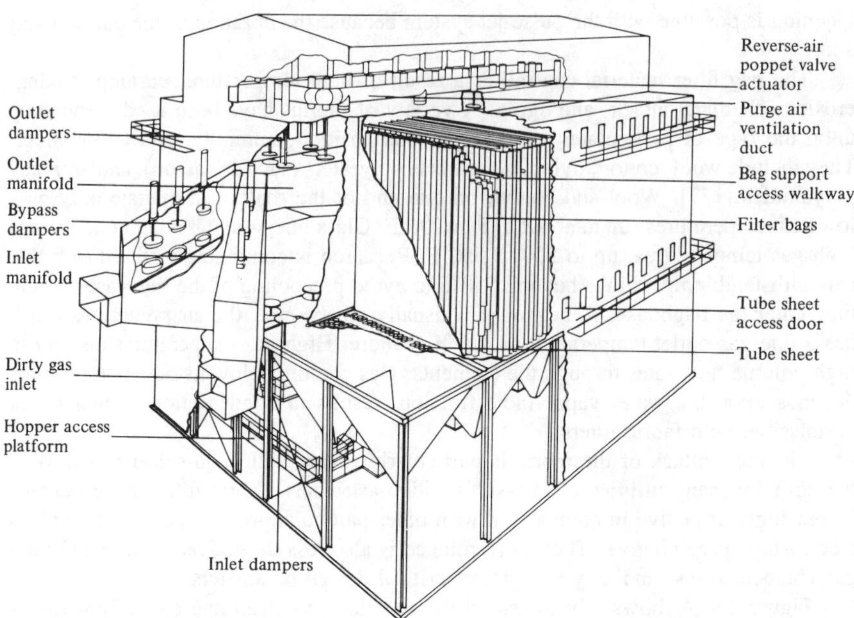

Figure 17-14 A baghouse. *(Courtesy Western Precipitation Division, Joy Manufacturing Company, Los Angeles, Calif.)*

from 2333 to 1500 MW. It can be seen that efficiency has a pronounced effect on heat rejected, greater than the change in its own value.

Because most modern thermal powerplants have efficiencies below 50 percent, actually in the 30 to 40 percent efficiency range, the amount of heat rejected is greater than the plant output. These very large amounts of heat are added to and affect the environment, and because they are mostly added to bodies of water they affect the aquatic ecological system.

The heat rejected in Rankine-cycle type powerplants is usually done via the cooling water of the condenser. The condenser is part of a once-through circulating-water system (the most common) or a closed system (Chap. 7). In the once-through system

Table 17-9 Heat added and rejected, MW, by plants producing 1000 MW, as a function of plant thermal efficiency

Thermal efficiency, %	10	20	25	30	33.3	40	50	60
Plant output W, MW	1,000	1000	1000	1000	1000	1000	1000	1000
Heat added Q_A, MW	10,000	5000	4000	3333	3000	2500	2000	1667
Heat rejected Q_R, MW	9,000	4000	3000	2333	2000	1500	1000	667

large amounts of relatively cool water are taken from the environment, passed through the condenser, and discharged at a higher temperature back to the environment. In the closed system only enough water is drawn from the environment to replace the condenser cooling water that is lost, usually by evaporation in a cooling tower or a cooling pond.

Once-Through Systems

In the case of the once-through system, two effects are of concern. The first is the effect of heat on the large volumes of water as it goes through the condenser cooling system. The second is the effect of discharging the same but now warm water on the lake or stream it originated from. A 1000-MW powerplant, depending upon its thermal efficiency, needs between 250×10^6 to 400×10^6 lb_m/h of condenser cooling water. This corresponds to about 65,000 to 100,000 ft^3/min, or about 30 to 50 m^3/s.

Water in a lake or a stream is a habitat for numerous species of animal and plant life. A good intake system must therefore be designed to screen out these organisms. Nevertheless, simple screens still let through small organisms such as fish eggs, larvae, and plankton. And some larger fish, unable to escape because of the speed and volume of intake water flow, impinge and accumulate on the screens along with other debris from the lake or stream, thus requiring frequent cleaning of the intake screens. A method conceived to minimize impingement and entrainment is the use of a *porous dike* upstream of the intake screen. Built of suitably sized rocks, it slows the speed of water at intake so that small fish are not easily drawn toward the intake screen.

The organisms that do filter through the screens pass through pumps, pipes, hot condenser tubes, and other cooling-water system components before being discharged back to the lake or stream. In this path, they are subjected to buffeting, pressure, and heat, resulting in a high mortality rate. The ecosystem is said to be "stressed." Some ecological studies, however, indicate that ecosystems, like populations, have a natural resilience to compensate for adverse impacts on small parts of the whole.

From an engineering point of view, these organisms, dead or alive, clog filters, pipes, and pumps. Furthermore, bacterial slime and algae grow rapidly on the inside surfaces of the tubes, a condition known as *bifouling*. This is usually accompanied by *scaling,* which is caused by corrosion and the deposition of suspended solids in the water on the same surfaces. Bifouling and scaling reduce the bore of the condenser tubes, reduce heat transfer, and thus reduce condenser effectiveness, which in turn raises the back pressure on the turbine and thus reduces plant efficiency. In addition, these conditions increase pressure drops and pumping power in the circulating-water system, further reducing plant net output.

Bifouling and scaling can be combatted by frequent applications of chemicals such as chlorine. But chlorine and its compounds are discharged with the cooling water and are potentially harmful to the ecosystem.* In severe cases, condenser tubes are cleaned by mechanical means that require costly plant shutdowns.

* The Ocean Water Act of 1977, which mandates that electric utilities change their cooling systems to the "best available technology," requires them to limit the use of chlorine.

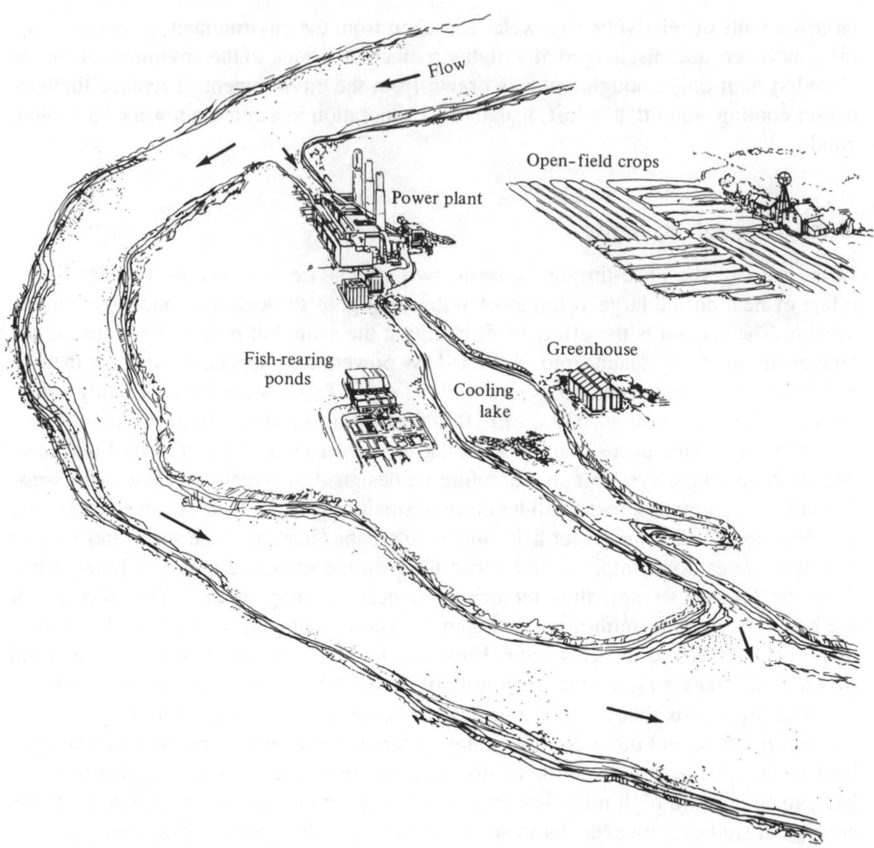

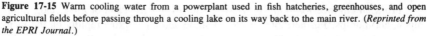

Figure 17-15 Warm cooling water from a powerplant used in fish hatcheries, greenhouses, and open agricultural fields before passing through a cooling lake on its way back to the main river. (*Reprinted from the EPRI Journal.*)

The warm water discharged in a once-through cooling system is some 20 to 25°F (11 to 14°C) above intake. The effect of this on the ecosystem depends upon the plant site, whether the plant is situated on a river, lake, estuary, or ocean, as well as on the workings of the aquatic world it is having an impact upon. Studies of flora and fauna,* necessary to understand the problem completely, are not simple and are time-consuming. Examples are fish spawning and migration patterns, especially in estuaries and sheltered coastal areas where environmental disturbances can cause great damage, the effect of small concentrations of chlorine on various species of aquatic life, etc.

* Plants and animals. In classic mythology: Flora is the goddess of flowers; Fauna is a Roman goddess, the sister of Faunus, the god of nature and patron of farming and animals.

Warm-water discharges in large volumes were once suspected of causing major ecological damage. It has, however, been recently demonstated that with good planning and management, warm water can be beneficial. It can for example be used to boost fish production in hatcheries (Fig. 17-15). (Commercial operations are already under way; an example is the Long Island Oyster Farm, operated in conjunction with the Long Island Lighting Company plant in Northport, New York.) It can also be used in agriculture, where it has been found to increase materially the production in greenhouses and open fields in cold climates. (An example is an agricultural facility operated in conjunction with the Northern States Power Company plant in Sherburne County, Minnesota.) Long-range economic questions, however, arise as to the relative gains from increased agricultural production versus costs of constructing the facilities, pumping the water, etc.

Closed Systems

Closed systems take in only sufficient water to make up for losses by evaporation from cooling towers, ponds, etc. They use corrosion-inhibiting additives with the cooling water. The systems are periodically flushed out, which releases these additives to the body of water on which the plant is situated, thus resulting in water pollution.

In general it is believed that fish mortality caused by powerplants is insignificant when compared with their natural mortality rate. It is also a fact that aquatic life is resilient and compensates naturally for the decrease in its population by producing more young or by increasing the survival rate of the normal number of young population, although this compensation is not always automatic and requires further study. Finally, an assessment should take into consideration the relative benefits of supplying the electric energy needed for the welfare of humankind and the loss, regrettable as it is, of an insignificant quantity of aquatic life, especially when the plant sites are chosen with care.

17-13 NATURAL AND ARTIFICIAL RADIOACTIVITY

Just like gaseous and particulate matter, which are caused by both natural and artificial events, radioactivity or radiation* in the environment is also caused by both natural and human-generated events. Of all environmental radioactivity, the portion contributed by nuclear powerplants in normal operation is miniscule. The history of nuclear-powerplant operation strongly suggests that even with abnormal operation (accidents), that portion is still insignificant. This suggests that the crisis atmosphere periodically generated by opposition to nuclear powerplants on this issue is largely unjustified.

Natural radioactivity has *always* been present. It comes from the sun and outer-

* The term *radiation* is a broad term that includes light (electromagnetic) and radio waves. However, it is often used to mean *radioactive* or *ionizing radiation,* which can present a health hazard. It will be used in the latter sense in this section.

space; it exists in numerous earth materials and in food, water, and air. It is generated by natural eruptions and volcanoes. It even exists within our own bodies; the human body contains about 0.35 percent (by mass) potassium, of which 0.0118 percent is radioactive potassium 40. The total of these is often called *natural background radiation*. Its level varies widely from location to location around the earth but averages about two-thirds of all radiation present.

It is instructive here to compare a natural occurrence with a human-made abnormal one. These are the eruptions from Mount St. Helens in the state of Washington, USA, that first erupted in May 1980, and the March 1979 accident at Three Mile Island nuclear powerplant unit No. 2 (TMI-2). The President's commission to investigate TMI-2, the Kemeny Commission, estimated in its report [178] that a total of about 2.5 million curies (Ci) of noble gases, mostly xenon, were released over the course of the accident. On the other hand, an extensive study by Battelle Pacific Northwest Laboratories of the ash from Mount St. Helens estimates that up to 3 million Ci of radon gas were released in the eruption of one day, May 18, alone.

Both xenon and radon are noble gases that are chemically and biologically inert. The difference, however, is that radon decays to a series of radioactive daughter elements that are chemically and biologically active. Thus the radioactive release from Mount St. Helens due to the gases alone is much more significant than that from TMI-2. In addition, the ash from the eruption at Mount St. Helens that spread over vast areas of the state of Washington and beyond was found to contain significant radioactivity from isotopes such as radium 226, potassium 40, thorium 232, polonium 210, and lead 214. Newly fallen ash was also found to contain rather high concentrations of the short-lived radon daughters lead 214, bismuth 214, and polonium 214. Note also that Mount St. Helens has erupted with varying intensity several times since May 1980.

Of the artificial sources of radioactivity we are subjected to, medical irradiations (x-rays, etc.) account for the largest portion, about 30 percent of the total. Occupational exposures, fallout from nuclear weapons tests, and miscellaneous sources (such as high-altitude flying) account for less than 1 percent each. Releases from the nuclear-power industry, including fuel manufacturing and reprocessing and powerplant emissions, account for some 0.15 percent of the total (Table 17-2).

Both natural and artificially generated radiations may be classified as particles and electromagnetic radiation. The particles include beta (β) or ${}_{-1}e^0$ (electrons in a free state); alpha (α) or ${}_2\mathrm{He}^4$ (helium nuclei); as well as positrons ${}_{+1}e^0$ (positively charged β); neutrons, n; protons, p or ${}_1\mathrm{H}^1$; tritons, ${}_1\mathrm{H}^3$; and fission products. The electromagnetic radiation includes gamma (γ) rays, x-rays, and Bremsstrahlung. Examples of radioactive reactions (or disintegrations) that cause these radiations are:

$$\text{Beta:} \quad {}_{19}\mathrm{K}^{40} \xrightarrow{1.28 \times 10^9 \text{ years}} {}_{20}\mathrm{C}^{40} + {}_{-1}e^0 \tag{17-18}$$

$${}_6\mathrm{C}^{14} \xrightarrow{5730 \text{ years}} {}_7\mathrm{N}^{14} + {}_{-1}e^0 \tag{17-19}$$

$$\text{Alpha:} \quad {}_{94}\mathrm{Pu}^{239} \xrightarrow{24{,}000 \text{ years}} {}_{92}\mathrm{U}^{235} + {}_2\mathrm{He}^4 \tag{17-20}$$

$$_{88}\mathrm{Ra}^{226} \xrightarrow{\text{1600 years}} {}_{86}\mathrm{Rn}^{222} + {}_{2}\mathrm{He}^{4} \tag{17-21}$$

$$\text{Positron: } {}_{15}\mathrm{P}^{30} \xrightarrow{\text{2.5 min}} {}_{14}\mathrm{Si}^{30} + {}_{+1}e^{0} \tag{17-22}$$

$$\text{Neutron: } {}_{54}\mathrm{Xe}^{137} \xrightarrow{\text{3.9 min}} {}_{54}\mathrm{Xe}^{136} + {}_{0}n^{1} \tag{17-23}$$

The original isotope is often called the *parent*. The first isotope on the right-hand side is called the *daughter*. The emitted particle is the radiation. (Gamma decay by itself does not alter the isotope, only its energy level, however γ rays often accompany other radiations.) The daughter may be stable or radioactive; if radioactive, it in turn decays, resulting in a radioactive chain, sometimes of considerable length. The half-life of each parent, shown above the arrows, is defined as the time during which one-half of the original parent nuclei decays and one-half is left. One-quarter is left after two half lives, one-eighth after three, etc. (Sec. 9-8).

The Curie

A radiation dose means that a person has been exposed to and has absorbed some radiaton energy. Radiation has both quality and quantity. The quantitative physical unit of radioactivity is the *curie,* Ci (and its fractions millicurie, mCi; picocurie, pCi, 10^{-12} Ci; etc.). 1 Ci = 3.70×10^{10} disintegrations per second. The number of curies emitted by a radioisotope depends upon both its mass and its half-life.

Example 17-2 Compute the activity in disintegrations per second of 1 g of radium 226. Ra^{226} has an atomic mass of 225.0245. It decays into radon gas with the emission of α particles, with a half-life $\theta_{1/2}$ = 1600 year = 5.049×10^{10} s.

SOLUTION Initial number of atoms of Ra^{226} in 1 g is

$$N_o = \frac{\text{Avogardro's number}}{\text{atomic mass}}$$

$$= \frac{6.0225 \times 10^{23}}{226.0245} = 2.6645 \times 10^{21}$$

$$\text{Decay constant } \lambda = \frac{0.6931}{\theta_{1/2}} = \frac{0.6931}{5.049 \times 10^{10}} = 1.3727 \times 10^{-11}\ \mathrm{s}^{-1}$$

$$\text{Activity} = \lambda N_o = \lambda\frac{0.6931}{\theta_{1/2}} = 1.3727 \times 10^{-11}$$

$$\times\ 2.6645 \times 10^{21} = 3.6476 \times 10^{10}\ \text{dis/s}$$

This activity is small compared with the initial number of atoms. The activity of radium may thus be considered constant, a true phenomenon for any radioactive species with a long half-life. Early measurements of radioactivity indicated that 1 g of radium had an activity of 3.70×10^{10} dis/s instead of the more accurate value given above. 3.7×10^{10} was, and still is, adopted as the numerical value for the curie. A curie is also used to indicate the quantity of any isotope having 1 Ci of radioactivity.

The Rad and the Gray

While the curie is a physical quantitative unit that indicates the number of radioactive events or disintegrations, it must be recognized that different radioisotopes emit different radiations with different energies. A more significant unit from the point of view of biological effect is the *rad*. The rad is a unit of radiation *energy* absorbed. 1 rad = 0.01 J/kg (4.3×10^{-6} Btu/lb_m), 1 millirad = mrad = 0.00001 J/kg. Some higher-than-average natural background levels, measured in mrad/year, emanate from monazite sand in Egypt (220 to 475), some beaches in Rio de Janeiro, Brazil (550 to 1250), the city of Kerala, India (800 to 8000), granite areas in Sri Lanka (3000 to 7000), and others. The rad is now being replaced by another unit in the SI system of units, called the *gray* (symbol Gy). 1 Gy = 1 J/kg = 100 rad.

The Rem and the Sievert

The biological effect of radiation on human beings must further take into account not only the total energy absorbed but also the relative biological effects of different types of ionizing radiation on people, such as the number of cells damaged by this radiation. Thus the rad is multiplied by a factor called the *relative biological effectiveness* (RBE), also called the *quality factor Q,* to obtain a more meaningful unit called the *rem,* for *roentgen equivalent man* (a millirem, *mrem,* is one-thousandth of a rem). Thus

$$1 \text{ rem} = 1 \text{ rad} \times \text{RBE} \tag{17-24}$$

some RBE values are 0.6 for 4-MeV gamma rays, 1.4 for 1-MeV electrons, 4 to 5 for thermal neutrons, and 2 to 10 for 1-MeV neutrons. These values are relative, based on an RBE for 200 kVp x-rays equal to 1. In the SI system of units, Gy replaces the rad, and therefore a new unit called the *sievert* (symbol Sv) replaces the rem. 1 Sv = 100 rem.

In addition, the rate at which the radiation is received is important. An analogy here may be made with drinking liquids. Some drinks are more harmful to health than others. A small quantity of a potentially harmful drink may not be harmful. A large quantity, say taken in the course of one evening, is. However, that same quantity would not be harmful if taken over a long period of time, say weeks. Thus the rate, mrem/h or mrem/year, is also important. Table 17-10 lists some examples of doses and dose rates we are exposed to in our normal lives.

Radiation effects are a complex subject. A complete treatment is beyond the scope of this text. They are, however, covered in the next sections where appropriate. For a fuller treatment, the reader is referred to specialized books on radiation and on health physics [179–184].

Table 17-10 Some common doses and dose rates of radioactivity

Source	Dose, mrem	Dose rate, mrem/year
1 diagnostic x-ray	20	
1 transatlantic flight	2	
Cosmic rays		45
Soil		15
Food, water, and air		25
Brick house		50–100
Concrete house		70–100
Wooden house		30–50
TV set		1–10
Living in the vicinity of a nuclear powerplant		1
Average background:		
New York		100
London		100
Paris		120
Denver		125
Kerala, India		400

17-14 NUCLEAR POWER AND THE ENVIRONMENT

In the United States, and doubtless in almost all countries constructing nuclear powerplants, federal licensing proceedings for each plant require the inclusion of detailed environmental statements to be issued as public documents. In the United States, these should be in accordance with the National Environmental Policy Act of 1969 (NEPA). Such statements must assess not only the impact upon the environment that is associated with the construction and the operation of the powerplant, but also the effect of the transportation of radioactive materials to and from that plant.

Besides thermal pollution (Sec. 17-12) which it shares with almost all types of powerplants, nuclear power's effects on the environment stem mainly from (1) the nuclear fuel cycle, (2) low-level dose radiations from nuclear-powerplant effluents, and (3) low- and high-level dose radiations from wastes.

The Fuel Cycle

Most nuclear powerplants in operation or under construction in the world today are using, and will continue to use for the near future, ordinary (light) water cooled and moderated reactors: the pressurized-water reactor (PWR) (Sec. 10-2) and the boiling-water reactor (BWR) (Sec. 10-7). A small number use the heavy water cooled and moderated reactor (PHWR) (Sec. 10-14). The expectations are that the fast-breeder reactor powerplant (Chap. 11) and perhaps an improved version of the gas-cooled reactor powerplant (Sec. 8-11) will come on line in increasing numbers in the twenty-first century. Almost all current water reactors use slightly enriched uranium dioxide, UO_2, fuel. The fuel has to go through a cycle that includes prereactor preparation, called the *front end,* in-reactor use, and postreactor management, called the *back end.* A typical fuel cycle may or may not incorporate fuel reprocessing in the back end (Fig. 17-16). The different processes are briefly explained below.

1. *Mining* of the uranium ore.
2. *Milling* and *refining* of the ore to produce uranium concentrates, U_3O_8.
3. *Processing* to produce of uranium hexafluoride, UF_6, from the uranium concentrates. This provides feed for isotopic (U^{235}) enrichment.
4. Isotopic *enrichment* of uranium hexafluoride to reach reactor enrichment requirements. This is done invariably now by the gaseous diffusion process.
5. *Fabrication* of the reactor fuel elements. This includes conversion of uranium hexafluoride to uranium dioxide UO_2, pelletizing, encapsulating in rods, and assembling the fuel rods into subassemblies.
6. *Power generation* in the reactor, resulting in *irradiated* or *spent* fuel.
7. *Short-term storage* of the spent fuel.
8. *Reprocessing* of the irradiated fuel and *conversion* of the residual uranium to uranium hexafluoride, UF_6 (for recycling through the gaseous diffusion plant for reenrichment) and/or extraction of Pu^{239} (converted from U^{238}) for recycling to the fuel-fabrication plant. Reprocessing can reuse up to 96 percent of the original material in the irradiated fuel with 4 percent actually becoming waste.
9. *Waste management,* which includes long-term storage of high-level wastes.

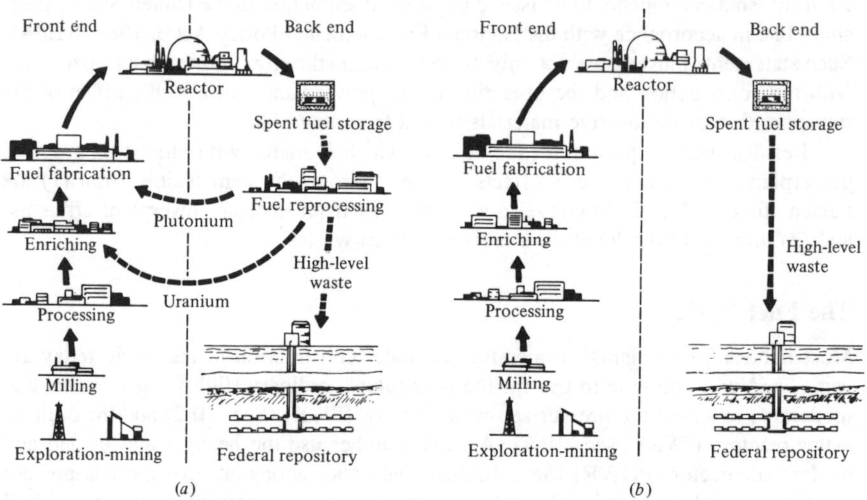

Figure 17-16 A typical nuclear fuel cycle (*a*) with reprocessing and (*b*) without reprocessing [186].

Step 8, reprocessing, may be bypassed, which results in disposal of both reusable fuel and wastes. This is the current (1982) U.S. Department of Energy process for dealing with irradiated fuel. The fuel assemblies are stored for at least 10 years and then buried. This is the so-called *throw-away* fuel cycle (Fig. 17-16*b*).

Wastes

The wastes associated with nuclear power can be summarized as:

1. *Gaseous effluents*. Under normal operation, these are released slowly from the powerplants into the biosphere and become diluted and dispersed harmlessly.
2. *Uranium mine and mill tailings*. *Tailings* are residues from uranium mining and milling operations. They contain low concentrations of naturally occurring radioactive materials. They are generated in large volumes and are stored at the mine or mill sites.
3. *Low-level wastes (LLW)*. These are classified as wastes that contain less than 10 nCi (nanocuries) per gram of transuranium contaminants and that have low but potentially hazardous concentrations of radioactive materials. They are generated in almost all activities (power generation, medical, industrial, etc.) that involve radioactive materials, require little or no shielding, and are usually disposed of in liquid form by shallow land burial (Fig. 17-17).
4. *High-level wastes (HLW)*. These are generated in the reprocessing of spent fuel. They contain essentially all the fission products and most of the transuranium elements not separated during reprocessing. Such wastes are to be disposed of carefully (Sec. 17-16).
5. *Spent fuel*. This is unreprocessed spent fuel that is removed from the reactor core after reaching its end-of-life core service. It is usually removed intact in its fuel-element structural form and then stored for 3 to 4 months under water on the plant

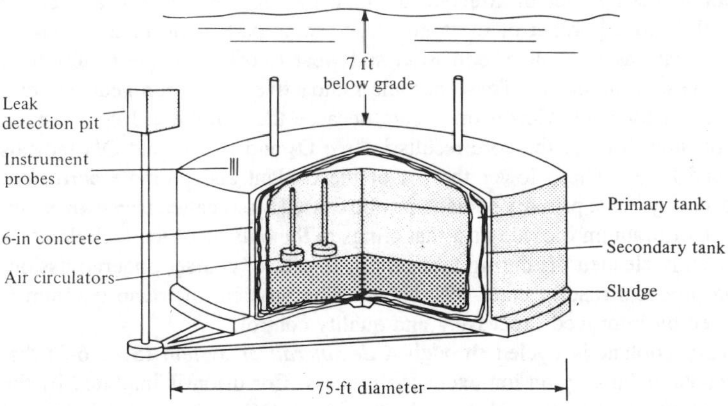

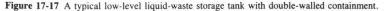

Figure 17-17 A typical low-level liquid-waste storage tank with double-walled containment.

site to give time for the most intense radioactive isotopes (which are the ones with shortest half-lives) to decay before shipment for reprocessing or disposal. Lack of a reprocessing capacity or a disposal policy has resulted in longer on-site storage, however. If the spent fuel is to be disposed of in a throw-away system (without reprocessing), it is treated as high-level waste.

17-15 RADIATIONS FROM NUCLEAR-POWERPLANT EFFLUENTS

Radiations from nuclear-powerplant effluents are low-dose-level types of radiations. The effluents are mainly gases and liquids. Environmental concerns about nuclear powerplants are prompted mainly by the effects of these radiations on the populations living near the plants. Sources of effluents vary with the type of reactor.

In both pressurized-water reactors (PWR) and boiling-water reactors (BWR), two important sources of effluents are (1) the condenser steam-jet air ejectors and (2) the turbine gland-seal system. The ejector uses high-pressure steam in a series of nozzles to create a vacuum, higher than that in the condenser, and thus draws air and other noncondensable gases from it. The mixture of steam and gases is collected, the steam portion condenses, and the gases are vented to the atmosphere (Sec. 6-3). In the gland seal, high-pressure steam is used to seal the turbine bearings by passing through a labyrinth from the outside in so that no turbine steam leaks out and, in the case of low-pressure turbines, no air leaks in. The escaping gland-seal steam is also collected and removed. In the BWR, the effluents come directly from the primary system. In the PWR, they come from the secondary system, so there is less likelihood of radioactive material being exhausted from a PWR than a BWR from these sources.

The primary-coolant radioactivity comes about mainly from fuel fission products that find their way into the coolant through the few small cracks that inevitably develop in the very thin cladding of some fuel elements. Such activity is readily detectable. However, to avoid frequent costly shutdowns and repairs, the system is designed to operate as long as the number of affected fuel elements does not exceed a tolerable limit, usually 0.25 to 1 percent of the total. Also, some particulate matter finds its way into the coolant as a result of corrosion and wear (erosion) of the materials of the primary system components. These become radioactive in the rich neutron environment of the reactor core. Corrosion occurs because the radiolytic decomposition of the water passing through the core results in free O_2 and free H and OH radicals as well as some H_2O_2. These lower the pH of the coolant and promote corrosion. Finally, radioactivity in the primary coolant may be caused by so-called *tramp uranium*. This is uranium or uranium dioxide dust that clings to the outside of the fuel elements and is insufficiently cleaned off during fabrication. It will, of course, undergo fission, and its fission products readily enter the coolant. The problem of tramp uranium is being minimized by improved processing and quality control.

The primary coolant is cycled through a *demineralizer system* (Sec. 6-8) that removes the contaminants by an ion-exchange process. Corrosion is inhibited by the slow addition of small amounts of lithium hydroxide, LiOH, to control acidity, and by the addition of hydrazine, NH_2NH_2, during shutdown to remove the radiolytic oxygen.

Of the above sources, the neutron activiation of the corrosion and wear products represents the major activity in the primary coolant. The materials irradiated include Zircaloy, Inconol, stainless steel, carbon steel, and other steel and copper alloys that may be rich in nickel, chromium, and cobalt. The principal "crud" activity in PWRs is due to Co^{58}, which results from the neutron irradiation of Ni^{58} with the release of a proton (hydrogen nucleus). In BWRs it is due to Co^{60}, which results from the irradiation of Co^{59}. Other activities in both reactor types are due to Fe^{59}, Cr^{51}, Mn^{54}, Zr^{95}. Zircaloy 4, used as cladding in most water reactors, contributes very little to the crud because the oxide film adheres well to the zircaloy.

Another form of activation in the primary coolant in water reactors is the result of the irradiation of the primary coolant itself. The main radioactivity is due to the neutron capture of oxygen in the water. (Neutrons captured by hydrogen convert it to nonradioactive deuterium.) The most important of the oxygen reactions, because of the high energy γ emitted, is the $O^{16}(n,p)N^{16}$ reaction. It has a microscopic cross section of 1.4×10^{-5} barn and is given by

$$_{8}O^{16} + {}_{0}n^{1} \rightarrow {}_{7}N^{16} + {}_{1}H^{1} \tag{17-25}$$

N^{16} is a radioactive β and γ emitter, reverting to O^{16}, with a half-life of 7.2 s. The β rays are mainly of 3.8, 4.3, and 0.5-MeV energy. Gamma rays are mainly of 6.13- and 7.10-MeV energy. Another reaction of somewhat less importance is caused by the neutron capture by O^{18}, which is present to the extent of 0.024 percent of all oxygen. The reaction is $O^{18}(n,\gamma)O^{19}$. O^{19} is also a β and γ emitter of 29-s half-life that converts to stable F^{19}. A third reaction of some importance is due to O^{17}, present to the extent of 0.037 percent of all oxygen. The reaction, $O^{17}(n,p)N^{17}$, results in N^{17}, also a β and γ emitter of 4.16-s half-life, which reverts back to O^{17}. Although the cross section for this reaction is greater than the first one by a factor of about 10^3, its effect is reduced by the small concentration of O^{17} in water. These and other resultant isotopes that yield only a few weak radiations are shown in Table 17-11.

Impurities other than corrosion and erosion products also exist in the primary water. The main one is argon 40, which enters into solution in water from the atmosphere, where it exists to the extent of 1 percent by volume. It undergoes the

Table 17-11 Principal activation of water in reactors

Isotope	Half-life	Concentrations	BWR release, μCi/s
N^{16}	7.20 s	1.0×10^2	1.7×10^8
O^{19}	26.80 s	8.0×10^{-1}	1.4×10^6
N^{17}	4.16 s	1.6×10^{-2}	2.6×10^4
N^{13}	9.9 min	6.5×10^{-3}	1.2×10^4
F^{18}	109.8 min	4.0×10^{-3}	7.2×10^3

$Ar^{40}(n,\gamma)Ar^{41}$ reaction. Ar^{41} is a β and γ emitter with a 1.83-h half-life that decays to stable potassium 41. With limited water exposure to the atmosphere and hold decay tanks used for effluents, the effect of this reaction is minimal.

Although all these activities can be released with effluents from water-reactor powerplants, the BWR is unique in that the primary coolant is also the working fluid and hence the gaseous isotopes, in particular, are more likely to be released to the atmosphere. The fourth column in Table 17-11 indicates the possible release rates in a 1000-MW BWR.

A problem that is more characteristic of PWRs than BWRs, on the other hand, is that of *tritium*. Tritium, a gaseous emitter of 12.25-year half-life, is primarily the result of using chemical shim with boric acid in the primary water in PWRs (Sec. 10-5). A little less than 20 percent of all boron is B^{10}. Upon neutron capture it undergoes the reaction $B^{10}(n,T)2\alpha$ given by

$$_{5}B^{10} + {_0}n^1 \rightarrow {_1}H^3 + 2{_2}He^4. \tag{17-26}$$

by which a tritium atom, $_1H^3$ or T, and two α particles (helium nuclei) are released. This reaction has a neutron-energy threshold of about 1.5 MeV, below which it will not take place. The tritium, like ordinary hydrogen, becomes a diatomic gas, $_1H_2^3$ or T_2, or combines with hydrogen to become HT.

Other tritium-producing reactions are due to lithium and deuterium. Lithium enters the primary coolant through the use of LiOH in the demineralizer or ion exchanger, which is used to help maintain the pH value of the primary water at an alkalinic 9.5. Lithium 6 constitutes about 7.4 percent of all lithium. Upon neutron capture, it undergoes the reaction $Li(n,\alpha)T$, given by

$$_{3}Li^6 + {_0}n^1 \rightarrow {_1}H^3 + {_2}He^4 \tag{17-27}$$

This reaction occurs at all neutron energies. Again the tritium becomes T_2 or HT gas. Tritium production from this source can be eliminated by substituting KOH or NH_4OH for LiOH in the demineralizer. It may be noted here that BWRs use no chemical shim and no lithium compounds, and, hence, the tritium in their liquid effluents is less than 10 percent of that in PWRs.

Deuterium dioxide or heavy water, H_2^2O or D_2O, is present to the extent of 0.015 percent of all water. Upon neutron capture one or both deuterium atoms in the heavy-water molecule undergoes the reaction $D(n,\gamma)T$, given by

$$_{1}H^2 + {_0}n^1 \rightarrow {_1}H^3 + \gamma \tag{17-28}$$

resulting in DTO or T_2O. This reaction contributes insignificantly to the production of tritium when compared with the preceding two. Tritium is also generated as a result of the use of boron-bearing control rods, which usually contain boron in the form of B_4C. (Reactors using Ag-In-Cd control material produce less tritium.) Tritium is also generated in *ternary fission*, i.e., fission that results in three, instead of two, fission fragments, one of which is tritium. A small fraction of this tritium, estimated at 1 percent, escapes the fuel through the Zircaloy cladding and enters the coolant. (Stainless steel cladding allows a much higher percentage of tritium to pass through.) Most of this tritium, therefore, is released in fuel-reprocessing plants rather than at the reactor site.

When tritium, generated as T_2 or HT gas, enters ordinary water H_2O, one T atom replaces one H atom in the water resulting in HTO and releasing hydrogen gas H_2 according to

$$\left.\begin{array}{l} T_2 + H_2O \rightarrow HTO + HT \\ \text{and } HT + H_2O \rightarrow HTO + H_2 \end{array}\right\} \quad (17\text{-}29)$$

and

HTO is physically and chemically similar to H_2O and becomes an integral part of it. Some of this "tritiated water" eventually finds its way into the environment.

In *heavy-water reactors,* as it is to be expected, the nearly 100 percent D_2O coolant accounts for particularly large tritium production, according to Eq. (17-28). As with light-water reactors, heavy-water reactors also suffer from N^{16} production, Eq. (17-25).

In *high-temperature gas-cooled reactors* that are helium-cooled (the HTGR, Sec. 10-12), tritium is generated by the neutron activation of He^3, which is present to the extent of 0.00013 percent of all helium. It is also generated by ternary fission to a greater degree than in water reactors. Tritium, however, can be removed from helium by absorption on a titanium sponge, which requires regeneration, normally every few months.

In *liquid-metal-cooled fast reactors,* the primary coolant, sodium, becomes intensely radioactive during plant operation. All naturally occuring sodium is made up of the isotope Na^{23}. When subjected to neutrons, it undergoes the reaction $Na^{23}(n,\gamma)Na^{24}$. Na^{24} is a radioisotope of about 15-h half-life that decays into stable Mg^{24} while emitting γ radiations, mainly of 2.76 and 1.38 MeV. The primary system, however, is separated from the steam cycle by a sodium intermediate system (Sec. 11-3) that does not become radioactive. The primary system operates at low pressures, below those of the intermediate and steam systems, so there is a very low likelihood of the radioactive sodium's leaking into the atmosphere and none of its leaking into the intermediate and steam systems.

One advantage of sodium is its ability to combine with or retain other elements. Some fast-reactor fuel elements are designed so that their gaseous fission products, such as iodine, xenon, and krypton, are vented into the coolant. Others may release fission products upon cladding failure. Sodium tends to retain some of these products, forming sodium iodide or holding cesium in solution, so that they do not escape to the gas blanket (usually argon) that separates sodium from air. Xe and Kr, on the other hand, are not retained by Na and would escape into the gas blanket.

A discussion of the biological effects of low-level radiation on human beings is a large subject and is beyond the scope of this text. It is, however, believed to be slight when compared with other hazards, such as toxic wastes [185].

17-16 HIGH-LEVEL WASTES

We will be concerned here with the high-level wastes that result from the spent fuel. High-level wastes, because of the intense exothermic activity (Sec. 9-7), generate too much heat for early passive burial and have to be cooled, usually by air circulation or other means, possibly for decades, before they can be permanently stored.

The level of activity is not as long as generally perceived. Although some fission products have extremely long half-lives, it should be recalled that the intensity of radiation is not a sole function of the half-life but is also a function of the energy generated in the reaction. Figure 17-18 [186,187] shows the relative energy generated by some of the more important fission products versus time. The level of activity is so high the first 10 years that storage can probably be best accomplished if the wastes

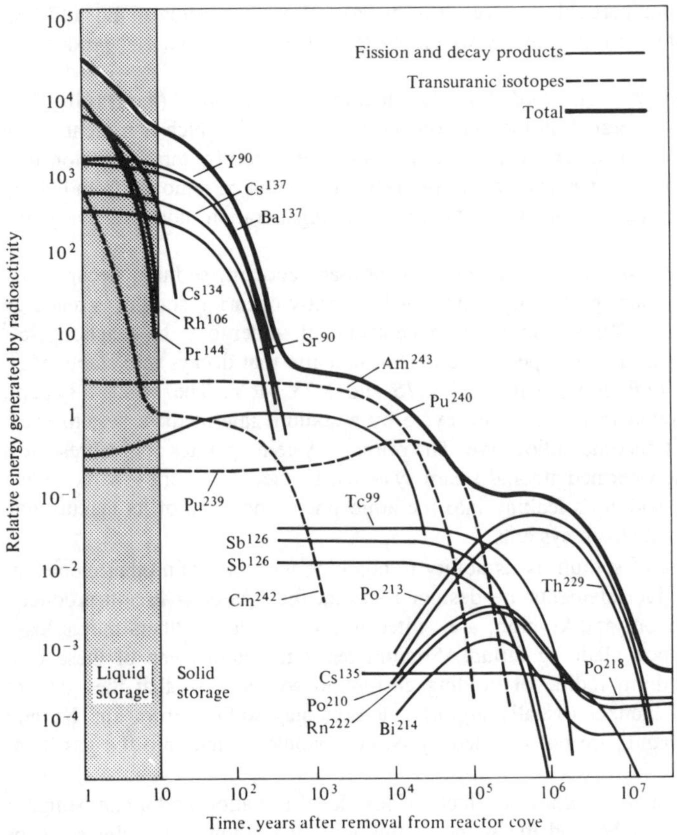

Figure 17-18 Relative energy generated by the more important fission products in high-level waste from spent nuclear fuel [186].

were in liquid form. After 10 years, the level drops by nearly an order of magnitude, thus permitting storage in solid form. It is interesting to note another and sharper drop in energy generated between 100 and 1000 years.

Figure 17-19 shows the relative "hazard" generated by the high-level waste and that generated by the original uranium ore as extracted from the ground. It shows that the high-level waste hazard drops below that of the ore it originally came from after about 800 years. The argument, based only on the half-lives of the isotopes, that stored nuclear waste would be a hazard to humanity for tens of thousands of years is, therefore, an inaccurate one.

Biological Effects of High-Level Wastes

Although all possible efforts are made to isolate high-level wastes from the biological system, it is instructive to study their effects. The principal effect is the destruction of body cells in the vicinity of the irradiated region. The effects are classified as somatic or genetic.

Somatic effects are limited to the exposed individuals. They are a direct result of the doses received by cells and manifest themselves in some form of malignancy. There is much information on high-dose irradiation effects on small animals that suggests that the frequency of effects within their population is proportional to the dose. There is, on the other hand, very little information on human beings. The only groups that have received high doses are the survivors of the atomic bombing of Hiroshima and Nagasaki, those receiving therapeutic radiation cancer treatment, and some occupationally exposed workers, such as those who work in underground uranium mines.

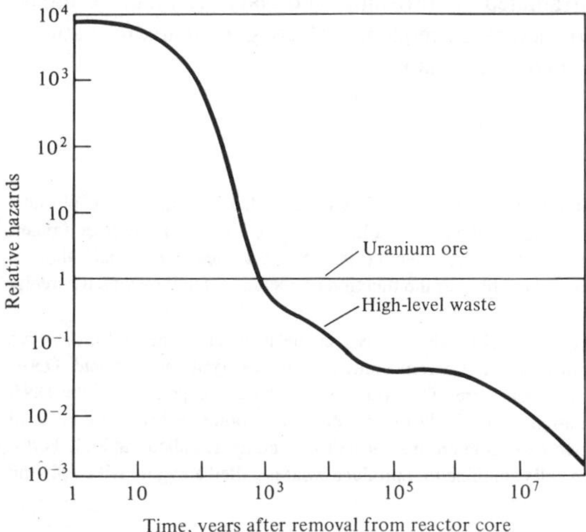

Figure 17-19 Ratio of hazards of high-level waste to those of original uranium ore versus time. *(From Ref. 188.)*

High-dose levels of radiation can result in leukemia, which occurs most frequently within a few years after doses of 50 to 500 rad. By 25 years, its frequency drops to levels that are normally encountered in the absence of radiation. Lung cancers have been observed in Hiroshima survivors who received 30 to 100 rad of gamma radiation. Estimates were that 10 cancer cases at 250 rad to 40 cases at 30 rad develop per rad for each million people during the first 25 years after exposure. Other information is available for the incidence of breast cancer (6 to 20 cases), thyroid cancer (40 cases), and other types of cancer (40 cases), all per rad for each million in the first 25 years after exposure to levels between 60 and 400 rad. Uranium miners exposed to at least a few hundred rads of alpha radiation from radon gas and its radioactive daughters show a high incidence of lung cancers. On the other hand, current evidence indicates few, if any, somatic effects can be detected at doses below 10 rad.

Genetic effects can be transmitted to the descendants of exposed individuals and thus affect unexposed generations. They are radiation-induced changes in the genetic materials of sex cells. They manifest themselves in different ways: (1) gene* mutations (changes) that result in changes in the functions of individual genes, (2) chromosome† aberrations due to breakage and reorganization of chromosomes, and (3) changes in the number of chromosomes. Such changes can result in offspring abnormalities, ranging from mild to lethal. Unfortunately (or rather fortunately), useful and adequate human data on the effect of high doses of radiation are not available and estimates of high doses to human reproductive cells are based on research on mice and other species. Such estimates yield ranges as wide as 30 to 1500 mutations per million babies per rad of acute exposures to males, and about half of that to females. Based on the response of mouse ovaries, it is expected that the effect on females may approach nil if conception occurs a sufficient time after exposure. Further, it is believed that dominant gene mutations are induced in the first-generation offspring of an irradiated population but that gene mutations appear at a higher frequency and last many more generations than chromosome mutations. Of the known human cases, however, about 75,000 children born to parents irradiated at Hiroshima and Nagasaki, and examined periodically since 1945, show no increased frequency of congenital malformations, stillbirths, or growth and development abnormalities.

* A *gene* is the carrier of information in the nucleus of a living cell that determines the physical characteristics of living things, such as the shape of the eyes in a human or the color of a flower or a plant. Genes are inherited, which is the reason why offspring resemble their parents. Genes were once thought to possess a physical structure. They are now thought of as the function or operational unit by which heredity is transmitted from parent to offspring.

† A chromosome is a threadlike microscopic part which exists in the nucleus of the cell and which carries hereditary information in the form of genes. The chromosome contains deoxyribonucleic acid, DNA, and ribonucleic acid, RNA, attached to a protein core. The arrangement of the components of the DNA molecules determines the genetic information. Each body or somatic cell contains a certain number of chromosomes, called diploid. Reproductive cells contain exactly half that number, called haploid. Fertilization occurs when two such reproductive cells combine to reproduce what is called a zygote with a diploid set of chromosomes.

Fuel Reprocessing

As indicated previously, spent fuel can either be stored directly or reprocessed. There are advantages and disadvantages to both schemes. The throw-away system (the former) avoids the costs and hazards associated with a reprocessing plant. The latter utilizes the unused uranium, converted plutonium, and other radioisotopes for use in a wide variety of services, such as isotopic generators, medicine, agriculture, and industry. There are also indications that the energy generated by the high-level waste produced after reprocessing is lower than that generated by the intact spent fuel by ratios of about 0.83 after 10 years, 0.38 after 100 years, and 0.06 after 1000 years [186].

Reprocessing of spent fuel is done by dissolving it, usually in nitric acid, then removing the converted plutonium and unspent uranium by solvent extraction. The remaining solution contains more than 99.9 percent of the nonvolatile fission products, plus some constituents of the cladding of the fuel elements, traces of plutonium, uranium, and others, and most of the resulting transuranium elements.

This remaining solution constitutes the high-level wastes. It is usually concentrated by evaporation and then stored as an aqueous nitric acid solution, usually in high-integrity stainless steel tanks. Permanent storage in liquid form, however, requires continual supervision and tank replacement over an indefinite period of time.

The experience with short-term tank storage has generally been good, though some leakage has been encountered, particularly in tanks used for storage of military wastes. It is now generally believed that, for the long term, a storage system based on solidification of the wastes would be more acceptable.

The conversion of the liquid wastes to a solid form has been studied since the 1960s, including large-scale pilot operations. Advanced processes are currently being developed. The aim is to convert the wastes to a solid product that is not liable to leakage, requires less supervision, and is more suited to final disposal. This solid product should maintain its mechanical strength. It should have good thermal conductivity, as it will continue to generate heat from radioactive decay for a long time, though at a decreasing rate. Ideally it should have a low leach rate.

Glasses and ceramics are now considered to be the most suitable forms for this final disposal. About 30 different processes have been developed over the past 20 years. The basic processes are shown in Fig. 17-20. The simplest one involves evaporation and denitration (or calcination) to form a granular or solid calcine. This is considered an interim product, since it does not meet all the above requirements; it is treated further by being mixed with additives and is then melted to form glasses or ceramics.

A second process involves mixing the additives with the original waste solution, then evaporating, denitrating, and melting this mixture to form the glasses or ceramics.

A third process uses an adsorption process and treatment at high temperature to produce the ceramics.

In recent years, attempts at improving the properties of the above glass or ceramic final products have been made. These essentially involve the further formation of either the calcine or glass into metal matrixes or coated particles to improve their leach resistance.

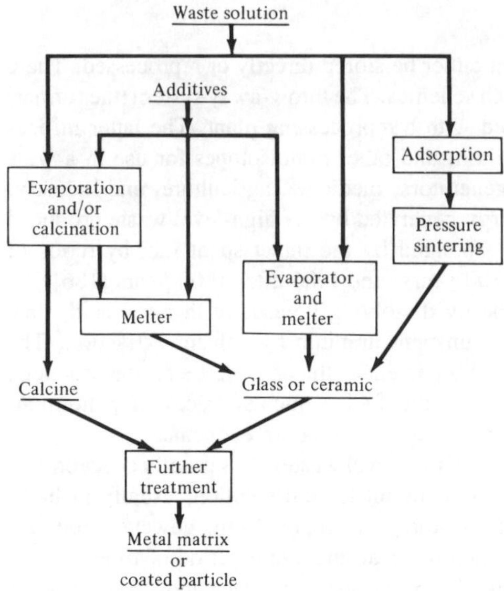

Figure 17-20 Basic high-level waste solidification processes.

Historically, the first attempt at converting wastes into glass was made at Chalk River, Canada, in 1960. This was closely followed by another developed at Harwell, England, and later by several attempts in the United States. In the United States a new plant was under construction in 1980. It is based on a 1963 process, used in a processing plant in Idaho, by which the waste is solidified by spraying it into a fluidized bed and then calcined at 400 to 500°C, thus producing a granular product that is then stored in air-cooled bins in underground vaults. Other promising plants are to be found in several countries, such as France, India, Germany, and the USSR.

Most solidification plants produce off-gases of steam and oxides of nitrogen that usually contain some fine particulate carryover and volatile radionuclides. These gases must be treated. All processes involve high temperatures and, of course, high levels of radioactivity. This combination imposes severe demands on plant design and operation. It is expected, however, that workable, commercially attractive solutions will emerge in the late 1980s. This is not late, as the problem of high-level waste is not expected to be of major effect for a few decades.

Disposal

The final disposition of the wastes, with or without the above treatments, is also of major concern. Many countries are undertaking activities involving underground disposal in deep geological formations. These activities include the investigation of suitable sites and suitable methods of storage in these sites. The main objectives, of course, are the protection of present and future populations from potential hazards. The suitable sites must be free of flowing groundwater, but the storage vessels must demonstrate reliability even in flowing-water conditions.

The disposal of low- and intermediate-level wastes (such as those used in research and medicine) has been done at relatively shallow depths in many countries. For example, since 1967 such wastes in solid form have been packaged in concrete or steel drums and buried in the old Asse Salt Mine in Germany.

If the spent fuel is to be disposed of in a throw-away process, it is buried intact.

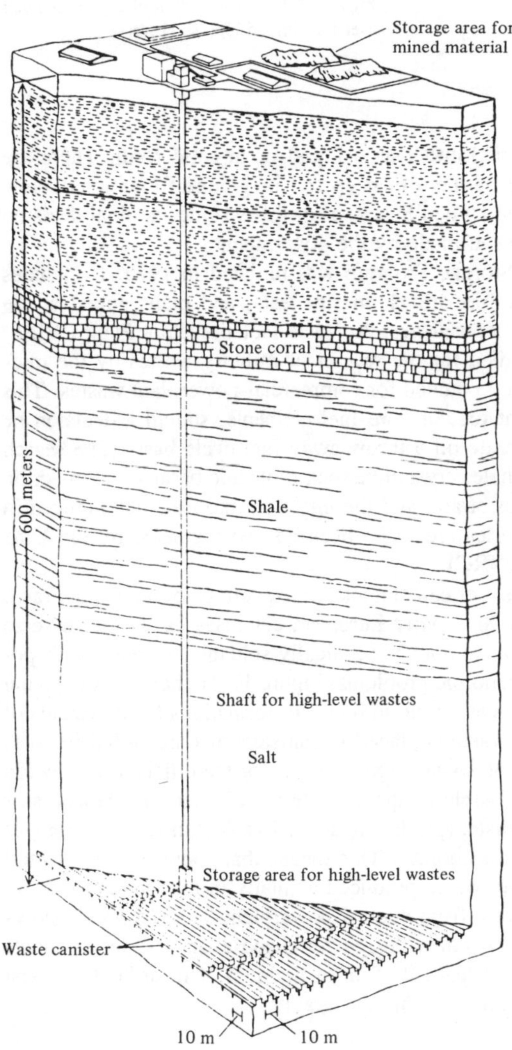

Figure 17-21 A conceptual depository of high-level waste in rock salt formations [186].

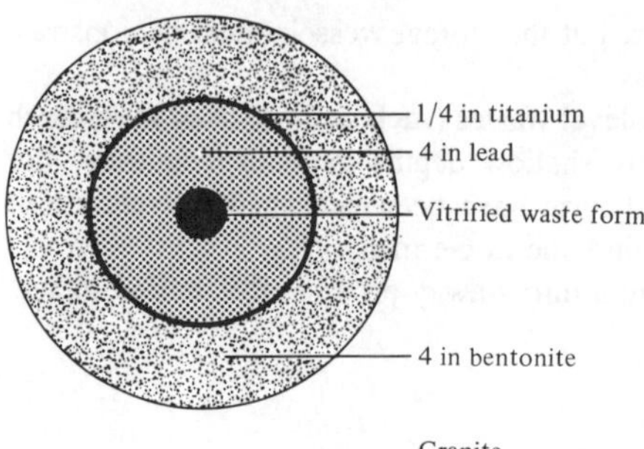

Figure 17-22 Cross section of a proposed canister for vitrified high-level waste.

If it is reprocessed, then it is buried either as a liquid or in solid form. There are several strategies for burial, most of them in deep salt or granite formations below ground or below the sea bed. In the United States, as of this writing, no final decision on high-level waste disposal has been made by the government.

A National Waste Terminal Storage Program was initiated in 1976. It includes the identification of suitable rock formations, in situ tests, and plans aimed at the establishment of up to six federal repositories. A Waste Isolation Pilot Plant (WIPP) in New Mexico is now being studied for the receipt of military wastes by the mid-1980s. Other federal repositories are planned for commercial-powerplant wastes. This also includes the disposal of spent reactor core fuel elements, should a decision be made not to treat them but to operate on a throw-away fuel cycle basis. The search includes sites in salt formations, shale, crystalline rock (volcanic basalt), the injection of liquid waste into isolated porous strata, and the injection into fractures induced in impermeable formations. Repositories are to be licensed and regulated by the U.S. Nuclear Regulatory Commission (NRC).

In France, which has the largest per capita nuclear program, plans are being drawn for long-range storage under the Alps. Other European countries have other diverse and innovative plans. The problem, although technically soluable, is laden with political, emotional, physical, and economic problems. Figure 17-21 shows a conceptual depository for the storage of high-level waste in rock salt formations for thousands of years. In the figure, the solidified waste is placed in canisters that are stored in holes drilled in rock salt with a spacing of about 10 m to allow for the efficient dissipation of energy without exceeding permissible temperature limits of either the canisters or the salt. It is estimated that each canister (producing about 10 kW at the time of storage) will require about 100 m^2 of salt for cooling. This means that about 0.5 km^2 would be needed to store all the high-level waste produced annually in the United States in 1980, if all electrical production were from nuclear powerplants. Figure 17-22 shows the cross section of a canister of Swedish design for disposal in granite. It shows vitrified waste surrounded by 4 in of lead, 0.25 in of titanium, 4 in of bentonite (an absorptive and colloidal clay mineral), and finally, granite.

PROBLEMS

17-1 A sample of particulate matter in the atmosphere has the following statistical count distribution. The particulate matter is assumed to have uniform density and the particles are approximated as spheres. d_i is the mean diameter, in micrometers, within each group i.

i	d_i	Count	i	d_i	Count
1	0.2	174	8	1.5	1602
2	0.3	347	9	2.0	1174
3	0.4	578	10	3.0	694
4	0.5	730	11	4.0	403
5	0.6	936	12	5.0	236
6	0.8	1271	13	6.0	150
7	1.0	1622	14	8.0	82

(*a*) Calculate the count, area, and mass mean diameters, and the standard deviation, in micrometers. (*b*) Plot on a semilogarithmic graph the count, area, and mass distributions, in percent, versus diameter. (*c*) Calculate the percent count and mass of particles that can be ingested with diameters of 1 μm and less, and (*d*) estimate the count modal diameter, in micrometers.

17-2 The particulate matter sample of Prob. 17-1 is acted upon by a collection device. Assuming that the device preferentially collects the larger particles, calculate the fraction of all the particles that escape collection, and the maximum diameter that escapes collection, in micrometers, if the device has an overall collection efficiency of (*a*) 95.16 percent or (*b*) 78.2 percent.

17-3 Flue gases from a powerplant contain 1 mass percent SO_2 · 1.2×10^4 kg/s of these gases are to be desulfurized by the Wellman-Lord process. In addition, elemental sulfur is to be generated by the Claus process. Estimate the various mass flow rates of the chemicals, in kilograms per second, and the amount of elemental sulfur produced, in tons per day.

17-4 A plate-type electrostatic precipitator is composed of 8-ft-high plates that are 9 in apart. The applied voltage is 50,000 V. The average flue gas velocity between the plates is 4.5 ft/s. Estimate (*a*) the minimum length of the precipitator, in feet, if the particles in the gas vary in size between 1 and 5 μm, and (*b*) the minimum and maximum fractional collection efficiencies using the Deutsch equation. Take gas viscosity as 2.95×10^5 kg/ms.

17-5 A plate-type electrostatic precipitator is composed of plates that are 8 ft high, 13.44 ft long, and 9 in apart. The average flue gas velocity at a given load is 4.5 ft/s. At these conditions the overall collection efficiency is 95 percent. Estimate (*a*) the change in that efficiency if the flue gas volume flow rate is increased by 15 percent, and (*b*) the change in the particulate matter mass removed, in percent. Assume that the flue gas temperature and particulate matter concentration and distribution in it remain unchanged.

17-6 A powerplant using a once-through cooling system is considered for construction on a river that has a water mass flow rate of 10^9 lb_m/h. The river water temperature at a given time of the year is 60°F. The plant is designed to have a net efficiency of 38 percent with a condenser cooling water temperature rise of 22°F. At the plant site, the river has a more-or-less uniform width of 1 mi. Environmental regulations do not permit plant operation to result in a mixed river water temperature rise of more than 2°F, 3 mi downstream of the plant. Assuming a water surface overall heat transfer coefficient of 10 Btu/h · ft^2 · °F (Sec. 7-10), calculate (*a*) the fraction of river water used for cooling, and (*b*) the maximum net power of the plant that can meet the above regulations.

17-7 Calculate the mass of carbon 14 and xenon 137, in kilograms, that would produce an activity of 1 Ci if their atomic masses are 14.00323 and 136.9080, respectively.

17-8 Calculate (*a*) the radioactivity of 1 g each of potassium 40 and plutonium 239 in curies if their atomic masses are 39.9740 and 239.0522, respectively, and (*b*) the activity of each after 1000 years.

17-9 Chlorine 36 has an atomic mass of 35.96852 and a half life of 3.1×10^5 years. It is a β emitter. The most probable energy of the emitted β particles is 0.25 MeV. For 1 kg of Cl^{36}, calculate (*a*) the activity, in curies and (*b*) the biological dose rate, in sieverts per second and millirems per second, assuming a quality factor (relative biological effectiveness) of 1.2, and that all radiation falls on a 75-kg body.

17-10 When an isotope *A* with a concentration N_A nuclei/m^3 is irradiated in a neutron flux ϕ with a microscopic absorption cross section σ_A and forms a daughter radioisotope *B* that decays with a half-life $\theta_{1/2}$, the concentration of the daughter N_B at any given time θ is equal to the neutron absorption reaction rate $N_A\sigma_A\ \phi$ (Sec. 9-12) times the quantity $(1\text{-}e^{-\lambda\theta})/\lambda$, where λ is the decay constant of *B*. Consider a water-cooled reactor and the reaction $0^{16}(n,p)N^{16}$. Water has a density of 800 kg/m^3. The average neutron flux is 10^{18} neutrons/s · m^2. Calculate the concentration of N^{16} and its activity, Ci/m^3, after 1 h of irradiation.

17-11 Repeat Prob. 17-10 but for a sodium-cooled reactor, considering the $Na^{23}(n,\gamma)N^{24}$ reaction for irradiations of 1 h, 30 d, and 300 d. Sodium is at a density of 760 kg/m^3. Na^{23} has a neutron microscopic absorption cross section of 2 barn and an atomic mass of 22.98977. The neutron flux is 10^{19} neutrons/s · m^2.

17-12 A pressurized-water reactor is chemically shimmed with boric acid (Sec. 10-5). The natural boron concentration is 1950 ppm. The water density is 800 kg/m^3. Consider the reaction $B^{10}(n,T)\alpha$. B^{10} constitutes 19.78 percent of all boron, has a molecular mass of 10.01294 and a thermal neutron microscopic cross section of 3,837 barn. The average neutron flux is 10^{18} neutrons/s · m^2. Calculate the tritium activities after 1 h, 30 d, and 300 d of irradiation.

APPENDIX

A

THERMODYNAMIC PROPERTIES OF WATER AND STEAM

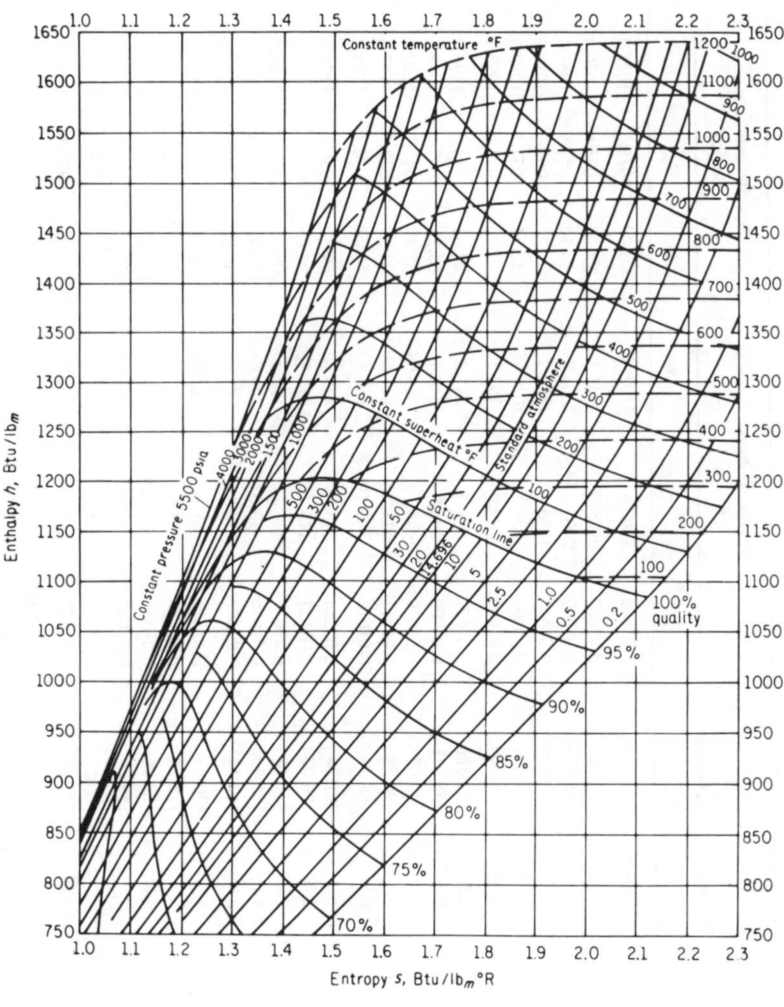

Figure A-1 Mollier diagram for steam. *(Based on data in Ref. 6.)*

Table A-1 Saturated steam: pressure table (English units)*

Pressure, psia	Temperature, °F	Specific volume, ft³/lb$_m$			Specific enthalpy, Btu/lb$_m$			Specific entropy, Btu/lb$_m$°F		
		v_f	v_{fg}	v_g	h_f	h_{fg}	h_g	s_f	s_{fg}	s_g
0.0886	32.018	0.01602	3302.4	3302.4	0.00	1075.5	1075.5	0	2.1872	2.1872
0.10	35.023	0.01602	2945.5	2945.5	3.03	1073.8	1076.8	0.0061	2.1705	2.1766
0.15	45.453	0.01602	2004.7	2004.7	13.50	1067.9	1081.4	0.0271	2.1140	2.1411
0.20	53.160	0.01603	1526.3	1526.3	21.22	1063.5	1084.7	0.0422	2.0738	2.1160
0.30	64.484	0.01604	1039.7	1039.7	32.54	1057.1	1089.7	0.0641	2.0168	2.0809
0.40	72.869	0.01606	792.0	792.1	40.92	1052.4	1093.3	0.0799	1.9762	2.0562
0.5	79.586	0.01607	641.5	641.5	47.62	1048.6	1096.3	0.0925	1.9446	2.0370
0.6	85.218	0.01609	540.0	540.1	53.25	1045.5	1098.7	0.1028	1.9186	2.0215
0.7	90.09	0.01610	446.93	466.94	58.10	1042.7	1100.8	0.3	1.8966	2.0083
0.8	94.38	0.01611	411.67	411.69	62.39	1040.3	1102.6	0.1117	1.8775	1.9970
0.9	98.24	0.01612	368.41	368.43	66.24	1038.1	1104.3	0.1264	1.8606	1.9870
1.0	101.74	0.01614	333.59	333.60	69.73	1036.1	1105.8	0.1326	1.8455	1.9781
2.0	126.07	0.01623	173.74	173.76	94.03	1022.1	1116.2	0.1750	1.7450	1.9200
3.0	141.47	0.01630	118.71	118.73	109.42	1013.2	1122.6	0.2009	1.6854	1.8864
4.0	152.96	0.01636	90.63	90.64	120.92	1006.4	1127.3	0.2199	1.6428	1.8626
5.0	162.24	0.01641	73.515	73.53	130.20	1000.9	1131.1	0.2349	1.6094	1.8443
6.0	170.05	0.01645	61.967	61.98	138.03	996.2	1134.2	0.2474	1.5820	1.8294
7.0	176.84	0.01649	53.634	53.65	144.83	992.1	1136.9	0.2581	1.5587	1.8168
8.0	182.86	0.01653	47.328	47.35	150.87	988.5	1139.3	0.2676	1.5384	1.8060
9.0	188.27	0.01656	42.385	42.40	156.30	985.1	1141.4	0.2760	1.5204	1.7964
10	193.21	0.01659	38.404	38.42	161.26	982.1	1143.3	0.2836	1.5043	1.7879
14.696	212.00	0.01672	26.782	26.80	180.17	970.3	1150.5	0.3121	1.4447	1.7568
15.0	213.03	0.016726	26.274	26.290	181.21	969.7	1150.9	0.3137	1.4415	1.7552
20.0	227.96	0.016834	20.070	20.087	196.27	960.1	1156.3	0.3358	1.3962	1.7320

30.0	250.34	0.017009	13.7266	13.7436	218.9	945.2	1164.1	0.3682	1.3313	1.6995
40.0	267.25	0.017151	10.4794	10.4965	236.1	933.6	1169.8	0.3921	1.2844	1.6765
50.0	281.02	0.017274	8.4967	8.5140	250.2	923.9	1174.1	0.4112	1.2474	1.6586
60.0	292.71	0.017383	7.1562	7.1736	262.2	915.4	1177.6	0.4273	1.2167	1.6440
70.0	302.93	0.017482	6.1875	6.2050	272.7	907.8	1180.6	0.4411	1.1905	1.6316
80.0	312.04	0.017573	5.4536	5.4711	282.1	900.9	1183.1	0.4534	1.1675	1.6208
90.0	320.28	0.017659	4.8779	4.8953	290.7	894.6	1185.3	0.4643	1.1470	1.6113
100.0	327.82	0.017740	4.4133	4.4310	298.5	888.6	1187.2	0.4743	1.1284	1.6027
110.0	334.79	0.01782	4.0306	4.0484	305.8	883.1	1188.9	0.4834	1.1115	1.5950
120.0	341.27	0.01789	3.7097	3.7275	312.6	877.8	1190.4	0.4919	1.0960	1.5879
130.0	347.33	0.01796	3.4364	3.4544	319.0	872.8	1191.7	0.4998	1.0815	1.5813
140.0	353.04	0.01803	3.2010	3.2190	325.0	868.0	1193.0	0.5071	1.0681	1.5752
150.0	358.43	0.01809	2.9958	3.0139	330.6	863.4	1194.1	0.5141	1.0554	1.5695
160.0	363.55	0.01815	2.8155	2.8336	336.1	859.0	1195.1	0.5206	1.0435	1.5641
170.0	368.42	0.01821	2.6556	2.6738	341.2	854.8	1196.0	0.5269	1.0322	1.5591
180.0	373.08	0.01827	2.5129	2.5312	346.2	850.7	1196.9	0.5328	1.0215	1.5543
190.0	377.53	0.01833	2.3847	2.4030	350.9	846.7	1197.6	0.5384	1.0113	1.5498
200.0	381.80	0.01839	2.2689	2.2873	355.5	842.8	1198.3	0.5438	1.0016	1.5454
210.0	385.91	0.01844	2.16373	2.18217	359.9	839.1	1199.0	0.5490	0.9923	1.5413
220.0	389.88	0.01850	2.06779	2.08629	364.2	835.4	1199.6	0.5540	0.9834	1.5374
230.0	393.70	0.01855	1.97991	1.99846	368.3	831.8	1200.1	0.5588	0.9748	1.5336
240.0	397.39	0.01860	1.89909	1.91769	372.3	828.4	1200.6	0.5634	0.9665	1.5299
250.0	400.97	0.01865	1.82452	1.84317	376.1	825.0	1201.1	0.5679	0.9585	1.5264
260.0	404.44	0.01870	1.75548	1.77418	379.9	821.6	1201.5	0.5722	0.9508	1.5230
270.0	407.80	0.01875	1.69137	1.71013	383.6	818.3	1201.9	0.5764	0.9433	1.5197
280.0	411.07	0.01880	1.63169	1.65049	387.1	815.1	1202.3	0.5805	0.9361	1.5166
290.0	414.25	0.01885	1.57597	1.59482	390.6	812.0	1202.6	0.5844	0.9291	1.5135
300.0	417.35	0.01889	1.52384	1.54274	394.0	808.9	1202.9	0.5882	0.9223	1.5105
350.0	431.73	0.01912	1.30642	1.32554	409.8	794.2	1204.0	0.6059	0.8909	1.4968
400.0	444.60	0.01934	1.14162	1.16095	424.2	780.4	1204.6	0.6217	0.8630	1.4847
450.0	456.28	0.01954	1.01224	1.03179	437.3	767.5	1204.8	0.6360	0.8378	1.4738

(continued)

Table A-1 Saturated steam: pressure table (English units)* (Continued)

Pressure, psia	Temperature, °F	Specific volume, ft^3/lb_m			Specific enthalpy, Btu/lb_m			Specific entropy, $Btu/lb_m{}^\circ F$		
		v_f	v_{fg}	v_g	h_f	h_{fg}	h_g	s_f	s_{fg}	s_g
500.0	467.01	0.01975	0.90787	0.92762	449.5	755.1	1204.7	0.6490	0.8148	1.4639
550.0	476.94	0.01994	0.82183	0.84177	460.9	743.3	1204.3	0.6611	0.7936	1.4547
600.0	486.20	0.02013	0.74962	0.76975	471.7	732.0	1203.7	0.6723	0.7738	1.4461
650.0	494.89	0.02032	0.68811	0.70843	481.9	720.9	1202.8	0.6828	0.7552	1.4381
700.0	503.08	0.02050	0.63505	0.65556	491.6	710.2	1201.8	0.6928	0.7377	1.4304
750.0	510.84	0.02069	0.58880	0.60949	500.9	699.8	1200.7	0.7022	0.7210	1.4232
800.0	518.21	0.02087	0.54809	0.56896	509.8	689.6	1199.4	0.7111	0.7051	1.4163
850.0	525.24	0.02105	0.51197	0.53302	518.4	679.5	1198.0	0.7197	0.6899	1.4096
900.0	531.95	0.02123	0.47968	0.50091	526.7	669.7	1196.4	0.7279	0.6753	1.4032
950.0	538.39	0.02141	0.45064	0.47205	534.7	660.0	1194.7	0.7358	0.6612	1.3970
1000.0	544.58	0.02159	0.42436	0.44596	542.6	650.4	1192.9	0.7434	0.6476	1.3910
1050.0	550.53	0.02177	0.40047	0.42224	550.1	640.9	1191.0	0.7507	0.6344	1.3851
1100.0	556.28	0.02195	0.37863	0.40058	557.5	631.5	1189.1	0.7578	0.6216	1.3794
1150.0	561.82	0.02214	0.35859	0.38073	564.8	622.2	1187.0	0.7647	0.6091	1.3738
1200.0	567.19	0.02232	0.34013	0.36245	571.9	613.0	1184.8	0.7714	0.5969	1.3683
1250.0	572.38	0.02250	0.32306	0.34556	578.8	603.8	1182.6	0.7780	0.5850	1.3630
1300.0	577.42	0.02269	0.30722	0.32991	585.6	594.6	1180.2	0.7843	0.5733	1.3577
1350.0	582.32	0.02288	0.29250	0.31537	592.3	585.4	1177.8	0.7906	0.5620	1.3525
1400.0	587.07	0.02307	0.27871	0.30178	598.8	576.5	1175.3	0.7966	0.5507	1.3474
1450.0	591.70	0.02327	0.26584	0.28911	605.3	567.4	1172.8	0.8026	0.5397	1.3423
1500.0	596.20	0.02346	0.25372	0.27719	611.7	558.4	1170.1	0.8085	0.5288	1.3373
1550.0	600.59	0.02366	0.24235	0.26601	618.0	549.4	1167.4	0.8142	0.5182	1.3324
1600.0	604.87	0.02387	0.23159	0.25545	624.2	540.3	1164.5	0.8199	0.5076	1.3274
1650.0	609.05	0.02407	0.22143	0.24551	630.4	531.3	1161.6	0.8254	0.4971	1.3225
1700.0	613.13	0.02428	0.21178	0.23607	636.5	522.2	1158.6	0.8309	0.4867	1.3176

1750.0	617.12	0.02450	0.20263	0.22713	642.5	513.1	1155.6	0.8363	0.4765	1.3128
1800.0	621.02	0.02472	0.19390	0.21861	648.5	503.8	1152.3	0.8417	0.4662	1.3079
1850.0	624.83	0.02495	0.18558	0.21052	654.5	494.6	1149.0	0.8470	0.4561	1.3030
1900.0	628.56	0.02517	0.17761	0.20278	660.4	485.2	1145.6	0.8522	0.4459	1.2981
1950.0	632.22	0.02541	0.16999	0.19540	666.3	475.8	1142.0	0.8574	0.4358	1.2931
2000.0	635.80	0.02565	0.16266	0.18831	672.1	466.2	1138.3	0.8625	0.4256	1.2881
2100.0	642.76	0.02615	0.14885	0.17501	683.8	446.7	1130.5	0.8727	0.4053	1.2780
2200.0	649.45	0.02669	0.13603	0.16272	695.5	426.7	1122.2	0.8828	0.3848	1.2676
2300.0	655.89	0.02727	0.12406	0.15133	707.2	406.0	1113.2	0.8929	0.3640	1.2569
2400.0	662.11	0.02790	0.11287	0.14076	719.0	384.8	1103.7	0.9031	0.3430	1.2460
2500.0	668.11	0.02859	0.10209	0.13068	731.7	361.6	1093.3	0.9139	0.3206	1.2345
2600.0	673.91	0.02938	0.09172	0.12110	744.5	337.6	1082.0	0.9247	0.2977	1.2225
2700.0	679.53	0.03029	0.08165	0.11194	757.3	312.3	1069.7	0.9356	0.2741	1.2097
2800.0	684.96	0.03134	0.07171	0.10305	770.7	285.1	1055.8	0.9468	0.2491	1.1958
2900.0	690.22	0.03262	0.06158	0.09420	785.1	254.7	1039.8	0.9588	0.2215	1.1803
3000.0	695.33	0.03428	0.05073	0.08500	801.8	218.4	1020.3	0.9728	0.1891	1.1619
3100.0	700.28	0.03681	0.03771	0.07452	824.0	169.3	993.3	0.9914	0.1460	1.1373
3200.0	705.08	0.04472	0.01191	0.05663	875.5	56.1	931.6	1.0351	0.0482	1.0832
3208.2†	705.47	0.05078	0.00000	0.05078	906.0	0.0	906.0	1.0612	0.0000	1.0612

* Tables A-1 through A-5 are abstracted from the ASME *Thermodynamic and Transport Properties of Steam*, Ref. 6.

† Critical point.

Table A-2 Saturated stream: temperature table (English units)

Temperature, °F	Pressure, psia	Specific volume, ft^3/lb_m			Specific enthalpy, Btu/lb_m			Specific entropy, $Btu/(lb_m)(°F)$		
		v_f	v_{fg}	v_g	h_f	h_{fg}	h_g	s_f	s_{fg}	s_g
32.018	0.08865	0.016022	3302.4	3302.4	0.0003	1075.5	1075.5	0.0000	2.1872	2.1872
33.0	0.09223	0.016021	3180.7	3180.7	0.989	1074.9	1075.9	0.0020	2.1817	2.1837
34.0	0.09600	0.016021	3061.9	3061.9	1.996	1074.4	1076.4	0.0041	2.1762	2.1802
35.0	0.09991	0.016020	2948.1	2948.1	3.002	1073.8	1076.8	0.0061	2.1706	2.1767
36.0	0.10395	0.016020	2839.0	2839.0	4.008	1073.2	1077.2	0.0081	2.1651	2.1732
37.0	0.10815	0.016019	2734.4	2734.4	5.013	1072.7	1077.7	0.0101	2.1596	2.1697
38.0	0.11249	0.016019	2634.1	2634.2	6.018	1072.1	1078.1	0.0122	2.1541	2.1663
39.0	0.11698	0.016019	2538.0	2538.0	7.023	1071.5	1078.5	0.0142	2.1487	2.1629
40.0	0.12163	0.016019	2445.8	2445.8	8.027	1071.0	1079.0	0.0162	2.1432	2.1594
41.0	0.12645	0.016019	2357.3	2357.3	9.031	1070.4	1079.4	0.0182	2.1325	2.1560
42.0	0.13143	0.016019	2272.4	2272.4	10.035	1069.8	1079.9	0.0202	2.1271	2.1527
43.0	0.13659	0.016019	2191.0	2191.0	11.038	1069.3	1080.3	0.0222	2.1217	2.1493
44.0	0.14192	0.016019	2112.8	2112.8	12.041	1068.7	1080.7	0.0242	2.1164	2.1459
45.0	0.14744	0.016020	2037.7	2037.8	13.044	1068.1	1081.2	0.0262	2.1111	2.1426
46.0	0.15314	0.016020	1965.7	1965.7	14.047	1067.6	1081.6	0.0282	2.1058	2.1493
47.0	0.15904	0.016021	1896.5	1896.5	15.049	1067.0	1082.1	0.0301	2.1006	2.1460
48.0	0.16514	0.016021	1830.0	1830.0	16.051	1066.4	1082.5	0.0321	2.1006	2.1327
49.0	0.17144	0.016022	1766.2	1766.2	17.053	1065.9	1082.9	0.0341	2.0953	2.1294
50.0	0.17796	0.016023	1704.8	1704.8	18.054	1065.3	1083.4	0.0361	2.0901	2.1262
51.0	0.18469	0.016023	1646.9	1645.9	19.056	1064.7	1083.8	0.0380	2.0849	2.1230
52.0	0.19165	0.016024	1589.2	1589.2	20.057	1064.2	1084.2	0.0400	2.0798	2.1197
53.0	0.19883	0.016025	1534.7	1534.8	21.058	1063.6	1084.7	0.0419	2.0746	2.1165
54.0	0.20625	0.016026	1482.4	1482.4	22.058	1063.1	1085.1	0.0439	2.0695	2.1134
55.0	0.21392	0.016027	1432.0	1432.0	23.059	1062.5	1085.6	0.0458	2.0644	2.1102
56.0	0.22183	0.016028	1383.6	1383.6	24.059	1061.9	1086.0	0.0478	2.0593	2.1070
57.0	0.23000	0.016029	1337.0	1337.0	25.060	1061.4	1086.4	0.0497	2.0542	2.1039
58.0	0.23843	0.016031	1292.2	1292.2	26.060	1060.8	1086.9	0.0516	2.0491	2.1008

59.0	0.24713	0.016032	1249.1	1249.1	27.060	1060.2	1087.3	0.0536	2.0441	2.0977
60.0	0.25611	0.016033	1207.6	1207.6	28.060	1059.7	1087.7	0.0555	2.0391	2.0946
61.0	0.26538	0.016035	1167.6	1167.6	29.059	1059.1	1088.2	0.0574	2.0341	2.0915
62.0	0.27494	0.016036	1129.2	1129.2	30.059	1058.5	1088.6	0.0593	2.0291	2.0885
63.0	0.28480	0.016038	1092.1	1092.1	31.058	1058.0	1089.0	0.0613	2.0242	2.0854
64.0	0.29497	0.016039	1056.5	1056.5	32.058	1057.4	1089.5	0.0632	2.0192	2.0824
65.0	0.30545	0.016041	1022.1	1022.1	33.057	1056.9	1089.9	0.0651	2.0143	2.0794
66.0	0.31626	0.016043	989.0	989.1	34.056	1056.3	1090.4	0.0670	2.0094	2.0764
67.0	0.32740	0.016044	957.2	957.2	35.055	1055.7	1090.8	0.0689	2.0045	2.0734
68.0	0.33889	0.016046	926.5	926.5	36.054	1055.2	1091.2	0.0708	1.9996	2.0704
69.0	0.35073	0.016048	896.9	896.9	37.053	1054.6	1091.7	0.0727	1.9948	2.0675
70.0	0.36292	0.016050	868.3	868.4	38.052	1054.0	1092.1	0.0745	1.9900	2.0645
71.0	0.37549	0.016052	840.8	840.9	39.050	1053.5	1092.5	0.0764	1.9852	2.0616
72.0	0.38844	0.016054	814.3	814.3	40.049	1052.9	1093.0	0.0783	1.9804	2.0587
73.0	0.40177	0.016056	788.8	788.8	41.048	1052.4	1093.4	0.0802	1.9756	2.0558
74.0	0.41550	0.016058	764.1	764.1	42.046	1051.8	1093.8	0.0821	1.9708	2.0529
75.0	0.42964	0.016060	740.3	740.3	43.045	1051.2	1094.3	0.0839	1.9961	2.0500
76.0	0.44420	0.016063	717.4	717.4	44.043	1050.7	1094.7	0.0858	1.9614	2.0472
77.0	0.45919	0.016065	695.2	695.2	45.042	1050.1	1095.1	0.0877	1.9567	2.0443
78.0	0.47461	0.016067	673.8	673.9	46.040	1049.5	1095.6	0.0895	1.9520	2.0415
79.0	0.49049	0.016070	653.2	653.2	47.038	1049.0	1096.0	0.0914	1.9473	2.0387
80.0	0.50683	0.016072	633.3	633.3	48.037	1048.4	1096.4	0.0932	1.9426	2.0359
81.0	0.52364	0.016074	614.1	614.1	49.035	1047.8	1096.9	0.0951	1.9380	2.0331
82.0	0.54093	0.016077	595.5	595.6	50.033	1047.3	1097.3	0.0969	1.9334	2.0303
83.0	0.55872	0.016079	577.6	577.6	51.031	1046.7	1097.7	0.0988	1.9288	2.0275
84.0	0.57702	0.016082	560.3	560.3	52.029	1046.1	1098.2	0.1006	1.9242	2.0248
85.0	0.59583	0.016085	543.6	543.6	53.027	1045.6	1098.6	0.1024	1.9196	2.0221
86.0	0.61518	0.016087	527.5	527.5	54.026	1045.0	1099.0	0.1043	1.9151	2.0193
87.0	0.63507	0.016090	511.9	511.9	55.024	1044.4	1099.5	0.1061	1.9105	2.0166
88.0	0.65551	0.016093	496.8	496.8	56.022	1043.9	1099.9	0.1079	1.9060	2.0139
89.0	0.67653	0.016096	482.2	482.2	57.020	1043.3	1100.3	0.1097	1.9015	2.0112
90.0	0.69813	0.016099	468.1	468.1	58.018	1042.7	1100.8	0.1115	1.8970	2.0086
91.0	0.72032	0.016102	454.5	454.5	59.016	1042.2	1101.2	0.1134	1.8926	2.0059

(*continued*)

Table A-2 Saturated stream: temperature table (English units) (Continued)

Temperature, °F	Pressure, psia	Specific volume, ft^3/lb$_m$			Specific enthalpy, Btu/lb$_m$			Specific entropy, Btu/(lb$_m$)(°F)		
		v_f	v_{fg}	v_g	h_f	h_{fg}	h_g	s_f	s_{fg}	s_g
92.0	0.74313	0.016105	441.3	441.3	60.014	1041.6	1101.6	0.1152	1.8881	2.0033
93.0	0.76655	0.016108	428.6	428.6	61.012	1041.0	1102.1	0.1170	1.8837	2.0006
94.0	0.79062	0.016111	416.3	416.3	62.010	1040.5	1102.5	0.1188	1.8792	1.9980
95.0	0.81534	0.016114	404.4	404.4	63.008	1039.9	1102.9	0.1206	1.8748	1.9954
96.0	0.84072	0.016117	392.8	392.9	64.006	1039.3	1103.3	0.1224	1.8704	1.9928
97.0	0.86679	0.16120	381.7	381.7	65.005	1038.8	1103.8	0.2342	1.8660	1.9902
98.0	0.89356	0.16123	370.9	370.9	66.003	1038.2	1104.2	0.1260	1.8617	1.9876
99.0	0.92103	0.16127	360.5	360.6	67.001	1037.6	1104.6	0.1278	1.8573	1.9851
100.0	0.94924	0.16130	350.4	360.4	67.999	1037.1	1105.1	0.1295	1.8530	1.9825
101.0	0.97818	0.16133	340.6	340.6	68.997	1036.5	1105.5	0.1313	1.8487	1.9800
102.0	1.00789	0.16137	331.1	331.1	69.995	1035.9	1105.9	0.1331	1.8444	1.9775
103.0	1.03838	0.016140	322.0	322.0	70.993	1035.4	1106.3	0.1349	1.8401	1.9750
104.0	1.06965	0.016144	313.1	313.1	71.992	1034.8	1106.8	0.1366	1.8358	1.9725
105.0	1.10174	0.016148	304.5	304.5	72.990	1034.2	1107.2	0.1384	1.8315	1.9700
110.0	1.2750	0.016165	265.37	265.39	77.98	1031.4	1109.3	0.1472	1.8105	1.9577
120.0	1.6927	0.016204	203.25	203.26	87.97	1025.6	1113.6	0.1646	1.7693	1.9339
130.0	2.2230	0.016247	157.32	157.33	97.96	1019.8	1117.8	0.1817	1.7295	1.9112
140.0	2.8892	0.016293	122.98	123.00	107.95	1014.0	1122.0	0.1985	1.6910	1.8895
150.0	3.7184	0.016343	97.05	97.07	117.95	1008.2	1126.1	0.2150	1.6536	1.8686
160.0	4.7414	0.016395	77.27	77.29	127.96	1002.2	1130.2	0.2313	1.6174	1.8487
170.0	5.9926	0.016451	62.04	62.06	137.97	995.2	1134.2	0.2473	1.5822	1.8295
180.0	7.5110	0.016510	50.21	50.22	148.00	990.2	1138.2	0.2631	1.5480	1.8111
190.0	9.340	0.016572	40.941	40.957	158.04	984.1	1142.1	0.2787	1.5148	1.7934
200.0	11.526	0.016637	33.622	33.639	168.09	977.9	1146.0	0.2940	1.4824	1.7764
220.0	17.186	0.016775	23.131	23.148	188.23	965.2	1153.4	0.3241	1.4201	1.7442
240.0	24.968	0.016926	16.304	16.321	208.45	952.1	1160.6	0.3533	1.3609	1.7142
260.0	35.427	0.017089	11.745	11.762	228.76	938.6	1167.4	0.3819	1.3043	1.6862

280.0	49.200	0.017264	8.627	8.644	249.17	924.6	1173.8	0.4098	1.2501	1.6599
300.0	67.005	0.01745	6.4483	6.4658	269.7	910.0	1179.7	0.4372	1.1979	1.6351
320.0	89.643	0.01766	4.8961	4.9138	290.4	894.8	1185.2	0.4640	1.1477	1.6116
340.0	117.992	0.01787	3.7699	3.7878	311.3	878.8	1190.1	0.4902	1.0990	1.5892
360.0	153.010	0.01811	2.9392	2.9573	332.3	862.1	1194.4	0.5161	1.0517	1.5678
380.0	195.729	0.01836	2.3170	2.3353	353.6	844.5	1198.0	0.5416	1.0057	1.5473
400.0	247.259	0.01864	1.8444	1.8630	375.1	825.9	1201.0	0.5667	0.9607	1.5274
420.0	308.780	0.01894	1.4808	1.4997	396.9	806.2	1203.1	0.5915	0.9165	1.5080
440.0	381.54	0.01926	1.19761	1.21687	419.0	785.4	1204.4	0.6161	0.8729	1.4890
460.0	466.87	0.01961	0.97463	0.99424	441.5	763.2	1204.8	0.6405	0.8299	1.4704
480.0	566.15	0.02000	0.79716	0.81717	464.5	739.6	1204.1	0.6648	0.7871	1.4518
500.0	680.86	0.02043	0.65448	0.67492	487.9	714.3	1202.2	0.6890	0.7443	1.4333
520.0	812.53	0.02091	0.53864	0.55956	512.0	687.0	1199.0	0.7133	0.7013	1.4146
540.0	962.79	0.02146	0.44367	0.46513	536.8	657.5	1194.3	0.7378	0.6577	1.3954
560.0	1133.38	0.02207	0.36507	0.38714	562.4	625.3	1187.7	0.7625	0.6132	1.3757
580.0	1326.17	0.02279	0.29937	0.32216	589.1	589.9	1179.0	0.7876	0.5673	1.3550
600.0	1543.2	0.02364	0.24384	0.26747	617.1	550.6	1167.7	0.8134	0.5196	1.3330
620.0	1786.9	0.02466	0.19615	0.22081	646.9	506.3	1153.2	0.8403	0.4689	1.3092
640.0	2059.9	0.02595	0.15427	0.18021	679.1	454.6	1133.7	0.8686	0.4134	1.2821
660.0	2365.7	0.02768	0.11663	0.14431	714.9	392.1	1107.0	0.8995	0.3502	1.2498
680.0	2708.6	0.03037	0.08080	0.11117	758.5	310.1	1068.5	0.9365	0.2720	1.2086
700.0	3094.3	0.03662	0.03857	0.07519	822.4	172.7	995.2	0.9901	0.1490	1.1390

Table A-3 Properties of superheated steam, compressed water, and supercritical steam (English units)*

Absolute pressure, psia (saturated temperature)		Temperature, °F														
		100	200	300	400	500	600	700	800	900	1000	1100	1200	1300	1400	1500
	v	0.0161	392.5	452.3	511.9	571.5	631.1	690.7								
1	h	68.00	1150.2	1195.7	1241.8	1288.6	1336.1	1384.5								
(101.74)	s	0.1295	2.0509	2.1152	2.1722	2.2237	2.2708	2.3144								
	v	0.0161	78.14	90.24	102.24	114.21	126.15	138.08	150.01	161.94	173.86	185.78	197.70	209.62	221.53	233.45
5	h	68.01	1148.6	1194.8	1241.3	1288.2	1335.9	1384.3	1433.6	1483.7	1534.7	1586.7	1639.6	1693.3	1748.0	1803.5
(162.24)	s	0.1295	1.8716	1.9369	1.9943	2.0460	2.0932	2.1369	2.1776	2.2159	2.2521	2.2866	2.3194	2.3509	2.3811	2.4101
	v	0.0161	38.84	44.98	51.03	57.04	63.03	69.00	74.98	80.94	86.91	92.87	98.84	104.80	110.76	116.72
10	h	68.02	1146.6	1193.7	1240.6	1287.8	1335.5	1384.0	1433.4	1483.5	1534.6	1586.6	1639.5	1693.3	1747.9	1803.4
(193.21)	s	0.1295	1.7928	1.8593	1.9173	1.9692	2.0166	2.0603	2.1011	2.1394	2.1757	2.2101	2.2430	2.2744	2.3046	2.3337
	v	0.0161	0.0166	29.899	33.963	37.985	41.986	45.978	49.964	53.946	57.926	61.905	65.882	69.858	73.833	77.807
15	h	68.04	168.09	1192.5	1239.9	1287.3	1335.2	1383.8	1433.2	1483.4	1534.5	1586.5	1639.4	1693.2	1747.8	1803.4
(213.03)	s	0.1295	0.2940	1.8134	1.8720	1.9242	1.9717	2.0155	2.0563	2.0946	2.1309	2.1653	2.1982	2.2297	2.2599	2.2890
	v	0.0161	0.0166	22.356	25.428	28.457	31.466	34.465	37.458	40.447	43.435	46.420	49.405	52.388	55.370	58.352
20	h	68.05	168.11	1191.4	1239.2	1286.9	1334.9	1383.5	1432.9	1483.2	1534.3	1586.3	1639.3	1693.1	1747.8	1803.3
(227.96)	s	0.1295	0.2940	1.7805	1.8397	1.8921	1.9397	1.9836	2.0244	2.0628	2.0991	2.1336	2.1665	2.1979	2.2282	2.2572
	v	0.0161	0.0166	11.036	12.624	14.165	15.685	17.195	18.699	20.199	21.697	23.194	24.689	26.183	27.676	29.168
40	h	68.10	168.15	1186.6	1236.4	1285.0	1333.6	1382.5	1432.1	1482.5	1533.7	1585.8	1638.8	1992.7	1747.5	1803.0
(267.25)	s	0.1295	0.2940	1.6992	1.7608	1.8143	1.8624	1.9065	1.9476	1.9860	2.0224	2.0569	2.0899	2.1224	2.1516	2.1807
	v	0.0161	0.0166	7.257	8.354	9.400	10.425	11.438	12.446	13.450	14.452	15.452	16.450	17.448	18.445	19.441
60	h	68.15	168.20	1181.6	1233.5	1283.2	1332.3	1381.5	1431.3	1481.8	1533.2	1585.3	1638.4	1692.4	1747.1	1802.8
(292.71)	s	0.1295	0.2939	1.6492	1.7134	1.7681	1.8168	1.8612	1.9024	1.9410	1.9774	2.0120	2.0450	2.0765	2.1068	2.1359
	v	0.0161	0.0166	0.0175	6.218	7.018	7.794	8.560	9.319	10.075	10.829	11.581	12.331	13.081	13.829	14.577
80	h	68.21	168.24	269.74	1230.5	1281.3	1330.9	1380.5	1430.5	1481.1	1532.6	1584.9	1638.0	1692.0	1746.8	1802.5
(312.04)	s	0.1295	0.2939	0.4371	1.6790	1.7349	1.7842	1.8289	1.8702	1.9089	1.9454	1.9800	2.0131	2.0446	2.0750	2.1041
	v	0.0161	0.0166	0.0175	4.935	5.588	6.216	6.833	7.443	8.050	8.655	9.258	9.860	10.460	11.060	11.659
100	h	68.26	168.29	269.77	1227.4	1279.3	1329.6	1379.5	1429.7	1480.4	1532.0	1584.4	1637.6	1691.6	1746.5	1802.2
(327.82)	s	0.1295	0.2939	0.4371	1.6516	1.7088	1.7586	1.8036	1.8451	1.8839	1.9205	1.9552	1.9883	2.0199	2.0502	2.0794

	v	0.0161	0.0166	0.0175	4.0786	4.6341	5.1637	5.6831	6.1928	6.7006	7.2060	7.7096	8.2119	8.7130	9.2134	9.7130
120	h	68.31	168.33	269.81	1224.1	1277.4	1328.1	1378.4	1428.8	1479.8	1531.4	1583.9	1637.1	1691.3	1746.2	1802.0
(341.27)	s	0.1295	0.2939	0.4371	1.6286	1.6872	1.7376	1.7829	1.8246	1.8635	1.9001	1.9349	1.9680	1.9996	2.0300	2.0592
	v	0.0161	0.0166	0.0175	3.4661	3.9526	4.4119	4.8585	5.2995	5.7364	6.1709	6.6036	7.0349	7.4652	7.8946	8.3233
140	h	68.37	168.38	269.85	1220.8	1275.3	1326.8	1377.4	1428.0	1479.1	1530.8	1583.4	1636.7	1690.9	1745.9	1801.7
(353.04)	s	0.1295	0.2939	0.4370	1.6085	1.6686	1.7196	1.7652	1.8071	1.8461	1.8828	1.9176	1.9508	1.9825	2.0129	2.0421
	v	0.0161	0.0166	0.0175	3.0060	3.4413	3.8480	4.2420	4.6295	5.0132	5.3945	5.7741	6.1522	6.5293	6.9055	7.2811
160	h	68.42	168.42	269.89	1217.4	1273.3	1325.4	1376.4	1427.2	1478.4	1530.3	1582.9	1636.3	1690.5	1745.6	1801.4
(363.55)	s	0.1294	0.2938	0.4370	1.5906	1.6522	1.7039	1.7499	1.7919	1.8310	1.8678	1.9027	1.9359	1.9676	1.9980	2.0273
	v	0.0161	0.0166	0.0174	2.6474	3.0433	3.4093	3.7621	4.1084	4.4505	4.7907	5.1289	5.4657	5.8014	6.1363	6.4704
180	h	68.47	168.47	269.92	1213.8	1271.2	1324.0	1375.3	1426.3	1477.7	1529.7	1582.4	1635.9	1690.2	1745.3	1801.2
(373.08)	s	0.1294	0.2938	0.4370	1.5743	1.6376	1.6900	1.7362	1.7784	1.8176	1.8545	1.8894	1.9227	1.9545	1.9849	2.0142
	v	0.0161	0.0166	0.0174	2.3598	2.7247	3.0583	3.3783	3.6915	4.0008	4.3077	4.6128	4.9165	5.2191	5.5209	5.8219
200	h	68.52	168.51	269.96	1210.1	1269.0	1322.6	1374.3	1425.5	1477.0	1529.1	1581.9	1635.4	1689.8	1745.0	1800.9
(381.80)	s	0.1294	0.2938	0.4369	1.5593	1.6242	1.6776	1.7239	1.7663	1.8057	1.8426	1.8776	1.9109	1.9427	1.9732	2.0025
	v	0.0161	0.0166	0.0174	0.0186	2.1504	2.4662	2.6872	2.9410	3.1909	3.4382	3.6837	3.9278	4.1709	4.4131	4.6546
250	h	68.66	168.63	270.05	375.10	1263.5	1319.0	1371.6	1423.4	1475.3	1527.6	1580.6	1634.4	1688.9	1744.2	1800.2
(400.97)	s	0.1294	0.2937	0.4368	0.5667	1.5951	1.6502	1.6976	1.7405	1.7801	1.8173	1.8524	1.8858	1.9177	1.9482	1.9776
	v	0.0161	0.0166	0.0174	0.0186	1.7665	2.0044	2.2263	2.4407	2.6509	2.8585	3.0643	3.2688	3.4721	3.6746	3.8764
300	h	68.79	168.74	270.14	375.15	1257.7	1315.2	1368.9	1421.3	1473.6	1526.2	1579.4	1633.3	1688.0	1743.4	1799.6
(417.35)	s	0.1294	0.2937	0.4307	0.5665	1.5703	1.6274	1.6758	1.7192	1.7591	1.7964	1.8317	1.8652	1.8972	1.9278	1.9572
	v	0.0161	0.0166	0.0174	0.0186	0.4913	1.7028	1.8970	2.0832	2.2652	2.4445	2.6219	2.7980	2.9730	3.1471	3.3205
350	h	68.92	168.85	270.24	375.21	1251.5	1311.4	1366.2	1419.2	1471.8	1524.7	1578.2	1632.3	1687.1	1742.6	1798.9
(431.73)	s	0.1293	0.2936	0.4367	0.5664	1.5483	1.6077	1.6571	1.7009	1.7411	1.7787	1.8141	1.8477	1.8798	1.9105	1.9400
	v	0.0161	0.0166	0.0174	0.0162	1.2841	1.4763	1.6499	1.8151	1.9759	2.1339	2.2901	2.4450	2.5987	2.7515	2.9037
400	h	69.05	168.97	270.33	375.27	1245.1	1307.4	1363.4	1417.0	1470.1	1523.3	1576.9	1631.2	1686.2	1741.9	1798.2
(444.60)	s	0.1293	0.2935	0.4366	0.5663	1.5282	1.5901	1.6406	1.6850	1.7255	1.7632	1.7988	1.8325	1.8647	1.8955	1.9250
	v	0.0161	0.0166	0.0174	0.0186	0.9919	1.1584	1.3037	1.4397	1.5708	1.6992	1.8256	1.9507	2.0746	2.1977	2.3200
500	h	69.32	169.19	270.51	375.38	1231.2	1299.1	1357.7	1412.7	1466.6	1520.3	1574.4	1629.1	1684.4	1740.3	1796.9
(467.01)	s	0.1292	0.2934	0.4364	0.5660	1.4921	1.5595	1.6123	1.6578	1.6990	1.7371	1.7730	1.8069	1.8393	1.8702	1.8998
	v	0.0161	0.0166	0.0174	0.0186	0.7944	0.9456	1.0726	1.1892	1.3008	1.4093	1.5160	1.6211	1.7252	1.8284	1.9309
600	h	69.58	169.42	270.70	375.49	1215.9	1290.3	1351.8	1408.3	1463.0	1517.4	1571.9	1627.0	1682.6	1738.8	1795.6
(486.20)	s	0.1292	0.2933	0.4362	0.5657	1.4590	1.5329	1.5844	1.6351	1.6769	1.7155	1.7517	1.7859	1.8184	1.8494	1.8792

(continued)

Table A-3 Properties of superheated steam, compressed water, and supercritical steam (English units)* (Continued)

Absolute pressure, psia (saturated temperature)		Temperature, °F														
		100	200	300	400	500	600	700	800	900	1000	1100	1200	1300	1400	1500
	v	0.0161	0.0166	0.0174	0.0186	0.0204	0.7928	0.9072	1.0102	1.1078	1.2023	1.2948	1.3858	1.4757	1.5647	1.6530
700	*h*	69.84	169.65	270.89	375.61	487.93	1281.0	1345.6	1403.7	1459.4	1514.4	1569.4	1624.8	1680.7	1737.2	1794.3
(503.08)	*s*	0.1291	0.2932	0.4360	0.5655	0.6889	1.5090	1.5673	1.6154	1.6580	1.6970	1.7335	1.7679	1.8006	1.8318	1.8617
	v	0.0161	0.0166	0.0174	0.0186	0.0204	0.6774	0.7828	0.8759	0.9631	1.0470	1.1289	1.2093	1.2885	1.3669	1.4446
800	*h*	70.11	169.88	271.07	375.73	487.88	1271.1	1339.2	1399.1	1455.8	1511.4	1566.9	1622.7	1678.9	1735.0	1792.9
(518.21)	*s*	0.1290	0.2930	0.4358	0.5652	0.6885	1.4869	1.5484	1.5980	1.6413	1.6807	1.7175	1.7522	1.7851	1.8164	1.8464
	v	0.0161	0.0166	0.0174	0.0186	0.0204	0.5869	0.6858	0.7713	0.8504	0.9262	0.9998	1.0720	1.1430	1.2131	1.2825
900	*h*	70.37	170.10	271.26	375.84	487.83	1260.6	1332.7	1394.4	1452.2	1508.5	1564.4	1620.6	1677.1	1734.1	1791.6
(531.95)	*s*	0.1290	0.2929	0.4357	0.5649	0.6881	1.4659	1.5311	1.5822	1.6263	1.6662	1.7033	1.7382	1.7713	1.8028	1.8329
	v	0.0161	0.0166	0.0174	0.0186	0.0204	0.5137	0.6080	0.6875	0.7603	0.8295	0.8966	0.9622	1.0266	1.0901	1.1529
1000	*h*	70.63	170.33	271.44	375.96	487.79	1249.3	1325.9	1389.6	1448.5	1504.4	1561.9	1618.4	1675.3	1732.5	1790.3
(544.58)	*s*	0.1289	0.2928	0.4355	0.5647	0.6876	1.4457	1.5149	1.5677	1.6126	1.6530	1.6905	1.7256	1.7589	1.7905	1.8207
	v	0.0161	0.0166	0.0174	0.0185	0.0203	0.4531	0.5440	0.6188	0.6865	0.7505	0.8121	0.8723	0.9313	0.9894	1.0468
1100	*h*	70.90	170.56	271.63	376.08	487.75	1237.3	1318.8	1384.7	1444.7	1502.4	1559.4	1616.3	1673.5	1731.0	1789.0
(556.28)	*s*	0.1289	0.2927	0.4353	0.5644	0.6872	1.4259	1.4996	1.5542	1.6000	1.6410	1.6787	1.7141	1.7475	1.7793	1.8097
	v	0.0161	0.0166	0.0174	0.0185	0.0203	0.4016	0.4905	0.5615	0.6250	0.6845	0.7418	0.7974	0.8519	0.9055	0.9584
1200	*h*	71.16	170.78	271.82	376.20	487.72	1224.2	1311.5	1379.7	1440.9	1499.4	1556.9	1614.2	1671.6	1729.4	1787.6
(567.19)	*s*	0.1288	0.2926	0.4351	0.5642	0.6868	1.4061	1.4851	1.5415	1.5883	1.6298	1.6679	1.7035	1.7371	1.7691	1.7996
	v	0.0161	0.0166	0.0174	0.0185	0.0203	0.3176	0.4059	0.4712	0.5282	0.5809	0.6311	0.6798	0.7272	0.7737	0.8195
1400	*h*	71.68	171.24	272.19	376.44	487.65	1194.1	1296.1	1369.3	1433.2	1493.2	1551.8	1609.9	1668.0	1726.3	1785.0
(587.07)	*s*	0.1287	0.2923	0.4348	0.5636	0.6859	1.3652	1.4575	1.5182	1.5670	1.6096	1.6484	1.6845	1.7185	1.7508	1.7815
	v	0.0161	0.0166	0.0173	0.0185	0.0202	0.0236	0.3415	0.4032	0.4555	0.5031	0.5482	0.5915	0.6336	0.6748	0.7153
1600	*h*	72.21	171.69	272.57	376.69	487.60	616.77	1279.4	1358.5	1425.2	1486.9	1546.6	1605.6	1664.3	1723.2	1782.3
(604.87)	*s*	0.1286	0.2921	0.4344	0.5631	0.6851	0.8129	1.4312	1.4968	1.5478	1.5916	1.6312	1.6678	1.7022	1.7344	1.7657
	v	0.0160	0.0165	0.0173	0.0185	0.0202	0.0235	0.2906	0.3500	0.3988	0.4426	0.4836	0.5229	0.5609	0.5980	0.6343
1800	*h*	72.7[illegible]	172.15	272.95	376.93	487.56	615.58	1261.1	1347.2	1417.1	1480.6	1541.1	1601.2	1660.7	1720.1	1779.7
(621.02)	*s*	0.1284	0.2918	0.4341	0.5626	0.6843	0.8109	1.4054	1.4768	1.5302	1.5753	1.6156	1.6528	1.6876	1.7204	1.7516

	v	0.0160	0.0165	0.0173	0.0184	0.0201	0.0233	0.2488	0.3072	0.3534	0.3942	0.4320	0.4680	0.5027	0.5365	0.5695
2000	h	73.26	172.60	273.32	377.19	487.53	614.48	1240.9	1353.4	1408.7	1474.1	1536.2	1596.9	1657.0	1717.0	1777.1
(635.80)	s	0.1283	0.2916	0.4337	0.5621	0.6834	0.8091	1.3794	1.4578	1.5138	1.5603	1.6014	1.6391	1.6743	1.7075	1.7389
	v	0.0160	0.0165	0.0173	0.0184	0.0200	0.0230	0.1681	0.2293	0.2712	0.3068	0.3390	0.3692	0.3980	0.4259	0.4529
2500	h	74.57	173.74	274.27	377.82	487.50	612.08	1176.7	1303.4	1386.7	1457.5	1522.9	1585.9	1647.8	1709.2	1770.4
(668.11)	s	0.1280	0.2910	0.4329	0.5609	0.6815	0.8048	1.3076	1.4129	1.4766	1.5269	1.5703	1.6094	1.6456	1.6796	1.7116
	v	0.0160	0.0165	0.0172	0.0183	0.0200	0.0228	0.0982	0.1759	0.2161	0.2484	0.2770	0.3033	0.3282	0.3522	0.3753
3000	h	75.88	174.88	275.22	378.47	487.52	610.08	1060.5	1267.0	1363.2	1440.2	1509.4	1574.8	1638.5	1701.4	1761.8
(695.33)	s	0.1277	0.2904	0.4320	0.5597	0.6796	0.8009	1.1966	1.3692	1.4429	1.4976	1.5434	1.5841	1.6214	1.6561	1.6888
	v	0.0160	0.0165	0.0172	0.0183	0.0199	0.0227	0.0335	0.1588	0.1987	0.2301	0.2576	0.2827	0.3065	0.3291	0.3510
3200	h	76.4	175.3	275.6	378.7	487.5	609.4	800.8	1250.9	1353.4	1433.1	1503.8	1570.3	1634.8	1698.3	1761.2
(705.08)	s	0.1276	0.2902	0.4317	0.5592	0.6788	0.7994	0.9708	1.3515	1.4300	1.4866	1.5335	1.5749	1.6126	1.6477	1.6806
	v	0.0160	0.0164	0.0172	0.0183	0.0199	0.0225	0.0307	0.1364	0.1764	0.2066	0.2326	0.2563	0.2784	0.2995	0.3198
3500	h	77.2	176.0	276.2	379.1	487.6	608.4	779.4	1224.6	1338.2	1422.2	1495.5	1563.3	1629.2	1693.6	1757.2
	s	0.1274	0.2899	0.4312	0.5585	0.6777	0.7973	0.9508	1.3242	1.4112	1.4709	1.5194	1.5618	1.6002	1.6358	1.6691
	v	0.0159	0.0164	0.0172	0.0182	0.0198	0.0223	0.0287	0.1052	0.1463	0.1752	0.1994	0.2210	0.2411	0.2601	0.2783
4000	h	78.5	177.2	277.1	379.8	487.7	606.9	763.0	1174.3	1311.6	1403.6	1481.3	1552.2	1619.8	1685.7	1750.6
	s	0.1271	0.2893	0.4304	0.5573	0.6760	0.7940	0.9343	1.2754	1.3807	1.4461	1.4976	1.5417	1.5812	1.6177	1.6516
	v	0.0159	0.0164	0.0171	0.0181	0.0196	0.0219	0.0268	0.0591	0.1038	0.1312	0.1529	0.1718	0.1890	0.2050	0.2203
5000	h	81.1	179.5	279.1	381.2	488.1	604.6	746.0	1042.9	1252.9	1364.6	1452.1	1529.1	1600.9	1670.0	1737.4
	s	0.1265	0.2881	0.4287	0.5550	0.6726	0.7880	0.9153	1.1593	1.3207	1.4001	1.4582	1.5061	1.5481	1.5863	1.6216
	v	0.0159	0.0163	0.0170	0.0180	0.0195	0.0216	0.0256	0.0397	0.0757	0.1020	0.1221	0.1391	0.1544	0.1684	0.1817
6000	h	83.7	181.7	281.0	382.7	488.6	602.9	736.1	945.1	1188.8	1323.6	1422.3	1505.9	1582.0	1654.2	1724.2
	s	0.1258	0.2870	0.4271	0.5528	0.6693	0.7826	0.9026	1.0176	1.2615	1.3574	1.4229	1.4748	1.5194	1.5593	1.5962
	v	0.0158	0.0163	0.0170	0.0180	0.0193	0.0213	0.0248	0.0334	0.0573	0.0816	0.1004	0.1160	0.1298	0.1424	0.1542
7000	h	86.2	184.4	283.0	384.2	489.3	601.7	729.3	901.8	1124.9	1281.7	1392.2	1482.6	1563.1	1638.6	1711.1
	s	0.1252	0.2859	0.4256	0.5507	0.6663	0.7777	0.8926	1.0350	1.2055	1.3171	1.3904	1.4466	1.4938	1.5355	1.5735

* Compressed (subcooled) water to the left of the vertical lines; superheated steam to the right of the vertical lines; supercritical steam below the horizontal line.

Table A-4*a* Saturated steam properties, temperature table (SI units)

Temperature, °C	Pressure, bar	Pressure, psia	Specific volume, m³/kg v_f	v_g	Specific enthalpy, kJ/kg h_f	h_{fg}	h_g	Specific entropy, kJ/(kg)(K) s_f	s_{fg}	s_g
0.01	0.00611	0.0886	0.0010002	206.3	0.00	2501	2501	0.0000	9.1544	9.1544
1	0.00657	0.0952	0.0010001	192.6	4.22	2498	2502	0.0154	9.1127	9.1281
2	0.00705	0.1023	0.0010001	179.9	8.42	2496	2504	0.0306	9.0712	9.1018
3	0.00758	0.1099	0.0010001	168.2	12.63	2493	2506	0.0458	9.0299	9.0757
4	0.00813	0.1179	0.0010001	157.3	16.84	2491	2508	0.0610	8.9888	9.0498
5	0.00872	0.1265	0.0010001	147.2	21.05	2489	2510	0.0762	8.9479	9.0241
6	0.00935	0.1356	0.0010001	137.8	25.25	2489	2512	0.0913	8.9065	8.9978
7	0.01001	0.1452	0.0010001	129.1	29.45	2485	2514	0.1063	8.8673	8.9736
8	0.01072	0.1555	0.0010002	121.0	33.55	2482	2516	0.1212	8.8273	8.9485
9	0.01147	0.1664	0.0010003	113.4	37.85	2479	2517	0.1361	8.7877	8.9238
10	0.01228	0.1781	0.0010004	106.42	42.04	2477	2519	0.1510	8.7484	8.8994
11	0.01312	0.1903	0.0010005	99.91	46.22	2475	2521	0.1658	8.7094	8.8752
12	0.01402	0.2033	0.0010006	93.84	50.41	2473	2523	0.1805	8.6708	8.8513
13	0.01497	0.2171	0.0010007	88.18	54.60	2470	2525	0.1952	8.6324	8.8276
14	0.01597	0.2316	0.0010008	82.90	58.78	2468	2527	0.2098	8.5942	8.8040
15	0.01704	0.2471	0.0010010	77.97	62.97	2465	2528	0.2244	8.5562	8.7806
16	0.01817	0.2635	0.0010011	73.39	67.16	2463	2530	0.2389	8.5185	8.7574
17	0.01936	0.2808	0.0010013	69.10	71.34	2461	2532	0.2534	8.4810	8.7344
18	0.02062	0.2991	0.0010015	65.09	75.53	2458	2534	0.2678	8.4438	8.7116
19	0.02196	0.3185	0.0010016	61.34	79.72	2456	2536	0.2821	8.4066	8.6890
20	0.02377	0.3390	0.0010018	57.84	83.90	2454	2537	0.2964	8.3701	8.6665
21	0.02486	0.3606	0.0010021	54.56	88.09	2451	2539	0.3107	8.3335	8.6442
22	0.02643	0.3833	0.0010023	51.50	92.27	2449	2541	0.3249	8.2971	8.6220
23	0.02808	0.4073	0.0010025	48.62	96.46	2447	2543	0.3391	8.2610	8.6001
24	0.02982	0.4325	0.0010028	45.93	100.63	2444	2545	0.3532	8.2253	8.5785
25	0.03166	0.4592	0.0010030	43.40	104.81	2442	2547	0.3672	8.1898	8.5570
26	0.03360	0.4873	0.0010033	41.04	108.99	2440	2548	0.3812	8.1546	8.5358
27	0.03564	0.5169	0.0010036	38.82	113.17	2437	2550	0.3951	8.1196	8.5147

28	0.03779	0.5481	0.0010038	36.73	117.35	2435	2552	0.4090	8.0848	8.4938
29	0.04004	0.5807	0.0010041	34.77	121.53	2432	2554	0.4228	8.0502	8.4730
30	0.04241	0.6151	0.0010044	32.93	125.71	2430	2556	0.4366	8.0157	8.4523
31	0.04491	0.6514	0.0010047	31.20	129.89	2428	2558	0.4503	7.9816	8.4319
32	0.04753	0.6894	0.0010051	29.57	134.07	2425	2559	0.4640	7.9477	8.4117
33	0.05029	0.7294	0.0010054	28.04	138.25	2423	2561	0.4777	7.9139	8.3916
34	0.05318	0.7713	0.0010057	26.60	142.42	2421	2563	0.4913	7.8803	8.3716
35	0.05622	0.8154	0.0010061	25.24	146.60	2418	2565	0.5049	7.8470	8.3519
36	0.05940	0.8615	0.0010064	23.97	150.78	2416	2567	0.5185	7.8138	8.3323
37	0.06274	0.9061	0.0010068	22.77	154.96	2414	2569	0.5320	7.7809	8.3129
38	0.06624	0.9607	0.0010071	21.63	159.14	2411	2570	0.5455	7.7483	8.2938
39	0.06991	1.0140	0.0010075	20.56	163.32	2409	2572	0.5589	7.7159	8.2748
40	0.07375	1.0697	0.0010079	19.55	167.50	2406	2574	0.5723	7.6836	8.2559
45	0.0958	1.3898	0.0010098	15.276	188.35	2394.9	2583.3	0.6383	7.5277	8.1661
50	0.1234	1.7890	0.0010121	12.046	209.26	2382.9	2592.2	0.7035	7.3741	8.0776
55	0.1574	2.2830	0.0010145	9.579	230.17	2370.8	2601.0	0.7677	7.2248	7.9925
60	0.1992	2.8892	0.0010171	7.679	251.09	2358.6	2609.7	0.8310	7.0798	7.9108
65	0.2501	3.6274	0.0010199	6.202	272.03	2346.3	2618.3	0.8933	6.9388	7.8322
70	0.3116	4.5194	0.0010228	5.946	292.97	2334.0	2626.9	0.9548	6.8017	7.7565
75	0.3855	5.5912	0.0010259	4.134	313.93	2321.5	2635.4	1.0154	6.6681	7.6835
80	0.4736	6.8690	0.0010292	3.409	334.92	2308.8	2643.8	1.0753	6.5380	7.6132
85	0.5780	8.3832	0.0010326	2.829	355.91	2296.1	2652.0	1.1343	6.4111	7.5454
90	0.7011	10.169	0.0010361	2.3613	376.94	2283.2	2660.1	1.1925	6.2873	7.4799
95	0.8453	12.280	0.0010398	1.9822	397.99	2270.2	2668.2	1.2501	6.1665	7.4166
100	1.0133	14.697	0.0010437	1.6730	419.06	2256.9	2676.0	1.3069	6.0485	7.3554
105	1.2080	17.521	0.0010477	1.4193	440.17	2243.6	2683.7	1.3630	5.9331	7.2962
110	1.4327	20.780	0.0010519	1.2099	461.32	2230.0	2691.3	1.4185	5.8203	7.2388
115	1.6906	24.520	0.0010562	1.0363	482.50	2216.2	2698.7	1.4733	5.7099	7.1832
120	1.9854	28.796	0.0010606	0.8915	503.72	2202.2	2706.0	1.5276	5.6017	7.1293
125	2.321	33.663	0.0010652	0.7702	524.99	2188.0	2713.0	1.5813	5.4957	7.0769
130	2.701	39.175	0.0010700	0.6681	546.31	2173.6	2719.9	1.6344	5.3917	7.0261
135	3.131	45.411	0.0010750	0.5818	567.68	2158.9	2726.6	1.6869	5.2897	6.9766
140	3.614	52.417	0.0010801	0.5085	589.10	2144.0	2733.1	1.7390	5.1894	6.9284
145	4.155	60.263	0.0010853	0.4460	610.59	2128.7	2739.3	1.7906	5.0910	6.8815
150	4.760	69.038	0.0010908	0.3924	632.15	2113.2	2745.4	1.8416	4.9941	6.8358
155	5.433	78.799	0.0010964	0.3464	653.77	2097.4	2751.2	1.8923	4.8989	6.7911

(continued)

Table A-4a Saturated steam properties, temperature table (SI units) (Continued)

Temperature, °C	Pressure, bar	Pressure, psia	Specific volume, m³/kg v_f	v_g	Specific enthalpy, kJ/kg h_f	h_{fg}	h_g	Specific entropy, kJ/(kg)(K) s_f	s_{fg}	s_g
160	6.181	89.648	0.0011022	0.3068	675.47	2081.3	2756.7	1.9425	4.8050	6.7475
165	7.008	101.64	0.0011082	0.2724	697.25	2064.8	2762.0	1.9923	4.7126	6.7048
170	7.920	114.87	0.0011145	0.2426	719.12	2047.9	2767.1	2.0416	4.6214	6.6630
175	8.924	129.43	0.0011209	0.21654	741.07	2030.7	2771.8	2.0906	4.5314	6.6221
180	10.027	145.43	0.0011275	0.19380	763.12	2013.2	2776.3	2.1393	4.4426	6.5819
185	11.233	162.92	0.0011344	0.17386	785.26	1995.2	2780.4	2.1876	4.3548	6.5424
190	12.551	182.04	0.00114415	0.15632	807.52	1976.7	2784.3	2.2356	4.2680	6.5036
195	13.987	202.86	0.0011489	0.14084	829.88	1957.9	2787.8	2.2833	4.1821	6.4654
200	15.549	225.52	0.0011565	0.12716	852.37	1938.6	2790.9	2.3307	4.0971	6.4278
210	19.077	276.89	0.0011726	0.10424	897.73	1898.5	2796.2	2.4247	3.9293	6.3539
220	23.198	336.45	0.0011900	0.08604	943.67	1856.2	2799.9	2.5178	3.7639	6.2817
230	27.98	405.82	0.0012087	0.07145	990.27	1811.7	2802.0	2.6102	3.6006	6.2107
240	33.48	485.59	0.0012291	0.05965	1037.60	1764.6	2802.2	2.7020	3.4386	6.1406
250	39.78	576.96	0.0012513	0.05004	1085.78	1714.7	2800.4	2.7935	3.2773	6.0708
260	46.94	680.81	0.0012756	0.04213	1134.94	1661.5	2796.4	2.8848	3.1161	6.0010
270	55.06	798.58	0.0013025	0.03559	1185.23	1604.6	2789.9	2.9763	2.9541	5.9304
280	64.20	931.14	0.0013324	0.03013	1236.84	1543.6	2780.4	3.0683	2.7903	5.8586
290	74.46	1079.95	0.0013659	0.02554	1290.01	1477.6	2767.6	3.1611	2.6237	5.7848
300	85.93	1246.31	0.0014041	0.021649	1345.05	1406.0	2751.0	3.2552	2.4529	5.7081
310	98.70	1431.52	0.0014480	0.018334	1402.39	1327.6	2730.0	3.3512	2.2766	5.6278
320	112.89	1637.33	0.0014995	0.015480	1462.60	1241.1	2703.7	3.4500	2.0923	5.5423
330	128.63	1865.62	0.0015615	0.012989	1526.52	1143.6	2670.2	3.5528	1.8962	5.4490
340	146.05	2118.28	0.0016387	0.010780	1595.47	1030.7	2626.2	3.6616	1.6811	5.3427
350	165.35	2398.20	0.0017411	0.008799	1671.94	895.7	2567.7	3.7800	1.4376	5.2177
360	186.75	2708.58	0.0018959	0.006940	1764.17	721.3	2485.4	3.9210	1.1390	5.0600
370	210.54	3053.62	0.0022136	0.004973	1890.21	452.6	2342.8	4.1108	0.7036	4.8144
374.15	221.20	3208.23	0.00317	0.00317	2107.37	0.0	2107.4	4.4429	0.0	4.4429

Table A-4*b* Saturated steam: pressure table (SI units)

Pressure, bar	Pressure, psia	Temperature, °C	Specific volume, m³/kg v_f	v_g	Specific enthalpy, kJ/kg h_f	h_{fg}	h_g	Specific entropy, kJ/(kg)(K) s_f	s_{fg}	s_g
0.010	0.1450	6.98	0.0010001	129.20	29.30	2484.9	2514.2	0.1034	8.8714	8.9748
0.015	0.2176	13.04	0.0010007	87.98	54.71	2470.6	2525.3	0.1958	8.6312	8.8270
0.020	0.2901	17.51	0.0010014	67.00	73.48	2460.0	2533.5	0.2569	8.4659	8.7228
0.025	0.3626	21.08	0.0010021	54.24	88.49	2451.6	2540.0	0.3083	8.3340	8.6423
0.030	0.4351	24.10	0.0010028	45.66	101.05	2444.5	2545.5	0.3510	8.2258	8.5768
0.040	0.5802	28.98	0.0010041	34.81	121.46	2432.9	2554.4	0.4197	8.0541	8.4738
0.050	0.7252	32.90	0.0010053	28.19	137.82	2423.7	2561.5	0.4740	7.9203	8.3943
0.060	0.8702	36.16	0.0010064	23.74	151.50	2415.0	2566.9	0.5191	7.8105	8.3296
0.070	1.0153	39.03	0.0010075	20.53	163.43	2409	2572.4	0.5591	7.7149	8.2740
0.080	1.1603	41.54	0.0010085	18.10	173.9	2402.6	2576.5	0.5915	7.6364	8.2279
0.090	1.3053	43.79	0.0010094	16.20	183.3	2396.7	2580.0	0.6225	7.5635	8.1860
0.10	1.4504	45.84	0.0010103	14.68	191.9	2392.3	2584.2	0.6488	7.5006	8.1494
0.11	1.5954	47.72	0.0010111	13.40	199.7	2388.3	2588.0	0.6740	7.4420	8.1160
0.12	1.7405	49.45	0.001012	12.36	207.1	2383.5	2590.6	0.6964	7.3891	8.0855
0.14	2.0305	52.58	0.001013	10.69	220.3	2375.8	2596.1	0.7371	7.2964	8.0317
0.16	2.3206	55.34	0.001015	9.433	231.9	2369.1	2601.0	0.7728	7.2124	7.9852
0.18	2.6107	57.82	0.001016	8.445	242.4	2362.9	2605.3	0.8045	7.1397	7.9442
0.20	2.9008	60.09	0.001017	7.649	251.9	2357.4	2609.3	0.8332	7.0745	7.9077
0.25	3.6259	64.99	0.001020	6.204	272.6	2345.1	2617.7	0.8947	6.9359	7.8306
0.30	4.3511	69.12	0.001022	5.229	289.9	2334.9	2624.8	0.9458	6.8220	7.7678
0.40	5.8015	75.88	0.001026	3.993	318.3	2318.0	2636.3	1.0279	6.6413	7.6692
0.50	7.2519	81.35	0.001030	3.240	341.3	2304.1	2645.4	1.0930	6.5001	7.5931
0.60	8.7023	85.95	0.001033	2.732	360.6	2292.4	2653.0	1.1471	6.3841	7.5312
0.80	11.6030	93.52	0.001038	2.087	392.3	2273.0	2665.3	1.2344	6.1994	7.4338
1.0	14.5038	99.64	0.001043	1.694	418.0	2257.0	2675.0	1.3038	6.0548	7.3580
1.013	14.696	100	0.001043	1.673	419.5	2256.1	2675.6	1.3079	6.0462	7.354
1.2	17.4045	104.81	0.001047	1.428	439.7	2243.4	2683.1	1.3617	5.9356	7.2973
1.4	20.305	109.3	0.001051	1.237	458.6	2231.4	2690.0	1.4115	5.8341	7.2456
1.6	23.206	113.3	0.001054	1.091	475.5	2220.5	2696.0	1.4553	5.7456	7.2009
1.8	26.107	116.9	0.001058	.9775	490.8	2210.6	2701.4	1.4945	5.6670	7.1615

(continued)

Table A-4*b* Saturated steam: pressure table (SI units) (Continued)

Pressure,		Temperature,	Specific volume, m^3/kg		Specific enthalpy, kJ/kg			Specific entropy, kJ/(kg)(K)		
bar	psia	°C	v_f	v_g	h_f	h_{fg}	h_g	s_f	s_{fg}	s_g
120.2	2.0	29.008	0.001061	.8857	504.7	2201.5	2706.2	1.5300	5.5963	7.1263
127.4	2.5	36.259	0.001067	.7187	535.2	2181.3	2716.5	1.6068	5.4451	7.0519
133.6	3.0	43.511	0.001073	.6058	561.2	2163.7	2724.9	1.6710	5.3201	6.9911
143.6	4.0	58.015	0.001084	.4625	604.3	2133.8	2738.1	1.7755	5.1196	6.8951
151.9	5.0	72.519	0.001093	.3749	639.8	2108.4	2748.2	1.8594	4.9611	6.8205
158.9	6.0	87.023	0.001101	.3157	670.1	2086.3	2756.4	1.9299	4.8293	6.7592
170.4	8.0	116.03	0.001115	.2404	720.7	2048.0	2768.7	2.0451	4.6169	6.6620
179.9	10.	145.04	0.001127	.1944	762.5	2015.1	2777.6	2.1378	4.4479	6.5857
188.0	12	174.05	0.001139	.1633	798.5	1985.9	2784.4	2.2160	4.3065	6.5225
195.1	14	203.05	0.001149	.1408	830.2	1959.4	2789.6	2.2838	4.1847	6.4685
201.4	16	232.06	0.001159	.1238	858.8	1934.8	2793.6	2.3440	4.0770	6.4210
207.1	18	261.07	0.001168	.1104	884.9	1911.8	2796.7	2.3981	3.9805	6.3786
212.4	20	290.08	0.001176	.0996	908.9	1890.2	2799.1	2.4474	3.8927	6.3401
224.0	25	362.59	0.001197	.0800	962.4	1840.2	2802.6	2.5549	3.7018	6.2567
233.9	30	435.11	0.001216	.0667	1008.7	1795.0	2803.7	2.6461	3.5400	6.1861
250.4	40	580.15	0.001252	.0498	1087.6	1713.4	2801.0	2.7968	3.2725	6.0693
264.0	50	725.19	0.001286	.0394	1154.5	1639.4	2793.9	2.9206	3.0520	5.9726
275.6	60	870.23	0.001319	.0324	1213.7	1570.2	2783.9	3.0271	2.8613	5.8884
285.9	70	1015.3	0.001352	.0274	1267.4	1504.3	2771.7	3.1216	2.6909	5.8125
295.1	80	1160.3	0.001385	.0235	1317.0	1440.5	2757.5	3.2073	2.5351	5.7424
303.3	90	1305.3	0.001417	.0205	1363.7	1379.3	2743.0	3.2870	2.3910	5.6780
311.1	100	1450.4	0.001453	.0180	1407.9	1316.4	2724.3	3.3600	2.2533	5.6133
318.2	110	1595.4	0.001489	.0160	1450.2	1255.0	2705.2	3.4296	2.1224	5.5520
324.8	120	1740.5	0.001527	.0143	1491.2	1193.2	2684.4	3.4960	1.9956	5.4916
330.9	130	1885.5	0.001567	.0128	1531.1	1130.7	2661.8	3.5599	1.8717	5.4316
336.8	140	2030.5	0.001610	.0115	1570.4	1066.8	2637.2	3.6220	1.7490	5.3710
347.4	160	2320.6	0.001710	.0093	1648.9	931.3	2580.2	3.7441	1.5007	5.2448
357.1	180	2610.7	0.001840	.0075	1731.4	777.4	2508.8	3.8703	1.2336	5.1039
365.8	200	2900.8	0.002041	.00584	1828.5	581.0	2409.5	4.0172	0.9093	4.9265
374.1	220.89	3203.7	0.003155	.003155	2098.8	0	2098.8	4.4289	0	4.4289

Table A-5 Properties of superheated steam (SI units)

Pressure, bar (saturated temperature, °C)		Temperature, °C									
		100	150	200	250	300	400	500	600	700	800
0.1	*v*	17.196	19.51	21.825	24.136	26.445	31.063	35.679	40.295	44.911	49.526
(45.81)	*h*	2867.5	2783.0	2879.5	2977.3	3076.5	3279.6	3489.1	3705.4	3928.7	4159.0
	s	8.4479	8.6882	8.9038	9.1002	9.2813	9.6077	9.8978	10.1608	10.4028	10.6281
0.5	*v*	3.418	3.889	4.356	4.820	5.284	6.209	7.134	8.057	8.981	9.904
(81.33)	*h*	2682.5	2780.1	2877.7	2976.0	3075.5	3278.9	3488.7	3705.1	3928.5	4158.9
	s	7.6947	7.9401	8.1580	8.3556	8.5373	8.8642	9.1546	9.4178	9.6599	9.8852
1.0	*v*	1.6958	1.9364	2.172	2.406	2.639	3.103	3.565	4.028	4.490	4.952
(99.63)	*h*	2676.2	2776.4	2875.3	2974.3	3074.3	3278.2	3488.1	3704.7	3928.2	4.158.6
	s	7.3614	7.6134	7.8343	8.0333	8.2158	8.5435	8.8342	9.0976	9.3398	9.5652
2.0	*v*		0.9596	1.0803	1.1988	1.3162	1.5493	1.7814	2.013	2.244	2.475
(120.23)	*h*		2768.8	2870.5	2971.0	3071.8	3276.6	3487.1	3704.0	3927.6	4.158.2
	s		7.2795	7.5066	7.7086	7.8926	8.2218	8.5133	8.7770	9.0194	9.2449
3.0	*v*		0.6339	0.7163	0.7964	0.8753	1.0315	1.1867	1.3414	1.4957	1.6499
(135.55)	*h*		2761.0	2865.6	2967.6	3069.3	3275.0	3486.0	3703.2	3927.1	4.157.8
	s		7.0778	7.3115	7.5166	7.7022	8.0330	8.3251	8.5892	8.8319	9.0576
4.0	*v*		0.4708	0.5342	0.5951	0.6548	0.7726	0.8893	1.0552	1.1215	1.2372
(143.63)	*h*		2752.8	2860.5	2964.2	3066.8	3273.4	3484.9	3702.4	3926.5	4.157.3
	s		6.9299	7.1706	7.5662	7.5662	7.8985	8.1913	8.4558	8.6987	8.9244
5	*v*			0.4249	0.4744	0.5226	0.6173	0.7109	0.8041	0.8969	0.9896
(151.86)	*h*			2855.4	2960.7	3064.2	3271.9	3483.9	3701.7	3925.9	4156.9
	s			7.0592	7.2709	7.4599	7.7938	8.0873	7.3522	8.5952	8.8211

(continued)

Table A-5 Properties of superheated steam (SI units) (Continued)

Pressure, bar (saturated temperature, °C)		Temperature, °C									
		100	150	200	250	300	400	500	600	700	800
6	v			0.3520	0.3938	0.4344	0.5137	0.5920	0.6697	0.7472	0.8245
(158.58)	h			2850.1	2957.2	3061.6	3270.3	3482.8	3700.9	3925.3	4156.5
	s			6.9665	7.1816	7.3724	7.7079	8.0021	8.2674	8.5107	8.7367
10	v			0.2060	0.2327	0.2579	0.3066	0.3541	0.4011	0.4478	0.4923
(179.91)	h			2827.9	2942.6	3051.2	3263.9	3478.5	3697.9	3923.1	4154.7
	s			6.6940	6.9247	7.1229	7.4651	7.7622	8.0290	8.2731	8.4996
20	v				0.1114	0.1255	0.1512	0.1757	0.1996	0.2232	0.2467
(212.42)	h				2902.5	3023.5	3247.6	3467.6	3690.1	3917.4	4150.3
	s				6.5453	6.7664	7.1271	7.4317	7.7024	7.9487	8.1765
30	v				0.0758	0.08114	0.0994	0.1162	0.1324	0.1484	0.1641
(233.9)	h				2855.8	2993.5	3230.9	3456.5	3682.3	3911.7	4145.9
	s				6.2872	6.5390	6.9212	7.2338	7.5085	7.7571	7.9862
40	v					0.0588	0.0734	0.0864	0.09885	0.11095	0.12287
(250.4)	h					2960.7	3213.6	3445.3	3674.4	3905.9	4141.5
	s					6.3615	6.7690	7.0901	7.3688	7.6198	7.8502
50	v					0.0453	0.0578	0.06857	0.07869	0.08849	0.0981
(263.99)	h					2924.5	3195.7	3433.8	3666.5	3900.1	4137.1
	s					6.2084	6.6459	6.9759	7.2589	7.5122	7.7440
60	v					0.0362	0.0474	0.0567	0.0653	0.0735	0.0816
(275.64)	h					2884.2	3177.2	3422.2	3658.4	3894.2	4132.7
	s					6.0674	6.5408	6.8803	7.1677	7.4234	7.6566
70	v					0.0295	0.0393	0.0481	0.0557	0.0628	0.0698
(285.88)	h					2838.4	3158.1	3410.3	3650.3	3888.3	4128.2
	s					5.9305	6.4478	6.7975	7.0894	7.3476	7.5822

80	*v*	0.0243	0.0343	0.0418	0.0486	0.0548	0.0610
(295.06)	*h*	2785.0	3138.3	3398.3	3642.0	3882.4	4123.8
	s	5.7906	6.3634	6.7240	7.0206	7.2812	7.5173
90	*v*		0.0299	0.0368	0.0429	0.0486	0.0541
(303.40)	*h*		3117.8	3386.1	3633.7	3876.5	4119.3
	s		6.2854	6.6576	6.9589	7.2221	7.4596
100	*v*		0.0264	0.0328	0.0384	0.0436	0.0486
(311.06)	*h*		3096.5	3373.7	3625.3	3870.5	4114.8
	s		6.2120	6.5966	6.9029	7.1687	7.4077
150	*v*		0.0157	0.0208	0.0249	0.0286	0.0321
(342.24)	*h*		2975.5	3308.6	3582.3	3840.1	4092.4
	s		5.8811	6.3443	6.6776	6.9572	7.2040
200	*v*		0.0099	0.0148	0.0182	0.0211	0.0239
(365.81)	*h*		2818.1	3238.2	3537.6	3809.0	4069.7
	s		5.5540	6.1401	6.5048	6.7993	7.0544
250	*v*		0.0060	0.0111	0.0141	0.0167	0.0189
	h		2580.2	3162.4	3491.4	3777.5	4047.1
	s		5.1418	5.9592	6.3602	6.6707	6.9345
300	*v*		0.0028	0.0087	0.0115	0.0137	0.0156
	h		2151.1	3081.1	3443.9	3745.6	4024.2
	s		4.4728	5.7905	6.2331	6.5606	6.8332
350	*v*		0.0021	0.0069	0.0095	0.0115	0.0313
	h		1987.6	2994.4	3395.5	3713.5	4001.5
	s		4.2126	5.6282	6.1179	6.4631	6.7450
400	*v*		0.0019	0.0056	0.0081	0.0099	0.0115
	h		1930.9	2903.3	3346.4	3681.2	3978.7
	s		4.1135	5.4700	6.0114	6.3750	6.6662
500	*v*		0.00173	0.0039	0.0061	0.0077	0.0091
	h		1874.6	2720.1	3247.6	3616.8	3933.6
	s		4.0031	5.1726	5.8178	6.2189	6.5290

APPENDIX

B

THERMODYNAMIC PROPERTIES OF FREON-12

Table B-1 Saturation properties of Freon-12 (English units)*

Temperature, °F	Pressure, psia	Specific volume, ft^3/lb_m		Specific enthalpy, Btu/lb_m			Specific entropy, $Btu/(lb_m)(°R)$	
		v_f	v_g	h_f	h_{fg}	h_g	s_f	s_g
60	72.433	0.011913	0.55839	21.766	61.643	83.409	0.046180	0.16479
62	74.807	0.011947	0.54112	22.221	61.380	83.601	0.047044	0.16470
64	77.239	0.011982	0.52450	22.676	61.116	83.792	0.047905	0.16460
66	79.729	0.012017	0.50848	23.133	60.849	83.982	0.048765	0.16451
68	82.279	0.012053	0.49305	23.591	60.580	84.171	0.049624	0.16442
70	84.888	0.012089	0.47818	24.050	60.309	84.359	0.050482	0.16434
72	87.559	0.012126	0.46383	24.511	60.035	84.546	0.051338	0.16425
74	90.292	0.012163	0.45000	24.973	59.759	84.732	0.052193	0.16417
76	93.087	0.012201	0.43666	25.435	59.481	84.916	0.053047	0.16408
78	95.946	0.012239	0.42378	25.899	59.201	85.100	0.053900	0.16400
80	98.870	0.012277	0.41135	26.365	58.917	85.282	0.054751	0.16392
82	101.86	0.012316	0.39935	26.832	58.631	85.463	0.055602	0.16384
84	104.92	0.012356	0.38776	27.300	58.343	85.643	0.056452	0.16376
86	108.04	0.012396	0.37657	27.769	58.052	85.821	0.057301	0.16368
88	111.23	0.012437	0.36575	28.241	57.757	85.998	0.058149	0.16360
90	114.49	0.012478	0.35529	28.713	57.461	86.174	0.058997	0.16353
92	117.82	0.012520	0.34518	29.187	57.161	86.348	0.059844	0.16345
94	121.22	0.012562	0.33540	29.663	56.585	86.521	0.060690	0.16338
96	124.70	0.012605	0.32594	30.140	56.551	86.691	0.061536	0.16330
98	128.24	0.012649	0.31679	30.619	56.242	86.861	0.062381	0.16323
100	131.86	0.012693	0.30794	31.100	55.929	87.029	0.063227	0.16315
102	135.56	0.012738	0.29937	31.583	55.613	87.196	0.064072	0.16308
104	139.33	0.012783	0.29106	32.067	55.293	87.360	0.064916	0.16301
106	143.18	0.012829	0.28303	32.553	54.970	87.523	0.065761	0.16293
108	147.11	0.012976	0.27524	33.041	54.643	87.684	0.066606	0.16282
110	151.11	0.012924	0.26769	33.531	54.313	87.844	0.067451	0.16279

Table B-1 Saturation properties of Freon-12 (English units)* (Continued)

Temperature, °F	Pressure, psia	Specific volume, ft³/lb$_m$		Specific enthalpy, Btu/lb$_m$			Specific entropy, Btu/(lb$_m$)(°R)	
		v_f	v_g	h_f	h_{fg}	h_g	s_f	s_g
112	155.19	0.012972	0.26037	34.023	53.978	88.001	0.068296	0.16271
114	159.36	0.013022	0.25328	34.517	53.639	88.156	0.069141	0.16264
116	163.61	0.013072	0.24641	35.014	53.296	88.310	0.069987	0.16256
118	167.94	0.013123	0.23974	35.512	52.949	88.461	0.070833	0.16249
120	172.35	0.013174	0.23326	36.013	52.597	88.610	0.071680	0.16241
122	176.85	0.013227	0.22698	36.516	52.241	88.757	0.072528	0.16234
124	181.43	0.013280	0.22089	37.021	51.881	88.902	0.073376	0.16226
126	186.10	0.013335	0.21497	37.529	51.515	89.044	0.074225	0.16218
128	190.86	0.013390	0.20922	38.040	51.144	89.184	0.075075	0.16210
130	195.71	0.013447	0.20364	38.553	50.768	89.321	0.075927	0.16202
132	200.64	0.013504	0.19821	39.069	50.387	89.456	0.076779	0.16194
134	205.67	0.013563	0.19294	39.588	50.000	89.588	0.077633	0.16185
136	210.79	0.013623	0.18782	40.110	49.608	89.718	0.078489	0.16177
138	216.01	0.013684	0.18283	40.634	49.210	89.844	0.079346	0.16168
140	221.32	0.013746	0.17799	41.162	48.805	89.967	0.080205	0.16159
142	226.72	0.013810	0.17327	41.693	48.394	90.087	0.081065	0.16150
144	232.22	0.013874	0.16868	42.227	47.977	90.204	0.81928	0.16150
146	237.82	0.013941	0.16422	42.765	47.553	90.318	0.082794	0.16130
148	243.51	0.014008	0.15987	43.306	47.122	90.428	0.083661	0.16120
150	249.31	0.014078	0.15564	43.850	46.684	90.534	0.084531	0.16110
152	255.20	0.014148	0.15151	44.399	46.238	90.637	0.085404	0.16099
154	261.20	0.014221	0.14750	44.951	45.784	90.735	0.086280	0.16088
156	267.30	0.014295	0.14358	45.508	45.322	90.830	0.087159	0.16077
158	273.51	0.014371	0.13976	46.068	44.852	90.920	0.088041	0.16065
160	279.82	0.014449	0.13604	46.633	44.373	90.006	0.088927	0.16055
162	286.24	0.014529	0.13241	47.202	43.885	91.087	0.089817	0.16040
164	292.77	0.014611	0.12886	47.777	43.386	91.163	0.090710	0.16027
166	299.40	0.014695	0.12540	48.355	42.879	91.234	0.091608	0.16014
168	306.15	0.014782	0.12202	48.939	42.360	91.299	0.092511	0.16000
170	313.00	0.014871	0.11873	49.529	41.830	91.359	0.093418	0.15985
172	319.97	0.014963	0.11550	50.123	41.290	91.413	0.094330	0.15969
174	327.06	0.015058	0.11235	50.724	40.736	91.460	0.095248	0.15953
176	334.25	0.015155	0.10927	51.330	40.171	91.501	0.096172	0.05936
178	341.57	0.015256	0.10625	51.943	39.592	91.535	0.097102	0.15919
180	349.00	0.015360	0.10330	52.562	38.999	91.561	0.098039	0.15900
182	356.55	0.015468	0.10041	53.188	38.391	91.579	0.098982	0.15881
184	364.23	0.015580	0.097584	53.822	37.767	91.589	0.099933	0.15861
186	372.02	0.015696	0.094810	54.463	37.127	91.590	0.10089	0.15839
188	379.94	0.015816	0.092089	55.111	36.469	91.580	0.10186	0.15817
190	387.98	0.015942	0.089418	55.769	35.792	91.561	0.10284	0.15793
192	396.14	0.016073	0.086796	56.435	35.096	91.531	0.10382	0.15768
194	404.44	0.016209	0.084218	57.111	34.377	91.488	0.10482	0.15741
196	412.86	0.016352	0.081683	57.797	33.636	91.433	0.10583	0.15713
198	421.41	0.016502	0.079188	58.494	32.869	91.363	0.10685	0.15683
200	430.09	0.016659	0.076728	59.203	32.075	91.278	0.10789	0.15651
202	438.91	0.016826	0.074301	59.924	31.252	91.176	0.10894	0.15617
204	447.85	0.017002	0.071903	60.659	30.396	91.055	0.11001	0.15580
206	456.94	0.017188	0.069531	61.409	29.505	90.914	0.11109	0.15541

(continued)

Table B-1 Saturation properties of Freon-12 (English units)* (Continued)

Temperature, °F	Pressure, psia	Specific volume, ft³/lb$_m$		Specific enthalpy, Btu/lb$_m$			Specific entropy, Btu/(lb$_m$)(°R)	
		v_f	v_g	h_f	h_{fg}	h_g	s_f	s_g
208	466.16	0.017387	0.067179	62.175	28.574	90.749	0.11220	0.15499
210	475.52	0.017601	0.064843	62.959	27.599	90.558	0.11332	0.15453
212	485.01	0.017713	0.062517	63.764	26.573	90.337	0.11448	0.15404
214	494.65	0.018079	0.060193	64.591	25.490	90.081	0.11566	0.15349
216	504.44	0.018351	0.057864	65.444	24.341	89.785	0.11687	0.15290
218	514.36	0.018651	0.055518	66.327	23.113	89.440	0.11813	0.15223
220	524.43	0.018986	0.053140	67.246	21.790	89.036	0.11943	0.15149
222	534.65	0.019365	0.050711	68.209	20.350	88.559	0.12079	0.15064
224	545.02	0.019804	0.048200	69.228	18.757	87.985	0.12223	0.14966
226	555.54	0.020327	0.045559	70.320	16.958	87.278	0.12377	0.14850
228	566.20	0.020978	0.042702	71.519	14.854	86.373	0.12545	0.14705
230	577.03	0.021854	0.039435	72.893	12.229	85.122	0.12739	0.14512
232	588.01	0.023262	0.035041	74.651	8.335	82.986	0.12987	0.14191
233.6 (Critical)	596.9	0.02870	0.02870	78.86	0	78.86	0.1359	0.1359

* Tables B-1 and B-2 from data of E. I. du Pont de Nemours and Company. Freon is a registered trademark of du Pont.

Table B-2 Properties of superheated Freon-12 (English units)

Pressure, psia (saturated temperature)		Temperature, °F											
		−40	−20	0	20	40	60	80	100	150	200	250	300
	v	7.363	7.726	8.088	8.450	8.812	9.173	9.533	9.893	10.79	11.69		
5	h	73.72	76.36	79.05	81.78	84.56	87.41	90.30	93.25	100.84	108.75		
(−62.5)	s	0.1859	0.1920	0.1979	0.2038	0.2095	0.2150	0.2205	0.2258	0.2388	0.2513		
	v	......	3.821	4.006	4.189	4.371	4.556	4.740	4.923	5.379	5.831	6.281	
10	h	......	76.11	78.81	81.56	84.35	87.19	90.11	93.05	100.66	108.63	116.88	
(−37.3)	s	......	0.1801	0.1861	0.1919	0.1977	0.2033	0.2087	0.2141	0.2271	0.2396	0.2517	
	v	......	2.521	2.646	2.771	2.895	3.019	3.143	3.266	3.571	3.877	4.191	
15	h	......	75.89	78.59	81.37	84.18	87.03	89.94	92.91	100.53	108.49	116.78	
(−20.8)	s	......	0.17307	0.17913	0.18499	0.19074	0.19635	0.20185	0.20723	0.22028	0.23282	0.24491	
	v	......	......	1.965	2.060	2.155	2.250	2.343	2.437	2.669	2.901	3.130	
20	h	......	......	78.39	81.14	83.97	86.85	89.78	92.75	100.40	108.38	116.67	
(−8.2)	s	......	......	0.17407	0.17996	0.18573	0.19138	0.19688	0.20229	0.21537	0.22794	0.24005	
	v	......	......	......	1.712	1.793	1.873	1.952	2.031	2.227	2.422	2.615	
25	h	......	......	......	80.95	83.78	86.67	89.61	92.56	100.26	108.26	116.56	
(2.2)	s	......	......	......	0.17637	0.18216	0.18783	0.19336	0.19748	0.21190	0.22450	0.23665	
	v	......	......	......	1.364	1.430	1.495	1.560	1.624	1.784	1.943	2.099	
30	h	......	......	......	80.75	83.59	86.49	89.43	92.42	100.12	108.13	116.45	
(11.1)	s	......	......	......	0.17278	0.17859	0.18429	0.18983	0.19527	0.20843	0.22105	0.23325	
	v	......	......	......	1.109	1.237	1.295	1.352	1.409	1.550	1.689	1.827	
35	h	......	......	......	80.49	83.40	86.30	89.26	92.26	99.98	108.01	116.33	
(18.9)	s	......	......	......	0.16963	0.17591	0.18162	0.18719	0.19266	0.20584	0.21849	0.23069	
	v	......	......	......	......	1.044	1.095	1.144	1.194	1.315	1.435	1.554	
40	h	......	......	......	......	83.20	86.11	89.09	92.09	99.83	107.88	116.21	
(25.9)	s	......	......	......	......	0.17322	0.17896	0.18455	0.19004	0.20325	0.21592	0.22813	
	v	......	......	......	......	0.821	0.863	0.904	0.944	1.044	1.142	1.239	1.332
50	h	......	......	......	......	82.76	85.72	88.72	91.75	99.54	107.62	116.00	124.69
(38.3)	s	......	......	......	......	0.16895	0.17475	0.19040	0.18591	0.19923	0.21196	0.22419	0.23600
	v	......	......	......	......	......	0.708	0.743	0.778	0.863	0.946	1.028	1.108
60	h	......	......	......	......	......	85.33	88.35	91.41	99.24	107.36	115.54	124.29
(48.7)	s	......	......	......	......	......	0.17120	0.17689	0.18246	0.19585	0.20865	0.22094	0.23280

(continued)

Table B-2 Properties of superheated Freon-12 (English units) (Continued)

Pressure, psia (saturated temperature)		Temperature, °F											
		−40	−20	0	20	40	60	80	100	150	200	250	300
70	v	......	......	......	......	......	0.553	0.642	0.673	0.750	0.824	0.896	0.967
	h	......	......	......	......	......	84.94	87.96	91.05	98.94	107.10	115.54	124.29
(57.9)	s	......	......	......	......	......	0.16765	0.17399	0.17961	0.19310	0.20597	0.21830	0.23020
80	v	......	......	......	......	......	......	0.540	0.568	0.636	0.701	0.764	0.826
	h	......	......	......	......	......	......	87.56	90.68	98.64	106.84	115.30	124.08
(66.3)	s	......	......	......	......	......	......	0.17108	0.17675	0.19035	0.20328	0.21566	0.22760
90	v	......	......	......	......	......	......	......	0.505	0.568	0.627	0.685	0.742
	h	......	......	......	......	......	......	......	90.31	98.32	106.56	115.07	123.88
(73.6)	s	......	......	......	......	......	......	......	0.17443	0.18813	0.20111	0.21356	0.22554
100	v	......	......	......	......	......	......	......	0.442	0.499	0.553	0.606	0.657
	h	......	......	......	......	......	......	......	89.93	97.99	106.29	114.84	123.67
(80.9)	s	......	......	......	......	......	......	......	0.17210	0.18590	0.19894	0.21145	0.22347
120	v	......	......	......	......	......	......	......	0.357	0.407	0.454	0.500	0.543
	h	......	......	......	......	......	......	......	89.13	97.30	105.70	114.35	123.25
(93.4)	s	......	......	......	......	......	......	......	0.16803	0.18207	0.19529	0.20792	0.22000
140	v	......	......	......	......	......	......	......	......	0.341	0.383	0.423	0.462
	h	......	......	......	......	......	......	......	......	96.65	105.14	113.85	122.85
(104.5)	s	......	......	......	......	......	......	......	......	0.17868	0.19205	0.20479	0.21701
160	v	......	......	......	......	......	......	......	......	0.318	0.335	0.372	0.408
	h	......	......	......	......	......	......	......	......	95.82	104.50	113.33	122.39
(114.5)	s	......	......	......	......	......	......	......	......	0.17561	0.18927	0.20213	0.21444
180	v	......	......	......	......	......	......	......	......	0.294	0.287	0.321	0.353
	h	......	......	......	......	......	......	......	......	94.99	103.85	112.81	121.92
(123.7)	s	......	......	......	......	......	......	......	......	0.17254	0.18648	0.19947	0.21187
200	v	......	......	......	......	......	......	......	......	0.241	0.255	0.288	0.317
	h	......	......	......	......	......	......	......	......	94.16	103.12	112.20	121.42
(132.1)	s	......	......	......	......	......	......	......	......	0.16970	0.18395	0.19717	0.20970
220	v	......	......	......	......	......	......	......	......	0.188	0.232	0.254	0.282
	h	......	......	......	......	......	......	......	......	93.32	102.39	111.59	120.91
(139.9)	s	......	......	......	......	......	......	......	......	0.16685	0.18142	0.19387	0.20753

APPENDIX

C

THERMODYNAMIC PROPERTIES OF AMMONIA

Table C-1 Saturated ammonia: pressure table (English units)*

Pressure, psia	Temperature, °F	Specific volume, ft^3/lb_m			Specific enthalpy, Btu/lb_m			Specific entropy, Btu $(lb_m)(°R)$		
		v_f	v_{fg}	v_g	h_f	h_{fg}	h_g	s_f	s_{fg}	s_g
80	44.40	0.02546	3.630	3.655	91.7	532.3	624.0	0.1982	1.0563	1.2545
82	45.66	0.02550	3.545	3.570	93.1	531.2	624.3	0.2010	1.5014	1.2524
84	46.89	0.02554	3.462	3.488	94.5	530.1	624.6	0.2037	1.0467	1.2504
86	48.11	0.03558	3.385	3.411	95.8	529.0	624.8	0.2064	1.0420	1.2484
88	49.30	0.02562	3.311	3.337	97.2	527.9	625.1	0.2090	1.0375	1.2465
90	50.47	0.02566	3.240	3.266	98.4	526.9	625.3	0.2115	1.0330	1.2445
92	51.62	0.02570	3.172	3.198	99.8	525.8	625.6	0.2141	1.0286	1.2427
94	52.76	0.02573	3.166	3.132	101.0	524.8	625.8	0.2165	1.0243	1.2408
96	53.87	0.02577	3.044	3.070	102.3	523.8	626.1	0.2190	1.0201	1.2391
98	54.97	0.02581	2.984	3.010	103.5	522.8	626.3	0.2213	1.0160	1.2373
100	56.05	0.02584	2.926	2.952	104.7	521.8	626.5	0.2237	1.0119	1.2356
102	57.11	0.02588	2.870	2.896	105.9	520.8	626.7	0.2260	1.0079	1.2339
104	58.16	0.02591	2.817	2.843	107.1	519.8	626.9	0.2282	1.0041	1.2323
106	59.19	0.02594	2.765	2.791	108.3	518.8	627.1	0.2305	1.0002	1.2307
108	60.21	0.02598	2.715	2.741	109.4	517.9	627.3	0.2327	0.9964	1.2291
110	61.21	0.02601	2.670	2.693	110.5	517.0	627.5	0.2348	0.9927	1.2275
116	64.13	0.02611	2.533	2.559	113.9	514.2	628.1	0.2411	0.9819	1.2230
118	65.08	0.02614	2.491	2.517	114.9	513.3	628.2	0.2431	0.9784	1.2215
120	66.92	0.02618	2.450	2.476	116.0	512.4	628.4	0.2452	0.9749	1.2201
122	66.94	0.02621	2.411	2.437	117.1	511.5	628.6	0.2471	0.9715	1.2186
124	67.86	0.02624	2.373	2.399	118.1	510.6	628.7	0.2491	0.9682	1.2173
126	68.76	0.02628	2.336	2.362	119.1	509.8	628.9	0.2510	0.9649	1.2159
128	69.65	0.02631	2.300	2.326	120.1	508.9	629.0	0.2529	0.9616	1.2145
130	70.53	0.02634	2.265	2.291	121.1	508.1	629.2	0.2548	0.9584	1.2132
132	71.40	0.02637	2.232	2.258	122.1	507.2	629.3	0.2567	0.9552	1.2119
134	72.26	0.02640	2.199	2.225	123.1	506.4	629.5	0.2585	0.9521	1.2106
136	73.11	0.02643	2.167	2.193	124.1	505.5	629.6	0.2603	0.9490	1.2093
138	73.95	0.02646	2.136	2.162	125.1	504.7	629.8	0.2621	0.9460	1.2081
140	74.79	0.02649	2.106	2.132	126.0	503.9	629.9	0.2638	0.9430	1.2068
142	75.61	0.02652	2.076	2.103	126.9	503.1	630.0	0.2656	0.9400	1.2056
144	76.42	0.02655	2.048	2.075	127.9	502.3	630.2	0.2673	0.9371	1.2044
146	77.23	0.02658	2.020	2.047	128.8	501.5	630.3	0.2690	0.9342	1.2032

148	78.03	0.02661	1.993	2.020	129.7	500.7	630.4	0.2707	0.9313	1.2020
150	78.81	0.02664	1.967	1.994	130.6	499.9	630.5	0.2724	0.9285	1.2009
152	79.60	0.02667	1.941	1.968	131.5	499.1	630.6	0.2740	0.9257	1.1997
154	80.37	0.02669	1.916	1.943	132.4	498.3	630.7	0.2756	0.9229	1.1985
156	81.13	0.02672	1.892	1.919	133.3	497.6	630.9	0.2772	0.9202	1.1974
158	81.89	0.02675	1.868	1.895	134.2	496.8	631.0	0.2788	0.9175	1.1963
160	82.64	0.02678	1.845	1.872	135.0	496.1	631.1	0.2804	0.9148	1.1952
162	83.39	0.02681	1.822	1.849	135.9	495.3	631.2	0.2820	0.9122	1.1942
164	84.12	0.02684	1.800	1.827	136.8	494.5	631.3	0.2835	0.9096	1.1931
166	84.85	0.02686	1.778	1.805	137.6	493.8	631.4	0.2850	0.9070	1.1920
168	85.57	0.02689	1.757	1.784	138.4	493.1	631.5	0.2866	0.9044	1.1910
170	86.29	0.02692	1.737	1.764	139.3	492.3	631.6	0.2881	0.9019	1.1900
172	87.00	0.02695	1.717	1.744	140.1	491.6	631.7	0.2895	0.8994	1.1889
174	87.71	0.02697	1.697	1.724	140.9	490.8	631.7	0.2910	0.8969	1.1879
176	88.40	0.02700	1.678	1.705	141.7	490.1	631.8	0.2925	0.8944	1.1869
178	89.10	0.02703	1.660	1.686	142.5	489.4	631.9	0.2939	0.8920	1.1859
180	89.78	0.02705	1.640	1.667	143.3	488.7	632.0	0.2954	0.8896	1.1850
182	90.46	0.02709	1.622	1.649	144.1	488.0	632.1	0.2968	0.8872	1.1840
184	91.80	0.02714	1.587	1.614	145.6	486.6	632.2	0.2996	0.8825	1.1821
186	92.47	0.02717	1.570	1.597	146.4	485.9	632.3	0.3010	0.8801	1.1811
188	93.13	0.02720	1.554	1.581	147.2	485.2	632.4	0.3024	0.8778	1.1802
190	93.78	0.02722	1.537	1.564	147.9	484.5	632.4	0.3037	0.8755	1.1792
192	94.43	0.02725	1.521	1.548	148.7	483.8	632.5	0.3050	0.8733	1.1783
194	95.07	0.02727	1.505	1.533	149.5	483.1	632.6	0.3064	0.8710	1.1774
196	95.71	0.02730	1.490	1.517	150.2	482.4	632.6	0.3077	0.8688	1.1765
198	96.34	0.02732	1.475	1.502	150.9	481.8	632.7	0.3090	0.8666	1.1756
200	99.43	0.02745	1.404	1.431	154.6	478.4	633.0	0.3154	0.8559	1.1713
202	102.42	0.02757	1.339	1.367	158.0	475.2	633.2	0.3216	0.8455	1.1671
204	105.30	0.02770	1.279	1.307	161.4	472.0	633.4	0.3275	0.8356	1.1631
206	108.09	0.02782	1.225	1.253	164.7	468.9	633.6	0.3332	0.8260	1.1592
208	110.80	0.02794	1.174	1.202	168.0	465.8	633.8	0.3388	0.8167	1.1555
210	113.42	0.02806	1.127	1.155	171.1	462.8	633.9	0.3441	0.8077	1.1518
212	115.97	0.02817	1.084	1.112	174.1	459.8	633.9	0.3494	0.7989	1.1483
214	118.45	0.02829	1.044	1.072	177.1	456.9	634.0	0.3545	0.7904	1.1449
216	120.86	0.02840	1.006	1.034	180.0	454.0	634.0	0.3594	0.7821	1.1415
218	123.21	0.02851	0.970	0.999	182.9	451.1	634.0	0.3642	0.7741	1.1383

* Table C-1 From Ref. 189.

Table C-2 Properties of superheated ammonia (English units)*

Pressure, psia (saturated temperature, °F)		Temperature of superheated vapor, °F																		
		40	60	80	100	120	140	160	180	200	220	240	260	280	300	320	340	360	380	400
	v	4.933	5.184	5.428	5.665	5.897	6.126	6.352	6.576	6.798	7.019	7.238	7.457	7.675	7.892					
60	*h*	626.8	639.0	650.7	662.1	673.3	684.4	695.5	706.5	717.5	728.6	739.7	750.9	762.1	773.3					
(30.21)	*s*	1.2913	1.3152	1.3373	1.3581	1.3778	1.3966	1.4148	1.4323	1.4493	1.4658	1.4819	1.4976	1.5130	1.5281					
	v	4.177	4.401	4.615	4.822	5.025	5.224	5.420	5.615	5.807	5.998	6.187	6.376	6.563	6.750					
70	*h*	623.9	636.6	648.7	660.4	671.8	683.1	694.3	705.5	716.6	727.7	738.9	750.1	761.4	772.7					
(37.70)	*s*	1.2688	1.2937	1.3166	1.3378	1.3579	1.3770	1.3954	1.4131	1.4302	1.4469	1.4631	1.4789	1.4943	1.5095					
	v		3.812	4.005	4.190	4.371	4.548	4.722	4.893	5.063	5.231	5.398	5.565	5.730	5.894					
80	*h*		634.3	646.7	658.7	670.4	681.8	693.2	704.4	715.6	726.9	738.1	749.4	760.7	772.1					
(53.10)	*s*		1.2745	1.2981	1.3199	1.3404	1.3598	1.3784	1.3693	1.4136	1.4304	1.4467	1.4626	1.4781	1.4933					
	v		3.353	3.529	3.698	3.862	4.021	4.178	4.332	4.484	4.635	4.785	4.933	5.081	5.228					
90	*h*		631.8	644.7	657.0	668.9	680.5	692.0	703.4	714.7	726.0	737.3	748.7	760.0	771.5					
(50.47)	*s*		1.2571	1.2814	1.3038	1.3247	1.3444	1.3633	1.3813	1.3988	1.4157	1.4321	1.4481	1.4637	1.4789					
	v			3.149	3.304	3.454	3.600	3.743	3.883	4.021	4.158	4.294	4.428	4.562	4.695					
100	*h*			642.6	655.2	667.3	679.2	690.8	702.3	713.7	725.1	736.5	747.9	759.4	770.8					
(56.05)	*s*			1.2661	1.2891	1.3104	1.3305	1.3495	1.3678	1.3854	1.4024	1.4190	1.4350	1.4507	1.4660					
	v			2.837	2.981	3.120	3.255	3.386	3.515	3.642	3.768	3.892	4.015	4.137	4.259					
110	*h*			642.6	653.4	665.8	677.8	689.6	701.2	712.8	724.3	735.7	747.2	758.7	770.2					
(61.21)	*s*			1.2661	1.2755	1.2972	1.3176	1.3370	1.3555	1.3732	1.3904	1.4070	1.4232	1.4389	1.4543					
	v			2.576	2.712	2.842	2.967	3.089	3.209	3.326	3.442	3.557	3.671	3.783	3.895					
120	*h*			638.3	651.6	664.2	676.5	688.4	700.2	711.8	723.4	734.9	746.5	758.0	769.6					
(66.02)	*s*			1.2519	1.2628	1.2850	1.3058	1.3254	1.3441	1.3620	1.3793	1.3960	1.4123	1.4281	1.4435					
	v			2.355	2.484	2.606	2.724	2.838	2.949	3.0590	3.167	3.273	3.379	3.483	3.587	3.690				
130	*h*			636.0	649.7	662.7	675.1	687.2	699.1	710.9	722.5	734.1	745.7	757.3	769.0	780.6				
(70.53)	*s*			1.2260	1.2509	1.2736	1.2947	1.3146	1.3335	1.3516	1.3690	1.3858	1.4022	1.4181	1.4336	1.4487				

	v	2.166	2.288	2.404	2.515	2.622	2.727	2.830	2.931	3.030	3.129	3.227	3.323	3.420			
140	h	633.8	647.8	661.1	673.7	686.0	698.0	709.9	721.6	733.3	745.0	756.7	768.3	780.0			
(74.79)	s	1.2140	1.2396	1.2628	1.2843	1.3045	1.3236	1.3418	1.3594	1.3763	1.3928	1.4088	1.4243	1.4395			
	v	2.001	2.118	2.228	2.334	2.435	2.534	2.631	2.726	2.820	2.912	3.004	3.095	3.185			
150	h	631.4	645.9	659.4	672.3	684.8	696.9	708.9	720.7	732.5	744.3	756.0	767.7	779.4			
(78.81)	s	1.2025	1.2289	1.2526	1.2745	1.2949	1.3142	1.3327	1.3504	1.3675	1.3840	1.4001	1.4157	1.4310			
	v		1.969	2.075	2.175	2.272	2.365	2.457	2.547	2.635	2.723	2.809	2.895	2.980	3.064		
160	h		643.9	657.8	670.9	683.5	695.8	707.9	719.9	731.7	743.5	755.3	767.1	778.9	790.7		
(82.64)	s		1.2186	1.2429	1.2652	1.2859	1.3054	1.3240	1.3419	1.3591	1.3757	1.3919	1.4076	1.4229	1.4379		
	v		1.837	1.939	2.035	2.127	2.216	2.303	2.389	2.473	2.555	2.637	2.718	2.798	2.878		
170	h		641.9	656.1	669.4	682.3	694.7	706.9	719.0	730.9	742.8	754.6	766.4	778.3	790.1		
(86.29)	s		1.2087	1.2336	1.2563	1.2773	1.2971	1.3159	1.3338	1.3512	1.3679	1.3841	1.3999	1.4153	1.4303		
	v		1.720	1.818	1.910	1.999	2.084	2.167	2.248	2.328	2.407	2.484	2.561	2.637	2.713		
180	h		639.9	654.4	668.0	681.0	693.6	705.9	718.1	730.1	742.0	753.9	765.8	777.7	789.6		
(89.78)	s		1.1992	1.2247	1.2477	1.2691	1.2891	1.3081	1.3262	1.3436	1.3605	1.3768	1.3926	1.4081	1.4231		
	v		1.615	1.710	1.799	1.884	1.966	2.045	2.123	2.199	2.274	2.348	2.421	2.493	2.565		
190	h		637.8	652.6	666.5	679.7	692.5	704.9	717.2	729.3	741.3	753.2	765.2	777.1	789.0		
(93.13)	s		1.1899	1.2160	1.2396	1.2612	1.2815	1.3007	1.3189	1.3365	1.3534	1.3698	1.3857	1.4012	1.4163		
	v		1.520	1.612	1.698	1.780	1.859	1.935	2.009	2.082	2.154	2.225	2.295	2.364	2.432	2.500	2.568
200	h		635.6	650.9	665.0	678.4	691.3	703.9	716.3	728.4	740.5	752.5	764.5	776.5	788.5	800.5	812.5
(96.34)	s		1.1947	1.2077	1.2317	1.2537	1.2742	1.2935	1.3120	1.3296	1.3467	1.3631	1.3791	1.3947	1.4099	1.4247	1.4392
	v			1.524	1.608	1.687	1.762	1.836	1.907	1.977	2.046	2.113	2.180	2.246	2.312	2.377	2.442
210	h			649.1	663.5	677.1	690.2	702.9	715.3	727.6	739.8	751.8	763.9	755.9	787.9	800.0	812.0
(99.43)	s			1.1996	1.2240	1.2464	1.2672	1.2867	1.3053	1.3231	1.3402	1.3568	1.3728	1.3884	1.4037	1.4186	1.4331
	v			1.443	1.525	1.601	1.675	1.745	1.814	1.881	1.947	2.012	2.076	2.140	2.203	2.265	2.327
220	h			647.3	662.0	675.8	689.1	701.9	714.4	726.8	739.0	751.1	763.2	775.3	787.4	799.5	811.6
(102.42)	s			1.1917	1.2167	1.2394	1.2604	1.2801	1.2989	1.3168	1.3340	1.3507	1.3668	1.3825	1.3978	1.4127	1.4273
	v			1.370	1.449	1.524	1.594	1.663	1.729	1.794	1.857	1.920	1.982	2.043	2.103	2.163	2.222
230	h			645.4	660.4	674.5	687.9	700.9	713.5	726.0	738.3	750.5	762.6	774.7	786.8	798.9	811.1
(105.30)	s			1.1840	1.2095	1.2325	1.2538	1.2738	1.2927	1.3107	1.3281	1.3448	1.3610	1.3767	1.3921	1.4070	1.4217

(continued)

Table C-2 Properties of superheated ammonia (English units)* (Continued)

Pressure, psia (saturated temperature, °F)		Temperature of superheated vapor, °F																		
		40	60	80	100	120	140	160	180	200	220	240	260	280	300	320	340	360	380	400
	v					1.302	1.380	1.452	1.521	1.587	1.651	1.714	1.775	1.835	1.895	1.954	2.012	2.069	2.126	
240	*h*					643.5	658.8	673.1	686.7	699.8	712.6	725.1	737.5	747.8	762.0	774.1	786.3	798.4	810.6	
(108.09)	*s*					1.1764	1.2025	1.2259	1.2475	1.2677	1.2867	1.3049	1.3224	1.3392	1.3554	1.3712	1.3866	1.4216	1.4163	
	v					1.240	1.316	1.386	1.453	1.518	1.580	1.640	1.699	1.758	1.815	1.872	1.928	1.983	2.038	2.093
250	*h*					641.5	657.2	671.8	685.5	698.8	711.7	724.3	736.7	749.1	761.3	773.5	785.7	797.9	810.1	822.3
(110.80)	*h*					1.1690	1.1956	1.2195	1.2414	1.2617	1.2810	1.2993	1.3168	1.3337	1.3501	1.3659	1.3814	1.3964	1.4111	1.4255
	v					1.182	1.257	1.326	1.391	1.453	1.514	1.572	1.630	1.686	1.741	1.796	1.850	1.904	1.957	2.009
260	*h*					639.5	655.6	670.4	684.4	697.7	710.7	723.4	736.0	748.4	760.7	772.9	785.2	797.4	809.6	821.9
(113.42)	*s*					1.1617	1.1889	1.2132	1.2354	1.2560	1.2754	1.2938	1.3115	1.3285	1.3449	1.3608	1.3763	1.3914	1.4062	1.4206
	v					1.128	1.202	1.269	1.333	1.394	1.452	1.509	1.565	1.620	1.673	1.726	1.778	1.830	1.881	1.932
270	*h*					637.5	653.9	669.0	683.2	696.7	709.8	722.6	735.2	747.7	760.0	772.3	784.6	796.9	809.1	821.4
(115.97)	*s*					1.1544	1.1823	1.2071	1.2296	1.2504	1.2700	1.2885	1.3063	1.3234	1.3399	1.3559	1.3714	1.3866	1.4014	1.4158
	v					1.078	1.151	1.217	1.279	1.339	1.396	1.451	1.505	1.558	1.610	1.661	1.712	1.762	1.811	1.861
280	*h*					635.4	652.2	667.6	681.9	695.6	708.8	721.8	734.4	747.0	759.4	771.7	784.0	796.3	808.7	821.0
(118.45)	*s*					1.1473	1.1759	1.2011	1.2239	1.2449	1.2647	1.2834	1.3013	1.3184	1.3350	1.3511	1.3367	1.3819	1.3967	1.4112
	v						1.103	1.168	1.229	1.287	1.343	1.397	1.449	1.501	1.551	1.601	1.650	1.698	1.747	1.794
290	*h*						650.5	666.1	680.7	694.6	707.9	720.9	733.7	746.3	758.7	771.1	783.5	795.8	808.2	820.5
(120.86)	*s*						1.695	1.1952	1.2183	1.2396	1.2596	1.2784	1.2964	1.3137	1.3303	1.3464	1.3621	1.3773	1.3922	1.4067
	v						1.058	1.123	1.183	1.239	1.294	1.346	1.397	1.447	1.496	1.544	1.592	1.639	1.686	1.732
300	*h*						648.7	664.7	679.5	693.5	706.9	720.0	732.9	745.5	758.1	770.5	782.9	795.3	807.7	820.1
(123.21)	*s*						1.1632	1.1894	1.2129	1.2344	1.2546	1.2736	1.2917	1.3090	1.3257	1.3419	1.3576	1.3729	1.3878	1.4024

* Abstracted from Ref. 189.

APPENDIX

D

THERMODYNAMIC PROPERTIES OF PROPANE

Table D Saturation properties of propane C_3H_8*

Temperature, °F	Pressure, psia	Specific volume, ft³/lb$_m$			Specific enthalpy, Btu/lb$_m$			Specific entropy, Btu/(lb$_m$)(°R)		
		v_f	v_{fg}	v_g	h_f	h_{fg}	h_g	s_f	s_{fg}	s_g
0	37.81	0.0289	2.711	2.740	205.0	172.2	377.2	0.9812	0.3743	1.3555
10	45.85	0.0293	2.271	2.300	210.7	169.3	380.0	0.9932	0.3599	1.3531
20	55.00	0.0297	1.900	1.930	216.6	166.0	382.6	1.0050	0.3460	1.3510
30	65.70	0.0301	1.570	1.600	222.3	162.8	385.1	1.0167	0.3324	1.3491
40	77.80	0.0306	1.299	1.330	227.9	159.6	387.5	1.0283	0.3190	1.3473
50	91.50	0.0310	1.109	1.140	233.8	156.1	389.9	1.0389	0.3067	1.3456
60	106.9	0.0315	0.953	0.984	239.6	152.6	392.2	1.0511	0.2930	1.3441
70	124.3	0.0321	0.822	0.854	245.7	148.7	394.4	1.0624	0.2803	1.3427
80	143.6	0.0327	0.712	0.745	251.9	144.5	396.4	1.0737	0.2680	1.3413
90	165.0	0.333	0.610	0.643	258.2	140.1	398.3	1.0850	0.2550	1.3400
100	188.7	0.0339	0.524	0.558	264.6	135.6	400.2	1.0963	0.2425	1.3388
110	214.8	0.0345	0.453	0.487	271.1	130.8	401.9	1.1080	0.2298	1.3378
120	243.4	0.0353	0.391	0.426	278.0	125.8	403.8	1.1195	0.2173	1.3368
130	274.5	0.0361	0.334	0.370	285.2	120.2	405.4	1.1310	0.2046	1.3356
140	308.4	0.0370	0.283	0.320	292.7	114.3	407.0	1.1430	0.1917	1.3347
150	354.4	0.0382	0.240	0.278	300.2	108.0	408.2	1.1552	0.1774	1.3326
160	385.0	0.0396	0.200	0.240	308.4	100.4	408.8	1.1680	0.1623	1.3303
170	426.0	0.0413	0.167	0.208	317.5	91.1	408.6	1.1816	0.1456	1.3272
180	473.2	0.0437	0.136	0.180	327.5	80.1	407.6	1.1970	0.1253	1.3223
190	523.4	0.0471	0.102	0.149	339.2	65.4	404.6	1.2140	0.1016	1.3156
200	575.0	0.0521	0.061	0.113	353.5	44.8	398.3	1.2360	0.0680	1.3040

* Abstracted from Ref. 190.

APPENDIX

E

THERMODYNAMIC PROPERTIES OF SODIUM

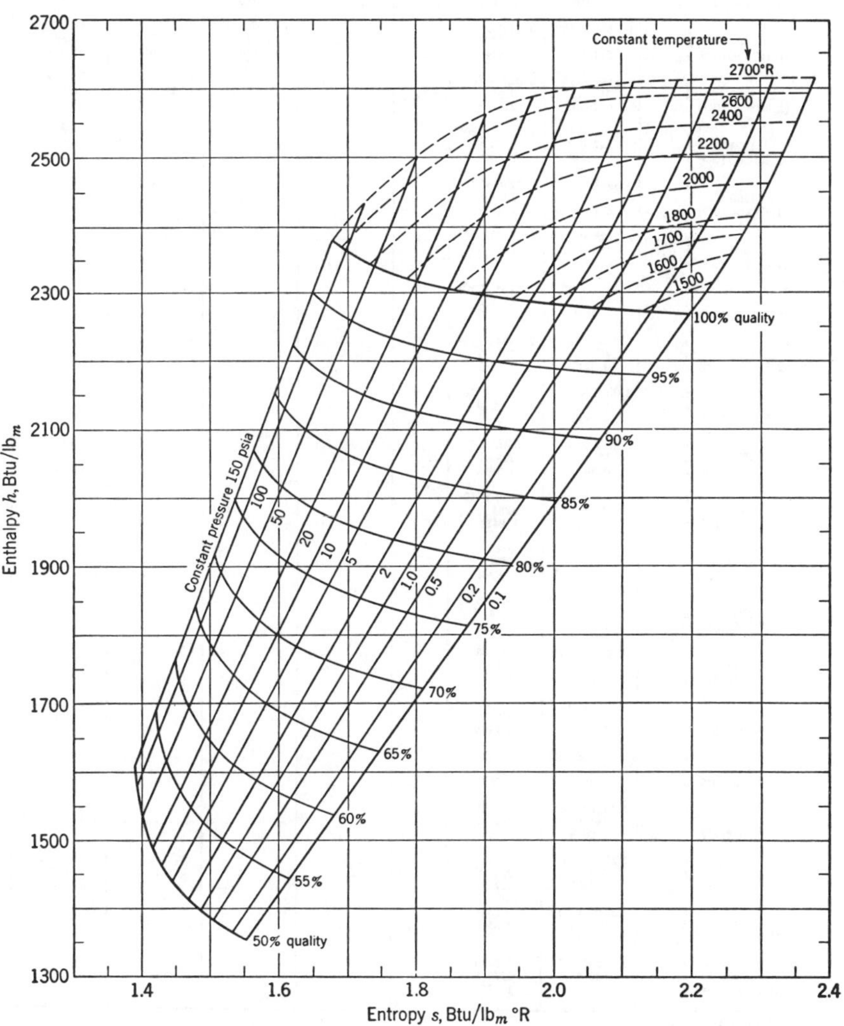

Figure E-1 Mollier diagram for sodium. *(Based on charts, courtesy Flight Propulsion Laboratory and the General Electric Company.)*

Table E Thermodynamic properties of sodium*

Temperature, °R (Sat. press., psia)		Sat. liquid	Sat. vapor	Temperature of				
				800	900	1000	1100	1200
700	v	1.7232×10^{-2}						
	h	219.7	2180.5	2203.1	2224.7	2246.4	2268.0	2289.5
(8.7472×10^{-9})	s	0.6854	3.4866	3.5169	3.5424	3.5652	3.5857	3.6043
800	v	1.7548×10^{-2}	7.4375×10^{8}	......	8.3835×10^{8}	9.3168×10^{8}	1.0249×10^{9}	1.1180×10^{9}
	h	252.3	2200.1	......	2224.4	2246.3	2267.9	2289.5
(5.0100×10^{-7})	s	0.7290	3.1637	......	3.1925	3.2155	3.2360	3.2546
900	v	1.7864×10^{-2}	3.6411×10^{7}	......	......	4.0267×10^{7}	4.4718×10^{7}	4.8789×10^{7}
	h	284.3	2217.6	......	......	2245.2	2267.7	2289.4
(1.1480×10^{-5})	s	0.7667	2.9148	......	......	2.9440	2.9653	2.9841
1000	v	1.8180×10^{-2}	3.323×10^{6}	......	......	......	3.6834×10^{6}	4.0245×10^{6}
	h	315.9	2232.7	......	......	......	2264.8	2288.6
(1.3909×10^{-4})	s	0.7999	2.7168	......	......	......	2.7474	2.7680
1100	v	1.8496×10^{-2}	4.7592×10^{5}	......	......	......	......	5.2512×10^{5}
	h	347.0	2245.1	......	......	......	......	2282.7
(1.0616×10^{-3})	s	0.8296	2.5551	......	......	......	......	2.5878
1200	v	1.8812×10^{-2}	9.5235×10^{4}	......	......	......	......	......
	h	377.7	2254.9	......	......	......	......	......
(5.7398×10^{-3})	s	0.8563	2.4207	......	......	......	......	......
1300	v	1.9128×10^{-2}	2.4520×10^{4}	......	......	......	......	......
	h	408.2	2262.8	......	......	......	......	......
(2.3916×10^{-2})	s	0.8807	2.3073	......	......	......	......	......
1400	v	1.9444×10^{-2}	7.6798×10^{3}	......	......	......	......	......
	h	438.4	2269.3	......	......	......	......	......
(8.1347×10^{-2})	s	0.9031	2.2109	......	......	......	......	......
1500	v	1.9760×10^{-2}	2.8334×10^{3}	......	......	......	......	......
	h	468.5	2274.9	......	......	......	......	......
(2.3351×10^{-1})	s	0.9239	2.1282	......	......	......	......	......
1600	v	2.0076×10^{-2}	1.1935×10^{3}	......	......	......	......	......
	h	498.5	2280.0	......	......	......	......	......
(5.8425×10^{-1})	s	0.9433	2.0567	......	......	......	......	......
1700	v	2.0392×10^{-2}	5.5585×10^{2}	......	......	......	......	......
	h	528.5	2285.3	......	......	......	......	......
(1.3170)	s	0.9615	1.9948	......	......	......	......	......
1800	v	2.0708×10^{-2}	2.8200×10^{2}	......	......	......	......	......
	h	558.6	2291.1	......	......	......	......	......
(2.7164)	s	0.9786	1.9411	......	......	......	......	......
1900	v	2.1024×10^{-2}	1.5512×10^{2}	......	......	......	......	......
	h	588.8	2297.2	......	......	......	......	......
(5.1529)	s	0.9949	1.8941	......	......	......	......	......
2000	v	2.1340×10^{-2}	90.914	......	......	......	......	......
	h	619.1	2304.1	......	......	......	......	......
(9.1533)	s	1.0105	1.8530	......	......	......	......	......
2100	v	2.1656×10^{-2}	56.185	......	......	......	......	......
	h	649.7	2312.1	......	......	......	......	......
(15.392)	s	1.0255	1.8171	......	......	......	......	......
2200	v	2.1972×10^{-2}	36.338	......	......	......	......	......
	h	680.7	2321.0	......	......	......	......	......
(24.692)	s	1.0399	1.7855	......	......	......	......	......
2300	v	2.2288×10^{-2}	24.454	......	......	......	......	......
	h	712.0	2330.7	......	......	......	......	......
(38.013)	s	1.0538	1.7576	......	......	......	......	......
2400	v	2.2604×10^{-2}	17.109	......	......	......	......	......
	h	743.8	2341.2	......	......	......	......	......
(56.212)	s	1.0673	1.7329	......	......	......	......	......
2500	v	2.2920×10^{-2}	12.388	......	......	......	......	......
	h	776.2	2352.6	......	......	......	......	......
(80.236)	s	1.0805	1.7111	......	......	......	......	......
2600	v	2.3236×10^{-2}	9.2328	......	......	......	......	......
	h	809.1	2365.1	......	......	......	......	......
(1.1116×10^{2})	s	1.0934	1.6919	......	......	......	......	......
2700	v	2.3552×10^{-2}	7.0380					
	h	842.7	2378.8					
(1.5052×10^{2})	s	1.1061	1.6751					

* From Ref. 191

Superheated Vapor, °R

1400	1600	1800	2000	2200	2400	2600	2700
2332.7 3.6381	2375.9 3.6665	2419.1 3.6924	2462.3 3.7148	2505.4 3.7354	2548.6 3.7545	2591.8 3.7713	2613.4 3.7796
1.3044×10^{9} 2332.7 3.2884	1.4907×10^{9} 2375.9 3.3169	1.6771×10^{9} 2419.1 3.3428	1.8634×10^{9} 2462.3 3.3652	2.0498×10^{9} 2505.4 3.3858	2.2361×10^{9} 2548.6 3.4048	2.4224×10^{9} 2591.8 3.5217	2.5156×10^{9} 2613.4 3.4299
5.6924×10^{7} 2332.7 3.0179	6.5056×10^{7} 2375.9 3.0464	7.3188×10^{7} 2419.1 3.0723	8.1320×10^{7} 2462.3 3.0947	8.9452×10^{7} 2505.4 3.1153	9.7584×10^{7} 2548.6 3.1343	1.0572×10^{8} 2591.8 3.1511	1.0978×10^{8} 2613.4 3.1594
4.6978×10^{6} 2332.6 2.8024	5.3693×10^{6} 2375.9 2.8309	6.0406×10^{6} 2419.1 2.8568	6.7118×10^{6} 2462.3 2.8792	7.383×10^{6} 2505.4 2.8998	8.0541×10^{6} 2548.6 2.9188	8.7253×10^{6} 2591.8 2.9357	9.0609×10^{6} 2613.4 2.9439
6.1515×10^{5} 2331.7 2.6263	7.0339×10^{5} 2375.7 2.6552	7.9139×10^{5} 2419.0 2.6812	8.7935×10^{5} 2462.3 2.7036	9.6729×10^{5} 2505.4 2.7243	1.0552×10^{6} 2548.6 2.7433	1.1432×10^{6} 2591.8 2.7601	1.1871×10^{6} 2613.4 2.7684
1.1345×10^{5} 2327.6 2.4778	1.3001×10^{5} 2374.7 2.5089	1.4635×10^{5} 2418.7 2.5353	1.6263×10^{5} 2462.2 2.5578	1.7891×10^{5} 2505.4 2.5785	1.9517×10^{5} 2548.6 2.5975	2.1144×10^{5} 2591.8 2.6143	2.1957×10^{5} 2613.4 2.6226
2.6936×10^{4} 2312.2 2.3445	3.1124×10^{4} 2371.1 2.3836	3.5095×10^{4} 2417.6 2.4115	3.9019×10^{4} 2461.7 2.4343	4.2931×10^{4} 2505.1 2.4551	4.6838×10^{4} 2548.5 2.4742	5.0743×10^{4} 2591.8 2.4911	5.2695×10^{4} 2613.4 2.4993
......	9.0793×10^{3} 2359.9 2.2715	1.0292×10^{4} 2414.0 2.3040	1.1460×10^{4} 2460.3 2.3280	1.2616×10^{4} 2504.5 2.3491	1.3767×10^{4} 2548.1 2.3683	1.4916×10^{4} 2591.6 2.3853	1.5491×10^{4} 2613.2 2.3935
......	3.1025×10^{3} 2332.6 2.1651	3.5625×10^{3} 2404.7 2.2083	3.9820×10^{3} 2456.5 2.2352	4.3896×10^{3} 2502.7 2.2573	4.7929×10^{3} 2547.2 2.2769	5.1944×10^{3} 2591.0 2.2940	5.3948×10^{3} 2612.8 2.3023
......	......	1.4040×10^{3} 2384.7 2.1192	1.5823×10^{3} 2448.0 2.1523	1.7496×10^{3} 2498.7 2.1765	1.9128×10^{3} 2545.1 2.1969	2.0743×10^{3} 2589.7 2.2144	2.1548×10^{3} 2611.8 2.2228
......	......	6.0659×10^{2} 2347.7 2.0309	6.9378×10^{2} 2431.3 2.0747	7.7180×10^{2} 2490.5 2.1031	8.4601×10^{2} 2540.7 2.1252	9.1858×10^{2} 2587.1 2.1433	9.5458×10^{2} 2609.8 2.1519
......	......	......	3.2952×10^{2} 2402.3 1.9996	3.7033×10^{2} 2475.6 2.0347	4.0785×10^{2} 2532.5 2.0597	4.4385×10^{2} 2582.2 2.0792	4.6158×10^{2} 2605.9 2.0882
......	......	......	1.6838×10^{2} 2359.5 1.9259	1.9197×10^{2} 2451.8 1.9701	2.1298×10^{2} 2518.8 1.9996	2.3265×10^{2} 2573.9 2.0212	2.4224×10^{2} 2599.3 2.0309
......	......	......	......	1.0543×10^{2} 2417.4 1.9072	1.1816×10^{2} 2498.0 1.9426	1.2980×10^{2} 2560.9 1.9673	1.3539×10^{2} 2588.9 1.9780
......	......	......	......	60.665 2372.9 1.8455	68.825 2469.0 1.8876	76.167 2451.9 1.9164	79.656 2573.5 1.9284
......	......	......	......	......	41.754 2431.8 1.8340	46.622 2516.2 1.8674	48.920 2552.3 1.8811
......	......	......	......	......	26.244 2388.2 1.7820	29.585 2484.0 1.8201	31.163 2525.0 1.8356
......	......	......	......	......	......	19.460 2446.8 1.7748	20.580 2492.6 1.7922
......	......	......	......	......	......	13.219 2406.5 1.7321	14.032 2456.5 1.7501
......	......	......	......	......	......	......	9.8326 2418.1 1.7120

APPENDIX

F

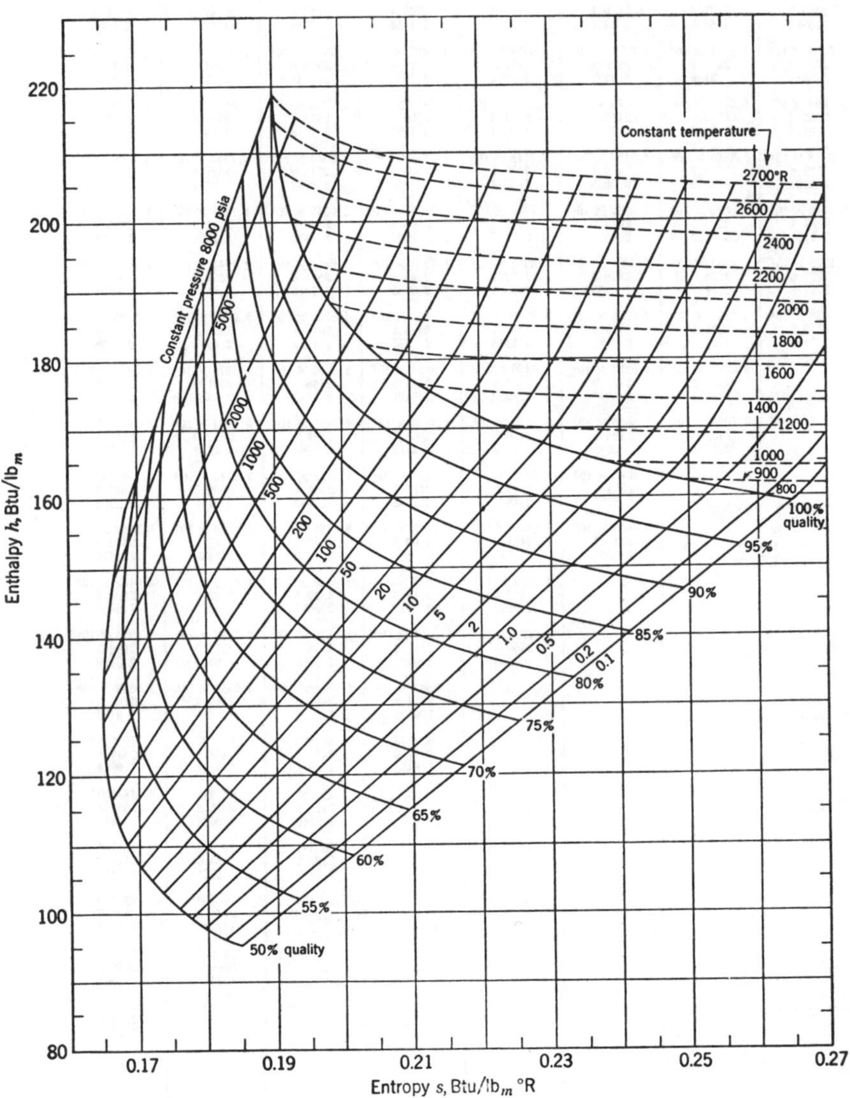

Figure F-1 Mollier diagram for mercury. *(Based on charts, courtesy Flight Propulsion Laboratory and the General Electric Company.)*

Table F Thermodynamic properties of mercury*

Temperature, °R (Sat. press., psia)		Sat. liquid	Sat. vapor	Temperature of superheated vapor, °R				
				800	900	1000	1100	1200
	v	0.0012	3.107×10^3	3.551×10^3	3.995×10^3	4.439×10^3	4.883×10^3	5.327×10^3
700	h	29.26	156.94	159.21	161.49	163.77	166.06	168.36
(1.2061×10^{-2})	s	0.09998	0.2824	0.2854	0.2882	0.2906	0.2928	0.2948
	v	0.00122	3.4150×10^2	...	38.42	42.69	46.96	51.23
800	h	32.51	159.75	...	162.03	164.31	166.60	168.89
(1.2541×10^{-1})	s	0.1043	0.2634	...	0.2661	0.2685	0.2707	0.2727
	v	0.00123	62.870	...	...	69.86	76.85	83.83
900	h	35.75	162.54	...	...	164.82	167.11	169.41
(0.76637)	s	0.1081	0.2490	...	...	0.2514	0.2536	0.2557
	v	0.00124	16.570	...	...	...	18.22	19.88
1000	h	38.98	165.32	...	...	...	167.61	169.91
(3.2319)	s	0.1115	0.2379	...	...	...	0.2041	0.2421
	v	0.00125	5.654	...	...	...	...	6.168
1100	h	42.21	168.10	...	...	...	...	170.40
(10.415)	s	0.1146	0.2291	...	...	...	...	0.2311
	v	0.00127	2.340	...	...	...	...	...
1200	h	45.45	170.88	...	...	...	...	...
(27.450)	s	0.1174	0.2220	...	...	...	...	...
	v	0.00128	1.122	...	...	...	...	...
1300	h	48.70	173.69	...	...	...	...	...
(62.007)	s	0.1200	0.2162	...	...	...	...	...
	v	0.00129	0.6038	...	...	...	...	...
1400	h	51.97	176.51	...	...	...	...	...
(1.2413×10^2)	s	0.1225	0.2114	...	...	...	...	...
	v	0.00131	0.3559	...	...	...	...	...
1500	h	55.27	179.36	...	...	...	...	...
(2.2566×10^2)	s	0.1247	0.2075	...	...	...	...	...
	v	0.00132	0.2258	...	...	...	...	...
1600	h	58.61	182.25	...	...	...	...	...
(3.7943×10^2)	s	0.1269	0.2042	...	...	...	...	...
	v	0.00133	0.1521	...	...	...	...	...
1700	h	62.00	185.19	...	...	...	...	...
(5.9840×10^2)	s	0.1289	0.2014	...	...	...	...	...
	v	0.00135	0.1077	...	...	...	...	...
1800	h	65.44	188.18	...	...	...	...	...
(8.9481×10^2)	s	0.1309	0.1991	...	...	...	...	...
	v	0.00136	0.0795	...	...	...	...	...
1900	h	68.94	191.23	...	...	...	...	...
(1.2795×10^3)	s	0.1328	0.1972	...	...	...	...	...
	v	0.00137	0.06078	...	...	...	...	...
2000	h	72.51	194.35	...	...	...	...	...
(1.7617×10^3)	s	0.1346	0.1956	...	...	...	...	...
	v	0.00139	0.04788	...	...	...	...	...
2100	h	76.16	197.55	...	...	...	...	...
(2.3483×10^3)	s	0.1364	0.1942	...	...	...	...	...
	v	0.00140	0.03869	...	...	...	...	...
2200	h	79.89	200.83	...	...	...	...	...
(3.0443×10^3)	s	0.1382	0.1931	...	...	...	...	...
	v	0.00141	0.03196	...	...	...	...	...
2300	h	83.71	204.21	...	...	...	...	...
(3.8523×10^3)	s	0.1399	0.1922	...	...	...	...	...
	v	0.00143	0.02692	...	...	...	...	...
2400	h	87.64	207.68	...	...	...	...	...
(4.7732×10^3)	s	0.1415	0.1915	...	...	...	...	...
	v	0.00144	0.02305	...	...	...	...	...
2500	h	91.67	211.27	...	...	...	...	...
(5.8059×10^3)	s	0.1432	0.1910	...	...	...	...	...
	v	0.00145	0.02003	...	...	...	...	...
2600	h	95.82	214.97	...	...	...	...	...
(6.9479×10^3)	s	0.1448	0.1906	...	...	...	...	...
	v	0.00147	0.01764					
2700	h	100.10	218.79					
(8.1953×10^3)	s	0.1464	0.1904					

* From Ref. 191.

(continued)

Table F Thermodynamic properties of mercury* (Continued)

Temperature, °R (Sat. press., psia)		Temperature of superheated vapor, °R							
		1400	1600	1800	2000	2200	2400	2600	2700
700	v	6.214×10^3	7.102×10^3	7.990×10^3	8.878×10^3	9.765×10^3	1.0650×10^4	1.1540×10^4	1.1980×10^4
	h	172.97	177.61	182.27	186.97	191.69	196.44	201.21	203.61
(1.2061×10^{-2})	s	0.2984	0.3015	0.3042	0.3067	0.3090	0.3111	0.3130	0.3139
800	v	5.977×10^2	6.831×10^2	7.684×10^2	8.538×10^2	9.392×10^2	1.025×10^3	1.110×10^3	1.153×10^3
	h	173.51	178.15	182.81	187.51	192.23	196.97	201.75	204.14
(1.2541×10^{-1})	s	0.2763	0.2794	0.2822	0.2847	0.2869	0.2890	0.2909	0.2918
900	v	97.80	1.118×10^2	1.257×10^2	1.397×10^2	1.537×10^2	1.677×10^2	1.816×10^2	1.886×10^2
	h	174.02	178.66	183.33	188.02	192.74	197.49	202.26	204.66
(0.76637)	s	0 2592	0.2633	0.2651	0.2676	0.2698	0.2719	0.2738	0.2747
1000	v	23.19	26.51	29.82	33.13	36.44	39.76	43.07	44.73
	h	174.52	179.16	183.82	188.52	193.24	197.98	202.76	205.14
(3.2319)	s	0.2457	0.2488	0.2515	0.2540	0.2563	0.2584	0.2603	0.2612
1100	v	7.197	8.225	9.253	10.28	11.31	12.34	13.37	13.88
	h	175.01	179.65	184.31	189.01	193.73	198.47	203.25	205.64
(10.415)	s	0.2347	0.2378	0.2405	0.2430	0.2453	0.2473	0.2493	0.2502
1200	v	2.731	3.121	3.511	3.901	4.291	4.681	5.071	5.266
	h	175.50	180.14	184.80	189.50	194.22	198.96	203.74	206.13
(27.450)	s	0.2255	0.2287	0.2314	0.2339	0.2362	0.2382	0.2402	0.2411
1300	v	1.209	1.381	1.554	1.727	1.900	2.072	2.245	2.331
	h	175.99	180.63	185.30	189.99	194.71	199.46	204.23	206.63
(62.007)	s	0.2179	0.2210	0.2238	0.2263	0.2285	0.2306	0.2325	0.2334
1400	v	....	0.6901	0.7764	0.8626	0.9489	1.035	1.121	1.165
	h	....	181.15	185.81	190.51	195.23	199.97	204.75	207.14
(1.2413×10^2)	s	....	0.2145	0.2173	0.2198	0.2220	0.2241	0.2260	0.2269
1500	v	....	0.3796	0.4271	0.4745	0.5220	0.5694	0.6169	0.6406
	h	....	181.68	186.35	191.04	195.76	200.51	205.28	207.68
(2.2566×10^2)	s	....	0.2090	0.2117	0.2142	0.2165	0.2185	0.2205	0.2214
1600	v	....	....	0.2540	0.2822	0.3104	0.3386	0.3669	0.3810
	h	....	....	186.92	191.61	196.33	201.08	205.85	208.25
(3.7943×10^2)	s	....	....	0.2069	0.2094	0.2117	0.2137	0.2157	0.2166
1700	v	....	....	0.1610	0.1789	0.1968	0.2147	0.2326	0.2416
	h	....	....	187.52	192.22	196.94	201.68	206.46	208.86
(5.9840×10^2)	s	....	....	0.2028	0.2052	0.2075	0.2096	0.2115	0.2124
1800	v	....	....	....	0.1197	0.1316	0.1436	0.1556	0.1615
	h	....	....	....	192.87	197.59	202.34	207.11	209.51
(8.9481×10^2)	s	....	....	....	0.2016	0.2038	0.2059	0.2078	0.2087
1900	v	....	....	....	0.08368	0.09205	0.1004	0.1088	0.1130
	h	....	....	....	193.58	198.30	203.05	207.82	210.22
(1.2795×10^3)	s	....	....	....	0.1984	0.2006	0.2027	0.2046	0.2055
2000	v	....	....	....	....	0.06686	0.07294	0.07901	0.08205
	h	....	....	....	....	199.07	203.82	208.59	210.99
(1.7617×10^3)	s	....	....	....	....	0.1978	0.1999	0.2018	0.2027
2100	v	....	....	....	....	0.05016	0.05472	0.05928	0.06156
	h	....	....	....	....	199.91	204.66	209.43	211.83
(2.3483×10^3)	s	....	....	....	....	0.1953	0.1974	0.1993	0.2002
2200	v	....	....	....	....	....	0.04221	0.04572	0.04748
	h	....	....	....	....	....	205.58	210.35	212.75
(3.0443×10^3)	s	....	....	....	....	....	0.1952	0.1971	0.1980
2300	v	....	....	....	....	....	0.03335	0.03613	0.03752
	h	....	....	....	....	....	206.58	211.36	213.75
(3.8523×10^3)	s	....	....	....	....	....	0.1933	0.1952	0.1961
2400	v	....	....	....	....	....	....	0.02804	0.03028
	h	....	....	....	....	....	....	210.07	214.85
(4.7732×10^3)	s	....	....	....	....	....	....	0.1925	0.1944
2500	v	....	....	....	....	....	....	0.02398	0.02490
	h	....	....	....	....	....	....	213.66	216.06
(5.8059×10^3)	s	....	....	....	....	....	....	0.1919	0.1929
2600	v	....	....	....	....	....	....	....	0.02081
	h	....	....	....	....	....	....	....	217.37
(6.9479×10^3)	s	....	....	....	....	....	....	....	0.1915
2700	v								
	h								
(8.1953×10^3)	s								

*From Ref. 191.

APPENDIX

G

THERMODYNAMIC PROPERTIES OF HELIUM

Table G Thermodynamic properties of helium*

Pressure, psia		Temperature, °F					
		100	200	300	400	500	600
14.696	v	102.23	120.487	138.743	157.00	175.258	193.515
	ρ	0.0097820	0.0082997	0.0072076	0.0063694	0.0057059	0.0051676
	h	707.73	827.56	952.38	1077.20	1202.02	1326.83
	s	6.8421	7.0472	7.2233	7.3776	7.5149	7.6386
50	v	30.085	35.451	40.817	46.183	51.549	56.915
	ρ	0.033239	0.028208	0.024500	0.021653	0.019399	0.017570
	h	703.08	827.90	952.72	1077.54	1202.36	1327.18
	s	6.2342	6.4393	6.6153	6.7697	6.9070	7.0307
150	v	10.063	11.8522	13.6407	15.4293	17.2183	19.008
	ρ	0.099372	0.084372	0.073310	0.064812	0.058078	0.052610
	h	704.08	828.91	953.73	1078.55	1203.37	1328.19
	s	5.6886	5.8937	6.0698	6.2241	6.3614	6.4852
400	v	3.8062	4.4775	5.1487	5.8197	6.4905	7.1616
	ρ	0.26273	0.22334	0.194225	0.171831	0.154072	0.139633
	h	706.58	831.42	956.24	1081.06	1205.88	1330.70
	s	5.2013	5.4065	5.5827	5.7371	5.8744	5.9981
600	v	2.5546	3.0023	3.44995	3.8973	4.3449	4.7923
	ρ	0.39146	0.33308	0.28986	0.25658	0.23016	0.20867
	h	708.49	833.33	958.15	1082.97	1207.79	1332.61
	s	4.9998	5.2050	5.3813	5.5357	5.6730	5.7968
900	v	1.7200	2.0187	2.3173	2.6157	2.91399	3.2124
	ρ	0.58139	0.49537	0.43154	0.38230	0.34317	0.31129
	h	710.29	835.38	960.40	1085.42	1210.42	1335.36
	s	4.7981	5.0035	5.1797	5.3342	5.4715	5.5953

(*continued*)

Table G Thermodynamic properties of helium* (Continued)

Pressure, psia		Temperature, °F					
		100	200	300	400	500	600
1,500	v	1.05192	1.2314	1.4108	1.58994	1.7690	1.9483
	ρ	0.95064	0.81207	0.70880	0.62897	0.56528	0.51328
	h	715.54	840.77	965.88	1090.92	1215.93	1340.97
	s	4.5437	4.7475	4.9257	5.0801	5.2176	5.3414
2,500	v	0.65044	0.75847	0.86635	0.97410	1.08176	1.18947
	ρ	1.53741	1.31845	1.15427	1.02659	0.92442	0.84071
	h	724.37	849.73	974.95	1100.10	1225.22	1350.29
	s	4.2887	4.4928	4.6712	4.8258	4.9634	5.0873
4,000	v	0.42377	0.49161	0.55932	0.62694	0.69444	0.76191
	ρ	2.3598	2.0341	1.78789	1.59503	1.44000	1.31248
	h	736.48	862.24	987.70	1113.12	1238.46	1363.73
	s	4.0531	4.2576	4.4363	4.5912	4.7287	4.8530

* From Ref. 177.

APPENDIX

H

THERMODYNAMIC PROPERTIES OF CARBON DIOXIDE

Table H Thermodynamic properties of carbon dioxide*

Pressure, psia		Temperature, °F												
		50	100	150	200	300	400	600	800	1000	1200	1400	1600	1800
	v	12.38	13.61	14.84	16.06	18.51	20.96	25.85	30.73	35.61	40.49	45.36	50.24	55.11
10.0	h	307.3	317.7	328.2	339.0	361.3	384.6	434.4	487.1	542.4	599.6	658.6	718.8	780.0
	s	1.4277	1.4467	1.4645	1.4813	1.5126	1.5412	1.5930	1.6384	1.6790	1.7158	1.7494	1.7799	1.8084
	v	6.119	6.778	7.407	8.016	9.247	10.47	12.92	15.36	17.80	20.24	22.68	25.11	27.55
20.0	h	306.8	317.3	327.9	338.8	361.1	384.5	434.3	487.1	542.4	599.6	658.6	718.8	780.0
	s	1.3964	1.4154	1.4332	1.4500	1.4813	1.5099	1.5617	1.6071	1.6477	1.6845	1.7181	1.7486	1.7771
	v	3.053	3.363	3.688	3.993	4.615	5.230	6.458	7.688	8.901	10.12	11.37	12.56	13.78
40.0	h	305.9	316.5	327.4	338.4	360.9	384.3	434.2	487.0	542.4	599.6	658.6	718.8	780.0
	s	1.3642	1.3834	1.4014	1.4184	1.4499	1.4787	1.5305	1.5759	1.6165	1.6533	1.6869	1.7174	1.7459
	v	1.498	1.657	1.828	1.982	2.298	2.608	3.226	3.839	4.448	5.060	5.670	6.281	6.887
80.0	h	304.1	315.1	326.4	337.7	360.2	383.9	434.0	486.9	542.3	599.5	658.6	718.8	780.0
	s	1.3284	1.3490	1.3679	1.3855	1.4177	1.4468	1.4991	1.5446	1.5852	1.6220	1.6556	1.6861	1.7146
	v	0.9799	1.088	1.208	1.311	1.525	1.734	2.148	1.559	2.966	3.373	3.781	4.188	4.592
120	h	302.2	313.6	325.4	337.0	359.7	383.5	433.8	486.8	542.3	599.5	658.6	718.8	780.0
	s	1.3086	1.3297	1.3488	1.3666	1.3993	1.4285	1.4808	1.5263	1.5669	1.6037	1.6373	1.6678	1.6963
	v	0.7207	0.8033	0.8986	0.9760	1.139	1.297	1.610	1.918	2.224	2.530	2.836	3.141	3.445
160	h	300.4	312.1	324.4	336.3	359.1	383.1	433.6	486.6	542.2	599.5	658.6	718.8	780.0
	s	1.2928	1.3154	1.3350	1.3529	1.3857	1.4151	1.4675	1.5133	1.5539	1.5907	1.6243	1.6548	1.6833
	v	0.5652	0.6376	0.7125	0.7748	0.9075	1.035	1.287	1.534	1.779	2.024	2.269	2.513	2.757
200	h	298.6	310.6	323.4	335.6	358.5	382.7	433.4	486.5	542.2	599.5	658.5	718.8	780.0
	s	1.2805	1.3038	1.3239	1.3421	1.3753	1.4049	1.4574	1.5033	1.5439	1.5807	1.6143	1.6448	1.6733
	v	0.4614	0.5237	0.5886	0.6407	0.7532	0.8604	1.071	1.273	1.482	1.687	1.891	2.095	2.297
240	h	296.7	309.1	322.4	334.9	358.0	382.3	433.1	486.4	542.1	599.5	658.5	718.8	780.0
	s	1.2694	1.2940	1.3145	1.3330	1.3671	1.3963	1.4490	1.4948	1.5356	1.5724	1.6060	1.6365	1.6650
	v	0.3563	0.4100	0.4636	0.5065	0.5985	0.6868	0.8556	1.021	1.186	1.349	1.513	1.676	1.838
300	h	294.0	306.9	320.9	333.9	357.1	381.6	432.8	486.2	542.0	599.4	658.5	718.7	780.0
	s	1.2562	1.2813	1.3029	1.3219	1.3560	1.3862	1.4389	1.4848	1.5256	1.5624	1.5960	1.6265	1.6550
	v	0.2858	0.3341	0.3780	0.4171	0.4958	0.5693	0.7212	0.8502	0.9874	1.125	1.261	1.397	1.533
360	h	291.2	304.6	319.4	332.8	356.3	381.0	432.5	486.0	541.9	599.4	658.5	718.7	779.9
	s	1.2436	1.2699	1.2925	1.3124	1.3475	1.3779	1.4307	1.4766	1.5174	1.5542	1.5878	1.6183	1.6468

(continued)

Table H Thermodynamic properties of carbon dioxide* (Continued)

Pressure, psia		Temperature, °F												
		50	100	150	200	300	400	600	800	1000	1200	1400	1600	1800
440	v......	0.2216	0.2652	0.3040	0.3358	0.4022	0.4633	0.5817	0.6950	0.8079	0.9201	1.032	1.142	1.255
	h......	287.6	301.6	317.4	331.4	355.1	380.2	432.1	485.8	541.7	599.3	658.4	718.6	779.9
	s......	1.2282	1.2559	1.2797	1.3006	1.3370	1.3681	1.4215	1.4675	1.5083	1.5451	1.5787	1.6092	1.6377
520	v......	0.1772	0.2174	0.2513	0.2795	0.3374	0.3901	0.4912	0.5881	0.6832	0.7785	0.8733	0.9672	1.062
	h......	283.9	298.7	315.4	330.0	354.0	379.4	431.7	485.5	541.5	599.2	658.3	718.6	779.9
	s......	1.2148	1.2438	1.2687	1.2905	1.3281	1.3599	1.4138	1.4599	1.5007	1.5375	1.5711	1.6010	1.6301
600	v......	0.1452	0.1823	0.2123	0.2383	0.2898	0.3363	0.4250	0.5093	0.5921	0.6747	0.7571	0.8385	0.9202
	h......	280.3	295.7	313.4	328.6	352.8	378.6	431.1	485.3	541.4	599.0	658.2	718.6	779.8
	s......	1.2020	1.2323	1.2583	1.2809	1.3198	1.3525	1.4071	1.4534	1.4942	1.5310	1.5646	1.5951	1.6236
800	v......	......	0.1196	0.1483	0.1712	0.2126	0.2489	0.3173	0.3812	0.4436	0.5060	0.5680	0.6292	0.6906
	h......	......	288.2	308.4	325.1	350.0	376.5	430.1	484.7	541.0	598.8	658.0	718.4	779.7
	s......	......	1.2111	1.2391	1.2631	1.3041	1.3380	1.3935	1.4404	1.4812	1.5180	1.5516	1.5821	1.6106
1000	v......	......	......	0.1101	0.1310	0.1663	0.1966	0.2526	0.3048	0.3547	0.4049	0.4545	0.5037	0.5526
	h......	......	......	303.4	321.6	347.1	374.5	429.1	484.0	540.6	598.5	657.8	718.3	779.6
	s......	......	......	1.2218	1.2472	1.2903	1.3258	1.3828	1.4302	1.4712	1.5080	1.5416	1.5721	1.6006

* From Ref. 193

APPENDIX

I

AIR TABLES

Table I-1 Enthalpy and isentropic pressure ratio for dry air*

T, °R	*h*, Btu/lb$_m$	Pr	*T*, °R	*h*, Btu/lb$_m$	Pr	*T*, °R	*h*, Btu/lb$_m$	Pr
450	107.50	0.7329	760	182.08	4.607	1070	258.47	15.734
460	109.90	0.7913	770	184.51	4.826	1080	260.97	16.278
470	112.30	0.8531	780	186.94	5.051	1090	263.48	16.838
480	114.69	0.9182	790	189.38	5.285	1100	265.99	17.413
490	117.08	0.9868	800	191.81	5.526	1110	268.52	18.000
500	119.48	1.0590	810	194.25	5.775	1120	271.03	18.604
510	121.87	1.1349	820	196.69	6.033	1130	273.56	19.223
520	124.27	1.2147	830	199.12	6.299	1140	276.08	19.858
530	126.66	1.2983	840	201.56	6.573	1150	278.61	20.15
540	129.06	1.3860	850	204.01	6.856	1160	281.14	21.18
550	131.46	1.4779	860	206.46	7.149	1170	283.68	21.86
560	133.86	1.5742	870	208.90	7.450	1180	286.2.	22.56
570	136.26	1.6748	880	211.35	7.761	1190	288.76	23.28
580	138.66	1.7800	890	213.80	8.081	1200	291.30	24.01
590	141.06	1.8899	900	216.26	8.411	1210	293.86	24.76
600	143.47	2.005	910	218.72	8.752	1220	296.41	25.53
610	145.88	2.124	920	221.18	9.102	1230	298.96	26.32
620	148.28	2.249	930	223.64	9.463	1240	301.52	27.13
630	150.68	2.379	940	226.11	9.834	1250	302.08	27.96
640	153.09	2.514	950	228.58	10.216	1260	306.65	28.80
650	155.50	2.655	960	231.06	10.255	1270	309.22	29.67
660	157.92	2.953	970	233.53	11.014	1280	311.79	30.55
670	160.33	2.953	980	236.02	11.430	1290	314.36	31.46
680	162.73	3.111	990	238.50	11.858	1300	316.94	32.39
690	165.15	3.276	1000	240.98	12.298	1310	319.53	33.34
700	167.56	3.446	1010	243.23	21.523	1320	322.11	34.31
710	169.98	3.623	1020	245.97	13.215	1330	324.69	35.30
720	172.39	3.806	1030	248.45	13.692	1340	327.29	36.31
730	174.82	3.996	1040	250.95	14.182	1350	329.88	37.35
740	177.23	4.193	1050	253.45	14.686	1360	332.48	38.41
750	179.66	4.396	1060	255.96	15.203	1370	335.09	39.49

(*continued*)

Table I-1 Enthalpy and isentropic pressure ratio for dry air* (Continued)

T, °R	h, Btu/lb$_m$	Pr	T, °R	h, Btu/lb$_m$	Pr	T, °R	h, Btu/lb$_m$	Pr
1380	337.39	40.59	1590	393.07	70.00	1800	449.71	114.03
1390	340.29	41.73	1600	395.74	71.73	1810	452.44	116.57
1400	342.90	42.88	1610	398.42	73.49	1820	455.17	119.16
1410	345.52	44.06	1620	401.09	75.29	1830	457.90	121.79
1420	348.14	45.26	1630	403.77	77.12	1840	460.63	124.47
1430	350.75	46.49	1640	406.45	78.99	1850	463.37	127.37
1440	353.37	47.75	1650	409.13	80.89	1860	466.12	129.56
1450	356.00	49.03	1660	411.82	82.83	1870	468.86	132.77
1460	358.63	50.34	1670	414.51	84.80	1880	471.60	135.64
1470	361.127	51.68	1680	417.20	86.82	1890	474.35	138.55
1480	363.89	53.04	1690	419.89	88.87	1900	477.09	141.51
1490	366.53	54.43	1700	422.59	90.95	1910	479.85	144.53
1500	369.17	55.86	1710	425.29	93.08	1920	482.60	147.59
1510	371.82	57.30	1720	428.00	95.24	1930	485.36	150.70
1520	374.47	58.78	1730	430.69	97.45	1940	488.12	153.87
1530	377.11	60.29	1740	433.41	99.69	1950	490.88	157.10
1540	379.77	61.83	1750	436.12	101.98	1960	493.64	160.37
1550	382.42	63.40	1760	438.83	104.30	1970	496.40	163.69
1560	385.08	65.00	1770	441.55	106.67	1980	499.17	167.07
1570	387.74	66.63	1780	444.26	109.08	1990	501.94	170.50
1580	390.40	68.30	1790	446.99	111.54	2000	504.71	174.00

* Abstracted from Ref. 8.

Table I-2 Enthalpy and isentropic pressure ratio for products of combustion with 200 percent theoretical air*

T, °R	h, Btu/lb$_m$ · mol	Pr	T, °R	h, Btu/lb$_m$ · mol	Pr	T, °R	h, Btu/lb$_m$ · mol	Pr	T, °R	h, Btu/lb$_m$ · mol	Pr
800	5676.3	5.690	1090	7840.3	18.173	1380	10091.9	45.68	1670	12432.1	99.41
810	5749.7	5.957	1100	7916.4	18.822	1390	10171.3	47.02	1680	12514.2	101.60
820	5823.1	6.234	1110	7993.0	19.486	1400	10250.7	48.38	1690	12596.3	104.13
830	5896.3	6.520	1120	8069.0	20.170	1410	10330.4	49.78	1700	12678.6	106.70
840	5969.9	6.815	1130	8145.7	20.873	1420	10410.0	51.21	1710	12761.0	109.33
850	6043.6	7.120	1140	8222.1	21.595	1430	10489.4	52.67	1720	12843.8	112.01
860	6117.5	7.437	1150	8298.7	22.34	1440	10569.3	54.17	1730	12926.1	114.75
870	6191.0	7.763	1160	8375.5	23.10	1450	10649.2	55.70	1740	13009.0	117.53
880	6264.9	8.101	1170	8452.5	23.88	1460	10729.3	57.26	1750	13091.7	120.38
890	6338.8	8.449	1180	8529.2	24.68	1470	10809.6	58.86	1760	13174.6	123.27
900	6413.0	8.808	1190	8606.5	25.50	1480	10889.5	60.49	1770	13257.5	126.22
910	6487.2	9.181	1200	8683.6	26.34	1490	10969.8	62.16	1780	13340.3	129.23
920	6561.5	9.564	1210	8761.1	27.20	1500	11050.2	63.88	1790	13423.7	132.31
930	6635.8	9.959	1220	8838.6	28.09	1510	11130.8	65.61	1800	13507.0	135.43
940	6710.4	10.366	1230	8915.9	29.00	1520	11211.4	67.40	1810	13590.4	138.61
950	6784.9	10.787	1240	8993.7	29.94	1530	11291.9	69.22	1820	13673.8	141.86
960	6859.8	11.221	1250	9071.4	30.90	1540	11372.8	71.08	1830	13757.0	145.17
970	6934.6	11.666	1260	9149.3	31.87	1550	11453.6	72.98	1840	13840.5	148.54
980	7009.7	12.126	1270	9227.4	32.88	1560	11534.7	74.91	1850	13924.4	151.96
990	7084.6	12.600	1280	9305.3	33.90	1570	11615.8	76.89	1860	14008.4	155.46
1000	7159.8	13.089	1290	9383.4	34.96	1580	11696.8	78.92	1870	14092.2	159.02
1010	7235.2	13.592	1300	9461.7	36.05	1590	11778.2	80.99	1880	14176.1	162.65
1020	7310.5	14.109	1310	9540.3	37.16	1600	11859.6	83.10	1890	14260.2	166.34
1030	7385.5	14.641	1320	9618.8	38.29	1610	11941.3	85.24	1900	14344.1	170.09
1040	7461.1	15.189	1330	9697.1	39.45	1620	12022.7	87.44	1910	14428.5	173.93
1050	7536.8	15.754	1340	9776.2	40.64	1630	12104.5	89.68	1920	14512.5	177.82
1060	7612.7	16.333	1350	9854.8	41.86	1640	12186.2	91.97	1930	14597.0	181.78
1070	7688.6	16.930	1360	9933.8	43.10	1650	12268.0	94.30	1940	14681.4	185.82
1080	7764.3	17.542	1370	10013.2	44.38	1660	12350.0	96.69	1950	14765.9	189.95

(continued)

Table I-2 Enthalpy and isentropic pressure ratio for products of combustion with 200 percent theoretical air* (Continued)

T, °R	h, Btu/lb$_m$ · mol	Pr	T, °R	h, Btu/lb$_m$ · mol	Pr	T, °R	h, Btu/lb$_m$ · mol	Pr	T, °R	h, Btu/lb$_m$ · mol	Pr
1960	14850.4	194.13	2370	18379.6	442.2	2780	22008.6	900.3	3190	25714.4	1683.3
1970	14935.0	198.38	2380	18467.0	450.4	2790	22097.8	914.9	3200	25805.6	1708.2
1980	15019.8	202.71	2390	18554.3	458.9	2800	22187.5	929.8	3210	25896.8	1732.8
1990	15104.4	207.11	2400	18642.1	467.4	2810	22277.2	945.0	3220	25988.1	1757.8
2000	15189.3	211.6	2410	18729.7	476.0	2820	22367.0	960.2	3230	26079.3	1783.2
2010	15274.5	216.2	2420	18817.4	484.8	2830	22456.6	975.6	3240	26170.6	1808.7
2020	15359.3	220.8	2430	18905.0	493.6	2840	22546.4	991.4	3250	26262.0	1834.4
2030	15444.5	225.5	2440	18992.8	502.7	2850	22636.3	1007.2	3260	26353.3	1860.6
2040	15529.7	230.4	2450	19080.7	511.9	2860	22726.1	1023.4	3270	26444.7	1887.0
2050	15651.1	235.2	2460	19168.6	521.1	2870	22816.2	1039.7	3280	26536.2	1913.8
2060	15700.3	240.2	2470	19256.6	530.6	2880	22905.8	1056.2	3290	26627.6	1940.9
2070	15785.8	245.3	2480	19344.3	540.2	2890	22995.8	1072.9	3300	26719.2	1968.3
2080	15871.3	250.4	2490	19432.3	550.0	2900	23086.0	1089.8	3310	26810.8	1995.9
2090	15956.5	255.6	2500	19520.7	559.8	2910	23176.2	1107.0	3320	26902.4	2023.8
2100	16042.4	260.9	2510	19608.9	569.8	2920	23266.2	1124.4	3330	26993.9	2052.2
2110	16128.3	266.4	2520	19697.0	580.0	2930	23356.5	1142.0	3340	27085.5	2080.7
2120	16214.1	271.9	2530	19785.3	590.3	2940	23446.7	1159.8	3350	27177.3	2109.5
2130	16299.7	277.5	2540	19873.7	600.7	2950	23536.7	1177.9	3360	27269.3	2138.7
2140	16385.6	283.1	2550	19962.0	611.3	2960	23626.9	1196.1	3370	27361.2	2168.4
2150	16471.5	288.9	2560	20050.4	622.0	2970	23717.3	1214.6	3380	27452.8	2198.4
2160	16557.7	294.8	2570	20138.8	633.0	2980	23807.8	1233.3	3390	27544.5	2228.6
2170	16643.8	300.8	2580	20227.5	644.0	2990	23897.9	1252.2	3400	27636.4	2259.1
2180	16730.0	306.8	2590	20315.9	655.3	3000	23988.5	1271.2	3410	27728.4	2290.1
2190	16816.2	313.0	2600	20404.6	666.6	3010	24079.0	1290.8	3420	27820.3	2321.4
2200	16902.5	319.2	2610	20493.3	678.1	3020	24260.1	1310.6	3430	27912.2	2353.0
2210	16989.1	325.6	2620	20582.0	689.9	3030	24350.7	1330.5	3440	28004.1	2385.0
2220	17075.6	332.0	2630	20670.7	701.8	3040	24441.3	1350.6	3450	28096.2	2417.3
2230	17161.9	338.6	2640	20759.5	713.7	3050	24532.2	1371.0	3460	28188.5	2450.0
2240	17248.6	345.2	2650	20848.4	725.9	3060	24623.0	1391.7	3470	28280.7	2483.0

2250	17335.3	352.0	2660	20937.5	738.2	3070	24713.7	1412.6	3480	28372.7	2516.4
2260	17422.0	358.9	2670	21026.6	750.8	3080	24713.7	1433.8	3490	28464.6	2550.0
2270	17508.7	365.9	2680	21115.4	763.4	3090	24804.5	1455.1	3500	28556.8	2584.0
2280	17595.3	373.0	2690	21204.5	776.3	3100	24895.3	1476.8			
2290	17682.4	380.3	2700	21293.8	789.4	3110	24986.1	1498.7			
2300	17769.3	387.5	2710	21382.9	802.5	3120	25077.0	1520.9			
2310	17856.5	395.0	2720	21472.0	815.9	3130	25168.0	1543.3			
2320	17943.4	402.6	2730	21561.5	829.6	3140	25259.0	1566.1			
2330	18030.5	410.2	2740	21650.9	843.4	3150	25350.1	1589.2			
2340	18117.8	418.0	2750	21740.3	857.2	3160	25441.1	1612.4			
2350	18204.9	425.9	2760	21829.4	871.4	3170	25532.2	1636.0			
2360	18292.2	434.0	2770	21918.8	885.8	3180	25623.4	1659.7			

* Abstracted from Ref. 8.

Table I-3 Enthalpy and isentropic pressure ratio for products of combustion with 400 percent theoretical air*

T,°R	h, Btu/lb$_m$·mol	Pr	T,°R	h, Btu/lb$_m$·mol	Pr	T,°R	h, Btu/lb$_m$·mol	Pr
1500	10875.6	59.80	1970	14662.6	180.50	2440	18608.9	445.8
1510	10954.4	61.39	1980	14745.3	184.33	2450	18694.1	453.7
1520	11033.1	63.01	1990	14827.7	188.23	2460	18779.5	461.7
1530	11111.6	64.67	2000	14910.3	192.21	2470	18864.9	469.9
1540	11190.7	66.37	2010	14993.2	196.24	2480	18950.2	478.1
1550	11269.6	68.10	2020	15075.8	200.35	2490	19035.7	486.5
1560	11348.6	69.86	2030	15158.7	204.51	2500	19121.4	494.9
1570	11427.6	71.66	2040	15241.6	208.76	2510	19207.2	503.5
1580	11506.8	73.51	2050	15324.7	213.06	2520	19292.8	512.3
1590	11586.3	75.39	2060	15407.6	217.45	2530	19378.4	521.1
1600	11665.6	77.30	2070	15490.9	221.90	2540	19464.2	530.0
1610	11745.2	79.25	2080	15574.0	226.42	2550	19550.1	539.1
1620	11824.6	81.24	2090	15656.9	231.04	2560	19635.9	548.3
1630	11904.4	83.27	2100	15740.5	235.7	2570	19721.8	557.7
1640	11984.1	85.35	2110	15740.5	240.5	2580	19807.8	567.1
1650	12063.8	87.46	2120	15907.5	245.3	2590	19893.7	576.7
1660	12143.8	89.61	2130	15990.7	250.2	2600	19979.7	586.4
1670	12223.8	91.80	2140	16074.2	255.1	2610	20065.9	596.2
1680	12303.9	94.05	2150	16157.9	260.2	2620	20152.0	606.3
1690	12383.9	96.33	2160	16241.7	265.4	2630	20238.1	616.5
1700	12464.3	98.64	2170	16325.5	270.6	2640	20324.3	626.6
1710	12544.6	101.02	2180	16409.4	275.9	2650	20410.6	636.9
1720	12625.3	103.43	2190	16493.3	281.3	2660	20497.0	647.4
1730	12705.5	105.89	2200	16577.1	286.7	2670	20583.5	658.1
1740	12786.4	108.40	2210	16661.3	292.3	2680	20670.0	668.9
1750	12867.0	110.96	2220	16745.4	297.9	2690	20756.4	679.8
1760	12947.8	113.55	2230	16829.4	303.7	2700	20842.8	690.9
1770	13028.6	116.20	2240	16913.6	309.4	2710	20929.2	702.1
1780	13109.4	118.90	2250	16998.0	315.3	2720	21015.7	713.5
1790	13190.5	121.66	2260	17082.3	321.3	2730	21102.4	725.1
1800	13271.7	124.45	2270	17166.6	327.5	2740	21189.2	736.8
1810	13353.0	127.30	2280	17250.7	333.6	2750	21276.0	748.5
1820	13434.3	130.21	2290	17335.4	340.0	2760	21362.6	760.5
1830	13515.5	133.17	2300	17419.8	346.2	2770	21449.4	772.6
1840	13597.0	136.18	2310	17504.6	352.7	2780	21536.2	784.9
1850	13678.5	139.23	2320	17589.0	359.3	2790	21622.7	797.3
1860	13760.3	142.35	2330	17673.6	365.9	2800	21709.8	809.9
1870	13842.0	145.53	2340	17758.4	372.6	2810	21796.9	822.6
1880	13923.6	148.76	2350	17843.2	379.5	2820	21884.0	835.5
1890	14005.6	152.05	2360	17928.1	386.5	2830	21970.9	848.5
1900	14087.2	155.39	2370	18013.0	393.6	2840	22058.0	861.8
1910	14169.5	158.80	2380	18098.0	400.7	2850	22145.3	875.2
1920	14251.3	162.26	2390	18182.9	408.0	2860	22232.5	888.8
1930	14333.4	165.78	2400	18268.0	415.3	2870	22319.8	902.6
1940	14415.7	169.37	2410	18353.2	422.8	2880	22406.6	916.4
1950	14498.0	173.02	2420	18438.4	430.2	2890	22494.1	930.4
1960	14580.3	176.73	2430	18523.5	438.0	2900	22581.4	944.7

Table I-3 Enthalpy and isentropic pressure ratio for products of combustion with 400 percent theoretical air* (Continued)

T,°R	h, Btu/lb$_m$·mol	Pr	T,°R	h, Btu/lb$_m$·mol	Pr	T,°R	h, Btu/lb$_m$·mol	Pr
2910	22668.8	959.1	3110	24423.6	1286.5	3310	26191.6	1698.1
2920	22756.3	973.7	3120	24511.8	1305.0	3320	26280.3	1721.1
2930	22843.7	988.5	3130	24600.1	1323.6	3330	26369.0	1744.5
2940	22931.2	1003.5	3140	24688.2	1342.5	3340	26457.7	1768.0
2950	23018.5	1018.6	3150	24776.5	1361.7	3350	26546.8	1791.6
2960	23105.9	1033.8	3160	24864.7	1381.0	3360	26635.9	1815.6
2970	23193.6	1049.4	3170	24953.0	1400.5	3370	26724.9	1840.0
2980	23281.3	1065.0	3180	25041.4	1420.2	3380	26813.6	1864.7
2990	23368.9	1080.9	3190	25129.6	1440.2	3390	26902.3	1889.5
3000	23456.6	1096.8	3200	25217.8	1460.4	3400	26991.4	1914.6
3010	23544.3	1113.2	3210	25306.3	1480.8	3410	27080.7	1940.0
3020	23632.0	1129.7	3220	25394.7	1501.5	3420	27169.8	1965.7
3030	23719.7	1146.3	3230	25483.1	1522.4	3430	27258.8	1991.6
3040	23807.7	1163.1	3240	25571.6	1543.5	3440	27347.9	2017.8
3050	23895.6	1180.1	3250	25660.1	1564.9	3439	27436.9	2044.3
3060	23983.6	1197.4	3260	25748.5	1586.5	3460	27526.2	2071.0
3070	24071.7	1214.8	3270	25837.1	1608.3	3470	27615.6	2098.0
3080	24159.6	1232.4	3280	25925.8	1630.4	3480	27704.7	2125.4
3090	24247.5	1250.2	3290	26014.3	1652.8	3490	27793.8	2152.9
3100	24335.5	1268.3	3300	26102.9	1675.3			

* Abstracted from Ref. 8.

APPENDIX

J

THE ELEMENTS

Table J Alphabetical list of the elements

Element	Symbol	Atomic number, Z	Element	Symbol	Atomic number, Z
Actinium	Ac	89	Mercury	Hg	80
Aluminum	Al	13	Molybdenum	Mo	42
Americium	Am	95	Neodymium	Nd	60
Antimony	Sb	51	Neon	Ne	10
Argon	A	18	Neptunium	Np	93
Arsenic	As	33	Nickel	Ni	28
Astatine	At	85	Niobium	Nb	41
Barium	Ba	56	Nitrogen	N	7
Berkelium	Bk	97	Nobelium	No	102
Beryllium	Be	4	Osmium	Os	76
Bismuth	Bi	83	Oxygen	O	8
Boron	B	5	Palladium	Pd	46
Bromine	Br	35	Phosphorus	P	15
Cadmium	Cd	48	Platinum	Pt	78
Calcium	Ca	20	Plutonium	Pu	94
Californium	Cf	98	Polonium	Po	84
Carbon	C	6	Potassium	K	19
Cerium	Ce	58	Praseodymium	Pr	59
Cesium	Cs	55	Promethium	Pm	61
Chlorine	Cl	17	Protactinium	Pa	91
Chromium	Cr	24	Radium	Ra	88
Cobalt	Co	27	Radon	Rn	86
Copper	Cu	29	Rhenium	Re	75
Curium	Cm	96	Rhodium	Rh	45
Dysprosium	Dy	66	Rubidium	Rb	37
Einsteinium	Es	99	Ruthenium	Ru	44
Erbium	Er	68	Samarium	Sm	62
Europium	Eu	63	Scandium	Sc	21
Fermium	Fm	100	Selenium	Se	34
Fluorine	F	9	Silicon	Si	14
Francium	Fr	87	Silver	Ag	47
Gadolinium	Gd	64	Sodium	Na	11
Gallium	Ga	31	Strontium	Sr	38
Germanium	Ge	32	Sulfur	S	16
Gold	Au	79	Tantalum	Ta	73
Hafnium	Hf	72	Technetium	Tc	43
Helium	He	2	Tellurium	Te	52
Holmium	Ho	67	Terbium	Tb	65
Hydrogen	H	1	Thallium	Tl	81
Indium	In	49	Thorium	Th	90
Iodine	I	53	Thulium	Tm	69
Iridium	Ir	77	Tin	Sn	50
Iron	Fe	26	Titanium	Ti	22
Krypton	Kr	36	Tungsten (Wolfram)	W	74
Lanthanum	La	57	Uranium	U	92
Lead	Pb	82	Vanadium	V	23
Lithium	Li	3	Xenon	Xe	54
Lutecium	Lu	71	Ytterbium	Yb	70
Magnesium	Mg	12	Yttrium	Y	39
Manganese	Mn	25	Zinc	Zn	30
Mendelevium	Md	101	Zirconium	Zr	40

APPENDIX

K

TUBES

Table K Properties of condenser and feedwater heater tubes

Outside diameter, in	Gage, BWG	Thickness, in	Inside diameter, in	Surface area, ft^2/ft	Waterflow, gpm/(ft)/(s)	Tube mass, lb_m/ft (Admiralty)
	16	0.065	0.495	0.1636	0.600	0.435
5/8	17	0.058	0.509	0.1636	0.630	0.393
	18	0.049	0.527	0.1636	0.680	0.337
	20	0.035	0.555	0.1636	0.750	0.247
	16	0.065	0.620	0.1963	0.942	0.532
3/4	17	0.058	0.634	0.1963	0.980	0.480
	18	0.049	0.652	0.1963	1.042	0.4110
	20	0.035	0.680	0.1963	1.130	0.299
	16	0.065	0.745	0.2291	1.360	0.630
7/8	17	0.058	0.759	0.2291	1.410	0.567
	18	0.049	0.777	0.2291	1.480	0.484
	20	0.035	0.805	0.2291	1.590	0.352
	16	0.065	0.870	0.2618	1.854	0.727
1	17	0.058	0.884	0.2618	1.910	0.653
	18	0.049	0.902	0.2618	1.994	0.557
	20	0.035	0.930	0.2618	2.120	0.404

APPENDIX

L

FRICTION FACTOR

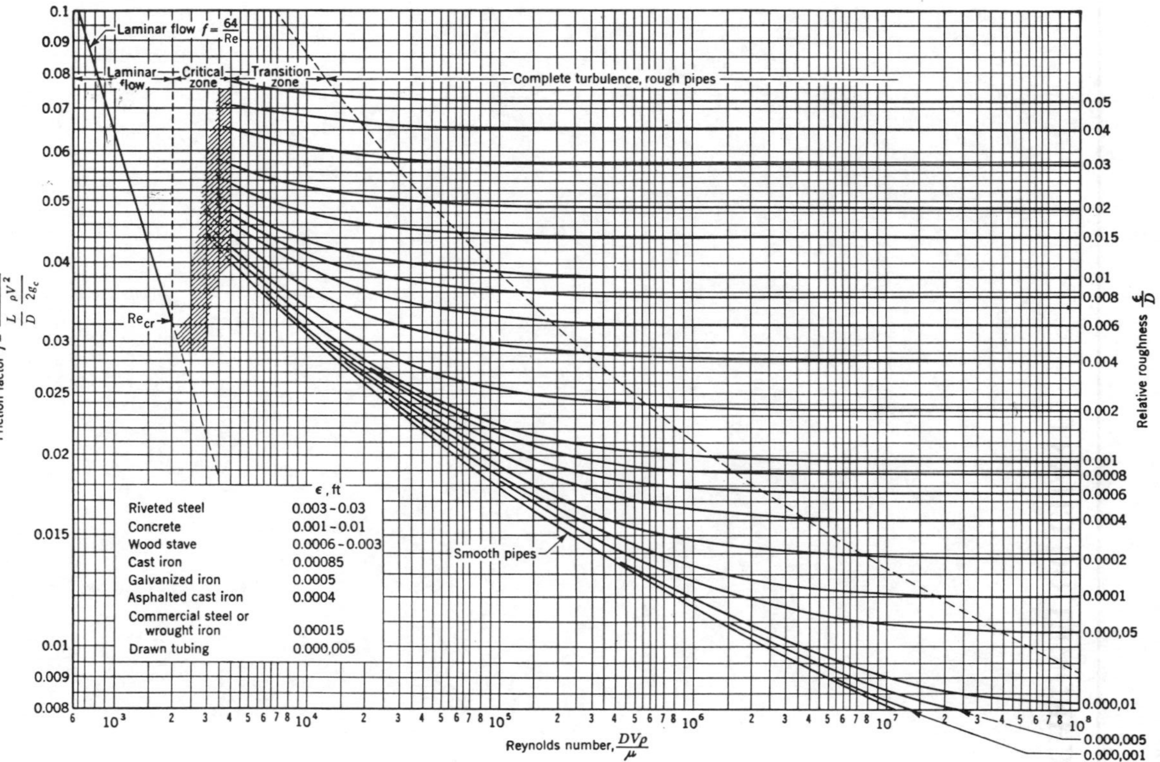

Figure L-1 Moody friction factor chart *(from Ref. 194)*.

APPENDIX

M

PSYCHROMETRIC CHART

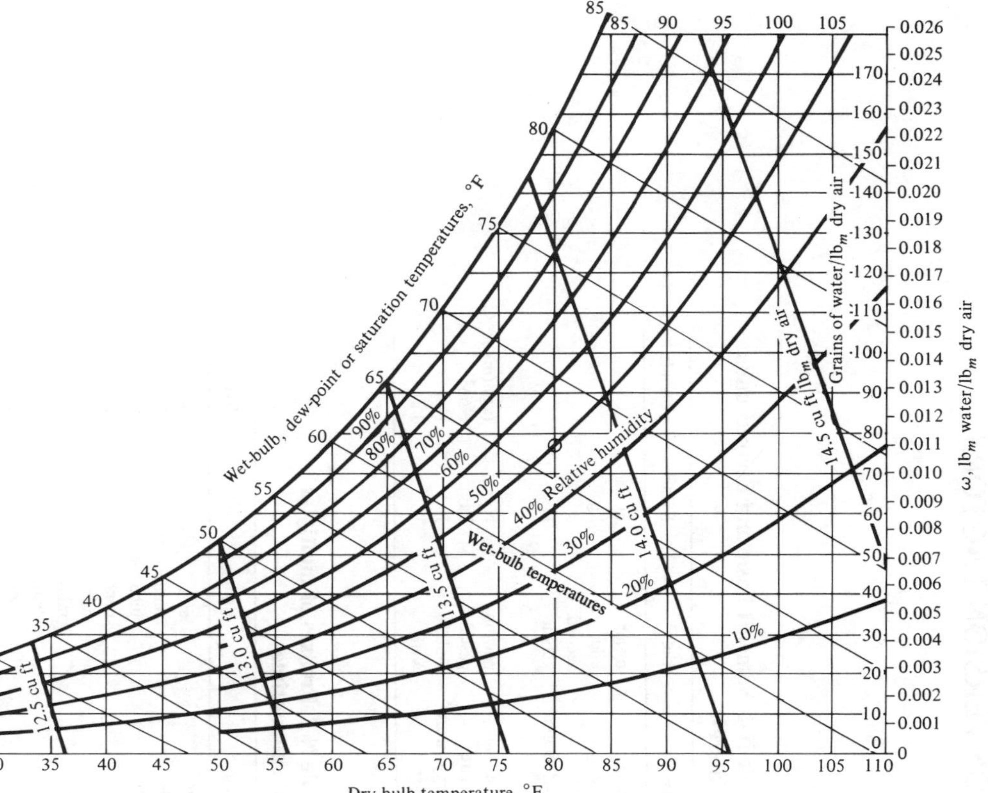

Figure M-1 Psychrometric chart at 1 standard atmosphere.

APPENDIX

N

CONVERSION FACTORS

Table N-1 Some SI basic and derived units of interest

Type	Unit	Symbol	Formula
Length	meter	m	
Mass	kilogram	kg	
Time	second	s	
Temperature	kelvin	K	
Force	newton	N	$(kg)\ (m)/s^2$
Pressure	pascal	Pa	N/m^2
Energy, work	joule	J	Nm
Power	watt	W	J/s

Table N-2 Prefixes of multiples and submultiples

Factor	Prefix	Symbol
10^{-18}	atto	a
10^{-15}	femto	f
10^{-12}	pico	p
10^{-9}	nano	n
10^{-6}	micro	μ
10^{-3}	milli	m
10^{-2}	centi	c
10^{-1}	deci	d
10	deka	da
10^{2}	hecto	h
10^{3}	kilo	k
10^{6}	mega	M
10^{9}	giga	G
10^{12}	tera	T
10^{15}	peta	P
10^{18}	exa	E

Table N-3 Length

Å	ft	in	light year	mil	micron	mi (statute)	m
1	3.280840 $E - 10$	3.937008 $E - 09$	1.057021 $E - 26$	3.937008 $E - 06$	1.000000 $E - 04$	6.213712 $E - 14$	1.000000 $E - 10$
3.048000 $E + 09$	1	1.200000 $E + 01$	3.221807 $E - 17$	1.200000 $E + 04$	3.048000 $E + 05$	1.893939 $E - 04$	3.048000 $E - 01$
2.540000 $E + 08$	8.333333 $E - 02$	1	2.684839 $E - 18$	1.000000 $E + 03$	2.540000 $E + 04$	1.578283 $E - 05$	2.540000 $E - 02$
9.460530 $E + 25$	3.103848 $E + 16$	3.724618 $E + 17$	1	3.724618 $E + 20$	9.460530 $E + 21$	5.878501 $E + 12$	9.460530 $E + 15$
2.540000 $E + 05$	8.333333 $E - 05$	1.000000 $E - 03$	2.684839 $E - 21$	1	2.540000 $E + 01$	1.578283 $E - 08$	2.540000 $E - 05$
1.000000 $E + 04$	3.280840 $E - 06$	3.937008 $E - 05$	1.057023 $E - 22$	3.937008 $E - 02$	1	6.213712 $E - 10$	1.000000 $E - 06$
1.609344 $E + 13$	5.280000 $E + 03$	6.336000 $E + 04$	1.701114 $E - 13$	6.336000 $E + 07$	1.609344 $E + 09$	1	1.609344 $E + 03$
1.000000 $E + 10$	3.280840 $E + 00$	3.937008 $E + 01$	1.057023 $E - 16$	3.937008 $E + 04$	1.000000 $E + 06$	6.213712 $E - 04$	1

Table N-4 Area

acre	barn	cm^2	ft^2	hectare	in^2	km^2	mi^2	m^2
1	4.046856 E + 31	4.046856 E + 07	4.356000 E + 04	4.046856 E − 01	6.272640 E + 06	4.046856 E − 03	1.562500 E − 03	4.046856 E + 03
2.471054 E − 32	1	1.000000 E − 24	1.076391 E − 27	1.000000 E − 32	1.550003 E − 25	1.000000 E − 34	3.861022 E − 35	1.000000 E − 28
1.940761 E − 08	1.000000 E + 24	1	1.076391 E − 03	1.000000 E − 08	1.550003 E − 01	1.000000 E − 10	3.861022 E − 11	1.000000 E − 04
2.295684 E − 05	9.290304 E + 26	9.290304 E + 02	1	9.290304 E − 06	1.440000 E + 02	9.290304 E − 08	3.587007 E − 08	9.290304 E − 02
2.471054 E + 00	1.000000 E + 32	1.000000 E + 08	1.076391 E + 05	1	1.550003 E + 07	1.000000 E − 02	3.861022 E − 03	1.000000 E + 04
1.594225 E − 07	6.451600 E + 24	6.451600 E + 00	6.944444 E − 03	6.451600 E − 08	1	6.451600 E − 10	2.490977 E − 10	6.451600 E − 04
2.471054 E + 02	1.000000 E + 34	1.000000 E + 10	1.076391 E + 07	1.000000 E + 02	1.550003 E + 09	1	3.861022 E − 01	1.000000 E + 06
6.400000 E + 02	2.589988 E + 34	2.589988 E + 10	2.787840 E + 07	2.589988 E + 02	4.014490 E + 09	2.589988 E + 00	1	2.589988 E + 06
2.471054 E − 04	1.000000 E + 28	1.000000 E + 04	1.076391 E + 01	1.000000 E − 04	1.550003 E + 03	1.000000 E − 06	3.861022 E − 07	1

Table N-5 Volume

acre-ft	cm^3	ft^3	gal (US liquid)	gal (Imperial liquid)	in^3	L	pt (US liquid)	qt (US liquid)	m^3
1	1.233482 E + 09	4.356000 E + 04	3.258514 E + 05	2.713283 E + 05	7.527168 E + 07	1.233482 E + 06	2.606811 E + 06	1.303406 E + 06	1.233482 E + 03
8.107132 E + 10	1	3.531467 E − 05	2.641721 E − 04	2.199694 E − 04	6.102374 E − 02	1.000000 E − 03	2.113376 E − 03	1.056688 E − 03	1.000000 E − 06
2.295684 E − 05	2.831685 E + 04	1	7.480520 E + 00	6.228841 E + 00	1.728000 E + 03	2.831685 E + 01	5.984416 E + 01	2.992208 E + 01	2.831685 E − 02
3.068883 E − 06	3.785412 E + 03	1.336806 E − 01	1	8.326748 E − 01	2.310000 E + 02	3.785412 E + 00	8.000000 E + 00	4.000000 E + 00	3.785412 E − 03
3.685572 E − 06	4.546087 E + 03	1.605435 E − 01	1.200949 E + 00	1	2.774192 E + 02	4.546086 E + 00	9.607592 E + 00	4.803796 E + 00	4.546086 E − 03
1.328521 E − 08	1.638706 E + 01	5.787037 E − 04	4.329004 E − 03	3.604653 E − 03	1	1.638706 E − 02	3.463204 E − 02	1.731602 E − 02	1.638706 E − 05
8.107132 E − 07	1.000000 E + 03	5.787037 E − 04	2.641721 E − 01	2.199694 E − 01	6.102375 E + 01	1	2.113376 E + 00	1.056688 E + 00	1.000000 E − 03

(continued)

Table N-5 Volume (Continued)

acre-ft	cm^3	ft^3	gal (US liquid)	gal (Imperial liquid)	in^3	L	pt (US liquid)	qt (US liquid)	m^3
3.836104	4.731765	1.671007	1.250000	1.040844	2.887500	4.731765	1	5.000000	4.731765
$E-07$	$E+02$	$E-02$	$E-01$	$E-01$	$E+01$	$E-01$		$E-01$	$E-04$
7.672208	9.463530	3.342014	2.500000	2.081687	5.775000	9.463530	2.000000	1	9.463530
$E-07$	$E+02$	$E-02$	$E-01$	$E-01$	$E+01$	$E-01$	$E+00$		$E-04$
8.107132	1.000000	3.531467	2.641721	2.199694	6.102375	1.000000	2.113376	1.056688	1
$E-04$	$E+06$	$E+01$	$E+02$	$E+02$	$E+04$	$E+03$	$E+03$	$E+03$	

Table N-6 Mass

amu	dram (avoirdupois)	gr	g	oz	lb_m	ton (long)	ton (metric)	ton (short)	kg
1	9.371957 *E* − 25	2.562645 *E* − 23	1.660566 *E* − 24	5.857476 *E* − 26	3.660921 *E* − 27	1.634343 *E* − 30	1.660566 *E* − 30	1.830460 *E* − 30	1.660566 *E* − 27
1.067013 *E* + 24	1	2.734375 *E* + 01	1.771845 *E* + 00	6.250000 *E* − 02	3.906250 *E* − 03	1.743862 *E* − 06	1.771845 *E* − 06	1.953125 *E* − 06	1.771845 *E* − 03
3.902218 *E* + 22	3.657143 *E* − 02	1	6.479891 *E* − 02	2.285714 *E* − 03	1.428571 *E* − 04	6.377551 *E* − 08	6.479891 *E* − 08	7.142857 *E* − 08	6.479891 *E* − 05
6.022043 *E* + 23	5.643834 *E* − 01	1.543236 *E* + 01	1	3.527396 *E* − 02	2.204623 *E* − 03	9.842065 *E* − 07	1.000000 *E* − 06	1.102311 *E* − 06	1.000000 *E* − 03
1.707220 *E* + 25	1.600000 *E* + 01	4.375000 *E* + 02	2.834952 *E* + 01	1	6.250000 *E* − 02	2.790179 *E* − 05	2.834952 *E* − 05	3.125000 *E* − 05	2.834952 *E* − 02
2.731553 *E* + 26	2.560000 *E* + 02	7.000000 *E* + 03	4.535924 *E* + 02	1.600000 *E* + 01	1	4.464286 *E* − 04	4.535924 *E* − 04	5.000000 *E* − 04	4.535924 *E* − 01
6.118679 *E* + 29	5.734400 *E* + 05	1.568000 *E* + 07	1.016047 *E* + 06	3.584000 *E* + 04	2.240000 *E* + 03	1	1.016047 *E* + 00	1.120000 *E* + 00	1.016047 *E* + 03

(continued)

Table N-6 Mass (Continued)

amu	dram (avoirdupois)	gr	g	oz	lb_m	ton (long)	ton (metric)	ton (short)	kg
6.022043 E + 29	5.643834 E + 05	1.543236 E + 07	1.000000 E + 06	3.527396 E + 04	2.204623 E + 03	9.842065 E − 01	1	1.102311 E + 00	1.000000 E + 03
5.463107 E + 29	5.120000 E + 05	1.400000 E + 07	9.071847 E + 05	3.200000 E + 04	2.000000 E + 03	8.928571 E − 01	9.071847 E − 01	1	9.071847 E + 02
6.022043 E + 26	5.643834 E + 02	1.543236 E + 04	1.000000 E + 03	3.527396 E + 01	2.204623 E + 00	9.842065 E − 04	1.000000 E − 03	1.102311 E − 03	1

Table N-7 Density

g/cm^3	g/L	lb_m/ft^3	lb_m/gal (US)	lb_m/gal (imperial)	lb_m/in^3	ton/yd^3	kg/m^3
1	1.000000 E + 03	6.242795 E + 01	8.345403 E + 00	1.002240 E + 01	3.612728 E − 02	8.427773 E − 01	1.000000 E + 03
1.000000 E − 03	1	6.242795 E − 02	8.345403 E − 03	1.002240 E − 02	3.612728 E − 05	8.427773 E − 04	1.000000 E + 00
1.601847 E − 02	1.601847 E + 01	1	1.336806 E − 01	1.605435 E − 01	5.787037 E − 04	1.350000 E − 02	1.601847 E + 01
1.198264 E − 01	1.198264 E + 02	7.480519 E + 00	1	1.200949 E + 00	4.329004 E − 03	1.009870 E − 01	1.198264 E + 02
9.977648 E − 02	9.977648 E + 01	6.228841 E + 00	8.326750 E − 01	1	3.604654 E − 03	8.408933 E − 02	9.977650 E + 01
2.767991 E + 01	2.767991 E + 04	1.728000 E + 03	2.310000 E + 02	2.774191 E + 02	1	2.332800 E + 01	2.767991 E + 04
1.186553 E + 00	1.186553 E + 03	7.407407 E + 01	9.902264 E + 00	1.189211 E + 01	4.286694 E − 02	1	1.186553 E + 03
1.000000 E − 03	1.000000 E + 00	6.242795 E − 02	8.345403 E − 03	1.002240 E − 02	3.612728 E − 05	8.427773 E − 04	1

Table N-8 Time (mean solar)

day	h	ms	min	lunar month	year	s
1	2.400000 E + 01	8.640000 E + 10	1.440000 E + 03	3.386319 E − 02	2.737909 E − 03	8.640000 E + 04
4.166667 E − 02	1	3.600000 E + 09	6.000000 E + 01	1.410966 E − 03	1.140796 E − 04	3.600000 E + 03
1.157407 E − 11	2.777778 E − 10	1	1.666667 E − 08	3.919351 E − 13	3.168876 E − 14	1.000000 E − 06

(continued)

Table N-8 Time (mean solar) (Continued)

day	h	ms	min	lunar month	year	s
6.944445 E − 04	1.666667 E − 02	6.000000 E + 07	1	2.351611 E − 05	1.901326 E − 06	6.000000 E + 01
2.953059 E + 01	7.087341 E + 02	2.551443 E + 12	4.252405 E + 04	1	8.085207 E − 02	2.551443 E + 06
3.652422 E + 02	8.765813 E + 03	3.155693 E + 13	5.259488 E + 05	1.236827 E + 01	1	3.155693 E + 07
1.157407 E − 05	2.777778 E − 04	1.000000 E + 06	1.666667 E − 02	3.919351 E − 07	3.168876 E − 08	1

Table N-9 Speed

ft/h	ft/s	in/s	km/h	kn	mi/h	mi/s	m/s
1	2.777778 E − 04	3.333333 E − 03	3.048000 E − 04	1.645789 E − 04	1.893939 E − 04	5.260943 E − 08	8.466667 E − 05
3.600000 E + 03	1	1.200000 E + 01	1.097280 E + 00	5.924841 E − 01	6.818182 E − 01	1.893939 E − 04	3.048000 E − 01
3.000000 E + 02	8.333333 E − 02	1	9.144000 E − 02	4.937367 E − 02	5.681818 E − 02	1.578283 E − 05	2.540000 E − 02
3.280840 E + 03	9.113444 E − 01	1.093613 E + 01	1	5.399570 E − 01	6.213712 E − 01	1.726031 E − 04	2.777778 E − 01
6.076112 E + 03	1.687809 E + 00	2.025371 E + 01	1.851999 E + 00	1	1.150779 E + 00	3.196608 E − 04	5.144444 E − 01
5.280000 E + 03	1.466667 E + 00	1.760000 E + 01	1.609344 E + 00	8.689766 E − 01	1	2.777778 E − 04	4.470400 E − 01
1.900800 E + 07	5.280000 E + 03	6.336000 E + 04	5.793639 E + 03	3.128316 E + 03	3.600000 E + 03	1	1.609344 E + 03
1.181102 E + 04	3.280840 E + 00	3.155693 E + 04	3.600000 E + 00	1.943845 E + 00	2.236936 E + 00	6.213712 E − 04	1

Table N-10 Acceleration

ft/h²	ft/min²	ft/s²	free fall	in/s²	km/h²	km/s²	mi/h²	m/s²
1	2.777778 E − 04	7.716049 E − 08	2.398221 E − 09	9.259259 E − 07	3.048000 E − 04	2.351852 E − 11	1.893939 E − 04	2.351852 E − 08
3.600000 E + 03	1	2.777778 E − 04	8.633597 E − 06	3.333333 E − 03	1.097280 E + 00	8.466666 E − 08	6.818182 E − 01	8.466666 E − 05
1.296000 E + 07	3.600000 E + 03	1	3.108095 E − 02	1.200000 E + 01	3.950208 E + 03	3.048000 E − 04	2.454546 E + 03	3.048000 E − 01
4.169757 E + 08	1.150266 E + 05	3.217405 E + 01	1	3.860886 E + 02	1.270942 E + 05	9.806650 E − 03	6.093570 E − 06	9.806650 E + 00
1.080000 E + 06	3.000000 E + 02	8.333333 E − 02	2.590079 E − 03	1	3.291840 E + 02	2.540000 E − 05	2.045455 E + 02	2.540000 E − 02
3.280840 E + 00	9.113444 E − 04	2.531512 E − 07	7.868180 E − 06	3.037815 E − 03	1	7.716049 E − 08	6.213712 E − 01	7.716049 E − 05
4.251969 E + 10	1.181102 E + 07	3.280840 E + 03	1.019716 E + 02	3.937008 E + 04	1.296000 E + 07	1	8.052971 E + 06	1.000000 E + 03
5.280000 E + 03	1.466667 E + 00	4.074074 E − 04	1.266261 E − 05	4.888889 E − 03	1.609344 E + 00	1.241778 E − 07	1	1.241778 E − 04
4.251969 E + 07	1.181102 E + 04	3.280840 E + 00	1.019716 E − 01	3.937008 E + 01	1.296000 E + 04	1.000000 E − 03	8.052971 E + 03	1

Table N-11 Volume flow rate

cm^3/s	in^3/s	ft^3/min	gal(US)/min	gal(imperial)/min	L/min	m^3/s
1	6.102374 E − 02	2.118880 E − 03	1.585032 E − 02	1.319816 E − 02	6.000000 E − 02	1.000000 E − 06
1.638706 E + 01	1	3.472222 E − 02	2.597403 E − 01	2.162792 E − 01	9.832238 E − 01	1.638707 E − 05
4.719474 E + 02	2.880000 E + 01	1	7.480520 E + 00	6.228841 E + 00	2.831685 E + 01	4.719474 E − 04
6.309020 E + 01	3.850000 E + 00	1.336806 E − 01	1	8.326748 E − 01	3.785412 E + 00	6.309020 E − 05
7.576811 E + 01	4.623654 E + 00	1.605436 E − 01	1.200949 E + 00	1	4.546087 E + 00	7.576811 E − 05
1.666667 E + 01	1.017062 E + 00	3.531467 E − 02	2.641721 E − 01	2.199695 E − 01	1	1.666667 E − 05
1.000000 E + 06	6.102374 E + 04	2.118880 E + 03	1.585032 E + 04	1.319816 E + 04	6.000000 E + 04	1

Table N-12 Mass flow rate

g/s	kg/h	lb_m/min	lb_m/h	ton(metric)/day	ton(short)/h	kg/s
1	3.600000 E + 00	1.322774 E − 01	7.936642 E + 00	8.640000 E − 02	3.968321 E − 03	1.000000 E − 03
2.777778 E − 01	1	3.674371 E − 02	2.204623 E + 00	2.400000 E − 02	1.102311 E − 03	2.777778 E − 04
7.559873 E + 00	2.721554 E + 01	1	6.000000 E + 01	6.531730 E − 01	3.000000 E − 02	7.559873 E − 03

(continued)

Table N-12 Mass flow rate (Continued)

g/s	kg/h	lb_m/min	lb_m/h	ton(metric)/day	ton(short)/h	kg/s
1.259979 $E-01$	4.535924 $E-01$	1.666667 $E-02$	1	1.088622 $E-02$	5.000000 $E-04$	1.259979 $E-04$
1.157407 $E+01$	4.166667 $E+01$	1.530988 $E+00$	9.185928 $E+01$	1	4.592964 $E-02$	1.157407 $E-02$
2.519958 $E+02$	9.071847 $E+02$	3.333333 $E+01$	2.000000 $E+03$	2.177243 $E+01$	1	2.519958 $E-01$
1.000000 $E+03$	3.600000 $E+03$	1.322774 $E+02$	7.936642 $E+03$	8.640000 $E+01$	3.968321 $E+00$	1

Table N-13 Force

dyn [g(m) · cm/s^2]	kg/force (or klb) kg_f	kip	lb-force lb_f	poundal ($lb_m \cdot ft/s^2$)	N [(kg)(m)/s^2]
1	1.019716 $E-06$	2.248089 $E-09$	2.248089 $E-06$	7.233014 $E-05$	1.000000 $E-05$
9.806650 $E+05$	1	2.204623 $E-03$	2.204623 $E+00$	7.093165 $E+01$	9.806650 $E+00$
4.448222 $E+08$	4.535924 $E+02$	1	1.000000 $E+03$	3.217405 $E+04$	4.448222 $E+03$
4.448222 $E+05$	4.535924 $E-01$	1.000000 $E-03$	1	3.217405 $E+01$	4.448222 $E+00$
1.382550 $E+04$	1.409808 $E-02$	3.108095 $E-05$	3.108095 $E-02$	1	1.382550 $E-01$
1.000000 $E+05$	1.019716 $E-01$	2.248089 $E-04$	2.248089 $E-01$	7.233014 $E+00$	1

Table N-14 Pressure

atm (normal)	bar	cm Hg (0°C)	dyn/cm^2	ft water (60°F)	in Hg (32°F)	in water (60°F)	kg$_f$/cm^2	lb$_f$/in^2	torr, mm Hg (0°C)	Pa, (N/m^2)
1	1.013250 $E+00$	7.600007 $E+01$	1.013250 $E+06$	3.393615 $E+01$	2.992129 $E+01$	4.072338 $E+02$	1.033227 $E+00$	1.469595 $E+01$	7.600007 $E+02$	1.013250 $E+05$
9.869233 $E-01$	1	7.500624 $E+01$	1.000000 $E+06$	3.349238 $E+01$	2.953001 $E+01$	4.019085 $E+02$	1.019716 $E+00$	1.450377 $E+01$	7.500624 $E+02$	1.000000 $E+05$
1.315788 $E-02$	1.333222 $E-02$	1	1.333222 $E+04$	4.465279 $E-01$	3.937008 $E-01$	5.358355 $E+00$	1.359509 $E-02$	1.933676 $E-01$	1.000000 $E+01$	1.333222 $E+03$
9.869233 $E-07$	1.000000 $E-06$	7.500624 $E-05$	1	3.349238 $E-05$	2.953001 $E-05$	4.019085 $E-04$	1.019716 $E-06$	1.450377 $E-05$	7.500624 $E-04$	1.000000 $E-01$
2.946710 $E-02$	2.985754 $E-02$	2.239502 $E+00$	2.985754 $E+04$	1	8.816936 $E-01$	1.200000 $E+01$	3.044622 $E-02$	4.330470 $E-01$	2.239502 $E+01$	2.985754 $E+03$

3.342102 E − 02	3.386385 E − 02	2.540000 E + 00	3.386385 E + 04	1.134181 E + 00	1	1.361017 E + 01	3.453152 E − 02	4.911537 E − 01	2.540000 E + 01	3.386385 E + 03
2.455592 E − 03	2.488128 E − 03	1.866251 E − 01	2.488128 E + 03	8.333333 E − 02	7.347447 E − 02	1	2.537185 E − 03	3.608725 E − 02	1.866251 E + 00	2.488126 E + 02
9.678411 E − 01	9.806650 E − 01	7.355599 E + 01	9.806650 E + 05	3.284480 E + 01	2.895905 E + 01	3.941376 E + 02	1	1.422334 E + 01	7.355599 E + 02	9.806650 E + 04
6.804596 E − 02	6.894757 E − 02	5.171498 E + 00	6.894757 E + 04	2.309218 E + 00	2.036023 E + 00	2.771062 E + 01	7.030696 E − 02	1	5.171498 E + 01	6.894757 $\times 10^3$
1.315788 E − 03	1.333222 E − 03	1.000000 E − 01	1.333222 E − 03	4.465279 E − 02	3.937008 E − 02	5.358335 E − 01	1.359509 E − 03	1.933676 E − 02	1	1.333222 E − 02
9.869233 E − 06	1.000000 E − 05	7.500624 E − 04	1.000000 E + 01	3.349238 E − 04	2.953001 E − 04	4.019085 E − 03	1.019716 E − 05	1.450377 E − 04	7.500624 E − 03	1

Table N-15 Energy

Btu	cal (Int. table)	EV	erg (dyn · cm)	ft · lb_f	hp · h	kw · h	MeV	J, W · s or N · m
1	2.521403 $E + 02$	6.585086 $E + 21$	1.055056 $E + 10$	7.781693 $E + 02$	3.930148 $E - 04$	2.930711 $E - 04$	6.585086 $E + 15$	1.055056 $E + 03$
3.968321 $E - 03$	1	2.613173 $E + 19$	4.186800 $E + 07$	3.088025 $E - 00$	1.559609 $E - 06$	1.163000 $E - 06$	2.613173 $E + 13$	4.186800 $E + 00$
1.518583 $E - 22$	3.826765 $E - 20$	1	1.602190 $E - 12$	1.181715 $E - 19$	5.968256 $E - 26$	4.450528 $E - 26$	1.000000 $E - 06$	1.602190 $E - 19$
9.478172 $E - 11$	2.388459 $E - 08$	6.241457 $E + 11$	1	7.375622 $E - 08$	3.725062 $E - 14$	2.777777 $E - 14$	6.241457 $E + 05$	1.000000 $E - 07$
1.285067 $E - 03$	3.238315 $E - 01$	8.462280 $E + 18$	1.355818 $E + 07$	1	5.050505 $E - 07$	3.766160 $E - 07$	8.462280 $E + 12$	1.355818 $E + 00$
2.544433 $E + 03$	6.411864 $E + 05$	1.675531 $E + 25$	2.684519 $E + 13$	1.980000 $E + 06$	1	7.456997 $E - 01$	1.675531 $E + 19$	2.684519 $E + 06$
3.412142 $E + 03$	8.598454 $E + 05$	2.246925 $E + 25$	3.600000 $E + 13$	2.655224 $E + 06$	1.341022 $E + 00$	1	2.246925 $E + 19$	3.600000 $E + 06$
1.518583 $E - 16$	3.826765 $E - 14$	1.000000 $E + 06$	1.602190 $E - 6$	1.181715 $E - 13$	5.968256 $E - 20$	4.450528 $E - 20$	1	1.602190 $E - 13$
9.478172 $E - 04$	2.388459 $E - 01$	6.241457 $E + 18$	1.000000 $E + 07$	7.375622 $E - 01$	3.725062 $E - 07$	2.777778 $E - 07$	6.241457 $E + 12$	1

Table N-16 Energy flux

Btu/(h)(ft^2)	cal/(s)(cm^2)	erg/(s)(cm^2)	ft lb_f/ (h)(ft^2)	hp/ft^2	W/cm^2	W/m^2
1	7.534608 E − 05	3.154589 E + 03	7.781696 E + 02	3.930150 E − 04	3.154589 E − 04	3.154589 E + 00
1.327209 E + 04	1	4.186800 E + 07	1.032794 E + 07	5.216129 E + 00	4.186800 E + 00	4.186800 E + 04
3.169984 E − 04	2.388459 E − 08	1	2.466785 E − 01	1.245851 E − 07	1.000000 E − 07	1.000000 E − 03
1.285067 E − 03	9.682478 E − 08	4.053860 E + 00	1	5.050505 E − 07	4.053859 E − 07	4.053859 E − 03
2.544432 E + 03	1.917130 E − 01	8.026641 E + 06	1.980000 E + 06	1	8.026641 E − 01	8.026641 E + 03
3.169984 E + 03	2.388459 E − 01	1.000000 E + 07	2.466785 E + 06	1.245851 E + 00	1	1.000000 E + 04
3.169984 E − 01	2.388459 E − 05	1.000000 E + 03	2.466785 E + 02	1.245851 E − 04	1.000000 E − 04	1

Table N-17 Specific energy

Btu/lb_m	cal/g	kwh/kg	MW · day/ton (metric)	J/kg, (W · s/kg)
1	5.555556 E − 01	6.461111 E − 04	2.692130 E − 05	2.326000 E + 03
1.800000 E + 00	1	1.163000 E − 03	4.845833 E − 05	4.186800 E + 03
1.547721 E + 03	8.598452 E + 02	1	4.166667 E − 02	3.600000 E + 06
3.714530 E + 04	2.063618 E + 04	2.400000 E + 01	1	8.640000 E + 07
4.299226 E − 04	2.308459 E − 04	2.777778 E − 07	1.157407 E − 08	1

Table N-18 Power

Btu/h	cal/s	erg/s	eV/s	ft · lb_f/min	hp	kw	W, (J/s)
1	6.999884 *E* − 02	2.930711 *E* + 06	1.829191 *E* + 18	1.296948 *E* + 01	3.930143 *E* − 04	2.930711 *E* − 04	2.930711 *E* − 01
1.428595 *E* + 01	1	4.186800 *E* + 07	2.613173 *E* + 19	1.852813 *E* + 02	5.614583 *E* − 03	4.186800 *E* − 03	4.186800 *E* + 00
3.412141 *E* − 07	2.380493 *E* − 08	1	6.241457 *E* + 11	4.425367 *E* − 06	1.341020 *E* − 10	1.000000 *E* − 10	1.000000 *E* − 07
5.466898 *E* − 19	3.814002 *E* − 20	1.602190 *E* − 12	1	7.090278 *E* − 18	2.148569 *E* − 22	1.602190 *E* − 22	1.602190 *E* − 19
7.710415 *E* − 02	5.397200 *E* − 03	2.259700 *E* + 05	1.410382 *E* + 17	1	3.030303 *E* − 05	2.259697 *E* − 01	2.259697 *E* − 02
2.544437 *E* + 03	1.781076 *E* + 02	7.457010 *E* + 09	4.654261 *E* + 21	3.300000 *E* + 04	1	7.456999 *E* − 05	7.456999 *E* + 02
3.412141 *E* + 03	2.388459 *E* + 02	1.000000 *E* + 10	6.241457 *E* + 21	4.425366 *E* + 04	1.341020 *E* + 00	1	1.000000 *E* + 03
3.412141 *E* + 00	2.388459 *E* − 01	1.000000 *E* + 07	6.241457 *E* + 18	4.425366 *E* + 01	1.341020 *E* − 03	1.000000 *E* − 03	1

Table N-19 Power density

Btu/(h)(ft^3)	cal/(s)(cm^3)	MeV/(s)(cm^3)	W/cm^3, (kW/L)	W/m^3
1	2.471986 $E-06$	6.459728 $E+07$	1.034971 $E-05$	1.034971 $E+01$
4.045330 $E+05$	1	2.613173 $E+13$	4.186800 $E+00$	4.186800 $E+06$
1.548053 $E-08$	3.826765 $E-14$	1	1.602190 $E-13$	1.602190 $E-07$
9.662106 $E+04$	2.388459 $E-01$	6.241457 $E+12$	1	1.000000 $E+06$
9.662106 $E-02$	2.388459 $E-07$	6.241457 $E+06$	1.000000 $E-06$	1

Table N-20 Specific power

Btu/(h)(lb$_m$)	cal/(s)(g$_m$)	ft lb$_f$/(h)(lb$_m$)	hp/lb$_m$	W/g, (kW/kg)	W/kg
1	1.543210 $E-04$	7.781693 $E+02$	3.930148 $E-04$	6.461113 $E-01$	6.461113 $E+02$
6.479999 $E+03$	1	5.042536 $E+06$	2.546735 $E+00$	4.186800 $E+03$	4.186800 $E+06$
1.285067 $E-03$	1.984456 $E-07$	1	5.050505 $E-07$	8.302965 $E-04$	8.302965 $E-01$
2.546136 $E+03$	3.929223 $E-01$	1.980000 $E+06$	1	1.643987 $E+03$	1.643987 $E+06$
1.547721 $E+00$	2.388459 $E-04$	1.204389 $E+03$	6.082773 $E-04$	1	1.000000 $E+03$
1.547721 $E-03$	2.388459 $E-07$	1.204389 $E+00$	6.082773 $E-07$	1.000000 $E-03$	1

Table N-21 Thermal conductivity

Btu/(h)(ft)(°F)	Btu · in/(h)(ft²)(°F)	cal/(s)(cm)(°C)	kcal/(h)(m)(°C)	W/(m)(°C)
1	1.200000 E + 01	4.133787 E − 05	1.488163 E + 00	1.730734 E + 00
8.333333 E − 02	1	3.444823 E − 06	1.240136 E − 01	1.442279 E − 01
2.419088 E + 04	2.902906 E + 05	1	3.600000 E + 04	4.186800 E + 04
6.719691 E − 01	8.063629 E + 00	2.777778 E − 05	1	1.163000 E + 00
5.777894 E − 01	6.933473 E + 00	2.388459 E − 05	8.598452 E − 01	1

Table N-22 Heat-transfer coefficient

Btu/(ft²)(h)(°F)	cal/(cm²)(s)(°C)	W/(cm²)(°C)	W/(m²)(°C)
1	1.356229 E − 04	5.678260 E − 04	5.678260 E + 00
7.373386 E + 03	1	4.186800 E + 00	4.186800 E + 04
1.761102 E + 03	2.388459 E − 01	1	1.000000 E + 04
1.761102 E − 01	2.388459 E − 05	1.000000 E − 04	1

Table N-23 Specific heat and specific gas constant

Btu/(lb_m)(°F)	cal/(g_m)(°C)	ft · lb_f/(lb_m)(°R)	kJ/(kg)(°C)
1	1.000000 $E+00$	7.835679 $E+02$	4.186800 $E+00$
1.000000 $E+00$	1	7.835679 $E+02$	4.186800 $E+00$
1.276214 $E-03$	1.276214 $E-04$	1	5.343251 $E-03$
2.388459 $E-01$	2.388459 $E-01$	1.871829 $E-02$	1

REFERENCES

1. El-Wakil, M. M.: "Nuclear Power Engineering," McGraw-Hill Book Company, New York, 1962.
2. El-Wakil, M. M.: "Nuclear Heat Transport," American Nuclear Society, LaGrange Park, Ill., 1981.
3. El-Wakil, M. M.: "Nuclear Energy Conversion," American Nuclear Society, LaGrange Park, Ill., 1982.
4. Obert, E. F.: "Concepts of Thermodynamics," McGraw-Hill Book Company, New York, 1960.
5. Sonntag, R. E., and G. J. Van Wylen: "Introduction to Thermodynamics," 2d ed., John Wiley & Sons, Inc., New York, 1982.
6. "Thermodynamic and Transport Properties of Steam," The American Society of Mechanical Engineers, New York, 1967.
7. Carnot, S.: "Reflections on the Motive Power of Heat," R. H. Thurston (trans., 1924), The American Society of Mechanical Engineers, New York, 1943.
8. Keenan, J. H., and J. Kaye: "Gas Tables," John Wiley & Sons, Inc., New York, 1948.
9. Franck, C. C.: Steam Turbine Developments, "Proceedings of the American Power Conference," vol. 16, pp. 130–147, 1954.
10. Bennett, S. B., and R. L. Bannister: Pulverized Coal Power Plants: The Next Logical Step, *Mech. Eng.*, vol. 103, no. 12, pp. 18–24, 1981.
11. Steam Generator, *Power and Steam,* vol. 6, no. 5, March 1886.
12. "Steam, Its Generation and Use," 37th ed. and 38th ed. (revised), Babcock and Wilcox Company, New York, 1963 and 1975.
13. Singer, J. G. (ed.): "Combustion Fossil Power Systems," Combustion Engineering, Inc., Windsor, Conn., 1981.
14. Hottel, H. C., and A. C. Sarafim: "Radiative Heat Transfer," McGraw-Hill Book Company, New York, 1967.
15. "ASME Boiler and Pressure Vessel Code," sec. 1, "Power Boilers Addenda," The American Society of Mechanical Engineers, New York, June 1970.
16. Spencer, R. C.: "Design of Double Reheat Turbines for Supercritical Pressures," presented at the 42d annual meeting of the American Power Conference, Chicago, Ill., April 1980.
17. "Fan Engineering: An Engineer's Handbook," 7th ed., Buffalo Forge Co., Buffalo, N.Y., 1970.

18. "Power Plant Fans: Specification Guidelines," Publication 801, Air Movement and Control Association (AMCA), Arlington Heights, Ill., 1977.
19. "Test Code for Sound Rating Air Moving Devices," Standard 300, Air Movement and Control Association, Arlington Heights, Ill., 1967.
20. Stern, A. C. (ed.): "Air Pollution," vol. 1, 2d ed., Academic Press, Inc., New York, 1968.
21. Carson, J. E., and H. Moses: The Validity of Several Plume Rise Formulas, *J. Air Pollut. Central Assoc.* vol. 26, no. 11, p. 1089, 1976.
22. Briggs, G. A.: "Plume Rise," Atomic Energy Commission Critical Review Series, T1D-25075, Washington, D.C., 1969.
23. Wark, K., and C. F. Warner: "Air Pollution, Its Origin and Control," 2d ed., Harper & Row, Publishers, Incorporated, New York, 1981.
24. Shade, D. H. (ed.): "Meteorology and Atomic Energy," Atomic Energy Commission Publication T1D-24190, Washington, D.C., 1968.
25. Cramer, H. E.: A Practical Method for Estimating the Dispersal of Atmospheric Contaminants, "Proceedings of the First National Conference on Applied Meteorology," Hartford, Conn., October 28 and 29, pp. C-33–C-55, The American Meteorological Society, Boston, Mass., 1957.
26. Lambers, W.S., and W. T. Reid: "A Graphical Form for Applying the Rosin and Hammler Equation for the Size Distribution of Broken Coal," U.S. Bureau of Mines Circular 7346, 1946.
27. Diesel, Rudolf: "Theory and Construction of a Rotational Heat Motor," Bryan Donkin (trans.), London, 1894.
28. Anderson, John: Pulverized Coal Under Central Station Boilers, *Power,* vol. 51, no. 9, pp. 336–339, March 1920. (Also, other articles on the topic in the same issue.)
29. Blizard, John: "Transportation and Combustion of Powdered Coal," U.S. Bureau of Mines Bulletin 217, Government Printing Office, Washington, D.C., 1923.
30. Nusselt, W.: The Combustion Process in Pulverized Coal Furnaces, *VDI Zeitschrift,* vol. 68, no. 6, pp. 124–128, February 1924.
31. Van Heerden, C., A. P. P. Nobel, and D. W. van Krenelen: Studies of Fluidization, I. The Critical Mass Velocity, *Chem. Eng. Sci.,* vol 1, pp. 37–49, 1952.
32. White, P. C., R. L. Zahradnik, and R. E. Vener: "Coal Gasification," Office of Fossil Energy, ERDA Quarterly Report, July–September 1975.
33. Kydd, P. H.: "Integrated Gasification-Gas Turbine Cycle Performance," General Electric Company Technical Information Series Report No. 75CRD021, March 1975.
34. Othmer, D. F.: Energy–Fluid from Solids, *Mech. Eng.,* pp. 29–35, November 1977.
35. "Environmental Development Plan Biomass Energy Systems," U.S. Department of Energy Report No. DOE/EDP-0032, September 1979.
36. "JANAF Thermochemical Tables," Thermal Research Laboratory, The Dow Chemical Company, Midland, Mich., issued between 1961 and 1966.
37. Church, E. F.: "Stream Turbines," 3d ed., McGraw-Hill Book Company, New York, 1950.
38. Fowler, J. E., and E. H. Miller: "Some Aspects of Efficient Last Stage Buckets for Steam Generators," General Electric Company Report No. GER-2630, presented at the ASME winter annual meeting, Los Angeles, Calif., November 1969.
39. Baily, F. G., and E. H. Miller: "Modern Turbine Designs for Water-Cooled Reactors," General Electric Company Report No. GER-2452, also in *Nucl. Eng.,* January 1967.
40. Cowgill, T., and K. Robbins: Understanding the Observed Effects of Erosion and Corrosion in Steam Turbines, *Power,* p. 28, September 1976.
41. Baily, F. G., K. C. Cotton, and R. C. Spencer: "Predicting the Performance of Large Steam Turbine-Generators Operating with Saturated and Low-Superheat Steam Conditions," General Electric Company Report No. GER-2454A, presented at 29th annual meeting of the American Power Conference, Chicago, Ill., April 1967.
42. Spencer, R. C., K. C. Cotton, and C. N. Cannon: A Method for Predicting the Performance of Steam Turbine-Generation—16,500 kW and Larger, "Transactions of the American Society of Mechanical Engineers," ser. A, *J. Eng. Power,* vol. 85, p. 249, 1963.
43. Hohn, A.: Rotors for Large Steam Turbines, *Brown Boveri Rev.,* vol. 60, no. 9, p. 404, September 1973.

44. Keller, C., and D. Schmidt: "The Helium Gas Turbine for Nuclear Power Plants," A report of the Brown Boveri-Sulzer Turbomachinery, Ltd., The American Society of Mechanical Engineers paper No. 67-GT-10, March 1967.
45. Cohen, H., G. F. C. Rogers, and H. I. H. Saravanamuttoo: "Gas Turbine Theory," 2d ed., John Wiley & Sons, Inc., New York, 1973.
46. Shepherd, D. G.: "Principles of Turbomachinery," Macmillan Publishing Co., Inc., New York, 1956.
47. Gebhart B., "Heat Transfer," McGraw-Hill Book Company, New York, 1961.
48. "Construction Standards for Surface Type Condensers for Ejector Service," The Heat Exchange Institute, New York, 1972.
49. "Standards for Closed Feedwater Heaters," The Heat Exchange Institute, New York, 1968.
50. "Standards of the Tubular Exchanger Manufacturers Association," Tubular Exchanger Manufacturers Association, Inc., New York, 1959.
51. Cooling Towers, A Special Report, *Power,* pp. S-1–S-24, March 1973.
52. Marks, R. H.: Cooling Towers, A Special Report, *Power,* pp. S-1–S-16, March 1963.
53. Kadel, J. O.: Cooling Towers—A Technological Tool to Increase Plant Site Potentials, "Proceedings of the American Power Conference," vol. 32, pp. 539–543, 1970.
54. "The Noise of Cooling Towers," Engineering Manual No. 251, prepared jointly by Baltimore Aircoil Company, Inc., subsidiary of Merck and Co., Inc., and Bolt Beranek and Neuman, Inc., acoustical consultants, 1971.
55. Sneck, H. J., and D. H. Brown: "Plume Rise from Large Thermal Sources such as Cooling Towers," General Electric Technical Information Series Report No. 72GEN025, November 1972.
56. Hollingshad, W. R.: "Chemical and Microbiological Corrosion, Cooling-Water Treatment, and Pollution Considerations," The Calgon Corporation, Pittsburgh, Penn., Cooling Tower Short Course, University of Wisconsin, Madison, Wis., October 1973.
57. Bartz, J. A., and J. S. Maulbetsch: Are Dry-Cooled Power Plants A Feasible Alternative?, *Mech. Eng.*, vol. 103, no. 10, pp. 34–41, October 1981.
58. Oplatka, G.: Prevention of Frost Damage in Dry Cooling Towers, *Brown Boveri Rev.*, vol. 65, pp. 555–564, August 1978.
59. Patterson, W. D., J. L. Leporati, and M. J. Scarpa: The Capacity of Cooling Ponds to Dissipate Heat, "Proceedings of the American Power Conference," vol. 33, pp. 446–456, 1971.
60. Rao, D. K., and R. W. Porter: Closed-Cycle Evaporative Cooling Systems: Their Off-Design Performance, *Mech. Eng.*, vol. 103, no. 12, pp. 40–45, December 1981.
61. Lorenzi, D. J., and R. W. Porter: Simplified Analysis of Surface Energy from Heated Bodies of Water, in S. Sengupta (ed.), "Advances in Heat and Mass Transfer at Air-Water Interfaces," The American Society of Mechanical Engineers, New York, pp. 93–101, 1978.
62. Edinger, J. E., and J. C. Geyer: "Heat Exchange in the Environment," Edison Electric Institute, New York, 1971.
63. "Spray Ponds, the Answer to Thermal Pollution Problems," a report by Spray Engineering Company, Burlington, Mass., updated.
64. Bartholomaei, N. T.: Aircraft Gas Turbines in Industrial Service, *Lubrication,* vol. 56, no. 7, pp. 89–100, 1970.
65. "Tables of Thermal Properties of Gases," U.S. Department of Commerce, National Bureau of Standards Circular 564, November 1955.
66. Gasparovic, N.: "Gas Turbines and Combined Cycles: Their Status Nowadays and Their Application for Power Generation," private correspondence, January 1982.
67. Gasparovic, N., and J. G. Hellemans: Gas Turbines with Heat Exchanger and Water Injection in the Compressed Air, "Proceedings of the Institution of Mechanical Engineers," vol. 185 66/71, pp. 953–961, 1970–1971.
68. Gasparovic, N., and D. Stapersma: Gas Turbines with Heat Exchangers and Water Injection in the Compressed Air, *Combustion,* pp. 6–16, December 1973.
69. Singh, P. P., "Formation and Control of Nitrogen Oxide Emissions from Gas Turbine Combustion Systems," The American Society of Mechanical Engineers paper No. 72-GT-22, 1972.
70. Sweigert, R. L., and N. W. Beardsley: Empirical Specific Heat Equations Based on Spectroscopic Data, *Ga. Sch. Technol. Tech. Bull.*, vol. 1, no. 3, June 1938.

71. Caruvana, A., and G. A. Cincotta: High Temperature Turbine Technology Advancements, "Proceedings of the American Power Conference," vol. 44, 1982.
72. Mukherjee, D. K.: The Cooling of Gas Turbine Blades, *Brown Boveri Rev.*, vol. 64, pp. 47–51, January 1977.
73. Alff, R. K., W. G. Taylor, R. C. Sheldon, and D. M. Todd: Coal-Derived Flexibility for Today's Combined Cycles, "Proceedings of the American Power Conference," vol. 42, pp. 101–111, 1980.
74. Felix, P. C.: Practical Experience with Crude and Heavy Oil in Stationary Gas Turbines, *Brown Boveri Rev.*, vol. 66, pp. 89–96, February 1979.
75. Scheirer, S. T., and H. L. Jaeger: The Combustion Turbine: Future Design and Fuel Flexibility, "Proceedings of the American Power Conference," vol. 43, pp. 325–334, 1981.
76. Felix, P. C.: Corrosion and Its Prevention in Modern Stationary Gas Turbines, *Brown Boveri Rev.*, vol. 64, pp. 40–46, January 1977.
77. Foster-Pegg, R. W.: Combined Cycles have Potential for Cleaner, More Economical Use of Coal, *Power Eng.*, vol. 86, no. 4, pp. 80–84, April 1982.
78. Wunsch, A.; Combined Gas/Steam Turbine Power Plants—The Present State of Progress and Future Developments, *Brown Boveri Rev.*, vol. 65, pp. 646–655, October 1978.
79. Sined, R. J., and R. G. Wood: The Selection and Design of a New Combined-Cycle Plant-STAG, "Proceedings of the American Power Conference," vol. 34, pp. 292–301, 1972.
80. Sokolowski, P. F., and D. A. Schwartz: Design and Development of High-Pressure Modular Heat Recovery Steam Generators for Gas Turbine-Steam Turbine Combined Cycles, "Proceedings of the American Power Conference," vol. 34, pp. 310–330, 1972.
81. Boland, C. R., and R. D. Patterson: A Unique Combined-Cycle System to Meet Utility Intermediate Cycling Loads, "Proceedings of the American Power Conference," vol. 34, pp. 302–309, 1972.
82. Kirparti, S. P., and M. M. Nagib: Combined Helium-Steam Cycle for Nuclear Power Plants, *Mech. Eng.*, vol. 93, no. 8, p. 14, August 1971.
83. Rose, D. and M. Clarke, "Plasmas and Controlled Fusion," John Wiley & Sons, Inc., New York, 1961.
84. Katcoff, S.: Fission Product Yields from U, Th, and Pu, *Nucleonics,* vol. 16, no. 4, p. 78, 1958.
85. Watt, B. E.: Energy Spectrum of Neutrons from Thermal Fission of U^{235}, *Phys. Rev.*, vol. 87, p. 1037, 1952.
86. Blatt, B. E., and V. F. Weisskoff: "Theoretical Nuclear Physics," John Wiley & Sons, Inc., New York, 1952.
87. Hughes, D. J., and R. B. Schwartz: "Neutron Cross Sections," U.S. Atomic Energy Commission Report ANL-325, 2d ed., 1958.
88. The World List of Nuclear Power Plants, *Nucl. News,* vol. 25, no. 2, p. 83, February 1982.
89. Glasser, T. H.: "Basic Equations for Predicting Performance of a Nuclear Power Plant Pressurized," presented at the Second Nuclear Engineering and Science Conference, Philadelphia, Penn., The American Society of Mechanical Engineers paper No. 57-NES-95, March 1975.
90. Lacey, P. G.: "Fine Structure Power Peaking in a Critical Experiment Mockup of a Chemical Shim Core," USAEC Report No. WCAP-3723, March 1963.
91. Aisu, H., R. F. Barry, and P. G. Lacey: "Load Variation Restrictions in a Chemically Poisoned Large PWR Core," USAEC Report No. WCAP-3724, March 1963.
92. Cohen, P. and H. W. Graves, Jr.: Chemical Shim Control for Power Reactors, *Nucleonics,* vol. 22, no. 5, p. 75, May 1964.
93. "Final Facility Description and Safety Analysis Report, Wisconsin Electric Power Company, and Wisconsin Michigan Power Company Point Beach Nuclear Plant, Units 1 and 2," U.S. Atomic Energy Commission, 1969.
94. "Brown's Ferry Nuclear Power Station," Tennessee Valley Authority, design and analysis report, 1968.
95. Commonwealth Edison Company, private correspondence.
96. Dahlberg, R. C., E. G. Beasley, T. K. DeBoor, T. C. Evans, D. F. Molino, W. S. Rothwell, and W. R. Sivka: "Gas-Cooled, Natural Uranium, D_2O-Moderated Power Reactor; Reactor Design and Feasibility Problem," U.S. Atomic Energy Commission Report CF-S6-8-207 (Del.), August 1956.

97. Alikhanov, A. I., V. V. Vladimirsky, P. A. Petrov, and P. I. Khristenko: A Heavy-Water Power Reactor with Gas-Cooling, *J. Nucl. Energy,* vol. 3, part 2, pp. 77–82, August 1956.
98. Calder Hall, A Special Report, *Nucleonics,* vol. 14, no. 12, December 1956.
99. Hinkley Point B, A Special Survey, *Nucl. Eng.*, vol. 13, no. 147, pp. 652–668, August 1968.
100. Jay, S., and W. V. Goeddel: "High-Temperature Gas-Cooled Reactor Fuels and Fueld Cycles—Their Progress and Promise," General Atomic Report No. GA-748, December 1966.
101. Ross-Ross, P. A., and K. L. Smith: "Pressure-Tube Development for Canada's Power Reactors," The American Society of Mechanical Engineers paper No. 64-WA/NE-5, 1964.
102. Civilian Power Reactor Program, Part III, TID-8515(1), Status Report on Fast Reactors as of 1959, U.S. Atomic Energy Commission, 1960.
103. Dounreay, *Nucl. Eng.*, vol. 2, no. 15, pp. 229–245, June 1957.
104. Glasstone, S., and A. Sesonske, "Nuclear Reactor Engineering," D. Van Nostrand Company, Inc., Princeton, N.J., 1963.
105. CRBR, Westinghouse Electric Corporation, Technical Progress Report No. CRBRP-ARD-0270, 1980.
106. Zaleski, C. P.: *Science,* vol. 208, no. 4440, pp. 137–144, April 1980.
107. The Creys-Malville Power Station, translated from *Bull. Inform. Sci. Tech.*, no. 227, January 1978, French Atomic Energy Commission, April 1980.
108. Potter, R. W., II: Geothermal Energy, an Assessment, *Mech. Eng.,* vol. 103, no. 5, pp. 20–23, May 1981.
109. United States Geological Survey, Circular 790, 1979. (See Muffler, L. J. P., and M. Guffanti: "Assessment of Geothermal Resources of the United States—1978," pp. 1–7; also see Brook, C. A., et al.: "Hydrothermal Convection Systems with Reservoir Temperatures—90°C," pp. 18–85.)
110. Ellis, A. J.: Geothermal Systems and Power Development, *Am. Sci.*, vol. 63, pp. 510–521, September-October 1975.
111. Pruce, L. M.: Using Geothermal Energy for Power, *Power,* vol. 123, no. 10, October 1979.
112. Nordyke, M. D.: "Peaceful Uses of Nuclear Explosions," International Atomic Energy Agency Report No. PL-388-12, 1970.
113. Charlson, M.: Geothermal: New Potential Underground, *EPRI J.*, vol. 6, no. 10, pp. 18–25, December 1981.
114. Austin, A. L., and A. W. Lundberg: Electric Power Generation from Geothermal Hot Water Deposits, *Mech. Eng.*, vol. 97, no. 12, pp. 18–25, December 1975.
115. Kestin, J., R. DiPippo, and H. E. Khalifa: Hybrid Geothermal-Fossil Power Plants, *Mech. Eng.,* vol. 100, no. 12, pp. 28–35, December 1978.
116. Duffie, J. A., and W. A. Beckman, "Solar Engineering of Thermal Processes," John Wiley & Sons, Inc., New York, 1980.
117. Thekaekara, M. P.: Data on Incident Solar Energy, "Supplement to the Proceedings of the 20th Annual Meeting of the Institute for Environmental Science," vol. 21, 1974.
118. Battleson, K. W.: "Solar Power Design Guide: Solar Thermal Central Receiver Power Systems. A Source of Electricity and/or Process Heat," Sandia National Laboratories Report No. SAND 81-8005, April 1981.
119. "Solar Central Receiver Hybrid Power Systems Sodium-Cooled Receiver Concept," Rockwell International Energy Systems Group Report No. DOE/ET-20567-1, January 1980.
120. "Solar Central Receiver Hybrid Power System (Nitrate Salt)," Martin-Marietta Company Report No. DOE/ET T-21038-1, September 1979.
121. "Solar Thermal Conversion to Electricity Utilizing a Central Receiver, Open Cycle Gas Turbine Design," Black and Veech Consulting Engineers, Electric Power Research Institute Report No. EPRI-ER-652, March 1978.
122. "Closed Brayton Cycle Advanced Central Receiver Solar Electric Power System," Boeing Engineering and Construction Co. Report No. SAN/1726-1, November 1978.
123. "Combined Cycle Solar Central Receiver Hybrid Power System," Bechtel National, Inc. Report No. DOE/ET-21050-1, November 1979.
124. "Central Receiver Solar Thermal Power System," McDonnell Douglas Astronautics Company Report No. MDC G6776, DOE/SAN-1108-76-8, October 1977.

125. Liebowitz, L., and Hanseth, E.: Solar Thermal Technology—Outlook for the 80s, *Mech. Eng.*, vol. 104, no. 1, pp. 30–35, January 1982.
126. Solar-Thermal Electric: Focal Point for the Desert Sun, *EPRI J.*, vol. 6, no. 10, pp. 36–43, December 1981.
127. Wolf, M.: "Methods for Low Cost Manufacture of Silicon Solar Arrays," The American Society of Mechanical Engineers paper No. 74-WA/Ener-4, presented at the ASME annual meeting, New York, November 17–22, 1974.
128. Kittel, C.: "Introduction to Solid State Physics," John Wiley & Sons, Inc., New York, 1956.
129. Loferski, J. J.: Theoretical Considerations Governing the Choice of the Optimum Semiconductor for Photovoltaic Solar Energy Conversions, *J. App. Phys.*, vol 27, pp. 777–784, 1956.
130. Glaser, P. E.: Power from the Sun, *Mech. Eng.*, vol. 91, no. 3, pp. 20–24, March 1969. (Also see *Science*, vol. 162, no. 3856, pp. 857–861, November 22, 1968.)
131. Heredeen, R. A., and T. K. Rebitzer: Energy Analysis of the Solar Power Satellite, *Science*, vol. 205, no. 4405, pp. 451–454, August 3, 1979.
132. Glaser, P. E.: Economic and Environmental Costs of Satellite Solar Power, *Mech. Eng.*, vol. 100, no. 1, pp. 32–37, January 1978.
133. Brunt, D.: "Physical and Dynamic Meteorology," Cambridge University Press, 1941.
134. Kung, E. C.: Large-Scale Balance of Kinetic Energy in the Atmosphere, *Mon. Weather Rev.*, vol. 94, no. 11, pp. 627–639, November 1966.
135. Hirschfeld, F.: Wind Power-Pipe Dream or Reality?, *Mech. Eng.*, vol. 99, no. 9, pp. 20–28, September 1977.
136. Wegley, H. L., M. M. Orgill, and R. L. Drake: "A Siting Handbook for Small Wind Energy Conversion Systems," Pacific Northwest Laboratories, Richland, Wash., Report No. PNL-2521, 1978.
137. Proceedings of the Conference and Workshop on Wind Energy Characteristics and Wind Energy Siting, 1979, Pacific Northwest Laboratories, Richland, Wash., Report No. PNL-3214, January 1980.
138. Justus, C. G., W. R. Hargraves, and A. Mikhail: "Reference Wind Speed Distributions and Height Profiles for Wind Turbine Design and Performance Evaluation Applications," Energy Research and Development Administration Report No. ORO/5108-76/4, August 1976.
139. Ramler, J. R., and R. M. Donovan: "Wind Turbines for Electric Utilities: Development Status and Economics," Report No. DOE/NASA 1028-79/23, 2d printing, presented at the Terrestrial Energy Systems Conference of the American Institute of Aeronautics and Astronautics, Orlando, Fla., June 1979.
140. Dodge, R. A., and M. J. Thompson: "Fluid Mechanics," McGraw-Hill Book Company, New York, 1937.
141. Davis, C. V., and K. E. Sorensen (eds.): "Handbook of Applied Hydraulics," 3d ed., McGraw-Hill Book Company, New York, 1969.
142. Walters, S.: Power in the Year 2001, Part 2—Thermal Sea Power, *Mech. Eng.*, vol. 93, no. 10, pp. 21–25, October 1971.
143. "World Atlas of Sea Surface Temperatures," H.O. Publication No. 225, 2d ed., U.S. Naval Hydrographic Office, Washington, D.C., 1944.
144. Claude, G.: Power from the Tropical Seas, *Mech. Eng.*, vol. 52, no. 12, pp. 1039–1044, December 1930.
145. Roe, R. C., and D. F. Othmer: Controlled Flash Evaporation, *Mech. Eng.*, vol. 93, no. 5, pp. 27–31, May 1971.
146. Anderson, J. H., and J. H. Anderson, Jr.: Thermal Power from Sea Water, *Mech. Eng.*, vol. 88, no. 4, pp. 41–46, April 1966. (See also Power from the Sun by Way of the Sea?, *Power*, January-February 1965.)
147. Hogben, N., and F. E. Lamb: "Ocean Wave Statistics," National Physical Laboratory, London, Her Majesty's Stationary Office, 1967.
148. Lamb, H.: "Hydrodynamics," 6th ed., Dover Publications, New York, 1932.
149. Milne-Thompson, L. M.: "Theoretical Hydrodynamics," 3d ed., Macmillan Publishing Co., Inc., New York, 1955.

150. Martin, M. D.: "Power from Ocean Waves," The American Society of Mechanical Engineers paper No. 74-WA/Pwr-5, presented at the ASME winter annual meeting, New York, November 1974.
151. Wave Motion Can Be Used to Tap Wind Energy, *Energy Int.*, pp. 19–20, April 1975.
152. Darwin, G. H.: "The Tides," W. H. Freeman and Company, San Francisco, 1962. Originally published in 1898.
153. Gray, T. J., and O. K. Gashus (eds.): "Tidal Power: Proceedings of the Tidal Power Conference, Nova Scotia, 1970," Plenum Press, New York, 1972.
154. Data supplied by Madison Gas and Electric Co., Madison, Wis., 1982.
155. CAES—It's More than Hot Air, Staff Report, *Mech. Eng.*, vol. 104, no. 1, pp. 20–23, January 1982.
156. Hoffeins, H., N. Romeyke, and D. Hebel: Commissioning the First Air Storage Gas Turbine Set, *Brown Boveri Rev.*, vol. 67, pp. 465–473, August 1980.
157. Fullman, R. L.: "Energy Storage by Flywheels," General Electric Company Report No. 75CRD051, April 1975.
158. Walters, S.: Briefing the Record, Low-Loss Energy Storage Flywheel, *Mech. Eng.*, vol. 99, no. 9, pp. 48–49, September 1977.
159. Boom, R. W. and H. A. Peterson: Superconductive Energy Storage for Power Systems, "Proceedings of the 1972 Intermag Conference," Kyoto, Japan, April 1972.
160. Grover, F. W.: Formulas for Mutual and Self Inductance, *Bull. U.S. Bur. Stand.*, vol. 8, no. 1, 1912.
161. Boom, R. W., and R. F. Bischke: Inductor-Converter Superconductive Magnetic Energy Storage for Electric Utility Usage, *Phys. Technol.*, vol. 13, pp. 18–27, 1982.
162. Gilli, P. V., and G. Beckmann: "The Nuclear Storage Plant—An Economic Method of Peak Power Generation," Paper No. 4.1.10, Ninth World Energy Conference, Detroit, 1974.
163. Gawron, V., and J. Schröder: "Proceedings of the Fourth European Symposium on Fluorine Chemistry," Ljublijana, Yugoslavia, 1972.
164. Bundy, F. G.: "Power Generating Plant with Nuclear Reactor-Heat Storage System Combination," U.S. Patent No. 3,848,416.
165. Schulten R., C. B. Van der Decken, K. Kugeler, and H. Barnert: Chemical Latent Heat for Transport of Nuclear Energy over Long Distances, "Proceedings of the British Nuclear Energy Society International Conference on The High Temperature Reactor and Process Applications," 1974.
166. Hanneman, R. E., H. B. Vakil, and R. H. Wentorf, Jr.: Closed Loop Chemical Systems for Energy Transmission, Conversion and Storage, "Proceedings of the Ninth Intersociety Energy Conversion Engineering Conference," 1974.
167. Van Wylen, G. J., and R. E. Sonntag: "Fundamentals of Classical Thermodynamics," 2d ed., John Wiley & Son, Inc., New York, 1973, chap. 12.
168. Golibersuch, D. G., F. P. Bundy, P. G. Kosky, and H. B. Vakil: "Thermal Energy Storage for Utility Applications," General Electric Company Report No. 75CRD256, December 1975.
169. Myers, P. S.: "Automobile Emissions—A Study in Environmental Benefits versus Technological Costs," Society of Automotive Engineers paper No. 700182, 1969.
170. Taylor, F. E., and G. A. M. Webb: "Radiation Exposure of the U.K. Population," National Radiological Protection Board Publication No. NRPB-R77, 1978.
171. Council on Environmental Control: "Global Energy Futures and the Carbon Dioxide Problem," U.S. Government Printing Office, January 1981.
172. Cowling, E. B., and J. N. Galloway: Effects of Precipitation on Aquatic and Terrestrial Ecosystems; A Proposed Precipitation Chemistry Network, *J. Air Pollut. Control Assoc.*, vol. 28, no. 3, pp. 228–235, March, 1978.
173. Wark, K., and C. F. Warner: "Air Pollution, its Origin and Control," 2d ed., Harper & Row, Publishers, Inc., New York, 1981.
174. White, H. J.: "Industrial Electrostatic Precipitation," Addison-Wesley Publishing Company, Inc., Reading, Mass., 1973.
175. White, H. J.: Modern Electrical Precipitation, *Ind. Eng. Chem.*, vol. 47, no. 2, pp. 932–939, 1955.
176. McCain, J. D., J. P. Gooch, and W. B. Smith: Results of Field Measurements of Industrial Particulate Sources and Electrostatic Precipitator Performance, *J. Air Pollut. Control Assoc.*, vol. 25, no. 2, pp. 117–121, 1975.

177. Fabric Filter Systems Study, "Handbook of Fabric Filter Technology," vol. 1, PB200-648, APTD-0690, National Technical Information Service, December 1970.
178. The President's Commission on the Accident at Three-Mile Island, J. G. Kemeny, Chairman, Washington, D.C., October 1979.
179. Johns, H. E., and J. R. Cunningham: "The Physics of Radiology," Charles C Thomas, Publisher, Springfield, Ill., 1969.
180. Attix, F. H., and W. C. Roesch (eds.): "Radiation Dosimetry," vol. 1, Academic Press, Inc., New York, 1968.
181. Blatz, H.: "Radiological Health," McGraw-Hill Book Company, New York, 1964.
182. Gloyna, E. F., and H. O. Ledbetter: "Principles of Radiological Health," Marcel Dekker, New York, 1969.
183. Duhamel, A. M. F. (ed.): "Health Physics," vol. 2, pt. 1, Progress in Nuclear Energy Series XII, Pergamon Press, Oxford, 1969.
184. Cember, H.: "Introduction to Health Physics," Pergamon Press, London, 1969.
185. Upton, A. C.: The Biological Effects of Low-Level Ionizing Radiation, *Sci. Am.*, vol. 246, no. 2, pp. 41–49, February 1982.
186. Gilbertson, J.: The Fraudulent Nuclear Waste Controversy, *Fusion*, vol. 2, no. 4, pp. 34–36, January 1979.
187. Cohen, B. L.: The Disposal of Radioactive Wastes from Fission Reactors, *Sci. Am.*, vol. 236, pp. 21–31, June 1977.
188. Sokol, J. P., and M. H. Cooper: Radioactive Waste in Perspective, *Nucl. Eng. Dig.*, Westinghouse Nuclear Energy Systems, pp. 15–21, 1979.
189. "Tables of Thermodynamic Properties of Ammonia," U.S. Bureau of Standards, Department of Commerce Circular No. 142, April 1923.
190. Perry, R. H., and C. H. Chilton (eds.): "Chemical Engineer's Handbook," 5th ed., McGraw-Hill Book Company, New York, 1973.
191. Meisel, C. J., and A. Shapiro: "Thermodynamic Properties of Alkali Metal Vapors and Mercury," 2d rev. ed., General Electric Flight Propulsion Laboratory Report No. R60FPD358-A, November 1960.
192. Hoegerton, J. F., and R. C. Grass (eds.): "Reactor Handbook," vol. 3, "Engineering, Selected Reference Material," U.S. Atomic Energy Commission, August 1955.
193. "Tables of the Thermodynamic Properties of Gases," U.S. National Bureau of Standards Circular No. 564, November 1955.
194. Moody, L. F., Friction Factors for Pipe Flows, "Transactions of the American Society of Mechanical Engineers," vol. 66, pp. 671–694, 1944.

INDEX